Subject	Symbol	Meaning	Page
Sequences	$\ldots$	and so forth	199
	$\displaystyle\sum_{k=m}^{n} a_k$	the summation from k equals m to n of a_k	202
	$\displaystyle\prod_{k=m}^{n} a_k$	the product from k equals m to n of a_k	205
	$n!$	n factorial	206
Set Theory	$a \in A$	a is an element of A	76
	$a \notin A$	a is not an element of A	76
	$\{a_1, a_2, \ldots, a_n\}$	the set with elements $a_1, a_2, \ldots, a_n$	76
	$\{x \in D \mid P(x)\}$	the set of all x in D for which $P(x)$ is true	77
	$\mathbf{R}, \mathbf{R}^-, \mathbf{R}^+, \mathbf{R}^{nonneg}$	the sets of all real numbers, negative real numbers, positive real numbers, and nonnegative real numbers	76, 77
	$\mathbf{Z}, \mathbf{Z}^-, \mathbf{Z}^+, \mathbf{Z}^{nonneg}$	the sets of all integers, negative integers, positive integers, and nonnegative integers	76, 77
	$\mathbf{Q}, \mathbf{Q}^-, \mathbf{Q}^+, \mathbf{Q}^{nonneg}$	the sets of all rational numbers, negative rational numbers, positive rational numbers, and nonnegative rational numbers	76, 77
	$\mathbf{N}$	the set of natural numbers	77
	$A \subseteq B$	A is a subset of B	256
	$A \nsubseteq B$	A is not a subset of B	257
	$A = B$	A equals B	258
	$A \cup B$	A union B	260
	$A \cap B$	A intersect B	260
	$B - A$	the difference of B minus A	260
	A^c	the complement of A	260
	(x, y)	ordered pair	264
	$(x_1, x_2, \ldots, x_n)$	ordered n-tuple	264
	$A \times B$	the Cartesian product of A and B	265
	$A_1 \times A_2 \times \cdots \times A_n$	the Cartesian product of $A_1, A_2, \ldots, A_n$	265
	$\emptyset$	the empty set	262
	$\mathscr{P}(A)$	the power set of A	264

List of Symbols

Subject	Symbol	Meaning	Page
Counting and Probability	$N(A)$	the number of elements in a set A	299
	$P(A)$	the probability of a set A	299
	$P(n, r)$	the number of r-permutations of a set of n elements	315
	$\binom{n}{r}$	n choose r, the number of r-combinations of a set of n elements, the number of r-element subsets of a set of n elements	334
	$[x_{i_1}, x_{i_2}, \ldots, x_{i_r}]$	multiset of size r	349
	$P(A \mid B)$	the probability of A given B	376
Functions	$f: X \to Y$	f is a function from X to Y	390
	$f(x)$	the value of f at x	390
	$x \xrightarrow{f} y$	f sends x to y	390
	$f(A)$	the image of A	402
	$f^{-1}(C)$	the inverse image of C	402
	i_x	the identity function on X	394
	b^x	b raised to the power x	411
	$\exp_b(x)$	b raised to the power x	411
	$\log_b(x)$	logarithm with base b of x	395
	F^{-1}	the inverse function of F	415
	$f \circ g$	the composition of g and f	432
Algorithm Efficiency	$x \cong y$	x is approximately equal to y	206
	$O(f(x))$	big-O of f of x	519
	$\Omega(f(x))$	big-Omega of f of x	519
	$\Theta(f(x))$	big-Theta of f of x	519
Relations	$x \, R \, y$	x is related to y by R	572
	R^{-1}	the inverse relation of R	578
	$m \equiv n \pmod{d}$	m is congruent to n modulo d	597
	$[a]$	the equivalence class of a	599
	$x \preceq y$	x is related to y by a partial order relation $\preceq$	635

Continued on first page of back endpapers.

DISCRETE MATHEMATICS
WITH APPLICATIONS

THIRD EDITION

SUSANNA S. EPP
DePaul University

THOMSON

BROOKS/COLE

Australia • Canada • Mexico • Singapore • Spain
United Kingdom • United States

THOMSON

BROOKS/COLE

Cover Photo: *The stones are discrete objects placed one on top of another like a chain of careful reasoning. A person who decides to build such a tower aspires to the heights and enjoys playing with a challenging problem. Choosing the stones takes both a scientific and an aesthetic sense. Getting them to balance requires patient effort and careful thought. And the tower that results is beautiful. A perfect metaphor for discrete mathematics!*

Publisher: *Bob Pirtle*
Assistant Editor: *Stacy Green*
Project Manager, Editorial Production: *Janet Hill*
Marketing Manager: *Tom Ziolkowski*
Marketing Assistant: *Jennifer Gee*
Print/Media Buyer: *Doreen Suruki*
Production Service: *Martha Emry Production Services*

Text Designer: *Kathleen Cunningham*
Art Editor: *Martha Emry*
Illustrator: *Techsetters, Inc.*
Cover Designer: *Jeanne Calabrese*
Cover Image: *Getty Images*
Cover/Interior Printer: *Quebecor World/Versailles*
Compositor: *Techsetters, Inc.*

For more information about our products, contact us at:
Thomson Learning Academic Resource Center
1-800-423-0563

For permission to use material from this text, contact us by:
Phone: 1-800-730-2214   **Fax:** 1-800-730-2215
Web: http://www.thomsonrights.com

COPYRIGHT 2004 Thomson Learning, Inc. All Rights Reserved. Thomson Learning WebTutor™ is a trademark of Thomson Learning, Inc.

Library of Congress Control Number: 2003107333

ISBN 0-534-35945-0

International Student Edition ISBN: 0-534-49096-4
(Not for sale in the United States)

Brooks/Cole–Thomson Learning
10 Davis Drive
Belmont, CA 94002
USA

Asia
Thomson Learning
5 Shenton Way #01-01
UIC Building
Singapore 068808

Australia/New Zealand
Thomson Learning
102 Dodds Street
Southbank, Victoria 3006
Australia

Canada
Nelson
1120 Birchmount Road
Toronto, Ontario M1K 5G4
Canada

Europe/Middle East/Africa
Thomson Learning
High Holborn House
50/51 Bedford Row
London WC1R 4LR
United Kingdom

Latin America
Thomson Learning
Seneca, 53
Colonia Polanco
11560 Mexico D.F.
Mexico

Spain/Portugal
Paraninfo
Calle Magallanes, 25
28015 Madrid, Spain

To Jayne and Ernest

CONTENTS

Chapter 3 **Elementary Number Theory and Methods of Proof**   *125*

PREFACE

My purpose in writing this book was to provide a clear, accessible treatment of discrete mathematics for students majoring or minoring in computer science, mathematics, mathematics education, and engineering. The goal of the book is to lay the mathematical foundation for computer science courses such as data structures, algorithms, relational database theory, automata theory and formal languages, compiler design, and cryptography, and for mathematics courses such as linear and abstract algebra, combinatorics, probability, logic and set theory, and number theory. By combining discussion of theory and practice, I have tried to show that mathematics has engaging and important applications as well as being interesting and beautiful in its own right.

A good background in algebra is the only prerequisite; the course may be taken by students either before or after a course in calculus. Previous editions of the book have been used successfully by students at hundreds of institutions in North and South America, Europe, the Middle East, Asia, and Australia.

Recent curricular recommendations from the Institute for Electrical and Electronic Engineers Computer Society (IEEE CS) and the Association for Computing Machinery (ACM) include discrete mathematics as the largest portion of "core knowledge" for computer science students and state that students should take at least a one-semester course in the subject as part of their first-year studies, with a two-semester course preferred when possible. This book includes all the topics recommended by those organizations and can be used effectively for either a one-semester or a two-semester course.

At one time, most of the topics in discrete mathematics were taught only to upper-level undergraduates. Discovering how to present these topics in ways that can be understood by first- and second-year students was the major and most interesting challenge of writing this book. The presentation was developed over a long period of experimentation during which my students were in many ways my teachers. Their questions, comments, and written work showed me what concepts and techniques caused them difficulty, and their reaction to my exposition showed me what worked to build their understanding and to encourage their interest. Many of the changes in this edition have resulted from continuing interaction with students.

Themes of a Discrete Mathematics Course

Discrete mathematics describes processes that consist of a sequence of individual steps. This contrasts with calculus, which describes processes that change in a continuous fashion. Whereas the ideas of calculus were fundamental to the science and technology of the industrial revolution, the ideas of discrete mathematics underlie the science and technology of the computer age. The main themes of a first course in discrete mathematics are logic and proof, induction and recursion, combinatorics and discrete probability, algorithms and their analysis, discrete structures, and applications and modeling.

Logic and Proof Probably the most important goal of a first course in discrete mathematics is to help students develop the ability to think abstractly. This means learning to use logically valid forms of argument and avoid common logical errors, appreciating what it means to reason from definitions, knowing how to use both direct and indirect argument to derive new results from those already known to be true, and being able to work with symbolic representations as if they were concrete objects.

Induction and Recursion   An exciting development of recent years has been the increased appreciation for the power and beauty of "recursive thinking." To think recursively means to address a problem by assuming that similar problems of a smaller nature have already been solved and figuring out how to put those solutions together to solve the larger problem. Such thinking is widely used in the analysis of algorithms, where recurrence relations that result from recursive thinking often give rise to formulas that are verified by mathematical induction.

Combinatorics and Discrete Probability   Combinatorics is the mathematics of counting and arranging objects, and probability is the study of laws concerning the measurement of random or chance events. Discrete probability focuses on situations involving discrete sets of objects, such as finding the likelihood of obtaining a certain number of heads when an unbiased coin is tossed a certain number of times. Skill in using combinatorics and probability is needed in almost every discipline where mathematics is applied, from economics to biology, to computer science, to chemistry and physics, to business management.

Algorithms and Their Analysis   The word *algorithm* was largely unknown in the middle of the twentieth century, yet now it is one of the first words encountered in the study of computer science. To solve a problem on a computer, it is necessary to find an algorithm or step-by-step sequence of instructions for the computer to follow. Designing an algorithm requires an understanding of the mathematics underlying the problem to be solved. Determining whether or not an algorithm is correct requires a sophisticated use of mathematical induction. Calculating the amount of time or memory space the algorithm will need in order to compare it to other algorithms that produce the same output requires knowledge of combinatorics, recurrence relations, functions, and O-, Ω-, and Θ-notations.

Discrete Structures   Discrete mathematical structures are the abstract structures that describe, categorize, and reveal the underlying relationships among discrete mathematical objects. Those studied in this book are the sets of integers and rational numbers, general sets, Boolean algebras, functions, relations, graphs and trees, formal languages and regular expressions, and finite-state automata.

Applications and Modeling   Mathematical topics are best understood when they are seen in a variety of contexts and used to solve problems in a broad range of applied situations. One of the profound lessons of mathematics is that the same mathematical model can be used to solve problems in situations that appear superficially to be totally dissimilar. A goal of this book is to show students the extraordinary practical utility of some very abstract mathematical ideas.

Special Features of This Book

Mathematical Reasoning   The feature that most distinguishes this book from other discrete mathematics texts is that it teaches—explicitly but in a way that is accessible to first- and second-year college and university students—the unspoken logic and reasoning that underlie mathematical thought. For many years I taught an intensively interactive transition-to-abstract-mathematics course to mathematics and computer science majors. This experience showed me that while it is possible to teach the majority of students to understand and construct straightforward mathematical arguments, the obstacles to doing so cannot be passed over lightly. To be successful, a text for such a course must address students' difficulties with logic and language directly and at some length. It must also include enough concrete examples and exercises to enable students to develop the mental models needed to conceptualize more abstract problems. The treatment of logic and proof in this book blends common sense and rigor in a way that explains the essentials, yet avoids overloading students with formal detail.

Spiral Approach to Concept Development A number of concepts in this book appear in increasingly more sophisticated forms in successive chapters to help students develop the ability to deal effectively with increasing levels of abstraction. For example, by the time students encounter the relatively advanced mathematics of Fermat's little theorem and the Chinese remainder theorem in the Section 10.4, they have been introduced to the logic of mathematical discourse in Chapters 1 and 2, learned the basic methods of proof and the concepts of *mod* and *div* in Chapter 3, studied partitions of the integers in Chapter 5, considered *mod* and *div* as functions in Chapter 7, and become familiar with equivalence relations in Sections 10.2 and 10.3. This approach builds in useful review and develops mathematical maturity in natural stages.

Support for the Student Students at colleges and universities inevitably have to learn a great deal on their own. Though it is often frustrating, learning to learn through self-study is a crucial step toward eventual success in a professional career. This book has a number of features to facilitate students' transition to independent learning.

Worked Examples
The book contains over 500 worked examples, which are written using a problem-solution format and are keyed in type and in difficulty to the exercises. Many solutions for the proof problems are developed in two stages: first a discussion of how one might come to think of the proof or disproof and then a summary of the solution, which is enclosed in a box. This format allows students to read the problem and skip immediately to the summary, if they wish, only going back to the discussion if they have trouble understanding the summary. The format also saves time for students who are rereading the text in preparation for an examination.

Exercises
The book contains almost 2,500 exercises. The sets at the end of each section have been designed so that students with widely varying backgrounds and ability levels will find some exercises they can be sure to do successfully and also some exercises that will challenge them.

Solutions for Exercises
To provide adequate feedback for students between class sessions, Appendix B contains a large number of complete solutions to exercises. Students are strongly urged not to consult solutions until they have tried their best to answer the questions on their own. Once they have done so, however, comparing their answers with those given can lead to significantly improved understanding. In addition, many problems, including some of the most challenging, have partial solutions or hints so that students can determine whether they are on the right track and make adjustments if necessary. There are also plenty of exercises without solutions to help students learn to grapple with mathematical problems in a realistic environment.

Figures and Tables
Figures and tables are included in every case where it seemed that doing so would help readers to a better understanding. In most, a second color is used to add meaning.

Reference Features
Many students have written me to say that the book helped them succeed in their advanced courses. One even wrote that he had used the first edition so extensively that it had fallen apart and he actually went out and bought a copy of the second edition,

which he was continuing to use in a master's program. My rationale for screening statements of definitions and theorems, for putting titles on exercises, and for giving the meaning of symbols and a list of reference formulas in the endpapers is to make it easier for students to use this book for review in a current course and as a reference in later ones.

Support for the Instructor I have received a great deal of valuable feedback from instructors who have used previous editions of this book. Many aspects of the book have been improved through their suggestions.

Exercises

The large variety of exercises at all levels of difficulty allows instructors great freedom to tailor a course to the abilities of their students. Exercises with solutions in the back of the book have numbers in blue and those whose solutions are given in a separate Student Solutions Manual/Study Guide have numbers that are a multiple of three. There are exercises of every type that are represented in this book which have no answer in either location to enable instructors to assign whatever mixture they prefer of exercises with and without answers. The ample number of exercises of all kinds gives instructors a significant choice of problems to use for review assignments and exams. Instructors are invited to use the many exercises stated as questions rather than in "prove that" form to stimulate class discussion on the role of proof and counterexample in problem solving.

Flexible Sections

Most sections are divided into subsections so that an instructor who is pressed for time can choose to cover certain subsections only and either omit the rest or leave them for the students to study on their own. The division into subsections also makes it easier for instructors to break up sections if they wish to spend more then one day on them.

Presentation of Proof Methods

It is inevitable that the proofs and disproofs in this book will seem easy to instructors. Many students, however, find them difficult. In showing students how to discover and construct proof and disproofs, I have tried to describe the kinds of approaches that mathematicians use when confronting challenging problems in their own research.

Instructor's Manual

An instructor's manual is available to anyone teaching a course from this book. It contains suggestions about how to approach the material of each chapter, solutions for all exercises not fully solved in Appendix B, transparency masters, review sheets, ideas for projects and writing assignments, and additional exercises.

Highlights of the Third Edition

The changes that have been made for this edition are based on suggestions from colleagues and other long-time users of the first and second editions, on continuing interactions with my students, and on developments within the evolving fields of computer science and mathematics.

Improved Pedagogy

- The number of exercises has been increased to almost 2,500. Approximately 980 new exercises have been added.
- Exercises have been added for topics where students seemed to need additional practice, and they have been modified, as needed, to address student difficulties.
- Additional full answers have been incorporated into Appendix B to give students more help for difficult topics.

- The exposition has been reexamined throughout and revised where needed.
- Careful work has been done to improve format and presentation.
- Discussion of historical background and recent results has been expanded and the number of photographs of mathematicians and computer scientists whose contributions are discussed in the book has been increased.

Logic

- The treatment of quantification has been significantly expanded, with a new section entirely devoted to multiple quantifiers.
- Exercises have been added using Tarski's World, an excellent pedagogical tool developed by Jon Barwise and John Etchemendy at Stanford University.
- Applications related to Internet searching are now included.
- Terms for various forms of argument have been simplified.

Introduction to Proof

- The directions for writing proofs have been expanded.
- The descriptions of methods of proof have been made clearer.
- Exercises have been revised and/or relocated to promote the development of student understanding.

Induction and Recursion

- The format for outlining proofs by mathematical induction has been improved.
- The subsections in the section on sequences have been reorganized.
- The sets of exercises for the sections on strong mathematical induction and the well-ordering principle and on recursive definitions have been significantly expanded.

Number Theory

- A subsection on open problems in number theory has been incorporated, and the discussion of recent mathematical discoveries in number theory has been expanded.
- A new section on modular arithmetic and cryptography has been added. It includes a discussion of RSA cryptography, Fermat's little theorem, and the Chinese remainder theorem.
- The discussion of testing for primality has been moved to later in Chapter 3 to make clear its dependence on indirect argument.

Set Theory

- The properties of the empty set are now introduced in the first section of Chapter 5.
- The second section of Chapter 5 is now entirely devoted to element proofs.
- Algebraic proofs of set properties and the use of counterexamples to disprove set properties have been moved to the third section of Chapter 5.
- The treatment of Boolean algebras has been expanded, and the relationship among logical equivalences, set properties, and Boolean algebras has been highlighted.

Combinatorics and Discrete Probability

- Exercises for the section on the binomial theorem has been significantly expanded.
- Two new sections have been added on probability, including expected value, conditional probability and independence, and Bayes' theorem.
- Combinatorial aspects of Internet protocol (IP) addresses are explained.

Functions

- Exercises about one-to-one and onto functions have been refined and improved.
- The set of exercises on cardinality with applications to computability has been significantly expanded.

Efficiency of Algorithms

- Sections 9.2 and 9.4 have been reworked to add Θ- and Ω-notations.
- Sections 9.3 and 9.5 have been revised correspondingly, with a clearer explanation of the meaning of order for an algorithm.
- The treatment of insertion sort and selection sort has been improved and expanded.

Regular Expressions and Finite-State Automata

- The previous disparate sections on formal languages and finite-state automata have been reassembled into a chapter of their own.
- A new section on regular expressions has been added, as well as discussion of the relationship between regular expressions and finite-state automata.

Website

A website has been developed for this book that contains information and materials for both students and instructors. It includes

- descriptions and links to many sites on the Internet with accessible information about discrete mathematical topics,
- links to applets that illustrate or provide practice in the concepts of discrete mathematics,
- additional examples and exercises with solutions,
- review guides for the chapters of the book.

A special section for instructors contains

- transparency masters and PowerPoint slides,
- additional exercises for quizzes and exams.

Student Solutions Manual/Study Guide

In writing this book, I strove to give sufficient help to students through the exposition in the text, the worked examples, and the exercise solutions, so that the book itself would provide all that a student would need to successfully master the material of the course. I believe that students who finish the study of this book with the ability to solve, on their own, all the exercises with full solutions in Appendix B will have developed an excellent command of the subject. Nonetheless, I have become aware that some students want the opportunity to obtain additional helpful materials. In response, I have developed a Student Solutions Manual/Study Guide, available separately from this book, which contains complete solutions to every exercise that is not completely answered in Appendix B and whose number is divisible by 3. The guide also includes alternative explanations for some of the concepts, and review questions for each chapter.

Organization

This book may be used effectively for a one- or two-semester course. Each chapter contains core sections, sections covering optional mathematical material, and sections covering optional applications. Instructors have the flexibility to choose whatever mixture will best serve the needs of their students. The following table shows a division of the sections into categories.

Chapter	Core Sections	Sections Containing Optional Mathematical Material	Sections Containing Optional Computer Science Applications
1	1.1–1.3		1.4, 1.5
2	2.1–2.4	2.2, 2.3	2.3
3	3.1–3.4, 3.6	3.5, 3.7	3.8
4	4.1–4.2	4.3–4.4	4.5
5	5.1	5.2–5.4	5.4
6	6.1–6.4	6.5–6.9	6.3
7	7.1–7.2	7.3–7.5	7.1, 7.2, 7.5
8	8.1, 8.2	8.3, 8.4	8.4
9	9.1, 9.2	9.4	9.3, 9.5
10	10.1–10.3	10.4, 10.5	10.4, 10.5
11	11.1, 11.5	11.2, 11.3, 11.4	11.1, 11.2, 11.5, 11.6
12	12.1, 12.2	12.3	12.1–12.3

The following tree diagram shows, approximately, how the chapters of this book depend on each other. Chapters on different branches of the tree are sufficiently independent that instructors need to make at most minor adjustments if they skip chapters but follow paths along branches of the tree.

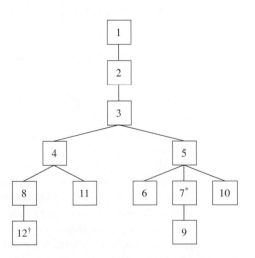

*Instructors who wish to define a function as a binary relation can cover Section 10.1 before Section 7.1.

†Section 10.3 is needed for Section 12.3 but not for Sections 12.1 and 12.2.

Acknowlegments

I owe a debt of gratitude to many people at DePaul University for their support and encouragement throughout the years I worked on editions of this book. A number of my colleagues used early versions and previous editions and provided many excellent suggestions for improvement. For this, I am thankful to Louis Aquila, J. Marshall Ash, Allan Berele, Jeffrey Bergen, William Chin, Barbara Cortzen, Constantine Georgakis, Sigrun Goes, Jerry Goldman, Lawrence Gluck, Leonid Krop, Carolyn Narasimhan, Walter Pranger, Eric Rieders, Ayse Sahin, Yuen-Fat Wong, and, most especially, Jeanne LaDuke. The thousands of students to whom I have taught discrete mathematics have had a profound influence on the book's form. By sharing their thoughts and thought processes with me, they taught me how to teach them better. I am very grateful for their help. I owe the DePaul University administration, especially my dean, Michael Mezey, and my former dean, Richard Meister, a special word of thanks for considering the writing of this book a worthwhile scholarly endeavor.

My thanks to the reviewers for their valuable suggestions for this edition of the book: Pablo Echeverria, Camden County College; William Gasarch, University of Maryland; Joseph Kolibal, University of Southern Mississippi; Benny Lo, International Technological University; George Luger, University of New Mexico; Norman Richert, University of Houston—Clear Lake; Peter Williams, California State University at San Bernardino; and Jay Zimmerman, Towson University. For their help with the first and second editions of the book, I am grateful to Itshak Borosh, Texas A & M University; Douglas M. Campbell, Brigham Young University; David G. Cantor, University of California at Los Angeles; C. Patrick Collier, University of Wisconsin-Oshkosh; Kevan H. Croteau, Francis Marion University; Irinel Drogan, University of Texas at Arlington; Henry A. Etlinger, Rochester Institute of Technology; Melvin J. Friske, Wisconsin Lutheran College; Ladnor Geissinger, University of North Carolina; Jerrold R. Griggs, University of South Carolina; Nancy Baxter Hastings, Dickinson College; Lillian Hupert, Loyola University Chicago; Leonard T. Malinowski, Finger Lakes Community College; John F. Morrison, Towson State Unviersity; Paul Pederson, University of Denver; George Peck, Arizona State University; Roxy Peck, California Polytechnic State University, San Luis Obispo; Dix Pettey, University of Missouri; Anthony Ralston, State University of New York at Buffalo; John Roberts, University of Louisville; and George Schultz, St. Petersburg Junior College, Clearwater. Special thanks are due John Carroll, San Diego State University; Dr. Joseph S. Fulda; and Porter G. Webster, University of Southern Mississippi for their unusual thoroughness and their encouragement.

I have also benefitted greatly from the suggestions of the many instructors who have generously offered me their ideas for improvement based on their experiences with previous editions of the book. I am especially grateful to Jonathan Goldstine, Pennsylvania State University; David Hecker, St. Joseph's University; Tom Jenkyns, Brock University; Robert Messer, Albion College; Piotr Rudnicki, University of Alberta; Anwar Shiek, Diné College; and Norton Starr, Amherst College. I also received excellent assistance from John Banks; Christopher Novak, DePaul University; and Ian Crewe, Ascension Collegiate School during the production of the book.

I am grateful to many people at the Wadsworth and Brooks/Cole Publishing Companies, especially my editor, Robert Pirtle, for his ability to make good things happen as if by magic, my previous editors, Heather Bennett and Barbara Holland, for their encouragement and enthusiasm, and my production supervisor, Janet Hill, for her understanding and her willingness to let me play an active role in all aspects of the production process. I cannot imagine a better production editor than Martha Emry, whose high standards and attention to detail were an inspiration. The design by Kathleen Cunningham and composition by Techsetters, Inc. will be appreciated by all readers of this book.

The older I get the more I realize the profound debt I owe my own mathematics teachers for shaping the way I perceive the subject. My first thanks must go to my husband, Helmut Epp, who, on a high school date (!), introduced me to the power and beauty of the field axioms and the view that mathematics is a subject with ideas as well as formulas and techniques. In my formal education, I am most grateful to Daniel Zelinsky and Ky Fan at Northwestern University and Izaak Wirszup, I. N. Herstein, and Irving Kaplansky at the University of Chicago, all of whom, in their own ways, helped lead me to appreciate the elegance, rigor, and excitement of mathematics.

To my family, I owe thanks beyond measure. I am grateful to my mother, whose keen interest in the workings of the human intellect started me many years ago on the track that led ultimately to this book, and to my late father, whose devotion to the written word has been a constant source of inspiration. I thank my children and grandchildren for their affection and cheerful acceptance of the demands this book has placed on my life. And, most of all, I am grateful to my husband, who for many years has encouraged me with his faith in the value of this project and supported me with his love and his wise advice.

Susanna Epp

CHAPTER 1

THE LOGIC OF COMPOUND STATEMENTS

Aristotle
(384 B.C.–322 B.C.)

The first great treatises on logic were written by the Greek philosopher Aristotle. They were a collection of rules for deductive reasoning that were intended to serve as a basis for the study of every branch of knowledge. In the seventeenth century, the German philosopher and mathematician Gottfried Leibniz conceived the idea of using symbols to mechanize the process of deductive reasoning in much the same way that algebraic notation had mechanized the process of reasoning about numbers and their relationships. Leibniz's idea was realized in the nineteenth century by the English mathematicians George Boole and Augustus De Morgan, who founded the modern subject of symbolic logic. With research continuing to the present day, symbolic logic has provided, among other things, the theoretical basis for many areas of computer science such as digital logic circuit design (see Sections 1.4 and 1.5), relational database theory (see Section 10.1), automata theory and computability (see Chapter 12), and artificial intelligence (see Sections 2.3, 11.1, and 11.5).

1.1 Logical Form and Logical Equivalence

Logic is a science of the necessary laws of thought, without which no employment of the understanding and the reason takes place. — Immanuel Kant, 1785

The central concept of deductive logic is the concept of argument form. An argument is a sequence of statements aimed at demonstrating the truth of an assertion. The assertion at the end of the sequence is called the *conclusion,* and the preceding statements are called *premises.* To have confidence in the conclusion that you draw from an argument, you must be sure that the premises are acceptable on their own merits or follow from other statements that are known to be true.

In logic, the form of an argument is distinguished from its content. Logical analysis won't help you determine the intrinsic merit of an argument's content, but it will help you analyze an argument's form to determine whether the truth of the conclusion follows *necessarily* from the truth of the premises. For this reason logic is sometimes defined as the science of necessary inference or the science of reasoning.

Consider the following two arguments, for example. Although their content is very different, their logical form is the same. Both arguments are *valid* in the sense that if their premises are true, then their conclusions must also be true. (In Section 1.3 you will learn how to test whether an argument is valid.)

1

Argument 1 If the program syntax is faulty or if program execution results in division by zero, then the computer will generate an error message. Therefore, if the computer does not generate an error message, then the program syntax is correct and program execution does not result in division by zero.

Argument 2 If x is a real number such that $x < -2$ or $x > 2$, then $x^2 > 4$. Therefore, if $x^2 \not> 4$, then $x \not< -2$ and $x \not> 2$.

To illustrate the logical form of these arguments, we use letters of the alphabet (such as p, q, and r) to represent the component sentences and the expression "not p" to refer to the sentence "It is not the case that p." Then the *common logical form* of both the arguments above is as follows:

> If p or q, then r.
>
> Therefore, if not r, then not p and not q.

Example 1.1.1 Identifying Logical Form

Fill in the blanks below so that argument (b) has the same form as argument (a). Then represent the common form of the arguments using letters to stand for component sentences.

a. If Jane is a math major or Jane is a computer science major, then Jane will take Math 150.
   Jane is a computer science major.
   Therefore, Jane will take Math 150.

b. If logic is easy or ___(1)___, then ___(2)___.
   I will study hard.
   Therefore, I will get an A in this course.

Solution

1. I (will) study hard.

2. I will get an A in this course.

   Common form: If p or q, then r.

   q.

   Therefore, r. ∎

Statements

Most of the definitions of formal logic have been developed so that they agree with the natural or intuitive logic used by people who have been educated to think clearly and use language carefully. The differences that exist between formal and intuitive logic are necessary to avoid ambiguity and obtain consistency.

In any mathematical theory, new terms are defined by using those that have been previously defined. However, this process has to start somewhere. A few initial terms necessarily remain undefined. In logic, the words *sentence, true,* and *false* are the initial undefined terms.

• Definition

A **statement** (or **proposition**) is a sentence that is true or false but not both.

For example, "Two plus two equals four" and "Two plus two equals five" are both statements, the first because it is true and the second because it is false. On the other hand, the truth or falsity of "He is a college student" depends on the reference for the pronoun *he*. For some values of *he* the sentence is true; for others it is false. If the sentence were preceded by other sentences that made the pronoun's reference clear, then the sentence would be a statement. Considered on its own, however, the sentence is neither true nor false, and so it is not a statement. We will discuss ways of transforming sentences of this form into statements in Section 2.1.

Similarly, $x + y > 0$ is not a statement because for some values of x and y the sentence is true, whereas for others it is false. For instance, if $x = 1$ and $y = 2$, the sentence is true; if $x = -1$ and $y = 0$, the sentence is false.

Compound Statements

We now introduce three symbols that are used to build more complicated logical expressions out of simpler ones. The symbol $\sim$ denotes *not,* $\wedge$ denotes *and,* and $\vee$ denotes *or.* Given a statement p, the sentence "$\sim p$" is read "not p" or "It is not the case that p" and is called the **negation of p.** In some computer languages the symbol $\neg$ is used in place of $\sim$. Given another statement q, the sentence "$p \wedge q$" is read "p and q" and is called **conjunction of p and q.** The sentence "$p \vee q$" is read "p or q" and is called the **disjunction of p and q.**

In expressions that include the symbol $\sim$ as well as $\wedge$ or $\vee$, the **order of operations** is that $\sim$ is performed first. For instance, $\sim p \wedge q = (\sim p) \wedge q$. In logical expressions, as in ordinary algebraic expressions, the order of operations can be overridden through the use of parentheses. Thus $\sim(p \wedge q)$ represents the negation of the conjunction of p and q. In this, as in most treatments of logic, the symbols $\wedge$ and $\vee$ are considered coequal in order of operation, and an expression such as $p \wedge q \vee r$ is considered ambiguous. This expression must be written as either $(p \wedge q) \vee r$ or $p \wedge (q \vee r)$ to have meaning.

A variety of English words translate into logic as $\wedge$, $\vee$, or $\sim$. For instance, the word *but* translates the same as *and* when it links two independent clauses, as in "Jim is tall but he is not heavy." Generally, the word *but* is used in place of *and* when the part of the sentence that follows is, in some way, unexpected. Another example involves the words *neither-nor*. When Shakespeare wrote, "Neither a borrower nor a lender be," he meant, "Do not be a borrower and do not be a lender." So if p and q are statements, then

p but q	means	p and q
neither p nor q	means	$\sim p$ and $\sim q$.

Example 1.1.2 Translating from English to Symbols: *But* and *Neither-Nor*

Write each of the following sentences symbolically, letting $h =$ "It is hot" and $s =$ "It is sunny."

a. It is not hot but it is sunny.

b. It is neither hot nor sunny.

Solution

a. The given sentence is equivalent to "It is not hot and it is sunny," which can be written symbolically as $\sim h \wedge s$.

b. To say it is neither hot nor sunny means that it is not hot and it is not sunny. Therefore, the given sentence can be written symbolically as $\sim h \wedge \sim s$. ∎

Example 1.1.3 Searching on the Internet

Advanced versions of many Internet search engines allow you to use some form of *and, or,* and *not* to refine the search process. For instance, imagine that you want to find web pages about careers in mathematics or computer science but not finance or marketing. With a search engine that uses quotation marks to enclose exact phrases and expresses *and* as AND, *or* as OR, and *not* as NOT, you would write

> Careers AND (mathematics OR "computer science")
> AND NOT (finance OR marketing). ∎

The notation for inequalities involves *and* and *or* statements. For instance, if x, a, and b are particular real numbers, then

$$
\begin{array}{lll}
x \leq a & \text{means} & x < a \quad \text{or} \quad x = a \\
a \leq x \leq b & \text{means} & a \leq x \quad \text{and} \quad x \leq b.
\end{array}
$$

Note that the inequality $2 \leq x \leq 1$ is not satisfied by any real numbers because

$$2 \leq x \leq 1 \quad \text{means} \quad 2 \leq x \quad \text{and} \quad x \leq 1,$$

and this is false no matter what number x happens to be. By the way, the point of specifying x, a, and b to be *particular* real numbers is to ensure that sentences such as "$x < a$" and "$x \geq b$" are either true or false and hence that they are statements.

Example 1.1.4 *And, Or,* and Inequalities

Suppose x is a particular real number. Let p, q, and r symbolize "$0 < x$," "$x < 3$," and "$x = 3$," respectively. Write the following inequalities symbolically:

a. $x \leq 3$

b. $0 < x < 3$

c. $0 < x \leq 3$

Solution

a. $q \vee r$

b. $p \wedge q$

c. $p \wedge (q \vee r)$ ∎

Truth Values

In Examples 1.1.2–1.1.4 we built compound sentences out of component statements and the terms *not, and,* and *or.* If such sentences are to be statements, however, they must have well-defined **truth values**—they must be either true or false. We now define such compound sentences as statements by specifying their truth values in terms of the statements that compose them.

The negation of a statement is a statement that exactly expresses what it would mean for the statement to be false. Therefore, the negation of a statement has opposite truth value from the statement.

> ### • Definition
>
> If p is a statement variable, the **negation** of p is "not p" or "It is not the case that p" and is denoted $\sim p$. It has opposite truth value from p: if p is true, $\sim p$ is false; if p is false, $\sim p$ is true.

The truth values for negation are summarized in a *truth table*.

Truth Table for $\sim p$

p	$\sim p$
T	F
F	T

In ordinary language the sentence "It is hot and it is sunny" is understood to be true when both conditions—being hot and being sunny—are satisfied. If it is hot but not sunny, or sunny but not hot, or neither hot nor sunny, the sentence is understood to be false. The formal definition of truth values for an *and* statement agrees with this general understanding.

> ### • Definition
>
> If p and q are statement variables, the **conjunction** of p and q is "p and q," denoted $p \wedge q$. It is true when, and only when, both p and q are true. If either p or q is false, or if both are false, $p \wedge q$ is false.

The truth values for conjunction can also be summarized in a truth table. The table is obtained by considering the four possible combinations of truth values for p and q. Each combination is displayed in one row of the table; the corresponding truth value for the whole statement is placed in the right-most column of that row. Note that the only row containing a T is the first one since the only way for an *and* statement to be true is for both component statements to be true.

Truth Table for $p \wedge q$

p	q	$p \wedge q$
T	T	T
T	F	F
F	T	F
F	F	F

By the way, the order of truth values for p and q in the table above is TT, TF, FT, FF. It is not necessary to write the truth values in this order, although it is customary to do so. We will use this order for all truth tables involving two statement variables. In Example 1.1.6 we will show the standard order for truth tables that involve three statement variables.

In the case of disjunction—statements of the form "p or q"—intuitive logic offers two alternative interpretations. In ordinary language *or* is sometimes used in an exclusive sense (p or q but not both) and sometimes in an inclusive sense (p or q or both). A waiter who says you may have "coffee, tea, or milk" uses the word *or* in an exclusive sense:

Extra payment is generally required if you want more than one beverage. On the other hand, a waiter who offers "cream or sugar" uses the word *or* in an inclusive sense: You are entitled to both cream and sugar if you wish to have them.

Mathematicians and logicians avoid possible ambiguity about the meaning of the word *or* by understanding it to mean the inclusive "and/or." The symbol ∨ comes from the Latin word *vel,* which means *or* in its inclusive sense. To express the exclusive *or,* the phrase *p or q but not both* is used.

• Definition

If p and q are statement variables, the **disjunction** of p and q is "p or q," denoted $p \vee q$. It is true when either p is true, or q is true, or both p and q are true; it is false only when both p and q are false.

Here is the truth table for disjunction:

Truth Table for $p \vee q$

p	q	$p \vee q$
T	T	T
T	F	T
F	T	T
F	F	F

Note that the statement "$2 \leq 2$" ("2 is less than 2 or 2 equals 2") is true because $2 = 2$.

Evaluating the Truth of More General Compound Statements

Now that truth values have been assigned to $\sim p$, $p \wedge q$, and $p \vee q$, consider the question of assigning truth values to more complicated expressions such as $\sim p \vee q$, $(p \vee q) \wedge \sim (p \wedge q)$, and $(p \wedge q) \vee r$. Such expressions are called *statement forms* (or *propositional forms*). The close relationship between statement forms and *Boolean expressions* is discussed in Section 1.4.

• Definition

A **statement form** (or **propositional form**) is an expression made up of statement variables (such as p, q, and r) and logical connectives (such as $\sim$, $\wedge$, and $\vee$) that becomes a statement when actual statements are substituted for the component statement variables. The **truth table** for a given statement form displays the truth values that correspond to all possible combinations of truth values for its component statement variables.

To compute the truth values for a statement form, follow rules similar to those used to evaluate algebraic expressions. For each combination of truth values for the statement variables, first evaluate the expressions within the innermost parentheses, then evaluate the expressions within the next innermost set of parentheses, and so forth until you have the truth values for the complete expression.

Example 1.1.5 Truth Table for *Exclusive Or*

Construct the truth table for the statement form $(p \vee q) \wedge \sim(p \wedge q)$. Note that when *or* is used in its exclusive sense, the statement "*p* or *q*" means "*p* or *q* but not both" or "*p* or *q* and not both *p* and *q*," which translates into symbols as $(p \vee q) \wedge \sim(p \wedge q)$. This is sometimes abbreviated $p \oplus q$ or p XOR q.

Solution   Set up columns labeled $p, q, p \vee q, p \wedge q, \sim(p \wedge q)$, and $(p \vee q) \wedge \sim(p \wedge q)$. Fill in the p and q columns with all the logically possible combinations of T's and F's. Then use the truth tables for $\vee$ and $\wedge$ to fill in the $p \vee q$ and $p \wedge q$ columns with the appropriate truth values. Next fill in the $\sim(p \wedge q)$ column by taking the opposites of the truth values for $p \wedge q$. For example, the entry for $\sim(p \wedge q)$ in the first row is F because in the first row the truth value of $p \wedge q$ is T. Finally, fill in the $(p \vee q) \wedge \sim(p \wedge q)$ column by considering the truth table for an *and* statement together with the computed truth values for $p \vee q$ and $\sim(p \wedge q)$. For example, the entry in the first row is F because the entry for $p \vee q$ is T, the entry for $\sim(p \wedge q)$ is F, and an *and* statement is false unless both components are true. The entry in the second row is T because both components are true in this row.

Truth Table for *Exclusive Or:* $(p \vee q) \wedge \sim(p \wedge q)$

p	q	$p \vee q$	$p \wedge q$	$\sim(p \wedge q)$	$(p \vee q) \wedge \sim(p \wedge q)$
T	T	T	T	F	F
T	F	T	F	T	T
F	T	T	F	T	T
F	F	F	F	T	F

Example 1.1.6 Truth Table for $(p \wedge q) \vee \sim r$

Construct a truth table for the statement form $(p \wedge q) \vee \sim r$.

Solution   Make columns headed $p, q, r, p \wedge q, \sim r$, and $(p \wedge q) \vee \sim r$. Since there are eight logically possible combinations of truth values for p, q, and r, enter these in the three left-most columns. Then fill in the truth values for $p \wedge q$ and for $\sim r$. Complete the table by considering the truth values for $(p \wedge q)$ and for $\sim r$ and the definition of an *or* statement. Since an *or* statement is false only when both components are false, the only rows in which the entry is F are the third, fifth, and seventh rows because those are the only rows in which the expressions $p \wedge q$ and $\sim r$ are both false. The entry for all the other rows is T.

p	q	r	$p \wedge q$	$\sim r$	$(p \wedge q) \vee \sim r$
T	T	T	T	F	T
T	T	F	T	T	T
T	F	T	F	F	F
T	F	F	F	T	T
F	T	T	F	F	F
F	T	F	F	T	T
F	F	T	F	F	F
F	F	F	F	T	T

The essential point about assigning truth values to compound statements is that it allows you—using logic alone—to judge the truth of a compound statement on the basis of your knowledge of the truth of its component parts. Logic does not help you determine the truth or falsity of the component statements. Rather, logic helps link these separate pieces of information together into a coherent whole.

Logical Equivalence

The statements

6 is greater than 2   and   2 is less than 6

are two different ways of saying the same thing. Why? Because of the definition of the phrases *greater than* and *less than.* By contrast, although the statements

(1) Dogs bark and cats meow   and   (2) Cats meow and dogs bark

are also two different ways of saying the same thing, the reason has nothing to do with the definition of the words. It has to do with the logical form of the statements. Any two statements whose logical forms are related in the same way as (1) and (2) would either both be true or both be false. You can see this by examining the following truth table, where the statement variables p and q are substituted for the component statements "Dogs bark" and "Cats meow," respectively. The table shows that for each combination of truth values for p and q, $p \wedge q$ is true when, and only when, $q \wedge p$ is true. In such a case, the statement forms are called *logically equivalent,* and we say that (1) and (2) are *logically equivalent statements.*

p	q	$p \wedge q$	$q \wedge p$
T	T	T	T
T	F	F	F
F	T	F	F
F	F	F	F

$p \wedge q$ and $q \wedge p$ always have the same truth values, so they are logically equivalent

• Definition

Two *statement forms* are called **logically equivalent** if, and only if, they have identical truth values for each possible substitution of statements for their statement variables. The logical equivalence of statement forms P and Q is denoted by writing $P \equiv Q$.

    Two *statements* are called **logically equivalent** if, and only if, they have logically equivalent forms when identical component statement variables are used to replace identical component statements.

Testing Whether Two Statement Forms P and Q Are Logically Equivalent

1. Construct a truth table for P with one column for the truth values of P and another column for the truth values of Q.

2. Check each combination of truth values of the statement variables to see whether the truth value of P is the same as the truth value of Q.

   a. If in each row the truth value of P is the same as the truth value of Q, then P and Q are logically equivalent.

   b. If in some row P has a different truth value from Q, then P and Q are not logically equivalent.

Example 1.1.7 Double Negative Property: $\sim(\sim p) \equiv p$

Check that the negation of the negation of a statement is logically equivalent to the statement.

Solution

p	$\sim p$	$\sim(\sim p)$
T	F	T
F	T	F

p and $\sim(\sim p)$ always have the same truth values, so they are logically equivalent

An alternate way to show that statement forms P and Q are not logically equivalent is to find concrete statements of each form, one of which is true and the other of which is false. This method is illustrated in part (b) of Example 1.1.8.

Example 1.1.8 Showing Nonequivalence

Show that the statement forms $\sim(p \wedge q)$ and $\sim p \wedge \sim q$ are not logically equivalent.

Solution

a. One way to show this is to use truth tables.

p	q	$\sim p$	$\sim q$	$p \wedge q$	$\sim(p \wedge q)$		$\sim p \wedge \sim q$
T	T	F	F	T	F		F
T	F	F	T	F	T	$\neq$	F
F	T	T	F	F	T	$\neq$	F
F	F	T	T	F	T		T

$\sim(p \wedge q)$ and $\sim p \wedge \sim q$ have different truth values in rows 2 and 3, so they are not logically equivalent

b. A second way to show that $\sim(p \wedge q)$ and $\sim p \wedge \sim q$ are not logically equivalent is by example. Let p be the statement "$0 < 1$" and let q be the statement "$1 < 0$." Then

$$\sim(p \wedge q) \quad \text{is} \quad \text{"It is not the case that both } 0 < 1 \text{ and } 1 < 0,\text{"}$$

which is true. On the other hand,

$$\sim p \wedge \sim q \quad \text{is} \quad \text{"} 0 \not< 1 \quad \text{and} \quad 1 \not< 0,\text{"}$$

which is false. This example shows that there are concrete statements you can substitute for p and q to make one of the statement forms true and the other false. Therefore, the statement forms are not logically equivalent. ∎

Example 1.1.9 Negations of *And* and *Or:* De Morgan's Laws

For the statement "John is tall and Jim is redheaded" to be true, both components must be true. So for the statement to be false, one or both components must be false. Thus the negation can be written as "John is not tall or Jim is not redheaded." In general, the negation of the conjunction of two statements is logically equivalent to the disjunction of their negations. That is, statements of the forms $\sim(p \wedge q)$ and $\sim p \vee \sim q$ are logically equivalent. Check this using truth tables.

Solution

p	q	$\sim p$	$\sim q$	$p \wedge q$	$\sim(p \wedge q)$	$\sim p \vee \sim q$
T	T	F	F	T	F	F
T	F	F	T	F	T	T
F	T	T	F	F	T	T
F	F	T	T	F	T	T

↑   ↑

$\sim(p \wedge q)$ and $\sim p \vee \sim q$ always have the same truth values, so they are logically equivalent

Symbolically,

$$\sim(p \wedge q) \equiv \sim p \vee \sim q.$$

In the exercises at the end of this section you are asked to show the analogous law that the negation of the disjunction of two statements is logically equivalent to the conjunction of their negations:

$$\sim(p \vee q) \equiv \sim p \wedge \sim q.$$

∎

The two logical equivalences of Example 1.1.9 are known as **De Morgan's laws** of logic in honor of Augustus De Morgan, who was the first to state them in formal mathematical terms.

Augustus De Morgan
(1806–1871)

Culver Pictures

> ### De Morgan's Laws
>
> The negation of an *and* statement is logically equivalent to the *or* statement in which each component is negated.
>
> The negation of an *or* statement is logically equivalent to the *and* statement in which each component is negated.

Example 1.1.10 Applying De Morgan's Laws

Write negations for each of the following statements:

a. John is 6 feet tall and he weighs at least 200 pounds.

b. The bus was late or Tom's watch was slow.

Solution

a. John is not 6 feet tall or he weighs less than 200 pounds.

b. The bus was not late and Tom's watch was not slow.

Since the statement "neither p nor q" means the same as "$\sim p$ and $\sim q$," an alternative answer for (b) is "Neither was the bus late nor was Tom's watch slow." ∎

If x is a particular real number, saying that x is not less than 2 ($x \not< 2$) means that x does not lie to the left of 2 on the number line. This is equivalent to saying that either $x = 2$ or x lies to the right of 2 on the number line ($x = 2$ or $x > 2$). Hence,

$$x \not< 2 \quad \text{is equivalent to} \quad x \geq 2.$$

Pictorially,

If $x \not< 2$, then x lies in here.

Similarly,

$$x \not> 2 \quad \text{is equivalent to} \quad x \leq 2,$$
$$x \not\leq 2 \quad \text{is equivalent to} \quad x > 2, \text{ and}$$
$$x \not\geq 2 \quad \text{is equivalent to} \quad x < 2.$$

Example 1.1.11 Inequalities and De Morgan's Laws

Use De Morgan's laws to write the negation of $-1 < x \leq 4$.

Solution   The given statement is equivalent to

$$-1 < x \quad \text{and} \quad x \leq 4.$$

By De Morgan's laws, the negation is

$$-1 \not< x \quad \text{or} \quad x \not\leq 4,$$

which is equivalent to

$$-1 \geq x \quad \text{or} \quad x > 4.$$

Pictorially, if $-1 \geq x$ or $x > 4$, then x lies in the shaded region of the number line, as shown below.

De Morgan's laws are frequently used in writing computer programs. For instance, suppose you want your program to delete all files modified outside a certain range of dates, say from date 1 through date 2 inclusive. You would use the fact that

$$\sim(date1 \leq file\_modification\_date \leq date2)$$

is equivalent to

$$(file\_modification\_date < date1) \quad \text{or} \quad (date2 < file\_modification\_date).$$

Example 1.1.12 A Cautionary Example

According to De Morgan's laws, the negation of

$$p:\text{ Jim is tall and Jim is thin}$$

is

$$\sim p:\text{ Jim is not tall or Jim is not thin}$$

because the negation of an *and* statement is the *or* statement in which the two components are negated.

Unfortunately, a potentially confusing aspect of the English language can arise when you are taking negations of this kind. Note that statement p can be written more compactly as

$$p':\text{ Jim is tall and thin.}$$

When it is so written, another way to negate it is

$$\sim(p'):\text{ Jim is not tall and thin.}$$

But in this form the negation looks like an *and* statement. Doesn't that violate De Morgan's laws?

Actually no violation occurs. The reason is that in formal logic the words *and* and *or* are allowed only between complete statements, not between sentence fragments.

One lesson to be learned from this example is that when you apply De Morgan's laws, you must have complete statements on either side of each *and* and on either side of each *or*. A deeper lesson is this:

> ⚠️ *Caution!*   Although the laws of logic are extremely useful, they should be used as an *aid* to thinking, not as a mechanical substitute for it.

∎

Tautologies and Contradictions

It has been said that all of mathematics reduces to tautologies. Although this is formally true, most working mathematicians think of their subject as having substance as well as form. Nonetheless, an intuitive grasp of basic logical tautologies is part of the equipment of anyone who reasons with mathematics.

> **• Definition**
>
> A **tautology** is a statement form that is always true regardless of the truth values of the individual statements substituted for its statement variables. A statement whose form is a tautology is a **tautological statement.**
>
> A **contradiction** is a statement form that is always false regardless of the truth values of the individual statements substituted for its statement variables. A statement whose form is a contradiction is a **contradictory statement.**

According to this definition, the truth of a tautological statement and the falsity of a contradictory statement are due to the logical structure of the statements themselves and are independent of the meanings of the statements.

Example 1.1.13 Tautologies and Contradictions

Show that the statement form $p \vee \sim p$ is a tautology and that the statement form $p \wedge \sim p$ is a contradiction.

Solution

p	$\sim p$	$p \vee \sim p$	$p \wedge \sim p$
T	F	T	F
F	T	T	F

                                    ↑   ↑

all T's so   all F's so
$p \vee \sim p$ is $p \wedge \sim p$ is a
a tautology   contradiction

Example 1.1.14 Logical Equivalence Involving Tautologies and Contradictions

If **t** is a tautology and **c** is a contradiction, show that $p \wedge \mathbf{t} \equiv p$ and $p \wedge \mathbf{c} \equiv \mathbf{c}$.

Solution

p	t	$p \wedge t$	p	c	$p \wedge c$
T	T	T	T	F	F
F	T	F	F	F	F

             ↑ ‾‾‾ ↑   ↑ ‾‾‾ ↑

same truth   same truth
values, so   values, so
$p \wedge t \equiv p$ $p \wedge \mathbf{c} \equiv \mathbf{c}$

Summary of Logical Equivalences

Knowledge of logically equivalent statements is very useful for constructing arguments. It often happens that it is difficult to see how a conclusion follows from one form of a statement, whereas it is easy to see how it follows from a logically equivalent form of the statement. A number of logical equivalences are summarized in Theorem 1.1.1 for future reference.

Theorem 1.1.1 Logical Equivalences

Given any statement variables p, q, and r, a tautology $\mathbf{t}$ and a contradiction $\mathbf{c}$, the following logical equivalences hold.

1. *Commutative laws:* $p \wedge q \equiv q \wedge p$ $p \vee q \equiv q \vee p$

2. *Associative laws:* $(p \wedge q) \wedge r \equiv p \wedge (q \wedge r)$ $(p \vee q) \vee r \equiv p \vee (q \vee r)$

3. *Distributive laws:* $p \wedge (q \vee r) \equiv (p \wedge q) \vee (p \wedge r)$ $p \vee (q \wedge r) \equiv (p \vee q) \wedge (p \vee r)$

4. *Identity laws:* $p \wedge \mathbf{t} \equiv p$ $p \vee \mathbf{c} \equiv p$

5. *Negation laws:* $p \vee {\sim}p \equiv \mathbf{t}$ $p \wedge {\sim}p \equiv \mathbf{c}$

6. *Double negative law:* ${\sim}({\sim}p) \equiv p$

7. *Idempotent laws:* $p \wedge p \equiv p$ $p \vee p \equiv p$

8. *Universal bound laws:* $p \vee \mathbf{t} \equiv \mathbf{t}$ $p \wedge \mathbf{c} \equiv \mathbf{c}$

9. *De Morgan's laws:* ${\sim}(p \wedge q) \equiv {\sim}p \vee {\sim}q$ ${\sim}(p \vee q) \equiv {\sim}p \wedge {\sim}q$

10. *Absorption laws:* $p \vee (p \wedge q) \equiv p$ $p \wedge (p \vee q) \equiv p$

11. *Negations of $\mathbf{t}$ and $\mathbf{c}$:* ${\sim}\mathbf{t} \equiv \mathbf{c}$ ${\sim}\mathbf{c} \equiv \mathbf{t}$

The proofs of laws 4 and 6, the first parts of laws 1 and 5, and the second part of law 9 have already been given as examples in the text. Proofs of the other parts of the theorem are left as exercises. In fact, it can be shown that the first five laws of Theorem 1.1.1 form a core from which the other laws can be derived. The first five laws are the axioms for a mathematical structure known as a Boolean algebra, which is discussed in Section 5.3.

The equivalences of Theorem 1.1.1 are general laws of thought that occur in all areas of human endeavor. They can also be used in a formal way to rewrite complicated statement forms more simply.

Example 1.1.15 Simplifying Statement Forms

Use Theorem 1.1.1 to verify the logical equivalence

$$ {\sim}({\sim}p \wedge q) \wedge (p \vee q) \equiv p. $$

Solution   Use the laws of Theorem 1.1.1 to replace sections of the statement form on the left by logically equivalent expressions. Each time you do this, you obtain a logically equivalent statement form. Continue making replacements until you obtain the statement form on the right.

$$
\begin{aligned}
{\sim}({\sim}p \wedge q) \wedge (p \vee q) &\equiv ({\sim}({\sim}p) \vee {\sim}q) \wedge (p \vee q) && \text{by De Morgan's laws} \\
&\equiv (p \vee {\sim}q) \wedge (p \vee q) && \text{by the double negative law} \\
&\equiv p \vee ({\sim}q \wedge q) && \text{by the distributive law} \\
&\equiv p \vee (q \wedge {\sim}q) && \text{by the commutative law for } \wedge \\
&\equiv p \vee \mathbf{c} && \text{by the negation law} \\
&\equiv p && \text{by the identity law} \quad \blacksquare
\end{aligned}
$$

Skill in simplifying statement forms is useful in constructing logically efficient computer programs and in designing digital logic circuits.

Although the properties in Theorem 1.1.1 can be used to prove the logical equivalence of two statement forms, they cannot be used to prove that statement forms are not logically equivalent. On the other hand, truth tables can always be used to determine both equivalence and nonequivalence, and truth tables are easy to program on a computer. When truth tables are used, however, checking for equivalence always requires 2^n steps, where n is the number of variables. Sometimes you can quickly see that two statement forms are equivalent by Theorem 1.1.1, whereas it would take quite a bit of calculating to show their equivalence using truth tables. For instance, it follows immediately from the associative law for $\wedge$ that $p \wedge (\sim q \wedge \sim r) \equiv (p \wedge \sim q) \wedge \sim r$, whereas a truth table verification requires constructing a table with eight rows.

Exercise Set 1.1

Appendix B contains either full or partial solutions to all exercises with blue numbers. When the solution is not complete, the exercise number has an **H** next to it. A ✶ next to an exercise number signals that the exercise is more challenging than usual. Be careful not to get into the habit of turning to the solutions too quickly. Make every effort to work exercises on your own before checking your answers. See the Preface for additional sources of assistance and further study.

In each of 1–4 represent the common form of each argument using letters to stand for component sentences, and fill in the blanks so that the argument in part (b) has the same logical form as the argument in part (a).

1. a. If all integers are rational, then the number 1 is rational.
   All integers are rational.
   Therefore, the number 1 is rational.
   b. If all algebraic expressions can be written in prefix notation, then _____.
   _____.
   Therefore, $(a + 2b)(a^2 - b)$ can be written in prefix notation.

2. a. If all computer programs contain errors, then this program contains an error.
   This program does not contain an error.
   Therefore, it is not the case that all computer programs contain errors.
   b. If _____, then _____.
   2 is not odd.
   Therefore, it is not the case that all prime numbers are odd.

3. a. This number is even or this number is odd.
   This number is not even.
   Therefore, this number is odd.
   b. _____ or logic is confusing.
   My mind is not shot.
   Therefore, _____.

4. a. If n is divisible by 6, then n is divisible by 3.
   If n is divisible by 3, then the sum of the digits of n is divisible by 3.
   Therefore, if n is divisible by 6, then the sum of the digits of n is divisible by 3.
   (Assume that n is a particular, fixed integer.)
   b. If _____,
   then the guard condition for the **while** loop is false.
   If _____,
   then program execution moves to the next instruction following the loop.

Therefore, if x equals 0, then _____.
(Assume that x is a particular variable in a particular computer program.)

5. Indicate which of the following sentences are statements.
   a. 1,024 is the smallest four-digit number that is a perfect square.
   b. She is a mathematics major.
   c. $128 = 2^6$ d. $x = 2^6$

Write the statements in 6–9 in symbolic form using the symbols $\sim$, $\vee$, and $\wedge$ and the indicated letters to represent component statements.

6. Let s = "stocks are increasing" and i = "interest rates are steady."
   a. Stocks are increasing but interest rates are steady.
   b. Neither are stocks increasing nor are interest rates steady.

7. Juan is a math major but not a computer science major. (m = "Juan is a math major," c = "Juan is a computer science major")

8. Let h = "John is healthy," w = "John is wealthy," and s = "John is wise."
   a. John is healthy and wealthy but not wise.
   b. John is not wealthy but he is healthy and wise.
   c. John is neither healthy, wealthy, nor wise.
   d. John is neither wealthy nor wise, but he is healthy.
   e. John is wealthy, but he is not both healthy and wise.

9. Either Olga will go out for tennis or she will go out for track but not both. (n = "Olga will go out for tennis," k = "Olga will go out for track")

10. Let p be the statement "DATAENDFLAG is off," q the statement "ERROR equals 0," and r the statement "SUM is less than 1,000." Express the following sentences in symbolic notation.
   a. DATAENDFLAG is off, ERROR equals 0, and SUM is less than 1,000.
   b. DATAENDFLAG is off but ERROR is not equal to 0.
   c. DATAENDFLAG is off; however ERROR is not 0 or SUM is greater than or equal to 1,000.

d. DATAENDFLAG is on and ERROR equals 0 but SUM is greater than or equal to 1,000.

e. Either DATAENDFLAG is on or it is the case that both ERROR equals 0 and SUM is less than 1,000.

11. In the following sentence, is the word *or* used in its inclusive or exclusive sense? A team wins the playoffs if it wins two games in a row or a total of three games.

In 12 and 13, imagine that you are searching the Internet using a search engine that uses AND for *and,* NOT for *not,* and OR for *or.*

12. You are trying to find the name of the fourteenth president of the United States of America. Write a logical expression to find Web pages containing the following: "United States president" and either "14th" or "fourteenth" but not "amendment" (to avoid pages about the Fourteenth Amendment to the United States Constitution).

13. You recall that the fastest mammal on earth is either a jaguar or a cheetah. To find a Web page to tell you which one is the fastest, write a logical expression containing "jaguar" and "cheetah,"and either "speed" or "fastest" but not "car," or "automobile," or "auto" (to avoid pages about the Jaguar automobile).

Write truth tables for the statement forms in 14–18.

14. $\sim p \wedge q$

15. $\sim(p \wedge q) \vee (p \vee q)$

16. $p \wedge (q \wedge r)$

17. $p \wedge (\sim q \vee r)$

H 18. $(p \vee (\sim p \vee q)) \wedge \sim(q \wedge \sim r)$

Determine which of the pairs of statement forms in 19–28 are logically equivalent. Justify your answers using truth tables and include a few words of explanation. Read **t** to be a tautology and **c** to be a contradiction.

19. $p \vee (p \wedge q)$ and p

20. $\sim(p \wedge q)$ and $\sim p \wedge \sim q$

21. $p \vee \mathbf{t}$ and $\mathbf{t}$

22. $p \wedge \mathbf{t}$ and p

23. $(p \wedge q) \wedge r$ and $p \wedge (q \wedge r)$

24. $p \wedge (q \vee r)$ and $(p \wedge q) \vee (p \wedge r)$

25. $(p \wedge q) \vee r$ and $p \wedge (q \vee r)$

26. $(p \vee q) \vee (p \wedge r)$ and $(p \vee q) \wedge r$

27. $((\sim p \vee q) \wedge (p \vee \sim r)) \wedge (\sim p \vee \sim q)$ and $\sim(p \vee r)$

28. $(r \vee p) \wedge ((\sim r \vee (p \wedge q)) \wedge (r \vee q))$ and $p \wedge q$

Use De Morgan's laws to write negations for the statements in 29–34.

29. Hal is a math major and Hal's sister is a computer science major.

30. Sam is an orange belt and Kate is a red belt.

31. The connector is loose or the machine is unplugged.

32. This computer program has a logical error in the first ten lines or it is being run with an incomplete data set.

33. The dollar is at an all-time high and the stock market is at a record low.

34. The train is late or my watch is fast.

Assume x is a particular real number and use De Morgan's laws to write negations for the statements in 35–38.

35. $-2 < x < 7$

36. $-10 < x < 2$

37. $1 > x \geq -3$

38. $0 > x \geq -7$

In 39 and 40, imagine that *num_orders* and *num_instock* are particular values, such as might occur during execution of a computer program. Write negations for the following statements.

39. (*num_orders* > 100 and *num_instock* ≤ 500) or *num_instock* < 200

40. (*num_orders* < 50 and *num_instock* > 300) or ($50 \leq$ *num_orders* < 75 and *num_instock* > 500)

Use truth tables to establish which of the statement forms in 41–44 are tautologies and which are contradictions.

41. $(p \wedge q) \vee (\sim p \vee (p \wedge \sim q))$

42. $(p \wedge \sim q) \wedge (\sim p \vee q)$

43. $((\sim p \wedge q) \wedge (q \wedge r)) \wedge \sim q$

44. $(\sim p \vee q) \vee (p \wedge \sim q)$

In 45 and 46 below, a logical equivalence is derived from Theorem 1.1.1. Supply a reason for each step.

45. $(p \wedge \sim q) \vee (p \wedge q) \equiv p \wedge (\sim q \vee q)$ by (a) _____
$\equiv p \wedge (q \vee \sim q)$ by (b) _____
$\equiv p \wedge \mathbf{t}$ by (c) _____
$\equiv p$ by (d) _____

Therefore, $(p \wedge \sim q) \vee (p \wedge q) \equiv p$.

46. $(p \vee \sim q) \wedge (\sim p \vee \sim q)$
$\equiv (\sim q \vee p) \wedge (\sim q \vee \sim p)$ by (a) _____
$\equiv \sim q \vee (p \wedge \sim p)$ by (b) _____
$\equiv \sim q \vee \mathbf{c}$ by (c) _____
$\equiv \sim q$ by (d) _____

Therefore, $(p \vee \sim q) \wedge (\sim p \vee \sim q) \equiv \sim q$.

Use Theorem 1.1.1 to verify the logical equivalences in 47–51. Supply a reason for each step.

47. $(p \wedge \sim q) \vee p \equiv p$

48. $p \wedge (\sim q \vee p) \equiv p$

49. $\sim(p \vee \sim q) \vee (\sim p \wedge \sim q) \equiv \sim p$

50. $\sim((\sim p \wedge q) \vee (\sim p \wedge \sim q)) \vee (p \wedge q) \equiv p$

51. $(p \wedge (\sim(\sim p \vee q))) \vee (p \wedge q) \equiv p$

✱ 52. In Example 1.1.5, the symbol ⊕ was introduced to denote *exclusive or*, so $p \oplus q \equiv (p \vee q) \wedge \sim(p \wedge q)$. Hence the truth table for *exclusive or* is as follows:

p	q	$p \oplus q$
T	T	F
T	F	T
F	T	T
F	F	F

  a. Find simpler statement forms that are logically equivalent to $p \oplus p$ and $(p \oplus p) \oplus p$.

  b. Is $(p \oplus q) \oplus r \equiv p \oplus (q \oplus r)$? Justify your answer.

  c. Is $(p \oplus q) \wedge r \equiv (p \wedge r) \oplus (q \wedge r)$? Justify your answer.

✱ **53.** In logic and in standard English, a double negative is equivalent to a positive. Is there any English usage in which a double positive is equivalent to a negative? Explain.

✱ 54. The rules for a certain frequent-flyer club include the following statements: "Any member who fails to earn any mileage during the first twelve months after enrollment in the program may be removed from the program. Except as otherwise provided, any member who fails at any time to earn mileage for a period of three consecutive years is subject to termination of his or her membership and forfeiture of all accrued mileage. Notwithstanding this provision, no pre-July 1, 2004, member who has earned mileage (other than enrollment bonus) prior to July 1, 2005, shall be subject under this provision to the termination of his or her membership and to the cancellation of mileage accrued prior to July 1, 2005, until the amount of such mileage falls below 10,000 miles (the amount necessary for the lowest available award under the structure in place as of June 30, 2004), or until December 15, 2015, whichever comes first."

Let x be a particular member of this club, and let

$p =$ "x fails to earn mileage during the first twelve months after enrollment,"

$q =$ "x fails to earn mileage for a period of three consecutive years,"

$r =$ "x became a member prior to July 1, 2004,"

$s =$ "x currently has at least 10,000 miles for pre-July 1, 2005, mileage (not including enrollment bonus miles),"

$t =$ "the current date is prior to December 15, 2015."

Use symbols to write the complete condition under which x's membership may be terminated.

1.2 Conditional Statements

. . . hypothetical reasoning implies the subordination of the real to the realm of the possible . . . — Jean Piaget, 1972

When you make a logical inference or deduction, you reason *from* a hypothesis *to* a conclusion. Your aim is to be able to say, "*If* such and such is known, *then* something or other must be the case."

Let p and q be statements. A sentence of the form "If p then q" is denoted symbolically by "$p \rightarrow q$"; p is called the *hypothesis* and q is called the *conclusion*. For instance, in

If 4,686 is divisible by 6, then 4,686 is divisible by 3

the hypothesis is "4,686 is divisible by 6" and the conclusion is "4,686 is divisible by 3." Such a sentence is called *conditional* because the truth of statement q is conditioned on the truth of statement p.

The notation $p \rightarrow q$ indicates that $\rightarrow$ is a connective, like $\wedge$ or $\vee$, that can be used to join statements to create new statements. To define $p \rightarrow q$ as a statement, therefore, we must specify the truth values for $p \rightarrow q$ as we specified truth values for $p \wedge q$ and for $p \vee q$. As is the case with the other connectives, the formal definition of truth values for $\rightarrow$ (if-then) is based on its everyday, intuitive meaning. Consider an example.

Suppose you go to interview for a job at a store and the owner of the store makes you the following promise:

If you show up for work Monday morning, then you will get the job.

Under what circumstances are you justified in saying the owner spoke falsely? That is, under what circumstances is the above sentence false? The answer is: You *do* show up for work Monday morning and you do *not* get the job.

After all, the owner's promise only says you will get the job *if* a certain condition (showing up for work Monday morning) is met; it says nothing about what will happen if the condition is *not* met. So if the condition is not met, you cannot in fairness say the promise is false regardless of whether or not you get the job.

The above example was intended to convince you that *the only combination of circumstances in which you would call a conditional sentence false occurs when the hypothesis is true and the conclusion is false.* In all other cases, you would not call the sentence false. This implies that the only row of the truth table for $p \rightarrow q$ that should be filled in with an F is the row where p is T and q is F. No other row should contain an F. But each row of a truth table must be filled in with either a T or an F. Thus all other rows of the truth table for $p \rightarrow q$ must be filled in with T's.

Truth Table for $p \rightarrow q$

p	q	$p \rightarrow q$
T	T	T
T	F	F
F	T	T
F	F	T

• Definition

If p and q are statement variables, the **conditional** of q by p is "If p then q" or "p implies q" and is denoted $p \rightarrow q$. It is false when p is true and q is false; otherwise it is true. We call p the **hypothesis** (or **antecedent**) of the conditional and q the **conclusion** (or **consequent**).

A conditional statement that is true by virtue of the fact that its hypothesis is false is often called **vacuously true** or **true by default.** Thus the statement "If you show up for work Monday morning, then you will get the job" is vacuously true if you do not show up for work Monday morning.

In expressions that include $\rightarrow$ as well as other logical operators such as $\wedge$, $\vee$, and $\sim$, the **order of operations** is that $\rightarrow$ is performed last. Thus, according to the specification of order of operations in Section 1.1, $\sim$ is performed first, then $\wedge$ and $\vee$, and finally $\rightarrow$.

The philosopher Willard VanOrman Quine advises against using the phrase "p implies q" to mean "$p \rightarrow q$" because the word *implies* suggests that q can be logically deduced from p and this is often not the case. Nonetheless, the phrase is used by many people, probably because it is a convenient replacement for the $\rightarrow$ symbol.

Example 1.2.1 Truth Table for $p \vee \sim q \rightarrow \sim p$

Construct a truth table for the statement form $p \vee \sim q \rightarrow \sim p$.

Solution By the order of operations given above, $p \vee \sim q \rightarrow \sim p$ means $(p \vee (\sim q)) \rightarrow (\sim p)$, and this order governs the construction of the truth table. First fill in the four possible combinations of truth values for p and q, and then enter the truth values for $\sim p$ and $\sim q$ using the definition of negation. Next fill in the $p \vee \sim q$ column using the definition of $\vee$. Finally, fill in the $p \vee \sim q \rightarrow \sim p$ column using the definition of $\rightarrow$. The only rows in which the hypothesis $p \vee \sim q$ is true and the conclusion $\sim p$ is false are the first and second rows. So you put F's in those two rows and T's in the other two rows.

		conclusion		hypothesis	
p	q	$\sim p$	$\sim q$	$p \vee \sim q$	$p \vee \sim q \rightarrow \sim p$
T	T	F	F	T	F
T	F	F	T	T	F
F	T	T	F	F	T
F	F	T	T	T	T

Logical Equivalences Involving $\rightarrow$

Imagine that you are trying to solve a problem involving three statements: p, q, and r. Suppose you know that the truth of r follows from the truth of p *and also* that the truth of r follows from the truth of q. Then no matter whether p or q is the case, the truth of r must follow. The division-into-cases method of analysis is based on this idea.

Example 1.2.2 Division into Cases: Showing that $p \vee q \rightarrow r \equiv (p \rightarrow r) \wedge (q \rightarrow r)$

Use truth tables to show the logical equivalence of the statement forms $p \vee q \rightarrow r$ and $(p \rightarrow r) \wedge (q \rightarrow r)$.

Solution   First fill in the eight possible combinations of truth values for p, q, and r. Then fill in the columns for $p \vee q$, $p \rightarrow r$, and $q \rightarrow r$ using the definitions of *or* and *if-then*. For instance, the $p \rightarrow r$ column has F's in the second and fourth rows because these are the rows in which p is true and q is false. Next fill in the $p \vee q \rightarrow r$ column using the definition of *if-then*. The rows in which the hypothesis $p \vee q$ is true and the conclusion r is false are the second, fourth, and sixth. So F's go in these rows and T's in all the others. The complete table shows that $p \vee q \rightarrow r$ and $(p \rightarrow r) \wedge (q \rightarrow r)$ have the same truth values for each combination of truth values of p, q, and r. Hence the two statement forms are logically equivalent.

p	q	r	$p \vee q$	$p \rightarrow r$	$q \rightarrow r$	$p \vee q \rightarrow r$	$(p \rightarrow r) \wedge (q \rightarrow r)$
T	T	T	T	T	T	T	T
T	T	F	T	F	F	F	F
T	F	T	T	T	T	T	T
T	F	F	T	F	T	F	F
F	T	T	T	T	T	T	T
F	T	F	T	T	F	F	F
F	F	T	F	T	T	T	T
F	F	F	F	T	T	T	T

$p \vee q \rightarrow r$ and $(p \rightarrow r) \wedge (q \rightarrow r)$ always have the same truth values, so they are logically equivalent

Representation of If-Then *As* Or

In exercise 13(a) at the end of this section you are asked to use truth tables to show that

$$p \to q \equiv \sim p \vee q.$$

The logical equivalence of "if p then q" and "not p or q" is occasionally used in everyday speech. Here is one instance.

Example 1.2.3 Application of the Equivalence between $\sim p \vee q$ and $p \to q$

Rewrite the following statement in if-then form.

> Either you get to work on time or you are fired.

Solution   Let $\sim p$ be

> You get to work on time.

and q be

> You are fired.

Then the given statement is $\sim p \vee q$. Also p is

> You do not get to work on time.

So the equivalent if-then version, $p \to q$, is

> If you do not get to work on time, then you are fired.   ■

The Negation of a Conditional Statement

By definition, $p \to q$ is false if, and only if, its hypothesis, p, is true and its conclusion, q, is false. It follows that

> The negation of "if p then q" is logically equivalent to "p and not q."

This can be restated symbolically as follows:

$$\sim(p \to q) \equiv p \wedge \sim q$$

You can also obtain this result by starting from the logical equivalence $p \to q \equiv \sim p \vee q$. Take the negation of both sides to obtain

$$
\begin{aligned}
\sim(p \to q) &\equiv \sim(\sim p \vee q) \\
&\equiv \sim(\sim p) \wedge (\sim q) &&\text{by De Morgan's laws} \\
&\equiv p \wedge \sim q &&\text{by the double negative law.}
\end{aligned}
$$

Yet another way to derive this result is to construct truth tables for $\sim(p \to q)$ and for $p \wedge \sim q$ and to check that they have the same truth values. (See exercise 13(b) at the end of this section.)

Example 1.2.4 Negations of *If-Then* Statements

Write negations for each of the following statements:

a. If my car is in the repair shop, then I cannot get to class.

b. If Sara lives in Athens, then she lives in Greece.

Solution

a. My car is in the repair shop and I can get to class.

b. Sara lives in Athens and she does not live in Greece. (Sara might live in Athens, Georgia; Athens, Ohio; or Athens, Wisconsin.) ■

> **Caution!**   It is tempting to write the negation of an if-then statement as another if-then statement. Please resist that temptation! Remember that **the negation of an if-then statement does not start with the word if.**

The Contrapositive of a Conditional Statement

One of the most fundamental laws of logic is the equivalence between a conditional statement and its contrapositive.

• Definition

The **contrapositive** of a conditional statement of the form "If p then q" is

$$\text{If } \sim q \text{ then } \sim p.$$

Symbolically,

$$\text{The contrapositive of } p \rightarrow q \text{ is } \sim q \rightarrow \sim p.$$

The fact is that

> A conditional statement is logically equivalent to its contrapositive.

You are asked to establish this equivalence in exercise 26 at the end of this section.

Example 1.2.5 Writing the Contrapositive

Write each of the following statements in its equivalent contrapositive form:

a. If Howard can swim across the lake, then Howard can swim to the island.

b. If today is Easter, then tomorrow is Monday.

Solution

a. If Howard cannot swim to the island, then Howard cannot swim across the lake.

b. If tomorrow is not Monday, then today is not Easter. ■

When you are trying to solve certain problems, you may find that the contrapositive form of a conditional statement is easier to work with than the original statement. Replacing a statement by its contrapositive may give the extra push that helps you over the top in your search for a solution. This logical equivalence is also the basis for one of the most important laws of deduction, modus tollens (to be explained in Section 1.3), and for the contrapositive method of proof (to be explained in Section 3.6).

The Converse and Inverse of a Conditional Statement

The fact that a conditional statement and its contrapositive are logically equivalent is very important and has wide application. Two other variants of a conditional statement are *not* logically equivalent to the statement.

• Definition

Suppose a conditional statement of the form "If p then q" is given.

1. The **converse** is "If q then p."

2. The **inverse** is "If $\sim p$ then $\sim q$."

Symbolically,

$$\text{The converse of } p \rightarrow q \text{ is } q \rightarrow p,$$

and

$$\text{The inverse of } p \rightarrow q \text{ is } \sim p \rightarrow \sim q.$$

Example 1.2.6 Writing the Converse and the Inverse

Write the converse and inverse of each of the following statements:

a. If Howard can swim across the lake, then Howard can swim to the island.

b. If today is Easter, then tomorrow is Monday.

Solution

a. *Converse:* If Howard can swim to the island, then Howard can swim across the lake.

   Inverse: If Howard cannot swim across the lake, then Howard cannot swim to the island.

b. *Converse:* If tomorrow is Monday, then today is Easter.

   Inverse: If today is not Easter, then tomorrow is not Monday. ∎

 Caution! Many people mistakenly believe that if a conditional statement is true, then its converse and inverse are also true. This is not so. If a conditional statement is true, then its converse and inverse may or may not be true. For instance, on any Sunday except Easter, the conditional statement in Example 1.2.6(b) is true; yet both its converse and its inverse are false.

> 1. A conditional statement and its converse are *not* logically equivalent.
>
> 2. A conditional statement and its inverse are *not* logically equivalent.
>
> 3. The converse and the inverse of a conditional statement are logically equivalent to each other.

In exercises 24, 25, and 27 at the end of this section, you are asked to use truth tables to verify the statements in the box above. Note that the truth of statement 3 also follows from the observation that the inverse of a conditional statement is the contrapositive of its converse.

Only If *and the Biconditional*

To say "p only if q" means that p can take place *only* if q takes place also. That is, if q does not take place, then p cannot take place. Another way to say this is that if p occurs, then q must also occur (by the logical equivalence between a statement and its contrapositive).

• Definition

If p and q are statements,

$$p \textbf{ only if } q \quad \text{means} \quad \text{"if not } q \text{ then not } p\text{,"}$$

or, equivalently,

$$\text{"if } p \text{ then } q\text{."}$$

Example 1.2.7 Converting *Only If* to *If-Then*

p only if *q*   if ~*q* → ~*p*
                  if *p* then *q*.

Use the contrapositive to rewrite the following statement in if-then form in two ways:

p
John will break world's record for the mile run only if
he runs the mile in under four minutes. *q*

Solution   *Version 1:* If John does not run the mile in under four minutes, then he will not break the world's record.

Version 2: If John breaks the world's record, then he will have run the mile in under four minutes.   ∎

 Caution!   "p only if q" does *not* mean "p if q." For instance, to say that John will break the world's record only if he runs the mile in under four minutes does not mean that John will break the world's record if he runs the mile in under four minutes. His time could be under four minutes but still not be fast enough to break the record.

• Definition

Given statement variables p and q, the **biconditional of p and q** is "p if, and only if, q" and is denoted $p \leftrightarrow q$. It is true if both p and q have the same truth values and is false if p and q have opposite truth values. The words *if and only if* are sometimes abbreviated **iff.**

The biconditional has the following truth table:

Truth Table for $p \leftrightarrow q$

p	q	$p \leftrightarrow q$
T	T	T
T	F	F
F	T	F
F	F	T

In order of operations $\leftrightarrow$ is coequal with $\rightarrow$. As with $\wedge$ and $\vee$, the only way to indicate precedence between them is to use parentheses. Here is the full hierarchy of operations for the five logical operators:

Order of Operations for Logical Operators

1. $\sim$	Evaluate negations first.
2. $\wedge, \vee$	Evaluate $\wedge$ and $\vee$ second. When both are present, parentheses may be needed.
3. $\rightarrow, \leftrightarrow$	Evaluate $\rightarrow$ and $\leftrightarrow$ third. When both are present, parentheses may be needed.

According to the separate definitions of *if* and *only if,* saying "p if, and only if, q" should mean the same as saying both "p if q" and "p only if q." The following truth table shows that this is the case:

Truth Table Showing that $p \leftrightarrow q \equiv (p \rightarrow q) \wedge (q \rightarrow p)$

p	q	$p \rightarrow q$	$q \rightarrow p$	$p \leftrightarrow q$	$(p \rightarrow q) \wedge (q \rightarrow p)$
T	T	T	T	T	T
T	F	F	T	F	F
F	T	T	F	F	F
F	F	T	T	T	T

$p \leftrightarrow q$ and $(p \rightarrow q) \wedge (q \rightarrow p)$ always have the same truth values, so they are logically equivalent

Example 1.2.8 *If* and *Only If*

Rewrite the following statement as a conjunction of two if-then statements:

This computer program is correct if, and only if, it produces the correct answer for all possible sets of input data.

Solution   If this program is correct, then it produces the correct answers for all possible sets of input data; and if this program produces the correct answers for all possible sets of input data, then it is correct.   ■

Earlier it was noted that $p \rightarrow q \equiv \sim p \vee q$. Since $p \leftrightarrow q \equiv (p \rightarrow q) \wedge (q \rightarrow p)$, it follows that

$$p \leftrightarrow q \equiv (\sim p \vee q) \wedge (\sim q \vee p).$$

Consequently, any statement form containing $\rightarrow$ or $\leftrightarrow$ is logically equivalent to one containing only $\sim$, $\wedge$, and $\vee$. (See exercises 38–41.)

Necessary and Sufficient Conditions

The phrases *necessary condition* and *sufficient condition,* as used in formal English, correspond exactly to their definitions in logic.

• Definition

If r and s are statements:

r is a **sufficient condition** for s means   "if r then s."

r is a **necessary condition** for s means   "if not r then not s."

In other words, to say "r is a sufficient condition for s" means that the occurrence of r is *sufficient* to guarantee the occurrence of s. On the other hand, to say "r is a necessary condition for s" means that if r does not occur, then s cannot occur either: The occurrence of r is *necessary* to obtain the occurrence of s. Note that because of the equivalence between a statement and its contrapositive,

---
r is a necessary condition for s also means   "if s then r."

Consequently,

---
r is a necessary and sufficient condition for s means   "r if, and only if, s."

Example 1.2.9 Interpreting Necessary and Sufficient Conditions

Consider the statement "If John is eligible to vote, then he is at least 18 years old." The truth of the condition "John is eligible to vote" is *sufficient* to ensure the truth of the condition "John is at least 18 years old." In addition, the condition "John is at least 18 years old" is *necessary* for the condition "John is eligible to vote" to be true. If John were younger than 18, then he would not be eligible to vote.   ■

Example 1.2.10 Converting a Sufficient Condition to If-Then Form

Rewrite the following statement in the form "If A then B":

Pia's birth on U.S soil is a sufficient condition for her to be a U.S. citizen.

Solution   If Pia was born on U.S. soil, then she is a U.S. citizen.   ■

Example 1.2.11 Converting a Necessary Condition to If-Then Form

Use the contrapositive to rewrite the following statement in two ways:

George's attaining age 35 is a necessary condition
for his being president of the United States.

Solution   *Version 1:*   If George has not attained the age of 35, then he cannot be president
of the United States.

       Version 2:   If George can be president of the United States, then he has attained
the age of 35. ∎

Remarks

1. *In logic, a hypothesis and conclusion are not required to have related subject matters.*

   In ordinary speech we never say things like "If computers are machines, then Babe
   Ruth was a baseball player" or "If $2 + 2 = 5$, then Mickey Mouse is president of the
   United States." We formulate a sentence like "If p then q" only if there is some
   connection of content between p and q.

   In logic, however, the two parts of a conditional statement need not have related
   meanings. The reason? If there were such a requirement, who would enforce it? What
   one person perceives as two unrelated clauses may seem related to someone else. There
   would have to be a central arbiter to check each conditional sentence before anyone
   could use it, to be sure its clauses were in proper relation. This is impractical, to say
   the least!

   Thus a statement like "if computers are machines, then Babe Ruth was a baseball
   player" is allowed, and it is even called true because both its hypothesis and its conclusion are true. Similarly, the statement "If $2 + 2 = 5$, then Mickey Mouse is president
   of the United States" is allowed and is called true because its hypothesis is false, even
   though doing so may seem ridiculous.

   In mathematics it often happens that a carefully formulated definition that successfully covers the situations for which it was primarily intended is later seen to be
   satisfied by some extreme cases that the formulator did not have in mind. But those
   are the breaks, and it is important to get into the habit of exploring definitions fully to
   seek out and understand *all* their instances, even the unusual ones.

2. *In informal language, simple conditionals are often used to mean biconditionals.*

   The formal statement "p if, and only if, q" is seldom used in ordinary language.
   Frequently, when people intend the biconditional they leave out either the *and only if*
   or the *if and.* That is, they say either "p if q" or "p only if q" when they really mean
   "p if, and only if, q." For example, consider the statement "You will get dessert if,
   and only if, you eat your dinner." Logically, this is equivalent to the conjunction of
   the following two statements.

   Statement 1: If you eat your dinner, then you will get dessert.

   Statement 2: You will get dessert only if you eat your dinner.

        or

            If you do not eat your dinner, then you will not get dessert.

   Now how many parents in the history of the world have said to their children "You
   will get dessert if, and only if, you eat your dinner"? Not many! Most say either "If you
   eat your dinner, you will get dessert" (these take the positive approach—they emphasize
   the reward) or "You will get dessert only if you eat your dinner" (these take the negative

approach—they emphasize the punishment). Yet the parents who promise the reward intend to suggest the punishment as well, and those who threaten the punishment will certainly give the reward if it is earned. Both sets of parents expect that their conditional statements will be interpreted as biconditionals.

Since we often (correctly) interpret conditional statements as biconditionals, it is not surprising that we may come to believe (mistakenly) that conditional statements are always logically equivalent to their inverses and converses. In formal settings, however, statements must have unambiguous interpretations. If-then statements can't sometimes mean "if-then" and other times mean "if and only if." When using language in mathematics, science, or other situations where precision is important, it is essential to interpret if-then statements according to the formal definition and not to confuse them with their converses and inverses.

Exercise Set 1.2

Rewrite the statements in 1–4 in if-then form.

1. This loop will repeat exactly *N* times if it does not contain a **stop** or a **go to.**

2. I am on time for work if I catch the 8:05 bus.

3. Freeze or I'll shoot.

4. Fix my ceiling or I won't pay my rent.

Construct truth tables for the statement forms in 5–11.

5. $\sim p \vee q \to \sim q$ 6. $(p \vee q) \vee (\sim p \wedge q) \to q$

7. $p \wedge \sim q \to r$ 8. $\sim p \vee q \to r$

9. $p \wedge \sim r \leftrightarrow q \vee r$ 10. $(p \to r) \leftrightarrow (q \to r)$

11. $(p \to (q \to r)) \leftrightarrow ((p \wedge q) \to r)$

12. Use the logical equivalence established in Example 1.2.3, $p \vee q \to r \equiv (p \to r) \wedge (q \to r)$, to rewrite the following statement. (Assume that *x* represents a fixed real number.)

If $x > 2$ or $x < -2$, then $x^2 > 4$.

13. Use truth tables to verify the following logical equivalences. Include a few words of explanation with your answers.

a. $p \to q = \sim p \vee q$ b. $\sim (p \to q) \equiv p \wedge \sim q$.

H 14. **a.** Show that the following statement forms are all logically equivalent.

$p \to q \vee r, \quad p \wedge \sim q \to r, \quad$ and $\quad p \wedge \sim r \to q$

b. Use the logical equivalences established in part (a) to rewrite the following sentence in two different ways. (Assume that *n* represents a fixed integer.)

If *n* is prime, then *n* is odd or *n* is 2.

15. Determine whether the following statement forms are logically equivalent:

$p \to (q \to r) \quad$ and $\quad (p \to q) \to r$

In 16 and 17, write each of the two statements in symbolic form and determine whether they are logically equivalent. Include a truth table and a few words of explanation.

16. If you paid full price, you didn't buy it at Crown Books. You didn't buy it at Crown Books or you paid full price.

17. If Rob is goalkeeper and Aaron plays forward, then Sam plays defense. Rob is not goalkeeper or Aaron does not play forward or Sam plays defense.

18. Write each of the following three statements in symbolic form and determine which pairs are logically equivalent. Include truth tables and a few words of explanation.

If it walks like a duck and it talks like a duck, then it is a duck.

Either it does not walk like a duck or it does not talk like a duck, or it is a duck.

If it does not walk like a duck and it does not talk like a duck, then it is not a duck.

19. True or false? The negation of "If Sue is Luiz's mother, then Deana is his cousin" is "If Sue is Luiz's mother, then Deana is not his cousin."

20. Write negations for each of the following statements. (Assume that all variables represent fixed quantities or entities, as appropriate.)

a. If *P* is a square, then *P* is a rectangle.

b. If today is New Year's Eve, then tomorrow is January.

c. If the decimal expansion of *r* is terminating, then *r* is rational.

d. If *n* is prime, then *n* is odd or *n* is 2.

e. If *x* is nonnegative, then *x* is positive or *x* is 0.

f. If Tom is Ann's father, then Jim is her uncle and Sue is her aunt.

g. If *n* is divisible by 6, then *n* is divisible by 2 and *n* is divisible by 3.

21. Suppose that p and q are statements so that $p \rightarrow q$ is false. Find the truth values of each of the following:

    a. $\sim p \rightarrow q$ b. $p \vee q$ c. $q \rightarrow p$

H 22. Write contrapositives for the statements of exercise 20.

H 23. Write the converse and inverse for each statement of exercise 20.

Use truth tables to establish the truth of each statement in 24–27.

24. A conditional statement is not logically equivalent to its converse.

25. A conditional statement is not logically equivalent to its inverse.

26. A conditional statement and its contrapositive are logically equivalent to each other.

27. The converse and inverse of a conditional statement are logically equivalent to each other.

H 28. "Do you mean that you think you can find out the answer to it?" said the March Hare.

    "Exactly so," said Alice.

    "Then you should say what you mean," the March Hare went on.

    "I do," Alice hastily replied; "at least—at least I mean what I say—that's the same thing, you know."

    "Not the same thing a bit!" said the Hatter. "Why, you might just as well say that 'I see what I eat' is the same thing as 'I eat what I see'!"

    — from "A Mad Tea-Party" in *Alice in Wonderland*, by Lewis Carroll

    The Hatter is right. "I say what I mean" is not the same thing as "I mean what I say." Rewrite each of these two sentences in if-then form and explain the logical relation between them. (This exercise is referred to in the introduction to Chapter 3.)

If statement forms P and Q are logically equivalent, then $P \leftrightarrow Q$ is a tautology. Conversely, if $P \leftrightarrow Q$ is a tautology, then P and Q are logically equivalent. Use $\leftrightarrow$ to convert each of the logical equivalences in 29–31 to a tautology. Then use a truth table to verify each tautology.

29. $p \rightarrow (q \vee r) \equiv (p \wedge \sim q) \rightarrow r$

30. $p \wedge (q \vee r) \equiv (p \wedge q) \vee (p \wedge r)$

31. $p \rightarrow (q \rightarrow r) \equiv (p \wedge q) \rightarrow r$

Use the contrapositive to rewrite the statements in 25 and 26 in if-then form in two ways. Assume that *only if* has its formal, logical meaning.

32. The Cubs will win the pennant only if they win tomorrow's game.

33. Sam will be allowed on Signe's racing boat only if he is an expert sailor.

34. Taking the long view on your education, you go to the Prestige Corporation and ask what you should do in college to be hired when you graduate. The personnel director replies that you will be hired *only if* you major in mathematics or computer science, get a B average or better, and take accounting. You do, in fact, become a math major, get a B$^+$ average, and take accounting. You return to Prestige Corporation, make a formal application, and are turned down. Did the personnel director lie to you?

Some programming languages use statements of the form "r unless s." This means that as long as s does not happen, then r will happen. More formally,

> **Definition:** If r and s are statements,
>
> > **r unless s**   means   if $\sim s$ then r.

In 35–37, rewrite the statements in if-then form.

35. Payment will be made on the fifth unless a new hearing is granted.

36. Ann will go unless it rains.

37. This door will not open unless a security code is entered.

In 38–41 (a) use the logical equivalences $p \rightarrow q \equiv \sim p \vee q$ and $p \leftrightarrow q \equiv (\sim p \vee q) \wedge (\sim q \vee p)$ to rewrite the given statement forms without using the symbol $\rightarrow$ or $\leftrightarrow$, and (b) use the logical equivalence $p \vee q \equiv \sim(\sim p \wedge \sim q)$ to rewrite each statement form using only $\wedge$ and $\sim$.

38. $p \wedge \sim q \rightarrow r$ 39. $p \vee \sim q \rightarrow r \vee q$

40. $(p \rightarrow r) \leftrightarrow (q \rightarrow r)$

41. $(p \rightarrow (q \rightarrow r)) \leftrightarrow ((p \wedge q) \rightarrow r)$

42. Given any statement form, is it possible to find a logically equivalent form that uses only $\sim$ and $\wedge$? Justify your answer.

Rewrite the statements in 43 and 44 in if-then form.

43. Catching the 8:05 bus is a sufficient condition for my being on time for work.

44. Having two 45° angles is a sufficient condition for this triangle to be a right triangle.

Use the contrapositive to rewrite the statements in 45 and 46 in if-then form in two ways.

45. Being divisible by 3 is a necessary condition for this number to be divisible by 9.

46. Doing homework regularly is a necessary condition for Jim to pass the course.

Note that "a sufficient condition for s is r" means r is a sufficient condition for s and that "a necessary condition for s is r" means r is a necessary condition for s. Rewrite the statements in 47 and 48 in if-then form.

47. A sufficient condition for Jon's team to win the championship is that it win the rest of its games.

48. A necessary condition for this computer program to be correct is that it not produce error messages during translation.

49. "If compound X is boiling, then its temperature must be at least 150°C." Assuming that this statement is true, which of the following must also be true?

    a. If the temperature of compound X is at least 150°C, then compound X is boiling.

    b. If the temperature of compound X is less than 150°C, then compound X is not boiling.

    c. Compound X will boil only if its temperature is at least 150°C.

    d. If compound X is not boiling, then its temperature is less than 150°C.

    e. A necessary condition for compound X to boil is that its temperature be at least 150°C.

    f. A sufficient condition for compound X to boil is that its temperature be at least 150°C.

1.3 Valid and Invalid Arguments

"Contrariwise," continued Tweedledee, "if it was so, it might be; and if it were so, it would be; but as it isn't, it ain't. That's logic." — Lewis Carroll, *Through the Looking Glass*

In mathematics and logic an argument is not a dispute. It is a sequence of statements ending in a conclusion. In this section we show how to determine whether an argument is valid—that is, whether the conclusion follows *necessarily* from the preceding statements. We will show that this determination depends only on the form of an argument, not on its content.

It was shown in Section 1.1 that the logical form of an argument can be abstracted from its content. For example, the argument

> If Socrates is a man, then Socrates is mortal.
>
> Socrates is a man.
>
> ∴ Socrates is mortal.

has the abstract form

> If p then q
> p
> ∴ q

When considering the abstract form of an argument, think of p and q as variables for which statements may be substituted. An argument form is called *valid* if, and only if, whenever statements are substituted that make all the premises true, the conclusion is also true.

• Definition

An **argument** is a sequence of statements, and an **argument form** is a sequence of statement forms. All statements in an argument and all statement forms in an argument form, except for the final one, are called **premises** (or **assumptions** or **hypotheses**). The final statement or statement form is called the **conclusion.** The symbol ∴, which is read "therefore," is normally placed just before the conclusion.

To say that an *argument form* is **valid** means that no matter what particular statements are substituted for the statement variables in its premises, if the resulting premises are all true, then the conclusion is also true. To say that an *argument* is **valid** means that its form is valid.

The crucial fact about a valid argument is that the truth of its conclusion follows *necessarily* or *inescapably* or *by logical form alone* from the truth of its premises. It is impossible to have a valid argument with true premises and a false conclusion. When an argument is valid and its premises are true, the truth of the conclusion is said to be *inferred* or *deduced* from the truth of the premises. If a conclusion "ain't necessarily so," then it isn't a valid deduction.

Testing an Argument Form for Validity

1. Identify the premises and conclusion of the argument form.

2. Construct a truth table showing the truth values of all the premises and the conclusion.

3. If the truth table contains a row in which all the premises are true and the conclusion is false, then it is possible for an argument of the given form to have true premises and a false conclusion, and so the argument form is invalid. Otherwise, in every case where all the premises are true, the conclusion is also true, and so the argument form is valid.

Example 1.3.1 An Invalid Argument Form

Show that the following argument form is invalid.

$$p \rightarrow q \vee \sim r$$
$$q \rightarrow p \wedge r$$
$$\therefore p \rightarrow r$$

Solution   Construct a truth table as shown below, and indicate which columns represent the premises and which the conclusion. Although there are several situations in which the premises and the conclusion are all true (rows 1, 7, and 8), there is one situation (shown in row 4) where the premises are true and the conclusion is false. This cannot occur when the argument form is valid, and so this argument form is invalid.

						premises		conclusion
p	q	r	$\sim r$	$q \vee \sim r$	$p \wedge r$	$p \rightarrow q \vee \sim r$	$q \rightarrow p \wedge r$	$p \rightarrow r$
T	T	T	F	T	T	T	T	T
T	T	F	T	T	F	T	F	F
T	F	T	F	F	T	F	T	F
T	F	F	T	T	F	T	T	F
F	T	T	F	T	F	T	F	F
F	T	F	T	T	F	T	F	F
F	F	T	F	F	F	T	T	T
F	F	F	T	T	F	T	T	T

This row shows it is possible for an argument of this form to have true premises and a false conclusion. Hence this form of argument is invalid.

Note that if you are in a hurry to check the validity of an argument, you need not fill in truth values for the conclusion except in the rows where all the premises are true. We call these the **critical rows.** The truth values in the other rows are irrelevant to the validity or invalidity of the argument. This is illustrated in the following example.

Example 1.3.2 A Valid Argument Form

Show that the following argument form is valid:

$$p \lor (q \lor r)$$
$$\sim r$$
$$\therefore p \lor q$$

Solution Construct a truth table as shown below, and indicate which columns represent the premises and which the conclusion.

				premises		conclusion
p	q	r	$q \lor r$	$p \lor (q \lor r)$	$\sim r$	$p \lor q$
T	T	T	T	T	F	
T	T	F	T	T	T	T
T	F	T	T	T	F	
T	F	F	F	T	T	T
F	T	T	T	T	F	
F	T	F	T	T	T	T
F	F	T	T	T	F	
F	F	F	F	F	T	

critical rows

In each situation where the premises are both true, the conclusion is also true, so the argument form is valid.

Modus Ponens and Modus Tollens

An argument form consisting of two premises and a conclusion is called a **syllogism.** The first and second premises are called the major and minor premises, respectively. The most famous form of syllogism in logic is called **modus ponens.** It has the following form:

$$\text{If } p \text{ then } q.$$
$$p$$
$$\therefore q$$

Here is an argument of this form:

> If the sum of the digits of 371,487 is divisible by 3,
> then 371,487 is divisible by 3.
>
> The sum of the digits of 371,487 is divisible by 3.
>
> $\therefore$ 371,487 is divisible by 3.

The term *modus ponens* is Latin meaning "method of affirming" (the conclusion is an affirmation). Long before you saw your first truth table, you were undoubtedly being convinced by arguments of this form. Nevertheless, it is instructive to prove that modus ponens is a valid form of argument, if for no other reason than to confirm the agreement between the formal definition of validity and the intuitive concept. To do so, we construct a truth table for the premises and conclusion.

		premises		conclusion	
p	q	$p \rightarrow q$	p	q	
T	T	T	T	T	⟵ critical row
T	F	F	T	F	
F	T	T	F	T	
F	F	T	F	F	

The first row is the only one in which both premises are true, and the conclusion in that row is also true. Hence the argument form is valid.

Now consider another valid argument form called **modus tollens.** It has the following form:

$$\text{If } p \text{ then } q.$$
$$\sim q$$
$$\therefore \sim p$$

Here is an example of modus tollens:

> If Zeus is human, then Zeus is mortal.
>
> Zeus is not mortal.
>
> $\therefore$ Zeus is not human.

An intuitive explanation for the validity of modus tollens uses proof by contradiction. It goes like this:

Suppose

(1) If Zeus is human, then Zeus is mortal; and

(2) Zeus is not mortal.

Must Zeus necessarily be nonhuman?

Yes!

Because, if Zeus were human, then by (1) he would be mortal.

But by (2) he is not mortal.

Hence, Zeus cannot be human.

Modus tollens is Latin meaning "method of denying" (the conclusion is a denial). The validity of modus tollens can be shown to follow from modus ponens together with the fact that a conditional statement is logically equivalent to its contrapositive. Or it can be established formally by using a truth table. (See exercise 11.)

Studies by cognitive psychologists have shown that although nearly 100% of college students have a solid, intuitive understanding of modus ponens, less than 60% are able to apply modus tollens correctly.* Yet in mathematical reasoning, modus tollens is used

---

Cognitive Psychology and Its Implications, 3d ed. by John R. Anderson (New York: Freeman, 1990), pp. 292–297.

almost as often as modus ponens. Thus it is important to study the form of modus tollens carefully to learn to use it effectively.

Example 1.3.3 Recognizing Modus Ponens and Modus Tollens

Use modus ponens or modus tollens to fill in the blanks of the following arguments so that they become valid inferences.

a. If there are more pigeons than there are pigeonholes, then two pigeons roost in the same hole.
   There are more pigeons than there are pigeonholes.
   ∴ _____.

b. If 870,232 is divisible by 6, then it is divisible by 3.
   870,232 is not divisible by 3.
   ∴ _____.

Solution

a. Two pigeons roost in the same hole.   by modus ponens

b. 870,232 is not divisible by 6.   by modus tollens   ■

Additional Valid Argument Forms: Rules of Inference

A **rule of inference** is a form of argument that is valid. Thus modus ponens and modus tollens are both rules of inference. The following are additional examples of rules of inference that are frequently used in deductive reasoning.

Example 1.3.4 Generalization

The following argument forms are valid:

$$\text{a.} \quad \begin{array}{l} p \\ \hline \therefore p \vee q \end{array} \qquad\qquad \text{b.} \quad \begin{array}{l} q \\ \hline \therefore p \vee q \end{array}$$

These argument forms are used for making generalizations. For instance, according to the first, if p is true, then, more generally, "p or q" is true for *any* other statement q. As an example, suppose you are given the job of counting the upperclassmen at your school. You ask what class Anton is in and are told he is a junior.
  You reason as follows:

Anton is a junior.
∴ (more generally) Anton is a junior or Anton is a senior.

Knowing that upperclassman means junior or senior, you add Anton to your list.   ■

Example 1.3.5 Specialization

The following argument forms are valid:

$$\text{a.} \quad \begin{array}{l} p \wedge q \\ \hline \therefore p \end{array} \qquad\qquad \text{b.} \quad \begin{array}{l} p \wedge q \\ \hline \therefore q \end{array}$$

These argument forms are used for specializing. When classifying objects according to some property, you often know much more about them than whether they do or do not have that property. When this happens, you discard extraneous information as you concentrate on the particular property of interest.

For instance, suppose you are looking for a person who knows graph algorithms to work with you on a project. You discover that Ana knows both numerical analysis and graph algorithms. You reason as follows:

Ana knows numerical analysis and Ana knows graph algorithms.

∴ (in particular) Ana knows graph algorithms.

Accordingly, you invite her to work with you on your project. ■

Both generalization and specialization are used frequently in mathematics to tailor facts to fit into hypotheses of known theorems in order to draw further conclusions. Elimination, transitivity, and proof by division into cases are also widely used tools.

Example 1.3.6 Elimination

The following argument forms are valid:

$$
\begin{array}{ll}
\text{a.} \quad p \vee q & \qquad \text{b.} \quad p \vee q \\
\qquad \sim q & \qquad \qquad \sim p \\
\qquad \therefore p & \qquad \qquad \therefore q
\end{array}
$$

These argument forms say that when you have only two possibilities and you can rule one out, the other must be the case. For instance, suppose you know that for a particular number x,

$$x - 3 = 0 \quad \text{or} \quad x + 2 = 0.$$

If you also know that x is not negative, then $x \neq -2$, so

$$x + 2 \neq 0.$$

By elimination, you can then conclude that

$$\therefore x - 3 = 0.$$ ■

Example 1.3.7 Transitivity

The following argument form is valid:

$$
\begin{array}{c}
p \rightarrow q \\
q \rightarrow r \\
\therefore p \rightarrow r
\end{array}
$$

Many arguments in mathematics contain chains of if-then statements. From the fact that one statement implies a second and the second implies a third, you can conclude that the first statement implies the third. Here is an example:

If 18,486 is divisible by 18, then 18,486 is divisible by 9.

If 18,486 is divisible by 9, then the sum of the digits of 18,486 is divisible by 9.

∴ If 18,486 is divisible by 18, then the sum of the digits of 18,486 is divisible by 9. ■

Example 1.3.8 Proof by Division into Cases

The following argument form is valid:

$$p \vee q$$
$$p \to r$$
$$q \to r$$
$$\therefore r$$

It often happens that you know one thing or another is true. If you can show that in either case a certain conclusion follows, then this conclusion must also be true. For instance, suppose you know that x is a particular nonzero real number. The trichotomy property of the real numbers says that any number is positive, negative, or zero. Thus (by elimination) you know that x is positive or x is negative. You can deduce that $x^2 > 0$ by arguing as follows:

x is positive or x is negative.

If x is positive, then $x^2 > 0$.

If x is negative, then $x^2 > 0$.

$\therefore x^2 > 0$. ∎

The rules of valid inference are used constantly in problem solving. Here is an example from everyday life.

Example 1.3.9 Application: A More Complex Deduction

You are about to leave for school in the morning and discover that you don't have your glasses. You know the following statements are true:

a. If my glasses are on the kitchen table, then I saw them at breakfast.

b. I was reading the newspaper in the living room or I was reading the newspaper in the kitchen.

c. If I was reading the newspaper in the living room, then my glasses are on the coffee table.

d. I did not see my glasses at breakfast.

e. If I was reading my book in bed, then my glasses are on the bed table.

f. If I was reading the newspaper in the kitchen, then my glasses are on the kitchen table.

Where are the glasses?

Note that (e) was not needed to derive the conclusion. In mathematics as in real life, we frequently deduce a conclusion from just a part of the information available to us. ■

The preceding example shows how to use rules of inferential logic to solve an ordinary problem that could occur in real life. Normally, of course, you use these rules unconsciously. Occasionally, however, problems are so complex that it is helpful to use symbolic logic explicitly. The next example shows how you could do this for the situation described in Example 1.3.9.

Example 1.3.10 Symbolizing a Situation to Find a Solution

Solve the problem of Example 1.3.9 symbolically.

Solution   Let $p =$ My glasses are on the kitchen table.

$q =$ I saw my glasses at breakfast.

$r =$ I was reading the newspaper in the living room.

$s =$ I was reading the newspaper in the kitchen.

$t =$ My glasses are on the coffee table.

$u =$ I was reading my book in bed.

$v =$ My glasses are on the bed table.

Then the statements of Example 1.3.9 translate into the following premises:

a. $p \rightarrow q$ b. $r \vee s$ c. $r \rightarrow t$
d. $\sim q$ e. $u \rightarrow v$ f. $s \rightarrow p$

The following deductions can be made:

1. $p \rightarrow q$ by (a)

     $\sim q$ by (d)

    $\therefore \sim p$ by modus tollens

2. $s \rightarrow p$ by (f)

     $\sim p$ by the conclusion of (1)

    $\therefore \sim s$ by modus tollens

3. $r \vee s$ by (b)

     $\sim s$ by the conclusion of (2)

    $\therefore r$ by elimination

4. $r \rightarrow t$ by (c)

     r by the conclusion of (3)

    $\therefore t$ by modus ponens

Hence t is true and the glasses are on the coffee table. ■

Fallacies

A **fallacy** is an error in reasoning that results in an invalid argument. Three common fallacies are **using ambiguous premises,** and treating them as if they were unambiguous,

begging the question (assuming what is to be proved without having derived it from the premises), and **jumping to a conclusion** (without adequate grounds). In this section we discuss two other fallacies, called *converse error* and *inverse error,* which give rise to arguments that superficially resemble those that are valid by modus ponens and modus tollens but are not, in fact, valid.

As in previous examples, you can show that an argument is invalid by constructing a truth table for the argument form and finding at least one critical row in which all the premises are true but the conclusion is false. Another way is to find an argument of the same form with true premises and a false conclusion. The reason is that for an argument to be valid, *any* argument of the same form that has true premises must have a true conclusion.

Example 1.3.11 Converse Error

Show that the following argument is invalid:

> If Zeke is a cheater, then Zeke sits in the back row.
>
> Zeke sits in the back row.
>
> ∴ Zeke is a cheater.

Solution   Many people recognize the invalidity of the above argument intuitively, reasoning something like this: The first premise gives information about Zeke *if* it is known he is a cheater. It doesn't give any information about him if it is not already known that he is a cheater. One can certainly imagine a person who is not a cheater but happens to sit in the back row. Then if that person's name is substituted for Zeke, the first premise is true by default and the second premise is also true but the conclusion is false.

The general form of the above argument is as follows:

$$p \rightarrow q$$
$$q$$
$$\therefore p$$

In exercise 13(a) at the end of this section you are asked to use a truth table to show that this form of argument is invalid.   ■

The fallacy underlying this invalid argument form is called the **converse error** because the conclusion of the argument would follow from the premises if the premise $p \rightarrow q$ were replaced by its converse. Such a replacement is not allowed, however, because a conditional statement is not logically equivalent to its converse. Converse error is also known as the *fallacy of affirming the consequent.*

Another common error in reasoning is called the *inverse error.*

Example 1.3.12 Inverse Error

Consider the following argument:

> If interest rates are going up, stock market prices will go down.
>
> Interest rates are not going up.
>
> ∴ Stock market prices will not go down.

Note that this argument has the following form:

$$p \rightarrow q$$
$$\sim p$$
$$\therefore \sim q$$

You are asked to give a truth table verification of the invalidity of this argument form in exercise 13(b) at the end of this section.

The fallacy underlying this invalid argument form is called the **inverse error** because the conclusion of the argument would follow from the premises if the premise $p \rightarrow q$ were replaced by its inverse. Such a replacement is not allowed, however, because a conditional statement is not logically equivalent to its inverse. Inverse error is also known as the *fallacy of denying the antecedent.* ■

 Caution! Sometimes people lump together the ideas of validity and truth. If an argument seems valid, they accept the conclusion as true. And if an argument seems fishy (really a slang expression for invalid), they think the conclusion must be false. This is not correct! In logic, the words *true* and *valid* have very different meanings.

A valid argument may have a false conclusion, and an invalid argument may have a true conclusion.

Example 1.3.13 A Valid Argument with a False Conclusion

The argument below is valid by modus ponens. But its major premise is false, and so is its conclusion.

If John Lennon was a rock star, then John Lennon had red hair.

John Lennon was a rock star.

∴ John Lennon had red hair. ■

Example 1.3.14 An Invalid Argument with a True Conclusion

The argument below is invalid by the converse error, but it has a true conclusion.

If New York is a big city, then New York has tall buildings.

New York has tall buildings.

∴ New York is a big city. ■

The important thing to note is that validity is a property of argument *forms:* If an argument is valid, then so is every other argument that has the same form. Similarly, if an argument is invalid, then so is every other argument that has the same form. What characterizes a valid argument is that no argument whose form is valid can have all true premises and a false conclusion. For each valid argument, there are arguments of that form with all true premises and a true conclusion, at least one false premise and a true conclusion, and at least one false premise and a false conclusion. On the other hand, for each invalid argument, there are arguments of that form with every combination of truth values for the premises and conclusion, including all true premises and a false conclusion.

Contradictions and Valid Arguments

The concept of logical contradiction can be used to make inferences through a technique of reasoning called the *contradiction rule.* Suppose p is some statement whose truth you wish to deduce.

Contradiction Rule

If you can show that the supposition that statement p is false leads logically to a contradiction, then you can conclude that p is true.

Example 1.3.15 Contradiction Rule

Show that the following argument form is valid:

$$\sim p \to \mathbf{c}, \text{ where } \mathbf{c} \text{ is a contradiction}$$

$$\therefore p$$

Solution   Construct a truth table for the premise and the conclusion of this argument.

			premises	conclusion
p	$\sim p$	$\mathbf{c}$	$\sim p \to \mathbf{c}$	p
T	F	F	T	T
F	T	F	F	F

← There is only one critical row in which the premise is true, and in this row the conclusion is also true. Hence this form of argument is valid.

The contradiction rule is the logical heart of the method of proof by contradiction. A slight variation also provides the basis for solving many logical puzzles by eliminating contradictory answers: *If an assumption leads to a contradiction, then that assumption must be false.*

Example 1.3.16 Knights and Knaves

Raymond Smullyan (born 1919)

The logician Raymond Smullyan describes an island containing two types of people: knights who always tell the truth and knaves who always lie.* You visit the island and are approached by two natives who speak to you as follows:

> *A* says: *B* is a knight.
>
> *B* says: *A* and I are of opposite type.

What are *A* and *B*?

---

*Raymond Smullyan has written a delightful series of whimsical yet profound books of logical puzzles starting with *What Is the Name of This Book?* (Englewood Cliffs, New Jersey: Prentice-Hall, 1978). Other good sources of logical puzzles are the many excellent books of Martin Gardner, such as *Aha! Insight* and *Aha! Gotcha* (New York: W. H. Freeman, 1978, 1982).

Solution   *A* and *B* are both knaves. To see this, reason as follows:

Suppose *A* is a knight.

∴ What *A* says is true.   by definition of *knight*

∴ *B* is also a knight.   That's what *A* said.

∴ What *B* says is true.   by definition of *knight*

∴ *A* and *B* are of opposite types.   That's what *B* said.

∴ We have arrived at the following contradiction: *A* and *B* are both knights and *A* and *B* are of opposite type.

∴ The supposition is false.   by the contradiction rule

∴ *A* is not a knight.   negation of supposition

∴ *A* is a knave.   by elimination: It's given that all inhabitants are knights or knaves, so since *A* is not a knight, *A* is a knave.

∴ What *A* says is false.

∴ *B* is not a knight.

∴ *B* is also a knave.   by elimination

This reasoning shows that if the problem has a solution at all, then *A* and *B* must both be knaves. It is conceivable, however, that the problem has no solution. The problem statement could be inherently contradictory. If you look back at the problem, though, you can see that it does work out for both *A* and *B* to be knaves. ∎

Summary of Rules of Inference

Table 1.3.1 summarizes some of the most important rules of inference.

Table 1.3.1 Valid Argument Forms

Modus Ponens	$p \rightarrow q$ p $\therefore q$	**Elimination**	**a.** $p \vee q$ **b.** $\quad p \vee q$ $\sim q \qquad\qquad\quad \sim p$ $\therefore p \qquad\qquad\quad \therefore q$
Modus Tollens	$p \rightarrow q$ $\sim q$ $\therefore \sim p$	**Transitivity**	$p \rightarrow q$ $q \rightarrow r$ $\therefore p \rightarrow r$
Generalization	**a.** $\quad p$ **b.** $\quad q$ $\therefore p \vee q \qquad\quad \therefore p \vee q$	**Proof by Division into Cases**	$p \vee q$ $p \rightarrow r$ $q \rightarrow r$ $\therefore r$
Specialization	**a.** $\quad p \wedge q$ **b.** $\quad p \wedge q$ $\therefore p \qquad\qquad\quad \therefore q$		
Conjunction	p q $\therefore p \wedge q$	**Contradiction Rule**	$\sim p \rightarrow c$ $\therefore p$

Exercise Set 1.3

Use modus ponens or modus tollens to fill in the blanks in the arguments of 1–5 so as to produce valid inferences.

1. If $\sqrt{2}$ is rational, then $\sqrt{2} = a/b$ for some integers a and b.
It is not true that $\sqrt{2} = a/b$ for some integers a and b.
$\therefore$ _____

2. If this is a **while** loop, then the body of the loop may never be executed.
_____
$\therefore$ The body of the loop may never be executed.

3. If logic is easy, then I am a monkey's uncle.
I am not a monkey's uncle.
$\therefore$ _____

4. If this figure is a quadrilateral, then the sum of its interior angles is $360°$.
The sum of the interior angles of this figure is not $360°$.
$\therefore$ _____

5. If they were unsure of the address, then they would have telephoned.
_____
$\therefore$ They were sure of the address.

Use truth tables to determine whether the argument forms in 6–10 are valid. Indicate which columns represent the premises and which represent the conclusion, and include a few words of explanation to support your answers.

6. $p \to q$
$q \to p$
$\therefore p \lor q$

7. p
$p \to q$
$\sim q \lor r$
$\therefore r$

8. $p \lor q$
$p \to \sim q$
$p \to r$
$\therefore r$

9. $p \land q \to \sim r$
$p \lor \sim q$
$\sim q \to p$
$\therefore \sim r$

10. $p \to r$
$q \to r$
$\therefore p \lor q \to r$

11. $p \to q \lor r$
$\sim q \lor \sim r$
$\therefore \sim p \lor \sim r$

12. Use a truth table to prove the validity of modus tollens.

$$p \to q$$
$$\sim q$$
$$\therefore \sim p$$

13. Use truth tables to show that the following forms of argument are invalid.

a. $p \to q$
q
$\therefore p$
(converse error)

b. $p \to q$
$\sim p$
$\therefore \sim q$
(inverse error)

Use truth tables to show that the argument forms referred to in 14–21 are valid. Indicate which columns represent the premises and which represent the conclusion, and include a few words of explanation to support your answers.

14. Example 1.3.4(a)

15. Example 1.3.4(b)

16. Example 1.3.5(a)

17. Example 1.3.5(b)

18. Example 1.3.6(a)

19. Example 1.3.6(b)

20. Example 1.3.7

21. Example 1.3.8

Use symbols to write the logical form of each argument in 22 and 23, and then use a truth table to test the argument for validity. Indicate which columns represent the premises and which represent the conclusion, and include a few words of explanation to support your answers.

22. If Tom is not on team A, then Hua is on team B.
If Hua is not on team B, then Tom is on team A.
$\therefore$ Tom is not on team A or Hua is not on team B.

23. Oleg is a math major or Oleg is an economics major
If Oleg is a math major, then Oleg is required to take Math 362.
$\therefore$ Oleg is an economics major or Oleg is not required to take Math 362.

Some of the arguments in 24–32 are valid, whereas others exhibit the converse or the inverse error. Use symbols to write the logical form of each argument. If the argument is valid, identify the rule of inference that guarantees its validity. Otherwise, state whether the converse or the inverse error is made.

24. If Jules solved this problem correctly, then Jules obtained the answer 2.
Jules obtained the answer 2.
$\therefore$ Jules solved this problem correctly.

25. This real number is rational or it is irrational.
This real number is not rational.
$\therefore$ This real number is irrational.

26. If I go to the movies, I won't finish my homework.
If I don't finish my homework, I won't do well on the exam tomorrow.
$\therefore$ If I go to the movies, I won't do well on the exam tomorrow.

27. If this number is larger than 2, then its square is larger than 4.
This number is not larger than 2.
$\therefore$ The square of this number is not larger than 4.

28. If there are as many rational numbers as there are irrational numbers, then the set of all irrational numbers is infinite.
The set of all irrational numbers is infinite.
$\therefore$ There are as many rational numbers as there are irrational numbers.

29. If at least one of these two numbers is divisible by 6, then the product of these two numbers is divisible by 6.
    Neither of these two numbers is divisible by 6.
    ∴ The product of these two numbers is not divisible by 6.

30. If this computer program is correct, then it produces the correct output when run with the test data my teacher gave me.
    This computer program produces the correct output when run with the test data my teacher gave me.
    ∴ This computer program is correct.

31. Sandra knows Java and Sandra knows C++.
    ∴ Sandra knows C++.

32. If I get a Christmas bonus, I'll buy a stereo.
    If I sell my motorcycle, I'll buy a stereo.
    ∴ If I get a Christmas bonus or I sell my motorcycle, then I'll buy a stereo.

33. Give an example (other than Example 1.3.13) of a valid argument with a false conclusion.

34. Give an example (other than Example 1.3.14) of an invalid argument with a true conclusion.

35. Explain in your own words what distinguishes a valid form of argument from an invalid one.

36. Given the following information about a computer program, find the mistake in the program.
   a. There is an undeclared variable or there is a syntax error in the first five lines.
   b. If there is a syntax error in the first five lines, then there is a missing semicolon or a variable name is misspelled.
   c. There is not a missing semicolon.
   d. There is not a misspelled variable name.

37. In the back of an old cupboard you discover a note signed by a pirate famous for his bizarre sense of humor and love of logical puzzles. In the note he wrote that he had hidden treasure somewhere on the property. He listed five true statements (a–e below) and challenged the reader to use them to figure out the location of the treasure.
   a. If this house is next to a lake, then the treasure is not in the kitchen.
   b. If the tree in the front yard is an elm, then the treasure is in the kitchen.
   c. This house is next to a lake.
   d. The tree in the front yard is an elm or the treasure is buried under the flagpole.
   e. If the tree in the back yard is an oak, then the treasure is in the garage.
   Where is the treasure hidden?

38. You are visiting the island described in Example 1.3.16 and have the following encounters with natives.
   a. Two natives A and B address you as follows:
   A says: Both of us are knights.
   B says: A is a knave.
   What are A and B?

   b. Another two natives C and D approach you but only C speaks.
   C says: Both of us are knaves.
   What are C and D?

   c. You then encounter natives E and F.
   E says: F is a knave.
   F says: E is a knave.
   How many knaves are there?

   H d. Finally, you meet a group of six natives, U, V, W, X, Y, and Z, who speak to you as follows:
   U says: None of us is a knight.
   V says: At least three of us are knights.
   W says: At most three of us are knights.
   X says: Exactly five of us are knights.
   Y says: Exactly two of us are knights.
   Z says: Exactly one of us is a knight.
   Which are knights and which are knaves?

39. The famous detective Percule Hoirot was called in to solve a baffling murder mystery. He determined the following facts:
   a. Lord Hazelton, the murdered man, was killed by a blow on the head with a brass candlestick.
   b. Either Lady Hazelton or a maid, Sara, was in the dining room at the time of the murder.
   c. If the cook was in the kitchen at the time of the murder, then the butler killed Lord Hazelton with a fatal dose of strychnine.
   d. If Lady Hazelton was in the dining room at the time of the murder, then the chauffeur killed Lord Hazelton.
   e. If the cook was not in the kitchen at the time of the murder, then Sara was not in the dining room when the murder was committed.
   f. If Sara was in the dining room at the time the murder was committed, then the wine steward killed Lord Hazelton.
   Is it possible for the detective to deduce the identity of the murderer from the above facts? If so, who did murder Lord Hazelton? (Assume there was only one cause of death.)

40. Sharky, a leader of the underworld, was killed by one of his own band of four henchmen. Detective Sharp interviewed the men and determined that all were lying except for one. He deduced who killed Sharky on the basis of the following statements:
   a. Socko: Lefty killed Sharky.
   b. Fats: Muscles didn't kill Sharky.
   c. Lefty: Muscles was shooting craps with Socko when Sharky was knocked off.
   d. Muscles: Lefty didn't kill Sharky.
   Who did kill Sharky?

In 41–44 a set of premises and a conclusion are given. Use the valid argument forms listed in Table 1.3.1 to deduce the conclusion from the premises, giving a reason for each step as in Example 1.3.10. Assume all variables are statement variables.

41. a. $\sim p \vee q \rightarrow r$

    b. $s \vee \sim q$

    c. $\sim t$

    d. $p \rightarrow t$

    e. $\sim p \wedge r \rightarrow \sim s$

    f. $\therefore \sim q$

42. a. $p \vee q$

    b. $q \rightarrow r$

    c. $p \wedge s \rightarrow t$

    d. $\sim r$

    e. $\sim q \rightarrow u \wedge s$

    f. $\therefore t$

43. a. $\sim p \rightarrow r \wedge \sim s$

    b. $t \rightarrow s$

    c. $u \rightarrow \sim p$

    d. $\sim w$

    e. $u \vee w$

    f. $\therefore \sim t$

44. a. $p \rightarrow q$

    b. $r \vee s$

    c. $\sim s \rightarrow \sim t$

    d. $\sim q \vee s$

    e. $\sim s$

    f. $\sim p \wedge r \rightarrow u$

    g. $w \vee t$

    h. $\therefore u \wedge w$

1.4 Application: Digital Logic Circuits

Only connect! — E. M. Forster, *Howards End*

*Claude Shannon
(1916–2001)*

M.I.T. Museum

In the late 1930s, a young M.I.T. graduate student named Claude Shannon noticed an analogy between the operations of switching devices, such as telephone switching circuits, and the operations of logical connectives. He used this analogy with striking success to solve problems of circuit design and wrote up his results in his master's thesis, which was published in 1938.

The drawing in Figure 1.4.1(a) shows the appearance of the two positions of a simple switch. When the switch is closed, current can flow from one terminal to the other; when it is open, current cannot flow. Imagine that such a switch is part of the circuit shown in Figure 1.4.1(b). The light bulb turns on if, and only if, current flows through it. And this happens if, and only if, the switch is closed.

Open   Closed

(a)

The symbol denotes a battery and the symbol denotes a light bulb.

(b)

Figure 1.4.1

Now consider the more complicated circuits of Figures 1.4.2(a) and 1.4.2(b).

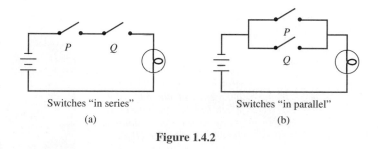

Switches "in series"

(a)

Switches "in parallel"

(b)

Figure 1.4.2

In the circuit of Figure 1.4.2(a) current flows and the light bulb turns on if, and only if, *both* switches P and Q are closed. The switches in this circuit are said to be **in series.** In the circuit of Figure 1.4.2(b) current flows and the light bulb turns on if, and only if, *at least one* of the switches P or Q is closed. The switches in this circuit are said to be **in parallel.** All possible behaviors of these circuits are discribed by Table 1.4.1.

Table 1.4.1

(a) Switches in Series		
Switches		Light Bulb
P	Q	State
closed	closed	on
closed	open	off
open	closed	off
open	open	off

(b) Switches in Parallel		
Switches		Light Bulb
P	Q	State
closed	closed	on
closed	open	on
open	closed	on
open	open	off

Observe that if the words *closed* and *on* are replaced by T and *open* and *off* are replaced by F, Table 1.4.1(a) becomes the truth table for *and* and Table 1.4.1(b) becomes the truth table for *or*. Consequently, the switching circuit of Figure 1.4.2(a) is said to correspond to the logical expression $P \wedge Q$, and that of Figure 1.4.2(b) is said to correspond to $P \vee Q$.

More complicated circuits correspond to more complicated logical expressions. This correspondence has been used extensively in the design and study of circuits.

In the 1940s and 1950s, switches were replaced by electronic devices, with the physical states of closed and open corresponding to electronic states such as high and low voltages. The new electronic technology led to the development of modern digital systems such as electronic computers, electronic telephone switching systems, traffic light controls, electronic calculators, and the control mechanisms used in hundreds of other types of electronic equipment. The basic electronic components of a digital system are called *digital logic circuits*. The word *logic* indicates the important role of logic in the design of such circuits, and the word *digital* indicates that the circuits process discrete, or separate, signals as opposed to continuous ones.

The INTEL Pentium integrated circuit, here shown enlarged, can function as the central processing unit of a powerful personal computer. It is a triumph of miniaturization, containing tens of millions of transistors that make up millions of digital logic circuits.

Courtesy of INTEL Corporation

Electrical engineers continue to use the language of logic when they refer to values of signals produced by an electronic switch as being "true" or "false." But they generally use the symbols 1 and 0 rather than T and F to denote these values. The symbols 0 and 1 are called **bits,** short for *b*inary dig*its*. This terminology was introduced in 1946 by the statistician John Tukey.

Black Boxes and Gates

John W. Tukey
(1915–2000)

Combinations of signal bits (1's and 0's) can be transformed into other combinations of signal bits (1's and 0's) by means of various circuits. Because a variety of different technologies are used in circuit construction, computer engineers and digital system designers find it useful to think of certain basic circuits as black boxes. The inside of a black box contains the detailed implementation of the circuit and is often ignored while attention is focused on the relation between the **input** and the **output** signals.

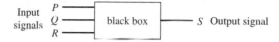

The operation of a black box is completely specified by constructing an **input/output table** that lists all its possible input signals together with their corresponding output signals. For example, the black box pictured above has three input signals. Since each of these signals can take the value 1 or 0, there are eight possible combinations of input signals. One possible correspondence of input to output signals is as follows:

An Input/Output Table

Input			Output
P	*Q*	*R*	*S*
1	1	1	1
1	1	0	0
1	0	1	0
1	0	0	1
0	1	1	0
0	1	0	1
0	0	1	1
0	0	0	0

The third row, for instance, indicates that for inputs $P = 1$, $Q = 0$, and $R = 1$, the output S is 0.

An efficient method for designing more complicated circuits is to build them by connecting less complicated black box circuits. Three such circuits are known as NOT-, AND-, and OR-gates.

A **NOT-gate** (or **inverter**) is a circuit with one input signal and one output signal. If the input signal is 1, the output signal is 0. Conversely, if the input signal is 0, then the output signal is 1. An **AND-gate** is a circuit with two input signals and one output signal. If both input signals are 1, then the output signal is 1. Otherwise, the output signal is 0. An **OR-gate** also has two input signals and one output signal. If both input signals are 0, then the output signal is 0. Otherwise, the output signal is 1.

The actions of NOT-, AND-, and OR-gates are summarized in Figure 1.4.3, where P and Q represent input signals and R represents the output signal. It should be clear from

Figure 1.4.3 that the actions of the NOT-, AND-, and OR-gates on signals correspond exactly to those of the logical connectives $\sim$, $\wedge$, and $\vee$ on statements, if the symbol 1 is identified with T and the symbol 0 is identified with F.

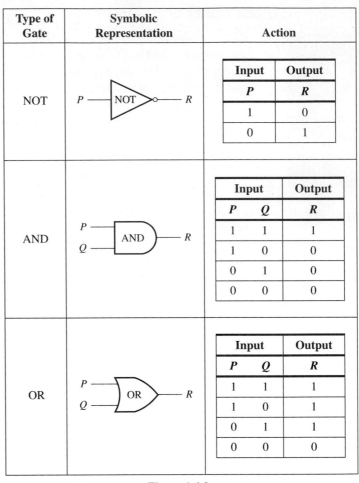

Type of Gate	Symbolic Representation	Action

NOT

Input	Output
P	R
1	0
0	1

AND

Input		Output
P	Q	R
1	1	1
1	0	0
0	1	0
0	0	0

OR

Input		Output
P	Q	R
1	1	1
1	0	1
0	1	1
0	0	0

Figure 1.4.3

Gates can be combined into circuits in a variety of ways. If the rules shown below are obeyed, the result is a **combinational circuit,** one whose output at any time is determined entirely by its input at that time without regard to previous inputs.

Rules for a Combinational Circuit

Never combine two input wires.   1.4.1

A single input wire can be split partway and used as input
for two separate gates.   1.4.2

An output wire can be used as input.   1.4.3

No output of a gate can eventually feed back into that gate.   1.4.4

Rule (1.4.4) is violated in more complex circuits, called **sequential circuits,** whose output at any given time depends both on the input at that time and also on previous inputs. These circuits are discussed in Section 12.2.

The Input/Output Table for a Circuit

If you are given a set of input signals for a circuit, you can find its output by tracing through the circuit gate by gate.

Example 1.4.1 Determining Output for a Given Input

Indicate the output of the circuits shown below for the given input signals.

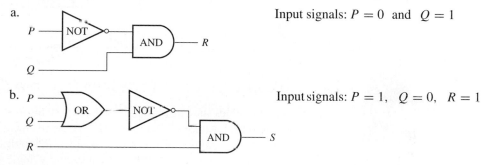

a. Input signals: $P = 0$ and $Q = 1$

b. Input signals: $P = 1$, $Q = 0$, $R = 1$

Solution

a. Move from left to right through the diagram, tracing the action of each gate on the input signals. The NOT-gate changes $P = 0$ to a 1, so both inputs to the AND-gate are 1; hence the output R is 1. This is illustrated by annotating the diagram as shown below.

b. The output of the OR-gate is 1 since one of the input signals, P, is 1. The NOT-gate changes this 1 into a 0, so the two inputs to the AND-gate are 0 and $R = 1$. Hence the output S is 0. The trace is shown below.

To construct the entire input/output table for a circuit, trace through the circuit to find the corresponding output signals for each possible combination of input signals.

Example 1.4.2 Constructing the Input/Output Table for a Circuit

Construct the input/output table for the following circuit.

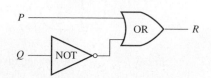

Solution   List the four possible combinations of input signals, and find the output for each by tracing through the circuit.

Input		Output
P	*Q*	*R*
1	1	1
1	0	1
0	1	0
0	0	1

George Boole (1815–1864)

The Boolean Expression Corresponding to a Circuit

In logic, variables such as p, q and r represent statements, and a statement can have one of only two truth values: T (true) or F (false). A statement form is an expression, such as $p \wedge (\sim q \vee r)$, composed of statement variables and logical connectives.

As noted earlier, one of the founders of symbolic logic was the English mathematician George Boole. In his honor, any variable, such as a statement variable or an input signal, that can take one of only two values is called a **Boolean variable.** An expression composed of Boolean variables and the connectives $\sim$, $\wedge$, and $\vee$ is called a **Boolean expression.***

Given a circuit consisting of combined NOT-, AND-, and OR-gates, a corresponding Boolean expression can be obtained by tracing the actions of the gates of the input variables.

Example 1.4.3 Finding a Boolean Expression for a Circuit

Find the Boolean expressions that correspond to the circuits shown below. A dot indicates a soldering of two wires; wires that cross without a dot are assumed not to touch.

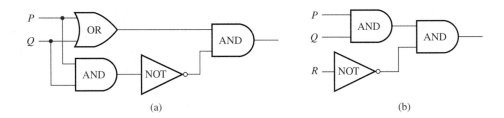

(a)   (b)

Solution

a. Trace through the circuit from left to right, indicating the output of each gate symbolically, as shown below.

*Strictly speaking, only meaningful expressions such as $(\sim p \wedge q) \vee (p \wedge r)$ and $\sim(\sim(p \wedge q) \vee r)$ are allowed, not meaningless ones like $p \sim q((rs \vee \wedge q \sim$. We use recursion to give a careful definition of Boolean expressions in Section 8.4.

The final expression obtained, $(P \lor Q) \land \sim(P \land Q)$, is the expression for exclusive or: P or Q but not both.

b. The Boolean expression corresponding to the circuit is $(P \land Q) \land \sim R$, as shown below.

Observe that the output of the circuit shown in Example 1.4.3(b) is 1 for exactly one combination of inputs ($P = 1$, $Q = 1$, and $R = 0$) and is 0 for all other combinations of inputs. For this reason, the circuit can be said to "recognize" one particular combination of inputs. The output column of the input/output table has a 1 in exactly one row and 0's in all other rows.

> **• Definition**
>
> A **recognizer** is a circuit that outputs a 1 for exactly one particular combination of input signals and outputs 0's for all other combinations.

Input/Output Table for a Recognizer

P	Q	R	$(P \land Q) \land \sim R$
1	1	1	0
1	1	0	1
1	0	1	0
1	0	0	0
0	1	1	0
0	1	0	0
0	0	1	0
0	0	0	0

The Circuit Corresponding to a Boolean Expression

The preceding examples showed how to find a Boolean expression corresponding to a circuit. The following example shows how to construct a circuit corresponding to a Boolean expression.

Example 1.4.4 Constructing Circuits for Boolean Expressions

Construct circuits for the following Boolean expressions.

a. $(\sim P \land Q) \lor \sim Q$ \qquad\qquad b. $((P \land Q) \land (R \land S)) \land T$

Solution

a. Write the input variables in a column on the left side of the diagram. Then go from the right side of the diagram to the left, working from the outermost part of the expression to the innermost part. Since the last operation executed when evaluating $(\sim P \wedge Q) \vee \sim Q$ is $\vee$, put an OR-gate at the extreme right of the diagram. One input to this gate is $\sim P \wedge Q$, so draw an AND-gate to the left of the OR-gate and show its output coming into the OR-gate. Since one input to the AND-gate is $\sim P$, draw a line from P to a NOT-gate and from there to the AND-gate. Since the other input to the AND-gate is Q, draw a line from Q directly to the AND-gate. The other input to the OR-gate is $\sim Q$, so draw a line from Q to a NOT-gate and from the NOT-gate to the OR-gate. The circuit you obtain is shown below.

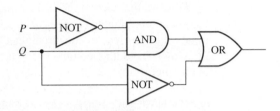

b. To start constructing this circuit, put one AND-gate at the extreme right for the $\wedge$ between $((P \wedge Q) \wedge (R \wedge S))$ and T. To the left of that put the AND-gate corresponding to the $\wedge$ between $P \wedge Q$ and $R \wedge S$. To the left of that put the AND-gates corresponding to the $\wedge$'s between P and Q and between R and S. The circuit is shown in Figure 1.4.4.

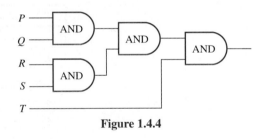

Figure 1.4.4

It follows from Theorem 1.1.1 that all the ways of adding parentheses to $P \wedge Q \wedge R \wedge S \wedge T$ are logically equivalent. Thus, for example,

$$((P \wedge Q) \wedge (R \wedge S)) \wedge T \equiv (P \wedge (Q \wedge R)) \wedge (S \wedge T).$$

It also follows that the circuit in Figure 1.4.5, which corresponds to $(P \wedge (Q \wedge R)) \wedge (S \wedge T)$, has the same input/output table as the circuit in Figure 1.4.4, which corresponds to $((P \wedge Q) \wedge (R \wedge S)) \wedge T$.

Figure 1.4.5

Each of the circuits in Figures 1.4.4 and 1.4.5 is, therefore, an implementation of the expression $P \wedge Q \wedge R \wedge S \wedge T$. Such a circuit is called a **multiple-input AND-gate** and is represented by the diagram shown in Figure 1.4.6. **Multiple-input OR-gates** are constructed similarly.

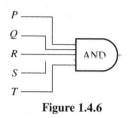

Figure 1.4.6

Finding a Circuit That Corresponds to a Given Input/Output Table

To this point, we have discussed how to construct the input/output table for a circuit, how to find the Boolean expression corresponding to a given circuit, and how to construct the circuit corresponding to a given Boolean expression. Now we address the question of how to design a circuit (or find a Boolean expression) corresponding to a given input/output table. The way to do this is to put several recognizers together in parallel.

Example 1.4.5   Designing a Circuit for a Given Input/Output Table

Design a circuit for the following input/output table:

Input			Output
P	*Q*	*R*	*S*
1	1	1	1
1	1	0	0
1	0	1	1
1	0	0	1
0	1	1	0
0	1	0	0
0	0	1	0
0	0	0	0

Solution   First construct a Boolean expression with this table as its truth table. To do this, identify each row for which the output is 1—in this case, the first, third, and fourth rows. For each such row, construct an *and* expression that produces a 1(or true) for the exact combination of input values for that row and a 0 (or false) for all other combinations of input values. For example, the expression for the first row is $P \wedge Q \wedge R$ because $P \wedge Q \wedge R$ is 1 if $P = 1$ and $Q = 1$ and $R = 1$, and it is 0 for all other values of P, Q, and R. The expression for the third row is $P \wedge {\sim}Q \wedge R$ because $P \wedge {\sim}Q \wedge R$ is 1 if $P = 1$ and $Q = 0$ and $R = 1$, and it is 0 for all other values of P, Q, and R. Similarly, the expression for the fourth row is $P \wedge {\sim}Q \wedge {\sim}R$.

Now any Boolean expression with the given table as its truth table has the value 1 in case $P \wedge Q \wedge R = 1$, or in case $P \wedge \sim Q \wedge R = 1$, or in case $P \wedge \sim Q \wedge \sim R = 1$, and in no other cases. It follows that a Boolean expression with the given truth table is

$$(P \wedge Q \wedge R) \vee (P \wedge \sim Q \wedge R) \vee (P \wedge \sim Q \wedge \sim R). \qquad 1.4.5$$

The circuit corresponding to this expression has the diagram shown in Figure 1.4.7. Observe that expression (1.4.5) is a disjunction of terms that are themselves conjunctions in which one of P or $\sim P$, one of Q or $\sim Q$, and one of R or $\sim R$ all appear. Such expressions are said to be in **disjunctive normal form** or **sum-of-products form.**

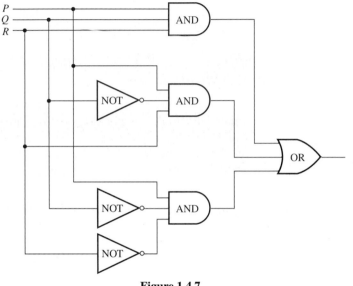

Figure 1.4.7 ■

Simplifying Combinational Circuits

Consider the two combinational circuits shown in Figure 1.4.8.

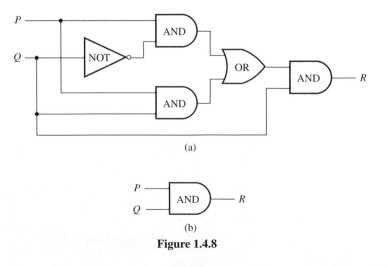

(a)

(b)

Figure 1.4.8

If you trace through circuit (a), you will find that its input/output table is

Input		Output
P	*Q*	*R*
1	1	1
1	0	0
0	1	0
0	0	0

which is the same as the input/output table for circuit (b). Thus these two circuits do the same job in the sense that they transform the same combinations of input signals into the same output signals. Yet circuit (b) is simpler than circuit (a) in that it contains many fewer logic gates. Thus, as part of an integrated circuit, it would take less space and require less power.

• Definition

Two digital logic circuits are **equivalent** if, and only if, their input/output tables are identical

---

Since logically equivalent statement forms have identical truth tables, you can determine that two circuits are equivalent by finding the Boolean expressions corresponding to the circuits and showing that these expressions, regarded as statement forms, are logically equivalent. Example 1.4.6 shows how this procedure works for circuits (a) and (b) in Figure 1.4.8.

Example 1.4.6 Showing That Two Circuits Are Equivalent

Find the Boolean expressions for each circuit in Figure 1.4.8. Use Theorem 1.1.1 to show that these expressions are logically equivalent when regarded as statement forms.

Solution   The Boolean expressions that correspond to circuits (a) and (b) are $((P \wedge \sim Q) \vee (P \wedge Q)) \wedge Q$ and $P \wedge Q$, respectively. By Theorem 1.1.1,

$$((P \wedge \sim Q) \vee (P \wedge Q)) \wedge Q$$

$\equiv (P \wedge (\sim Q \vee Q)) \wedge Q$ by the distributive law (Theorem 1.1.1 (3))

$\equiv (P \wedge (Q \vee \sim Q)) \wedge Q$ by the commutative law for $\vee$ (Theorem 1.1.1(1))

$\equiv (P \wedge \mathbf{t}) \wedge Q$ by the negation law (Theorem 1.1.1(5))

$\equiv P \wedge Q$ by the identity law (Theorem 1.1.1(4)).

It follows that the truth tables for $((P \wedge \sim Q) \vee (P \wedge Q)) \wedge Q$ and $P \wedge Q$ are the same. Hence the input/output tables for the circuits corresponding to these expressions are also the same, and so the circuits are equivalent. ■

In general, you can simplify a combinational circuit by finding the corresponding Boolean expression, using the properties listed in Theorem 1.1.1 to find a Boolean expression that is simpler and logically equivalent to it (when both are regarded as statement forms), and constructing the circuit corresponding to this simpler Boolean expression.

NAND and NOR Gates

Another way to simplify a circuit is to find an equivalent circuit that uses the least number of different kinds of logic gates. Two gates not previously introduced are particularly useful for this: NAND-gates and NOR-gates. A NAND-gate is a single gate that acts like an AND-gate followed by a NOT-gate. A NOR-gate acts like an OR-gate followed by a NOT-gate. Thus the output signal of a NAND-gate is 0 when, and only when, both input signals are 1, and the output signal for a NOR-gate is 1 when, and only when, both input signals are 0. The logical symbols corresponding to these gates are $|$ (for NAND) and $\downarrow$ (for NOR), where $|$ is called a **Sheffer stroke** (after H.M. Sheffer 1882–1964) and $\downarrow$ is called a **Peirce arrow** (after C.S. Peirce, 1839–1914; see page 78). Thus

$$P \mid Q \equiv \sim(P \wedge Q) \quad \text{and} \quad P \downarrow Q \equiv \sim(P \vee Q).$$

The table below summarizes the actions of NAND and NOR gates.

H. M. Sheffer
(1882–1964)

Harvard University Archives

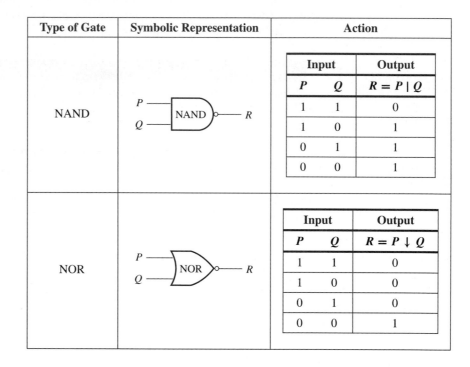

Type of Gate	Symbolic Representation	Action		

NAND:

Input		Output
P	**Q**	**R = P | Q**
1	1	0
1	0	1
0	1	1
0	0	1

NOR:

Input		Output
P	**Q**	**R = P ↓ Q**
1	1	0
1	0	0
0	1	0
0	0	1

It can be shown that any Boolean expression is equivalent to one written entirely with Sheffer strokes or entirely with Peirce arrows. Thus any digital logic circuit is equivalent to one that uses only NAND-gates or only NOR-gates. Example 1.4.7 develops part of the derivation of this result; the rest is left for the exercises.

Example 1.4.7 Rewriting Expressions Using the Sheffer Stroke

Show that

a. $\sim P \equiv P \mid P$ and   b. $P \vee Q \equiv (P \mid P) \mid (Q \mid Q)$.

Solution

a. $\sim P \equiv \sim(P \wedge P)$ by the idempotent law for $\wedge$ (Theorem 1.1.1(7))

$\equiv P \mid P$ by definition of $|$.

b. $P \vee Q \equiv \sim(\sim(P \vee Q))$ by the double negative law (Theorem 1.1.1(6))

$\equiv \sim(\sim P \wedge \sim Q)$ by De Morgan's laws (Theorem 1.1.1(9))

$\equiv \sim((P \mid P) \wedge (Q \mid Q))$ by part (a)

$\equiv (P \mid P) \mid (Q \mid Q)$ by definition of $\mid$. ∎ ■

Exercise Set 1.4

Give the output signals for the circuits in 1–4 if the input signals are as indicated.

1.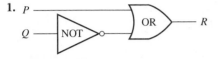

input signals: $P = 1$ and $Q = 1$

2.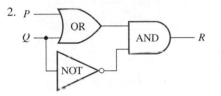

input signals: $P = 1$ and $Q = 0$

3.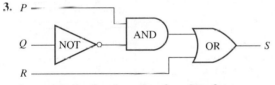

input signals; $P = 1$, $Q = 0$, $R = 0$

4.

input signals: $P = 0$, $Q = 0$, $R = 0$

In 5–8, write an input/output table for the circuit in the referenced exercise.

5. Exercise 1   **6.** Exercise 2

7. Exercise 3   **8.** Exercise 4

In 9–12, find the Boolean expression that corresponds to the circuit in the referenced exercise.

9. Exercise 1   **10.** Exercise 2

11. Exercise 3   **12.** Exercise 4

Construct circuits for the Boolean expressions in 13–17.

13. $\sim P \vee Q$ **14.** $\sim(P \vee Q)$

15. $P \vee (\sim P \wedge \sim Q)$ **16.** $(P \wedge Q) \vee \sim R$

17. $(P \wedge \sim Q) \vee (\sim P \wedge R)$

For each of the tables in 18–21, construct (a) a Boolean expression having the given table as its truth table and (b) a circuit having the given table as its input/output table.

18.

P	Q	R	S
1	1	1	0
1	1	0	1
1	0	1	0
1	0	0	0
0	1	1	1
0	1	0	0
0	0	1	0
0	0	0	0

19.

P	Q	R	S
1	1	1	0
1	1	0	1
1	0	1	0
1	0	0	1
0	1	1	0
0	1	0	1
0	0	1	0
0	0	0	0

20.

P	Q	R	S
1	1	1	1
1	1	0	0
1	0	1	1
1	0	0	0
0	1	1	0
0	1	0	0
0	0	1	0
0	0	0	1

21.

P	Q	R	S
1	1	1	0
1	1	0	1
1	0	1	0
1	0	0	0
0	1	1	1
0	1	0	1
0	0	1	0
0	0	0	0

22. Design a circuit to take input signals P, Q, and R and output a 1 if, and only if, P and Q have the same value and Q and R have opposite values.

23. Design a circuit to take input signals P, Q, and R and output a 1 if, and only if, all three of P, Q, and R have the same value.

24. The lights in a classroom are controlled by two switches: one at the back and one at the front of the room. Moving either switch to the opposite position turns the lights off if they are on and on if they are off. Assume the lights have been installed so that when both switches are in the down position, the lights are off. Design a circuit to control the switches.

25. An alarm system has three different control panels in three different locations. To enable the system, switches in at least two of the panels must be in the on position. If fewer than two are in the on position, the system is disabled. Design a circuit to control the switches.

Use the properties listed in Theorem 1.1.1 to show that each pair of circuits in 26–29 have the same input/output table. (Find the Boolean expressions for the circuits and show that they are logically equivalent when regarded as statement forms.)

26. a.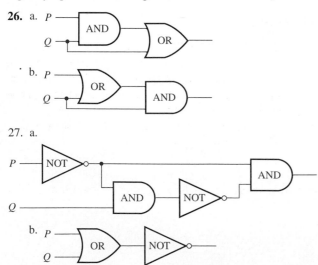

27. a.

For the circuits corresponding to the Boolean expressions in each of 30 and 31 there is an equivalent circuit with at most two logic gates. Find such a circuit.

30. $(P \wedge Q) \vee (\sim P \wedge Q) \vee (\sim P \wedge \sim Q)$

31. $(\sim P \wedge \sim Q) \vee (\sim P \wedge Q) \vee (P \wedge \sim Q)$

32. The Boolean expression for the circuit in Example 1.4.5 is

$$(P \wedge Q \wedge R) \vee (P \wedge \sim Q \wedge R) \vee (P \wedge \sim Q \wedge \sim R)$$

(a disjunctive normal form). Find a circuit with at most three logic gates that is equivalent to this circuit.

33. a. Show that for the Sheffer stroke $|$,

$$P \wedge Q \equiv (P \mid Q) \mid (P \mid Q).$$

b. Use the results of Example 1.4.7 and part (a) above to write $P \wedge (\sim Q \vee R)$ using only Sheffer strokes.

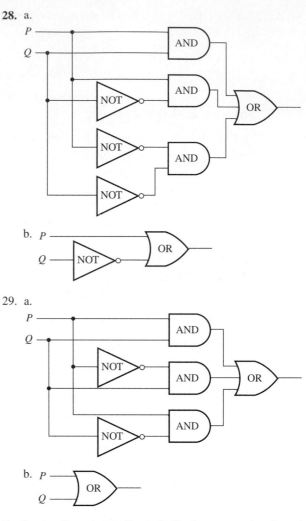

28. a.

b.

29. a.

b.

34. Show that the following logical equivalences hold for the Peirce arrow $\downarrow$, where $P \downarrow Q \equiv \sim(P \vee Q)$.
    a. $\sim P \equiv P \downarrow P$
    b. $P \vee Q \equiv (P \downarrow Q) \downarrow (P \downarrow Q)$
    c. $P \wedge Q \equiv (P \downarrow P) \downarrow (Q \downarrow Q)$
    H d. Write $P \rightarrow Q$ using Peirce arrows only.
    e. Write $P \leftrightarrow Q$ using Peirce arrows only.

1.5 Application: Number Systems and Circuits for Addition

Counting in binary is just like counting in decimal if you are all thumbs. — Glaser and Way

In elementary school, you learned the meaning of decimal notation: that to interpret a string of decimal digits as a number, you mentally multiply each digit by its place value. For instance, 5,049 has a 5 in the thousands place, a 0 in the hundreds place, a 4 in the tens place, and a 9 in the ones place. Thus

$$5,049 = 5 \cdot (1,000) + 0 \cdot (100) + 4 \cdot (10) + 9 \cdot (1).$$

Using exponential notation, this equation can be rewritten as

$$5,049 = 5 \cdot 10^3 + 0 \cdot 10^2 + 4 \cdot 10^1 + 9 \cdot 10^0.$$

More generally, decimal notation is based on the fact that any positive integer can be written uniquely as a sum of products of the form

$$d \cdot 10^n,$$

where each n is a nonnegative integer and each d is one of the decimal digits 0, 1, 2, 3, 4, 5, 6, 7, 8, or 9. The word *decimal* comes from the Latin root *deci,* meaning "ten." Decimal (or base 10) notation expresses a number as a string of digits in which each digit's position indicates the power of 10 by which it is multiplied. The right-most position is the ones place (or 10^0 place), to the left of that is the tens place (or 10^1 place), to the left of that is the hundreds place (or 10^2 place), and so forth, as illustrated below.

Place	10^3 thousands	10^2 hundreds	10^1 tens	10^0 ones
Decimal Digit	5	0	4	9

Binary Representation of Numbers

There is nothing sacred about the number 10; we use 10 as a base for our usual number system because we happen to have ten fingers. In fact, any integer greater than 1 can serve as a base for a number system. In computer science, **base 2 notation,** or **binary notation,** is of special importance because the signals used in modern electronics are always in one of only two states. (The Latin root *bi* means "two.")

In Section 4.4, we show that any integer can be represented uniquely as a sum of powers of the form

$$d \cdot 2^n,$$

where each n is an integer and each d is one of the binary digits (or bits) 0 or 1. For example,

$$27 = 16 + 8 + 2 + 1$$
$$= 1 \cdot 2^4 + 1 \cdot 2^3 + 0 \cdot 2^2 + 1 \cdot 2^1 + 1 \cdot 2^0.$$

In binary notation, as in decimal notation, we write just the binary digits, and not the powers of the base. In binary notation, then,

$$1 \cdot 2^4 + 1 \cdot 2^3 + 0 \cdot 2^2 + 1 \cdot 2^1 + 1 \cdot 2^0$$

$$27_{10} \qquad = \qquad 1\,1\,0\,1\,1_2$$

where the subscripts indicate the base, whether 10 or 2, in which the number is written. The places in binary notation correspond to the various powers of 2. The right-most position is the one's place (or 2^0 place), to the left of that is the twos place (or 2^1 place), to the left of that is the fours place (or 2^2 place), and so forth, as illustrated below.

Place	2^4 sixteens	2^3 eights	2^2 fours	2^1 twos	2^0 ones
Binary Digit	1	1	0	1	1

As in the decimal notation, leading zeros may be added or dropped as desired. For example,

$$003_{10} = 3_{10} = 1 \cdot 2^1 + 1 \cdot 2^0 = 11_2 = 011_2.$$

Example 1.5.1 Binary Notation for Integers from 1 to 9

Derive the binary notation for the integers from 1 to 9.

Solution
$$
\begin{aligned}
1_{10} &= & 1 \cdot 2^0 &= & 1_2 \\
2_{10} &= & 1 \cdot 2^1 + 0 \cdot 2^0 &= & 10_2 \\
3_{10} &= & 1 \cdot 2^1 + 1 \cdot 2^0 &= & 11_2 \\
4_{10} &= & 1 \cdot 2^2 + 0 \cdot 2^1 + 0 \cdot 2^0 &= & 100_2 \\
5_{10} &= & 1 \cdot 2^2 + 0 \cdot 2^1 + 1 \cdot 2^0 &= & 101_2 \\
6_{10} &= & 1 \cdot 2^2 + 1 \cdot 2^1 + 0 \cdot 2^0 &= & 110_2 \\
7_{10} &= & 1 \cdot 2^2 + 1 \cdot 2^1 + 1 \cdot 2^0 &= & 111_2 \\
8_{10} &= 1 \cdot 2^3 + 0 \cdot 2^2 + 0 \cdot 2^1 + 0 \cdot 2^0 &= & 1000_2 \\
9_{10} &= 1 \cdot 2^3 + 0 \cdot 2^2 + 0 \cdot 2^1 + 1 \cdot 2^0 &= & 1001_2
\end{aligned}
$$
∎

A list of powers of 2 is useful for doing binary-to-decimal and decimal-to-binary conversions. See Table 1.5.1.

Table 1.5.1 Powers of 2

Power of 2	2^{10}	2^9	2^8	2^7	2^6	2^5	2^4	2^3	2^2	2^1	2^0
Decimal Form	1024	512	256	128	64	32	16	8	4	2	1

Example 1.5.2 Converting a Binary to a Decimal Number

Represent 110101_2 in decimal notation.

Solution
$$110101_2 = 1 \cdot 2^5 + 1 \cdot 2^4 + 0 \cdot 2^3 + 1 \cdot 2^2 + 0 \cdot 2^1 + 1 \cdot 2^0$$
$$= 32 + 16 + 4 + 1$$
$$= 53_{10}$$

Alternatively, the schema below may be used.

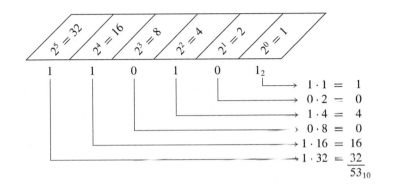

Example 1.5.3 Converting a Decimal to a Binary Number

Represent 209 in binary notation.

Solution   Use Table 1.5.1 to write 209 as a sum of powers of 2, starting with the highest power of 2 that is less than 209 and continuing to lower powers.

Since 209 is between 128 and 256, the highest power of 2 that is less than 209 is 128. Hence

$$209_{10} = 128 + \text{ a smaller number.}$$

Now $209 - 128 = 81$, and 81 is between 64 and 128, so the highest power of 2 that is less than 81 is 64. Hence

$$209_{10} = 128 + 64 + \text{ a smaller number.}$$

Continuing in this way, you obtain

$$209_{10} = 128 + 64 + 16 + 1$$
$$= 1 \cdot 2^7 + 1 \cdot 2^6 + 0 \cdot 2^5 + 1 \cdot 2^4 + 0 \cdot 2^3 + 0 \cdot 2^2 + 0 \cdot 2^1 + 1 \cdot 2^0.$$

For each power of 2 that occurs in the sum, there is a 1 in the corresponding position of the binary number. For each power of 2 that is missing from the sum, there is a 0 in the corresponding position of the binary number. Thus

$$209_{10} = 11010001_2$$

Another procedure for converting from decimal to binary notation is discussed in Section 4.1.

Binary Addition and Subtraction

The computational methods of binary arithmetic are analogous to those of decimal arithmetic. In binary arithmetic the number 2 ($= 10_2$ in binary notation) plays a role similar to that of the number 10 in decimal arithmetic. But do not read 10_2 as "ten"; it is the number two. Read 10_2 as "one oh base two."

Example 1.5.4 Addition in Binary Notation

Add 1101_2 and 111_2 using binary notation.

Solution   Because $2_{10} = 10_2$ and $1_{10} = 1_2$, the translation of $1_{10} + 1_{10} = 2_{10}$ to binary notation is

$$
\begin{array}{r}
1_2 \\
+\ 1_2 \\
\hline
10_2
\end{array}
$$

It follows that adding two 1's together results in a carry of 1 when binary notation is used. Adding three 1's together also results in a carry of 1 since $3_{10} = 11_2$ ("one one base two").

$$
\begin{array}{r}
1_2 \\
+\ 1_2 \\
+\ 1_2 \\
\hline
11_2
\end{array}
$$

Thus the addition can be performed as follows:

$$
\begin{array}{r}
1 \quad 1 \quad 1 \qquad \leftarrow \text{carry row} \\
1 \quad 1 \quad 0 \quad 1_2 \\
+ \qquad 1 \quad 1 \quad 1_2 \\
\hline
1 \quad 0 \quad 1 \quad 0 \quad 0_2
\end{array}
$$

Example 1.5.5 Subtraction in Binary Notation

Subtract 1011_2 from 11000_2 using binary notation.

Solution   In decimal subtraction the fact that $10_{10} - 1_{10} = 9_{10}$ is used to borrow across several columns. For example, consider the following:

$$
\begin{array}{r}
9 \quad 9 \\
\not{1} \quad 1 \qquad \leftarrow \text{borrow row} \\
\not{1}\not{0} \quad \not{0} \quad 0_{10} \\
- \qquad 5 \quad 8_{10} \\
\hline
9 \quad 4 \quad 2_{10}
\end{array}
$$

In binary subtraction it may also be necessary to borrow across more than one column. But when you borrow a 1_2 from 10_2, what remains is 1_2.

$$
\begin{array}{r}
10_2 \\
-\ 1_2 \\
\hline
1_2
\end{array}
$$

Thus the subtraction can be performed as follows:

$$
\begin{array}{r}
0 \quad 1 \quad 1 \\
\not{1} \quad \not{1} \quad 1 \qquad \leftarrow \text{borrow row} \\
1 \quad \not{1} \quad \not{0} \quad \not{0} \quad 0_2 \\
- \qquad 1 \quad 0 \quad 1 \quad 1_2 \\
\hline
1 \quad 1 \quad 0 \quad 1_2
\end{array}
$$

Circuits for Computer Addition

Consider the question of designing a circuit to produce the sum of two binary digits P and Q. Both P and Q can be either 0 or 1. And the following facts are known:

$$1_2 + 1_2 \quad\;\; = 10_2,$$
$$1_2 + 0_2 = 1_2 = 01_2,$$
$$0_2 + 1_2 = 1_2 = 01_2,$$
$$0_2 + 0_2 = 0_2 = 00_2.$$

It follows that the circuit to be designed must have two outputs—one for the left binary digit (this is called the **carry**) and one for the right binary digit (this is called the **sum**). The carry output is 1 if both P and Q are 1; it is 0 otherwise. Thus the carry can be produced using the AND-gate circuit that corresponds to the Boolean expression $P \wedge Q$. The sum output is 1 if either P or Q, but not both, is 1. The sum can, therefore, be produced using a circuit that corresponds to the Boolean expression for *exclusive or:* $(P \vee Q) \wedge \sim(P \wedge Q)$. (See Example 1.4.3(a).) Hence, a circuit to add two binary digits P and Q can be constructed as in Figure 1.5.1. This circuit is called a **half-adder.**

HALF-ADDER

Circuit

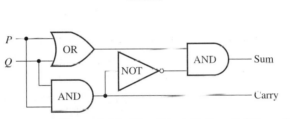

Input/Output Table

P	Q	Carry	Sum
1	1	1	0
1	0	0	1
0	1	0	1
0	0	0	0

Figure 1.5.1 Circuit to Add $P + Q$, Where P and Q Are Binary Digits

Now consider the question of how to construct a circuit to add two binary integers, each with more than one digit. Because the addition of two binary digits may result in a carry to the next column to the left, it may be necessary to add three binary digits at certain points. In the following example, the sum in the right column is the sum of two binary digits, and, because of the carry, the sum in the left column is the sum of three binary digits.

$$
\begin{array}{r}
1 \quad\quad \leftarrow \text{carry row}\\
1 \quad 1_2\\
+ \quad 1 \quad 1_2\\
\hline
1 \quad 1 \quad 0_2
\end{array}
$$

Thus, in order to construct a circuit that will add multidigit binary numbers, it is necessary to incorporate a circuit that will compute the sum of three binary digits. Such a circuit is called a **full-adder.** Consider a general addition of three binary digits P, Q, and R that results in a carry (or left-most digit) C and a sum (or right-most digit) S.

$$
\begin{array}{r}
P\\
+ \quad Q\\
+ \quad R\\
\hline
C\,S
\end{array}
$$

The operation of the full-adder is based on the fact that addition is a binary operation: Only two numbers can be added at one time. Thus P is first added to Q and then the result is added to R. For instance, consider the following addition:

$$\left.\begin{array}{r} 1_2 \\ +\ 0_2 \\ +\ 1_2 \\ \hline 10_2 \end{array}\right\} \begin{array}{l} 1_2 + 0_2 = 01_2 \\ \end{array}\right\} 1_2 + 1_2 = 10_2$$

The process illustrated here can be broken down into steps that use half-adder circuits.

Step 1:   Add P and Q using a half-adder to obtain a binary number with two digits.

$$\begin{array}{r} P \\ +\quad Q \\ \hline C_1 S_1 \end{array}$$

Step 2:   Add R to the sum $C_1 S_1$ of P and Q.

$$\begin{array}{r} C_1 S_1 \\ +\quad R \\ \hline \end{array}$$

To do this, proceed as follows:

Step 2a: Add R to S_1 using a half-adder to obtain the two-digit number $C_2 S$.

$$\begin{array}{r} S_1 \\ +\quad R \\ \hline C_2 S \end{array}$$

Then S is the right-most digit of the entire sum of P, Q, and R.

Step 2b: Determine the left-most digit, C, of the entire sum as follows: First note that it is impossible for both C_1 and C_2 to be 1's. For if $C_1 = 1$, then P and Q are both 1, and so $S_1 = 0$. Consequently, the addition of S_1 and R gives a binary number $C_2 S_1$ where $C_2 = 0$. Next observe that C will be a 1 in the case that the addition of P and Q gives a carry of 1 or in the case that the addition of S_1 (the right-most digit of $P + Q$) and R gives a carry of 1. In other words, $C = 1$ if, and only if, $C_1 = 1$ or $C_2 = 1$. It follows that the circuit shown in Figure 1.5.2 will compute the sum of three binary digits.

FULL-ADDER

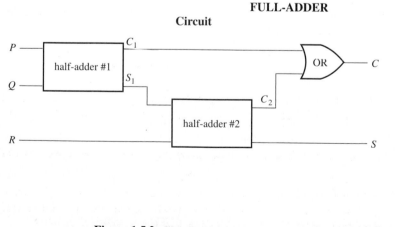

Circuit

Input/Output Table

P	Q	R	C	S
1	1	1	1	1
1	1	0	1	0
1	0	1	1	0
1	0	0	0	1
0	1	1	1	0
0	1	0	0	1
0	0	1	0	1
0	0	0	0	0

Figure 1.5.2 Circuit to Add $P + Q + R$, Where P, Q and R Are Binary Digits

Two full-adders and one half-adder can be used together to build a circuit that will add two three-digit binary numbers PQR and STU to obtain the sum $WXYZ$. This is illustrated in Figure 1.5.3. Such a circuit is called a **parallel adder.** Parallel adders can be constructed to add binary numbers of any finite length.

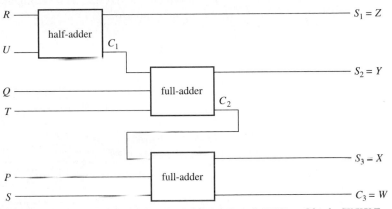

Figure 1.5.3 A Parallel Adder to Add PQR and STU to Obtain $WXYZ$

Two's Complements and the Computer Representation of Negative Integers

Typically, a fixed number of bits is used to represent integers on a computer, and these are required to represent negative as well as nonnegative integers. Sometimes a particular bit, normally the left-most, is used as a sign indicator, and the remaining bits are taken to be the absolute value of the number in binary notation. The problem with this approach is that the procedures for adding the resulting numbers are somewhat complicated and the representation of 0 is not unique. A more common approach, using *two's complements,* makes it possible to add integers quite easily and results in a unique representation for 0. The two's complement of an integer relative to a fixed bit length is defined as follows:

> • **Definition**
>
> Given a positive integer a, the **two's complement of a relative to a fixed bit length** n is the n-bit binary representation of
> $$2^n - a.$$

Bit lengths of 16 and 32 are the most commonly used in practice. However, because the principles are the same for all bit lengths, we use a bit length of 8 for simplicity in this discussion. For instance, because

$$(2^8 - 27)_{10} = (256 - 27)_{10} = 229_{10} = (128 + 64 + 32 + 4 + 1)_{10} = 11100101_2,$$

the 8-bit two's complement of 27 is 11100101_2.

It turns out that there is a convenient way to compute two's complements that involves less arithmetic than direct application of the definition. For an 8-bit representation, it is based on three facts:

1. $2^8 - a = [(2^8 - 1) - a] + 1.$

2. The binary representation of $2^8 - 1$ is 11111111_2.

3. Subtracting an 8-bit binary number a from 11111111_2 just switches all the 0's in a to 1's and all the 1's to 0's. (The resulting number is called the **one's complement** of the given number.)

For instance, by (2) and (3), with $a = 27$,

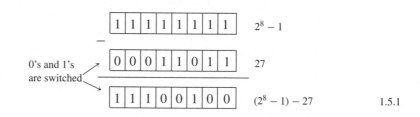

and so in binary notation the difference $(2^8 - 1) - 27$ is 11100100_2. But by (1) with $a = 27, 2^8 - 27 = [(2^8 - 1) - 27] + 1$, and so if we add 1 to (1.5.1), we obtain the 8-bit binary representation of $2^8 - 27$, which is the 8-bit two's complement of 27:

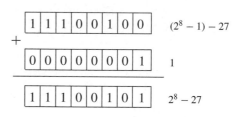

In general,

> To find the 8-bit two's complement of a positive integer a that is at most 255:
>
> - Write the 8-bit binary representation for a.
> - Flip the bits (that is, switch all the 1's to 0's and all the 0's to 1's).
> - Add 1 in binary notation.

Example 1.5.6   Finding a Two's Complement

Find the 8-bit two's complement of 19.

Solution   Write the 8-bit binary representation for 19, switch all the 0's to 1's and all the 1's to 0's, and add 1.

$$19_{10} = (16 + 2 + 1)_{10} = 00010011_2 \xrightarrow{\text{flip the bits}} 11101100 \xrightarrow{\text{add 1}} 11101101$$

To check this result, note that

$$11101101_2 = (128 + 64 + 32 + 8 + 4 + 1)_{10} = 237_{10} = (256 - 19)_{10}$$
$$= (2^8 - 19)_{10},$$

which *is* the two's complement of 19.   ∎

Observe that because

$$2^8 - (2^8 - a) = a$$

the *two's complement of the two's complement of a number is the number itself,* and therefore,

> To find the decimal representation of the integer with a given 8-bit two's complement:
>
> • Find the two's complement of the given two's complement.
> • Write the decimal equivalent of the result.

Example 1.5.7 Finding a Number with a Given Two's Complement

What is the decimal representation for the integer with two's complement 10101001?

Solution

$$10101001_2 \xrightarrow{\text{flip the bits}} 01010110$$
$$\xrightarrow{\text{add 1}} 01010111_2 = (64 + 16 + 4 + 2 + 1)_{10} = 87_{10}$$

To check this result, note that the given number is

$$10101001_2 = (128 + 32 + 8 + 1)_{10} = 169_{10} = (256 - 87)_{10} = (2^8 - 87)_{10},$$

which is the two's complement of 87. ∎

8-Bit Representation of a Number

Now consider the two's complement of an integer n that satisfies the inequality $1 \leq n \leq 128$. Then

$$-1 \geq -n \geq -128 \qquad \text{because multiplying by } -1 \text{ reverses the direction of the inequality}$$

and

$$2^8 - 1 \geq 2^8 - n \geq 2^8 - 128 \qquad \text{by adding } 2^8 \text{ to all parts of the inequality.}$$

But $2^8 - 128 = 256 - 128 = 128 = 2^7$. Hence

$$2^7 \leq \text{the decimal form of the two's complement of } n < 2^8.$$

It follows that the 8-bit two's complement of an integer from 1 through 128 has a leading bit of 1. Note also that the ordinary 8-bit representation of an integer from 0 through 127 has a leading bit of 0. Consequently, eight bits can be used to represent both nonnegative and negative integers by representing each nonnegative integer up through 127 using ordinary 8-bit binary notation and representing each negative integer from -1 through -128 as the two's complement of its absolute value. That is, for any integer a from -128 through 127,

> The 8-bit representation of a
>
> $$= \begin{cases} \text{the 8-bit binary representation of } a & \text{if } a \geq 0 \\ \text{the 8-bit binary representation of } 2^8 - |a| & \text{if } a < 0 \end{cases}.$$

The representations are illustrated in Table 1.5.2.

Table 1.5.2

Integer	8-Bit Representation (ordinary 8-bit binary notation if nonnegative or 8-bit two's complement of absolute value if negative)	Decimal Form of Two's Complement for Negative Integers
127	01111111	
126	01111110	
$\vdots$	$\vdots$	
2	00000010	
1	00000001	
0	00000000	
−1	11111111	$2^8 - 1$
−2	11111110	$2^8 - 2$
−3	11111101	$2^8 - 3$
$\vdots$	$\vdots$	$\vdots$
−127	10000001	$2^8 - 127$
−128	10000000	$2^8 - 128$

Computer Addition with Negative Integers

Here is an example of how two's complements enable addition circuits to perform subtraction. Suppose you want to compute $72 - 54$. First note that this is the same as $72 + (-54)$, and the 8-bit binary representations of 72 and −54 are 01001000 and 11001010, respectively. So if you add the 8-bit binary representations for both numbers, you get

$$
\begin{array}{r}
0\ 1\ 0\ 0\ 1\ 0\ 0\ 0 \\
+\quad 1\ 1\ 0\ 0\ 1\ 0\ 1\ 0 \\
\hline
1\ 0\ 0\ 0\ 1\ 0\ 0\ 1\ 0
\end{array}
$$

And if you truncate the leading 1, you get 00010010. This is the 8-bit binary representation for 18, which is the right answer!

The description below explains how to use this method to add any two integers between −128 and 127. It is easily generalized to apply to 16-bit and 32-bit representations in order to add integers between about −2,000,000 and 2,000,000.

> To add two integers in the range −128 through 127 whose sum is also in the range −128 through 127:
>
> • Convert both integers to their 8-bit representations (representing negative integers by using the two's complements of their absolute values).
>
> • Add the resulting integers using ordinary binary addition.
>
> • Truncate any leading 1 (overflow) that occurs in the 2^8th position.
>
> • Convert the result back to decimal form (interpreting 8-bit integers with leading 0's as nonnegative and 8-bit integers with leading 1's as negative).

To see why this result is true, consider four cases: (1) both integers are nonnegative, (2) one integer is nonnegative and the other is negative and the absolute value of the negative integer is greater than that of the nonnegative one, (3) one integer is nonnegative and the other is negative and the absolute value of the negative integer is less than or equal to that of the nonnegative one, and (4) both integers are negative.

Case (1), where both integers are nonnegative, is easy because if two nonnegative integers from 0 through 127 are written in their 8-bit representation and if their sum is also in the range 0 through 127, then the 8-bit representation of their sum has a leading 0 and is therefore interpreted correctly as a nonnegative integer. The example below illustrates what happens when 38 and 69 are added.

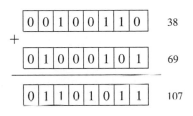

Both cases (2) and (3) involve adding a negative and a nonnegative integer. To be concrete, let the nonnegative integer be a and the negative integer be $-b$ and suppose both a and $-b$ are in the range -128 through 127. The crucial observation is that adding the 8-bit representations of a and $-b$ is equivalent to computing

$$a + (2^8 - b)$$

because the 8-bit representation of $-b$ is the binary representation of $2^8 - b$.

In case $|a| < |b|$, observe that

$$a + (2^8 - b) = 2^8 - (b - a),$$

and the binary representation of this number is the 8-bit representation of $-(b - a) = a + (-b)$. We must be careful to check that $2^8 - (b - a)$ is between 2^7 and 2^8. But it *is* because

$$2^7 = 2^8 - 2^7 \leq 2^8 - (b - a) < 2^8 \qquad \text{since } 0 < b - a \leq 128 = 2^7.$$

Hence in case $|a| < |b|$, adding the 8-bit representations of a and $-b$ gives the 8-bit representation of $a + (-b)$.

In case $|a| \geq |b|$, observe that

$$a + (2^8 - b) = 2^8 + (a - b).$$

Also

$$2^8 \leq 2^8 + (a - b) < 2^8 + 2^7 \qquad \text{because } 0 \leq a - b < 128.$$

So the binary representation of $a + (2^8 - b) = 2^8 + (a - b)$ has a leading 1 in the ninth (2^8th) position. This leading 1 is often called "overflow" because it does not fit in the 8-bit integer format. Now subtracting 2^8 from $2^8 + (a - b)$ is equivalent to truncating the leading 1 in the 2^8th position of the binary representation of the number. But

$$[a + (2^8 - b)] - 2^8 = 2^8 + (a - b) - 2^8 = a - b = a + (-b).$$

Hence in case $|a| \geq |b|$, adding the 8-bit representations of a and $-b$ and truncating the leading 1 (which is sure to be present) gives the 8-bit representation of $a + (-b)$.

Example 1.5.8 Computing $a + (-b)$ Where $0 \leq a < b \leq 128$

Use 8-bit representations to compute $39 + (-89)$.

Solution

Step 1: Change from decimal to 8-bit representations using the two's complement to represent -89.

Since $39_{10} = (32 + 4 + 2 + 1)_{10} = 100111_2$, the 8-bit representation of 39 is 00100111. Now the 8-bit representation of -89 is the two's complement of 89. This is obtained as follows:

$$89_{10} = (64 + 16 + 8 + 1)_{10} = 01011001_2 \xrightarrow{\text{flip the bits}}$$
$$10100110 \xrightarrow{\text{add 1}} 10100111$$

So the 8-bit representation of -89 is 10100111.

Step 2: Add the 8-bit representations in binary notation and truncate the 1 in the 2^8th position if there is one:

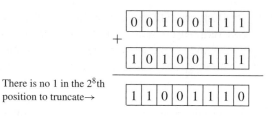

$$
\begin{array}{c}
\;\;0\;0\;1\;0\;0\;1\;1\;1 \\
+\;\;1\;0\;1\;0\;0\;1\;1\;1 \\
\hline
\end{array}
$$

There is no 1 in the 2^8th position to truncate→ $\boxed{1\;1\;0\;0\;1\;1\;1\;0}$

Step 3: Find the decimal equivalent of the result. Since its leading bit is 1, this number is the 8-bit representation of a negative integer.

$$11001110 \xrightarrow{\text{flip the bits}} 00110001 \xrightarrow{\text{add 1}} 00110010$$
$$\leftrightarrow -(32 + 16 + 2)_{10} = -50_{10}$$

Note that since $39 - 89 = -50$, this procedure gives the correct answer. ∎

Example 1.5.9 Computing $a + (-b)$ Where $1 \leq b \leq a \leq 127$

Use 8-bit representations to compute $39 + (-25)$.

Solution

Step 1: Change from decimal to 8-bit representations using the two's complement to represent -25

As in Example 1.5.8, the 8-bit representation of 39 is 00100111. Now the 8-bit representation of -25 is the two's complement of 25, which is obtained as follows:

$$25_{10} = (16 + 8 + 1)_{10} = 00011001_2 \xrightarrow{\text{flip the bits}}$$
$$11100110 \xrightarrow{\text{add 1}} 11100111$$

So the 8-bit representation of -25 is 11100111.

Step 2: Add the 8-bit representations in binary notation and truncate the 1 in the 2^8th position if there is one:

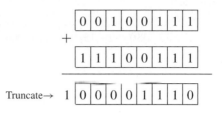

Step 3: Find the decimal equivalent of the result:

$$00001110_2 = (8 + 4 + 2)_{10} = 14_{10}.$$

Since $39 - 25 = 14$, this is the correct answer. ■

Case (4) involves adding two negative integers in the range -1 through -128 whose sum is also in this range. To be specific, consider the sum $(-a) + (-b)$ where a, b, and $a + b$ are all in the range 1 through 128. In this case, the 8-bit representations of $-a$ and $-b$ are the 8-bit representations of $2^8 - a$ and $2^8 - b$. So if the 8-bit representations of $-a$ and $-b$ are added, the result is

$$(2^8 - a) + (2^8 - b) = [2^8 - (a + b)] + 2^8.$$

Recall that truncating a leading 1 in the ninth (2^8th) position of a binary number is equivalent to subtracting 2^8. So when the leading 1 is truncated from the 8-bit representation of $(2^8 - a) + (2^8 - b)$, the result is $2^8 - (a + b)$, which is the 8-bit representation of $-(a + b) = (-a) + (-b)$. (In exercise 37 you are asked to show that the sum $(2^8 - a) + (2^8 - b)$ *has* a leading 1 in the ninth (2^8th) position.)

Example 1.5.10 Computing $(-a) + (-b)$ Where $1 \le a, b \le 128$, and $1 \le a + b \le 128$

Use 8-bit representations to compute $(-89) + (-25)$.

Solution

Step 1: Change from decimal to 8-bit representations using the two's complements to represent -89 and -25.

   The 8-bit representations of -89 and -25 were shown in Examples 1.5.8 and 1.5.9 to be 10100111 and 11100111, respectively.

Step 2: Add the 8-bit representations in binary notation and truncate the 1 in the 2^8th position if there is one:

Step 3: Find the decimal equivalent of the result. Because its leading bit is 1, this number is the 8-bit representation of a negative integer.

$$10001110 \xrightarrow{\text{flip the bits}} 01110001 \xrightarrow{\text{add 1}} 01110010_2$$

$$\leftrightarrow -(64 + 32 + 16 + 2)_{10} = -114_{10}$$

Since $(-89) + (-25) = -114$, that is the correct answer. ■

Hexadecimal Notation

It should now be obvious that numbers written in binary notation take up much more space than numbers written in decimal notation. Yet many aspects of computer operation can best be analyzed using binary numbers. **Hexadecimal notation** is even more compact than decimal notation, and it is much easier to convert back and forth between hexadecimal and binary notation than it is between binary and decimal notation. The word *hexadecimal* comes from the Greek root *hex-,* meaning "six," and the Latin root *deci-,* meaning "ten." Hence *hexadecimal* refers to "sixteen," and hexadecimal notation is also called **base 16 notation.** Hexadecimal notation is based on the fact that any integer can be uniquely expressed as a sum of numbers of the form

$$d \cdot 16^n,$$

where each n is a nonnegative integer and each d is one of the integers from 0 to 15. In order to avoid ambiguity, each hexadecimal digit must be represented by a single symbol. So digits 10 through 15 are represented by the first six letters of the alphabet. The sixteen hexadecimal digits are shown in Table 1.5.3, together with their decimal equivalents and, for future reference, their 4-bit binary equivalents.

Table 1.5.3

Decimal	Hexadecimal	4-Bit Binary Equivalent
0	0	0000
1	1	0001
2	2	0010
3	3	0011
4	4	0100
5	5	0101
6	6	0110
7	7	0111
8	8	1000
9	9	1001
10	A	1010
11	B	1011
12	C	1100
13	D	1101
14	E	1110
15	F	1111

Example 1.5.11 Converting from Hexadecimal to Decimal Notation

Convert $3CF_{16}$ to decimal notation.

Solution   A schema similar to the one introduced in Example 1.5.2 can be used here.

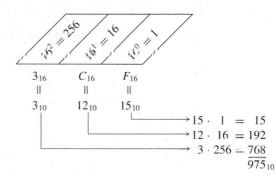

So $3CF_{16} = 975_{10}$.   ■

Now consider how to convert from hexadecimal to binary notation. In the example below the numbers are rewritten using powers of 2, and the laws of exponents are applied. The result suggests a general procedure.

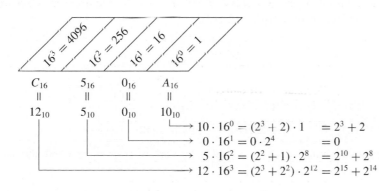

$$10 \cdot 16^0 = (2^3 + 2) \cdot 1 \quad = 2^3 + 2 \qquad \text{since } 10 = 2^3 + 2$$
$$0 \cdot 16^1 = 0 \cdot 2^4 \qquad = 0 \qquad \text{since } 16^1 = 2^4$$
$$5 \cdot 16^2 = (2^2 + 1) \cdot 2^8 \quad = 2^{10} + 2^8 \qquad \text{since } 5 = 2^2 + 1,\ 16^2 = (2^4)^2 = 8 \text{ and } 2^2 \cdot 2^8 = 2^{10}$$
$$12 \cdot 16^3 = (2^3 + 2^2) \cdot 2^{12} = 2^{15} + 2^{14} \qquad \text{since } 12 = 2^3 + 2^2,\ 16^2 = (2^4)^3 = 2^{12},$$
$$2^3 \cdot 2^{12} = 2^{15}, \text{ and } 2^2 \cdot 2^{12} = 2^{14}$$

But

$$(2^{15} + 2^{14}) + (2^{10} + 2^8) + 0 + (2^3 + 2)$$

$$= 1100\ 0000\ 0000\ 0000_2 + 0101\ 0000\ 0000_2 \qquad \text{by the rules for writing binary numbers.}$$
$$+ 0000\ 0000_2 + 1010_2$$

So

$$C50A_{16} = \underbrace{1100}_{C_{16}}\ \underbrace{0101}_{5_{16}}\ \underbrace{0000}_{0_{16}}\ \underbrace{1010}_{A_{16}}{}_2 \qquad \text{by the rules for adding binary numbers.}$$

The procedure illustrated in this example can be generalized. In fact, the following sequence of steps will always give the correct answer:

> To convert an integer from hexadecimal to binary notation:
>
> - Write each hexadecimal digit of the integer in fixed 4-bit binary notation.
> - Juxtapose the results.

Example 1.5.12 Converting from Hexadecimal to Binary Notation

Convert $B09F_{16}$ to binary notation.

Solution $B_{16} = 11_{10} = 1011_2$, $0_{16} = 0_{10} = 0000_2$, $9_{16} = 9_{10} = 1001_2$, and $F_{16} = 15_{10} = 1111_2$. Consequently,

$$
\begin{array}{cccc}
B & 0 & 9 & F \\
\updownarrow & \updownarrow & \updownarrow & \updownarrow \\
1011 & 0000 & 1001 & 1111
\end{array}
$$

and the answer is 1011000010011111_2. ■

To convert integers written in binary notation into hexadecimal notation, reverse the steps of the previous procedure.

> To convert an integer from binary to hexadecimal notation:
>
> - Group the digits of the binary number into sets of four, starting from the right and adding leading zeros as needed.
> - Convert the binary numbers in each set of four into hexadecimal digits. Juxtapose those hexadecimal digits.

Example 1.5.13 Converting from Binary to Hexadecimal Notation

Convert 100110110101001_2 to hexadecimal notation.

Solution   First group the binary digits in sets of four, working from right to left and adding leading 0's if necessary.

$$0100 \quad 1101 \quad 1010 \quad 1001.$$

Convert each group of four binary digits into a hexadecimal digit.

$$
\begin{array}{cccc}
0100 & 1101 & 1010 & 1001 \\
\updownarrow & \updownarrow & \updownarrow & \updownarrow \\
4 & D & A & 9
\end{array}
$$

Then juxtapose the hexadecimal digits.

$$4DA9_{16}$$ ■

Example 1.5.14 Reading a Memory Dump

The smallest addressable memory unit on most computers is one byte, or eight bits. In some debugging operations a dump is made of memory contents; that is, the contents of each memory location are displayed or printed out in order. To save space and make the output easier on the eye, the hexadecimal versions of the memory contents are given, rather than the binary versions. Suppose, for example, that a segment of the memory dump looks like

$$A3 \ BB \ 59 \ 2E.$$

What is the actual content of the four memory locations?

Solution

$$A3_{16} = 10100011_2$$
$$BB_{16} = 10111011_2$$
$$59_{16} = 01011001_2$$
$$2E_{16} = 00101110_2$$ ■

Exercise Set 1.5

Represent the decimal integers in 1–6 in binary notation.

1. 19   **2.** 55   **3.** 287

4. 458   **5.** 1609   **6.** 1424

Represent the integers in 7–12 in decimal notation.

7. 1110_2 **8.** 10111_2 **9.** 110110_2

10. 1100101_2 **11.** 1000111_2 **12.** 1011011_2

Perform the arithmetic in 13–20 using binary notation.

13. 1011_2
    $+ \ \ 101_2$

14. 1001_2
    $+ 1011_2$

15. 101101_2
    $+ \ 11101_2$

16. 110111011_2
    $+ \ 1001011010_2$

17. 10100_2
    $- \ \ 1101_2$

18. 11010_2
    $- \ \ 1101_2$

19. 101101_2
    $- \ \ 10011_2$

20. 1010100_2
    $- \ \ 10111_2$

21. Give the output signals S and T for the circuit below if the input signals P, Q, and R are as specified. Note that this is *not* the circuit for a full-adder.
   a. $P = 1, Q = 1, R = 1$
   b. $P = 0, Q = 1, R = 0$
   c. $P = 1, Q = 0, R = 1$

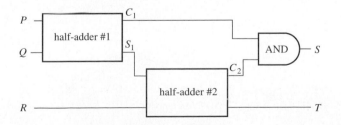

22. Add $11111111_2 + 1_2$ and convert the result to decimal notation, to verify that $11111111_2 = (2^8 - 1)_{10}$.

Find the 8-bit two's complements for the integers in 23–26.

23. 23   **24.** 67   **25.** 4   **26.** 115

Find the decimal representations for the integers with the 8-bit representations given in 27–30.

27. 11010011   **28.** 10011001

29. 11110010   **30.** 10111010

Use 8-bit representations to compute the sums in 31–36.

31. $57 + (-118)$ **32.** $62 + (-18)$

33. $(-6) + (-73)$ **34.** $89 + (-55)$

35. $(-15) + (-46)$ **36.** $123 + (-94)$

✶ 37. Show that if a and b are integers in the range 1 through 128, and the sum of a and b is also in this range, then $2^8 \le (2^8 - a) + (2^8 - b) < 2^9$. Explain why it follows that the binary representation of $(2^8 - a) + (2^8 - b)$ has a leading 1 in the 2^8th position.

Convert the integers in 38–40 from hexadecimal to decimal notation.

38. $A2BC_{16}$ **39.** $E0D_{16}$ **40.** $39EB_{16}$

Convert the integers 41–43 from hexadecimal to binary notation.

41. $1C0ABE_{16}$ **42.** $B53DF8_{16}$ **43.** $4ADF83_{16}$

Convert the integers in 44–46 from binary to hexadecimal notation.

44. 00101110_2 **45.** 1011011111000101_2

46. 11001001011100_2

47. **Octal Notation:** In addition to binary and hexadecimal, computer scientists also use *octal notation* (base 8) to represent numbers. Octal notation is based on the fact that any integer can be uniquely represented as a sum of numbers of the form $d \cdot 8^n$, where each n is a nonnegative integer and each d is one of the integers from 0 to 7. Thus, for example, $5073_8 = 5 \cdot 8^3 + 0 \cdot 8^2 + 7 \cdot 8^1 + 3 \cdot 8^0 = 2619_{10}$.

   a. Convert 61502_8 to decimal notation.

   b. Convert 20763_8 to decimal notation.

   c. Describe methods for converting integers from octal to binary notation and the reverse that are similar to the methods used in Examples 1.5.12 and 1.5.13 for converting back and forth from hexadecimal to binary notation. Give examples showing that these methods result in correct answers.

CHAPTER 2

THE LOGIC OF QUANTIFIED STATEMENTS

In Chapter 1 we discussed the logical analysis of compound statements—those made of simple statements joined by the connectives $\sim$, $\wedge$, $\vee$, $\rightarrow$, and $\leftrightarrow$. Such analysis casts light on many aspects of human reasoning, but it cannot be used to determine validity in the majority of everyday and mathematical situations. For example, the argument

> All men are mortal.
>
> Socrates is a man.
>
> $\therefore$ Socrates is mortal.

is intuitively perceived as correct. Yet its validity cannot be derived using the methods outlined in Section 1.3. To determine validity in examples like this, it is necessary to separate the statements into parts in much the same way that you separate declarative sentences into subjects and predicates. And you must analyze and understand the special role played by words that denote quantities such as "all" or "some." The symbolic analysis of predicates and quantified statements is called the **predicate calculus.** The symbolic analysis of ordinary compound statements (as outlined in Sections 1.1–1.3) is called the **statement calculus** (or the **propositional calculus**).

2.1 Introduction to Predicates and Quantified Statements I

People who call this "instinct" are merely giving the phenomenon a name, not explaining anything. — Douglas Adams, *Dirk Gently's Holistic Detective Agency,* 1987

As noted in Section 1.1, the sentence "He is a college student" is not a statement because it may be either true or false depending on the value of the pronoun *he*. Similarly, the sentence "$x + y$ is greater than 0" is not a statement because its truth value depends on the values of the variables x and y.

In grammar, the word *predicate* refers to the part of a sentence that gives information about the subject. In the sentence "James is a student at Bedford College," the word *James*

75

is the subject and the phrase *is a student at Bedford College* is the predicate. The predicate is the part of the sentence from which the subject has been removed.

In logic, predicates can be obtained by removing some or all of the nouns from a statement. For instance, let P stand for "is a student at Bedford College" and let Q stand for "is a student at." Then both P and Q are *predicate symbols*. The sentences "x is a student at Bedford College" and "x is a student at y" are symbolized as $P(x)$ and as $Q(x, y)$ respectively, where x and y are *predicate variables* that take values in appropriate sets. When concrete values are substituted in place of predicate variables, a statement results. For simplicity, we define a *predicate* to be a predicate symbol together with suitable predicate variables. In some other treatments of logic, such objects are referred to as **propositional functions** or **open sentences.**

• Definition

A **predicate** is a sentence that contains a finite number of variables and becomes a statement when specific values are substituted for the variables. The **domain** of a predicate variable is the set of all values that may be substituted in place of the variable.

Example 2.1.1 Finding Truth Values of a Predicate

Let $P(x)$ be the predicate "$x^2 > x$" with domain the set $\mathbf{R}$ of all real numbers. Write $P(2)$, $P(\frac{1}{2})$, and $P(-\frac{1}{2})$, and indicate which of these statements are true and which are false.

Solution

$$P(2): 2^2 > 2, \quad \text{or} \quad 4 > 2. \quad \text{True.}$$

$$P\left(\tfrac{1}{2}\right): \left(\tfrac{1}{2}\right)^2 > \tfrac{1}{2}, \quad \text{or} \quad \tfrac{1}{4} > \tfrac{1}{2}. \quad \text{False.}$$

$$P\left(-\tfrac{1}{2}\right): \left(-\tfrac{1}{2}\right)^2 > -\tfrac{1}{2}, \quad \text{or} \quad \tfrac{1}{4} > -\tfrac{1}{2}. \quad \text{True.} \qquad ■$$

The sets in which predicate variables take their values may be described either in words or in symbols. When symbols are used, sets are normally denoted by uppercase letters and elements of sets by lower-case letters. The notation $x \in A$ indicates that x is an element of the set A, or, more briefly, x is in A. Then $x \notin A$ means that x is not in A. One way to define a set is simply to indicate its elements between a pair of braces. For instance, $\{1, 2, 3\}$ refers to the set whose elements are 1, 2, and 3, and $\{1, 2, 3, \ldots\}$ indicates the set of all positive integers. (The symbol "$\ldots$" is called an **ellipsis** and is read "and so forth.") Two sets are equal if, and only if, they have exactly the same elements.

Certain sets of numbers are so frequently referred to that they are given special symbolic names. These are summarized in the table below.

Symbol	Set
$\mathbf{R}$	set of all real numbers
$\mathbf{Z}$	set of all integers*
$\mathbf{Q}$	set of all rational numbers, or quotients of integers

*The $\mathbf{Z}$ stands for the first letter of the German word for integers, *Zahlen*.

Addition of a superscript $+$ or $-$ or the letters *nonneg* indicates that only the positive or negative or nonnegative elements of the set, respectively, are to be included. Thus $\mathbf{R}^+$ denotes the set of positive real numbers, and $\mathbf{Z}^{nonneg}$ refers to the set of nonnegative integers: 0, 1, 2, 3, 4, and so forth. Some authors refer to the set of nonnegative integers as the set of **natural numbers** and denote it as **N**. Other authors call only the positive integers natural numbers. To avoid confusion, we simply avoid using the phrase *natural numbers* in this book.

The set of real numbers is usually pictured as the set of all points on a line, as shown below. The number 0 corresponds to a middle point, called the *origin*. A unit of distance is marked off, and each point to the right of the origin corresponds to a positive real number found by computing its distance from the origin. Each point to the left of the origin corresponds to a negative real number, which is denoted by computing its distance from the origin and putting a minus sign in front of the resulting number. The set of real numbers is therefore divided into three parts: the set of positive real numbers, the set of negative real numbers, and the number 0. *Note that 0 is neither positive nor negative.* Labels are given for a few real numbers corresponding to points on the line.

The real number line is called *continuous* because it is imagined to have no holes. The set of integers corresponds to a collection of points located at fixed intervals along the real number line. Thus every integer is a real number, and because the integers are all separated from each other, the set of integers is called *discrete*. The name *discrete mathematics* comes from the distinction between continuous and discrete mathematical objects.

When an element in the domain of the variable of a one-variable predicate is substituted for the variable, the resulting statement is either true or false. The set of all such elements that make the predicate true is called the *truth set* of the predicate.

• Definition

If $P(x)$ is a predicate and x has domain D, the **truth set** of $P(x)$ is the set of all elements of D that make $P(x)$ true when they are substituted for x. The truth set of $P(x)$ is denoted

$$\{x \in D \mid P(x)\}$$

the set of all   such that

which is read "the set of all x in D such that $P(x)$."

Example 2.1.2 Finding the Truth Set of a Predicate

Let $Q(n)$ be the predicate "n is a factor of 8." Find the truth set of $Q(n)$ if

a. the domain of n is the set $\mathbf{Z}^+$ of all positive integers

b. the domain of n is the set $\mathbf{Z}$ of all integers.

Solution

a. The truth set is {1, 2, 4, 8} because these are exactly the positive integers that divide 8 evenly.

b. The truth set is {1, 2, 4, 8, −1, −2, −4, −8} because the negative integers −1, −2, −4, and −8 also divide into 8 without leaving a remainder. ∎

The Universal Quantifier: ∀

*Charles Sanders Peirce
(1839–1914)*

One sure way to change predicates into statements is to assign specific values to all their variables. For example, if x represents the number 35, the sentence "x is (evenly) divisible by 5" is a true statement since $35 = 5 \cdot 7$. Another way to obtain statements from predicates is to add **quantifiers.** Quantifiers are words that refer to quantities such as "some" or "all" and tell for how many elements a given predicate is true. The formal concept of quantifier was introduced into symbolic logic in the late nineteenth century by the American philosopher, logician, and engineer Charles Sanders Peirce and, independently, by the German logician Gottlob Frege.

The symbol ∀ denotes "for all" and is called the **universal quantifier.** For example, another way to express the sentence "All human beings are mortal" is to write

$$\forall \text{ human beings } x, x \text{ is mortal,}$$

or, more formally,

$$\forall x \in S, x \text{ is mortal,}$$

where S denotes the set of all human beings. (Think "for all" when you see the symbol ∀.) The domain of the predicate variable is generally indicated between the ∀ symbol and the variable name (as in ∀ human beings x) or immediately following the variable name (as in $\forall x \in S$). Some other expressions that can be used instead of *for all* are *for every, for arbitrary, for any, for each,* and *given any*. In a sentence such as "∀ real numbers x and $y, x + y = y + x$," the ∀ symbol is understood to refer to both x and y.*

Sentences that are quantified universally are defined as statements by giving them the truth values specified in the following definition:

*Gottlob Frege
(1848–1925)*

> **• Definition**
>
> Let $Q(x)$ be a predicate and D the domain of x. A **universal statement** is a statement of the form "$\forall x \in D, Q(x)$." It is defined to be true if, and only if, $Q(x)$ is true for every x in D. It is defined to be false if, and only if, $Q(x)$ is false for at least one x in D. A value for x for which $Q(x)$ is false is called a **counterexample** to the universal statement.

Example 2.1.3 Truth and Falsity of Universal Statements

a. Let $D = \{1, 2, 3, 4, 5\}$, and consider the statement

$$\forall x \in D, x^2 \geq x.$$

Show that this statement is true.

---

*More formal versions of symbolic logic would require writing a separate ∀ for each variable: "$\forall x \in \mathbf{R}(\forall y \in \mathbf{R}(x + y = y + x))$."

b. Consider the statement

$$\forall x \in \mathbf{R}, x^2 \geq x.$$

Find a counterexample to show that this statement is false.

Solution

a. Check that "$x^2 \geq x$" is true for each individual x in D.

$$1^2 \geq 1, \qquad 2^2 > 2, \qquad 3^2 \geq 3, \qquad 4^2 \geq 4, \qquad 5^2 \geq 5.$$

Hence "$\forall x \in D, x^2 \geq x$" is true.

b. *Counterexample:* Take $x = \frac{1}{2}$. Then x is in $\mathbf{R}$ (since $\frac{1}{2}$ is a real number) and

$$\left(\frac{1}{2}\right)^2 = \frac{1}{4} \ngeq \frac{1}{2}.$$

Hence "$\forall x \in \mathbf{R}, x^2 \geq x$" is false. ■

The technique used to show the truth of the universal statement in Example 2.1.3(a) is called the **method of exhaustion.** It consists of showing the truth of the predicate separately for each individual element of the domain. (The idea is to exhaust the possibilities before you exhaust yourself!) This method can, in theory, be used whenever the domain of the predicate variable is finite. In recent years the prevalence of digital computers has greatly increased the convenience of using the method of exhaustion. Computer expert systems, or knowledge-based systems, use this method to arrive at answers to many of the questions posed to them. Because most mathematical sets are infinite, however, the method of exhaustion can rarely be used to derive general mathematical results.

The Existential Quantifier: ∃

The symbol ∃ denotes "there exists" and is called the **existential quantifier.** For example, the sentence "There is a student in Math 140" can be written as

∃ a person s such that s is a student in Math 140,

or, more formally,

∃$s \in S$ such that s is a student in Math 140,

where S is the set of all people. (Think "there exists" when you see the symbol ∃.) The domain of the predicate variable is generally indicated either between the ∃ symbol and the variable name or immediately following the variable name. The words *such that* are inserted just before the predicate. Some other expressions that can be used in place of *there exists* are *there is a, we can find a, there is at least one, for some,* and *for at least one.* In a sentence such as "∃ integers m and n such that $m + n = m \cdot n$," the ∃ symbol is understood to refer to both m and n.*

Sentences that are quantified existentially are defined as statements by giving them the truth values specified in the following definition.

---

*In more formal versions of symbolic logic, the words *such that* are not written out (although they are understood) and a separate ∃ symbol is used for each variable: "∃$m \in \mathbf{Z}(\exists n \in \mathbf{Z}(m + n = m \cdot n))$."

> • **Definition**
>
> Let $Q(x)$ be a predicate and D the domain of x. An **existential statement** is a statement of the form "$\exists x \in D$ such that $Q(x)$." It is defined to be true if, and only if, $Q(x)$ is true for at least one x in D. It is false if, and only if, $Q(x)$ is false for all x in D.

Example 2.1.4 Truth and Falsity of Existential Statements

a. Consider the statement

$$\exists m \in \mathbf{Z} \text{ such that } m^2 = m.$$

Show that this statement is true.

b. Let $E = \{5, 6, 7, 8, 9, 10\}$ and consider the statement

$$\exists m \in E \text{ such that } m^2 = m.$$

Show that this statement is false.

Solution

a. Observe that $1^2 = 1$. Thus "$m^2 = m$" is true for at least one integer m. Hence "$\exists m \in \mathbf{Z}$ such that $m^2 = m$" is true.

b. Note that $m^2 = m$ is not true for any integers m from 5 to 10:

$$5^2 = 25 \neq 5, \qquad 6^2 = 36 \neq 6, \qquad 7^2 = 49 \neq 7, \qquad 8^2 = 64 \neq 8,$$
$$9^2 = 81 \neq 9, \qquad 10^2 = 100 \neq 10.$$

Thus "$\exists m \in E$ such that $m^2 = m$" is false. ∎

Formal Versus Informal Language

It is important to be able to translate from formal to informal language when trying to make sense of mathematical concepts that are new to you. It is equally important to be able to translate from informal to formal language when thinking out a complicated problem.

Example 2.1.5 Translating from Formal to Informal Language

Rewrite the following formal statements in a variety of equivalent but more informal ways. Do not use the symbol ∀ or ∃.

a. $\forall x \in \mathbf{R}, x^2 \geq 0$.

b. $\forall x \in \mathbf{R}, x^2 \neq -1$.

c. $\exists m \in \mathbf{Z}$ such that $m^2 = m$.

Solution

a. All real numbers have nonnegative squares.
   Every real number has a nonnegative square.
   Any real number has a nonnegative square.
   x has a nonnegative square, for each real number x.
   The square of any real number is nonnegative.
   (Note that the singular noun is used to refer to the domain when the ∀ symbol is translated as *every, any,* or *each*.)

b. All real numbers have squares not equal to -1.
No real numbers have squares equal to -1.
(The words *none are* or *no . . . are* are equivalent to the words *all are not*.)

c. There is an integer whose square is equal to itself.
We can find at least one integer equal to its own square.
$m^2 = m$, for some integer m.
Some integer equals its own square.
Some integers equal their own squares.
(In ordinary English, this last statement might be taken to be true only if there are at least two integers equal to their own squares. In mathematics, we understand the last two statements to mean the same thing.) ■

Example 2.1.6 Translating from Informal to Formal Language

Rewrite each of the following statements formally. Use quantifiers and variables.

a. All triangles have three sides.

b. No dogs have wings.

c. Some programs are structured.

Solution

a. ∀ triangles t, t has three sides, or
$\forall t \in T$, t has three sides (where T is the set of all triangles).

b. ∀ dogs d, d does not have wings, or
$\forall d \in D$, d does not have wings (where D is the set of all dogs).

c. ∃ a program p such that p is structured, or
$\exists p \in P$ such that p is structured (where P is the set of all programs). ■

Universal Conditional Statements

A reasonable argument can be made that the most important form of statement in mathematics is the **universal conditional statement:**

$$\forall x, \text{ if } P(x) \text{ then } Q(x).$$

Familiarity with statements of this form is essential if you are to learn to speak mathematics.

Example 2.1.7 Writing Universal Conditional Statements Informally

Rewrite the following formal statement in a variety of informal ways. Do not use quantifiers or variables.

$$\forall x \in \mathbf{R}, \text{ if } x > 2 \text{ then } x^2 > 4.$$

Solution   If a real number is greater than 2 then its square is greater than 4.

The square of any real number that is greater than 2, its square is greater than 4.

The square of any real number that is greater than 2 is greater than 4.

The squares of all real numbers greater than 2 are greater than 4. ■

Example 2.1.8 Writing Universal Conditional Statements Formally

Rewrite each of the following statements in the form

$$\forall \underline{\qquad}, \text{ if } \underline{\qquad} \text{ then } \underline{\qquad}.$$

a. If a real number is an integer, then it is a rational number.

b. All bytes have eight bits.

c. No fire trucks are green.

Solution

a. $\forall$ real numbers x, if x is an integer, then x is a rational number, or
   $\forall x \in \mathbf{R}$, if $x \in \mathbf{Z}$ then $x \in \mathbf{Q}$.

b. $\forall x$, if x is a byte, then x has eight bits.

c. $\forall x$, if x is a fire truck, then x is not green.

It is common, as in (b) and (c) above, to omit explicit identification of the domain of predicate variables in universal conditional statements. ■

Careful thought about the meaning of universal conditional statements leads to another level of understanding for why the truth table for an if-then statement must be defined as it is. Consider again the statement

$$\forall \text{ real numbers } x, \text{ if } x > 2 \text{ then } x^2 > 4.$$

Your experience and intuition tell you that this statement is true. But that means that

$$\text{If } x > 2 \text{ then } x^2 > 4$$

must be true for every single real number x. Consequently, it must be true even for x that make its hypothesis "$x > 2$" false. In particular, both statements

$$\text{If } 1 > 2 \text{ then } 1^2 > 4 \quad \text{and} \quad \text{If } -3 > 2 \text{ then} (-3)^2 > 4$$

must be true. In both cases the hypothesis is false, but in the first case the conclusion "$1^2 > 4$" is false, and in the second case the conclusion "$(-3)^2 > 4$" is true. Hence, regardless of whether its conclusion is true or false, an if-then statement with a false hypothesis must be true.

Note also that the definition of valid argument is a universal conditional statement:

$$\forall \text{ combinations of truth values for the component statements,}$$
if the premises are all true then the conclusion is also true.

Equivalent Forms of Universal and Existential Statements

Observe that the two statements "$\forall$ real numbers x, if x is an integer then x is rational" and "$\forall$ integers x, x is rational" mean the same thing. Both have informal translations "All integers are rational." In fact, a statement of the form

$$\forall x \in U, \text{ if } P(x) \text{ then } Q(x)$$

can always be rewritten in the form

$$\forall x \in D, Q(x)$$

by narrowing U to be the domain D consisting of all values of the variable x that make $P(x)$ true. Conversely, a statement of the form

$$\forall x \in D, \, Q(x)$$

can be rewritten as

$$\forall x, \text{ if } x \text{ is in } D \text{ then } Q(x).$$

Example 2.1.9 Equivalent Forms for Universal Statements

Rewrite the following statement in the two forms "$\forall x$, if _____ then _____" and "$\forall$ _____ x, _____": All squares are rectangles.

Solution $\forall x$, if x is a square then x is a rectangle.

                                $\forall$ squares x, x is a rectangle. ■

Example 2.1.10 Equivalent Forms for Existential Statements

A **prime number** is an integer greater than 1 whose only positive integer factors are itself and 1. Consider the statement "There is an integer that is both prime and even." Let Prime(n) be "n is prime" and Even(n) be "n is even." Use the notation Prime(n) and Even(n) to rewrite this statement in the following two forms:

a. $\exists n$ such that _____ $\wedge$ _____.

b. $\exists$ _____ n such that _____.

Solution

a. $\exists n$ such that Prime(n) $\wedge$ Even(n).

b. Two answers: $\exists$ a prime number n such that Even(n).
                      $\exists$ an even number n such that Prime(n).

Implicit Quantification

Consider the statement

                 If a number is an integer, then it is a rational number.

As shown earlier, this statement is equivalent to a universal statement. However, it does not contain the telltale word *all* or *every* or *any* or *each*. The only clue to indicate its universal quantification comes from the presence of the indefinite article *a*. This is an example of *implicit* universal quantification.

Existential quantification can also be implicit. For instance, the statement "The number 24 can be written as a sum of two even integers" can be expressed formally as "$\exists$ even integers m and n such that $24 = m + n$."

Mathematical writing contains many examples of implicitly quantified statements. Some occur, as in the first example above, through the presence of the word *a* or *an*. Others occur in cases where the general context of a sentence supplies part of its meaning. For example, in an algebra course in which the letter x is always used to indicate a real number, the predicate

$$\text{If } x > 2 \text{ then } x^2 > 4$$

is interpreted to mean the same as the statement

$$\forall \text{ real numbers } x, \text{ if } x > 2 \text{ then } x^2 > 4.$$

Mathematicians often use a double arrow to indicate implicit quantification symbolically. For instance, they might express the above statement as

$$x > 2 \quad \Rightarrow \quad x^2 > 4.$$

• Notation

Let $P(x)$ and $Q(x)$ be predicates and suppose the common domain of x is D. The notation $P(x) \Rightarrow Q(x)$ means that every element in the truth set of $P(x)$ is in the truth set of $Q(x)$, or, equivalently, $\forall x, P(x) \rightarrow Q(x)$. The notation $P(x) \Leftrightarrow Q(x)$ means that $P(x)$ and $Q(x)$ have identical truth sets, or, equivalently, $\forall x, P(x) \leftrightarrow Q(x)$.

Example 2.1.11 Using $\Rightarrow$ and $\Leftrightarrow$

Let

$$Q(n) \text{ be "}n \text{ is a factor of 8,"}$$
$$R(n) \text{ be "}n \text{ is a factor of 4,"}$$
$$S(n) \text{ be "}n < 5 \text{ and } n \neq 3,\text{"}$$

and suppose the domain of n is $\mathbf{Z}^+$, the set of positive integers. Use the $\Rightarrow$ and $\Leftrightarrow$ symbols to indicate true relationships among $P(n)$, $Q(n)$, and $R(n)$.

Solution

1. As noted in Example 2.1.2, the truth set of $Q(n)$ is $\{1, 2, 4, 8\}$ when the domain of n is $\mathbf{Z}^+$. By similar reasoning the truth set of $R(n)$ is $\{1, 2, 4\}$. Thus it is true that every element in the truth set of $R(n)$ is in the truth set of $Q(n)$, or, equivalently, $\forall n$ in $\mathbf{Z}^+$, $R(n) \rightarrow Q(n)$. So $R(n) \Rightarrow Q(n)$, or, equivalently

$$n \text{ is a factor of 4} \quad \Rightarrow \quad n \text{ is a factor of 8.}$$

2. The truth set of $S(n)$ is $\{1, 2, 4\}$, which is identical to the truth set of $R(n)$, or, equivalently, $\forall n$ in $\mathbf{Z}^+$, $R(n) \leftrightarrow S(n)$. So $R(n) \Leftrightarrow S(n)$, or, equivalently,

$$n \text{ is a factor of 4} \quad \Leftrightarrow \quad n < 5 \text{ and } n \neq 3.$$

   Moreover, since every element in the truth set of $S(n)$ is in the truth set of $Q(n)$, or, equivalently, $\forall n$ in $\mathbf{Z}^+$, $S(n) \rightarrow Q(n)$, then $S(n) \Rightarrow Q(n)$, or, equivalently,

$$n < 5 \text{ and } n \neq 3 \quad \Rightarrow \quad n \text{ is a factor of 8.} \qquad \blacksquare$$

Some questions of quantification can be quite subtle. For instance, a mathematics text might contain the following:

a. $(x + 1)^2 = x^2 + 2x + 1$.   b. Solve $(x + 2)^2 = 25$.

Although neither (a) nor (b) contains explicit quantification, the reader is supposed to understand that the x in (a) is universally quantified whereas the x in (b) is existentially quantified. When the quantification is made explicit, (a) and (b) become

a. $\forall$ real numbers x, $(x + 1)^2 = x^2 + 2x + 1$.

b. Show (by finding a value) that $\exists$ a real number x such that $(x + 2)^2 = 25$.

The quantification of a statement—whether universal or existential—crucially determines both how the statement can be applied and what method must be used to establish its truth. Thus it is important to be alert to the presence of hidden quantifiers when you read mathematics so that you will interpret statements in a logically correct way.

Tarski's World

Tarski's World is a computer program developed by information scientists Jon Barwise and John Etchemendy to help teach the principles of logic. It is described in their book *The Language of First-Order Logic,* which is accompanied by a CD-Rom containing the program Tarski's World, named after the great logician Alfred Tarski.

Example 2.1.12 Investigating Tarski's World

*Alfred Tarski
(1902–1983)*

The program for Tarski's World provides pictures of blocks of various sizes, shapes, and colors, which are located on a grid. Shown in Figure 2.1.1 is a picture of an arrangement of objects in a two-dimensional Tarski world. The configuration can be described using logical operators and—for the two-dimensional version—notation such as Triangle(x), meaning "x is a triangle," Blue(y), meaning "y is blue," and RightOf(x, y), meaning "x is to the right of y (but possibly in a different row)." Individual objects can be given names such as a, b, or c.

Figure 2.1.1

Determine the truth or falsity of each of the following statements. The domain for all variables is the set of objects in the Tarski world shown above.

a. $\forall t$, Triangle(t) $\rightarrow$ Blue(t).

b. $\forall x$, Blue(x) $\rightarrow$ Triangle(x).

c. $\exists y$ such that Square(y) $\wedge$ RightOf(d, y).

d. $\exists z$ such that Square(z) $\wedge$ Gray(z).

Solution

a. This statement is true: all the triangles are blue.

b. This statement is false. As a counterexample, note that e is blue and it is not a triangle.

c. This statement is true because e and h are both square and d is to their right.

d. This statement is false: all the squares are either blue or black.

Exercise Set 2.1*

1. A menagerie consists of seven brown dogs, two black dogs, six gray cats, ten black cats, five blue birds, six yellow birds, and one black bird. Determine which of the following statements are true and which are false.
   a. There is an animal in the menagerie that is red.
   b. Every animal in the menagerie is a bird or a mammal.
   c. Every animal in the menagerie is brown or gray or black.
   d. There is an animal in the menagerie that is neither a cat nor a dog.
   e. No animal in the menagerie is blue.
   f. There are in the menagerie a dog, a cat, and a bird that all have the same color.

2. Indicate which of the following statements are true and which are false. Justify your answers as best as you can.
   a. Every integer is a real number.
   b. 0 is a positive real number.
   c. For all real numbers r, $-r$ is a negative real number.
   d. Every real number is an integer.

3. Let $P(x)$ be the predicate "$x > 1/x$."
   a. Write $P(2)$, $P(\frac{1}{2})$, $P(-1)$, $P(-\frac{1}{2})$, and $P(-8)$, and indicate which of these statements are true and which are false.
   b. Find the truth set of $P(x)$ if the domain of x is $\mathbf{R}$, the set of all real numbers.
   c. If the domain is the set $\mathbf{R}^+$ of all positive real numbers, what is the truth set of $P(x)$?

4. Let $Q(n)$ be the predicate "$n^2 \leq 30$."
   a. Write $Q(2)$, $Q(-2)$, $Q(7)$, and $Q(-7)$, and indicate which of these statements are true and which are false.
   b. Find the truth set of $Q(n)$ if the domain of n is $\mathbf{Z}$, the set of all integers.
   c. If the domain is the set $\mathbf{Z}^+$ of all positive integers, what is the truth set of $Q(n)$?

5. Let $Q(x, y)$ be the predicate "If $x < y$ then $x^2 < y^2$" with domain for both x and y being the set $\mathbf{R}$ of real numbers.
   a. Explain why $Q(x, y)$ is false if $x = -2$ and $y = 1$.
   b. Give values different from those in part (a) for which $Q(x, y)$ is false.
   c. Explain why $Q(x, y)$ is true if $x = 3$ and $y = 8$.
   d. Give values different from those in part (c) for which $Q(x, y)$ is true.

6. Let $R(m, n)$ be the predicate "If m is a factor of n^2 then m is a factor of n," with domain for both m and n being the set $\mathbf{Z}$ of integers.
   a. Explain why $R(m, n)$ is false if $m = 25$ and $n = 10$.
   b. Give values different from those in part (a) for which $R(m, n)$ is false.
   c. Explain why $R(m, n)$ is true if $m = 5$ and $n = 10$.

d. Give values different from those in part (c) for which $R(m, n)$ is true.

7. Find the truth set of each predicate.
   a. predicate: $6/d$ is an integer, domain: $\mathbf{Z}$
   b. predicate: $6/d$ is an integer, domain: $\mathbf{Z}^+$
   c. predicate: $1 \leq x^2 \leq 4$, domain: $\mathbf{R}$
   d. predicate: $1 \leq x^2 \leq 4$, domain: $\mathbf{Z}$

8. Let $B(x)$ be "$-10 < x < 10$." Find the truth set of $B(x)$ for each of the following domains.
   a. $\mathbf{Z}$ b. $\mathbf{Z}^+$ c. The set of all even integers

Find counterexamples to show that the statements in 9–12 are false.

9. $\forall x \in \mathbf{R}, x > 1/x$.

10. $\forall a \in \mathbf{Z}, (a - 1)/a$ is not an integer.

11. $\forall$ positive integers m and n, $m \cdot n \geq m + n$.

12. $\forall$ real numbers x and y, $\sqrt{x + y} = \sqrt{x} + \sqrt{y}$.

13. Consider the following statement:

    $\forall$ basketball players x, x is tall.

    Which of the following are equivalent ways of expressing this statement?
    a. Every basketball player is tall.
    b. Among all the basketball players, some are tall.
    c. Some of all the tall people are basketball players.
    d. Anyone who is tall is a basketball player.
    e. All people who are basketball players are tall.
    f. Anyone who is a basketball player is a tall person.

14. Consider the following statement:

    $\exists x \in \mathbf{R}$ such that $x^2 = 2$.

    Which of the following are equivalent ways of expressing this statement?
    a. The square of each real number is 2.
    b. Some real numbers have square 2.
    c. The number x has square 2, for some real number x.
    d. If x is a real number, then $x^2 = 2$.
    e. Some real number has square 2.
    f. There is at least one real number whose square is 2.

15. Rewrite the following statements informally in at least two different ways without using variables or the symbol $\forall$ or $\exists$.
    a. $\forall$ squares x, x is a rectangle.
    b. $\exists$ a set A such that A has 16 subsets.

16. Rewrite each of the following statements in the form "$\forall _____ x, _____.$"
    a. All dinosaurs are extinct.
    b. Every real number is positive, negative, or zero.

---

*For exercises with blue numbers or letters, solutions are given in Appendix B. The symbol *H* indicates that only a hint or a partial solution is given. The symbol ✴ signals that an exercise is more challenging than usual.

   c. No irrational numbers are integers.

   d. No logicians are lazy.

   e. The number 2,147,581,953 is not equal to the square of any integer.

   f. The number -1 is not equal to the square of any real number.

17. Rewrite each of the following in the form "∃ _____ x such that _____."

   a. Some exercises have answers.

   b. Some real numbers are rational.

18. Let D be the set of all students at your school, and let $M(s)$ be "s is a math major," let $C(s)$ be "s is a computer science student," and let $E(s)$ be "s is an engineering student." Express each of the following statements using quantifiers, variables, and the predicates $M(s)$, $C(s)$, and $E(s)$.

   a. There is an engineering student who is a math major.

   b. Every computer science student is an engineering student.

   c. No computer science students are engineering students.

   d. Some computer science students are also math majors.

   e. Some computer science students are engineering students and some are not.

19. Consider the following statement:

$$\forall \text{ integers } n, \text{ if } n^2 \text{ is even then } n \text{ is even.}$$

Which of the following are equivalent ways of expressing this statement?

   a. All integers have even squares and are even.

   b. Given any integer whose square is even, that integer is itself even.

   c. For all integers, there are some whose square is even.

   d. Any integer with an even square is even.

   e. If the square of an integer is even, then that integer is even.

   f. All even integers have even squares.

20. Rewrite the following statement informally in at least two different ways without using variables or the symbol ∀ or ∃.

$$\forall \text{ students } S, \text{ if } S \text{ is in CSC 321}$$
$$\text{then } S \text{ has taken MAT 140.}$$

21. Rewrite each of the following statements in the form "∀ _____ x, if _____ then _____" or "∀ _____ x and y, if _____ then _____."

   a. All Java programs have at least 5 lines.

   b. Any valid argument with true premises has a true conclusion.

   c. The sum of any two even integers is even.

   d. The product of any two odd integers is odd.

22. Rewrite each of the following statements in the two forms "∀x, if _____ then _____" and "∀ _____ x, _____" (without an if-then).

   a. The square of any even integer is even.

   b. Every computer science student needs to take data structures.

23. Rewrite the following statements in the two forms "∃ _____ x such that _____" and "∃x such that _____ and _____."

   a. Some hatters are mad.   **b.** Some questions are easy.

24. Consider the statement "All integers are rational numbers but some rational numbers are not integers."

   a. Write this statement in the form "∀x, if _____ then _____, but ∃ _____ x such that _____."

   b. Let Ratl(x) be "x is a rational number" and Int(x) be "x is an integer." Write the given statement formally using only the symbols Ratl(x), Int(x), ∀, ∃, ∧, ∨, ~, and →.

25. Refer to the picture of Tarski's world given in Example 2.1.12. Let Above(x, y) mean that x is above y (but possibly in a different column). Determine the truth or falsity of each of the following statements. Give reasons for your answers.

   a. $\forall u$, Circle(u) → Gray(u).

   b. $\forall u$, Gray(u) → Circle(u).

   c. ∃y such that Square(y) ∧ Above(y, d).

   d. ∃z such that Triangle(z) ∧ Above(f, z).

In 26–28, rewrite each statement without using quantifiers or variables. Indicate which are true and which are false, and justify your answers as best as you can.

26. Let the domain of x be the set D of objects discussed in mathematics courses, and let Real(x) be "x is a real number," Pos(x) be "x is a positive real number," Neg(x) be "x is a negative real number," and Int(x) be "x is an integer."

   a. Pos(0)

   b. $\forall x$, Real(x) ∧ Neg(x) → Pos($-x$).

   c. $\forall x$, Int(x) → Real(x).

   d. ∃x such that Real(x) ∧ ~Int(x).

27. Let the domain of x be the set of geometric figures in the plane, and let Square(x) be "x is a square" and Rect(x) be "x is a rectangle."

   a. ∃x such that Rect(x) ∧ Square(x).

   b. ∃x such that Rect(x) ∧ ~Square(x).

   c. $\forall x$, Square(x) → Rect(x).

28. Let the domain of x be the set **Z** of integers, and let Odd(x) be "x is odd," Prime(x) be "x is prime," and Square(x) be "x is a perfect square." (An integer n is said to be a **perfect square** if, and only if, it equals the square of some integer. For example, 25 is a perfect square because $25 = 5^2$.)

   a. ∃x such that Prime(x) ∧ ~Odd(x).

   b. $\forall x$, Prime(x) → ~Square(x).

   c. ∃x such that Odd(x) ∧ Square(x).

H 29. In any mathematics or computer science text other than this book, find an example of a statement that is universal but is implicitly quantified. Copy the statement as it appears and rewrite it making the quantification explicit. Give a complete citation for your example, including title, author, publisher, year, and page number.

30. Let **R** be the domain of the predicate variable x. Which of the following are true and which are false? Give counterexamples for the statements that are false.
    a. $x > 2 \Rightarrow x > 1$
    b. $x > 2 \Rightarrow x^2 > 4$
    c. $x^2 > 4 \Rightarrow x > 2$
    d. $x^2 > 4 \Leftrightarrow |x| > 2$

31. Let **R** be the domain of the predicate variables a, b, c, and d. Which of the following are true and which are false? Give counterexamples for the statements that are false.
    a. $a > 0$ and $b > 0 \Rightarrow ab > 0$
    b. $a < 0$ and $b < 0 \Rightarrow ab < 0$
    c. $ab = 0 \Rightarrow a = 0$ or $b = 0$
    d. $a < b$ and $c < d \Rightarrow ac < bd$

2.2 Introduction to Predicates and Quantified Statements II

TOUCHSTONE: *Stand you both forth now: stroke your chins, and swear by your beards that I am a knave.*
CELIA: *By our beards—if we had them—thou art.*
TOUCHSTONE: *By my knavery—if I had it—then I were; but if you swear by that that is not, you are not forsworn.* — William Shakespeare, *As You Like It*

This section continues the discussion of predicates and quantified statements begun in Section 2.1. It contains the rules for negating quantified statements; an exploration of the relation among ∀, ∃, ∧, and ∨; an introduction to the concept of vacuous truth of universal statements; examples of variants of universal conditional statements; and an extension of the meaning of *necessary, sufficient,* and *only if* to quantified statements.

Negations of Quantified Statements

Consider the statement "All mathematicians wear glasses." Many people would say that its negation is "No mathematicians wear glasses." In fact, the negation is "One or more mathematicians do not wear glasses" or "Some mathematicians do not wear glasses." After all, if even one mathematician does not wear glasses, the sweeping statement that all mathematicians wear glasses must be false.

The general form of the negation of a universal statement follows immediately from the definitions of negation and of the truth values for universal and existential statements.

Theorem 2.2.1 Negation of a Universal Statement

The negation of a statement of the form

$$\forall x \text{ in } D, Q(x)$$

is logically equivalent to a statement of the form

$$\exists x \text{ in } D \text{ such that } \sim Q(x).$$

Symbolically,

$$\sim(\forall x \in D, Q(x)) \equiv \exists x \in D \text{ such that } \sim Q(x).$$

Thus

The negation of a universal statement ("all are") is logically equivalent to an existential statement ("some are not").

Note that when we speak of **logical equivalence for quantified statements,** we mean that the statements always have identical truth values no matter what predicates are substituted for the predicate variables and no matter what sets are used for the domains of the predicate variables.

Now consider the statement "Some fish breathe air." What is its negation? Many people would answer that it is "Some fish do not breathe air." Actually, the negation is "*No* fish breathe air." After all, if it is not true that some fish breathe air, then not a single fish breathes air. "That is, no fish breathe air, or all fish are non–air-breathers.

The general form for the negation of an existential statement follows immediately from the definitions of negation and of the truth values for existential and universal statements.

Theorem 2.2.2 Negation of an Existential Statement

The negation of a statement of the form

$$\exists x \text{ in } D \text{ such that } Q(x)$$

is logically equivalent to a statement of the form

$$\forall x \text{ in } D, \sim Q(x).$$

Symbolically,

$$\sim(\exists x \in D \text{ such that } Q(x)) \equiv \forall x \in D, \sim Q(x).$$

Thus

The negation of an existential statement ("some are") is logically equivalent to a universal statement ("all are not").

Example 2.2.1 Negating Quantified Statements

Write formal negations for the following statements:

a. $\forall$ primes p, p is odd.

b. $\exists$ a triangle T such that the sum of the angles of T equals $200°$.

Solution

a. By applying the rule for the negation of a $\forall$ statement, you can see that the answer is

$$\exists \text{ a prime } p \text{ such that } p \text{ is not odd.}$$

b. By applying the rule for the negation of a $\exists$ statement, you can see that the answer is

$$\forall \text{ triangles } T, \text{ the sum of the angles of } T \text{ does not equal } 200°. \qquad \blacksquare$$

You need to exercise special care to avoid mistakes when writing negations of statements that are given informally. One way to avoid error is to rewrite the statement formally and take the negation using the formal rule.

Example 2.2.2 More Negations

Rewrite the following statement formally. Then write formal and informal negations.

No politicians are honest.

Solution   *Formal version:* $\forall$ politicians x, x is not honest.

Formal negation: $\exists$ a politician x such that x is honest.

Informal negation: Some politicians are honest. $\qquad \blacksquare$

Another way to avoid error when taking negations of statements that are given in informal language is to ask yourself, "What *exactly* would it mean for the given statement to be false? What statement, if true, would be equivalent to saying that the given statement is false?"

Example 2.2.3 Still More Negations

Write informal negations for the following statements:

a. All computer programs are finite.

b. Some computer hackers are over 40.

Solution

a. What exactly would it mean for this statement to be false? The statement asserts a property for all computer programs. So for it to be false, there would simply have to be some computer program that does not have the property. Thus the answer is

There is a computer program that is not finite.

Or:

Some computer programs are not finite.

b. This statement is equivalent to saying that there is at least one computer hacker with a certain property. So for it to be false, not a single computer hacker can have that property. Thus the negation is

No computer hackers are over 40.

Or:

All computer hackers are 40 or under. ∎

Caution! Informal negations of many universal statements can be constructed simply by inserting the word *not* or the words *do not* at an appropriate place. However, the resulting statements may be ambiguous. For example, a possible negation of "All mathematicians wear glasses" is "All mathematicians do not wear glasses." The problem is that this sentence has two meanings. With the proper verbal stress on the word *not*, it could be interpreted as the logical negation. (What! You say that all mathematicians wear glasses? Nonsense! All mathematicians do *not* wear glasses.) On the other hand, stated in a flat tone of voice (try it!), it would mean that all mathematicians are nonwearers of glasses; that is, not a single mathematician wears glasses. This is a much stronger statement than the logical negation: It implies the negation but is not equivalent to it.

Negations of Universal Conditional Statements

Negations of universal conditional statements are of special importance in mathematics. The form of such negations can be derived from facts that have already been established. By definition of the negation of a *for all* statement,

$$\sim(\forall x,\, P(x) \rightarrow Q(x)) \;\equiv\; \exists x \text{ such that } \sim(P(x) \rightarrow Q(x)). \qquad \text{2.2.1}$$

But the negation of an if-then statement is logically equivalent to an *and* statement. More precisely,

$$\sim(P(x) \rightarrow Q(x)) \equiv P(x) \wedge \sim Q(x).$$ 2.2.2

Substituting (2.2.2) into (2.2.1) gives

$$\sim(\forall x,\ P(x) \rightarrow Q(x)) \equiv \exists x \text{ such that } (P(x) \wedge \sim Q(x)).$$

Written less symbolically, this becomes

$$\sim(\forall x, \text{ if } P(x) \text{ then } Q(x)) \equiv \exists x \text{ such that } P(x) \text{ and } \sim Q(x).$$

Example 2.2.4 Negating Universal Conditional Statements

Write a formal negation for statement (a) and an informal negation for statement (b).

a. $\forall$ people p, if p is blond then p has blue eyes.

b. If a computer program has more than 100,000 lines, then it contains a bug.

Solution

a. $\exists$ a person p such that p is blond and p does not have blue eyes.

b. There is at least one computer program that has more than 100,000 lines and does not contain a bug. ∎

The Relation among ∀, ∃, ∧, and ∨

The negation of a *for all* statement is a *there exists* statement, and the negation of a *there exists* statement is a *for all* statement. These facts are analogous to De Morgan's laws, which state that the negation of an *and* statement is an *or* statement and that the negation of an *or* statement is an *and* statement. This similarity is not accidental. In a sense, universal statements are generalizations of *and* statements, and existential statements are generalizations of *or* statements.

If $Q(x)$ is a predicate and the domain D of x is the set $\{x_1, x_2, \ldots, x_n\}$, then the statements

$$\forall x \in D,\ Q(x)$$

and

$$Q(x_1) \wedge Q(x_2) \wedge \cdots \wedge Q(x_n)$$

are logically equivalent. For example, let $Q(x)$ be "$x \cdot x = x$" and suppose $D = \{0, 1\}$. Then

$$\forall x \in D,\ Q(x)$$

can be rewritten as

$$\forall \text{ binary digits } x,\ x \cdot x = x.$$

This is equivalent to

$$0 \cdot 0 = 0 \quad \text{and} \quad 1 \cdot 1 = 1,$$

which can be rewritten in symbols as

$$Q(0) \wedge Q(1).$$

Similarly, if $Q(x)$ is a predicate and $D = \{x_1, x_2, \ldots, x_n\}$, then the statements

$$\exists x \in D \text{ such that } Q(x)$$

and

$$Q(x_1) \vee Q(x_2) \vee \cdots \vee Q(x_n)$$

are logically equivalent. For example, let $Q(x)$ be "$x + x = x$" and suppose $D = \{0, 1\}$. Then

$$\exists x \in D \text{ such that } Q(x)$$

can be rewritten as

$$\exists \text{ a binary digit } x \text{ such that } x + x = x.$$

This is equivalent to

$$0 + 0 = 0 \quad \text{or} \quad 1 + 1 = 1,$$

which can be rewritten in symbols as

$$Q(0) \vee Q(1).$$

Vacuous Truth of Universal Statements

Suppose a bowl sits on a table and next to the bowl is a pile of five blue and five gray balls, any of which may be placed in the bowl. If three blue balls and one gray ball are placed in the bowl, as shown in Figure 2.2.1(a), the statement "All the balls in the bowl are blue" would be false (since one of the balls in the bowl is gray).

Now suppose that no balls at all are placed in the bowl, as shown in Figure 2.2.1(b). Consider the statement

All the balls in the bowl are blue.

Is this statement true or false? The statement is false if, and only if, its negation is true. And its negation is

There exists a ball in the bowl that is not blue.

But the only way this negation can be true is for there actually to be a nonblue ball in the bowl. And there is not! Hence the negation is false, and so the statement is true "by default."

(a)   (b)

Figure 2.2.1

In general, a statement of the form

$$\forall x \text{ in } D, \text{ if } P(x) \text{ then } Q(x)$$

is called **vacuously true** or **true by default** if, and only if, $P(x)$ is false for every x in D.

By the way, in ordinary language the words *in general* mean that something is usually, but not always, the case. (In general, I take the bus home, but today I walked.) In mathematics, the words *in general* are used quite differently. When they occur just after discussion of a particular example (as in the preceding paragraph), they are a signal that what is to follow is a generalization of some aspect of the example that always holds true.

Variants of Universal Conditional Statements

Recall from Section 1.2 that a conditional statement has a contrapositive, a converse, and an inverse. The definitions of these terms can be extended to universal conditional statements.

• Definition

Consider a statement of the form

$$\forall x \in D, \text{ if } P(x) \text{ then } Q(x).$$

1. Its **contrapositive** is the statement

$$\forall x \in D, \text{ if } \sim Q(x) \text{ then } \sim P(x).$$

2. Its **converse** is the statement

$$\forall x \in D, \text{ if } Q(x) \text{ then } P(x).$$

3. Its **inverse** is the statement

$$\forall x \in D, \text{ if } \sim P(x) \text{ then } \sim Q(x).$$

Example 2.2.5 Contrapositive, Converse, and Inverse of a Universal Conditional Statement

Write the contrapositive, converse, and inverse for the following statement:

If a real number is greater than 2, then its square is greater than 4.

Solution   The formal version of this statement is $\forall x \in \mathbf{R}$, if $x > 2$ then $x^2 > 4$.

Contrapositive: $\forall x \in \mathbf{R}$, if $x^2 \leq 4$ then $x \leq 2$; or,
  If the square of a real number is less than or equal to 4,
  then the number is less than or equal to 2.

Converse: $\forall x \in \mathbf{R}$, if $x^2 > 4$ then $x > 2$; or,
  If the square of a real number is greater than 4, then the
  number is greater than 2.

Inverse: $\forall x \in \mathbf{R}$, if $x \leq 2$ then $x^2 \leq 4$; or
  If a real number is less that or equal to 2, then the square
  of the number is less than or equal to 4.

Note that in solving this example, we have used the equivalence of "$x \not> a$" and "$x \leq a$" for all real numbers x and a. (See page 11.) ∎

In Section 1.2 we showed that a conditional statement is logically equivalent to its contrapositive and that it is not logically equivalent to either its converse or its inverse. The following discussion shows that these facts generalize to the case of universal conditional statements and their contrapositives, converses, and inverses.

Let $P(x)$ and $Q(x)$ be any predicates, let D be the domain of x, and consider the statement

$$\forall x \in D, \text{ if } P(x) \text{ then } Q(x)$$

and its contrapositive

$$\forall x \in D, \text{if } \sim Q(x) \text{ then } \sim P(x).$$

Any particular x in D that makes "if $P(x)$ then $Q(x)$" true also makes "if $\sim Q(x)$ then $\sim P(x)$" true (by the logical equivalence between $p \to q$ and $\sim q \to \sim p$). It follows that the sentence "If $P(x)$ then $Q(x)$" is true for all x in D if, and only if, the sentence "If $\sim Q(x)$ then $\sim P(x)$" is true for all x in D. This is what is meant, in the predicate calculus, by the statements

$$\forall x \in D, \text{if } P(x) \text{ then } Q(x) \quad \text{and} \quad \forall x \in D, \text{if } \sim Q(x) \text{ then } \sim P(x)$$

being logically equivalent to each other. Thus we write the following and say that a universal conditional statement is logically equivalent to its contrapositive:

$$\boxed{\forall x \in D, \text{if } P(x) \text{ then } Q(x) \equiv \forall x \in D, \text{if } \sim Q(x) \text{ then } \sim P(x)}$$

In Example 2.2.5 we noted that the statement

$$\forall x \in \mathbf{R}, \text{if } x > 2 \text{ then } x^2 > 4$$

has the converse

$$\forall x \in \mathbf{R}, \text{if } x^2 > 4 \text{ then } x > 2.$$

Observe that the statement is true whereas its converse is false (since, for instance, $(-3)^2 = 9 > 4$ but $-3 \not> 2$). This shows that a universal conditional statement may have a different truth value from its converse. Hence a universal conditional statement is not logically equivalent to its converse. This is written in symbols as follows:

$$\boxed{\forall x \in D, \text{if } P(x) \text{ then } Q(x) \not\equiv \forall x \in D, \text{if } Q(x) \text{ then } P(x).}$$

In the exercises at the end of this section, you are asked to show similarly that a universal conditional statement is not logically equivalent to its inverse.

$$\boxed{\forall x \in D, \text{if } P(x) \text{ then } Q(x) \not\equiv \forall x \in D, \text{if } \sim P(x) \text{ then } \sim Q(x).}$$

Necessary and Sufficient Conditions, Only If

The definitions of *necessary, sufficient,* and *only if* can also be extended to apply to universal conditional statements.

• Definition

1. "$\forall x, r(x)$ is a **sufficient condition** for $s(x)$" means "$\forall x, \text{if } r(x) \text{ then } s(x)$."

2. "$\forall x, r(x)$ is a **necessary condition** for $s(x)$" means "$\forall x, \text{if } \sim r(x) \text{ then } \sim s(x)$" or, equivalently, "$\forall x, \text{if } s(x) \text{ then } r(x)$."

3. "$\forall x, r(x)$ **only if** $s(x)$" means "$\forall x, \text{if } \sim s(x) \text{ then } \sim r(x)$" or, equivalently, "$\forall x, \text{if } r(x) \text{ then } s(x)$."

Example 2.2.6 Necessary and Sufficient Conditions

Rewrite the following statements as quantified conditional statements. Do not use the word *necessary* or *sufficient*.

a. Squareness is a sufficient condition for rectangularity.

b. Being at least 35 years old is a necessary condition for being President of the United States.

Solution

a. A formal version of the statement is

$$\forall x, \text{ if } x \text{ is a square, then } x \text{ is a rectangle.}$$

Or, in informal language:

$$\text{If a figure is a square, then it is a rectangle.}$$

b. Using formal language, you could write the answer as

$$\forall \text{ people } x, \text{ if } x \text{ is younger than 35, then } x$$
$$\text{cannot be President of the United States.}$$

Or, by the equivalence between a statement and its contrapositive:

$$\forall \text{ people } x, \text{ if } x \text{ is President of the United States,}$$
$$\text{then } x \text{ is at least 35 years old.} \qquad \blacksquare$$

Example 2.2.7 Only If

Rewrite the following as a universal conditional statement:

$$\text{A product of two numbers is 0 only if one of the numbers is 0.}$$

Solution   Using informal language, you could write the answer as

$$\text{If neither of two numbers is 0, then the product of the numbers is not 0.}$$

Or, by the equivalence between a statement and its contrapositive,

$$\text{If a product of two numbers is 0, then one of the numbers is 0.} \qquad \blacksquare$$

Exercise Set 2.2

1. Which of the following is a negation for "All discrete mathematics students are athletic." More than one answer may be correct.
a. There is a discrete mathematics student who is nonathletic.
b. All discrete mathematics students are nonathletic.
c. There is an athletic person who is a discrete mathematics student.
d. No discrete mathematics students are athletic.
e. Some discrete mathematics students are nonathletic.
f. Some nonathletic people are not discrete mathematics students.

2. Which of the following is a negation for "All dogs are loyal"? More than one answer may be correct.
a. All dogs are disloyal.   b. No dogs are loyal.
c. Some dogs are disloyal.   d. Some dogs are loyal.

e. There is a disloyal animal that is not a dog.
f. There is a dog that is disloyal.
g. No animals that are not dogs are loyal.
h. Some animals that are not dogs are loyal.

3. Write a formal negation for each of the following statements:
a. $\forall$ fish x, x has gills.
b. $\forall$ computers c, c has a CPU.
c. $\exists$ a movie m such that m is over 6 hours long.
d. $\exists$ a band b such that b has won at least 10 Grammy awards.

4. Write an informal negation for each of the following statements:
a. All pots have lids.   b. All birds can fly.
c. Some pigs can fly.   d. Some dogs have spots.

In 5 and 6, write a formal and an informal negation for each statement in the referenced exercise.

H **5.** Section 2.1, exercise 16

H **6.** Section 2.1, exercise 17

7. Informal language is actually more complex than formal language. That is what makes the job of a systems analyst so challenging. A systems analyst works as an intermediary between a client who uses informal language and a programmer who needs precise specifications in order to produce code. For instance, the sentence "There are no orders from store *A* for item *B*" contains the words *there are*. Is the statement existential? Write an informal negation for the statement, and then write the statement formally using quantifiers and variables.

8. Consider the statement "There are no simple solutions to life's problems." Write an informal negation for the statement, and then write the statement formally using quantifiers and variables.

Write a negation for each statement in 9 and 10.

9. ∀ real numbers x, if $x > 3$ then $x^2 > 9$.

10. ∀ computer programs P, if P compiles without error messages, then P is correct.

In each of 11–14 determine whether the proposed negation is correct. If it is not, write a correct negation.

11. *Statement:* The sum of any two irrational numbers is irrational.
Proposed negation: The sum of any two irrational numbers is rational.

12. *Statement:* The product of any irrational number and any rational number is irrational.
Proposed negation: The product of any irrational number and any rational number is rational.

13. *Statement:* For all integers n, if n^2 is even then n is even.
Proposed negation: For all integers n, if n^2 is even then n is not even.

14. *Statement:* For all real numbers x_1 and x_2, if $x_1^2 = x_2^2$ then $x_1 = x_2$.
Proposed negation: For all real numbers x_1 and x_2, if $x_1^2 = x_2^2$ then $x_1 \neq x_2$.

15. Let $D = \{-48, -14, -8, 0, 1, 3, 16, 23, 26, 32, 36\}$. Determine which of the following statements are true and which are false. Provide counterexamples for those statements that are false.
   a. $\forall x \in D$, if x is odd then $x > 0$.
   b. $\forall x \in D$, if x is less than 0 then x is even.
   c. $\forall x \in D$, if x is even then $x \leq 0$.
   d. $\forall x \in D$, if the ones digit of x is 2, then the tens digit is 3 or 4.
   e. $\forall x \in D$, if the ones digit of x is 6, then the tens digit is 1 or 2.

In 16 and 17, write a negation for each statement in the referenced exercise.

H **16.** Section 2.1, exercise 21

H **17.** Section 2.1, exercise 22

In 18–25, write a negation for each statement.

18. ∀ real numbers x, if $x^2 \geq 1$ then $x > 0$.

19. ∀ integers d, if $6/d$ is an integer then $d = 3$.

20. $\forall x \in \mathbf{R}$, if $x(x + 1) > 0$ then $x > 0$ or $x < -1$.

21. $\forall n \in \mathbf{Z}$, if n is prime then n is odd or $n = 2$.

22. ∀ integers a, b and c, if $a - b$ is even and $b - c$ is even, then $a - c$ is even.

23. ∀ animals x, if x is a dog then x has paws and x has a tail.

24. If an integer is divisible by 2, then it is even.

25. If the square of an integer is odd, then the integer is odd.

✶ 26. If $P(x)$ is a predicate and the domain of x is the set of all real numbers, let R be "$\forall x \in \mathbf{Z}, P(x)$," let S be "$\forall x \in \mathbf{Q}, P(x)$," and let T be "$\forall x \in \mathbf{R}, P(x)$."
   a. Find a definition for $P(x)$ (but do not use "$x \in \mathbf{Z}$") so that R is true and both S and T are false.
   b. Find a definition for $P(x)$ (but do not use "$x \in \mathbf{Q}$") so that both R and S are true and T is false.

27. Consider the following string of numbers: 0204. A person claims that all the 1's in the string are to the left of all the 0's in the string. Is this true? Justify your answer. (*Hint:* Write the claim formally and write a formal negation for it. Is the negation true or false?)

28. True or false? All the occurrences of the letter u in the title of this book are lower case. Justify your answer.

In 29–36, give the contrapositive, converse, and inverse of each statement in the referenced exercise.

29. Exercise 18   30. Exercise 19

31. Exercise 20   32. Exercise 21

33. Exercise 22   34. Exercise 23

35. Exercise 24   36. Exercise 25

37. Give an example to show that a universal conditional statement is not logically equivalent to its inverse.

Rewrite each statement of 38–41 in if-then form.

38. Earning a grade of C− in this course is a sufficient condition for it to count toward graduation.

39. Being divisible by 8 is a sufficient condition for being divisible by 4.

40. Being on time each day is a necessary condition for keeping this job.

41. Passing a comprehensive exam is a necessary condition for obtaining a master's degree.

Use the facts that the negation of a ∀ statement is a ∃ statement and that the negation of an if-then statement is an *and* statement to rewrite each of the statements 42–45 without using the word *sufficient* or *necessary*.

42. Being divisible by 8 is not a necessary condition for being divisible by 4.

43 Having a large income is not a necessary condition for a person to be happy.

44. Having a large income is not a sufficient condition for a person to be happy.

45. Being a polynomial is not a sufficient condition for a function to have a real root.

46. The computer scientists Richard Conway and David Gries once wrote:

> The absence of error messages during translation of a computer program is only a necessary and not a sufficient condition for reasonable [program] correctness.

Rewrite this statement without using the words *necessary* or *sufficient*.

47. A frequent-flyer club brochure states, "You may select among carriers only if they offer the same lowest fare." Assuming that "only if" has its formal, logical meaning, does this statement guarantee that if two carriers offer the same lowest fare, the customer will be free to choose between them? Explain.

2.3 Statements Containing Multiple Quantifiers

It is not enough to have a good mind. The main thing is to use it well. — René Descartes

Imagine you are visiting a factory that manufactures computer microchips. The factory guide tells you,

> There is a person supervising every detail of the production process.

Note that this statement contains informal versions of both the existential quantifier *there is* and the universal quantifier *every*. Which of the following best describes its meaning?

- There is one single person who supervises all the details of the production process.
- For any particular production detail, there is a person who supervises that detail, but there might be different supervisors for different details.

As it happens, either interpretation could be what the guide meant. (Reread the sentence to be sure you agree!) Taken by itself, his statement is genuinely ambiguous, although other things he may have said (the context for his statement) might have clarified it. In our ordinary lives, we deal with this kind of ambiguity all the time. Usually context helps resolve it, but sometimes we simply misunderstand each other.

In mathematics, formal logic, and computer science, by contrast, it is essential that we all interpret statements in exactly the same way. For instance, the initial stage of software development typically involves careful discussion between a programmer analyst and a client to turn vague descriptions of what the client wants into unambiguous program specifications that client and programmer can mutually agree on.

Because many important technical statements contain both ∃ and ∀, a convention has developed for interpreting them uniformly. When a statement contains more than one quantifier, we imagine the actions suggested by the quantifiers as being performed in the order in which the quantifiers occur. For instance, consider a statement of the form

$$\forall x \text{ in set } D, \exists y \text{ in set } E \text{ such that } x \text{ and } y \text{ satisfy property } P(x, y).$$

To show that such a statement is true, you must be able to meet the following challenge:

- Imagine that someone is allowed to choose any element whatsoever from the set D, and imagine that the person gives you that element. Call it x.
- The challenge for you is to find an element y in E so that the person's x and your y, taken together, satisfy property $P(x, y)$.

Note that *because you do not have to specify the y until after the other person has specified the x, you are allowed to find a different value of y for each different x you are given.*

Example 2.3.1 Truth of a ∀∃ Statement in a Tarski World

Consider the Tarski world shown in Figure 2.3.1.

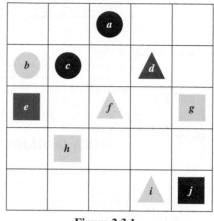

Figure 2.3.1

Show that the following statement is true in this world:

For all triangles x, there is a square y such that x and y have the same color.

Solution   The statement says that no matter which triangle someone gives you, you will be able to find a square of the same color. There are only three triangles, d, f, and i. The following table shows that for each of these triangles a square of the same color can be found.

Given $x =$	choose $y =$	and check that y is the same color as x.
d	e	yes ✓
f or i	h or g	yes ✓

Now consider a statement containing both ∀ and ∃, where the ∃ comes before the ∀:

∃ an x in D such that ∀y in E, x and y satisfy property $P(x, y)$.

To show that a statement of this form is true:

You must find one single element (call it x) in D with the following property:

- After you have found your x, someone is allowed to choose any element whatsoever from E. The person challenges you by giving you that element. Call it y.
- Your job is to show that your x together with the person's y satisfy property $P(x, y)$.

Note that your x has to work for *any y* the person gives you; *you are not allowed to change your x once you have specified it initially.*

Example 2.3.2 Truth of a ∃∀ Statement in a Tarski World

Consider again the Tarski world in Figure 2.3.1. Show that the following statement is true: There is a triangle x such that for all circles y, x is to the right of y.

Solution   The statement says that you can find a triangle that is to the right of all the circles. Actually, either d or i would work, as you can see in the following table.

Choose x =	Then, given y =	check that x is to the right of y.
d or i	a	yes ✓
	b	yes ✓
	c	yes ✓

■

Here is a summary of the convention for interpreting statements with two different quantifiers:

Interpreting Statements with Two Different Quantifiers

If you want to establish the truth of a statement of the form

forall x in D, therexists y in E such that $P(x, y)$

your challenge is to allow someone else to pick whatever element x in D they wish and then you must find an element y in E that "works" for that particular x.

If you want to establish the truth of a statement of the form

$\exists x$ in D such that $\forall y$ in E, $P(x, y)$

your job is to find one particular x in D that will "work" no matter what y in E anyone might choose to challenge you with.

Example 2.3.3 Interpreting Multiply-Quantified Statements

A college cafeteria line has four stations: salads, main courses, desserts, and beverages. The salad station offers a choice of green salad or fruit salad; the main course station offers spaghetti or fish; the dessert station offers pie or cake; and the beverage station offers milk, soda, or coffee. Three students, Uta, Tim, and Yuen, go through the line and make the following choices:

Uta: green salad, spaghetti, pie, milk

Tim: fruit salad, fish, pie, cake, milk, coffee

Yuen: spaghetti, fish, pie, soda

These choices are illustrated in Figure 2.3.2.

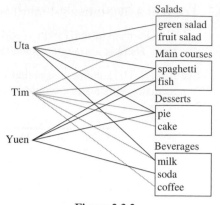

Figure 2.3.2

Write each of following statements informally and find its truth value.

a. $\exists$ an item I such that $\forall$ students S, S chose I.

b. $\exists$ a student S such that $\forall$ items I, S chose I.

c. $\exists$ a student S such that $\forall$ stations Z, $\exists$ an item I in Z such that S chose I.

d. $\forall$ students S and $\forall$ stations Z, $\exists$ an item I in Z such that S chose I.

Solution

a. There is an item that was chosen by every student. This is true; every student chose pie.

b. There is a student who chose every available item. This is false; no student chose all nine items.

c. There is a student who chose at least one item from every station. This is true; both Uta and Tim chose at least one item from every station.

d. Every student chose at least one item from every station. This is false; Yuen did not choose a salad. ■

Translating from Informal to Formal Language

Most problems are stated in informal language, but solving them often requires translating them into more formal terms.

Example 2.3.4 Translating Multiply-Quantified Statements from Informal to Formal Language

The **reciprocal** of a real number a is a real number b such that $ab = 1$. The following two statements are true. Rewrite them formally using quantifiers and variables:

a. Every nonzero real number has a reciprocal.

b. There is a real number with no reciprocal.   The number 0 has no reciprocal.

Solution

a. $\forall$ nonzero real numbers u, $\exists$ a real number v such that $uv = 1$.

b. $\exists$ a real number c such that $\forall$ real numbers d, $cd \neq 1$. ■

Example 2.3.5 There Is a Smallest Positive Integer

Recall that every integer is a real number and that real numbers are of three types: positive, negative, and zero (zero being neither positive nor negative). Consider the statement "There is a smallest positive integer." Write this statement formally using both symbols ∃ and ∀.

Solution   To say that there is a smallest positive integer means that there is one, single positive integer m with the property that no matter what positive integer n a person might pick, m will be less than or equal to n.

$$∃ \text{ a positive integer } m \text{ such that } ∀ \text{ positive integers } n, m \leq n.$$

Note that this statement is true because 1 is a positive integer that is less than or equal to every positive integer.

Example 2.3.6 There Is No Smallest Positive Real Number

Imagine any positive real number x on the real number line. These numbers correspond to all the points to the right of 0. Observe that no matter how small x is, the number $x/2$ will be both positive and less than x.*

Thus the following statement is true: "There is no smallest positive real number" Write this statement formally using both symbols ∀ and ∃.

Solution   ∀ positive real numbers x, ∃ a positive real number y such that $y < x$.

Example 2.3.7 The Definition of Limit

The definition of limit of a sequence, studied in calculus, uses both quantifiers ∀ and ∃ and also if-then. We say that the limit of the sequence a_n as n goes to infinity equals L and write

$$\lim_{n \to \infty} a_n = L$$

if, and only if, the values of a_n become *arbitrarily* close to L as n gets larger and larger without bound. More precisely, this means that given any positive number ε, we can find an integer N such that whenever n is larger than N, the number a_n sits between $L - \varepsilon$ and $L + \varepsilon$ on the number line.

a_n must lie in here when $n > N$

---

*This can be deduced from the properties of the real numbers given in Appendix A. Because x is positive, $0 < x$. Add x to both sides to obtain $x < 2x$. Then $0 < x < 2x$. Now multiply all parts of the inequality by the positive number $1/2$. This does not change the direction of the inequality, so $0 < x/2 < x$.

Symbolically:

$$\forall \varepsilon > 0, \exists \text{ an integer } N \text{ such that } \forall \text{ integers } n,$$
$$\text{if } n > N \text{ then } L - \varepsilon < a_n < L + \varepsilon.$$

Considering the logical complexity of this definition, it is no wonder that many students find it hard to understand. ∎

Ambiguous Language

The drawing in Figure 2.3.3 is a famous example of visual ambiguity. When you look at it for a while, you will probably see either a silhouette of a young woman wearing a large hat or an elderly woman with a large nose. Whichever image first pops into your mind, try to see how the drawing can be interpreted in the other way. (*Hint:* The mouth of the elderly woman is the necklace on the young woman.)

Figure 2.3.3

Once most people see one of the images, it is difficult for them to perceive the other. So it is with ambiguous language. Once you interpreted the sentence at the beginning of this section in one way, it may have been hard for you to see that it could be understood in the other way. Perhaps you had difficulty even though the two possible meanings were explained, just as many people have difficulty seeing the second interpretation for the drawing even when they are told what to look for.

Although statements written informally may be open to multiple interpretations, we cannot determine their truth or falsity without interpreting them one way or another. Therefore, we have to use context to try to ascertain their meaning as best we can.

Negations of Multiply-Quantified Statements

You can use the same rules to negate multiply-quantified statements that you used to negate simpler quantified statements. Recall that

$$\sim(\forall x \text{ in } D, P(x)) \equiv \exists x \text{ in } D \text{ such that } \sim P(x).$$

and

$$\sim(\exists x \text{ in } D \text{ such that } P(x)) \equiv \forall x \text{ in } D, \sim P(x).$$

We apply these laws to negate a statement of the form

$$\sim(\forall x \text{ in } D, \exists x \text{ in } E \text{ such that } P(x, y))$$

by moving in stages from left to right along the sentence.

First version of negation: $\exists x$ in D such that $\sim(\exists y$ in E such that $P(x, y))$.

Final version of negation: $\exists x$ in D such that $\forall y$ in E, $\sim P(x, y)$.

Similarly, to find

$$\sim(\exists x \text{ in } D \text{ such that } \forall y \text{ in } E, P(x, y)),$$

we have

First version of negation: $\forall x$ in D, $\sim(\forall y$ in E, $P(x, y))$.

Final version of negation: $\forall x$ in D, $\exists y$ in E such that $\sim P(x, y)$.

These facts can be summarized as follows:

Negations of Multiply-Quantified Statements

$\sim(\forall x \text{ in } D, \exists x \text{ in } E \text{ such that } P(x, y)) \equiv \exists x \text{ in } D \text{ such that } \forall y \text{ in } E, \sim P(x, y).$

$\sim(\exists x \text{ in } D \text{ such that } \forall y \text{ in } E, P(x, y)) \equiv \forall x \text{ in } D, \exists y \text{ in } E \text{ such that } \sim P(x, y).$

Example 2.3.8 Negating Statements in a Tarski World

Refer to the Tarski world of Example 2.3.1, which is reprinted here for reference.

Write a negation for each of the following statements, and determine which is true, the given statement or its negation.

a. For all squares x, there is a circle y such that x and y have the same color.

b. There is a triangle x such that for all squares y, x is to the right of y.

Solution

a.   *First version of negation:* ∃ a square x such that ∼(∃ a circle y such that x and y have the same color).

    Final version of negation: ∃ a square x such that ∀ circles y, x and y do not have the same color.

The negation is true. Square e is black and no circle is black, so there is a square that does not have the same color as any circle.

b.   *First version of negation:* ∀ triangles x, ∼ (∀ squares y, x is to the right of y).

    Final version of negation: ∀ triangles x, ∃ a square y such that x is not to the right of y.

The negation is true because no matter what triangle is chosen, it is not to the right of square g (or square j).   ■

Order of Quantifiers

Consider the following two statements:

   ∀ people x, ∃ a person y such that x loves y.

   ∃ a person y such that ∀ people x, x loves y.

Note that except for the order of the quantifiers, these statements are identical. However, the first means that given any person, it is possible to find someone who loves that person, whereas the second means that there is one amazing individual who is loved by all people. (Reread the statements carefully to verify these interpretations!) The two sentences illustrate an extremely important property about multiply-quantified statements:

 Caution! If a statement contains two different quantifiers, reversing the order of the quantifiers can change the truth value of the statement to its opposite.

Interestingly, however, if one quantifier immediately follows another quantifier *of the same type,* then the order of the quantifiers does not affect the meaning. Consider the commutative property of addition of real numbers, for example:

   ∀ real numbers x and ∀ real numbers y, $x + y = y + x$.

This means the same as

   ∀ real numbers y and ∀ real numbers x, $x + y = y + x$.

Thus the property can be expressed more briefly as

   ∀ real numbers x and y, $x + y = y + x$.

Example 2.3.9 Quantifier Order in a Tarski World

Consider again the Tarski world of Example 2.3.1. Do the following two statements have the same truth value?

a. For every square x there is a triangle y such that x and y have different colors.

b. There exists a triangle y such that for every square x, x and y have different colors.

Solution Statement (a) says that if someone gives you one of the squares from the Tarski world, you can find a triangle that has a different color. This is true. If someone gives you square *g* or *h* (which are gray), you can use triangle *d* (which is black); if someone gives you square *e* (which is black), you can use either triangle *f* or triangle *i* (which are both gray); and if someone gives you square *j* (which is blue), you can use triangle *d* (which is black) or triangle *f* or *i* (which are both gray).

Statement (b) says that there is one particular triangle in the Tarski world that has a different color from every one of the squares in the world. This is false. Two of the triangles are gray, but they cannot be used to show the truth of the statement because the Tarski world contains gray squares. The only other triangle is black, but it cannot be used either because there is a black square in the Tarski world.

Thus one of the statements is true and the other is false, and so they have opposite truth values. ∎

Formal Logical Notation

In many areas of computer science, logical statements are expressed in purely symbolic notation. The notation involves using predicates to describe all properties of variables and omitting the words *such that* in existential statements. (When you try to figure out the meaning of a formal statement, however, it is helpful to think the words *such that* to yourself each time they are appropriate.) The formalism also depends on ideas introduced in Examples 2.1.10 and 2.1.12 —namely, that

"$\forall x$ in D, $P(x)$" can be written as "$\forall x (x$ in $D \rightarrow P(x))$," and

"$\exists x$ in D such that $P(x)$" can be written as "$\exists x (x$ in $D \wedge P(x))$."

We illustrate these ideas in Example 2.3.10.

Example 2.3.10 Formalizing Statements in a Tarski World

Consider once more the Tarski world of Example 2.3.1:

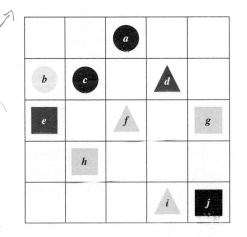

Let Triangle(x), Circle(x), and Square(x) mean "x is a triangle," "x is a circle," and "x is a square"; let Blue(x), Gray(x), and Black(x) mean "x is blue," "x is gray," and "x is black"; let RightOf(x, y), Above(x, y), and SameColorAs(x, y) mean "x is to the right of y," "x is above y," and "x has the same color as y"; and use the notation $x = y$ to denote the predicate "x is equal to y". Let the common domain D of all variables be the set of all the objects in the Tarski world. Use formal, logical notation to write each of the following statements, and write a formal negation for each statement.

a. For all circles x, x is above f.

b. There is a square x such that x is black.

c. For all circles x, there is a square y such that x and y have the same color.

d. There is a square x such that for all triangles y, x is to right of y.

Solution

a. *Statement:* $\forall x(\text{Circle}(x) \rightarrow \text{Above}(x, f))$.
   Negation: $\sim(\forall x(\text{Circle}(x) \rightarrow \text{Above}(x, f)))$

$$\equiv \exists x \sim (\text{Circle}(x) \rightarrow \text{Above}(x, f))$$
by the law for negating a ∀ statement

$$\equiv \exists x(\text{Circle}(x) \wedge \sim\text{Above}(x, f))$$
by the law of negating an if-then statement

b. *Statement:* $\exists x(\text{Square}(x) \wedge \text{Black}(x))$.
   Negation: $\sim(\exists x(\text{Square}(x) \wedge \text{Black}(x)))$

$$\equiv \forall x \sim (\text{Square}(x) \wedge \text{Black}(x))$$
by the law for negating a ∃ statement

$$\equiv \forall x(\sim\text{Square}(x) \vee \sim\text{Black}(x))$$
by De Morgan's law

c. *Statement:* $\forall x(\text{Circle}(x) \rightarrow \exists y(\text{Square}(y) \wedge \text{SameColor}(x, y)))$.
   Negation: $\sim(\forall x(\text{Circle}(x) \rightarrow \exists y(\text{Square}(y) \wedge \text{SameColor}(x, y))))$

$$\equiv \exists x \sim (\text{Circle}(x) \rightarrow \exists y(\text{Square}(y) \wedge \text{SameColor}(x, y)))$$
by the law for negating a ∀ statement

$$\equiv \exists x(\text{Circle}(x) \wedge \sim(\exists y(\text{Square}(y) \wedge \text{SameColor}(x, y))))$$
by the law for negating an if-then statement

$$\equiv \exists x(\text{Circle}(x) \wedge \forall y(\sim(\text{Square}(y) \wedge \text{SameColor}(x, y))))$$
by the law for negating a ∃ statement

$$\equiv \exists x(\text{Circle}(x) \wedge \forall y(\sim\text{Square}(y) \vee \sim\text{SameColor}(x, y)))$$
by De Morgan's law

d. *Statement:* $\exists x(\text{Square}(x) \wedge \forall y(\text{Triangle}(y) \rightarrow \text{RightOf}(x, y)))$.
   Negation: $\sim(\exists x(\text{Square}(x) \wedge \forall y(\text{Triangle}(y) \rightarrow \text{RightOf}(x, y))))$

$$\equiv \forall x \sim (\text{Square}(x) \wedge \forall y(\text{Triangle}(x) \rightarrow \text{RightOf}(x, y)))$$
by the law for negating a ∃ statement

$$\equiv \forall x(\sim\text{Square}(x) \vee \sim(\forall y(\text{Triangle}(y) \rightarrow \text{RightOf}(x, y))))$$
by De Morgan's law

$$\equiv \forall x(\sim\text{Square}(x) \vee \exists y(\sim(\text{Triangle}(y) \rightarrow \text{RightOf}(x, y))))$$
by the law for negating a ∀ statement

$$\equiv \forall x(\sim\text{Square}(x) \vee \exists y(\text{Triangle}(y) \wedge \sim\text{RightOf}(x, y)))$$
by the law for negating an if-then statement

■

The disadvantage of the fully formal notation is that because it is complex and somewhat remote from intuitive understanding, when we use it, we may make errors that go unrecognized. The advantage, however, is that operations, such as taking negations, can be made completely mechanical and programmed on a computer. Also, when we become comfortable with formal manipulations, we can use them to check our intuition, and then we can use our intuition to check our formal manipulations. Formal logical notation is used in branches of computer science such as artificial intelligence, program verification, and automata theory and formal languages.

Taken together, the symbols for quantifiers, variables, predicates, and logical connectives make up what is known as the **language of first-order logic.** Even though this language is simpler in many respects than the language we use every day, learning it requires the same kind of practice needed to acquire any foreign language.

Prolog

The programming language Prolog (short for *pro*gramming in *log*ic) was developed in France in the 1970s by A. Colmerauer and P. Roussel to help programmers working in the field of artificial intelligence. A simple Prolog program consists of a set of statements describing some situation together with questions about the situation. Built into the language are search and inference techniques needed to answer the questions by deriving the answers from the given statements. This frees the programmer from the necessity of having to write separate programs to answer each type of question. Example 2.3.11 gives a very simple example of a Prolog program.

Example 2.3.11 A Prolog Program

Consider the following picture, which shows colored blocks stacked on a table.

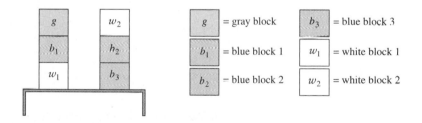

The following are statements in Prolog that describe this picture and ask two questions about it.*

isabove(g, b_1)	color(g, gray)	color(b_3, blue)
isabove(b_1, w_1)	color(b_1, blue)	color(w_1, white)
isabove(w_2, b_2)	color(b_2, blue)	color(w_2, white)
isabove(b_2, b_3)	isabove(X, Z) if isabove(X, Y) and isabove(Y, Z)	
?color(b_1, blue)	?isabove(X, w_1)	

The statements "isabove(g, b_1)" and "color(g, gray)" are to be interpreted as "g is above b_1" and "g is colored gray". The statement "isabove(X, Z) if isabove(X, Y) and isabove(Y, Z)" is to be interpreted as "For all X, Y, and Z, if X is above Y and Y is above Z, then X is above Z." The program statement

$$?color(b_1, blue)$$

---

*Different Prolog implementations follow different conventions as to how to represent constant, variable, and predicate names and forms of questions and answers. The conventions used here are similar to those of Edinburgh Prolog.

is a question asking whether block b_1 is colored blue. Prolog answers this by writing

Yes.

The statement

$$?\text{isabove}(X, w_1)$$

is a question asking for which blocks X the predicate "X is above w_1" is true. Prolog answers by giving a list of all such blocks. In this case, the answer is

$$X = b_1, X = g.$$

Note that Prolog can find the solution $X = b_1$ by merely searching the original set of given facts. However, Prolog must *infer* the solution $X = g$ from the following statements:

$$\text{isabove}(g, b_1),$$

$$\text{isabove}(b_1, w_1),$$

$$\text{isabove}(X, Z) \text{ if } \text{isabove}(X, Y) \text{ and } \text{isabove}(Y, Z).$$

Write the answers Prolog would give if the following questions were added to the program above.

a. $?\text{isabove}(b_2, w_1)$ b. $?\text{color}(w_1, X)$ c. $?\text{color}(X, \text{blue})$

Solution

a. The question means "Is b_2 above w_1?"; so the answer is "No."

b. The question means "For what colors X is the predicate 'w_1 is colored X' true?"; so the answer is "$X = \text{white}$."

c. The question means "For what blocks is the predicate 'X is colored blue' true?"; so the answer is "$X = b_1$," "$X = b_2$," and "$X = b_3$." ∎

Exercise Set 2.3

1. Let C be the set of cities in the world, let N be the set of nations in the world, and let $P(c, n)$ be "c is the capital city of n." Determine the truth values of the following statements.
   a. $P(\text{Tokyo, Japan})$ **b.** $P(\text{Athens, Egypt})$
   c. $P(\text{Paris, France})$ **d.** $P(\text{Miami, Brazil})$

2. Let $G(x, y)$ be "$x^2 > y$." Indicate which of the following statements are true and which are false.
   a. $G(2, 3)$ **b.** $G(1, 1)$
   c. $G(\frac{1}{2}, \frac{1}{2})$ **d.** $G(-2, 2)$

3. The following statement is true: "$\forall$ nonzero numbers x, $\exists$ a real number y such that $xy = 1$." For each x given below, find a y to make the predicate "$xy = 1$" true.
   a. $x = 2$ **b.** $x = -1$ **c.** $x = 3/4$

4. The following statement is true: "$\forall$ real numbers x, $\exists$ an integer n such that $n > x$."* For each x given below, find an n to make the predicate "$n > x$" true.
   a. $x = 15.83$ **b.** $x = 10^8$ **c.** $x = 10^{10^{10}}$

*This is called the Archimedean principle because it was first formulated (in geometric terms) by the great Greek mathematician Archimedes of Syracuse, who lived from about 287 to 212 B.C.

The statements in exercises 5–8 refer to the Tarski world given in Example 2.3.1. Explain why each is true.

5. For all circles x there is a square y such that x and y have the same color.

6. For all squares x there is a circle y such that x and y have different colors and y is above x.

7. There is a triangle x such that for all squares y, x is above y.

8. There is a triangle x such that for all circles y, y is above x.

9. Let $D = E = \{-2, -1, 0, 1, 2\}$. Explain why the following statements are true.
   a. $\forall x$ in D, $\exists y$ in E such that $x + y = 0$.
   b. $\exists x$ in D such that $\forall y$ in E, $x + y = y$.

10. This exercise refers to Example 2.3.3. Determine whether each of the following statements is true or false.
    a. $\forall$ students S, $\exists$ a dessert D such that S chose D.
    b. $\forall$ students S, $\exists$ a salad T such that S chose T.
    c. $\exists$ a dessert D such that $\forall$ students S, S chose D.

d. ∃ a beverage B such that ∀ students D, D chose B.

e. ∃ an item I such that ∀ students S, S did not choose I.

f. ∃ a station Z such that ∀ students S, ∃ an item I such that S chose I from Z.

11. How could you determine the truth or falsity of the following statements for the students in your discrete mathematics class? Assume that students will respond truthfully to questions that are asked of them.

a. There is a student in this class who has dated at least one person from every residence hall at this school.

b. There is a residence hall at this school with the property that every student in this class has dated at least one person from that residence hall.

c. Every residence hall at this school has the property that if a student from this class has dated at least one person from that hall, then that student has dated at least two people from that hall.

12. Let S be the set of students at your school, let M be the set of movies that have ever been released, and let $V(s, m)$ be "student s has seen movie m." Rewrite each of the following statements without using the symbol ∀, the symbol ∃, or variables.

a. ∃$s \subset S$ such that $V(s,$ Casablanca$)$.

b. ∀$s \in S$, $V(s,$ Star Wars$)$.

c. ∀$s \in S$, ∃$m \in M$ such that $V(s, m)$.

d. ∃$m \in M$ such that ∀$s \in S$, $V(s, m)$.

e. ∃$s \in S$, ∃$t \in S$, and ∃$m \in M$ such that $s \neq t$ and $V(s, m) \wedge V(t, m)$.

f. ∃$s \subset S$ and ∃$t \in S$ such that $s \neq t$ and ∀$m \in M$, $V(s, m) \rightarrow V(t, m)$.

13. Let $D = E = \{-2, -1, 0, 1, 2\}$. Write negations for each of the following statements and determine which is true, the given statement or its negation.

a. ∀x in D, ∃y in E such that $x + y = 1$.

b. ∃x in D such that ∀y in E, $x + y = -y$.

In each of 14–19, (a) rewrite the statement in English without using the symbol ∀ or ∃ but expressing your answer as simply as possible, and (b) write a negation for the statement.

14. ∀ colors C, ∃ an animal A such that A is colored C.

15. ∃ a book b such that ∀ people p, p has read b.

16. ∀ odd integers n, ∃ an integer k such that $n = 2k + 1$.

17. ∀$r \in \mathbf{Q}$, ∃ integers a and b such that $r = a/b$.

18. ∀$x \in \mathbf{R}$, ∃ a real number y such that $x + y = 0$.

19. ∃$x \in \mathbf{R}$ such that for all real numbers y, $x + y = 0$.

20. Recall that reversing the order of the quantifiers in a statement with two different quantifiers may change the truth value of the statement—but it does not necessarily do so. All the statements in the pairs below refer to the Tarski world of Example 2.3.1. In each pair, the order of the quantifiers is reversed but everything else is the same. For each pair,

determine whether the statements have the same or opposite truth values. Justify your answers.

a. (1) For all circles x there is a triangle y such that x and y do not have the same color.

(2) There is a triangle x such that for all circles y, x and y have different colors.

b. (1) For all circles x there is a square y such that x and y have the same color.

(2) There is a square x such that for all circles y, x and y have the same color.

In 21 and 22, rewrite each statement without using variables or the symbol ∀ or ∃. Indicate whether the statement is true or false.

21. a. ∀ real numbers x, ∃ a real number y such that $x + y = 0$.

b. ∃ a real number y such that ∀ real numbers x, $x + y = 0$.

22. a. ∀ nonzero real numbers r, ∃ a real number s such that $rs = 1$.

b. ∃ a real number s such that ∀ real numbers r, $rs = 1$.

23. Use the laws for negating universal and existential statements to derive the following rules:

a. $\sim(\forall x \in D(\forall y \in E(P(x, y))))$
$\equiv ∃x \in D(∃y \in E(\sim P(x, y)))$

b. $\sim(∃x \in D(∃y \in E(P(x, y))))$
$\equiv \forall x \in D(\forall y \in E(\sim P(x, y)))$

Each statement in 24–27 refers to the Tarski world of Example 2.3.1. For each, (a) determine whether the statement is true or false and justify your answer, (b) write a negation for the statement (referring, if you wish, to the result in exercise 23).

24. ∀ circles x and ∀ squares y, x is above y.

25. ∀ circles x and ∀ triangles y, x is above y.

26. ∃ a circle x and ∃ a square y such that x is above y and x and y have different colors.

27. ∃ a circle x and ∃ a square y such that x is above y and x and y have the same color.

For each of the statements in 28 and 29, (a) write a new statement by interchanging the symbols ∀ and ∃, and (b) state which is true: the given statement, the version with interchanged quantifiers, neither, or both.

28. ∀$x \in \mathbf{R}$, ∃$y \in \mathbf{R}$ such that $x < y$.

29. ∃$x \in \mathbf{R}$ such that ∀$y \in \mathbf{R}^-$ (the set of negative real numbers), $x > y$.

30. Consider the statement "Everybody is older than somebody." Rewrite this statement in the form "∀ people x, ∃ _____."

31. Consider the statement "Somebody is older than everybody." Rewrite this statement in the form "∃ a person x such that ∀ _____."

In 32–38, (a) rewrite the statement formally using quantifiers and variables, and (b) write a negation for the statement.

32. Everybody loves somebody.

33. Somebody loves everybody.

34. Everybody trusts somebody.

35. Somebody trusts everybody.

36. Any even integer equals twice some integer.

37. Every action has an equal and opposite reaction.

38. There is a program that gives the correct answer to every question that is posed to it.

39. In informal speech most sentences of the form "There is _____ every _____" are intended to be understood as meaning "∀ _____ ∃ _____," even though the existential quantifier *there is* comes before the universal quantifier *every*. Note that this interpretation applies to the following well-known sentences. Rewrite them using quantifiers and variables.
a. There is a sucker born every minute.
b. There is a time for every purpose under heaven.

40. Indicate which of the following statements are true and which are false. Justify your answers as best you can.
a. $\forall x \in \mathbf{Z}^+, \exists y \in \mathbf{Z}^+$ such that $x = y + 1$.
b. $\forall x \in \mathbf{Z}, \exists y \in \mathbf{Z}$ such that $x = y + 1$.
c. $\exists x \in \mathbf{R}$ such that $\forall y \in \mathbf{R}, x = y + 1$.
d. $\forall x \in \mathbf{R}^+, \exists y \in \mathbf{R}^+$ such that $xy = 1$.
e. $\forall x \in \mathbf{R}, \exists y \in \mathbf{R}$ such that $xy = 1$.
f. $\forall x \in \mathbf{Z}^+$ and $\forall y \in \mathbf{Z}^+, \exists z \in \mathbf{Z}^+$ such that $z = x - y$.
g. $\forall x \in \mathbf{Z}$ and $\forall y \in \mathbf{Z}, \exists z \in \mathbf{Z}$ such that $z = x - y$.
h. $\exists u \in \mathbf{R}^+$ such that $\forall v \in \mathbf{R}^+, uv < v$.
i. $\forall v \in \mathbf{R}^+, \exists u \in \mathbf{R}^+$ such that $uv < v$.

41. Write the negation of the definition of limit of a sequence given in Example 2.3.7.

42. Write a negation for the following statement (which is the definition of $\lim_{x \to a} f(x) = L$):

> For all real numbers $\varepsilon > 0$, there exists a real number $\delta > 0$ such that if $a - \delta < x < a + \delta$ then $L - \varepsilon < f(x) < L + \varepsilon$.

43. The notation $\exists!$ stands for the words "there exists a unique." Thus, for instance, "$\exists! x$ such that x is prime and x is even" means that there is one and only one even prime number. Which of the following statements are true and which are false? Explain.
a. $\exists!$ real number x such that $\forall$ real numbers y, $xy = y$.
b. $\exists!$ integer x such that $1/x$ is an integer.
c. $\forall$ real numbers x, $\exists!$ real number y such that $x + y = 0$.

∗ 44. Suppose that $P(x)$ is a predicate and D is the domain of x. Rewrite the statement "$\exists! x \in D$ such that $P(x)$" without using the symbol $\exists!$. (See exercise 43 for the meaning of $\exists!$.)

In 45–52, refer to the Tarski world given in Example 2.1.1, which is printed again here for reference. The domains of all variables consist of all the objects in the Tarski world. For each statement, (a) indicate whether the statement is true or false and justify your answer, (b) write the given statement using the formal logical notation illustrated in Example 2.3.10, and (c) write the negation of the given statement using the formal logical notation of Example 2.3.10.

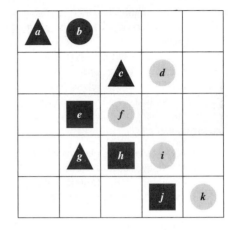

45. There is a triangle x such that for all squares y, x is above y.

46. There is a triangle x such that for all circles y, x is above y.

47. For all circles x, there is a square y such that y is to the right of x.

48. For every object x, there is an object y such that if $x \neq y$ then x and y have different colors.

49. There is an object y such that for all objects x, if $x \neq y$ then x and y have different colors.

50. For all circles x and for all triangles y, x is to the right of y.

51. There is a circle x and there is a square y such that x and y have the same color.

52. There is a circle x and there is a triangle y such that x and y have the same color.

Let $P(x)$ and $Q(x)$ be predicates and suppose D is the domain of x. In 53–56, for the statement forms in each pair, determine whether (a) they have the same truth value for every choice of $P(x)$, $Q(x)$, and D, or (b) there is a choice of $P(x)$, $Q(x)$, and D for which they have opposite truth values.

53. $\forall x \in D, (P(x) \wedge Q(x))$, and
$(\forall x \in D, P(x)) \wedge (\forall x \in D, Q(x))$

54. $\exists x \in D, (P(x) \wedge Q(x))$, and
$(\exists x \in D, P(x)) \wedge (\exists x \in D, Q(x))$

55. $\forall x \in D, (P(x) \vee Q(x))$, and
$(\forall x \in D, P(x)) \vee (\forall x \in D, Q(x))$

56. $\exists x \in D, (P(x) \vee Q(x))$, and
$(\exists x \in D, P(x)) \vee (\exists x \in D, Q(x))$

In 57–59, find the answers Prolog would give if the following
questions were added to the program given in Example 2.3.11.

57. a. ?isabove(b_1, w_1)
    b. ?color(X, white)
    c. ?isabove(X, b_3)

58. a. ?isabove(w_1, g)
    b. ?color(w_2, blue)
    c. ?isabove(X, b_1)

59. a. ?isabove(w_2, b_3)
    b. ?color(X, gray)
    c. ?isabove(g, X)

2.4 Arguments with Quantified Statements

*The only complete safeguard against reasoning ill, is the habit of reasoning well;
familiarity with the principles of correct reasoning; and practice in applying those
principles.* — John Stuart Mill

The rule of **universal instantiation** (in-stan-she-AY-shun) says that

> If some property is true of *everything* in a domain, then
> it is true of *any particular* thing in the domain.

Use of the words *universal instantiation* indicates that the truth of a property in a particular
case follows as a special instance of its more general or universal truth. The validity of
this argument form follows immediately from the definition of truth values for a universal
statement. One of the most famous examples of universal instantiation is the following:

> All men are mortal.
>
> Socrates is a man.
>
> ∴ Socrates is mortal.

Universal instantiation is *the* fundamental tool of deductive reasoning. Mathematical
formulas, definitions, and theorems are like general templates that are used over and over
in a wide variety of particular situations. A given theorem says that such and such is true
for all things of a certain type. If, in a given situation, you have a particular object of
that type, then by universal instantiation, you conclude that such and such is true for that
particular object. You may repeat this process 10, 20, or more times in a single proof or
problem solution.

As an example of universal instantiation, suppose you are doing a problem that requires
you to simplify

$$r^{k+1} \cdot r,$$

where r is a particular real number and k is a particular integer. You know from your
study of algebra that the following universal statements are true:

1. For all real numbers x and all integers m and n, $x^m \cdot x^n = x^{m+n}$.

2. For all real numbers x, $x^1 = x$.

So you proceed as follows:

$$r^{k+1} \cdot r = r^{k+1} \cdot r^1 \quad \text{Step 1}$$
$$= r^{(k+1)+1} \quad \text{Step 2}$$
$$= r^{k+2} \quad \text{by basic algebra.}$$

The reasoning behind step 1 and step 2 is outlined as follows.

Step 1:   For all real numbers x, $x^1 = x$.   universal truth
           r is a particular real number.   particular instance
           $\therefore r^1 = r$.   conclusion

Step 2:   For all real numbers x and all integers
           m and n, $x^m \cdot x^n = x^{m+n}$.   universal truth
           r is a particular real number and $k + 1$
           and 1 are particular integers.   particular instance
           $\therefore r^{k+1} \cdot r^1 = r^{(k+1)+1}$.   conclusion

Both arguments are examples of universal instantiation.

Universal Modus Ponens

The rule of universal instantiation can be combined with modus ponens to obtain the valid form of argument called *universal modus ponens.*

Universal Modus Ponens	
Formal Version	*Informal Version*
$\forall x$, if $P(x)$ then $Q(x)$.	If x makes $P(x)$ true, then x makes $Q(x)$ true.
$P(a)$ for a particular a.	a makes $P(x)$ true.
$\therefore Q(a)$.	$\therefore a$ makes $Q(x)$ true.

Note that the first, or major, premise of universal modus ponens could be written "All things that make $P(x)$ true make $Q(x)$ true," in which case the conclusion would follow by universal instantiation alone. However, the if-then form is more natural to use in the majority of mathematical situations.

Example 2.4.1 Recognizing Universal Modus Ponens

Rewrite the following argument using quantifiers, variables, and predicate symbols. Is this argument valid? Why?

> If a number is even, then its square is even.
>
> k is a particular number that is even.
>
> $\therefore k^2$ is even.

Solution   The major premise of this argument can be rewritten as

$$\forall x, \text{ if } x \text{ is even then } x^2 \text{ is even.}$$

Let $E(x)$ be "x is even," let $S(x)$ be "x^2 is even," and let k stand for a particular number that is even. Then the argument has the following form:

$$\forall x, \text{ if } E(x) \text{ then } S(x).$$
$$E(k), \text{ for a particular } k.$$
$$\therefore S(k).$$

This argument has the form of universal modus ponens and is therefore valid.   ■

Example 2.4.2 Drawing Conclusions Using Universal Modus Ponens

Write the conclusion that can be inferred using universal modus ponens.

If T is any right triangle with hypotenuse c and legs a and b, then $c^2 = a^2 + b^2$.   Pythagorean theorem

The triangle shown at the right is a right triangle with both legs equal to 1 and hypotenuse c.

∴ _____.

Solution $\qquad\qquad\qquad\qquad c^2 = 1^2 + 1^2 = 2$

Note that if you take the nonnegative square root of both sides of this equation, you obtain $c = \sqrt{2}$. This shows that there is a line segment whose length is $\sqrt{2}$. Section 3.7 contains a proof that $\sqrt{2}$ is not a rational number. ∎

Use of Universal Modus Ponens in a Proof

In Chapter 3 we discuss methods of proving quantified statements. Here is a proof that the sum of any two even integers is even. It makes use of the definition of even integer, namely, that an integer is *even* if, and only if, it equals twice some integer. (Or, more formally: ∀ integers x, x is even if, and only if, ∃ an integer k such that $x = 2k$.)

Suppose m and n are particular but arbitrarily chosen even integers. Then $m = 2r$ for some integer r,[1] and $n = 2s$ for some integer s.[2] Hence

$$m + n = 2r + 2s \qquad \text{by substitution}$$
$$= 2(r + s)^{(3)} \qquad \text{by factoring out the 2.}$$

Now $r + s$ is an integer,[4] and so $2(r + s)$ is even.[5] Thus $m + n$ is even.

The following expansion of the proof shows how each of the numbered steps is justified by arguments that are valid by universal modus ponens.

(1)   If an integer is even, then it equals twice some integer.
   m is a particular even integer.
   ∴ m equals twice some integer r.

(2)   If an integer is even, then it equals twice some integer.
   n is a particular even integer.
   ∴ n equals twice some integer s.

(3)   If a quantity is an integer, then it is a real number.
   r and s are particular integers.
   ∴ r and s are real numbers.

   For all a, b, and c, if a, b, and c are real numbers, then $ab + ac = a(b + c)$.
   2, r, and s are particular real numbers.
   ∴ $2r + 2s = 2(r + s)$.

(4)   For all u and v, if u and v are integers, then $u + v$ is an integer.
   r and s are two particular integers.
   ∴ $r + s$ is an integer.

(5)   If a number equals twice some integer, then that number is even.
   $2(r + s)$ equals twice the integer $r + s$.
   ∴ $2(r + s)$ is even.

Of course, the actual proof that the sum of even integers is even does not explicitly contain the sequence of arguments given above. (Heaven forbid!) And, in fact, people who are good at analytical thinking are normally not even conscious that they are reasoning in this way. But that is because they have absorbed the method so completely that it has become almost as automatic as breathing.

Universal Modus Tollens

Another crucially important rule of inference is *universal modus tollens.* Its validity results from combining universal instantiation with modus tollens. Universal modus tollens is the heart of proof of contradiction, which is one of the most important methods of mathematical argument.

Universal Modus Tollens

Formal Version	*Informal Version*
$\forall x$, if $P(x)$ then $Q(x)$.	If x makes $P(x)$ true, then x makes $Q(x)$ true.
$\sim Q(a)$, for a particular a.	a does not make $Q(x)$ true.
$\therefore \sim P(a)$.	$\therefore a$ does not make $P(x)$ true.

Example 2.4.3 Recognizing the Form of Universal Modus Tollens

Rewrite the following argument using quantifiers, variables, and predicate symbols. Write the major premise in conditional form. Is this argument valid? Why?

> All human beings are mortal.
>
> Zeus is not mortal.
>
> $\therefore$ Zeus is not human.

Solution   The major premise can be rewritten as

$$\forall x, \text{ if } x \text{ is human then } x \text{ is mortal.}$$

Let $H(x)$ be "x is human," let $M(x)$ be "x is mortal," and let Z stand for Zeus. The argument becomes

$$\forall x, \text{ if } H(x) \text{ then } M(x)$$
$$\sim M(Z)$$
$$\therefore \sim H(Z).$$

This argument has the form of universal modus tollens and is therefore valid.   ∎

Example 2.4.4 Drawing Conclusions Using Universal Modus Tollens

Write the conclusion that can be inferred using universal modus tollens.

> All professors are absent-minded.
>
> Tom Hutchins is not absent-minded.
>
> $\therefore$ _____.

Solution   Tom Hutchins is not a professor.   ∎

Proving Validity of Arguments with Quantified Statements

The intuitive definition of validity for arguments with quantified statements is the same as for arguments with compound statements. An argument is valid if, and only if, the truth of its conclusion follows *necessarily* from the truth of its premises. The formal definition is as follows:

• Definition

To say that an *argument form* is **valid** means the following: No matter what particular predicates are substituted for the predicate symbols in its premises, if the resulting premise statements are all true, then the conclusion is also true.

  An *argument* is called **valid** if, and only if, its form is valid.

As already noted, the validity of universal instantiation follows immediately from the definition of truth values of a universal statement. General formal proofs of validity of arguments in the predicate calculus are beyond the scope of this book. We give the proof of the validity of universal modus ponens as an example to show that such proofs are possible and to give an idea of how they look.

Universal modus ponens asserts that

$$\forall x, \text{if } P(x) \text{ then } Q(x).$$
$$P(a) \text{ for a particular } a.$$
$$\therefore Q(a).$$

To prove that this form of argument is valid, suppose the major and minor premises are both true. [*We must show that the conclusion "$Q(a)$" is also true.*] By the minor premise, $P(a)$ is true for a particular value of a. By the major premise and universal instantiation, the statement "If $P(a)$ then $Q(a)$" is true for that particular a. But by modus ponens, since the statements "If $P(a)$ then $Q(a)$" and "$P(a)$" are both true, it follows that $Q(a)$ is true also. [*This is what was to be shown.*]

The proof of validity given above is abstract and somewhat subtle. We include the proof not because we expect that you will be able to make up such proofs yourself at this stage of your study. Rather, it is intended as a glimpse of a more advanced treatment of the subject, which you can try your hand at in exercises 35 and 36 at the end of this section if you wish.

One of the paradoxes of the formal study of logic is that the laws of logic are used to prove that the laws of logic are valid!

In the next part of this section we show how you can use diagrams to analyze the validity or invalidity of arguments that contain quantified statements. Diagrams do not provide totally rigorous proofs of validity and invalidity, and in some complex settings they may even be confusing, but in many situations they are helpful and convincing.

Using Diagrams to Test for Validity

Consider the statement

All integers are rational numbers.

Or, formally,

$\forall$ integers n, n is a rational number.

Picture the set of all integers and the set of all rational numbers as disks. The truth of the given statement is represented by placing the integers disk entirely inside the rationals disk, as shown in Figure 2.4.1.

Figure 2.4.1

G. W. Leibniz
(1646–1716)

Because the two statements "$\forall x \in D,\ Q(x)$" and "$\forall x$, if x is in D then $Q(x)$" are logically equivalent, both can be represented by diagrams like the foregoing.

Perhaps the first person to use diagrams like these to analyze arguments was the German mathematician and philosopher Gottfried Wilhelm Leibniz. Leibniz (LIPE-nits) was far ahead of his time in anticipating modern symbolic logic. He also developed the main ideas of the differential and integral calculus at approximately the same time as (and independently of) Isaac Newton (1642–1727).

To test the validity of an argument diagrammatically, represent the truth of both premises with diagrams. Then analyze the diagrams to see whether they necessarily represent the truth of the conclusion as well.

Example 2.4.5 Using a Diagram to Show Validity

Use diagrams to show the validity of the following syllogism:

All human beings are mortal.

Zeus is not mortal.

∴ Zeus is not a human being.

Solution   The major premise is pictured on the left in Figure 2.4.2 by placing a disk labeled "human beings" inside a disk labeled "mortals." The minor premise is pictured on the right in Figure 2.4.2 by placing a dot labeled "Zeus" outside the disk labeled "mortals."

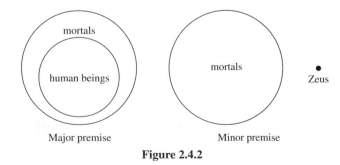

Figure 2.4.2

The two diagrams fit together in only one way, as shown in Figure 2.4.3.

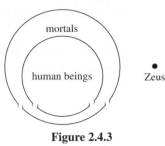

Figure 2.4.3

Since the Zeus dot is outside the mortals disk, it is necessarily outside the human beings disk. Thus the truth of the conclusion follows necessarily from the truth of the premises. It is impossible for the premises of this argument to be true and the conclusion false; hence the argument is valid. ∎

Example 2.4.6 Using Diagrams to Show *In*validity

Use a diagram to show the invalidity of the following argument:

All human beings are mortal.

Felix is mortal.

∴ Felix is a human being.

Solution   The major and minor premises are represented diagrammatically in Figure 2.4.4.

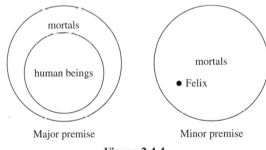

Major premise   Minor premise

Figure 2.4.4

All that is known is that the Felix dot is located *somewhere* inside the mortals disk. Where it is located with respect to the human beings disk cannot be determined. Either one of the situations shown in Figure 2.4.5 might be the case.

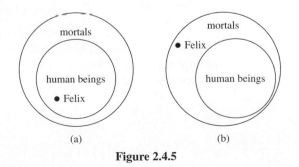

(a)   (b)

Figure 2.4.5

The conclusion "Felix is a human being" is true in the first case but not in the second (Felix might, for example, be a cat). Because the conclusion does not necessarily follow from the premises, the argument is invalid. ∎

The argument of Example 2.4.6 would be valid if the major premise were replaced by its converse. But since a universal conditional statement is not logically equivalent to its converse, such a replacement cannot, in general, be made. We say that this argument exhibits the converse error.

Converse Error (Quantified Form)

Formal Version	*Informal Version*
$\forall x$, if $P(x)$ then $Q(x)$.	If x makes $P(x)$ true, then x makes $Q(x)$ true.
$Q(a)$ for a particular a.	a makes $Q(x)$ true.
$\therefore P(a)$. ← invalid conclusion	$\therefore a$ makes $P(x)$ true. ← invalid conclusion

---

The following form of argument would be valid if a conditional statement were logically equivalent to its inverse. But it is not, and the argument form is invalid. We say that it exhibits the inverse error. You are asked to show the invalidity of this argument form in the exercises at the end of this section.

Inverse Error (Quantified Form)

Formal Version	*Informal Version*
$\forall x$, if $P(x)$ then $Q(x)$.	If x makes $P(x)$ true, then x makes $Q(x)$ true.
$\sim P(a)$, for a particular a.	a does not make $P(x)$ true.
$\therefore \sim Q(a)$. ← invalid conclusion	$\therefore a$ does not make $Q(x)$ true. ← invalid conclusion

---

Example 2.4.7 An Argument with "No"

Use diagrams to test the following argument for validity:

> No polynomial functions have horizontal asymptotes.
>
> This function has a horizontal asymptote.
>
> ∴ This function is not a polynomial function.

Solution A good way to represent the major premise diagrammatically is shown in Figure 2.4.6, two disks—a disk for polynomial functions and a disk for functions with horizontal asymptotes—that do not overlap at all. The minor premise is represented by placing a dot labeled "this function" inside the disk for functions with horizontal asymptotes.

$\sim \exists x (P(x) \wedge HA(x))$

$\forall x (P(x) \to \sim HA(x).)$

$\sim HA(x)$

$\therefore \sim P(x)$

Figure 2.4.6

The diagram shows that "this function" must lie outside the polynomial functions disk, and so the truth of the conclusion necessarily follows from the truth of the premises. Hence the argument is valid. ∎

An alternative approach to this example is to transform the statement "No polynomial functions have horizontal asymptotes" into the equivalent form "$\forall x$, if x is a polynomial function, then x does not have a horizontal asymptote." If this is done, the argument can be seen to have the form

$$\forall x, \text{if } P(x) \text{ then } Q(x).$$
$$\sim Q(a), \text{for a particular } a.$$
$$\therefore \sim P(a)$$

where $P(x)$ is "x is a polynomial function" and $Q(x)$ is "x does not have a horizontal asymptote." This is valid by universal modus tollens.

> ⚠ **Caution!** You need to be careful when using diagrams to test for validity, because you may miss seeing one or more of the ways the diagrams fit together. For instance, in Example 2.4.6, if you saw only Figure 2.4.5(a) and not Figure 2.4.5(b), you would conclude erroneously that the argument was valid.

Creating Additional Forms of Argument

Universal modus ponens and modus tollens were obtained by combining universal instantiation with modus ponens and modus tollens. In the same way, additional forms of arguments involving universally quantified statements can be obtained by combining universal instantiation with other of the valid argument forms given in Section 1.3. For instance, in Section 1.3 the argument form called transitivity was introduced:

$$p \to q$$
$$q \to r$$
$$\therefore p \to r$$

This argument form can be combined with universal instantiation to obtain the following valid argument form.

	Universal Transitivity	
Formal Version		*Informal Version*
$\forall x\, P(x) \rightarrow Q(x).$		Anything that x makes $P(x)$ true makes $Q(x)$ true.
$\forall x\, Q(x) \rightarrow R(x).$		Anything that x makes $Q(x)$ true makes $R(x)$ true.
$\therefore \forall x\, P(x) \rightarrow R(x).$		$\therefore$ Anything that x makes $P(x)$ true makes $R(x)$ true.

Example 2.4.8 Evaluating an Argument for Tarski's World

The following argument refers to the kind of arrangement of objects of various types and colors described in Examples 2.1.12 and 2.3.1. Reorder and rewrite the premises to show that the conclusion follows as a valid consequence from the premises.

1. All the triangles are blue.

2. If an object is to the right of all the squares, then it is above all the circles.

3. If an object is not to the right of all the squares, then it is not blue.

$\therefore$ All the triangles are above all the circles.

Solution   It is helpful to begin by rewriting the premises and the conclusion in if-then form:

1. $\forall x$, if x is a triangle, then x is blue.

2. $\forall x$, if x is to the right of all the squares, then x is above all the circles.

3. $\forall x$, if x is not to the right of all the squares, then x is not blue.

$\therefore$ $\forall x$, if x is a triangle, then x is above all the circles.

The goal is to reorder the premises so that the conclusion of each is the same as the hypothesis of the next. Also, the hypothesis of the argument's conclusion should be the same as the hypothesis of the first premise, and the conclusion of the argument's conclusion should be the same as the conclusion of the last premise. To achieve this goal, it may be necessary to rewrite some of the statements in contrapositive form.

In this example you can see that the first premise should remain where it is, but the second and third premises should be interchanged. Then the hypothesis of the argument is the same as the hypothesis of the first premise, and the conclusion of the argument's conclusion is the same as the conclusion of the third premise. But the hypotheses and conclusions of the premises do not quite line up. This is remedied by rewriting the third premise in contrapositive form.

Thus the premises and conclusion of the argument can be rewritten as follows:

1. $\forall x$, if x is a triangle, then x is blue.

3. $\forall x$, if x is blue, then x is to the right of all the squares.

2. $\forall x$, if x is to the right of all the squares, then x is above all the circles.

$\therefore$ $\forall x$, if x is a triangle, then x is above all the circles.

The validity of this argument follows easily from the validity of universal transitivity. Putting 1 and 2 together and using universal transitivity gives that

4. $\forall x$, if x is a triangle, then x is to the right of all the squares.

And putting 4 together with 3 and using universal transitivity gives that

$$\forall x, \text{ if } x \text{ is a triangle, then } x \text{ is above all the circles,}$$

which is the conclusion of the argument. ∎

Remark on the Converse and Inverse Errors

One reason why so many people make converse and inverse errors is that the forms of the resulting arguments would be valid if the major premise were a biconditional rather than a simple conditional. And, as we noted in Section 1.2, many people tend to confuse biconditionals and conditionals.

Consider, for example, the following argument:

> All the town criminals frequent the Den of Iniquity bar.
>
> John frequents the Den of Iniquity bar.
>
> ∴ John is one of the town criminals.

The conclusion of this argument is invalid—it results from making the converse error. Therefore, it may be false even when the premises of the argument are true. This type of argument attempts unfairly to establish guilt by association.

The closer, however, the major premise comes to being a biconditional, the more likely the conclusion is to be true. If hardly anyone but criminals frequents the bar and John also frequents the bar, then it is likely (though not certain) that John is a criminal. On the basis of the given premises, it might be sensible to be suspicious of John, but it would be wrong to convict him.

A variation of the converse error is a very useful reasoning tool, provided that it is used with caution. It is the type of reasoning that is used by doctors to make medical diagnoses and by auto mechanics to repair cars. It is the type of reasoning used to generate explanations for phenomena. It goes like this: If a statement of the form

$$\text{For all } x, \text{ if } P(x) \text{ then } Q(x)$$

is true, and if

$$Q(a) \text{ is true, for a particular } a,$$

then check out the statement $P(a)$; it just might be true. For instance, suppose a doctor knows that

> For all x, if x has pneumonia, then x has a fever and chills,
> coughs deeply, and feels exceptionally tired and miserable.

And suppose the doctor also knows that

> John has a fever and chills, coughs deeply,
> and feels exceptionally tired and miserable.

On the basis of these data, the doctor concludes that a diagnosis of pneumonia is a strong possibility, though not a certainty. The doctor will probably attempt to gain further support for this diagnosis through laboratory testing that is specifically designed to detect pneumonia. Note that the closer a set of symptoms comes to being a necessary and sufficient condition for an illness, the more nearly certain the doctor can be of his or her diagnosis.

This form of reasoning has been named **abduction** by researchers working in artificial intelligence. It is used in certain computer programs, called expert systems, that attempt to duplicate the functioning of an expert in some field of knowledge.

Exercise Set 2.4

1. Let the following law of algebra be the first statement of an argument:

$$\text{For all real numbers } a \text{ and } b,$$
$$(a + b)^2 = a^2 + 2ab + b^2.$$

Suppose each of the following statements is, in turn, the second statement of the argument. Use universal instantiation or universal modus ponens to write the conclusion that follows in each case.

   a. $a = x$ and $b = y$ are particular real numbers.
   b. $a = f_i$ and $b = f_j$ are particular real numbers.
   c. $a = 3u$ and $b = 5v$ are particular real numbers.
   d. $a = g(r)$ and $b = g(s)$ are particular real numbers.
   e. $a = \log(t_1)$ and $b = \log(t_2)$ are particular real numbers.

Use universal instantiation or universal modus ponens to fill in valid conclusions for the arguments in 2–4.

2. If an integer n equals $2 \cdot k$ and k is an integer, then n is even.
   0 equals $2 \cdot 0$ and 0 is an integer.
   ∴ _____

3. For all real numbers $a, b, c,$ and d, if $b \neq 0$ and $d \neq 0$, then $a/b + c/d = (ad + bc)/bd$.
   $a = 2, b = 3, c = 4$ and $d = 5$ are particular real numbers such that $b \neq 0$ and $d \neq 0$.
   ∴ _____

4. ∀ real numbers $r, a,$ and b, if r is positive, then $(r^a)^b = r^{ab}$.
   $r = 3, a = 1/2,$ and $b = 6$ are particular real numbers such that r is positive.
   ∴ _____

Use universal modus tollens to fill in valid conclusions for the arguments in 5 and 6.

5. All healthy people eat an apple a day.
   Adster does not eat an apple a day.
   ∴ _____

6. If a computer program is correct, then compilation of the program does not produce error messages.
   Compilation of this program produces error messages.
   ∴ _____

Some of the arguments in 7–18 are valid by universal modus ponens or universal modus tollens; others are invalid and exhibit the converse or the inverse error. State which are valid and which are invalid. Justify your answers.

7. All healthy people eat an apple a day.
   Keisha eats an apple a day.
   ∴ Keisha is a healthy person.

8. All freshmen must take writing.
   Caroline is a freshman.
   ∴ Caroline must take writing.

9. All healthy people eat an apple a day.
   Herbert is not a healthy person.
   ∴ Herbert does not eat an apple a day.

10. If a product of two numbers is 0, then at least one of the numbers is 0.
    For a particular number x, neither $(2x + 1)$ nor $(x - 7)$ equals 0.
    ∴ The product $(2x + 1)(x - 7)$ is not 0.

11. All cheaters sit in the back row.
    Monty sits in the back row.
    ∴ Monty is a cheater.

12. All honest people pay their taxes.
    Darth is not honest.
    ∴ Darth does not pay his taxes.

13. For all students x, if x studies discrete mathematics, then x is good at logic.
    Tarik studies discrete mathematics.
    ∴ Tarik is good at logic.

14. If compilation of a computer program produces error messages, then the program is not correct.
    Compilation of this program does not produce error messages.
    ∴ This program is correct.

15. Any sum of two rational numbers is rational.
    The sum $r + s$ is rational.
    ∴ The numbers r and s are both rational.

16. If a number is even, then twice that number is even.
    The number $2n$ is even, for a particular number n.
    ∴ The particular number n is even.

17. If an infinite series converges, then the terms go to 0.
    The terms of the infinite series $\sum_{n=1}^{\infty} \frac{1}{n}$ go to 0.
    ∴ The infinite series $\sum_{n=1}^{\infty} \frac{1}{n}$ converges.

18. If an infinite series converges, then its terms go to 0.
    The terms of the infinite series $\sum_{n=1}^{\infty} \frac{n}{n + 1}$ do not go to 0.
    ∴ The infinite series $\sum_{n=1}^{\infty} \frac{n}{n + 1}$ does not converge.

19. Rewrite the statement "No good cars are cheap" in the form "$\forall x$, if $P(x)$ then $\sim Q(x)$." Indicate whether each of the following arguments is valid or invalid, and justify your answers.
    a. No good car is cheap.
    A Rimbaud is a good car.
    ∴ A Rimbaud is not cheap.

b. No good car is cheap.
A Simbaru is not cheap.
∴ A Simbaru is a good car.

c. No good car is cheap.
A VX Roadster is cheap.
∴ A VX Roadster is not good.

d. No good car is cheap.
An Omnex is not a good car.
∴ An Omnex is cheap.

20. **a.** Use a diagram to show that the following argument can have true premises and a false conclusion.

All dogs are carnivorous.

Aaron is not a dog.

∴ Aaron is not carnivorous.

b. What can you conclude about the validity or invalidity of the following argument form? Explain how the result from part (a) leads to this conclusion.

$\forall x$, if $P(x)$ then $Q(x)$.

$\sim P(a)$ for a particular a.

∴ $\sim Q(a)$.

Indicate whether the arguments in 21–26 are valid or invalid. Support your answers by drawing diagrams.

21. All people are mice.
All mice are mortal.
∴ All people are mortal.

22. All discrete mathematics students can tell a valid argument from an invalid one.
All thoughtful people can tell a valid argument from an invalid one.
∴ All discrete mathematics students are thoughtful.

23. All teachers occasionally make mistakes.
No gods ever make mistakes.
∴ No teachers are gods.

24. No vegetarians eat meat.
All vegans are vegetarian.
∴ No vegans eat meat.

25. No college cafeteria food is good.
No good food is wasted.
∴ No college cafeteria food is wasted.

26. All polynomial functions are differentiable.
All differentiable functions are continuous.
∴ All polynomial functions are continuous.

27. [Adapted from Lewis Carroll.]
Nothing intelligible ever puzzles *me*.
Logic puzzles me.
∴ Logic is unintelligible.

In exercises 28–33, reorder the premises in each of the arguments to show that the conclusion follows as a valid consequence from the premises. It may be helpful to rewrite the statements in if-then form and replace some statements by their contrapositives. Exercises 28–30 refer to the kinds of Tarski worlds discussed in Example 2.1.12 and 2.3.1. Exercises 31 and 32 are adapted from *Symbolic Logic* by Lewis Carroll.*

28. 1. Every object that is to the right of all the blue objects is above all the triangles.
2. If an object is a circle, then it is to the right of all the blue objects.
3. If an object is not a circle, then it is not gray.
∴ All the gray objects are above all the triangles.

29. 1. All the objects that are to the right of all the triangles are above all the circles.
2. If an object is not above all the black objects, then it is not a square.
3. All the objects that are above all the black objects are to the right of all the triangles.
∴ All the squares are above all the circles.

30. 1. If an object is not blue, then it is not a triangle.
2. If an object is not above all the gray objects, then it is not a square.
3. Every black object is a square.
4. Every object that is above all the gray objects is above all the triangles.
∴ If an object is black, then it is above all the blue objects.

31. 1. I trust every animal that belongs to me.
2. Dogs gnaw bones.
3. I admit no animals into my study unless they will beg when told to do so.
4. All the animals in the yard are mine.
5. I admit every animal that I trust into my study.
6. The only animals that are really willing to beg when told to do so are dogs.
∴ All the animals in the yard gnaw bones.

32. 1. When I work a logic example without grumbling, you may be sure it is one I understand.
2. The arguments in these examples are not arranged in regular order like the ones I am used to.
3. No easy examples make my head ache.
4. I can't understand examples if the arguments are not arranged in regular order like the ones I am used to.
5. I never grumble at an example unless it gives me a headache.
∴ These examples are not easy.

*Lewis Carroll, *Symbolic Logic* (New York: Dover, 1958), pp. 118, 120, 123.

In 33 and 34 a conclusion follows from the given premises, but it is difficult to see because the premises are jumbled up. Reorder the premises to make it clear that a conclusion follows logically, and state the valid conclusion that can be drawn. (It may be helpful to rewrite some of the statements in if-then form and to replace some statements by their contrapositives.)

33. 1. No birds except ostriches are at least 9 feet tall.
2. There are no birds in this aviary that belong to anyone but me.
3. No ostrich lives on mince pies.
4. I have no birds less than 9 feet high.

34. 1. All writers who understand human nature are clever.
2. No one is a true poet unless he can stir the human heart.
3. Shakespeare wrote *Hamlet*.
4. No writer who does not understand human nature can stir the human heart.
5. None but a true poet could have written *Hamlet*.

✳ 35. Derive the validity of universal modus tollens from the validity of universal instantiation and modus tollens.

✳ 36. Derive the validity of universal transitivity from the validity of universal instantiation and the valid argument called transitivity in Section 1.3.

CHAPTER 3

ELEMENTARY NUMBER THEORY AND METHODS OF PROOF

The underlying content of this chapter is likely to be familiar to you. It consists of properties of integers (whole numbers), rational numbers (integer fractions), and real numbers. The underlying theme of this chapter is the question of how to determine the truth or falsity of a mathematical statement.

Here is an example involving a concept used frequently in computer science. Given any real number x, the floor of x, or greatest integer in x, denoted $\lfloor x \rfloor$, is the largest integer that is less than or equal to x. On the number line, $\lfloor x \rfloor$ is the integer immediately to the left of x (or equal to x if x is, itself, an integer). Thus $\lfloor 2.3 \rfloor = 2$, $\lfloor 12.99999 \rfloor = 12$, and $\lfloor -1.5 \rfloor = -2$. Consider the following two questions:

1. For any real number x, is $\lfloor x - 1 \rfloor = \lfloor x \rfloor - 1$?

2. For any real numbers x and y, is $\lfloor x - y \rfloor = \lfloor x \rfloor - \lfloor y \rfloor$?

Take a few minutes to try to answer these questions for yourself.

It turns out that the answer to (1) is yes, whereas the answer to (2) is no. Are these the answers you got? If not, don't worry. In Section 3.5 you will learn the techniques you need to answer these questions and more. If you did get the correct answers, congratulations! You have excellent mathematical intuition. Now ask yourself, "How sure am I of my answers? Were they plausible guesses or absolute certainties? Was there any difference in certainty between my answers to (1) and (2)? Would I have been willing to bet a large sum of money on the correctness of my answers?"

One of the best ways to think of a mathematical proof is as a carefully reasoned argument to convince a skeptical listener (often yourself) that a given statement is true. Imagine the listener challenging your reasoning every step of the way, constantly asking, "Why is that so?" If you can counter every possible challenge, then your proof as a whole will be correct.

As an example, imagine proving to someone not very familiar with mathematical notation that if x is a number with $5x + 3 = 33$, then $x = 6$. You could argue as follows:

> If $5x + 3 = 33$, then $5x + 3$ minus 3 will equal $33 - 3$ since subtracting the same number from two equal quantities gives equal results. But $5x + 3$ minus 3 equals $5x$ because adding 3 to $5x$ and then subtracting 3 just leaves $5x$. Also, $33 - 3 = 30$. Hence $5x = 30$. This means that x is a number which when multiplied by 5 equals 30. But the only number with this property is 6. Therefore, $x = 6$.

Of course there are other ways to phrase this proof, depending on the level of mathematical sophistication of the intended reader. In practice, mathematicians often omit reasons for certain steps of an argument when they are confident that the reader can easily supply them. When you are first learning to write proofs, however, it is better to err on the side of supplying too many reasons rather than too few. All too frequently, when even the best mathematicians carefully examine some "details" in their arguments, they discover that those details are actually false. Probably the most important reason for requiring proof in mathematics is that writing a proof forces us to become aware of weaknesses in our arguments and in the unconscious assumptions we have made.

Sometimes correctness of a mathematical argument can be a matter of life or death. Suppose, for example, that a mathematician is part of a team charged with designing a new type of airplane engine, and suppose that the mathematician is given the job of determining whether the thrust delivered by various engine types is adequate. If you knew that the mathematician was only fairly sure, but not positive, of the correctness of his analysis, you would probably not want to ride in the resulting aircraft.

At a certain point in Lewis Carroll's *Alice in Wonderland* (see exercise 28 in Section 1.2), the March Hare tells Alice to "say what you mean." In other words, she should be precise in her use of language: If she means a thing, then that is exactly what she should say. In this chapter, perhaps more than in any other mathematics course you have ever taken, you will find it necessary to say what you mean. Precision of thought and language is essential to achieve the mathematical certainty that is needed if you are to have complete confidence in your solutions to mathematical problems.

3.1 Direct Proof and Counterexample I: Introduction

Mathematics, as a science, commenced when first someone, probably a Greek, proved propositions about "any" things or about "some" things without specification of definite particular things. — Alfred North Whitehead, 1861–1947

Both discovery and proof are integral parts of problem solving. When you think you have discovered that a certain statement is true, try to figure out why it is true. If you succeed, you will know that your discovery is genuine. Even if you fail, the process of trying will give you insight into the nature of the problem and may lead to the discovery that the statement is false. For complex problems, the interplay between discovery and proof is not reserved to the end of the problem-solving process but, rather, is an important part of each step.

In this text we assume a familiarity with the laws of basic algebra, which are listed in Appendix A. We also use the three properties of equality: For all objects A, B, and C, (1) $A = A$, (2) if $A = B$ then $B = A$, and (3) if $A = B$ and $B = C$, then $A = C$. In addition, we assume that the set of integers is closed under addition, subtraction, and multiplication. This means that sums, differences, and products of integers are integers. Of course, most quotients of integers are not integers. For example, $3 \div 2$, which equals $3/2$, is not an integer, and $3 \div 0$ is not even a number.

The mathematical content of this section primarily concerns even and odd integers and prime and composite numbers.

Definitions

In order to evaluate the truth or falsity of a statement, you must understand what the statement is about. In other words, you must know the meanings of all terms that occur in the statement. Mathematicians define terms very carefully and precisely and consider it important to learn definitions virtually word for word.

• Definitions

An integer n is **even** if, and only if, n equals twice some integer. An integer n is **odd** if, and only if, n equals twice some integer plus 1.

Symbolically, if n is an integer, then

$$n \text{ is even} \quad \Leftrightarrow \quad \exists \text{ an integer } k \text{ such that } n = 2k.$$
$$n \text{ is odd} \quad \Leftrightarrow \quad \exists \text{ an integer } k \text{ such that } n = 2k + 1.$$

It follows from the definition that if you are doing a problem in which you happen to know that a certain integer is even, you can deduce that it has the form $2 \cdot$ (some integer). Conversely, if you know in some situation that a particular integer equals $2 \cdot$ (some integer), then you can deduce that the integer is even.

Know a particular integer n is even.	$\xrightarrow{\text{deduce}}$	n has the form $2 \cdot$ (some integer).
Know n has the form $2 \cdot$ (some integer).	$\xrightarrow{\text{deduce}}$	n is even.

Example 3.1.1 Even and Odd Integers

Use the definitions of *even* and *odd* to justify your answers to the following questions.

a. Is 0 even?

b. Is -301 odd?

c. If a and b are integers, is $6a^2b$ even?

d. If a and b are integers, is $10a + 8b + 1$ odd?

e. Is every integer either even or odd?

Solution

a. Yes, $0 = 2 \cdot 0$.

b. Yes, $-301 = 2(-151) + 1$.

c. Yes, $6a^2b = 2(3a^2b)$, and since a and b are integers, so is $3a^2b$ (being a product of integers).

d. Yes, $10a + 8b + 1 = 2(5a + 4b) + 1$, and since a and b are integers, so is $5a + 4b$ (being a sum of products of integers).

e. The answer is yes, although the proof is not obvious. (Try giving a reason yourself.) We will show in Section 3.4 that this fact results from another fact known as the quotient-remainder theorem. ∎

The integer 6, which equals $2 \cdot 3$, is a product of two smaller positive integers. On the other hand, 7 cannot be written as a product of two smaller positive integers; its only positive factors are 1 and 7. A positive integer, such as 7, that cannot be written as a product of two smaller positive integers is called *prime*.

> **• Definition**
>
> An integer n is **prime** if, and only if, $n > 1$ and for all positive integers r and s, if $n = r \cdot s$, then $r = 1$ or $s = 1$. An integer n is **composite** if, and only if, $n > 1$ and $n = r \cdot s$ for some positive integers r and s with $r \neq 1$ and $s \neq 1$.
>
> Symbolically, if n is an integer that is greater than 1, then
>
> n is prime $\quad \Leftrightarrow \quad \forall$ positive integers r and s, if $n = r \cdot s$
> $\qquad\qquad\qquad\qquad$ then $r = 1$ or $s = 1$.
>
> n is composite $\quad \Leftrightarrow \quad \exists$ positive integers r and s such that $n = r \cdot s$
> $\qquad\qquad\qquad\qquad$ and $r \neq 1$ and $s \neq 1$.

Example 3.1.2 Prime and Composite Numbers

a. Is 1 prime?

b. Is it true that every integer greater than 1 is either prime or composite?

c. Write the first six prime numbers.

d. Write the first six composite numbers.

Solution

a. No. A prime number is required to be greater than 1.

b. Yes. For any integer greater than 1, the two definitions are negations of each other.

c. 2, 3, 5, 7, 11, 13

d. 4, 6, 8, 9, 10, 12　　　　　　　　　　　　　　　　　　　　■

Proving Existential Statements

According to the definition given in Section 2.1, a statement in the form

$$\exists x \in D \text{ such that } Q(x)$$

is true if, and only if,

$$Q(x) \text{ is true for at least one } x \text{ in } D.$$

One way to prove this is to find an x in D that makes $Q(x)$ true. Another way is to give a set of directions for finding such an x. Both of these methods are called **constructive proofs of existence.**

Example 3.1.3 Constructive Proofs of Existence

a. Prove the following: $\exists$ an even integer n that can be written in two ways as a sum of two prime numbers.

b. Suppose that r and s are integers. Prove the following: $\exists$ an integer k such that $22r + 18s = 2k$.

Solution

a. Let $n = 10$. Then $10 = 5 + 5 = 3 + 7$ and 3, 5, and 7 are all prime numbers.

b. Let $k = 11r + 9s$. Then k is an integer because it is a sum of products of integers; and by substitution, $2k = 2(11r + 9s)$, which equals $22r + 18s$ by the distributive law of algebra.　　　　　　　　　　　　　　　　　　　　■

A **nonconstructive proof of existence** involves showing either (a) that the existence of a value of x that makes $Q(x)$ true is guaranteed by an axiom or a previously proved theorem or (b) that the assumption that there is no such x leads to a contradiction. The disadvantage of a nonconstructive proof is that it may give virtually no clue about where or how x may be found. The widespread use of digital computers in recent years has led to some dissatisfaction with this aspect of nonconstructive proofs and to increased efforts to produce constructive proofs containing directions for computer calculation of the quantity in question.

Disproving Universal Statements by Counterexample

To disprove a statement means to show that it is false. Consider the question of disproving a statement of the form

$$\forall x \text{ in } D, \text{ if } P(x) \text{ then } Q(x).$$

Showing that this statement is false is equivalent to showing that its negation is true. The negation of the statement is existential:

$$\exists x \text{ in } D \text{ such that } P(x) \text{ and not } Q(x).$$

But to show that an existential statement is true, we generally give an example, and because the example is used to show that the original statement is false, we call it a *counterexample*. Thus the method of disproof by *counterexample* can be written as follows:

Disproof by Counterexample

To disprove a statement of the form "$\forall x \in D$, if $P(x)$ then $Q(x)$," find a value of x in D for which $P(x)$ is true and $Q(x)$ is false. Such an x is called a **counterexample.**

Example 3.1.4 Disproof by Counterexample

Disprove the following statement by finding a counterexample:

$$\forall \text{ real numbers } a \text{ and } b, \text{ if } a^2 = b^2 \text{ then } a = b.$$

Solution   To disprove this statement, you need to find real numbers a and b such that $a^2 = b^2$ and $a \neq b$. The fact that both positive and negative integers have positive squares helps in the search. If you flip through some possibilities in your mind, you will quickly see that 1 and -1 will work (or 2 and -2, or 0.5 and -0.5, and so forth).

Statement: $\forall$ real numbers a and b, if $a^2 = b^2$, then $a = b$.

Counterexample: Let $a = 1$ and $b = -1$. Then $a^2 = 1^2 = 1$ and $b^2 = (-1)^2 = 1$, and so $a^2 = b^2$. But $a \neq b$ since $1 \neq -1$.

■

It is a sign of intelligence to make generalizations. Frequently, after observing a property to hold in a large number of cases, you may guess that it holds in all cases. You may, however, run into difficulty when you try to prove your guess. Perhaps you just have not figured out the key to the proof. But perhaps your guess is false. Consequently, when you are having serious difficulty proving a general statement, you should interrupt your efforts to look for a counterexample. Analyzing the kinds of problems you are encountering in your proof efforts may help in the search. It may even happen that if you find a counterexample and therefore prove the statement false, your understanding may be sufficiently clarified that you can formulate a more limited but true version of the statement. For instance, Example 3.1.4 shows that it is not always true that if the squares of two numbers are equal, then the numbers are equal. However, it is true that if the squares of two *positive* numbers are equal, then the numbers are equal.

Proving Universal Statements

The vast majority of mathematical statements to be proved are universal. In discussing how to prove such statements, it is helpful to imagine them in a standard form:

$$\forall x \in D, \text{ if } P(x) \text{ then } Q(x).$$

In Section 2.1 we showed that any universal statement can be written in this form and that when D is finite, such a statement can be proved by the method of exhaustion. This method can also be used when there are only a finite number of elements that satisfy the condition $P(x)$.

Example 3.1.5 The Method of Exhaustion

Use the method of exhaustion to prove the following statement:

> $\forall n \in \mathbf{Z}$, if n is even and $4 \leq n \leq 30$, then n can be written as a sum of two prime numbers.

Solution

$$
\begin{array}{llll}
4 = 2 + 2 & 6 = 3 + 3 & 8 = 3 + 5 & 10 = 5 + 5 \\
12 = 5 + 7 & 14 = 11 + 3 & 16 = 5 + 11 & 18 = 7 + 11 \\
20 = 7 + 13 & 22 = 5 + 17 & 24 = 5 + 19 & 26 = 7 + 19 \\
28 = 11 + 17 & 30 = 11 + 19
\end{array}
$$

■

In most cases in mathematics, however, the method of exhaustion cannot be used. For instance, can you prove by exhaustion that *every* even integer greater than 2 can be written as a sum of two prime numbers? No. To do that you would have to check every even integer, and because there are infinitely many such numbers, this is an impossible task.

Even when the domain is finite, it may be infeasible to use the method of exhaustion. Imagine, for example, trying to check by exhaustion that the multiplication circuitry of a particular computer gives the correct result for every pair of numbers in the computer's range. Since a typical computer would require thousands of years just to compute all possible products of all numbers in its range (not to mention the time it would take to check the accuracy of the answers), checking correctness by the method of exhaustion is obviously impractical.

The most powerful technique for proving a universal statement is one that works regardless of the size of the domain over which the statement is quantified. It is called the *method of generalizing from the generic particular*. Here is the idea underlying the method:

Method of Generalizing from the Generic Particular

To show that every element of a domain satisfies a certain property, suppose x is a *particular* but *arbitrarily chosen* element of the domain, and show that x satisfies the property.

Example 3.1.6 Generalizing from the Generic Particular

At some time you may have been shown a "mathematical trick" like the following. You ask a person to pick any number, add 5, multiply by 4, subtract 6, divide by 2, and subtract twice the original number. Then you astound the person by announcing that their final result was 7. How does this "trick" work? Let x stand for the number the person picks. Here is what happens when the person follows your directions:

Step	Result
Pick a number.	x
Add 5.	$x + 5$
Multiply by 4.	$(x + 5) \cdot 4 = 4x + 20$
Subtract 6.	$(4x + 20) - 6 = 4x + 14$
Divide by 2.	$\dfrac{4x + 14}{2} = 2x + 7$
Subtract twice the original number.	$(2x + 7) - 2x = 7$

Thus no matter what number the person starts with, the result will always be 7. Note that the x in the analysis above is *particular* (because it represents a single quantity), but it is also *arbitrarily chosen* or *generic* (because it can represent any number whatsoever). This illustrates the process of drawing a general conclusion from a particular but generic object. ∎

The point of having x be arbitrarily chosen (or generic) is to make a proof that can be generalized to all elements of the domain. By choosing x arbitrarily, you are making no special assumptions about x that are not also true of all other elements of the domain. The word *generic* means "sharing all the common characteristics of a group or class." Thus everything you deduce about a generic element x of the domain is equally true of any other element of the domain.

When the method of generalizing from the generic particular is applied to a property of the form "If $P(x)$ then $Q(x)$," the result is the method of *direct proof*. Recall that the only way "If $P(x)$ then $Q(x)$" can be false is for $P(x)$ to be true and $Q(x)$ to be false. Thus to show that "If $P(x)$ then $Q(x)$" is true, suppose $P(x)$ is true and show that $Q(x)$ must also be true. It follows by the method of generalizing from the generic particular that to prove a statement of the form "$\forall x \in D$, if $P(x)$ then $Q(x)$," you suppose x is a particular but arbitrarily chosen element of D that satisfies $P(x)$, and then you show that x satisfies $Q(x)$.

Method of Direct Proof

1. Express the statement to be proved in the form "$\forall x \in D$, if $P(x)$ then $Q(x)$." (This step is often done mentally.)

2. Start the proof by supposing x is a particular but arbitrarily chosen element of D for which the hypothesis $P(x)$ is true. (This step is often abbreviated "Suppose $x \in D$ and $P(x)$.")

3. Show that the conclusion $Q(x)$ is true by using definitions, previously established results, and the rules for logical inference.

Example 3.1.7 A Direct Proof of a Theorem

Prove that the sum of any two even integers is even.

 Caution! The word two in this statement does not necessarily refer to two distinct integers. If a choice of integers is made arbitrarily, the integers are very likely to be distinct, but they might be the same.

Solution   Whenever you are presented with a statement to be proved, it is a good idea to ask yourself whether you believe it to be true. In this case you might imagine some pairs of even integers, say $2 + 4, 6 + 10, 12 + 12, 28 + 54$, and mentally check that their sums are even. However, since you cannot possibly check all pairs of even numbers, you cannot know for sure that the statement is true in general by checking its truth in these particular instances. Many properties hold for a large number of examples and yet fail to be true in general.

To prove this statement in general, you need to show that no matter what even integers are given, their sum is even. But given any two even integers, it is possible to represent them as $2r$ and $2s$ for some integers r and s. And by the distributive law of algebra, $2r + 2s = 2(r + s)$, which is even. Thus the statement is true in general.

Suppose the statement to be proved were much more complicated than this. What is the statement method you could use to derive a proof?

Formal Restatement: $\forall$ integers m and n, if m and n are even then $m + n$ is even.

This statement is universally quantified over an infinite domain. Thus to prove it in general, you need to show that no matter what two integers you might be given, if both of them are even then their sum will also be even.

Next ask yourself, "Where am I starting from?" or "What am I supposing?" The answer to such a question gives you the starting point, or first sentence, of the proof.

Starting Point: Suppose m and n are particular but arbitrarily chosen integers that are even.

Or, in abbreviated form:

Suppose m and n are any even integers.

Then ask yourself, "What conclusion do I need to show in order to complete the proof?"

To Show: $m + n$ is even.

At this point you need to ask yourself, "How do I get from the starting point to the conclusion?" Since both involve the term *even integer,* you must use the definition of this term—and thus you must know what it means for an integer to be even. It follows from the definition that since m and n are even,

$$m = 2r, \text{ for some integer } r \quad \text{and} \quad n = 2s, \text{ for some integer } s.$$

(The reason you have to use two different letters r and s is that m and n are arbitrarily chosen—they could be any pair of even integers whatsoever. If you had set $m = 2r$ and $n = 2r$, then m would equal n, which need not be the case.)

Now what you want to show is that $m + n$ is even. In other words, you want to show something about the expression $m + n$. Having just found alternate representations for m (as $2r$) and n (as $2s$), it seems reasonable to substitute these representations in place of m and n:

$$m + n = 2r + 2s.$$

Your goal is to show that $m + n$ is even. By definition of even, this means that $m + n$ can be written in the form

$$2 \cdot (\text{some integer}).$$

This analysis narrows the gap between the starting point and what is to be shown to showing that

$$2r + 2s = 2 \cdot (\text{some integer}).$$

Why is this true? First, because of the distributive law from algebra, which says that

$$2r + 2s = 2(r + s),$$

and, second, because the sum of any two integers is an integer, which implies that $r + s$ is an integer.

This discussion is summarized by rewriting the statement as a theorem and giving a formal proof of it. (In mathematics, the word *theorem* refers to a statement that is known to be true because it has been proved.) The formal proof, as well as many others in this text, includes explanatory notes to make its logical flow apparent. Such comments are purely a convenience for the reader and could be omitted entirely. For this reason they are italicized and enclosed in square brackets: [ ].

Donald Knuth, one of the pioneers of the science of computing, has compared constructing a computer program from a set of specifications to writing a mathematical proof based on a set of axioms.[*] In keeping with this analogy, the bracketed comments can be thought of as similar to the explanatory documentation provided by a good programmer. Documentation is not necessary for a program to run, but it helps a human reader understand what is going on.

Theorem 3.1.1

The sum of any two even integers is even.

Proof:

Suppose m and n are [*particular but arbitrarily chosen*] even integers. [*We must show that $m + n$ is even.*] By definition of even, $m = 2r$ and $n = 2s$ for some integers r and s. Then

$$m + n = 2r + 2s \qquad \text{by substitution}$$
$$= 2(r + s) \qquad \text{by factoring out a 2.}$$

Let $k = r + s$. Note that k is an integer because it is a sum of integers. Hence

$$m + n = 2k \quad \text{where } k \text{ is an integer.}$$

It follows by definition of even that $m + n$ is even. [*This is what we needed to show.*][†]

∎

Most theorems, like the one above, can be analyzed to a point where you realize that as soon as a certain thing is shown, the theorem will be proved. When that thing has been

---

[*]Donald E. Knuth, *The Art of Computer Programming,* 2nd ed., Vol. I (Reading, MA: Addison-Wesley, 1973), p. ix.
[†]See page 113 for a discussion of the role of universal modus ponens in this proof.

shown, it is natural to end the proof with the words "this is what we needed to show." The Latin words for this are *quod erat demonstrandum,* or Q.E.D. for short. Proofs in older mathematics books end with these initials.

Note that both the *if* and the *only if* parts of the definition of even were used in the proof of Theorem 3.1.1. Since m and n were known to be even, the *only if* ($\Rightarrow$) part of the definition was used to deduce that m and n had a certain general form. Then, after some algebraic substitution and manipulation, the *if* ($\Leftarrow$) part of the definition was used to deduce that $m + n$ was even.

Directions for Writing Proofs of Universal Statements

Think of a proof as a way to communicate a convincing argument for the truth of a mathematical statement. When you write a proof, imagine that you will be sending it to a capable classmate who has had to miss the last week or two of your course. Try to be clear and complete. Keep in mind that your classmate will see only what you actually write down, not any unexpressed thoughts behind it.

Over the years, the following rules of style have become fairly standard for writing the final versions of proofs:

1. **Copy the statement of the theorem to be proved on your paper.**

2. **Clearly mark the beginning of your proof with the word <u>Proof</u>.**

3. **Make your proof self-contained.**

   This means that you should identify each variable used in your proof in the body of the proof. Thus you will begin proofs by introducing the initial variables and stating what kind of objects they are. The first sentence of your proof would be something like "Suppose m and n are integers" or "Let x be a real number that is greater than 2." This is similar to declaring variables and their data types at the beginning of a computer program.

   At a later point in your proof, you may introduce a new variable to represent a quantity that is known to exist. For example, if you know that a particular integer n is even, then you know that n equals 2 times some integer. It is usually convenient to give this integer a name so that you can work with it concretely later in the proof. Thus if you decide to call the integer, say, s, you would write, "Since n is even, $n = 2s$ for some integer s."

4. **Write your proof in complete sentences.**

   This does not mean that you should avoid using symbols and shorthand abbreviations, just that you should incorporate them into sentences. For example, the proof of Theorem 3.1.1 contains the sentence

   $$\text{Then } m + n = 2r + 2s$$
   $$= 2(r + s).$$

   To read this as a sentence, read the first equals sign as "equals" and each subsequent equals sign as "which equals."

5. **Give a reason for each assertion you make in your proof.**

   Each assertion in a proof should come directly from the hypothesis of the theorem, or follow from the definition of one of the terms in the theorem, or be a result obtained earlier in the proof, or be a mathematical result that has previously been established or is agreed to be assumed. Indicate the reason for each step of your proof using phrases such as *by hypothesis, by definition of . . .* , and *by theorem . . . .*

6. **Include the "little words" that make the logic of your arguments clear.**

When writing a mathematical argument, especially a proof, indicate how each sentence is related to the previous one. Does it follow from the previous sentence or from a combination of the previous sentence and earlier ones? If so, start the sentence by stating the reason why it follows or by writing *Then,* or *Thus,* or *So,* or *Hence,* or *Therefore,* or *It follows that,* and include the reason at the end of the sentence. For instance, in the proof of Theorem 3.1.1, once you know that m is even, you can write: "By definition of even, $m = 2k$ for some integer k," or you can write, "Then $m = 2k$ for some integer k by definition of even."

If a sentence expresses a new thought or fact that does not follow as an immediate consequence of the preceding statement but is needed for a later part of a proof, introduce it by writing *Observe that,* or *Note that,* or *But,* or *Now.*

Sometimes in a proof it is desirable to define a new variable in terms of previous variables. In such a case, introduce the new variable with the word *Let.* For instance, in the proof of Theorem 3.1.1, once it is known that $m + n = 2(r + s)$, where r and s are integers, a new variable k is introduced to represent $r + s$. The proof goes on to say, "Let $k = r + s$. Then k is an integer because it is a sum of two integers."

Variations among Proofs

It is rare that two proofs of a given statement, written by two different people, are identical. Even when the basic mathematical steps are the same, the two people may use different notation or may give differing amounts of explanation for their steps, or may choose different words to link the steps together into paragraph form. An important question is how detailed to make the explanations for the steps of a proof. This must ultimately be worked out between the writer of a proof and the intended reader, whether they be student and teacher, teacher and student, student and fellow student, or mathematician and colleague. Your teacher may provide explicit guidelines for you to use in your course. Or you may follow the example of the proofs in this book (which are generally explained rather fully in order to be understood by students at various stages of mathematical development). Remember that the phrases written inside brackets [ ] are intended to elucidate the logical flow or underlying assumptions of the proof and need not be written down at all. It is entirely your decision whether to include such phrases in your own proofs.

Common Mistakes

The following are some of the most common mistakes people make when writing mathematical proofs.

1. **Arguing from examples.**

Looking at examples is one of the most helpful practices a problem solver can engage in and is encouraged by all good mathematics teachers. However, it is a mistake to think that a general statement can be proved by showing it to be true for some special cases. A universal statement may be true in many instances without being true in general.

Here is an example of this mistake. It is an incorrect "proof" of the fact that the sum of any two even integers is even. (Theorem 3.1.1).

> This is true because if $m = 14$ and $n = 6$, which are both even,
> then $m + n = 20$, which is also even.

Some people find this kind of argument convincing because it does, after all, consist of evidence in support of a true conclusion. But remember that when we discussed valid arguments, we pointed out that an argument may be invalid and yet have a true conclusion. In the same way, an argument from examples may be mistakenly used to "prove" a true statement. In the example above, it is not sufficient to show that the conclusion "$m + n$ is even" is true for $m = 14$ and $n = 6$. You must give an argument to show that the conclusion is true for any even integers m and n.

2. **Using the same letter to mean two different things.**

    Some beginning theorem provers give a new variable quantity the same letter name as a previously introduced variable. Consider the following "proof" fragment:

    > Suppose m and n are odd integers. Then by definition of odd, $m = 2k + 1$ and $n = 2k + 1$ for some integer k.

    This is incorrect. Using the same symbol, k, in the expressions for both m and n implies that $m = 2k + 1 = n$. It follows that the rest of the proof applies only to integers m and n that equal each other. This is inconsistent with the supposition that m and n are arbitrarily chosen odd integers. For instance, the proof would not show that the sum of 3 and 5 is even.

3. **Jumping to a conclusion.**

    To jump to a conclusion means to allege the truth of something without giving an adequate reason. Consider the following "proof" that the sum of any two even integers is even.

    > Suppose m and n are any even integers. By definition of even, $m = 2r$ and $n = 2s$ for some integers r and s. Then $m + n = 2r + 2s$. So $m + n$ is even.

    The problem with this "proof" is that the crucial calculation

    $$2r + 2s = 2(r + s)$$

    is missing. The author of the "proof" has jumped prematurely to a conclusion.

4. **Begging the question.**

    To beg the question means to assume what is to be proved; it is a variation of jumping to a conclusion. As an example, consider the following "proof" of the fact that the product of any two odd integers is odd:

    > Suppose m and n are any odd integers. When any odd integers are multiplied, their product is odd. Hence mn is odd.

    Here is another, more subtle example of the same mistake.

    > Suppose m and n are odd integers. If mn is odd, then $mn = 2k + 1$ for some integer k. Also by definition of odd, $m = 2a + 1$ and $n = 2b + 1$ for some integers a and b. Then $mn = (2a + 1)(2b + 1) = 2k + 1$, which is odd by definition of odd. This is what was to be shown.

    The problem with this "proof" is that the author first states what it means for the conclusion to be true (that $m \cdot n$ can be expressed as $2k + 1$ for some integer k) and later just assumes it to be true (by setting $(2a + 1) \cdot (2b + 1)$ equal to $2k + 1$). Thus the author of the "proof" begs the question.

5. **Misuse of the word *if*.**

    Another common error is not serious in itself, but it reflects imprecise thinking that sometimes leads to problems later in a proof. This error involves using the word *if* when the word *because* is really meant. Consider the following proof fragment:

> Suppose p is a prime number. If p is prime, then p cannot be written as a product of two smaller positive integers.

The use of the word *if* in the second sentence is inappropriate. It suggests that the primeness of p is in doubt. But p is known to be prime by the first sentence. It cannot be written as a product of two smaller positive integers *because* it is prime. Here is a correct version of the fragment:

> Suppose p is a prime number. Because p is prime, p cannot be written as a product of two smaller positive integers.

Getting Proofs Started

Believe it or not, once you understand the idea of generalizing from the generic particular and the method of direct proof, you can write the beginnings of proofs even for theorems you do not understand. The reason is that the starting point and what is to be shown in a proof depend only on the linguistic form of the statement to be proved, not on the content of the statement.

Example 3.1.8 Identifying the "Starting Point" and the "Conclusion to Be Shown"

Write the first sentence of a proof of the following statement (the "starting point") and the last sentence of a proof (the "conclusion to be shown"):

> Every complete, bipartite graph is connected.   You are not expected to
> understand this statement.

Solution   It is helpful to rewrite the statement formally using a quantifier and a variable:

$$\overbrace{\text{domain}}\qquad\overbrace{\text{hypothesis}}\qquad\overbrace{\text{conclusion}}$$

Formal Restatement: $\forall$ graphs G, if G is complete and bipartite, then G is connected.

The first sentence, or starting point, of a proof supposes the existence of an object (in this case G) in the domain (in this case the set of all graphs) that satisfies the hypothesis of the if-then part of the statement (in this case that G is complete and bipartite). The conclusion to be shown is just the conclusion of the if-then part of the statement (in this case that G is connected).

Starting Point: Suppose G is a [*particular but arbitrarily chosen*] graph such that G is complete and bipartite.

Conclusion to Be Shown: G is connected.

Thus the proof has the following first and last sentences:

First sentence of proof: Suppose G is a [*particular but arbitrarily chosen*] graph such that G is complete and bipartite.

Last sentence of proof: Therefore G is connected.

Of course, to reach the last sentence of the proof, the definitions of the terms will have to be used.   ∎

Showing That an Existential Statement Is False

Recall that the negation of an existential statement is universal. It follows that to prove an existential statement is false, you must prove a universal statement (its negation) is true.

Example 3.1.9 Disproving an Existential Statement

Show that the following statement is false:

There is a positive integer n such that $n^2 + 3n + 2$ is prime.

Solution   Proving that the given statement is false is equivalent to proving its negation is true. The negation is

For all positive integers n, $n^2 + 3n + 2$ is not prime.

Because the negation is universal, it is proved by generalizing from the generic particular.

Claim: The statement "There is a positive integer n such that $n^2 + 3n + 2$ is prime" is false.

Proof:

Suppose n is any [*particular but arbitrarily chosen*] positive integer. [*We will show that $n^2 + 3n + 2$ is not prime.*] We can factor $n^2 + 3n + 2$ to obtain $n^2 + 3n + 2 = (n + 1)(n + 2)$. We also note that $n + 1$ and $n + 2$ are integers (because they are sums of integers) and that both $n + 1 > 1$ and $n + 2 > 1$ (because $n \geq 1$). Thus $n^2 + 3n + 2$ is a product of two integers each greater than 1, and so $n^2 + 3n + 2$ is not prime. ∎

Conjecture, Proof, and Disproof

Pierre de Fermat
(1601–1665)

Andrew Wiles
(born 1953)

More than 350 years ago, the French mathematician Pierre de Fermat claimed that it is impossible to find positive integers x, y, and z with $x^n + y^n = z^n$ if n is an integer that is at least 3. (For $n = 2$, the equation has many integer solutions, such as $3^2 + 4^2 = 5^2$ and $5^2 + 12^2 = 13^2$.) Fermat wrote his claim in the margin of a book, along with the comment "I have discovered a truly remarkable proof of this theorem which this margin is too small to contain." No proof, however, was found among his papers, and over the years some of the greatest mathematical minds tried and failed to discover a proof or a counterexample, for what came to be known as Fermat's last theorem.

In 1986 Kenneth Ribet of the University of California at Berkeley showed that if a certain other statement, the Taniyama–Shimura conjecture, could be proved, then Fermat's theorem would follow. Andrew Wiles, an English mathematician and faculty member at Princeton University, had become intrigued by Fermat's claim while still a child and, as an adult, had come to work in the branch of mathematics to which the Taniyama–Shimura conjecture belonged. As soon as he heard of Ribet's result, Wiles immediately set to work to prove the conjecture. In June of 1993, after 7 years of concentrated effort, he presented a proof to worldwide acclaim.

During the summer of 1993, however, while every part of the proof was being carefully checked to prepare for formal publication, Wiles found that he could not justify one step and that that step might actually be wrong. He worked unceasingly for another year to resolve the problem, finally realizing that the gap in the proof was a genuine error but that an approach he had worked on years earlier and abandoned provided a way around the difficulty. By the end of 1994, the revised proof had been thoroughly checked and pronounced correct in every detail by experts in the field. It was published in the *Annals of Mathematics* in 1995. Several books and an excellent documentary television show have been produced that convey the drama and excitement of Wiles's discovery.*

---

*"The Proof," produced in 1997, for the series *Nova* on the Public Broadcasting System; *Fermat's Enigma: The Epic Quest to Solve the World's Greatest Mathematical Problem,* by Simon Singh and John Lynch (New York: Bantam Books, 1998); *Fermat's Last Theorem: Unlocking the Secret of an Ancient Mathematical Problem* by Amir D. Aczel (New York: Delacorte Press, 1997).

One of the oldest problems in mathematics that remain unsolved is the Goldbach conjecture. In Example 3.1.5 it was shown that every even integer from 4 to 30 can be represented as a sum of two prime numbers. More than 250 years ago, Christian Goldbach (1690–1764) conjectured that every even integer greater than 2 can be so represented. Explicit computer-aided calculations have shown the conjecture to be true up to at least 10^{16}. But there is a huge chasm between 10^{16} and infinity. As pointed out by James Gleick of the *New York Times,* many other plausible conjectures in number theory have proved false. Leonhard Euler (1707–1783), for example, proposed in the eighteenth century that $a^4 + b^4 + c^4 = d^4$ had no nontrivial whole number solutions. In other words, no three perfect fourth powers add up to another perfect fourth power. For small numbers, Euler's conjecture looked good. But in 1987 a Harvard mathematician, Noam Elkies, proved it wrong. One counterexample, found by Roger Frye of Thinking Machines Corporation in a long computer search, is $95{,}800^4 + 217{,}519^4 + 414{,}560^4 = 422{,}481^4$.*

In May 2000, "to celebrate mathematics in the new millennium," the Clay Mathematics Institute of Cambridge, Massachusetts, announced that it would award prizes of $1 million each for the solutions to seven longstanding, classical mathematical questions. One of them, "P vs. NP," asks whether problems belonging to a certain class can be solved on a computer using more efficient methods than the very inefficient methods that are presently known to work for them. This question is discussed briefly at the end of Chapter 9.

Exercise Set 3.1†

In 1–3, use the definitions of even, odd, prime, and composite to justify each of your answers.

1. Assume that k is a particular integer.
   a. Is -17 an odd integer?   b. Is 0 an even integer?
   c. Is $2k - 1$ odd?

2. Assume that m and n are particular integers.
   a. Is $6m + 8n$ even?   b. Is $10mn + 7$ odd?
   c. If $m > n > 0$, is $m^2 - n^2$ composite?

3. Assume that r and s are particular integers.
   a. Is $4rs$ even?   b. Is $6r + 4s^2 + 3$ odd?
   c. If r and s are both positive, is $r^2 + 2rs + s^2$ composite?

Prove the statements in 4–10.

4. There are integers m and n such that $m > 1$ and $n > 1$ and $\dfrac{1}{m} + \dfrac{1}{n}$ is an integer.

5. There are distinct integers m and n such that $\dfrac{1}{m} + \dfrac{1}{n}$ is an integer.

6. There are real numbers a and b such that
$$\sqrt{a + b} = \sqrt{a} + \sqrt{b}.$$

7. There is an integer $n > 5$ such that $2^n - 1$ is prime.

8. There is a real number x such that $x > 1$ and $2^x > x^{10}$.

Definition: An integer n is called a **perfect square** if, and only if, $n = k^2$ for some integer k.

9. There is a perfect square that can be written as a sum of two other perfect squares.

10. There is an integer n such that $2n^2 - 5n + 2$ is prime.

Disprove the statements in 11–13 by giving a counterexample.

11. For all real numbers a and b, if $a < b$ then $a^2 < b^2$.

12. For all integers n, if n is odd then $\dfrac{n-1}{2}$ is odd.

13. For all integers m and n, if $2m + n$ is odd then m and n are both odd.

In 14–16, determine whether the property is true for all integers, true for no integers, or true for some integers and false for other integers. Justify your answers.

14. $(a + b)^2 = a^2 + b^2$ 15. $3n^2 - 4n + 1$ is prime.

16. The average of any two odd integers is odd.

Prove the statements in 17 and 18 by the method of exhaustion.

17. Every positive even integer less than 26 can be expressed as a sum of three or fewer perfect squares. (For instance, $10 = 1^2 + 3^2$ and $16 = 4^2$.)

---

*James Gleick, "Fermat's Last Theorem Still Has Zero Solutions," *New York Times,* 17 April 1988.

†For exercises with blue numbers, solutions are given in Appendix B. The symbol **H** indicates that only a hint or partial solution is given. The symbol ✶ signals that an exercise is more challenging than usual.

18. For each integer n with $1 \le n \le 10$, $n^2 - n + 11$ is a prime number.

19. a. Rewrite the following theorem in the form $\forall$ _____, if _____ then _____.

    b. Fill in the blanks in the proof.

    Theorem: The sum of any even integer and any odd integer is odd.

    Proof: Suppose m is any even integer and n is _(a)_. By definition of even, $m = 2r$ for some _(b)_, and by definition of odd, $n = 2s + 1$ for some integer s. By substitution and algebra, $m + n = $ _(c)_ $ = 2(r + s) + 1$. Since r and s are both integers, so is their sum $r + s$. Hence $m + n$ has the form $2 \cdot$ (some integer) $+ 1$, and so _(d)_ by definition of odd.

Each of the statements in 20–23 is true. For each, (a) rewrite the statement using a variable or variables and the form $\forall$ _____, if _____ then _____, and (b) write the first sentence of a proof (the "starting point") and the last sentence of a proof (the "conclusion to be shown"). Note that you do not need to understand the statements in order to be able to do these exercises.

20. For all integers m, if $m > 1$ then $0 < \dfrac{1}{m} < 1$.

21. For all real numbers x, if $x > 1$ then $x^2 > x$.

22. For all integers m and n, if $mn = 1$ then $m = n = 1$ or $m = n = -1$.

23. For all real numbers x, if $0 < x < 1$ then $x^2 < x$.

Prove the statements in 24–30. Follow the directions given in this section for writing proofs of universal statements.

24. The negative of any even integer is even.

25. The difference of any even integer minus any odd integer is odd.

26. The difference of any odd integer minus any even integer is odd. (*Note:* The "proof" shown in exericse 35 contains an error. Can you spot it?)

27. The sum of any two odd integers is even.

28. For all integers n, if n is odd then n^2 is odd.

29. If n is any even integer, then $(-1)^n = 1$.

30. If n is any odd integer, then $(-1)^n = -1$.

Prove that the statements in 31–33 are false.

31. There exists an integer $m \ge 3$ such that $m^2 - 1$ is prime.

32. There exists an integer n such that $6n^2 + 27$ is prime.

33. There exists an integer k such that $k \ge 4$ and $2k^2 - 5k + 2$ is prime.

Find the mistakes in the "proofs" shown in 34–38.

34. **Theorem:** For all integers k, if $k > 0$ then $k^2 + 2k + 1$ is composite.

    "**Proof:** For $k = 2$, $k^2 + 2k + 1 = 2^2 + 2 \cdot 2 + 1 = 9$. But $9 = 3 \cdot 3$, and so 9 is composite. Hence the theorem is true."

35. **Theorem:** The difference between any odd integer and any even integer is odd.

    "**Proof:** Suppose n is any odd integer, and m is any even integer. By definition of odd, $n = 2k + 1$ where k is an integer, and by definition of even, $m = 2k$ where k is an integer. Then $n - m = (2k + 1) - 2k = 1$. But 1 is odd. Therefore, the difference between any odd integer and any even integer is odd."

36. **Theorem:** For all integers k, if $k > 0$ then $k^2 + 2k + 1$ is composite.

    "**Proof:** Suppose k is any integer such that $k > 0$. If $k^2 + 2k + 1$ is composite, then $k^2 + 2k + 1 = r \cdot s$ for some integers r and s such that $1 < r < (k^2 + 2k + 1)$ and $1 < s < (k^2 + 2k + 1)$. Since $k^2 + 2k + 1 = r \cdot s$ and both r and s are strictly between 1 and $k^2 + 2k + 1$, then $k^2 + 2k + 1$ is not prime. Hence $k^2 + 2k + 1$ is composite as was to be shown."

37. **Theorem:** The product of an even integer and an odd integer is even.

    "**Proof:** Suppose m is an even integer and n is an odd integer. If $m \cdot n$ is even, then by definition of even there exists an integer r such that $m \cdot n = 2r$. Also since m is even, there exists an integer p such that $m = 2p$, and since n is odd there exists an integer q such that $n = 2q + 1$. Thus

    $$m \cdot n = (2p) \cdot (2q + 1) = 2r,$$

    where r is an integer. By definition of even, then, $m \cdot n$ is even, as was to be shown."

38. **Theorem:** The sum of any two even integers equals $4k$ for some integer k.

    "**Proof:** Suppose m and n are any two even integers. By definition of even, $m = 2k$ for some integer k and $n = 2k$ for some integer k. By substitution, $m + n = 2k + 2k = 4k$. This is what was to be shown."

In 39–56 determine whether the statement is true or false. Justify your answer with a proof or a counterexample, as appropriate.

39. The product of any two odd integers is odd.

40. The negative of any odd integer is odd.

41. The difference of any two odd integers is odd.

42. The product of any even integer and any integer is even.

43. If a sum of two integers is even, then one of the summands is even. (In the expression $a + b$, a and b are called **summands.**)

44. The difference of any two even integers is even.

45. The difference of any two odd integers is even.

46. For all integers n and m, if $n - m$ is even then $n^3 - m^3$ is even.

47. For all integers n, if n is prime then $(-1)^n = -1$.

48. For all integers m, if $m > 2$ then $m^2 - 4$ is composite.

49. For all integers n, $n^2 - n + 11$ is a prime number.

50. For all integers n, $4(n^2 + n + 1) - 3n^2$ is a perfect square.

51. Every positive integer can be expressed as a sum of three or fewer perfect squares.

H ✱ 52. (Two integers are **consecutive** if, and only if, one is one more than the other.) Any product of four consecutive integers is one less than a perfect square.

53. If m and n are positive integers and mn is a perfect square, then m and n are perfect squares.

54. The difference of the squares of any two consecutive integers is odd.

55. For all nonnegative real numbers a and b, $\sqrt{ab} = \sqrt{a}\sqrt{b}$. (Note that if x is a nonnegative real number, then there is a unique nonnegative real number y, denoted $\sqrt{x}$, such that $y^2 = x$.)

56. For all nonnegative real numbers a and b,
$$\sqrt{a + b} = \sqrt{a} + \sqrt{b}.$$

57. If m and n are perfect squares, then $m + n + 2\sqrt{mn}$ is also a perfect square. Why?

H ✱ 58. If p is a prime number, must $2^p - 1$ also be prime? Prove or give a counterexample.

✱ 59. If n is a nonnegative integer, must $2^{2^n} + 1$ be prime? Prove or give a counterexample.

60. When expressions of the form $(x - r)(x - s)$ are multiplied out, a quadratic polynomial is obtained. For instance, $(x - 2)(x - (-7)) = (x - 2)(x + 7) = x^2 + 5x - 14$.

   H a. What can be said about the coefficients of the polynomial obtained by multiplying out $(x - r)(x - s)$ when both r and s are odd integers? when both r and s are even integers? when one of r and s is even and the other is odd?

   b. It follows from part (a) that $x^3 - 1253x + 255$ cannot be written as a product of two polynomials with integer coefficients. Explain why this is so.

✱ 61. Observe that $(x - r)(x - s)(x - t)$
$$= x^3 - (r + s + t)x^2 + (rs + rt + st)x - rst.$$

   a. Derive a result for cubic polynomials similar to the result in part (a) of exercise 60 for quadratic polynomials.

   b. Can $15x^3 + 7x^2 - 8x - 27$ be written as a product of three polynomials with integer coefficients? Explain.

3.2 Direct Proof and Counterexample II: Rational Numbers

Such, then, is the whole art of convincing. It is contained in two principles: to define all notations used, and to prove everything by replacing mentally the defined terms by their definitions. — Blaise Pascal, 1623–1662

Sums, differences, and products of integers are integers. But most quotients of integers are not integers. Quotients of integers are, however, important; they are known as *rational numbers*.

• Definition

A real number r is **rational** if, and only if, it can be expressed as a quotient of two integers with a nonzero denominator. A real number that is not rational is **irrational.** More formally, if r is a real number, then

$$r \text{ is rational } \Leftrightarrow \exists \text{ integers } a \text{ and } b \text{ such that } r = \frac{a}{b} \text{ and } b \neq 0.$$

The word *rational* contains the word *ratio,* which is another word for quotient. A rational number is a fraction or ratio of integers.

Example 3.2.1 Determining Whether Numbers Are Rational or Irrational

  a. Is 10/3 a rational number?

  b. Is $-(5/39)$ a rational number?

  c. Is 0.281 a rational number?

  d. Is 7 a rational number?

  e. Is 0 a rational number?

  f. Is 2/0 a rational number?

  g. Is 2/0 an irrational number?

  h. Is 0.12121212... a rational number (where the digits 12 are assumed to repeat forever)?

  i. If m and n are integers and neither m nor n is zero, is $(m + n)/mn$ a rational number?

Solution

  a. Yes, 10/3 is a quotient of the integers 10 and 3 and hence is rational.

  b. Yes, $-(5/39) = -5/39$, which is a quotient of the integers -5 and 39 and hence is rational.

  c. Yes, $0.281 = 281/1000$. Note that the real numbers represented on a typical calculator display are all finite decimals. An explanation similar to the one in this example shows that any such number is rational. It follows that a calculator with such a display can represent only rational numbers.

  d. Yes, $7 = 7/1$.

  e. Yes, $0 = 0/1$.

  f. No, 2/0 is not a number (division by 0 is not allowed).

  g. No, because every irrational number is a number, and 2/0 is not a number. We discuss additional techniques for determining whether numbers are irrational in Sections 3.6, 3.7, and 7.4.

  h. Yes. Let $x = 0.12121212...$. Then $100x = 12.12121212...$. Hence

$$100x - x = 12.12121212... - 0.12121212... = 12.$$

But also

$$100x - x = 99x \qquad \text{by basic algebra}$$

Hence

$$99x = 12,$$

and so

$$x = \frac{12}{99}.$$

Therefore, $0.12121212... = 12/99$, which is a ratio of two nonzero integers and thus is a rational number.

Note that you can use an argument similar to this one to show that any repeating decimal is a rational number. In Section 7.3 we show that any rational number can be written as a repeating or terminating decimal.

i. Yes, since m and n are integers, so are $m + n$ and mn (because sums and products of integers are integers). Also $mn \neq 0$ by the **zero product property.** (One version of this property says that if neither of two real numbers is 0, then their product is also not 0. See exercise 8 at the end of this section.) It follows that $(m + n)/mn$ is a quotient of two integers with a nonzero denominator and hence is a rational number. ∎

More on Generalizing from the Generic Particular

Some people like to think of the method of generalizing from the generic particular as a challenge process. If you claim a property holds for all elements in a domain, then someone can challenge your claim by picking any element in the domain whatsoever and asking you to prove that that element satisfies the property. To prove your claim, you must be able to meet all such challenges. That is, you must have a way to convince the challenger that the property is true for an *arbitrarily chosen* element in the domain.

For example, suppose "A" claims that every integer is a rational number. "B" challenges this claim by asking "A" to prove it for $n = 7$. "A" observes that

$$7 = \frac{7}{1} \quad \text{which is a quotient of integers and hence rational.}$$

"B" accepts this explanation but challenges again with $n = -12$. "A" responds that

$$-12 = \frac{-12}{1} \quad \text{which is a quotient of integers and hence rational.}$$

Next "B" tries to trip up "A" by challenging with $n = 0$, but "A" answers that

$$0 = \frac{0}{1} \quad \text{which is a quotient of integers and hence rational.}$$

As you can see, "A" is able to respond effectively to all "B"s challenges because "A" has a general procedure for putting integers into the form of rational numbers: "A" just divides whatever integer "B" gives by 1. That is, no matter what integer n "B" gives "A", "A" writes

$$n = \frac{n}{1} \quad \text{which is a quotient of integers and hence rational.}$$

This discussion proves the following theorem.

Theorem 3.2.1

Every integer is a rational number.

In exercise 11 at the end of this section you are asked to condense the above discussion into a formal proof.

Proving Properties of Rational Numbers

The next example shows how to use the method of generalizing from the generic particular to prove a property of rational numbers.

Example 3.2.2   A Sum of Rationals Is Rational

Prove that the sum of any two rational numbers is rational.

Solution   Begin by mentally or explicitly rewriting the statement to be proved in the form "∀ _____, if _____ then _____."

Formal Restatement: ∀ real numbers r and s, if r and s are rational then $r + s$ is rational.

Next ask yourself, "Where am I starting from?" or "What am I supposing?" The answer gives you the starting point, or first sentence, of the proof.

Starting Point: Suppose r and s are particular but arbitrarily chosen real numbers such that r and s are rational; or, more simply,

Suppose r and s are rational numbers.

Then ask yourself, "What must I show to complete the proof?"

To Show: $r + s$ is rational.

Finally you ask, "How do I get from the starting point to the conclusion?" or "Why must $r + s$ be rational if both r and s are rational?" The answer depends in an essential way on the definition of rational.

Rational numbers are quotients of integers, so to say that r and s are rational means that

$$r = \frac{a}{b} \quad \text{and} \quad s = \frac{c}{d} \quad \text{for some integers } a, b, c, \text{ and } d \\ \text{where } b \neq 0 \text{ and } d \neq 0.$$

It follows by substitution that

$$r + s = \frac{a}{b} + \frac{c}{d}.$$

Hence you must show that the right-hand sum can be written as a single fraction or ratio of two integers with a nonzero denominator. But

$$\frac{a}{b} + \frac{c}{d} = \frac{ad}{bd} + \frac{bc}{bd} \qquad \text{rewriting the fraction with a common denominator}$$

$$= \frac{ad + bc}{bd} \qquad \text{adding fractions with a common denominator.}$$

Is this fraction a ratio of integers? Yes. Because products and sums of integers are integers, $ad + bc$ and bd are both integers. Is the denominator $bd \neq 0$? Yes, by the zero product property (since $b \neq 0$ and $d \neq 0$). Thus $r + s$ is a rational number.

This discussion is summarized as follows:

Theorem 3.2.2

The sum of any two rational numbers is rational.

Proof:

Suppose r and s are rational numbers. [*We must show that $r + s$ is rational.*] Then, by definition of rational, $r = a/b$ and $s = c/d$ for some integers a, b, c, and d with $b \neq 0$ and $d \neq 0$. Thus

$$r + s = \frac{a}{b} + \frac{c}{d} \qquad \text{by substitution}$$

$$= \frac{ad + bc}{bd} \qquad \text{by basic algebra.}$$

Let $p = ad + bc$ and $q = bd$. Then p and q are integers because products and sums of integers are integers and because a, b, c, and d are all integers. Also $q \neq 0$ by the zero product property. Thus

$$r + s = \frac{p}{q} \text{ where } p \text{ and } q \text{ are integers and } q \neq 0.$$

Therefore, $r + s$ is rational by definition of a rational number. [*This is what was to be shown.*]

■

Deriving New Mathematics from Old

Section 3.1 focused on establishing truth and falsity of mathematical theorems using only the basic algebra normally taught in secondary school; the fact that the integers are closed under addition, subtraction, and multiplication; and the definitions of the terms in the theorems themselves. In the future, when we ask you to **prove something directly from the definitions,** we will mean that you should restrict yourself to this approach. However, once a collection of statements has been proved directly from the definitions, another method of proof becomes possible. The statements in the collection can be used to derive additional results.

Example 3.2.3 Deriving Additional Results about Even and Odd Integers

Suppose that you have already proved the following properties of even and odd integers:

1. The sum, product, and difference of any two even integers are even.

2. The sum and difference of any two odd integers are even.

3. The product of any two odd integers is odd.

4. The product of any even integer and any odd integer is even.

5. The sum of any odd integer and any even integer is odd.

6. The difference of any odd integer minus any even integer is odd.

7. The difference of any even integer minus any odd integer is odd.

Use the properties listed above to prove that if a is any odd integer and b is any given integer, then $\dfrac{a^2 + b^2 + 1}{2}$ is an integer.

Solution   Suppose a is any odd integer and b is any even integer. By property 3, b^2 is odd, and by property 1, a^2 is even. Then by property 5, $a^2 + b^2$ is odd, and because 1 is also odd, the sum $(a^2 + b^2) + 1 = a^2 + b^2 + 1$ is even by property 2. Hence, by definition of even, there exists an integer k such that $a^2 + b^2 + 1 = 2k$. Dividing both sides by 2 gives $\dfrac{a^2 + b^2 + 1}{2} = k$, which is an integer. Thus $\dfrac{a^2 + b^2 + 1}{2}$ is an integer [*as was to be shown*]. ∎

A **corollary** is a statement whose truth can be immediately deduced from a theorem that has already been proved.

Example 3.2.4 The Double of a Rational Number

Derive the following as a corollary of Theorem 3.2.2.

> **Corollary 3.2.3**
>
> The double of a rational number is rational.

Solution   The double of a number is just its sum with itself. But since the sum of any two rational numbers is rational (Theorem 3.2.2), the sum of a rational number with itself is rational. Hence the double of a rational number is rational. Here is a formal version of this argument:

Proof:

Suppose r is any rational number. Then $2r = r + r$ is a sum of two rational numbers. So, by Theorem 3.2.2, $2r$ is rational. ∎

Exercise Set 3.2

The numbers in 1–7 are all rational. Write each number as a ratio of two integers.

1. $-\dfrac{35}{6}$

2. 4.6037

3. $\dfrac{4}{5} + \dfrac{2}{9}$

4. $0.37373737\ldots$

5. $0.56565656\ldots$

6. $320.5492492492\ldots$

7. $52.4672167216721\ldots$

8. The zero product property says that if a product of two real numbers is 0, then one of the numbers must be 0.
   a. Write this property formally using quantifiers and variables.
   b. Write the contrapositive of your answer to part (a).
   c. Write an informal version (without quantifier symbols or variables) for your answer to part (b).

9. Assume that a and b are both integers and that $a \neq 0$ and $b \neq 0$. Explain why $(b - a)/(ab^2)$ must be a rational number.

10. Assume that m and n are both integers and that $n \neq 0$. Explain why $(5m + 12n)/(4n)$ must be a rational number.

11. Prove that every integer is a rational number.

12. Fill in the blanks in the following proof that the square of any rational number is rational:

   Proof: Suppose that r is __(a)__. By definition of rational, $r = a/b$ for some __(b)__ with $b \neq 0$. By substitution, $r^2 = $ __(c)__ $= a^2/b^2$. Since a and b are both integers, so are the products a^2 and __(d)__. Also $b^2 \neq 0$ by the __(e)__. Hence r^2 is a ratio of two integers with a nonzero denominator, and so __(f)__ by definition of rational.

Determine which of the statements in 13–19 are true and which are false. Prove each true statement directly from the definitions, and give a counterexample for each false statement. In case the statement is false, determine whether a small change would make it true. If so, make the change and prove the new statement.

13. The product of any two rational numbers is a rational number.

H 14. The quotient of any two rational numbers is a rational number.

15. The difference of any two rational numbers is a rational number.

16. Given any rational number r, $-r$ is also a rational number.

H 17. If r and s are any two rational numbers with $r < s$, then $\dfrac{r+s}{2}$ is rational.

H 18. For all real numbers a and b, if $a < b$ then $a < \dfrac{a+b}{2} < b$. (You may use the properties of inequalities in T16–T25 of Appendix A.)

19. Given any two rational numbers r and s with $r < s$, there is another rational number between r and s. (*Hint:* Use the results of exercises 17 and 18.)

Use the properties of even and odd integers that are listed in Example 3.2.3 to do exercises 20–22. Indicate which properties you use to justify your reasoning.

20. True or false? If m is any even integer and n is any odd integer, then $m^2 + 3n$ is odd. Explain.

21. True or false? If a is any odd integer, then $a^2 + a$ is even. Explain.

22. True or false? If k is any even integer and m is any odd integer, then $(k + 2)^2 - (m - 1)^2$ is even. Explain.

Derive the statements in 23–25 as corollaries of Theorems 3.2.1, 3.2.2, and the results of exercises 12, 13, 15, and 16.

23. For any rational numbers r and s, $2r + 3s$ is rational.

24. If r is any rational number, then $3r^2 - 2r + 4$ is rational.

25. For any rational number s, $5s^3 + 8s^2 - 7$ is rational.

26. It is a fact that if n is any nonnegative integer, then
$$1 + \frac{1}{2} + \frac{1}{2^2} + \frac{1}{2^3} + \cdots + \frac{1}{2^n} = \frac{1 - (1/2^{n+1})}{1 - (1/2)}.$$

(A more general form of this statement is proved in Section 4.2). Is a number of this form rational? If so, express it as a ratio of two integers.

27. Suppose a, b, c, and d are integers and $a \neq c$. Suppose also that x is a real number that satisfies the equation
$$\frac{ax + b}{cx + d} = 1.$$

Must x be rational? If so, express x as a ratio of two integers.

★ 28. Suppose a, b, and c are integers and x, y, and z are nonzero real numbers that satisfy the following equations:
$$\frac{xy}{x + y} = a \quad \text{and} \quad \frac{xz}{x + z} = b \quad \text{and} \quad \frac{yz}{y + z} = c.$$

Is x rational? If so, express it as a ratio of two integers.

29. Prove that if one solution for a quadratic equation of the form $x^2 + bx + c = 0$ is rational (where b and c are ra-

tional), then the other solution is also rational. (Use the fact that if the solutions of the equation are r and s, then $x^2 + bx + c = (x - r)(x - s)$.)

30. Prove that if a real number c satisfies a polynomial equation of the form
$$r_3 x^3 + r_2 x^2 + r_1 x + r_0 = 0,$$

where r_0, r_1, r_2, and r_3 are rational numbers, then c satisfies an equation of the form
$$n_3 x^3 + n_2 x^2 + n_1 x + n_0 = 0,$$

where n_0, n_1, n_2, and n_3 are integers.

> **Definition:** A number c is called a **root** of a polynomial $p(x)$ if, and only if, $p(c) = 0$.

★ 31. Prove that for all real numbers c, if c is a root of a polynomial with rational coefficients, then c is a root of a polynomial with integer coefficients.

In 32–36 find the mistakes in the "proofs" that the sum of any two rational numbers is a rational number.

32. "**Proof:** Let rational numbers $r = \frac{1}{4}$ and $s = \frac{1}{2}$ be given. Then $r + s = \frac{1}{4} + \frac{1}{2} = \frac{3}{4}$, which is a rational number. This is what was to be shown."

33. "**Proof:** Any two rational numbers produce a rational number when added together. So if r and s are particular but arbitrarily chosen rational numbers, then $r + s$ is rational."

34. "**Proof:** Suppose r and s are rational numbers. By definition of rational, $r = a/b$ for some integers a and b with $b \neq 0$, and $s = a/b$ for some integers a and b with $b \neq 0$. Then $r + s = a/b + a/b = 2a/b$. Let $p = 2a$. Then p is an integer since it is a product of integers. Hence $r + s = p/b$, where p and b are integers and $b \neq 0$. Thus $r + s$ is a rational number by definition of rational. This is what was to be shown."

35. "**Proof:** Suppose r and s are rational numbers. Then $r = a/b$ and $s = c/d$ for some integers a, b, c, and d with $b \neq 0$ and $d \neq 0$ (by definition of rational). Then $r + s = a/b + c/d$. But this is a sum of two fractions, which is a fraction. So $r + s$ is a rational number since a rational number is a fraction."

36. "**Proof:** Suppose r and s are rational numbers. If $r + s$ is rational, then by definition of rational $r + s = a/b$ for some integers a and b with $b \neq 0$. Also since r and s are rational, $r = i/j$ and $s = m/n$ for some integers i, j, m, and n with $j \neq 0$ and $n \neq 0$. It follows that $r + s = i/j + m/n = a/b$, which is a quotient of two integers with a nonzero denominator. Hence it is a rational number. This is what was to be shown."

3.3 Direct Proof and Counterexample III: Divisibility

The essential quality of a proof is to compel belief. — Pierre de Fermat

When you were first introduced to the concept of division in elementary school, you were probably taught that 12 divided by 3 is 4 because if you separate 12 objects into groups of 3, you get 4 groups with nothing left over.

You may also have been taught to describe this fact by saying that "12 is evenly divisible by 3" or "3 divides 12 evenly."

The notion of divisibility is the central concept of one of the most beautiful subjects in advanced mathematics: **number theory,** the study of properties of integers.

• Definition

If n and d are integers, then

n is **divisible by** d if, and only if, $n = dk$ for some integer k.

Alternatively, we say that

n **is a multiple of** d, or
d **is a factor of** n, or
d **is a divisor of** n, or
d **divides** n.

The notation **d | n** is read "d divides n." Symbolically, if n and d are integers and $d \neq 0$,

$$d \mid n \quad \Leftrightarrow \quad \exists \text{ an integer } k \text{ such that } n = dk.$$

Example 3.3.1 Divisibility

a. Is 21 divisible by 3?   b. Does 5 divide 40?   c. Does 7 | 42?

d. Is 32 a multiple of -16?   e. Is 6 a factor of 54?   f. Is 7 a factor of -7?

Solution

a. Yes, $21 = 3 \cdot 7$.   b. Yes, $40 = 5 \cdot 8$.   c. Yes, $42 = 7 \cdot 6$.

d. Yes, $32 = (-16) \cdot (-2)$.   e. Yes, $54 = 6 \cdot 9$.   f. Yes, $-7 = 7 \cdot (-1)$. ■

Example 3.3.2 Divisors of Zero

If k is any integer, does k divide 0?

Solution   Yes, because $0 = k \cdot 0$. ■

Example 3.3.3 The Positive Divisors of a Positive Number

Suppose a and b are positive integers and $a \mid b$. Is $a \leq b$?

Solution Yes. To say that $a \mid b$ means that $b = ka$ for some integer k. Now k must be a positive integer because both a and b are positive. It follows that

$$1 \leq k$$

because every positive integer is greater than or equal to 1. Multiplying both sides by a gives

$$a \leq k \cdot a = b$$

(since multiplying both sides of an inequality by a positive number preserves the inequality—property T19 of Appendix A). ∎

Example 3.3.4 Divisors of 1

Which integers divide 1?

Solution By Example 3.3.3 any positive integer that divides 1 is less than or equal to 1. Since $1 = 1 \cdot 1$, 1 divides 1, and there are no positive integers that are less than 1. So the only positive divisor of 1 is 1.

On the other hand, if d is a negative integer that divides 1, then $1 = dk$, and so $1 = |d| \cdot |k|$. Hence $|d|$ is a positive integer that divides 1. Thus $|d| = 1$, and so $d = -1$. It follows that the only divisors of 1 are 1 and   1. ∎

Example 3.3.5 Divisibility of Algebraic Expressions

a. If a and b are integers, is $3a + 3b$ divisible by 3?

b. If k and m are integers, is $10km$ divisible by 5?

Solution

a. Yes. By the distributive law of algebra, $3a + 3b = 3(a + b)$ and $a + b$ is an integer because it is a sum of two integers.

b. Yes. By the associative law of algebra, $10km = 5 \cdot (2km)$ and $2km$ is an integer because it is a product of three integers. ∎

When the definition of divides is rewritten formally using the existential quantifier, the result is

$$d \mid n \quad \Leftrightarrow \quad \exists \text{ an integer } k \text{ such that } n = dk.$$

Since the negation of an existential statement is universal, it follows that d does not divide n (denoted $d \nmid n$) if, and only if, $\forall$ integers k, $n \neq dk$, or, in other words, the quotient n/d is not an integer.

> For all integers n and d, $d \nmid n$ $\Leftrightarrow$ $\dfrac{n}{d}$ is not an integer.

Example 3.3.6 Checking Nondivisibility

Does $4 \mid 15$?

Solution No, $\frac{15}{4} = 3.75$, which is not an integer. ∎

> ⚠️ **Caution!** Be careful to distinguish between the notation $a \mid b$ and the notation a/b. The notation $a \mid b$ stands for the sentence "a divides b," which means that there is an integer k such that $b = a \cdot k$. Dividing both sides by a gives $b/a = k$, an integer. Thus, when $d \neq 0$, $a \mid b$ if, and only if, b/a is an integer. On the other hand, the notation a/b stands for the fractional number a/b (the inverse fraction!), which may or may not be an integer.

Example 3.3.7 Prime Numbers and Divisibility

An alternative way to define a prime number is to say that an integer $n > 1$ is prime if, and only if, its only positive integer divisors are 1 and itself. ∎

Proving Properties of Divisibility

One of the most useful properties of divisibility is that it is transitive. If one number divides a second and the second number divides a third, then the first number divides the third.

Example 3.3.8 Transitivity of Divisibility

Prove that for all integers a, b, and c, if $a \mid b$ and $b \mid c$, then $a \mid c$.

Solution   Since the statement to be proved is already written formally, you can immediately pick out the starting point, or first sentence of the proof, and the conclusion that must be shown.

Starting Point: Suppose a, b, and c are particular but arbitrarily chosen integers such that $a \mid b$ and $b \mid c$.

To Show: $a \mid c$.

You need to show that $a \mid c$, or, in other words, that

$$c = a \cdot \text{(some integer)}.$$

But since $a \mid b$,

$$b = ar \quad \text{for some integer } r. \tag{3.3.1}$$

And since $b \mid c$,

$$c = bs \quad \text{for some integer } s. \tag{3.3.2}$$

Equation 3.3.2 expresses c in terms of b, and equation 3.3.1 expresses b in terms of a. Thus if you substitute 3.3.1 into 3.3.2, you will have an equation that expresses c in terms of a.

$$c = bs \qquad \text{by equation 3.3.2}$$
$$= (ar)s \qquad \text{by equation 3.3.1.}$$

But $(ar)s = a(rs)$ by the associative law for multiplication. Hence

$$c = a(rs).$$

Now you are almost finished. You have expressed c as $a \cdot$ (something). It remains only to verify that that something is an integer. But of course it is, because it is a product of two integers.

This discussion is summarized as follows:

Theorem 3.3.1 Transitivity of Divisibility

For all integers a, b, and c, if a divides b and b divides c, then a divides c.

Proof:

Suppose a, b, and c are [*particular but arbitrarily chosen*] integers such that a divides b and b divides c [*We must show that a divides c.*] By definition of divisibility,

$$b = ar \quad \text{and} \quad c = bs \quad \text{for some integers } r \text{ and } s.$$

By substitution

$$c = bs$$
$$= (ar)s$$
$$= a(rs) \qquad \text{by basic algebra.}$$

Let $k = rs$. Then k is an integer since it is a product of integers, and therefore

$$c = ak \quad \text{where } k \text{ is an integer.}$$

Thus a divides c by definition of divisibility. [*This is what was to be shown.*]

It would appear from the definition of prime that to show that an integer is prime you would need to show that it is not divisible by any integer greater than 1 and less than itself. In fact, you need only check divisibility by prime numbers. This follows from Theorem 3.3.1, Example 3.3.3, and the following theorem, which says that any integer greater than 1 is divisible by a prime number. The idea of the proof is quite simple. You start with a positive integer. If it is prime, you are done; if not, it is a product of two smaller positive factors. If one of these is prime, you are done; if not, you can pick one of the factors and write it as a product of still smaller positive factors. You can continue in this way, factoring the factors of the number you started with, until one of them turns out to be prime. This must happen eventually because all the factors can be chosen to be positive and each is smaller than the preceding one.

Theorem 3.3.2 Divisibility by a Prime

Any integer $n > 1$ is divisible by a prime number.

Proof:

Suppose n is a [*particular but arbitrarily chosen*] integer that is greater than 1. [*We must show that there is a prime number that divides n.*] If n is prime, then n is divisible by a prime number (namely itself), and we are done. If n is not prime, then, as discussed in Example 3.1.2b,

$$n = r_0 s_0 \quad \text{where } r_0 \text{ and } s_0 \text{ are integers and}$$
$$1 < r_0 < n \text{ and } 1 < s_0 < n.$$

It follows by definition of divisibility that $r_0 \mid n$.

continued on page 152

If r_0 is prime, then r_0 is a prime number that divides n, and we are done. If r_0 is not prime, then

$$r_0 = r_1 s_1 \quad \text{where } r_1 \text{ and } s_1 \text{ are integers and} \\ 1 < r_1 < r_0 \text{ and } 1 < s_1 < r_0.$$

It follows by the definition of divisibility that $r_1 \mid r_0$. But we already know that $r_0 \mid n$. Consequently, by transitivity of divisibility, $r_1 \mid n$.

If r_1 is prime, then r_1 is a prime number that divides n, and we are done. If r_1 is not prime, then

$$r_1 = r_2 s_2 \quad \text{where } r_2 \text{ and } s_2 \text{ are integers and} \\ 1 < r_2 < r_1 \text{ and } 1 < s_2 < r_1.$$

It follows by definition of divisibility that $r_2 \mid r_1$. But we already know that $r_1 \mid n$. Consequently, by transitivity of divisibility, $r_2 \mid n$.

If r_2 is prime, then r_2 is a prime number that divides n, and we are done. If r_2 is not prime, then we may repeat the above process by factoring r_2 as $r_3 s_3$.

We may continue in this way, factoring successive factors of n until we find a prime factor. We must succeed in a finite number of steps because each new factor is both less than the previous one (which is less than n) and greater than 1, and there are fewer than n integers strictly between 1 and n.* Thus we obtain a sequence

$$r_0, r_1, r_2, \ldots, r_k,$$

where $k \geq 0$, $1 < r_k < r_{k-1} < \cdots < r_2 < r_1 < r_0 < n$, and $r_i \mid n$ for each $i = 0, 1, 2, \ldots, k$. The condition for termination is that r_k should be prime. Hence r_k is a prime number that divides n. [*This is what we were to show.*]

Counterexamples and Divisibility

To show that a proposed divisibility property is not universally true, you need to find some integers for which it is false.

Example 3.3.9 Checking a Proposed Divisibility Property

Is it true or false that for all integers a and b, if $a \mid b$ and $b \mid a$ then $a = b$?

Solution   This proposed property is false. Can you think of a counterexample just by concentrating for a minute or so?

The following discussion describes a mental process that may take just a few seconds. It is helpful to be able to use it consciously, however, to solve more difficult problems.

To discover the truth or falsity of a statement such as the one given above, start off much as you would if you were trying to prove it.

Starting Point: Suppose a and b are integers such that $a \mid b$ and $b \mid a$.

Ask yourself, "*Must* it follow that $a = b$, or *could* it happen that $a \neq b$ for some a and b?" Focus on the supposition. What does it mean? By definition of divisibility, the conditions $a \mid b$ and $b \mid a$ mean that

$$b = ka \quad \text{and} \quad a = lb \quad \text{for some integers } k \text{ and } l.$$

---

*Strictly speaking, this statement is justified by an axiom for the integers called the well-ordering principle, which is discussed in Section 4.4. Theorem 3.3.2 can also be proved using strong mathematical induction, as shown in Example 4.4.1.

Must it follow that $a = b$, or can you find integers a and b that satisfy these equations for which $a \neq b$? The equations imply that

$$b = ka = k(lb) = (kl)b.$$

Since $b \mid a$, $b \neq 0$, and so you can cancel b from the extreme left and right sides to obtain

$$1 = kl.$$

In other words, k and l are divisors of 1. But the only divisors of 1 are 1 and -1 (see Example 3.3.1). Thus k and l are both 1 or -1. If $k = l = 1$, then $b = a$. But if $k = l = -1$, then $b = -a$ and so $a \neq b$. This analysis suggests that you can find a counterexample by taking $b = -a$. Here is a formal answer:

Statement: For all integers a and b, if $a \mid b$ and $b \mid a$ then $a = b$.

Counterexample: Let $a = 2$ and $b = -2$. Then

$a \mid b$ since $2 \mid (-2)$ and $b \mid a$ since $(-2) \mid 2$, but $a \neq b$ since $2 \neq -2$.

Therefore, the proposed divisibility property is false.

The search for a proof will frequently help you discover a counterexample (provided the statement you are trying to prove is, in fact, false). Conversely, in trying to find a counterexample for a statement, you may come to realize the reason why it is true (if it is, in fact, true). The important thing is to keep an open mind until you are convinced by the evidence of your own careful reasoning.

The Unique Factorization Theorem

The most comprehensive statement about divisibility of integers is contained in a theorem known as the *unique factorization theorem* for the integers. Because of its importance, this theorem is also called the *fundamental theorem of arithmetic.* Although Euclid, who lived about 300 B.C., seems to have been acquainted with the theorem, it was first stated precisely by the great German mathematician Carl Friedrich Gauss (rhymes with *house*) in 1801.

The unique factorization theorem says that any integer greater than 1 either is prime or can be written as a product of prime numbers in a way that is unique except, perhaps, for the order in which the primes are written. For example,

$$72 = 2 \cdot 2 \cdot 2 \cdot 3 \cdot 3 = 2 \cdot 3 \cdot 3 \cdot 2 \cdot 2 = 3 \cdot 2 \cdot 2 \cdot 3 \cdot 2$$

and so forth. The three 2's and two 3's may be written in any order, but any factorization of 72 as a product of primes must contain exactly three 2's and two 3's—no other collection of prime numbers besides three 2's and two 3's multiplies out to 72.

Theorem 3.3.3 Unique Factorization Theorem for the Integers
(Fundamental Theorem of Arithmetic)

Given any integer $n > 1$, there exist a positive integer k, distinct prime numbers $p_1, p_2, \ldots, p_k$, and positive integers $e_1, e_2, \ldots, e_k$ such that

$$n = p_1^{e_1} p_2^{e_2} p_3^{e_3} \cdots p_k^{e_k},$$

and any other expression of n as a product of prime numbers is identical to this except, perhaps, for the order in which the factors are written.

The proof of the unique factorization theorem is included in Section 10.4.

Because of the unique factorization theorem, any integer $n > 1$ can be put into a *standard factored form* in which the prime factors are written in ascending order from left to right.

• Definition

Given any integer $n > 1$, the **standard factored form** of n is an expression of the form

$$n = p_1^{e_1} p_2^{e_2} p_3^{e_3} \cdots p_k^{e_k},$$

where k is a positive integer; $p_1, p_2, \ldots, p_k$ are prime numbers; $e_1, e_2, \ldots, e_k$ are positive integers; and $p_1 < p_2 < \cdots < p_k$.

Example 3.3.10 Writing Integers in Standard Factored Form

Write 3,300 in standard factored form.

Solution   First find all the factors of 3,300. Then write them in ascending order:

$$3,300 = 100 \cdot 33 = 4 \cdot 25 \cdot 3 \cdot 11$$
$$= 2 \cdot 2 \cdot 5 \cdot 5 \cdot 3 \cdot 11 = 2^2 \cdot 3^1 \cdot 5^2 \cdot 11^1.$$ ∎

Example 3.3.11 Using Unique Factorization to Solve a Problem

Suppose m is an integer such that

$$8 \cdot 7 \cdot 6 \cdot 5 \cdot 4 \cdot 3 \cdot 2 \cdot m = 17 \cdot 16 \cdot 15 \cdot 14 \cdot 13 \cdot 12 \cdot 11 \cdot 10.$$

Does $17 \mid m$?

Solution   Since 17 is one of the prime factors of the right-hand side of the equation, it is also a prime factor of the left-hand side (by the unique factorization theorem). But 17 does not equal any prime factor of 8, 7, 6, 5, 4, 3, or 2 (because it is too large). Hence 17 must occur as one of the prime factors of m, and so $17 \mid m$.   ∎

Exercise Set 3.3

Give a reason for your answer in each of 1–13. Assume that all variables represent integers.

1. Is 52 divisible by 13?   **2.** Is 54 divisible by 18?

3. Does $5 \mid 0$?

4. Is $(3k + 1)(3k + 2)(3k + 3)$ divisible by 3?

5. Is $6m(2m + 10)$ divisible by 4?

6. Is 29 a multiple of 3?   **7.** Is -3 a factor of 66?

8. Is $6a(a + b)$ a multiple of $3a$?

9. Is 4 a factor of $2a \cdot 34b$?

10. Does $7 \mid 34$?   **11.** Does $13 \mid 73$?

12. If $n = 4k + 1$, does 8 divide $n^2 - 1$?

13. If $n = 4k + 3$, does 8 divide $n^2 - 1$?

14. Fill in the blanks in the following proof that for all integers a and b, if $a \mid b$ then $a \mid (-b)$.

Proof: Suppose a and b are any integers such that ___(a)___. By definition of divisibility, $b =$ ___(b)___ for some ___(c)___ k. By substitution, $-b =$ ___(d)___ $= a \cdot (-k)$. But $-k = (-1) \cdot k$ is an integer since -1 and k are integers. Hence, by definition of divisibility, ___(e)___, as was to be shown.

Prove statements 15 and 16 directly from the definition of divisibility.

15. For all integers a, b, and c, if $a \mid b$ and $a \mid c$ then $a \mid (b + c)$.

16. For all integers a, b, and c, if $a \mid b$ and $a \mid c$ then $a \mid (b - c)$.

For each statement in 17–28, determine whether the statement is true or false. Prove the statement directly from the definitions if it is true, and give a counterexample if it is false.

17. The sum of any three consecutive integers is divisible by 3. (Two integers are **consecutive** if, and only if, one is one more than the other.)

18. The product of any two even integers is a multiple of 4.

19. A necessary condition for an integer to be divisible by 6 is that it be divisible by 2.

20. A sufficient condition for an integer to be divisible by 8 is that it be divisible by 16.

21. For all integers a, b, and c, if $a \mid b$ and $a \mid c$ then $a \mid (2b - 3c)$.

H 22. For all integers a, b, and c, if $ab \mid c$ then $a \mid c$ and $b \mid c$.

23. For all integers a, b, and c, if a is a factor of c then ab is a factor of c.

H 24. For all integers a, b, and c, if $a \mid (b + c)$ then $a \mid b$ or $a \mid c$.

25. For all integers a, b, and c, if $a \mid bc$ then $a \mid b$ or $a \mid c$.

26. For all integers a and b, if $a \mid b$ then $a^2 \mid b^2$.

27. For all integers a and n, if $a \mid n^2$ and $a \le n$ then $a \mid n$.

28. For all integers a and b, if $a \mid 10b$ then $a \mid 10$ or $a \mid b$.

29. A fast-food chain has a contest in which a card with numbers on it is given to each customer who makes a purchase. If some of the numbers on the card add up to 100, then the customer wins $100. A certain customer receives a card containing the numbers

    72, 21, 15, 36, 69, 81, 9, 27, 42, and 63.

    Will the customer win $100? Why or why not?

30. Is it possible to have a combination of nickels, dimes, and quarters that add up to $4.72? Explain.

31. Is it possible to have 50 coins, made up of pennies, dimes, and quarters, that add up to $3? Explain.

32. Two athletes run a circular track at a steady pace so that the first completes one round in 8 minutes and the second in 10 minutes. If they both start from the same spot at 4 P.M., when will be the first time they return to the start together?

33. It can be shown (see exercises 41–45) that an integer is divisible by 3 if, and only if, the sum of its digits is divisible by 3. An integer is divisible by 9 if, and only if, the sum of its digits is divisible by 9. An integer is divisible by 5 if, and only if, its right-most digit is a 5 or a 0. And an integer is divisible by 4 if, and only if, the number formed by its right-most two digits is divisible by 4. Check the following integers for divisibility by 3, 4, 5 and 9.
    a. 637,425,403,705,125　　**b.** 12,858,306,120,312
    c. 517,924,440,926,512　　**d.** 14,328,083,360,232

34. Use the unique factorization theorem to write the following integers in standard factored form.
    a. 1176　　**b.** 5377　　**c.** 3675

35. Suppose that in standard factored form $a = p_1^{e_1} p_2^{e_2} \cdots p_k^{e_k}$, where k is a positive integer; $p_1, p_2, \ldots, p_k$ are prime numbers; and $e_1, e_2, \ldots, e_k$ are positive integers.
    a. What is the standard factored form for a^2?
    b. Find the least positive integer n such that $2^5 \cdot 3 \cdot 5^2 \cdot 7^3 \cdot n$ is a perfect square. Write the resulting product as a perfect square.
    c. Find the least positive integer m such that $2^2 \cdot 3^5 \cdot 7 \cdot 11 \cdot m$ is a perfect square. Write the resulting product as a perfect square.

36. Suppose that in standard factored form $a = p_1^{e_1} p_2^{e_2} \cdots p_k^{e_k}$, where k is a positive integer; $p_1, p_2, \ldots, p_k$ are prime numbers; and $e_1, e_2, \ldots, e_k$ are positive integers.
    a. What is the standard factored form for a^3?
    b. Find the least positive integer k such that $2^4 \cdot 3^5 \cdot 7 \cdot 11^2 \cdot k$ is a perfect cube (i.e., equals an integer to the third power). Write the resulting product as a perfect cube.

37. **a.** If a and b are integers and $12a = 25b$, does $12 \mid b$? does $25 \mid a$? Explain.
    b. If x and y are integers and $10x = 9y$, does $10 \mid y$? does $9 \mid x$? Explain.

38. How many zeros are at the end of $45^8 \cdot 88^5$? Explain how you can answer this question without actually computing the number. (*Hint:* $10 = 2 \cdot 5$.)

39. If n is an integer and $n > 1$, then $n!$ is the product of n and every other positive integer that is less than n. For example, $5! = 5 \cdot 4 \cdot 3 \cdot 2 \cdot 1$.
    a. Write 6! in standard factored form.
    b. Write 20! in standard factored form.
    c. Without computing the value of $(20!)^2$ determine how many zeros are at the end of this number when it is written in decimal form. Justify your answer.

✱ 40. In a certain town 2/3 of the adult men are married to 3/5 of the adult women. Assume that all marriages are monogamous (no one is married to more than one other person). Also assume that there are at least 100 adult men in the town. What is the least possible number of adult men in the town? of adult women in the town?

Definition: Given any nonnegative integer n, the **decimal representation** of n is an expression of the form

$$d_k d_{k-1} \cdots d_2 d_1 d_0,$$

where k is a nonnegative integer; $d_0, d_1, d_2, \ldots, d_k$ (called the **decimal digits** of n) are integers from 0 to 9 inclusive; $d_k \ne 0$ unless $n = 0$ and $k = 0$; and

$$n = d_k \cdot 10^k + d_{k-1} \cdot 10^{k-1} + \cdots + d_2 \cdot 10^2 + d_1 \cdot 10 + d_0.$$

(For example, $2{,}503 = 2 \cdot 10^3 + 5 \cdot 10^2 + 0 \cdot 10 + 3$.)

---

41. Prove that if n is any nonnegative integer whose decimal representation ends in 0, then $5 \mid n$. (*Hint:* If the decimal representation of a nonnegative integer n ends in d_0, then $n = 10m + d_0$ for some integer m.)

42. Prove that if n is any nonnegative integer whose decimal representation ends in 5, then $5 \mid n$.

43. Prove that if the decimal representation of a nonnegative integer n ends in $d_1 d_0$ and if $4 \mid (10d_1 + d_0)$, then $4 \mid n$. (*Hint:* If the decimal representation of a nonnegative integer n ends in $d_1 d_0$, then there is an integer s such that $n = 100s + 10d_1 + d_0$.)

H $\ast$ **44.** Observe that

$$7524 = 7 \cdot 1000 + 5 \cdot 100 + 2 \cdot 10 + 4$$
$$= 7(999 + 1) + 5(99 + 1) + 2(9 + 1) + 4$$
$$= (7 \cdot 999 + 7) + (5 \cdot 99 + 5) + (2 \cdot 9 + 2) + 4$$
$$= (7 \cdot 999 + 5 \cdot 99 + 2 \cdot 9) + (7 + 5 + 2 + 4)$$
$$= (7 \cdot 111 \cdot 9 + 5 \cdot 11 \cdot 9 + 2 \cdot 9) + (7 + 5 + 2 + 4)$$
$$= (7 \cdot 111 + 5 \cdot 11 + 2) \cdot 9 + (7 + 5 + 2 + 4)$$
$$= (\text{an integer divisible by } 9)$$
$$\qquad + (\text{the sum of the digits of } 7524).$$

Since the sum of the digits of 7524 is divisible by 9, 7524 can be written as a sum of two integers each of which is divisible by 9. It follows from exercise 15 that 7524 is divisible by 9.

Generalize the argument given in this example to any nonnegative integer n. In other words, prove that for any nonnegative integer n, if the sum of the digits of n is divisible by 9, then n is divisible by 9.

$\ast$ **45.** Prove that for any nonnegative integer n, if the sum of the digits of n is divisible by 3, then n is divisible by 3.

$\ast$ **46.** Given a positive integer n written in decimal form, the alternating sum of the digits of n is obtained by starting with the right-most digit, subtracting the digit immediately to its left, adding the next digit to the left, subtracting the next digit, and so forth. For example, the alternating sum of the digits of 180,928 is $8 - 2 + 9 - 0 + 8 - 1 = 22$. Justify the fact that for any nonnegative integer n, if the alternating sum of the digits of n is divisible by 11, then n is divisible by 11.

3.4 *Direct Proof and Counterexample IV: Division into Cases and the Quotient-Remainder Theorem*

Be especially critical of any statement following the word "obviously."
— Anna Pell Wheeler 1883–1966

When you divide 11 by 4, you get a quotient of 2 and a remainder of 3.

$$\begin{array}{r} 2 \ \leftarrow \text{quotient} \\ 4 \overline{) 11} \\ \underline{8} \\ 3 \ \leftarrow \text{remainder} \end{array}$$

Another way to say this is that 11 equals 2 groups of 4 with 3 left over:

| XXXX | XXXX | XXX |

↑   ↑

2 groups of 4   3 left over

Or,

$$11 = 2 \cdot 4 + 3.$$

↑   ↑

2 groups of 4   3 left over

Of course, the number left over (3) is less than the size of the groups (4) because if more than 4 were left over, another group of 4 could be separated off.

The quotient-remainder theorem says that when any integer n is divided by any positive integer d, the result is a quotient q and a nonnegative remainder r that is smaller than d.

Theorem 3.4.1 The Quotient-Remainder Theorem

Given any integer n and positive integer d, there exist unique integers q and r such that

$$n = dq + r \quad \text{and} \quad 0 \le r < d.$$

We give a proof of the quotient-remainder theorem in Section 4.4.

If n is positive, the quotient-remainder theorem can be illustrated on the number line as follows:

If n is negative, the picture changes. Since $n = dq + r$, where r is nonnegative, d must be multiplied by a negative integer q to go below n. Then the nonnegative integer r is added to come back up to n. This is illustrated as follows:

Example 3.4.1 The Quotient-Remainder Theorem

For each of the following values of n and d, find integers q and r such that $n = dq + r$ and $0 \le r < d$.

a. $n = 54, d = 4$ b. $n = -54, d = 4$ c. $n = 54, d = 70$

Solution

a. $54 = 4 \cdot 13 + 2$; hence $q = 13$ and $r = 2$.

b. $-54 = (-14) \cdot 4 + 2$; hence $q = -14$ and $r = 2$.

c. $54 = 70 \cdot 0 + 54$; hence $q = 0$ and $r = 54$.   ■

div and mod

A number of computer languages have built-in functions that enable you to compute many values of q and r for the quotient-remainder theorem. These functions are called **div** and **mod** in Pascal, are called / and % in C and C++, are called / and % in Java, and are called / (or \) and **mod** in .NET. The functions give the values that satisfy the quotient-remainder theorem when a *nonnegative* integer n is divided by a positive integer d and the result is assigned to an integer variable. However, they do not give the values that satisfy the quotient-remainder theorem when a negative integer n is divided by a positive integer d (see exercise 16 at the end of this section). So we restrict our definitions for *div* (short for "divided by") and *mod* (short for "modulo") to division of a nonnegative integer. The modulo concept is discussed in greater detail in Sections 10.3 and 10.4.

> ° **Definition**
>
> Given a nonnegative integer n and a positive integer d,
>
> $$n \ div \ d = \text{the integer quotient obtained}$$
> $$\text{when } n \text{ is divided by } d, \text{ and}$$
>
> $$n \ mod \ d = \text{the integer remainder obtained}$$
> $$\text{when } n \text{ is divided by } d.$$
>
> Symbolically, if n and d are positive integers, then
>
> $$n \ div \ d = q \quad \text{and} \quad n \ mod \ d = r \quad \Leftrightarrow \quad n = dq + r$$
>
> where q and r are integers and $0 \le r < d$.

Note that it follows from the quotient-remainder theorem that $n \ mod \ d$ equals one of the integers from 0 through $d - 1$ (since the remainder of the division of n by d must be one of these integers). Note also that a necessary and sufficient condition for an integer n to be divisible by an integer d is that $n \ mod \ d = 0$. You are asked to prove this in the exercises at the end of this section.

You can also use a calculator to compute values of *div* and *mod*. To compute $n \ div \ d$ for a nonnegative integer n and a positive integer d, you just divide n by d and ignore the fractional part of the answer. To find $n \ mod \ d$, you can use the fact that if $n = dq + r$, then $r = n - dq$. Thus since $n = d \cdot (n \ div \ d) + n \ mod \ d$, we have that

$$n \ mod \ d = n - d \cdot (n \ div \ d).$$

So you can compute $n \ div \ d$, multiply by d, and subtract the result from n, to obtain $n \ mod \ d$.

Example 3.4.2 *div* and *mod*

Compute 32 *div* 9 and 32 *mod* 9.

Solution

$$
\begin{array}{r}
3 \quad \leftarrow 32 \ div \ 9 \\
9\overline{)32} \\
\underline{27} \\
5 \quad \leftarrow 32 \ mod \ 9
\end{array}
$$

Thus 32 *div* 9 = 3 and 32 *mod* 9 = 5. ∎

Example 3.4.3 Computing the Day of the Week

Suppose today is Tuesday, and neither this year nor next year is a leap year. What day of the week will it be 1 year from today?

Solution   There are 365 days in a year that is not a leap year, and each week has 7 days. Now

$$365 \ div \ 7 = 52 \quad \text{and} \quad 365 \ mod \ 7 = 1$$

because $365 = 52 \cdot 7 + 1$. Thus 52 weeks, or 364 days, from today will be a Tuesday, and so 365 days from today will be 1 day later, namely Wednesday.

More generally, if *DayT* is the day of the week today and *DayN* is the day of the week in N days, then

$$DayN = (DayT + N) \ mod \ 7,$$

3.4.1

where Sunday $= 0$, Monday $= 1, \ldots,$ Saturday $= 6$. ∎

Representations of Integers

In Section 3.1 we defined an even integer to have the form $2k$ for some integer k. At that time we could have defined an odd integer to be one that was not even. Instead, because it was more useful for proving theorems, we specified that an odd integer has the form $2k + 1$ for some integer k. The quotient-remainder theorem brings these two ways of describing odd integers together by guaranteeing that any integer is either even or odd. To see why, let n be any integer, and consider what happens when n is divided by 2. By the quotient-remainder theorem (with $d = 2$), there exist integers q and r such that

$$n = 2q + r \quad \text{and} \quad 0 \leq r < 2.$$

But the only integers that satisfy $0 \leq r < 2$ are $r = 0$ and $r = 1$. It follows that given any integer n, there exists an integer q with

$$n = 2q + 0 \quad \text{or} \quad n = 2q + 1.$$

In the case that $n = 2q + 0 = 2q$, n is even. In the case that $n = 2q + 1$, n is odd. Hence n is either even or odd.

The *parity* of an integer refers to whether the integer is even or odd. For instance, 5 has odd parity and 28 has even parity. We call the fact that any integer is either even or odd the **parity property.**

Example 3.4.4 Consecutive Integers Have Opposite Parity

Prove that given any two consecutive integers, one is even and the other is odd.

Solution   Two integers are called *consecutive* if, and only if, one is one more than the other. So if one integer is m, the next consecutive integer is $m + 1$.

To prove the given statement, start by supposing that you have two particular but arbitrarily chosen consecutive integers. If the smaller is m, then the larger will be $m + 1$. How do you know for sure that one of these is even and the other is odd? You might imagine some examples: 4, 5; 12, 13; 1,073, 1,074. In the first two examples, the smaller of the two integers is even and the larger is odd; in the last example, it is the reverse. These observations suggest dividing the analysis into two cases.

Case 1: The smaller of the two integers is even.

Case 2: The smaller of the two integers is odd.

In the first case, when m is even, it appears that the next consecutive integer is odd. Is this always true? If an integer m is even, must $m + 1$ necessarily be odd? Of course the answer is yes. Because if m is even, then $m = 2k$ for some integer k, and so $m + 1 = 2k + 1$, which is odd.

In the second case, when m is odd, it appears that the next consecutive integer is even. Is this always true? If an integer m is odd, must $m + 1$ necessarily be even? Again, the answer is yes. For if m is odd, then $m = 2k + 1$ for some integer k, and so $m + 1 = (2k + 1) + 1 = 2k + 2 = 2(k + 1)$, which is even.

This discussion is summarized as follows.

Theorem 3.4.2

Any two consecutive integers have opposite parity.

Proof:

Suppose that two [*particular but arbitrarily chosen*] consecutive integers are given; call them m and $m + 1$. [*We must show that one of m and $m + 1$ is even and that the other is odd.*] By the parity property, either m is even or m is odd. [*We break the proof into two cases depending on whether m is even or odd.*]

Case 1 (m is even): In this case, $m = 2k$ for some integer k, and so $m + 1 = 2k + 1$, which is odd [*by definition of odd*]. Hence in this case, one of m and $m + 1$ is even and the other is odd.

Case 2 (m is odd): In this case, $m = 2k + 1$ for some integer k, and so $m + 1 = (2k + 1) + 1 = 2k + 2 = 2(k + 1)$. But $k + 1$ is an integer because it is a sum of two integers. Therefore, $m + 1$ equals twice some integer, and thus $m + 1$ is even. Hence in this case also, one of m and $m + 1$ is even and the other is odd.

It follows that regardless of which case actually occurs for the particular m and $m + 1$ that are chosen, one of m and $m + 1$ is even and the other is odd. [*This is what was to be shown.*]

■

The division into cases in a proof is like the transfer of control for an **if-then-else** statement in a computer program. If m is even, control transfers to case 1; if not, control transfers to case 2. For any given integer, only one of the cases will apply. You must consider both cases, however, to obtain a proof that is valid for an arbitrarily given integer whether even or not.

There are times when division into more than two cases is called for. Suppose that at some stage of developing a proof, you know that a statement of the form

$$A_1 \text{ or } A_2 \text{ or } A_3 \text{ or} \ldots \text{or } A_n$$

is true, and suppose you want to deduce a conclusion C. By definition of *or*, you know that at least one of the statements A_i is true (although you may not know which). In this situation, you should use the method of division into cases. First assume A_1 is true and deduce C; next assume A_2 is true and deduce C; and so forth until you have assumed A_n is true and deduced C. At that point, you can conclude that regardless of which statement A_i happens to be true, the truth of C follows. In symbols,

Given that A_1 or A_2 or A_3 or $\cdots$ or A_n, to show that
$(A_1 \text{ or } A_2 \text{ or } A_3 \text{ or } \cdots \text{ or } A_n) \to C$, show all the implications

$$A_1 \to C,$$
$$A_2 \to C,$$
$$A_3 \to C,$$
$$\vdots$$
$$A_n \to C.$$

Note that this form of argument is a generalization of the one given in Example 1.3.8. Its validity was proved in exercise 21 of Section 1.3.

The procedure used to derive the parity property can be applied with other values of d to obtain a variety of alternative representations of integers.

Example 3.4.5 Representations of Integers Modulo 4

Show that any integer can be written in one of the four forms

$$n = 4q \quad \text{or} \quad n = 4q + 1 \quad \text{or} \quad n = 4q + 2 \quad \text{or} \quad n = 4q + 3$$

for some integer q.

Solution   Given any integer n, apply the quotient-remainder theorem to n with $d = 4$. This implies that there exist an integer quotient q and a remainder r such that

$$n = 4q + r \quad \text{and} \quad 0 \leq r < 4.$$

But the only nonnegative remainders r that are less than 4 are 0, 1, 2, and 3. Hence

$$n = 4q \quad \text{or} \quad n = 4q + 1 \quad \text{or} \quad n = 4q + 2 \quad \text{or} \quad n = 4q + 3$$

for some integer q. ∎

The next example illustrates how alternative representations for integers can help establish results in number theory. The solution is broken into two parts: a discussion and a formal proof. These correspond to the stages of actual proof development. Very few people, when asked to prove an unfamiliar theorem, immediately write down the kind of formal proof you find in a mathematics text. Most need to experiment with several possible approaches before they find one that works. A formal proof is much like the ending of a mystery story—the part in which the action of the story is systematically reviewed and all the loose ends are carefully tied together.

Example 3.4.6 The Square of an Odd Integer

Prove that the square of any odd integer has the form $8m + 1$ for some integer m.

Solution   Begin by asking yourself, "Where am I starting from?" and "What do I need to show?" To help answer these questions, introduce variables to represent the quantities in the statement to be proved.

Formal Restatement: ∀ odd integers n, ∃ an integer m such that $n^2 = 8m + 1$.

From this, you can immediately identify the starting point and what is to be shown.

Starting Point: Suppose n is a particular but arbitrarily chosen odd integer.

To Show: ∃ an integer m such that $n^2 = 8m + 1$.

This looks tough. Why should there be an integer m with the property that $n^2 = 8m + 1$? That would say that $(n^2 - 1)/8$ is an integer, or that 8 divides $n^2 - 1$. Perhaps you could make use of the fact that $n^2 - 1 = (n - 1)(n + 1)$. Does 8 divide $(n - 1)(n + 1)$? Since n is odd, both $(n - 1)$ and $(n + 1)$ are even. That means that their product is divisible by 4. But that's not enough. You need to show that the product is divisible by 8. This seems to be a blind alley.

You could try another tack. Since n is odd, you could represent n as $2q + 1$ for some integer q. Then $n^2 = (2q + 1)^2 = 4q^2 + 4q + 1 = 4(q^2 + q) + 1$. It is clear from this analysis that n^2 can be written in the form $4m + 1$, but it may not be clear that it can be written as $8m + 1$. This also seems to be a blind alley.*

Yet another possibility is to use the result of Example 3.4.5. That example showed that any integer can be written in one of the four forms $4q, 4q + 1, 4q + 2$, or $4q + 3$. Two of these, $4q + 1$ and $4q + 3$, are odd. Thus any odd integer can be written in the

---

*See exercise 25 for a different perspective.

form $4q + 1$ or $4q + 3$ for some integer q. You could try breaking into cases based on these two different forms.*

It turns out that this last possibility works! In each of the two cases, the conclusion follows readily by direct calculation. The details are shown in the following formal proof:

Theorem 3.4.3

The square of any odd integer has the form $8m + 1$ for some integer m.

Proof:

Suppose n is a [*particular but arbitrarily chosen*] odd integer. By the quotient-remainder theorem, n can be written in one of the forms

$$4q \quad \text{or} \quad 4q + 1 \quad \text{or} \quad 4q + 2 \quad \text{or} \quad 4q + 3$$

for some integer q. In fact, since n is odd and $4q$ and $4q + 2$ are even, n must have one of the forms

$$4q + 1 \quad \text{or} \quad 4q + 3.$$

Case 1 ($n = 4q + 1$ for some integer q): [*We must find an integer m such that $n^2 = 8m + 1$.*] Since $n = 4q + 1$,

$$
\begin{aligned}
n^2 &= (4q + 1)^2 && \text{by substitution} \\
&= (4q + 1)(4q + 1) && \text{by definition of square} \\
&= 16q^2 + 8q + 1 \\
&= 8(2q^2 + q) + 1 && \text{by the laws of algebra.}
\end{aligned}
$$

Let $m = 2q^2 + q$. Then m is an integer since 2 and q are integers and sums and products of integers are integers. Thus, substituting,

$$n^2 = 8m + 1 \quad \text{where } m \text{ is an integer.}$$

Case 2 ($n = 4q + 3$ for some integer q): [*We must find an integer m such that $n^2 = 8m + 1$.*] Since $n = 4q + 3$,

$$
\begin{aligned}
n^2 &= (4q + 3)^2 && \text{by substitution} \\
&= (4q + 3)(4q + 3) && \text{by definition of square} \\
&= 16q^2 + 24q + 9 \\
&= 16q^2 + 24q + (8 + 1) \\
&= 8(2q^2 + 3q + 1) + 1 && \text{by the laws of algebra.}
\end{aligned}
$$

[*The motivation for the choice of algebra steps was the desire to write the expression in the form $8 \cdot$ (some integer) $+ 1$.*]

Let $m = 2q^2 + 3q + 1$. Then m is an integer since 1, 2, 3, and q are integers and sums and products of integers are integers. Thus, substituting,

$$n^2 = 8m + 1 \quad \text{where } m \text{ is an integer.}$$

Cases 1 and 2 show that given any odd integer, whether of the form $4q + 1$ or $4q + 3$, $n^2 = 8m + 1$ for some integer m. [*This is what we needed to show.*]

■

---

*Desperation can spur creativity. When you have tried all the obvious approaches without success and you really care about solving a problem, you reach into the odd corners of your memory for *anything* that may help.

Note that the result of Theorem 3.4.3 can also be written, "For any odd integer n, $n^2 \bmod 8 = 1$."

Exercise Set 3.4

For each of the values of n and d given in 1–6, find integers q and r such that $n = dq + r$ and $0 \le r < d$.

1. $n = 70, d = 9$ 2. $n = 62, d = 7$

3. $n = 36, d = 40$ 4. $n = 3, d = 11$

5. $n = -45, d = 11$ 6. $n = -27, d = 8$

Evaluate the expressions in 7–10.

7. a. $43\ div\ 9$ b. $43\ mod\ 9$

8. a. $50\ div\ 7$ b. $50\ mod\ 7$

9. a. $28\ div\ 5$ b. $28\ mod\ 5$

10. a. $30\ div\ 2$ b. $30\ mod\ 2$

11. Check the correctness of formula (3.4.1) given in Example 3.4.3 for the following values of $DayT$ and N.
   a. $DayT = 6$ (Saturday) and $N = 15$
   b. $DayT = 0$ (Sunday) and $N = 7$
   c. $DayT = 4$ (Thursday) and $N = 12$

✳ 12. Justify formula (3.4.1) for general values of $DayT$ and N.

13. On a Monday a friend says he will meet you again in 30 days. What day of the week will that be?

H 14. If today is Tuesday, what day of the week will it be 1,000 days from today?

15. January 1, 2000 was a Saturday, and 2000 was a leap year. What day of the week will January 1, 2050 be?

H 16. The / and % functions in Java operate as follows: If q and r are the integers obtained from the quotient-remainder theorem when a negative integer n is divided by a positive integer d, then n/d is $q + 1$ and $n\%d$ is $r - d$, provided that these values are assigned to an integer variable. Show that n/d and $n\%d$ satisfy one of the conclusions of the quotient-remainder theorem but not the other. To be specific, show that the equation $n = d \cdot n/d + n\%d$ is true but the condition $0 \le n\%d < d$ is false. (The functions **div** and **mod** in Pascal, / and % in C and C++, and / (or \ ) and **mod** in .NET operate similarly to / and % in Java.)

17. When an integer a is divided by 7, the remainder is 4. What is the remainder when $5a$ is divided by 7?

18. When an integer b is divided by 12, the remainder is 5. What is the remainder when $8b$ is divided by 12?

19. When an integer c is divided by 15, the remainder is 3. What is the remainder when $10c$ is divided by 15?

20. Suppose d is a positive integer and n is any integer. If $d \mid n$, what is the remainder obtained when the quotient-remainder theorem is applied to n with divisor d?

H 21. Prove that a necessary and sufficient condition for a non-negative integer n to be divisible by a positive integer d is that $n \bmod d = 0$.

22. A matrix **M** has 3 rows and 4 columns.

$$\begin{bmatrix} a_{11} & a_{12} & a_{13} & a_{14} \\ a_{21} & a_{22} & a_{23} & a_{24} \\ a_{31} & a_{32} & a_{33} & a_{34} \end{bmatrix}$$

The 12 entries in the matrix are to be stored in *row major* form in locations 7,609 to 7,620 in a computer's memory. This means that the entries in the first row (reading left to right) are stored first, then the entries in the second row, and finally the entries in the third row.
   a. Which location will a_{22} be stored in?
   b. Write a formula (in i and j) that gives the integer n so that a_{ij} is stored in location $7{,}609 + n$.
   c. Find formulas (in n) for r and s so that a_{rs} is stored in location $7{,}609 + n$.

23. Let **M** be a matrix with m rows and n columns, and suppose that the entries of **M** are stored in a computer's memory in row major form (see exercise 22) in locations $N, N + 1, N + 2, \ldots, N + mn - 1$. Find formulas in k for r and s so that a_{rs} is stored in location $N + k$.

24. Prove that the product of any two consecutive integers is even.

25. The result of exercise 24 suggests that the second apparent blind alley in the discussion of Example 3.4.6 might not be a blind alley after all. Write a new proof of Theorem 3.4.3 based on this observation.

26. Prove that for all integers n, $n^2 - n + 3$ is odd.

27. Show that any integer n can be written in one of the three forms

$$n = 3q \quad \text{or} \quad n = 3q + 1 \quad \text{or} \quad n = 3q + 2$$

for some integer q.

28. a. Use the quotient-remainder theorem with $d = 3$ to prove that the product of any three consecutive integers is divisible by 3.
   b. Use the *mod* notation to rewrite the result of part (a).

H 29. Use the quotient-remainder theorem with $d = 3$ to prove that the square of any integer has the form $3k$ or $3k + 1$ for some integer k.

30. Use the quotient-remainder theorem with $d = 3$ to prove that the product of any two consecutive integers has the form $3k$ or $3k + 2$ for some integer k.

31. a. Prove that for all integers m and n, $m + n$ and $m - n$ are either both odd or both even.
   b. Find all solutions to the equation $m^2 - n^2 = 56$ for which both m and n are positive integers.
   c. Find all solutions to the equation $m^2 - n^2 = 88$ for which both m and n are positive integers.

32. Given any integers a, b, and c, if $a - b$ is even and $b - c$ is even, what can you say about the parity of $2a - (b + c)$? Prove your answer. You may use the properties listed in Example 3.2.3.

33. Given any integers a, b, and c, if $a - b$ is odd and $b - c$ is even, what can you say about the parity of $a - c$? Prove your answer.

H 34. Given any integer n, if $n > 3$, could n, $n + 2$, and $n + 4$ all be prime? Prove or give a counterexample.

35. The fourth power of any integer has the form $8m$ or $8m + 1$ for some integer m.

Prove each of the statements in 36–43.

H 36. The product of any four consecutive integers is divisible by 8.

37. The square of any integer has the form $4k$ or $4k + 1$ for some integer k.

H 38. For any integer $n \geq 1$, $n^2 + 1$ has the form $4k + 1$ or $4k + 2$ for some integer k.

H 39. The sum of any four consecutive integers has the form $4k + 2$ for some integer k.

40. For any integer n, $n(n^2 - 1)(n + 2)$ is divisible by 4.

41. For all integers m, $m^2 = 5k$, or $m^2 = 5k + 1$, or $m^2 = 5k + 4$ for some integer k.

H 42. Every prime number except 2 and 3 has the form $6q + 1$ or $6q + 5$ for some integer q.

43. If n is an odd integer, then $n^4 \bmod 16 = 1$.

✶ 44. If m, n, and d are integers and $m \bmod d = n \bmod d$, does it necessarily follow that $m = n$? That $m - n$ is divisible by d? Prove your answers.

✶ 45. If m, n, and d are integers and $d \mid (m - n)$, what is the relation between $m \bmod d$ and $n \bmod d$? Prove your answer.

✶ 46. If m, n, a, b, and d are integers and $m \bmod d = a$ and $n \bmod d = b$, is $(m + n) \bmod d = a + b$? Is $(m + n) \bmod d = (a + b) \bmod d$? Prove your answers.

✶ 47. If m, n, a, b, and d are integers and $m \bmod d = a$ and $n \bmod d = b$, is $(mn) \bmod d = ab$? Is $(mn) \bmod d = ab \bmod d$? Prove your answers.

48. Prove that if m, d, and k are integers and $d \neq 0$, then $(m + dk) \bmod d = m \bmod d$.

Use the following definition to prove each statement in 49–53.

Definition: For any real number x, the **absolute value of x**, denoted $|x|$, is defined as follows:

$$|x| = \begin{cases} x & \text{if } x \geq 0 \\ -x & \text{if } x < 0 \end{cases}.$$

---

49. For all real numbers x, $|-x| = |x|$.

50. For all real numbers x and y, $|x| \cdot |y| = |xy|$.

51. For all real numbers x, $-|x| \leq x \leq |x|$.

52. If c is a positive real number and x is any real number, then $-c \leq x \leq c$ if, and only if, $|x| \leq c$. (To prove a statement of the form "A if, and only if, B," you must prove "if A then B" and "if B then A.")

53. For all real numbers x and y, $|x + y| \leq |x| + |y|$. This result is called the **triangle inequality.** (*Hint:* Use 51 and 52 above.)

3.5 Direct Proof and Counterexample V: Floor and Ceiling

Proof serves many purposes simultaneously. In being exposed to the scrutiny and judgment of a new audience, [a] proof is subject to a constant process of criticism and revalidation. Errors, ambiguities, and misunderstandings are cleared up by constant exposure. Proof is respectability. Proof is the seal of authority.

Proof, in its best instances, increases understanding by revealing the heart of the matter. Proof suggests new mathematics. The novice who studies proofs gets closer to the creation of new mathematics. Proof is mathematical power, the electric voltage of the subject which vitalizes the static assertions of the theorems.

Finally, proof is ritual, and a celebration of the power of pure reason.

— Philip J. Davis and Reuben Hersh, *The Mathematical Experience*, 1981

Imagine a real number sitting on a number line. The *floor* and *ceiling* of the number are the integers to the immediate left and to the immediate right of the number (unless the number is, itself, an integer, in which case its floor and ceiling both equal the number itself). Many computer languages have built-in functions that compute floor and ceiling automatically. These functions are very convenient to use when writing certain kinds of computer programs. In addition, the concepts of floor and ceiling are important in analyzing the efficiency of many computer algorithms.

• Definition

Given any real number x, the **floor of x,** denoted $\lfloor x \rfloor$, is defined as follows:

$$\lfloor x \rfloor = \text{that unique integer } n \text{ such that } n \leq x < n+1.$$

Symbolically, if x is a real number and n is an integer, then

$$\lfloor x \rfloor = n \quad \Leftrightarrow \quad n \leq x < n+1.$$

---

floor of $x = \lfloor x \rfloor$

• Definition

Given any real number x, the **ceiling of x,** denoted $\lceil x \rceil$, is defined as follows:

$$\lceil x \rceil = \text{that unique integer } n \text{ such that } n-1 < x \leq n.$$

Symbolically, if x is a real number and n is an integer, then

$$\lceil x \rceil = n \quad \Leftrightarrow \quad n-1 < x \leq n.$$

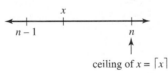

ceiling of $x = \lceil x \rceil$

Example 3.5.1 Computing Floors and Ceilings

Compute $\lfloor x \rfloor$ and $\lceil x \rceil$ for each of the following values of x:

a. $25/4$ b. 0.999 c. -2.01

Solution

a. $25/4 = 6.25$ and $6 < 6.25 < 7$; hence $\lfloor 25/4 \rfloor = 6$ and $\lceil 25/4 \rceil = 7$.

b. $0 < 0.999 < 1$; hence $\lfloor 0.999 \rfloor = 0$ and $\lceil 0.999 \rceil = 1$.

c. $-3 < -2.01 < -2$; hence $\lfloor -2.01 \rfloor = -3$ and $\lceil -2.01 \rceil = -2$.

Note that on some calculators $\lfloor x \rfloor$ is denoted INT (x). ■

Example 3.5.2 An Application

The 1,370 soldiers at a military base are given the opportunity to take buses into town for an evening out. Each bus holds a maximum of 40 passengers.

a. For reasons of economy, the base commander will send only full buses. What is the maximum number of buses the base commander will send?

b. If the base commander is willing to send a partially filled bus, how many buses will the commander need to allow all the soldiers to take the trip?

Solution

a. $\lfloor 1370/40 \rfloor = \lfloor 34.25 \rfloor = 34$ b. $\lceil 1370/40 \rceil = \lceil 34.25 \rceil = 35$ ∎

Example 3.5.3 Some General Values of Floor

If k is an integer, what are $\lfloor k \rfloor$ and $\lfloor k + 1/2 \rfloor$? Why?

Solution   Suppose k is an integer. Then

$$\lfloor k \rfloor = k \text{ because } k \text{ is an integer and } k \leq k < k + 1,$$

and

$$\left\lfloor k + \frac{1}{2} \right\rfloor = k \text{ because } k \text{ is an integer and } k \leq k + \frac{1}{2} < k + 1.$$ ∎

Example 3.5.4 Disproving an Alleged Property of Floor

Is the following statement true or false?

For all real numbers x and y, $\lfloor x + y \rfloor = \lfloor x \rfloor + \lfloor y \rfloor$.

Solution   The statement is false. As a counterexample, take $x = y = \frac{1}{2}$. Then

$$\lfloor x \rfloor + \lfloor y \rfloor = \left\lfloor \frac{1}{2} \right\rfloor + \left\lfloor \frac{1}{2} \right\rfloor = 0 + 0 = 0,$$

whereas

$$\lfloor x + y \rfloor = \left\lfloor \frac{1}{2} + \frac{1}{2} \right\rfloor = \lfloor 1 \rfloor = 1.$$

Hence $\lfloor x + y \rfloor \neq \lfloor x \rfloor + \lfloor y \rfloor$.

To arrive at this counterexample, you could have reasoned as follows: Suppose x and y are real numbers. Must it necessarily be the case that $\lfloor x + y \rfloor = \lfloor x \rfloor + \lfloor y \rfloor$, or could x and y be such that $\lfloor x + y \rfloor \neq \lfloor x \rfloor + \lfloor y \rfloor$? Imagine values that the various quantities could take. For instance, if both x and y are positive, then $\lfloor x \rfloor$ and $\lfloor y \rfloor$ are the integer parts of $\lfloor x \rfloor$ and $\lfloor y \rfloor$ respectively; just as

$$2\frac{3}{5} = 2 + \frac{3}{5}$$

$$\text{integer part} \qquad \text{fractional part}$$

so is

$$x = \lfloor x \rfloor + \text{fractional part of } x$$

and

$$y = \lfloor y \rfloor + \text{fractional part of } y.$$

Thus if x and y are positive,

$$x + y = \lfloor x \rfloor + \lfloor y \rfloor + \textit{the sum of the fractional parts of x and y.}$$

But also

$$x + y = \lfloor x + y \rfloor + \textit{the fractional part of } (x + y).$$

These equations show that if there exist numbers x and y such that the sum of the fractional parts of x and y is at least 1, then a counterexample can be found. But there do exist such x and y; for instance, $x = \frac{1}{2}$ and $y = \frac{1}{2}$ as before. ∎

The analysis of Example 3.5.4 indicates that if x and y are positive and the sum of their fractional parts is less than 1, then $\lfloor x + y \rfloor = \lfloor x \rfloor + \lfloor y \rfloor$. In particular, if x is positive and m is a positive integer, then $\lfloor x + m \rfloor = \lfloor x \rfloor + \lfloor m \rfloor = \lfloor x \rfloor + m$. (The fractional part of m is 0; hence the sum of the fractional parts of x and m equals the fractional part of x, which is less than 1.) It turns out that you can use the definition of floor to show that this equation holds for all real numbers x and for all integers m.

Example 3.5.5 Proving a Property of Floor

Prove that for all real numbers x and for all integers m, $\lfloor x + m \rfloor = \lfloor x \rfloor + m$.

Solution   Begin by supposing that x is a particular but arbitrarily chosen real number and that m is a particular but arbitrarily chosen integer. You must show that $\lfloor x + m \rfloor = \lfloor x \rfloor + m$. Since this is an equation involving $\lfloor x \rfloor$ and $\lfloor x + m \rfloor$, it is reasonable to give one of these quantities a name: Let $n = \lfloor x \rfloor$. By definition of floor,

$$n \text{ is an integer} \quad \text{and} \quad n \leq x < n + 1.$$

This double inequality enables you to compute the value of $\lfloor x + m \rfloor$ in terms of n by adding m to all sides:

$$n + m \leq x + m < n + m + 1.$$

Thus the left-hand side of the equation to be shown is

$$\lfloor x + m \rfloor = n + m.$$

On the other hand, since $n = \lfloor x \rfloor$, the right-hand side of the equation to be shown is

$$\lfloor x \rfloor + m = n + m$$

also. Thus $\lfloor x + m \rfloor = \lfloor x \rfloor + m$. This discussion is summarized as follows:

Theorem 3.5.1

For all real numbers x and all integers m, $\lfloor x + m \rfloor = \lfloor x \rfloor + m$.

Proof:

Suppose a real number x and an integer m are given. [*We must show that* $\lfloor x + m \rfloor = \lfloor x \rfloor + m$.] Let $n = \lfloor x \rfloor$. By definition of floor, n is an integer and

$$n \leq x < n + 1.$$

Add m to all sides to obtain

$$n + m \leq x + m < n + m + 1$$

continued on page 168

[*since adding a number to both sides of an inequality does not change the direction of the inequality*].

Now $n + m$ is an integer [*since n and m are integers and a sum of integers is an integer,*] and so, by definition of floor, the left-hand side of the equation to be shown is

$$\lfloor x + m \rfloor = n + m.$$

But $n = \lfloor x \rfloor$. Hence, by substitution,

$$n + m = \lfloor x \rfloor + m,$$

which is the right-hand side of the equation to be shown. Thus $\lfloor x + m \rfloor = \lfloor x \rfloor + m$ [*as was to be shown*].

∎

The analysis of a number of computer algorithms, such as the binary search and merge sort algorithms, requires that you know the value of $\lfloor n/2 \rfloor$, where n is an integer. The formula for computing this value depends on whether n is even or odd.

Theorem 3.5.2 The Floor of $n/2$

For any integer n,

$$\left\lfloor \frac{n}{2} \right\rfloor = \begin{cases} \dfrac{n}{2} & \text{if } n \text{ is even} \\[2mm] \dfrac{n-1}{2} & \text{if } n \text{ is odd} \end{cases}.$$

Proof:

Suppose n is a [*particular but arbitrarily chosen*] integer. By the quotient-remainder theorem, n is odd or n is even.

Case 1 (n is odd): In this case, $n = 2k + 1$ for some integer k. [*We must show that* $\lfloor n/2 \rfloor = (n-1)/2$.] But the left-hand side of the equation to be shown is

$$\left\lfloor \frac{n}{2} \right\rfloor = \left\lfloor \frac{2k+1}{2} \right\rfloor = \left\lfloor \frac{2k}{2} + \frac{1}{2} \right\rfloor = \left\lfloor k + \frac{1}{2} \right\rfloor = k$$

because k is an integer and $k \leq k + 1/2 < k + 1$. And the right-hand side of the equation to be shown is

$$\frac{n-1}{2} = \frac{(2k+1) - 1}{2} = \frac{2k}{2} = k$$

also. So since both the left-hand and right-hand sides equal k, they are equal to each other. That is, $\left\lfloor \dfrac{n}{2} \right\rfloor = \dfrac{n-1}{2}$ [*as was to be shown*].

Case 2 (n is even): In this case, $n = 2k$ for some integer k. [*We must show that* $\lfloor n/2 \rfloor = n/2$]. The rest of the proof of this case is left as an exercise.

Given a nonnegative integer n and a positive integer d, the quotient-remainder theorem guarantees the existence of unique integers q and r such that

$$n = dq + r \quad \text{and} \quad 0 \leq r < d.$$

The following theorem states that the floor notation can be used to describe q and r as follows:

$$q = \left\lfloor \frac{n}{d} \right\rfloor \quad \text{and} \quad r = n - d \left\lfloor \frac{n}{d} \right\rfloor.$$

Thus if, on a calculator or in a computer language, floor is built in but *div* and *mod* are not, *div* and *mod* can be defined as follows: For a nonnegative integer n and a positive integer d,

$$n \ div \ d = \left\lfloor \frac{n}{d} \right\rfloor \quad \text{and} \quad n \ mod \ d = n - d \left\lfloor \frac{n}{d} \right\rfloor. \qquad \text{3.5.1}$$

Note that d divides n if, and only if, $n \ mod \ d = 0$, or, in other words, $n = d\lfloor n/d \rfloor$. You are asked to prove this in exercise 13.

Theorem 3.5.3

If n is a nonnegative integer and d is a positive integer, and if $q = \lfloor n/d \rfloor$ and $r = n - d\lfloor n/d \rfloor$, then

$$n = dq + r \quad \text{and} \quad 0 \leq r < d.$$

Proof:

Suppose n is a nonnegative integer, d is a positive integer, $q = \lfloor n/d \rfloor$, and $r = n - d\lfloor n/d \rfloor$. [*We must show that $n = dq + r$ and $0 \leq r < d$.*] By substitution,

$$dq + r = d \left\lfloor \frac{n}{d} \right\rfloor + \left(n - d \left\lfloor \frac{n}{d} \right\rfloor \right) = n.$$

So it remains only to show that $0 \leq r < d$. But $q = \lfloor n/d \rfloor$. Thus, by definition of floor,

$$q \leq \frac{n}{d} < q + 1.$$

Then

$$dq \leq n < dq + d \quad \text{by multiplying all parts by } d$$

and so

$$0 \leq n - dq < d \quad \text{by subtracting } dq \text{ from all parts.}$$

But

$$r = n - d \left\lfloor \frac{n}{d} \right\rfloor = n - dq.$$

Hence

$$0 \leq r < d \quad \text{by substitution.}$$

[*This is what was to be shown.*]

Example 3.5.6 Computing *div* and *mod*

Use the floor notation to compute 3850 *div* 17 and 3850 *mod* 17.

Solution By formula (3.5.1),

$$3850 \ div \ 17 = \lfloor 3850/17 \rfloor = \lfloor 226.47 \rfloor = 226$$

$$3850 \ mod \ 17 = 3850 - 17 \cdot \lfloor 3850/17 \rfloor$$
$$= 3850 - 17 \cdot 226$$
$$= 3850 - 3842 = 8. \qquad \blacksquare$$

Exercise Set 3.5

Compute $\lfloor x \rfloor$ and $\lceil x \rceil$ for each of the values of x in 1–4.

1. 37.999

2. 17/4

3. −14.00001

4. −32/5

5. Use the floor notation to express 259 *div* 11 and 259 *mod* 11.

6. If k is an integer, what is $\lceil k \rceil$? Why?

7. If k is an integer, what is $\lceil k + \frac{1}{2} \rceil$? Why?

8. Seven pounds of raw material are needed to manufacture each unit of a certain product. Express the number of units that can be produced from n pounds of raw material using either the floor or the ceiling notation. Which notation is more appropriate?

9. Boxes, each capable of holding 36 units, are used to ship a product from the manufacturer to a wholesaler. Express the number of boxes that would be required to ship n units of the product using either the floor or the ceiling notation. Which notation is more appropriate?

10. If 0 = Sunday, 1 = Monday, 2 = Tuesday, ... , 6 = Saturday, then January 1 of year n occurs on the day of the week given by the following formula:

$$\left(n + \left\lfloor \frac{n-1}{4} \right\rfloor - \left\lfloor \frac{n-1}{100} \right\rfloor + \left\lfloor \frac{n-1}{400} \right\rfloor \right) mod \ 7.$$

a. Use this formula to find January 1 of
   i. 2050   ii. 2100   iii. the year of your birth.
H b. Interpret the different components of this formula.

11. State a necessary and sufficient condition for the floor of a real number to equal that number.

12. Prove that if n is any even integer, then $\lfloor n/2 \rfloor = n/2$.

13. Suppose n and d are integers and $d \neq 0$. Prove each of the following.
   a. If $d \mid n$, then $n = \lfloor n/d \rfloor \cdot d$.
   b. If $n = \lfloor n/d \rfloor \cdot d$, then $d \mid n$.
   c. Use the floor notation to state a necessary and sufficient condition for an integer n to be divisible by an integer d.

Some of the statements in 14–22 are true and some are false. Prove each true statement and find a counterexample for each false statement.

14. For all real numbers x and y, $\lfloor x - y \rfloor = \lfloor x \rfloor - \lfloor y \rfloor$.

15. For all real numbers x, $\lfloor x - 1 \rfloor = \lfloor x \rfloor - 1$.

16. For all real numbers x, $\lfloor x^2 \rfloor = \lfloor x \rfloor^2$.

H **17.** For all integers n,

$$\lfloor n/3 \rfloor = \begin{cases} n/3 & \text{if } n \ mod \ 3 = 0 \\ (n-1)/3 & \text{if } n \ mod \ 3 = 1 \\ (n-2)/3 & \text{if } n \ mod \ 3 = 2 \end{cases}.$$

H **18.** For all real numbers x and y, $\lceil x + y \rceil = \lceil x \rceil + \lceil y \rceil$.

H **19.** For all real numbers x, $\lceil x + 1 \rceil = \lceil x \rceil + 1$.

20. For all real numbers x and y, $\lceil xy \rceil = \lceil x \rceil \cdot \lceil y \rceil$.

21. For all odd integers n, $\lceil n/2 \rceil = (n + 1)/2$.

22. For all real numbers x and y, $\lceil xy \rceil = \lceil x \rceil \cdot \lfloor y \rfloor$.

Prove each of the statements in 23–29.

23. For any real number x, if x is not an integer, then $\lfloor x \rfloor + \lfloor -x \rfloor = -1$.

24. For any integer m and any real number x, if x is not an integer, then $\lfloor x \rfloor + \lfloor m - x \rfloor = m - 1$.

H **25.** For all real numbers x, $\lfloor \lfloor x/2 \rfloor /2 \rfloor = \lfloor x/4 \rfloor$.

26. For all real numbers x, if $x - \lfloor x \rfloor < 1/2$ then $\lfloor 2x \rfloor = 2 \lfloor x \rfloor$.

27. For all real numbers x, if $x - \lfloor x \rfloor \geq 1/2$ then $\lfloor 2x \rfloor = 2 \lfloor x \rfloor + 1$.

28. For any odd integer n,

$$\left\lfloor \frac{n^2}{4} \right\rfloor = \left(\frac{n-1}{2} \right) \left(\frac{n+1}{2} \right).$$

29. For any odd integer n,

$$\left\lceil \frac{n^2}{4} \right\rceil = \frac{n^2 + 3}{4}.$$

30. Find the mistake in the following "proof" that $\lfloor n/2 \rfloor = (n-1)/2$ if n is an odd integer.

"**Proof:** Suppose n is any odd integer. Then $n = 2k + 1$ for some integer k. Consequently,

$$\left\lfloor \frac{2k+1}{2} \right\rfloor = \frac{(2k+1)-1}{2} = \frac{2k}{2} = k.$$

But $n = 2k + 1$. Solving for k gives $k = (n-1)/2$. Hence, by substitution, $\lfloor n/2 \rfloor = (n-1)/2$."

3.6 Indirect Argument: Contradiction and Contraposition

Reductio ad absurdum is one of a mathematician's finest weapons. It is a far finer gambit than any chess gambit: a chess player may offer the sacrifice of a pawn or even a piece, but the mathematician offers the game. — G. H. Hardy, 1877–1947

In a direct proof you start with the hypothesis of a statement and make one deduction after another until you reach the conclusion. Indirect proofs are more roundabout. One kind of indirect proof, *argument by contradiction,* is based on the fact that either a statement is true or it is false but not both. Suppose you can show that the assumption that a given statement is not true leads logically to a contradiction, impossibility, or absurdity. Then that assumption must be false; hence, the given statement must be true. This method of proof is also known as *reductio ad impossible* or *reductio ad absurdum* because it relies on reducing a given assumption to an impossibility or absurdity.

Argument by contradiction occurs in many different settings. For example, if a man accused of holding up a bank can prove that he was some place else at the time the crime was committed, he will certainly be acquitted. The logic of his defense is as follows:

Suppose I did commit the crime. Then at the time of the crime, I would have had to be at the scene of the crime. In fact, at the time of the crime I was in a meeting with 20 people far from the crime scene, as they will testify. This contradicts the assumption that I committed the crime, since it is impossible to be in two places at one time. Hence that assumption is false.

Another example occurs in debate. One technique of debate is to say, "Suppose for a moment that what my opponent says is correct." Starting from this supposition, the debater then deduces one statement after another until finally arriving at a statement that is completely ridiculous and unacceptable to the audience. By this means the debater shows the opponent's statement to be false.

The point of departure for a proof by contradiction is the supposition that the statement to be proved is false. The goal is to reason to a contradiction. Thus proof by contradiction has the following outline:

Method of Proof by Contradiction

1. Suppose the statement to be proved is false. That is, suppose that the negation of the statement is true. (Be very careful when writing the negation!)

2. Show that this supposition leads logically to a contradiction.

3. Conclude that the statement to be proved is true.

There are no clear-cut rules for when to try a direct proof and when to try a proof by contradiction. There are some general guidelines, however. Proof by contradiction is indicated if you want to show that there is no object with a certain property, or if you want to show that a certain object does not have a certain property. The next two examples illustrate these situations.

Example 3.6.1 There Is No Greatest Integer

Use proof by contradiction to show that there is no greatest integer.

Solution   Most small children believe there is a greatest integer—they often call it a "zillion." But with age and experience, they change their belief. At some point they realize that if there were a greatest integer, they could add 1 to it to obtain an integer that was greater still. Since that is a contradiction, no greatest integer can exist. This line of reasoning is the heart of the formal proof.

For the proof, the certain property is the property of being the greatest integer. To prove that there is no object with this property, begin by supposing the negation: that there is an object with the property.

Starting Point: Suppose not. Suppose there is a greatest integer; call it N.

This means that $N \geq n$ for all integers n.

To Show: This supposition leads logically to a contradiction.

Theorem 3.6.1

There is no greatest integer.

Proof:

[*We take the negation of the theorem and suppose it to be true.*] Suppose not. That is, suppose there is a greatest integer N. [*We must deduce a contradiction.*] Then $N \geq n$ for every integer n. Let $M = N + 1$. Now M is an integer since it is a sum of integers. Also $M > N$ since $M = N + 1$.

Thus M is an integer that is greater than N. So N is the greatest integer and N is not the greatest integer, which is a contradiction. [*This contradiction shows that the supposition is false and, hence, that the theorem is true.*]

After a contradiction has been reached, the logic of the argument is always the same: "This is a contradiction. Hence the supposition is false and the theorem is true." Because of this, most mathematics texts end proofs by contradiction at the point at which the contradiction has been obtained.

The contradiction in the next example is based on the fact that $1/2$ is not an integer.

Example 3.6.2 No Integer Can Be Both Even and Odd

Is it possible for an integer to be both even and odd? The intuitive answer is "Of course not." What justifies this certainty? A proof by contradiction!

Theorem 3.6.2

There is no integer that is both even and odd.

Proof:

[*We take the negation of the theorem and suppose it to be true.*] Suppose not. That is, suppose there is an integer n that is both even and odd. [*We must deduce a contradiction.*] By definition of even, $n = 2a$ for some integer a and by definition of odd, $n = 2b + 1$ for some integer b. Consequently,

$$2a = 2b + 1 \quad \text{by equating the two expressions for } n$$

and so

$$2a - 2b = 1$$
$$2(a - b) = 1$$
$$(a - b) = 1/2 \quad \text{by algebra.}$$

Now since a and b are integers, the difference $a - b$ must also be an integer. But $a - b = 1/2$, and $1/2$ is not an integer. Thus $a - b$ is an integer and $a - b$ is not an integer, which is a contradiction. [*This contradiction shows that the supposition is false and, hence, that the theorem is true.*]

■

The next example asks you to show that the sum of any rational number and any irrational number is irrational. One way to think of this is in terms of a certain object (the sum of a rational and an irrational) not having a certain property (the property of being rational). This suggests trying a proof by contradiction: suppose the object has the property and deduce a contradiction.

Example 3.6.3 The Sum of a Rational Number and an Irrational Number

Use proof by contradiction to show that the sum of any rational number and any irrational number is irrational.

Solution   Begin by supposing the negation of what you are to prove. Be very careful when writing down what this means. If you take the negation incorrectly, the entire rest of the proof will be flawed. In this example, the statement to be proved can be written formally as

∀ real numbers r and s, if r is rational and
s is irrational, then $r + s$ is irrational.

From this you can see that the negation is

∃ a rational number r and an irrational
number s such that $r + s$ is rational.

 Caution!   The negation of "The sum of any irrational number and any rational number is irrational" is NOT "The sum of any irrational number and any rational number is rational."

It follows that the starting point and what is to be shown are as follows:

Starting Point: Suppose not. That is, suppose there is a rational number r and an irrational number s such that $r + s$ is rational.

To Show: This supposition leads to a contradiction.

To derive a contradiction, you need to understand what you are supposing: There are numbers r and s such that r is rational, s is irrational, and $r + s$ is rational. By definition of rational and irrational, this means that s cannot be written as a quotient of any two integers but that r and $r + s$ can:

$$r = \frac{a}{b} \quad \text{for some integers } a \text{ and } b \text{ with } b \neq 0, \text{ and} \qquad 3.6.1$$

$$r + s = \frac{c}{d} \quad \text{for some integers } c \text{ and } d \text{ with } d \neq 0. \qquad 3.6.2$$

If you substitute (3.6.1) into (3.6.2), you obtain

$$\frac{a}{b} + s = \frac{c}{d}.$$

Subtracting a/b from both sides gives

$$
\begin{aligned}
s &= \frac{c}{d} - \frac{a}{b} \\
&= \frac{bc}{bd} - \frac{ad}{bd} \quad \text{by rewriting } c/d \text{ and } a/b \text{ as equivalent fractions} \\
&= \frac{bc - ad}{bd} \quad \begin{array}{l}\text{by the rule for subtracting fractions} \\ \text{with the same denominator.}\end{array}
\end{aligned}
$$

But both $bc - ad$ and bd are integers because products and differences of integers are integers, and $bd \neq 0$ by the zero product property. Hence s can be expressed as a quotient of two integers with a nonzero denominator, and so s is rational, which contradicts the supposition that it is irrational.

This discussion is summarized in a formal proof.

Theorem 3.6.3

The sum of any rational number and any irrational number is irrational.

Proof:

[*We take the negation of the theorem and suppose it to be true.*] Suppose not. That is, suppose there is a rational number r and an irrational number s such that $r + s$ is rational. [*We must deduce a contradiction.*] By definition of rational, $r = a/b$ and $r + s = c/d$ for some integers $a, b, c,$ and d with $b \neq 0$ and $d \neq 0$. By substitution,

$$\frac{a}{b} + s = \frac{c}{d},$$

and so

$$
\begin{aligned}
s &= \frac{c}{d} - \frac{a}{b} \quad \text{by subtracting } a/b \text{ from both sides} \\
&= \frac{bc - ad}{bd} \quad \text{by the laws of algebra.}
\end{aligned}
$$

Now $bc - ad$ and bd are both integers [*since a, b, c, and d are, and since products and differences of integers are integers*], and $bd \neq 0$ [*by the zero product property*]. Hence s is a quotient of the two integers $bc - ad$ and bd with $bd \neq 0$. Thus, by definition of rational, s is rational, which contradicts the supposition that s is irrational. [*Hence the supposition is false and the theorem is true.*]

■

Argument by Contraposition

A second form of indirect argument, *argument by contraposition,* is based on the logical equivalence between a statement and its contrapositive. To prove a statement by contraposition, you take the contrapositive of the statement, prove the contrapositive by a direct proof, and conclude that the original statement is true. The underlying reasoning is that since a conditional statement is logically equivalent to its contrapositive, if the contrapositive is true then the statement must also be true.

Method of Proof by Contraposition

1. Express the statement to be proved in the form

$$\forall x \text{ in } D, \text{ if } P(x) \text{ then } Q(x).$$

   (This step may be done mentally.)

2. Rewrite this statement in the contrapositive form

$$\forall x \text{ in } D, \text{ if } Q(x) \text{ is false then } P(x) \text{ is false.}$$

   (This step may also be done mentally.)

3. Prove the contrapositive by a direct proof.

   a. Suppose x is a (particular but arbitrarily chosen) element of D such that $Q(x)$ is false.

   b. Show that $P(x)$ is false.

Example 3.6.4 If the Square of an Integer Is Even, Then the Integer Is Even

Prove that for all integers n, if n^2 is even then n is even.

Solution   First form the contrapositive of the statement to be proved.

Contrapositive: For all integers n, if n is not even then n^2 is not even.

By the quotient-remainder theorem with $d = 2$, any integer is even or odd, so any integer that is not even is odd. Also by Theorem 3.6.2, no integer can be both even and odd. So if an integer is odd, then it is not even. Thus the contrapositive can be restated as follows:

Contrapositive: For all integers n, if n is odd then n^2 is odd.

A straightforward computation is the heart of a direct proof for this statement, as shown below.

Proposition 3.6.4

For all integers n, if n^2 is even then n is even.

Proof (by contraposition):

Suppose n is any odd integer. [*We must show that n^2 is odd.*] By definition of odd, $n = 2k + 1$ for some integer k. By substitution and algebra, $n^2 = (2k + 1)^2 = 4k^2 + 4k + 1 = 2(2k^2 + 2k) + 1$. But $2k^2 + 2k$ is an integer because products and sums of integers are integers. So $n^2 = 2 \cdot (\text{an integer}) + 1$, and thus, by definition of odd, n^2 is odd [*as was to be shown*].

We used the word *proposition* here rather than *theorem* because although the word *theorem* can refer to any statement that has been proved, mathematicians often restrict it to especially important statements that have many and varied consequences. Then they use the word **proposition** to refer to a statement that is somewhat less consequential but nonetheless worth writing down. We will use Proposition 3.6.4 in Section 3.7 to prove that $\sqrt{2}$ is irrational. ∎

Relation between Proof by Contradiction and Proof by Contraposition

Observe that any proof by contraposition can be recast in the language of proof by contradiction. In a proof by contraposition, the statement

$$\forall x \text{ in } D, \text{ if } P(x) \text{ then } Q(x)$$

is proved by giving a direct proof of the equivalent statement

$$\forall x \text{ in } D, \text{ if } \sim Q(x) \text{ then } \sim P(x).$$

To do this, you suppose you are given an arbitrary element x of D such that $\sim Q(x)$. You then show that $\sim P(x)$. This is illustrated in Figure 3.6.1.

Figure 3.6.1 Proof by Contraposition

Exactly the same sequence of steps can be used as the heart of a proof by contradiction for the given statement. The only thing that changes is the context in which the steps are written down.

To rewrite the proof as a proof by contradiction, you suppose there is an x in D such that $P(x)$ and $\sim Q(x)$. You then follow the steps of the proof by contraposition to deduce the statement $\sim P(x)$. But $\sim P(x)$ is a contradiction to the supposition that $P(x)$ and $\sim Q(x)$. (Because to contradict a conjunction of statement, it is only necessary to contradict one component.) This process is illustrated in Figure 3.6.2.

Figure 3.6.2 Proof by Contradiction

As an example, here is a proof by contradiction of Proposition 3.6.4, namely that for any integer n, if n^2 is even then n is even.

Proposition 3.6.4

For all integers n, if n^2 is even then n is even.

Proof (by contradiction):

[*We take the negation of the theorem and suppose it to be true.*] Suppose not. That is, suppose there is an integer n such that n^2 is even and n is not even. [*We must deduce a contradiction.*] By the quotient-remainder theorem with $d = 2$, any integer is even or odd. Hence, since n is not even it is odd, and thus, by definition of odd, $n = 2k + 1$ for some integer k. By substitution and algebra: $n^2 = (2k + 1)^2 = 4k^2 + 4k + 1 = 2(2k^2 + 2k) + 1$. But $2k^2 + 2k$ is an integer because products and sums of integers are integers. So $n^2 = 2 \cdot$ (an integer) $+ 1$, and so, by definition of odd, n^2 is odd. Therefore, n^2 is both even and odd. This contradicts Theorem 3.6.2, which states that no integer can be both even and odd. [*This contradiction shows that the supposition is false and, hence, that the proposition is true.*]

Note that when you use proof by contraposition, you know exactly what conclusion you need to show, namely the negation of the hypothesis; whereas in proof by contradiction, it may be difficult to know what contradiction to head for. On the other hand, when you use proof by contradiction, once you have deduced any contradiction whatsoever, you are done. The main advantage of contraposition over contradiction is that you avoid having to take (possibly incorrectly) the negation of a complicated statement. The disadvantage of contraposition as compared with contradiction is that you can use contraposition only for a specific class of statements—those that are universal and conditional. The discussion above shows that any statement that can be proved by contraposition can be proved by contradiction. But the converse is not true. Statements such as "$\sqrt{2}$ is irrational" (discussed in the next section) can be proved by contradiction but not by contraposition.

Proof as a Problem-Solving Tool

Direct proof, disproof by counterexample, proof by contradiction, and proof by contraposition are all tools that may be used to help determine whether statements are true or false. Given a statement of the form

For all elements in a domain, if (hypothesis) then (conclusion),

imagine elements in the domain that satisfy the hypothesis. Ask yourself: Must they satisfy the conclusion? If you can see that the answer is "yes" in all cases, then the statement is true and your insight will form the basis for a direct proof. If after some thought it is not clear that the answer is "yes," try to think whether there are elements of the domain that satisfy the hypothesis and *not* the conclusion. If you are successful in finding some, then the statement is false and you have a counterexample. On the other hand, if you are not successful in finding such elements, perhaps none exist. Perhaps you can show that assuming the existence of elements in the domain that satisfy the hypothesis and not the conclusion leads logically to a contradiction. If so, then the given statement is true and you have the basis for a proof by contradiction. Alternatively, you could imagine elements of the domain for which the conclusion is false and ask whether such elements also fail to satisfy the hypothesis. If the answer in all cases is "yes," then you have a basis for a proof by contraposition.

Solving problems, especially difficult problems, is rarely a straightforward process. At any stage of following the guidelines above, you might want to try the method of a previous stage again. If, for example, you fail to find a counterexample for a certain statement, your experience in trying to find it might help you decide to reattempt a direct argument rather than trying an indirect one. Psychologists who have studied problem solving have found that the most successful problem solvers are those who are flexible and willing to use a variety of approaches without getting stuck in any one of them for very long. Mathematicians sometimes work for months (or longer) on difficult problems. Don't be discouraged if some problems in this book take you quite a while to solve.

Learning the skills of proof and disproof is much like learning other skills, such as those used in swimming, tennis, or playing a musical instrument. When you first start out, you may feel bewildered by all the rules, and you may not feel confident as you attempt new things. But with practice the rules become internalized and you can use them in conjunction with all your other powers—of balance, coordination, judgment, aesthetic sense—to concentrate on winning a meet, winning a match, or playing a concert successfully.

Now that you have worked through the first six sections of this chapter, return to the idea that, above all, a proof or disproof should be a convincing argument. You need to know how direct and indirect proofs and counterexamples are structured. But to use this knowledge effectively, you must use it in conjunction with your imaginative powers, your intuition, and especially your common sense.

Exercise Set 3.6

1. Fill in the blanks in the following proof that there is no least positive real number.

   Proof: [*We take the negation of the statement and suppose it to be true.*] Suppose not. That is, suppose that there is a real number x such that x is positive and __(a)__ for all positive real numbers y. [*We must deduce __(b)__.*] Consider the number $x/2$. Then __(c)__ because x is positive, and $x/2 < x$ because __(d)__. Hence __(e)__, which is a contradiction. [*Thus the supposition is false, and so there is no least positive real number.*]

2. Is $\dfrac{1}{0}$ an irrational number? Explain.

3. Use proof by contradiction to show that for all integers n, $3n + 2$ is not divisible by 3.

4. Use proof by contradiction to show that for all integers m, $7m + 4$ is not divisible by 7.

Carefully formulate the negations of each of the statements in 5–8. Then prove each statement by contradiction.

5. There is no greatest even integer.

6. There is no greatest negative real number.

7. There is no least positive rational number.

8. a. When asked to prove that the difference of any rational number and any irrational number is irrational, a student begins, "Suppose not. Suppose the difference of any rational number and any irrational number is rational." Comment.

   b. Prove by contradiction that the difference of any rational number and any irrational numbr is irrational.

Prove each statement in 9–15 by contradiction.

9. For all real numbers x and y, if x is irrational and y is rational then $x - y$ is irrational.

10. The product of any nonzero rational number and any irrational number is irrational.

11. If a and b are rational numbers, $b \neq 0$, and r is an irrational number, then $a + br$ is irrational.

H 12. For any integer n, $n^2 - 2$ is not divisible by 4.

H 13. For all prime numbers a, b, and c, $a^2 + b^2 \neq c^2$.

H 14. If a, b, and c are integers and $a^2 + b^2 = c^2$, then at least one of a and b is even.

H ✱ 15. For all odd integers a, b, and c, if z is a solution of $ax^2 + bx + c = 0$ then z is irrational.

16. Fill in the blanks in the following proof by contraposition that for all integers n, if $5 \nmid n^2$ then $5 \nmid n$.

    Proof (by contraposition): [*The contrapositive is: For all integers n, if $5 \mid n$ then $5 \mid n^2$.*] Suppose n is any integer such that __(a)__. [*We must show that __(b)__.*] By definition of divisibility, $n = $ __(c)__ for some integer k. By substitution, $n^2 = $ __(d)__ $= 5(5k^2)$. But $5k^2$ is an integer because it is a product of integers. Hence $n^2 = 5 \cdot$ (an integer), and so __(e)__ [*as was to be shown*].

Prove the statements in 17 and 18 by contraposition.

17. If a product of two positive real numbers is greater than 100, then at least one of the numbers is greater than 10.

18. If a sum of two real numbers is less than 50, then at least one of the numbers is less than 25.

19. Consider the statement "For all integers n, if n^2 is odd then n is odd."
   a. Write what you would suppose and what you would need to show to prove this statement by contradiction.
   b. Write what you would suppose and what you would need to show to prove this statement by contraposition.

20. Consider the statement "For all real numbers r, if r^2 is irrational then r is irrational."
   a. Write what you would suppose and what you would need to show to prove this statement by contradiction.
   b. Write what you would suppose and what you would need to show to prove this statement by contraposition.

Prove each of the statements in 21–26 in two ways: (a) by contraposition and (b) by contradiction.

21. The negative of any irrational number is irrational.

22. The reciprocal of any irrational number is irrational. (The **reciprocal** of a nonzero real number x is $1/x$.)

H **23.** For all integers n, if n^2 is odd then n is odd.

24. For all integers a, b, and c, if $a \nmid bc$ then $a \nmid b$. (Recall that the symbol $\nmid$ means "does not divide.")

H **25.** For all integers m and n, if $m + n$ is even then m and n are both even or m and n are both odd.

26. For all integers a, b, and c, if $a \mid b$ and $a \nmid c$, then $a \nmid (b + c)$. (*Hint:* To prove $p \rightarrow q \vee r$, it suffices to prove either $p \wedge \sim q \rightarrow r$ or $p \wedge \sim r \rightarrow q$. See exercise 14 in Section 1.2.)

27. The following "proof" that every integer is rational is incorrect. Find the mistake.

"**Proof (by contradiction):** Suppose not. Suppose every integer is irrational. Then the integer 1 is irrational. But

$1 = 1/1$, which is rational. This is a contradiction. [*Hence the supposition is false and the theorem is true.*]"

28. a. Use the properties of inequalities in Appendix A to prove that for all integers r, s, and n, if $r > \sqrt{n}$ and $s > \sqrt{n}$ then $rs > n$.

H **b.** Use proof by contraposition and the result of part (a) to show that for all integers $n > 1$, if n is not divisible by any positive integer that is greater than 1 and less than or equal to $\sqrt{n}$, then n is prime.

   c. Use proof by contraposition and the result of part (b) to show that for all integers $n > 1$, if n is not divisible by any prime number less than or equal to $\sqrt{n}$, then n is prime.

29. Use the result of exercise 28 to determine whether the following numbers are prime.
   a. 667   **b.** 557   c. 527   d. 613

30. The sieve of Eratosthenes, named after its inventor, the Greek scholar Eratosthenes (276–194 B.C.), provides a way to find all prime numbers less than or equal to some fixed number n. To construct it, write out all the integers from 2 to n. Cross out all multiples of 2 except 2 itself, then all multiples of 3 except 3 itself, then all multiples of 5 except 5 itself, and so forth. Continue crossing out the multiples of each successive prime number up to $\sqrt{n}$. The numbers that are not crossed out are all the prime numbers from 2 to n. Here is a sieve of Eratosthenes that includes the numbers from 2 to 27. The multiples of 2 are crossed out with a /, the multiples of 3 with a \, and the multiples of 5 with a —.

$$2 \quad 3 \quad \not{4} \quad 5 \quad \not{6} \quad 7 \quad \not{8} \quad \not{9} \quad \not{10} \quad 11 \quad \not{12} \quad 13 \quad \not{14}$$
$$\not{15} \quad \not{16} \quad 17 \quad \not{18} \quad 19 \quad \not{20} \quad \not{21} \quad \not{22} \quad 23 \quad \not{24} \quad \not{25} \quad \not{26} \quad \not{27}$$

Use the sieve of Eratosthenes to find all prime numbers less than 100.

31. Use the results of exercises 28 and 30 to determine whether the following numbers are prime.
   a. 9,269   **b.** 9,103   c. 8,623   d. 7,917

H ✶ **32.** Use proof by contradiction to show that every integer greater than 11 is a sum of two composite numbers.

3.7 Two Classical Theorems

How flat and dead would be a mind that saw nothing in a negation but an opaque barrier! A live mind can see a window onto a world of possibilities.
— Douglas Hofstadter, *Gödel, Escher, Bach*, 1979

This section contains proofs of two of the most famous theorems in mathematics: that $\sqrt{2}$ is irrational and that there are infinitely many prime numbers. Both proofs are examples of indirect arguments and were well known more than 2,000 years ago, but they remain exemplary models of mathematical argument to this day.

The Irrationality of $\sqrt{2}$

When mathematics flourished at the time of the ancient Greeks, mathematicians believed that given any two line segments, say A: ——— and B: ———, two integers, say a and b, could be found so that the ratio of the lengths of A and B would be in the same proportion as the ratio of a and b. Symbolically:

$$\frac{\text{length } A}{\text{length } B} = \frac{a}{b}.$$

Now it is easy to find a line segment of length $\sqrt{2}$; just take the diagonal of the unit square:

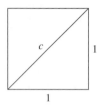

By the Pythagorean theorem, $c^2 = 1^2 + 1^2 = 2$, and so $c = \sqrt{2}$. If the belief of the ancient Greeks were correct, there would be integers a and b such that

$$\frac{\text{length (diagonal)}}{\text{length (side)}} = \frac{a}{b}.$$

And this would imply that

$$\frac{c}{1} = \frac{\sqrt{2}}{1} = \sqrt{2} = \frac{a}{b}.$$

But then $\sqrt{2}$ would be a ratio of two integers, or, in other words, $\sqrt{2}$ would be rational.

In the fourth or fifth century B.C., the followers of the Greek mathematician and philosopher Pythagoras discovered that $\sqrt{2}$ was not rational. This discovery was very upsetting to them, for it undermined their deep, quasi-religious belief in the power of whole numbers to describe phenomena.

The following proof of the irrationality of $\sqrt{2}$ was known to Aristotle and is similar to that in the tenth book of Euclid's *Elements of Geometry*. The Greek mathematician Euclid is best known as a geometer. In fact, knowledge of the geometry in the first six books of his *Elements* has been considered an essential part of a liberal education for more than 2,000 years. Books 7–10 of his *Elements*, however, contain much that we would now call number theory.

The proof begins by supposing the negation: $\sqrt{2}$ is rational. This means that there exist integers m and n such that $\sqrt{2} = m/n$. Now if m and n have any common factors, these may be factored out to obtain a new fraction, equal to m/n, in which the numerator and denominator have no common factors. (For example, $18/12 = (6 \cdot 3)/(6 \cdot 2) = 3/2$, which is a fraction whose numerator and denominator have no common factors.) Thus, without loss of generality, we may assume that m and n had no common factors in the first place.* We will then derive the contradiction that m and n *do* have a common factor of 2. The argument makes use of Proposition 3.6.4: If the square of an integer is even, then that integer is even.

Euclid
(fl. 300 B.C.)

CORBIS

---

*Strictly speaking, this deduction is a consequence of an axiom called the "well-ordering principle," which is discussed in Section 4.4.

Theorem 3.7.1 Irrationality of $\sqrt{2}$

$\sqrt{2}$ is irrational.

Proof:

[*We take the negation and suppose it to be true.*] Suppose not. That is, suppose $\sqrt{2}$ is rational. Then there are integers m and n with no common factors such that

$$\sqrt{2} = \frac{m}{n} \qquad 3.7.1$$

[*by dividing m and n by any common factors if necessary*]. [*We must derive a contradiction.*] Squaring both sides of equation (3.7.1) gives

$$2 = \frac{m^2}{n^2}.$$

Or, equivalently,

$$m^2 = 2n^2. \qquad 3.7.2$$

Note that equation (3.7.2) implies that m^2 is even (by definition of even). It follows that m is even (by Proposition 3.6.4). We file this fact away for future reference and also deduce (by definition of even) that

$$m = 2k \quad \text{for some integer } k. \qquad 3.7.3$$

Substituting equation (3.7.3) into equation (3.7.2), we see that

$$m^2 = (2k)^2 = 4k^2 = 2n^2.$$

Dividing both sides of the right-most equation by 2 gives

$$n^2 = 2k^2.$$

Consequently, n^2 is even, and so n is even (by Proposition 3.6.4). But we also know that m is even. [*This is the fact we filed away.*] Hence both m and n have a common factor of 2. But this contradicts the supposition that m and n have no common factors. [*Hence the supposition is false and so the theorem is true.*]

Now that you have seen the proof that $\sqrt{2}$ is irrational, you can easily derive the irrationality of certain other real numbers.

Example 3.7.1 Irrationality of $1 + 3\sqrt{2}$

Prove by contradiction that $1 + 3\sqrt{2}$ is irrational.

Solution The essence of the argument is the observation that if $1 + 3\sqrt{2}$ could be written as a fraction, then so could $\sqrt{2}$. But by Theorem 3.7.1, we know that to be impossible.

Proposition 3.7.2

$1 + 3\sqrt{2}$ is irrational.

Proof:

Suppose not. Suppose $1 + 3\sqrt{2}$ is rational. [*We must derive a contradiction.*] Then by definition of rational,

$$1 + 3\sqrt{2} = \frac{a}{b} \quad \text{for some integers } a \text{ and } b \text{ with } b \neq 0.$$

It follows that

$$3\sqrt{2} = \frac{a}{b} - 1 \qquad \text{by subtracting 1 from both sides}$$

$$= \frac{a}{b} - \frac{b}{b} \qquad \text{by substitution}$$

$$= \frac{a - b}{b} \qquad \begin{array}{l}\text{by the rule for subtracting fractions} \\ \text{with a common denominator.}\end{array}$$

Hence

$$\sqrt{2} = \frac{a - b}{3b} \qquad \text{by dividing both sides by 3.}$$

But $a - b$ and $3b$ are integers (since a and b are integers and differences and products of integers are integers), and $3b \neq 0$ by the zero product property. Hence $\sqrt{2}$ is a quotient of the two integers $a - b$ and $3b$ with $3b \neq 0$, and so $\sqrt{2}$ is rational (by definition of rational.) This contradicts the fact that $\sqrt{2}$ is irrational. [*This contradiction shows that the supposition is false.*] Hence $1 + 3\sqrt{2}$ is irrational.

∎

The Infinitude of the Set of Prime Numbers

You know that a prime number is a positive integer that cannot be factored as a product of two smaller positive integers. Is the set of all such numbers infinite, or is there a largest prime number? The answer was known to Euclid, and a proof that the set of all prime numbers is infinite appears in Book 9 of his *Elements of Geometry*.

Euclid's proof requires one additional fact we have not yet established: If a prime number divides an integer a, then it does not divide $a + 1$.

Proposition 3.7.3

For any integer a and any prime number p, if $p \mid a$ then $p \nmid (a + 1)$.

Proof:

Suppose not. That is, suppose there exists an integer a and a prime number p such that $p \mid a$ and $p \mid (a + 1)$. Then, by definition of divisibility, there exist integers r and s such that $a = pr$ and $a + 1 = ps$. It follows that $1 = (a + 1) - a = ps - pr = p(s - r)$, and so (since $s - r$ is an integer) $p \mid 1$. But the only integer divisors of 1 are 1 and -1 (see Example 3.3.4), and since p is prime, $p > 1$. Thus $p \leq 1$ and $p > 1$, which is a contradiction. [*Hence the supposition is false, and the proposition is true.*]

The idea of Euclid's proof is this: Suppose the set of prime numbers were finite. Then you could take the product of all the prime numbers and add one. By Theorem 3.3.2 this number must be divisible by some prime number. But by Proposition 3.7.3, this number is not divisible by any of the prime numbers in the set. Hence there must be a prime number that is not in the set of all prime numbers, which is impossible.

The following formal proof fills in the details of this outline.

Theorem 3.7.4 Infinitude of the Primes

The set of prime numbers is infinite.

Proof (by contradiction):

Suppose not. Suppose the set of prime numbers is finite. [*We must deduce a contradiction.*] Then all the prime numbers can be listed, say, in ascending order:

$$p_1 = 2, \ p_2 = 3, \ p_3 = 5, \ p_4 = 7, \ p_5 = 11, \ldots, \ p_n.$$

Consider the integer

$$N = p_1 p_2 p_3 \cdots p_n + 1.$$

Then $N > 1$, and so, by Theorem 3.3.2, N is divisible by some prime number p. Also, since p is prime, p must equal one of the prime numbers $p_1, p_2, p_3, \ldots, p_n$. Thus $p \mid (p_1 p_2 p_3 \cdots p_n)$. By Proposition 3.7.3, $p \nmid (p_1 p_2 p_3 \cdots p_n + 1)$, and so $p \nmid N$. Hence $p \mid N$ and $p \nmid N$, which is a contradiction. [*Hence the supposition is false and the theorem is true.*]

The proof of Theorem 3.7.4 shows that if you form the product of all prime numbers up to a certain point and add one, the result, N, is divisible by a prime number not on the list. The proof does not show that N is, itself, prime. In the exercises at the end of this section you are asked to find an example of an integer N constructed in this way that is not prime.

When to Use Indirect Proof

The examples in this section and Section 3.6 have not provided a definitive answer to the question of when to prove a statement directly and when to prove it indirectly. Many theorems can be proved either way. Usually, however, when both types of proof are possible, indirect proof is clumsier than direct proof. In the absence of obvious clues suggesting indirect argument, try first to prove a statement directly. Then, if that does not succeed, look for a counterexample. If the search for a counterexample is unsuccessful, look for a proof by contradiction or contraposition.

Open Questions in Number Theory

In this section we proved that there are infinitely many prime numbers. There is no known formula for obtaining primes, but a few formulas have been found to be more successful at producing them than other formulas. One such is due to Marin Mersenne, a French monk who lived from 1588–1648. *Mersenne primes* have the form $2^p - 1$, where p is prime. Not all numbers of this form are prime, but because of the greater likelihood of finding primes among them, those seeking large prime numbers often test these for primality. As a result, many of the largest known prime numbers are Mersenne primes.

An interesting question is whether there are infinitely many Mersenne primes. As of the date of publication of this book, the answer is not known, but new mathematical discoveries are being made every day and by the time you read this someone may have discovered the answer. Another formula that seems to produce a relatively large number of prime numbers is due to Fermat. *Fermat primes* are prime numbers of the form $2^{2^n} + 1$, where n is a positive integer. Are there infinitely many Fermat primes? Again, as of now, no one knows. Similarly unknown are whether there are infinitely many primes of the form $n^2 + 1$, where n is a positive integer, and whether there is always a prime number between integers n^2 and $(n + 1)^2$.

Another famous open question involving primes is the *twin primes conjecture,* which states that there are infinitely many pairs of prime numbers of the form p and $p + 2$. As with other well-known problems in number theory, this conjecture has withstood computer testing up to extremely large numbers. But compared with infinity, any number, no matter how large, is less than a drop in the ocean.

In 1844, the Belgian mathematician Eugène Catalan conjectured that the only solutions to the equation $x^n - y^m = 1$, where x, y, n, and m are all integers greater than 1, is $3^2 - 2^3 = 1$. This conjecture also remains unresolved to this day.

In 1993, while trying to prove Fermat's last theorem, an amateur number theorist, Andrew Beal, became intrigued by the equation $x^m + y^n = z^k$, where no two of x, y, or z have any common factor other than ± 1. When diligent effort, first by hand and then by computer, failed to reveal any solutions, Beal conjectured that no solutions exist. His conjecture has become known as *Beal's conjecture,* and he has offered a prize of $100,000 to anyone who can either prove or disprove it.

These are just a few of a large number of open questions in number theory. Many people believe that mathematics is a fixed subject that changes very little from one century to the next. In fact, more mathematical questions are being raised and more results are being discovered now than ever before in history.

Exercise Set 3.7

1. A calculator display shows that $\sqrt{2} = 1.414213562$, and $1.414213562 = \dfrac{1414213562}{1000000000}$. This suggests that $\sqrt{2}$ is a rational number, which contradicts Theorem 3.7.1. Explain the discrepancy.

2. Example 3.2.1(h) illustrates a technique for showing that any repeating decimal number is rational. A calculator display shows the result of a certain calculation as 40.72727272727. Can you be sure that the result of the calculation is a rational number? Explain.

Determine which statements in 3–13 are true and which are false. Prove those that are true and disprove those that are false.

3. $6 - 7\sqrt{2}$ is irrational.

4. $3\sqrt{2} - 7$ is irrational.

5. $\sqrt{4}$ is irrational.

6. $\sqrt{2}/6$ is rational.

7. The sum of any two irrational numbers is irrational.

8. The difference of any two irrational numbers is irrational.

9. The square root of an irrational number is irrational.

10. If r is any rational number and s is any irrational number, then r/s is irrational.

11. The sum of any two positive irrational numbers is irrational.

12. The product of any two irrational numbers is irrational.

H 13. If an integer greater than 1 is a perfect square, then its cube root is irrational.

14. Consider the following sentence: If x is rational then $\sqrt{x}$ is irrational. Is this sentence always true, sometimes true and sometimes false, or always false? Justify your answer.

15. a. Prove that for all integers a, if a^3 is even then a is even.
    b. Prove that $\sqrt[3]{2}$ is irrational.

16. a. Use proof by contradiction to show that for any integer n, it is impossible for n to equal both $3q_1 + r_1$ and $3q_2 + r_2$, where q_1, q_2, r_1, and r_2, are integers, $0 \le r_1 < 3, 0 \le r_2 < 3$, and $r_1 \ne r_2$.
    b. Use proof by contradiction, the quotient-remainder theorem, division into cases, and the result of part (a) to prove that for all integers n, if n^2 is divisible by 3 then n is divisible by 3.
    c. Prove that $\sqrt{3}$ is irrational.

17. Give an example to show that if d is not prime and n^2 is divisible by d, then n need not be divisible by d.

H **18.** The quotient-remainder theorem says not only that there exist quotients and remainders but also that the quotient and remainder of a division are unique. Prove the uniqueness. That is, prove that if a and d are integers with $d > 0$ and if $q_1, r_1, q_2,$ and r_2 are integers such that

$$a = dq_1 + r_1 \quad \text{where } 0 \leq r_1 < d$$

and

$$a = dq_2 + r_2 \quad \text{where } 0 \leq r_2 < d,$$

then

$$q_1 = q_2 \quad \text{and} \quad r_1 = r_2.$$

H **19.** Prove that $\sqrt{5}$ is irrational.

H **20.** Prove that for any integer a, $9 \nmid (a^2 - 3)$.

21. a. Use the unique factorization theorem to answer the following question: If the prime factorization of an integer n contains k occurrences of a prime number p, how many occurrences of p are contained in the prime factorization of n^2?

  b. An alternative proof of the irrationality of $\sqrt{2}$ counts the number of 2's on the two sides of the equation $2n^2 = m^2$ and deduces a contradiction. Write a proof that uses this approach.

22. Use the proof technique illustrated in exercise 21(b) to prove that if n is any integer that is not a perfect square, then $\sqrt{n}$ is irrational.

H **23.** Prove that $\sqrt{2} + \sqrt{3}$ is irrational.

✱ **24.** Prove that $\log_5(2)$ is irrational.

H **25.** Let $N = 2 \cdot 3 \cdot 5 \cdot 7 + 1$. What remainder is obtained when N is divided by 2? 3? 5? 7? Is N prime? Justify your answer.

H **26.** Suppose a is an integer and p is a prime number such that $p \mid a$ and $p \mid (a + 3)$. What can you deduce about p? Why?

27. Let $p_1, p_2, p_3, \ldots$ be a list of all prime numbers in ascending order. Here is a table of the first six:

p_1	p_2	p_3	p_4	p_5	p_6
2	3	5	7	11	13

  H **a.** For each $i = 1, 2, 3, 4, 5, 6$, let $N_i = p_1 p_2 \cdots p_i + 1$. Calculate $N_1, N_2, N_3, N_4, N_5,$ and N_6.

  b. For each $i = 1, 2, 3, 4, 5, 6$, find the smallest prime number q_i such that q_i divides N_i.

28. An alternative proof of the infinitude of the prime numbers begins as follows:

Proof: Suppose there are only finitely many prime numbers. Then one is the largest. Call it p. Let $M = p! + 1$. We will show that there is a prime number q such that $q > p$.

Complete this proof.

H ✱ **29.** Prove that if $p_1, p_2, \ldots,$ and p_n are distinct prime numbers with $p_1 = 2$ and $n > 1$, then $p_1 p_2 \cdots p_n + 1$ can be written in the form $4k + 3$ for some integer k.

H ✱ **30.** Prove that for all integers n, if $n > 2$ then there is a prime number p such that $n < p < n!$.

$$(n! = n(n - 1) \cdots 3 \cdot 2 \cdot 1)$$

H **31. a.** Fermat's last theorem says that for all integers $n > 2$, the equation $x^n + y^n = z^n$ has no positive integer solution (solution for which x, y, and z are positive integers). Prove the following: If for all prime numbers $p > 2$, $x^p + y^p = z^p$ has no positive integer solution, then for any integer $n > 2$ that is not a power of 2, $x^n + y^n = z^n$ has no positive integer solution.

  b Fermat proved that there are no integers x, y, and z such that $x^4 + y^4 = z^4$. Use this result to remove the restriction in part (a) that n not be a power of 2. That is, prove that if n is a power of 2 and $n > 4$, then $x^n + y^n = z^n$ has no positive integer solution.

For exercises 32–35 note that to show there is a unique object with a certain property, show that (1) there is an object with the property and (2) if objects A and B have the property, then $A = B$.

32. Prove that there exists a unique prime number of the form $n^2 - 1$, where n is an integer that is greater than or equal to 2.

33. Prove that there exists a unique prime number of the form $n^2 + 2n - 3$, where n is a positive integer.

34. Prove that there is at most one real number a with the property that $a + r = r$ for all real numbers r. (Such a number is called an *additive identity*.)

35. Prove that there is at most one real number b with the property that $br = r$ for all real numbers r. (Such a number is called a *multiplicative identity*.)

3.8 Application: Algorithms

Begin at the beginning . . . and go on till you come to the end: then stop.
— Lewis Carroll, *Alice's Adventures in Wonderland,* 1865

Lady Lovelace
(1815–1852)

Getty Images

In this section we will show how the number theory facts developed in this chapter form the basis for some useful computer algorithms.

The word *algorithm* refers to a step-by-step method for performing some action. Some examples of algorithms in everyday life are food preparation recipes, directions for assembling equipment or hobby kits, sewing pattern instructions, and instructions for filling out income tax forms. Much of elementary school mathematics is devoted to learning algorithms for doing arithmetic such as multidigit addition and subtraction, multidigit (or long) multiplication, and long division.

The idea of a computer algorithm is credited to Ada Augusta, Countess of Lovelace. Trained as a mathematician, she became very interested in Charles Babbage's design for an "Analytical Engine," a machine similar in concept to a modern computer. Lady Lovelace extended Babbage's explorations of how such a machine would operate, recognizing that its importance lay "in the possibility of using a given sequence of instructions repeatedly, the number of times being either preassigned or dependent on the results of the computation." This is the essence of a modern computer algorithm.

An Algorithmic Language

The algorithmic language used in this book is a kind of pseudocode, combining elements of Pascal, C, Java, and VB.NET, and ordinary, but fairly precise, English. We will use some of the formal constructs of computer languages—such as assignment statements, loops, and so forth—but we will ignore the more technical details, such as the requirement for explicit end-of-statement delimiters, the range of integer values available on a particular installation, and so forth. The algorithms presented in this text are intended to be precise enough to be easily translated into virtually any high-level computer language.

In high-level computer languages, the term **variable** is used to refer to a specific storage location in a computer's memory. To say that the variable x has the value 3 means that the memory location corresponding to x contains the number 3. A given storage location can hold only one value at a time. So if a variable is given a new value during program execution, then the old value is erased. The **data type** of a variable indicates the set in which the variable takes its values, whether the set of integers, or real numbers, or character strings, or the set $\{0, 1\}$ (for a Boolean variable), and so forth.

An **assignment statement** gives a value to a variable. It has the form

$$x := e,$$

where x is a variable and e is an expression. This is read "x is assigned the value e" or "let x be e." When an assignment statement is executed, the expression e is evaluated (using the current values of all the variables in the expression), and then its value is placed in the memory location corresponding to x (replacing any previous contents of this location).

Ordinarily, algorithm statements are executed one after another in the order in which they are written. **Conditional statements** allow this natural order to be overridden by using the current values of program variables to determine which algorithm statement will be executed next. Conditional statements are denoted either

a. **if** (*condition*)   or   b. **if** (*condition*) **then** s_1

   then s_1

   else s_2

where *condition* is a predicate involving algorithm variables and where s_1 and s_2 are algorithm statements or groups of algorithm statements. We generally use indentation to indicate that statements belong together as a unit. When ambiguity is possible, however, we may explicitly bind a group of statements together into a unit by preceding the group with the word **do** and following it with the words **end do.**

Execution of an **if-then-else** statement occurs as follows:

1. The *condition* is evaluated by substituting the current values of all algorithm variables appearing in it and evaluating the truth or falsity of the resulting statement

2. If *condition* is true, then s_1 is executed and execution moves to the next algorithm statement following the **if-then-else** statement.

3. If *condition* is false, then s_2 is executed and moves to the next algorithm statement following the **if-then-else** statement.

Execution of an **if-then** statement is similar to execution of an **if-then-else** statement, except that if *condition* is false, execution passes immediately to the next algorithm statement following the **if-then** statement.

Often *condition* is called a **guard** because it is stationed before s_1 and s_2 and restricts access to them.

Example 3.8.1 Execution of if-then-else and if-then Statements

Consider the following algorithm segments:

a. **if** $x > 2$
    then $y := x + 1$
    else do $x := x - 1$
        $y := 3 \cdot x$ **end do**

b. $y := 0$
    if $x > 2$ **then** $y := 2^x$

What is the value of y after execution of these segments for the following values of x?

i. $x = 5$ ii. $x = 2$

Solution

a. (i) Because the value of x is 5 before execution, the guard condition $x > 2$ is true at the time it is evaluated. Hence the statement following **then** is executed, and so the value of $x + 1 = 5 + 1$ is computed and placed in the storage location corresponding to y. So after execution, $y = 6$.

   (ii) Because the value of x is 2 before execution, the guard condition $x > 2$ is false at the time it is evaluated. Hence the statement following **else** is executed. The value of $x - 1 = 2 - 1$ is computed and placed in the storage location corresponding to x, and the value of $3 \cdot x = 3 \cdot 1$ is computed and placed in the storage location corresponding to y. So after execution, $y = 3$.

b. (i) Since $x = 5$ initially, the condition $x > 2$ is true at the time it is evaluated. So the statement following **then** is executed, and y obtains the value $2^5 = 32$.

   (ii) Since $x = 2$ initially, the condition $x > 2$ is false at the time it is evaluated. Execution, therefore, moves to the next statement following the if-then statement, and the value of y does not change from its initial value of 0. ∎

Iterative statements are used when a sequence of algorithm statements is to be executed over and over again. We will use two types of iterative statements: **while** loops and **for-next** loops.

A **while** loop has the form

while (*condition*)

[*statements that make up
the body of the loop*]

end while

where *condition* is a predicate involving algorithm variables. The word **while** marks the beginning of the loop, and the words **end while** mark its end. Execution of a **while** loop occurs as follows:

1. The *condition* is evaluated by substituting the current values of all the algorithm variables and evaluating the truth or falsity of the resulting statement.

2. If *condition* is true, all the statements in the body of the loop are executed in order. Then execution moves back to the beginning of the loop and the process repeats.

3. If *condition* is false, execution passes to the next algorithm statement following the loop.

The loop is said to be **iterated** (IT-a-rate-ed) each time the statements in the body of the loop are executed. Each execution of the body of the loop is called an **iteration** (it-er-AY-shun) of the loop.

Example 3.8.2 Tracing Execution of a **while** Loop

Trace the execution of the following algorithm segment by finding the values of all the algorithm variables each time they are changed during execution:

$$i := 1, s := 0$$

while ($i \leq 2$)

$$s := s + i$$

$$i := i + 1$$

end while

Solution   Since i is given an initial value of 1, the condition $i \leq 2$ is true when the **while** loop is entered. So the statements within the loop are executed in order:

$$s = 0 + 1 = 1 \quad \text{and} \quad i = 1 + 1 = 2.$$

Then execution passes back to the beginning of the loop.

The condition $i \leq 2$ is evaluated using the current value of i, which is 2. The condition is true, and so the statements within the loop are executed again:

$$s = 1 + 2 = 3 \quad \text{and} \quad i = 2 + 1 = 3.$$

Then execution passes back to the beginning of the loop.

The condition $i \leq 2$ is evaluated using the current value of i, which is 3. This time the condition is false, and so execution passes beyond the loop to the next statement of the algorithm.

This discussion can be summarized in a table, called a **trace table,** that shows the current values of algorithm variables at various points during execution. The trace table for a **while** loop generally gives all values immediately following each iteration of the loop. ("After the zeroth iteration" means the same as "before the first iteration.")

Trace Table

	Iteration Number		
Variable Name	0	1	2
i	1	2	3
s	0	1	3

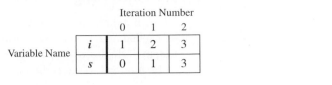

The second form of iteration we will use is a **for-next** loop. A **for-next** loop has the following form:

> **for** *variable* := *initial expression* **to** *final expression*
>
> [*statements that make up the body of the loop*]
>
> **next** *(same) variable*

A **for-next** loop is executed as follows:

1. The **for-next** loop *variable* is set equal to the value of *initial expression*.

2. A check is made to determine whether the value of *variable* is less than or equal to the value of *final expression*.

3. If the value of *variable* is less than or equal to the value of *final expression*, then the statements in the body of the loop are executed, *variable* is increased by 1, and execution returns back to step 2.

4. If the value of *variable* is greater than the value of *final expression*, then execution passes to the next algorithm statement following the loop.

Example 3.8.3 Trace Table for a **for-next** Loop

Convert the **for-next** loop shown below into a **while** loop. Construct a trace table for the loop.

> **for** $i := 1$ **to** 4
>
> $x := i^2$
>
> **next** i

Solution   The given **for-next** loop is equivalent to the following:

> $i := 1$
>
> **while** $(i \leq 4)$
>
> $x := i^2$
>
> $i := i + 1$
>
> **end while**

Its trace table is as follows:

Trace Table

	Iteration Number				
Variable Name	0	1	2	3	4
x		1	4	9	16
i	1	2	3	4	5

A Notation for Algorithms

We will express algorithms as subroutines that can be called upon by other algorithms as needed and used to transform a set of input variables with given values into a set of output variables with specific values. The output variables and their values are assumed to be returned to the calling algorithm. For example, the division algorithm specifies a procedure for taking any two positive integers as input and producing the quotient and remainder of the division of one number by the other as output. Whenever an algorithm requires such a computation, the algorithm can just "call" the division algorithm to do the job.

We generally include the following information when describing algorithms formally:

1. The name of the algorithm, together with a list of input and output variables.

2. A brief description of how the algorithm works.

3. The input variable names, labeled by data type (whether integer, real number, and so forth).

4. The statements that make up the body of the algorithm, possibly with explanatory comments.

5. The output variable names, labeled by data type.

al-Khowârizmî
(ca. 780–850)

You may wonder where the word *algorithm* came from. It evolved from the last part of the name of the Persian mathematician Abu Ja'far Mohammed ibn Mûsâ al-Khowârizmî. During Europe's Dark Ages, the Arabic world enjoyed a period of intense intellectual activity. One of the great mathematical works of that period was a book written by al-Khowârizmî that contained foundational ideas for the subject of algebra. The translation of this book into Latin in the thirteenth century had a profound influence on the development of mathematics during the European Renaissance.

The Division Algorithm

For an integer a and a positive integer d, the quotient-remainder theorem guarantees the existence of integers q and r such that

$$a = dq + r \quad \text{and} \quad 0 \leq r < d.$$

In this section, we give an algorithm to calculate q and r for given a and d where a is nonnegative. (The extension to negative a is left to the exercises at the end of this section.) The following example illustrates the idea behind the algorithm. Consider trying to find the quotient and the remainder of the division of 32 by 9, but suppose that you do not remember your multiplication table and have to figure out the answer from basic principles. The quotient represents that number of 9's that are contained in 32. The remainder is the number left over when all possible groups of 9 are subtracted. Thus you can calculate the quotient and remainder by repeatedly subtracting 9 from 32 until you obtain a number less than 9:

$$32 - 9 = 23 \geq 9, \text{ and}$$
$$32 - 9 - 9 = 14 \geq 9, \text{ and}$$
$$32 - 9 - 9 - 9 = 5 < 9.$$

This shows that 3 groups of 9 can be subtracted from 32 with 5 left over. Thus the quotient is 3 and the remainder is 5.

Algorithm 3.8.1 Division Algorithm

[*Given a nonnegative integer a and a positive integer d, the aim of the algorithm is to find integers q and r that satisfy the conditions $a = dq + r$ and $0 \leq r < d$. This is done by subtracting d repeatedly from a until the result is less than d but is still nonnegative.*

$$0 \leq a - d - d - d - \cdots - d = a - dq < d.$$

The total number of d's that are subtracted is the quotient q. The quantity $a - dq$ equals the remainder r.]

Input: a [*a nonnegative integer*], d [*a positive integer*]

Algorithm Body:

$r := a, q := 0$

[*Repeatedly subtract d from r until a number less than d is obtained. Add 1 to q each time d is subtracted.*]

while $(r \geq d)$

$r := r - d$

$q := q + 1$

end while

[*After execution of the* **while** *loop, $a = dq + r$.*]

Output: q, r [*nonnegative integers*]

Note that the values of q and r obtained from the division algorithm are the same as those computed by the *div* and *mod* functions built into a number of computer languages. That is, if q and r are the quotient and remainder obtained from the division algorithm with input a and d, then the output variables q and r satisfy

$$q = a \; div \; d \quad \text{and} \quad r = a \; mod \; d.$$

The next example asks for a trace of the division algorithm.

Example 3.8.4 Tracing the Division Algorithm

Trace the action of Algorithm 3.8.1 on the input variables $a = 19$ and $d = 4$.

Solution   Make a trace table as shown below. The column under the kth iteration gives the states of the variables after the kth iteration of the loop.

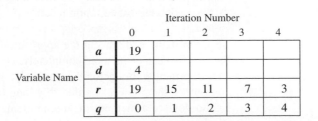

	Iteration Number				
	0	1	2	3	4
a	19				
d	4				
r	19	15	11	7	3
q	0	1	2	3	4

Variable Name

The Euclidean Algorithm

The greatest common divisor of two integers a and b is the largest integer that divides both a and b. For example, the greatest common divisor of 12 and 30 is 6. The Euclidean algorithm provides a very efficient way to compute the greatest common divisor of two integers.

• Definition

Let a and b be integers that are not both zero. The **greatest common divisor** of a and b, denoted **gcd**(a, b), is that integer d with the following properties:

1. d is a common divisor of both a and b. In other words,

$$d \mid a \quad \text{and} \quad d \mid b.$$

2. For all integers c, if c is a common divisor of both a and b, then c is less than or equal to d. In other words,

$$\text{for all integers } c, \text{ if } c \mid a \text{ and } c \mid b, \text{ then } c \leq d.$$

Example 3.8.5 Calculating Some gcd's

　　a. Find gcd(72, 63).

　　b. Find gcd(10^{20}, 6^{30}).

　　c. In the definition of greatest common divisor, gcd(0, 0) is not allowed. Why not? What would gcd(0, 0) equal if it were found in the same way as the greatest common divisors for other pairs of numbers?

Solution

　　a. $72 = 9 \cdot 8$ and $63 = 9 \cdot 7$. So $9 \mid 72$ and $9 \mid 63$, and no integer larger than 9 divides both 72 and 63. Hence gcd(72, 63) $= 9$.

　　b. By the laws of exponents, $10^{20} = 2^{20} \cdot 5^{20}$ and $6^{30} = 2^{30} \cdot 3^{30} = 2^{20} \cdot 2^{10} \cdot 3^{30}$. It follows that

$$2^{20} \mid 10^{20} \quad \text{and} \quad 2^{20} \mid 6^{30},$$

and by the unique factorization theorem, no integer larger than 2^{20} divides both 10^{20} and 6^{30} (because no more than twenty 2's divide 10^{20}, no 3's divide 10^{20}, and no 5's divide 6^{30}). Hence gcd(10^{20}, 6^{30}) $= 2^{20}$.

　　c. Suppose gcd(0, 0) were defined to be the largest common factor that divides 0 and 0. The problem is that *every* positive integer divides 0 and there is no largest integer. So there is no largest common factor! ■

　　Calculating gcd's using the approach illustrated in Example 3.8.5 works only when the numbers can be factored completely. By the unique factorization theorem, all numbers can, in principle, be factored completely. But, in practice, even using the highest-speed computers, the process is unfeasibly long for very large integers. Over 2,000 years ago, Euclid devised a method for finding greatest common divisors that is easy to use and is much more efficient than either factoring the numbers or repeatedly testing both numbers for divisibility by successively larger integers.

The Euclidean algorithm is based on the following two facts, which are stated as lemmas. A lemma is a statement that does not have much intrinsic interest but helps prove a theorem.

Lemma 3.8.1

If r is a positive integer, then $\gcd(r, 0) = r$.

Proof:

Suppose r is a positive integer. [*We must show that the greatest common divisor of both r and 0 is r.*] Certainly, r is a common divisor of both r and 0 because r divides itself and also r divides 0 (since every positive integer divides 0). Also no integer larger than r can be a common divisor of r and 0 (since no integer larger than r can divide r). Hence r is the greatest common divisor of r and 0.

The proof of the second lemma is based on a clever pattern of argument that is used in many different areas of mathematics: To prove that $A = B$, prove that $A \leq B$ and that $B \leq A$.

Lemma 3.8.2

If a and b are any integers with $b \neq 0$ and q and r are nonnegative integers such that

$$a = bq + r,$$

then

$$\gcd(a, b) = \gcd(b, r).$$

Proof:

[*The proof is divided into two sections: (1) proof that $\gcd(a, b) \leq \gcd(b, r)$, and (2) proof that $\gcd(b, r) \leq \gcd(a, b)$. Since each gcd is less than or equal to the other, the two must be equal.*]

1. **$\gcd(a, b) \leq \gcd(b, r)$:**

   a. [*We will first show that any common divisor of a and b is also a common divisor of b and r.*]

   Let c be a common divisor of a and b. Then $c \mid a$ and $c \mid b$, and so, by definition of divisibility, $a = nc$ and $b = mc$, for some integers n and m. Now substitute into the equation

   $$a = bq + r$$

   to obtain

   $$nc = (mc)q + r.$$

   Then solve for r:

   $$r = nc - (mc)q = (n - mq)c.$$

   But $n - mq$ is an integer, and so, by definition of divisibility, $c \mid r$. Because we already know that $c \mid b$, we can conclude that c is a common divisor of b and r [*as was to be shown*].

continued on page 194

b. [*Next we show that* $\gcd(a, b) \leq \gcd(b, r)$.]

By part (a), every common divisor of a and b is a common divisor of b and r. It follows that the greatest common divisor of a and b is a common divisor of b and r. But then $\gcd(a, b)$ (being one of the common divisors of b and r) is less than or equal to the greatest common divisor of b and r:

$$\gcd(a, b) \leq \gcd(b, r).$$

2. **$\gcd(b, r) \leq \gcd(a, b)$:**

The second part of the proof is very similar to the first part. It is left as an exercise.

The Euclidean algorithm can be described as follows:

1. Let A and B be integers with $A > B \geq 0$.

2. To find the greatest common divisor of A and B, first check whether $B = 0$. If it is, then $\gcd(A, B) = A$ by Lemma 3.8.1. If it isn't, then $B > 0$ and the quotient-remainder theorem can be used to divide A by B to obtain a quotient q and a remainder r:

$$A = Bq + r \quad \text{where } 0 \leq r < B.$$

By Lemma 3.8.2, $\gcd(A, B) = \gcd(B, r)$. Thus the problem of finding the greatest common divisor of A and B is reduced to the problem of finding the greatest common divisor of B and r.

What makes this piece of information useful is that B and r are smaller numbers than A and B. To see this, recall that we assumed

$$A > B \geq 0.$$

Also the r found by the quotient-remainder theorem satisfies

$$0 \leq r < B.$$

Putting these two inequalities together gives

$$0 \leq r < B < A.$$

So the larger number of the pair (B, r) is smaller than the larger number of the pair (A, B).

3. Now just repeat the process, starting àgain at (2), but use B instead of A and r instead of B. The repetitions are guaranteed to terminate eventually with $r = 0$ because each new remainder is less than the preceding one and all are nonnegative.*

By the way, it is always the case that the number of steps required in the Euclidean algorithm is at most five times the number of digits in the smaller integer. This was proved by the French mathematician Gabriel Lamé (1795–1870).

---

*Strictly speaking, this statement is justified by an axiom for the integers called the well-ordering principle, which is discussed in Section 4.4.

The following example illustrates how to use the Euclidean algorithm.

Example 3.8.6 Hand-Calculation of gcd's Using the Euclidean Algorithm

Use the Euclidean algorithm to find gcd(330, 156).

Solution

1. Divide 330 by 156:

Thus $330 = 156 \cdot 2 + 18$ and hence $\gcd(330, 156) = \gcd(156, 18)$ by Lemma 3.8.2.

2. Divide 156 by 18:

Thus $156 = 18 \cdot 8 + 12$ and hence $\gcd(156, 18) = \gcd(18, 12)$ by Lemma 3.8.2.

3. Divide 18 by 12:

Thus $18 = 12 \cdot 1 + 6$ and hence $\gcd(18, 12) = \gcd(12, 6)$ by Lemma 3.8.2.

4. Divide 12 by 6:

$$6 \overline{)12} \quad \begin{array}{l} 2 \leftarrow \text{quotient} \\ 12 \\ \overline{0} \leftarrow \text{remainder} \end{array}$$

Thus $12 = 6 \cdot 2 + 0$ and hence $\gcd(12, 6) = \gcd(6, 0)$ by Lemma 3.8.2.

Putting all the equations above together gives

$$\begin{aligned}
\gcd(330, 156) &= \gcd(156, 18) \\
&= \gcd(18, 12) \\
&= \gcd(12, 6) \\
&= \gcd(6, 0) \\
&= 6 \qquad \text{by Lemma 3.8.1.}
\end{aligned}$$

Therefore, $\gcd(330, 156) = 6$. ∎

The following is a version of the Euclidean algorithm written using formal algorithm notation.

Algorithm 3.8.2 Euclidean Algorithm

[*Given two integers A and B with $A > B \geq 0$, this algorithm computes $\gcd(A, B)$. It is based on two facts:*

1. $\gcd(a, b) = \gcd(b, r)$ *if a, b, q, and r are integers with $a = b \cdot q + r$ and $0 \leq r < b$.*

2. $\gcd(a, 0) = a$.]

Input: A, B [*integers with $A > B \geq 0$*]

Algorithm Body:

$a := A, b := B, r := B$
[*If $b \neq 0$, compute $a \bmod b$, the remainder of the integer division of a by b, and set r equal to this value. Then repeat the process using b in place of a and r in place of b.*]

while ($b \neq 0$)

$r := a \bmod b$
[*The value of $a \bmod b$ can be obtained by calling the division algorithm.*]

$a := b$

$b := r$

end while

[*After execution of the **while** loop, $\gcd(A, B) = a$.*]
$\gcd := a$

Output: $\gcd$ [*a positive integer*]

Exercise Set 3.8

Find the value of z when each of the algorithm segments in 1 and 2 is executed.

1. $i := 2$

if ($i > 3$ or $i \leq 0$)

then $z := 1$

else $z := 0$

2. $i := 3$

if ($i \leq 3$ or $i > 6$)

then $z := 2$

else $z := 0$

3. Consider the following algorithm segment:

if $x \cdot y > 0$ **then do** $y := 3 \cdot x$

$x := x + 1$ **end do**

$z := x \cdot y$

Find the value of z if prior to execution x and y have the values given below.

a. $x = 2, y = 3$ b. $x = 1, y = 1$

Find the values of a and e after execution of the loops in 4 and 5:

4. $a := 2$

for $i := 1$ **to** 2

$a := \dfrac{a}{2} + \dfrac{1}{a}$

next i

5. $e := 0, f := 2$

for $j := 1$ **to** 4

$f := f \cdot j$

$e := e + \dfrac{1}{f}$

next j

Make a trace table to trace the action of Algorithm 3.8.1 for the input variables given in 6 and 7.

6. $a = 26, d = 7$ 7. $a = 59, d = 13$

8. The following algorithm segment makes change; given an amount of money A between 1¢ and 99¢, it determines a breakdown of A into quarters (q), dimes (d), nickels (n), and pennies (p).

$$q := A \ div \ 25$$
$$A := A \ mod \ 25$$
$$d := A \ div \ 10$$
$$A := A \ mod \ 10$$
$$n := A \ div \ 5$$
$$p := A \ mod \ 5$$

   a. Trace this algorithm segment for $A = 69$.
   b. Trace this algorithm segment for $A = 87$.

Find the greatest common divisor of each of the pairs of integers in 9–12. (Use any method you wish.)

   9. 27 and 72   **10.** 5 and 9

   11. 7 and 21   12. 48 and 54

Use the Euclidean algorithm to hand-calculate the greatest common divisors of each of the pairs of integers in 13–16.

   13. 1,188 and 385   **14.** 509 and 1,177

   15. 832 and 10,933   16. 4,131 and 2,431

Make a trace table to trace the action of Algorithm 3.8.2 for the input variables given in 17 and 18.

   17. 1,001 and 871   18. 5,859 and 1,232

H 19. Prove that for all positive integers a and b, $a \mid b$ if, and only if, $\gcd(a, b) = a$. (Note that to prove "A if, and only if, B," you need to prove "if A then B" and "if B then A.")

20. Write an algorithm that accepts the numerator and denominator of a fraction as input and produces as output the numerator and denominator of that fraction written in lowest terms. (The algorithm may call upon the Euclidean algorithm as needed.)

21. Complete the proof of Lemma 3.8.2 by proving the following: If a and b are any positive integers and q and r are any integers such that

$$a = bq + r \quad \text{and} \quad 0 \leq r < b,$$

   then

$$\gcd(b, r) \leq \gcd(a, b).$$

H 22. a. Prove: If a and d are positive integers and q and r are integers such that $a = dq + r$ and $0 < r < d$, then

$$-a = d(-(q + 1)) + (d - r)$$

   and

$$0 < d - r < d.$$

   b. Indicate how to modify Algorithm 3.8.1 to allow for the input a to be negative.

23. **a.** Prove that if a, d, q, and r are integers such that $a = dq + r$ and $0 \leq r < d$, then

$$q = \lfloor a/d \rfloor \quad \text{and} \quad r = a - \lfloor a/d \rfloor \cdot d.$$

   b. In a computer language with a built-in-floor function, div and mod can be calculated as follows:

$$a \ div \ d = \lfloor a/d \rfloor \quad \text{and} \quad a \ mod \ d = a - \lfloor a/d \rfloor \cdot d.$$

   Rewrite the steps of Algorithm 3.8.2 for a computer language with a built-in floor function but without div and mod.

24. An alternative to the Euclidean algorithm uses subtraction rather than division to compute greatest common divisors. (After all, division is repeated subtraction.) It is based on the following lemma:

Lemma 3.8.3

If $a \geq b > 0$, then $\gcd(a, b) = \gcd(b, a - b)$.

Algorithm 3.8.3 Computing gcd's by Subtraction

[*Given two positive integers A and B, variables a and b are set equal to A and B. Then a repetitive process begins. If $a \neq 0$, and $b \neq 0$, then the larger of a and b is set equal to $a - b$ (if $a \geq b$) or to $b - a$ (if $a < b$), and the smaller of a and b is left unchanged. This process is repeated over and over until eventually a or b becomes 0. By Lemma 3.8.3, after each repetition of the process,*

$$\gcd(A, B) = \gcd(a, b).$$

After the last repetition,

$$\gcd(A, B) = \gcd(a, 0) \quad or \quad \gcd(A, B) = \gcd(0, b)$$

depending on whether a or b is nonzero. But by Lemma 3.8.1,

$$\gcd(a, 0) = a \quad and \quad \gcd(0, b) = b.$$

Hence, after the last repetition,

$$\gcd(A, B) = a \ if \ a \neq 0 \quad or \quad \gcd(A, B) = b \ if \ b \neq 0.]$$

Input: A, B [*positive integers*]
Algorithm Body:

   $a := A, b := B$
   while $(a \neq 0$ and $b \neq 0)$
       if $a \geq b$ **then** $a := a - b$
           else $b := b - a$
   end while
       if $a = 0$ **then** $\gcd := b$
       else $\gcd := a$
       [*After execution of the* **if-then-else** *statement,* $\gcd = \gcd(A, B)$.]

Output: gcd [*a positive integer*]

---

a. Prove Lemma 3.8.3.

b. Trace the execution of Algorithm 3.8.3 for $A = 630$ and $B = 336$.

c. Trace the execution of Algorithm 3.8.3 for $A = 768$ and $B = 348$.

Exercises 25–29 refer to the following definition.

Definition: The **least common multiple** of two nonzero integers a and b, denoted $\mathbf{lcm}(a, b)$, is the positive integer c such that

a. $a \mid c$ and $b \mid c$

b. for all integers m, if $a \mid m$ and $b \mid m$, then $c \mid m$.

---

25. Find
    a. lcm(12, 18) b. lcm($2^2 \cdot 3 \cdot 5, 2^3 \cdot 3^2$)
    c. lcm(2800, 6125)

H 26. Prove that for all positive integers a and b, $\gcd(a, b) = \text{lcm}(a, b)$ if, and only if $a = b$.

27. Prove that for all positive integers a and b, $a \mid b$ if, and only if, $\text{lcm}(a, b) = b$.

H 28. Prove that for all integers a and b, $\gcd(a, b) \mid \text{lcm}(a, b)$.

29. Prove that for all positive integers a and b, $\gcd(a, b) \cdot \text{lcm}(a, b) = ab$.

CHAPTER 4

SEQUENCES AND MATHEMATICAL INDUCTION

One of the most important tasks of mathematics is to discover and characterize regular patterns, such as those associated with processes that are repeated. The main mathematical structure used to study repeated processes is the *sequence,* and the main mathematical tool used to verify conjectures about patterns governing the arrangement of terms in sequences is *mathematical induction.* In this chapter we introduce the notation and terminology of sequences, show how to use both the ordinary and the strong forms of mathematical induction, and give an application showing how to prove the correctness of computer algorithms.

4.1 Sequences

A mathematician, like a painter or poet, is a maker of patterns.
— G. H. Hardy, *A Mathematician's Apology,* 1940

Imagine that a person decides to count his ancestors. He has two parents, four grandparents, eight great-grandparents, and so forth, These numbers can be written in a row as

$$2, 4, 8, 16, 32, 64, 128, \ldots$$

The symbol ".." is called an *ellipsis.* It is shorthand for "and so forth."

To express the pattern of the numbers, suppose that each is labeled by an integer giving its position in the row.

Position in the row	1	2	3	4	5	6	7...
Number of ancestors	2	4	8	16	32	64	128...

The number corresponding to position 1 is 2, which equals 2^1. The number corresponding to position 2 is 4, which equals 2^2. For positions 3, 4, 5, 6, and 7, the corresponding numbers are 8, 16, 32, 64, and 128, which equal $2^3, 2^4, 2^5, 2^6,$ and 2^7, respectively. For a general value of k, let A_k be the number of ancestors in the kth generation back. The pattern of computed values strongly suggests the following for each k:

$$A_k = 2^k.^*$$

---

*Strictly speaking, the true value of A_k is probably less than 2^k when k is large, because ancestors from one branch of the family tree may also appear on other branches of the tree.

In this section we define the term **sequence** informally as a set of elements written in a row. (We give a more formal definition of sequence in terms of functions in Section 7.1.) In the sequence denoted

$$a_m, a_{m+1}, a_{m+2}, \ldots, a_n,$$

each individual element a_k (read "a sub k") is called a **term.** The k in a_k is called a **subscript** or **index,** m (which may be any integer) is the subscript of the **initial term,** and n (which must be greater than or equal to m) is the subscript of the **final term.** The notation

$$a_m, a_{m+1}, a_{m+2}, \ldots$$

denotes an **infinite sequence.** An **explicit formula** or **general formula** for a sequence is a rule that shows how the values of a_k depend on k.

The following example shows that it is possible for two different formulas to give sequences with the same terms.

Example 4.1.1 Finding Terms of Sequences Given by Explicit Formulas

Define sequences $a_1, a_2, a_3, \ldots$ and $b_2, b_3, b_4, \ldots$ by the following explicit formulas:

$$a_k = \frac{k}{k+1} \quad \text{for all integers } k \geq 1,$$

$$b_i = \frac{i-1}{i} \quad \text{for all integers } i \geq 2.$$

Compute the first five terms of both sequences.

Solution

$$a_1 = \frac{1}{1+1} = \frac{1}{2} \qquad b_2 = \frac{2-1}{2} = \frac{1}{2}$$

$$a_2 = \frac{2}{2+1} = \frac{2}{3} \qquad b_3 = \frac{3-1}{3} = \frac{2}{3}$$

$$a_3 = \frac{3}{3+1} = \frac{3}{4} \qquad b_4 = \frac{4-1}{4} = \frac{3}{4}$$

$$a_4 = \frac{4}{4+1} = \frac{4}{5} \qquad b_5 = \frac{5-1}{5} = \frac{4}{5}$$

$$a_5 = \frac{5}{5+1} = \frac{5}{6} \qquad b_6 = \frac{6-1}{6} = \frac{5}{6}$$

As you can see, the first terms of both sequences are $\frac{1}{2}, \frac{2}{3}, \frac{3}{4}, \frac{4}{5}, \frac{5}{6}$; in fact, it can be shown that all terms of both sequences are identical. ∎

The next example shows that an infinite sequence may have only a finite number of values.

Example 4.1.2 An Alternating Sequence

Compute the first six terms of the sequence $c_0, c_1, c_2, \ldots$ defined as follows:

$$c_j = (-1)^j \quad \text{for all integers } j \geq 0.$$

Solution
$$c_0 = (-1)^0 = 1$$
$$c_1 = (-1)^1 = -1$$
$$c_2 = (-1)^2 = 1$$
$$c_3 = (-1)^3 = -1$$
$$c_4 = (-1)^4 = 1$$
$$c_5 = (-1)^5 = -1$$

Thus the first six terms are $1, -1, 1, -1, 1, -1$. By exercises 29 and 30 of Section 3.1, even powers of -1 equal 1 and odd powers of -1 equal -1. It follows that the sequence oscillates endlessly between 1 and -1. ∎

In Examples 4.1.1 and 4.1.2 the task was to compute initial values of a sequence given by an explicit formula. The next example treats the question of how to find an explicit formula for a sequence with given initial terms. Any such formula is a guess, but it is very useful to be able to make such guesses.

Example 4.1.3 Finding an Explicit Formula to Fit Given Initial Terms

Find an explicit formula for a sequence that has the following initial terms:

$$1, \quad -\frac{1}{4}, \quad \frac{1}{9}, \quad -\frac{1}{16}, \quad \frac{1}{25}, \quad -\frac{1}{36}, \dots$$

Solution   Denote the general term of the sequence by a_k and suppose the first term is a_1. Then observe that the denominator of each term is a perfect square. Thus the terms can be rewritten as

$$\frac{1}{1^2}, \quad \frac{(-1)}{2^2}, \quad \frac{1}{3^2}, \quad \frac{(-1)}{4^2}, \quad \frac{1}{5^2}, \quad \frac{(-1)}{6^2}.$$
$$\updownarrow \qquad \updownarrow \qquad \updownarrow \qquad \updownarrow \qquad \updownarrow \qquad \updownarrow$$
$$a_1 \qquad a_2 \qquad a_3 \qquad a_4 \qquad a_5 \qquad a_6$$

Note that the denominator of each term equals the square of the subscript of that term, and that the numerator equals ± 1. Hence

$$a_k = \frac{\pm 1}{k^2}.$$

Also the numerator oscillates back and forth between $+1$ and -1; it is $+1$ when k is odd and -1 when k is even. To achieve this oscillation, insert a factor of $(-1)^{k+1}$ (or $(-1)^{k-1}$) into the formula for a_k. [*For when k is odd, $k + 1$ is even and thus $(-1)^{k+1} = +1$; and when k is even, $k + 1$ is odd and thus $(-1)^{k+1} = -1$.*] Consequently, an explicit formula that gives the correct first six terms is

$$a_k = \frac{(-1)^{k+1}}{k^2} \quad \text{for all integers } k \geq 1.$$

Note that making the first term a_0 would have led to the alternative formula

$$a_k = \frac{(-1)^k}{(k+1)^2} \quad \text{for all integers } k \geq 0.$$

You should check that this formula also gives the correct first six terms. ∎

⚠ **Caution!** Two sequences may start off with the same initial values but diverge later on. See exercise 7 at the end of this section.

Summation Notation

Joseph Louis Lagrange
(1736–1813)

CORBIS

Consider again the example in which $A_k = 2^k$ represented the number of ancestors a person has in the kth generation back. What is the total number of ancestors for the past six generations? The answer is

$$A_1 + A_2 + A_3 + A_4 + A_5 + A_6 = 2^1 + 2^2 + 2^3 + 2^4 + 2^5 + 2^6 = 126.$$

It is convenient to use a shorthand notation to write such sums. In 1772 the French mathematician Joseph Louis Lagrange introduced the capital Greek letter sigma, Σ, to denote the word *sum* (or *summation*), and the notation

$$\sum_{k=1}^{n} a_k$$

to represent the sum given in **expanded form** by

$$a_1 + a_2 + a_3 + \cdots + a_n.$$

More generally, if m and n are integers and $m \le n$, then the **summation from k equals m to n of a_k** is the sum of all the terms $a_m, a_{m+1}, a_{m+2}, \ldots, a_n$. We write

$$\sum_{k=m}^{n} a_k = a_m + a_{m+1} + a_{m+2} + \cdots + a_n$$

and call k the **index** of the summation, m the **lower limit** of the summation, and n the **upper limit** of the summation.

Example 4.1.4 Computing Summations

Let $a_1 = -2$, $a_2 = -1$, $a_3 = 0$, $a_4 = 1$, and $a_5 = 2$. Compute the following:

a. $\displaystyle\sum_{k=1}^{5} a_k$ b. $\displaystyle\sum_{k=2}^{2} a_k$ c. $\displaystyle\sum_{k=1}^{2} a_{2k}$

Solution

a. $\displaystyle\sum_{k=1}^{5} a_k = a_1 + a_2 + a_3 + a_4 + a_5 = (-2) + (-1) + 0 + 1 + 2 = 0$

b. $\displaystyle\sum_{k=2}^{2} a_k = a_2 = -1$

c. $\displaystyle\sum_{k=1}^{2} a_{2k} = a_{2\cdot 1} + a_{2\cdot 2} = a_2 + a_4 = -1 + 1 = 0$ ∎

Oftentimes, the terms of a summation are expressed using an explicit formula. For instance, it is common to see summations such as

$$\sum_{k=1}^{5} k^2 \quad \text{or} \quad \sum_{i=0}^{8} \frac{(-1)^i}{i+1}.$$

Example 4.1.5 When the Terms of a Summation Are Given by a Formula

Compute the following summation:

$$\sum_{k=1}^{5} k^2.$$

Solution

$$\sum_{k=1}^{5} k^2 = 1^2 + 2^2 + 3^2 + 4^2 + 5^2 = 55.$$ ∎

When the upper limit of a summation is a variable, an ellipsis is used to write the summation in expanded form.

Example 4.1.6 Changing from Summation Notation to Expanded Form

Write the following summation in expanded form:

$$\sum_{i=0}^{n} \frac{(-1)^i}{i+1}.$$

Solution

$$\sum_{i=0}^{n} \frac{(-1)^i}{i+1} = \frac{(-1)^0}{0+1} + \frac{(-1)^1}{1+1} + \frac{(-1)^2}{2+1} + \frac{(-1)^3}{3+1} + \cdots + \frac{(-1)^n}{n+1}$$

$$= \frac{1}{1} + \frac{(-1)}{2} + \frac{1}{3} + \frac{(-1)}{4} + \cdots + \frac{(-1)^n}{n+1}$$

$$= 1 - \frac{1}{2} + \frac{1}{3} - \frac{1}{4} + \cdots + \frac{(-1)^n}{n+1}$$ ∎

Example 4.1.7 Changing from Expanded Form to Summation Notation

Express the following using summation notation:

$$\frac{1}{n} + \frac{2}{n+1} + \frac{3}{n+2} + \cdots + \frac{n+1}{2n}.$$

Solution The general term of this summation can be expressed as $\dfrac{k+1}{n+k}$ for integers k from 0 to n. Hence

$$\frac{1}{n} + \frac{2}{n+1} + \frac{3}{n+2} + \cdots + \frac{n+1}{2n} = \sum_{k=0}^{n} \frac{k+1}{n+k}.$$ ∎

Caution! The expanded form of a sum may appear ambiguous for small values of n. For instance, consider

$$1^2 + 2^2 + 3^2 + \cdots + n^2.$$

This expression is intended to represent the sum of squares of consecutive integers starting with 1^2 and ending with n^2. Thus, if $n = 1$ the sum is just 1^2, if $n = 2$ the sum is $1^2 + 2^2$, and if $n = 3$ the sum is $1^2 + 2^2 + 3^2$.

Example 4.1.8 Evaluating $a_1, a_2, a_3, \ldots, a_n$ for Small n

What is the value of the expression $\dfrac{1}{1 \cdot 2} + \dfrac{1}{2 \cdot 3} + \dfrac{1}{3 \cdot 4} + \cdots + \dfrac{1}{n \cdot (n+1)}$ when $n = 1$? $n = 2$? $n = 3$?

Solution

When $n = 1$, the expression equals $\dfrac{1}{1 \cdot 2} = \dfrac{1}{2}$.

When $n = 2$, it equals $\dfrac{1}{1 \cdot 2} + \dfrac{1}{2 \cdot 3} = \dfrac{1}{2} + \dfrac{1}{6} = \dfrac{3}{6} + \dfrac{1}{6} = \dfrac{4}{6} = \dfrac{2}{3}$.

When $n = 3$, it is $\dfrac{1}{1 \cdot 2} + \dfrac{1}{2 \cdot 3} + \dfrac{1}{3 \cdot 4} = \dfrac{1}{2} + \dfrac{1}{6} + \dfrac{1}{12} = \dfrac{6}{12} + \dfrac{2}{12} + \dfrac{1}{12} = \dfrac{9}{12} = \dfrac{3}{4}$. ∎

A more mathematically precise definition of summation, called a *recursive definition*, is the following:* If m and n are any integers with $m < n$, then

$$\sum_{k=m}^{m} a_k = a_m \qquad \text{and} \qquad \sum_{k=m}^{n} a_k = \sum_{k=m}^{n-1} a_k + a_n \quad \text{for all integers } n > m.$$

When solving problems, it is often useful to rewrite a summation using the recursive form of the definition, either by separating off the final term of a summation or by adding a final term to a summation.

Example 4.1.9 Separating Off a Final Term and Adding On a Final Term

a. Rewrite $\displaystyle\sum_{i=1}^{n} \frac{1}{k^2}$ by separating off the final term.

b. Write $\displaystyle\sum_{i=0}^{n-1} 2^k + 2^n$ as a single summation.

Solution

a. $\displaystyle\sum_{i=1}^{n} \frac{1}{i^2} = \sum_{i=1}^{n-1} \frac{1}{i^2} + \frac{1}{n^2}$ \qquad b. $\displaystyle\sum_{k=0}^{n-1} 2^k + 2^n = \sum_{k=0}^{n} 2^k$ ∎

In certain sums each term is a difference of two quantities. When you write such sums in expanded form, you sometimes see that all the terms cancel except the first and the last. Successive cancellation of terms collapses the sum like a telescope.

*Recursive definitions are discussed in Section 8.4.

Example 4.1.10 A Telescoping Sum

Some sums can be transformed into telescoping sums, which then can be rewritten as a simple expression. For instance, observe that

$$\frac{1}{k} - \frac{1}{k+1} = \frac{(k+1) - k}{k(k+1)} = \frac{1}{k(k+1)}.$$

Use this identity to find a simple expression for $\displaystyle\sum_{k=1}^{n} \frac{1}{k(k+1)}$.

Solution

$$\sum_{k=1}^{n} \frac{1}{k(k+1)}$$

$$= \sum_{k=1}^{n} \left(\frac{1}{k} - \frac{1}{k+1} \right)$$

$$= \left(\frac{1}{1} - \frac{1}{2} \right) + \left(\frac{1}{2} - \frac{1}{3} \right) + \left(\frac{1}{3} - \frac{1}{4} \right) + \cdots + \left(\frac{1}{n-1} - \frac{1}{n} \right) + \left(\frac{1}{n} - \frac{1}{n+1} \right)$$

$$= 1 - \frac{1}{n+1}.$$ ∎

Product Notation

The notation for the product of a sequence of numbers is analogous to the notation for their sum. The Greek capital letter pi, Π, denotes a product. For example,

$$\prod_{k=1}^{5} a_k = a_1 a_2 a_3 a_4 a_5.$$

More generally, the **product from k equals m to n of a_k** is the product of all the terms $a_m, a_{m+1}, a_{m+2}, \ldots, a_n$. That is,

$$\prod_{k=m}^{n} a_k = a_m \cdot a_{m+1} \cdot a_{m+2} \cdots a_n.$$

A recursive definition for the product notation is the following: If m and n are any integers with $m < n$, then

$$\prod_{k=m}^{m} a_k = a_m \quad \text{and} \quad \prod_{k=m}^{n} a_k = \left(\prod_{k=m}^{n-1} a_k \right) \cdot a_n \quad \text{for all integers } n > m.$$

Example 4.1.11 Computing Products

Compute the following products:

a. $\displaystyle\prod_{k=1}^{5} k$

b. $\displaystyle\prod_{k=1}^{1} \frac{k}{k+1}$

Solution

a. $\displaystyle\prod_{k=1}^{5} k = 1 \cdot 2 \cdot 3 \cdot 4 \cdot 5 = 120$ b. $\displaystyle\prod_{k=1}^{1} \frac{k}{k+1} = \frac{1}{1+1} = \frac{1}{2}$ ∎

Factorial Notation

The product of all consecutive integers up to a given integer occurs so often in mathematics that it is given a special notation—*factorial* notation.

• Definition

For each positive integer n, the quantity n **factorial** denoted $n!$, is defined to be the product of all the integers from 1 to n:

$$n! = n \cdot (n-1) \cdots 3 \cdot 2 \cdot 1.$$

Zero factorial, denoted $0!$, is defined to be 1:

$$0! = 1.$$

The definition of zero factorial as 1 may seem odd, but, as you will see when you read Chapter 6, it is convenient for many mathematical formulas.

Example 4.1.12 The First Ten Factorials

$$0! = 1 \qquad\qquad 1! = 1$$
$$2! = 2 \cdot 1 = 2 \qquad\qquad 3! = 3 \cdot 2 \cdot 1 = 6$$
$$4! = 4 \cdot 3 \cdot 2 \cdot 1 = 24 \qquad\qquad 5! = 5 \cdot 4 \cdot 3 \cdot 2 \cdot 1 = 120$$
$$6! = 6 \cdot 5 \cdot 4 \cdot 3 \cdot 2 \cdot 1 = 720 \qquad\qquad 7! = 7 \cdot 6 \cdot 5 \cdot 4 \cdot 3 \cdot 2 \cdot 1 = 5{,}040$$
$$8! = 8 \cdot 7 \cdot 6 \cdot 5 \cdot 4 \cdot 3 \cdot 2 \cdot 1 \qquad\qquad 9! = 9 \cdot 8 \cdot 7 \cdot 6 \cdot 5 \cdot 4 \cdot 3 \cdot 2 \cdot 1$$
$$= 40{,}320 \qquad\qquad\qquad = 362{,}880 \qquad\blacksquare$$

As you can see from the example above, the values of $n!$ grow very rapidly. For instance, $40! \cong 8.16 \times 10^{47}$, which is a number that is too large to be computed exactly using the standard integer arithmetic of the machine-specific implementations of many computer languages. (The symbol $\cong$ means "is approximately equal to.")

A recursive definition for factorial is the following: Given any nonnegative integer n,

$$n! = \begin{cases} 1 & \text{if } n = 0 \\ n \cdot (n-1)! & \text{if } n \geq 1 \end{cases}.$$

Example 4.1.13 illustrates the usefulness of the recursive definition for making computations.

Example 4.1.13 Computing with Factorials

Simplify the following expressions:

a. $\dfrac{8!}{7!}$ b. $\dfrac{5!}{2! \cdot 3!}$ c. $\dfrac{1}{2! \cdot 4!} + \dfrac{1}{3! \cdot 3!}$ d. $\dfrac{(n+1)!}{n!}$ e. $\dfrac{n!}{(n-3)!}$

Solution

a. $\dfrac{8!}{7!} = \dfrac{8 \cdot \cancel{7!}}{\cancel{7!}} = 8$

b. $\dfrac{5!}{2! \cdot 3!} = \dfrac{5 \cdot 4 \cdot \cancel{3!}}{2! \cdot \cancel{3!}} = \dfrac{5 \cdot 4}{2 \cdot 1} = 10$

c. $\dfrac{1}{2! \cdot 4!} + \dfrac{1}{3! \cdot 3!} = \dfrac{1}{2! \cdot 4!} \cdot \dfrac{3}{3} + \dfrac{1}{3! \cdot 3!} \cdot \dfrac{4}{4}$ by multiplying each numerator and denominator by just what is necessary to obtain a common denominator

$= \dfrac{3}{3 \cdot 2! \cdot 4!} + \dfrac{4}{3! \cdot 4 \cdot 3!}$ by rearranging factors

$= \dfrac{3}{3! \cdot 4!} + \dfrac{4}{3! \cdot 4!}$ because $3 \cdot 2! = 3!$ and $4 \cdot 3! = 4!$

$= \dfrac{7}{3! \cdot 4!}$ by the rule for adding fractions with a common denominator

$= \dfrac{7}{144}$

d. $\dfrac{(n+1)!}{n!} = \dfrac{(n+1) \cdot \cancel{n!}}{\cancel{n!}} = n + 1$

e. $\dfrac{n!}{(n-3)!} = \dfrac{n \cdot (n-1) \cdot (n-2) \cdot \cancel{(n-3)!}}{\cancel{(n-3)!}} = n \cdot (n-1) \cdot (n-2)$

$= n^3 - 3n^2 + 2n$ ∎

Properties of Summations and Products

The following theorem states general properties of summations and products. The proof of the theorem is discussed in Section 8.4.

Theorem 4.1.1

If $a_m, a_{m+1}, a_{m+2}, \ldots$ and $b_m, b_{m+1}, b_{m+2}, \ldots$ are sequences of real numbers and c is any real number, then the following equations hold for any integer $n \geq m$:

1. $\displaystyle\sum_{k=m}^{n} a_k + \sum_{k=m}^{n} b_k = \sum_{k=m}^{n} (a_k + b_k)$

2. $c \cdot \displaystyle\sum_{k=m}^{n} a_k = \sum_{k=m}^{n} c \cdot a_k$ generalized distributive law

3. $\left(\displaystyle\prod_{k=m}^{n} a_k \right) \cdot \left(\displaystyle\prod_{k=m}^{n} b_k \right) = \prod_{k=m}^{n} (a_k \cdot b_k).$

Example 4.1.14 Using Properties of Summation and Product

Let $a_k = k + 1$ and $b_k = k - 1$ for all integers k. Write each of the following expressions as a single summation or product:

a. $\displaystyle\sum_{k=m}^{n} a_k + 2 \cdot \sum_{k=m}^{n} b_k$ b. $\left(\displaystyle\prod_{k=m}^{n} a_k \right) \cdot \left(\displaystyle\prod_{k=m}^{n} b_k \right)$

Solution

a. $\displaystyle\sum_{k=m}^{n} a_k + 2 \cdot \sum_{k=m}^{n} b_k = \sum_{k=m}^{n}(k+1) + 2 \cdot \sum_{k=m}^{n}(k-1)$ by substitution

$\displaystyle = \sum_{k=m}^{n}(k+1) + \sum_{k=m}^{n} 2 \cdot (k-1)$ by Theorem 4.1.1 (2)

$\displaystyle = \sum_{k=m}^{n}((k+1) + 2 \cdot (k-1))$ by Theorem 4.1.1 (1)

$\displaystyle = \sum_{k=m}^{n}(3k-1)$ by algebraic simplification

b. $\displaystyle\left(\prod_{k=m}^{n} a_k\right) \cdot \left(\prod_{k=m}^{n} b_k\right) = \left(\prod_{k=m}^{n}(k+1)\right) \cdot \left(\prod_{k=m}^{n}(k-1)\right)$ by substitution

$\displaystyle = \prod_{k=m}^{n}(k+1) \cdot (k-1)$ by Theorem 4.1.1 (3)

$\displaystyle = \prod_{k=m}^{n}(k^2-1)$ by algebraic simplification ∎

Change of Variable

Observe that

$$\sum_{k=1}^{3} k^2 = 1^2 + 2^2 + 3^2$$

and also that

$$\sum_{i=1}^{3} i^2 = 1^2 + 2^2 + 3^2.$$

Hence

$$\sum_{k=1}^{3} k^2 = \sum_{i=1}^{3} i^2.$$

This equation illustrates the fact that the symbol used to represent the index of a summation can be replaced by any other symbol as long as the replacement is made in each location where the symbol occurs. As a consequence, the index of a summation is called a *dummy variable*. A dummy variable is a symbol that derives its entire meaning from its local context. Outside of that context (both before and after), the symbol may have another meaning entirely.

The appearance of a summation can be altered by more complicated changes of variable as well. For example, observe that

$$\sum_{j=2}^{4} (j-1)^2 = (2-1)^2 + (3-1)^2 + (4-1)^2$$

$$= 1^2 + 2^2 + 3^2$$

$$= \sum_{k=1}^{3} k^2.$$

A general procedure to transform the first summation into the second is illustrated in Example 4.1.15.

Example 4.1.15 Transforming a Sum by a Change of Variable

Transform the following summation by making the specified change of variable.

$$summation: \sum_{k=0}^{6} \frac{1}{k+1} \qquad change\ of\ variable:\ j = k+1$$

Solution First calculate the lower and upper limits of the new summation:

When $k = 0$, $j = k + 1 = 0 + 1 = 1$.
When $k = 6$, $j = k + 1 = 6 + 1 = 7$.

Thus the new sum goes from $j = 1$ to $j = 7$.

Next calculate the general term of the new summation. You will need to replace each occurrence of k by an expression in j:

Since $j = k + 1$, then $k = j - 1$.

Hence $\dfrac{1}{k+1} = \dfrac{1}{(j-1)+1} = \dfrac{1}{j}$.

Finally, put the steps together to obtain

$$\sum_{k=0}^{6} \frac{1}{k+1} = \sum_{j=1}^{7} \frac{1}{j}. \qquad\qquad 4.1.1$$

∎

Equation (4.1.1) can be given an additional twist by noting that because the j in the right-hand summation is a dummy variable, it may be replaced by any other variable name, as long as the substitution is made in every location where j occurs. In particular, it is legal to substitute k in place of j to obtain

$$\sum_{j=1}^{7} \frac{1}{j} = \sum_{k=1}^{7} \frac{1}{k}. \qquad\qquad 4.1.2$$

Putting equations (4.1.1) and (4.1.2) together gives

$$\sum_{k=0}^{6} \frac{1}{k+1} = \sum_{k=1}^{7} \frac{1}{k}.$$

Sometimes it is necessary to shift the limits of one summation in order to add it to another. An example is the algebraic proof of the binomial theorem, given in Section 6.7. A general procedure for making such a shift when the upper limit is part of the summand is illustrated in Example 4.1.16.

Example 4.1.16 When the Upper Limit Appears in the Expression to Be Summed

a. Transform the following summation by making the specified change of variable.

$$\text{summation: } \sum_{k=1}^{n+1} \left(\frac{k}{n+k} \right) \qquad \text{change of variable: } j = k - 1$$

b. Transform the summation obtained in part (a) by changing all j's to k's.

Solution

a. When $k = 1$, then $j = k - 1 = 1 - 1 = 0$. (So the new lower limit is 0.) When $k = n + 1$, then $j = k - 1 = (n + 1) - 1 = n$. (So the new upper limit is n.)

Since $j = k - 1$, then $k = j + 1$. Also note that n is a constant as far as the terms of the sum are concerned. It follows that

$$\frac{k}{n+k} = \frac{j+1}{n+(j+1)}$$

and so the general term of the new summation is

$$\frac{j+1}{n+(j+1)}.$$

Therefore,

$$\sum_{k=1}^{n+1} \frac{k}{n+k} = \sum_{j=0}^{n} \frac{j+1}{n+(j+1)}. \qquad 4.1.3$$

b. Changing all the j's to k's in the right-hand side of equation (4.1.3) gives

$$\sum_{j=0}^{n} \frac{j+1}{n+(j+1)} = \sum_{k=0}^{n} \frac{k+1}{n+(k+1)} \qquad 4.1.4$$

Combining equations (4.1.3) and (4.1.4) results in

$$\sum_{k=1}^{n+1} \frac{k}{n+k} = \sum_{k=0}^{n} \frac{k+1}{n+(k+1)}. \qquad ■$$

Sequences In Computer Programming

An important data type in computer programming consists of finite sequences. In computer programming contexts, these are usually referred to as *one-dimensional arrays*. For example, consider a program that analyzes the wages paid to a sample of 50 workers. Such a program might compute the average wage and the difference between each individual wage and the average. This would require that each wage be stored in memory for retrieval later in the calculation. To avoid the use of entirely separate variable names for all of the 50 wages, each is written as a term of a one-dimensional array:

$$W[1], W[2], W[3], \ldots, W[50].$$

Note that the subscript labels are written inside square brackets. The reason is that until relatively recently, it was impossible to type actual dropped subscripts on most computer keyboards.

The main difficulty programmers have when using one-dimensional arrays is keeping the labels straight.

Example 4.1.17 Dummy Variable in a Loop

The index variable for a **for-next** loop is a dummy variable. For example, the following three algorithm segments all produce the same output:

1. **for** $i := 1$ **to** n 2. **for** $j := 0$ **to** $n - 1$ 3. **for** $k := 2$ **to** $n + 1$

       print $a[i]$ **print** $a[j + 1]$ **print** $a[k - 1]$

    next i **next** j **next** k ■

The recursive definitions for summation, product, and factorial lead naturally to computational algorithms. For instance, here are two sets of pseudocode to find the sum of $a[1], a[2], \ldots, a[n]$. The one on the left exactly mimics the recursive definition by initializing the sum to equal $a[1]$; the one on the right initializes the sum to equal 0. In both cases the output is $\sum_{k=1}^{n} a[k]$.

$s := a[1]$ $s := 0$

for $k := 2$ **to** n **for** $k := 1$ **to** n

     $s := s + a[k]$ $s := s + a[k]$

next k **next** k

Application: Algorithm to Convert from Base 10 to Base 2 Using Repeated Division by 2

Section 1.5 contains some examples of converting integers from decimal to binary notation. The method shown there, however, is only convenient to use with small numbers. A systematic algorithm to convert any nonnegative integer to binary notation uses repeated division by 2.

Suppose a is a nonnegative integer. Divide a by 2 using the quotient-remainder theorem to obtain a quotient $q[0]$ and a remainder $r[0]$. If the quotient is nonzero, divide by 2 again to obtain a quotient $q[1]$ and a remainder $r[1]$. Continue this process until a quotient of 0 is obtained. At each stage, the remainder must be less than the divisor, which is 2. Thus each remainder is either 0 or 1. The process is illustrated below for $a = 38$. (Read the divisions from the bottom up.)

The results of all these divisions can be written as a sequence of equations:

$$38 = 19 \cdot 2 + 0,$$
$$19 = 9 \cdot 2 + 1,$$
$$9 = 4 \cdot 2 + 1,$$
$$4 = 2 \cdot 2 + 0,$$
$$2 = 1 \cdot 2 + 0,$$
$$1 = 0 \cdot 2 + 1.$$

By repeated substitution, then,

$$
\begin{aligned}
38 &= 19 \cdot 2 + 0 \\
&= (9 \cdot 2 + 1) \cdot 2 + 0 \ = \ 9 \cdot 2^2 + 1 \cdot 2 + 0 \\
&= (4 \cdot 2 + 1) \cdot 2^2 + 1 \cdot 2 + 0 \ = \ 4 \cdot 2^3 + 1 \cdot 2^2 + 1 \cdot 2 + 0 \\
&= (2 \cdot 2 + 0) \cdot 2^3 + 1 \cdot 2^2 + 1 \cdot 2 + 0 \\
&= 2 \cdot 2^4 + 0 \cdot 2^3 + 1 \cdot 2^2 + 1 \cdot 2 + 0 \\
&= (1 \cdot 2 + 0) \cdot 2^4 + 0 \cdot 2^3 + 1 \cdot 2^2 + 1 \cdot 2 + 0 \\
&= 1 \cdot 2^5 + 0 \cdot 2^4 + 0 \cdot 2^3 + 1 \cdot 2^2 + 1 \cdot 2 + 0.
\end{aligned}
$$

Note that each coefficient of a power of 2 on the right-hand side above is one of the remainders obtained in the repeated division of 38 by 2. This is true for the left-most 1 as well, because $1 = 0 \cdot 2 + 1$. Thus

$$38_{10} = 100110_2 = (r[5]r[4]r[3]r[2]r[1]r[0])_2.$$

In general, if a nonnegative integer a is repeatedly divided by 2 until a quotient of zero is obtained and the remainders are found to be $r[0], r[1], \ldots, r[k]$, then by the quotient-remainder theorem each $r[i]$ equals 0 or 1, and by repeated substitution from the theorem,

$$a = 2^k \cdot r[k] + 2^{k-1} \cdot r[k-1] + \cdots + 2^2 \cdot r[2] + 2^1 \cdot r[1] + 2^0 \cdot r[0]. \qquad 4.1.5$$

Thus the binary representation for a can be read from equation (4.1.5):

$$a_{10} = (r[k]r[k-1] \cdots r[2]r[1]r[0])_2.$$

Example 4.1.18 Converting from Decimal to Binary Notation Using Repeated Division by 2

Use repeated division by 2 to write the number 29_{10} in binary notation.

Solution

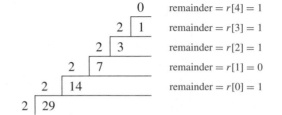

Hence $29_{10} = (r[4]r[3]r[2]r[1]r[0])_2 = 11101_2$. ∎

The procedure we have described for converting from base 10 to base 2 is formalized in the following algorithm:

Algorithm 4.1.1 Decimal to Binary Conversion Using Repeated Division by 2

[In Algorithm 4.1.1 the input is a nonnegative integer a. The aim of the algorithm is to produce a sequence of binary digits $r[0], r[1], r[2], \ldots, r[k]$ so that the binary representation of a is

$$(r[k]r[k-1]\cdots r[2]r[1]r[0])_2.$$

That is,

$$a = 2^k \cdot r[k] + 2^{k-1} \cdot r[k-1] + \cdots + 2^2 \cdot r[2] + 2^1 \cdot r[1] + 2^0 \cdot r[0].]$$

Input: a [*a nonnegative integer*]

Algorithm Body:

$q := a, i := 0$

*[Repeatedly perform the integer division of q by 2 until q becomes 0. Store successive remainders in a one-dimensional array $r[0], r[1], r[2], \ldots, r[k]$. Even if the initial value of q equals 0, the loop should execute one time (so that $r[0]$ is computed). Thus the guard condition for the **while** loop is $i = 0$ or $q \neq 0$.]*

while ($i = 0$ or $q \neq 0$)

$\quad r[i] := q \bmod 2$

$\quad q := q \operatorname{div} 2$

[$r[i]$ and q can be obtained by calling the division algorithm.]

$\quad i := i + 1$

end while

[After execution of this step, the values of $r[0], r[1], \ldots, r[i-1]$ are all 0's and 1's, and $a = (r[i-1]r[i-2]\cdots r[2]r[1]r[0])_2.$]

Output: $r[0], r[1], r[2], \ldots, r[i-1]$ [*a sequence of integers*]

Exercise Set 4.1*

Write the first four terms of the sequences defined by the formulas in 1–6.

1. $a_k = \dfrac{k}{10+k}$, for all integers $k \geq 1$.

2. $b_j = \dfrac{5-k}{5+k}$, for all integers $j \geq 1$.

3. $c_i = \dfrac{(-1)^i}{3^i}$, for all integers $i \geq 0$.

4. $d_m = 1 + \left(\dfrac{1}{2}\right)^m$ for all integers $m \geq 0$.

5. $e_n = \left\lfloor \dfrac{n}{2} \right\rfloor \cdot 2$, for all integers $n \geq 0$.

6. $f_n = \left\lfloor \dfrac{n}{4} \right\rfloor \cdot 4$, for all integers $n \geq 1$.

7. Let $a_k = 2k+1$ and $b_k = (k-1)^3 + k + 2$ for all integers $k \geq 0$. Show that the first three terms of these sequences are identical but that their fourth terms differ.

Compute the first fifteen terms of each of the sequences in 8 and 9, and describe the general behavior of these sequences in words. (A definition of logarithm is given in Section 7.1.)

8. $g_n = \lfloor \log_2 n \rfloor$ for all integers $n \geq 1$.

9. $h_n = n \lfloor \log_2 n \rfloor$ for all integers $n \geq 1$.

Find explicit formulas for sequences of the form $a_1, a_2, a_3, \ldots$ with the initial terms given in 10–16.

10. $-1, 1, -1, 1, -1, 1$ **11.** $0, 1, -2, 3, -4, 5$

12. $\dfrac{1}{4}, \dfrac{2}{9}, \dfrac{3}{16}, \dfrac{4}{25}, \dfrac{5}{36}, \dfrac{6}{49}$

---

*For exercises with blue numbers or letters, solutions are given in Appendix B. The symbol **H** indicates that only a hint or a partial solution is given. The symbol ✳ signals that an exercise is more challenging than usual.

13. $1 - \dfrac{1}{2}, \dfrac{1}{2} - \dfrac{1}{3}, \dfrac{1}{3} - \dfrac{1}{4}, \dfrac{1}{4} - \dfrac{1}{5}, \dfrac{1}{5} - \dfrac{1}{6}, \dfrac{1}{6} - \dfrac{1}{7}$

14. $\dfrac{1}{3}, \dfrac{2}{9}, \dfrac{3}{27}, \dfrac{4}{81}, \dfrac{5}{243}, \dfrac{6}{729}$

15. $1, -\dfrac{1}{2}, \dfrac{2}{3}, -\dfrac{3}{4}, \dfrac{4}{5}, -\dfrac{5}{6}, \dfrac{6}{7}$

16. $3, 6, 12, 24, 48, 96$

✶ 17. Consider the sequence defined by $a_n = \dfrac{2n + (-1)^n - 1}{4}$ for all integers $n \geq 0$. Find an alternative explicit formula for a_n that uses the floor notation.

18. Let $a_0 = 2, a_1 = 3, a_2 = -2, a_3 = 1, a_4 = 0, a_5 = -1$, and $a_6 = -2$. Compute each of the summations and products below.

    a. $\displaystyle\sum_{i=0}^{6} a_i$ **b.** $\displaystyle\sum_{i=0}^{0} a_i$ **c.** $\displaystyle\sum_{j=1}^{3} a_{2j}$ **d.** $\displaystyle\prod_{k=0}^{6} a_k$ **e.** $\displaystyle\prod_{k=2}^{2} a_k$

Compute the summations and products in 19–28

19. $\displaystyle\sum_{k=1}^{5}(k+1)$ 20. $\displaystyle\prod_{k=2}^{4} k^2$ 21. $\displaystyle\sum_{m=0}^{3} \dfrac{1}{2^m}$

22. $\displaystyle\prod_{j=0}^{4}(-1)^j$ 23. $\displaystyle\sum_{i=1}^{1} i(i+1)$ 24. $\displaystyle\sum_{j=0}^{0}(j+1)\cdot 2^j$

25. $\displaystyle\prod_{k=2}^{2}\left(1 - \dfrac{1}{k}\right)$ 26. $\displaystyle\sum_{k=-1}^{1}(k^2 + 3)$

27. $\displaystyle\sum_{n=1}^{10}\left(\dfrac{1}{n} - \dfrac{1}{n+1}\right)$ 28. $\displaystyle\prod_{i=2}^{5} \dfrac{i(i+2)}{(i-1)\cdot(i+1)}$

Write the summations in 29–31 in expanded form.

29. $\displaystyle\sum_{i=1}^{n}(-2)^i$ 30. $\displaystyle\sum_{j=1}^{n} j(j+1)$ 31. $\displaystyle\sum_{k=0}^{n} \dfrac{1}{k!}$

Write each of 32–41 using summation or product notation.

32. $1^2 - 2^2 + 3^2 - 4^2 + 5^2 - 6^2 + 7^2$

33. $(1^3 - 1) - (2^3 - 1) + (3^3 - 1) - (4^3 - 1) + (5^3 - 1)$

34. $(2^2 - 1)\cdot(3^2 - 1)\cdot(4^2 - 1)$

35. $\dfrac{2}{3\cdot 4} - \dfrac{3}{4\cdot 5} + \dfrac{4}{5\cdot 6} - \dfrac{5}{6\cdot 7} + \dfrac{6}{7\cdot 8}$

36. $1 - r + r^2 - r^3 + r^4 - r^5$

37. $(1 - t)\cdot(1 - t^2)\cdot(1 - t^3)\cdot(1 - t^4)$

38. $1^3 + 2^3 + 3^3 + \cdots + n^3$

39. $\dfrac{1}{2!} + \dfrac{2}{3!} + \dfrac{3}{4!} + \cdots + \dfrac{n}{(n+1)!}$

40. $n + (n-1) + (n-2) + \cdots + 1$

41. $n + \dfrac{n-1}{2!} + \dfrac{n-2}{3!} + \dfrac{n-3}{4!} + \cdots + \dfrac{1}{n!}$

Compute each of 42–50.

42. $\dfrac{4!}{3!}$ 43. $\dfrac{6!}{8!}$ 44. $\dfrac{4!}{0!}$

45. $\dfrac{n!}{(n-1)!}$ 46. $\dfrac{(n-1)!}{(n+1)!}$ 47. $\dfrac{n!}{(n-2)!}$

48. $\dfrac{((n+1)!)^2}{(n!)^2}$ 49. $\dfrac{n!}{(n-k)!}$ 50. $\dfrac{n!}{(n-k+1)!}$

51. **a.** Prove that $n! + 2$ is divisible by 2, for all integers $n \geq 2$.

    b. Prove that $n! + k$ is divisible by k, for all integers $n \geq 2$ and $k = 2, 3, \ldots, n$.

  H **c.** Given any integer $m \geq 2$, is it possible to find a sequence of $m - 1$ consecutive positive integers none of which is prime? Explain your answer.

Transform each of 52 and 53 by making the change of variable $i = k + 1$.

52. $\displaystyle\sum_{k=0}^{5} k(k-1)$ 53. $\displaystyle\prod_{k=1}^{n} \dfrac{k}{k^2 + 4}$

Transform each of 54–57 by making the change of variable $j = i - 1$.

54. $\displaystyle\sum_{i=1}^{n+1} \dfrac{(i-1)^2}{i\cdot n}$ 55. $\displaystyle\sum_{i=3}^{n} \dfrac{i}{i + n - 1}$

56. $\displaystyle\sum_{i=1}^{n-1} \dfrac{i}{(n-i)^2}$ 57. $\displaystyle\prod_{i=n}^{2n} \dfrac{n - i + 1}{n + i}$

Write each of 58–60 as a single summation or product.

58. $3 \cdot \displaystyle\sum_{k=1}^{n}(2k - 3) + \sum_{k=1}^{n}(4 - 5k)$

59. $2 \cdot \displaystyle\sum_{k=1}^{n}(3k^2 + 4) + 5 \cdot \sum_{k=1}^{n}(2k^2 - 1)$

60. $\left(\displaystyle\prod_{k=1}^{n} \dfrac{k}{k+1}\right) \cdot \left(\prod_{k=1}^{n} \dfrac{k+1}{k+2}\right)$

61. Check Theorem 4.1.1 for $m = 1$ and $n = 4$ by writing out the left-hand and right-hand sides of the equations in expanded form. The two sides are equal by repeated application of certain laws. What are these laws?

62. Suppose $a[1], a[2], a[3], \ldots, a[m]$ is a one-dimensional array and consider the following algorithm segment:

    sum := 0
    for $k := 1$ **to** m
        sum := *sum* + *a*[*k*]
    next k

Fill in the blanks below so that each algorithm segment performs the same job as the one given above.

Use repeated division by 2 to convert (by hand) the integers in 63–65 from base 10 to base 2.

63. 90   **64.** 98   **65.** 205

Make a trace table to trace the action of Algorithm 4.1.1 on the input in 66–68.

66. 23   **67.** 28   **68.** 44

69. Write an informal description of an algorithm (using repeated division by 16) to convert a nonnegative integer from decimal notation to hexadecimal notation (base 16).

Use the algorithm you developed for exercise 69 to convert the integers in 70–72 to hexadecimal notation.

70. 287   **71.** 693   **72.** 2,301

73. Write a formal version of the algorithm you developed for exercise 69.

4.2 Mathematical Induction I

[Mathematical induction is] the standard proof technique in computer science.
— Anthony Ralston, 1984

Mathematical induction is one of the more recently developed techniques of proof in the history of mathematics. It is used to check conjectures about the outcomes of processes that occur repeatedly and according to definite patterns. We introduce the technique with an example.

Some people claim that the United States penny is such a small coin that it should be abolished. They point out that frequently a person who drops a penny on the ground does not even bother to pick it up. Other people argue that abolishing the penny would not give enough flexibility for pricing merchandise. What prices could still be paid with exact change if the penny were abolished and another coin worth 3¢ were introduced? The answer is that the only prices that could not be paid with exact change would be 1¢, 2¢, 4¢, and 7¢. In other words,

Any whole number of cents of at least 8¢ can be obtained using 3¢ and 5¢ coins.

More formally:

For all integers $n \geq 8$, n cents can be obtained using 3¢ and 5¢ coins.

Even more formally:

For all integers $n \geq 8$, $P(n)$ is true, where $P(n)$ is the sentence
"n cents can be obtained using 3¢ and 5¢ coins."

You could check that $P(n)$ is true for a few particular values of n, as is done in the table below.

Number of Cents	How to Obtain It
8¢	3¢ + 5¢
9¢	3¢ + 3¢ + 3¢
10¢	5¢ + 5¢
11¢	3¢ + 3¢ + 5¢
12¢	3¢ + 3¢ + 3¢ + 3¢
13¢	3¢ + 5¢ + 5¢
14¢	3¢ + 3¢ + 3¢ + 5¢
15¢	5¢ + 5¢ + 5¢
16¢	3¢ + 3¢ + 5¢ + 5¢
17¢	3¢ + 3¢ + 3¢ + 3¢ + 5¢

The cases shown in the table provide inductive evidence to support the claim that $P(n)$ is true for general n. Indeed, *$P(n)$ is true for all $n \geq 8$ if, and only if, it is possible to continue filling in the table for arbitrarily large values of n.*

The kth line of the table gives information about how to obtain $k\cent$ using $3\cent$ and $5\cent$ coins. To continue the table to the next row, directions must be given for how to obtain $(k+1)\cent$ using $3\cent$ and $5\cent$ coins. The secret is to observe first that if $k\cent$ can be obtained using at least one $5\cent$ coin, then $(k+1)\cent$ can be obtained by replacing the $5\cent$ coin by two $3\cent$ coins, as shown in Figure 4.2.1.

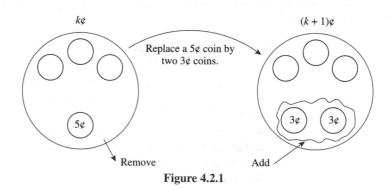

Figure 4.2.1

If, on the other hand, $k\cent$ is obtained without using a $5\cent$ coin, then $3\cent$ coins are used exclusively. And since the total is at least $8\cent$, three or more $3\cent$ coins must be included. Three of the $3\cent$ coins can be replaced by two $5\cent$ coins to obtain a total of $(k+1)\cent$, as shown in Figure 4.2.2.

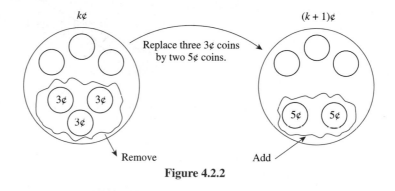

Figure 4.2.2

The structure of the argument above can be summarized as follows: To show that $P(n)$ is true for all integers $n \geq 8$, (1) show that $P(8)$ is true, and (2) show that the truth of $P(k+1)$ follows necessarily from the truth of $P(k)$ for each $k \geq 8$. Any argument of this form is called an argument by *mathematical induction*.

> **Principle of Mathematical Induction**
>
> Let $P(n)$ be a property that is defined for integers n, and let a be a fixed integer. Suppose the following two statements are true:
>
> 1. $P(a)$ is true.
>
> 2. For all integers $k \geq a$, if $P(k)$ is true then $P(k+1)$ is true.
>
> Then the statement
>
> $$\text{for all integers } n \geq a, \ P(n)$$
>
> is true.

The first known use of mathematical induction occurs in the work of the Italian scientist Francesco Maurolico in 1575. In the seventeenth century both Pierre de Fermat and Blaise Pascal used the technique, Fermat calling it the "method of infinite descent." In 1883 Augustus De Morgan (best known for De Morgan's laws) described the process carefully and gave it the name *mathematical induction*.

To visualize the idea of mathematical induction, imagine an infinite collection of dominoes positioned one behind the other in such a way that if any given domino falls backward, it makes the one behind it fall backward also. (See Figure 4.2.3) Then imagine that the first domino falls backward. What happens? ... They all fall down!

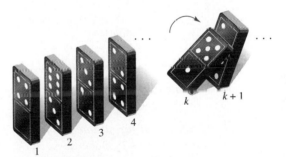

Figure 4.2.3 If the kth domino falls backward, it pushes the $(k+1)$st domino backward also.

To see the connection between this image and the principle of mathematical induction, let $P(n)$ be the sentence "The nth domino falls backward." It is given that for each $k \geq 1$, if $P(k)$ is true (the kth domino falls backward), then $P(k+1)$ is also true (the $(k+1)$st domino falls backward). It is also given that $P(1)$ is true (the first domino falls backward). Thus by the principle of mathematical induction, $P(n)$ (the nth domino falls backward) is true for every integer $n \geq 1$.

The validity of proof by mathematical induction is generally taken as an axiom. That is why it is referred to as the *principle* of mathematical induction rather than as a theorem. It is equivalent to the following property of the integers, which is easy to accept on intuitive grounds:

> Suppose S is any set of integers satisfying (1) $a \in S$, and (2) for all integers k, if $k \in S$ then $k+1 \in S$. Then S must contain every integer greater than or equal to a.

To understand the equivalence of this formulation and the one given earlier, just let S be the set of all integers for which $P(n)$ is true.

Proving a statement by mathematical induction is a two-step process. The first step is called the *basis step,* and the second step is called the *inductive step.*

Method of Proof by Mathematical Induction

Consider a statement of the form, "For all integers $n \geq a$, a property $P(n)$ is true." To prove such a statement, perform the following two steps:

Step 1 (basis step): Show that the property is true for $n = a$.

Step 2 (inductive step): Show that for all integers $k \geq a$, if the property is true for $n = k$ then it is true for $n = k + 1$. To perform this step,

> **suppose** that the property is true for $n = k$, where k is any particular but arbitrarily chosen integer with $k \geq a$.
>> [*This supposition is called the* **inductive hypothesis.**]

Then

> **show** that the property is true for $n = k + 1$.

Here is a formal version of the proof about coins developed informally above.

Proposition 4.2.1

Let $P(n)$ be the property "n¢ can be obtained using 3¢ and 5¢ coins." Then $P(n)$ is true for all integers $n \geq 8$.

Proof:

Show that the property is true for $n = 8$: The property is true for $n = 8$ because 8¢ $= 3$¢ $+ 5$¢.

Show that for all integers $k \geq 8$, if the property is true for $n = k$, then it is true for $n = k + 1$: Suppose k¢ can be obtained using 3¢ and 5¢ coins for some integer $k \geq 8$. [*This is the inductive hypothesis.*] We must show that $(k + 1)$¢ can be obtained using 3¢ and 5¢ coins. In case there is a 5¢ coin among those used to make up the k¢, replace it by two 3¢ coins; the result will be $(k + 1)$¢. In case no 5¢ coin is used to make up the k¢, then at least three 3¢ coins must be used because $k \geq 8$. Remove three 3¢ coins and replace them by two 5¢ coins; the result will be $(k + 1)$¢. Thus in either case $(k + 1)$¢ can be obtained using 3¢ and 5¢ coins [*as was to be shown*].

The following example shows how to use mathematical induction to prove a formula for the sum of the first n integers.

Example 4.2.1 Sum of the First n Integers

Use mathematical induction to prove that

$$1 + 2 + \cdots + n = \frac{n(n + 1)}{2} \quad \text{for all integers } n \geq 1.$$

Solution   To construct a proof by induction, you must first identify the property $P(n)$. In this case, $P(n)$ is

$$1 + 2 + \cdots + n = \frac{n(n+1)}{2}.$$ ← the property ($P(n)$)

[*To see that $P(n)$ is a sentence, note that its subject is "the sum of the integers from 1 to n" and its verb is "equals."*]

In the basis step of the proof, you must show that the property is true for $n = 1$, or, in other words that $P(1)$ is true. Now $P(1)$ is obtained by substituting 1 in place of n in $P(n)$. The left-hand side of $P(1)$ is the sum of all the successive integers starting at 1 and ending at 1. This is just 1. Thus $P(1)$ is

$$1 = \frac{1(1+1)}{2}.$$ ← basis ($P(1)$)

Of course, this equation is true because the right-hand side is

$$\frac{1(1+1)}{2} = \frac{1 \cdot 2}{2} = 1,$$

which equals the left-hand side.

In the inductive step, you assume that $P(k)$ is true, for some integer k with $k \geq 1$. [*This assumption is the inductive hypothesis.*] You must then show that $P(k+1)$ is true. What are $P(k)$ and $P(k+1)$? $P(k)$ is obtained by substituting k for every n in $P(n)$. Thus $P(k)$ is

$$1 + 2 + \cdots + k = \frac{k(k+1)}{2}.$$ ← inductive hypothesis ($P(k)$)

Similarly, $P(k+1)$ is obtained by substituting the quantity $(k+1)$ for every n that appears in $P(n)$. Thus $P(k+1)$ is

$$1 + 2 + \cdots + (k+1) = \frac{(k+1)((k+1)+1)}{2},$$

or, equivalently,

$$1 + 2 + \cdots + (k+1) = \frac{(k+1)(k+2)}{2}.$$ ← to show ($P(k+1)$)

Now the inductive hypothesis is the supposition that $P(k)$ is true. How can this supposition be used to show that $P(k+1)$ is true? $P(k+1)$ is an equation, and the truth of an equation can be shown in a variety of ways. One of the most straightforward is to transform the left-hand side into the right-hand side using algebra and other known facts and legal assumptions (such as the inductive hypothesis). In this case, the left-hand side of $P(k+1)$ is

$$1 + 2 + \cdots + (k+1),$$

which equals

$$(1 + 2 + \cdots + k) + (k + 1) \qquad \text{by explicitly identifying the next-to-last term and regrouping.}$$

But by substitution from the inductive hypothesis,

$$(1 + 2 + \cdots + k) + (k + 1)$$
$$= \frac{k(k + 1)}{2} + (k + 1) \qquad \text{since the inductive hypothesis says that } 1 + 2 + \cdots + k = \frac{k(k + 1)}{2}.$$

Now use algebra to show that this expression equals the right-hand side of $P(k + 1)$:

$$\frac{k(k + 1)}{2} + (k + 1)$$

$$= \frac{k(k + 1)}{2} + \frac{2(k + 1)}{2} \qquad \text{multiply numerator and denominator of the second term by 2 to obtain a common denominator}$$

$$= \frac{k(k + 1) + 2(k + 1)}{2} \qquad \text{by adding fractions}$$

$$= \frac{(k + 2)(k + 1)}{2} \qquad \text{by factoring out } (k + 1)$$

$$= \frac{(k + 1)(k + 2)}{2} \qquad \text{by commuting the factors } (k + 1) \text{ and } (k + 2)$$

which equals the right-hand side of $P(k + 1)$.

This discussion is summarized as follows:

Theorem 4.2.2 Sum of the First n Integers

For all integers $n \geq 1$,

$$1 + 2 + \cdots + n = \frac{n(n + 1)}{2}.$$

Proof (by mathematical induction):

Let the property $P(n)$ be the equation $1 + 2 + \cdots + n = \frac{n(n + 1)}{2}$.

Show that the property is true for $n = 1$: To establish the property for $n = 1$, We must show that $1 = \frac{1(1 + 1)}{2}$. But the left-hand side of this equation is 1, and the right-hand side is $\frac{1(1 + 1)}{2} = \frac{2}{2} = 1$ also. Hence the property is true for $n = 1$.

Show that for all integers $k \geq 1$, if the property is true for $n = k$ then it is true for $n = k + 1$:

[*Suppose the property $1 + 2 + \cdots + n = \frac{n(n + 1)}{2}$ is true when an integer $k \geq 1$ is substituted for n.*]

Suppose $1 + 2 + \cdots + k = \frac{k(k + 1)}{2}$, for some integer $k \geq 1$.

[*This is the inductive hypothesis.*]

[*We must show that the property* $1 + 2 + \cdots + n = \dfrac{n(n+1)}{2}$ *is true when* $k + 1$ *is substituted for* n].

We must show that $1 + 2 + \cdots + (k + 1) = \dfrac{(k+1)((k+1)+1)}{2}$, or, equivalently

that $1 + 2 + \cdots + (k + 1) = \dfrac{(k+1)(k+2)}{2}$.   4.2.1

[*We will show that the left-hand side of equation (4.2.1) equals the right-hand side.*]

But the left-hand side of equation (4.2.1) is

$1 + 2 + \cdots + (k + 1)$

$\qquad = (1 + 2 + \cdots + k) + (k + 1)$ The next-to-last term is k because the terms are successive integers and the last term is $k + 1$.

$\qquad = \dfrac{k(k+1)}{2} + (k + 1)$ by substitution from the inductive hypothesis

$\qquad = \dfrac{k(k+1)}{2} + \dfrac{(k+1) \cdot 2}{2}$

$\qquad = \dfrac{(k+1)(k+2)}{2},$

which is the right-hand side of equation (4.2.1) [*as was to be shown*].

[*Since we have proved both the basis step and the inductive step, we conclude that the theorem is true.*]

■

The story is told that one of the greatest mathematicians of all time, Carl Friedrich Gauss (1777–1855), was given the problem of adding the numbers from 1 to 100 by his teacher when he was a young child. The teacher had asked his students to compute the sum, supposedly to gain himself some time to grade papers. But after just a few moments, Gauss produced the correct answer. Needless to say, the teacher was dumbfounded. How could young Gauss have calculated the quantity so rapidly? In his later years, Gauss explained that he had imagined the numbers paired according to the following schema.

The sum of the numbers in each pair is 101, and there are 50 pairs in all; hence the total sum is $50 \cdot 101 = 5{,}050$.

Example 4.2.2 Applying the Formula for the Sum of the First n Integers

a. Find $2 + 4 + 6 + \cdots + 500$.

b. Find $5 + 6 + 7 + 8 + \cdots + 50$.

c. For an integer $h \geq 2$, find $1 + 2 + 3 + \cdots + (h - 1)$.

Solution

a. $2 + 4 + 6 + \cdots + 500 = 2 \cdot (1 + 2 + 3 + \cdots + 250)$

$$= 2 \cdot \left(\frac{250 \cdot 251}{2} \right) \quad \begin{array}{l} \text{by applying the formula for the sum} \\ \text{of the first } n \text{ integers with } n = 250 \end{array}$$

$$= 62{,}750.$$

b. $5 + 6 + 7 + 8 + \cdots + 50 = (1 + 2 + 3 + \cdots + 50) - (1 + 2 + 3 + 4)$

$$= \frac{50 \cdot 51}{2} - 10 \quad \begin{array}{l} \text{by applying the formula for the sum} \\ \text{of the first } n \text{ integers with } n = 50 \end{array}$$

$$= 1{,}265$$

c. $1 + 2 + 3 + \cdots + (h - 1) = \dfrac{(h - 1) \cdot [(h - 1) + 1]}{2} \quad \begin{array}{l} \text{by applying the formula for the sum} \\ \text{of the first } n \text{ integers with } n = h - 1 \end{array}$

$$= \frac{(h - 1) \cdot h}{2} \quad \text{since } (h - 1) + 1 = h. \quad \blacksquare$$

The next example asks for a proof of another famous and important formula in mathematics—the formula for the sum of a geometric sequence. In a **geometric sequence,** each term is obtained from the preceding one by multiplying by a constant factor. If the first term is 1 and the constant factor is r, then the sequence is $1, r, r^2, r^3, \ldots, r^n, \ldots$. The sum of the first n terms of this sequence is given by the formula

$$\sum_{i=0}^{n} r^i = \frac{r^{n+1} - 1}{r - 1}$$

for all integers $n \geq 0$ and real numbers r not equal to 1. The expanded form of this formula is

$$r^0 + r^1 + r^2 + \cdots + r^n = \frac{r^{n+1} - 1}{r - 1},$$

and because $r^0 = 1$ and $r^1 = r$, the formula for $n \geq 1$ can be rewritten as

$$1 + r + r^2 + \cdots + r^n = \frac{r^{n+1} - 1}{r - 1}.$$

Example 4.2.3 Sum of a Geometric Sequence

Prove that $\displaystyle\sum_{i=0}^{n} r^i = \frac{r^{n+1} - 1}{r - 1}$, for all integers $n \geq 0$ and all real numbers r except 1.

Solution   In this example the property $P(n)$ is again an equation, although in this case it contains a real variable r:

$$\boxed{\sum_{i=0}^{n} r^i = \frac{r^{n+1} - 1}{r - 1}.} \quad \leftarrow \text{the property } (P(n))$$

Because r can be any real number other than 1, the proof begins by supposing that r is a particular but arbitrarily chosen real number not equal to 1. Then the proof continues by mathematical induction on n, starting with $n = 0$. In the basis step, you must show that $P(0)$ is true; that is, you show the property is true for $n = 0$. So you substitute 0 for each n in $P(n)$:

$$\sum_{0}^{0} r^i = \frac{r^{0+1} - 1}{r - 1}. \qquad \leftarrow \text{basis } (P(0))$$

In the inductive step, you suppose $P(k)$ is true; that is, you suppose the property is true for $n = k$. So you substitute k for each n in $P(n)$:

$$\sum_{i=0}^{k} r^i = \frac{r^{k+1} - 1}{r - 1}. \qquad \leftarrow \text{inductive hypothesis } (P(k))$$

Then you show that $P(k + 1)$ is true; that is, you show the property is true for $n = k + 1$. So you substitute $k + 1$ for each n in $P(n)$:

$$\sum_{i=0}^{k+1} r^i = \frac{r^{(k+1)+1} - 1}{r - 1},$$

or, equivalently,

$$\sum_{i=0}^{k+1} r^i = \frac{r^{k+2} - 1}{r - 1}. \qquad \leftarrow \text{to show } (P(k+1))$$

Theorem 4.2.3 Sum of a Geometric Sequence

For any real number r except 1, and any integer $n \geq 0$,

$$\sum_{i=0}^{n} r^i = \frac{r^{n+1} - 1}{r - 1}.$$

Proof:

Let the property $P(n)$ be the equation $\sum_{i=0}^{n} r^i = \frac{r^{n+1} - 1}{r - 1}$.

Suppose r is a particular but arbitrarily chosen real number that is not equal to 1. We must show that for all integers $n \geq 0$,

$$\sum_{i=0}^{n} r^i = \frac{r^{n+1} - 1}{r - 1}.$$

We show this by mathematical induction on n.

continued on page 224

Show that the property is true for n = 0: For $n = 0$ we must show that

$$\sum_{i=0}^{0} r^i = \frac{r^{0+1} - 1}{r - 1}.$$

The left-hand side of this equation is $r^0 = 1$. The right-hand side is

$$\frac{r^1 - 1}{r - 1} = \frac{r - 1}{r - 1} = 1$$

also because $r^1 = r$ and $r \neq 1$. [*So the property is true for n = 0.*]

Show that for all integers k ≥ 0, if the property is true for n = k then it is true for n = k + 1:

[*Suppose the property* $\displaystyle\sum_{i=0}^{n} r^i = \frac{r^{n+1} - 1}{r - 1}$ *is true when an integer k ≥ 0 is substituted in place of n.*]

Suppose $\displaystyle\sum_{i=0}^{k} r^i = \frac{r^{k+1} - 1}{r - 1}$, for $k \geq 0$.   [*This is the inductive hypothesis.*]

[*We must show that the property is true when k + 1 is substituted in place of n.*]

We must show that

$$\sum_{i=0}^{k+1} r^i = \frac{r^{(k+1)+1} - 1}{r - 1}.$$ 4.2.2

[*We will show that the left-hand side of this equation equals the right-hand side.*]

But the left-hand side of equation (4.2.2) is

$$\sum_{i=0}^{k+1} r^i = \sum_{i=0}^{k} r^i + r^{k+1} \qquad \text{by writing the } (k+1)\text{st term separately from the first } k \text{ terms}$$

$$= \frac{r^{k+1} - 1}{r - 1} + r^{k+1} \qquad \text{by substitution from the inductive hypothesis}$$

$$= \frac{r^{k+1} - 1}{r - 1} + \frac{r^{k+1}(r - 1)}{r - 1} \qquad \text{by multiplying the numerator and denominator of the second term by } (r - 1) \text{ to obtain a common denominator}$$

$$= \frac{(r^{k+1} - 1) + r^{k+1}(r - 1)}{r - 1} \qquad \text{by adding fractions}$$

$$= \frac{r^{k+1} - 1 + r^{k+2} - r^{k+1}}{r - 1} \qquad \text{by multiplying out and using the fact that } r^{k+1} \cdot r = r^{k+1} \cdot r^1 = r^{k+2}.$$

$$= \frac{r^{k+2} - 1}{r - 1} \qquad \text{by canceling the } r^{k+1}\text{'s.}$$

which is the right-hand side of equation (4. 2. 2) [*as was to be shown*].

[*Since we have proved the basis step and the inductive step, we conclude that the theorem is true.*]

■

Note that the formula for the sum of a geometric sequence can be thought of as a family of different formulas in r, one for each real number r except 1.

Example 4.2.4 Applying the Formula for the Sum of a Geometric Sequence

In each of (a) and (b) below, assume that m is an integer that is greater than or equal to 3.

a. Find $1 + 3 + 3^2 + \cdots + 3^{m-2}$.

b. Find $3^2 + 3^3 + 3^4 + \cdots + 3^m$.

Solution

a. $1 + 3 + 3^2 + \cdots + 3^{m-2} = \dfrac{3^{(m-2)+1} - 1}{3 - 1}$ by applying the formula for the sum of a geometric sequence with $r = 3$ and $n = m - 2$

$$= \dfrac{3^{m-1} - 1}{2}.$$

b. $3^2 + 3^3 + 3^4 + \cdots + 3^m = 3^2 \cdot (1 + 3 + 3^2 + \cdots + 3^{m-2})$ by factoring out 3^2

$$= 9 \cdot \left(\dfrac{3^{m-1} - 1}{2} \right) \qquad \text{by part (a).} \qquad \blacksquare$$

As with the formula for the sum of the first n integers, there is a way to think of the formula for the sum of the terms of a geometric sequence that makes it seem simple and intuitive. Let

$$S_n = 1 + r + r^2 + \cdots + r^n.$$

Then

$$rS_n = r + r^2 + r^3 + \cdots + r^{n+1},$$

and so

$$rS_n - S_n = (r + r^2 + r^3 + \cdots + r^{n+1}) - (1 + r + r^2 + \cdots + r^n)$$
$$= r^{n+1} - 1. \qquad\qquad 4.2.3$$

But

$$rS_n - S_n = (r - 1)S_n. \qquad\qquad 4.2.4$$

Equating the right-hand sides of equations (4.2.3) and (4.2.4) and dividing by $r - 1$ gives

$$S_n = \dfrac{r^{n+1} - 1}{r - 1}.$$

This derivation of the formula is attractive and is quite convincing. However, it is not as logically airtight as the proof by mathematical induction. To go from one step to another in the calculations above, the argument is made that each term among those indicated by the ellipsis (...) has such-and-such an appearance and when these are canceled such-and-such occurs. But it is impossible actually to see each such term and each such calculation, and so the accuracy of these claims cannot be fully checked. With mathematical induction it is possible to focus exactly on what happens in the middle of the ellipsis and verify without doubt that the calculations are correct.

Exercise Set 4.2

1. Use mathematical induction (and the proof of Proposition 4.2.1 as a model) to show that any amount of money of at least 14¢ can be made up using 3¢ and 8¢ coins.

2. Use mathematical induction to show that any postage of at least 12¢ can be obtained using 3¢ and 7¢ stamps.

3. For each positive integer n, let $P(n)$ be the formula

$$1^2 + 2^2 + \cdots + n^2 = \frac{n(n+1)(2n+1)}{6}.$$

   a. Write $P(1)$. Is $P(1)$ true?
   b. Write $P(k)$.
   c. Write $P(k+1)$.
   d. In a proof by mathematical induction that the formula holds for all integers $n \geq 1$, what must be shown in the inductive step?

4. For each integer n with $n \geq 2$, let $P(n)$ be the formula

$$\sum_{i=1}^{n-1} i(i+1) = \frac{n(n-1)(n+1)}{3}.$$

   a. Write $P(2)$. Is $P(2)$ true?
   b. Write $P(k)$.
   c. Write $P(k+1)$.
   d. In a proof by mathematical induction that the formula holds for all integers $n \geq 2$, what must be shown in the inductive step?

5. Fill in the missing pieces in the following proof that

$$1 + 3 + 5 + \cdots + (2n-1) = n^2$$

for all integers $n \geq 1$.

   Proof: Let the property $P(n)$ be the equation

$$1 + 3 + 5 + \cdots + (2n-1) = n.$$

   Show that the property is true for $n = 1$: To establish the formula for $n = 1$, we must show that when 1 is substituted in place of n, the left-hand side equals the right-hand side. But when $n = 1$, the left-hand side is the sum of all the odd integers from 1 to $2 \cdot 1 - 1$, which is the sum of the odd integers from 1 to 1, which is just 1. The right-hand side is __(a)__, which also equals 1. So the property is true for $n = 1$.

   Show that for all integers $k \geq 1$, if the property is true for $n = k$ then it is true for $n = k + 1$: Let k be any integer with $k \geq 1$.

   [*Suppose the property $1 + 3 + 5 + \cdots + (2n-1) = n^2$ is true when k is substituted for n.*]

   Suppose $1 + 3 + 5 + \cdots + (2k-1) =$ __(b)__.
   [*This is the inductive hypothesis.*]

   [*We must show that the property is true when $k + 1$ is substituted for n*].

   We must show that

$$\underline{(c)} = \underline{(d)}.$$

4.2.5

But the left-hand side of equation (4.2.5) is

$$1 + 3 + 5 + \cdots + (2(k+1) - 1)$$
$$= 1 + 3 + 5 + \cdots + (2k+1) \quad \text{by algebra}$$
$$= [1 + 3 + 5 + \cdots + (2k-1)] + (2k+1)$$
$$\text{the next-to-last term is } 2k - 1 \text{ because } \underline{(e)}$$
$$= k^2 + (2k+1) \quad \text{by } \underline{(f)}$$
$$= (k+1)^2 \quad \text{by algebra}$$

which is the right-hand side of equation (4.2.5) [*as was to be shown*].

   [*Since we have proved the basis step and the inductive step, we conclude that the given statement is true.*]

The proof above was heavily annotated to help make its logical flow more obvious. In standard mathematical writing, such annotation is omitted.

Prove each statement in 6–9 using mathematical induction. Do not derive them from Theorem 4.2.2 or Theorem 4.2.3.

6. For all integers $n \geq 1$, $2 + 4 + 6 + \cdots + 2n = n^2 + n$.

7. For all integers $n \geq 1$,

$$1 + 6 + 11 + 16 + \cdots + (5n - 4) = \frac{n(5n-3)}{2}.$$

8. For all integers $n \geq 0$, $1 + 2 + 2^2 + \cdots + 2^n = 2^{n+1} - 1$.

9. For all integers $n \geq 3$,

$$4^3 + 4^4 + 4^5 + \cdots + 4^n = \frac{4(4^n - 16)}{3}.$$

Prove each of the statements in 10–17 by mathematical induction.

10. $1^2 + 2^2 + \cdots + n^2 = \dfrac{n(n+1)(2n+1)}{6}$, for all integers $n \geq 1$.

11. $1^3 + 2^3 + \cdots + n^3 = \left[\dfrac{n(n+1)}{2}\right]^2$, for all integers $n \geq 1$.

12. $\dfrac{1}{1 \cdot 2} + \dfrac{1}{2 \cdot 3} + \cdots + \dfrac{1}{n(n+1)} = \dfrac{n}{n+1}$, for all integers $n \geq 1$.

13. $\displaystyle\sum_{i=1}^{n-1} i(i+1) = \dfrac{n(n-1)(n+1)}{3}$, for all integers $n \geq 2$.

14. $\displaystyle\sum_{i=1}^{n+1} i \cdot 2^i = n \cdot 2^{n+2} + 2$, for all integers $n \geq 0$.

H 15. $\displaystyle\sum_{i=1}^{n} i(i!) = (n+1)! - 1$, for all integers $n \geq 1$.

16. $\left(1 - \dfrac{1}{2^2}\right) \cdot \left(1 - \dfrac{1}{3^2}\right) \cdots \left(1 - \dfrac{1}{n^2}\right) = \dfrac{n+1}{2n}$, for all integers $n \geq 2$.

17. $\displaystyle\prod_{i=0}^{n}\left(\frac{1}{2i+1}\cdot\frac{1}{2i+2}\right)=\frac{1}{(2n+2)!}$, for all integers $n \geq 0$.

H ✶ **18.** If x is a real number not divisible by π, then for all integers $n \geq 1$,

$$\sin x + \sin 3x + \sin 5x + \cdots + \sin (2n-1)x$$
$$= \frac{1 - \cos 2nx}{2 \sin x}.$$

Use the formula for the sum of the first n integers and/or the formula for the sum of a geometric sequence to find the sums in 19–28.

19. $4 + 8 + 12 + 16 + \cdots + 200$

20. $5 + 10 + 15 + 20 + \cdots + 300$

21. $3 + 4 + 5 + 6 + \cdots + 1000$

22. $7 + 8 + 9 + 10 + \cdots + 600$

23. $1 + 2 + 3 + \cdots + (k-1)$, where k is a positive integer

24. a. $1 + 2 + 2^2 + \cdots + 2^{25}$
    b. $2 + 2^2 + 2^3 + \cdots + 2^{26}$

25. $3 + 3^2 + 3^3 + \cdots + 3^n$, where n is an integer with $n \geq 2$

26. $5^3 + 5^4 + 5^5 + \cdots + 5^k$, where k is any integer with $k > 3$.

27. $1 + \dfrac{1}{2} + \dfrac{1}{2^2} + \cdots + \dfrac{1}{2^n}$, where n is a positive integer

28. $1 - 2 + 2^2 - 2^3 + \cdots + (-1)^n 2^n$, where n is a positive integer

H **29.** Find a formula in n, a, m, and d for the sum $(a+md) + (a+(m+1)d) + (a+(m+2)d) + \cdots + (a+(m+n)d)$, where m and n are integers, $n \geq 0$, and a and d are real numbers. Justify your answer.

30. Find a formula in a, r, m, and n for the sum $ar^m + ar^{m+1} + ar^{m+2} + \cdots + ar^{m+n}$, where m and n are integers, $n \geq 0$, and a and r are real numbers. Justify your answer.

31. You have two parents, four grandparents, eight great-grandparents, and so forth.
    a. If all your ancestors were distinct, what would be the total number of your ancestors for the past 40 generations (counting your parents' generation as number one)? (*Hint:* Use the formula for the sum of a geometric sequence.)
    b. Assuming that each generation represents 25 years, how long is 40 generations?
    c. The total number of people who have ever lived is approximately 10 billion, which equals 10^{10} people. Compare this fact with the answer to part (a). What do you deduce?

32. Find the mistake in the following proof fragment.

Theorem: For any integer $n \geq 1$,

$$1^2 + 2^2 + \cdots + n^2 = \frac{n(n+1)(2n+1)}{6}.$$

"**Proof (by mathematical induction):** Certainly the theorem is true for $n = 1$ because $1^2 = 1$ and $\dfrac{1(1+1)(2\cdot 1 + 1)}{6} = 1$. So the basis step is true. For the inductive step, suppose that for some integer $k \geq 1$, $k^2 = \dfrac{k(k+1)(2k+1)}{6}$. We must show that $(k+1)^2 = \dfrac{(k+1)((k+1)+1)(2(k+1)+1)}{6} \cdots$."

✶ **33.** Use Theorem 4.2.2 to prove that if m is any odd integer and n is any integer, then $\sum_{k=0}^{m-1}(n+k)$ is divisible by m. Does the conclusion hold if m is even? Justify your answer.

H ✶ **34.** Use Theorem 4.2.2 and the result of exercise 10 to prove that if p is any prime number with $p \geq 5$, then the sum of squares of any p consecutive integers is divisible by p.

4.3 Mathematical Induction II

A good proof is one which makes us wiser. — I. Manin, *A Course in Mathematical Logic,* 1977

In natural science courses, deduction and induction are presented as alternative modes of thought—deduction being to infer a conclusion from general principles using the laws of logical reasoning, and induction being to enunciate a general principle after observing it to hold in a large number of specific instances. In this sense, then, *mathematical induction* is not inductive but deductive. Once proved by mathematical induction, a theorem is known just as certainly as if it were proved by any other mathematical method. Inductive reasoning, in the natural sciences sense, *is* used in mathematics, but only to make conjectures, not to prove them. For example, observe that

$$1 - \frac{1}{2} = \frac{1}{2}$$

$$\left(1 - \frac{1}{2}\right)\left(1 - \frac{1}{3}\right) = \frac{1}{3}$$

$$\left(1 - \frac{1}{2}\right)\left(1 - \frac{1}{3}\right)\left(1 - \frac{1}{4}\right) = \frac{1}{4}$$

This pattern seems so unlikely to occur by pure chance that it is reasonable to conjecture (though it is by no means certain) that the pattern holds true in general. In a case like this, a proof by mathematical induction (which you are asked to write in exercise 1 at the end of this section) gets to the essence of why the pattern holds in general. It reveals the mathematical mechanism that necessitates the truth of each successive case from the previous one. For instance, in this example observe that if

$$\left(1 - \frac{1}{2}\right)\left(1 - \frac{1}{3}\right) \cdots \left(1 - \frac{1}{k}\right) = \frac{1}{k},$$

then by substitution

$$\left(1 - \frac{1}{2}\right)\left(1 - \frac{1}{3}\right) \cdots \left(1 - \frac{1}{k}\right)\left(1 - \frac{1}{k+1}\right)$$

$$= \frac{1}{k}\left(1 - \frac{1}{k+1}\right) = \frac{1}{k}\left(\frac{k+1-1}{k+1}\right) = \frac{1}{k}\left(\frac{k}{k+1}\right) = \frac{1}{k+1}.$$

Thus mathematical induction makes knowledge of the general pattern a matter of mathematical certainty rather than vague conjecture.

In the remainder of this section we show how to use mathematical induction to prove additional kinds of statements such as divisibility properties of the integers and inequalities. The basic outlines of the proofs are the same in all cases, but the details of the basis and inductive steps differ from one to another.

In the example below, mathematical induction is used to establish a divisibility property.

Example 4.3.1 Proving a Divisibility Property

Use mathematical induction to prove that for all integers $n \geq 1$, $2^{2n} - 1$ is divisible by 3.

Solution   As in the previous proofs by mathematical induction, you need to identify the property $P(n)$. In this example, $P(n)$ is

$$\boxed{2^{2n} - 1 \text{ is divisible by 3.}} \quad \leftarrow \text{the property } (P(n))$$

By substitution, the statement for the basis step, $P(1)$, is

$$\boxed{2^{2 \cdot 1} - 1 \text{ is divisible by 3.}} \quad \leftarrow \text{basis } (P(1))$$

The supposition for the inductive step, $P(k)$, is

$$\boxed{2^{2k} - 1 \text{ is divisible by 3,}} \quad \leftarrow \text{inductive hypothesis } (P(k))$$

and the conclusion to be shown, $P(k + 1)$, is

$$2^{2(k+1)} - 1 \text{ is divisible by 3.} \qquad \leftarrow \text{to show } (P(k+1))$$

Recall that an integer m is divisible by 3 if, and only if, $m = 3r$ for some integer r. Now the statement $P(1)$ is true because $2^{2 \cdot 1} - 1 = 2^2 - 1 = 4 - 1 = 3$, which is divisible by 3 because $3 = 3 \cdot 1$.

To prove the inductive step, you suppose that k is an integer greater than or equal to 1 such that $P(k)$ is true. This means that $2^{2k} - 1$ is divisible by 3. You must then prove the truth of $P(k + 1)$. Or, in other words, you must show that $2^{2(k+1)} - 1$ is divisible by 3. But

$$\begin{aligned}
2^{2(k+1)} - 1 &= 2^{2k+2} - 1 \\
&= 2^{2k} \cdot 2^2 - 1 \qquad \text{by the laws of exponents} \\
&= 2^{2k} \cdot 4 - 1.
\end{aligned}$$

The aim is to show that this quantity, $2^{2k} \cdot 4 - 1$, is divisible by 3. Why should that be so? By the inductive hypothesis, $2^{2k} - 1$ is divisible by 3, and $2^{2k} \cdot 4 - 1$ resembles $2^{2k} - 1$. Indeed, if you subtract $2^{2k} - 1$ from $2^{2k} \cdot 4 - 1$, you obtain $2^{2k} \cdot 3$, which is divisible by 3:

$$\underbrace{2^{2k} \cdot 4 - 1}_{\uparrow} - \underbrace{(2^{2k} - 1)}_{\uparrow} = \underbrace{2^{2k} \cdot 3}_{\uparrow}$$

divisible by 3?   divisible by 3   divisible by 3

Adding $2^{2k} - 1$ to both sides gives

$$\underbrace{2^{2k} \cdot 4 - 1}_{\uparrow} = \underbrace{2^{2k} \cdot 3}_{\uparrow} + \underbrace{2^{2k} - 1}_{\uparrow}.$$

divisible by 3?   divisible by 3   divisible by 3

Both terms of the sum on the right-hand side of this equation are divisible by 3; hence the sum is divisible by 3. (See exercise 15 of Section 3.3.) Therefore, the left-hand side of the equation is also divisible by 3, which is what was to be shown.

This discussion is summarized as follows:

Proposition 4.3.1

For all integers $n \geq 1$, $2^{2n} - 1$ is divisible by 3.

Proof (by mathematical induction):

Let the property $P(n)$ be the sentence "$2^{2n} - 1$ is divisible by 3."

Show that the property is true for $n = 1$: To show the property is true for $n = 1$, we must show that $2^{2 \cdot 1} - 1 = 2^2 - 1 = 3$ is divisible by 3. But this is true because $3 = 3 \cdot 1$.

Show that for all integers $k \geq 1$, if the property is true for $n = k$ then it is true for $n = k + 1$:

[*Suppose the property "$2^{2n} - 1$ is divisible by 3" is true when an integer $k \geq 1$ is substituted for n.*]

Suppose $2^{2k} - 1$ is divisible by 3, for some integer $k \geq 1$.

[*Inductive hypothesis*]

continued on page 230

By definition of divisibility, this means that

$$2^{2k} - 1 = 3r \quad \text{for some integer } r.$$

[*We must show that the property "$2^{2n} - 1$ is divisible by 3" is true when $k + 1$ is substituted for n.*]

We must show that $2^{2(k+1)} - 1$ is divisible by 3.

But

$$
\begin{aligned}
2^{2(k+1)} - 1 &= 2^{2k+2} - 1 \\
&= 2^{2k} \cdot 2^2 - 1 && \text{by the laws of exponents} \\
&= 2^{2k} \cdot 4 - 1 \\
&= 2^{2k}(3 + 1) - 1 \\
&= 2^{2k} \cdot 3 + (2^{2k} - 1) && \text{by the laws of algebra} \\
&= 2^{2k} \cdot 3 + 3r && \text{by inductive hypothesis} \\
&= 3(2^{2k} + r) && \text{by factoring out the 3.}
\end{aligned}
$$

But $2^{2k} + r$ is an integer because it is a sum of products of integers, and so, by definition of divisibility, $2^{2(k+1)} - 1$ is divisible by 3 [*as was to be shown*].

[*Since we have proved the basis step and the inductive step, we conclude that the proposition is true.*]

∎

The next example illustrates the use of mathematical induction to prove an inequality.

Example 4.3.2 Proving an Inequality

Use mathematical induction to prove that for all integers $n \geq 3$,

$$2n + 1 < 2^n.$$

Solution   In this example the property $P(n)$ is the inequality

$$\boxed{2n + 1 < 2^n.}$$ ← the property $(P(n))$

By substitution, the statement for the basis step, $P(3)$, is

$$\boxed{2 \cdot 3 + 1 < 2^3.}$$ ← basis $(P(3))$

The supposition for the inductive step, $P(k)$, is

$$\boxed{2k + 1 < 2^k,}$$ ← inductive hypothesis $(P(k))$

and the conclusion to be shown is

$$\boxed{2(k + 1) + 1 < 2^{k+1}.}$$ ← to show $(P(k + 1))$

To prove the basis step, observe that the statement $P(3)$ is true because $2 \cdot 3 + 1 = 7$, $2^3 = 8$, and $7 < 8$.

To prove the inductive step, suppose $P(k)$ is true for an integer $k \geq 3$. [*This is the inductive hypothesis.*] This means that $2k + 1 < 2^k$ is assumed to be true for an integer $k \geq 3$. Then derive the truth of $P(k + 1)$. Or, in other words, show that the inequality $2(k + 1) + 1 < 2^{k+1}$ is true. But by multiplying out and regrouping,

$$2(k + 1) + 1 = 2k + 3 = (2k + 1) + 2, \qquad \text{4.3.1}$$

and by substitution from the inductive hypothesis,

$$(2k + 1) + 2 < 2^k + 2. \qquad \text{4.3.2}$$

Hence

$$2(k + 1) + 1 < 2^k + 2 \qquad \text{The left-most part of equation (4.3.1) is less than the right-most part of inequality (4.3.2).}$$

If it can be shown that $2^k + 2$ is less than 2^{k+1}, then the desired inequality will have been proved. But since the quantity 2^k can be added to or subtracted from an inequality without changing its direction,

$$2^k + 2 < 2^{k+1} \quad \Leftrightarrow \quad 2 < 2^{k+1} - 2^k = 2^k(2 - 1) = 2^k.$$

And since multiplying or dividing an inequality by 2 does not change its direction,

$$2 < 2^k \quad \Leftrightarrow \quad 1 = \frac{2}{2} < \frac{2^k}{2} = 2^{k-1} \qquad \text{by the laws of exponents.}$$

This last inequality is clearly true for all $k \geq 2$. Hence it is true that $2(k + 1) + 1 < 2^{k+1}$.

This discussion is made more flowing (but less intuitive) in the following formal proof:

Proposition 4.3.2

For all integers $n \geq 3$, $2n + 1 < 2^n$.

Proof (by mathematical induction):

Let the property $P(n)$ be the inequality

$$2n + 1 < n.$$

Show that the property is true for $n = 3$: To prove the property for $n = 3$, we must show that $2 \cdot 3 + 1 < 2^3$. But $2 \cdot 3 + 1 = 7$, $2^3 = 8$, and $7 < 8$. Hence the property is true for $n = 3$.

Show that for all integers $k \geq 3$, if the property is true for $n = k$ then it is true for $n = k + 1$:

[*Suppose "$2n + 1 < 2^n$" is true when an integer $k \geq 3$ is substituted for n.*]

Suppose $2k + 1 < 2^k$, for some integer k such that $k \geq 3$.

[*Inductive hypothesis*]

[*We must show that "$2n + 1 < 2^n$" is true when $k + 1$ is substituted for n.*]

We must show that $2(k + 1) + 1 < 2^{k+1}$, or, equivalently, $2k + 3 < 2^{k+1}$.

continued on page 232

But

$$2k + 3 = (2k + 1) + 2 \quad \text{by algebra}$$

$$< 2^k + 2^k \quad \text{because } 2k + 1 < 2^k \text{ by the inductive hypothesis}$$
$$\text{and because } 2 < 2^k \text{ for all integers } k \geq 2$$

$$\therefore 2k + 3 < 2 \cdot 2^k = 2^{k+1} \quad \text{by the laws of exponents.}$$

[*This is what we needed to show.*]

[*Since we have proved the basis step and the inductive step, we conclude that the proposition is true.*]

∎

The last example of this section demonstrates how to use mathematical induction to show that the terms of a sequence satisfy a certain explicit formula.

Example 4.3.3 Proving a Property of a Sequence

Define a sequence $a_1, a_2, a_3, \ldots$ as follows.*

$$a_1 = 2$$
$$a_k = 5a_{k-1} \quad \text{for all integers } k \geq 2.$$

a. Write the first four terms of the sequence.

b. Use mathematical induction to show that the terms of the sequence satisfy the property

$$a_n = 2 \cdot 5^{n-1} \quad \text{for all integers } n \geq 1.$$

Solution

a. $a_1 = 2$.

$a_2 = 5a_{2-1} = 5a_1 = 5 \cdot 2 = 10$

$a_3 = 5a_{3-1} = 5a_2 = 5 \cdot 10 = 50$

$a_4 = 5a_{4-1} = 5a_3 = 5 \cdot 50 = 250$.

b. To use mathematical induction to show that the property is true in general, you begin by showing that the first term of the sequence satisfies the property. Then you suppose that the kth term of the sequence (for some integer $k \geq 1$) satisfies the property and show that the $(k + 1)$st term also satisfies the property.

Show that the property is true for $n = 1$: For $n = 1$, the property states that $2 \cdot 5^{1-1} = 2 \cdot 5^0 = 2 \cdot 1 = 2$. But by definition of the sequence, $a_1 = 2$. Hence the property is true for $n = 1$.

Show that for all integers $n \geq 1$, if the property is true for $n = k$ then it is true for $n = k + 1$: Let k be an integer with $k \geq 1$ and suppose that $a_k = 2 \cdot 5^{k-1}$. [*This is the inductive hypothesis.*]

*This is an example of a recursive definition. The general subject of recursion is discussed in Chapter 8.

We must show that $a_{k+1} = 2 \cdot 5^{(k+1)-1} = 2 \cdot 5^k$.

But the left-hand side of the equation is

$$
\begin{aligned}
a_{k+1} &= 5a_{(k+1)-1} && \text{by definition of } a_1, a_2, a_3, \ldots \\
&= 5a_k && \text{since } (k+1) - 1 = k \\
&= 5 \cdot (2 \cdot 5^{k-1}) && \text{by inductive hypothesis} \\
&= 2 \cdot (5 \cdot 5^{k-1}) && \text{by regrouping} \\
&= 2 \cdot 5^k && \text{by the laws of exponents}
\end{aligned}
$$

which is the right-hand side of the equation [*as was to be shown*].

[*Since we have proved the basis step and the inductive step, we conclude that the formula holds for all terms of the sequence.*] ∎

Exercise Set 4.3

1. Based on the discussion of the product $(1 - \frac{1}{2})(1 - \frac{1}{3})$ $(1 - \frac{1}{4}) \cdots (1 - \frac{1}{n})$ at the beginning of this section, conjecture a formula for general n. Prove your conjecture by mathematical induction.

2. Experiment with computing values of the product $(1 + \frac{1}{1})(1 + \frac{1}{2})(1 + \frac{1}{3}) \cdots (1 + \frac{1}{n})$ for small values of n to conjecture a formula for this product for general n. Prove your conjecture by mathematical induction.

3. Observe that

$$
\frac{1}{1 \cdot 3} = \frac{1}{3}
$$

$$
\frac{1}{1 \cdot 3} + \frac{1}{3 \cdot 5} = \frac{2}{5}
$$

$$
\frac{1}{1 \cdot 3} + \frac{1}{3 \cdot 5} + \frac{1}{5 \cdot 7} = \frac{3}{7}
$$

$$
\frac{1}{1 \cdot 3} + \frac{1}{3 \cdot 5} + \frac{1}{5 \cdot 7} + \frac{1}{7 \cdot 9} = \frac{4}{9}
$$

Guess a general formula and prove it by mathematical induction.

H 4. Observe that

$$
\begin{aligned}
1 &= 1, \\
1 - 4 &= -(1 + 2), \\
1 - 4 + 9 &= 1 + 2 + 3, \\
1 - 4 + 9 - 16 &= -(1 + 2 + 3 + 4), \\
1 - 4 + 9 - 16 + 25 &= 1 + 2 + 3 + 4 + 5.
\end{aligned}
$$

Guess a general formula and prove it by mathematical induction.

5. Evaluate the sum $\sum_{k=1}^{n} \frac{k}{(k+1)!}$ for $n = 1, 2, 3, 4,$ and 5. Make a conjecture about a formula for this sum for general n, and prove your conjecture by mathematical induction.

6. For each positive integer n, let $P(n)$ be the property

$$5^n - 1 \text{ is divisible by } 4.$$

   a. Write $P(0)$. Is $P(0)$ true?
   b. Write $P(k)$.
   c. Write $P(k + 0)$.
   d. In a proof by mathematical induction that this divisibility property holds for all integers $n \geq 0$, what must be shown in the inductive step?

7. For each positive integer n, let $P(n)$ be the property

$$2^n < (n + 1)!.$$

   a. Write $P(2)$. Is $P(2)$ true?
   b. Write $P(k)$.
   c. Write $P(k + 1)$.
   d. In a proof by mathematical induction that this inequality holds for all integers $n \geq 2$, what must be shown in the inductive step?

Prove each statement in 8–23 by mathematical induction.

8. $5^n - 1$ is divisible by 4, for each integer $n \geq 0$.

9. $7^n - 1$ is divisible by 6, for each integer $n \geq 0$.

10. $n^3 - 7n + 3$ is divisible by 3, for each integer $n \geq 0$.

11. $3^{2n} - 1$ is divisible by 8, for each integer $n \geq 0$.

12. For any integer $n \geq 1$, $7^n - 2^n$ is divisible by 5.

H 13. For any integer $n \geq 1$, $x^n - y^n$ is divisible by $x - y$, where x and y are any integers with $x \neq y$.

H 14. $n^3 - n$ is divisible by 6, for each integer $n \geq 2$.

15. $n(n^2 + 5)$ is divisible by 6, for each integer $n \geq 1$.

16. $2^n < (n + 1)!$, for all integers $n \geq 2$.

17. $1 + 3n \leq 4^n$, for every integer $n \geq 0$.

18. $5^n + 9 < 6^n$, for all integers $n \geq 2$.

19. $n^2 < 2^n$, for all integers $n \geq 5$.

20. $2^n < (n+2)!$, for all integers $n \geq 0$.

21. $\sqrt{n} < \dfrac{1}{\sqrt{1}} + \dfrac{1}{\sqrt{2}} + \cdots + \dfrac{1}{\sqrt{n}}$, for all integers $n \geq 2$.

22. $1 + nx \leq (1+x)^n$, for all real numbers $x > -1$ and integers $n \geq 2$.

23. a. $n^3 > 2n + 1$, for all integers $n \geq 2$.
 b. $n! > n^2$, for all integers $n \geq 4$.

24. A sequence $a_1, a_2, a_3, \ldots$ is defined by letting $a_1 = 3$ and $a_k = 7a_{k-1}$ for all integers $k \geq 2$. Show that $a_n = 3 \cdot 7^{n-1}$ for all integers $n \geq 1$.

25. A sequence $b_0, b_1, b_2, \ldots$ is defined by letting $b_0 = 5$ and $b_k = 4 + b_{k-1}$ for all integers $k \geq 1$. Show that $(b_n)^2 > 16n^2$ for all integers $n \geq 0$.

26. A sequence $c_0, c_1, c_2, \ldots$ is defined by letting $c_0 = 3$ and $c_k = (c_{k-1})^2$ for all integers $k \geq 1$. Show that $c_n = 3^{2^n}$ for all integers $n \geq 0$.

27. A sequence $d_1, d_2, d_3, \ldots$ is defined by letting $d_1 = 2$ and $d_k = \dfrac{d_{k-1}}{k}$ for all integers $k \geq 2$. Show that for all integers $n \geq 1$, $d_n = \dfrac{2}{n!}$.

28. Prove that for all integers $n \geq 1$,

$$\frac{1}{3} = \frac{1+3}{5+7} = \frac{1+3+5}{7+9+11} = \cdots$$
$$= \frac{1+3+\cdots+(2n-1)}{(2n+1)+\cdots+(4n-1)}.$$

29. As each of a group of business people arrives at a meeting, each shakes hands with all the other people present. Use mathematical induction to show that if n people come to the meeting then $[n(n-1)]/2$ handshakes occur.

In order for a proof by mathematical induction to be valid, the basis statement must be true for $n = a$ and the argument of the inductive step must be correct for every integer $k \geq a$. In 30 and 31 find the mistakes in the "proofs" by mathematical induction.

30. **"Theorem:"** For any integer $n \geq 1$, all the numbers in a set of n numbers are equal to each other.

"**Proof (by mathematical induction):** It is obviously true that all the numbers in a set consisting of just one number are equal to each other, so the basis step is true. For the inductive step, let $A = \{a_1, a_2, \ldots, a_k, a_{k+1}\}$ be any set of $k+1$ numbers. Form two subsets each of size k:

$$B = \{a_1, a_2, a_3, \ldots, a_k\} \quad \text{and}$$
$$C = \{a_1, a_3, a_4, \ldots, a_{k+1}\}.$$

(B consists of all the numbers in A except a_{k+1}, and C consists of all the numbers in A except a_2.) By inductive hypothesis, all the numbers in B equal a_1 and all the numbers in C equal a_1 (since both sets have only k numbers).

But every number in A is in B or C, so all the numbers in A equal a_1; hence all are equal to each other."

31. **"Theorem:"** For all integers $n \geq 1$, $3^n - 2$ is even.

"**Proof (by mathematical induction):** Suppose the theorem is true for an integer k, where $k \geq 1$. That is, suppose that $3^k - 2$ is even. We must show that $3^{k+1} - 2$ is even. But

$$3^{k+1} - 2 = 3^k \cdot 3 - 2 = 3^k(1+2) - 2$$
$$= (3^k - 2) + 3^k \cdot 2.$$

Now $3^k - 2$ is even by inductive hypothesis and $3^k \cdot 2$ is even by inspection. Hence the sum of the two quantities is even (by Theorem 3.1.1). It follows that $3^{k+1} - 2$ is even, which is what we needed to show."

32. An L-tromino, or tromino for short, is similar to a domino but is shaped like an L: ⌐. Call a checkerboard that is formed using m squares on a side an $m \times m$ checkerboard. If one square is removed from a 4×4 checkerboard, the remaining squares can be completely covered by trominos. For instance, a covering for one such board is the following:

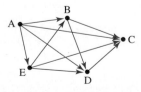

Use mathematical induction to prove that for any integer $n \geq 1$, if one square is removed from a $2^n \times 2^n$ checkerboard, the remaining squares can be completely covered by trominos.

33. In a round-robin tournament each team plays every other team exactly once. If the teams are labeled $T_1, T_2, \ldots, T_n$, then the outcome of such a tournament can be represented by a drawing, called a *directed graph,* in which the teams are represented as dots and an arrow is drawn from one dot to another if, and only if, the team represented by the first dot beats the team represented by the second dot. For example, the directed graph below shows one outcome of a round-robin tournament involving five teams, A, B, C, D, and E.

Use mathematical induction to show that in any round-robin tournament involving n teams, where $n \geq 2$, it is possible to label the teams $T_1, T_2, \ldots, T_n$ so that T_i beats T_{i+1} for all $i = 1, 2, \ldots, n - 1$. (For instance, one such labeling in the example above is $T_1 = A, T_2 = B, T_3 = C, T_4 = E, T_5 = D$. (*Hint:* Given $k + 1$ teams, pick one—say T'—and apply the inductive hypothesis to the remaining teams to obtain an ordering $T_1, T_2, \ldots, T_k$. Consider three cases: T' beats T_1, T' loses to the first m teams (where $1 \le m \le k - 1$) and beats the $(m + 1)$st team, and T' loses to all the other teams.)

H ✱ **34.** On the outside rim of a circular disk the integers from 1 through 30 are painted in random order. Show that no matter what this order is, there must be three successive integers whose sum is at least 45.

4.4 Strong Mathematical Induction and the Well-Ordering Principle

Mathematics takes us still further from what is human into the region of absolute necessity, to which not only the actual world, but every possible world, must conform.
— Bertrand Russell, 1902

Strong mathematical induction is similar to ordinary mathematical induction in that it is a technique for establishing the truth of a sequence of statements about integers. Also, a proof by strong mathematical induction consists of a basis step and an inductive step. However, the basis step may contain proofs for several initial values, and in the inductive step the truth of the predicate $P(n)$ is assumed not just for one value of n but for *all* values through $k - 1$, and then the truth of $P(k)$ is proved.

Principle of Strong Mathematical Induction

Let $P(n)$ be a property that is defined for integers n, and let a and b be fixed integers with $a \le b$. Suppose the following two statements are true:

1. $P(a)$, $P(a + 1)$, ..., and $P(b)$ are all true. (**basis step**)

2. For any integer $k > b$, if $P(i)$ is true for all integers i with $a \le i < k$, then $P(k)$ is true. (**inductive step**)

Then the statement

$$\text{for all integers } n \ge a, \ P(n)$$

is true. (The supposition that $P(i)$ is true for all integers i with $a \le i < k$ is called the **inductive hypothesis.**)

Any statement that can be proved with ordinary mathematical induction can be proved with strong mathematical induction. The reason is that given any integer $k \ge b$, if the truth of $P(k)$ alone implies the truth of $P(k + 1)$, then certainly the truth of $P(a)$, $P(a + 1), \ldots$, and $P(k)$ implies the truth of $P(k + 1)$. It is also the case that any statement that can be proved with strong mathematical induction can be proved with ordinary mathematical induction. A proof is sketched in exercise 25 at the end of this section.

Strictly speaking, the principle of strong mathematical induction can be written without a basis step if the inductive step is changed to "$\forall k \ge a$, if $P(k)$ is true then $P(k + 1)$ is true." The reason for this is that the statement "$P(i)$ is true for all integers i with $a \le i < k$" is vacuously true for $k = a$. Hence, if the implication in the inductive step is true, then the conclusion $P(a)$ must also be true,* which proves the basis step. However, in many cases

---

*If you have proved that a certain if-then statement is true and if you also know that the hypothesis is true, then the conclusion must be true.

the proof of the implication for $k > b$ does not work for $a \leq k \leq b$. So it is a good idea to get into the habit of thinking separately about the cases where $a \leq k \leq b$ by explicitly including a basis step.

The principle of strong mathematical induction is known under a variety of different names including the second principle of induction, the second principle of finite induction, and the principle of complete induction.

Applying Strong Mathematical Induction

The divisibility by a prime theorem (Theorem 3.3.2) states that any integer greater than 1 is divisible by a prime number. We prove this theorem below using strong mathematical induction.

Example 4.4.1 Divisibility by a Prime

Prove that any integer greater than 1 is divisible by a prime number.

Solution   Let the property $P(n)$ be the sentence "n is divisible by a prime number." We use strong mathematical induction to prove that the property is true for every integer $n \geq 2$.

Show that the property is true for $n = 2$: The property is true for $n = 2$ because 2 is a prime number and 2 is divisible by 2.

Show that for all integers $k > 2$, if the property is true for all i with $2 \leq i < k$, then it is true for k: Let k be an integer with $k > 2$. Suppose that

For all integers i with $2 \leq i < k$, i is divisible by a prime number.
[*This is the inductive hypothesis.*]

[*We must show that k is divisible by a prime number.*] Either k is prime or k is not prime. If k is prime, then k is divisible by a prime number, namely itself. If k is not prime, then $k = ab$, where a and b are integers with $2 \leq a < k$ and $2 \leq b < k$. By the inductive hypothesis, a is divisible by a prime number p, and so by transitivity of divisibility, k is also divisible by p. Hence, regardless of whether k is prime or not, k is divisible by a prime number [*as was to be shown*].

[*Since we have proved the basis step and the inductive step of the strong mathematical induction, we conclude that the given statement is true.*] ■

Sometimes strong mathematical induction must be used to show that the terms of certain sequences satisfy a certain property.

Example 4.4.2 Proving a Property of a Sequence

Define a sequence $a_1, a_2, a_3, \ldots$ as follows:

$$a_1 = 0,$$
$$a_2 = 2,$$
$$a_k = 3a_{\lfloor k/2 \rfloor} + 2 \quad \text{for all integers } k \geq 3.$$

a. Find the first seven terms of the sequence.

b. Prove that a_n is even for each integer $n \geq 1$.

Solution

a. $a_1 = 0,$

   $a_2 = 2,$

   $a_3 = 3a_{\lfloor 3/2 \rfloor} + 2 = 3a_1 + 2 = 3 \cdot 0 + 2 = 2,$

   $a_4 = 3a_{\lfloor 4/2 \rfloor} + 2 = 3a_2 + 2 = 3 \cdot 2 + 2 = 8,$

   $a_5 = 3a_{\lfloor 5/2 \rfloor} + 2 = 3a_2 + 2 = 3 \cdot 2 + 2 = 8,$

   $a_6 = 3a_{\lfloor 6/2 \rfloor} + 2 = 3a_3 + 2 = 3 \cdot 2 + 2 = 8$

   $a_7 = 3a_{\lfloor 7/2 \rfloor} + 2 = 3a_3 + 2 = 3 \cdot 2 + 2 = 8.$

b. Let the property $P(n)$ be the sentence "a_n is even." We use strong mathematical induction to show that the property holds for all integers $n \geq 1$.

Show that the property is true for $n = 1$ and $n = 2$: The property is true for $n = 1$ and $n = 2$ because $a_1 = 0$ and $a_2 = 2$ and both 0 and 2 are even integers.

Show that for all integers $k > 2$, if the property is true for all i with $1 \leq i < k$, then it is true for k: Let k be an integer with $k > 2$ and suppose that

   a_i is even for all integers i with $1 \leq i < k$.

   [*This is the inductive hypothesis.*]

[*We must show that a_k is even.*] By definition of $a_1, a_2, a_3, \ldots,$

$$a_k = 3a_{\lfloor k/2 \rfloor} + 2 \quad \text{for all integers } k \geq 3.$$

Now $a_{\lfloor k/2 \rfloor}$ is even by the inductive hypothesis [*because $k > 2$ and so $1 \leq \lfloor k/2 \rfloor < k$*]. Thus $3a_{\lfloor k/2 \rfloor}$ is even [*because odd · even = even*], and hence $3a_{\lfloor k/2 \rfloor} + 2$ is even [*because even + even = even—see Section 3.1*]. Consequently, a_k, which equals $3a_{\lfloor k/2 \rfloor} + 2$, is even [*as was to be shown*].

[*Since we have proved the basis step and the inductive step of the strong mathematical induction, we conclude that the given statement is true.*] ■

Another use of strong induction concerns the computation of products. A product of four numbers may be computed in a variety of different ways as indicated by the placing of parentheses. For instance,

$((x_1 x_2)x_3)x_4$ means multiply x_1 and x_2, multiply the result by x_3, and then multiply that number by x_4.

And

$(x_1 x_2)(x_3 x_4)$ means multiply x_1 and x_2, multiply x_3 and x_4, and then take the product of the two.

Note that in both examples above, although the factors are multiplied in a different order, the number of multiplications—three—is the same. Strong mathematical induction is used to prove a generalization of this fact.

Convention

Let us agree to say that a single number x_1 is a product with one factor and can be computed with zero multiplications.

Example 4.4.3 The Number of Multiplications Needed to Multiply n Distinct Numbers

Prove that for any integer $n \geq 1$, if $x_1, x_2, \ldots, x_n$ are n distinct real numbers, then no matter how the parentheses are inserted into their product, the number of multiplications used to compute the product is $n - 1$.

Solution Let the property $P(n)$ be the sentence "If $x_1, x_2, \ldots, x_n$ are n distinct real numbers, then no matter how the parentheses are inserted into their product, the number of multiplications used to compute the product is $n - 1$." We use strong mathematical induction to show that the property is true for all integers $n \geq 1$.

Show that the property is true for $n = 1$: The property is true for $n = 1$ because by agreement, x_1 is a product with one factor and is computed using $1 - 1$, or 0, multiplications.

Show that for all integers $k > 1$, if the property is true for all i with $1 \leq i < k$, then it is true for k: Let $k > 1$ be an integer and suppose that

For all i with $1 \leq i < k$, if $x_1, x_2, \ldots, x_i$ are i distinct real numbers, then no matter how the parentheses are inserted into their product, the number of multiplications used to compute the product is $i - 1$.

[*This is the inductive hypothesis.*]

Consider a product of k distinct factors: $x_1, x_2, \ldots, x_k$. [*We must show that no matter how parentheses are inserted into the product of these factors, the number of multiplications is $k - 1$.*] When parentheses are inserted in order to compute the product of the factors $x_1, x_2, \ldots, x_n$, some multiplication must be the final one. (For instance, in the product $((x_1 x_2)x_3)((x_4(x_5 x_6))x_7)$, the final multiplication is between $((x_1 x_2)x_3)$ and $((x_4(x_5 x_6))x_7)$.) Consider the two factors in this final multiplication. Each is itself a product of fewer than k factors. Say the left-hand product consists of r_k and the right-hand product of s_k factors. Then $1 \leq r_k < k$ and $1 \leq s_k < k$, and so by the inductive hypothesis, the number of multiplications for the left-hand product is $r_k - 1$ and the number of multiplications for the right-hand product is $s_k - 1$. It follows that the number of multiplications to compute the product of all the factors $x_1, x_2, \ldots, x_k$ is

$$(r_k - 1) + (s_k - 1) + 1,$$

where the $+1$ at the end represents the final multiplication between the left-hand and right-hand products. But the sum of the factors in the left-hand product plus those in the right-hand product is the total number of factors in the product. Hence $r_k + s_k = k$, and the number of multiplications equals

$$(r_k - 1) + (s_k - 1) + 1 = (r_k + s_k) - 1 = k - 1.$$

[*This is what was to be shown.*]

[*Since we have proved the basis step and the inductive step of the strong mathematical induction, we conclude that the given statement is true.*] ∎

Binary Representation of Integers

Strong mathematical induction makes possible a proof of the frequently used fact that every positive integer n has a unique binary integer representation. The proof looks complicated because of all the notation needed to write down the various steps. But the idea of the proof is simple. It is that if smaller integers than n have unique representations as sums of powers of 2, then the unique representation for n as a sum of powers of 2 can be found by taking the representation for $n/2$ (or for $(n-1)/2$ if n is odd) and multiplying it by 2.

Theorem 4.4.1 Existence and Uniqueness of Binary Integer Representations

Given any positive integer n, n has a unique representation in the form

$$n = c_r \cdot 2^r + c_{r-1} \cdot 2^{r-1} + \cdots + c_2 \cdot 2^2 + c_1 \cdot 2 + c_0,$$

where r is a nonnegative integer, $c_r = 1$, and $c_j = 1$ or 0 for all $j = 0, 1, 2, \ldots, r - 1$.

Proof:

We give separate proofs by strong mathematical induction to show first the existence and second the uniqueness of the binary representation.

Existence (proof by strong mathematical induction): Consider the formula

$$n = c_r \cdot 2^r + c_{r-1} \cdot 2^{r-1} + \cdots + c_2 \cdot 2^2 + c_1 \cdot 2 + c_0,$$

where r is a nonnegative integer, $c_r = 1$, and $c_j = 1$ or 0 for all $j = 0, 1, 2, \ldots, r - 1$.

[*This is the property $P(n)$.*]

Show that the property is true for $n = 1$: Let $r = 0$ and $c_0 = 1$. Then $1 = c_r \cdot 2^r$, and so $n = 1$ can be written in the required form.

Show that for all integers $k > 1$, if the property is true for all i with $1 \leq i < k$, then it is true for k: Let k be an integer with $k > 1$. Suppose that for all integers i with $1 \leq i < k$, i can be written in the required form

$$i = c_r \cdot 2^r + c_{r-1} \cdot 2^{r-1} + \cdots + c_2 \cdot 2^2 + c_1 \cdot 2 + c_0,$$

where r is a nonnegative integer, $c_r = 1$, and $c_j = 1$ or 0 for all $j = 0, 1, 2, \ldots, r - 1$.

[*This is the inductive hypothesis.*]

We must show that k can be written as a sum of powers of 2 in the required form:

Case 1 (k is even): In this case $k/2$ is an integer, and since $1 \leq k/2 < k$, then by inductive hypothesis,

$$\frac{k}{2} = c_r \cdot 2^r + c_{r-1} \cdot 2^{r-1} + \cdots + c_2 \cdot 2^2 + c_1 \cdot 2 + c_0,$$

where r is a nonnegative integer, $c_r = 1$, and $c_j = 1$ or 0 for all $j = 0, 1, 2, \ldots, r - 1$. Multiplying both sides of the equation by 2 gives

$$k = c_r \cdot 2^{r+1} + c_{r-1} \cdot 2^r + \cdots + c_2 \cdot 2^3 + c_1 \cdot 2^2 + c_0 \cdot 2,$$

which is a sum of powers of 2 of the required form.

Case 2 (k is odd): In this case $(k - 1)/2$ is an integer, and since $1 \leq (k - 1)/2 < k$, then by inductive hypothesis,

$$\frac{k - 1}{2} = c_r \cdot 2^r + c_{r-1} \cdot 2^{r-1} + \cdots + c_2 \cdot 2^2 + c_1 \cdot 2 + c_0,$$

where r is a nonnegative integer, $c_r = 1$, and $c_j = 1$ or 0 for all $j = 0, 1, 2, \ldots, r - 1$. Multiplying both sides of the equation by 2 and adding 1 gives

$$k = c_r \cdot 2^{r+1} + c_{r-1} \cdot 2^r + \cdots + c_2 \cdot 2^3 + c_1 \cdot 2^2 + c_0 \cdot 2 + 1,$$

which is the sum of powers of 2 of the required form.

continued on page 240

The preceding arguments show that regardless of whether k is even or odd, k has a representation of the required form. [*Or, in other words, $P(k)$ is true as was to be shown*].

[*Since we have proved the basis step and the inductive step of the strong mathematical induction, the existence half of the theorem is true.*]

***Uniqueness*:** To prove uniqueness, suppose that there is an integer n with two different representations as a sum of nonnegative integer powers of 2. Equating the two representations and canceling all identical terms gives

$$2^r + c_{r-1} \cdot 2^{r-1} + \cdots + c_1 \cdot 2 + c_0 \ = \ 2^s + d_{s-1} \cdot 2^{s-1} + \cdots + d_1 \cdot 2 + d_0 \qquad \text{4.4.1}$$

where r and s are nonnegative integers, $r < s$, and each c_i and each d_i equal 0 or 1. But by the formula for the sum of a geometric sequence (Theorem 4.2.3),

$$
\begin{aligned}
2^r + c_{r-1} \cdot 2^{r-1} + \cdots + c_1 \cdot 2 + c_0 &\leq \ 2^r + 2^{r-1} + \cdots + 2 + 1 \ = \ 2^{r+1} - 1 \\
&< \ 2^s \\
&\leq \ 2^s + d_{s-1} \cdot 2^{s-1} + \cdots + d_1 \cdot 2 + d_0,
\end{aligned}
$$

which contradicts equation (4.4.1). Hence the supposition is false, so any integer n has only one representation as a sum of nonnegative integer powers of 2.

The Well-Ordering Principle for the Integers

The well-ordering principle for the integers looks very different from both the ordinary and the strong principles of mathematical induction, but it can be shown that all three principles are equivalent. That is, if any one of the three is true, then so are both of the others.

Well-Ordering Principle for the Integers

Let S be a set containing one or more integers all of which are greater than some fixed integer. Then S has a least element.

Note that when the context makes the reference clear, we will write simply "the well-ordering principle" rather than "the well-ordering principle for integers."

Example 4.4.4 Finding Least Elements

In each case, if the set has a least element, state what it is. If not, explain why the well-ordering principle is not violated.

a. The set of all positive real numbers.

b. The set of all nonnegative integers n such that $n^2 < n$.

c. The set of all nonnegative integers of the form $46 - 7k$, where k is an integer.

Solution

a. There is no least positive real number. For if x is any positive real number, then $x/2$ is a positive real number that is less than x. No violation of the well-ordering principle occurs because the well-ordering principle refers only to sets of integers, and this set is not a set of integers.

b. There is no least nonnegative integer n such that $n^2 < n$ because there is *no* nonnegative integer that satisfies this inequality. The well-ordering principle is not violated because the well-ordering principle refers only to sets that contain at least one element.

c. The following table shows values of $46 - 7k$ for various values of k.

k	0	1	2	3	4	5	6	7	$\cdots$	-1	-2	-3	$\cdots$
46 − 7k	46	39	32	25	18	11	4	-3	$\cdots$	53	60	67	

The table suggests, and you can easily confirm, that $46 - 7k < 0$ for $k \geq 7$ and that $46 - 7k \geq 46$ for $k \leq 0$. Therefore, from the other values in the table it is clear that 4 is the least nonnegative integer of the form $46 - 7k$. This corresponds to $k = 6$. ∎

Another way to look at the analysis of Example 4.4.4(c) is to observe that subtracting six 7's from 46 leaves 4 left over and this is the least nonnegative integer obtained by repeated subtraction of 7's from 46. In other words, 6 is the quotient and 4 is the remainder for the division of 46 by 7. More generally, in the division of any integer n by any positive integer d, the remainder r is the least nonnegative integer of the form $n - dk$. This is the heart of the following proof of the existence part of the quotient-remainder theorem (the part that guarantees the existence of a quotient and a remainder of the division of an integer by a positive integer). For a proof of the uniqueness of the quotient and remainder, see exercise 18 of Section 3.7.

Quotient-Remainder Theorem (Existence Part)

Given any integer n and any positive integer d, there exist integers q and r such that

$$n = dq + r \quad \text{and} \quad 0 \leq r < d.$$

Proof:

Let S be the set of all nonnegative integers of the form

$$n - dk,$$

where k is an integer. This set has at least one element. [*For if n is nonnegative, then*

$$n - 0 \cdot d = n \geq 0,$$

and so n − 0 · d is in S. And if n is negative, then

$$n - nd = n\underbrace{(1 - d)}_{} \geq 0,$$

$\underset{\leq 0}{\uparrow} \qquad \overset{}{\underset{\leq 0 \text{ since } d \text{ is a positive integer}}{\nwarrow}}$

and so n − nd is in S.] It follows by the well-ordering principle that S contains a least element r. Then, for some specific integer $k = q$,

$$n - dq = r$$

[*because every integer in S can be written in this form*]. Adding dq to both sides gives

$$n = dq + r.$$

continued on page 242

Furthermore, $r < d$. [*For suppose $r \geq d$. Then*

$$n - d(q + 1) = n - dq - d = r - d \geq 0,$$

and so $n - d(q + 1)$ would be a nonnegative integer in S that would be smaller than r. But r is the smallest integer in S. This contradiction shows that the supposition $r \geq d$ must be false.] The preceding arguments prove that there exist integers r and q for which

$$n = dq + r \quad \text{and} \quad 0 \leq r < d.$$

[*This is what was to be shown.*]

Another consequence of the well-ordering principle is the fact that any strictly decreasing sequence of nonnegative integers is finite. That is, if $r_1, r_2, r_3, \ldots$ is a sequence of nonnegative integers satisfying

$$r_i > r_{i+1}$$

for all $i \geq 1$, then $r_1, r_2, r_3, \ldots$ is a finite sequence. [*For by the well-ordering principle such a sequence would have to have a least element r_k. It follows that r_k must be the final term of the sequence because if there were a term r_{k+1}, then since the sequence is strictly decreasing, $r_{k+1} < r_k$, which would be a contradiction.*] This fact is frequently used in computer science to prove that algorithms terminate after a finite number of steps and to prove that the guard conditions for loops eventually become false. It was also used implicitly in the proof of Theorem 3.3.2 and to justify the claim in Section 3.8 that the Euclidean algorithm eventually terminates.

Exercise Set 4.4

1. Suppose $a_1, a_2, a_3, \ldots$ is a sequence defined as follows:

$$a_1 = 1, a_2 = 3,$$
$$a_k = a_{k-2} + 2a_{k-1} \quad \text{for all integers } k \geq 3.$$

Prove that a_n is odd for all integers $n \geq 1$.

2. Suppose $b_1, b_2, b_3, \ldots$ is a sequence defined as follows:

$$b_1 = 4, b_2 = 12$$
$$b_k = b_{k-2} + b_{k-1} \quad \text{for all integers } k \geq 3.$$

Prove that b_n is divisible by 4 for all integers $n \geq 1$.

3. Suppose that $c_0, c_1, c_2, \ldots$ is a sequence defined as follows:

$$c_0 = 2, c_1 = 2, c_2 = 6,$$
$$c_k = 3c_{k-3} \quad \text{for all integers } k \geq 3.$$

Prove that c_n is even for all integers $n \geq 0$.

4. Suppose that $d_1, d_2, d_3, \ldots$ is a sequence defined as follows:

$$d_1 = \frac{9}{10}, d_2 = \frac{10}{11},$$
$$d_k = d_{k-1} \cdot d_{k-2} \quad \text{for all integers } k \geq 3.$$

Prove that $d_n \leq 1$ for all integers $n \geq 0$.

5. Suppose that $e_0, e_1, e_2, \ldots$ is a sequence defined as follows:

$$e_0 = 1, e_1 = 2, e_2 = 3,$$
$$e_k = e_{k-1} + e_{k-2} + e_{k-3} \quad \text{for all integers } k \geq 3.$$

Prove that $e_n \leq 3^n$ for all integers $n \geq 0$.

6. Suppose that $f_1, f_2, f_3, \ldots$ is a sequence defined as follows:

$$f_1 = 1, f_k = 2 \cdot f_{\lfloor k/2 \rfloor} \quad \text{for all integers } k \geq 2.$$

Prove that $f_n \leq n$ for all integers $n \geq 1$.

7. Suppose that $g_0, g_1, g_2, \ldots$ is a sequence defined as follows:

$$g_0 = 12, g_1 = 29,$$
$$g_k = 5g_{k-1} - 6g_{k-2} \quad \text{for all integers } k \geq 2.$$

Prove that $g_n = 5 \cdot 3^n + 7 \cdot 2^n$ for all integers $n \geq 0$.

8. Suppose that $h_0, h_1, h_2, \ldots$ is a sequence defined as follows:

$$h_0 = 1, h_1 = 2, h_2 = 3,$$
$$h_k = h_{k-1} + h_{k-2} + h_{k-3} \quad \text{for all integers } k \geq 3.$$

a. Prove that $h_n \leq 3^n$ for all integers $n \geq 0$.
b. Suppose that s is any real number such that $s^3 \geq s^2 + s + 1$. (This implies that $s > 1.83$.) Prove that $h_n \leq s^n$ for all $n \geq 2$.

9. Define a sequence $a_1, a_2, a_3, \ldots$ as follows: $a_1 = 1$, $a_2 = 3$, and $a_k = a_{k-1} + a_{k-2}$ for all integers $k \geq 3$. (This sequence is known as the Lucas sequence.) Use strong mathematical induction to prove that $a_n \leq \left(\dfrac{7}{4}\right)^n$ for all integers $n \geq 1$.

10. You begin solving a jigsaw puzzle by finding two pieces that match and fitting them together. Each subsequent step of the solution consists of fitting together two blocks made up of one or more pieces that have previously been assembled. Use strong mathematical induction to prove that the number of steps required to put together all n pieces of a jigsaw puzzle is $n - 1$.

H 11. Use strong mathematical induction to prove the existence part of the unique factorization theorem: Every integer greater than or equal to 2 is either a prime number or a product of prime numbers.

12. Any product of two or more integers is a result of successive multiplications of two integers at a time. For instance, here are a few of the ways in which $a_1 a_2 a_3 a_4$ might be computed: $(a_1 a_2)(a_3 a_4)$ or $((a_1 a_2)a_3)a_4)$ or $a_1((a_2 a_3)a_4)$. Use strong mathematical induction to prove that any product of two or more odd integers is odd.

13. Any sum of two or more integers is a result of successive additions of two integers at a time. For instance, here are a few of the ways in which $a_1 + a_2 + a_3 + a_4$ might be computed: $(a_1 + a_2) + (a_3 + a_4)$ or $((a_1 + a_2) + a_3) + a_4$ or $a_1 + ((a_2 + a_3) + a_4)$. Use strong mathematical induction to prove that any sum of two or more even integers is even.

H 14. Use strong mathematical induction to prove that for any integer $n > 2$, if n is even, then any sum of n odd integers is even, and if n is odd, then any sum of n odd integers is odd.

15. Compute $4^1, 4^2, 4^3, 4^4, 4^5, 4^6, 4^7$, and 4^8. Make a conjecture about the units digit of 4^n where n is a positive integer. Use strong mathematical induction to prove your conjecture.

16. Compute $3^0, 3^1, 3^2, 3^3, 3^4, 3^5, 3^6, 3^7, 3^8, 3^9$, and 3^{10}. Make a conjecture about the units digit of 3^n where n is a positive integer. Use strong mathematical induction to prove your conjecture.

17. Find the mistake in the following "proof" that purports to show that every nonnegative integer power of every nonzero real number is 1.

"**Proof:** Let r be any nonzero real number and let the property $P(n)$ be the equation "$r^n = 1$."

Show that the property is true for $n = 0$: The property is true for $n = 0$ because $r^0 = 1$ by definition of zeroth power.

Show that for all integers $k > 0$, if the property is true for all integers i with $0 \leq i < k$, then it is true for k: Let $k > 0$ be an integer, and suppose that $r^i = 1$ for all integers

i with $0 \leq i < k$. [*We must show that $r^k = 1$.*] Now

$$\begin{aligned} r^k &= r^{(k-1)+(k-1)-(k-2)} && \text{because } (k-1) + \\ & && (k-1) - (k-2) = k \\ &= \frac{r^{k-1} \cdot r^{k-1}}{r^{k-2}} && \text{by the laws of exponents} \\ &= \frac{1 \cdot 1}{1} && \text{by inductive hypothesis} \\ &= 1. \end{aligned}$$

Thus $r^k = 1$ [*as was to be shown*].

[*Since we have proved the basis step and the inductive step, we conclude that $r^n = 1$ for all integers $n \geq 0$.*]"

✳ 18. Use the well-ordering principle to prove Theorem 3.3.2: Every integer greater than 1 is divisible by a prime number.

19. Use the well-ordering principle to prove that every integer n greater than 1 is either a prime number or a product of prime numbers.

20. The Archimedean property for the rational numbers states that for all rational numbers r, there is an integer n such that $n > r$. Prove this property.

21. Use the result of exercise 20 and the well-ordering principle for the integers to show that given any rational number r, there is an integer m such that $m \leq r < m + 1$.

H 22. Use the well-ordering principle to prove that given any integer $n \geq 1$, there exists an odd integer m and a nonnegative integer k such that $n = 2^k \cdot m$.

✳ 23. Use the well-ordering principle to prove that if a and b are any integers not both zero, then there exist integers u and v such that $\gcd(a, b) = ua + vb$. (*Hint:* Let S be the set of all positive integers of the form $ua + vb$ for some integers u and v.)

24. Suppose $P(n)$ is a property such that
   1. $P(0), P(1), P(2)$ are all true,
   2. for all integers $k \geq 0$, if $P(k)$ is true, then $P(3k)$ is true.
   Must it follow that $P(n)$ is true for all integers $n \geq 0$? If yes, explain why; if no, give a counterexample.

25. Prove that if a statement can be proved by strong mathematical induction, then it can be proved by ordinary mathematical induction. To do this, let $P(n)$ be a property that is defined for integers n, and suppose the following two statements are true:
   1. $P(a), P(a + 1), P(a + 2), \ldots, P(b)$.
   2. For any integer $k > b$, if $P(i)$ is true for all integers i with $a \leq i < k$, then $P(k)$ is true.
   The principle of strong mathematical induction would allow us to conclude immediately that $P(n)$ is true for all integers $n \geq a$. Can we reach the same conclusion using the principle of ordinary mathematical induction? Yes! To see

this, let $Q(n)$ be the property

$P(j)$ is true for all integers j with $a \leq j \leq n$.

Then use ordinary mathematical induction to show that $Q(n)$ is true for all integers $n \geq b$. That is, prove

1. $Q(b)$ is true.
2. For any integer $k \geq b$, if $Q(k)$ is true the $Q(k+1)$ is true.

26. Give examples to illustrate the proof of Theorem 4.4.1.

H 27. It is a fact that every integer $n \geq 1$ can be written in the form

$$c_r \cdot 3^r + c_{r-1} \cdot 3^{r-1} + \cdots + c_2 \cdot 3^2 + c_1 \cdot 3 + c_0,$$

where $c_r = 1$ or 2 and $c_i = 0, 1,$ or 2 for all integers $i = 0, 1, 2, \ldots, r - 1$. Sketch a proof of this fact.

H ✱ 28. Use mathematical induction to prove the existence part of the quotient-remainder theorem for integers $n \geq 0$.

H ✱ 29. Prove that if a statement can be proved by ordinary mathematical induction, then it can be proved by the well-ordering principle.

✱ 30. Prove that if a statement can be proved by the well-ordering principle, then it can be proved by ordinary mathematical induction.

4.5 Application: Correctness of Algorithms

[P]rogramming reliably—must be an activity of an undeniably mathematical nature .... You see, mathematics is about thinking, and doing mathematics is always trying to think as well as possible. — Edsger W. Dijkstra (1981)

Robert W. Floyd (1936–2002)

Edsger W. Dijkstra (1930–2002)

What does it mean for a computer program to be correct? Each program is designed to do a specific task—calculate the mean or median of a set of numbers, compute the size of the paychecks for a company payroll, rearrange names in alphabetical order, and so forth. We will say that a program is correct if it produces the output specified in its accompanying documentation for each set of input data of the type specified in the documentation.*

Most computer programmers write their programs using a combination of logical analysis and trial and error. In order to get a program to run at all, the programmer must first fix all syntax errors (such as writing **ik** instead of **if,** failing to declare a variable, or using a restricted keyword for a variable name). When the syntax errors have been removed, however, the program may still contain logical errors that prevent it from producing correct output. Frequently, programs are tested using sets of sample data for which the correct output is known in advance. And often the sample data are deliberately chosen to test the correctness of the program under extreme circumstances. But for most programs the number of possible sets of input data is either infinite or unmanageably large, and so no amount of program testing can give perfect confidence that the program will be correct for all possible sets of legal input data.

Since 1967, with the publication of a paper by Robert W. Floyd,[†] considerable effort has gone into developing methods for proving programs correct at the time they are composed. One of the pioneers in this effort, Edsger W. Dijkstra, asserts that "we now take the position that it is not only the programmer's task to produce a correct program but also to demonstrate its correctness in a convincing manner."[‡] Another leader in the

---

*Consumers of computer programs want an even more stringent definition of correctness. If a user puts in data of the wrong type, the user wants a decent error message, not a system crash.

[†]R. W. Floyd, "Assigning meanings to programs," *Proc. Symp. Appl. Math.,* Amer. Math. Soc. **19** (1967), 19–32.

[‡]Edsger Dijkstra in O. J. Dahl, E. W. Dijkstra, and C. A. R. Hoare, *Structured Programming* (London: Academic Press, 1972), p. 5.

field, David Gries, goes so far as to say that "a program and its proof should be developed hand-in-hand, with the *proof* usually leading the way."* If such methods can eventually be used to write large scientific and commercial programs, the benefits to society will be enormous.

As with most techniques that are still in the process of development, methods for proving program correctness are somewhat awkward and unwieldy. In this section we give an overview of the general format of correctness proofs and the details of one crucial technique, the *loop invariant procedure*. At this point, we switch from using the term *program,* which refers to a particular programming language, to the more general term *algorithm*.

Assertions

Consider an algorithm that is designed to produce a certain final state from a certain initial state. Both the initial and final states can be expressed as predicates involving the input and output variables. Often the predicate describing the initial state is called the **pre-condition for the algorithm,** and the predicate describing the final state is called the **post-condition for the algorithm.**

Example 4.5.1 Algorithm Pre-Conditions and Post-Conditions

Here are pre- and post-conditions for some typical algorithms.

a. Algorithm to compute a product of nonnegative integers

Pre-condition: The input variables m and n are nonnegative integers.

Post-condition: The output variable p equals mn.

b. Algorithm to find quotient and remainder of the division of one positive integer by another

Pre-condition: The input variables a and b are positive integers.

Post-condition: The output variables q and r are integers such that
$a = bq + r$ and $0 \leq r < b$.

c. Algorithm to sort a one-dimensional array of real numbers

Pre-condition: The input variable $A[1], A[2], \ldots, A[n]$ is a one-dimensional array of real numbers.

Post-condition: The output variable $B[1], B[2], \ldots, B[n]$ is a one-dimensional array of real numbers with same elements as $A[1], A[2], \ldots, A[n]$ but with the property that $B[i] \leq B[j]$ whenever $i \leq j$. ■

A proof of algorithm correctness consists of showing that if the pre-condition for the algorithm is true for a collection of values for the input variables and if the statements of the algorithms are executed, then the post-condition is also true.

The divide-and-conquer principle has been useful in many aspects of computer programming, and proving algorithm correctness is no exception. The steps of an algorithm

*David Gries, *The Science of Programming* (New York: Springer-Verlag, 1981), p. 164.

are divided into sections with assertions about the current state of algorithm variables inserted at strategically chosen points:

[*Assertion 1: pre-condition for the algorithm*]

{Algorithm statements}

[*Assertion 2*]

{Algorithm statements}

$$\vdots$$

[*Assertion k − 1*]

{Algorithm statements}

[*Assertion k: post-condition for the algorithm*]

Successive pairs of assertions are then treated as pre- and post-conditions for the algorithm statements between them. For each $i = 1, 2, \ldots, k - 1$, one proves that if Assertion i is true and all the algorithm statements between Assertion i and Assertion $(i + 1)$ are executed, then Assertion $(i + 1)$ is true. Once all these individual proofs have been completed, one knows that Assertion k is true. And since Assertion 1 is the same as the pre-condition for the algorithm and Assertion k is the same as the post-condition for the algorithm, one concludes that the entire algorithm is correct with respect to its pre- and post-conditions.

Loop Invariants

The method of loop invariants is used to prove correctness of a loop with respect to certain pre- and post-conditions. It is based on the principle of mathematical induction. Suppose that an algorithm contains a **while** loop and that entry to this loop is restricted by a condition G, called the **guard.** Suppose also that assertions describing the current states of algorithm variables have been placed immediately preceding and immediately following the loop. The assertion just preceding the loop is called the **pre-condition for the loop** and the one just following is called the **post-condition for the loop.** The annotated loop has the following appearance:

[*Pre-condition for the loop*]

while (G)

  [*Statements in the body of the loop.
  None contain branching statements
  that lead outside the loop.*]

end while

[*Post-condition for the loop*]

• Definition

A loop is defined as **correct with respect to its pre- and post-conditions** if, and only if, whenever the algorithm variables satisfy the pre-condition for the loop and the loop terminates after a finite number of steps, the algorithm variables satisfy the post-condition for the loop.

Establishing the correctness of a loop uses the concept of loop invariant. **A loop invariant** is a predicate with domain a set of integers, which is chosen to satisfy three conditions:

1. It is true before the first iteration of the loop.

2. For each iteration of the loop, if it is true before the iteration, then it is true after the iteration.

3. If the loop terminates after a finite number of iterations, the truth of the loop invariant ensures the truth of the post condition of the loop.

A version of the following theorem, called the *loop invariant theorem,* was first formulated by C. A. R. Hoare in 1969.

*C. A. R. Hoare
(born 1934)*

Courtesy of Tony Hoare

Theorem 4.5.1 Loop Invariant Theorem

Let a **while** loop with guard G be given, together with pre- and post-conditions that are predicates in the algorithm variables. Also let a predicate $I(n)$, called the **loop invariant,** be given. If the following four properties are true, then the loop is correct with respect to its pre- and post-conditions.

I. Basis Property: The pre-condition for the loop implies that $I(0)$ is true before the first iteration of the loop.

II. Inductive Property: For all integers $k \geq 0$, if the guard G and the loop invariant $I(k)$ are both true before an iteration of the loop, then $I(k+1)$ is true after iteration of the loop.

III. Eventual Falsity of Guard: After a finite number of iterations of the loop, the guard G becomes false.

IV. Correctness of the Post-Condition: If N is the least number of iterations after which G is false and $I(N)$ is true, then the values of the algorithm variables will be as specified in the post-condition of the loop.

Proof:

The loop invariant theorem follows easily from the principle of mathematical induction. Assume that $I(n)$ is a predicate that satisfies properties I–IV of the loop invariant theorem. [*We will prove that the loop is correct with respect to its pre- and post-conditions.*] Properties I and II are the basis and inductive steps needed to prove the truth of the following statement:

> For all integers $n \geq 0$, if the **while** loop
> iterates n times, then $I(n)$ is true. 4.5.1

Thus, by the principle of mathematical induction, since both I and II are true, statement (4.5.1) is also true.

Property III says that the guard G eventually becomes false. At that point the loop will have been iterated some number, say N, of times. Since $I(n)$ is true after the nth iteration for every $n \geq 0$, then $I(n)$ is true after the Nth iteration. That is, after the Nth iteration the guard is false and $I(N)$ is true. But this is the hypothesis of property IV, which is an if-then statement. Since statement IV is true (by assumption) and its hypothesis is true (by the argument just given), it follows (by modus ponens) that its conclusion is also true. That is, the values of all algorithm variables after execution of the loop are as specified in the post-condition for the loop.

The loop invariant in the procedure for proving loop correctness may seem like a rabbit in a hat. Where does it come from? The fact is that developing a good loop invariant is a tricky process. Although learning how to do it is beyond the scope of this book, it is worth pursuing in a more advanced course. Many people who have become good at the process claim it has significantly altered their outlook on programming and has greatly improved their ability to write good code.

Another tricky aspect of handling correctness proofs arises from the fact that execution of an algorithm is a dynamic process—it takes place in time. As execution progresses, the values of variables keep changing, yet often their names stay the same. In the following discussion, when we need to make a distinction between the values of a variable just before execution of an algorithm statement and just after execution of the statement, we will attach the subscripts *old* and *new* to the variable name.

Example 4.5.2 Correctness of a Loop to Compute a Product

The following loop is designed to compute the product mx for a nonnegative integer m and a real number x, without using a built-in multiplication operation. Prior to the loop, variables i and *product* have been introduced and given initial values $i = 0$ and *product* $= 0$.

> [*Pre-condition: m is a nonnegative integer,*
> *x is a real number, i = 0, and product = 0.*]
>
> **while** $(i \neq m)$
>
> > 1. *product* := *product* + x
> > 2. $i := i + 1$
>
> **end while**
>
> [*Post-condition: product = mx*]

Let the loop invariant be

$$\boxed{I(n): \ i = n \quad \text{and} \quad product = nx}$$

The guard condition G of the **while** loop is

$$\boxed{G: \ i \neq m}$$

Use the loop invariant theorem to prove that the **while** loop is correct with respect to the given pre- and post-conditions.

Solution

I. Basis Property: [$I(0)$ *is true before the first iteration of the loop.*]

$I(0)$ is "$i = 0$ and *product* $= 0 \cdot x$", which is true before the first iteration of the loop because $0 \cdot x = 0$.

II. Inductive Property: [*If* $G \wedge I(k)$ *is true before a loop iteration (where* $k \geq 0$*), then* $I(k + 1)$ *is true after the loop iteration.*]

Suppose k is a nonnegative integer such that $G \wedge I(k)$ is true before an iteration of the loop. Then as execution reaches the top of the loop, $i \neq m$, *product* $= kx$, and $i = k$. Since $i \neq m$, the guard is passed and statement 1 is executed. Before execution of statement 1,

$$product_{\text{old}} = kx.$$

Thus execution of statement 1 has the following effect:

$$product_{\text{new}} = product_{\text{old}} + x = kx + x = (k+1)x.$$

Similarly, before statement 2 is executed,

$$i_{\text{old}} = k,$$

so after execution of statement 2,

$$i_{\text{new}} = i_{\text{old}} + 1 = k + 1.$$

Hence after the loop iteration, the statement $I(k+1)$ ($i = k+1$ and $product = (k+1)x$) is true. This is what we needed to show.

III. Eventual Falsity of Guard: [*After a finite number of iterations of the loop, G becomes false.*]

The guard G is the condition $i \neq m$, and m is a nonnegative integer. By I and II, it is known that

for all integers $n \geq 0$, if the loop is iterated
n times, then $i = n$ and $product = nx$.

So after m iterations of the loop, $i = m$. Thus G becomes false after m iterations of the loop.

IV. Correctness of the Post-Condition: [*If N is the least number of iterations after which G is false and $I(N)$ is true, then the value of the algorithm variables will be as specified in the post-condition of the loop.*]

According to the post-condition, the value of $product$ after execution of the loop should be mx. But if G becomes false after N iterations, $i = m$. And if $I(N)$ is true, $i = N$ and $product = Nx$. Since both conditions (G false and $I(N)$ true) are satisfied, $m = i = N$ and $product = mx$ as required. ∎

In the remainder of this section, we present proofs of the correctness of the crucial loops in the division algorithm and the Euclidean algorithm. (These algorithms were given in Section 3.8.)

Correctness of the Division Algorithm

The division algorithm is supposed to take a nonnegative integer a and a positive integer d and compute nonnegative integers q and r such that $a = dq + r$ and $0 \leq r < d$. Initially, the variables r and q are introduced and given the values $r = a$ and $q = 0$. The crucial loop, annotated with pre- and post-conditions, is the following:

[*Pre-condition: a is a nonnegative integer
and d is a positive integer, $r = a$, and $q = 0$.*]
    while $(r \geq d)$
        1. $r := r - d$
        2. $q := q + 1$
    end while
[*Post-condition: q and r are nonnegative integers
with the property that $a = qd + r$ and $0 \leq r < d$.*]

Proof:

To prove the correctness of the loop, let the loop invariant be

$$I(n): r = a - nd \geq 0 \quad \text{and} \quad n = q.$$

The guard of the **while** loop is

$$G: r \geq d$$

I. **Basis property:** [$I(0)$ *is true before the first iteration of the loop.*]

$I(0)$ is "$r = a - 0 \cdot d$ and $q = 0$." But by the pre-condition, $r = a$. So since $a = a - 0 \cdot d$, then $r = a - 0 \cdot d$. Also $q = 0$ by the pre-condition. Hence $I(0)$ is true before the first iteration of the loop.

II. **Inductive Property:** [*If $G \wedge I(k)$ is true before an iteration of the loop (where $k \geq 0$), then $I(k+1)$ is true after iteration of the loop.*]

Suppose k is a nonnegative integer such that $G \wedge I(k)$ is true before an iteration of the loop. Since G is true, $r \geq d$ and the loop is entered. Also since $I(k)$ is true, $r = a - kd \geq 0$ and $k = q$. Hence, before execution of statements 1 and 2,

$$r_{\text{old}} \geq d \quad \text{and} \quad r_{\text{old}} = a - kd \quad \text{and} \quad q_{\text{old}} = k.$$

When statements 1 and 2 are executed, then,

$$r_{\text{new}} = r_{\text{old}} - d = (a - kd) - d = a - (k+1)d \qquad 4.5.2$$

and

$$q_{\text{new}} = q_{\text{old}} + 1 = k + 1 \qquad 4.5.3$$

In addition, since $r_{\text{old}} \geq d$ before execution of statements 1 and 2, after execution of these statements,

$$r_{\text{new}} = r_{\text{old}} - d \geq d - d \geq 0. \qquad 4.5.4$$

Putting equations (4.5.2), (4.5.3), and (4.5.4) together shows that after iteration of the loop,

$$r_{\text{new}} \geq 0 \quad \text{and} \quad r_{\text{new}} = a - (k+1)d \quad \text{and} \quad q_{\text{new}} = k + 1.$$

Hence $I(k+1)$ is true.

III. **Eventual Falsity of the Guard:** [*After a finite number of iterations of the loop, G becomes false.*]

The guard G is the condition $r \geq d$. Each iteration of the loop reduces the value of r by d and yet leaves r nonnegative. Thus the values of r form a decreasing sequence of nonnegative integers, and so (by the well-ordering principle) there must be a smallest such r, say $r_{\min}$. Then $r_{\min} < d$. [*If $r_{\min}$ were greater than d, the loop would iterate another time, and a new value of r equal to $r_{\min} - d$ would be obtained. But this new value would be smaller than $r_{\min}$ which would contradict the fact that $r_{\min}$ is the smallest remainder obtained by repeated iteration of the loop.*] Hence as soon as the value $r = r_{\min}$ is computed, the value of r becomes less than d, and so the guard G is false.

IV. **Correctness of the Post-Condition:** [*If N is the least number of iterations after which G is false and $I(N)$ is true, then the values of the algorithm variables will be as specified in the post-condition of the loop.*]

Suppose that for some nonnegative integer N, G is false and $I(N)$ is true. Then $r < d, r = a - Nd, r \geq 0$, and $q = N$. Since $q = N$, by substitution,

$$r = a - qd.$$

Or, adding qd to both sides,

$$a = qd + r.$$

Combining the two inequalities involving r gives

$$0 \leq r < d.$$

But these are the values of q and r specified in the post-condition, so the proof is complete.

Correctness of the Euclidean Algorithm

The Euclidean algorithm is supposed to take integers A and B with $A > B \geq 0$ and compute their greatest common divisor. Just before the crucial loop, variables a, b, and r have been introduced with $a = A, b = B$, and $r = B$. The crucial loop, annotated with pre- and post-conditions, is the following:

[*Pre-condition: A and B are integers*
with $A > B \geq 0, a = A, b = B, r = B$.]

 while $(b \neq 0)$

    1. $r := a \bmod b$
    2. $a := b$
    3. $b := r$

 end while

[*Post-condition: $a = \gcd(A, B)$*]

Proof:

To prove the correctness of the loop, let the invariant be

$$\boxed{I(n): \gcd(a, b) = \gcd(A, B) \quad \text{and} \quad 0 \leq b < a.}$$

The guard of the **while** loop is

$$\boxed{G: b \neq 0.}$$

I. Basis Property: [$I(0)$ *is true before the first iteration of the loop.*]
    $I(0)$ is

$$\gcd(A, B) = \gcd(a, b) \quad \text{and} \quad 0 \leq b < a.$$

According to the pre-condition,

$$a = A, \quad b = B, \quad r = B, \quad \text{and} \quad 0 \leq B < A.$$

Hence $\gcd(A, B) = \gcd(a, b)$. Since $0 \leq B < A$, $b = B$, and $a = A$ then $0 \leq b < a$. Hence $I(0)$ is true.

II. **Inductive Property:** [*If $G \wedge I(k)$ is true before an iteration of the loop (where $k \geq 0$), then $I(k+1)$ is true after iteration of the loop.*]

Suppose k is a nonnegative integer such that $G \wedge I(k)$ is true before an iteration of the loop. [*We must show that $I(k+1)$ is true after iteration of the loop.*] Since G is true, $b_{\text{old}} \neq 0$ and the loop is entered. And since $I(k)$ is true, immediately before statement 1 is executed,

$$\gcd(a_{\text{old}}, b_{\text{old}}) = \gcd(A, B) \quad \text{and} \quad 0 \leq b_{\text{old}} < a_{\text{old}}. \qquad 4.5.5$$

After execution of statement 1,

$$r_{\text{new}} = a_{\text{old}} \bmod b_{\text{old}}.$$

Thus, by the quotient-remainder theorem,

$$a_{\text{old}} = b_{\text{old}} \cdot q + r_{\text{new}} \quad \text{for some integer } q$$

and r_{new} has the property that

$$0 \leq r_{\text{new}} < b_{\text{old}}. \qquad 4.5.6$$

By Lemma 3.8.2,

$$\gcd(a_{\text{old}}, b_{\text{old}}) = \gcd(b_{\text{old}}, r_{\text{new}}),$$

and by the equation of (4.5.5),

$$\gcd(a_{\text{old}}, b_{\text{old}}) = \gcd(A, B).$$

Hence

$$\gcd(b_{\text{old}}, r_{\text{new}}) = \gcd(A, B). \qquad 4.5.7$$

When statements 2 and 3 are executed,

$$a_{\text{new}} = b_{\text{old}} \quad \text{and} \quad b_{\text{new}} = r_{\text{new}}. \qquad 4.5.8$$

Substituting equations (4.5.8) into equation (4.5.7) yields

$$\gcd(a_{\text{new}}, b_{\text{new}}) = \gcd(A, B). \qquad 4.5.9$$

By inequality (4.5.6),

$$0 \leq r_{\text{new}} < b_{\text{old}}.$$

So substituting the values from equations (4.5.8) gives

$$0 \leq b_{\text{new}} < a_{\text{new}}. \qquad 4.5.10$$

Hence after the iteration of the loop, by equation (4.5.9) and inequality (4.5.10),

$$\gcd(a, b) = \gcd(A, B) \quad \text{and} \quad 0 \leq b < a,$$

which is $I(k+1)$. [*This is what we needed to show.*]

III. **Eventual Falsity of the Guard:** [*After a finite number of iterations of the loop, G becomes false.*]

Each value of b obtained by repeated iteration of the loop is nonnegative and less than the previous value of b. Thus, by the well-ordering principle, there is a least value $b_{\min}$. The fact is that $b_{\min} = 0$. [*If $b_{\min}$ were not 0, then since r is given the value of $b_{\min}$ in statement 3, r would not be 0 either. But $r \neq 0$ means that the guard is true, and so the loop is iterated another time. In this iteration a value of r is calculated that is less than the previous value of b, $b_{\min}$. Then the value of b is changed to r, which is less than $b_{\min}$. This contradicts the fact that $b_{\min}$ is the least value of b obtained by repeated iteration of the loop. Hence $b_{\min} = 0$.*] Since

$b_{\min} = 0$, the guard is false immediately following the loop interaction in which $b_{\min}$ is calculated.

IV. Correctness of the Post-Condition: [*If N is the least number of iterations after which G is false and I(N) is true, then the values of the algorithm variables will be as specified in the post-condition.*]

Suppose that for some nonnegative integer N, G is false and $I(N)$ is true. [*We must show the truth of the post-condition: $a = gcd(A, B)$.*] Since G is false, $b = 0$, and since $I(N)$ is true,

$$\gcd(a, b) = \gcd(A, B).$$ 4.5.11

Substituting $b = 0$ into equation (4.5.11) gives

$$\gcd(a, 0) = \gcd(A, B).$$

But by Lemma 3.8.1,

$$\gcd(a, 0) = a.$$

Hence $a = \gcd(A, B)$ [*as was to be shown*].

Exercise Set 4.5

Exercises 1–5 contain a while loop and a predicate. In each case show that if the predicate is true before entry to the loop, then it is also true after exit from the loop.

1. loop: **while** ($m \geq 0$ and $m \leq 100$)
      $m := m + 1$
      $n := n - 1$
   end while

   predicate: $m + n = 100$

2. loop: **while** ($m \geq 0$ and $m \leq 100$)
      $m := m + 4$
      $n := n - 2$
   end while

   predicate: $m + n$ is odd

3. loop: **while** ($m \geq 0$ and $m \leq 100$)
      $m := 3 \cdot m$
      $n := 5 \cdot n$
   end while

   predicate: $m^3 > n^2$

4. loop: **while** ($n \geq 0$ and $n \leq 100$)
      $n := n + 1$
   end while

   predicate: $2^n < (n + 2)!$

5. loop: **while** ($n \geq 3$ and $n \leq 100$)
      $n := n + 1$
   end while

   predicate: $2n + 1 \leq 2^n$

Exercises 6–9 each contain a while loop annotated with a pre- and a post-condition and also a loop invariant. In each case, use the loop invariant theorem to prove the correctness of the loop with respect to the pre- and post-conditions.

6. [*Pre-condition: m is a nonnegative integer, x is a real number, i = 0, and exp = 1.*]

   while ($i \neq m$)
      1. $exp := exp \cdot x$
      2. $i := i + 1$
   end while

   [*Post-condition: $exp = exp \cdot x^n$ and $i = n$*]
   loop invariant: $I(n)$ is "$exp = x^n$ and $i = n$."

7. [*Pre-condition: largest = A[1] and i = 1*]

   while ($i \neq m$)
      1. $i := i + 1$
      2. **if** $A[i] >$ largest **then** largest $:= A[i]$
   end while

   [*Post-condition: largest = maximum value of A[1], A[2], ..., A[m]*]

   loop invariant: $I(n)$ is "largest = maximum value of $A[1]$, $A[2]$, ..., $A[n + 1]$ and $i = n + 1$."

8. [*Pre-condition: sum = A[1] and i = 1*]

> **while** ($i \neq m$)
>> 1. $i := i + 1$
>> 2. $sum := sum + A[i]$
>
> **end while**

[*Post-condition: sum = A[1] + A[2] + $\cdots$ + A[m]*]

loop invariant: $I(n)$ is "$i = n + 1$ and $sum = A[1] + A[2] + \cdots + A[n + 1]$."

9. [*Pre-condition: a = A and A is a positive integer.*]

> **while** ($a > 0$)
>> 1. $a := a - 2$
>
> **end while**

[*Post-condition: a = 0 if A is even and a = −1 if A is odd.*]

loop invariant: $I(n)$ is "both a and A are even integers or both are odd integers and $a \geq -1$."

H ✱ **10.** Prove correctness of the **while** loop of Algorithm 3.8.3 (in exercise 24 of Exercise Set 3.8) with respect to the following pre- and post-conditions:

Pre-condition: A and B are positive integers, $a = A$, and $b = B$.

Post-condition: One of a or b is zero and the other is nonzero. Whichever is nonzero equals gcd(A, B).

Use the loop invariant

$I(n)$ "(1) a and b are nonnegative integers with gcd(a, b) = gcd(A, B).
(2) at most one of a and b equals 0,
(3) $0 \leq a + b \leq A + B - n$."

11. The following **while** loop implements a way to multiply two numbers that was developed by the ancient Egyptians.

[*Pre-condition: A and B are positive integers, $x = A$, $y = B$, and product = 0.*]

> **while** ($y \neq 0$)
>> $r := y \bmod 2$
>>
>> **if** $r = 0$
>>> **then do** $x := 2 \cdot x$
>>>> $y := y \ div \ 2$
>>>
>>> **end do**
>>
>> **if** $r = 1$
>>> **then do** $product := product + x$
>>>
>>> **end do**
>
> **end while**

[*Post-condition: product = A $\cdot$ B*]

Prove the correctness of this loop with respect to its pre- and post-conditions by using the loop invariant

$$I(n): \quad \text{"} xy + product = A \cdot B \text{."}$$

✱ **12.** The following sentence could be added to the loop invariant for the Euclidean algorithm:

> There exist integers u, v, s, and t such that
> $$a = uA + vB \quad \text{and} \quad b = sA + tB. \qquad 4.5.12$$

a. Show that this sentence is a loop invariant for

> **while** ($b \neq 0$)
>> $r := a \bmod b$
>>
>> $a := b$
>>
>> $b := r$
>
> **end while**

b. Show that if initially $a = A$ and $b = B$, then sentence (4.5.12) is true before the first iteration of the loop.

c. Explain how the correctness proof for the Euclidean algorithm together with the results of (a) and (b) above allow you to conclude that given any integers A and B with $A > B \geq 0$, there exist integers u and v so that gcd(A, B) = $uA + vB$.

d. By actually calculating u, v, s, and t at each stage of execution of the Euclidean algorithm, find integers u and v so that gcd(330, 156) = $330u + 156v$.

CHAPTER 5

SET THEORY

*Georg Cantor
(1845–1918)*

In the late nineteenth century, Georg Cantor was the first to realize the potential usefulness of investigating properties of sets in general as distinct from properties of the elements that comprise them. Many mathematicians of his time resisted accepting the validity of Cantor's work. Now, however, abstract set theory is regarded as the foundation of mathematical thought. All mathematical objects (even numbers!) can be defined in terms of sets, and the language of set theory is used in every mathematical subject.

In this chapter we introduce the basic definitions and notation of set theory and show how to establish properties of sets through the use of proofs and counterexamples. We also introduce the notion of a Boolean algebra, explain how to derive its properties, and discuss their relationships to logical equivalencies and set identities. The chapter ends with a discussion of a famous "paradox" of set theory and its relation to computer science.

5.1 Basic Definitions of Set Theory

The introduction of suitable abstractions is our only mental aid to organize and master complexity. — E. W. Dijkstra, 1930–2002

The words *set* and *element* are undefined terms of set theory just as *sentence, true,* and *false* are undefined terms of logic. The founder of set theory, Georg Cantor, suggested imagining a set as a "collection into a whole M of definite and separate objects of our intuition or our thought. These objects are called the elements of M." Cantor used the letter M because it is the first letter of the German word for set: *Menge.* Following the spirit of his notation (though not the letter), let S denote a set and a an element of S. Then, as indicated in Section 2.1, $a \in S$ means that a is an element of S, $a \notin S$ means that a is not an element of S, $\{1, 2, 3\}$ refers to the set whose elements are 1, 2, and 3, and $\{1, 2, 3, \ldots\}$ refers to the set of all positive integers. The **axiom of extension** says that a set is completely determined by its elements; the order in which the elements are listed is irrelevant, as is the fact that some elements may be listed more than once.

Example 5.1.1 The { } Notation for Sets

a. Suppose that Ann, Bob, and Cal are three students in a discrete mathematics class. Since the following sets all have the same elements—namely Ann, Bob, and Cal—they all represent the same set:

$$\{\text{Ann, Bob, Cal}\}, \{\text{Bob, Cal, Ann}\}, \{\text{Bob, Bob, Ann, Cal, Ann}\}$$

b. {Ann} denotes the set whose only element is Ann, whereas the word *Ann* denotes Ann herself. Since these are different {Ann} ≠ Ann.

c. Sets can themselves be elements of other sets. For example, $\{1, \{1\}\}$ has two elements: the number 1 and the set $\{1\}$.

d. Sometimes a set may appear to have more elements than it really has. For every nonnegative integer n, let $U_n = \{-n, n\}$. Then $U_2 = \{-2, 2\}$ and $U_1 = \{-1, 1\}$ both have two elements, but

$$U_0 = \{-0, 0\} = \{0\}$$

has only one element since $-0 = 0$. ∎

As noted in Section 2.1, a set may also be defined by writing $A = \{x \in S \mid P(x)\}$, where the left-hand brace is read "the set of all," the vertical bar is read "such that," and $P(x)$ is a property. An element x is in A if, and only if, x is in S and $P(x)$ is true.

Occasionally we will write $\{x \mid P(x)\}$ without being specific about where the element x comes from. It turns out that unrestricted use of this notation can lead to genuine contradictions in set theory. We will discuss one of these in Section 5.4 and will be careful to use this notation purely as a convenience in cases where the set S could be specified if necessary.

Example 5.1.2 Sets Given by a Defining Property

Recall that **R** denotes the set of all real numbers, **Z** the set of all integers, and $\mathbf{Z}^+$ the set of all positive integers. Describe each of the following sets.

a. $\{x \in \mathbf{R} \mid -2 < x < 5\}$

b. $\{x \in \mathbf{Z} \mid -2 < x < 5\}$

c. $\{x \in \mathbf{Z}^+ \mid -2 < x < 5\}$

Solution

a. $\{x \in \mathbf{R} \mid -2 < x < 5\}$ is the open interval of real numbers strictly between -2 and 5. It is pictured as follows:

b. $\{x \in \mathbf{Z} \mid -2 < x < 5\}$ is the set of all integers between -2 and 5. It is equal to the set $\{-1, 0, 1, 2, 3, 4\}$.

c. Since all the integers in $\mathbf{Z}^+$ are positive, $\{x \in \mathbf{Z}^+ \mid -2 < x < 5\} = \{1, 2, 3, 4\}$. ∎

Subsets

A basic relation between sets is that of subset.

• Definition

If A and B are sets, then A is called a **subset** of B, written $A \subseteq B$, if, and only if, every element of A is also an element of B.

Symbolically:

$$A \subseteq B \quad \Leftrightarrow \quad \forall x, \text{ if } x \in A \text{ then } x \in B.$$

The phrases A *is contained in* B and B *contains* A are alternative ways of saying that A is a subset of B.

It follows from the definition of subset that a set A is not a subset of a set B, written $A \nsubseteq B$, if, and only if, there is at least one element of A that is not an element of B. Symbolically:

$$A \nsubseteq B \quad \Leftrightarrow \quad \exists x \text{ such that } x \in A \text{ and } x \notin B.$$

Example 5.1.3 Subsets

Suppose B76, XR3, D54, ES2, and XL5 are the model numbers of certain pieces of equipment. Let $A = \{B76, XR3, D54, XL5\}$, $B = \{B76, D54\}$, and $C = \{ES2, XL5\}$.

a. Is $B \subseteq A$?   b. Is $C \subseteq A$?   c. Is $B \subseteq B$?

Solution

a. Yes. Both elements of B are in A.

b. No; ES2 is in C but not in A.

c. Yes. Both elements of B are in B. (The definition of subset implies that any set is a subset of itself.) ∎

> • **Definition**
>
> Let A and B be sets. A is a **proper subset** of B if, and only if, every element of A is in B but there is at least one element of B that is not in A.

John Venn
(1834–1923)

Royal Society of London

If sets A and B are represented as regions in the plane, relationships between A and B can be represented by pictures, called **Venn diagrams,** that were introduced by the British mathematician John Venn in 1881. For instance, the relationship $A \subseteq B$ can be pictured in one of two ways, as shown in Figure 5.1.1.

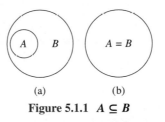

(a)   (b)

Figure 5.1.1 $A \subseteq B$

The relationship $A \nsubseteq B$ can be represented in three different ways with Venn diagrams, as shown in Figure 5.1.2. If we allow the possibility that some subregions of Venn diagrams do not contain any points, then in Figure 5.1.1 diagram (b) can be viewed as a special case of diagram (a) by imagining that the part of B outside A does not contain any points. Similarly, diagrams (a) and (c) of Figure 5.1.2 can be viewed as special cases of diagram (b). To obtain (a) from (b), imagine that the region of overlap between A and B does not contain any points. To obtain (c), imagine that the part of B that lies outside

(a)   (b)   (c)

Figure 5.1.2 $A \nsubseteq B$

A does not contain any points. However, in all three diagrams it would be necessary to specify that there is a point in *A* that is not in *B*.

Venn diagrams are useful for exploring properties that involve only two or three sets, but they are not very helpful when the number of sets is four or more. For instance, if the requirement is made that a Venn diagram must show every possible intersection of the sets, it is impossible to draw a symmetric Venn diagram for four sets, or, in fact, for any nonprime number of sets. In 2002, computer scientists/mathematicians Carla Savage and Jerrold Griggs and undergraduate student Charles Killian solved a longstanding open problem by proving that it *is* possible to draw such a symmetric Venn diagram for any prime number of sets. For $n > 5$, however, the resulting pictures are very complicated! The existence of such symmetric diagrams has applications in the area of computer science called coding theory.

Example 5.1.4 Relations among Sets of Numbers

Figure 5.1.3

Since **Z**, **Q**, and **R** denote the sets of integers, rational numbers, and real numbers, respectively, **Z** is a subset of **Q** because every integer is rational (any integer n can be written in the form $\frac{n}{1}$), and **Q** is a subset of **R** because every rational number is real (any rational number can be represented as a length on the number line). **Z** is a proper subset of **Q** because there are rational numbers that are not integers (for example, $\frac{1}{2}$), and **Q** is a proper subset of **R** because there are real numbers that are not rational (for example, $\sqrt{2}$). This is shown diagrammatically in Figure 5.1.3. ∎

It is important to distinguish clearly between the concepts of set membership ($\in$) and set containment ($\subseteq$). The following example illustrates some distinctions between them.

Example 5.1.5 Distinction between ∈ and ⊆

Which of the following are true statements?

a. $2 \in \{1, 2, 3\}$ b. $\{2\} \in \{1, 2, 3\}$ c. $2 \subseteq \{1, 2, 3\}$

d. $\{2\} \subseteq \{1, 2, 3\}$ e. $\{2\} \subseteq \{\{1\}, \{2\}\}$ f. $\{2\} \in \{\{1\}, \{2\}\}$

Solution Only (a), (d), and (f) are true.

For (b) to be true, the set $\{1, 2, 3\}$ would have to contain the element $\{2\}$. But the only elements of $\{1, 2, 3\}$ are 1, 2, and 3, and 2 is not equal to $\{2\}$. Hence (b) is false.

For (c) to be true, the number 2 would have to be a set and every element in the set 2 would have to be an element of $\{1, 2, 3\}$. This is not the case, so (c) is false.

For (e) to be true, every element in the set containing only the number 2 would have to be an element of the set whose elements are $\{1\}$ and $\{2\}$. But 2 is not equal to $\{1\}$ or $\{2\}$, and so (e) is false. ∎

Set Equality

Recall that by the principle of extension, sets *A* and *B* are equal if, and only if, they have exactly the same elements. We restate this as a definition using the language of subsets.

> **• Definition**
>
> Given sets *A* and *B*, *A* **equals** *B*, written $A = B$, if, and only if, every element of *A* is in *B* and every element of *B* is in *A*.
> Symbolically:
> $$A = B \iff A \subseteq B \text{ and } B \subseteq A.$$

This version of the definition of equality implies the following:

> To know that a set A equals a set B, you must know
> that $A \subseteq B$ and you must also know that $B \subseteq A$.

Example 5.1.6 Set Equality

Let sets A, B, C, and D be defined as follows:

$$A = \{n \in \mathbf{Z} \mid n = 2p, \text{ for some integer } p\},$$
$$B = \text{the set of all even integers,}$$
$$C = \{m \in \mathbf{Z} \mid m = 2q - 2, \text{ for some integer } q\},$$
$$D = \{k \in \mathbf{Z} \mid k = 3r + 1, \text{ for some integer } r\}.$$

a. Is $A = B$?   b. Is $A = D$?   c. Is $A = C$?

Solution

a. Yes. $A = B$ because every integer of the form $2p$, for some integer p, is even (so $A \subseteq B$), and every integer that is even can be written in the form $2p$, for some integer p (so $B \subseteq A$).

b. No. $A \neq D$ for the following reason: $2 \in A$ since $2 = 2 \cdot 1$; but $2 \notin D$. For if 2 were an element of D, then 2 would equal $3r + 1$, for some integer r. Solving for r would give

$$3r + 1 = 2$$
$$3r = 2 - 1$$
$$3r = 1$$
$$r = \frac{1}{3}.$$

This argument shows that if 2 were an element of D, then there would be an integer r such that $r = 1/3$. But $1/3$ is not an integer, and so $2 \notin D$. Since there is an element in A that is not in D, $A \neq D$.

c. Yes. $A = C$ if, and only if, every element of A is in C and every element of C is in A. Considering the definitions of A and C, deciding whether $A = C$ involves deciding whether both of the following questions can be answered yes:

1. Can any integer that can be written in the form $2p$, for some integer p, also be written in the form $2q - 2$, for some integer q?

2. Can any integer that can be written in the form $2q - 2$, for some integer q, also be written in the form $2p$, for some integer p?

To answer question (1), suppose an integer n equals $2p$, for some integer p. Can you find an integer q so that n equals $2q - 2$? If so, then

$$2q - 2 = 2p$$
$$2q = 2p + 2 = 2(p + 1)$$

and thus
$$q = p + 1.$$

Now, go backwards through the steps: If $n = 2p$, where p is an integer, let $q = p + 1$. Then q is an integer (since it is a sum of integers) and

$$2q - 2 = 2(p + 1) - 2 = 2p - 2 + 2 = 2p.$$

Hence the answer to question (1) is yes: $A \subseteq C$.

To answer question (2), suppose an integer m equals $2q - 2$, for some integer q. Can you find an integer p such that m equals $2p$? If so, then

$$2p = 2q - 2 = 2(q - 1)$$

and thus
$$p = q - 1.$$

Again go backwards through the steps: If $m = 2q - 2$, where q is an integer, let $p = q - 1$. Then p is an integer (since it is a difference of two integers), and

$$2p = 2(q - 1) = 2q - 2.$$

Hence the answer to (2) is yes: $C \subseteq A$.

Since $A \subseteq C$ and $C \subseteq A$, then $A = C$ by definition of set equality.

The type of explanation used in this example is called an *element argument*. We discuss element arguments further in the next section.   ∎

Operations on Sets

Most mathematical discussions are carried on within some context. For example, in a certain situation all sets being considered might be sets of real numbers. In such a situation, the set of real numbers would be called a **universal set** or a **universe of discourse** for the discussion.

• Definition

Let A and B be subsets of a universal set U.

1. The **union** of A and B, denoted $A \cup B$, is the set of all elements x in U such that x is in A or x is in B.

2. The **intersection** of A and B, denoted $A \cap B$, is the set of all elements x in U such that x is in A and x is in B.

3. The **difference** of B minus A (or **relative complement** of A in B), denoted $B - A$, is the set of all elements x in U such that x is in B and x is not in A.

4. The **complement** of A, denoted A^c, is the set of all elements x in U such that x is not in A.

Symbolically:

$$A \cup B = \{x \in U \mid x \in A \text{ or } x \in B\},$$
$$A \cap B = \{x \in U \mid x \in A \text{ and } x \in B\},$$
$$B - A = \{x \in U \mid x \in B \text{ and } x \notin A\},$$
$$A^c = \{x \in U \mid x \notin A\}.$$

Giuseppe Peano
(1858–1932)

Stock Montage

Thus the union of A and B is the set of elements in U that are in at least one of the sets A and B. The intersection of A and B is the set of elements common to both sets A and B. The difference of B minus A is the set of elements in B that are not in A. And the complement of A is the set of elements in the universal set U that are not in A. The symbols $\in$, $\cup$, and $\cap$ were introduced in 1889 by the Italian mathematician Giuseppe Peano.

Example 5.1.7 Unions, Intersections, Differences, and Complements

Let the universal set be the set $U = \{a, b, c, d, e, f, g\}$ and let $A = \{a, c, e, g\}$ and $B = \{d, e, f, g\}$. Find $A \cup B, A \cap B, B - A,$ and A^c.

Solution

$$A \cup B = \{a, c, d, e, f, g\} \qquad A \cap B = \{e, g\}$$
$$B - A = \{d, f\} \qquad A^c = \{b, d, f\}$$ ■

Example 5.1.8 An Example with Intervals

Let the universal set be the set **R** of all real numbers and let $A = \{x \in \mathbf{R} \mid -1 < x \leq 0\}$ and $B = \{x \in \mathbf{R} \mid 0 \leq x < 1\}$. These sets are shown on the number lines below.

Find $A \cup B, A \cap B,$ and A^c.

Solution $A \cup B = \{x \in \mathbf{R} \mid -1 < x \leq 0 \text{ or } 0 \leq x < 1\} = \{x \in \mathbf{R} \mid -1 < x < 1\}$.

$A \cap B = \{x \in \mathbf{R} \mid -1 < x < 0 \text{ and } 0 \leq x < 1\} = \{0\}$.

$A^c = \{x \in \mathbf{R} \mid \text{ it is not the case that } -1 < x \leq 0\}$

$\quad = \{x \in \mathbf{R} \mid \text{ it is not the case that } (-1 < x \text{ and } x \leq 0)\}$ by definition of the double inequality

$\quad = \{x \in \mathbf{R} \mid x \leq -1 \text{ or } x > 0\}$ by De Morgan's law

The Venn diagram representations for union, intersection, difference, and complement are shown in Figure 5.1.4.

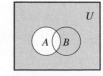

Shaded region represents $A \cup B$.   Shaded region represents $A \cap B$.   Shaded region represents $B - A$.   Shaded region represents A^c.

Figure 5.1.4

The Empty Set

We have stated that a set is defined by the elements that compose it. This being so, can there be a set that has no elements? It turns out that it is convenient to allow such a set. Otherwise, every time we wanted to take the intersection of two sets or to define a set by specifying a property, we would have to check that the result had elements and hence qualified for "sethood." For example, if $A = \{1, 3\}$ and $B = \{2, 4\}$, then $A \cap B$ has no elements. Neither does $\{x \in \mathbf{R} \mid x^2 = -1\}$ because no real numbers have negative squares.

It is somewhat unsettling to talk about a set with no elements, but it often happens in mathematics that the definitions formulated to fit one set of circumstances are satisfied by some extreme cases not originally anticipated. Yet changing the definitions to exclude those cases would seriously undermine the simplicity and elegance of the theory taken as a whole.

In Section 5.2 we will show that there is only one set with no elements. Because it is unique, we can give it a special name. We call it the **empty set** (or **null set**) and denote it by the symbol $\emptyset$. Thus $\{1, 3\} \cap \{2, 4\} = \emptyset$ and $\{x \in \mathbf{R} \mid x^2 = -1\} = \emptyset$.

Example 5.1.9 A Set with No Elements

Describe the set $D = \{x \in \mathbf{R} \mid 3 < x < 2\}$.

Solution   Recall that $a < x < b$ means that $a < x$ and $x < b$. So X consists of all real numbers that are both greater than 3 and less than 2. Since there are no such numbers, D has no elements and so $D = \emptyset$. ∎

Partitions of Sets

In many applications of set theory, sets are divided up into nonoverlapping (or *disjoint*) pieces. Such a division is called a *partition*.

> **• Definition**
>
> Two sets are called **disjoint** if, and only if, they have no elements in common. Symbolically:
>
> $$A \text{ and } B \text{ are disjoint} \quad \Leftrightarrow \quad A \cap B = \emptyset.$$

Example 5.1.10 Disjoint Sets

Let $A = \{1, 3, 5\}$ and $B = \{2, 4, 6\}$. Are A and B disjoint?

Solution   Yes. By inspection A and B have no elements in common, or, in other words, $\{1, 3, 5\} \cap \{2, 4, 6\} = \emptyset$. ∎

> **• Definition**
>
> Sets $A_1, A_2, \ldots, A_n$ are **mutually disjoint** (or **pairwise disjoint** or **nonoverlapping**) if, and only if, no two sets A_i and A_j with distinct subscripts have any elements in common. More precisely, for all $i, j = 1, 2, \ldots, n$,
>
> $$A_i \cap A_j = \emptyset \quad \text{whenever } i \neq j.$$

Example 5.1.11 Mutually Disjoint Sets

    a. Let $A_1 = \{3, 5\}$, $A_2 = \{1, 4, 6\}$, and $A_3 = \{2\}$. Are A_1, A_2, and A_3 mutually disjoint?

    b. Let $B_1 = \{2, 4, 6\}$, $B_2 = \{3, 7\}$, and $B_3 = \{4, 5\}$. Are B_1, B_2, and B_3 mutually disjoint?

Solution

    a. Yes. A_1 and A_2 have no elements in common, A_1 and A_3 have no elements in common, and A_2 and A_3 have no elements in common.

    b. No. B_1 and B_3 both contain 4.   ■

Suppose A, A_1, A_2, A_3, and A_4 are the sets of points represented by the regions shown in Figure 5.1.5. Then A_1, A_2, A_3, and A_4 are subsets of A, and $A = A_1 \cup A_2 \cup A_3 \cup A_4$. Suppose further that boundaries are assigned to the regions representing A_2, A_3, and A_4 in such a way that these sets are mutually disjoint. Then A is called a *union of mutually disjoint subsets,* and the collection of sets $\{A_1, A_2, A_3, A_4\}$ is said to be a *partition* of A.

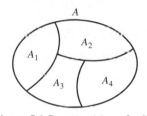

Figure 5.1.5 A Partition of a Set

> **• Definition**
>
> A collection of nonempty sets $\{A_1, A_2, \ldots, A_n\}$ is a **partition** of a set A if, and only if,
>
> 1. $A = A_1 \cup A_2 \cup \cdots \cup A_n$;
>
> 2. $A_1, A_2, \ldots, A_n$ are mutually disjoint.

Example 5.1.12 Partitions of Sets

    a. Let $A = \{1, 2, 3, 4, 5, 6\}$, $A_1 = \{1, 2\}$, $A_2 = \{3, 4\}$, and $A_3 = \{5, 6\}$. Is $\{A_1, A_2, A_3\}$ a partition of A?

    b. Let **Z** be the set of all integers and let

$$T_0 = \{n \in \mathbf{Z} \mid n = 3k, \text{ for some integer } k\},$$
$$T_1 = \{n \in \mathbf{Z} \mid n = 3k + 1, \text{ for some integer } k\}, \text{ and}$$
$$T_2 = \{n \in \mathbf{Z} \mid n = 3k + 2, \text{ for some integer } k\}.$$

Is $\{T_0, T_1, T_2\}$ a partition of **Z**?

Solution

    a. Yes. By inspection, $A = A_1 \cup A_2 \cup A_3$ and the sets A_1, A_2, and A_3 are mutually disjoint.

    b. Yes. By the quotient-remainder theorem and exercise 18 of Section 3.7, every integer n can be represented in exactly one of the three forms

$$n = 3k \quad \text{or} \quad n = 3k + 1 \quad \text{or} \quad n = 3k + 2,$$

for some integer k. This implies that no integer can be in any two of the sets T_0, T_1, or T_2. So T_0, T_1, and T_2 are mutually disjoint. It also implies that every integer is in one of the sets T_0, T_1, or T_2. So $\mathbf{Z} = T_0 \cup T_1 \cup T_2$.   ∎

Power Sets

There are a variety of situations in which it is useful to consider the set of all subsets of a particular set. The **power set axiom** guarantees that this is a set.

> **• Definition**
>
> Given a set A, the **power set** of A, denoted $\mathscr{P}(A)$, is the set of all subsets of A.

Example 5.1.13 Power Set of a Set

Find the power set of the set $\{x, y\}$. That is, find $\mathscr{P}(\{x, y\})$.

Solution $\mathscr{P}(\{x, y\})$ is the set of all subsets of $\{x, y\}$. Now since $\emptyset$ is a subset of every set, $\emptyset \in \mathscr{P}(\{x, y\})$. Also any set is a subset of itself, so $\{x, y\} \in \mathscr{P}(\{x, y\})$. The only other subsets of $\{x, y\}$ are $\{x\}$ and $\{y\}$, so

$$\mathscr{P}(\{x, y\}) = \{\emptyset, \{x\}, \{y\}, \{x, y\}\}.$$ ∎

Cartesian Products

Recall that the definition of a set is unaffected by the order in which its elements are listed or the fact that some elements may be listed more than once. Thus $\{a, b\}$, $\{b, a\}$, and $\{a, a, b\}$ all represent the same set. The notation for an *ordered n-tuple* takes both order and multiplicity into account.

> **• Definition**
>
> Let n be a positive integer and let $x_1, x_2, \ldots, x_n$ be (not necessarily distinct) elements. The **ordered n-tuple, $(x_1, x_2, \ldots, x_n)$,** consists of $x_1, x_2, \ldots, x_n$ together with the ordering: first x_1, then x_2, and so forth up to x_n. An ordered 2-tuple is called an **ordered pair,** and an ordered 3-tuple is called an **ordered triple.**
>
> Two ordered n-tuples $(x_1, x_2, \ldots, x_n)$ and $(y_1, y_2, \ldots, y_n)$ are **equal** if, and only if, $x_1 = y_1, x_2 = y_2, \ldots, x_n = y_n$.
>
> Symbolically:
>
> $$(x_1, x_2, \ldots, x_n) = (y_1, y_2, \ldots, y_n) \quad \Leftrightarrow \quad x_1 = y_1, x_2 = y_2, \ldots, x_n = y_n.$$
>
> In particular,
>
> $$(a, b) = (c, d) \quad \Leftrightarrow \quad a = c \text{ and } b = d.$$

Example 5.1.14 Ordered n-tuples

a. Is $(1, 2) = (2, 1)$?

b. Is $\left(3, (-2)^2, \frac{1}{2}\right) = \left(\sqrt{9}, 4, \frac{3}{6}\right)$?

Solution

a. No. By definition of equality of ordered pairs,

$$(1, 2) = (2, 1) \quad \Leftrightarrow \quad 1 = 2 \text{ and } 2 = 1.$$

But $1 \neq 2$, and so the ordered pairs are not equal.

b. Yes. By definition of equality of ordered triples,

$$\left(3, (-2)^7, \tfrac{1}{2}\right) = \left(\sqrt{9}, 4, \tfrac{1}{6}\right) \quad \Leftrightarrow \quad 3 = \sqrt{9} \text{ and } (-2)^7 = 4 \text{ and } \tfrac{1}{2} = \tfrac{7}{6}.$$

Because these equations are all true, the two ordered triples are equal. ∎

> **• Definition**
>
> Given two sets A and B, the **Cartesian product** of A and B, denoted $A \times B$ (read "A cross B"), is the set of all ordered pairs (a, b), where a is in A and b is in B.
>
> Given sets $A_1, A_2, \ldots, A_n$, the **Cartesian product** of $A_1, A_2, \ldots, A_n$ denoted $A_1 \times A_2 \times \cdots \times A_n$, is the set of all ordered n-tuples $(a_1, a_2, \ldots, a_n)$ where $a_1 \in A_1, a_2 \in A_2, \ldots, a_n \in A_n$.
>
> Symbolically:
>
> $$A \times B = \{(a, b) \mid a \in A \text{ and } b \in B\},$$
>
> $$A_1 \times A_2 \times \cdots \times A_n = \{(a_1, a_2, \ldots, a_n) \mid a_1 \in A_1, a_2 \in A_2, \ldots, a_n \in A_n\}.$$

Example 5.1.15 Cartesian Products

Let $A = \{x, y\}$, $B = \{1, 2, 3\}$, and $C = \{a, b\}$.

a. Find $A \times B$.   b. Find $(A \times B) \times C$.   c. Find $A \times B \times C$.

Solution

a. $A \times B = \{(x, 1), (x, 2), (x, 3), (y, 1), (y, 2), (y, 3)\}$

b. The Cartesian product of A and B is a set, so it may be used as one of the sets making up another Cartesian product. This is the case for $(A \times B) \times C$:

$$(A \times B) \times C = \{(u, v) \mid u \in A \times B \text{ and } v \in C\} \quad \text{by definition of Cartesian product}$$
$$= \{((x, 1), a), ((x, 2), a), ((x, 3), a), ((y, 1), a),$$
$$((y, 2), a), ((y, 3), a), ((x, 1), b), ((x, 2), b), ((x, 3), b),$$
$$((y, 1), b), ((y, 2), b), ((y, 3), b)\}$$

c. The Cartesian product $A \times B \times C$ is superficially similar to, but is not quite the same mathematical object as, $(A \times B) \times C$. $(A \times B) \times C$ is a set of ordered pairs of which one element is itself an ordered pair, whereas $A \times B \times C$ is a set of ordered triples. By definition of Cartesian product,

$$A \times B \times C = \{(u, v, w) \mid u \in A, v \in B, \text{ and } w \in C\}$$
$$= \{(x, 1, a), (x, 2, a), (x, 3, a), (y, 1, a), (y, 2, a),$$
$$(y, 3, a), (x, 1, b), (x, 2, b), (x, 3, b), (y, 1, b),$$
$$(y, 2, b), (y, 3, b)\}. \quad ∎$$

An Algorithm to Check Whether One Set Is a Subset of Another (Optional)

You may get some additional insight into the concept of subset by considering an algorithm for checking whether one finite set is a subset of another. Order the elements of both sets and successively compare each element of the first set with each element of the second set. If some element of the first set is not found to equal any element of the second, then the first set is not a subset of the second. But if each element of the first set is found to equal an element of the second set, then the first set is a subset of the second. The following algorithm formalizes this reasoning.

Algorithm 5.1.1 Testing Whether $A \subseteq B$

[*Input sets A and B are represented as one-dimensional arrays $a[1], a[2], \ldots, a[m]$ and $b[1], b[2], \ldots, b[n]$, respectively. Starting with $a[1]$ and for each successive $a[i]$ in A, a check is made to see whether $a[i]$ is in B. To do this, $a[i]$ is compared to successive elements of B. If $a[i]$ is not equal to any element of B, then answer is given the value "$A \nsubseteq B$." If $a[i]$ equals some element of B, the next successive element in A is checked to see whether it is in B. If every successive element of A is found to be in B, then answer never changes from its initial value "$A \subseteq B$."*]

Input: m [*a positive integer*], $a[1], a[2], \ldots, a[m]$ [*a one-dimensional array representing the set A*], n [*a positive integer*], $b[1], b[2], \ldots, b[n]$ [*a one-dimensional array representing the set B*]

Algorithm Body:

$i := 1, answer := $ "$A \subseteq B$"

while ($i \leq m$ and $answer = $ "$A \subseteq B$")

$\quad j := 1, found := $ "no"

$\quad$ **while** ($j \leq n$ and $found = $ "no")

$\quad\quad$ **if** $a[i] = b[j]$ **then** $found := $ "yes"

$\quad\quad j := j + 1$

$\quad$ **end while**

$\quad$ [*If found has not been given the value "yes" when execution reaches this point, then $a[i] \notin B$.*]

$\quad$ **if** $found = $ "no" **then** $answer := $ "$A \nsubseteq B$"

$\quad i := i + 1$

end while

Output: $answer$ [*a string*]

Example 5.1.16 Tracing Algorithm 5.1.1

Trace the action of Algorithm 5.1.1 on the variables i, j, *found*, and *answer* for $m = 3, n = 4$, and sets A and B represented as the arrays $a[1] = u, a[2] = v, a[3] = w$, $b[1] = w, b[2] = x, b[3] = y$, and $b[4] = u$.

Solution

i	1					2					3
j	1	2	3	4	5	1	2	3	4	5	
found	no			yes		no					
answer	$A \subseteq B$									$A \nsubseteq B$	

In the exercises at the end of this section, you are asked to write an algorithm to check whether a given element is in a given set. To do this, you can represent the set as a one-dimensional array and compare the given element with successive elements of the array to determine whether the two elements are equal. If they are, then the element is in the set; if the given element does not equal any element of the array, then the element is not in the set.

Exercise Set 5.1*

1. Which of the following sets are equal?

    $A - \{a, b, c, d\}$ $B = \{d, e, a, c\}$
    $C = \{d, b, a, c\}$ $D = \{a, a, d, e, c, e\}$

2. Is $4 = \{4\}$? Explain.

3. Which of the following sets are equal?

    $$A = \{0, 1, 2\}$$
    $$B = \{x \in \mathbf{R} \mid -1 \le x < 3\}$$
    $$C = \{x \in \mathbf{R} \mid -1 < x < 3\}$$
    $$D = \{x \in \mathbf{Z} \mid -1 < x < 3\}$$
    $$E = \{x \in \mathbf{Z}^{+} \mid -1 < x < 3\}$$

4. Indicate the elements in each set defined in (a)–(f).
    a. $S = \{n \in \mathbf{Z} \mid n = (-1)^k, \text{ for some integer } k\}$.
    b. $T = \{m \in \mathbf{Z} \mid m = 1 + (-1)^i, \text{ for some integer } i\}$.
    c. $U = \{r \in \mathbf{Z} \mid 2 \le r \le -2\}$
    d. $V = \{s \in \mathbf{Z} \mid s > 2 \text{ or } s < 3\}$
    e. $W = \{t \in \mathbf{Z} \mid -1 < t < -3\}$
    f. $X = \{u \in \mathbf{Z} \mid u \le 4 \text{ or } u \ge 1\}$

5. a. Is the number 0 in $\emptyset$? Why?   b. Is $\emptyset = \{\emptyset\}$? Why?
    c. Is $\emptyset \in \{\emptyset\}$? Why?   d. Is $\emptyset \in \emptyset$? Why?

6. Write in words how to read each of the following out loud. Then write the shorthand notation for each set.
    a. $\{x \in U \mid x \in A \text{ and } x \in B\}$
    b. $\{x \in U \mid x \in A \text{ or } x \in B\}$
    c. $\{x \in U \mid x \in A \text{ and } x \notin B\}$
    d. $\{x \in U \mid x \notin A\}$

7. Let $A = \{c, d, f, g\}$, $B = \{f, j\}$, and $C = \{d, g\}$. Answer each of the following questions. Give reasons for your answers.
    a. Is $B \subseteq A$?   b. Is $C \subseteq A$?

c. Is $C \subseteq C$?   **d.** Is C a proper subset of A?

8. **a.** Is $3 \in \{1, 2, 3\}$?   **b.** Is $1 \subseteq \{1\}$?
    c. Is $\{2\} \subset \{1, 2\}$?   d. Is $\{3\} \in \{1, \{2\}, \{3\}\}$?
    e. Is $1 \in \{1\}$?   **f.** Is $\{2\} \subseteq \{1, \{2\}, \{3\}\}$?
    g. Is $\{1\} \subseteq \{1, 2\}$?   h. Is $1 \in \{\{1\}, 2\}$?
    i. Is $\{1\} \subseteq \{1, \{2\}\}$?   j. Is $\{1\} \subseteq \{1\}$?

9. Let $A = \{1, 3, 5, 7, 9\}$, $B = \{3, 6, 9\}$, and $C = \{2, 4, 6, 8\}$. Find each of the following:
    a. $A \cup B$ b. $A \cap B$ c. $A \cup C$ d. $A \cap C$
    e. $A - B$ f. $B - A$ g. $B \cup C$ h. $B \cap C$

10. Let the universal set be the set $\mathbf{R}$ of all real numbers and let $A = \{x \in \mathbf{R} \mid 0 < x \le 2\}$, $B = \{x \in \mathbf{R} \mid 1 \le x < 4\}$ and $C = \{x \in \mathbf{R} \mid 3 \le x < 9\}$. Find each of the following:
    a. $A \cup B$ b. $A \cap B$ c. A^c d. $A \cup C$
    e. $A \cap C$ f. B^c g. $A^c \cap B^c$
    h. $A^c \cup B^c$ i. $(A \cap B)^c$ j. $(A \cup B)^c$

11. Let the universal set be the set $\mathbf{R}$ of all real numbers and let $A = \{x \in \mathbf{R} \mid -3 \le x \le 0\}$, $B = \{x \in \mathbf{R} \mid -1 < x < 2\}$, and $C = \{x \in \mathbf{R} \mid 6 < x \le 8\}$. Find each of the following:
    a. $A \cup B$ b. $A \cap B$ c. A^c d. $A \cup C$
    e. $A \cap C$ f. B^c g. $A^c \cap B^c$
    h. $A^c \cup B^c$ i. $(A \cap B)^c$ j. $(A \cup B)^c$

12. Indicate which of the following relationships are true and which are false:
    a. $\mathbf{Z}^{+} \subseteq \mathbf{Q}$ **b.** $\mathbf{R}^{-} \subseteq \mathbf{Q}$
    c. $\mathbf{Q} \subseteq \mathbf{Z}$ **d.** $\mathbf{Z}^{-} \cup \mathbf{Z}^{+} = \mathbf{Z}$
    e. $\mathbf{Z}^{-} \cap \mathbf{Z}^{+} = \emptyset$ f. $\mathbf{Q} \cap \mathbf{R} = \mathbf{Q}$
    g. $\mathbf{Q} \cup \mathbf{Z} = \mathbf{Q}$ h. $\mathbf{Z}^{+} \cap \mathbf{R} = \mathbf{Z}^{+}$
    i. $\mathbf{Z} \cup \mathbf{Q} = \mathbf{Z}$

13. **a.** Write a negation for the following statement: $\forall$ sets A, if $A \subseteq \mathbf{R}$ then $A \subseteq \mathbf{Z}$. Which is true, the statement or its negation? Explain.

  b. Write a negation for the following statement: $\forall$ sets S, if $S \subseteq Q^+$ then $S \subseteq Q^-$. Which is true, the statement or its negation? Explain.

14. Let sets R, S, and T be defined as follows:

$$R = \{x \in \mathbf{Z} \mid x \text{ is divisible by } 2\}$$
$$S = \{y \in \mathbf{Z} \mid y \text{ is divisible by } 3\}$$
$$T = \{z \in \mathbf{Z} \mid z \text{ is divisible by } 6\}$$

  a. Is $R \subseteq T$? Explain.
  b. Is $T \subseteq R$? Explain.
  c. Is $T \subseteq S$? Explain.
  d. Find $R \cap S$. Explain.

15. Let $A = \{n \in \mathbf{Z} \mid n = 5r \text{ for some integer } r\}$ and $B = \{m \in \mathbf{Z} \mid m = 20s \text{ for some integer } s\}$.
  a. Is $A \subseteq B$?
  b. Is $B \subseteq A$?

16. Let $C = \{n \in \mathbf{Z} \mid n = 6r - 5 \text{ for some integer } r\}$ and $D = \{m \in \mathbf{Z} \mid m = 3s + 1 \text{ for some integer } s\}$.
  a. Is $C \subseteq D$?
  b. Is $D \subseteq C$?

17. Let $A = \{m \in \mathbf{Z} \mid m = 5i - 1, \text{ for some integer } i\}$,
  $B = \{n \in \mathbf{Z} \mid n = 3j + 2, \text{ for some integer } j\}$,
  $C = \{p \in \mathbf{Z} \mid p = 5r + 4, \text{ for some integer } r\}$, and
  $D = \{q \in \mathbf{Z} \mid q = 3s - 1, \text{ for some integer } s\}$.
  a. Is $A = B$? Explain.
  b. Is $A = C$? Explain.
  c. Is $A = D$? Explain.
  d. Is $B = D$? Explain.

18. In each of the following, draw a Venn diagram for sets A, B, and C that satisfy the given conditions:
  a. $A \subseteq B$; $C \subseteq B$; $A \cap C = \emptyset$
  b. $C \subseteq A$; $A \cap C = \emptyset$

19. Draw Venn diagrams to describe sets A, B, and C that satisfy the given conditions.
  a. $A \cap B = \emptyset$, $A \subseteq C$, $C \cap B \neq \emptyset$
  b. $A \subseteq B$, $C \subseteq B$, $A \cap C \neq \emptyset$
  c. $A \cap B \neq \emptyset$, $B \cap C \neq \emptyset$, $A \cap C = \emptyset$, $A \nsubseteq B$, $C \nsubseteq B$

20. Let $A = \{a, b, c\}$, $B = \{b, c, d\}$, and $C = \{b, c, e\}$.
  a. Find $A \cup (B \cap C)$, $(A \cup B) \cap C$, and $(A \cup B) \cap (A \cup C)$. Which of these sets are equal?
  b. Find $A \cap (B \cup C)$, $(A \cap B) \cup C$, and $(A \cap B) \cup (A \cap C)$. Which of these sets are equal?
  c. Find $(A - B) - C$ and $A - (B - C)$. Are these sets equal?

21. Consider the Venn diagram shown in the next column. For each of (a)–(f), copy the diagram and shade the region corresponding to the indicated set.

  a. $A \cap B$ **b.** $B \cup C$ **c.** A^c
  d. $A - (B \cup C)$ **e.** $(A \cup B)^c$ **f.** $A^c \cap B^c$

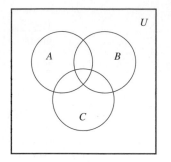

22. **a.** Is $\{\{a, d, e\}, \{b, c\}, \{d, f\}\}$ a partition of $\{a, b, c, d, e, f\}$?
  b. Is $\{\{w, x, v\}, \{u, y, q\}, \{p, z\}\}$ a partition of $\{p, q, u, v, w, x, y, z\}$?
  c. Is $\{\{5, 4\}, \{7, 2\}, \{1, 3, 4\}, \{6, 8\}\}$ a partition of $\{1, 2, 3, 4, 5, 6, 7, 8\}$?
  d. Is $\{\{3, 7, 8\}, \{2, 9\}, \{1, 4, 5\}\}$ a partition of $\{1, 2, 3, 4, 5, 6, 7, 8, 9\}$?
  e. Is $\{\{1, 5\}, \{4, 7\}, \{2, 8, 6, 3\}\}$ a partition of $\{1, 2, 3, 4, 5, 6, 7, 8\}$?

23. Let E be the set of all even integers and O the set of all odd integers. Is $\{E, O\}$ a partition of $\mathbf{Z}$, the set of all integers? Explain your answer.

24. Let $\mathbf{R}$ be the set of all real numbers. Is $\{\mathbf{R}^+, \mathbf{R}^-, \{0\}\}$ a partition of $\mathbf{R}$? Explain your answer.

25. Let $\mathbf{Z}$ be the set of all integers and let

$$A_0 = \{n \in \mathbf{Z} \mid n = 4k, \text{ for some integer } k\},$$
$$A_1 = \{n \in \mathbf{Z} \mid n = 4k + 1, \text{ for some integer } k\},$$
$$A_2 = \{n \in \mathbf{Z} \mid n = 4k + 2, \text{ for some integer } k\}, \text{ and}$$
$$A_3 = \{n \in \mathbf{Z} \mid n = 4k + 3, \text{ for some integer } k\}.$$

  Is $\{A_0, A_1, A_2, A_3\}$ a partition of $\mathbf{Z}$? Explain your answer.

26. Suppose $A = \{1, 2\}$ and $B = \{2, 3\}$. Find each of the following:
  a. $\mathscr{P}(A \cap B)$ b. $\mathscr{P}(A)$
  c. $\mathscr{P}(A \cup B)$ d. $\mathscr{P}(A \times B)$

27. **a.** Suppose $A = \{1\}$ and $B = \{u, v\}$. Find $\mathscr{P}(A \times B)$.
  b. Suppose $X = \{a, b\}$ and $Y = \{x, y\}$. Find $\mathscr{P}(X \times Y)$.

28. a. Find $\mathscr{P}(\emptyset)$. **b.** Find $\mathscr{P}(\mathscr{P}(\emptyset))$.
  c. Find $\mathscr{P}(\mathscr{P}(\mathscr{P}(\emptyset)))$.

29. Let $A = \{x, y, z, w\}$ and $B = \{a, b\}$. List the elements of each of the following sets:
  a. $A \times B$ b. $B \times A$
  c. $A \times A$ d. $B \times B$

30. Let $A = \{1, 2, 3\}$, $B = \{u, v\}$, and $C = \{m, n\}$. List the elements of each of the following sets:
    a. $A \times (B \times C)$ b. $(A \times B) \times C$ c. $A \times B \times C$

31. Trace the action of Algorithm 5.1.1 on the variables i, j, *found*, and *answer* for $m = 3$, $n = 3$, and sets A and B represented as the arrays $a[1] = u, a[2] = v, a[3] = w$, $b[1] = w$, $b[2] = u$, and $b[3] = v$.

32. Trace the action of Algorithm 5.1.1 on the variables i, j, *found*, and *answer* for $m = 4$, $n = 4$, and sets A and B represented as the arrays $a[1] = u, a[2] = v, a[3] = w$, $a[4] = x, b[1] = r, b[2] = u, b[3] = y, b[4] = z$.

33. Write an algorithm to determine whether a given element x belongs to a given set, which is represented as an array $a[1], a[2], \ldots, a[n]$.

5.2 Properties of Sets

. . . only the last line is a genuine theorem here—everything else is in the fantasy.
— Douglas Hofstadter, *Gödel, Escher, Bach*, 1979

It is possible to list many relations involving unions, intersections, complements, and differences of sets. Some of these are true for all sets, whereas others fail to hold in some cases. In this section we show how to establish basic set properties using *element arguments*, the most basic method used for proofs involving sets, and we discuss a variation used to prove that a set is empty. In the next section we will show how to disprove a proposed set property by constructing a counterexample and how to use an algebraic technique to derive new set properties from set properties already known to be true.

We begin by listing some set properties that involve subset relations. As you read them, keep in mind that the operations of union, intersection, and difference take precedence over set inclusion. Thus, for example, $A \cap B \subseteq C$ means $(A \cap B) \subseteq C$.

Theorem 5.2.1 Some Subset Relations

1. *Inclusion of Intersection:* For all sets A and B,

    (a) $A \cap B \subseteq A$ and   (b) $A \cap B \subseteq B$.

2. *Inclusion in Union:* For all sets A and B,

    (a) $A \subseteq A \cup B$ and   (b) $B \subseteq A \cup B$.

3. *Transitive Property of Subsets:* For all sets A, B, and C,

    if $A \subseteq B$ and $B \subseteq C$,   then $A \subseteq C$.

The conclusion of each part of Theorem 5.2.1 states that one set is a subset of another. Recall that by definition of subset, if X and Y are sets, then

$$X \subseteq Y \quad \Leftrightarrow \quad \forall x, \text{if } x \in X \text{ then } x \in Y.$$

Since the definition of subset is a universal conditional statement, the most basic way to prove that one set is a subset of another is as follows:

Element Argument: The Basic Method for Proving That One Set Is a Subset of Another

Let sets X and Y be given. To prove that $X \subseteq Y$,

1. suppose that x is a particular but arbitrarily chosen element of X,

2. show that x is an element of Y.

In most set theoretic proofs, the secret of getting from the assumption that x is in X to the conclusion that x is in Y is to think of the definitions of basic set operations in procedural terms. For example, the union of sets X and Y, $X \cup Y$, is defined as

$$X \cup Y = \{x \mid x \in X \text{ or } x \in Y\}.$$

This means that any time you know an element x is in $X \cup Y$, you can conclude that x must be in X or x must be in Y. Conversely, any time you know that a particular x is in some set X or is in some set Y, you can conclude that x is in $X \cup Y$. Thus, for any sets X and Y and any element x,

$$x \in X \cup Y \quad \text{if, and only if,} \quad x \in X \text{ or } x \in Y.$$

Procedural versions of the definitions of the other set operations are derived similarly and are summarized below.

Procedural Versions of Set Definitions

Let X and Y be subsets of a universal set U and suppose x and y are elements of U.

1. $x \in X \cup Y \quad \Leftrightarrow \quad x \in X \text{ or } x \in Y$

2. $x \in X \cap Y \quad \Leftrightarrow \quad x \in X \text{ and } x \in Y$

3. $x \in X - Y \quad \Leftrightarrow \quad x \in X \text{ and } x \notin Y$

4. $x \in X^c \quad \Leftrightarrow \quad x \notin X$

5. $(x, y) \in X \times Y \quad \Leftrightarrow \quad x \in X \text{ and } y \in Y$

Example 5.2.1 Proof of a Subset Relation

Prove Theorem 5.2.1(1)(a): For all sets A and B, $A \cap B \subseteq A$.

Solution We start by giving a proof of the statement and then explain how you can obtain such a proof yourself.

Proof:

> Suppose A and B are any sets and suppose x is any element of $A \cap B$.
> Then $x \in A$ and $x \in B$ by definition of intersection. In particular, $x \in A$.

The underlying structure of this proof is not difficult, but it is more complicated than the brevity of the proof suggests. The first important thing to realize is that the statement to be proved is universal (it says that for *all* sets A and B, $A \cap B \subseteq A$). The proof, therefore, has the following outline:

Starting Point: Suppose A and B are any (particular but arbitrarily chosen) sets.

To Show: $A \cap B \subseteq A$

Now to prove that $A \cap B \subseteq A$, you must show that

$$\forall x, \text{ if } x \in A \cap B \text{ then } x \in A.$$

But this statement also is universal. So to prove it, you

> **suppose** x is an element in $A \cap B$

and then you

> **show** that x is in A.

Filling in the gap between the "suppose" and the "show" is easy if you use the procedural version of the definition of intersection: To say that x is in $A \cap B$ means that

$$x \text{ is in } A \quad \text{and} \quad x \text{ is in } B.$$

This allows you to complete the proof by deducing that, in particular,

$$x \text{ is in } A,$$

as was to be shown. Note that this deduction is just a special case of the valid argument form

$$p \wedge q$$
$$\therefore p.$$

■

In his book *Gödel, Escher, Bach,** Douglas Hofstadter introduces the fantasy rule for mathematical proof. Hofstadter points out that when you start a mathematical argument with *if, let,* or *suppose,* you are stepping into a fantasy world where not only are all the facts of the real world true but whatever you are supposing is also true. Once you are in that world, you can suppose something else. That sends you into a subfantasy world where not only is everything in the fantasy world true but also the new thing you are supposing. Of course you can continue stepping into new subfantasy worlds in this way indefinitely. You return one level closer to the real world each time you derive a conclusion that makes a whole if-then or universal statement true. Your aim in a proof is to continue deriving such conclusions until you return to the world from which you made your first supposition.

Occasionally, mathematical problems are stated in the following form:

Suppose (*statement* 1). Prove that (*statement* 2).

When this phrasing is used, the author intends the reader to add statement 1 to his or her general mathematical knowledge and not to make explicit reference to it in the proof. In Hofstadter's terms, the author invites the reader to enter a fantasy world where statement 1 is known to be true and to prove statement 2 in this fantasy world. Thus the solver of such a problem would begin a proof with the starting point for a proof of statement 2. Consider, for instance, the following restatement of Example 5.2.1:

Suppose A and B are arbitrarily chosen sets.
Prove that $A \cap B \subseteq A$.

The proof would begin "Suppose $x \in A \cap B$," it being *understood* that sets A and B have already been chosen arbitrarily.

The proof of Example 5.2.1 is called an **element argument** because it shows one set to be a subset of another by demonstrating that every element in the one set is also an element in the other. In higher mathematics, element arguments are the standard method of establishing relations among sets. High school students are often allowed to justify set properties by using Venn diagrams. This method is appealing, but for it to be mathematically rigorous may be more complicated than you might expect. For instance, it is impossible to draw a single Venn diagram in which four circular disks represent sets in such a way that all 16 subsets appear as regions of the diagram. (It *is* possible to represent all the subsets if noncircular regions are used.) Even when appropriate Venn diagrams can be drawn, the verbal explanations needed to justify conclusions inferred from them are normally as long as a straightforward element proof.

*Gödel, Escher, Bach: An Eternal Golden Braid (New York: Basic Books, 1979).

Set Identities

An **identity** is an equation that is universally true for all elements in some set. For example, the equation $a + b = b + a$ is an identity for real numbers because it is true for all real numbers a and b. The collection of set properties in the next theorem consists entirely of set identities. That is, they are equations that are true for all sets in some universal set.

Theorem 5.2.2 Set Identities

Let all sets referred to below be subsets of a universal set U.

1. *Commutative Laws:* For all sets A and B,

$$\text{(a) } A \cup B = B \cup A \quad \text{and} \quad \text{(b) } A \cap B = B \cap A.$$

2. *Associative Laws:* For all sets A, B, and C,

$$\text{(a) } (A \cup B) \cup C = A \cup (B \cup C) \quad \text{and}$$
$$\text{(b) } (A \cap B) \cap C = A \cap (B \cap C).$$

3. *Distributive Laws:* For all sets, A, B, and C,

$$\text{(a) } A \cup (B \cap C) = (A \cup B) \cap (A \cup C) \quad \text{and}$$
$$\text{(b) } A \cap (B \cup C) = (A \cap B) \cup (A \cap C).$$

4. *Identity Laws:* For all sets A,

$$\text{(a) } A \cup \emptyset = A \quad \text{and} \quad \text{(b) } A \cap U = A.$$

5. *Complement Laws:*

$$\text{(a) } A \cup A^c = U \quad \text{and} \quad \text{(b) } A \cap A^c = \emptyset.$$

6. *Double Complement Law:* For all sets A,

$$(A^c)^c = A.$$

7. *Idempotent Laws:* For all sets A,

$$\text{(a) } A \cup A = A \quad \text{and} \quad \text{(b) } A \cap A = A.$$

8. *Universal Bound Laws:* For all sets A,

$$\text{(a) } A \cup U = U \quad \text{and} \quad \text{(b) } A \cap \emptyset = \emptyset.$$

9. *De Morgan's Laws:* For all sets A and B,

$$\text{(a) } (A \cup B)^c = A^c \cap B^c \quad \text{and} \quad \text{(b) } (A \cap B)^c = A^c \cup B^c.$$

10. *Absorption Laws:* For all sets A and B,

$$\text{(a) } A \cup (A \cap B) = A \quad \text{and} \quad \text{(b) } A \cap (A \cup B) = A.$$

11. *Complements of U and $\emptyset$:*

$$\text{(a) } U^c = \emptyset \quad \text{and} \quad \text{(b) } \emptyset^c = U.$$

12. *Set Difference Law:* For all sets A and B,

$$A - B = A \cap B^c.$$

Proving Set Identities

The conclusion of each part of Theorem 5.2.2 is that one set equals another set. As we noted in Section 5.1,

Two sets are equal   ⇔   each is a subset of the other.

The method derived from this fact is the most basic way to prove equality of sets.

Basic Method for Proving That Sets Are Equal

Let sets X and Y be given. To prove that $X = Y$:

1. Prove that $X \subseteq Y$.

2. Prove that $Y \subseteq X$.

Example 5.2.2 Proof of a Distributive Law

Prove that for all sets A, B, and C,

$$A \cup (B \cap C) = (A \cup B) \cap (A \cup C).$$

Solution The proof of this fact is somewhat more complicated than the proof in Example 5.2.1, so we first derive its logical structure, then find the core arguments, and end with a formal proof as a summary. As in Example 5.2.1, the statement to be proved is universal, and so, by the method of generalizing from the generic particular, the proof has the following outline:

Starting Point: Suppose A, B, and C are arbitrarily chosen sets.

To Show: $A \cup (B \cap C) = (A \cup B) \cap (A \cup C)$.

Now two sets are equal if, and only if, each is a subset of the other. Hence, the following two statements must be proved:

$$A \cup (B \cap C) \subseteq (A \cup B) \cap (A \cup C)$$

and

$$(A \cup B) \cap (A \cup C) \subseteq A \cup (B \cap C).$$

Showing the first containment requires showing that

$$\forall x, \text{ if } x \in A \cup (B \cap C) \text{ then } x \in (A \cup B) \cap (A \cup C).$$

Showing the second containment requires showing that

$$\forall x, \text{ if } x \in (A \cup B) \cap (A \cup C) \text{ then } x \in A \cup (B \cap C).$$

Note that both of these statements are universal. So to prove the first containment, you

suppose you have any element x in $A \cup (B \cap C)$,

and then you

show that $x \in (A \cup B) \cap (A \cup C)$.

And to prove the second containment, you

suppose you have any element x in $(A \cup B) \cap (A \cup C)$,

and then you

show that $x \in A \cup (B \cap C)$.

In Figure 5.2.1, the structure of the proof is illustrated by the kind of diagram that is often used in connection with structured programs. The analysis in the diagram reduces the proof to two concrete tasks: filling in the steps indicated by dots in the two center boxes of Figure 5.2.1.

Suppose A, B, and C are sets. [*Show $A \cup (B \cap C) = (A \cup B) \cap (A \cup C)$. That is, show $A \cup (B \cap C) \subseteq (A \cup B) \cap (A \cup C)$ and $(A \cup B) \cap (A \cup C) \subseteq A \cup (B \cap C)$.*]

Show $A \cup (B \cap C) \subseteq (A \cup B) \cap (A \cup C)$. [*That is, show $\forall x$, if $x \in A \cup (B \cap C)$ then $x \in (A \cup B) \cap (A \cup C)$.*]

Suppose $x \in A \cup (B \cap C)$. [*Show $x \in (A \cup B) \cap (A \cup C)$.*]
$\vdots$
Thus $x \in (A \cup B) \cap (A \cup C)$.

Hence $A \cup (B \cap C) \subseteq (A \cup B) \cap (A \cup C)$.

Show $(A \cup B) \cap (A \cup C) \subseteq A \cup (B \cap C)$. [*That is, show $\forall x$, if $x \in (A \cup B) \cap (A \cup C)$ then $x \in A \cup (B \cap C)$.*]

Suppose $x \in (A \cup B) \cap (A \cup C)$. [*Show $x \in A \cup (B \cap C)$.*]
$\vdots$
Thus $x \in A \cup (B \cap C)$.

Hence $(A \cup B) \cap (A \cup C) \subseteq A \cup (B \cap C)$.

Thus $(A \cup B) \cap (A \cup C) = A \cup (B \cap C)$.

Figure 5.2.1

The top inner box goes from the supposition that $x \in A \cup (B \cap C)$ to the conclusion that $x \in (A \cup B) \cap (A \cup C)$.

Now when $x \in A \cup (B \cap C)$, then by definition of union, $x \in A$ or $x \in B \cap C$. But either of these possibilities might be the case because x is assumed to be chosen arbitrarily from the set $A \cup (B \cap C)$. So you have to show you can reach the conclusion that $x \in (A \cup B) \cap (A \cup C)$ regardless of whether x happens to be in A or x happens to be in $B \cap C$. This leads you to break your analysis into two cases: $x \in A$ and $x \in B \cap C$.

In case $x \in A$, your goal is to show that $x \in (A \cup B) \cap (A \cup C)$, which means that $x \in A \cup B$ and $x \in A \cup C$ (by definition of intersection). But when $x \in A$, both statements $x \in A \cup B$ and $x \in A \cup C$ are true by virtue of x's being in A.

In case $x \in B \cap C$, your goal is also to show that $x \in (A \cup B) \cap (A \cup C)$, which means that $x \in A \cup B$ and $x \in A \cup C$. But when $x \in B \cap C$, then $x \in B$ and $x \in C$ (by definition of intersection), and so $x \in A \cup B$ (by virtue of being in B) and $x \in A \cup C$ (by virtue of being in C).

This analysis shows that regardless of whether $x \in A$ or $x \in B \cap C$, the conclusion $x \in (A \cup B) \cap (A \cup C)$ follows. So you can fill in the steps in the top inner box.

To fill in the steps of the bottom inner box, you need to go from the supposition that $x \in (A \cup B) \cap (A \cup C)$ to the conclusion that $x \in A \cup (B \cap C)$.

Now when $x \in (A \cup B) \cap (A \cup C)$ and when x happens to be in A, then the statement "$x \in A$ or $x \in B \cap C$" is certainly true, and so x is in $A \cup (B \cap C)$ by definition of union. But either x is in A or x is not in A. So it remains only to be shown that even in the case when x is not in A, if $x \in (A \cup B) \cap (A \cup C)$, then $x \in A \cup (B \cap C)$.

Now to say that $x \in (A \cup B) \cap (A \cup C)$ means that $x \in A \cup B$ and $x \in A \cup C$ (by definition of union). But if $x \in A \cup B$, then x is in at least one of A or B, so if x is not in A, then x must be in B. Similarly, if $x \in A \cup C$, then x is in at least one of A or C, so if x is not in A, then x must be in C. Thus, when x is not in A and $x \in (A \cup B) \cap (A \cup C)$, then x is in both B and C, which means that $x \in B \cap C$. It follows that the statement "$x \in A$ or $x \in B \cap C$" is true, and so $x \in A \cup (B \cap C)$ by definition of union.

This analysis shows that if $x \in (A \cup B) \cap (A \cup C)$, then regardless of whether $x \in A$ or $x \notin A$, you can conclude that $x \in A \cup (B \cap C)$. Hence you can fill in the steps of the bottom inner box.

A formal proof is shown below.

Theorem 5.2.2(3)(a) A Distributive Law for Sets

For all sets A, B, and C,

$$A \cup (B \cap C) = (A \cup B) \cap (A \cup C).$$

Proof:

Suppose A and B are sets.

$A \cup (B \cap C) \subseteq (A \cup B) \cap (A \cup C)$:

Suppose $x \in A \cup (B \cap C)$. By definition of union, $x \in A$ or $x \in B \cap C$.

Case 1 ($x \in A$): Since $x \in A$, then $x \in A \cup B$ by definition of union and also $x \in A \cup C$ by definition of union. Hence $x \in (A \cup B) \cap (A \cup C)$ by definition of intersection.

Case 2 ($x \in B \cap C$): Since $x \in B \cap C$, then $x \in B$ and $x \in C$ by definition of intersection. Since $x \in B$, $x \in A \cup B$ and since $x \in C$, $x \in A \cup C$, both by definition of union. Hence $x \in (A \cup B) \cap (A \cup C)$ by definition of intersection.

In both cases, $x \in (A \cup B) \cap (A \cup C)$. Hence $A \cup (B \cap C) \subseteq (A \cup B) \cap (A \cup C)$ by definition of subset.

$(A \cup B) \cap (A \cup C) \subseteq A \cup (B \cap C)$:

Suppose $x \in (A \cup B) \cap (A \cup C)$. By definition of intersection, $x \in A \cup B$ and $x \in A \cup C$. Consider the two cases $x \in A$ and $x \notin A$.

Case 1 ($x \in A$): Since $x \in A$, we can immediately conclude that $x \in A \cup (B \cap C)$ by definition of union.

Case 2 ($x \notin A$): Since $x \in A \cup B$, x is in at least one of A or B. But x is not in A; hence x is in B. Similarly, since $x \in A \cup C$, x is in at least one of A or C. But x is not in A; hence x is in C. We have shown that both $x \in B$ and $x \in C$, and so by definition of intersection, $x \in B \cap C$. It follows by definition of union that $x \in A \cup (B \cap C)$.

In both cases $x \in A \cup (B \cap C)$. Hence, by definition of subset, $(A \cup B) \cap (A \cup C) \subseteq A \cup (B \cap C)$.

Since both subset relations have been proved, it follows by definition of set equality that $A \cup (B \cap C) = (A \cup B) \cap (A \cup C)$.

In the study of artificial intelligence, the types of reasoning used above to derive the proof of the distributive law are called *forward chaining* and *backward chaining*. First what is to be shown is viewed as a goal to be reached starting from a certain initial position: the starting point. Analysis of this goal leads to the realization that if a certain job is accomplished, then the goal will be reached. Call this job subgoal 1: SG_1. (For instance, if the goal is to show that $A \cup (B \cap C) = (A \cup B) \cap (A \cup C)$, then SG_1 would be to show that each set is a subset of the other.) Analysis of SG_1 shows that when yet another job is completed, SG_1 will be reached. Call this job subgoal 2: SG_2. Continuing in this way, a chain of argument leading backward from the goal is constructed.

$$\boxed{\text{starting point}} \qquad\qquad\qquad \to SG_3 \to SG_2 \to SG_1 \to \boxed{\text{goal}}$$

At a certain point, backward chaining becomes difficult, but analysis of the current subgoal suggests it may be reachable by a direct line of argument, called forward chaining, beginning at the starting point. Using the information contained in the starting point, another piece of information, I_1, is deduced; from that another piece of information, I_2, is deduced; and so forth until finally one of the subgoals is reached. This completes the chain and proves the theorem. A completed chain is illustrated below.

$$\boxed{\text{starting point}} \to I_1 \to I_2 \to I_3 \to I_4 \to SG_3 \to SG_2 \to SG_1 \to \boxed{\text{goal}}$$

Since set complement is defined in terms of *not,* and since unions and intersections are defined in terms of *or* and *and,* it is not surprising that there are analogues of De Morgan's laws of logic for sets.

Example 5.2.3 Proof of a De Morgan's Law for Sets

Prove that for all sets A and B, $(A \cup B)^c = A^c \cap B^c$.

Solution As in previous examples, the statement to be proved is universal, and so the starting point of the proof and the conclusion to be shown are as follows:

Starting Point: Suppose A and B are arbitrarily chosen sets.

To Show: $(A \cup B)^c = A^c \cap B^c$

To do this, you must show that $(A \cup B)^c \subseteq A^c \cap B^c$ and that $A^c \cap B^c \subseteq (A \cup B)^c$. To show the first containment means to show that

$$\forall x, \text{ if } x \in (A \cup B)^c \text{ then } x \in A^c \cap B^c.$$

And to show the second containment means to show that

$$\forall x, \text{ if } x \in A^c \cap B^c \text{ then } x \in (A \cup B)^c.$$

Since each of these statements is universal and conditional, for the first containment, you

$$\textbf{suppose } x \in (A \cup B)^c,$$

and then you

$$\textbf{show that } x \in A^c \cap B^c.$$

And for the second containment, you

$$\textbf{suppose } x \in A^c \cap B^c,$$

and then you

$$\textbf{show that } x \in (A \cup B)^c.$$

To fill in the steps of these arguments, you use the procedural versions of the definitions of complement, union, and intersection, and at crucial points you use De Morgan's laws of logic.

Theorem 5.2.2(9)(a) A De Morgan's Law for Sets

For all sets A and B, $(A \cup B)^c = A^c \cap B^c$.

Proof:

Suppose A and B are sets.

$(A \cup B)^c \subseteq A^c \cap B^c$:

[*We must show that* $\forall x$, *if* $x \in (A \cup B)^c$ *then* $x \in A^c \cap B^c$.]

Suppose $x \in (A \cup B)^c$. [*We must show that* $x \in A^c \cap B^c$.] By definition of complement,

$$x \notin A \cup B.$$

But to say that $x \notin A \cup B$ means that

$$\text{it is false that } (x \text{ is in } A \text{ or } x \text{ is in } B).$$

By De Morgan's laws of logic, this implies that

$$x \text{ is not in } A \text{ and } x \text{ is not in } B,$$

which can be written

$$x \notin A \quad \text{and} \quad x \notin B.$$

Hence $x \in A^c$ and $x \in B^c$ by definition of complement. It follows, by definition of intersection, that $x \in A^c \cap B^c$ [*as was to be shown*]. So $(A \cup B)^c \subseteq A^c \cap B^c$ by definition of subset.

$A^c \cap B^c \subseteq (A \cup B)^c$:

[*We must show that* $\forall x$, *if* $x \in A^c \cap B^c$ *then* $x \in (A \cup B)^c$.]

Suppose $x \in A^c \cap B^c$. [*We must show that* $x \in (A \cup B)^c$.] By definition of intersection, $x \in A^c$ and $x \in B^c$, and by definition of complement,

$$x \notin A \quad \text{and} \quad x \notin B.$$

In other words,

$$x \text{ is not in } A \text{ and } x \text{ is not in } B.$$

By De Morgan's laws of logic this implies that

$$\text{it is false that } (x \text{ is in } A \text{ or } x \text{ is in } B),$$

which can be written

$$x \notin A \cup B$$

by definition of union. Hence, by definition of complement, $x \in (A \cup B)^c$ [*as was to be shown*]. It follows that $A^c \cap B^c \subseteq (A \cup B)^c$ by definition of subset.

Since both set containments have been proved, $(A \cup B)^c = A^c \cap B^c$ by definition of set equality.

The set property given in the next theorem says that if one set is a subset of another, then their intersection is the smaller of the two sets and their union is the larger of the two sets.

Theorem 5.2.3 Intersection and Union with a Subset

For any sets A and B, if $A \subseteq B$, then

$$\text{(a) } A \cap B = A \quad \text{and} \quad \text{(b) } A \cup B = B.$$

Proof:

Part (a): Suppose A and B are sets with $A \subseteq B$. To show part (a) we must show both that $A \cap B \subseteq A$ and that $A \subseteq A \cap B$. We already know that $A \cap B \subseteq A$ by the inclusion of intersection property. To show that $A \subseteq A \cap B$, let $x \in A$. [*We must show that $x \in A \cap B$.*] Since $A \subseteq B$, then $x \in B$ also. Hence

$$x \in A \quad \text{and} \quad x \in B,$$

and thus

$$x \in A \cap B$$

by definition of intersection [*as was to be shown*].

Proof:

Part (b): The proof of part (b) is left as an exercise.

The Empty Set

In Section 5.1 we introduced the concept of a set with no elements and promised that in this section we would show that there is only one such set. To do so, we start with the most basic—and strangest—property of a set with no elements: It is a subset of every set. To see why this is true, just ask yourself, "Could it possibly be false? Could there be a set without elements that is *not* a subset of some given set?" The crucial fact is that the negation of a universal statement is existential: If a set B is not a subset of a set A, then there exists x in B such that x is not in A. But if B has no elements, then no such x can exist.

Theorem 5.2.4 A Set with No Elements Is a Subset of Every Set

If E is a set with no elements and A is any set, then $E \subseteq A$.

Proof (by contradiction):

Suppose not. [*We take the negation of the theorem and suppose it to be true.*] Suppose there exists a set E with no elements and a set A such that $E \nsubseteq A$. [*We must deduce a contradiction.*] Then there would be an element of E that is not an element of A [*by definition of subset*]. But there can be no such element since E has no elements. This is a contradiction. [*Hence the supposition that there are sets E and A, where E has no elements and $E \nsubseteq A$, is false, and so the theorem is true.*]

The truth of Theorem 5.2.4 can also be understood by appeal to the notion of vacuous truth. If E is a set with no elements and A is any set, then to say that $E \subseteq A$ is the same as saying that

$$\forall x \in E, x \in A.$$

But since E has no elements, this statement is vacuously true.

How many sets with no elements are there? Only one.

Corollary 5.2.5 Uniqueness of the Empty Set

There is only one set with no elements.

Proof:

Suppose E_1 and E_2 are both sets with no elements. By Theorem 5.2.4, $E_1 \subseteq E_2$ since E_1 has no elements. Also $E_2 \subseteq E_1$ since E_2 has no elements. Thus $E_1 = E_2$ by definition of set equality.

It follows from Corollary 5.2.5 that the set of pink elephants is equal to the set of all real numbers whose square is -1 because each set has no elements! Since there is only one set with no elements, we are justified in calling it by a special name, the empty set (or null set) and in denoting it by the special symbol $\varnothing$.

Note that whereas $\varnothing$ is the set with no elements, the set $\{\varnothing\}$ has one element, the empty set. This is similar to the convention in the computer programming languages LISP and Scheme, in which ( ) denotes the empty list and (( )) denotes the list whose one element is the empty list.

Suppose you need to show that a certain set equals the empty set. By Corollary 5.2.5 it suffices to show that the set has no elements. For since there is only one set with no elements (namely $\varnothing$), if the given set has no elements, then it must equal $\varnothing$.

Element Method for Proving a Set Equals the Empty Set

To prove that a set X is equal to the empty set $\varnothing$, prove that X has no elements. To do this, suppose X has an element and derive a contradiction.

Example 5.2.4 Proving That a Set Is Empty

Prove Theorem 5.2.2(8)(b). That is, prove that for any set A, $A \cap \varnothing = \varnothing$.

Solution Let A be a [*particular, but arbitrarily chosen*] set. To show that $A \cap \varnothing = \varnothing$, it suffices to show that $A \cap \varnothing$ has no elements [*by the element method for proving a set equals the empty set*]. Suppose not. That is, suppose there is an element x such that $x \in A \cap \varnothing$. Then, by definition of intersection, $x \in A$ and $x \in \varnothing$. In particular, $x \in \varnothing$. But this is impossible since $\varnothing$ has no elements. [*This contradiction shows that the supposition that there is an element x in $A \cap \varnothing$ is false. So $A \cap \varnothing$ has no elements, as was to be shown.*] Thus $A \cap \varnothing = \varnothing$. ■

Example 5.2.5 A Proof for a Conditional Statement

Prove that for all sets A, B, and C, if $A \subseteq B$ and $B \subseteq C^c$, then $A \cap C = \emptyset$.

Solution   Since the statement to be proved is both universal and conditional, you start with the method of direct proof:

> **Suppose** A, B, and C are arbitrarily chosen sets
> that satisfy the condition: $A \subseteq B$ and $B \subseteq C^c$.
>
> **Show** that $A \cap C = \emptyset$.

Since the conclusion to be shown is that a certain set is empty, you can use the principle for proving that a set equals the empty set. A complete proof is shown below.

Proposition 5.2.6

For all sets A, B, and C, if $A \subseteq B$ and $B \subseteq C^c$, then $A \cap C = \emptyset$.

Proof:

Suppose A, B, and C are any sets such that $A \subseteq B$ and $B \subseteq C^c$. We must show that $A \cap C = \emptyset$. Suppose not. That is, suppose there is an element x in $A \cap C$. By definition of intersection, $x \in A$ and $x \in C$. Then, since $A \subseteq B$, $x \in B$ by definition of subset. Also, since $B \subseteq C^c$, then $x \in C^c$ by definition of subset again. It follows by definition of complement that $x \notin C$. Thus $x \in C$ and $x \notin C$, which is a contradiction. So the supposition that there is an element x in $A \cap C$ is false, and thus $A \cap C = \emptyset$ [*as was to be shown*].

Exercise Set 5.2

1. **a.** To say that an element is in $A \cap (B \cup C)$ means that it is in __(1)__ and in __(2)__.
    b. To say that an element is in $(A \cap B) \cup C$ means that it is in __(1)__ or in __(2)__.
    c. To say that an element is in $A - (B \cap C)$ means that it is in __(1)__ and not in __(2)__.

2. The following are two proofs that for all sets A and B, $A - B \subseteq A$. The first is less formal, and the second is more formal. Fill in the blanks.

    a. **Proof:** Suppose A and B are any sets. To show that $A - B \subseteq A$, we must show that every element in __(1)__ is in __(2)__. But any element in $A - B$ is in __(3)__ and not in __(4)__ (by definition of $A - B$). In particular, such an element is in A.

    b. **Proof:** Suppose A and B are any sets and $x \in A - B$. [*We must show that __(1)__.*] By definition of set difference, $x \in$ __(2)__ and $x \notin$ __(3)__. In particular, $x \in$ __(4)__ [*which is what was to be shown*].

3. The following is a proof that for all sets A, B, and C, if $A \subseteq B$ and $B \subseteq C$, then $A \subseteq C$. Fill in the blanks.

    Proof: Suppose A, B, and C are sets and $A \subseteq B$ and $B \subseteq C$. To show that $A \subseteq C$, we must show that every element in __(1)__ is in __(2)__. But given any element in A, that element is in __(3)__ (because $A \subseteq B$), and so that element is also in __(4)__ (because __(5)__). Hence $A \subseteq C$.

4. The following is a proof that for all sets A and B, if $A \subseteq B$, then $A \cup B \subseteq B$. Fill in the blanks.

    Proof: Suppose A and B are any sets and $A \subseteq B$. [*We must show that __(a)__.*] Let $x \in$ __(b)__. [*We must show that __(c)__.*] By definition of union, $x \in$ __(d)__ __(e)__ $x \in$ __(f)__. In case $x \in$ __(g)__, then since $A \subseteq B$, $x \in$ __(h)__. In case $x \in B$, then clearly $x \in B$. So in either case, $x \in$ __(i)__ [*as was to be shown*].

5. Prove that for all sets A and B, $B - A = B \cap A^c$.

6. The following is a proof that for any sets A, B, and C, $A \cap (B \cup C) = (A \cap B) \cup (A \cap C)$. Fill in the blanks.

Proof: Suppose A, B, and C are any sets.

(1) $A \cap (B \cup C) \subseteq (A \cap B) \cup (A \cap C)$:

Let $x \in A \cap (B \cup C)$. [*We must show that $x \in$ (a) .*] By definition of intersection, $x \in$ (b) and $x \in$ (c) . Thus $x \in A$ and, by definition of union, $x \in B$ or (d) .

Case 1 ($x \in A$ and $x \in B$): In this case, by definition of intersection, $x \in$ (e) , and so, by definition of union, $x \in (A \cap B) \cup (A \cap C)$.

Case 2 ($x \in A$ and $x \in C$): In this case, (f) .

Hence in either case, $x \in (A \cap B) \cup (A \cap C)$ [*as was to be shown*].

[*So $A \cap (B \cup C) \subseteq (A \cap B) \cup (A \cap C)$ by definition of subset.*]

(2) $(A \cap B) \cup (A \cap C) \subseteq A \cap (B \cup C)$:

Let $x \in (A \cap B) \cup (A \cap C)$. [*We must show that* (a) .] By definition of union, $x \in$ (b) or $x \in$ (c) .

Case 1 ($x \in A \cap B$): In this case, by definition of intersection, (d) and (e) . Since $x \in B$, then by definition of union, $x \in B \cup C$. Hence $x \in A$ and $x \in B \cup C$, and so, by definition of intersection, $x \in$ (f) .

Case 2 ($x \in A \cap C$): In this case, (g) .

In either case, $x \in A \cap (B \cup C)$ [*as was to be shown*]. [*Thus $(A \cap B) \cup (A \cap C) \subseteq A \cap (B \cup C)$ by definition of subset.*]

[*Since both subset relations have been proved, it follows, by definition of set equality, that* (h) .]

H 7. Prove that for all sets A and B, $(A \cap B)^c = A^c \cup B^c$.

Use an element argument to prove each statement in 8–17. Assume that all sets are subsets of a universal set U.

8. For all sets A, B, and C,
$$(A - B) \cup (C - B) = (A \cup C) - B.$$

9. For all sets A, B, and C,
$$(A - B) \cap (C - B) = (A \cap C) - B.$$

H 10. For all sets A and B, $A \cup (A \cap B) = A$.

11. For all sets A, $A \cup \emptyset = A$.

12. For all sets A, B, and C, if $A \subseteq B$ then $A \cap C \subseteq B \cap C$.

13. For all sets A, B, and C, if $A \subseteq B$ then $A \cup C \subseteq B \cup C$.

14. For all sets A and B, if $A \subseteq B$ then $B^c \subseteq A^c$.

H 15. For all sets A, B, and C, if $A \subseteq B$ and $A \subseteq C$ then
$$A \subseteq B \cap C.$$

16. For all sets A, B, and C,
$$A \times (B \cup C) = (A \times B) \cup (A \times C).$$

17. For all sets A, B, and C,
$$A \times (B \cap C) = (A \times B) \cap (A \times C).$$

18. Find the mistake in the following "proof" that for all sets A, B, and C, if $A \subseteq B$ and $B \subseteq C$ then $A \subseteq C$.

"**Proof:** Suppose A, B, and C are sets such that $A \subseteq B$ and $B \subseteq C$. Since $A \subseteq B$, there is an element x such that $x \in A$ and $x \in B$. Since $B \subseteq C$, there is an element x such that $x \in B$ and $x \in C$. Hence there is a element x such that $x \in A$ and $x \in C$ and so $A \subseteq C$."

H 19. Find the mistake in the following "proof."

"**Theorem:**" For all sets A and B, $A^c \cup B^c \subseteq (A \cup B)^c$.

"**Proof:** Suppose A and B are sets, and $x \in A^c \cup B^c$. Then $x \in A^c$ or $x \in B^c$ by definition of union. It follows that $x \notin A$ or $x \notin B$ by definition of complement, and so $x \notin A \cup B$ by definition of union. Thus $x \in (A \cup B)^c$ by definition of complement, and hence $A^c \cup B^c \subset (A \cup B)^c$."

20. Find the mistake in the following "proof" that for all sets A and B, $(A - B) \cup (A \cap B) \subseteq A$.

"**Proof:** Suppose A and B are sets, and suppose $x \in (A - B) \cup (A \cap B)$. If $x \in A$ then $x \in A - B$. Then, by definition of difference, $x \in A$ and $x \notin B$. Hence $x \in A$, and so $(A - B) \cup (A \cap B) \subseteq A$ by definition of subset."

21. Consider the Venn diagram below.

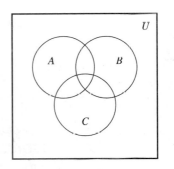

a. Illustrate one of the distributive laws by shading in the region corresponding to $A \cup (B \cap C)$ on one copy of the diagram and $(A \cup B) \cap (A \cup C)$ on another.

b. Illustrate the other distributive law by shading in the region corresponding to $A \cap (B \cup C)$ on one copy of the diagram and $(A \cap B) \cup (A \cap C)$ on another.

c. Illustrate one of De Morgan's laws by shading in the region corresponding to $(A \cup B)^c$ on one copy of the diagram and $A^c \cap B^c$ on the other. (Leave the set C out of your diagrams.)

d. Illustrate the other De Morgan's law by shading in the region corresponding to $(A \cap B)^c$ on one copy of the diagram and $A^c \cup B^c$ on the other. (Leave the set C out of your diagrams.)

22. Fill in the blanks in the following proof that for all sets A and B, $(A - B) \cap (B - A) = \emptyset$.

   Proof: Let A and B be any sets and supppose $(A - B) \cap (B - A) \neq \emptyset$. That is, suppose there were an element x in __(a)__. By definition of __(b)__, $x \in A - B$ and $x \in$ __(c)__. Then by definition of set difference, $x \in A$ and $x \notin B$ and $x \in$ __(d)__ and $x \notin$ __(e)__. In particular $x \in A$ and $x \notin$ __(f)__, which is a contradiction. Hence [*the supposition that* $(A - B) \cap (B - A) \neq \emptyset$ *is false, and so*] __(g)__.

Use the element method for proving a set equals the empty set to prove each statement in 23–34. Assume that all sets are subsets of a universal set U.

23. For all sets A and B, $(A - B) \cap (A \cap B) = \emptyset$.

24. For all sets A, B, and C,
$$(A - C) \cap (B - C) \cap (A - B) = \emptyset.$$

25. For all subsets A of a universal set U, $A \cap A^c = \emptyset$.

26. If U denotes a universal set, then $U^c = \emptyset$.

27. For all sets A, $A \times \emptyset = \emptyset$.

28. For all sets A and B, if $A \subseteq B$ then $A \cap B^c = \emptyset$.

29. For all sets A and B, if $B \subseteq A^c$ then $A \cap B = \emptyset$.

30. For all sets A, B, and C, if $A \subseteq B$ and $B \cap C = \emptyset$ then $A \cap C = \emptyset$.

31. For all sets A, B, and C, if $B \subseteq C$ and $A \cap C = \emptyset$, then $A \cap B = \emptyset$.

32. For all sets A, B, and C, if $C \subseteq B - A$, then $A \cap C = \emptyset$.

33. For all sets A, B, and C,
$$\text{if } B \cap C \subseteq A, \text{ then } (C - A) \cap (B - A) = \emptyset.$$

34. For all sets A, B, C, and D,
$$\text{if } A \cap C = \emptyset \text{ then } (A \times B) \cap (C \times D) = \emptyset.$$

Use mathematical induction and the following definitions to prove each statement in 35–37. If n is an integer with $n \geq 3$ and if $C_1, C_2, C_3, \ldots, C_n$ are any sets,

$$C_1 \cup C_2 \cup C_3 \cup \cdots \cup C_n = (C_1 \cup C_2 \cup C_3 \cup \cdots \cup C_{n-1}) \cup C_n,$$

and

$$C_1 \cap C_2 \cap C_3 \cap \cdots \cap C_n = (C_1 \cap C_2 \cap C_3 \cap \cdots \cap C_{n-1}) \cap C_n.$$

(More rigorous versions of the definitions are given in Section 8.4.)

35. *Generalized Distributive Law for Sets:* For any integer $n \geq 1$, if A and $B_1, B_2, B_3, \ldots, B_n$ are any sets, then
$$(A \cap B_1) \cup (A \cap B_2) \cup \cdots \cup (A \cap B_n)$$
$$= A \cap (B_1 \cup B_2 \cup B_3 \cup \cdots \cup B_n).$$

36. For any integer $n \geq 1$, if $A_1, A_2, A_3, \ldots, A_n$ and B are any sets, then
$$(A_1 - B) \cup (A_2 - B) \cup \cdots \cup (A_n - B)$$
$$= (A_1 \cup A_2 \cup A_3 \cup \cdots \cup A_n) - B.$$

(To prove the inductive step, you can use the result of exercise 8.)

37. For any integer $n \geq 1$, if $A_1, A_2, A_3, \ldots, A_n$ and B are any sets, then
$$(A_1 - B) \cap (A_2 - B) \cap \cdots \cap (A_n - B)$$
$$= (A_1 \cap A_2 \cap A_3 \cap \cdots \cap A_n) - B.$$

(To prove the inductive step, you can use the result of exercise 9.)

5.3 Disproofs, Algebraic Proofs, and Boolean Algebras

If a fact goes against common sense, and we are nevertheless compelled to accept and deal with this fact, we learn to alter our notion of common sense.
— Phillip J. Davis and Reuben Hersh, *The Mathematical Experience*, 1981

In Section 5.2 we gave examples only of set properties that were true. Occasionally, however, a proposed set property is false. We begin this section by discussing how to disprove such a proposed property. Then we prove an important theorem about the power set of a set and go on to discuss an "algebraic" method for deriving new set properties from set properties already known to be true. We finish the section with an introduction to Boolean algebras.

Disproving an Alleged Set Property

Recall that to show a universal statement is false, it suffices to find one example (called a counterexample) for which it is false.

Example 5.3.1 Finding a Counterexample for a Set Identity

Is the following set property true?

For all sets A, B, and C, $(A - B) \cup (B - C) = A - C$.

Solution   Observe that the property is true if, and only if,

the given equality holds for *all* sets A, B, and C.

So it is false if, and only if,

there are sets A, B, and C for which the equality does *not* hold.

Thus one way to analyze the question is to ask yourself what elements are in each of the sets $(A - B) \cup (B - C)$ and $A - C$ and then try to find a condition that A, B, and C could satisfy so that one of the sets would contain an element that is not in the other set.

Now $A - C$ consists of all the elements that are in A but not in C, and $(A - B) \cup (B - C)$ consists of all the elements that are in A but not in B, together with all those that are in B but not in C. Think, for instance, about an element that is in $(A - B) \cup (B - C)$ by virtue of being in A but not in B. Could such an element fail to be in $A - C$? The answer is yes, provided that the element is in C. In other words, an element that is in both A and C but is not in B will be in $(A - B) \cup (B - C)$, but it will not be in $A - C$. From this analysis, you can construct a counterexample to the proposed set property such as the following. Because you might have made a mistake in your analysis, always be sure to check that your counterexample works.

Counterexample: Let $A = \{1, 2\}$, $B = \{2\}$, and $C = \{1\}$. Then

$A - C = \{2\}$, $A - B = \{1\}$, $B - C = \{2\}$, and $(A - B) \cup (B - C) = \{1, 2\}$.

Since $1 \in (A - B) \cup (B - C)$ but $1 \notin A - C$, then $(A - B) \cup (B - C) \neq A - C$.

Alternatively, you could think about an element that is in $(A - B) \cup (B - C)$ by virtue of being in B but not in C. Could such an element fail to be in $A - C$? This time the answer is yes, provided that the element is not in A. So an element that is in B but not in either A or C will be in $(A - B) \cup (B - C)$ but not in $A - C$. Can you think of sets A, B, and C that satisfy this condition and confirm that they produce a counterexample to the proposed property?

A different approach to solving this problem is to picture sets A, B, and C by drawing a Venn diagram such as that shown in Figure 5.3.1. If you assume that any of the eight regions of the diagram may be empty of points, then the diagram is quite general.

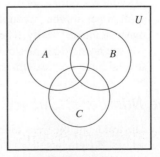

Figure 5.3.1

Find and shade the region corresponding to $(A - B) \cup (B - C)$. Then shade the region corresponding to $A - C$. These are shown in Figure 5.3.2

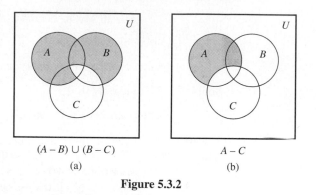

$(A - B) \cup (B - C)$

(a)

$A - C$

(b)

Figure 5.3.2

When you compare the shaded regions, you can see that there may be points in $(A - B) \cup (B - C)$ that are not in $A - C$. The property is therefore false, and a concrete counterexample consists of any sets A, B, and C with points inside regions shaded in one diagram but not the other. For example, A, B, and C could be taken to be the sets of *all* points inside each of the disks shown.

Another approach is to use the diagrams to help construct a discrete counterexample. The shading of the diagrams shows that for sets A, B, and C to be a counterexample, B must contain points that are not in either A or C, or there must be points in both A and C that are not in B. For example, you could take

$$A = \{a, b\}, \quad B = \{b, c\}, \quad \text{and} \quad C = \{a, d\}.$$

Then

$$A - B = \{a\}, \quad B - C = \{b, c\}, \quad \text{and} \quad A - C = \{b\}.$$

Hence

$$(A - B) \cup (B - C) = \{a, b, c\} \quad \text{whereas} \quad A - C = \{b\}.$$

So $(A - B) \cup (B - C) \neq A - C$. ∎

Problem-Solving Strategy

How can you discover whether a given universal statement about sets is true or false? There are two basic approaches: the optimistic and the pessimistic. In the optimistic approach, you simply plunge in and start trying to prove the statement, asking yourself, "What do I need to show?" and "How do I show it?" In the pessimistic approach, you start by searching your mind for a set of conditions that must be fulfilled to construct a counterexample. With either approach you may have clear sailing and be immediately successful or you may run into difficulty. The trick is to be ready to switch to the other approach if the one you are trying does not look promising. For more difficult questions, you may alternate several times between the two approaches before arriving at the correct answer.

The Number of Subsets of a Set

The following theorem states the important fact that if a set has n elements, then its power set has 2^n elements. The proof uses mathematical induction and is based on the following observations. Suppose X is a set and z is an element of X.

1. The subsets of X can be split into two groups: those that do not contain z and those that do contain z.

2. The subsets of X that do not contain z are the same as the subsets of $X - \{z\}$.

3. The subsets of X that do not contain z can be matched up one for one with the subsets of X that do contain z by matching each subset A that does not contain z to the subset $A \cup \{z\}$ that contains z. Thus there are as many subsets of X that contain z as there are subsets of X that do not contain z. For instance, if $X = \{x, y, z\}$, the following table shows the correspondence between subsets of X that do not contain z and subsets of X that contain z.

Subsets of X That Do Not Contain z		Subsets of X That Contain z
$\emptyset$	$\longleftrightarrow$	$\emptyset \cup \{z\} = \{z\}$
$\{x\}$	$\longleftrightarrow$	$\{x\} \cup \{z\} = \{x, z\}$
$\{y\}$	$\longleftrightarrow$	$\{y\} \cup \{z\} = \{y, z\}$
$\{x, y\}$	$\longleftrightarrow$	$\{x, y\} \cup \{z\} = \{x, y, z\}$

Theorem 5.3.1

For all integers $n \geq 0$, if a set X has n elements, then $\mathscr{P}(X)$ has 2^n elements.

Proof (by mathematical induction):

Consider the property "Any set with n elements has 2^n subsets."

Show that the property is true for $n = 0$: We must show that a set with zero elements has 2^0 subsets. But the only set with zero elements is the empty set, and the only subset of the empty set is itself. Thus a set with zero elements has one subset. Since $1 = 2^0$, the theorem is true for $n = 0$.

Show that for all integers $k \geq 1$, if the property is true for $n = k$, then it is true for $n = k + 1$: Let k be any integer with $k \geq 0$ and suppose that any set with k elements has 2^k subsets. [*This is the inductive hypothesis.*] We must show that any set with $k + 1$ elements has 2^{k+1} subsets.

Let X be a set with $k + 1$ elements and pick an element z in X. Observe that any subset of X either contains z or not. Furthermore, any subset of X that does not contain z is a subset of $X - \{z\}$. And any subset A of $X - \{z\}$ can be matched up with a subset B, equal to $A \cup \{z\}$, of X that contains z. Consequently, there are as many subsets of X that contain z as do not, and thus there are twice as many subsets of X as there are subsets of $X - \{z\}$. But $X - \{z\}$ has k elements, and so

$$\text{the number of subsets of } X - \{z\} = 2^k \quad \text{by inductive hypothesis.}$$

Therefore,

$$\text{the number of subsets of } X = 2 \cdot (\text{the number of subsets of } X - \{z\})$$
$$= 2 \cdot (2^k) \quad \text{by substitution}$$
$$= 2^{k+1} \quad \text{by basic algebra.}$$

[*This is what was to be shown.*]

[*Since we have proved both the basis step and the inductive step, we conclude that the theorem is true.*]

"Algebraic" Proofs of Set Identities

Let U be a universal set and consider the power set of U, $\mathcal{P}(U)$. To prepare the way for a discussion of Boolean algebras, let $S = \mathcal{P}(U)$. The set identities given in Theorem 5.2.2 hold for all elements of S. Once a certain number of identities and other properties have been established, new properties can be derived from them algebraically. It turns out that only identities (1–5) of Theorem 5.2.2 are needed to prove any other identity involving only unions, intersections, and complements. With the addition of identity (12), the set difference law, any set identity involving unions, intersections, complements, and set differences can be established.

To use known properties to derive new ones, you need to use the fact that such properties are universal statements. Like the laws of algebra for real numbers, they apply to a wide variety of different situations. For instance, one of the distributive laws states that

for all sets A, B, and C, $\quad A \cap (B \cup C) = (A \cap B) \cup (A \cap C)$.

This law can be viewed as a general template into which *any* three particular sets can be placed. Thus, for example, if A_1, A_2, and A_3 represent particular sets, then

$$\underbrace{A_1} \cap (\underbrace{A_2} \cup \underbrace{A_3}) = (\underbrace{A_1} \cap \underbrace{A_2}) \cup (\underbrace{A_1} \cap \underbrace{A_3}),$$
$$A \quad \cap \quad (B \quad \cup \quad C) \; = \; (A \quad \cap \quad B) \; \cup \; (A \quad \cap \quad C)$$

where A_1 plays the role of A, A_2 plays the role of B, and A_3 plays the role of C. Similarly, if W, X, Y, and Z are any particular sets, then, by the distributive law,

$$(\underbrace{W \cap X}) \cap (\underbrace{Y} \cup \underbrace{Z}) = ((\underbrace{W \cap X}) \cap \underbrace{Y}) \cup ((\underbrace{W \cap X}) \cap \underbrace{Z}),$$
$$A \quad\quad \cap (B \cup C) = \quad (A \quad\quad \cap B) \cup \quad (A \quad\quad \cap C)$$

where $W \cap X$ plays the role of A, Y plays the role of B, and Z plays the role of C.

Example 5.3.2 Deriving a Set Difference Property

Use the properties in Theorem 5.2.2 to construct an algebraic proof that for all sets A, B, and C,

$$(A \cup B) - C = (A - C) \cup (B - C).$$

Solution   Let sets A, B, and C be given. Then

$(A \cup B) - C = (A \cup B) \cap C^c$	by the set difference law
$= C^c \cap (A \cup B)$	by the commutative law for $\cap$
$= (C^c \cap A) \cup (C^c \cap B)$	by the distributive law
$= (A \cap C^c) \cup (B \cap C^c)$	by the commutative law for $\cap$
$= (A - C) \cup (B - C)$	by the set difference law. ∎

Example 5.3.3 Deriving a Set Identity Using Properties of ∅

Use the properties in Theorem 5.2.2 to construct an algebraic proof that for all sets A and B,

$$A - (A \cap B) = A - B.$$

Solution   Suppose A and B are sets. Then

$$\begin{aligned}
A - (A \cap B) &= A \cap (A \cap B)^c && \text{by the set difference law} \\
&= A \cap (A^c \cup B^c) && \text{by De Morgan's laws} \\
&= (A \cap A^c) \cup (A \cap B^c) && \text{by the distributive law} \\
&= \emptyset \cup (A \cap B^c) && \text{by the complement law} \\
&= (A \cap B^c) \cup \emptyset && \text{by the commutative law for } \cup \\
&= A \cap B^c && \text{by the identity law for } \cup \\
&= A - B && \text{by the set difference law.}
\end{aligned}$$
∎

To many people an algebraic proof seems more attractive than an element proof. But often an element proof is actually simpler. For instance, in Example 5.3.3 above, you could see immediately that $A - (A \cap B) = A - B$ because for an element to be in $A - (A \cap B)$ means that it is in A and not in both A and B, and this is equivalent to saying that it is in A and not in B.

Example 5.3.4 Deriving a Generalized Associative Law

Prove that for any sets A_1, A_2, A_3, and A_4,

$$((A_1 \cup A_2) \cup A_3) \cup A_4 = A_1 \cup ((A_2 \cup A_3) \cup A_4).$$

Solution   Let sets A_1, A_2, A_3, and A_4 be given. Then

$$\begin{aligned}
((A_1 \cup A_2) \cup A_3) \cup A_4 &= (A_1 \cup (A_2 \cup A_3)) \cup A_4 && \text{by the associative law for } \cup \text{ with } A_1 \\
& && \text{playing the role of } A, A_2 \text{ playing the role} \\
& && \text{of } B, \text{ and } A_3 \text{ playing the role of } C \\
&= A_1 \cup ((A_2 \cup A_3) \cup A_4) && \text{by the associative law for } \cup \text{ with } A_1 \\
& && \text{playing the role of } A, A_2 \cup A_3 \text{ playing} \\
& && \text{the role of } B, \text{ and } A_4 \text{ playing the role of } C.
\end{aligned}$$
∎

 Caution! When doing problems similar to Examples 5.3.2–5.3.4, be sure to use the set properties exactly as they are stated.

Boolean Algebras

If you look back at the logical equivalences of Theorem 1.1.1 and compare them to the set identities of Theorem 5.2.2, you will notice that numbers 1–11 are very similar. This reflects a similarity of underlying structure between the set of all statement forms in a finite number of variables with the operations of $\vee$ and $\wedge$ and the set of all subsets of a set together with the operations of $\cup$ and $\cap$. Both are special cases of a general algebraic structure known as a Boolean algebra. Properties of Boolean algebras are used extensively in the simplification of digital logic circuits.

• Definition: Boolean Algebra

A **Boolean algebra** is a set B together with two operations, generally denoted $+$ and $\cdot$, such that for all a and b in B both $a + b$ and $a \cdot b$ are in B and the following properties hold:

1. *Commutative Laws:* For all a and b in B,

$$\text{(a) } a + b = b + a \quad \text{and} \quad \text{(b) } a \cdot b = b \cdot a.$$

2. *Associative Laws:* For all a, b, and c in B,

$$\text{(a) } (a + b) + c = a + (b + c) \quad \text{and} \quad \text{(b) } (a \cdot b) \cdot c = a \cdot (b \cdot c).$$

3. *Distributive Laws:* For all a, b, and c in B,

$$\text{(a) } a + (b \cdot c) = (a + b) \cdot (a + c) \quad \text{and} \quad \text{(b) } a \cdot (b + c) = (a \cdot b) + (a \cdot c).$$

4. *Identity Laws:* There exist distinct elements 0 and 1 in B such that for all a in B,

$$\text{(a) } a + 0 = a \quad \text{and} \quad \text{(b) } a \cdot 1 = a.$$

5. *Complement Laws:* For each a in B, there exists an element in B, denoted $\overline{a}$ and called the **complement** or **negation** of a, such that

$$\text{(a) } a + \overline{a} = 1 \quad \text{and} \quad \text{(b) } a \cdot \overline{a} = 0.$$

For the set of statement forms in a finite number of variables, $\vee$ and $\wedge$ play the roles of $+$ and $\cdot$, the tautology **t** and contradiction **c** play the roles of 1 and 0, and $\sim$ plays the role of $-$. For a set of subsets of a nonempty set U, $\cup$ and $\cap$ play the roles of $+$ and $\cdot$, U and $\emptyset$ play the roles of 1 and 0, and complementation ^{c} plays the role of $-$. Thus the set of all subsets of a universal set U is a Boolean algebra with operations $\cup$ and $\cap$, and the set of all statement forms in a finite number of variables is a Boolean algebra with operations $\vee$ and $\wedge$. It turns out that in any Boolean algebra, the complement of each element is unique, the quantities 0 and 1 are unique, and identities analogous to numbers 6–11 in Theorem 1.1.1 and Theorem 5.2.2 can be deduced.

Theorem 5.3.2 Properties of a Boolean Algebra

Let B be any Boolean Algebra.

1. *Uniqueness of the Complement Law:* For all a and x in B, if $a + x = 1$ and $a \cdot x = 0$ then $x = \overline{a}$.

2. *Uniqueness of 0 and 1:* If there exists x in B such that $a + x = a$ for all a in B, then $x = 0$, and if there exists y in B such that $a \cdot y = a$ for all a in B, then $y = 1$.

3. *Double Complement Law:* For all $a \in B$, $\overline{(\overline{a})} = a$.

4. *Idempotent Law:* For all $a \in B$,

$$\text{(a) } a + a = a \quad \text{and} \quad \text{(b) } a \cdot a = a.$$

5. *Universal Bound Law:* For all $a \in B$,

$$\text{(a) } a + 1 = 1 \quad \text{and} \quad \text{(b) } a \cdot 0 = 0.$$

6. *De Morgan's Laws:* For all a and $b \in B$,

$$\text{(a) } \overline{a + b} = \overline{a} \cdot \overline{b} \quad \text{and} \quad \text{(b) } \overline{a \cdot b} = \overline{a} + \overline{b}.$$

7. *Absorption Laws:* For all a and $b \in B$,

$$\text{(a) } (a + b) \cdot a = a \quad \text{and} \quad \text{(b) } (a \cdot b) + a = a.$$

8. *Complements of* 0 *and* 1:

$$\text{(a) } \overline{0} = 1 \quad \text{and} \quad \text{(b) } \overline{1} = 0.$$

Proof:

Part 1: Uniqueness of the Complement Law

Suppose a and x are particular, but arbitrarily chosen, elements of B that satisfy the following hypothesis: $a + x = 1$ and $a \cdot x = 0$. Then

$$
\begin{aligned}
x &= x \cdot 1 & & \text{because 1 is an identity for } \cdot \\
&= x \cdot (a + \overline{a}) & & \text{by the complement law for } + \\
&= x \cdot a + x \cdot \overline{a} & & \text{by the distributive law for } \cdot \text{ over } + \\
&= a \cdot x + x \cdot \overline{a} & & \text{by the commutative law for } \cdot \\
&= 0 + x \cdot \overline{a} & & \text{by hypothesis} \\
&= a \cdot \overline{a} + x \cdot \overline{a} & & \text{by the complement law for } \cdot \\
&= (\overline{a} \cdot a) + (\overline{a} \cdot x) & & \text{by the commutative law for } \cdot \\
&= \overline{a} \cdot (a + x) & & \text{by the distributive law for } \cdot \text{ over } + \\
&= \overline{a} \cdot 1 & & \text{by hypothesis} \\
&= \overline{a} & & \text{because 1 is an identity for } \cdot.
\end{aligned}
$$

Proofs of the other parts of the theorem are discussed in the examples below and in the exercises.

You may notice that all parts of the definition of a Boolean algebra and most parts of Theorem 5.3.2 contain paired statements. For instance, the distributive laws state that for all a, b, and c in B,

$$\text{(a) } a + (b \cdot c) = (a + b) \cdot (a + c) \quad \text{and} \quad \text{(b) } a \cdot (b + c) = (a \cdot b) + (a \cdot c),$$

and the identity laws state that for all a in B,

$$\text{(a) } a + 0 = a \quad \text{and} \quad \text{(b) } a \cdot 1 = a.$$

Note that each of the paired statements can be obtained from the other by interchanging all the $+$ and $\cdot$ signs and interchanging 1 and 0. Such interchanges transform any Boolean identity into its **dual** identity. It can be proved that the dual of any Boolean identity is also an identity. This fact is often called the **duality principle** for a Boolean algebra.

Example 5.3.5 Proof of the Double Complement Law

Prove that for all elements a in a Boolean algebra B, $\overline{(\overline{a})} = a$.

Solution   Start by supposing that B is a Boolean algebra and a is any element of B. The basis for the proof is the uniqueness of the complement law: that each element in B has a unique complement that satisfies certain equations with respect to it. So if a can be shown to satisfy those equations with respect to $\overline{a}$, then a must be the complement of $\overline{a}$.

> **Theorem 5.3.2(3) Double Complement Law**
>
> For all elements a in a Boolean algebra B, $\overline{(\overline{a})} = a$.
>
> **Proof:**
>
> Suppose B is a Boolean algebra and a is any element of B. Then
>
> $$\overline{a} + a = a + \overline{a} \qquad \text{by the commutative law}$$
> $$= 1 \qquad \text{by the complement law for 1}$$
>
> and
>
> $$\overline{a} \cdot a = a \cdot \overline{a} \qquad \text{by the commutative law}$$
> $$= 0 \qquad \text{by the complement law for 0.}$$
>
> Thus a satisfies the two equations with respect to $\overline{a}$ that are satisfied by the complement of $\overline{a}$. From the fact that the complement of a is unique, we conclude that $\overline{(\overline{a})} = a$. ∎

Example 5.3.6 Proof of an Idempotent Law

Fill in the blanks in the following proof that for all elements a in a Boolean algebra B, $a + a = a$.

Proof:

Suppose B is a Boolean algebra and a is any element of B. Then

$$a = a + 0 \qquad \underline{\text{(a)}}$$
$$= a + (a \cdot \overline{a}) \qquad \underline{\text{(b)}}$$
$$= (a + a) \cdot (a + \overline{a}) \qquad \underline{\text{(c)}}$$
$$= (a + a) \cdot 1 \qquad \underline{\text{(d)}}$$
$$= a + a \qquad \underline{\text{(e)}}.$$

Solution

  a. because 0 is an identity for $+$

  b. by the complement law for $\cdot$

  c. by the distributive law for $+$ over $\cdot$

  d. by the complement law for $+$

  e. because 1 is an identity for $\cdot$ ∎

Exercise Set 5.3

For each of 1–4 find a counterexample to show that the statement is false. Assume all sets are subsets of a universal set U.

1. For all sets A, B, and C, $(A \cap B) \cup C = A \cap (B \cup C)$.

2. For all sets A, B, and C, if $A \subseteq B$ then $A \cap (B \cap C)^c = \emptyset$.

3. For all sets A, B, and C, if $A \nsubseteq B$ and $B \nsubseteq C$ then $A \nsubseteq C$.

4. For all sets A, B, and C, if $B \cap C \subseteq A$ then $(A - B) \cap (A - C) = \emptyset$.

For each of 5–17 prove each statement that is true and find a counterexample for each statement that is false. Assume all sets are subsets of a universal set U.

5. For all sets A, B, and C, $A - (B - C) = (A - B) - C$.

6. For all sets A and B, $A \cap (A \cup B) = A$.

7. For all sets A, B, and C,

$$(A - B) \cap (C - B) = A - (B \cup C).$$

8. For all sets A and B, if $A^c \subseteq B$ then $A \cup B = U$.

9. For all sets A, B, and C, if $A \subseteq C$ and $B \subseteq C$ then $A \cup B \subseteq C$.

10. For all sets A and B, if $A \subseteq B$ then $A \cap B^c = \emptyset$.

II 11. For all sets A, B, and C, if $A \subseteq B$ and $B \cap C = \emptyset$ then $A \cap C = \emptyset$.

12. For all sets A and B, if $A \cap B = \emptyset$ then $A \times B = \emptyset$.

13. For all sets A and B, if $A \subseteq B$ then $\mathcal{P}(A) \subseteq \mathcal{P}(B)$.

14. For all sets A and B, $\mathcal{P}(A \cup B) \subseteq \mathcal{P}(A) \cup \mathcal{P}(B)$.

H 15. For all sets A and B, $\mathcal{P}(A) \cup \mathcal{P}(B) \subseteq \mathcal{P}(A \cup B)$.

16. For all sets A and B, $\mathcal{P}(A \cap B) = \mathcal{P}(A) \cap \mathcal{P}(B)$.

17. For all sets A and B, $\mathcal{P}(A \times B) = \mathcal{P}(A) \times \mathcal{P}(B)$.

18. Write a negation for each of the following statements. Indicate which is true, the statement or its negation. Justify your answers.
    a. $\forall$ sets S, $\exists$ a set T such that $S \cap T = \emptyset$.
    b. $\exists$ a set S such that $\forall$ sets T, $S \cup T = \emptyset$.

19. Let $S = \{a, b, c\}$ and for each integer $i = 0, 1, 2, 3$, let S_i be the set of all subsets of S that have i elements. List the elements in S_0, S_1, S_2, and S_3. Is $\{S_0, S_1, S_2, S_3\}$ a partition of $\mathcal{P}(S)$?

20. Let $S = \{a, b, c\}$ and let S_a be the set of all subsets of S that contain a, let S_b be the set of all subsets of S that contain b, let S_c be the set of all subsets of S that contain c, and let $S_\emptyset$ be the set whose only element is $\emptyset$. Is $\{S_a, S_b, S_c, S_\emptyset\}$ a partition of $\mathcal{P}(S)$?

21. Let $A = \{t, u, v, w\}$ and let S_1 be the set of all subsets of A that do not contain w and S_2 the set of all subsets of A that contain w.
    a. Find S_1.
    b. Find S_2.
    c. Are S_1 and S_2 disjoint?
    d. Compare the sizes of S_1 and S_2.
    e. How many elements are in $S_1 \cup S_2$?
    f. What is the relation between $S_1 \cup S_2$ and $\mathcal{P}(A)$?

H ✳ 22. The following problem, devised by Ginger Bolton, appeared in the January 1989 issue of the *College Mathematics Journal* (Vol. 20, No. 1, p. 68): Given a positive integer $n \geq 2$, let S be the set of all nonempty subsets of $\{2, 3, \ldots, n\}$. For each $S_i \in S$, let P_i be the product of the elements of S_i. Prove or disprove that

$$\sum_{i=1}^{2^{n-1}-1} P_i = \frac{(n+1)!}{2} - 1.$$

In 23 and 24 supply a reason for each step in the derivation.

23. For all sets A, B, and C,

$$(A \cup B) \cap C = (A \cap C) \cup (B \cap C).$$

Proof: Suppose A, B, and C are any sets. Then

$$\begin{aligned}
(A \cup B) \cap C &= C \cap (A \cup B) & \text{by } \underline{\text{(a)}} \\
&= (C \cap A) \cup (C \cap B) & \text{by } \underline{\text{(b)}} \\
&= (A \cap C) \cup (B \cap C) & \text{by } \underline{\text{(c)}}.
\end{aligned}$$

H 24. For all sets A, B, and C,
$$(A \cup B) - (C - A) = A \cup (B - C).$$

Proof: Suppose A, B, and C are any sets. Then

$$\begin{aligned}
(A \cup B) - (C - A) &= (A \cup B) \cap (C - A)^c & \text{by } \underline{\text{(a)}} \\
&= (A \cup B) \cap (C \cap A^c)^c & \text{by } \underline{\text{(b)}} \\
&= (A \cup B) \cap (A^c \cap C)^c & \text{by } \underline{\text{(c)}} \\
&= (A \cup B) \cap ((A^c)^c \cup C^c) & \text{by } \underline{\text{(d)}} \\
&= (A \cup B) \cap (A \cup C^c) & \text{by } \underline{\text{(e)}} \\
&= A \cup (B \cap C^c) & \text{by } \underline{\text{(f)}} \\
&= A \cup (B - C) & \text{by } \underline{\text{(g)}}.
\end{aligned}$$

In 25–33 use the properties in Theorem 5.2.2 to construct an algebraic proof for the given statement.

25. For all sets A, B, and C,

$$(A \cap B) \cup C = (A \cup C) \cap (B \cup C).$$

26. For all sets A and B, $A \cup (B - A) = A \cup B$.

27. For all sets A, B, and C,

$$(A - B) - C = A - (B \cup C).$$

28. For all sets A and B, $A - (A - B) = A \cap B$.

29. For all sets A and B, $((A^c \cup B^c) - A)^c = A$.

30. For all sets A and B, $(B^c \cup (B^c - A))^c = B$.

31. For all sets A and B, $A - (A \cap B) = A - B$.

H 32. For all sets A and B,

$$(A - B) \cup (B - A) = (A \cup B) - (A \cap B).$$

33. For all sets A, B, and C,

$$(A - B) - (B - C) = A - B.$$

In 34–36 use the properties in Theorem 5.2.2 to simplify the given expression.

H 34. $A \cap ((B \cup A^c) \cap B^c)$

35. $(A - (A \cap B)) \cap (B - (A \cap B))$

36. $((A \cap (B \cup C)) \cap (A - B)) \cap (B \cup C^c)$

37. Consider the following set property: For all sets A and B, $A - B$ and B are disjoint.
    a. Use an element argument to derive the property.
    b. Use an algebraic argument to derive the property (by applying properties from Theorem 5.2.2).
    c. Comment on which method you found easier.

38. Consider the following set property: For all sets A, B, and C, $(A - B) \cup (B - C) = (A \cup B) - (B \cap C)$.
    a. Use an element argument to derive the property.
    b. Use an algebraic argument to derive the property (by applying properties from Theorem 5.2.2).
    c. Comment on which method you found easier.

Definition: Given sets A and B, the **symmetric difference of A and B,** denoted $A \triangle B$, is

$$A \triangle B = (A - B) \cup (B - A).$$

---

39. Let $A = \{1, 2, 3, 4\}$, $B = \{3, 4, 5, 6\}$, and $C = \{5, 6, 7, 8\}$. Find each of the following sets:
    a. $A \triangle B$ b. $B \triangle C$
    c. $A \triangle C$ d. $(A \triangle B) \triangle C$

Refer to the definition of symmetric difference given above. Prove each of 40–45, assuming that A, B, and C are all subsets of a universal set U.

40. $A \triangle B = B \triangle A$

41. $A \triangle \emptyset = A$

42. $A \triangle A^c = U$

43. $A \triangle A = \emptyset$

H 44. If $A \triangle C = B \triangle C$, then $A = B$.

H 45. $(A \triangle B) \triangle C = A \triangle (B \triangle C)$.

46. Derive the set identity $A \cup (A \cap B) = A$ from the properties listed in Theorem 5.2.2(1)–(5). Start by showing that for all subsets B of a universal set U, $U \cup B = U$. Then intersect both sides with A and deduce the identity.

47. Derive the set identity $A \cap (A \cup B) = A$ from the properties listed in Theorem 5.2.2(1)–(5). Start by showing that for all subsets B of a universal set U, $\emptyset = \emptyset \cap B$. Then take the union of both sides with A and deduce the identity.

In 48–50 assume that B is a Boolean algebra with operations $+$ and $\cdot$. Give the reasons needed to fill in the blanks in the proofs, but do not use any parts of Theorem 5.3.2 unless they have already been proved. You may use any part of the definition of a Boolean algebra, and the results of previous exercises, however.

48. For all a in B, $a \cdot a = a$.

    Proof: Let a be any element of B. Then

$$\begin{aligned} a &= a \cdot 1 & \text{(a)} \\ &= a \cdot (a + \overline{a}) & \text{(b)} \\ &= (a \cdot a) + (a \cdot \overline{a}) & \text{(c)} \\ &= (a \cdot a) + 0 & \text{(d)} \\ &= a \cdot a & \text{(e)} \end{aligned}$$

49. For all a in B, $a + 1 = 1$.

    Proof: Let a be any element of B. Then

$$\begin{aligned} a + 1 &= a + (a + \overline{a}) & \text{(a)} \\ &= (a + a) + \overline{a} & \text{(b)} \\ &= a + \overline{a} & \text{by Example 5.3.6} \\ &= 1 & \text{(d)} \end{aligned}$$

50. For all a and b in B, $(a + b) \cdot a = a$.

    Proof: Let a and b be any elements of B. Then

$$\begin{aligned} (a + b) \cdot a &= a \cdot (a + b) & \text{(a)} \\ &= a \cdot a + a \cdot b & \text{(b)} \\ &= a + a \cdot b & \text{(c)} \\ &= a \cdot 1 + a \cdot b & \text{(d)} \\ &= a \cdot (1 + b) & \text{(d)} \\ &= a \cdot (b + 1) & \text{(e)} \\ &= a \cdot 1 & \text{by exercise 49} \\ &= a & \text{(f)} \end{aligned}$$

In 51–57 assume that B is a Boolean algebra with operations $+$ and $\cdot$. Prove each statement without using any parts of Theorem 5.3.2 unless they have already been proved. You may use any part of the definition of a Boolean algebra and the results of previous exercises, however.

51. For all a in B, $a \cdot 0 = 0$.

52. For all a and b in B, $(a \cdot b) + a = a$.

53. $\overline{0} = 1$.   54. $\overline{1} = 0$

55. For all a and b in B, $\overline{a \cdot b} = \overline{a} + \overline{b}$. (*Hint:* Prove that $(a \cdot b) + (\overline{a} + \overline{b}) = 1$ and that $(a \cdot b) \cdot (\overline{a} + \overline{b}) = 0$, and use the fact that $a \cdot b$ has a unique complement.)

56. For all a and b in B, $\overline{a + b} = \overline{a} \cdot \overline{b}$.

H 57. For all x, y, and z in B, if $x + y = x + z$ and $x \cdot y = x \cdot z$, then $y = z$.

58. Let $S = \{0, 1\}$, and define operations $+$ and $\cdot$ on S by the following tables:

+	0	1		·	0	1
0	0	1		0	0	0
1	1	1		1	0	1

    a. Show that the elements of S satisfy the following properties:
       (i) the commutative law for $+$
       (ii) the commutative law for $\cdot$
       (iii) the associative law for $+$
       (iv) the associative law for $\cdot$
    H (v) the distributive law for $+$ over $\cdot$
       (vi) the distributive law for $\cdot$ over $+$

H **b.** Show that 0 is an identity element for $+$ and that 1 is an identity element for $\cdot$.

c. Define $\overline{0} = 1$ and $\overline{1} = 0$. Show that for all a in S, $a + \overline{a} = 1$ and $a \cdot \overline{a} = 0$. It follows from parts (a)–(c) that S is a Boolean algebra with the operations $+$ and $\cdot$.

H ✳ **59.** Prove that the associative laws for a Boolean algebra can be omitted from the definition. That is, prove that the associative laws can be derived from the other laws in the definition.

5.4 Russell's Paradox and the Halting Problem

From the paradise created for us by Cantor, no one will drive us out.
— David Hilbert 1862–1943

Bertrand Russell
(1872–1970)

By the beginning of the twentieth century, abstract set theory had gained such wide acceptance that a number of mathematicians were working hard to show that all of mathematics could be built upon a foundation of set theory. In the midst of this activity, the English mathematician and philosopher Bertrand Russell discovered a "paradox" (really a genuine contradiction) that seemed to shake the very core of the foundation. The paradox assumes Cantor's definition of set as "any collection into a whole of definite and separate objects of our intuition or our thought."

Russell's Paradox: Most sets are not elements of themselves. For instance, the set of all integers is not an integer and the set of all horses is not a horse. However, we can imagine the possibility of a set's being an element of itself. For instance, the set of all abstract ideas might be considered an abstract idea. If we are allowed to use any description of a property as the defining property of a set, we can let S be the set of all sets that are not elements of themselves:

$$S = \{A \mid A \text{ is a set and } A \notin A\}.$$

Is S an element of itself?

The answer is neither yes nor no. For if $S \in S$, then S satisfies the defining property for S, and hence $S \notin S$. But if $S \notin S$, then S is a set such that $S \notin S$ and so S satisfies the defining property for S, which implies that $S \in S$. Thus neither is $S \in S$ nor is $S \notin S$, which is a contradiction.

To help explain his discovery to lay people, Russell devised a puzzle, the barber puzzle, whose solution exhibits the same logic as his paradox.

Example 5.4.1 The Barber Puzzle

In a certain town there is a male barber who shaves all those men, and only those men, who do not shave themselves. *Question:* Does the barber shave himself?

Solution   Neither yes nor no. If the barber shaves himself, he is a member of the class of men who shave themselves. But no member of this class is shaved by the barber, and so the barber does *not* shave himself. On the other hand, if the barber does not shave himself, he belongs to the class of men who do not shave themselves. But the barber shaves every man in this class, so the barber *does* shave himself. ■

But how can the answer be neither yes nor no? Surely any barber either does or does not shave himself. You might try to think of circumstances that would make the paradox disappear. For instance, maybe the barber happens to have no beard and never shaves. But a condition of the puzzle is that the barber is a man who shaves *all* those men who do not shave themselves. If he does not shave, then he does not shave himself, in which case he is shaved by the barber and the contradiction is as present as ever. Other attempts at resolving the paradox by considering details of the barber's situation are similarly doomed to failure.

So let's accept the fact that the paradox has no easy resolution and see where that thought leads. Since the barber neither shaves himself nor doesn't shave himself, the sentence "The barber shaves himself" is neither true nor false. But the sentence arose in a natural way from a description of a situation. If the situation actually existed, then the sentence would have to be true or false. Thus we are forced to conclude that the situation described in the puzzle simply cannot exist in the world as we know it.

In a similar way, the conclusion to be drawn from Russell's paradox itself is that the object S is not a set. Because if it actually were a set, in the sense of satisfying the general properties of sets that we have been assuming, then it either would be an element of itself or not.

In the years following Russell's discovery, several ways were found to define the basic concepts of set theory so as to avoid his contradiction. The way used in this text requires that, except for the power set whose existence is guaranteed by an axiom, whenever a set is defined using a predicate as a defining property, the stipulation must also be made that the set is a subset of a known set. This method does not allow us to talk about "the set of all sets that are not elements of themselves." We can speak only of "the set of all sets that are subsets of some known set and that are not elements of themselves." When this restriction is made, Russell's paradox ceases to be contradictory. Here is what happens:

Let U be a universal set and suppose that all sets under discussion are subsets of U. Let

$$S = \{A \mid A \subseteq U \text{ and } A \notin A\}.$$

In Russell's paradox, both implications

$$S \in S \rightarrow S \notin S \quad \text{and} \quad S \notin S \rightarrow S \in S$$

are proved, and the contradictory conclusion

$$\text{neither } S \in S \quad \text{nor} \quad S \notin S$$

is therefore deduced. In the situation in which all sets under discussion are subsets of U, the implication $S \in S \rightarrow S \notin S$ is proved in almost the same way as it is for Russell's paradox: (Suppose $S \in S$. Then by definition of S, $S \subseteq U$ and $S \notin S$. In particular, $S \notin S$.) On the other hand, from the supposition that $S \notin S$ we can only deduce that the statement $S \subseteq U$ and $S \notin S$ is false. By one of De Morgan's laws, this means that $S \nsubseteq U$ or $S \in S$. Since $S \in S$ would contradict the supposition that $S \notin S$, we eliminate it and conclude that $S \nsubseteq U$. In other words, the only conclusion we can draw is that the seeming "definition" of S is faulty—that is, that S is not a set in U.

Russell's discovery had a profound impact on mathematics because even though his contradiction could be made to disappear by more careful definitions, its existence caused people to wonder whether other contradictions remained. In 1931 Kurt Gödel showed that it is not possible to prove, in a mathematically rigorous way, that mathematics is free of contradictions. You might think that Gödel's result would have caused mathematicians to give up their work in despair, but that has not happened. On the contrary, there has been more mathematical activity since 1931 than in any other period in history.

Kurt Gödel
(1906–1978)

The Halting Problem

Well before the actual construction of an electronic computer, Alan M. Turing (1912–1954) deduced a profound theorem about how such computers would have to work. The argument he used is similar to that in Russell's paradox. It is also related to those used by Gödel to prove his theorem and by Cantor to prove that it is impossible to write all the real numbers in an infinitely long list, even given an infinitely long period of time (see Section 7.5 and Chapter 12).

If you have some experience programming computers, you know how badly an infinite loop can tie up a computer system. It would be useful to be able to preprocess a program and its data set by running it through a checking program that determines whether execution of the given program with the given data set would result in an infinite loop. Can an algorithm for such a program be written? In other words, can an algorithm be written that will accept any algorithm X and any data set D as input and will then print "halts" or "loops forever" to indicate whether X terminates in a finite number of steps or loops forever when run with data set D? In the 1930s, Turing proved that the answer to this question is no.

Theorem 5.4.1

There is no computer algorithm that will accept any algorithm X and data set D as input and then will output, "halts" or "loops forever" to indicate whether X terminates in a finite number of steps when X is run with data set D.

Proof (by contradiction):

Suppose there is an algorithm, CheckHalt, such that if an algorithm X and a data set D are input, then

CheckHalt(X, D) prints

    "halts"   if X terminates in a finite number of steps when run with data set D

or

    "loops forever"   if X does not terminate in a finite number of steps when run with data set D.

[*To show that no algorithm such as CheckHalt can exist, we will deduce a contradiction.*]

Observe that the sequence of characters making up an algorithm X can be regarded as a data set itself. Thus it is possible to consider running CheckHalt with input (X, X). Define a new algorithm, Test, as follows: For any input algorithm X,

    Test(X)

        loops forever if CheckHalt(X, X) prints "halts"

or

        stops if CheckHalt(X, X) prints "loops forever".

Now run algorithm Test with input Test. If Test(Test) terminates after a finite number of steps, then the value of CheckHalt(Test, Test) is "halts" and so Test(Test) loops forever.

On the other hand, if Test(Test) does not terminate after a finite number of steps, then CheckHalt(Test, Test) prints "loops forever" and so Test(Test) terminates.

The two paragraphs above show that Test(Test) loops forever and also that it terminates. This is a contradiction. But the existence of Test follows logically from the supposition of the existence of an algorithm CheckHalt that can check any algorithm and data set for termination. [*Hence the supposition must be false, and there is no such algorithm.*]

In recent years, the axioms for set theory that guarantee that Russell's paradox will not arise have been found inadequate to deal with the full range of recursively defined objects in computer science, and a new theory of "non-well-founded" sets has been developed. In addition, computer scientists and logicians working on programs to enable computers to process natural language have seen the importance of exploring further the kinds of semantic issues raised by the barber puzzle and are developing new theories of logic to deal with them.

Exercise Set 5.4

In 1–6 determine whether each sentence is a statement. Explain your answers.

1. This sentence is false.

2. If $1 + 1 = 3$, then $1 = 0$.

3. ⟨The sentence in this box is a lie.⟩

4. All real numbers with negative squares are prime.

5. This sentence is false or $1 + 1 = 3$.

6. This sentence is false and $1 + 1 = 2$.

7. a. Assuming that the following sentence is a statement, prove that $1 + 1 = 3$:

    If this sentence is true, then $1 + 1 = 3$.

   b. What can you deduce from part (a) about the status of "This sentence is true"? Why? (This example is known as **Löb's paradox.**)

H 8. The following two sentences were devised by the logician Saul Kripke. While not intrinsically paradoxical, they could be paradoxical under certain circumstances. Describe such circumstances.

   (i) Most of Nixon's assertions about Watergate are false.

   (ii) Everything Jones says about Watergate is true.

   (*Hint:* Suppose Nixon says (ii) and the only utterance Jones makes about Watergate is (i).)

9. Can there exist a computer program that has as output a list of all the computer programs that do not list themselves in their output? Explain your answer.

10. Can there exist a book that refers to all those books and only those books that do not refer to themselves? Explain your answer.

11. Some English adjectives are descriptive of themselves (for instance, the word *polysyllabic* is polysyllabic) whereas others are not (for instance, the word *monosyllabic* is not monosyllabic). The word *heterological* refers to an adjective that does not describe itself. Is *heterological* heterological? Explain your answer.

12. As strange as it may seem, it is possible to give a precise-looking verbal definition of an integer that, in fact, is not a definition at all. The following was devised by an English librarian, G. G. Berry, and reported by Bertrand Russell. Explain how it leads to a contradiction. Let n be "the smallest integer not describable in fewer than 12 English words." (Note that the total number of strings consisting of 11 or fewer English words is finite.)

H 13. Is there an algorithm which, for a fixed quantity a and any input algorithm X and data set D, can determine whether X prints a when run with data set D? Explain. (This problem is called the **printing problem.**)

14. Use a technique similar to that used to derive Russell's paradox to prove that for any set A, $\mathcal{P}(A) \nsubseteq A$.

CHAPTER 6

COUNTING AND PROBABILITY

"It's as easy as 1–2–3."

That's the saying. And in certain ways, counting *is* easy. But other aspects of counting aren't so simple. Have you ever agreed to meet a friend "in three days" and then realized that you and your friend might mean different things? For example, on the European continent, to meet in eight days means to meet on the same day as today one week hence; on the other hand, in English-speaking countries, to meet in seven days means to meet one week hence. The difference is that on the continent, all days including the first and the last are counted. In the English-speaking world, it's the number of 24-hour periods that are counted.

Continental countries	1	2	3	4	5	6	7	8
	↕	↕	↕	↕	↓	↕	↕	↕
	Sun	Mon	Tue	Wed	Thu	Fri	Sat	Sun
English-speaking countries	1	2	3	4	5	6	7	

The English convention for counting days follows the almost universal convention for counting hours. If it is 9 A.M. and two people anywhere in the world agree to meet in three hours, they mean that they will get back together again at 12 noon.

Musical intervals, on the other hand, are universally reckoned the way the Continentals count the days of a week. An interval of a third consists of two tones with a single tone in between, and an interval of a second consists of two adjacent tones. (See Figure 6.1.1.)

C   E   C   D

Interval of a third   Interval of a second

Figure 6.1.1

Of course, the complicating factor in all these examples is not how to count but rather what to count. And, indeed, in the more complex mathematical counting problems discussed in this chapter, it is what to count that is the central issue. Once one knows exactly what to count, the counting itself is as easy as 1–2–3.

6.1 Introduction

Imagine tossing two coins and observing whether 0, 1, or 2 heads are obtained. It would be natural to guess that each of these events occurs about one-third of the time, but in fact this is not the case. Table 6.1.1 below shows actual data obtained from tossing two quarters 50 times.

Table 6.1.1 Experimental Data Obtained from Tossing Two Quarters 50 Times

Event	Tally	Frequency (Number of times the event occurred)	Relative Frequency (Fraction of times the event occurred)
2 heads obtained	ЖЖ ЖЖ I	11	22%
1 head obtained	ЖЖ ЖЖ ЖЖ ЖЖ ЖЖ II	27	54%
0 heads obtained	ЖЖ ЖЖ II	12	24%

As you can see, the relative frequency of obtaining exactly 1 head was roughly twice as great as that of obtaining either 2 heads or 0 heads. It turns out that the mathematical theory of probability can be used to predict that a result like this will almost always occur. To see how, call the two coins *A* and *B*, and suppose that each is perfectly balanced. Then each has an equal chance of coming up heads or tails, and when the two are tossed together, the four outcomes pictured in Figure 6.1.2 are all equally likely.

2 heads obtained   1 head obtained   0 heads obtained
Figure 6.1.2 Equally Likely Outcomes from Tossing Two Balanced Coins

Figure 6.1.2 shows that there is a 1 in 4 chance of obtaining two heads and a 1 in 4 chance of obtaining no heads. The chance of obtaining one head, however, is 2 in 4 because either *A* could come up heads and *B* tails or *B* could come up heads and *A* tails. So if you repeatedly toss two balanced coins and record the number of heads, you should expect relative frequencies similar to those shown in Table 6.1.1.

To formalize this analysis and extend it to more complex situations, we introduce the notions of random process, sample space, event and probability. To say that a process is **random** means that when it takes place, one outcome from some set of outcomes is sure to occur, but it is impossible to predict with certainty which outcome that will be. For instance, if an ordinary person performs the experiment of tossing an ordinary coin into the air and allowing it to fall flat on the ground, it can be predicted with certainty that the coin will land either heads up or tails up (so the set of outcomes can be denoted {heads, tails}), but it is not known for sure whether heads or tails will occur. We restricted this experiment to ordinary people because a skilled magician can toss a coin in a way that appears random but is not, and a physicist equipped with first-rate measuring devices may be able to analyze all the forces on the coin and correctly predict its landing position. Just a few of many examples of random processes or experiments are choosing winners in state lotteries, selecting respondents in public opinion polls, and choosing subjects to receive treatments or serve as controls in medical experiments. The set of outcomes that can result from a random process or experiment is called a *sample space*.

• Definition

A **sample space** is the set of all possible outcomes of a random process or experiment. An **event** is a subset of a sample space.

In case an experiment has finitely many* outcomes and all outcomes are equally likely to occur, the *probability* of an event (set of outcomes) is just the ratio of the number of outcomes in the event to the total number of outcomes. Strictly speaking, this result can be deduced from a set of axioms for probability formulated in 1933 by the Russian mathematician A. N. Kolmogorov. In Section 6.8 we discuss the axioms and show how to derive their consequences formally. At present, we take a naïve approach to probability and simply state the result as a principle.

Equally Likely Probability Formula

If S is a finite sample space in which all outcomes are equally likely and E is an event in S, then the **probability of E,** denoted $P(E)$, is

$$P(E) = \frac{\text{the number of outcomes in } E}{\text{the total number of outcomes in } S}.$$

• Notation

For any finite set, $N(A)$ denotes the number of elements in A.

With this notation, the equally likely probability formula becomes

$$P(E) = \frac{N(E)}{N(S)}.$$

Example 6.1.1 Probabilities for a Deck of Cards

An ordinary deck of cards contains 52 cards divided into four *suits*. The *red suits* are diamonds (♦) and hearts (♥) and the *black suits* are clubs (♣) and spades (♠). Each suit contains 13 cards of the following *denominations:* 2, 3, 4, 5, 6, 7, 8, 9, 10, J (jack), Q (queen), K (king), and A (ace). The cards J, Q, and K are called *face cards.*

---

*In Section 7.3 the concepts of finite and infinite are defined formally.

Mathematician Persi Diaconis, working with David Aldous in 1986 and Dave Bayer in 1992, showed that seven shuffles are needed to "thoroughly mix up" the cards in an ordinary deck. In 2000 mathematician Nick Trefethen, working with his father, Lloyd Trefethen, a mechanical engineer, used a somewhat different definition of "thoroughly mix up" to show that six shuffles will nearly always suffice. Imagine that the cards in a deck have become—by some method—so thoroughly mixed up that if you spread them out face down and pick one at random, you are as likely to get any one card as any other.

a. What is the sample space of outcomes?

b. What is the event that the chosen card is a black face card?

c. What is the probability that the chosen card is a black face card?

Solution

a. The outcomes in the sample space S are the 52 cards in the deck.

b. Let E be the event that a black face card is chosen. The outcomes in E are the jack, queen, and king of clubs and the jack, queen, and king of spades. Symbolically,

$$E = \{J\clubsuit, Q\clubsuit, K\clubsuit, J\spadesuit, Q\spadesuit, K\spadesuit\}.$$

c. By part (b), $N(E) = 6$, and according to the description of the situation, all 52 outcomes in the sample space are equally likely. Therefore, by the equally likely probability formula, the probability that the chosen card is a black face card is

$$P(E) = \frac{N(E)}{N(S)} = \frac{6}{52} \cong 11.5\%.$$ ■

Example 6.1.2 Rolling a Pair of Dice

A die is one of a pair of dice. It is a cube with six sides, each containing from one to six dots, called *pips*. Suppose a blue die and a gray die are rolled together, and the numbers of dots that occur face up on each are recorded. The possible outcomes can be listed as follows, where in each case the die on the left is blue and the one on the right is gray.

A more compact notation identifies, say, with the string 24, with 53, and so forth.

a. Use the compact notation to write the sample space S of possible outcomes.

b. Use set notation to write the event E that the numbers showing face up have a sum of 6 and find the probability of this event.

Solution

a. $S = \{11, 12, 13, 14, 15, 16, 21, 22, 23, 24, 25, 26, 31, 32, 33, 34, 35, 36, 41, 42, 43,$
$44, 45, 46, 51, 52, 53, 54, 55, 56, 61, 62, 63, 64, 65, 66\}.$

b. $E = \{15, 24, 33, 42, 51\}.$

The probability that the sum of the numbers is $6 = P(E) = \dfrac{N(E)}{N(S)} = \dfrac{5}{36}.$ ∎

The next example is called the Monty Hall problem after the host of an old game show, "Let's Make A Deal." When it was originally publicized in a newspaper column and on a radio show, it created tremendous controversy. Many highly educated people, even some with Ph.D.'s, submitted incorrect solutions or argued vociferously against the correct solution. Before you read the answer given below, think about what your own response to the situation would be.

Example 6.1.3 The Monty Hall Problem

There are three doors on the set for a game show. Let's call them A, B, and C. If you pick the right door you win the prize. You pick door A. The host of the show, Monty Hall, then opens one of the other doors and reveals that there is no prize behind it. Keeping the remaining two doors closed, he asks you whether you want to switch your choice to the other closed door or stay with your original choice of door A. What should you do if you want to maximize your chance of winning the prize: stay with door A or switch—or would the likelihood of winning be the same either way?

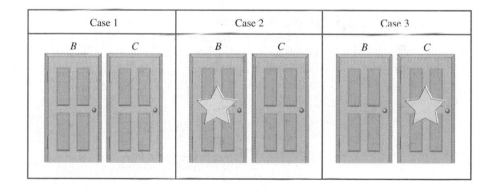

Solution   At the point just before the host opens one of the closed doors, there is no information about the location of the prize. Thus there are three equally likely possibilities for what lies behind the doors: (Case 1) the prize is behind A (i.e., it is not behind either B or C), (Case 2) the prize is behind B; (Case 3) the prize is behind C.

Since there is no prize behind the door the host opens, in Case 1 the host could open either door and you would win by staying with your original choice: door A. In Case 2 the host must open door C, and so you would win by switching to door B. In Case 3 the host must open door B, and so you would win by switching to door C. Thus, in two of the three equally likely cases, you would win by switching from A to the other closed door. In only one of the three equally likely cases would you win by staying with your original choice. Therefore, you should switch.

A reality note: The analysis used for this solution applies only if the host *always* opens one of the closed doors and offers the contestant the choice of staying with the original choice or switching. In the original show, Monty Hall made this offer only occasionally—most often when he knew the contestant had already chosen the correct door. ∎

Pierre-Simon Laplace
(1749–1827)

Many of the fundamental principles of probability were formulated in the mid-1600s in an exchange of letters between Pierre de Fermat and Blaise Pascal in response to questions posed by a French nobleman interested in games of chance. In 1812, Pierre-Simon Laplace published the first general mathematical treatise on the subject and extended the range of applications to a variety of scientific and practical problems.

Counting the Elements of a List

Some counting problems are as simple as counting the elements of a list. For instance, how many integers are there from 5 through 12? To answer this question, imagine going along the list of integers from 5 to 12, counting each in turn.

$$
\begin{array}{ccccccccc}
\text{list:} & 5 & 6 & 7 & 8 & 9 & 10 & 11 & 12 \\
 & \updownarrow & \updownarrow & \updownarrow & \updownarrow & \updownarrow & \updownarrow & \updownarrow & \updownarrow \\
\text{count:} & 1 & 2 & 3 & 4 & 5 & 6 & 7 & 8
\end{array}
$$

So the answer is 8.

More generally, if m and n are integers and $m \leq n$, how many integers are there from m through n? To answer this question, note that $n = m + (n - m)$, where $n - m \geq 0$ [*since* $n \geq m$]. Note also that the element $m + 0$ is the first element of the list, the element $m + 1$ is the second element, the element $m + 2$ is the third, and so forth. In general, the element $m + i$ is the $(i + 1)$st element of the list.

$$
\begin{array}{cccccc}
\text{list:} & m(= m + 0) & m + 1 & m + 2 & \ldots, & n\,(= m + (n - m)) \\
 & \updownarrow & \updownarrow & \updownarrow & & \updownarrow \\
\text{count:} & 1 & 2 & 3 & \ldots & (n - m) + 1
\end{array}
$$

And so the number of elements in the list is $n - m + 1$.

This general result is important enough to be restated as a theorem, the formal proof of which uses mathematical induction. (See exercise 33 at the end of this section.) The heart of the proof is the observation that if the list $m, m + 1, \ldots, k$ has $k - m + 1$ numbers, then the list $m, m + 1, \ldots, k, k + 1$ has $(k - m + 1) + 1 = (k + 1) - m + 1$ numbers.

> **Theorem 6.1.1 The Number of Elements in a List**
>
> If m and n are integers and $m \leq n$, then there are $n - m + 1$ integers from m to n inclusive.

Example 6.1.4 Counting the Elements of a Sublist

a. How many three-digit integers (integers from 100 to 999 inclusive) are divisible by 5?

b. What is the probability that a randomly chosen three-digit integer is divisible by 5?

Solution

a. Imagine writing the three-digit integers in a row, noting those that are multiples of 5 and drawing arrows between each such integer and its corresponding multiple of 5.

$$
\begin{array}{cccccccccccccccccc}
100 & 101 & 102 & 103 & 104 & 105 & 106 & 107 & 108 & 109 & 110 & \cdots & 994 & 995 & 996 & 997 & 998 & 999 \\
\updownarrow & & & & & \updownarrow & & & & & \updownarrow & & & \updownarrow & & & & \\
5 \cdot 20 & & & & & 5 \cdot 21 & & & & & 5 \cdot 22 & & & 5 \cdot 199 & & & &
\end{array}
$$

From the sketch it is clear that there are as many three-digit integers that are multiples of 5 as there are integers from 20 to 199 inclusive. By Theorem 6.1.1, there are $199 - 20 + 1$, or 180, such integers. Hence there are 180 three-digit integers that are divisible by 5.

b. By Theorem 6.1.1 the total number of integers from 100 through 999 is $999 - 100 + 1 = 900$. By part (a), 180 of these are divisible by 5. Hence the probability that a randomly chosen three-digit integer is divisible by 5 is $180/900 = 1/5$. ■

Example 6.1.5 Application: Counting Elements of a One-Dimensional Array

Analysis of many computer algorithms requires skill at counting the elements of a one-dimensional array. Let $A[1], A[2], \ldots, A[n]$ be a one-dimensional array, where n is a positive integer.

a. Suppose the array is cut at a middle value $A[m]$ so that two subarrays are formed:

(1) $A[1], A[2], \ldots, A[m]$ and (2) $A[m+1], A[m+2], \ldots, A[n]$.

How many elements does each subarray have?

b. What is the probability that a randomly chosen element of the array has an even subscript

(i) if n is even? (ii) if n is odd?

Solution

a. Array (1) has the same number of elements as the list of integers from 1 through m. So by Theorem 6.1.1, it has m, or $m - 1 + 1$, elements. Array (2) has the same number of elements as the list of integers from $m + 1$ through n. So by Theorem 6.1.1, it has $n - m$, or $n - (m + 1) + 1$, elements.

b. (i) If n is even, each even subscript starting with 2 and ending with n can be matched up with an integer from 1 to $n/2$.

$$
\begin{array}{cccccccccccc}
1 & 2 & 3 & 4 & 5 & 6 & 7 & 8 & 9 & 10 & \cdots & n \\
& \updownarrow & & \updownarrow & & \updownarrow & & \updownarrow & & \updownarrow & & \updownarrow \\
& 2 \cdot 1 & & 2 \cdot 2 & & 2 \cdot 3 & & 2 \cdot 4 & & 2 \cdot 5 & & 2 \cdot n/2
\end{array}
$$

So there are $n/2$ array elements with even subscripts. Since the entire array has n elements, the probability that a randomly chosen element has an even subscript is $\dfrac{n/2}{n} = \dfrac{1}{2}$.

(ii) If n is odd, then the greatest even subscript of the array is $n - 1$. So there are as many even subscripts between 1 and n as there are from 2 through $n - 1$. Then the reasoning of (i) can be used to conclude that there are $(n - 1)/2$ array elements with even subscripts.

$$
\begin{array}{cccccccc}
1 & 2 & 3 & 4 & 5 & 6 & \cdots & n - 1 & n \\
& \updownarrow & & \updownarrow & & \updownarrow & & \updownarrow \\
& 2 \cdot 1 & & 2 \cdot 2 & & 2 \cdot 3 & \cdots & 2 \cdot (n - 1)/2
\end{array}
$$

Since the entire array has n elements, the probability that a randomly chosen element has an even subscript is $\dfrac{(n - 1)/2}{n} = \dfrac{n - 1}{2n}$. Observe that as n gets larger and larger, this probability gets closer and closer to $1/2$.

Note that the answers to (i) and (ii) can be combined using the floor notation. By Theorem 3.5.2, the number of array elements with even subscripts is $\lfloor n/2 \rfloor$, so the probability that a randomly chosen element has an even subscript is $\dfrac{\lfloor n/2 \rfloor}{n}$. ∎

Exercise Set 6.1*

1. Toss two coins 30 times and make a table showing the relative frequencies of 0, 1, and 2 heads. How do your values compare with those shown in Table 6.1.1?

2. In the example of tossing two quarters, what is the probability that at least one head is obtained? that coin A is a head? that coins A and B are either both heads or both tails?

In 3–6 use the sample space given in Example 6.1.1. Write each event as a set, and compute its probability.

3. The event that the chosen card is red and is not a face card.

4. The event that the chosen card is black and has an even number on it.

5. The event that the denomination of the chosen card is at least 10 (counting aces high).

6. The event that the denomination of the chosen card is at most 4 (counting aces high).

In 7–10, use the sample space given in Example 6.1.2. Write each of the following events as a set and compute its probability.

7. The event that the sum of the numbers showing face up is 8.

8. The event that the numbers showing face up are the same.

9. The event that the sum of the numbers showing face up is at most 6.

10. The event that the sum of the numbers showing face up is at least 9.

11. Suppose that a coin is tossed three times and the side showing face up on each toss is noted. Suppose also that on each toss heads and tails are equally likely. Let *HHT* indicate the outcome heads on the first two tosses and tails on the third, *THT* the outcome tails on the first and third tosses and heads on the second, and so forth.
    a. List the eight elements in the sample space whose outcomes are all the possible head–tail sequences obtained in the three tosses.
    b. Write each of the following events as a set and find its probability:
       (i) The event that exactly one toss results in a head.
       (ii) The event that at least two tosses result in a head.
       (iii) The event that no head is obtained.

12. Suppose that each child born is equally likely to be a boy or a girl. Consider a family with exactly three children. Let

BBG indicate that the first two children born are boys and the third child is a girl, let *GBG* indicate that the first and third children born are girls and the second is a boy, and so forth.
    a. List the eight elements in the sample space whose outcomes are all possible genders of the three children.
    b. Write each of the following events as a set and find its probability.
       (i) The event that exactly one child is a girl.
       (ii) The event that at least two children are girls.
       (iii) The event that no child is a girl.

13. Suppose that on a true/false exam you have no idea at all about the answers to three questions. You choose answers randomly and therefore have a 50–50 chance of being correct on any one question. Let *CCW* indicate that you were correct on the first two questions and wrong on the third, let *WCW* indicate that you were wrong on the first and third questions and correct on the second, and so forth.
    a. List the elements in the sample space whose outcomes are all possible sequences of correct and incorrect responses on your part.
    b. Write each of the following events as a set and find its probability:
       (i) The event that exactly one answer is correct.
       (ii) The event that at least two answers are correct.
       (iii) The event that no answer is correct.

14. Three people have been exposed to a certain illness. Once exposed, a person has a 50–50 chance of actually becoming ill.
    a. What is the probability that exactly one of the people becomes ill?
    b. What is the probability that at least two of the people become ill?
    c. What is the probability that none of the three people becomes ill?

15. When discussing counting and probability, we often consider situations that may appear frivolous or of little practical value, such as tossing coins, choosing cards, or rolling dice. The reason is that these relatively simple examples serve as models for a wide variety of more complex situations in the real world. In light of this remark, comment on the relationship between your answer to exercise 11 and your answers to exercises 12–14.

---

*For exercises with blue numbers or letters, solutions are given in Appendix B. The symbol *H* indicates that only a hint or a partial solution is given. The symbol ✶ signals that an exercise is more challenging than usual.

16. Two faces of a six-sided die are painted red, two are painted blue, and two are painted yellow. The die is rolled three times, and the colors that appear face up on the first, second, and third rolls are recorded.
    a. Let *BBR* denote the outcome where the color appearing face up on the first and second rolls is blue and the color appearing face up on the third roll is red. Because there are as many faces of one color as of any other, the outcomes of this experiment are equally likely. List all 27 possible outcomes.
    b. Consider the event that all three rolls produce different colors. One outcome in this event is *RBY* and another *RYB*. List all outcomes in the event. What is the probability of the event?
    c. Consider the event that two of the colors that appear face up are the same. One outcome in this event is *RRB* and another is *RBR*. List all outcomes in the event. What is the probability of the event?

17. Consider the situation described in exercise 16.
    a. Find the probability of the event that exactly one of the colors that appears face up is red.
    b. Find the probability of the event that at least one of the colors that appears face up is red.

18. An urn contains two blue balls (denoted B_1 and B_2) and one white ball (denoted W). One ball is drawn, its color is recorded, and it is replaced in the urn. Then another ball is drawn, and its color is recorded.
    a. Let $B_1 W$ denote the outcome that the first ball drawn is B_1 and the second ball drawn is W. Because the first ball is replaced before the second ball is drawn, the outcomes of the experiment are equally likely. List all nine possible outcomes of the experiment.
    b. Consider the event that the two balls that are drawn are both blue. List all outcomes in the event. What is the probability of the event?
    c. Consider the event that the two balls that are drawn are of different colors. List all outcomes in the event. What is the probability of the event?

19. An urn contains two blue balls (denoted B_1 and B_2) and three white balls (denoted W_1, W_2, and W_3). One ball is drawn, its color is recorded, and it is replaced in the urn. Then another ball is drawn and its color is recorded.
    a. Let $B_1 W_2$ denote the outcome that the first ball drawn is B_1 and the second ball drawn is W_2. Because the first ball is replaced before the second ball is drawn, the outcomes of the experiment are equally likely. List all 25 possible outcomes of the experiment.
    b. Consider the event that the first ball that is drawn is blue. List all outcomes in the event. What is the probability of the event?
    c. Consider the event that only white balls are drawn. List all outcomes in the event. What is the probability of the event?

20. Refer to Example 6.1.3. Suppose you are appearing on a game show with a prize behind one of five closed doors: A, B, C, D, and E. If you pick the right door, you win the prize. You pick door A. The game show host then opens one of the other doors and reveals that there is no prize behind it. Then the host gives you the option of staying with your original choice of door A or switching to one of the other doors that is still closed.
    a. If you stick with your original choice, what is the probability that you will win the prize?
    b. If you switch to another door, what is the probability that you will win the prize?

21. a. How many positive two-digit integers are multiples of 3?
    b. What is the probability that a randomly chosen positive two-digit integer is a multiple of 3?

22. a. How many positive three-digit integers are multiples of 6?
    b. What is the probability that a randomly chosen positive three-digit integer is a multiple of 6?

23. Suppose $A[1], A[2], A[3], \ldots, A[n]$ is a one-dimensional array and $n \geq 50$.
    a. How many elements are in the array?
    b. How many elements are in the subarray

    $$A[4], A[5], \ldots, A[39]?$$

    c. If $3 \leq m \leq n$, what is the probability that a randomly chosen array element is in the subarray

    $$A[3], A[4], \ldots, A[m]?$$

    d. What is the probability that a randomly chosen array element is in the subarray shown below if $n = 39$?

    $$A[\lfloor n/2 \rfloor], A[\lfloor n/2 \rfloor + 1], \ldots, A[n]$$

24. Suppose $A[1], A[2], \ldots, A[n]$ is a one-dimensional array and $n \geq 2$. Consider the subarray

    $$A[1], A[2], \ldots, A[\lfloor n/2 \rfloor].$$

    a. How many elements are in the subarray (i) if n is even? and (ii) if n is odd?
    b. What is the probability that a randomly chosen array element is in the subarray (i) if n is even? and (ii) if n is odd?

25. Suppose $A[1], A[2], \ldots, A[n]$ is a one-dimensional array and $n \geq 2$. Consider the subarray

    $$A[\lfloor n/2 \rfloor], A[\lfloor n/2 \rfloor + 1], \ldots, A[n].$$

    a. How many elements are in the subarray (i) if n is even? and (ii) if n is odd?
    b. What is the probability that a randomly chosen array element is in the subarray (i) if n is even? and (ii) if n is odd?

26. What is the 27th element in the one-dimensional array $A[42], A[43], \ldots, A[100]$?

27. What is the 62nd element in the one-dimensional array $B[29], B[30], \ldots, B[100]$?

28. If the largest of 56 consecutive integers is 279, what is the smallest?

29. If the largest of 87 consecutive integers is 326, what is the smallest?

30. How many even integers are there between 1 and 1,001?

31. How many integers are there between 1 and 1,001 that are multiples of 3?

32. A non-leap year has 365 days. Assume that January 1 is a Monday.
   a. How many Sundays are there in the year?
   b. How many Mondays are there in the year?

✱ 33. Prove Theorem 6.1.1. (Let *m* be any integer and prove the theorem by mathematical induction on *n*.)

6.2 Possibility Trees and the Multiplication Rule

Don't believe anything unless you have thought it through for yourself.
— Anna Pell Wheeler, 1883–1966

A tree structure is a useful tool for keeping systematic track of all possibilities in situations in which events happen in order. The following example shows how to use such a structure to count the number of different outcomes of a tournament.

Example 6.2.1 Possibilities for Tournament Play

Teams *A* and *B* are to play each other repeatedly until one wins two games in a row or a total of three games. One way in which this tournament can be played is for *A* to win the first game, *B* to win the second, and *A* to win the third and fourth games. Denote this by writing *A–B–A–A*.

a. How many ways can the tournament be played?

b. Assuming that all the ways of playing the tournament are equally likely, what is the probability that five games are needed to determine the tournament winner?

Solution

a. The possible ways for the tournament to be played are represented by the distinct paths from "root" (the start) to "leaf" (a terminal point) in the tree shown side-ways in Figure 6.2.1. The label on each branching point indicates the winner of the game. The notations in parentheses indicate the winner of the tournament.

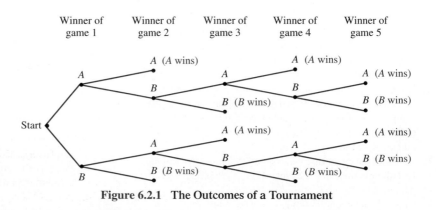

Figure 6.2.1   The Outcomes of a Tournament

The fact that there are ten paths from the root of the tree to its leaves shows that there are ten possible ways for the tournament to be played. They are (moving from the top down): *A–A, A–B–A–A, A–B–A–B–A, A–B–A–B–B, A–B–B, B–A–A,*

B–A–B–A–A, B–A–B–A–B, B–A–B–B, and B–B. In five cases A wins, and in the other five B wins. The least number of games that must be played to determine a winner is two, and the most that will need to be played is five.

b. Since all the possible ways of playing the tournament listed in part (a) are assumed to be equally likely, and the listing shows that five games are needed in four different cases (A–B–A–B–A, A–B–A–B–B, B–A–B–A–B, and B–A–B–A–A), the probability that five games are needed is $4/10 = 2/5 = 40\%$. ∎

The Multiplication Rule

Consider the following example. Suppose a computer installation has four input/output units (A, B, C, and D) and three central processing units (X, Y, and Z). Any input/output unit can be paired with any central processing unit. How many ways are there to pair an input/output unit with a central processing unit?

To answer this question, imagine the pairing of the two types of units as a two-step operation:

Step 1: Choose the input/output unit.

Step 2: Choose the central processing unit.

The possible outcomes of this operation are illustrated in the possibility tree of Figure 6.2.2.

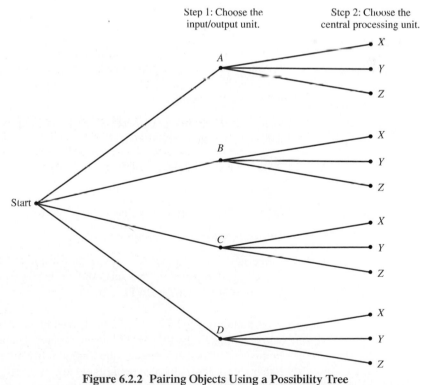

Figure 6.2.2 Pairing Objects Using a Possibility Tree

The top-most path from "root" to "leaf" indicates that input/output unit A is to be paired with central processing unit X. The next lower branch indicates that input/output unit A is to be paired with central processing unit Y. And so forth.

Thus the total number of ways to pair the two types of units is the same as the number of branches of the tree, which is

$$3 + 3 + 3 + 3 = 4 \cdot 3 = 12.$$

The idea behind this example can be used to prove the following rule. A formal proof uses mathematical induction and is left to the exercises.

Theorem 6.2.1 The Multiplication Rule

If an operation consists of k steps and

the first step can be performed in n_1 ways,

the second step can be performed in n_2 ways [*regardless of how the first step was performed*],

$$\vdots$$

the kth step can be performed in n_k ways [*regardless of how the preceding steps were performed*],

then the entire operation can be performed in $n_1 n_2 \cdots n_k$ ways.

To apply the multiplication rule, think of the objects you are trying to count as the output of a multistep operation. The possible ways to perform a step may depend on how preceding steps were performed, but the *number* of ways to perform each step must be constant regardless of the action taken in prior steps.

Example 6.2.2 Number of Personal Identification Numbers (PINs)

A typical PIN (personal identification number) is a sequence of any four symbols chosen from the 26 letters in the alphabet and the ten digits, with repetition allowed. How many different PINs are possible?

Solution   Typical PINs are CARE, 3387, B32B, and so forth. You can think of forming a PIN as a four-step operation.

Step 1: Choose the first symbol.

Step 2: Choose the second symbol.

Step 3: Choose the third symbol.

Step 4: Choose the fourth symbol.

There is a fixed number of ways to perform each step, namely 36, regardless of how preceding steps were performed. And so, by the multiplication rule, there are $36 \cdot 36 \cdot 36 \cdot 36 = 36^4 = 1,679,616$ PINs in all. ■

Another way to look at the PINs of Example 6.2.2 is as ordered 4-tuples. For example, you can think of the PIN M2ZM as the ordered 4-tuple (M, 2, Z, M). Therefore, the total number of PINs is the same as the total number of ordered 4-tuples whose elements are either letters of the alphabet or digits. One of the most important uses of the multiplication rule is to derive a general formula for the number of elements in any Cartesian product of a finite number of finite sets. In Example 6.2.3, this is done for a Cartesian product of four sets.

Example 6.2.3 The Number of Elements in a Cartesian Product

Suppose A_1, A_2, A_3, and A_4 are sets with n_1, n_2, n_3, and n_4 elements, respectively. Show that the set $A_1 \times A_2 \times A_3 \times A_4$ has $n_1 n_2 n_3 n_4$ elements.

Solution   Each element in $A_1 \times A_2 \times A_3 \times A_4$ is an ordered 4-tuple of the form (a_1, a_2, a_3, a_4), where $a_1 \in A_1, a_2 \in A_2, a_3 \in A_3$, and $a_4 \in A_4$. Imagine the process of constructing these ordered tuples as a four-step operation:

Step 1: Choose the first element of the 4-tuple.

Step 2: Choose the second element of the 4-tuple.

Step 3: Choose the third element of the 4-tuple.

Step 4: Choose the fourth element of the 4-tuple.

There are n_1 ways to perform step 1, n_2 ways to perform step 2, n_3 ways to perform step 3, and n_4 ways to perform step 4. Hence, by the multiplication rule, there are $n_1 n_2 n_3 n_4$ ways to perform the entire operation. Therefore, there are $n_1 n_2 n_3 n_4$ distinct 4-tuples in $A_1 \times A_2 \times A_3 \times A_4$. ■

Example 6.2.4 Number of PINs without Repetition

In Example 6.2.2 we formed PINs using four symbols, either letters of the alphabet or digits, and supposing that letters could be repeated. Now suppose that repetition is not allowed.

a. How many different PINs are there?

b. If all PINs are equally likely, what is the probability that a PIN chosen at random contains no repeated symbol?

Solution

a. Again think of forming a PIN as a four-step operation: Choose the first symbol, then the second, then the third, and then the fourth. There are 36 ways to choose the first symbol, 35 ways to choose the second (since the first symbol cannot be used again), 34 ways to choose the third (since the first two symbols cannot be reused), and 33 ways to choose the fourth (since the first three symbols cannot be reused). Thus, the multiplication rule can be applied to conclude that there are $36 \cdot 35 \cdot 34 \cdot 33 = 1,413,720$ different PINs with no repeated symbol.

b. By part (a) there are 1,413,720 PINs with no repeated symbol, and by Example 6.2.2 there are 1,679,616 PINs in all. Thus the probability that a PIN chosen at random contains no repeated symbol is $\dfrac{1,413,720}{1,679,616} \cong .8417$. In other words, approximately 84% of PINs have no repeated symbol. ■

Any circuit with two input signals P and Q has an input/output table consisting of four rows corresponding to the four possible assignments of values to P and Q: 11, 10, 01, and 00. The next example shows that there are only 16 distinct ways in which such a circuit can function.

Example 6.2.5 Number of Input/Output Tables for a Circuit with Two Input Signals

Consider the set of all circuits with two input signals P and Q. For each such circuit an input/output table can be constructed, but, as shown in Section 1.4, two such input/output tables may have the same values. How many distinct input/output tables can be constructed for circuits with input/output signals P and Q?

Solution   Fix the order of the input values for P and Q. Then two input/output tables are distinct if their output values differ in at least one row. For example, the input/output tables shown below are distinct, because their output values differ in the first row.

P	Q	Output
1	1	1
1	0	0
0	1	1
0	0	0

P	Q	Output
1	1	0
1	0	0
0	1	1
0	0	0

For a fixed ordering of input values, you can obtain a complete input/output table by filling in the entries in the output column. You can think of this as a four-step operation:

Step 1: Fill in the output value for the first row.

Step 2: Fill in the output value for the second row.

Step 3: Fill in the output value for the third row.

Step 4: Fill in the output value for the fourth row.

Each step can be performed in exactly two ways: either a 1 or a 0 can be filled in. Hence, by the multiplication rule, there are

$$2 \cdot 2 \cdot 2 \cdot 2 = 16$$

ways to perform the entire operation. It follows that there are $2^4 = 16$ distinct input/output tables for a circuit with two input signals P and Q. This means that such a circuit can function in only 16 distinct ways. ■

• Definition

Let n be a positive integer. Given a finite set S, a **string of length n over S** is an ordered n-tuple of elements of S written without parentheses or commas. The elements of S are called the **characters** of the string. The **null string** over S is defined to be the "string" with no characters. It is usually denoted ϵ and is said to have length 0. If $S = \{0, 1\}$, then a string over S is called a **bit string.**

Observe that in Examples 6.2.2 and 6.2.4, the set of all PINs of length 4 is the same as the set of all strings of length 4 over the set

$$S = \{x \mid x \text{ is a letter of the alphabet or } x \text{ is a digit}\}.$$

Also observe that another way to think of Example 6.2.5 is to realize that there are as many input/output tables for a circuit with two input signals as there are bit strings of length 4 (written vertically) that can be used to fill in the output values. As another example, here is a listing of all bit strings of length 3:

$$000, \quad 001, \quad 010, \quad 100, \quad 011, \quad 101, \quad 110, \quad 111.$$

Example 6.2.6 Counting the Number of Iterations of a Nested Loop

Consider the following nested loop:

> **for** $i := 1$ **to** 4
>> **for** $j := 1$ **to** 3
>>> [*Statements in body of inner loop.*
>>> *None contain branching statements*
>>> *that lead out of the inner loop.*]
>> **next** j
> **next** i

How many times will the inner loop be iterated when the algorithm is implemented and run?

Solution The outer loop is iterated four times, and during each iteration of the outer loop, there are three iterations of the inner loop. Hence by the multiplication rule, the total number of iterations of the inner loop is $4 \cdot 3 = 12$. This is illustrated by the trace table below.

When the Multiplication Rule is Difficult or Impossible to Apply

Consider the following problem:

> Three officers—a president, a treasurer, and a secretary—are to be chosen from among four people: Ann, Bob, Cyd, and Dan. Suppose that, for various reasons, Ann cannot be president and either Cyd or Dan must be secretary. How many ways can the officers be chosen?

It is natural to try to solve this problem using the multiplication rule. A person might answer as follows:

> There are three choices for president (all except Ann), three choices for treasurer (all except the one chosen as president), and two choices for secretary (Cyd or Dan). Therefore, by the multiplication rule, there are $3 \cdot 3 \cdot 2 = 18$ choices in all.

Unfortunately, this analysis is incorrect. The number of ways to choose the secretary varies depending on who is chosen for president and treasurer. For instance, if Bob is chosen for president and Ann for treasurer, then there are two choices for secretary: Cyd and Dan. But if Bob is chosen for president and Cyd for treasurer, then there is just one choice for secretary: Dan. The clearest way to see all the possible choices is to construct the possibility tree, as is shown in Figure 6.2.3.

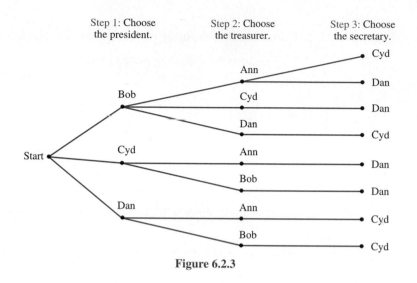

Figure 6.2.3

From the tree it is easy to see that there are only eight ways to choose a president, treasurer, and secretary so as to satisfy the given conditions.

Another way to solve this problem is somewhat surprising. It turns out that the steps can be reordered in a slightly different way so that the number of ways to perform each step is constant regardless of the way previous steps were performed.

Example 6.2.7 A More Subtle Use of the Multiplication Rule

Reorder the steps for choosing the officers in the example above so that the total number of ways to choose officers can be computed using the multiplication rule.

Solution

Step 1: Choose the secretary.

Step 2: Choose the president.

Step 3: Choose the treasurer.

There are exactly two ways to perform step 1 (either Cyd or Dan may be chosen), two ways to perform step 2 (neither Ann nor the person chosen in step 1 may be chosen but either of the other two may), and two ways to perform step 3 (either of the two people not chosen as secretary or president may be chosen as treasurer). Thus, by the multiplication rule, the total number of ways to choose officers is $2 \cdot 2 \cdot 2 = 8$. A possibility tree illustrating this sequence of choices is shown in Figure 6.2.4. Note how balanced this tree is compared with the one in Figure 6.2.3.

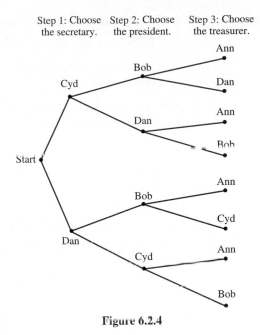

Figure 6.2.4

Permutations

A **permutation** of a set of objects is an ordering of the objects in a row. For example, the set of elements a, b, and c has six permutations.

$$abc \quad acb \quad cba \quad bac \quad bca \quad cab$$

In general, given a set of n objects, how many permutations does the set have? Imagine forming a permutation as an n-step operation:

Step 1: Choose an element to write first.

Step 2: Choose an element to write second.

$$\vdots \qquad \vdots$$

Step n: Choose an element to write nth.

Any element of the set can be chosen in step 1, so there are n ways to perform step 1. Any element except that chosen in step 1 can be chosen in step 2, so there are $n - 1$ ways to perform step 2. In general, the number of ways to perform each successive step is one less than the number of ways to perform the preceding step. At the point when the nth element is chosen, there is only one element left, so there is only one way to perform step n. Hence, by the multiplication rule, there are

$$n(n - 1)(n - 2) \cdots 2 \cdot 1 = n!$$

ways to perform the entire operation. In other words, there are $n!$ permutations of a set of n elements. This reasoning is summarized in the following theorem. A formal proof uses mathematical induction and is left as an exercise.

Theorem 6.2.2

For any integer n with $n \geq 1$, the number of permutations of a set with n elements is $n!$.

Example 6.2.8 Permutations of the Letters in a Word

   a. How many ways can the letters in the word *COMPUTER* be arranged in a row?

   b. How many ways can the letters in the word *COMPUTER* be arranged if the letters *CO* must remain next to each other (in order) as a unit?

   c. If letters of the word *COMPUTER* are randomly arranged in a row, what is the probability that the letters *CO* remain next to each other (in order) as a unit?

Solution

   a. All the eight letters in the word *COMPUTER* are distinct, so the number of ways in which we can arrange the letters equals the number of permutations of a set of eight elements. This equals $8! = 40,320$.

   b. If the letter group *CO* is treated as a unit, then there are effectively only seven objects that are to be arranged in a row.

Hence there are as many ways to write the letters as there are permutations of a set of seven elements, namely $7! = 5,040$.

   c. When the letters are arranged randomly in a row, the total number of arrangements is 40,320 by part (a), and the number of arrangements with the letters *CO* next to each other (in order) as a unit is 5,040. Thus the probability is

$$\frac{5,040}{40,320} = \frac{1}{8}$$ ∎

Example 6.2.9 Permutations of Objects Around a Circle

At a meeting of diplomats, the six participants are to be seated around a circular table. Since the table has no ends to confer particular status, it doesn't matter who sits in which chair. But it does matter how the diplomats are seated relative to each other. In other words, two seatings are considered the same if one is a rotation of the other. How many different ways can the diplomats be seated?

Solution   Call the diplomats by the letters A, B, C, D, E, and F. Since only relative position matters, you can start with any diplomat (say A), place that diplomat anywhere (say in the top seat of the diagram shown in Figure 6.2.5), and then consider all arrangements of the other diplomats around that one. B through F can be arranged in the seats around diplomat A in all possible orders. So there are $5! = 120$ ways to seat the group.

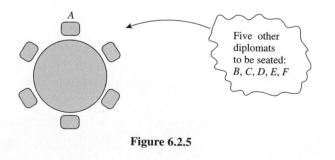

Figure 6.2.5 ∎

Permutations of Selected Elements

Given the set $\{a, b, c\}$, there are six ways to select two letters from the set and write them in order.

$$ab \quad ac \quad ba \quad bc \quad ca \quad cb$$

Each such ordering of two elements of $\{a, b, c\}$ is called a 2-*permutation* of $\{a, b, c\}$.

> **• Definition**
>
> An ***r*-permutation** of a set of n elements is an ordered selection of r elements taken from the set of n elements. The number of r-permutations of a set of n elements is denoted $P(n, r)$.

> **Theorem 6.2.3**
>
> If n and r are integers and $1 \le r \le n$, then the number of r-permutations of a set of n elements is given by the formula
>
> $$P(n, r) = n(n-1)(n-2)\cdots(n-r+1) \qquad \text{first version}$$
>
> or, equivalently,
>
> $$P(n, r) = \frac{n!}{(n-r)!} \qquad \text{second version.}$$

A formal proof of this theorem uses mathematical induction and is based on the multiplication rule. The idea of the proof is the following.

Suppose a set of n elements is given. Formation of an r-permutation can be thought of as an r-step process. Step 1 is to choose the element to be first. Since the set has n elements, there are n ways to perform step 1. Step 2 is to choose the element to be second. Since the element chosen in step 1 is no longer available, there are $n-1$ ways to perform step 2. Step 3 is to choose the element to be third. Since neither of the two elements chosen in the first two steps is available, there are $n-2$ choices for step 3. This process is repeated r times, as shown below.

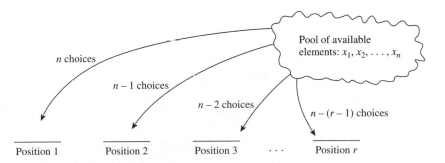

The number of ways to perform each successive step is one less than the number of ways to perform the preceding step. Step r is to choose the element to be rth. At the point just before step r is performed, $r-1$ elements have already been chosen, and so there are

$$n - (r - 1) = n - r + 1$$

left to choose from. Hence there are $n - r + 1$ ways to perform step r. It follows by the multiplication rule that the number of ways to form an r-permutation is

$$P(n, r) = n(n - 1)(n - 2) \cdots (n - r + 1).$$

Note that

$$\frac{n!}{(n-r)!} = \frac{n(n-1)(n-2)\cdots(n-r+1)(n-r)(n-r-1)\cdots 3 \cdot 2 \cdot 1}{(n-r)(n-r-1)\cdots 3 \cdot 2 \cdot 1}$$

$$= n(n-1)(n-2)\cdots(n-r+1).$$

Thus the formula can be written as

$$P(n, r) = \frac{n!}{(n - r)!}.$$

The second version of the formula is easier to remember. When you actually use it, however, first substitute the values of n and r and then immediately cancel the numerical value of $(n - r)!$ from the numerator and denominator. Because factorials become so large so fast, direct use of the first version of the formula without cancellation can overload your calculator's capacity for exact arithmetic even when n and r are quite small. For instance, if $n = 15$ and $r = 2$, then

$$\frac{n!}{(n - r)!} = \frac{15!}{13!} = \frac{1{,}307{,}674{,}368{,}000}{6{,}227{,}020{,}800}.$$

But if you cancel $(n - r)! = 13!$ from numerator and denominator before multiplying out, you obtain

$$\frac{n!}{(n - r)!} = \frac{15!}{13!} = \frac{15 \cdot 14 \cdot 13!}{13!} = 15 \cdot 14 = 210.$$

In fact, many scientific calculators allow you to compute $P(n, r)$ simply by entering the values of n and r and pressing a key or making a menu choice. Alternative notations for $P(n, r)$ that you may see in your calculator manual are $_nP_r$, $P_{n,r}$ and ^nP_r.

Example 6.2.10 Evaluating r-Permutations

    a. Evaluate $P(5, 2)$.

    b. How many 4-permutations are there of a set of seven objects?

    c. How many 5-permutations are there of a set of five objects?

Solution

    a. $P(5, 2) = \dfrac{5!}{(5 - 2)!} = \dfrac{5 \cdot 4 \cdot 3 \cdot 2 \cdot 1}{3 \cdot 2 \cdot 1} = 20$

    b. The number of 4-permutations of a set of seven objects is

$$P(7, 4) = \frac{7!}{(7 - 4)!} = \frac{7 \cdot 6 \cdot 5 \cdot 4 \cdot 3 \cdot 2 \cdot 1}{3 \cdot 2 \cdot 1} = 7 \cdot 6 \cdot 5 \cdot 4 = 840.$$

    c. The number of 5-permutations of a set of five objects is

$$P(5, 5) = \frac{5!}{(5 - 5)!} = \frac{5!}{0!} = \frac{5!}{1} = 5! = 120.$$

Note that the definition of 0! as 1 makes this calculation come out as it should, for the number of 5-permutations of a set of five objects is certainly equal to the number of permutations of the set. ∎

Example 6.2.11 Permutations of Selected Letters of a Word

a. How many different ways can three of the letters of the word *BYTES* be chosen and written in a row?

b. How many different ways can this be done if the first letter must be *B*?

Solution

a. The answer equals the number of 3-permutations of a set of five elements. This equals

$$P(5, 3) = \frac{5!}{(5-3)!} = \frac{5 \cdot 4 \cdot 3 \cdot \cancel{2} \cdot \cancel{1}}{\cancel{2} \cdot \cancel{1}} = 5 \cdot 4 \cdot 3 = 60.$$

b. Since the first letter must be *B*, there are effectively only two letters to be chosen and placed in the other two positions. And since the *B* is used in the first position, there are four letters available to fill the remaining two positions.

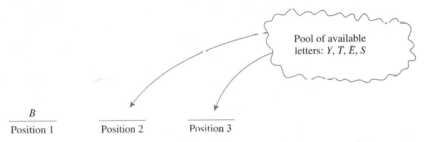

Pool of available letters: *Y, T, E, S*

$\underline{\quad B \quad}$	$\underline{\qquad\qquad}$	$\underline{\qquad\qquad}$
Position 1	Position 2	Position 3

Hence the answer is the number of 2-permutations of a set of four elements, which is

$$P(4, 2) = \frac{4!}{(4-2)!} = \frac{4 \cdot 3 \cdot \cancel{2} \cdot \cancel{1}}{\cancel{2} \cdot \cancel{1}} = 4 \cdot 3 = 12.$$ ∎

In many applications of the mathematics of counting, it is necessary to be skillful in working algebraically with quantities of the form $P(n, r)$. The next example shows a kind of problem that gives practice in developing such skill.

Example 6.2.12 Proving a Property of $P(n, r)$

Prove that for all integers $n \geq 2$,

$$P(n, 2) + P(n, 1) = n^2.$$

Solution Suppose n is an integer that is greater than or equal to 2. By Theorem 6.2.3,

$$P(n, 2) = \frac{n!}{(n-2)!} = \frac{n(n-1)\cancel{(n-2)!}}{\cancel{(n-2)!}} = n(n-1)$$

and

$$P(n, 1) = \frac{n!}{(n-1)!} = \frac{n \cdot \cancel{(n-1)!}}{\cancel{(n-1)!}} = n.$$

Hence

$$P(n, 2) + P(n, 1) = n \cdot (n-1) + n = n^2 - n + n = n^2,$$

which is what we needed to show. ∎

Exercise Set 6.2

In 1–4, use the fact that in baseball's World Series, the first team to win four games wins the series.

1. Suppose team A wins the first three games. How many ways can the series be completed? (Draw a tree.)

2. Suppose team A wins the first two games. How many ways can the series be completed? (Draw a tree.)

3. How many ways can a World Series be played if team A wins four games in a row?

4. How many ways can a World Series be played if no team wins two games in a row?

5. In a competition between players X and Y, the first player to win three games in a row or a total of four games wins. How many ways can the competition be played if X wins the first game and Y wins the second and third games? (Draw a tree.)

6. One urn contains two black balls (labeled B_1 and B_2) and one white ball. A second urn contains one black ball and two white balls (labeled W_1 and W_2). Suppose the following experiment is performed: One of the two urns is chosen at random. Next a ball is randomly chosen from the urn. Then a second ball is chosen at random from the same urn without replacing the first ball.
   a. Construct the possibility tree showing all possible outcomes of this experiment.
   b. What is the total number of outcomes of this experiment?
   c. What is the probability that two black balls are chosen?
   d. What is the probability that two balls of opposite color are chosen?

7. One urn contains one blue ball (labeled B_1) and three red balls (labeled R_1, R_2, and R_3). A second urn contains two red balls (R_4 and R_5) and two blue balls (B_2 and B_3). An experiment is performed in which one of the two urns is chosen at random and then two balls are randomly chosen from it, one after the other without replacement.
   a. Construct the possibility tree showing all possible outcomes of this experiment.
   b. What is the total number of outcomes of this experiment?
   c. What is the probability that two red balls are chosen?

8. A person buying a personal computer system is offered a choice of three models of the basic unit, two models of keyboard, and two models of printer. How many distinct systems can be purchased?

9. Suppose there are three roads from city A to city B and five roads from city B to city C.
   a. How many ways is it possible to travel from city A to city C via city B?
   b. How many different round-trip routes are there from city A to B to C to B and back to A?

   c. How many different routes are there from city A to B to C to B and back to A in which no road is traversed twice?

10. Suppose there are three routes from North Point to Boulder Creek, two routes from Boulder Creek to Beaver Dam, two routes from Beaver Dam to Star Lake, and four routes directly from Boulder Creek to Star Lake. (Draw a sketch.)
    a. How many routes from North Point to Star Lake pass through Beaver Dam?
    b. How many routes from North Point to Star Lake bypass Beaver Dam?

11. a. A bit string is a finite sequence of 0's and 1's. How many bit strings have length 8?
    b. How many bit strings of length 8 begin with three 0's?
    c. How many bit strings of length 8 begin and end with a 1?
    d. In Section 1.5 we showed how integers can be represented by strings of 0's and 1's inside a digital computer. In fact, through various coding schemes, strings of 0's and 1's can be used to represent all kinds of symbols. One commonly used code is the Extended Binary-Coded Decimal Interchange Code (EBCDIC) in which each symbol has an 8-bit representation. How many distinct symbols can be represented by this code?

12. Hexadecimal numbers are made using the sixteen digits 0, 1, 2, 3, 4, 5, 6, 7, 8, 9, A, B, C, D, E, F. They are denoted by the subscript 16.
    a. How many hexadecimal numbers begin with one of the digits 3 through B, end with one of the digits 5 through F and are 5 digits long?
    b. How many hexadecimal numbers begin with one of the digits 4 through D, end with one of the digits 2 through E and are 6 digits long?

13. A coin is tossed four times. Each time the result H for heads or T for tails is recorded. An outcome of $HHTT$ means that heads were obtained on the first two tosses and tails on the second two. Assume that heads and tails are equally likely on each toss.
    a. How many distinct outcomes are possible?
    b. What is the probability that exactly two heads occur?
    c. What is the probability that exactly one head occurs?

14. Suppose that in a certain state, all automobile license plates have four letters followed by three digits.
    a. How many different license plates are possible?
    b. How many license plates could begin with A and end in 0?
    c. How many license plates could begin with $TGIF$?
    d. How many license plates are possible in which all the letters and digits are distinct?
    e. How many license plates could begin with AB and have all letters and digits distinct?

15. A combination lock requires three selections of numbers, each from 1 through 30.
    a. How many different combinations are possible?
    b. Suppose the locks are constructed in such a way that no number may be used twice. How many different combinations are possible?

16. The diagram below shows the keypad for an automatic teller machine. As you can see, the same sequence of keys represents a variety of different PINs. For instance, 2133, AZDE, and BQ3F are all keyed in exactly the same way.

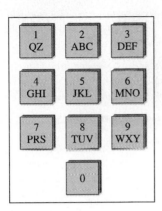

    a. How many different PINs are represented by the same sequence of keys as 2133?
    b. How many different PINs are represented by the same sequence of keys as 5031?
    c. At an automatic teller machine, each PIN corresponds to a four-digit numeric sequence. For instance, TWJM corresponds to 8956. How many such numeric sequences contain no repeated digit?

17. Three officers—a president, a treasurer, and a secretary—are to be chosen from among four people: Ann, Bob, Cyd, and Dan. Suppose that Bob is not qualified to be treasurer and Cyd's other commitments make it impossible for her to be secretary. How many ways can the officers be chosen? Can the multiplication rule be used to solve this problem?

18. Modify Example 6.2.4 by supposing that a PIN must not begin with any of the letters A–M and must end with a digit. Continue to assume that no symbol may be used more than once and that the total number of PINs is to be determined.
    a. Find the error in the following "solution."

    "Constructing a PIN is a four-step process.

    Step 1: Choose the left-most symbol.

    Step 2: Choose the second symbol from the left.

    Step 3: Choose the third symbol from the left.

    Step 4: Choose the right-most symbol.

    Because none of the thirteen letters from A through M may be chosen in step 1, there are $36 - 13 = 23$ ways

to perform step 1. There are 35 ways to perform step 2 and 34 ways to perform step 3 because previously used symbols may not be used. Since the symbol chosen in step 4 must be a previously unused digit, there are $10 - 3 = 7$ ways to perform step 4. Thus there are $23 \cdot 35 \cdot 34 \cdot 7 = 191,590$ different PINs that satisfy the given conditions."
    b. Reorder steps 1–4 in part (a) as follows:

    Step 1: Choose the right-most symbol.

    Step 2: Choose the left-most symbol.

    Step 3: Choose the second symbol from the left.

    Step 4: Choose the third symbol from the left.

    Use the multiplication rule to find the number of PINs that satisfy the given conditions.

19. a. How many integers are there from 10 through 99?
    b. How many odd integers are there from 10 through 99?
    c. How many integers from 10 through 99 have distinct digits?
    d. How many odd integers from 10 through 99 have distinct digits?
    e. What is the probability that a randomly chosen two-digit integer has distinct digits? has distinct digits and is odd?

20. a. How many integers are there from 1000 through 9999?
    b. How many odd integers are there from 1000 through 9999?
    c. How many integers from 1000 through 9999 have distinct digits?
    d. How many odd integers from 1000 through 9999 have distinct digits?
    e. What is the probability that a randomly chosen four-digit integer has distinct digits? has distinct digits and is odd?

In each of 21–25, determine how many times the innermost loop will be iterated when the algorithm segment is implemented and run. (Assume that $m, n, p, a, b, c,$ and d are all positive integers.)

21. **for** $i := 1$ **to** 30
        for $j := 1$ **to** 15
            [*Statements in body of inner loop.*
            None contain branching statements that
            lead outside the loop.]
        next j
      next i

22. **for** $j := 1$ **to** m
        for $k := 1$ **to** n
            [*Statements in body of inner loop.*
            None contain branching statements that
            lead outside the loop.]
        next k
      next j

23. **for** $i := 1$ **to** m
    for $j := 1$ **to** n
        for $k := 1$ **to** p
            [*Statements in body of inner loop.*
            None contain branching statements that
            lead outside the loop.]
        next k
    next j
  next i

24. **for** $i := 5$ **to** 50
    for $j := 10$ **to** 20
        [*Statements in body of inner loop.*
        None contain branching statements that
        lead outside the loop.]
    next j
  next i

25. Assume $a \le b$ and $c \le d$.

  for $i := a$ **to** b
    for $j := c$ **to** d
        [*Statements in body of inner loop.*
        None contain branching statements that
        lead outside the loop.]
    next j
  next i

H ✻ 26. Consider the numbers 1 through 99,999 in their ordinary decimal representations. How many contain exactly one of each of the digits 2, 3, 4, and 5?

✻ 27. Let $n = p_1^{k_1} p_2^{k_2} \cdots p_m^{k_m}$ where $p_1, p_2, \ldots, p_m$ are distinct prime numbers and $k_1, k_2, \ldots, k_m$ are positive integers. How many ways can n be written as a product of two positive integers that have no common factors
  a. assuming that order matters (i.e., $8 \cdot 15$ and $15 \cdot 8$ are regarded as different)?
  b. assuming that order does not matter (i.e., $8 \cdot 15$ and $15 \cdot 8$ are regarded as the same)?

✻ 28. a. If p is a prime number and a is a positive integer, how many divisors does p^a have?
  b. If p and q are prime numbers and a and b are positive integers, how many possible divisors does $p^a q^b$ have?
  c. If p, q, and r are prime numbers and a, b, and c are positive integers, how many possible divisors does $p^a q^b r^c$ have?
  d. If $p_1, p_2, \ldots, p_m$ are prime numbers and $a_1, a_2, \ldots, a_m$ are positive integers, how many possible divisors does $p_1^{a_1} p_2^{a_2} \cdots p_m^{a_m}$ have?
  e. What is the smallest positive integer with exactly 12 divisors?

29. a. How many ways can the letters of the word *ALGORITHM* be arranged in a row?
  b. How many ways can the letters of the word *ALGORITHM* be arranged in a row if A and L must remain together (in order) as a unit?

c. How many ways can the letters of the word *ALGORITHM* be arranged in a row if the letters *GOR* must remain together (in order) as a unit?

30. Six people attend the theater together and sit in a row with exactly six seats.
  a. How many ways can they be seated together in the row?
  b. Suppose one of the six is a doctor who must sit on the aisle in case she is paged. How many ways can the people be seated together in the row with the doctor in an aisle seat?
  c. Suppose the six people consist of three married couples and each couple wants to sit together with the husband on the left. How many ways can the six be seated together in the row?

31. Five people are to be seated around a circular table. Two seatings are considered the same if one is a rotation of the other. How many different seatings are possible?

32. Write all the 2-permutations of $\{W, X, Y, Z\}$.

33. Write all the 3-permutations of $\{s, t, u, v\}$.

34. Evaluate the following quantities.
  a. $P(6, 4)$ **b.** $P(6, 6)$ **c.** $P(6, 3)$ **d.** $P(6, 1)$

35. **a.** How many 3-permutations are there of a set of five objects?
  b. How many 2-permutations are there of a set of eight objects?

36. **a.** How many ways can three of the letters of the word *ALGORITHM* be selected and written in a row?
  b. How many ways can six of the letters of the word *ALGORITHM* be selected and written in a row?
  c. How many ways can six of the letters of the word *ALGORITHM* be selected and written in a row if the first letter must be A?
  d. How many ways can six of the letters of the word *ALGORITIIM* be selected and written in a row if the first two letters must be *OR*?

37. Prove that for all integers $n \ge 2$,
$$P(n + 1, 3) = n^3 - n.$$

38. Prove that for all integers $n \ge 2$,
$$P(n + 1, 2) - P(n, 2) = 2P(n, 1).$$

39. Prove that for all integers $n \ge 3$,
$$P(n + 1, 3) - P(n, 3) = 3P(n, 2).$$

40. Prove that for all integers $n \ge 2$,
$$P(n, n) = P(n, n - 1).$$

41. Prove Theorem 6.2.1 by mathematical induction.

H 42. Prove Theorem 6.2.2 by mathematical induction.

✻ 43. Prove Theorem 6.2.3 by mathematical induction.

6.3 *Counting Elements of Disjoint Sets: The Addition Rule*

The whole of science is nothing more than a refinement of everyday thinking.
— Albert Einstein, 1879–1955

In the last section we discussed counting problems that can be solved using possibility trees. In this section we look at counting problems that can be solved by counting the number of elements in the union of two sets, the difference of two sets, or the intersection of two sets.

The basic rule underlying the calculation of the number of elements in a union or difference or intersection is the addition rule. This rule states that the number of elements in a union of mutually disjoint finite sets equals the sum of the number of elements in each of the component sets.

Theorem 6.3.1 The Addition Rule

Suppose a finite set A equals the union of k distinct mutually disjoint subsets A_1, $A_2, \ldots, A_k$. Then

$$N(A) = N(A_1) + N(A_2) + \cdots + N(A_k).$$

A formal proof of this theorem uses mathematical induction and is left to the exercises.

Example 6.3.1 Counting Passwords with Three or Fewer Letters

A computer access password consists of from one to three letters chosen from the 26 in the alphabet with repetitions allowed. How many different passwords are possible?

Solution   The set of all passwords can be partitioned into subsets consisting of those of length 1, those of length 2, and those of length 3 as shown in Figure 6.3.1.

Set of All Passwords of Length ≤ 3

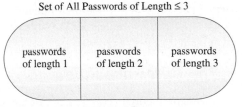

Figure 6.3.1

By the addition rule, the total number of passwords equals the number of passwords of length 1, plus the number of passwords of length 2, plus the number of passwords of length 3. Now the

number of passwords of length $1 = 26$	because there are 26 letters in the alphabet
number of passwords of length $2 = 26^2$	because forming such a word can be thought of as a two-step process in which there are 26 ways to perform each step
number of passwords of length $3 = 26^3$	because forming such a word can be thought of as a three-step process in which there are 26 ways to perform each step.

Hence the

total number of passwords $= 26 + 26^2 + 26^3 = 18{,}278.$ ∎

Example 6.3.2 Counting the Number of Integers Divisible by 5

How many three-digit integers (integers from 100 to 999 inclusive) are divisible by 5?

Solution   One solution to this problem was discussed in Example 6.1.4. Another approach uses the addition rule. Integers that are divisible by 5 end either in 5 or in 0. Thus the set of all three-digit integers that are divisible by 5 can be split into two mutually disjoint subsets A_1 and A_2 as shown in Figure 6.3.2.

Figure 6.3.2

Now there are as many three-digit integers that end in 0 as there are possible choices for the left-most and middle digits (because the right-most digit must be a 0). As illustrated below, there are nine choices for the left-most digit (the digits 1 through 9) and ten choices for the middle digit (the digits 0 through 9). Hence $N(A_1) = 9 \cdot 10 = 90$.

Similar reasoning (using 5 instead of 0) shows that $N(A_2) = 90$ also. So

$$\begin{bmatrix} \text{the number of} \\ \text{three-digit integers} \\ \text{that are divisible by 5} \end{bmatrix} = N(A_1) + N(A_2) = 90 + 90 = 180. \quad \blacksquare$$

The Difference Rule

An important consequence of the addition rule is the fact that if the number of elements in a set A and the number in a subset B of A are both known, then the number of elements that are in A and not in B can be computed.

Theorem 6.3.2 The Difference Rule

If A is a finite set and B is a subset of A, then

$$N(A - B) = N(A) - N(B).$$

The difference rule is illustrated in Figure 6.3.3.

Figure 6.3.3 The Difference Rule

The difference rule holds for the following reason: If B is a subset of A, then the two sets B and $A - B$ have no elements in common and $B \cup (A - B) = A$. Hence, by the addition rule,

$$N(B) + N(A - B) = N(A).$$

Subtracting $N(B)$ from both sides gives the equation

$$N(A - B) = N(A) - N(B).$$

Example 6.3.3 Counting PINs with Repeated Symbols

The PINs discussed in Examples 6.2.2 and 6.2.4 are made from exactly four symbols chosen from the 26 letters of the alphabet and the ten digits, with repetitions allowed.

a. How many PINs contain repeated symbols?

b. If all PINs are equally likely, what is the probability that a randomly chosen PIN contains a repeated symbol?

Solution

a. According to Example 6.2.2, there are $36^4 = 1,679,616$ PINs when repetition is allowed, and by Example 6.2.4, there are 1,413,720 PINs when repetition is not allowed. Thus, by the difference rule, there are

$$1,679,616 - 1,413,720 = 265,896$$

PINs that contain at least one repeated symbol.

b. By Example 6.2.2 there are 1,679,616 PINs in all, and by part (a) 265,896 of these contain at least one repeated symbol. Thus, by the equally likely probability formula, the probability that a randomly chosen PIN contains a repeated symbol is $\dfrac{265,896}{1,679,616} \cong$ $0.158 = 15.8\%$. ∎

An alternative solution to Example 6.3.3(b) is based on the observation that if S is the set of all PINs and A is the set of all PINs with no repeated symbol, then $S - A$ is the set of all PINs with at least one repeated symbol. It follows that

$$P(S - A) = \frac{N(S - A)}{N(S)} \qquad \text{by definition of probability in the equally likely case}$$

$$= \frac{N(S) - N(A)}{N(S)} \qquad \text{by the difference rule}$$

$$= \frac{N(S)}{N(S)} - \frac{N(A)}{N(S)} \qquad \text{by the laws of fractions}$$

$$= 1 - P(A) \qquad \text{by definition of probability in the equally likely case}$$

$$\cong 1 - 0.842 \qquad \text{by Example 6.2.4}$$

$$\cong 0.158 = 15.8\%$$

This solution illustrates a more general property of probabilities: that the probability of the complement of an event is obtained by subtracting the probability of the event from the number 1. In Section 6.8 we derive this formula from the axioms for probability.

Formula for the Probability of the Complement of an Event

If S is a finite sample space and A is an event in S, then

$$P(A^c) = 1 - P(A).$$

Example 6.3.4 Number of Python Identifiers of Eight or Fewer Characters

In the computer language Python, identifiers must start with one of 53 symbols: either one of the 52 letters of the upper- and lower-case Roman alphabet or a hyphen (-). The initial character may stand alone, or it may be followed by any number of additional characters chosen from a set of 63 symbols: the 53 symbols allowed as an initial character plus the ten digits. Certain keywords, however, such as *and, if, print,* and so forth, are set aside and may not be used as identifiers. In one implementation of Python there are 29 such reserved keywords, none of which has more than eight characters. How many Python identifiers are there that are less than or equal to eight characters in length?

Solution   The set of all Python identifiers with eight or fewer characters can be partitioned into eight subsets—identifiers of length 1, identifiers of length 2, and so on—as shown in Figure 6.3.4. The reserved words have various lengths (all less than or equal to 8), so the set of reserved words is shown overlapping the various subsets.

Set of Python Identifiers with Eight or Fewer Characters

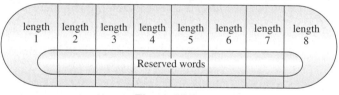

Figure 6.3.4

According to the rules for creating Python identifiers, there are

53 potential identifiers of length 1 — because there are 53 choices for the first character

$53 \cdot 63$ potential identifiers of length 2 — because the first character can be any one of 53 symbols, and the second character can be any one of 63 symbols

$53 \cdot 63^2$ potential identifiers of length 3 — because the first character can be any one of 53 symbols, and each of the next two characters can be any one of 63 symbols

$\vdots$

$53 \cdot 63^7$ potential identifiers of length 8 — because the first character can be any one of 53 symbols, and each of the next seven characters can be any one of 63 symbols.

Thus, by the addition rule, the number of potential Python identifiers with eight or fewer characters is

$$53 + 53 \cdot 63 + 53 \cdot 63^2 + 53 \cdot 63^3 + 53 \cdot 63^4 + 53 \cdot 63^5 + 53 \cdot 63^6 + 53 \cdot 63^7$$

$$= 53 \left(\frac{63^8 - 1}{63 - 1} \right)$$

$$= 212,133,167,002,880.$$

Now 29 of these potential identifiers are reserved, so by the difference rule, the actual number of Python identifiers with eight or fewer characters is

$$212,133,167,002,880 - 29 = 212,133,167,002,851. \qquad \blacksquare$$

Example 6.3.5 Internet Addresses

In order to communicate effectively, each computer in a network needs a distinguishing name called an address. For the Internet this address is currently a 32-bit number called the Internet Protocol (IP) address (although 128-bit addresses are being phased in to accommodate the growth of the Internet). For technical reasons some computers have more than one address, whereas other sets of computers, which use the Internet only sporadically, may share a pool of addresses that are assigned on a temporary basis. Like telephone numbers, IP addresses are divided into parts: one, the network ID, specifies the local network to which a given computer belongs, and the other, the host ID, specifies the particular computer.

An example of an IP address is 10001100 11000000 00100000 10001000, where the 32 bits have been divided into four groups of 8 for easier reading. To make the reading even easier, IP addresses are normally written as "dotted decimals," in which each group of 8 bits is converted into a decimal number between 0 and 255. For instance, the IP address above converts into 140.192.32.136.

In order to accommodate the various sizes of the local networks connected through the Internet, the network IDs are divided into several classes, the most important of which are called A, B, and C. In every class, a host ID may not consist of either all 0's or all 1's.

Class A network IDs are used for very large local networks. The left-most bit is set to 0, and the left-most 8 bits give the full network ID. The remaining 24 bits are used for individual host IDs. However, neither 00000000 nor 01111111 is allowed as a network ID for a class A IP address.

Class B network IDs are used for medium to large local networks. The two left-most bits are set to 10, and the left-most 16 bits give the full network ID. The remaining 16 bits are used for individual host IDs.

Class C network IDs are used for small local networks. The three left-most bits are set to 110, and the left-most 24 bits give the full network ID. The remaining 8 bits are used for individual host IDs.

a. Check that the dotted decimal form of 10001100 11000000 00100000 10001000 is 140.192.32.136.

b. How many Class B networks can there be?

c. What is the dotted decimal form of the IP address for a computer in a Class B network?

d. How many host IDs can there be for a Class B network?

Solution

a. $10001100 = 1 \cdot 2^7 + 1 \cdot 2^3 + 1 \cdot 2^2 = 128 + 8 + 4 = 140$

$11000000 = 1 \cdot 2^7 + 1 \cdot 2^6 = 128 + 64 = 192$

$00100000 = 1 \cdot 2^5 = 32$

$10001000 = 1 \cdot 2^7 + 1 \cdot 2^3 = 128 + 8 = 136$

b. The network ID for a Class B network consists of 16 bits and begins with 10. Because there are two choices for each of the remaining 14 positions (either 0 or 1), the total number of possible network IDs is 2^{14}, or 16,384.

c. The network ID part of a Class B IP address goes from

$$10000000\ 00000000 \quad \text{to} \quad 10111111\ 11111111.$$

As dotted decimals, these numbers range from 128.0 to 191.255 because $10000000_2 = 128_{10}$, $00000000_2 = 0_{10}$, $10111111_2 = 191_{10}$, and $11111111_2 = 255_{10}$. Thus the dotted decimal form of the IP address of a computer in a Class B network is $w.x.y.z$, where $128 \le w \le 191, 0 \le x \le 255, 0 \le y \le 255$, and $0 \le z \le 255$. However, y and z are not allowed both to be 0 or both to be 255 because host IDs may not consist of either all 0's or all 1's.

d. For a class B network, 16 bits are used for host IDs. Having two choices (either 0 or 1) for each of 16 positions gives a potential total of 2^{16}, or 65,536, host IDs. But because two of these are not allowed (all 0's and all 1's), the total number of host IDs is 65,534. ∎

The Inclusion/Exclusion Rule

The addition rule says how many elements are in a union of sets if the sets are mutually disjoint. Now consider the question of how to determine the number of elements in a union of sets when some of the sets overlap. For simplicity, begin by looking at a union of two sets A and B, as shown in Figure 6.3.5.

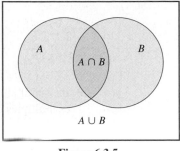

Figure 6.3.5

First observe that the number of elements in $A \cup B$ varies according to the number of elements the two sets have in common. If A and B have no elements in common, then $N(A \cup B) = N(A) + N(B)$. If A and B coincide, then $N(A \cup B) = N(A)$. Thus any general formula for $N(A \cup B)$ must contain a reference to the number of elements the two sets have in common, $N(A \cap B)$, as well as to $N(A)$ and $N(B)$.

The simplest way to derive a formula for $N(A \cup B)$ is to reason as follows:* The number $N(A)$ counts the elements that are in A and not in B and also the elements that are in both A and B. Similarly, the number $N(B)$ counts the elements that are in B and not in A and also the elements that are in both A and B. Hence when the two numbers $N(A)$ and $N(B)$ are added, the elements that are in both A and B are counted twice. To get an accurate count of the elements in $A \cup B$, it is necessary to subtract the number of elements that are in both A and B. Because these are the elements in $A \cap B$,

$$N(A \cup B) = N(A) + N(B) - N(A \cap B).$$

A similar analysis gives a formula for the number of elements in a union of three sets, as shown in Theorem 6.3.3.

Theorem 6.3.3 The Inclusion/Exclusion Rule for Two or Three Sets

If A, B, and C are any finite sets, then

$$N(A \cup B) = N(A) + N(B) - N(A \cap B)$$

and

$$N(A \cup B \cup C) = N(A) + N(B) + N(C) - N(A \cap B) - N(A \cap C)$$
$$-N(B \cap C) + N(A \cap B \cap C).$$

It can be shown using mathematical induction (see exercise 36 at the end of this section) that formulas analogous to those of Theorem 6.3.3 hold for unions of any finite number of sets.

Example 6.3.6 Counting Elements of a General Union

a. How many integers from 1 through 1,000 are multiples of 3 or multiples of 5?

b. How many integers from 1 through 1,000 are neither multiples of 3 nor multiples of 5?

*An alternative proof is outlined in exercise 34 at the end of this section.

Solution

a. Let A = the set of all integers from 1 through 1,000 that are multiples of 3.
   Let B = the set of all integers from 1 through 1,000 that are multiples of 5.

Then

$A \cup B$ = the set of all integers from 1 through 1,000 that are multiples of 3 or multiples of 5

and

$A \cap B$ = the set of all integers from 1 through 1,000 that are multiples of both 3 and 5

= the set of all integers from 1 through 1,000 that are multiples of 15.

[*Now calculate $N(A)$, $N(B)$, and $N(A \cap B)$ and use the inclusion/exclusion rule to solve for $N(A \cup B)$.*]

Because every third integer from 3 through 999 is a multiple of 3, each can be represented in the form $3k$, for some integer k from 1 through 333. Hence there are 333 multiples of 3 from 1 through 1,000, and so $N(A) = 333$.

$$
\begin{array}{ccccccccccc}
1 & 2 & 3 & 4 & 5 & 6 & \ldots & 996 & 997 & 998 & 999 \\
  &   & \updownarrow &   &   & \updownarrow &   & \updownarrow &   &   & \updownarrow \\
  &   & 3 \cdot 1 &   &   & 3 \cdot 2 &   & 3 \cdot 332 &   &   & 3 \cdot 333
\end{array}
$$

Similarly, each multiple of 5 from 1 through 1,000 has the form $5k$, for some integer k from 1 through 200.

$$
\begin{array}{cccccccccccccc}
1 & 2 & 3 & 4 & 5 & 6 & 7 & 8 & 9 & 10 & \ldots & 995 & 996 & 997 & 998 & 999 & 1,000 \\
  &   &   &   & \updownarrow &   &   &   &   & \updownarrow &   & \updownarrow &   &   &   &   & \updownarrow \\
  &   &   &   & 5 \cdot 1 &   &   &   &   & 5 \cdot 2 &   & 5 \cdot 199 &   &   &   &   & 5 \cdot 200
\end{array}
$$

Thus there are 200 multiples of 5 from 1 through 1,000 and $N(B) = 200$.
Finally, each multiple of 15 from 1 through 1,000 has the form $15k$, for some integer k from 1 through 66 (since $990 = 66 \cdot 15$).

$$
\begin{array}{ccccccccccccc}
1 & 2 & \ldots & 15 & \ldots & 30 & \ldots & 975 & \ldots & 990 & \ldots & 999 & 1,000 \\
  &   &   & \updownarrow &   & \updownarrow &   & \updownarrow &   & \updownarrow &   &   & \\
  &   &   & 15 \cdot 1 &   & 15 \cdot 2 &   & 15 \cdot 65 &   & 15 \cdot 66 &   &   &
\end{array}
$$

Hence there are 66 multiples of 15 from 1 through 1,000, and $N(A \cap B) = 66$.
It follows by the inclusion/exclusion rule that

$$
\begin{aligned}
N(A \cup B) &= N(A) + N(B) - N(A \cap B) \\
&= 333 + 200 - 66 \\
&= 467.
\end{aligned}
$$

Thus there are 467 integers from 1 through 1,000 that are multiples of 3 or multiples of 5.

b. There are 1,000 integers from 1 through 1,000, and by part (a), 467 of these are multiples of 3 or multiples of 5. Thus, by the set difference rule, there are $1,000 - 467 = 533$ that are neither multiples of 3 nor multiples of 5. ∎

Note that the solution to part (b) of Example 6.3.6 hid a use of De Morgan's law. The number of elements that are neither in A nor in B is $N(A^c \cap B^c)$, and by De Morgan's law, $A^c \cap B^c = (A \cup B)^c$. So $N((A \cup B)^c)$ was then calculated using the set difference rule: $N((A \cup B)^c) = N(U) - N(A \cup B)$, where the universe U was the set of all integers from 1 through 1,000. Exercises 30–32 at the end of this section explore this technique further.

Example 6.3.7   Counting the Number of Elements in an Intersection

A professor in an advanced computer course takes a survey on the first day of class to determine how many students know certain computer languages. The finding is that out of a total of 50 students in the class,

30 know Java;

18 know C++;

26 know C#;

9 know both Java and C++;

16 know both Java and C#;

8 know both C++ and C#;

47 know at least one of the three languages.

Note that when we write "30 students know Java," we mean that the total number of students who know Java is 30, and we allow for the possibility that some of these students may know one or both of the other languages. If we want to say that 30 students know Java *only* (and not either of the other languages), we will say so explicitly.

a. How many students know none of the three languages?

b. How many students know all three languages?

c. How many students know Java and C++ but not C#? How many students know Java but neither C++ nor C#?

Solution

a. By the difference rule, the number of students who know none of the three languages equals the number in the class minus the number who know at least one language. Thus the number of students who know none of the three languages is

$$50 - 47 = 3.$$

b. Let

$$J = \text{the set of students who know Java}$$
$$P = \text{the set of students who know C++}$$
$$S = \text{the set of students who know C\#.}$$

Then, by the inclusion/exclusion rule,

$$N(J \cup P \cup S) = N(J) + N(P) + N(S) - N(J \cap P) - N(J \cap S)$$
$$- N(P \cap S) + N(J \cap P \cap S)$$

Substituting known values, we get

$$47 = 30 + 26 + 18 - 9 - 16 - 8 + N(J \cap P \cap S).$$

Solving for $N(J \cap P \cap S)$ gives

$$N(J \cap P \cap S) = 6.$$

Hence there are six students who know all three languages. In general, if you know any seven of the eight terms in the inclusion/exclusion formula for three sets, you can solve for the eighth term.

c. To answer the questions of part (c), look at the diagram in Figure 6.3.6.

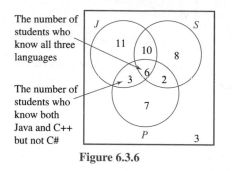

The number of students who know all three languages

The number of students who know both Java and C++ but not C#

Figure 6.3.6

Since $N(J \cap P \cap S) = 6$, put the number 6 inside the innermost region. Then work outward to find the numbers of students represented by the other regions of the diagram. For example, since nine students know both Java and C++ and six know all three languages, $9 - 6 = 3$ students know Java and C++ but not C#. Similarly, since 16 students know Java and C# and six know all three languages, $16 - 6 = 10$ students know Java and C# but not C++. Now the total number of students who know Java is 30. Of these 30, three also know C++ but not C#, ten know C# but not C++, and six know both C++ and C#. That leaves 11 students who know Java but neither of the other two languages.

A similar analysis can be used to fill in the numbers for the other regions of the diagram. ■

Exercise Set 6.3

1. a. How many bit strings consist of from one through four digits? (Strings of different lengths are considered distinct. Thus 10 and 0010 are distinct strings.)
   b. How many bit strings consist of from five through eight digits?

2. a. How many strings of hexadecimal digits consist of from one through three digits? (Recall that hexadecimal numbers are constructed using the 16 digits 0, 1, 2, 3, 4, 5, 6, 7, 8, 9, A, B, C, D, E, F.)
   b. How many strings of hexadecimal digits consist of from two through five digits?

3. a. How many integers from 1 through 999 do not have any repeated digits?
   b. What is the probability that an integer chosen at random from 1 through 999 has at least one repeated digit?

4. How many arrangements in a row of no more than three letters can be formed using the letters of the word *NETWORK* (with no repetitions allowed)?

5. a. How many five-digit integers (integers from 10,000 through 99,999) are divisible by 5?
   b. What is the probability that a five-digit integer chosen at random is divisible by 5?

6. In a certain state, license plates consist of from zero to three letters followed by from zero to four digits, with the provision, however, that a blank plate is not allowed.
   a. How many different license plates can the state produce?
   b. Suppose 85 letter combinations are not allowed because of their potential for giving offense. How many different license plates can the state produce?

✶ **7.** A calculator has an eight-digit display and a decimal point that is located at the extreme right of the number displayed, at the extreme left, or between any pair of digits. The calculator can also display a minus sign at the extreme left of the number. How many distinct numbers can the calculator display? (Note that certain numbers are equal, such as 1.9, 1.90, and 01.900, and should, therefore, not be counted twice.)

8. **a.** Consider the following algorithm segment:

> **for** $i := 1$ **to** 4
>> **for** $j := 1$ **to** i
>> [*Statements in body of inner loop.*
>> *None contain branching statements*
>> *that lead outside the loop.*]
>> **next** j
>
>> **next** i

How many times will the inner loop be iterated when the algorithm is implemented and run?

b. Let n be a positive integer, and consider the following algorithm segment:

> **for** $i := 1$ **to** n
>> **for** $j := 1$ **to** i
>> [*Statements in body of inner loop.*
>> *None contain branching statements*
>> *that lead outside the loop.*]
>> **next** j
>
>> **next** i

How many times will the inner loop be iterated when the algorithm is implemented and run?

9. **a.** How many ways can the letters of the word *QUICK* be arranged in a row?
   b. How many ways can the letters of the word *QUICK* be arranged in a row if the Q and the U must remain next to each other in the order QU?
   c. How many ways can the letters of the word *QUICK* be arranged in a row if the letters QU must remain together but may be in either the order QU or the order UQ?

10. **a.** How many ways can the letters of the word *THEORY* be arranged in a row?
    b. How many ways can the letters of the word *THEORY* be arranged in a row if T and H must remain next to each other as either TH or HT.

11. A group of eight people are attending the movies together.
    a. Two of the eight insist on sitting side-by-side. In how many ways can the eight be seated together in a row?
    b. Two of the people do not like each other and do not want to sit side-by-side. Now how many ways can the eight be seated together in a row?

12. An early compiler recognized variable names according to the following rules: Numeric variable names had to begin with a letter, and then the letter could be followed by another letter or a digit or by nothing at all. String variable names had to begin with the symbol $ followed by a letter, which could then be followed by another letter or a digit or by nothing at all. How many distinct variable names were recognized by this compiler?

H 13. Identifiers in a certain database language must begin with a letter, and then the letter may be followed by other characters, which can be letters, digits, or underscores (_). However, 82 keywords (all consisting of 15 or fewer characters) are reserved and cannot be used as identifiers. How many identifiers with 30 or fewer characters are possible? (Write the answer using summation notation and evaluate it using a formula from Section 4.2.)

14. **a.** If any seven digits could be used to form a telephone number, how many seven-digit telephone numbers would not have any repeated digits?
    b. How many seven-digit telephone numbers would have at least one repeated digit?
    c. What is the probability that a randomly chosen seven-digit telephone number would have at least one repeated digit?

15. **a.** How many strings of four hexadecimal digits do not have any repeated digits?
    b. How many strings of four hexadecimal digits have at least one repeated digit?
    c. What is the probability that a randomly chosen string of four hexadecimal digits has at least one repeated digit?

16. Just as the difference rule gives rise to a formula for the probability of the complement of an event, so the addition and inclusion/exclusion rules give rise to formulas for the probability of the union of mutually disjoint events and for a general union of (not necessarily mutually exclusive) events.
    a. Prove that for mutually disjoint events A and B,
    $$P(A \cup B) = P(A) + P(B).$$
    b. Prove that for any events A and B.
    $$P(A \cup B) = P(A) + P(B) - P(A \cap B).$$

H 17. A combination lock requires three selections of numbers, each from 1 through 39. Suppose the lock is constructed in such a way that no number can be used twice in a row but the same number may occur both first and third. How many different combinations are possible?

✱18. **a.** How many integers from 1 through 100,000 contain the digit 6 exactly once?
    b. How many integers from 1 through 100,000 contain the digit 6 at least once?
    c. If an integer is chosen at random from 1 through 100,000, what is the probability that it contains two or more occurrences of the digit 6?

H ✱ 19. Six new employees, two of whom are married to each other, are to be assigned six desks that are lined up in a row. If the assignment of employees to desks is made randomly, what is the probability that the married couple will have nonadjacent desks? (*Hint:* First find the probability that the couple will have adjacent desks, and then subtract this number from 1.)

✱ 20. Consider strings of length n over the set $\{a, b, c, d\}$.
 a. How many such strings contain at least one pair of adjacent characters that are the same?
 b. If a string of length ten over $\{a, b, c, d\}$ is chosen at random, what is the probability that it contains at least one pair of adjacent characters that are the same?

21. a. How many integers from 1 through 1,000 are multiples of 4 or multiples of 7?
 b. Suppose an integer from 1 through 1,000 is chosen at random. Use the result of part (a) to find the probability that the integer is a multiple of 4 or a multiple of 7.
 c. How many integers from 1 through 1,000 are neither multiples of 4 nor multiples of 7?

22. a. How many integers from 1 through 1,000 are multiples of 2 or multiples of 9?
 b. Suppose an integer from 1 through 1,000 is chosen at random. Use the result of part (a) to find the probability that the integer is a multiple of 2 or a multiple of 9.
 c. How many integers from 1 through 1,000 are neither multiples of 2 nor multiples of 9?

23. Refer to Example 6.3.5.
 a. Write the following IP address in dotted decimal form:

    11001010 00111000 01101011 11101110

 b. How many Class A networks can there be?
 c. What is the dotted decimal form of the IP address for a computer in a Class A network?
 d. How many host IDs can there be for a Class A network?
 e. How many Class C networks can there be?
 f. What is the dotted decimal form of the IP address for a computer in a Class C network?
 g. How many host IDs can there be for a Class C network?
 h. How can you tell, by looking at the first of the four numbers in the dotted decimal form of an IP address, what kind of network the address is from? Explain.
 i. An IP address is 140.192.32.136. What class of network does it come from?
 j. An IP address is 202.56.107.238. What class of network does it come from?

24. Assume that birthdays are equally likely to occur in any one of the 12 months of the year.
 a. Given a group of four people, A, B, C, and D. What is the total number of ways in which birth months could be associated with A, B, C, and D? (For instance, A and B might have been born in May, C in September, and D in February. As another example, A might have been born in January, B in June, C in March, and D in October.)
 b. How many ways could birth months be associated with A, B, C, and D so that no two people would share the same birth month?
 c. How many ways could birth months be associated with A, B, C, and D so that at least two people would share the same birth month?

 d. What is the probability that at least two people out of A, B, C, and D share the same birth month?
 e. How large must n be so that in any group of n people, the probability that two or more share the same birth month is at least 50%?

H 25. Assuming that all years have 365 days and all birthdays occur with equal probability, how large must n be so that in any randomly chosen group of n people, the probability that two or more have the same birthday is at least 1/2? (This is called the **birthday problem.** Many people find the answer surprising.)

26. A market research project studied student readership of certain news magazines by asking students to place checks underneath the names of all news magazines they read occasionally. Out of a sample of 100 students, it was found that 28 checked *Time*, 26 checked *Newsweek*, 14 checked *U.S. News and World Report*, 8 checked both *Time* and *Newsweek*, 4 checked both *Time* and *U.S. News*, 3 checked both *Newsweek* and *U.S. News*, and 2 checked all three. Note that some students who checked *Time* may also have checked one or both of the other magazines. A similar occurrence may be true for the other data.
 a. How many students checked at least one of the magazines?
 b. How many students checked none of the magazines?
 c. Let T be the set of students who checked *Time*, N the set of students who checked *Newsweek*, and U the set of students who checked *U.S. News*. Fill in the numbers for all eight regions of the diagram below.

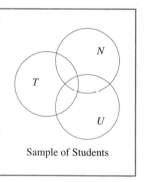

Sample of Students

 d. How many students read *Time* and *Newsweek* but not *U.S. News*?
 e. How many students read *Newsweek* and *U.S. News* but not *Time*?
 f. How many students read *Newsweek* but neither of the other two?

27. A study was done to determine the efficacy of three different drugs—A, B, and C—in relieving headache pain. Over the period covered by the study, 50 subjects were given the

chance to use all three drugs. The following results were obtained:

21 reported relief from drug A.

21 reported relief from drug B.

31 reported relief from drug C.

9 reported relief from both drugs A and B.

14 reported relief from both drugs A and C.

15 reported relief from both drugs B and C.

41 reported relief from at least one of the drugs.

Note that some of the 21 subjects who reported relief from drug A may also have reported relief from drugs B or C. A similar occurrence may be true for the other data.

a. How many people got relief from none of the drugs?

b. How many people got relief from all three drugs?

c. Let A be the set of all subjects who got relief from drug A, B the set of all subjects who got relief from drug B, and C the set of all subjects who got relief from drug C. Fill in the numbers for all eight regions of the diagram below.

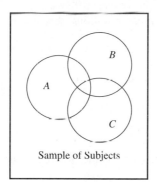

Sample of Subjects

d. How many subjects got relief from A only?

28. An interesting use of the inclusion/exclusion rule is to check survey numbers for consistency. For example, suppose a public opinion polltaker reports that out of a national sample of 1,200 adults, 675 are married, 682 are from 20 to 30 years old, 684 are female, 195 are married and are from 20 to 30 years old, 467 are married females, 318 are females from 20 to 30 years old, and 165 are married females from 20 to 30 years old. Are the polltaker's figures consistent? Could they have occurred as a result of an actual sample survey?

29. Fill in the reasons for each step below. If A and B are sets in a finite universe U, then

$$N(A \cap B) = N(U) - N((A \cap B)^c) \qquad \text{(a)}$$
$$= N(U) - N(A^c \cup B^c) \qquad \text{(b)}$$
$$= N(U) - (N(A^c) + N(B^c) - N(A^c \cap B^c)) \quad \text{(c)}$$

For each of exercises 30–32 below, the number of elements in a certain set can be found by computing the number in some larger universe that are not in the set and subtracting this from the total. In each case, as indicated by exercise 29, De Morgan's laws and the inclusion/exclusion rule can be used to compute the number that are not in the set.

30. How many positive integers less than 1,000 have no common factors with 1,000?

✳ 31. How many permutations of $abcde$ are there in which the first character is a, b, or c and the last character is c, d, or e?

✳ 32. How many integers from 1 through 999,999 contain each of the digits 1, 2, and 3 at least once? (*Hint:* For each $i = 1, 2,$ and 3, let A_i be the set of all integers from 1 through 999,999 that do not contain the digit i.)

33. Use mathematical induction to prove Theorem 6.3.1.

34. Prove the inclusion/exclusion rule for two sets A and B by showing that $A \cup B$ can be partitioned into $A \cap B$, $A - (A \cap B)$, and $B - (A \cap B)$, and then using the addition and difference rules.

35. Prove the inclusion/exclusion rule for three sets.

H ✳ 36. Use mathematical induction to prove the general inclusion/exclusion rule:

If $A_1, A_2, \ldots, A_n$ are finite sets, then
$$N(A_1 \cup A_2 \cup \cdots \cup A_n)$$

$$= \sum_{1 \le i \le n} N(A_i) - \sum_{1 \le i < j \le n} N(A_i \cap A_j)$$

$$+ \sum_{1 \le i < j < k \le n} N(A_i \cap A_j \cap A_k)$$

$$- \cdots + (-1)^{n+1} N(A_1 \cap A_2 \cap \cdots \cap A_n).$$

(The notation $\sum_{1 \le i < j \le n} N(A_i \cap A_j)$ means that quantities of the form $N(A_i \cap A_j)$ are to be added together for all integers i and j with $1 \le i < j \le n$.)

6.4 Counting Subsets of a Set: Combinations

"But 'glory' doesn't mean 'a nice knock-down argument,'" Alice objected. "When I use a word," Humpty Dumpty said, in rather a scornful tone, "it means just what I choose it to mean—neither more nor less." — Lewis Carroll, *Through the Looking Glass*, 1872

Consider the following question:

Suppose five members of a group of twelve are to be chosen to work as a team on a special project. How many distinct five-person teams can be selected?

This question is answered in Example 6.4.5. It is a special case of the following more general question:

Given a set S with n elements, how many subsets of size r can be chosen from S?

The number of subsets of size r that can be chosen from S equals the number of subsets of size r that S has. Each individual subset of size r is called an *r-combination* of the set.

• Definition

Let n and r be nonnegative integers with $r \leq n$. An **r-combination** of a set of n elements is a subset of r of the n elements. The symbol $\binom{n}{r}$, which is read "*n* choose *r*," denotes the number of subsets of size r (r-combinations) that can be chosen from a set of n elements.

Note that on a calculator the symbol $C(n, r)$, $_nC_r$, $C_{n,r}$, or ^nC_r is sometimes used instead of $\binom{n}{r}$.

Example 6.4.1 3-Combinations

Let $S = \{$Ann, Bob, Cyd, Dan$\}$. Each committee consisting of three of the four people in S is a 3-combination of S.

a. List all such 3-combinations of S.   b. What is $\binom{4}{3}$?

Solution

a. Each 3-combination of S is a subset of S of size 3. But each subset of size 3 can be obtained by leaving out one of the elements of S. The 3-combinations are

$\{$Bob, Cyd, Dan$\}$ leave out Ann
$\{$Ann, Cyd, Dan$\}$ leave out Bob
$\{$Ann, Bob, Dan$\}$ leave out Cyd
$\{$Ann, Bob, Cyd$\}$ leave out Dan.

b. Because $\binom{4}{3}$ is the number of 3-combinations of a set with four elements, by part (a),
$\binom{4}{3} = 4$. ∎

There are two distinct methods that can be used to select r objects from a set of n elements. In an **ordered selection,** it is not only what elements are chosen but also the order in which they are chosen that matters. Two ordered selections are said to be the same if the elements chosen are the same and also if the elements are chosen in the same order. An ordered selection of r elements from a set of n elements is an r-permutation of the set.

In an **unordered selection,** on the other hand, it is only the identity of the chosen elements that matters. Two unordered selections are said to be the same if they consist of the same elements, regardless of the order in which the elements are chosen. An unordered selection of r elements from a set of n elements is the same as a subset of size r or an r-combination of the set.

Example 6.4.2 Unordered Selections

How many unordered selections of two elements can be made from the set $\{0, 1, 2, 3\}$?

Solution   An unordered selection of two elements from $\{0, 1, 2, 3\}$ is the same as a 2-combination, or subset of size 2, taken from the set. These can be listed systematically as follows:

$\{0, 1\}, \{0, 2\}, \{0, 3\}$ subsets containing 0

$\{1, 2\}, \{1, 3\}$ subsets containing 1 but not already listed

$\{2, 3\}$ subsets containing 2 but not already listed.

Since this listing exhausts all possibilities, there are six subsets in all. Thus $\binom{4}{2} = 6$, which is the number of unordered selections of two elements from a set of four.   ∎

When the values of n and r are small, it is reasonable to calculate values of $\binom{n}{r}$ using the method of **complete enumeration** (listing all possibilities) illustrated in Examples 6.4.1 and 6.4.2. But when n and r are large, it is not feasible to compute these numbers by listing and counting all possibilities.

The general values of $\binom{n}{r}$ can be found by a somewhat indirect but simple method. An equation is derived that contains $\binom{n}{r}$ as a factor. Then this equation is solved to obtain a formula for $\binom{n}{r}$. The method is illustrated by Example 6.4.3.

Example 6.4.3 Relation between Permutations and Combinations

Write all 2-permutations of the set $\{0, 1, 2, 3\}$. Find an equation relating the number of 2-permutations, $P(4, 2)$, and the number of 2-combinations, $\binom{4}{2}$, and solve this equation for $\binom{4}{2}$.

Solution   According to Theorem 6.2.3, the number of 2-permutations of the set $\{0, 1, 2, 3\}$ is $P(4, 2)$, which equals

$$\frac{4!}{(4-2)!} = \frac{4 \cdot 3 \cdot \cancel{2} \cdot \cancel{1}}{\cancel{2} \cdot \cancel{1}} = 12.$$

Now the act of constructing a 2-permutation of $\{0, 1, 2, 3\}$ can be thought of as a two-step process:

Step 1: Choose a subset of two elements from $\{0, 1, 2, 3\}$.

Step 2: Choose an ordering for the two-element subset.

This process can be illustrated by the possibility tree shown in Figure 6.4.1.

Figure 6.4.1 Relation between Permutations and Combinations

The number of ways to perform step 1 is $\binom{4}{2}$, the same as the number of subsets of size 2 that can be chosen from $\{0, 1, 2, 3\}$. The number of ways to perform step 2 is $2!$, the number of ways to order the elements in a subset of size 2. Because the number of ways of performing the whole process is the number of 2-permutations of the set $\{0, 1, 2, 3\}$, which equals $P(4, 2)$, it follows from the product rule that

$$P(4, 2) = \binom{4}{2} \cdot 2!. \quad \text{\small This is an equation that relates } P(4, 2) \text{ and } \binom{4}{2}.$$

Solving the equation for $\binom{4}{2}$ gives

$$\binom{4}{2} = \frac{P(4, 2)}{2!}$$

Recall that $P(4, 2) = \dfrac{4!}{(4-2)!}$. Hence, substituting yields

$$\binom{4}{2} = \frac{\dfrac{4!}{(4-2)!}}{2!} = \frac{4!}{2!(4-2)!} = 6. \qquad \blacksquare$$

The reasoning used in Example 6.4.3 applies in the general case as well. To form an r-permutation of a set of n elements, first choose a subset of r of the n elements (there are $\binom{n}{r}$ ways to perform this step), and then choose an ordering for the r elements (there are $r!$ ways to perform this step). Thus the number of r-permutations is

$$P(n, r) = \binom{n}{r} \cdot r!.$$

Now solve for $\binom{n}{r}$ to obtain the formula

$$\binom{n}{r} = \frac{P(n, r)}{r!}.$$

Since $P(n, r) = \dfrac{n!}{(n-r)!}$, substitution gives

$$\binom{n}{r} = \frac{\dfrac{n!}{(n-r)!}}{r!} = \frac{n!}{r!(n-r)!}.$$

The result of this discussion is summarized and extended in Theorem 6.4.1.

Theorem 6.4.1

The number of subsets of size r (or r-combinations) that can be chosen from a set of n elements, $\binom{n}{r}$, is given by the formula

$$\binom{n}{r} = \frac{P(n, r)}{r!} \qquad \text{first version}$$

or, equivalently,

$$\binom{n}{r} = \frac{n!}{r!(n - r)!} \qquad \text{second version}$$

where n and r are nonnegative integers with $r \leq n$.

Note that the analysis presented before the theorem proves the theorem in all cases where n and r are positive. If r is zero and n is any nonnegative integer, then $\binom{n}{0}$ is the number of subsets of size zero of a set with n elements. But you know from Section 5.3 that there is only one set that does not have any elements. Consequently, $\binom{n}{0} = 1$. Also

$$\frac{n!}{0!(n - 0)!} = \frac{n!}{1 \cdot n!} = 1$$

since $0! = 1$ by definition. (Remember we said that definition would turn out to be convenient!) Hence the formula

$$\binom{n}{0} = \frac{n!}{0!(n - 0)!}$$

holds for all integers $n \geq 0$, and so the theorem is true for all nonnegative integers n and r with $r \leq n$.

Many electronic calculators have keys for computing values of $\binom{n}{r}$. Theorem 6.4.1 enables you to compute these by hand as well.

Example 6.4.4 Computing $\binom{n}{r}$ by Hand

Compute $\binom{8}{5}$.

Solution By Theorem 6.4.1,

$$\binom{8}{5} = \frac{8!}{5!(8 - 5)!}$$
$$= \frac{8 \cdot 7 \cdot \cancel{6} \cdot \cancel{5} \cdot \cancel{4} \cdot \cancel{3 \cdot 2 \cdot 1}}{(\cancel{5} \cdot \cancel{4} \cdot \cancel{3} \cdot \cancel{2} \cdot 1) \cdot (\cancel{3 \cdot 2 \cdot 1})} \qquad \text{always cancel common factors} \atop \text{before multiplying}$$
$$= 56. \qquad\qquad\qquad\qquad\qquad\qquad ■$$

Example 6.4.5 Calculating the Number of Teams

Consider again the problem of choosing five members from a group of twelve to work as a team on a special project. How many distinct five-person teams can be chosen?

Solution The number of distinct five-person teams is the same as the number of subsets of size 5 (or 5-combinations) that can be chosen from the set of twelve. This number is $\binom{12}{5}$. By Theorem 6.4.1,

$$\binom{12}{5} = \frac{12!}{5!(12 - 5)!} = \frac{\cancel{12} \cdot 11 \cdot \cancel{10} \cdot 9 \cdot 8 \cdot \cancel{7!}}{(\cancel{5} \cdot \cancel{4} \cdot \cancel{3} \cdot \cancel{2} \cdot 1) \cdot \cancel{7!}} = 11 \cdot 9 \cdot 8 = 792.$$

Thus there are 792 distinct five-person teams. ■

The formula for the number of r-combinations of a set can be applied in a wide variety of situations. Some of these are illustrated in the following examples.

Example 6.4.6 Teams That Contain Both or Neither

Suppose two members of the group of twelve insist on working as a pair—any team must contain either both or neither. How many five-person teams can be formed?

Solution   Call the two members of the group that insist on working as a pair A and B. Then any team formed must contain both A and B or neither A nor B. The set of all possible teams can be partitioned into two subsets as shown in Figure 6.4.2 below.

Because a team that contains both A and B contains exactly three other people from the remaining ten in the group, there are as many such teams as there are subsets of three people that can be chosen from the remaining ten. By Theorem 6.4.1, this number is

$$\binom{10}{3} = \frac{10!}{3! \cdot 7!} = \frac{10 \cdot \overset{3}{\cancel{9}} \cdot \overset{4}{\cancel{8}} \cdot \cancel{7!}}{\cancel{3} \cdot \cancel{2} \cdot 1 \cdot \cancel{7!}} = 120.$$

Because a team that contains neither A nor B contains exactly five people from the remaining ten, there are as many such teams as there are subsets of five people that can be chosen from the remaining ten. By Theorem 6.4.1, this number is

$$\binom{10}{5} = \frac{10!}{5! \cdot 5!} = \frac{\cancel{10} \cdot 9 \cdot \overset{2}{\cancel{8}} \cdot 7 \cdot \overset{2}{\cancel{6}} \cdot \cancel{5!}}{\cancel{5} \cdot \cancel{4} \cdot \cancel{3} \cdot \cancel{2} \cdot 1 \cdot \cancel{5!}} = 252.$$

Because the set of teams that contain both A and B is disjoint from the set of teams that contain neither A nor B, by the addition rule,

$$= 120 + 252 = 372.$$

This reasoning is summarized in Figure 6.4.2.

All Possible Five-Person Teams
Containing Both or Neither

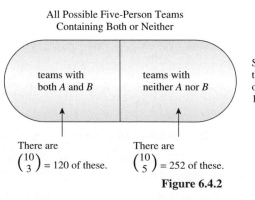

So the total number of teams that contain either both A and B or neither A nor B is $120 + 252 = 372$.

Figure 6.4.2

Example 6.4.7 Teams That Do Not Contain Both

Suppose two members of the group don't get along and refuse to work together on a team. How many five-person teams can be formed?

Solution   Call the two people who refuse to work together C and D. There are two different ways to answer the given question: One uses the addition rule and the other uses the difference rule.

To use the addition rule, partition the set of all teams that don't contain both C and D into three subsets as shown in Figure 6.4.3 below.

Because any team that contains C but not D contains exactly four other people from the remaining ten in the group, by Theorem 6.4.1 the number of such teams is

$$\binom{10}{4} = \frac{10!}{4!(10-4)!} = \frac{10 \cdot \overset{3}{\cancel{9}} \cdot \cancel{8} \cdot 7 \cdot \cancel{6!}}{\cancel{4} \cdot \cancel{3} \cdot \cancel{2} \cdot 1 \cdot \cancel{6!}} = 210.$$

Similarly, there are $\binom{10}{4} = 210$ teams that contain D but not C. Finally, by the same reasoning as in Example 6.4.6, there are 252 teams that contain neither C nor D. Thus, by the addition rule,

$$\begin{bmatrix} \text{number of teams that do} \\ \text{not contain both } C \text{ and } D \end{bmatrix} = 210 + 210 + 252 = 672.$$

This reasoning is summarized in Figure 6.4.3.

All Possible Five-Person Teams
That Do Not Contain Both C and D

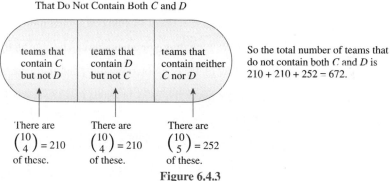

So the total number of teams that do not contain both C and D is $210 + 210 + 252 = 672$.

Figure 6.4.3

The alternative solution by the difference rule is based on the following observation: The set of all five-person teams that don't contain both C and D equals the set difference between the set of all five-person teams and the set of all five-person teams that contain both C and D. By Example 6.4.5, the total number of five-person teams is $\binom{12}{5} = 792$. Thus, by the difference rule,

$$\begin{bmatrix} \text{number of teams that don't} \\ \text{contain both } C \text{ and } D \end{bmatrix} = \begin{bmatrix} \text{total number of} \\ \text{teams of five} \end{bmatrix} - \begin{bmatrix} \text{number of teams that} \\ \text{contain both } C \text{ and } D \end{bmatrix}$$

$$= \binom{12}{5} - \binom{10}{3} = 792 - 120 = 672.$$

This reasoning is summarized in Figure 6.4.4.

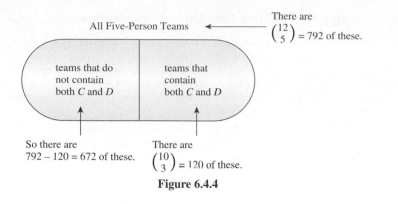

Figure 6.4.4

Before we begin the next example, a remark on the phrases *at least* and *at most* is in order:

> The phrase **at least** *n* means "*n* or more."
> The phrase **at most** *n* means "*n* or fewer."

For instance, if a set consists of three elements and you are to choose at least two, you will choose two or three; if you are to choose at most two, you will choose none, or one, or two.

Example 6.4.8 Teams with Members of Two Types

Suppose the group of twelve consists of five men and seven women.

a. How many five-person teams can be chosen that consist of three men and two women?

b. How many five-person teams contain at least one man?

c. How many five-person teams contain at most one man?

Solution

a. To answer this question, think of forming a team as a two-step process:

Step 1: Choose the men.

Step 2: Choose the women.

There are $\binom{5}{3}$ ways to choose the three men out of the five and $\binom{7}{2}$ ways to choose the two women out of the seven. Hence, by the product rule,

$$\begin{bmatrix} \text{number of teams of five that} \\ \text{contain three men and two women} \end{bmatrix} = \binom{5}{3}\binom{7}{2} = \frac{5!}{3!2!} \cdot \frac{7!}{2!5!}$$

$$= \frac{7 \cdot 6 \cdot 5 \cdot \cancel{4} \cdot \cancel{3} \cdot \cancel{2} \cdot 1}{\cancel{3} \cdot \cancel{2} \cdot 1 \cdot \cancel{2} \cdot 1 \cdot \cancel{2} \cdot 1}$$

$$= 210.$$

b. This question can also be answered either by the addition rule or by the difference rule. The solution by the difference rule is shorter and is shown first.

Observe that the set of five-person teams containing at least one man equals the set difference between the set of all five-person teams and the set of five-person teams that do not contain any men. See Figure 6.4.5 below.

Now a team with no men consists entirely of five women chosen from the seven women in the group, so there are $\binom{7}{5}$ such teams. Also, by Example 6.4.5, the total number of five-person teams is $\binom{12}{5} = 792$. Hence, by the difference rule,

$$
\begin{bmatrix} \text{number of teams} \\ \text{with at least} \\ \text{one man} \end{bmatrix} = \begin{bmatrix} \text{total number} \\ \text{of teams} \\ \text{of five} \end{bmatrix} - \begin{bmatrix} \text{number of teams} \\ \text{of five that do not} \\ \text{contain any men} \end{bmatrix}
$$

$$
= \binom{12}{5} - \binom{7}{5} = 792 - \frac{7!}{5! \cdot 2!}
$$

$$
= 792 - \frac{7 \cdot \overset{3}{\cancel{6}} \cdot \cancel{5!}}{\cancel{5!} \cdot \cancel{2} \cdot 1} = 792 - 21 = 771.
$$

This reasoning is summarized in Figure 6.4.5.

Figure 6.4.5

Alternatively, to use the addition rule, observe that the set of teams containing at least one man can be partitioned as shown in Figure 6.4.6 on page 342. The number of teams in each subset of the partition is calculated using the method illustrated in part (a). There are

$\binom{5}{1}\binom{7}{4}$ teams with one man and four women

$\binom{5}{2}\binom{7}{3}$ teams with two men and three women

$\binom{5}{3}\binom{7}{2}$ teams with three men and two women

$\binom{5}{4}\binom{7}{1}$ teams with four men and one woman

$\binom{5}{5}\binom{7}{0}$ teams with five men and no women.

Hence, by the addition rule,

$$\begin{bmatrix} \text{number of teams with} \\ \text{at least one man} \end{bmatrix}$$

$$= \binom{5}{1}\binom{7}{4} + \binom{5}{2}\binom{7}{3} + \binom{5}{3}\binom{7}{2} + \binom{5}{4}\binom{7}{1} + \binom{5}{5}\binom{7}{0}$$

$$= \frac{5!}{1!4!} \cdot \frac{7!}{4!3!} + \frac{5!}{2!3!} \cdot \frac{7!}{3!4!} + \frac{5!}{3!2!} \cdot \frac{7!}{2!5!} + \frac{5!}{4!1!} \cdot \frac{7!}{1!6!} + \frac{5!}{5!0!} \cdot \frac{7!}{0!7!}$$

$$= \frac{5 \cdot \cancel{4!} \cdot 7 \cdot \cancel{6} \cdot 5 \cdot \cancel{4!}}{\cancel{4!} \cdot 3 \cdot 2 \cdot \cancel{4!}} + \frac{5 \cdot \overset{2}{\cancel{4}} \cdot \cancel{3!} \cdot 7 \cdot \cancel{6} \cdot 5 \cdot \cancel{4!}}{\cancel{3!} \cdot \cancel{2} \cdot \cancel{4!} \cdot 3 \cdot 2} + \frac{5 \cdot \overset{2}{\cancel{4}} \cdot \cancel{3!} \cdot 7 \cdot \overset{3}{\cancel{6}} \cdot \cancel{5!}}{\cancel{2} \cdot \cancel{3!} \cdot \cancel{5!} \cdot 2}$$

$$+ \frac{5 \cdot \cancel{4!} \cdot 7 \cdot \cancel{6!}}{\cancel{4!} \cdot \cancel{6!}} + \frac{\cancel{5!} \cdot \cancel{7!}}{\cancel{5!} \cdot \cancel{7!}}$$

$$= 175 + 350 + 210 + 35 + 1 = 771.$$

This reasoning is summarized in Figure 6.4.6.

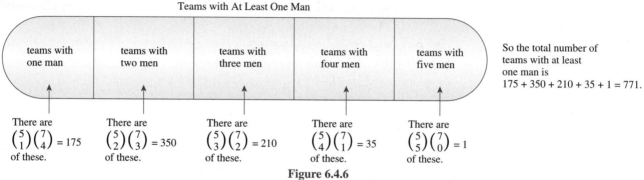

Teams with At Least One Man

| teams with one man | teams with two men | teams with three men | teams with four men | teams with five men |

So the total number of teams with at least one man is $175 + 350 + 210 + 35 + 1 = 771$.

There are $\binom{5}{1}\binom{7}{4} = 175$ of these.

There are $\binom{5}{2}\binom{7}{3} = 350$ of these.

There are $\binom{5}{3}\binom{7}{2} = 210$ of these.

There are $\binom{5}{4}\binom{7}{1} = 35$ of these.

There are $\binom{5}{5}\binom{7}{0} = 1$ of these.

Figure 6.4.6

c. As shown in Figure 6.4.7 below, the set of teams containing at most one man can be partitioned into the set that does not contain any men and the set that contains exactly one man. Hence, by the addition rule,

$$\begin{bmatrix} \text{number of teams} \\ \text{with at} \\ \text{most one man} \end{bmatrix} = \begin{bmatrix} \text{number of} \\ \text{teams without} \\ \text{any men} \end{bmatrix} + \begin{bmatrix} \text{number of} \\ \text{teams with} \\ \text{one man} \end{bmatrix}$$

$$= \binom{5}{0}\binom{7}{5} + \binom{5}{1}\binom{7}{4} = 21 + 175 = 196.$$

This reasoning is summarized in Figure 6.4.7.

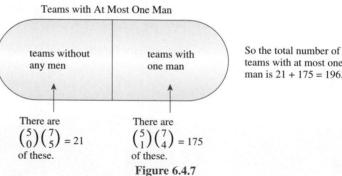

Teams with At Most One Man

| teams without any men | teams with one man |

So the total number of teams with at most one man is $21 + 175 = 196$.

There are $\binom{5}{0}\binom{7}{5} = 21$ of these.

There are $\binom{5}{1}\binom{7}{4} = 175$ of these.

Figure 6.4.7

Example 6.4.9 Poker Hand Problems

The game of poker is played with an ordinary deck of cards (see Example 6.1.1). Various five-card holdings are given special names, and certain holdings beat certain other holdings. The named holdings are listed from highest to lowest below.

Royal flush: 10, J, Q, K, A of the same suit

Straight flush: five adjacent denominations of the same suit but not a royal flush—aces can be high or low, so A, 2, 3, 4, 5 of the same suit is a straight flush.

Four of a kind: four cards of one denomination—the fifth card can be any other in the deck

Full house: three cards of one denomination, two cards of another denomination

Flush: five cards of the same suit but not a straight or a royal flush

Straight: five cards of adjacent denominations but not all of the same suit—aces can be high or low

Three of a kind: three cards of the same denomination and two other cards of different denominations

Two pairs: two cards of one denomination, two cards of a second denomination, and a fifth card of a third denomination

One pair: two cards of one denomination and three other cards all of different denominations

No pairs: all cards of different denominations but not a straight or straight flush or flush

a. How many five-card poker hands contain two pairs?

b. If a five-card hand is dealt at random from an ordinary deck of cards, what is the probability that the hand contains two pairs?

Solution

a. Consider forming a hand with two pairs as a four-step process:

Step 1: Choose the two denominations for the pairs.

Step 2: Choose two cards from the smaller denomination.

Step 3: Choose two cards from the larger denomination.

Step 4: Choose one card from those remaining.

The number of ways to perform step 1 is $\binom{13}{2}$ because there are 13 denominations in all. The number of ways to perform steps 2 and 3 is $\binom{4}{2}$ because there are four cards of each denomination, one in each suit. The number of ways to perform step 4 is $\binom{44}{1}$ because removing the eight cards in the two chosen denominations from the 52 in the deck leaves 44 from which to choose the fifth card. Thus

$$\begin{bmatrix} \text{the total number of} \\ \text{hands with two pairs} \end{bmatrix} = \binom{13}{2}\binom{4}{2}\binom{4}{2}\binom{44}{1}$$

$$= \frac{13!}{2!(13-2)!} \cdot \frac{4!}{2!(4-2)!} \cdot \frac{4!}{2!(4-2)!} \cdot \frac{44!}{1!(44-1)!}$$

$$= \frac{13 \cdot 12 \cdot 11!}{(2 \cdot 1) \cdot 11!} \cdot \frac{4 \cdot 3 \cdot 2!}{(2 \cdot 1) \cdot 2!} \cdot \frac{4 \cdot 3 \cdot 2!}{(2 \cdot 1) \cdot 2!} \cdot \frac{44 \cdot 43!}{1 \cdot 43!}$$

$$= 78 \cdot 6 \cdot 6 \cdot 44 = 123{,}552.$$

b. The total number of five-card hands from an ordinary deck of cards is $\binom{52}{5} = 2{,}598{,}960$. Thus if all hands are equally likely, the probability of obtaining a hand with two pairs is $\dfrac{123{,}552}{2{,}598{,}960} \cong 4.75\%$. ∎

Example 6.4.10 Number of Bit Strings with Fixed Number of 1's

How many eight-bit strings have exactly three 1's?

Solution   To solve this problem, imagine eight empty positions into which the 0's and 1's of the bit string will be placed.

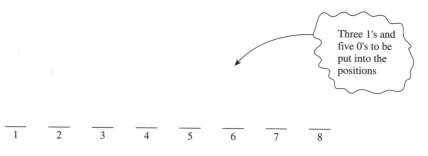

Three 1's and five 0's to be put into the positions

Once a subset of three positions has been chosen from the eight to contain 1's, then the remaining five positions must all contain 0's (since the string is to have exactly three 1's). It follows that the number of ways to construct an eight-bit string with exactly three 1's is the same as the number of subsets of three positions that can be chosen from the eight into which to place the 1's. By Theorem 6.4.1, this equals

$$\binom{8}{3} = \frac{8!}{3! \cdot 5!} = \frac{8 \cdot 7 \cdot 6 \cdot 5!}{3 \cdot 2 \cdot 5!} = 56.$$ ∎

Example 6.4.11 Permutations of a Set with Repeated Elements

Consider various ways of ordering the letters in the word *MISSISSIPPI*:

IIMSSPISSIP,   ISSSPMIIPIS,   PIMISSSSIIP,   and so on.

How many distinguishable orderings are there?

Solution   This example generalizes Example 6.4.10. Imagine placing the 11 letters of *MISSISSIPPI* one after another into 11 positions.

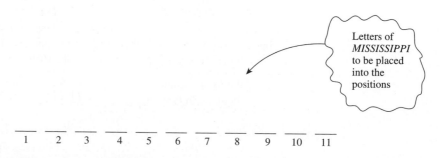

Letters of *MISSISSIPPI* to be placed into the positions

Because copies of the same letter cannot be distinguished from one another, once the positions for a certain letter are known, then all copies of the letter can go into the positions in any order. It follows that constructing an ordering for the letters can be thought of as a four-step process:

Step 1: Choose a subset of four positions for the S's.

Step 2: Choose a subset of four positions for the I's.

Step 3: Choose a subset of two positions for the P's.

Step 4: Choose a subset of one position for the M.

Since there are 11 positions in all, there are $\binom{11}{4}$ subsets of four positions for the S's. Once the four S's are in place, there are seven positions that remain empty, so there are $\binom{7}{4}$ subsets of four positions for the I's. After the I's are in place, there are three positions left empty, so there are $\binom{3}{2}$ subsets of two positions for the P's. That leaves just one position for the M. But $1 = \binom{1}{1}$. Hence by the multiplication rule,

$$\begin{bmatrix} \text{number of ways to} \\ \text{position all the letters} \end{bmatrix} = \binom{11}{4}\binom{7}{4}\binom{3}{2}\binom{1}{1}$$

$$= \frac{11!}{4!\,7!} \cdot \frac{7!}{4!\,3!} \cdot \frac{3!}{2!\,1!} \cdot \frac{1!}{1!\,0!}$$

$$= \frac{11!}{4! \cdot 4! \cdot 2! \cdot 1!} = 34,650. \qquad \blacksquare$$

In exercise 18 at the end of the section, you are asked to show that changing the order in which the letters are placed into the positions does not change the answer to this example.

The same reasoning used in this example can be used to derive the following general theorem.

Theorem 6.4.2

Suppose a collection consists of n objects of which

$\qquad n_1$ are of type 1 and are indistinguishable from each other

$\qquad n_2$ are of type 2 and are indistinguishable from each other

$\qquad \vdots$

$\qquad n_k$ are of type k and are indistinguishable from each other,

and suppose that $n_1 + n_2 + \cdots + n_k = n$. Then the number of distinct permutations of the n objects is

$$\binom{n}{n_1}\binom{n-n_1}{n_2}\binom{n-n_1-n_2}{n_3}\cdots\binom{n-n_1-n_2-\cdots-n_{k-1}}{n_k}$$

$$= \frac{n!}{n_1!\,n_2!\,n_3!\cdots n_k!}.$$

Some Advice about Counting

Students learning counting techniques often ask, "How do I know what to multiply and what to add? When do I use the multiplication rule and when do I use the addition rule?" Unfortunately, these questions have no easy answers. You need to imagine, as vividly as possible, the objects you are to count. You might even start to make an actual list of the items you are trying to count to get a sense for how to obtain them in a systematic way.

You should then construct a model that would allow you to continue counting the objects one by one if you had enough time. If you can imagine the elements to be counted as being obtained through a multistep process (in which each step is performed in a fixed number of ways regardless of how preceding steps were performed), then you can use the multiplication rule. The total number of elements will be the product of the number of ways to perform each step. If, however, you can imagine the set of elements to be counted as being broken up into disjoint subsets, then you can use the addition rule. The total number of elements in the set will be the sum of the number of elements in each subset.

One of the most common mistakes students make is to count certain possibilities more than once.

Example 6.4.12 Double Counting

Consider again the problem of Example 6.4.8(b). A group consists of five men and seven women. How many teams of five contain at least one man?

Caution! The following is a *false solution.* Imagine constructing the team as a two-step process:

Step 1: Choose a subset of one man from the five men.

Step 2: Choose a subset of four others from the remaining eleven people.

Hence, by the multiplication rule, there are $\binom{5}{1} \cdot \binom{11}{4} = 1,650$ five-person teams that contain at least one man.

Analysis of the False Solution: The problem with the solution above is that some teams are counted more than once. Suppose the men are Anwar, Ben, Carlos, Dwayne, and Ed and the women are Fumiko, Gail, Hui-Fan, Inez, Jill, Kim, and Laura. According to the method described above, one possible outcome of the two-step process is as follows:

Outcome of step 1: Anwar

Outcome of step 2: Ben, Gail, Inez, and Jill.

In this case the team would be {Anwar, Ben, Gail, Inez, Jill}. But another possible outcome is

Outcome of step 1: Ben

Outcome of step 2: Anwar, Gail, Inez, and Jill,

which also gives the team {Anwar, Ben, Gail, Inez, Jill}. Thus this one team is given by two different branches of the possibility tree, and so it is counted twice. ∎

The best way to avoid mistakes such as the one described above is to visualize the possibility tree that corresponds to any use of the multiplication rule and the set partition that corresponds to a use of the addition rule. Check how your division into steps works by applying it to some actual data—as was done in the analysis above—and try to pick data that are as typical or generic as possible.

It often helps to ask yourself (1) "Am I counting everything?" and (2) "Am I counting anything twice?" When using the multiplication rule, these questions become (1) "Does every outcome appear as some branch of the tree?" and (2) "Does any outcome appear on more than one branch of the tree?" When using the addition rule, the questions become (1) "Does every outcome appear in some subset of the diagram?" and (2) "Do any two subsets in the diagram share common elements?"

Exercise Set 6.4

1. a. List all 2-combinations for the set $\{x_1, x_2, x_3\}$. Deduce the value of $\binom{3}{2}$.
   b. List all unordered selections of four elements from the set $\{a, b, c, d, e\}$. Deduce the value of $\binom{5}{4}$.

2. a. List all 3-combinations for the set $\{x_1, x_2, x_3, x_4, x_5\}$. Deduce the value of $\binom{5}{3}$.
   b. List all unordered selections of two elements from the set $\{x_1, x_2, x_3, x_4, x_5, x_6\}$. Deduce the value of $\binom{6}{2}$.

3. Write an equation relating $P(7, 2)$ and $\binom{7}{2}$.

4. Write an equation relating $P(8, 3)$ and $\binom{8}{3}$.

5. Compute each of the following.
   a. $\binom{5}{0}$ b. $\binom{5}{1}$ c. $\binom{5}{2}$
   d. $\binom{5}{3}$ e. $\binom{5}{4}$ f. $\binom{5}{5}$

6. A student council consists of 15 students.
   a. In how many ways can a committee of six be selected from the membership of the council?
   b. Two council members have the same major and are not permitted to serve together on a committee. How many ways can a committee of six be selected from the membership of the council?
   c. Two council members always insist on serving on committees together. If they can't serve together, they won't serve at all. How many ways can a committee of six be selected from the council membership?
   d. Suppose the council contains eight men and seven women.
      (i) How many committees of six contain three men and three women?
      (ii) How many committees of six contain at least one woman?
   e. Suppose the council consists of three freshmen, four sophomores, three juniors, and five seniors. How many committees of eight contain two representatives from each class?

7. A computer programming team has 13 members.
   a. How many ways can a group of seven be chosen to work on a project?
   b. Suppose seven team members are women and six are men.
      (i) How many groups of seven can be chosen that contain four women and three men?
      (ii) How many groups of seven can be chosen that contain at least one man?
      (iii) How many groups of seven can be chosen that contain at most three women?
   c. Suppose two team members refuse to work together on projects. How many groups of seven can be chosen to work on a project?

   d. Suppose two team members insist on either working together or not at all on projects. How many groups of seven can be chosen to work on a project?

H 8. An instructor gives an exam with twelve questions. Students are allowed to choose any ten to answer.
   a. How many different choices of ten questions are there?
   b. Suppose five questions require proof and seven do not.
      (i) How many groups of ten questions contain four that require proof and six that do not?
      (ii) How many groups of ten questions contain at least one that requires proof?
      (iii) How many groups of ten questions contain at most three that require proof?
   c. Suppose the exam instructions specify that at most one of questions 1 and 2 may be included among the ten. How many different choices of ten questions are there?
   d. Suppose the exam instructions specify that either both questions 1 and 2 are to be included among the ten or neither is to be included. How many different choices of ten questions are there?

9. A club is considering changing its by-laws. In an initial straw vote on the issue, 24 of the 40 members of the club favored the change and 16 did not. A committee of six is to be chosen from the 40 club members to devote further study to the issue.
   a. How many committees of six can be formed from the club membership?
   b. How many of the committees will contain at least three club members who, in the preliminary survey, favored the change in the by-laws?
   (If you do not have a calculator that computes values of $\binom{n}{r}$, write your answers as numeric expressions using the symbol $\binom{n}{r}$ for some particular values of n and r.)

10. Two new drugs are to be tested using a group of 60 laboratory mice, each tagged with a number for identification purposes. Drug A is to be given to 22 mice, drug B is to be given to another 22 mice, and the remaining 16 mice are to be used as controls. How many ways can the assignment of treatments to mice be made? (A single assignment involves specifying the treatment for each mouse—whether drug A, drug B, or no drug.)

★ 11. Refer to Example 6.4.9. For each poker holding below, (1) find the number of five-card poker hands with that holding; (2) find the probability that a randomly chosen set of five cards has that holding.
   a. royal flush   b. straight flush   c. four of a kind
   d. full house   e. flush   f. straight
   g. three of a kind   h. one pair
   i. no repeated denomination and not of five adjacent denominations

12. How many pairs of two distinct integers chosen from the set $\{1, 2, 3, \ldots, 101\}$ have a sum that is even?

13. A coin is tossed ten times. In each case the outcome H (for heads) or T (for tails) is recorded. (One possible outcome of the ten tossings is denoted $THHTTTHTTH$.)
    a. What is the total number of possible outcomes of the coin-tossing experiment?
    b. In how many of the possible outcomes are exactly five heads obtained?
    c. In how many of the possible outcomes are at least eight heads obtained?
    d. In how many of the possible outcomes is at least one head obtained?
    e. In how many of the possible outcomes is at most one head obtained?

14. a. How many 16-bit strings contain exactly seven 1's?
    b. How many 16-bit strings contain at least thirteen 1's?
    c. How many 16-bit strings contain at least one 1?
    d. How many 16-bit strings contain at most one 1?

15. a. How many even integers are in the set

    $$\{1, 2, 3, \ldots, 100\}?$$

    b. How many odd integers are in the set

    $$\{1, 2, 3, \ldots, 100\}?$$

    c. How many ways can two integers be selected from the set $\{1, 2, 3, \ldots, 100\}$ so that their sum is even?
    d. How many ways can two integers be selected from the set $\{1, 2, 3, \ldots, 100\}$ so that their sum is odd?

16. Suppose that three computer boards in a production run of forty are defective. A sample of five is to be selected to be checked for defects.
    a. How many different samples can be chosen?
    b. How many samples will contain at least one defective board?
    c. What is the probability that a randomly chosen sample of five contains at least one defective board?

17. Ten points labeled $A, B, C, D, E, F, G, H, I, J$ are arranged in a plane in such a way that no three lie on the same straight line.
    a. How many straight lines are determined by the ten points?
    b. How many of these straight lines do not pass through point A?
    c. How many triangles have three of the ten points as vertices?
    d. How many of these triangles do not have A as a vertex?

18. Suppose that you placed the letters in Example 6.4.11 into positions in the following order: first the M, then the I's, then the S's, and then the P's. Show that you would obtain the same answer for the number of distinguishable orderings.

19. a. How many distinguishable ways can the letters of the word $HULLABALOO$ be arranged?
    b. How many distinguishable arrangements of the letters of $HULLABALOO$ begin with U and end with L?

c. How many distinguishable arrangements of the letters of $HULLABALOO$ contain the two letters HU next to each other in order?

20. a. How many distinguishable ways can the letters of the word $MILLIMICRON$ be arranged?
    b. How many distinguishable arrangements of the letters of $MILLIMICRON$ begin with M and end with N?
    c. How many distinguishable arrangements of the letters of $MILLIMICRON$ contain the letters CR next to each other in order and also the letters ON next to each other in order?

21. When the expression $(a + b)^4$ is multiplied out, terms of the form $aaaa, abaa, baba, bbba,$ and so on are obtained. Consider the set S of all strings of length 4 over $\{a, b\}$.
    a. What is $N(S)$? In other words, how many strings of length 4 can be constructed using a's and b's?
    b. How many strings of length 4 over $\{a, b\}$ have three a's and one b?
    c. How many strings of length 4 over $\{a, b\}$ have two a's and two b's?

22. In Morse code, symbols are represented by variable-length sequences of dots and dashes. (For example, $A = \cdot -$, $1 = \cdot - - - -$, $? = \cdot \cdot - - \cdot \cdot$.) How many different symbols can be represented by sequences of seven or fewer dots and dashes?

23. Each symbol in the Braille code is represented by a rectangular arrangement of six dots, each of which may be raised or flat against a smooth background. For instance, when the word Braille is spelled out, it looks like this:

    Given that at least one of the six dots must be raised, how many symbols can be represented in the Braille code?

24. On an 8×8 chessboard, a rook is allowed to move any number of squares either horizontally or vertically. How many different paths can a rook follow from the bottom-left square of the board to the top-right square of the board if all moves are to the right or upward?

25. The number 42 has the prime factorization $2 \cdot 3 \cdot 7$. Thus 42 can be written in four ways as a product of two positive integer factors: $1 \cdot 42, 6 \cdot 7, 14 \cdot 3,$ and $2 \cdot 21$.
    a. List the distinct ways the number 210 can be written as a product of two positive integer factors.
    b. If $n = p_1 p_2 p_3 p_4$, where the p_i are distinct prime numbers, how many ways can n be written as a product of two positive integer factors?
    c. If $n = p_1 p_2 p_3 p_4 p_5$, where the p_i are distinct prime numbers, how many ways can n be written as a product of two positive integer factors?
    d. If $n = p_1 p_2 \cdots p_k$, where the p_i are distinct prime numbers, how many ways can n be written as a product of two positive integer factors?

H ✶ **26.** A student council consists of three freshmen, four sophomores, four juniors, and five seniors. How many committees of eight members of the council contain at least one member from each class?

✶ **27.** An alternative way to derive Theorem 6.4.1 uses the following *division rule:* Let *n* and *k* be integers so that *k* divides *n*. If a set consisting of *n* elements is divided into subsets that each contain *k* elements, then the number of such subsets is *n*/*k*. Explain how Theorem 6.4.1 can be derived using the division rule

28. Find the error in the following reasoning: "Consider forming a poker hand with two pairs as a five-step process.

Step 1: Choose the denomination of one of the pairs.

Step 2: Choose the two cards of that denomination.

Step 3: Choose the denomination of the other of the pairs.

Step 4: Choose the two cards of that second denomination.

Step 5: Choose the fifth card from the remaining denominations.

There are $\binom{13}{1}$ ways to perform step 1, $\binom{4}{2}$ ways to perform step 2, $\binom{12}{1}$ ways to perform step 3, $\binom{4}{2}$ ways to perform step 4, and $\binom{44}{1}$ ways to perform step 5. Therefore, the total number of five-card poker hands with two pairs is $13 \cdot 6 \cdot 12 \cdot 6 \cdot 44 = 247{,}104$."

6.5 *r*-Combinations with Repetition Allowed

The value of mathematics in any science lies more in disciplined analysis and abstract thinking than in particular theories and techniques. — Alan Tucker, 1982

In Section 6.4 we showed that there are $\binom{n}{r}$ *r*-combinations, or subsets of size *r*, of a set of *n* elements. In other words, there are $\binom{n}{r}$ ways to choose *r* distinct elements without regard to order from a set of *n* elements. For instance, there are $\binom{4}{3} = 4$ ways to choose three elements out of a set of four: {1, 2, 3}, {1, 2, 4}, {1, 3, 4}, {2, 3, 4}.

In this section we ask: How many ways are there to choose *r* elements without regard to order from a set of *n* elements *if repetition is allowed?* A good way to imagine this is to visualize the *n* elements as categories of objects from which multiple selections may be made. For instance, if the categories are labeled 1, 2, 3, and 4 and three elements are chosen, it is possible to choose two elements of type 3 and one of type 1, or all three of type 2, or one each of types 1, 2 and 4. We denote such choices by [3, 3, 1], [2, 2, 2], and [1, 2, 4], respectively. Note that because order does not matter, [3, 3, 1] = [3, 1, 3] = [1, 3, 3], for example.

> **• Definition**
>
> An ***r*-combination with repetition allowed**, or **multiset of size *r*,** chosen from a set *X* of *n* elements is an unordered selection of elements taken from *X* with repetition allowed. If $X = \{x_1, x_2, \ldots, x_n\}$, we write an *r*-combination with repetition allowed, or multiset of size *r*, as $[x_{i_1}, x_{i_2}, \ldots, x_{i_r}]$ where each x_{i_j} is in *X* and some of the x_{i_j} may equal each other.

Example 6.5.1   *r*-Combinations with Repetition Allowed

Write a complete list to find the number of 3-combinations with repetition allowed, or multisets of size 3, that can be selected from {1, 2, 3, 4}. Observe that because the order in which the elements are chosen does not matter, the elements of each selection may be written in increasing order, and writing the elements in increasing order will ensure that no combinations are overlooked.

Solution [1, 1, 1] ; [1, 1, 2]; [1, 1, 3]; [1, 1, 4]   all combinations with 1, 1

[1, 2, 2] ; [1, 2, 3]; [1, 2, 4];   all additional combinations with 1, 2

[1, 3, 3] ; [1, 3, 4]; [1, 4, 4];   all additional combinations with 1, 3 or 1, 4

[2, 2, 2] ; [2, 2, 3]; [2, 2, 4];   all additional combinations with 2, 2

[2, 3, 3] ; [2, 3, 4]; [2, 4, 4];   all additional combinations with 2, 3 or 2, 4

[3, 3, 3] ; [3, 3, 4]; [3, 4, 4];   all additional combinations with 3, 3 or 3, 4

[4, 4, 4]   the only additional combination with 4, 4

Thus there are twenty 3-combinations with repetition allowed. ∎

How could the number twenty have been predicted other than by making a complete list? Consider the numbers 1, 2, 3, and 4 as categories and imagine choosing a total of three numbers from the categories with multiple selections from any category allowed. The results of several such selections are represented by the table below.

Category 1	Category 2	Category 3	Category 4	Result of the Selection
	| × |	| × ×		1 from category 2 2 from category 4
× |	| × |	×		1 each from categories 1, 3, and 4
× × × |	|	|		3 from category 1

As you can see, each selection of three numbers from the four categories can be represented by a string of vertical bars and crosses. Three vertical bars are used to separate the four categories, and three crosses are used to indicate how many items from each category are chosen. Each distinct string of three vertical bars and three crosses represents a distinct selection. For instance, the string

$$\times \times \mid \ \mid \times \mid$$

represents the selection: two from category 1, none from category 2, one from category 3, and none from category 4. Thus the number of distinct selections of three elements that can be formed from the set $\{1, 2, 3, 4\}$ with repetition allowed equals the number of distinct strings of six symbols consisting of three $\mid$'s and three $\times$'s. But this equals the number of ways to select three positions out of six because once three positions have been chosen for the $\times$'s, the $\mid$'s are placed in the remaining three positions. Thus the answer is

$$\binom{6}{3} = \frac{6!}{3!(6-3)!} = \frac{6 \cdot 5 \cdot 4 \cdot 3!}{3 \cdot 2 \cdot 1 \cdot 3!} = 20,$$

as was obtained earlier by a careful listing.

The analysis of this example extends to the general case. To count the number of r-combinations with repetition allowed, or multisets of size r, that can be selected from a set of n elements, think of the elements of the set as categories. Then each r-combination with repetition allowed can be represented as a string of $n - 1$ vertical bars (to separate the n categories) and r crosses (to represent the r elements to be chosen). The number of $\times$'s in each category represents the number of times the element represented by that category is repeated.

The number of strings of $n-1$ vertical bars and r crosses is the number of ways to choose r positions, into which to place the r crosses, out of a total of $r+(n-1)$ positions, leaving the remaining positions for the vertical bars. But by Theorem 6.4.1, this number is $\binom{r+n-1}{r}$.

This discussion proves the following theorem.

Theorem 6.5.1

The number of r-combinations with repetition allowed (multisets of size r) that can be selected from a set of n elements is

$$\binom{r+n-1}{r}.$$

This equals the number of ways r objects can be selected from n categories of objects with repetition allowed.

Example 6.5.2 Selecting 15 Cans of Soft Drinks of Five Different Types

A person giving a party wants to set out 15 assorted cans of soft drinks for his guests. He shops at a store that sells five different types of soft drinks.

a. How many different selections of cans of 15 soft drinks can he make?

b. If root beer is one of the types of soft drink, how many different selections include at least six cans of root beer?

c. What is the probability that a randomly chosen selection of 15 soft drinks includes at least six cans of root beer?

Solution

a. Think of the five different types of soft drinks as the n categories and the 15 cans of soft drinks to be chosen as the r objects (so $n=5$ and $r=15$). Each selection of cans of soft drinks is represented by a string of $5-1=4$ vertical bars (to separate the categories of soft drinks) and 15 crosses (to represent the cans selected). For instance, the string

$$\times\times\times\mid\times\times\times\times\times\times\times\mid\ \mid\times\times\times\mid\times\times$$

represents a selection of three cans of soft drinks of type 1, seven of type 2, none of type 3, three of type 4, and two of type 5. The total number of selections of 15 cans of soft drinks of the five types is the number of strings of 19 symbols, $5-1=4$ of them $\mid$ and 15 of them $\times$:

$$\binom{15+5-1}{15}=\binom{19}{15}=\frac{19\cdot\overset{6}{\cancel{18}}\cdot17\cdot\overset{2}{\cancel{16}}\cdot\cancel{15!}}{\cancel{15!}\cdot\cancel{4}\cdot\cancel{3}\cdot\cancel{2}\cdot1}=3{,}876.$$

b. If at least six cans of root beer are included, we can imagine choosing six such cans first and then choosing 9 additional cans. The choice of the nine additional cans can be represented as a string of 9 ×'s and 4 |'s. For example, if root beer is type 1, then the string × × × | | × × | × × × × | represents a selection of three cans of root beer (in addition to the six chosen initially), none of type 2, two of type 3, four of type 4, and none of type 5. Thus the total number of selections of 15 cans of soft drinks of the five types, including at least six cans of root beer, is the number of strings of 13 symbols, 4 (= 5 − 1) of them | and 9 of them ×:

$$\binom{9+4}{9} = \binom{13}{9} = \frac{13 \cdot \cancel{12} \cdot 11 \cdot \overset{5}{\cancel{10}} \cdot \cancel{9!}}{\cancel{9!} \cdot \cancel{4} \cdot \cancel{3} \cdot \cancel{2} \cdot 1} = 715.$$

c. The probability that a randomly chosen selection of cans will include at least six of root beer is the ratio of the number of selections that contain at least six cans of root beer (the answer to (b)) to the total number of selections (the answer to (a)). Therefore, the probability is $715/3,876 \cong 18.45\%$. ∎

Example 6.5.3 Counting Triples (i, j, k) with $1 \le i \le j \le k \le n$

If n is a positive integer, how many triples of integers from 1 through n can be formed in which the elements of the triple are written in increasing order but are not necessarily distinct? In other words, how many triples of integers (i, j, k) are there with $1 \le i \le j \le k \le n$?

Solution   Any triple of integers (i, j, k) with $1 \le i \le j \le k \le n$ can be represented as a string of $n − 1$ vertical bars and three crosses, with the positions of the crosses indicating which three integers from 1 to n are included in the triple. The table below illustrates this for $n = 5$.

1	2	Category 3	4	5	Result of the Selection
	|	| × ×	|	| ×	(3, 3, 5)
×	| ×	|	| ×	|	(1, 2, 4)

Thus the number of such triples is the same as the number of strings of $(n − 1)$ |'s and 3 ×'s, which is

$$\binom{3 + (n-1)}{3} = \binom{n+2}{3} = \frac{(n+2)!}{3!(n+2-3)!}$$

$$= \frac{(n+2)(n+1)n\cancel{(n-1)!}}{3!\cancel{(n-1)!}} = \frac{n(n+1)(n+2)}{6}.$$ ∎

Note that in Examples 6.5.2 and 6.5.3 the reasoning behind Theorem 6.5.1 was used rather than the statement of the theorem itself. Alternatively, in either example we could invoke Theorem 6.5.1 directly by recognizing that the items to be counted either are r-combinations with repetition allowed or are the same in number as such combinations. For instance, in Example 6.5.3 we might observe that there are exactly as many triples of integers (i, j, k) with $1 \le i \le j \le k \le n$ as there are 3-combinations of integers from 1 through n with repetition allowed because the elements of any such 3-combination can be written in increasing order in only one way.

Example 6.5.4 Counting Iterations of a Loop

How many times will the innermost loop be iterated when the algorithm segment below is implemented and run? (Assume n is a positive integer.)

> **for** $k := 1$ **to** n
>> **for** $j := 1$ **to** k
>>> **for** $i := 1$ **to** j
>>>> [*Statements in the body of the inner loop,*
>>>> *none containing branching statements that lead*
>>>> *outside the loop*]
>>> **next** i
>> **next** j
> **next** k

Solution Construct a trace table for the values of k, j, and i for which the statements in the body of the innermost loop are executed. (See the table that follows.) Because i goes from 1 to j, it is always the case that $i \leq j$. Similarly, because j goes from 1 to k, it is always the case that $j \leq k$. To focus on the details of the table construction, consider what happens when $k = 3$. In this case, j takes each value 1, 2, and 3. When $j = 1$, i can only take the value 1 (because $i \leq j$). When $j = 2$, i takes each value 1 and 2 (again because $i \leq j$). When $j = 3$, i takes each value 1, 2, and 3 (yet again because $i \leq j$).

k	1	2 →			3 →						$\cdots$	n →						
j	1	1	2 →		1	2 →	3 →				$\cdots$	1	2 →		$\cdots$	n →		
i	1	1	1	2	1	1	2	1	2	3	$\cdots$	1	1	2	$\cdots$	1	$\cdots$	n

Observe that there is one iteration of the innermost loop for each column of this table, and there is one column of the table for each triple of integers (i, j, k) with $1 \leq i \leq j \leq k \leq n$. But Example 6.5.3 showed that the number of such triples is $[n(n + 1)(n + 2)]/6$. Thus there are $[n(n + 1)(n + 2)]/6$ iterations of the innermost loop. ∎

This solution in Example 6.5.4 is the most elegant and generalizable (see exercises 8 and 9) to the given problem. An alternative solution using summations is outlined in exercise 20.

Example 6.5.5 The Number of Integral Solutions of an Equation

How many solutions are there to the equation $x_1 + x_2 + x_3 + x_4 = 10$ if x_1, x_2, x_3, and x_4 are nonnegative integers?

Solution Think of the number 10 as divided into ten individual units and the variables x_1, x_2, x_3, and x_4 as four categories into which these units are placed. The number of units in each category x_i indicates the value of x_i in a solution of the equation. Each solution can, then, be represented by a string of three vertical bars (to separate the four categories) and ten crosses (to represent the ten individual units). For example, in the following table, the two crosses under x_1, five crosses under x_2, and three crosses under x_4 represent the solution $x_1 = 2$, $x_2 = 5$, $x_3 = 0$, and $x_4 = 3$.

	Categories			Solution to the equation $x_1 + x_2 + x_3 + x_4 = 10$
x_1	x_2	x_3	x_4	
× ×	× × × × × ×		× × ×	$x_1 = 2$, $x_2 = 5$, $x_3 = 0$, and $x_4 = 3$
× × × ×	× × × × × ×			$x_1 = 4$, $x_2 = 6$, $x_3 = 0$, and $x_4 = 0$

Therefore, there are as many solutions to the equation as there are strings of ten crosses and three vertical bars, namely

$$\binom{10 + 3}{10} = \binom{13}{10} = \frac{13!}{10!(13 - 10)!} = \frac{13 \cdot 12 \cdot 11 \cdot 10!}{10! \cdot 3 \cdot 2 \cdot 1} = 286. \qquad \blacksquare$$

Example 6.5.6 illustrates a variation on Example 6.5.5.

Example 6.5.6 Additional Constraints on the Number of Solutions

How many integer solutions are there to the equation $x_1 + x_2 + x_3 + x_4 = 10$ if each $x_i \geq 1$?

Solution   In this case imagine starting by putting one cross in each of the four categories. Then distribute the remaining six crosses among the categories. Such a distribution can be represented by a string of three vertical bars and six crosses. For example, the string

$$\times \times \times \mid \quad \mid \times \times \mid \times$$

indicates that there are three more crosses in category x_1 in addition to the one cross already there (so $x_1 = 4$), no more crosses in category x_2 in addition to the one already there (so $x_2 = 1$), two more crosses in category x_3 in addition to the one already there (so $x_3 = 3$), and one more cross in category x_4 in addition to the one already there (so $x_4 = 2$). It follows that the number of solutions to the equation that satisfy the given condition is the same as the number of strings of three vertical bars and six crosses, namely

$$\binom{6 + 3}{6} = \binom{9}{6} = \frac{9!}{6!(9 - 6)!} = \frac{9 \cdot 8 \cdot 7 \cdot 6!}{6! \cdot 3 \cdot 2 \cdot 1} = 84.$$

An alternative solution to this example is based on the observation that since each $x_i \geq 1$, we may introduce new variables $y_i = x_i - 1$ for each $i = 1, 2, 3, 4$. Then each $y_i \geq 0$, and $y_1 + y_2 + y_3 + y_4 = 6$. Thus the number of solutions of $y_1 + y_2 + y_3 + y_4 = 6$ in nonnegative integers is the same as the number of solutions of $x_1 + x_2 + x_3 + x_4 = 10$ in positive integers. $\qquad \blacksquare$

Which Formula To Use?

Sections 6.2–6.5 have discussed four different ways of choosing k elements from n. The order in which the choices are made may or may not matter, and repetition may or may not be allowed. The following table summarizes which formula to use in which situation.

	Order Matters	Order Does Not Matter
Repetition Is Allowed	n^k	$\binom{n + k - 1}{k}$
Repetition Is Not Allowed	$P(n, k)$	$\binom{n}{k}$

Exercise Set 6.5

1. a. According to Theorem 6.5.1, how many 5-combinations with repetition allowed can be chosen from a set of three elements?

   b. List all of the 5-combinations that can be chosen with repetition allowed from $\{1, 2, 3\}$.

2. a. According to Theorem 6.5.1, how many multisets of size four can be chosen from a set of three elements?

   b. List all of the multisets of size four that can be chosen from the set $\{x, y, z\}$.

3. A bakery produces six different kinds of pastry.

   a. How many different selections of twenty pastries are there?

   b. Assuming that eclairs are one kind of pastry produced, how many different selections of twenty pastries are there if at least three must be eclairs?

   c. If a selection of twenty pastries is chosen randomly, what is the probability that at least three are eclairs?

   d. If a selection of twenty pastries is chosen randomly, what is the probability that exactly three are eclairs?

4. A camera shop stocks eight different types of batteries.

   a. How many ways can a total inventory of 30 batteries be distributed among the eight different types?

   b. Assuming that one of the types of batteries is A76, how many ways can a total inventory of 30 batteries be distributed among the eight different types if the inventory must include at least four A76 batteries?

   c. If an inventory of 30 batteries is selected at random from the eight different types, what is the probability that at least four A76 batteries will be included?

   d. If an inventory of 30 batteries is selected at random from the eight different types, what is the probability that exactly four A76 batteries will be included?

5. If n is a positive integer, how many 4-tuples of integers from 1 through n can be formed in which the elements of the 4-tuple are written in increasing order but are not necessarily distinct? In other words, how many 4-tuples of integers (i, j, k, m) are there with $1 \le i \le j \le k \le m \le n$?

6. If n is a positive integer, how many 5-tuples of integers from 1 through n can be formed in which the elements of the 5-tuple are written in decreasing order but are not necessarily distinct? In other words, how many 5-tuples of integers (h, i, j, k, m) are there with $n \ge h \ge i \ge j \ge k \ge m \ge 1$?

7. Another way to count the number of nonnegative integral solutions to an equation of the form $x_1 + x_2 + \cdots + x_n = m$ is to reduce the problem to one of finding the number of n-tuples $(y_1, y_2, \ldots, y_n)$ with $0 \le y_1 \le y_2 \le \cdots \le y_n \le m$. The reduction results from letting $y_i = x_1 + x_2 + \cdots + x_i$ for each $i = 1, 2, \ldots, n$. Use this approach to derive a general formula for the number of nonnegative integral solutions to $x_1 + x_2 + \cdots + x_n = m$.

In 8 and 9, how many times will the innermost loop be iterated when the algorithm segment is implemented and run? Assume $n, m, k,$ and j are positive integers.

8. for $m := 1$ to n
       for $k := 1$ to m
           for $j := 1$ to k
               for $i := 1$ to j
                   [*Statements in the body of the inner loop, none containing branching statements that lead outside the loop*]
               next i
           next j
       next k
   next m

9. for $k := 1$ to n
       for $j := k$ to n
           for $i := j$ to n
               [*Statements in the body of the inner loop, none containing branching statements that lead outside the loop*]
           next i
       next j
   next k

In 10–14, find how many solutions there are to the given equation that satisfy the given condition.

10. $x_1 + x_2 + x_3 = 20$, each x_i is a nonnegative integer.

11. $x_1 + x_2 + x_3 = 20$, each x_i is a positive integer.

12. $y_1 + y_2 + y_3 + y_4 = 30$, each y_i is a nonnegative integer.

13. $y_1 + y_2 + y_3 + y_4 = 30$, each y_i is an integer that is at least 2.

14. $a + b + c + d + e = 500$, each of $a, b, c, d,$ and e is an integer that is at least 10.

15. a. A store sells 30 kinds of balloons. How many different combinations of 12 balloons can be chosen?

   b. What is the probability that a combination of 12 balloons chosen at random will contain at least one balloon of each kind?

16. A large pile of coins consists of pennies, nickels, dimes, and quarters (at least 30 of each).

   a. How many different collections of 30 coins can be chosen?

   b. What is the probability that a collection of 30 coins chosen at random will contain at least four coins of each type?

✱ 17. For how many integers from 1 through 99,999 is the sum of their digits equal to 9?

✱ 18. Suppose the bakery in exercise 3 has only ten eclairs but has at least twenty of each of the other kinds of pastry.
   a. How many different selections of twenty pastries are there?
   b. Suppose in addition to having only ten eclairs, the bakery has only eight napoleon slices. How many different selections of twenty pastries are there?

✱ 19. Suppose the camera shop in exercise 4 can obtain at most ten A76 batteries but can get at least 30 of each of the other types.

   a. How many ways can a total inventory of 30 batteries be distributed among the eight different types?
   b. Suppose that in addition to being able to obtain only ten A76 batteries, the store can get only six of type D303. How many ways can a total inventory of 30 batteries be distributed among the eight different types?

20. Observe that the number of columns in the trace table for Example 6.5.4 can be expressed as the sum

$$1 + (1 + 2) + (1 + 2 + 3) + \cdots + (1 + 2 + \cdots + n).$$

Explain why this is so, and show how this sum simplifies to the same expression given in the solution of Example 6.5.4.

6.6 The Algebra of Combinations

Let us grant that the pursuit of mathematics is a divine madness of the human spirit, a refuge from the goading urgency of contingent happenings.
— Alfred North Whitehead, 1861–1947

In this section we derive a number of useful formulas that give values of $\binom{n}{r}$ in special cases and explore relations among different values of $\binom{n}{r}$.

Example 6.6.1 Values of $\binom{n}{n}, \binom{n}{n-1}, \binom{n}{n-2}$

Show that for all integers $n \geq 0$,

$$\binom{n}{n} = 1 \qquad\qquad 6.6.1$$

$$\binom{n}{n-1} = n, \quad \text{if } n \geq 1 \qquad\qquad 6.6.2$$

$$\binom{n}{n-2} = \frac{n(n-1)}{2}, \quad \text{if } n \geq 2. \qquad\qquad 6.6.3$$

Solution

$$\binom{n}{n} = \frac{n!}{n!(n-n)!} = \frac{1}{0!} = 1 \qquad \text{since } 0! = 1 \text{ by definition}$$

$$\binom{n}{n-1} = \frac{n!}{(n-1)!(n-(n-1))!}$$

$$= \frac{n \cdot (n-1)!}{(n-1)!(n-n+1)!} = \frac{n}{1} = n$$

$$\binom{n}{n-2} = \frac{n!}{(n-2)!(n-(n-2))!}$$

$$= \frac{n \cdot (n-1) \cdot (n-2)!}{(n-2)!2!} = \frac{n(n-1)}{2} \qquad ■$$

Note that the result derived algebraically above, that $\binom{n}{n}$ equals 1, agrees with the fact that a set with n elements has just one subset of size n, namely itself. Similarly, exercise 1 at the end of the section asks you to show algebraically that $\binom{n}{0} = 1$, which agrees with the fact that a set with n elements has one subset, the empty set, of size 0. In exercise 2 you are also asked to show algebraically that $\binom{n}{1} = n$. This result agrees with the fact that there are n subsets of size 1 that can be chosen from a set with n elements, namely the subsets consisting of each element taken alone.

Example 6.6.2 $\binom{n}{r} = \binom{n}{n-r}$

In exercise 5 at the end of the section you are asked to verify algebraically that

$$\binom{n}{r} = \binom{n}{n-r}$$

for all nonnegative integers n and r with $r \leq n$.

An alternative way to deduce this formula is to interpret it as saying that a set A with n elements has exactly as many subsets of size r as it has subsets of size $n - r$. Derive the formula using this reasoning.

Solution   Observe that any subset of size r can be specified either by saying which r elements lie in the subset or by saying which $n - r$ elements lie outside the subset.

A, A Set with n Elements

Any subset B with r elements completely determines a subset, $A - B$, with $n - r$ elements.

Suppose A has k subsets of size r: $B_1, B_2, \ldots, B_k$. Then each B_i can be paired up with exactly one set of size $n - r$, namely its complement $A - B_i$ as shown below.

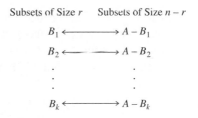

All subsets of size r are listed in the left-hand column, and all subsets of size $n - r$ are listed in the right-hand column. The number of subsets of size r equals the number of subsets of size $n - r$, and so $\binom{n}{r} = \binom{n}{n-r}$. ∎

The type of reasoning used in this example is called *combinatorial,* which means that a result is obtained by counting things that are combined in different ways. A number of theorems have both combinatorial proofs and proofs that are purely algebraic.

Example 6.6.3 New Formulas from Old by Substitution

The formulas established in Example 6.6.1 are true for all integers n in some specified range. For example, formula (6.6.3) states that

$$\binom{n}{n-2} = \frac{n(n-1)}{2} \quad \text{for all integers } n \geq 2. \qquad 6.6.4$$

The letter n in this formula is a dummy variable; it can be replaced by any other symbol or expression as long as each occurrence is replaced and the new symbol or expression represents an integer that is at least 2. Write the formulas obtained by substituting each of the following for n: $m+1, s-1$, and $n+2$. Simplify the result, and give the range of values of each variable for which the formula holds.

Solution

a. $\binom{m+1}{(m+1)-2} = \frac{(m+1)((m+1)-1)}{2}$ for all integers $(m+1) \geq 2$.

So $\binom{m+1}{m-1} = \frac{m(m+1)}{2}$ for all integers $m \geq 1$.

b. $\binom{s-1}{(s-1)-2} = \frac{(s-1)((s-1)-1)}{2}$ for all integers $(s-1) \geq 2$.

So $\binom{s-1}{s-3} = \frac{(s-1)(s-2)}{2}$ for all integers $s \geq 3$.

c. $\binom{n+2}{(n+2)-2} = \frac{(n+2)((n+2)-1)}{2}$ for all integers $n+2 \geq 2$.

So $\binom{n+2}{n} = \frac{(n+1)(n+2)}{2}$ for all integers $n \geq 0$. ■

Pascal's Formula

Blaise Pascal
(1623–1662)

Hulton-Deutch Collection/CORBIS

Pascal's formula, named after the seventeenth-century French mathematician and philosopher Blaise Pascal, is one of the most famous and useful in combinatorics (which is the formal term for the study of counting and listing problems). It relates the value of $\binom{n+1}{r}$ to the values of $\binom{n}{r-1}$ and $\binom{n}{r}$. Specifically, it says that

$$\binom{n+1}{r} = \binom{n}{r-1} + \binom{n}{r}$$

whenever n and r are positive integers with $r \leq n$. This formula makes it easy to compute higher combinations in terms of lower ones: If all the values of $\binom{n}{r}$ are known, then the values of $\binom{n+1}{r}$ can be computed for all r such that $0 < r \leq n$.

Pascal's triangle, shown in Table 6.1.1, is a geometric version of Pascal's formula. Sometimes it is simply called the arithmetic triangle because it was used centuries before Pascal by Chinese and Persian mathematicians. But Pascal discovered it independently, and ever since 1654, when he published a treatise that explored many of its features, it has generally been known as Pascal's triangle.

Table 6.6.1 Pascal's Triangle $\left(\text{Values of }\binom{n}{r}\right)$

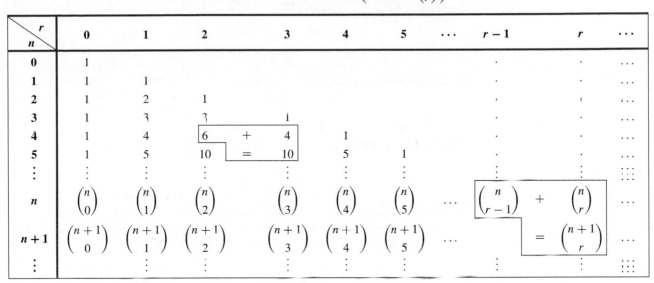

Each entry in the triangle is a value of $\binom{n}{r}$. Pascal's formula translates into the fact that the entry in row $n + 1$, column r equals the sum of the entry in row n, column $r - 1$ plus the entry in row n, column r. That is, the entry in a given interior position equals the sum of the two entries directly above and to the above left. The left-most and right-most entries in each row are 1 because $\binom{n}{n} = 1$ by Example 6.6.1 and $\binom{n}{0} = 1$ by exercise 1 at the end of this section.

Example 6.6.4 Calculating $\binom{n}{r}$ Using Pascal's Triangle

Use Pascal's triangle to compute the values of

$$\binom{6}{2} \quad \text{and} \quad \binom{6}{3}.$$

Solution By construction, the value in row n, column r of Pascal's triangle is the value of $\binom{n}{r}$, for every pair of positive integers n and r with $r \leq n$. By Pascal's formula, $\binom{n+1}{r}$ can be computed by adding together $\binom{n}{r-1}$ and $\binom{n}{r}$, which are located directly above and above left of $\binom{n+1}{r}$. Thus,

$$\binom{6}{2} = \binom{5}{1} + \binom{5}{2} = 5 + 10 = 15 \quad \text{and}$$

$$\binom{6}{3} = \binom{5}{2} + \binom{5}{3} = 10 + 10 = 20. \qquad \blacksquare$$

Pascal's formula can be derived by two entirely different arguments. One is algebraic; it uses the formula for the number of r-combinations obtained in Theorem 6.4.1. The other is combinatorial; it uses the definition of the number of r-combinations as the number of subsets of size r taken from a set with a certain number of elements. We give both proofs since both approaches have applications in many other situations.

Theorem 6.6.1 Pascal's Formula

Let n and r be positive integers and suppose $r \leq n$. Then

$$\binom{n+1}{r} = \binom{n}{r-1} + \binom{n}{r}.$$

Proof (algebraic version):

Let n and r be positive integers with $r \leq n$. By Theorem 6.4.1,

$$\binom{n}{r-1} + \binom{n}{r} = \frac{n!}{(r-1)!(n-(r-1))!} + \frac{n!}{r!(n-r)!}$$

$$= \frac{n!}{(r-1)!(n-r+1)!} + \frac{n!}{r!(n-r)!}.$$

To add these fractions, a common denominator, is needed, so multiply the numerator and denominator of the left-hand fraction by r and multiply the numerator and denominator of the right-hand fraction by $(n-r+1)$. Then

$$\binom{n}{r-1} + \binom{n}{r} = \frac{n!}{(r-1)!(n-r+1)!} \cdot \frac{r}{r} + \frac{n!}{r!(n-r)!} \cdot \frac{(n-r+1)}{(n-r+1)}$$

$$= \frac{r\,n!}{(n-r+1)!r(r-1)!} + \frac{n \cdot n! - r \cdot n! + n!}{(n-r+1)(n-r)!r!}$$

$$= \frac{n! \cdot r + n! \cdot n - n! \cdot r + n!}{(n-r+1)!r!} = \frac{n!(n+1)}{(n+1-r)!r!}$$

$$= \frac{(n+1)!}{((n+1)-r)!r!} = \binom{n+1}{r}.$$

Proof (combinatorial version):

Let n and r be positive integers with $r \leq n$. Suppose S is a set with $n+1$ elements. The number of subsets of S of size r can be calculated by thinking of S as consisting of two pieces: one with n elements $\{x_1, x_2, \ldots, x_n\}$ and the other with one element $\{x_{n+1}\}$.

Any subset of S with r elements either contains x_{n+1} or it does not. If it contains x_{n+1}, then it contains $r-1$ elements from the set $\{x_1, x_2, \ldots, x_n\}$. If it does not contain x_{n+1}, then it contains r elements from the set $\{x_1, x_2, \ldots, x_n\}$.

Subsets of Size r of $\{x_1, x_2, \ldots, x_{n+1}\}$

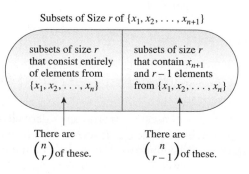

By the addition rule,

$$\begin{bmatrix} \text{number of subsets of} \\ \{x_1, x_2, \ldots, x_n, x_{n+1}\} \\ \text{of size } r \end{bmatrix} = \begin{bmatrix} \text{number of subsets of} \\ \{x_1, x_2, \ldots, x_n\} \\ \text{of size } r-1 \end{bmatrix} + \begin{bmatrix} \text{number of subsets of} \\ \{x_1, x_2, \ldots, x_n\} \\ \text{of size } r \end{bmatrix}.$$

By Theorem 6.4.1, the set $\{x_1, x_2, \ldots, x_n, x_{n+1}\}$ has $\binom{n+1}{r}$ subsets of size r, the set $\{x_1, x_2, \ldots, x_n\}$ has $\binom{n}{r-1}$ subsets of size $r-1$, and the set $\{x_1, x_2, \ldots, x_n\}$ has $\binom{n}{r}$ subsets of size r. Thus

$$\binom{n+1}{r} = \binom{n}{r-1} + \binom{n}{r},$$

as was to be shown.

Example 6.6.5 Deriving New Formulas from Pascal's Formula

Use Pascal's formula to derive a formula for $\binom{n+2}{r}$ in terms of values of $\binom{n}{r}$, $\binom{n}{r-1}$, and $\binom{n}{r-2}$. Assume n and r are nonnegative integers and $2 \le r \le n$.

Solution   By Pascal's formula,

$$\binom{n+2}{r} = \binom{n+1}{r-1} + \binom{n+1}{r}.$$

Now apply Pascal's formula to $\binom{n+1}{r-1}$ and $\binom{n+1}{r}$ and substitute into the above to obtain

$$\binom{n+2}{r} = \left[\binom{n}{r-2} + \binom{n}{r-1}\right] + \left[\binom{n}{r-1} + \binom{n}{r}\right].$$

Combining the two middle terms gives

$$\binom{n+2}{r} = \binom{n}{r-2} + 2\binom{n}{r-1} + \binom{n}{r}$$

for all nonnegative integers n and r such that $2 \le r \le n+2$. ∎

Exercise Set 6.6

In 1–4, use Theorem 6.4.1 to compute the values of the indicated quantities. (Assume n is an integer.)

1. $\binom{n}{0}$, for $n \ge 0$

2. $\binom{n}{1}$, for $n \ge 1$

3. $\binom{n}{2}$, for $n \ge 2$

4. $\binom{n}{3}$, for $n \ge 3$

5. Use Theorem 6.4.1 to prove algebraically that $\binom{n}{r} = \binom{n}{n-r}$, for integers n and r with $0 \le r \le n$. (This can be done by direct calculation; it is not necessary to use mathematical induction.)

Apply substitution to the formulas of Example 6.6.1 to derive the formulas in 6–8. (Assume n, k, and r are integers.)

6. $\binom{n+k}{n+k-1} = n+k$, for $n+k \ge 1$

7. $\binom{n+3}{n+1} = \dfrac{(n+3)(n+2)}{2}$, for $n \ge -1$

8. $\binom{k-r}{k-r} = 1$, for $k-r \ge 0$

9. Use Pascal's triangle given in Table 6.6.1 to compute the values of $\binom{6}{4}$ and $\binom{6}{5}$.

10. Complete the row of Pascal's triangle that corresponds to $n = 7$.

11. The row of Pascal's triangle that corresponds to $n = 8$ is as follows:

    1   8   28   56   70   56   28   8   1.

    What is the row that corresponds to $n = 9$?

12. Use Pascal's formula repeatedly to derive a formula for $\binom{n+3}{r}$ in terms of values of $\binom{n}{k}$ with $k \le r$. (Assume n and r are integers with $n + 3 \ge r \ge 0$.)

13. Prove that for all nonnegative integers n and r with $r + 1 \leq n$,

$$\binom{n}{r+1} = \frac{n-r}{r+1}\binom{n}{r}.$$

14. Prove by mathematical induction that if n is an integer and $n \geq 1$, then

$$\sum_{i=2}^{n+1}\binom{i}{2} = \binom{2}{2} + \binom{3}{2} + \cdots + \binom{n+1}{2}$$
$$= \binom{n+2}{3}.$$

H 15. Prove that if n is an integer and $n \geq 1$, then

$$1\cdot 2 + 2\cdot 3 + \cdots + n(n+1) = 2\binom{n+2}{3}.$$

16. Prove the following generalization of exercise 14: Let r be a fixed nonnegative integer. For all integers n with $n \geq r$,

$$\sum_{i=r}^{n}\binom{i}{r} = \binom{n+1}{r+1}.$$

17. The sequence of **Catalan numbers,** named after the Belgian mathematician Eugène Catalan (1814–1894), arises in a variety of different contexts. It is defined as follows: For each integer $n \geq 1$,

$$C_n = \frac{1}{n+1}\binom{2n}{n}.$$

  a. Find C_1, C_2, and C_3.
  b. Prove that $C_n = \frac{1}{4n+2}\binom{2n+2}{n+1}$, for any integer $n \geq 1$.

18. Think of a set with $m + n$ elements as composed of two parts, one with m elements and the other with n elements. Give a combinatorial argument to show that

$$\binom{m+n}{r} = \binom{m}{0}\binom{n}{r} + \binom{m}{1}\binom{n}{r-1} + \cdots + \binom{m}{r}\binom{n}{0},$$

where m and n are positive integers and r is an integer that is less than or equal to both m and n.

This identity gives rise to many useful additional identities involving the quantities $\binom{n}{k}$. Because Alexander Vandermonde published an influential article about it in 1772, it is generally called the *Vandermonde convolution.* However, it was known at least in the 1300s in China by Chu Shih-chieh.

H 19. Prove that for all integers $n \geq 0$,

$$\binom{n}{0}^2 + \binom{n}{1}^2 + \cdots + \binom{n}{n}^2 = \binom{2n}{n}.$$

20. Let m be any nonnegative integer. Use mathematical induction and Pascal's formula to prove that for all integers $n \geq 0$,

$$\binom{m}{0} + \binom{m+1}{1} + \cdots + \binom{m+n}{n} = \binom{m+n+1}{n}.$$

21. Prove that if p is a prime number and r is an integer with $0 < r < p$, then $\binom{p}{r}$ is divisible by p.

Eugène Catalan
(1814–1894)

Academic Royale de Belgique

6.7 The Binomial Theorem

I'm very well acquainted, too, with matters mathematical, I understand equations both
the simple and quadratical. About binomial theorem I am teaming with a lot of news,
With many cheerful facts about the square of the hypotenuse.
— William S. Gilbert, *The Pirates of Penzance*, 1880

In algebra a sum of two terms, such as $a + b$, is called a **binomial.** The *binomial theorem* gives an expression for the powers of a binomial $(a + b)^n$, for each positive integer n and all real numbers a and b.

Consider what happens when you calculate the first few powers of $a + b$. According to the distributive law of algebra, you take the sum of the products of all combinations of individual terms:

$$(a + b)^2 = (a + b) \cdot (a + b) = aa + ab + ba + bb,$$
$$(a + b)^3 = (a + b) \cdot (a + b) \cdot (a + b)$$
$$= aaa + aab + aba + abb + baa + bab + bba + bbb,$$
$$(a + b)^4 = \underbrace{(a + b)}_{\substack{1\text{st} \\ \text{factor}}} \cdot \underbrace{(a + b)}_{\substack{2\text{nd} \\ \text{factor}}} \cdot \underbrace{(a + b)}_{\substack{3\text{rd} \\ \text{factor}}} \cdot \underbrace{(a + b)}_{\substack{4\text{th} \\ \text{factor}}}$$

$$= aaaa + aaab + aaba + aabb + abaa + abab + abba + abbb$$
$$+ baaa + baab + baba + babb + bbaa + bbab + bbba + bbbb.$$

Now focus on the expansion of $(a + b)^4$. (It is concrete, and yet it has all the features of the general case.) A typical term of this expansion is obtained by multiplying one of the two terms from the first factor times one of the two terms from the second factor times one of the two terms from the third factor times one of the two terms from the fourth factor. For example, the term $abab$ is obtained by multiplying the a's and b's marked with arrows below.

$$\downarrow \qquad\qquad \downarrow \quad \downarrow \qquad\qquad \downarrow$$
$$(a + b) \cdot (a + b) \cdot (a + b) \cdot (a + b)$$

Since there are two possible values—a or b—for each term selected from one of the four factors, there are $2^4 = 16$ terms in the expansion of $(a + b)^4$.

Now some terms in the expansion are "like terms" and can be combined. Consider all possible orderings of three a's and one b, for example. By the techniques of Section 6.4, there are $\binom{4}{1} = 4$ of them. And each of the four occurs as a term in the expansion of $(a + b)^4$:

$$aaab \quad aaba \quad abaa \quad baaa.$$

By the commutative and associative laws of algebra, each such term equals a^3b, so all four are "like terms." When the like terms are combined, therefore, the coefficient of a^3b equals $\binom{4}{1}$.

Similarly, the expansion of $(a + b)^4$ contains the $\binom{4}{2} = 6$ different orderings of two a's and two b's,

$$aabb \quad abab \quad abba \quad baab \quad baba \quad bbaa,$$

all of which equal a^2b^2, so the coefficient of a^2b^2 equals $\binom{4}{2}$. By a similar analysis, the coefficient of ab^3 equals $\binom{4}{3}$. Also, since there is only one way to order four a's, the coefficient of a^4 is 1 (which equals $\binom{4}{0}$, and since there is only one way to order four b's, the coefficient of b^4 is 1 (which equals $\binom{4}{4}$). Thus, when all of the like terms are combined,

$$(a + b)^4 = \binom{4}{0} a^4 + \binom{4}{1} a^3b + \binom{4}{2} a^2b^2 + \binom{4}{3} ab^3 + \binom{4}{4} b^4$$
$$= a^4 + 4a^3b + 6a^2b^2 + 4ab^3 + b^4.$$

The binomial theorem generalizes this formula to an arbitrary nonnegative integer n.

Theorem 6.7.1 Binomial Theorem

Given any real numbers a and b and any nonnegative integer n,

$$(a+b)^n = \sum_{k=0}^{n} \binom{n}{k} a^{n-k} b^k$$

$$= a^n + \binom{n}{1} a^{n-1} b^1 + \binom{n}{2} a^{n-2} b^2 + \cdots + \binom{n}{n-1} a^1 b^{n-1} + b^n.$$

Note that the second expression equals the first because $\binom{n}{0} = 1$ and $\binom{n}{n} = 1$, for all nonnegative integers n.

It is instructive to see two proofs of the binomial theorem: an algebraic proof and a combinatorial proof. Both require a precise definition of integer power.

• Definition

For any real number a and any nonnegative integer n, the **nonnegative integer powers of a** are defined as follows:

$$a^n = \begin{cases} 1 & \text{if } n = 0 \\ a \cdot a^{n-1} & \text{if } n > 0 \end{cases}$$

In some mathematical subjects, 0^0 is left undefined. Defining it to be 1, as is done here, makes it possible to write general formulas such as $\sum_{i=0}^{n} x^i = \dfrac{1}{1-x}$ without having to exclude values of the variables that result in the expression 0^0.*

The algebraic version of the binomial theorem uses mathematical induction and calls upon Pascal's formula at a crucial point.

Proof of the Binomial Theorem (algebraic version):

Suppose a and b are real numbers. We use mathematical induction and let the property be the equation

$$(a+b)^n = \sum_{k=0}^{n} \binom{n}{k} a^{n-k} b^k.$$

Show that the property is true for $n = 0$: When $n = 0$, the binomial theorem states that

$$(a+b)^0 = \sum_{k=0}^{0} \binom{0}{k} a^{0-k} b^k.$$

---

*See *The Art of Computer Programming, Volume 1: Fundamental Algorithms,* Second Edition, by Donald E. Knuth (Reading, Mass.: Addison-Wesley, 1973), p. 56.

But the left-hand side is $(a+b)^0 = 1$ [*by definition of power*], and the right-hand side is

$$\sum_{k=0}^{0} \binom{0}{k} a^{0-k} b^k = \binom{0}{0} a^{0-0} b^0$$

$$= \frac{0!}{0! \cdot (0-0)!} \cdot 1 \cdot 1 = \frac{1}{1 \cdot 1} = 1$$

also [*since* $0! = 1$, $a^0 = 1$, *and* $b^0 = 1$], Hence the binomial theorem is true for $n = 0$.

Show that for all integers $m \geq 0$, if the property is true for $n = m$ then it is true for $n = m + 1$: Let an integer $m \geq 0$ be given, and suppose the property holds for m. That is, suppose

$$(a+b)^m = \sum_{k=0}^{m} \binom{m}{k} a^{m-k} b^k. \qquad [\textit{This is the inductive hypothesis.}]$$

We need to show that the property holds for $n = m + 1$:

$$(a+b)^{m+1} = \sum_{k=0}^{m+1} \binom{m+1}{k} a^{(m+1)-k} b^k.$$

Now, by definition of the $(m+1)$st power,

$$(a+b)^{m+1} = (a+b) \cdot (a+b)^m,$$

so by substitution from the inductive hypothesis,

$$(a+b)^{m+1} = (a+b) \cdot \sum_{k=0}^{m} \binom{m}{k} a^{m-k} b^k$$

$$= a \cdot \sum_{k=0}^{m} \binom{m}{k} a^{m-k} b^k + b \cdot \sum_{k=0}^{m} \binom{m}{k} a^{m-k} b^k$$

$$= \sum_{k=0}^{m} \binom{m}{k} a^{m+1-k} b^k + \sum_{k=0}^{m} \binom{m}{k} a^{m-k} b^{k+1}$$

by the generalized distributive law and the facts that $a \cdot a^{m-k} = a^{1+m-k} = a^{m+1-k}$ and $b \cdot b^k = b^{1+k} = b^{k+1}$.

We transform the second summation on the right-hand side by making the change of variable $j = k + 1$. When $k = 0$, then $j = 1$. When $k = m$, then $j = m + 1$. And since $k = j - 1$, the general term is

$$\binom{m}{k} a^{m-k} b^{k+1} = \binom{m}{j-1} a^{m-(j-1)} b^j = \binom{m}{j-1} a^{m+1-j} b^j.$$

Hence the second summation on the right-hand side above is

$$\sum_{j=1}^{m+1} \binom{m}{j-1} a^{m+1-j} b^j.$$

continued on page 366

But the j in this summation is a dummy variable; it can be replaced by the letter k, as long as the replacement is made everywhere the j occurs:

$$\sum_{j=1}^{m+1} \binom{m}{j-1} a^{m+1-j} b^j = \sum_{k=1}^{m+1} \binom{m}{k-1} a^{m+1-k} b^k.$$

Substituting back, we get

$$(a+b)^{m+1} = \sum_{k=0}^{m} \binom{m}{k} a^{m+1-k} b^k + \sum_{k=1}^{m+1} \binom{m}{k-1} a^{m+1-k} b^k.$$

[*The reason for the above maneuvers was to make the powers of a and b agree so that we can add the summations together term by term, except for the first and the last terms, which we must write separately.*]

Thus

$$(a+b)^{m+1} = \binom{m}{0} a^{m+1-0} b^0 + \sum_{k=1}^{m} \left[\binom{m}{k} + \binom{m}{k-1} \right] a^{m+1-k} b^k$$

$$+ \binom{m}{(m+1)-1} a^{m+1-(m+1)} b^{m+1}$$

$$= a^{m+1} + \sum_{k=1}^{m} \left[\binom{m}{k} + \binom{m}{k-1} \right] a^{m+1-k} b^k + b^{m+1}$$

since $a^0 = b^0 = 1$ and $\binom{m}{0} = \binom{m}{m} = 1.$

But

$$\left[\binom{m}{k} + \binom{m}{k-1} \right] = \binom{m+1}{k} \qquad \text{by Pascal's formula.}$$

Hence

$$(a+b)^{m+1} = a^{m+1} + \sum_{k=1}^{m} \binom{m+1}{k} a^{(m+1)-k} b^k + b^{m+1}$$

$$= \sum_{k=0}^{m+1} \binom{m+1}{k} a^{(m+1)-k} b^k \qquad \begin{array}{l} \text{because } \binom{m+1}{0} = \binom{m+1}{m+1} = 1 \\ \text{and } \binom{m+1}{0} = \binom{m+1}{m+1} = 1, \end{array}$$

which is what we needed to show.

It is instructive to write out the product $(a+b) \cdot (a+b)^m$ without using the summation notation but using the inductive hypothesis about $(a+b)^m$:

$$(a+b)^{m+1} = (a+b) \cdot \left[a^m + \binom{m}{1} a^{m-1} b + \cdots + \binom{m}{k-1} a^{m-(k-1)} b^{k-1} \right.$$

$$\left. + \binom{m}{k} a^{m-k} b^k + \cdots + \binom{m}{m-1} ab^{m-1} + b^m \right].$$

You will see that the first and last coefficients are clearly 1 and that the term containing $a^{m+1-k} b^k$ is obtained from multiplying $a^{m-k} b^k$ by a and $a^{m-(k-1)} b^{k-1}$ by b [*because* $m+1-k = m-(k-1)$]. Hence the coefficient of $a^{m+1-k} b^k$ equals the sum of $\binom{m}{k}$ and $\binom{m}{k-1}$. This is the crux of the algebraic proof.

If n and r are nonnegative integers and $r \leq n$, then $\binom{n}{r}$ is called a **binomial coefficient** because it is one of the coefficients in the expansion of the binomial expression $(a + b)^n$. The combinatorial proof of the binomial theorem follows.

Proof of Binomial Theorem (combinatorial version):

[*The combinatorial argument used here to prove the binomial theorem works only for $n \geq 1$. If we were giving only this combinatorial proof, we would have to prove the case $n = 0$ separately. Since we have already given a complete algebraic proof that includes the case $n = 0$, we do not prove it again here.*]

Let a and b be real numbers and n an integer that is at least 1. The expression $(a + b)^n$ can be expanded into products of n letters, where each letter is either a or b. For each $k = 0, 1, 2, \ldots, n$, the product

$$a^{n-k}b^k = \underbrace{a \cdot a \cdot a \cdots a}_{n - k \text{ factors}} \cdot \underbrace{b \cdot b \cdot b \cdots b}_{k \text{ factors}}$$

occurs as a term in the sum the same number of times as there are orderings of $(n - k)$ a's and k b's. But this number is $\binom{n}{k}$, the number of ways to choose k positions into which to place the b's. [*The other $n - k$ positions will be filled by a's.*] Hence, when the like terms are combined, the coefficient of $a^{n-k}b^k$ in the sum is $\binom{n}{k}$. Thus

$$(a + b)^n = \sum_{k=0}^{n} \binom{n}{k} a^{n-k}b^k.$$

This is what was to be proved.

Example 6.7.1 Substituting into the Binomial Theorem

Expand the following expressions using the binomial theorem:

a. $(a + b)^5$ b. $(x - 4y)^4$

Solution

a. $(a + b)^5 = \displaystyle\sum_{k=0}^{5} \binom{5}{k} a^{5-k}b^k$

$= a^5 + \binom{5}{1} a^{5-1}b^1 + \binom{5}{2} a^{5-2}b^2 + \binom{5}{3} a^{5-3}b^3 + \binom{5}{4} a^{5-4}b^4 + b^5$

$= a^5 + 5a^4b + 10a^3b^2 + 10a^2b^3 + 5ab^4 + b^5$

b. Observe that $(x - 4y)^4 = (x + (-4y))^4$. So let $a = x$ and $b = (-4y)$, and substitute into the binomial theorem.

$(x - 4y)^4 = \displaystyle\sum_{k=0}^{4} \binom{4}{k} x^{4-k}(-4y)^k$

$= x^4 + \binom{4}{1} x^{4-1}(-4y)^1 + \binom{4}{2} x^{4-2}(-4y)^2 + \binom{4}{3} x^{4-3}(-4y)^3 + (-4y)^4$

$= x^4 + 4x^3(-4y) + 6x^2(16y^2) + 4x^1(-64y^3) + (256y^4)$

$= x^4 - 16x^3y + 96x^2y^2 - 256xy^3 + 256y^4$ ∎

Example 6.7.2 Deriving Another Combinatorial Identity from the Binomial Theorem

Use the binomial theorem to show that

$$2^n = \sum_{k=0}^{n} \binom{n}{k} = \binom{n}{0} + \binom{n}{1} + \binom{n}{2} + \cdots + \binom{n}{n}$$

for all integers $n \geq 0$.

Solution Since $2 = 1 + 1$, $2^n = (1 + 1)^n$. Apply the binomial theorem to this expression by letting $a = 1$ and $b = 1$. Then

$$2^n = \sum_{k=0}^{n} \binom{n}{k} \cdot 1^{n-k} \cdot 1^k = \sum_{k=0}^{n} \binom{n}{k} \cdot 1 \cdot 1$$

since $1^{n-k} = 1$ and $1^k = 1$. Consequently,

$$2^n = \sum_{k=0}^{n} \binom{n}{k} = \binom{n}{0} + \binom{n}{1} + \binom{n}{2} + \cdots + \binom{n}{n}.$$ ∎

Example 6.7.3 Using a Combinatorial Argument to Derive the Identity

According to Theorem 5.3.5, a set with n elements has 2^n subsets. Apply this fact to give a combinatorial argument to justify the identity

$$\binom{n}{0} + \binom{n}{1} + \binom{n}{2} + \binom{n}{3} + \cdots + \binom{n}{n} = 2^n.$$

Solution Suppose S is a set with n elements. Then every subset of S has some number of elements k, where k is between 0 and n. It follows that the total number of subsets of S, $N(\mathscr{P}(S))$, can be expressed as the following sum:

$$\begin{bmatrix} \text{number of} \\ \text{subsets} \\ \text{of } S \end{bmatrix} = \begin{bmatrix} \text{number of} \\ \text{subsets of} \\ \text{size } 0 \end{bmatrix} + \begin{bmatrix} \text{number of} \\ \text{subsets of} \\ \text{size } 1 \end{bmatrix} + \cdots + \begin{bmatrix} \text{number of} \\ \text{subsets of} \\ \text{size } n \end{bmatrix}.$$

Now the number of subsets of size k of a set with n elements is $\binom{n}{k}$. Hence the

$$\text{number of subsets of } S = \binom{n}{0} + \binom{n}{1} + \binom{n}{2} + \cdots + \binom{n}{n}$$

But by Theorem 5.3.5, S has 2^n subsets. Hence

$$\binom{n}{0} + \binom{n}{1} + \binom{n}{2} + \binom{n}{3} + \cdots + \binom{n}{n} = 2^n.$$ ∎

Example 6.7.4 Using the Binomial Theorem to Simplify a Sum

Express the following sum in **closed form** (without using a summation symbol and without using an ellipsis $\cdots$):

$$\sum_{k=0}^{n} \binom{n}{k} 9^k$$

Solution   When the number 1 is raised to any power, the result is still 1. Thus

$$\sum_{k=0}^{n} \binom{n}{k} 9^k = \sum_{k=0}^{n} \binom{n}{k} 1^{n-k} 9^k$$

$$= (1+9)^n \qquad \text{by the binomial theorem with } a = 1 \text{ and } b = 9$$

$$= 10^n.$$

∎

Exercise Set 6.7

Expand the expressions in 1–9 using the binomial theorem.

1. $(1+x)^7$ 2. $(p+q)^6$ 3. $(1-x)^6$

4. $(u-v)^5$ 5. $(p-2q)^4$ 6. $(u^2-3v)^4$

7. $\left(x + \dfrac{1}{x}\right)^5$ 8. $\left(\dfrac{3}{a} - \dfrac{a}{3}\right)^5$ 9. $\left(x^2 + \dfrac{1}{x}\right)^5$

10. In Example 6.7.1 it was shown that

$$(a+b)^5 = a^5 + 5a^4b + 10a^3b^2 + 10a^2b^3 + 5ab^4 + b^5.$$

Evaluate $(a+b)^6$ by substituting the expression above into the equation

$$(a+b)^6 = (a+b) \cdot (a+b)^5$$

and then multiplying out and combining like terms.

In 11–16, find the coefficient of the given term when the expression is expanded by the binomial theorem.

11. $x^6 y^3$ in $(x+y)^9$ 12. x^7 in $(2x+3)^{10}$

13. $a^5 b^7$ in $(a-2b)^{12}$ 14. $u^{16}v^4$ in $(u^2-v^2)^{10}$

15. $p^{16}q^7$ in $(3p^2-2q)^{15}$ 16. $x^9 y^{10}$ in $(2x-3y^2)^{14}$

Use the binomial theorem to prove each statement in 17–22.

17. For all integers $n \geq 1$,

$$\binom{n}{0} - \binom{n}{1} + \binom{n}{2} - \cdots + (-1)^n \binom{n}{n} = 0.$$

(Use the fact that $1 + (-1) = 0$.)

H 18. For all integers $n \geq 0$,

$$3^n = \binom{n}{0} + 2\binom{n}{1} + 2^2\binom{n}{2} + \cdots + 2^n\binom{n}{n}.$$

19. For all integers $m \geq 0$, $\displaystyle\sum_{i=0}^{m} (-1)^i \binom{m}{i} 2^{m-i} = 1$.

20. For all integers $n \geq 0$, $\displaystyle\sum_{i=0}^{n} (-1)^i \binom{n}{i} 3^{n-i} = 2^n$.

21. For all integers $n \geq 0$ and for all positive real numbers x, $1 + nx \leq (1+x)^n$.

H 22. For all integers $n \geq 1$,

$$\binom{n}{0} - \frac{1}{2}\binom{n}{1} + \frac{1}{2^2}\binom{n}{2} - \frac{1}{2^3}\binom{n}{3}$$

$$+ \cdots + (-1)^{n-1}\frac{1}{2^{n-1}}\binom{n}{n-1} = \begin{cases} 0 & \text{if } n \text{ is even} \\ \dfrac{1}{2^{n-1}} & \text{if } n \text{ is odd} \end{cases}.$$

23. Use mathematical induction to prove that for all integers $n \geq 1$, if S is a set with n elements, then S has the same number of subsets with an even number of elements as with an odd number of elements. Use this fact to give a combinatorial argument to justify the identity of exercise 17.

Express each of the sums in 24–35 in closed form (without using a summation symbol and without using an ellipsis $\cdots$).

24. $\displaystyle\sum_{k=0}^{n} \binom{n}{k} 5^k$

25. $\displaystyle\sum_{i=0}^{m} \binom{m}{i} 4^i$

26. $\displaystyle\sum_{i=0}^{n} \binom{n}{i} x^i$

27. $\displaystyle\sum_{k=0}^{m} \binom{m}{k} 2^{m-k} x^k$

28. $\displaystyle\sum_{j=0}^{2n} (-1)^j \binom{2n}{j} x^j$

29. $\displaystyle\sum_{r=0}^{n} \binom{n}{r} x^{2r}$

30. $\displaystyle\sum_{i=0}^{m} \binom{m}{i} p^{m-i} q^{2i}$

31. $\displaystyle\sum_{k=0}^{n} \binom{n}{k} \frac{1}{2^k}$

32. $\displaystyle\sum_{i=0}^{m} (-1)^i \binom{m}{i} \frac{1}{2^i}$

33. $\displaystyle\sum_{k=0}^{n} (-1)^k \binom{n}{k} 3^{2n-2k} 2^{2k}$

34. $\displaystyle\sum_{i=0}^{n} (-1)^i \binom{n}{i} 5^{n-i} 2^i$

35. $\displaystyle\sum_{k=0}^{n} (-1)^k \binom{n}{k} 3^{2n-2k} 2^{2k}$

★ 36. (For students who have studied calculus)
   a. Explain how the equation below follows from the binomial theorem:

$$(1+x)^n = \sum_{k=0}^{n} \binom{n}{k} x^k.$$

b. Write the formula obtained by taking the derivative of both sides of the equation in part (a) with respect to x.

c. Use the result of part (b) to derive the formulas below.

(i) $2^{n-1} = \dfrac{1}{n}\left[\dbinom{n}{1} + 2\dbinom{n}{2} + 3\dbinom{n}{3} + \cdots + n\dbinom{n}{n}\right]$

(ii) $\displaystyle\sum_{k=1}^{n} k\dbinom{n}{k}(-1)^k = 0$

d. Express $\displaystyle\sum_{k=1}^{n} k\dbinom{n}{k} 3^k$ in closed form (without using a summation sign or $\cdots$).

6.8 Probability Axioms and Expected Value

The theory of probability is at bottom nothing but common sense reduced to a calculus.
— Pierre-Simon Laplace (1749–1827)

Andrei Nikolaevich Kolmogorov (1903–1987)

Up to this point, you have calculated probabilities only for situations, such as tossing a fair coin or rolling a pair of balanced dice, where the outcomes in the sample space are all equally likely. But coins are not always fair and dice are not always balanced. How is it possible to calculate probabilities for these more general situations?

The following axioms were formulated by A. N. Kolmogorov in 1933 to provide a theoretical foundation for a far-ranging theory of probability. In this section we state the axioms, derive a few consequences, and introduce the notion of expected value.

Recall that a sample space is a set of all outcomes of a random process or experiment and that an event is a subset of a sample space.

Probability Axioms

Let S be a sample space, and let A and B be any events in S. Then

1. $0 \le P(A) \le 1$

2. $P(\emptyset) = 0$ and $P(S) = 1$

3. If A and B are mutually disjoint (that is, if $A \cap B = \emptyset$), then the probability of the union of A and B is

$$P(A \cup B) = P(A) + P(B).$$

Example 6.8.1 Applying the Probability Axioms

Suppose that A and B are events in a sample space S. If A and B are mutually disjoint, could $P(A) = 0.6$ and $P(B) = 0.8$?

Solution No. Probability axiom 3 would imply that $P(A \cup B) = P(A) + P(B) = 0.6 + 0.8 = 1.4$, and since $1.4 > 1$, this result would violate probability axiom 1. ■

Example 6.8.2 The Probability of the Complement of an Event

Suppose that A is an event in a sample space S. Deduce that $P(A^c) = 1 - P(A)$.

Solution By Theorem 5.2.2(5), with S playing the role of the universal set U,

$$A \cap A^c = \emptyset \quad \text{and} \quad A \cup A^c = S.$$

Thus S is the disjoint union of A and A^c, and so

$$P(A \cup A^c) = P(A) + P(A^c) = P(S) = 1.$$

Subtracting $P(A)$ from both sides gives the result that $P(A^c) = 1 - P(A)$. ■

Probability of the Complement of an Event

If A is any event in a sample space S, then

$$P(A^c) = 1 - P(A). \qquad\qquad 6.8.1$$

It is important to check that Kolmogorov's probability axioms are consistent with the results obtained using the equally likely probability formula. To see that this is the case, let S be a finite sample space with outcomes $a_1, a_2, a_3, \ldots, a_n$. It is clear that all the singleton sets $\{a_1\}, \{a_2\}, \{a_3\}, \ldots, \{a_n\}$ are mutually disjoint and that their union is S. Since $P(S) = 1$, probability axiom 3 can be applied multiple times (see exercise 13 at the end of this section) to obtain

$$P(\{a_1\} \cup \{a_2\} \cup \{a_3\} \cup \cdots \cup \{a_n\}) = \sum_{k=1}^{n} P(\{a_k\}) = 1.$$

If, in addition, all the outcomes are equally likely, there is a positive real number c so that

$$P(\{a_1\}) = P(\{a_2\}) = P(\{a_3\}) = \cdots = P(\{a_n\}) = c.$$

Hence

$$1 = \sum_{k=1}^{n} c = \underbrace{c + c + \cdots + c}_{n \text{ terms}} = nc,$$

and thus

$$c = \frac{1}{n}.$$

It follows that if A is any event with outcomes $a_{i_1}, a_{i_2}, a_{i_3}, \ldots, a_{i_m}$, then

$$P(A) = \sum_{k=1}^{m} P(\{a_{i_k}\}) = \sum_{k=1}^{m} \frac{1}{n} = \frac{m}{n} = \frac{N(A)}{N(S)},$$

which is the result given by the equally likely probability formula.

Example 6.8.3 The Probability of a General Union of Two Events

Follow the steps outlined in parts (a) and (b) below to prove the following formula:

Probability of a General Union of Two Events

If S is any sample space and A and B are any events in S, then

$$P(A \cup B) = P(A) + P(B) - P(A \cap B). \qquad\qquad 6.8.2$$

In both steps, suppose that A and B are any events in a sample space S.

a. Show that $A \cup B$ is a disjoint union of the following sets: $A - (A \cap B)$, $B - (A \cap B)$, and $A \cap B$.

b. In exercise 12 at the end of the section, you are asked to prove that for any events U and V in a sample space S, if $U \subseteq V$ then $P(V - U) = P(V) - P(U)$. Use this result and the result of part (a) to finish the proof of the formula.

Solution

a. Refer to Figure 6.8.1 as you read the following explanation. Elements in the set $A - (A \cap B)$ are in the region shaded blue, elements in $B - (A \cap B)$ are in the region shaded gray, and elements in $A \cap B$ are in the white region.

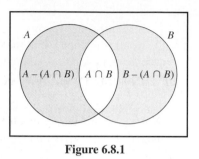

Figure 6.8.1

Part 1: Show that $A \cup B \subseteq (A - (A \cap B)) \cup (B - (A \cap B)) \cup (A \cap B)$: Given any element x in $A \cup B$, x satisfies exactly one of the following three conditions:

$$(1) \ \ x \in A \text{ and } x \in B$$
$$(2) \ \ x \in A \text{ and } x \notin B$$
$$(3) \ \ x \in B \text{ and } x \notin A$$

1. In the first case, $x \in A \cap B$, and so $x \in (A - (A \cap B)) \cup (B - (A \cap B)) \cup (A \cap B)$ by definition of union.

2. In the second case, $x \notin A \cap B$ (because $x \notin B$), and so $x \in A - (A \cap B)$. Therefore $x \in (A - (A \cap B)) \cup (B - (A \cap B)) \cup (A \cap B)$ by definition of union.

3. In the third case, $x \notin A \cap B$ (because $x \notin A$), and hence $x \in B - (A \cap B)$. So, again, $x \in (A - (A \cap B)) \cup (B - (A \cap B)) \cup (A \cap B)$ by definition of union.

Hence, in all three cases, $x \in (A - (A \cap B)) \cup (B - (A \cap B)) \cup (A \cap B)$, which completes the proof of part 1.

Moreover, since the three conditions are mutually exclusive, the three sets $A - (A \cap B)$, $B - (A \cap B)$, and $A \cap B$ are mutually disjoint.

Part 2: Show that $(A - (A \cap B)) \cup (B - (A \cap B)) \cup (A \cap B) \subseteq A \cup B$: Suppose x is any element in $(A - (A \cap B)) \cup (B - (A \cap B)) \cup (A \cap B)$. By definition of union, $x \in A - (A \cap B)$ or $x \in B - (A \cap B)$ or $x \in A \cap B$.

1. In case $x \in A - (A \cap B)$, then $x \in A$ and $x \notin A \cap B$ by definition of set difference. In particular, $x \in A$ and so $x \in A \cup B$.

2. In case $x \in B - (A \cap B)$, then $x \in B$ and $x \notin A \cap B$ by definition of set difference. In particular, $x \in B$ and so $x \in A \cup B$.

3. In case $x \in A \cap B$, then in particular, $x \in A$ and so $x \in A \cup B$.

Hence, in all three cases, $x \in A \cup B$, which completes the proof of part 2.

b. $P(A \cup B) = P((A - (A \cap B)) \cup (B - (A \cap B)) \cup (A \cap B))$ by part (a)

$\qquad\qquad = P(A - (A \cap B)) + P(B - (A \cap B)) + P(A \cap B)$

by exercise 13 at the end of the section and the fact that
$A - (A \cap B)$, $B - (A \cap B)$, and $A \cap B$ are mutually disjoint

$\qquad\qquad = P(A) - P(A \cap B) + P(B) - P(A \cap B) + P(A \cap B)$

by exercise 12 at the end of the section
because $A \cap B \subseteq A$ and $A \cap B \subseteq B$

$\qquad\qquad = P(A) + P(B) - P(A \cap B)$ by algebra. ∎

Example 6.8.4 Computing the Probability of a General Union of Two Events

Suppose a card is chosen at random from an ordinary 52-card deck (see Section 6.1). What is the probability that the card is a face card (jack, queen, or king) or is from one of the red suits (hearts or diamonds)?

Solution   Let A be the event that the chosen card is a face card, and let B be the event that the chosen card is from one of the red suits. The event that the card is a face card or is from one of the red suits is $A \cup B$. Now $N(A) = 4 \cdot 3 = 12$ (because each of the four suits has three face cards), and so $P(A) = 12/52$. Also $N(B) = 26$ (because half the cards are red), and so $P(B) = 26/52$. Finally, $N(A \cap B) = 6$ (because there are three face cards in hearts and another three in diamonds), and so $P(A \cap B) = 6/52$. It follows from the formula for the probability of a union of any two events that

$$P(A \cup B) = P(A) + P(B) - P(A \cap B) = \frac{12}{52} + \frac{26}{52} - \frac{6}{52} = \frac{32}{52} \cong 61.5\%.$$

Thus the probability that the chosen card is a face card or is from one of the red suits is approximately 61.5%. ∎

Expected Value

People who buy lottery tickets regularly often justify the practice by saying that, even though they know that on average they will lose money, they are hoping for one significant gain, after which they believe they will quit playing. Unfortunately, when people who have lost money on a string of losing lottery tickets win some or all of it back, they generally decide to keep trying their luck instead of quitting.

The technical way to say that on average a person will lose money on the lottery is to say that the *expected value* of playing the lottery is negative.

• Definition

Suppose the possible outcomes of an experiment, or random process, are real numbers $a_1, a_2, a_3, \ldots, a_n$, which occur with probabilities $p_1, p_2, p_3, \ldots, p_n$. The **expected value** of the process is

$$\sum_{k=1}^{n} a_k p_k = a_1 p_1 + a_2 p_2 + a_3 p_3 + \cdots + a_n p_n.$$

Example 6.8.5 Expected Value of a Lottery

Suppose that 500,000 people pay $5 each to play a lottery game with the following prizes: a grand prize of $1,000,000, 10 second prizes of $1,000 each, 1,000 third prizes of $500 each, and 10,000 fourth prizes of $10 each. What is the expected value of the game?

Solution   Each of the 500,000 people has the same chance as any other of picking a winning lottery number, and so $p_k = \dfrac{1}{500000}$ for all $k = 1, 2, 3, \ldots, 500000$. Let $a_1, a_2,$ $a_3, \ldots, a_{500000}$ be the net gains of the people playing the lottery. Let $a_1 = 999995$ (the net gain for the grand prize winner, which is one million dollars minus the $5 cost of the winning ticket), $a_2 = a_3 = \cdots = a_{11} = 995$ (the net gain for each of the 10 second prize winners), $a_{12} = a_{13} = \cdots = a_{1011} = 495$ (the net gain for each of the 1,000 third prize winners), and $a_{1012} = a_{1013} = \cdots = a_{11011} = 5$ (the net gain for each of the 10,000 fourth prize winners). Since the remaining 488,989 people just lose their $5, $a_{11012} = a_{11013} = \cdots = a_{500000} = -5$. The expected value of the game is therefore

$$
\sum_{k=1}^{500000} a_k p_k = \sum_{k=1}^{500000} \left(a_k \cdot \frac{1}{500000} \right) \qquad \text{because each } p_k = 1/500000
$$

$$
= \frac{1}{500000} \sum_{k=1}^{500000} a_k \qquad \text{by Theorem 4.1.1(2)}
$$

$$
= \frac{1}{500000}(999995 + 10 \cdot 995 + 1000 \cdot 495 + 10000 \cdot 5 + (-5) \cdot 488989)
$$

$$
= \frac{1}{500000}(999995 + 9950 + 495000 + 50000 - 2444945)
$$

$$
= -1.78.
$$

In other words, a person who continues to play this lottery for a very long time will probably win some money occasionally but on average will lose $1.78 per game.   ■

Exercise Set 6.8

1. In any sample space S, what is $P(\emptyset)$?

2. Suppose A, B, and C are mutually exclusive events in a sample space S, $A \cup B \cup C = S$, and A and B have probabilities 0.3 and 0.5, respectively.
a. What is $P(A \cup B)$?   b. What is $P(C)$?

3. Suppose A and B are mutually exclusive events in a sample space S, C is another event in S, $A \cup B \cup C = S$, and A and B have probabilities 0.4 and 0.2 respectively.
a. What is $P(A \cup B)$?
b. Is it possible that $P(C) = 0.2$? Explain.

4. Suppose A and B are events in a sample space S with probabilities 0.8 and 0.7, respectively. Suppose also that $P(A \cap B) = 0.6$. What is $P(A \cup B)$?

5. Suppose A and B are events in a sample space S and suppose that $P(A) = 0.6$, $P(B^c) = 0.4$, and $P(A \cap B) = 0.2$. What is $P(A \cup B)$?

6. Suppose U and V are events in a sample space S and suppose that $P(U^c) = 0.3$, $P(V) = 0.6$, and $P(U^c \cup V^c) = 0.4$. What is $P(U \cup V)$?

7. Suppose a sample space S consists of three outcomes: 0, 1, and 2. Let $A = \{0\}$, $B = \{1\}$, and $C = \{2\}$, and suppose $P(A) = 0.4$, and $P(B) = 0.3$. Find each of the following:
a. $P(A \cup B)$ b. $P(C)$ c. $P(A \cup C)$
d. $P(A^c)$ e. $P(A^c \cap B^c)$ f. $P(A^c \cup B^c)$

8. Redo exercise 7 assuming that $P(A) = 0.5$ and $P(B) = 0.4$.

9. Let A and B be events in a sample space S, and let $C = S - (A \cup B)$. Suppose $P(A) = 0.4$, $P(B) = 0.5$, and $P(A \cap B) = 0.2$. Find each of the following:
a. $P(A \cup B)$ b. $P(C)$ c. $P(A^c)$
d. $P(A^c \cap B^c)$ e. $P(A^c \cup B^c)$ f. $P(B^c \cap C)$

10. Redo exercise 9 assuming that $P(A) = 0.7$, $P(B) = 0.3$, and $P(A \cap B) = 0.1$.

H 11. Prove that if S is any sample space and U and V are events in S with $U \subseteq V$, then $P(U) \leq P(V)$.

H 12. Prove that if S is any sample space and U and V are any events in S, then $P(V - U) = P(V) - P(U \cap V)$.

H 13. Use the axioms for probability and mathematical induction to prove that for all integers $n \geq 2$, if $A_1, A_2, A_3, \ldots, A_n$ are any mutually disjoint events in a sample space S, then

$$P(A_1 \cup A_2 \cup A_3 \cup \cdots \cup A_n) = \sum_{k=1}^{n} P(A_k).$$

14. A lottery game offers $2 million to the grand prize winner, $20 to each of 10,000 second prize winners, and $4 to each of 50,000 third prize winners. The cost of the lottery is $2 per ticket. Suppose that 1.5 million tickets are sold. What is the expected gain or loss of a ticket?

15. A company sends millions of people an entry form for a sweepstakes accompanied by an order form for magazine subscriptions. The first, second, and third prizes are $10,000,000, $1,000,000, and $50,000, respectively. In order to qualify for a prize, a person is not required to order any magazines but has to spend 60 cents to mail back the entry form. If 30 million people qualify by sending back their entry forms, what is a person's expected gain or loss?

16. An urn contains four balls numbered 2, 2, 5, and 6. If a person selects a set of two balls at random, what is the expected value of the sum of the numbers on the balls?

17. An urn contains five balls numbered 1, 2, 2, 8, and 8. If a person selects a set of two balls at random, what is the expected value of the sum of the numbers on the balls?

18. An urn contains five balls numbered 1, 2, 2, 8, and 8. If a person selects a set of three balls at random, what is the expected value of the sum of the numbers on the balls?

19. When a pair of balanced dice are rolled and the sum of the numbers showing face up is computed, the result can be any number from 2 to 12, inclusive. What is the expected value of the sum?

H 20. Suppose a person offers to play a game with you. In this game, when you draw a card from a standard 52-card deck, if the card is a face card you win $3, and if the card is anything else you lose $1. If you agree to play the game, what is your expected gain or loss?

21. A person pays $1 to play the following game: The person tosses a fair coin four times. If no heads occur, the person pays an additional $2, if one head occurs, the person pays an additional $1, if two heads occur, the person just loses the initial dollar, if three heads occur, the person wins $3, and if four heads occur, the person wins $4. What is the person's expected gain or loss?

H 22. A fair coin is tossed until either a head comes up or four tails are obtained. What is the expected number of tosses?

6.9 Conditional Probability, Bayes' Formula, and Independent Events

It is remarkable that a science which began with the consideration of games of chance should have become the most important object of human knowledge . . . . The most important questions of life are, for the most part, really only problems of probability.
— Pierre-Simon Laplace 1749–1827

In this section we introduce the notion of conditional probability and discuss Bayes' Theorem and the kind of interesting results to which it leads. We then define the concept of independent events and give some applications.

Conditional Probability

Imagine a couple with two children, each of whom is equally likely to be a boy or a girl. Now suppose you are given the information that one is a boy. What is the probability that the other child is a boy?

Figure 6.9.1 shows the four equally likely combinations of gender for the children. You can imagine that the first letter refers to the older child and the second letter to the younger. Thus the combination BG indicates that the older child is a boy and the younger is a girl.

Figure 6.9.1

The combinations where one of the children is a boy are shaded gray, and the combination where the other child is also a boy is shaded blue. Given that you know one child is a boy, only the three combinations in the gray region could be the case, so you can think of the set of those outcomes as a new sample space with three elements, all of which are equally likely. Within the new sample space, there is one combination where the other child is a boy (in the region shaded blue-gray). Thus it would be reasonable to say that the likelihood that the other child is a boy, given that at least one is a boy, is $1/3 = 33\frac{1}{3}\%$. Note that because the original sample space contained four outcomes,

$$\frac{P(\text{at least one child is a boy and the other child is also a boy})}{P(\text{at least one child is a boy})} = \frac{\frac{1}{4}}{\frac{3}{4}} = \frac{1}{3}$$

also. A generalization of this observation forms the basis for the following definition.

• **Definition**

Let A and B be events in a sample space S. If $P(A) \neq 0$, then the **conditional probability of B given A,** denoted $P(B \mid A)$, is

$$P(B \mid A) = \frac{P(A \cap B)}{P(A)}. \qquad \text{6.9.1}$$

Example 6.9.1 Computing a Conditional Probability

A pair of fair dice, one blue and the other gray, are rolled. What is the probability that the sum of the numbers showing face up is 8, given that both of the numbers are even?

Solution The sample space is the set of all 36 outcomes obtained from rolling the two dice and noting the numbers showing face up on each. As in Section 6.1, denote by ab the outcome that the number showing face up on the blue die is a and the one on the gray die is b. Let A be the event that both numbers are even and B the event that the sum of the numbers is 8. Then $A = \{22, 24, 26, 42, 44, 46, 62, 64, 66\}$, $B = \{26, 35, 44, 53, 62\}$, and $A \cap B = \{26, 44, 62\}$. Because the dice are fair (so all outcomes are equally likely), $P(A) = 9/36$, $P(B) = 5/36$ and $P(A \cap B) = 3/36$. By definition of conditional probability,

$$P(B \mid A) = \frac{P(A \cap B)}{P(A)} = \frac{\frac{3}{36}}{\frac{9}{36}} = \frac{3}{9} = \frac{1}{3}. \qquad \blacksquare$$

Note that when both sides of the formula for conditional probability (formula 6.9.1) are multiplied by $P(A)$, a formula for $P(A \cap B)$ is obtained:

$$P(A \cap B) = P(B \mid A) \cdot P(A). \qquad \text{6.9.2}$$

Dividing both sides of formula (6.9.2) by $P(B \mid A)$ gives a formula for $P(A)$:

$$P(A) = \frac{P(A \cap B)}{P(B \mid A)}. \qquad \text{6.9.3}$$

Example 6.9.2 Further Applications of the Conditional Probability Formula

An urn contains 5 blue and 7 gray balls. Let us say that 2 are chosen at random, one after the other, without replacement.

a. What is the probability that both balls are blue?

b. What is the probability that the second ball is blue but the first ball is not?

c. What is the probability that the second ball is blue?

d. What is the probability that at least one of the balls is blue?

e. If the experiment of choosing 2 balls from the urn were repeated many times over, what would be the expected value of the number of blue balls?

Solution   Let S denote the sample space of all possible choices of two balls from the urn, let E be the event that the first ball is blue, and let F be the event that the second ball is blue.

a. The probability that both balls are blue is $P(E \cap F)$. Because there are 12 balls of which 5 are blue, the probability that the first ball is blue is

$$P(E) = \frac{5}{12}.$$

If the first ball is blue, then when the second ball is chosen the urn will contain 4 blue and 7 gray balls. Thus $P(F \mid E) = 4/11$, so by formula (6.9.2),

$$P(E \cap F) = P(F \mid E) \cdot P(E) = \frac{4}{11} \cdot \frac{5}{12} = \frac{20}{132}.$$

b. The probability that the second ball is blue but the first ball is not is $P(F \cap E^c)$. To compute this number, note that because there are 12 balls of which 7 are not blue, $P(E^c) = 7/12$. Also if the first ball is not blue, then when the second ball is chosen, the urn will contain 5 blue and 6 gray balls, and thus $P(F \mid E^c) = 5/11$. Hence, by formula 6.9.2,

$$P(F \cap E^c) = P(F \mid E^c) \cdot P(E^c) = \frac{5}{11} \cdot \frac{7}{12} = \frac{35}{132}.$$

c. The event that the second ball is blue can occur in one of two mutually exclusive ways: Either the first ball is blue and the second is also blue, or the first ball is gray and the second is blue. In other words, F is the disjoint union of $F \cap E$ and $F \cap E^c$. Hence

$$
\begin{aligned}
P(F) &= P((F \cap E) \cup (F \cap E^c)) \\
&= P(F \cap E) + P(F \cap E^c) \quad \text{by probability axiom 3} \\
&= \frac{20}{132} + \frac{35}{132} \quad\quad\quad \text{by parts (a) and (b)} \\
&= \frac{55}{132} = \frac{5}{12}.
\end{aligned}
$$

Thus the probability that the second ball is blue is 5/12, the same as the probability that the first ball is blue.

d. By formula 6.8.2, for the union of any two events,

$$P(E \cup F) = P(E) + P(F) - P(E \cap F)$$

$$= \frac{5}{12} + \frac{5}{12} - \frac{20}{132} \quad \text{by parts (a) and (c)}$$

$$= \frac{90}{132} = \frac{15}{22}.$$

Thus the probability is 15/22, or approximately 68.2%, that at least one of the balls is blue.

e. The event that neither ball is blue is the complement of the event that at least one of the balls is blue, so

$$P(0 \text{ blue balls}) = 1 - P \text{ (at least one ball is blue)} \quad \text{by formula 6.8.1}$$

$$= 1 - \frac{15}{22} \quad \text{by part (d)}$$

$$= \frac{7}{22}.$$

The event that one ball is blue can occur in one of two mutually exclusive ways: Either the second ball is blue and the first is not, or the first ball is blue and the second is not. Part (b) showed that the probability of the first way is $\frac{35}{132}$, and the same technique shows that the probability of the second way is also $\frac{35}{132}$. Thus, by probability axiom 3,

$$P(1 \text{ blue ball}) = \frac{35}{132} + \frac{35}{132} = \frac{70}{132}.$$

Finally, by part (a),

$$P(2 \text{ blue balls}) = \frac{20}{132}.$$

Therefore,

$$\begin{bmatrix} \text{the expected value of} \\ \text{the number of blue balls} \end{bmatrix} = 0 \cdot P(0 \text{ blue balls}) + 1 \cdot P(1 \text{ blue ball})$$

$$+ 2 \cdot P(2 \text{ blue balls})$$

$$= 0 \cdot \frac{7}{22} + 1 \cdot \frac{70}{132} + 2 \cdot \frac{20}{132}$$

$$= \frac{110}{132} \cong 0.8. \qquad \blacksquare$$

Bayes' Theorem

Suppose that one urn contains 3 blue and 4 gray balls and a second urn contains 5 blue and 3 gray balls. A ball is selected by choosing one of the urns at random and then picking a ball at random from that urn. If the chosen ball is blue, what is the probability that it came from the first urn?

This problem can be solved by carefully interpreting all the information that is known and putting it together in just the right way. Let A be the event that the chosen ball is blue, B_1 the event that the ball came from the first urn, and B_2 the event that the ball came from the second urn. Because 3 of the 7 balls in urn one are blue, and 5 of the 8 balls in urn two are blue,

$$P(A \mid B_1) = \frac{3}{7} \quad \text{and} \quad P(A \mid B_2) = \frac{5}{8}.$$

And because the urns are equally likely to be chosen,

$$P(B_1) = P(B_2) = \frac{1}{2}.$$

Moreover, by formula (6.9.2),

$$P(A \cap B_1) = P(A \mid B_1) \cdot P(B_1) = \frac{3}{7} \cdot \frac{1}{2} = \frac{3}{14}, \quad \text{and}$$

$$P(A \cap B_2) = P(A \mid B_2) \cdot P(B_2) = \frac{5}{8} \cdot \frac{1}{2} = \frac{5}{16}.$$

But A is the disjoint union of $(A \cap B_1)$ and $(A \cap B_2)$, so by probability axiom 3,

$$P(A) = P((A \cap B_1) \cup (A \cap B_2)) = P(A \cap B_1) + P(A \cap B_2) = \frac{3}{14} + \frac{5}{16} = \frac{59}{112}.$$

Finally, by definition of conditional probability,

$$P(B_1 \mid A) = \frac{P(B_1 \cap A)}{P(A)} = \frac{\frac{3}{14}}{\frac{59}{112}} = \frac{336}{826} \cong 40.7\%.$$

Thus, if the chosen ball is blue, the probability is approximately 40.7% that it came from the first urn.

The steps used to derive the answer in the example above can be generalized to prove Bayes' Theorem. (See exercises 8 and 9 at the end of this section.) Thomas Bayes was an English Presbyterian minister who devoted much of his energies to mathematics. The theorem that bears his name was published posthumously in 1763. The portrait at the left is the only one attributed to him, but its authenticity has recently come into question.

Thomas Bayes
(1702–1761)

Theorem 6.9.1 Bayes' Theorem

Suppose that a sample space S is a union of mutually disjoint events $B_1, B_2, B_3, \ldots, B_n$, and suppose A is an event in S with $P(A) \neq 0$. If k is an integer with $1 \leq k \leq n$, then

$$P(B_k \mid A) = \frac{P(A \mid B_k) \cdot P(B_k)}{P(A \mid B_1) \cdot P(B_1) + P(A \mid B_2) \cdot P(B_2) + \cdots + P(A \mid B_n) \cdot P(B_n)}$$

Example 6.9.3 Applying Bayes' Theorem

Most medical tests occasionally produce incorrect results, called false positives and false negatives. When a test is designed to determine whether a patient has a certain disease, a **false positive** result indicates that a patient has the disease when the patient does not have it. A **false negative** result indicates that a patient does not have the disease when the patient does have it.

When large-scale health screenings are performed for diseases with relatively low incidence, those who develop the screening procedures have to balance several considerations: the per-person cost of the screening, follow-up costs for further testing of false positives, and the possibility that people who have the disease will develop unwarranted confidence in the state of their health.

Consider a medical test that screens for a disease found in 5 people in 1,000. Suppose that the false positive rate is 3% and the false negative rate is 1%. Then 99% of the time a person who has the condition tests positive for it, and 97% of the time a person who does not have the condition tests negative for it. (See exercise 3 at the end of this section.)

a. What is the probability that a randomly chosen person who tests positive for the disease actually has the disease?

b. What is the probability that a randomly chosen person who tests negative for the disease does not indeed have the disease?

Solution   Consider a person chosen at random from among those screened. Let A be the event that the person tests positive for the disease, B_1 the event that the person actually has the disease, and B_2 the event that the person does not have the disease. Then

$$P(A \mid B_1) = 0.99, \quad P(A^c \mid B_1) = 0.01, \quad P(A^c \mid B_2) = 0.97, \quad \text{and} \quad P(A \mid B_2) = 0.03.$$

Also, because 5 people in 1,000 have the disease,

$$P(B_1) = 0.005 \quad \text{and} \quad P(B_2) = 0.995.$$

a. By Bayes' Theorem,

$$
\begin{aligned}
P(B_1 \mid A) &= \frac{P(A \mid B_1) \cdot P(B_1)}{P(A \mid B_1) \cdot P(B_1) + P(A \mid B_2) \cdot P(B_2)} \\
&= \frac{(0.99) \cdot (0.005)}{(0.99) \cdot (0.005) + (0.03) \cdot (0.995)} \\
&\cong 0.1422 \cong 14.2\%.
\end{aligned}
$$

Thus the probability that a person with a positive test result actually has the disease is approximately 14.2%.

b. By Bayes' Theorem,

$$
\begin{aligned}
P(B_2 \mid A^c) &= \frac{P(A^c \mid B_2) \cdot P(B_2)}{P(A^c \mid B_1) \cdot P(B_1) + P(A^c \mid B_2) \cdot P(B_2)} \\
&= \frac{(0.97) \cdot (0.995)}{(0.01) \cdot (0.005) + (0.97) \cdot (0.995)} \\
&\cong 0.999948 \cong 99.995\%.
\end{aligned}
$$

Thus the probability that a person with a negative test result does not have the disease is approximately 99.995%.

You might be surprised by these numbers, but they are fairly typical of the situation where the screening test is significantly less expensive than a more accurate test for the same disease yet produces positive results for nearly all people with the disease. Using the screening test limits the expense of unnecessarily using the more costly test to a relatively small percentage of the population being screened, while only rarely indicating that a person who has the disease is free of it. ∎

Independent Events

Suppose a coin is tossed twice. It seems intuitively clear that the outcome of the first toss does not depend in any way on the outcome of the second toss, and conversely. In other words, if, for instance, A is the event that a head is obtained on the first toss and B is the event that a head is obtained on the second toss, then if the coin is tossed randomly both times, events A and B should be *independent* in the sense that $P(A \mid B) = P(A)$ and $P(B \mid A) = P(B)$. This intuitive idea of independence is supported by the following analysis. If the coin is fair, then the four outcomes $HH, HT, TH,$ and TT are equally likely, and

$$A = \{HH, HT\}, \quad B = \{TH, HH\}, \quad A \cap B = \{HH\}.$$

Hence

$$P(A) = P(B) = \frac{2}{4} = \frac{1}{2}.$$

But also

$$P(A \mid B) = \frac{P(A \cap B)}{P(B)} = \frac{\frac{1}{4}}{\frac{1}{2}} = \frac{1}{2} \quad \text{and} \quad P(B \mid A) = \frac{P(A \cap B)}{P(A)} = \frac{\frac{1}{4}}{\frac{1}{2}} = \frac{1}{2},$$

and thus $P(A \mid B) = P(A)$ and $P(B \mid A) = P(B)$.

To obtain the final form for definition of independence, observe that

if $P(B) \neq 0$ and $P(A \mid B) = P(A)$, then $P(A \cap B) = P(A \mid B) \cdot P(B) = P(A) \cdot P(B)$.

By the same argument,

if $P(A) \neq 0$ and $P(B \mid A) = P(B)$, then $P(A \cap B) = P(A) \cdot P(B)$.

Conversely (see exercise 17 at the end of this section),

if $P(A \cap B) = P(A) \cdot P(B)$ and $P(A) \neq 0$, then $P(B \mid A) = P(B)$,

and

if $P(A \cap B) = P(A) \cdot P(B)$ and $P(B) \neq 0$, then $P(A \mid B) = P(A)$.

Thus, for convenience and to eliminate the requirement that the probabilities be nonzero, we use the following product formula to define independent events.

• Definition

If A and B are events in a sample space S, then A and B are **independent** if, and only if,

$$P(A \cap B) = P(A) \cdot P(B).$$

It would be natural to think that mutually disjoint events would be independent, but in fact almost the opposite is true: Mutually disjoint events with nonzero probabilities are dependent.

Example 6.9.4 Mutually Disjoint Events and Independence

Let A and B be events in a sample space S, and suppose $A \cap B = \emptyset$, $P(A) \neq 0$, and $P(B) \neq 0$. Show that $P(A \cap B) \neq P(A) \cdot P(B)$.

Solution Because $A \cap B = \emptyset$, $P(A \cap B) = 0$ by probability axiom 2. But $P(A) \cdot P(B) \neq 0$ because neither $P(A)$ nor $P(B)$ equals zero. Thus $P(A \cap B) \neq P(A) \cdot P(B)$. ∎

The following example, and its immediate consequence, show how the independence of two events extends to their complements.

Example 6.9.5 The Probability of $A \cap B^c$ When A and B Are Independent Events

Suppose A and B are independent events in a sample space S. Show that A and B^c are also independent.

Solution Observe that

$$
\begin{aligned}
(A \cap B) \cup (A \cap B^c) &= A \cap (B \cup B^c) && \text{by the distributive law for sets} \\
&= A \cap S && \text{by the complement law for union} \\
&= A && \text{by the identity law for intersection}
\end{aligned}
$$

Also $(A \cap B) \cap (A \cap B^c) = \emptyset$ because $B \cap B^c = \emptyset$. Apply probability axiom 3 to the above equality to obtain

$$P((A \cap B) \cup (A \cap B^c)) = P(A \cap B) + P(A \cap B^c) = P(A).$$

Solving for $P(A \cap B^c)$ gives that

$$
\begin{aligned}
P(A \cap B^c) &= P(A) - P(A \cap B) \\
&= P(A) - P(A) \cdot P(B) && \text{because } A \text{ and } B \text{ are independent} \\
&= P(A)(1 - P(B)) && \text{by factoring out } P(A) \\
&= P(A) \cdot P(B^c) && \text{by formula 6.8.1.}
\end{aligned}
$$

Thus A and B^c are independent events. ∎

It follows immediately from Example 6.9.5 that if A and B are independent, then A^c and B are also independent and so are A^c and B^c. (See exercises 21 and 22 at the end of this section.) These results are applied in Example 6.9.6.

Example 6.9.6 Computing Probabilities of Intersections of Independent Events

A coin is loaded so that the probability of heads is 0.6. Suppose the coin is tossed twice. Although the probability of heads is greater than the probability of tails, there is no reason to believe that whether the coin lands heads or tails on one toss will affect whether it lands heads or tails on the other toss. Thus it is reasonable to assume that the results of the tosses are independent.

a. What is the probability of obtaining two heads?

b. What is the probability of obtaining one head?

c. What is the probability of obtaining no heads?

d. What is the probability of obtaining at least one head?

Solution   The sample space S consists of the four outcomes $\{HH, HT, TH, TT\}$, which are not equally likely. Let E be the event that a head is obtained on the first toss, and let F be the event that a head is obtained on the second toss. Then $P(E) = P(F) = 0.6$, and it is to be assumed that E and F are independent.

a. The probability of obtaining two heads is $P(E \cap F)$. Because E and F are independent,

$$P(\text{two heads}) = P(E \cap F) = P(E) \cdot P(F) = (0.6)(0.6) = 0.36 = 36\%.$$

b. One head can be obtained in two mutually exclusive ways: head on the first toss and tail on the second, or tail on the first toss and head on the second. Thus, the event of obtaining exactly one head is $(E \cap F^c) \cup (E^c \cap F)$. Also $(E \cap F^c) \cap (E^c \cap F) = \emptyset$, and, moreover, by the formula for the probability of the complement of an event, $P(E^c) = P(F^c) = 1 - 0.6 = 0.4$. Hence

$$
\begin{aligned}
P(\text{one head}) &= P((E \cap F^c) \cup (E^c \cap F)) \\
&= P(E) \cdot P(F^c) + P(E^c) \cdot P(F) \quad \text{by Example 6.9.5 and exercise 21} \\
&= (0.6)(0.4) + (0.4)(0.6) \\
&= 0.48 = 48\%.
\end{aligned}
$$

c. The probability of obtaining no heads is $P(E^c \cap F^c)$. By exercise 22,

$$P(\text{no heads}) = P(E^c \cap F^c) = P(E^c) \cdot P(F^c) = (0.4)(0.4) = 0.16 = 16\%.$$

d. There are two ways to solve this problem. One is to observe that because the event of obtaining one head and the event of obtaining two heads are mutually disjoint,

$$
\begin{aligned}
P(\text{at least one head}) &= P(\text{one head}) + P(\text{two heads}) \\
&= 0.48 + 0.36 \quad \text{by parts (a) and (b)} \\
&= 0.84 = 84\%.
\end{aligned}
$$

The second way is to use the fact that the event of obtaining at least one head is the complement of the event of obtaining no heads. So

$$
\begin{aligned}
P(\text{at least one head}) &= 1 - P(\text{no heads}) \\
&= 1 - 0.16 \quad \text{by part (c)} \\
&= 0.84 = 84\%. \quad\blacksquare
\end{aligned}
$$

Example 6.9.7 Expected Value of Tossing a Loaded Coin Twice

Suppose that a coin is loaded so that the probability of heads is 0.6, and suppose the coin is tossed twice. If this experiment is repeated many times, what is the expected value of the number of heads?

Solution   Think of the outcomes of the coin tossings as just 0, 1, or 2 heads. Example 6.9.6 showed that the probabilities of these outcomes are 0.16, 0.48, and 0.36, respectively. Thus, by definition of expected value, the

$$\text{expected number of heads} = 0 \cdot (0.16) + 1 \cdot (0.48) + 2 \cdot (0.36) = 1.2. \quad\blacksquare$$

What if a loaded coin is tossed more than twice? Suppose it is tossed ten times, or a hundred times. What are the probabilities of various numbers of heads? To answer this question, it is necessary to expand the notion of independence to more than two events. For instance, we say three events A, B, and C are *pairwise independent* if, and only if,

$$P(A \cap B) = P(A) \cdot P(B), \, P(A \cap C) = P(A) \cdot P(C), \text{ and } P(B \cap C) = P(B) \cdot P(C).$$

The next example shows that events can be pairwise independent without satisfying the condition $P(A \cap B \cap C) = P(A) \cdot P(B) \cdot P(C)$. Conversely, they can satisfy the condition $P(A \cap B \cap C) = P(A) \cdot P(B) \cdot P(C)$ without being pairwise independent (see exercise 26 at the end of this section).

Example 6.9.8 Exploring Independence for Three Events

Suppose that a fair coin is tossed twice. Let A be the event that a head is obtained on the first toss, B the event that a head is obtained on the second toss, and C the event that either two heads or two tails are obtained. Show that A, B, and C are pairwise independent but do not satisfy the condition $P(A \cap B \cap C) = P(A) \cdot P(B) \cdot P(C)$.

Solution Because there are four equally likely outcomes—HH, HT, TH, and TT—it is clear that $P(A) = P(B) = P(C) = \frac{1}{2}$. You can also see that $A \cap B = \{HH\}$, $A \cap C = \{HH\}$, $B \cap C = \{HH\}$, and $A \cap B \cap C = \{HH\}$. Hence $P(A \cap B) = P(A \cap C) = P(B \cap C) = \frac{1}{4}$, and so $P(A \cap B) = P(A) \cdot P(B)$, $P(A \cap C) = P(A) \cdot P(C)$, and $P(B \cap C) = P(B) \cdot P(C)$. Thus A, B, and C are pairwise independent. But

$$P(A \cap B \cap C) = P(\{HH\}) = \frac{1}{4} \neq \left(\frac{1}{2}\right)^3 = P(A) \cdot P(B) \cdot P(C). \qquad \blacksquare$$

Because of situations like that in Example 6.9.6, four conditions must be included in the definition of independence for three events.

> **• Definition**
>
> Let A, B, and C be events in a sample space S. A, B, and C are **pairwise independent** if, and only if, they satisfy conditions 1–3 below. They are **mutually independent** if, and only if, they satisfy all four conditions below.
>
> 1. $P(A \cap B) = P(A) \cdot P(B)$
> 2. $P(A \cap C) = P(A) \cdot P(C)$
> 3. $P(B \cap C) = P(B) \cdot P(C)$
> 4. $P(A \cap B \cap C) = P(A) \cdot P(B) \cdot P(C)$

The definition of mutual independence for any collection of n events with $n \geq 2$ generalizes the two definitions given previously.

> **• Definition**
>
> Events $A_1, A_2, A_3, \ldots, A_n$ in a sample space S are **mutually independent** if, and only if, the probability of the intersection of any subset of the events is the product of the probabilities of the events in the subset.

Example 6.9.9 Tossing a Loaded Coin Ten Times

A coin is loaded so that the probability of heads is 0.6 (and thus the probability of tails is 0.4). Suppose the coin is tossed ten times. As in Example 6.9.6, it is reasonable to assume that the results of the tosses are mutually independent.

a. What is the probability of obtaining eight heads?

b. What is the probability of obtaining at least eight heads?

Solution

a. For each $i = 1, 2, \ldots, 10$, let H_i be the event that a head is obtained on the ith toss, and let T_i be the event that a tail is obtained on the ith toss. Suppose that the eight heads occur on the first eight tosses and that the remaining two tosses are tails. This is the event $H_1 \cap H_2 \cap H_3 \cap H_4 \cap H_5 \cap H_6 \cap H_7 \cap H_8 \cap T_9 \cap T_{10}$. For simplicity we denote it as $HHHHHHHHTT$. By definition of mutually independent events,

$$P(HHHHHHHHTT) = (0.6)^8(0.4)^2.$$

Because of the commutative law for multiplication, if the eight heads occur on any other of the ten tosses, the same number is obtained. For instance, if we denote the event $H_1 \cap H_2 \cap T_3 \cap H_4 \cap H_5 \cap H_6 \cap H_7 \cap H_8 \cap T_9 \cap H_{10}$ by $HHTHHHHHTH$, then

$$P(HHTHHHHHTH) = (0.6)^2(0.4)(0.6)^5(0.4)(0.6) = (0.6)^8(0.4)^2.$$

Now there are as many different ways to obtain eight heads in ten tosses as there are subsets of eight elements (the toss numbers on which heads are obtained) that can be chosen from a set of ten elements. This number is $\binom{10}{8}$. It follows that, because the different ways of obtaining eight heads are all mutually exclusive,

$$P(\text{eight heads}) = \binom{10}{8}(0.6)^8(0.4)^2.$$

b. By reasoning similar to that in part (a),

$$P(\text{nine heads}) = \begin{bmatrix} \text{the number of different} \\ \text{ways nine heads can be} \\ \text{obtained in ten tosses} \end{bmatrix} \cdot (0.6)^9(0.4)^1 = \binom{10}{9}(0.6)^9(0.4),$$

and

$$P(\text{ten heads}) = \begin{bmatrix} \text{the number of different} \\ \text{ways ten heads can be} \\ \text{obtained in ten tosses} \end{bmatrix} \cdot (0.6)^{10}(0.4)^0 = \binom{10}{10}(0.6)^{10}.$$

Because obtaining eight, obtaining nine, and obtaining ten heads are mutually disjoint events,

$$P(\text{at least eight heads}) = P(\text{eight heads}) + P(\text{nine heads}) + P(\text{ten heads})$$

$$= \binom{10}{8}(0.6)^8(0.4)^2 + \binom{10}{9}(0.6)^9(0.4) + \binom{10}{10}(0.6)^{10}$$

$$\cong 0.167 = 16.7\%. \qquad \blacksquare$$

Note the occurrence of the binomial coefficients $\binom{n}{k}$ in solutions to problems like the one in Example 6.9.9. For that reason, probabilities of the form

$$\binom{n}{k}p^{n-k}(1 - p)^k,$$

where $0 \le p \le 1$, are called **binomial probabilities.** Binomial probabilities occur in situations with multiple, mutually independent repetitions of a random process, with the same two possible outcomes that have the same probabilities on each repetition.

Exercise Set 6.9

1. Suppose $P(A \mid B) = 1/2$ and $P(A \cap B) = 1/6$. What is $P(B)$?

2. Suppose $P(X \mid Y) = 1/3$ and $P(Y) = 1/4$. What is $P(X \cap Y)$?

3. **a.** Prove that if A and B are any events in a sample space S, with $P(B) \neq 0$, then $P(A^c \mid B) = 1 - P(A \mid B)$.
   b. Explain how this result justifies the following statements: (1) If the probability of a false negative on a test for a condition is 4%, then there is a 96% probability that a person who does not have the condition will have a negative test result. (2) If the probability of a false positive on a test for a condition is 1%, then there is a 99% probability that a person who does have the condition will test positive for it.

H 4. Suppose that A and B are events in a sample space S and that $P(A)$, $P(B)$, and $P(A \mid B)$ are known. Derive a formula for $P(A \mid B^c)$.

5. An urn contains 25 red balls and 15 blue balls. Two are chosen at random, one after the other, without replacement.
   a. What is the probability that both balls are red?
   b. What is the probability that the second ball is red but the first ball is not?
   c. What is the probability that the second ball is red?
   d. What is the probability that at least one of the balls is red?

6. Redo exercise 5 assuming that the urn contains 30 red balls and 40 blue balls.

7. A pool of 10 semifinalists for a job consists of 7 men and 3 women. Because all are considered equally qualified, the names of two of the semifinalists are drawn, one after the other, at random, to become finalists for the job.
   a. What is the probability that both finalists are women?
   b. What is the probability that both finalists are men?
   H c. What is the probability that one finalist is a woman and the other is a man?

H 8. Prove Bayes' Theorem for $n = 2$. That is, prove that if a sample space S is a union of mutually disjoint events B_1 and B_2, if A is an event in S with $P(A) \neq 0$, and if $k = 1$ or $k = 2$, then
$$P(B_k \mid A) = \frac{P(A \mid B_k) \cdot P(B_k)}{P(A \mid B_1) \cdot P(B_1) + P(A \mid B_2) \cdot P(B_2)}.$$

H 9. Use the result of exercise 8 and mathematical induction to prove Bayes' Theorem.

10. One urn contains 12 blue balls and 7 white balls, and a second urn contains 8 blue balls and 19 white balls. An urn is selected at random, and a ball is chosen from the urn.
    a. What is the probability that the chosen ball is blue?
    b. If the chosen ball is blue, what is the probability that it came from the first urn?

11. Redo exercise 10 assumuing that the first urn contains 4 blue balls and 16 white balls and the second urn contains 10 blue balls and 9 white balls.

H 12. One urn contains 10 red balls and 25 green balls, and a second urn contains 22 red balls and 15 green balls. A ball is chosen as follows: First an urn is selected by tossing a loaded coin with probability 0.4 of landing heads up and probability 0.6 of landing tails up. If the coin lands heads up, the first urn is chosen; otherwise, the second urn is chosen. Then a ball is picked at random from the chosen urn.
    a. What is the probability that the chosen ball is green?
    b. If the chosen ball is green, what is the probability that it was picked from the first urn?

13. A drug-screening test is used in a large population of people of whom 4% actually use drugs. Suppose that the false positive rate is 3% and the false negative rate is 2%. Thus a person who uses drugs tests positive for them 97% of the time, and a person who does not use drugs tests negative for them 98% of the time.
    a. What is the probability that a randomly chosen person who tests positive for drugs actually uses drugs?
    b. What is the probability that a randomly chosen person who tests negative for drugs does not use drugs?

14. Two different factories both produce a certain automobile part. The probability that a component from the first factory is defective is 2%, and the probability that a component from the second factory is defective is 5%. In a supply of 180 of the parts, 100 were obtained from the first factory and 80 from the second factory.
    a. What is the probability that a part chosen at random from the 180 is from the first factory?
    b. What is the probability that a part chosen at random from the 180 is from the second factory?
    c. What is the probability that a part chosen at random from the 180 is defective?
    d. If the chosen part is defective, what is the probability that it came from the first factory?

H 15. Three different suppliers—X, Y, and Z—provide produce for a grocery store. Twelve percent of produce from X is superior grade, 8% of produce from Y is superior grade and 15% of produce from Z is superior grade. The store obtains 20% of its produce from X, 45% from Y, and 35% from Z.
    a. If a piece of produce is purchased, what is the probability that it is superior grade?
    b. If a piece of produce in the store is superior grade, what is the probability that it is from X?

16. Prove that if A and B are events in a sample space S with the property that $P(A \mid B) = P(A)$ and $P(A) \neq 0$, then $P(B \mid A) = P(B)$.

17. Prove that if $P(A \cap B) = P(A) \cdot P(B)$, $P(A) \neq 0$, and $P(B) \neq 0$, then $P(A \mid B) = P(A)$ and $P(B \mid A) = P(B)$.

18. A pair of fair dice, one blue and the other gray, are rolled. Let A be the event that the number face up on the blue die is 2, and let B be the event that the number face up on the gray die is 4 or 5. Show that $P(A \mid B) = P(A)$ and $P(B \mid A) = P(B)$.

19. Suppose a fair coin is tossed three times. Let A be the event that a head appears on the first toss, and let B be the event that an even number of heads is obtained. Show that $P(A \mid B) = P(A)$ and $P(B \mid A) = P(B)$.

20. If A and B are events in a sample space S and both $A \cap B = \emptyset$, what must be true in order for A and B to be independent? Explain.

21. Prove that if A and B are independent events in a sample space S, then A^c and B are also independent.

22. Prove that if A and B are independent events in a sample space S, then A^c and B^c are also independent.

23. A student taking a multiple-choice exam does not know the answers to two questions. All have five choices for the answer. For one of the two questions, the student can eliminate two answer choices as incorrect but has no idea about the other answer choices. For the other question, the student has no clue about the correct answer at all. Assume that whether the student chooses the correct answer on one of the questions does not affect whether the student chooses the correct answer on the other question.
    a. What is the probability that the student will answer both questions correctly?
    b. What is the probability that the student will answer exactly one of the questions correctly?
    c. What is the probability that the student will answer neither question correctly?

24. A company uses two proofreaders X and Y to check a certain manuscript. X misses 12% of typographical errors and Y misses 15%. Assume that the proofreaders work independently.
    a. What is the probability that a randomly chosen typographical error will be missed by both proofreaders?
    b. If the manuscript contains 1,000 typographical errors, what number can be expected to be missed?

25. A coin is loaded so that the probability of heads is 0.7 and the probability of tails is 0.3. Suppose that the coin is tossed twice and that the results of the tosses are independent.
    a. What is the probability of obtaining exactly two heads?
    b. What is the probability of obtaining exactly one head?
    c. What is the probability of obtaining no heads?
    d. What is the probability of obtaining at least one head?

✱26. Describe a sample space and events A, B, and C, where $P(A \cap B \cap C) = P(A) \cdot P(B) \cdot P(C)$ but A, B, and C are not pairwise independent.

H 27. The example used to introduce conditional probability described a family with two children each of whom was equally likely to be a boy or a girl. The example showed that if it is known that one child is a boy, the probability that the other child is a boy is 1/3. Now imagine the same kind of family—two children each of whom is equally likely to be a boy or a girl. Suppose you meet one of the children and see that it is a boy. What is the probability that the other child is a boy? Explain. (Be careful. The answer may surprise you.)

28. A coin is loaded so that the probability of heads is 0.7 and the probability of tails is 0.3. Suppose that the coin is tossed ten times and that the results of the tosses are mutually independent.
    a. What is the probability of obtaining exactly seven heads?
    b. What is the probability of obtaining exactly ten heads?
    c. What is the probability of obtaining no heads?
    d. What is the probability of obtaining at least one head?

29. Suppose that ten items are chosen at random from a large batch delivered to a company. The manufacturer claims that just 3% of the items in the batch are defective. Assume that the batch is large enough so that even though the selection is made without replacement, the number 0.03 can be used to approximate the probability that any one of the ten items is defective. In addition, assume that because the items are chosen at random, the outcomes of the choices are mutually independent. Finally, assume that the manufacturer's claim is correct.
    a. What is the probability that none of the ten is defective?
    b. What is the probability that at least one of the ten is defective?
    c. What is the probability that exactly four of the ten are defective?
    d. What is the probability that at most two of the ten are defective?

30. Suppose the probability of a false positive result on a mammogram is 4%, and suppose that the radiologists' interpretations of mammograms are mutually independent in the sense that whether or not a radiologist finds a positive result on one mammogram does not influence whether or not the radiologist finds a positive result on another mammogram. Assume that a woman without breast cancer has a mammogram every year for ten years.
    a. What is the probability that she will have no false positive results during that time?
    b. What is the probability that she will have at least one false positive result during that time?
    c. What is the probability that she will have exactly two false positive results during that time?
    d. Suppose that the probability of a false negative result on a mammogram is 2%.
        (i) If a woman has a positive test result one year, what is the probability that she actually has breast cancer?
        (ii) If a woman has a negative test result one year, what is the probability that she actually has breast cancer?

31. Empirical data indicate that approximately 103 out of every 200 children born are male. Hence the probability of a newborn being male is about 51.5%. Suppose that a family has six children, and suppose that the genders of all the children are mutually independent.

H a. What is the probability that none of the children is male?
  b. What is the probability that at least one of the children is male?
  c. What is the probability that exactly five of the children are male?

32. A person takes a multiple-choice exam in which each question has four possible answers. Suppose that the person has no idea about the answers to three of the questions and simply chooses randomly for each one.
  a. What is the probability that the person will answer all three questions correctly?
  b. What is the probability that the person will answer exactly two questions correctly?
  c. What is the probability that the person will answer exactly one question correctly?
  d. What is the probability that the person will answer no questions correctly?
  e. Suppose that the person gets one point of credit for each correct answer and that 1/3 point is deducted for each incorrect answer. What is the expected value of the person's score for the three questions?

CHAPTER 7

FUNCTIONS

Functions are ubiquitous in mathematics and computer science. That means you can hardly take two steps in these subjects without running into one. In this book we have already referred to truth tables and input/output tables (which are really Boolean functions), sequences (which are really functions defined on sets of integers), *mod* and *div* (which are really functions defined on Cartesian products of integers), and floor and ceiling (which are really functions from **R** to **Z**).

In this chapter we consider a wide variety of functions, focusing on those defined on discrete sets (such as finite sets or sets of integers). We then look at properties of functions such as one-to-one and onto, existence of inverse functions, and the interaction of composition of functions and the properties of one-to-one and onto. We end the chapter with a discussion of sizes of infinite sets and an application to computability.

7.1 Functions Defined on General Sets

The theory that has had the greatest development in recent times is without any doubt the theory of functions. — Vito Volterra, 1888

As used in ordinary language, the word *function* indicates dependence of one varying quantity on another. If your teacher tells you that your grade in a course will be a function of your performance on the exams, you interpret this to mean that the teacher has some rule for translating exam scores into grades. To each collection of exam scores there corresponds a certain grade.

More generally, suppose two sets of objects are given—a first set and a second set—and suppose that with each element of the first set is associated a particular element of the second set. The relationship between the elements of the sets is called a *function*. Functions are generally denoted by single letters such as f, g, h, F, G, and so forth, although special functions are denoted by strings of letters or other symbols such as log, exp, and *mod*.

• Definition

A **function f from a set X to a set Y** is a relation* between elements of X, called **inputs,** and elements of Y, called **outputs,** with the property that each input is related to one and only one output. The notation $f: X \rightarrow Y$ means that f is a function from X to Y. X is called the **domain** of f, and Y is called the **co-domain** of f.

Given an input element x in X, there is a unique output element y in Y that is related to x by f. We say that "f sends x to y" and write $x \xrightarrow{f} y$ or $f: x \rightarrow y$. The unique element y to which f sends x is denoted

$f(x)$ and is called   **f of x,** or
                       the output of f for the input x, or
                       the value of f at x, or
                       the image of x under f.

The set of all values of f taken together is called the *range of f* or the *image* of X *under f*. Symbolically,

range of f = image of X under f $= \{y \in Y \mid y = f(x), \text{for some } x \text{ in } X\}$.

Given an element y in Y, there may exist elements in X with y as their image. If $f(x) = y$, then x is called **a preimage of y** or **an inverse image of y.** The set of all inverse images of y is called *the inverse image of y*. Symbolically,

inverse image of y $= \{x \in X \mid f(x) = y\}$.

The concept of function was developed over a period of centuries. The definition given above was first formulated for sets of numbers by the German mathematician Lejeune Dirichlet (DEER-ish-lay) in 1837.

Johann Peter Gustav Lejeune Dirichlet (1805–1859)

Arrow Diagrams

If X and Y are finite sets, you can define a function f from X to Y by making a list of elements in X and a list of elements in Y and drawing an arrow from each element in X to the corresponding element in Y. Such a drawing is called an **arrow diagram.** An example of an arrow diagram is shown in Figure 7.1.1.

Figure 7.1.1

The definition of function implies that the arrow diagram for a function f has the following two properties:

1. Every element of X has an arrow coming out of it.

2. No element of X has two arrows coming out of it that point to two different elements of Y.

---

*In Chapter 10 we give a precise definition of the term *relation*.

Property 1 holds because the definition of function says that *each* element of X is sent to a unique element of Y. Property 2 holds because the definition of function says that each element of X is sent to a *unique* element of Y.

The range of f consists of all points in Y that have arrows pointing to them. The inverse image of an element y consists of all points in X that have arrows pointing from them to y.

Note that once X and Y have been given, the arrow diagram can also be specified by writing the set of all ordered pairs (x, y) for which there is an arrow from x to y. For instance, instead of drawing the arrows in Figure 7.1.1, we could write the set $\{(x_1, y_3), (x_2, y_1), (x_3, y_3), (x_4, y_4)\}$. In Chapter 10 we will discuss the formal definition of function, which specifies that a function from a set X to a set Y is a subset of $X \times Y$ satisfying certain properties.

Example 7.1.1 A Function Defined by an Arrow Diagram

Let $X = \{a, b, c\}$ and $Y = \{1, 2, 3, 4\}$. Define a function f from X to Y by the arrow diagram in Figure 7.1.2.

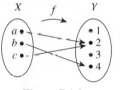

Figure 7.1.2

a. Write the domain and co-domain of f.

b. Find $f(a)$, $f(b)$, and $f(c)$.

c. What is the range of f?

d. Is c an inverse image of 2? Is b an inverse image of 3?

e. Find the inverse images of 2, 4, and 1.

f. Represent f as a set of ordered pairs.

Solution

a. domain of $f = \{a, b, c\}$, co-domain of $f = \{1, 2, 3, 4\}$

b. $f(a) = 2$, $f(b) = 4$, $f(c) = 2$

c. range of $f = \{2, 4\}$

d. Yes, No

e. inverse image of $2 = \{a, c\}$
   inverse image of $4 = \{b\}$
   inverse image of $1 = \emptyset$ (*since no arrows point to* 1)

f. $\{(a, 2), (b, 4), (c, 2)\}$ ∎

In Example 7.1.1 there are no arrows pointing to the 1 or the 3. This illustrates the fact that although each element of the domain of a function must have an arrow pointing out from it, there can be elements of the co-domain to which no arrows point. Note also that there are two arrows pointing to the 2—one coming from a and the other from c.

This illustrates the fact that although no two arrows can start from the same element of the domain, there can be two or more arrows pointing to the same element of the co-domain.

Example 7.1.2 Functions and Nonfunctions

Which of the arrow diagrams in Figure 7.1.3 define functions from $X = \{a, b, c\}$ to $Y = \{1, 2, 3, 4\}$?

(a) (b) (c)

Figure 7.1.3

Solution   Only (c) defines a function. In (a) there is an element of X, namely b, that is not sent to any element of Y; that is, there is no arrow coming out of b. And in (b) the element c is not sent to a *unique* element of Y; that is, there are two arrows coming out of c, one pointing to 2 and the other to 3. ∎

Function Machines

Another useful way to think of a function is as a machine. Suppose f is a function from X to Y and an input x of X is given. Imagine f to be a machine that processes x in a certain way to produce the output $f(x)$. This is illustrated in Figure 7.1.4.

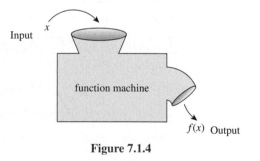

Input

x

function machine

$f(x)$ Output

Figure 7.1.4

Example 7.1.3 Functions Defined by Formulas

The **squaring function** $f: \mathbf{R} \to \mathbf{R}$ is defined by the formula $f(x) = x^2$ for all real numbers x. This means that no matter what real number input is substituted for x, the output of f will be the square of that number. This idea can be represented by writing $f(\square) = \square^2$. In other words, f sends each real number x to x^2, or, symbolically, $f: x \to x^2$. Note that the variable x is a dummy variable; any other symbol could replace it, as long as the replacement is made everywhere the x appears.

The **successor function** $g: \mathbf{Z} \to \mathbf{Z}$ is defined by the formula $g(n) = n + 1$. Thus, no matter what integer is substituted for n, the output of g will be that number plus one: $g(\square) = \square + 1$. In other words, g sends each integer n to $n + 1$, or, symbolically, $g: n \to n + 1$.

An example of a **constant function** is the function $h: \mathbf{Q} \to \mathbf{Z}$ defined by the formula $h(r) = 2$ for all rational numbers r. This function sends each rational number r to 2. In other words, no matter what the input, the output is always 2: $h(\square) = 2$ or $h: r \to 2$.

The functions f, g, and h are represented by the function machines in Figure 7.1.5.

$f(x) = x^2$

(a)

$g(n) = n + 1$

(b)

$h(r) = 2$

(c)

Figure 7.1.5

A function is an entity in its own right. It can be thought of as a certain relationship between sets or as an input/output machine that operates according to a certain rule. This is the reason why a function is generally denoted by a single symbol or string of symbols, such as f, G, or log.

> ⚠️ **Caution!** In some mathematical contexts, the notation $f(x)$ is used to refer both to the value of f at x and to the function f itself. Because using the notation this way can lead to confusion, we avoid it whenever possible. In this book, unless explicitly stated otherwise, the symbol $f(x)$ always refers to the value of the function f at x and not to the function f itself.

• Definition

Suppose f and g are functions from X to Y. Then f **equals** g, written $f = g$, if, and only if,

$$f(x) = g(x) \quad \text{for all } x \in X.$$

Note that if functions are defined formally as in Chapter 10, the definition given above is an immediate consequence of the function definition.

Example 7.1.4 Equality of Functions

a. Define $f: \mathbf{R} \to \mathbf{R}$ and $g: \mathbf{R} \to \mathbf{R}$ by the following formulas:

$$f(x) = |x| \quad \text{for all } x \in \mathbf{R},$$
$$g(x) = \sqrt{x^2} \quad \text{for all } x \in \mathbf{R}.$$

Does $f = g$?

b. Let $F: \mathbf{R} \to \mathbf{R}$ and $G: \mathbf{R} \to \mathbf{R}$ be functions. Define new functions $F + G: \mathbf{R} \to \mathbf{R}$ and $G + F: \mathbf{R} \to \mathbf{R}$ as follows:

$$(F + G)(x) = F(x) + G(x) \quad \text{for all } x \in \mathbf{R},$$
$$(G + F)(x) = G(x) + F(x) \quad \text{for all } x \in \mathbf{R}.$$

Does $F + G = G + F$?

Solution

a. Yes. Because the absolute value of a number equals the square root of its square, $|x| = \sqrt{x^2}$ for all $x \in \mathbf{R}$.

Hence $f = g$.

b. Again the answer is yes. For all real numbers x,

$$
\begin{aligned}
(F + G)(x) &= F(x) + G(x) && \text{by definition of } F + G \\
&= G(x) + F(x) && \text{by the commutative law for addition of real numbers} \\
&= (G + F)(x) && \text{by definition of } G + F
\end{aligned}
$$

Hence $F + G = G + F$. ∎

Examples of Functions

The following examples illustrate some of the wide variety of different types of functions.

Example 7.1.5 The Identity Function on a Set

Given a set X, define a function i_X from X to X by

$$i_X(x) = x \quad \text{for all } x \text{ in } X.$$

The function i_X is called the **identity function on X** because it sends each element of X to the element that is identical to it. Thus the identity function can be pictured as a machine that sends each piece of input directly to the output chute without changing it in any way.

Let X be any set and suppose that a_{ij}^k and $\phi(z)$ are elements of X. Find $i_X(a_{ij}^k)$ and $i_X(\phi(z))$.

Solution   Whatever is input to the identity function comes out unchanged, so $i_X(a_{ij}^k) = a_{ij}^k$ and $i_X(\phi(z)) = \phi(z)$. ∎

Example 7.1.6 Sequences

The formal definition of sequence specifies that a sequence is a function defined on the set of integers that are greater than or equal to a particular integer. For example, the sequence denoted

$$1, -\frac{1}{2}, \frac{1}{3}, -\frac{1}{4}, \frac{1}{5}, \ldots, \frac{(-1)^n}{n+1}, \ldots$$

can be thought of as the function f from the nonnegative integers to the real numbers that associates $0 \rightarrow 1$, $1 \rightarrow -\frac{1}{2}$, $2 \rightarrow \frac{1}{3}$, $3 \rightarrow -\frac{1}{4}$, $4 \rightarrow \frac{1}{5}$, and, in general, $n \rightarrow \frac{-1^n}{n+1}$. In other words, $f \colon \mathbf{Z}^{nonneg} \rightarrow \mathbf{R}$ is the function defined as follows:

$$\text{Send each integer } n \geq 0 \text{ to } f(n) = \frac{(-1)^n}{n+1}.$$

In fact, there are many functions that can be used to define a given sequence. For instance, express the sequence above as a function from the set of *positive* integers to the set of real numbers.

Solution   Define $g \colon \mathbf{Z}^+ \rightarrow \mathbf{R}$ by $g(n) = \frac{-1^{n+1}}{n}$, for each $n \in \mathbf{Z}^+$. Then $g(1) = 1$, $g(2) = -\frac{1}{2}$, $g(3) = \frac{1}{3}$, and in general

$$g(n+1) = \frac{(-1)^{n+2}}{n+1} = \frac{(-1)^n}{n+1} = f(n).$$ ∎

Example 7.1.7 A Function Defined on a Power Set

Recall from Section 5.3 that $\mathscr{P}(A)$ denotes the set of all subsets of the set A. Define a function $F\colon \mathscr{P}(\{a, b, c\}) \to \mathbf{Z}^{nonneg}$ as follows: For each $X \in \mathscr{P}(\{a, b, c\})$,

$$F(X) = \text{the number of elements in } X.$$

Draw an arrow diagram for F.

Solution

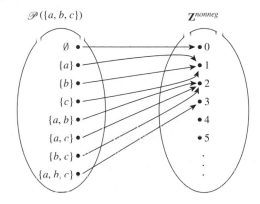

Example 7.1.8 Function Defined on a Set of Strings

In automata theory, the fundamental objects are sets of strings. Let S be the set of all strings of a's and b's, and let ϵ represent the null string (the "string" with no characters). Define a function $g\colon S \to \mathbf{Z}$ as follows: For each string $s \in S$,

$$g(s) = \text{the number of } a\text{'s in } s.$$

Find the following.

a. $g(\epsilon)$ b. $g(bb)$ c. $g(ababb)$ d. $g(bbbaa)$

Solution

a. 0   b. 0   c. 2   d. 2

Example 7.1.9 The Logarithmic Function

Let b be a positive real number. For each positive real number x, the **logarithm with base b of x,** written $\log_b x$, is the exponent to which b must be raised to obtain x.*Symbolically,

$$\log_b x = y \quad \Leftrightarrow \quad b^y = x.$$

The **logarithmic function with base b** is the function from $\mathbf{R}^+$ to $\mathbf{R}$ that takes each positive real number x to $\log_b x$. Find the following:

a. $\log_3 9$ b. $\log_2\left(\frac{1}{2}\right)$ c. $\log_{10}(1)$ d. $\log_2(2^m)$

Solution

a. $\log_3 9 = 2$ because $3^2 = 9$.   b. $\log_2(\frac{1}{2}) = -1$ because $2^{-1} = \frac{1}{2}$.

c. $\log_{10}(1) = 0$ because $10^0 = 1$.

d. $\log_2(2^m) = m$ because the exponent to which 2 must be raised to obtain 2^m is m. ■

---

*It is not obvious, but it is true, that for any positive real number x there is a unique real number y such that $b^y = x$. Most calculus books contain a discussion of this result.

Example 7.1.10 Encoding and Decoding Functions

Digital messages consist of finite sequences of 0's and 1's. When they are communicated across a transmission channel, they are frequently coded in special ways to reduce the possibility that they will be garbled by interfering noise in the transmission lines. For example, suppose a message consists of a sequence of 0's and 1's. A simple way to encode the message is to write each bit three times. Thus the message

$$00101111$$

would be encoded as

$$000000111000111111111111.$$

The receiver of the message decodes it by replacing each section of three identical bits by the one bit to which all three are equal.

Let A be the set of all strings of 0's and 1's, and let T be the set of all strings of 0's and 1's that consist of consecutive triples of identical bits. The encoding and decoding processes described above are actually functions from A to T and from T to A. The encoding function E is the function from A to T defined as follows: For each string $s \in A$,

$$E(s) = \text{the string obtained from } s \text{ by replacing each}$$
$$\text{bit of } s \text{ by the same bit written three times.}$$

The decoding function D is defined as follows: For each string $t \in T$,

$$D(t) = \text{the string obtained from } t \text{ by replacing each consecutive}$$
$$\text{triple of three identical bits of } t \text{ by a single copy of that bit.}$$

The advantage of this particular coding scheme is that it makes it possible to do a certain amount of error correction when interference in the transmission channels has introduced errors into the stream of bits. If the receiver of the coded message observes that one of the sections of three consecutive bits that should be identical does not consist of identical bits, then one bit differs from the other two. In this case, if errors are rare, it is likely that the single bit that is different is the one in error, and this bit is changed to agree with the other two before decoding. ∎

Example 7.1.11 The Hamming Distance Function

Richard Hamming (1915–1998)

Courtesy of the U.S. Naval Academy

The Hamming distance function, named after the computer scientist Richard W. Hamming, is very important in coding theory. It gives a measure of the "difference" between two strings of 0's and 1's that have the same length. Let S_n be the set of all strings of 0's and 1's of length n. Define a function $H: S_n \times S_n \to \mathbf{Z}^{nonneg}$ as follows: For each pair of strings $(s, t) \in S_n \times S_n$,

$$H(s, t) = \text{the number of positions in which } s \text{ and } t \text{ have different values.}$$

Thus, letting $n = 5$,

$$H(11111, 00000) = 5$$

because 11111 and 00000 differ in all five positions, whereas

$$H(11000, 00000) = 2$$

because 11000 and 00000 differ only in the first two positions.

a. Find $H(00101, 01110)$.   b. Find $H(10001, 01111)$.

Solution

a. 3   b. 4   ∎

Boolean Functions

In Section 1.4 we showed how to find input/output tables for certain digital logic circuits. Any such input/output table defines a function in the following way: The elements in the input column can be regarded as ordered tuples of 0's and 1's; the set of all such ordered tuples is the domain of the function. The elements in the output column are all either 0 or 1; thus $\{0, 1\}$ is taken to be the co-domain of the function. The relationship is that which sends each input element to the output element in the same row. Thus, for instance, the input/output table of Figure 7.1.6(a) defines the function with the arrow diagram shown in Figure 7.1.6(b).

More generally, the input/output table corresponding to a circuit with n input wires has n input columns. Such a table defines a function from the set of all n-tuples of 0's and 1's to the set $\{0, 1\}$.

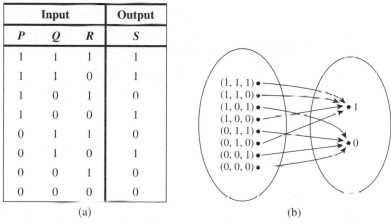

Input			Output
P	Q	R	S
1	1	1	1
1	1	0	1
1	0	1	0
1	0	0	1
0	1	1	0
0	1	0	1
0	0	1	0
0	0	0	0

(a)   (b)

Figure 7.1.6 Two Representations of a Boolean Function

• Definition

An (*n*-place) **Boolean function** f is a function whose domain is the set of all ordered n-tuples of 0's and 1's and whose co-domain is the set $\{0, 1\}$. More formally, the domain of a Boolean function can be described as the Cartesian product of n copies of the set $\{0, 1\}$, which is denoted $\{0, 1\}^n$. Thus $f: \{0, 1\}^n \to \{0, 1\}$.

It is customary to omit one set of parentheses when referring to functions defined on Cartesian products. For example, we write $f(1, 0, 1)$ rather than $f((1, 0, 1))$.

Example 7.1.12 A Boolean Function

Consider the three-place Boolean function defined from the set of all 3-tuples of 0's and 1's to $\{0, 1\}$ as follows: For each triple (x_1, x_2, x_3) of 0's and 1's,

$$f(x_1, x_2, x_3) = (x_1 + x_2 + x_3) \bmod 2.$$

Describe f using an input/output table.

Solution
$$f(1, 1, 1) = (1 + 1 + 1) \bmod 2 = 3 \bmod 2 = 1$$
$$f(1, 1, 0) = (1 + 1 + 0) \bmod 2 = 2 \bmod 2 = 0$$

The rest of the values of f can be calculated similarly to obtain the following table.

Input			Output
x_1	x_2	x_3	$(x_1 + x_2 + x_3) \bmod 2$
1	1	1	1
1	1	0	0
1	0	1	0
1	0	0	1
0	1	1	0
0	1	0	1
0	0	1	1
0	0	0	0

■

Checking Whether a Function Is Well Defined

It can sometimes happen that what appears to be a function defined by a rule is not really a function at all. To give an example, suppose we wrote, "Define a function $f : \mathbf{R} \to \mathbf{R}$ by the formula

$$f(x) = \sqrt{-x^2} \quad \text{for all real numbers } x.\text{"}$$

This definition is contradictory: On the one hand, f is supposed to be a function from the real numbers to the real numbers, but on the other hand, $\sqrt{-x^2}$ is a real number only when $x = 0$. In a situation like this we say that f is **not well defined** because the formula does not define a function.

Example 7.1.13 A Function That Is Not Well Defined

Recall that $\mathbf{Q}$ represents the set of all rational numbers. Suppose you read that a function $f : \mathbf{Q} \to \mathbf{Z}$ is to be defined by the formula

$$f\left(\frac{m}{n}\right) = m \quad \text{for all integers } m \text{ and } n \text{ with } n \neq 0.$$

That is, the integer associated by f to the number $\frac{m}{n}$ is m. Is f well defined? Why?

Solution   The function f is not well defined. The reason is that fractions have more than one representation as quotients of integers. For instance, $\frac{1}{2} = \frac{3}{6}$. Now if f were a function, then the definition of a function would imply that $f(\frac{1}{2}) = f(\frac{3}{6})$ since $\frac{1}{2} = \frac{3}{6}$. But applying the formula for f, you find that

$$f\left(\frac{1}{2}\right) = 1 \quad \text{and} \quad f\left(\frac{3}{6}\right) = 3,$$

and so

$$f\left(\frac{1}{2}\right) \neq f\left(\frac{3}{6}\right).$$

This contradiction shows that f is not well defined and is, therefore, not a function. ■

Note that the phrase *well-defined function* is actually redundant; for a function to be well defined really means that it is worthy of being called a function.

Exercise Set 7.1*

1. Let $X = \{1, 3, 5\}$ and $Y = \{s, t, u, v\}$. Define $f: X \to Y$ by the following arrow diagram.

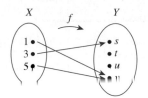

a. Write the domain of f and the co-domain of f.
b. Find $f(1)$, $f(3)$, and $f(5)$.
c. What is the range of f?
d. Is 3 an inverse image of s? Is 1 an inverse image of u?
e. What is the inverse image of s? of u? of v?
f. Represent f as a set of ordered pairs.

2. Let $X = \{1, 3, 5\}$ and $Y = \{a, b, c, d\}$. Define $g: X \to Y$ by the following arrow diagram.

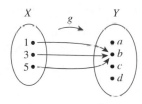

a. Write the domain of g and the co-domain of g.
b. Find $g(1)$, $g(3)$, and $g(5)$.
c. What is the range of g?
d. Is 3 an inverse image of a? Is 1 an inverse image of b?
e. What is the inverse image of b? of c?
f. Represent g as a set of ordered pairs.

3. Let $X = \{2, 4, 5\}$ and $Y = \{1, 2, 4, 6\}$. Which of the following arrow diagrams determine functions from X to Y?

a.

b.

c.

u.

e.

4. Indicate whether the statements in parts (a)–(d) are true or false. Justify your answers.

a. If two elements in the domain of a function are equal, then their images in the co-domain are equal.
b. If two elements in the co-domain of a function are equal, then their preimages in the domain are also equal.
c. A function can have the same output for more than one input.
d. A function can have the same input for more than one output.

5. a. Find all functions from $X = \{a, b\}$ to $Y = \{u, v\}$.
b. Find all functions from $X = \{a, b, c\}$ to $Y = \{u\}$.
c. Find all functions from $X = \{a, b, c\}$ to $Y = \{u, v\}$.

6. a. How many functions are there from a set with three elements to a set with four elements?
b. How many functions are there from a set with five elements to a set with two elements?
c. How many functions are there from a set with m elements to a set with n elements, where m and n are positive integers?

7. Define functions f and g from $\mathbf{R}$ to $\mathbf{R}$ by the following formulas:

For all $x \in \mathbf{R}$,

$$f(x) = 2x \quad \text{and} \quad g(x) = \frac{2x^3 + 2x}{x^2 + 1}.$$

Show that $f = g$.

8. Define functions H and K from $\mathbf{R}$ to $\mathbf{R}$ by the following formulas:

For all $x \in \mathbf{R}$,

$$H(x) = \lfloor x \rfloor + 1 \quad \text{and} \quad K(x) = \lceil x \rceil.$$

Does $H = K$? Explain.

9. Let F and G be functions from the set of all real numbers to itself. Define the product functions $F \cdot G: \mathbf{R} \to \mathbf{R}$ and $G \cdot F: \mathbf{R} \to \mathbf{R}$ as follows:

$$(F \cdot G)(x) = F(x) \cdot G(x) \quad \text{for all } x \in \mathbf{R},$$
$$(G \cdot F)(x) = G(x) \cdot F(x) \quad \text{for all } x \in \mathbf{R}.$$

Does $F \cdot G = G \cdot F$? Explain.

10. Let F and G be functions from the set of all real numbers to itself. Define new functions $F - G: \mathbf{R} \to \mathbf{R}$ and $G - F: \mathbf{R} \to \mathbf{R}$ as follows:

$$(F - G)(x) = F(x) - G(x) \quad \text{for all } x \in \mathbf{R},$$
$$(G - F)(x) = G(x) - F(x) \quad \text{for all } x \in \mathbf{R}.$$

Does $F - G = G - F$? Explain.

11. Let $i_{\mathbf{Z}}$ be the identity function defined on the set of all integers, and suppose that $e, b_i^{jk}, K(t)$, and u_{kj} all represent integers. Find
    a. $i_{\mathbf{Z}}(e)$ **b.** $i_{\mathbf{Z}}(b_i^{jk})$ **c.** $i_{\mathbf{Z}}(K(t))$ **d.** $i_{\mathbf{Z}}(U_{kj})$

12. Find functions defined on the set of nonnegative integers that define the sequences whose first six terms are given below.
    a. $1, -\dfrac{1}{3}, \dfrac{1}{5}, -\dfrac{1}{7}, \dfrac{1}{9}, -\dfrac{1}{11}$ **b.** $0, -2, 4, -6, 8, -10$

13. Let $A = \{1, 2, 3, 4, 5\}$ and define a function $F: \mathscr{P}(A) \to \mathbf{Z}$ as follows: For all sets X in $\mathscr{P}(A)$,

$$F(X) = \begin{cases} 0 & \text{if } X \text{ has an even} \\ & \text{number of elements} \\ 1 & \text{if } X \text{ has an odd} \\ & \text{number of elements} \end{cases}.$$

Find the following:
    a. $F(\{1, 3, 4\})$ **b.** $F(\emptyset)$
    c. $F(\{2, 3\})$ **d.** $F(\{2, 3, 4, 5\})$

14. Let S be the set of all strings of a's and b's.
    a. Define $f: S \to \mathbf{Z}$ as follows: For each string s in

$$f(s) = \begin{cases} \text{the number of } b\text{'s to the left} \\ \text{of the left-most } a \text{ in } s \\ 0 \quad \text{if } s \text{ contains no } a\text{'s} \end{cases}.$$

Find $f(aba)$, $f(bbab)$, and $f(b)$. What is the range of f?
    b. Define $g: S \to S$ as follows: For each string s in S,

$$g(s) = \text{the string obtained by writing the}$$
$$\text{characters of } s \text{ in reverse order.}$$

Find $g(aba)$, $g(bbab)$, and $g(b)$. What is the range of g?

15. Use the definition of logarithm to fill in the blanks below.
    a. $\log_2 8 = 3$ because _____.
    b. $\log_5(\frac{1}{25}) - 2$ because _____.
    c. $\log_4 4 = 1$ because _____.
    d. $\log_3(3^n) = n$ because _____.
    e. $\log_4 1 = 0$ because _____.

16. Find exact values for each of the following quantities. Do not use a calculator.
    a. $\log_3 81$ **b.** $\log_2 1024$ **c.** $\log_3(\frac{1}{27})$
    d. $\log_2 1$ **e.** $\log_{10}(\frac{1}{10})$ **f.** $\log_3 3$
    g. $\log_2(2^k)$

17. Use the definition of logarithm to prove that for any positive real number b with $b \neq 1$, $\log_b b = 1$.

18. Use the definition of logarithm to prove that for any positive real number b with $b \neq 1$, $\log_b 1 = 0$.

19. If b is any positive real number and x is any real number, b^{-x} is defined as follows: $b^{-x} = \dfrac{1}{b^x}$. Use this definition and the definition of logarithm to prove that $\log_b\left(\dfrac{1}{u}\right) = -\log_b(u)$ for all positive real numbers u and b.

H 20. Use the unique factorization theorem (Section 3.3) and the definition of logarithm to prove that $\log_3(7)$ is irrational.

21. If b and y are positive real numbers such that $\log_b y = 3$, what is $\log_{1/b}(y)$? Why?

22. If b and y are positive real numbers such that $\log_b y = 2$, what is $\log_{b^2}(y)$? Why?

23. Let $A = \{2, 3, 5\}$ and $B = \{x, y\}$. Let p_1 and p_2 be the **projections of $A \times B$ onto the first and second coordinates.** That is, for each pair $(a, b) \in A \times B$, $p_1(a, b) = a$ and $p_2(a, b) = b$.
    a. Find $p_1(2, y)$ and $p_1(5, x)$. What is the range of p_1?
    b. Find $p_2(2, y)$ and $p_2(5, x)$. What is the range of p_2?

24. Observe that *mod* and *div* can be defined as functions from $\mathbf{Z}^{nonneg} \times \mathbf{Z}^+$ to $\mathbf{Z}$. For each ordered pair (n, d) consisting of a nonnegative integer n and a positive integer d, let

$$mod(n, d) = n \bmod d \text{ (the nonnegative remainder} \\ \text{obtained when } n \text{ is divided by } d).$$

$$div(n, d) = n \operatorname{div} d \text{ (the integer quotient} \\ \text{obtained when } n \text{ is divided by } d).$$

Find each of the following:
    a. $mod(67, 10)$ and $div(67, 10)$
    b. $mod(59, 8)$ and $div(59, 8)$
    c. $mod(30, 5)$ and $div(30, 5)$

25. Consider the coding and decoding functions E and D defined in Example 7.1.10.
    a. Find $E(0110)$ and $D(111111000111)$.
    b. Find $E(1010)$ and $D(000000111111)$.

26. Consider the Hamming distance function defined in Example 7.1.11.
    a. Find $H(10101, 00011)$.
    b. Find $H(00110, 10111)$.

27. A permutation on a set can be regarded as a function from the set to itself. For instance, one permutation of $\{1, 2, 3, 4\}$ is 2341. It can be identified with the function that sends each position number to the number occupying that position. Since position 1 is occupied by 2, 1 is sent to 2 or $1 \rightarrow 2$; since position 2 is occupied by 3, 2 is sent to 3 or $2 \rightarrow 3$; and so forth. The entire permutation can be written using arrows as follows:

$$
\begin{array}{cccc}
1 & 2 & 3 & 4 \\
\downarrow & \downarrow & \downarrow & \downarrow \\
2 & 3 & 4 & 1
\end{array}
$$

a. Use arrows to write each of the six permutations of $\{1, 2, 3\}$.

b. Use arrows to write each of the permutations of $\{1, 2, 3, 4\}$ that keep 2 and 4 fixed.

c. Which permutations of $\{1, 2, 3\}$ keep no elements fixed?

d. Use arrows to write all permutations of $\{1, 2, 3, 4\}$ that keep no elements fixed.

28. Draw arrow diagrams for the Boolean functions defined by the following input/output tables.

a.

Input		Output
P	*Q*	*R*
1	1	0
1	0	1
0	1	0
0	0	1

b.

Input			Output
P	*Q*	*R*	*S*
1	1	1	1
1	1	0	0
1	0	1	1
1	0	0	1
0	1	1	0
0	1	0	0
0	0	1	0
0	0	0	1

29. Fill in the following table to show the values of all possible two-place Boolean functions.

Input	f_1	f_2	f_3	f_4	f_5	f_6	f_7	f_8	f_9	f_{10}	f_{11}	f_{12}	f_{13}	f_{14}	f_{15}	f_{16}
1 1																
1 0																
0 1																
0 0																

30. Consider the three-place Boolean function f defined by the following rule: For each triple (x_1, x_2, x_3) of 0's and 1's,

$$f(x_1, x_2, x_3) = (4x_1 + 3x_2 + 2x_3) \bmod 2.$$

a. Find $f(1, 1, 1)$ and $f(0, 0, 1)$.

b. Describe f using an input/output table.

31. Student A tries to define a function $g: \mathbf{Q} \rightarrow \mathbf{Z}$ by the rule

$$g\left(\frac{m}{n}\right) = m, \quad \text{for all integers } m \text{ and } n \text{ with } n \neq 0.$$

Student B claims that g is not well defined. Justify student B's claim.

32. Student C tries to define a function $h: \mathbf{Q} \rightarrow \mathbf{Q}$ by the rule

$$h\left(\frac{m}{n}\right) = \frac{m^2}{n}, \text{ for all integers } m \text{ and } n \text{ with } n \neq 0.$$

Student D claims that h is not well defined. Justify student D's claim.

33. On certain computers the integer data type goes from $-2,147,483,648$ through $2,147,483,647$. Let S be the set of all integers from $-2,147,483,648$ through $2,147,483,647$. Try to define a function $f: S \rightarrow S$ by the rule $f(n) = n^2$ for each n in S. Is f well defined? Why?

34. Given a set S and a subset A, the **characteristic function of A**, denoted χ_A, is the function defined from S to $\mathbf{Z}$ with the property that for all $u \in S$,

$$\chi_A(u) = \begin{cases} 1 & \text{if } u \in A \\ 0 & \text{if } u \notin A \end{cases}.$$

Show that each of the following holds for all subsets A and B of S and all $u \in S$.

a. $\chi_{A \cap B}(u) = \chi_A(u) \cdot \chi_B(u)$

b. $\chi_{A \cup B}(u) = \chi_A(u) + \chi_B(u) - \chi_A(u) \cdot \chi_B(u)$

Each of exercises 35–39 refers to the Euler phi function, denoted ϕ, which is defined as follows: For each integer $n \geq 1$, $\phi(n)$ is the number of positive integers less than or equal to n that have no common factors with n except ± 1. For example, $\phi(10) = 4$ because there are four positive integers less than or equal to 10 that have no common factors with 10 except ± 1; namely, 1, 3, 7, and 9.

35. Find each of the following:

a. $\phi(15)$ b. $\phi(2)$ c. $\phi(5)$

d. $\phi(12)$ e. $\phi(11)$ f. $\phi(1)$

✱ 36. Prove that if p is a prime number and n is an integer with $n \geq 1$, then $\phi(p^n) = p^n - p^{n-1}$.

H 37. Prove that there are infinitely many integers n for which $\phi(n)$ is a perfect square.

H 38. Use the inclusion/exclusion principle to prove the following: If $n = pq$, where p and q are distinct prime numbers, then $\phi(n) = (p - 1)(q - 1)$.

39. Use the inclusion/exclusion principle to prove the following: If $n = pqr$, where p, q, and r are distinct prime numbers, then $\phi(n) = (p-1)(q-1)(r-1)$.

Exercises 40–47 refer to the following definition:

Definition: If $f: X \to Y$ is a function and $A \subseteq X$ and $C \subseteq Y$, then

$$f(A) = \{y \in Y \mid y = f(x) \text{ for some } x \text{ in } A\}$$

and

$$f^{-1}(C) = \{x \in X \mid f(x) \in C\}.$$

$f(A)$ is called the **image of A,** and $f^{-1}(C)$ is called the **inverse image of C.**

Determine which of the properties in 40–47 are true for all functions f from a set X to a set Y and which are false for some function f. Justify your answers.

40. For all subsets A and B of X, if $A \subseteq B$, then $f(A) \subseteq f(B)$.

41. For all subsets A and B of X, $f(A \cup B) = f(A) \cup f(B)$.

42. For all subsets A and B of X, $f(A \cap B) = f(A) \cap f(B)$.

43. For all subsets A and B of X, $f(A - B) = f(A) - f(B)$.

44. For all subsets C and D of Y, if $C \subseteq D$, then

$$f^{-1}(C) \subseteq f^{-1}(D).$$

H 45. For all subsets C and D of Y,

$$f^{-1}(C \cup D) = f^{-1}(C) \cup f^{-1}(D).$$

46. For all subsets C and D of Y,

$$f^{-1}(C \cap D) = f^{-1}(C) \cap f^{-1}(D).$$

47. For all subsets C and D of Y,

$$f^{-1}(C - D) = f^{-1}(C) - f^{-1}(D).$$

7.2 One-to-One and Onto, Inverse Functions

Don't accept a statement just because it is printed. — Anna Pell Wheeler, 1883–1966

In this section we discuss two important properties that functions may satisfy: the property of being *one-to-one* and the property of being *onto*. Functions that satisfy both properties are called *one-to-one correspondences* or *one-to-one onto functions*. When a function is a one-to-one correspondence, the elements of its domain and co-domain match up perfectly, and we can define an *inverse function* from the co-domain to the domain that "undoes" the action of the function.

One-to-One Functions

In Section 7.1 we noted that a function may send several elements of its domain to the same element of its co-domain. In terms of arrow diagrams, this means that two or more arrows that start in the domain can point to the same element in the co-domain. On the other hand, a function may associate a different element of its co-domain to each element of its domain, which would mean that no two arrows that start in the domain would point to the same element of its co-domain. A function with this property is called *one-to-one* or *injective*. For a one-to-one function, each element of the range is the image of at most one element of the domain.

• Definition

Let F be a function from a set X to a set Y. F is **one-to-one** (or **injective**) if, and only if, for all elements x_1 and x_2 in X,

$$\text{if } F(x_1) = F(x_2), \text{ then } x_1 = x_2.$$

Or, equivalently,

$$\text{if } x_1 \neq x_2, \text{ then } F(x_1) \neq F(x_2).$$

Symbolically,

$$F: X \to Y \text{ is one-to-one} \quad \Leftrightarrow \quad \forall x_1, x_2 \in X, \text{ if } F(x_1) = F(x_2) \text{ then } x_1 = x_2.$$

To obtain a precise statement of what it means for a function *not* to be one-to-one, take the negation of one of the equivalent versions of the definition above. Thus:

A function $F: X \to Y$ is *not* one-to-one $\Leftrightarrow$ $\exists$ elements x_1 and x_2 in X with $F(x_1) = F(x_2)$ and $x_1 \neq x_2$.

That is, if elements x_1 and x_2 can be found that have the same function value but are not equal, then F is not one-to-one.

In terms of arrow diagrams, a one-to-one function can be thought of as a function that separates points. That is, it takes distinct points of the domain to distinct points of the co-domain. A function that is not one-to-one fails to separate points. That is, at least two points of the domain are taken to the same point of the co-domain. This is illustrated in Figure 7.2.1.

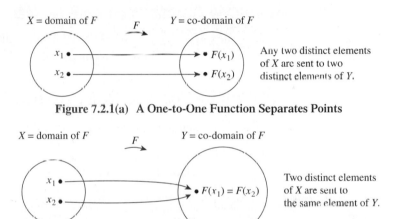

Figure 7.2.1(a) **A One-to-One Function Separates Points**

Figure 7.2.1(b) **A Function That Is Not One-to-One Collapses Points Together**

Example 7.2.1 Identifying One-to-One Functions Defined on Finite Sets

a. Do either of the arrow diagrams in Figure 7.2.2 define one-to-one functions?

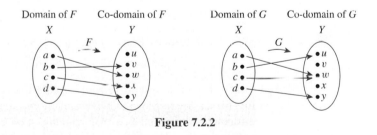

Figure 7.2.2

b. Let $X = \{1, 2, 3\}$ and $Y = \{a, b, c, d\}$. Define $H: X \to Y$ as follows: $H(1) = c$, $H(2) = a$, and $H(3) = d$. Define $K: X \to Y$ as follows: $K(1) = d$, $K(2) = b$, and $K(3) = d$. Is either H or K one-to-one?

Solution

a. F is one-to-one but G is not. F is one-to-one because no two different elements of X are sent by F to the same element of Y. G is not one-to-one because the elements a and c are both sent by G to the same element of Y: $G(a) = G(c) = w$ but $a \neq c$.

b. *H* is one-to-one but *K* is not. H is one-to-one because each of the three elements of the domain of *H* is sent by *H* to a different element of the co-domain: $H(1) \neq H(2)$, $H(1) \neq H(3)$, and $H(2) \neq H(3)$. *K*, however, is not one-to-one because $K(1) = K(3) = d$ but $1 \neq 3$. ∎

Consider the problem of writing a computer algorithm to check whether a function *F* is one-to-one. If *F* is defined on a finite set and there is an independent algorithm to compute values of *F*, then an algorithm to check whether *F* is one-to-one can be written as follows: Represent the domain of *F* as a one-dimensional array $a[1], a[2], \ldots, a[n]$ and use a nested loop to examine all possible pairs $(a[i], a[j])$, where $i < j$. If there is a pair $(a[i], a[j])$ for which $F(a[i]) = F(a[j])$ and $a[i] \neq a[j]$, then *F* is not one-to-one. If, however, all pairs have been examined without finding such a pair, then *F* is one-to-one. You are asked to write such an algorithm in exercise 53 at the end of this section.

One-to-One Functions on Infinite Sets

Now suppose *f* is a function defined on an infinite set *X*. By definition, *f* is one-to-one if, and only if, the following universal statement is true:

$$\forall x_1, x_2 \in X, \text{if } f(x_1) = f(x_2) \text{ then } x_1 = x_2.$$

Thus, to prove *f* is one-to-one, you will generally use the method of direct proof:

suppose x_1 and x_2 are elements of *X* such that $f(x_1) = f(x_2)$

and

show that $x_1 = x_2$.

To show that *f* is *not* one-to-one, you will ordinarily

find elements x_1 and x_2 in *X* so that $f(x_1) = f(x_2)$ but $x_1 \neq x_2$.

Example 7.2.2 Proving or Disproving That Functions Are One-to-One

Define $f: \mathbf{R} \to \mathbf{R}$ and $g: \mathbf{Z} \to \mathbf{Z}$ by the rules

$$f(x) = 4x - 1 \quad \text{for all} \quad x \in \mathbf{R}$$

and

$$g(n) = n^2 \quad \text{for all} \quad n \in \mathbf{Z}.$$

a. Is *f* one-to-one? Prove or give a counterexample.

b. Is *g* one-to-one? Prove or give a counterexample.

Solution   It is usually best to start by taking a positive approach to answering questions like these. Try to prove the given functions are one-to-one and see whether you run into difficulty. If you finish without running into any problems, then you have a proof. If you do encounter a problem, then analyzing the problem may lead you to discover a counterexample.

a. The function $f: \mathbf{R} \to \mathbf{R}$ is defined by the rule

$$f(x) = 4x - 1 \quad \text{for all real numbers } x.$$

To prove that f is one-to-one, it is necessary to prove that

$$\forall \text{ real numbers } x_1 \text{ and } x_2, \text{ if } f(x_1) = f(x_2) \text{ then } x_1 = x_2.$$

Substituting the definition of f into the outline of a direct proof, you

suppose x_1 and x_2 are any real numbers such that $4x_1 - 1 = 4x_2 - 1$,

and

show that $x_1 = x_2$,

Can you reach what is to be shown from the supposition? Of course. Just add 1 to both sides of the equation in the supposition and then divide both sides by 4.

This discussion is summarized in the following formal answer.

Answer to (a):

If the function $f: \mathbf{R} \to \mathbf{R}$ is defined by the rule $f(x) = 4x - 1$, for all real numbers x, then f is one-to-one.

Proof:

Suppose x_1 and x_2 are real numbers such that $f(x_1) = f(x_2)$. [*We must show that* $x_1 = x_2$.] By definition of f,

$$4x_1 - 1 = 4x_2 - 1.$$

Adding 1 to both sides gives

$$4x_1 = 4x_2,$$

and dividing both sides by 4 gives

$$x_1 = x_2,$$

which is what was to be shown.

---

b. The function $g: \mathbf{Z} \to \mathbf{Z}$ is defined by the rule

$$g(n) = n^2 \quad \text{for all integers } n.$$

As above, you start as though you were going to prove that g is one-to-one. Substituting the definition of g into the outline of a direct proof, you

suppose n_1 and n_2 are integers such that $n_1^2 = n_2^2$,

and

try to show that $n_1 = n_2$.

Can you reach what is to be shown from the supposition? No! It is quite possible for two numbers to have the same squares and yet be different. For example, $2^2 = (-2)^2$ but $2 \neq -2$.

Thus, in trying to prove that g is one-to-one, you run into difficulty. But analyzing this difficulty leads to the discovery of a counterexample, which shows that g is not one-to-one.

This discussion is summarized as follows:

Answer to (b):

If the function $g: \mathbf{Z} \to \mathbf{Z}$ is defined by the rule $g(n) = n^2$, for all $n \in \mathbf{Z}$, then g is not one-to-one.

Counterexample:

Let $n_1 = 2$ and $n_2 = -2$. Then by definition of g,

$$g(n_1) = g(2) = 2^2 = 4 \quad \text{and also}$$
$$g(n_2) = g(-2) = (-2)^2 = 4.$$

Hence

$$g(n_1) = g(n_2) \quad \text{but} \quad n_1 \neq n_2,$$

and so g is not one-to-one.

■

Application: Hash Functions

Imagine a set of student records, each of which includes the student's social security number, and suppose the records are to be stored in a table in which a record can be located if the social security number is known. One way to do this would be to place the record with social security number n into position n of the table. However, since social security numbers have nine digits, this method would require a table with 999,999,999 positions. The problem is that creating such a table for a small set of records would be very wasteful of computer memory space. **Hash functions** are functions defined from larger to smaller sets of integers, frequently using the *mod* function, which provide part of the solution to this problem. We illustrate how to define and use a hash function with a very simple example.

Example 7.2.3 A Hash Function

Suppose there are no more than seven student records. Define a function h from the set of all social security numbers (ignoring hyphens) to the set $\{0, 1, 2, 3, 4, 5, 6\}$ as follows:

$$h(n) = n \bmod 7 \quad \text{for all social security numbers } n.$$

To use your calculator to find $n \bmod 7$, use the formula $n \bmod 7 = n - 7 \cdot (n \ div \ 7)$. (See Section 3.4.) In other words, divide n by 7, multiply the integer part of the result by 7, and subtract that number from n. For instance, since $328343419/7 = 46906202.71\ldots,$

$$h(328\text{-}34\text{-}3419) = 328343419 - (7 \cdot 46906202) = 5.$$

As a first approximation to solving the problem of storing the records, try to place the record with social security number n in position $h(n)$. For instance, if the social security numbers are 328-34-3419, 356-63-3102, 223-79-9061, and 513-40-8716, the positions of the records are as shown in Table 7.2.1.

The problem with this approach is that h may not be one-to-one; h might assign the same position in the table to records with different social security numbers. Such an assignment is called a **collision.** When collisions occur, various **collision resolution**

Table 7.2.1

0	356-63-3102
1	
2	513-40-8716
3	223-79-9061
4	
5	328-34-3419
6	

methods are used. One of the simplest is the following: If, when the record with social security number n is to be placed, position $h(n)$ is already occupied, start from that position and search downward to place the record in the first empty position that occurs, going back up to the beginning of the table if necessary. To locate a record in the table from its social security number, n, you compute $h(n)$ and search downward from that position to find the record with social security number n. If there are not too many collisions, this is a very efficient way to store and locate records.

Suppose the social security number for another record to be stored is 908-37-1011. Find the position in Table 7.2.1 into which this record would be placed.

Solution   When you compute h you find that h(908-37-1011) $= 2$, which is already occupied by the record with social security number 513-40-8716. Searching downward from position 2, you find that position 3 is also occupied but position 4 is free.

$$908\text{-}37\text{-}1011 \;\; \xrightarrow{\;h\;} \;\; \underset{\substack{\uparrow \\ \text{occupied}}}{2} \;\; \rightarrow \;\; \underset{\substack{\uparrow \\ \text{occupied}}}{3} \;\; \rightarrow \;\; \underset{\substack{\uparrow \\ \text{free}}}{4}$$

Therefore, you place the record with social security number n into position 4. ∎

Onto Functions

It was noted in Section 7.1 that there may be an element of the co-domain of a function that is not the image of any element in the domain. On the other hand, a function may have the property that *every* element of its co-domain is the image of some element of its domain. Such a function is called *onto* or *surjective*. When a function is onto, its range is equal to its co-domain.

• Definition

Let F be a function from a set X to a set Y. F is **onto** (or **surjective**) if, and only if, given any element y in Y, it is possible to find an element x in X with the property that $y = F(x)$.

Symbolically:

$$F: X \to Y \text{ is onto} \;\; \Leftrightarrow \;\; \forall y \in Y, \exists x \in X \text{ such that } F(x) = y.$$

To obtain a precise statement of what it means for a function *not* to be onto, take the negation of the definition of onto:

$$\boxed{\; F: X \to Y \text{ is } not \text{ onto} \;\; \Leftrightarrow \;\; \exists y \text{ in } Y \text{ such that } \forall x \in X, F(x) \neq y. \;}$$

That is, there is some element in Y that is *not* the image of *any* element in X.

In terms of arrow diagrams, a function is onto if each element of the co-domain has an arrow pointing to it from some element of the domain. A function is not onto if at least one element in its co-domain does not have an arrow pointing to it. This is illustrated in Figure 7.2.3.

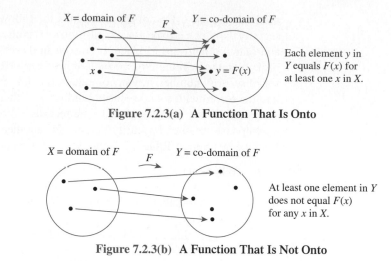

Figure 7.2.3(a) A Function That Is Onto

Each element y in Y equals $F(x)$ for at least one x in X.

Figure 7.2.3(b) A Function That Is Not Onto

At least one element in Y does not equal $F(x)$ for any x in X.

Example 7.2.4 Identifying Onto Functions Defined on Finite Sets

a. Do either of the arrow diagrams in Figure 7.2.4 define onto functions?

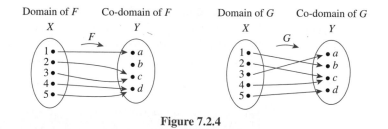

Figure 7.2.4

b. Let $X = \{1, 2, 3, 4\}$ and $Y = \{a, b, c\}$. Define $H: X \to Y$ as follows: $H(1) = c$, $H(2) = a, H(3) = c, H(4) = b$. Define $K: X \to Y$ as follows: $K(1) = c, K(2) = b$, $K(3) = b$, and $K(4) = c$. Is either H or K onto?

Solution

a. F is not onto because $b \neq F(x)$ for any x in X. G is onto because each element of Y equals $G(x)$ for some x in X: $a = G(3), b = G(1), c = G(2) = G(4)$, and $d = G(5)$.

b. H is onto but K is not. H is onto because each of the three elements of the co-domain of H is the image of some element of the domain of H: $a = H(2), b = H(4)$, and $c = H(1) = H(3)$. K, however, is not onto because $a \neq K(x)$ for any x in $\{1, 2, 3, 4\}$. ∎

It is possible to write a computer algorithm to check whether a function F is onto, provided F is defined from a finite set X to a finite set Y and there is an independent algorithm to compute values of F. Represent X and Y as one-dimensional arrays $a[1], a[2], \ldots, a[n]$ and $b[1], b[2], \ldots, b[m]$, respectively, and use a nested loop to pick each element y of Y in turn and search through the elements of X to find an x such that y is the image of x. If any search is unsuccessful, then F is not onto. If each such search is successful, then F is onto. You are asked to write such an algorithm in exercise 54 at the end of this section.

Onto Functions on Infinite Sets

Now suppose F is a function from a set X to a set Y, and suppose Y is infinite. By definition, F is onto if, and only if, the following universal statement is true:

$$\forall y \in Y, \exists x \in X \text{ such that } F(x) = y.$$

Thus to prove F is onto, you will ordinarily use the method of generalizing from the generic particular:

> **suppose** that y is any element of Y

and

> **show** that there is an element of X with $F(x) = y$.

To prove F is *not* onto, you will usually

> **find** an element y of Y such that $y \neq F(x)$ for *any* x in X.

Example 7.2.5 Proving or Disproving That Functions Are Onto

Define $f\colon \mathbf{R} \to \mathbf{R}$ and $h\colon \mathbf{Z} \to \mathbf{Z}$ by the rules

$$f(x) = 4x - 1 \quad \text{for all } x \in \mathbf{R}$$

and

$$h(n) = 4n - 1 \quad \text{for all } n \in \mathbf{Z}.$$

a. Is f onto? Prove or give a counterexample.

b. Is h onto? Prove or give a counterexample.

Solution

a. The best approach is to start trying to prove that f is onto and be alert for difficulties that might indicate that it is not. Now $f\colon \mathbf{R} \to \mathbf{R}$ is the function defined by the rule

$$\boxed{f(x) = 4x - 1 \quad \text{for all real numbers } x.}$$

To prove that f is onto, you must prove

$$\forall y \in Y, \exists x \in X \text{ such that } f(x) = y.$$

Substituting the definition of f into the outline of a proof by the method of generalizing from the generic particular, you

> **suppose** y is a real number

and

> **show** that there exists a real number x such that $y = 4x - 1$.

Scratch Work: If such a real number x exists, then

$$4x - 1 = y$$
$$4x = y + 1 \qquad \text{by adding 1 to both sides}$$
$$x = \frac{y + 1}{4} \qquad \text{by dividing both sides by 4.}$$

Thus *if* such a number x exists, it must equal $(y + 1)/4$. Does such a number exist? Yes. To show this, let $x = (y + 1)/4$, and then check that (1) x is a real number and (2) the steps above are valid, if followed in reverse order, to conclude $y = 4x - 1$. The following formal answer summarizes this process.

Answer to (a):

If $f: \mathbf{R} \to \mathbf{R}$ is the function defined by the rule $f(x) = 4x - 1$ for all real numbers x, then f is onto.

Proof:

Let $y \in \mathbf{R}$. [*We must show that* $\exists x$ *in* $\mathbf{R}$ *such that* $f(x) = y$.] Let $x = (y + 1)/4$. Then x is a real number since sums and quotients (other than by 0) of real numbers are real numbers. It follows that

$$f(x) = f\left(\frac{y + 1}{4}\right) \qquad \text{by substitution}$$

$$= 4 \cdot \left(\frac{y + 1}{4}\right) - 1 \qquad \text{by definition of } f$$

$$= (y + 1) - 1 = y \qquad \text{by basic algebra.}$$

This is what was to be shown.

---

b. The function $h: \mathbf{Z} \to \mathbf{Z}$ is defined by the rule

$$h(n) = 4n - 1 \quad \text{for all integers } n.$$

To prove that h is onto, it would be necessary to prove that

$$\forall \text{ integers } m, \exists \text{ an integer } n \text{ such that } h(n) = m.$$

Substituting the definition of h into the outline of a proof by the method of generalizing from the generic particular, you

suppose m is any integer

and

try to show that there is an integer n with $4n - 1 = m$.

Can you reach what is to be shown from the supposition? No! If $4n - 1 = m$, then

$$n = \frac{m + 1}{4} \qquad \text{by adding 1 and dividing by 4.}$$

But n must be an integer. And when, for example, $m = 0$, then

$$n = \frac{0 + 1}{4} = \frac{1}{4},$$

which is *not* an integer.

Thus, in trying to prove that h is onto, you run into difficulty, and this difficulty reveals a counterexample that shows h is not onto.

This discussion is summarized in the following formal answer.

Answer to (b):

If the function $h \colon \mathbf{Z} \to \mathbf{Z}$ is defined by the rule $h(n) = 4n - 1$ for all integers n, then h is not onto.

Counterexample:

The co-domain of h is $\mathbf{Z}$ and $0 \in \mathbf{Z}$. But $h(n) \neq 0$ for any integer n. For if $h(n) = 0$, then

$$4n - 1 = 0 \qquad \text{by definition of } h$$
$$4n = 1 \qquad \text{by adding 1 to both sides}$$
$$n = \frac{1}{4} \qquad \text{by dividing both sides by 4.}$$

But $1/4$ is not an integer. Hence there is no integer n for which $f(n) = 0$, and so f is not onto.

■

The Exponential and Logarithmic Functions

The **exponential function with base b,** denoted $\exp_b$, is the function from $\mathbf{R}$ to $\mathbf{R}^+$ defined as follows: For all real numbers x,

$$\exp_b(x) = b^x$$

where $b^0 = 1$ and $b^{-x} = 1/b^x$.*

When working with the exponential function, it is useful to recall the laws of exponents from elementary algebra.

Laws of Exponents

If b and c are any positive real numbers and u and v are any real numbers, the following laws of exponents hold true:

$$b^u b^v = b^{u+v} \qquad\qquad 7.2.1$$

$$(b^u)^v = b^{uv} \qquad\qquad 7.2.2$$

$$\frac{b^u}{b^v} = b^{u-v} \qquad\qquad 7.2.3$$

$$(bc)^u = b^u c^u \qquad\qquad 7.2.4$$

The logarithmic function with base b was defined in Example 7.1.9 to be the function from $\mathbf{R}^+$ to $\mathbf{R}$ with the property that for each positive real number x,

$$\log_b(x) = \text{ the exponent to which } b \text{ must be raised to obtain } x.$$

---

*That the quantity b^x is a real number for any real number x follows from the least-upper-bound property of the real number system. (See Appendix A.)

Or, equivalently, for each positive real number x and real number y,

$$\log_b x = y \iff b^y = x.$$

It can be shown using calculus that both the exponential and logarithmic functions are one-to-one and onto. Therefore, by definition of one-to-one, the following properties hold true:

For any positive real number b,

$$\text{if } b^u = b^v \text{ then } u = v \quad \text{for all numbers } u \text{ and } v, \qquad 7.2.5$$

and

$$\text{if } \log_b u = \log_b v \text{ then } u = v \quad \text{for all positive real numbers } u \text{ and } v. \qquad 7.2.6$$

These properties are used to derive many additional facts about exponents and logarithms. One example is given below.

Example 7.2.6 Using the One-to-Oneness of the Exponential Function

Use the definition of logarithm, the laws of exponents, and the one-to-oneness of the exponential function (property 7.2.5) to show that for any positive real numbers b, c, and x, with $c \neq 1$,

$$\log_c x = \frac{\log_b x}{\log_b c}. \qquad 7.2.7$$

Solution Suppose positive real numbers b, c, and x are given. Let

$$(1) \ \ u = \log_b c \qquad (2) \ \ v = \log_c x \qquad (3) \ \ w = \log_b x.$$

Then, by definition of logarithm,

$$(1') \ \ c = b^u \qquad (2') \ \ x = c^v \qquad (3') \ \ x = b^w.$$

Substituting $(1')$ into $(2')$ gives

$$x = c^v = (b^u)^v = b^{uv} \quad \text{by 7.2.2}$$

But by (3), $x = b^w$ also. Hence

$$b^{uv} = b^w,$$

and so by the one-to-oneness of the exponential function (property 7.2.5),

$$uv = w.$$

Substituting from (1), (2), and (3) gives that

$$(\log_b c)(\log_c x) = \log_b x.$$

And dividing both sides by $\log_b c$ (which is nonzero because $c \neq 1$) results in

$$\log_c x = \frac{\log_b x}{\log_b c}. \qquad \blacksquare$$

Example 7.2.7 Computing Logarithms with Base 2 on a Calculator

In computer science it is often necessary to compute logarithms with base 2. Most calculators do not have keys to compute logarithms with base 2 but do have keys to compute logarithms with base 10 (called **common logarithms** and often denoted simply log) and logarithms with base e (called **natural logarithms** and usually denoted ln). Suppose your calculator shows that $\ln 5 \cong 1.609437912$ and $\ln 2 \cong 0.6931471806$. Use formula (7.2.7) from Example 7.2.6 to find an approximate value for $\log_2 5$.

Solution By formula (7.2.7),

$$\log_2 5 = \frac{\ln 5}{\ln 2} \cong \frac{1.609437912}{0.6931471806} \cong 2.321928095. \qquad \blacksquare$$

One-to-One Correspondences

Consider a function $F\colon X \to Y$ that is both one-to-one and onto. Given any element x in X, there is a unique corresponding element $y = F(x)$ in Y (since F is a function). Also given any element y in Y, there is an element x in X such that $F(x) = y$ (since F is onto) and there is only one such x (since F is one-to-one). Thus, a function that is one to one and onto sets up a pairing between the elements of X and the elements of Y that matches each element of X with exactly one element of Y and each element of Y with exactly one element of X. Such a pairing is called a *one-to-one correspondence* or *bijection* and is illustrated by the arrow diagram in Figure 7.2.5. In Chapter 6 we frequently used one-to-one correspondences to count the number of elements in a set. The pairing of Figure 7.2.5, for example, shows that there are five elements in the set X.

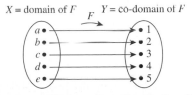

Figure 7.2.5 An Arrow Diagram for a One-to-One Correspondence

> **• Definition**
>
> A **one-to-one correspondence** (or **bijection**) from a set X to a set Y is a function $F\colon X \to Y$ that is both one-to-one and onto.

Example 7.2.8 A Function from a Power Set to a Set of Strings

Let $\mathscr{P}(\{a, b\})$ be the set of all subsets of $\{a, b\}$ and let S be the set of all strings of length 2 made up of 0's and 1's. Then $\mathscr{P}(\{a, b\}) = \{\emptyset, \{a\}, \{b\}, \{a, b\}\}$ and $S = \{00, 01, 10, 11\}$. Define a function h from $\mathscr{P}(\{a, b\})$ to S as follows: Given any subset A of $\{a, b\}$, a is either in A or not in A, and b is either in A or not in A. If a is in A, write a 1 in the first position of the string $h(A)$. If a is not in A, write a 0 in the first position of the string $h(A)$. Similarly, if b is in A, write a 1 in the second position of the string $h(A)$. If b is not in A, write a 0 in the second position of the string $h(A)$. This definition is summarized in the following table.

h

Subset of $\{a, b\}$	Status of a	Status of b	String in S
$\emptyset$	not in	not in	00
$\{a\}$	in	not in	10
$\{b\}$	not in	in	01
$\{a, b\}$	in	in	11

Is h a one-to-one correspondence?

Solution   The arrow diagram shown in Figure 7.2.6 shows clearly that h is a one-to-one correspondence. It is onto because each element of S has an arrow pointing to it. It is one-to-one because each element of S has no more than one arrow pointing to it.

Figure 7.2.6

Example 7.2.9 A String-Reversing Function

Let T be the set of all finite strings of x's and y's. Define $g: T \to T$ by the rule

For all strings $s \in T$,

$$g(s) = \text{the string obtained by writing the}$$
$$\text{characters of } s \text{ in reverse order.}$$

Is g a one-to-one correspondence from T to itself?

Solution   The answer is yes. To show that g is a one-to-one correspondence, it is necessary to show that g is one-to-one and onto.

To see that g is one-to-one, suppose that for some strings s_1 and s_2 in T, $g(s_1) = g(s_2)$. [*We must show that $s_1 = s_2$.*] Now to say that $g(s_1) = g(s_2)$ is the same as saying that the string obtained by writing the characters of s_1 in reverse order equals the string obtained by writing the characters of s_2 in reverse order. But if s_1 and s_2 are equal when written in reverse order, then they must be equal to start with. In other words, $s_1 = s_2$ [*as was to be shown*].

To show that g is onto, suppose t is a string in T. [*We must find a string s in T such that $g(s) = t$.*] Let $s = g(t)$. By definition of g, $s = g(t)$ is the string in T obtained by writing the characters of t in reverse order. But when the order of the characters of a string is reversed once and then reversed again, the original string is recovered. Thus

$$g(s) = g(g(t)) = \text{the string obtained by writing the characters}$$
$$\text{of } t \text{ in reverse order and then writing those}$$
$$\text{characters in reverse order again}$$
$$= t.$$

This is what was to be shown.

Inverse Functions

If F is a one-to-one correspondence from a set X to a set Y, then there is a function from Y to X that "undoes" the action of F; that is, it sends each element of Y back to the element of X that it came from. This function is called the *inverse function* for F.

Theorem 7.2.1

Suppose $F: X \to Y$ is a one-to-one correspondence, that is, suppose F is one-to-one and onto. Then there is a function $F^{-1}: Y \to X$ that is defined as follows:

Given any element y in Y,

$$F^{-1}(y) = \text{that unique element } x \text{ in } X \text{ such that } F(x) \text{ equals } y.$$

In other words,

$$F^{-1}(y) = x \quad \Leftrightarrow \quad y = F(x).$$

The proof of Theorem 7.2.1 follows immediately from the definition of one-to-one and onto. Given an element y in Y, there is an element x in X with $F(x) = y$ because F is onto; x is unique because F is one-to-one.

• Definition

The function F^{-1} of Theorem 7.2.1 is called the **inverse function** for F.

Note that according to this definition, the logarithmic function with base $b > 0$ is the inverse of the exponential function with base b.

The diagram that follows illustrates the fact that an inverse function sends each element back to where it came from.

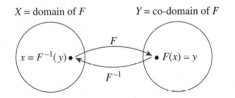

$X = $ domain of F $Y = $ co-domain of F

Example 7.2.10 Finding an Inverse Function for a Function Given by an Arrow Diagram

Define the inverse function for the one-to-one correspondence h given in Example 7.2.8.

Solution   The arrow diagram for h^{-1} is obtained by tracing the h-arrows back from S to $\mathscr{P}(\{a, b\})$ as shown below.

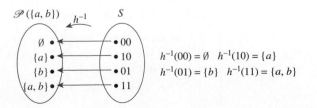

$h^{-1}(00) = \emptyset \quad h^{-1}(10) = \{a\}$

$h^{-1}(01) = \{b\} \quad h^{-1}(11) = \{a, b\}$

Example 7.2.11 Finding an Inverse Function for a Function Given in Words

Define the inverse function for the one-to-one correspondence g given in Example 7.2.9.

Solution The function $g: T \rightarrow T$ is defined by the rule

For all strings t in T,

$$g(t) = \text{the string obtained by writing the characters of } t \text{ in reverse order.}$$

Now if the characters of t are written in reverse order and then written in reverse order again, the original string is recovered. Thus given any string t in T,

$$g^{-1}(t) = \text{the unique string that, when written in reverse order, equals } t$$

$$= \text{the string obtained by writing the characters of } t \text{ in reverse order}$$

$$= g(t).$$

Hence $g^{-1}: T \rightarrow T$ is the same as g, or, in other words, $g^{-1} = g$. ∎

Example 7.2.12 Finding an Inverse Function for a Function Given by a Formula

The function $f: \mathbf{R} \rightarrow \mathbf{R}$ defined by the formula

$$f(x) = 4x - 1 \quad \text{for all real numbers } x$$

was shown to be one-to-one in Example 7.2.2 and onto in Example 7.2.5. Find its inverse function.

Solution By definition of f^{-1},

$$f^{-1}(y) = \text{that unique real number } y \text{ such that } f(x) = y.$$

But

$$f(x) = y$$
$$\Leftrightarrow \quad 4x - 1 = y \qquad \text{by definition of } f$$
$$\Leftrightarrow \qquad x = \frac{y+1}{4} \qquad \text{by adding 1 and dividing both sides by 4.}$$

Hence $f^{-1}(y) = \dfrac{y+1}{4}$. ∎

The following theorem follows easily from the definitions.

Theorem 7.2.2

If X and Y are sets and $F: X \rightarrow Y$ is one-to-one and onto, then $F^{-1}: Y \rightarrow X$ is also one-to-one and onto.

Proof:

F^{-1} *is one-to-one:* Suppose y_1 and y_2 are elements of Y such that $F^{-1}(y_1) = F^{-1}(y_2)$. [*We must show that* $y_1 = y_2$.] Let $x = F^{-1}(y_1) = F^{-1}(y_2)$. Then $x \in X$, and by definition of F^{-1},

$$F(x) = y_1 \quad \text{since } x = F^{-1}(y_1)$$

and

$$F(x) = y_2 \quad \text{since } x = F^{-1}(y_2).$$

Consequently, $y_1 = y_2$ since each is equal to $F(x)$. This is what was to be shown.

F^{-1} is onto: Suppose $x \in X$. [*We must show that there exists an element y in Y such that $F^{-1}(y) = x$.*] Let $y = F(x)$. Then $y \in Y$, and by definition of F^{-1}, $F^{-1}(y) = x$. This is what was to be shown.

Exercise Set 7.2

1. The definition of one-to-one is stated in two ways:

$$\forall x_1, x_2 \in X, \text{ if } F(x_1) = F(x_2) \text{ then } x_1 = x_2$$

and

$$\forall x_1, x_2 \in X, \text{ if } x_1 \neq x_2 \text{ then } F(x_1) \neq F(x_2).$$

Why are these two statements logically equivalent?

2. Fill in each blank with the word *most* or *least.*
   a. A function F is one-to-one if, and only if, each element in the co-domain of F is the image of at ___most___ one element in the domain of F.
   b. A function F is onto if, and only if, each element in the co-domain of F is the image of at _____ one element in the domain of F.

3. When asked to state the definition of one-to-one, a student replies, "A function f is one-to-one if, and only if, every element of X is sent by f to exactly one element of Y. Give a counterexample to show that the student's reply is incorrect.

H 4. Let $f: X \to Y$ be a function. True or false? A sufficient condition for f to be one-to-one is that for all elements y in Y, there is at most one x in X with $f(x) = y$.

H 5. All but two of the following statements are correct ways to express the fact that a function f is onto. Find the two that are incorrect.
   a. f is onto $\Leftrightarrow$ every element in its co-domain is the image of some element in its domain.
   b. f is onto $\Leftrightarrow$ every element in its domain has a corresponding image in its co-domain.
   c. f is onto $\Leftrightarrow \forall y \in Y, \exists x \in X$ such that $f(x) = y$.
   d. f is onto $\Leftrightarrow \forall x \in X, \exists y \in Y$ such that $f(x) = y$.
   e. f is onto $\Leftrightarrow$ the range of f is the same as the co-domain of f.

6. Let $X = \{1, 5, 9\}$ and $Y = \{3, 4, 7\}$.
   a. Define $f: X \to Y$ by specifying that
   $$f(1) = 4, \quad f(5) = 7, \quad f(9) = 4.$$
   Is f one-to-one? Is f onto? Explain your answers.
   b. Define $g: X \to Y$ by specifying that
   $$g(1) = 7, \quad g(5) = 3, \quad g(9) = 4.$$
   Is g one-to-one? Is g onto? Explain your answers.

7. Let $X = \{a, b, c, d\}$ and $Y = \{x, y, z\}$. Define functions F and G by the arrow diagrams below.

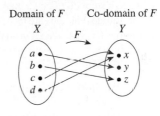

Domain of F Co-domain of F

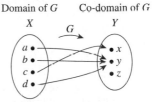

Domain of G Co-domain of G

   a. Is F one-to-one? Why or why not? Is it onto? Why or why not?
   b. Is G one-to-one? Why or why not? Is it onto? Why or why not?

8. Let $X = \{a, b, c\}$ and $Y = \{w, x, y, z\}$. Define functions H and K by the arrow diagrams below.

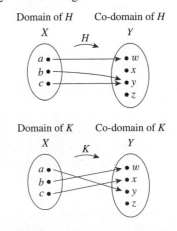

Domain of H Co-domain of H

Domain of K Co-domain of K

   a. Is H one-to-one? Why or why not? Is it onto? Why or why not?
   b. Is K one-to-one? Why or why not? Is it onto? Why or why not?

9. Let $X = \{1, 2, 3\}$, $Y = \{1, 2, 3, 4\}$, and $Z = \{1, 2\}$.
   a. Define a function $f: X \to Y$ that is one-to-one but not onto.
   b. Define a function $g: X \to Z$ that is onto but not one-to-one.
   c. Define a function $h: X \to X$ that is neither one-to-one nor onto.
   d. Define a function $k: X \to X$ that is one-to-one and onto but is not the identity function on X.

10. a. How many one-to-one functions are there from a set with three elements to a set with four elements?
    b. How many one-to-one functions are there from a set with three elements to a set with two elements?
    c. How many one-to-one functions are there from a set with three elements to a set with three elements?
    d. How many one-to-one functions are there from a set with three elements to a set with five elements?
  H e. How many one-to-one functions are there from a set with m elements to a set with n elements, where $m \leq n$?

11. a. How many onto functions are there from a set with three elements to a set with two elements?
    b. How many onto functions are there from a set with three elements to a set with five elements?
  H c. How many onto functions are there from a set with three elements to a set with three elements?
    d. How many onto functions are there from a set with four elements to a set with two elements?
    e. How many onto functions are there from a set with four elements to a set with three elements?
  H ✳ f. Let $c_{m,n}$ be the number of onto functions from a set of m elements to a set of n elements, where $m \geq n \geq 1$. Find a formula relating $c_{m,n}$ to $c_{m-1,n}$ and $c_{m-1,n-1}$.

12. a. Define $f: \mathbf{Z} \to \mathbf{Z}$ by the rule $f(n) = 2n$, for all integers n.
      (i) Is f one-to-one? Prove or give a counterexample.
      (ii) Is f onto? Prove or give a counterexample.
    b. Let $2\mathbf{Z}$ denote the set of all even integers. That is, $2\mathbf{Z} = \{n \in \mathbf{Z} \mid n = 2k, \text{ for some integer } k\}$. Define $h: \mathbf{Z} \to 2\mathbf{Z}$ by the rule $h(n) = 2n$, for all integers n. Is h onto? Prove or give a counterexample.

13. a. Define $g: \mathbf{Z} \to \mathbf{Z}$ by the rule $g(n) = 4n - 5$, for all integers n.
      (i) Is g one-to-one? Prove or give a counterexample.
      (ii) Is g onto? Prove or give a counterexample.
    b. Define $G: \mathbf{R} \to \mathbf{R}$ by the rule $G(x) = 4x - 5$ for all real numbers x. Is G onto? Prove or give a counterexample.

14. a. Define $H: \mathbf{R} \to \mathbf{R}$ by the rule $H(x) = x^2$, for all real numbers x.
      (i) Is H one-to-one? Prove or give a counterexample.
      (ii) Is H onto? Prove or give a counterexample.
    b. Define $K: \mathbf{R}^{nonneg} \to \mathbf{R}^{nonneg}$ by the rule $K(x) = x^2$, for all nonnegative real numbers x. Is K onto? Prove or give a counterexample.

15. Explain the mistake in the following "proof."

    Theorem: The function $f: \mathbf{Z} \to \mathbf{Z}$ defined by the formula $f(n) = 4n + 3$, for all integers n, is one-to-one.

    "**Proof:** Suppose any integer n is given. Then by definition of f, there is only one possible value for $f(n)$, namely, $4n + 3$. Hence f is one-to-one."

In each of 16–19 a function f is defined on a set of real numbers. Determine whether or not f is one-to-one and justify your answer.

16. $f(x) = \dfrac{x + 1}{x}$, for all real numbers $x \neq 0$

17. $f(x) = \dfrac{x}{x^2 + 1}$, for all real numbers x

18. $f(x) = \dfrac{3x - 1}{x}$, for all real numbers $x \neq 0$

19. $f(x) = \dfrac{x + 1}{x - 1}$, for all real numbers $x \neq 1$

20. Referring to Example 7.2.3, assume that records with the following social security numbers are to be placed in sequence into Table 7.2.1. Find the position into which each record is placed.
    a. 417-30-2072   b. 364-98-1703   c. 283-09-0787

21. Define Floor: $\mathbf{R} \to \mathbf{Z}$ by the formula $\text{Floor}(x) = \lfloor x \rfloor$, for all real numbers x.
    a. Is Floor one-to-one? Prove or give a counterexample.
    b. Is Floor onto? Prove or give a counterexample.

22. Let S be the set of all strings of 0's and 1's, and define $l: S \to \mathbf{Z}^{nonneg}$ by
    $$l(s) = \text{the length of } s, \quad \text{for all strings } s \text{ in } S.$$
    a. Is l one-to-one? Prove or give a counterexample.
    b. Is l onto? Prove or give a counterexample.

23. Let S be the set of all strings of 0's and 1's, and define $D: S \to \mathbf{Z}$ as follows: For all $s \in S$,
    $$D(s) = \text{the number of 1's in } s \text{ minus the number of 0's in } s.$$
    a. Is D one-to-one? Prove or give a counterexample.
    b. Is D onto? Prove or give a counterexample.

24. Define $F: \mathscr{P}(\{a, b, c\}) \to \mathbf{Z}$ as follows: For all A in $\mathscr{P}(\{a, b, c\})$,
    $$F(A) = \text{the number of elements in } A.$$
    a. Is F one-to-one? Prove or give a counterexample.
    b. Is F onto? Prove or give a counterexample.

25. Let S be the set of all strings of a's and b's, and define $N: S \to \mathbf{Z}$ by
    $$N(s) = \text{the number of } a\text{'s in } s, \quad \text{for all } s \in S.$$
    a. Is N one-to-one? Prove or give a counterexample.
    b. Is N onto? Prove or give a counterexample.

26. Let S be the set of all strings in a's and b's, and define $C: S \rightarrow S$ by

$$C(s) = as, \quad \text{for all } s \in S.$$

(C is called **concatenation** by a on the left.)
   a. Is C one-to-one? Prove or give a counterexample.
   b. Is C onto? Prove or give a counterexample.

✱ 27. Define $F: \mathbf{Z}^+ \times \mathbf{Z}^+ \rightarrow \mathbf{Z}^+$ and $G: \mathbf{Z}^+ \times \mathbf{Z}^+ \rightarrow \mathbf{Z}^+$ as follows: for all $(n, m) \in \mathbf{Z}^+ \times \mathbf{Z}^+$,

$$F(n, m) = 3^n 5^m \quad \text{and} \quad G(n, m) = 3^n 6^m.$$

   a. Is F one-to-one? Prove or give a counterexample.
   b. Is G one-to-one? Prove or give a counterexample.

28. a. Is $\log_8 27 = \log_2 3$? Why or why not?
   b. Is $\log_{16} 9 = \log_4 3$? Why or why not?

The properties of logarithms established in 29 and 30 are used in Sections 9.4 and 9.5.

29. Prove that for all positive real numbers $b, x,$ and y with $b \neq 1$,

$$\log_b \left(\frac{x}{y} \right) = \log_b x - \log_b y.$$

30. Prove that for all positive real numbers $b, x,$ and y with $b \neq 1$,

$$\log_b(xy) = \log_b x + \log_b y.$$

31. Prove that for all real numbers $a, b,$ and x with b and x positive and $b \neq 1$,

$$\log_b(x^a) = a \log_b x.$$

Exercises 32 and 33 use the following definition: If $f: \mathbf{R} \rightarrow \mathbf{R}$ and $g: \mathbf{R} \rightarrow \mathbf{R}$ are functions, then the function $(f + g): \mathbf{R} \rightarrow \mathbf{R}$ is defined by the formula $(f + g)(x) = f(x) + g(x)$ for all real numbers x.

32. If $f: \mathbf{R} \rightarrow \mathbf{R}$ and $g: \mathbf{R} \rightarrow \mathbf{R}$ are both one-to-one, is $f + g$ also one-to-one? Justify your answer.

33. If $f: \mathbf{R} \rightarrow \mathbf{R}$ and $g: \mathbf{R} \rightarrow \mathbf{R}$ are both onto, is $f + g$ also onto? Justify your answer.

Exercises 34 and 35 use the following definition: If $f: \mathbf{R} \rightarrow \mathbf{R}$ is a function and c is a nonzero real number, the function $(c \cdot f): \mathbf{R} \rightarrow \mathbf{R}$ is defined by the formula $(c \cdot f)(x) = c \cdot f(x)$ for all real numbers x.

34. Let $f: \mathbf{R} \rightarrow \mathbf{R}$ be a function and c a nonzero real number. If f is one-to-one, is $c \cdot f$ also one-to-one? Justify your answer.

35. Let $f: \mathbf{R} \rightarrow \mathbf{R}$ be a function and c a nonzero real number. If f is onto, is $c \cdot f$ also onto? Justify your answer.

Let $X = \{a, b, c, d, e\}$ and $Y = \{s, t, u, v, w\}$. In each of 36 and 37 a one-to-one correspondence $F: X \rightarrow Y$ is defined by an arrow diagram. In each case draw an arrow diagram for F^{-1}.

36.

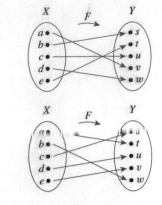

37.

In 38–51 indicate which of the functions in the referenced exercise are one-to-one correspondences. For each function that is a one-to-one correspondence, find the inverse function.

38. Exercise 12a   **39.** Exercise 12b

40. Exercise 13a   **41.** Exercise 13b

42. Exercise 14b   **43.** Exercise 21

44. Exercise 22   45. Exercise 23

46. Exercise 24   47. Exercise 25

48. Exercise 16 with the co-domain taken to be the set of all real numbers not equal to 1.

H **49.** Exercise 17 with the co-domain taken to be the set of all real numbers.

50. Exercise 18 with the co-domain taken to be the set of all real numbers not equal to 3.

51. Exercise 19 with the co-domain taken to be the set of all real numbers not equal to 1.

52. In Example 7.2.8 a one-to-one correspondence was defined from the power set of $\{a, b\}$ to the set of all strings of 0's and 1's that have length 2. Thus the elements of these two sets can be matched up exactly, and so the two sets have the same number of elements.
   a. Let $X = \{x_1, x_2, \ldots, x_n\}$ be a set with n elements. Use Example 7.2.8 as a model to define a one-to-one correspondence from $\mathscr{P}(X)$, the set of all subsets of X, to the set of all strings of 0's and 1's that have length n.
   b. Use the one-to-one correspondence of part (a) to deduce that a set with n elements has 2^n subsets. (This provides an alternative proof of Theorem 5.3.5.)

H **53.** Write a computer algorithm to check whether a function from one finite set to another is one-to-one. Assume the existence of an independent algorithm to compute values of the function.

H **54.** Write a computer algorithm to check whether a function from one finite set to another is onto. Assume the existence of an independent algorithm to compute values of the function.

7.3 Application: The Pigeonhole Principle

The shrewd guess, the fertile hypothesis, the courageous leap to a tentative
conclusion—these are the most valuable coin of the thinker at work.
— Jerome S. Bruner, 1960

The pigeonhole principle states that if n pigeons fly into m pigeonholes and $n > m$, then at least one hole must contain two or more pigeons. This principle is illustrated in Figure 7.3.1 for $n = 5$ and $m = 4$. Illustration (a) shows the pigeons perched next to their holes, and (b) shows the correspondence from pigeons to pigeonholes. The pigeonhole principle is sometimes called the *Dirichlet box principle* because it was first stated formally by J. P. G. L. Dirichlet (1805–1859).

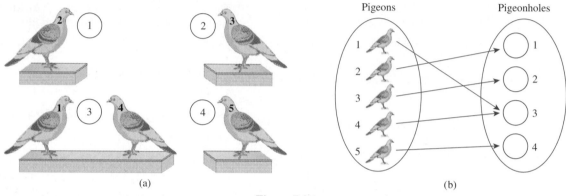

(a) (b)

Figure 7.3.1

Illustration (b) suggests the following mathematical way to phrase the principle.

Pigeonhole Principle

A function from one finite set to a smaller finite set cannot be one-to-one: There must be a least two elements in the domain that have the same image in the co-domain.

Thus an arrow diagram for a function from a finite set to a smaller finite set must have at least two arrows from the domain that point to the same element of the co-domain. In Figure 7.3.1(b), arrows from pigeons 1 and 4 both point to pigeonhole 3.

Since the truth of the pigeonhole principle is easy to accept on an intuitive basis, we move immediately to applications, leaving a formal proof to the end of the section. Applications of the pigeonhole principle range from the totally obvious to the extremely subtle. A representative sample is given in the examples and exercises that follow.

Example 7.3.1 Applying the Pigeonhole Principle

a. In a group of six people, must there be at least two who were born in the same month? In a group of thirteen people, must there be at least two who were born in the same month? Why?

b. Among the residents of New York City, must there be at least two people with the same number of hairs on their heads? Why?

Solution

a. A group of six people need not contain two who were born in the same month. For instance, the six people could have birthdays in each of the six months January through June.

A group of thirteen people, however, must contain at least two who were born in the same month, for there are only twelve months in a year and $13 > 12$. To get at the essence of this reasoning, think of the thirteen people as the pigeons and the twelve months of the year as the pigeonholes. Denote the thirteen people by the symbols $x_1, x_2, \ldots, x_{13}$ and define a function B from the set of people to the set of twelve months as shown in the following arrow diagram.

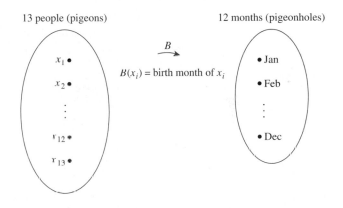

The pigeonhole principle says that no matter what the particular assignment of months to people, there must be at least two arrows pointing to the same month. Thus at least two people must have been born in the same month.

b. The answer is yes. In this example the pigeons are the people of New York City and the pigeonholes are all possible numbers of hairs on any individual's head. Call the population of New York City P. It is known that P is at least 5,000,000. Also the maximum number of hairs on any person's head is known to be no more than 300,000. Define a function H from the set of people in New York City $\{x_1, x_2, \ldots, x_p\}$ to the set $\{0, 1, 2, 3, \ldots, 300\,000\}$, as shown below.

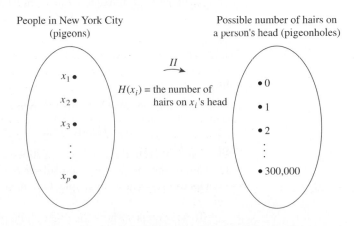

Since the number of people in New York City is larger than the number of possible hairs on their heads, the function H is not one-to-one; at least two arrows point to the same number. But that means that at least two people have the same number of hairs on their heads. ∎

Example 7.3.2 Finding the Number to Pick to Ensure a Result

A drawer contains ten black and ten white socks. You reach in and pull some out without looking at them. What is the *least* number of socks you must pull out to be sure to get a matched pair? Explain how the answer follows from the pigeonhole principle.

Solution   If you pick just two socks, they may have different colors. But when you pick a third sock, it must be the same color as one of the socks already chosen. Hence the answer is three.

This answer could be phrased more formally as follows: Let the socks pulled out be denoted $s_1, s_2, s_3, \ldots, s_n$ and consider the function C that sends each sock to its color, as shown below.

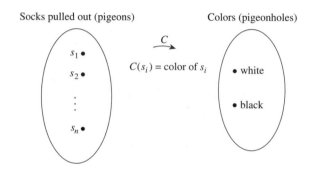

Socks pulled out (pigeons)   Colors (pigeonholes)

If $n = 2$, C could be a one-to-one correspondence (if the two socks pulled out were of different colors). But if $n > 2$, then the number of elements in the domain of C is larger than the number of elements in the co-domain of C. Thus by the pigeonhole principle, C is not one-to-one: $C(s_i) = C(s_j)$ for some $s_i \neq s_j$. This means that if at least three socks are pulled out, then at least two of them have the same color. ■

Example 7.3.3 Selecting a Pair of Integers with a Certain Sum

Let $A = \{1, 2, 3, 4, 5, 6, 7, 8\}$.

a. If five integers are selected from A, must at least one pair of the integers have a sum of 9?

b. If four integers are selected from A, must at least one pair of the integers have a sum of 9?

Solution

a. Yes. Partition the set A into the following four disjoint subsets:

$$\{1, 8\}, \quad \{2, 7\}, \quad \{3, 6\}, \quad \text{and} \quad \{4, 5\}$$

Observe that each of the integers in A occurs in exactly one of the four subsets and that the sum of the integers in each subset is 9. Thus if five integers from A are chosen, then by the pigeonhole principle, two must be from the same subset. It follows that the sum of these two integers is 9.

To see precisely how the pigeonhole principle applies, let the pigeons be the five selected integers (call them $a_1, a_2, a_3, a_4,$ and a_5) and let the pigeonholes be the subsets of the partition. The function P from pigeons to pigeonholes is defined by letting $P(a_i)$ be the subset that contains a_i.

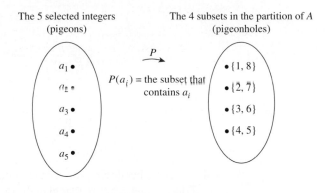

The function P is well defined because for each integer a_i in the domain, a_i belongs to one of the subsets (since the union of the subsets is A) and a_i does not belong to more than one subset (since the subsets are disjoint).

Because there are more pigeons than pigeonholes, at least two pigeons must go to the same hole. Thus two distinct integers are sent to the same set. But that implies that those two integers are the two distinct elements of the set, so their sum is 9. More formally, by the pigeonhole principle, since P is not one-to-one, there are integers a_i and a_j such that

$$P(a_i) = P(a_j) \quad \text{and} \quad a_i \neq a_j.$$

But then, by definition of P, a_i and a_j belong to the same subset. Since the elements in each subset add up to 9, $a_i + a_j = 9$.

b. The answer is no. This is a case where the pigeonhole principle does not apply; the number of pigeons is not larger than the number of pigeonholes. For instance, if you select the numbers 1, 2, 3, and 4, then since the largest sum of any two of these numbers is 7, no two of them add up to 9. ∎

Application to Decimal Expansions of Fractions

One important consequence of the piegonhole principle is the fact that

the decimal expansion of any rational number either terminates or repeats.

A terminating decimal is one like

$$3.625,$$

and a repeating decimal is one like

$$2.38\overline{246},$$

where the bar over the digits 246 means that these digits are repeated forever.*

Recall that a rational number is one that can be written as a ratio of integers—in other words, as a fraction. Recall also that the decimal expansion of a fraction is obtained by

---

*Strictly speaking, a terminating decimal like 3.625 can be regarded as a repeating decimal by adding trailing zeros: $3.625 = 3.625\overline{0}$. This can also be written as $3.624\overline{9}$.

dividing its numerator by its denominator using long division. For example, the decimal expansion of 4/33 is obtained as follows:

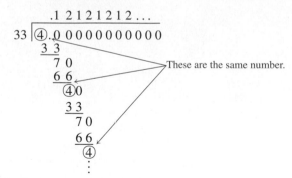

Because the number 4 reappears as a remainder in the long-division process, the sequence of quotients and remainders that give the digits of the decimal expansion repeats forever; hence the digits of the decimal expansion repeat forever.

In general, when one integer is divided by another, it is the pigeonhole principle (together with the quotient-remainder theorem) that guarantees that such a repetition of remainders and hence decimal digits must always occur. This is explained in the following example. The analysis in the example uses an obvious generalization of the pigeonhole principle, namely that a function from an infinite set to a finite set cannot be one-to-one.

Example 7.3.4 The Decimal Expansion of a Fraction

Consider a fraction a/b, where for simplicity a and b are both assumed to be positive. The decimal expansion of a/b is obtained by dividing the a by the b as illustrated here for $a = 3$ and $b = 14$.

$$
\begin{array}{r}
.2\,1\,4\,2\,8\,5\,7\,1\,4\,2\,8\,5\,7\ldots \\
14\,\overline{)\,3.0\,0\,0\,0\,0\,0\,0\,0\,0\,0\,0\,0\,0\,0\,0\,0\,0\,0} \\
\end{array}
$$

$$r_0 = 3$$
$$r_1 = 2$$
$$r_2 = 6$$
$$r_3 = 4$$
$$r_4 = 12$$
$$r_5 = 8$$
$$r_6 = 10$$
$$r_7 = 2 = r_1$$
$$r_8 = 6 = r_2$$
$$r_9 = 4 = r_3$$

Let $r_0 = a$ and let $r_1, r_2, r_3, \ldots$ be the successive remainders obtained in the long division of a by b. By the quotient-remainder theorem, each remainder must be between 0 and $b - 1$. (In this example, a is 3 and b is 14, and so the remainders are from 0 to 13.) If some remainder $r_i = 0$, then the division terminates and a/b has a terminating decimal expansion. If no $r_i = 0$, then the division process and hence the sequence of remainders continues forever. By the pigeonhole principle, since there are more remainders than values that the remainders can take, some remainder value must repeat: $r_j = r_k$, for some indices j and k with $j < k$. This is illustrated below for $a = 3$ and $b = 14$.

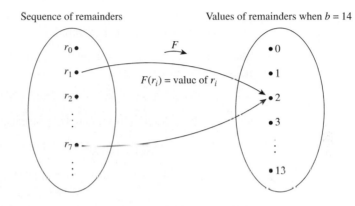

If follows that the decimal digits obtained from the divisions between r_j and r_{k-1} repeat forever. In the case of 3/14, the repetition begins with $r_7 = 2 = r_1$ and the decimal expansion repeats the quotients obtained from the divisions from r_1 through r_6 forever: $3/14 = 0.2\overline{142857}$. ∎

Note that since the decimal expansion of any rational number either terminates or repeats, if a number has a decimal expansion that neither terminates nor repeats, then it cannot be rational. Thus, for example, the following number cannot be rational: 0.01011011101111011111 ... (where each string of 1's is one longer than the previous string).

Generalized Pigeonhole Principle

A generalization of the pigeonhole principle states that if n pigeons fly into m pigeonholes and, for some positive integer $k, n > km$, then at least one pigeonhole contains $k + 1$ or more pigeons. This is illustrated in Figure 7.3.2 for $m = 4, n = 9$, and $k = 2$. Since $9 > 2 \cdot 4$, at least one pigeonhole contains three $(2 + 1)$ or more pigeons. (In this example, it is pigeonhole 3 that contains three pigeons.)

Figure 7.3.2

Generalized Pigeonhole Principle

For any function f from a finite set X to a finite set Y and for any positive integer k, if $N(X) > k \cdot N(Y)$, then there is some $y \in Y$ such that y is the image of at least $k + 1$ distinct elements of X.

Example 7.3.5 Applying the Generalized Pigeonhole Principle

Show how the generalized pigeonhole principle implies that in a group of 85 people, at least 4 must have the same last initial.

Solution   In this example the pigeons are the 85 people and the pigeonholes are the 26 possible last initials of their names. Note that

$$85 > 3 \cdot 26 = 78.$$

Consider the function I from people to initials defined by the following arrow diagram.

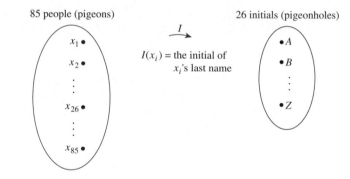

Since $85 > 3 \cdot 26$, the generalized pigeonhole states that some initial must be the image of at least four $(3 + 1)$ people. Thus at least four people have the same last initial. ∎

Consider the following contrapositive form of the generalized pigeonhole principle.

Generalized Pigenohole Principle (Contrapositive Form)

For any function f from a finite set X to a finite set Y and for any positive integer k, if for each $y \in Y$, $f^{-1}(y)$ has at most k elements, then X has at most $k \cdot N(Y)$ elements.

You may find it natural to use the contrapositive form of the generalized pigeonhole principle in certain situations. For instance, the result of Example 7.3.5 can be explained as follows:

Suppose no 4 people out of the 85 had the same last initial. Then at most 3 would share any particular one. By the generalized pigeonhole principle (contrapositive form), this would imply that the total number of people is at most $3 \cdot 26 = 78$. But this contradicts the fact that there are 85 people in all. Hence at least 4 people share a last initial.

Example 7.3.6 Using the Contrapositive Form of the Generalized Pigeonhole Principle

There are 42 students who are to share 12 computers. Each students uses exactly 1 computer, and no computer is used by more than 6 students. Show that at least 5 computers are used by 3 or more students.

Solution

a. ***Using an Argument by Contradiction:*** Suppose not. Suppose that 4 or fewer computers are used by 3 or more students. [*A contradiction will be derived.*] Then 8 or more computers are used by 2 or fewer students. Divide the set of computers into two subsets: C_1 and C_2. Into C_1 place 8 of the computers used by 2 or fewer students; into C_2 place the computers used by 3 or more students plus any remaining computers (to make a total of 4 computers in C_2). (See Figure 7.3.3.)

The Set of 12 Computers

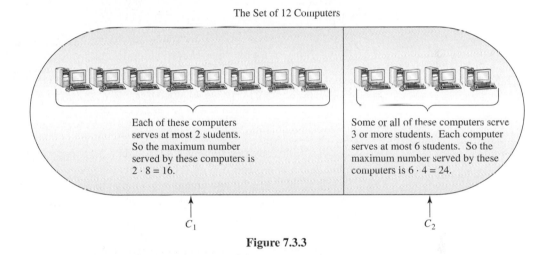

Each of these computers serves at most 2 students. So the maximum number served by these computers is $2 \cdot 8 = 16$.

Some or all of these computers serve 3 or more students. Each computer serves at most 6 students. So the maximum number served by these computers is $6 \cdot 4 = 24$.

C_1 C_2

Figure 7.3.3

Since at most 6 students are served by any one computer, by the contrapositive form of the generalized pigeonhole principle, the computers in set C_2 serve at most $6 \cdot 4 = 24$ students. Since at most 2 students are served by any one computer in C_1, by the generalized pigeonhole principle (contrapositive form), the computers in set C_1 serve at most $2 \cdot 8 = 16$ students. Hence the total number of students served by the computers is $24 + 16 = 40$. But this contradicts the fact that each of the 42 students is served by a computer. Therefore, the supposition is false: At least 5 computers are used by 3 or more students.

b. ***Using a Direct Argument:*** Let k be the number of computers used by 3 or more students. [*We must show that $k \geq 5$.*] Because each computer is used by at most 6 students, these computers are used by at most $6k$ students (by the contrapositive form of the generalized pigeonhole principle). Each of the remaining $12 - k$ computers is used by at most 2 students. Hence, taken together, they are used by at most $2(12 - k) = 24 - 2k$ students (again, by the contrapositive form of the generalized pigeonhole principle). Thus the maximum number of students served by the computers is $6k + (24 - 2k) = 4k + 24$. Because 42 students are served by the computers, $4k + 24 \geq 42$. Solving for k gives that $k \geq 4.5$, and since k is an integer, this implies that $k \geq 5$ [*as was to be shown*]. ∎

Proof of the Pigeonhole Principle

The truth of the pigeonhole principle depends essentially on the sets involved being finite. Formal definitions of finite and infinite can be stated as follows:

> **• Definition**
>
> A set is called **finite** if, and only if, it is the empty set or there is a one-to-one correspondence from $\{1, 2, \ldots, n\}$ to it, where n is a positive integer. In the first case the **number of elements** in the set is said to be 0, and in the second case it is said to be n. A set that is not finite is called **infinite.**

Note that it follows immediately from the definition that for a set to be finite means that it either is empty or can be written in the form $\{x_1, x_2, \ldots, x_n\}$ where n is a positive integer.

> **Theorem 7.3.1 The Pigeonhole Principle**
>
> For any function f from a finite set X to a finite set Y, if $N(X) > N(Y)$, then f is not one-to-one.
>
> **Proof:**
>
> Suppose f is any function from a finite set X to a finite set Y, where $N(X) > N(Y)$. Let $N(Y) = m$, and denote the elements of Y by $y_1, y_2, \ldots, y_m$. Recall that for each y_i in Y, the inverse image set $f^{-1}(y_i) = \{x \in X \mid f(x) = y_i\}$. Now consider the collection of all the inverse image sets for all the elements of Y:
>
> $$f^{-1}(y_1), f^{-1}(y_2), \ldots, f^{-1}(y_m).$$
>
> By definition of function, each element of X is sent by f to some element of Y. Hence each element of X is in one of the inverse image sets, and so the union of all these sets equals X. But also, by definition of function, no element of X is sent by f to more than one element of Y. Thus each element of X is in only one of the inverse image sets, and so the inverse image sets are mutually disjoint. By the addition rule, therefore,
>
> $$N(X) = N(f^{-1}(y_1)) + N(f^{-1}(y_2)) + \cdots + N(f^{-1}(y_m)). \qquad 7.3.1$$
>
> Now suppose that f *is* one-to-one [*which is the opposite of what we want to prove*]. Then each set $f^{-1}(y_i)$ has at most one element, and so
>
> $$N(f^{-1}(y_1)) + N(f^{-1}(y_2)) + \cdots + N(f^{-1}(y_m)) \leq \underbrace{1 + 1 + \cdots + 1}_{m \text{ terms}} = m \qquad 7.3.2$$
>
> Putting equations (7.3.1) and (7.3.2) together gives that
>
> $$N(X) \leq m = N(Y).$$
>
> This contradicts the fact that $N(X) > N(Y)$, and so the supposition that f is one-to-one must be false. Hence f is not one-to-one [*as was to be shown*].

An important theorem that follows from the pigeonhole principle states that a function from one finite set to another finite set of the same size is one-to-one if, and only if, it is onto. We will show in Section 7.5 that this result does not hold for infinite sets.

Theorem 7.3.2 One-to-One and Onto for Finite Sets

Let X and Y be finite sets with the same number of elements and suppose f is a function from X to Y. Then f is one-to-one if, and only if, f is onto.

Proof:

Suppose f is a function from X to Y, where X and Y are finite sets each with m elements. Let $X = \{x_1, x_2, \ldots, x_m\}$ and $Y = \{y_1, y_2, \ldots, y_m\}$

If f is one-to-one, then f is onto: Suppose f is one-to-one. Then $f(x_1), f(x_2), \ldots, f(x_m)$ are all distinct. Consider the set S of all elements of Y that are not the image of any element of X:
   Then the sets

$$\{f(x_1)\}, \{f(x_2)\}, \ldots, \{f(x_m)\} \quad \text{and} \quad S$$

are mutually disjoint. By the addition rule,

$$N(Y) = N(\{f(x_1)\}) + N(\{f(x_2)\}) + \cdots + N(\{f(x_m)\}) + N(S)$$
$$= \underbrace{1 + 1 + \cdots + 1}_{m \text{ terms}} + N(S) \quad \text{because each } \{f(x_i)\} \text{ is a singleton set}$$
$$= m + N(S).$$

Thus

$$m = m + N(S) \quad \text{because } N(Y) = m,$$
$$\Rightarrow \quad N(S) = 0 \quad \text{by subtracting } m \text{ from both sides.}$$

Hence S is empty, and so there is no element of Y that is not the image of some element of X. Consequently, f is onto.

If f is onto, then f is one-to-one: Suppose f is onto. Then $f^{-1}(y_i) \neq \emptyset$ and so $N(f^{-1}(y_i)) \geq 1$ for all $i = 1, 2, \ldots, m$. As in the proof of the pigeonhole principle (Theorem 7.3.1), X is the union of the mutually disjoint sets $f^{-1}(y_1), f^{-1}(y_2), \ldots, f^{-1}(y_m)$. By the addition principle,

$$N(X) = \underbrace{N(f^{-1}(y_1)) + N(f^{-1}(y_2)) + \cdots + N(f^{-1}(y_m))}_{m \text{ terms, each } \geq 1} \geq m. \qquad 7.3.3$$

Now if any one of the sets $f^{-1}(y_i)$ has more than one element, then the sum in equation (7.3.3) is greater than m. But we know this is not the case because $N(X) = m$. Hence each set $f^{-1}(y_i)$ has exactly one element, and thus f is one-to-one [*as was to be shown*].

Note that Theorem 7.3.2 applies in particular to the case $X = Y$. Thus a one-to-one function from a finite set to itself is onto, and an onto function from a finite set to itself is one-to-one. Such functions can be identified with permutations of the sets on which they are defined. For instance, the function defined by the diagram below can be identified with the permutation *cdba* obtained by listing the images of a, b, c, and d in order.

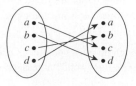

Exercise Set 7.3

1. a. If 4 cards are selected from a standard 52-card deck, must at least 2 be of the same suit? Why?

   b. If 5 cards are selected from a standard 52-card deck, must at least 2 be of the same suit? Why?

2. a. If 13 cards are selected from a standard 52-card deck, must at least 2 be of the same denomination? Why?

   b. If 20 cards are selected from a standard 52-card deck, must at least 2 be of the same denomination? Why?

3. A small town has only 500 residents. Must there be 2 residents who have the same birthday? Why?

4. In a group of 700 people, must there be 2 who have the same first and last initials? Why?

5. a. Given any set of four integers, must there be two that have the same remainder when divided by 3? Why?

   b. Given any set of three integers, must there be two that have the same remainder when divided by 3? Why?

6. a. Given any set of seven integers, must there be two that have the same remainder when divided by 6? Why?

   b. Given any set of seven integers, must there be two that have the same remainder when divided by 8? Why?

H 7. Let $S = \{3, 4, 5, 6, 7, 8, 9, 10, 11, 12\}$. Suppose six integers are chosen from S. Must there be two integers whose sum is 15? Why?

8. Let $T = \{1, 2, 3, 4, 5, 6, 7, 8, 9\}$. Suppose five integers are chosen from T. Must there be two integers whose sum is 10? Why?

9. a. If seven integers are chosen from between 1 and 12 inclusive, must at least one of them be odd? Why?

   b. If ten integers are chosen from between 1 and 20 inclusive, must at least one of them be even? Why?

10. If $n + 1$ integers are chosen from the set

    $$\{1, 2, 3, \ldots, 2n\},$$

    where n is a positive integer, must at least one of them be odd? Why?

11. If $n + 1$ integers are chosen from the set

    $$\{1, 2, 3, \ldots, 2n\},$$

    where n is a positive integer, must at least one of them be even? Why?

12. How many cards must you pick from a standard 52-card deck to be sure of getting at least 1 red card? Why?

13. Suppose six pairs of similar-looking boots are thrown together in a pile. How many individual boots must you pick to be sure of getting a matched pair? Why?

14. How many integers from 0 through 60 must you pick in order to be sure of getting at least one that is odd? at least one that is even?

15. If n is a positive integer, how many integers from 0 through $2n$ must you pick in order to be sure of getting at least one that is odd? at least one that is even?

16. How many integers from 1 through 100 must you pick in order to be sure of getting one that is divisible by 5?

17. How many integers must you pick in order to be sure that at least two of them have the same remainder when divided by 7?

18. How many integers must you pick in order to be sure that at least two of them have the same remainder when divided by 15?

19. How many integers from 100 through 999 must you pick in order to be sure that at least two of them have a digit in common? (For example, 256 and 530 have the common digit 5.)

20. If repeated divisions by 20,483 are performed, how many distinct remainders can be obtained?

21. When 5/20483 is written as a decimal, what is the maximum length of the repeating section of the representation?

22. Is 0.101001000100001000001 . . . (where each string of 0's is one longer than the previous one) rational or irrational?

23. Is 56.556655566655556666 . . . (where the strings of 5's and 6's become longer in each repetition) rational or irrational?

24. Show that within any set of thirteen integers chosen from 2 through 40, there are at least two integers with a common divisor greater than 1.

25. In a group of 30 people, must at least 3 have been born in the same month? Why?

26. In a group of 30 people, must at least 4 have been born in the same month? Why?

27. In a group of 2,000 people, must at least 5 have the same birthday? Why?

28. A programmer writes 500 lines of computer code in 17 days. Must there have been at least 1 day when the programmer wrote 30 or more lines of code? Why?

29. A certain college class has 40 students. All the students in the class are known to be from 17 through 34 years of age. You want to make a bet that the class contains at least x students of the same age. How large can you make x and yet be sure to win your bet?

30. A penny collection contains twelve 1967 pennies, seven 1968 pennies, and eleven 1971 pennies. If you are to pick some pennies without looking at the dates, how many must you pick to be sure of getting at least five pennies from the same year?

H **31.** A group of 15 executives are to share 5 secretaries. Each executive is assigned exactly 1 secretary, and no secretary is assigned to more than 4 executives. Show that at least 3 secretaries are assigned to 3 or more executives.

H ✶ **32.** Let A be a set of six positive integers each of which is less than 13. Show that there must be two distinct subsets of A whose elements when added up give the same sum. (For example, if $A = \{5, 12, 10, 1, 3, 4\}$, then the elements of the subsets $S_1 = \{1, 4, 10\}$ and $S_2 = \{5, 10\}$ both add up to 15.)

33. Let S be a set of ten integers chosen from 1 through 50. Show that the set contains at least two different (but not necessarily disjoint) subsets of four integers that add up to the same number. (For instance, if the ten numbers are $\{3, 8, 9, 18, 24, 34, 35, 41, 44, 50\}$, the subsets can be taken to be $\{8, 24, 34, 35\}$ and $\{9, 18, 24, 50\}$. The numbers in both of these add up to 101.)

H ✶ **34.** Given a set of 52 distinct integers, show that there must be 2 whose sum or difference is divisible by 100.

H ✶ **35.** Show that if 101 integers are chosen from 1 to 200 inclusive, there must be 2 with the property that one is divisible by the other.

✶ **36.** a. Suppose $u_1, u_2, \ldots, a_n$ is a sequence of n integers none of which is divisible by n. Show that at least one of the differences $a_i - a_j$ (for $i \neq j$) must be divisible by n.

   H b. Show that every finite sequence $x_1, x_2, \ldots, x_n$ of n integers has a consecutive subsequence $x_{i+1}, x_{i+2}, \ldots, x_j$

whose sum is divisible by n. (For instance, the sequence 3, 4, 17, 7, 16 has the consecutive subsequence 17, 7, 16 whose sum is divisible by 5.)*

H ✶ **37.** Observe that the sequence 12, 15, 8, 13, 7, 18, 19, 11, 14, 10 has three increasing subsequences of length four: 12, 15, 18, 19; 12, 13, 18, 19; and 8, 13, 18, 19. It also has one decreasing subsequence of length four: 15, 13, 11, 10. Show that in any sequence of $n^2 + 1$ distinct real numbers, there must be a sequence of length $n + 1$ that is either strictly increasing or strictly decreasing.

✶ **38.** What is the largest number of elements that a set of integers from 1 through 100 can have so that no one element in the set is divisible by another? (*Hint:* Imagine writing all the numbers from 1 through 100 in the form $2^k \cdot m$, where $k \geq 0$ and m is odd.)

39. Suppose X and Y are finite sets, X has more elements than Y, and $F: X \rightarrow Y$ is a function. By the pigeonhole principle, there exist elements a and b in X such that $a \neq b$ and $F(a) = F(b)$. Write a computer algorithm to find such a pair of elements a and b.

---

*James E. Schultz and William F. Burger, "An Approach to Problem-Solving Using Equivalence Classes Modulo n," *College Mathematics Journal (15)*, No. 5, 1984, 401–405.

7.4 *Composition of Functions*

It is no paradox to say that in our most theoretical moods we may be nearest to our most practical applications. — Alfred North Whitehead

Consider two functions, the successor function and the squaring function, defined from **Z** (the set of integers) to **Z**, and imagine that each is represented by a machine. If the two machines are hooked up so that the output from the successor function is used as input to the squaring function, then they work together to operate as one larger machine. In this larger machine, an integer n is first increased by 1 to obtain $n + 1$; then the quantity $n + 1$ is squared to obtain $(n + 1)^2$. This is illustrated in the following drawing.

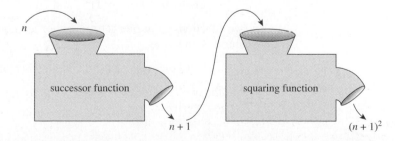

Combining functions in this way is called *composing* them; the resulting function is called the *composition* of the two functions. Note that the composition can be formed

only if the output of the first function is acceptable input to the second function. That is, the range of the first function must be contained in the domain of the second function.

• Definition

Let $f: X \to Y'$ and $g: Y \to Z$ be functions with the property that the range of f is a subset of the domain of g. Define a new function $g \circ f: X \to Z$ as follows:

$$(g \circ f)(x) = g(f(x)) \quad \text{for all } x \in X,$$

where $g \circ f$ is read "g circle f" and $g(f(x))$ is read "g of f of x." The function $g \circ f$ is called the **composition of f and g.** (We put the f first when we say "the composition of f and g" because an element x is acted upon first by f and then by g.)

This definition is shown schematically below.

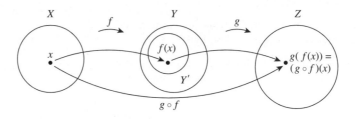

Example 7.4.1 Composition of Functions Defined by Formulas

Let $f: \mathbf{Z} \to \mathbf{Z}$ be the successor function and let $g: \mathbf{Z} \to \mathbf{Z}$ be the squaring function. Then $f(n) = n + 1$ for all $n \in \mathbf{Z}$ and $g(n) = n^2$ for all $n \in \mathbf{Z}$.

a. Find the compositions $g \circ f$ and $f \circ g$.

b. Is $g \circ f = f \circ g$? Explain.

Solution

a. The functions $g \circ f$ and $f \circ g$ are defined as follows:

$$(g \circ f)(n) = g(f(n)) = g(n + 1) = (n + 1)^2 \quad \text{for all } n \in \mathbf{Z},$$

and

$$(f \circ g)(n) = f(g(n)) = f(n^2) = n^2 + 1 \quad \text{for all } n \in \mathbf{Z}.$$

Thus

$$(g \circ f)(n) = (n + 1)^2 \quad \text{and} \quad (f \circ g)(n) = n^2 + 1 \quad \text{for all } n \in \mathbf{Z}.$$

b. Two functions from one set to another are equal if, and only if, they take the same values. In this case,

$$(g \circ f)(1) = (1 + 1)^2 = 4, \text{ whereas } (f \circ g)(1) = 1^2 + 1 = 2.$$

Thus the two functions $g \circ f$ and $f \circ g$ are not equal:

$$g \circ f \neq f \circ g. \qquad \blacksquare$$

Example 7.4.1 illustrates the important fact that composition of functions is not a commutative operation: *For general functions F and G, $F \circ G$ need not necessarily equal $G \circ F$* (although the two *may* be equal).

Example 7.4.2 Composition of Functions Defined on Finite Sets

Let $X = \{1, 2, 3\}$, $Y' = \{a, b, c, d\}$, $Y = \{a, b, c, d, e\}$, and $Z = \{x, y, z\}$. Define functions $f: X \to Y'$ and $g: Y \to Z$ by the arrow diagrams below.

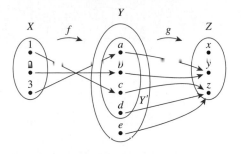

That is, $f(1) = c$, $f(2) = b$, $f(3) = a$, and $g(a) = y$, $g(b) = y$, $g(c) = z$, $g(d) = z$, and $g(e) = z$. Find the arrow diagram for $g \circ f$. What is the range of $g \circ f$?

Solution To find the arrow diagram for $g \circ f$, just trace the arrows all the way across from X to Z through Y. The result is shown below.

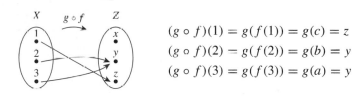

$$(g \circ f)(1) = g(f(1)) = g(c) = z$$
$$(g \circ f)(2) = g(f(2)) = g(b) = y$$
$$(g \circ f)(3) = g(f(3)) = g(a) = y$$

The range of $g \circ f$ is $\{y, z\}$. ■

Recall that the identity function on a set X, i_X, is the function from X to X defined by the formula

$$i_X(x) = x \quad \text{for all } x \in X.$$

That is, the identity function on X sends each element of X to itself. What happens when an identity function is composed with another function?

Example 7.4.3 Composition with the Identity Function

Let $X = \{a, b, c, d\}$ and $Y = \{u, v, w\}$, and suppose $f: X \to Y$ is given by the arrow diagram shown below.

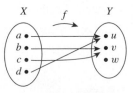

Find $f \circ i_X$ and $i_Y \circ f$.

Solution   The values of $f \circ i_X$ are obtained by tracing through the arrow diagram shown below.

$$(f \circ i_X)(a) = f(i_X(a)) = f(a) = u$$
$$(f \circ i_X)(b) = f(i_X(b)) = f(b) = v$$
$$(f \circ i_X)(c) = f(i_X(c)) = f(c) = v$$
$$(f \circ i_X)(d) = f(i_X(d)) = f(d) = u$$

Note that for all elements x in X,

$$(f \circ i_X)(x) = f(x).$$

By definition of equality of functions, this means that $f \circ i_X = f$.

Similarly, the equality $i_Y \circ f = f$ can be verified by tracing through the arrow diagram below for each x in X and noting that in each case, $(i_Y \circ f)(x) = f(x)$.

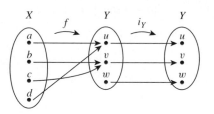

More generally, the composition of any function with an identity function equals the function.

Theorem 7.4.1 Composition with an Identity Function

If f is a function from a set X to a set Y, and i_X is the identity function on X, and i_Y is the identity function on Y, then

$$\text{(a)} \quad f \circ i_X = f \quad \text{and} \quad \text{(b)} \quad i_Y \circ f = f.$$

Proof:

Part (a): Suppose f is a function from a set X to a set Y and i_X is the identity function on X. Then, for all x in X,

$$(f \circ i_X)(x) = f(i_X(x)) = f(x).$$

Hence, by definition of equality of functions, $f \circ i_X = f$, as was to be shown.

Part (b): This is exercise 13 at the end of this section.

Now let f be a function from a set X to a set Y, and suppose f has an inverse function f^{-1}. Recall that f^{-1} is the function from Y to X with the property that

$$f^{-1}(y) = x \quad \Leftrightarrow \quad f(x) = y.$$

What happens when f is composed with f^{-1}? Or when f^{-1} is composed with f?

Example 7.4.4 Composing a Function with Its Inverse

Let $X = \{a, b, c\}$ and $Y = \{x, y, z\}$. Define $f: X \to Y$ by the following arrow diagram.

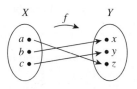

Then f is one-to-one and onto. Thus f^{-1} exists and is found by tracing the arrows backwards, as shown below.

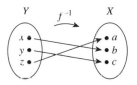

Now $f^{-1} \circ f$ is found by following the arrows from X to Y by f and back to X by f^{-1}. If you do this, you will see that

$$(f^{-1} \circ f)(a) = f^{-1}(f(a)) = f^{-1}(z) = a$$
$$(f^{-1} \circ f)(b) = f^{-1}(f(b)) = f^{-1}(x) = b$$

and

$$(f^{-1} \circ f)(c) = f^{-1}(f(c)) = f^{-1}(y) = c.$$

Thus the composition of f and f^{-1} sends each element to itself. So by definition of the identity function,

$$f^{-1} \circ f = i_X.$$

In a similar way, you can see that

$$f \circ f^{-1} = i_Y. \qquad \blacksquare$$

More generally, the composition of any function with its inverse (if it has one) is an identity function. Intuitively, the function sends an element in its domain to an element in its co-domain and the inverse function sends it back again, so the composition of the two sends each element to itself. This reasoning is formalized in Theorem 7.4.2.

Theorem 7.4.2 Composition of a Function with Its Inverse

If $f: X \to Y$ is a one-to-one and onto function with inverse function $f^{-1}: Y \to X$, then

$$\text{(a)} \quad f^{-1} \circ f = i_X \quad \text{and} \quad \text{(b)} \quad f \circ f^{-1} = i_Y.$$

Proof:

Part (a): Suppose $f: X \to Y$ is a one-to-one and onto function with inverse function $f^{-1}: Y \to X$. [*To show that $f^{-1} \circ f = i_X$, we must show that for all $x \in X$, $(f^{-1} \circ f)(x) = x$.*] Let x be an element in X. Then

$$(f^{-1} \circ f)(x) = f^{-1}(f(x))$$

by definition of composition of functions. Now the inverse function f^{-1} satisfies the condition

$$f^{-1}(b) = a \quad \Leftrightarrow \quad f(a) = b \quad \text{for all } a \in X \text{ and } b \in Y. \qquad 7.4.1$$

Let

$$x' = f^{-1}(f(x)). \qquad 7.4.2$$

Apply property (7.4.1) with x' playing the role of a and $f(x)$ playing the role of b. Then

$$f(x') = f(x).$$

But since f is one-to-one, this implies that $x' = x$. Substituting x for x' in equation (7.4.2) gives

$$x = f^{-1}(f(x)).$$

Then by definition of composition of functions,

$$(f^{-1} \circ f)(x) = x,$$

as was to be shown.

Part (b): This is exercise 14 at the end of this section.

Composition of One-to-One Functions

The composition of functions interacts in interesting ways with the properties of being one-to-one and onto. What happens, for instance, when two one-to-one functions are composed? Must their composition be one-to-one? For example, let $X = \{a, b, c\}$, $Y = \{w, x, y, z\}$, and $Z = \{1, 2, 3, 4, 5\}$, and define one-to-one functions $f: X \to Y$ and $g: Y \to Z$ as shown in the arrow diagrams of Figure 7.4.1.

Figure 7.4.1

Then $g \circ f$ is the function with the arrow diagram shown in Figure 7.4.2.

Figure 7.4.2

From the diagram it is clear that for these particular functions, the composition is one-to-one. This result is no accident. It turns out that the compositions of two one-to-one functions is always one-to-one.

Theorem 7.4.3

If $f: X \to Y$ and $g: Y \to Z$ are both one-to-one functions, then $g \circ f$ is one-to-one.

By the method of direct proof, the proof of Theorem 7.4.3 has the following starting point and conclusion to be shown.

Starting Point: Suppose f is a one-to-one function from X to Y and g is a one-to-one function from Y to Z.

To Show: $g \circ f$ is a one-to-one function from X to Z.

The conclusion to be shown says that a certain function is one-to-one. How do you show that? The crucial step is to realize that if you substitute $g \circ f$ into the definition of one-to-one, you see that

$$g \circ f \text{ is one-to-one} \quad \Leftrightarrow \quad \forall x_1, x_2 \in X, \text{if } (g \circ f)(x_1) = (g \circ f)(x_2) \text{ then } x_1 = x_2.$$

By the method of direct proof, then, to show $g \circ f$ is one-to-one, you

suppose x_1 and x_2 are elements of X such that $(g \circ f)(x_1) = (g \circ f)(x_2)$,

and you

show that $x_1 = x_2$.

Now the heart of the proof begins. To show that $x_1 = x_2$, you work forward from the supposition that $(g \circ f)(x_1) = (g \circ f)(x_2)$, using the fact that f and g are both one-to-one. By definition of composition,

$$(g \circ f)(x_1) = g(f(x_1)) \quad \text{and} \quad (g \circ f)(x_2) = g(f(x_2)).$$

Since the left-hand sides of the equations are equal, so are the right-hand sides. Thus

$$g(f(x_1)) = g(f(x_2)).$$

Now just stare at the above equation for a moment. It says that

$$g(\text{something}) = g(\text{something else}).$$

Because g is a one-to-one function, any time g of one thing equals g of another thing, those two things are equal. Hence

$$f(x_1) = f(x_2).$$

But f is also a one-to-one function. Any time f of one thing equals f of another thing, those two things are equal. Therefore,

$$x_1 = x_2.$$

This is what was to be shown!

This discussion is summarized in the following formal proof.

Proof of Theorem 7.4.3:

Suppose $f: X \to Y$ and $g: Y \to Z$ are both one-to-one functions. [*We must show that $g \circ f$ is one-to-one.*] Suppose x_1 and x_2 are elements of X such that

$$(g \circ f)(x_1) = (g \circ f)(x_2).$$

[*We must show that $x_1 = x_2$.*] By definition of composition of functions,

$$g(f(x_1)) = g(f(x_2)).$$

Since g is one-to-one,

$$f(x_1) = f(x_2).$$

And since f is one-to-one,

$$x_1 = x_2.$$

This is what was to be shown. Hence $g \circ f$ is one-to-one.

Composition of Onto Functions

Now consider what happens when two onto functions are composed. For example, let $X = \{a, b, c, d, e\}$, $Y = \{w, x, y, z\}$, and $Z = \{1, 2, 3\}$. Define onto functions f and g by the following arrow diagrams.

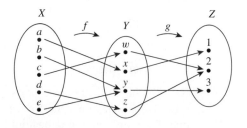

Then $g \circ f$ is the function with the arrow diagram shown below.

It is clear from the diagram that $g \circ f$ is onto.

It turns out that the composition of any two onto functions (that can be composed) is onto.

Theorem 7.4.4

If $f: X \to Y$ and $g: Y \to Z$ are both onto functions, then $g \circ f$ is onto.

By the method of direct proof, the proof of Theorem 7.4.4 has the following starting point and conclusion to be shown:

Starting Point: Suppose f is an onto function from X to Y, and g is an onto function from Y to Z.

To Show: $g \circ f$ is an onto function from X to Z.

The conclusion to be shown says that a certain function is onto. How do you show that? The crucial step is to realize that if you substitute $g \circ f$ into the definition of onto, you see that

---

$g \circ f: X \to Z$ is onto $\Leftrightarrow$ given any element of Z, it is possible to find an element of X such that $(g \circ f)(x) = z$.

Since this statement is universal, to prove it you

> **suppose** z is a [*particular but arbitrarily chosen*] element of Z

and

> **show** that there is an element x in X such that $(g \circ f)(x) = z$.

Hence you must start the proof by supposing you are given a particular but arbitrarily chosen element in Z. Let us call it z. Your job is to find an element x in X such that $(g \circ f)(x) = z$.

To find x, reason from the supposition that z is in Z, using the fact that both g and f are onto. Imagine arrow diagrams for the functions f and g.

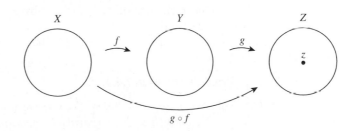

You have a particular element z in Z, and you need to find an element x in X such that when x is sent over to Z by $g \circ f$, its image will be z. Since g is onto, z is at the tip of some arrow coming from Y. That is, there is an element y in Y such that

$$g(y) = z.$$ 7.4.3

This means that the arrow diagrams can be drawn as follows:

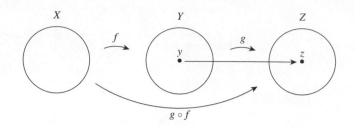

But f also is onto, so every element in Y is at the tip of an arrow coming from X. In particular, y is at the tip of some arrow. That is, there is an element x in X such that

$$f(x) = y. \qquad\qquad 7.4.4$$

The diagram, therefore, can be drawn as shown below.

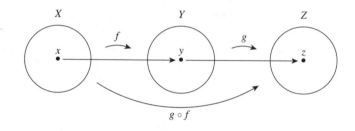

Now just substitute equation (7.4.4) into equation (7.4.3) to obtain

$$g(f(x)) = z.$$

But by definition of $g \circ f$,

$$g(f(x)) = (g \circ f)(x).$$

Hence

$$(g \circ f)(x) = z.$$

Thus x is an element of X that is sent by $g \circ f$ to z, and so x is the element you were supposed to find.

This discussion is summarized in the following formal proof.

Proof of Theorem 7.4.4:

Suppose $f: X \to Y$ and $g: Y \to Z$ are both onto functions. [*We must show that $g \circ f$ is onto*]. Let z be a [*particular but arbitrarily chosen*] element of Z. [*We must show the existence of an element x in X such that $(g \circ f)(x) = z$.*] Since g is onto, there is an element y in Y such that $g(y) = z$. And since f is onto, there is an element x in X such that $f(x) = y$. Hence there exists an element x in X such that

$$(g \circ f)(x) = g(f(x)) = g(y) = z.$$

If follows that $g \circ f$ is onto.

Example 7.4.5 An Incorrect "Proof" That a Function Is Onto

To prove that a composition of onto functions is onto, a student wrote,

"Suppose $f: X \to Y$ and $g: Y \to Z$ are both onto. Then

$$\forall y \in Y, \exists x \in X \text{ such that } f(x) = y \quad (*)$$

and

$$\forall z \in Z, \exists y \in Y \text{ such that } f(y) = z.$$

So

$$(g \circ f)(x) = g(f(x)) = g(y) = z,$$

and thus $g \circ f$ is onto."

Explain the mistakes in this "proof."

Solution   To show that $g \circ f$ is onto, you must be able to meet the following challenge: If someone gives you an element z in Z (over which you have no control), you must be able to explain how to find an element x in X such that $(g \circ f)(x) = z$. Thus a proof that $g \circ f$ is onto must start with the assumption that you have been given a particular but arbitrarily chosen element of Z. This proof does not do that.

Moreover, note that statement (*) simply asserts that f is onto. An informal version of (*) is the following: Given any element in the co-domain of f, there is an element in the domain of f that is sent by f to the given element. Use of the symbols x and y to denote these elements is arbitrary. Any other two symbols could equally well have been used. Thus, if we replace the x and y in (*) by u and v, we obtain a logically equivalent statement, and the "proof" becomes the following:

"Suppose $f: X \to Y$ and $g: Y \to Z$ are both onto. Then

$$\forall v \subset Y, \exists u \in X \text{ such that } f(u) = v$$

and

$$\forall z \in Z, \exists y \in Y \text{ such that } f(y) = z.$$

So (??!)

$$(g \circ f)(x) = g(f(x)) = g(y) = z,$$

and thus $g \circ f$ is onto."

From this logically equivalent version of the "proof," you can see that the statements leading up to the word *So* do not provide a rationale for the statement that follows it. The original reason for writing *So* was based on a misinterpretation of the meaning of the notation. ∎

Exercise Set 7.4

In each of 1 and 2, functions f and g are defined by arrow diagrams. Find $g \circ f$ and $f \circ g$ and determine whether $g \circ f$ equals $f \circ g$.

1.

2.

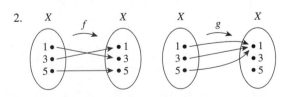

In each of 3–6, functions F and G are defined by formulas. Find $G \circ F$ and $F \circ G$ and determine whether $G \circ F$ equals $F \circ G$.

3. $F(x) = x^3$ and $G(x) = x - 1$, for all real numbers x.

4. $F(x) = x^5$ and $G(x) = x^{1/5}$ for all real numbers x.

5. $F(n) = 2n$ and $G(n) = \lfloor n/2 \rfloor$, for all integers n.

6. $F(x) = 3x$, $G(x) = \lfloor x/3 \rfloor$ for all real numbers x.

7. Let S be the set of all strings in a's and b's and let $L \colon S \to \mathbf{Z}$ be the length function:

   For all strings $s \in S$,

   $$L(s) = \text{ the number of characters in } s.$$

   Let $T \colon \mathbf{Z} \to \{0, 1, 2\}$ be the *mod* 3 function:

   For all integers n, $\quad T(n) = n \bmod 3$.

   What is $(T \circ L)(abaa)$? $(T \circ L)(baaab)$? $(T \circ L)(aaa)$?

8. Define $F \colon \mathbf{R} \to \mathbf{R}$ and $G \colon \mathbf{R} \to \mathbf{Z}$ by the following formulas: $F(x) = x^2/3$ and $G(x) = \lfloor x \rfloor$ for all $x \in \mathbf{R}$. What is $(G \circ F)(2)$? $(G \circ F)(-3)$? $(G \circ F)(5)$?

The functions of each pair in 9–11 are inverse to each other. For each pair, check that both compositions give the identity function.

9. $F \colon \mathbf{R} \to \mathbf{R}$ and $F^{-1} \colon \mathbf{R} \to \mathbf{R}$ are defined by

   $$F(x) = 3x + 2, \quad \text{for all } x \in \mathbf{R}$$

   and

   $$F^{-1}(y) = \frac{y - 2}{3}, \quad \text{for all } y \in \mathbf{R}.$$

10. $G \colon \mathbf{R}^+ \to \mathbf{R}^+$ and $G^{-1} \colon \mathbf{R}^+ \to \mathbf{R}^+$ are defined by

    $$G(x) = x^2, \quad \text{for all } x \in \mathbf{R}^+$$

    and

    $$G^{-1}(x) = \sqrt{x}, \quad \text{for all } x \in \mathbf{R}^+.$$

11. H and H^{-1} are both defined from $\mathbf{R} - \{1\}$ to $\mathbf{R} - \{1\}$ by the formula

    $$H(x) = H^{-1}(x) = \frac{x + 1}{x - 1}, \quad \text{for all } x \in \mathbf{R} - \{1\}.$$

12. Explain how it follows from the definition of logarithm that
    a. $\log_b(b^x) = x$, for all real numbers x.
    b. $b^{\log_b x} = x$, for all positive real numbers x.

H 13. Prove Theorem 7.4.1(b): If f is any function from a set X to a set Y, then $i_Y \circ f = f$, where i_Y is the identity function on Y.

14. Prove Theorem 7.4.2(b): If $f \colon X \to Y$ is a one-to-one and onto function with inverse function $f^{-1} \colon Y \to X$, then $f \circ f^{-1} = i_Y$, where i_Y is the identity function on Y.

15. Suppose Y and Z are sets and $g \colon Y \to Z$ is a one-to-one function. This means that if g takes the same value on any two elements of Y, then those elements are equal. Thus, for example, if a and b are elements of Y and $g(a) = g(b)$, then

it can be inferred that $a = b$. What can be inferred in the following situations?
a. s_k and s_m are elements of Y and $g(s_k) = g(s_m)$.
b. $z/2$ and $t/2$ are elements of Y and $g(z/2) = g(t/2)$.
c. $f(x_1)$ and $f(x_2)$ are elements of Y and $g(f(x_1)) = g(f(x_2))$.

16. If $f \colon X \to Y$ and $g \colon Y \to Z$ are functions and $g \circ f$ is one-to-one, must both f and g be one-to-one? Prove or give a counterexample.

17. If $f \colon X \to Y$ and $g \colon Y \to Z$ are functions and $g \circ f$ is onto, must both f and g be onto? Prove or give a counterexample.

H 18. If $f \colon X \to Y$ and $g \colon Y \to Z$ are functions and $g \circ f$ is one-to-one, must f be one-to-one? Prove or give a counterexample.

H 19. If $f \colon X \to Y$ and $g \colon Y \to Z$ are functions and $g \circ f$ is onto, must g be onto? Prove or give a counterexample.

20. Let $f \colon W \to X$, $g \colon X \to Y$, and $h \colon Y \to Z$ be functions. Must $h \circ (g \circ f) = (h \circ g) \circ f$? Prove or give a counterexample.

21. True or False? Given any set X and given any functions $f \colon X \to X$, $g \colon X \to X$, and $h \colon X \to X$, if h is one-to-one and $h \circ f = h \circ g$, then $f = g$. Justify your answer.

22. True or False? Given any set X and given any functions $f \colon X \to X$, $g \colon X \to X$, and $h \colon X \to X$, if h is one-to-one and $f \circ h = g \circ h$, then $f = g$. Justify your answer.

In 23 and 24 find $g \circ f$, $(g \circ f)^{-1}$, g^{-1}, f^{-1}, and $f^{-1} \circ g^{-1}$, and state how $(g \circ f)^{-1}$ and $f^{-1} \circ g^{-1}$ are related.

23. Let $X = \{a, c, b\}$, $Y = \{x, y, z\}$, and $Z = \{u, v, w\}$. Define $f \colon X \to Y$ and $g \colon Y \to Z$ by the arrow diagrams below.

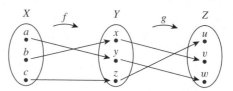

24. Define $f \colon \mathbf{R} \to \mathbf{R}$ and $g \colon \mathbf{R} \to \mathbf{R}$ by the formulas

    $$f(x) = x + 3 \quad \text{and} \quad g(x) = -x \quad \text{for all } x \in \mathbf{R}.$$

25. Prove or give a counterexample: If $f \colon X \to Y$ and $g \colon Y \to X$ are functions such that $g \circ f = i_X$ and $f \circ g = i_Y$, then f and g are both one-to-one and onto and $g = f^{-1}$.

H 26. Suppose $f \colon X \to Y$ and $g \colon Y \to Z$ are both one-to-one and onto. Prove that $(g \circ f)^{-1}$ exists and that $(g \circ f)^{-1} = f^{-1} \circ g^{-1}$.

Exercises 27–31 refer to the definition given immediately before exercise 40 in Section 7.1. Determine which of the properties in 27–30 are true for all functions $f \colon X \to Y$ and which are false for some function f. Also determine whether the property in 31 is true for all functions $f: X \rightarrow Y$ and $g: Y \rightarrow Z$ or false for some functions f and g. Justify your answers.

27. For all subsets A of X, $f^{-1}(f(A)) \subseteq A$.

28. For all subsets A of X, $A \subseteq f^{-1}(f(A))$.

29. For all subsets C of Y, $f(f^{-1}(C)) \subseteq C$.

30. For all subsets C of Y, $C \subseteq f(f^{-1}(C))$.

31. For all subsets E of Z, $(g \circ f)^{-1}(E) = f^{-1}(g^{-1}(E))$.

7.5 Cardinality with Applications to Computability

There are as many squares as there are numbers because they are just as numerous as their roots. — Galileo Galilei, 1632

Historically, the term *cardinal number* was introduced to describe the size of a set ("This set has *eight* elements") as distinguished from an *ordinal number* that refers to the order of an element in a sequence ("This is the *eighth* element in the row"). The definition of cardinal number derives from the primitive technique of representing numbers by fingers or tally marks. Small children, when asked how old they are, will usually answer by holding up a certain number of fingers, each finger being paired with a year of their life. As was discussed in Section 7.2, a pairing of the elements of two sets is called a one-to-one correspondence. We say that two finite sets whose elements can be paired by a one-to-one correspondence have the *same size*. This is illustrated by the following diagram.

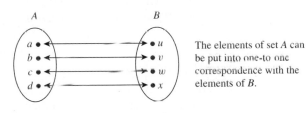

The elements of set A can be put into one-to-one correspondence with the elements of B.

Now a **finite set** is one that has no elements at all or that can be put into one-to-one correspondence with a set of the form $\{1, 2, \ldots, n\}$ for some positive integer n. By contrast, an **infinite set** is a nonempty set that cannot be put into one-to-one correspondence with $\{1, 2, \ldots, n\}$ for any positive integer n. Suppose that, as suggested by the quote from Galileo at the beginning of this section, we extend the concept of size to infinite sets by saying that one infinite set has the same size as another if, and only if, the first set can be put into one-to-one correspondence with the second. What consequences follow from such a definition? Do all infinite sets have the same size, or are some infinite sets larger than others? These are the questions we address in this section. The answers are sometimes surprising and have interesting applications to determining what can and what cannot be computed on a computer.

Galileo Galilei (1564–1642)

• Definition

Let A and B be any sets. **A has the same cardinality as B** if, and only if, there is a one-to-one correspondence from A to B. In other words, A has the same cardinality as B if, and only if, there is a function f from A to B that is one-to-one and onto.

The following theorem gives some basic properties of cardinality, most of which follow from statements proved earlier about one-to-one and onto functions.

Theorem 7.5.1

For all sets A, B, and C,

a. **(Reflexive property of cardinality)** A has the same cardinality as A.

b. **(Symmetric property of cardinality)** If A has the same cardinality as B, then B has the same cardinality as A.

c. **(Transitive property of cardinality)** If A has the same cardinality as B and B has the same cardinality as C, then A has the same cardinality as C.

Proof:

Part (a), Reflexivity: Suppose A is any set. [*To show that A has the same cardinality as A, we must show there is a one-to-one correspondence from A to A.*] Consider the identity function i_A from A to A. This function is one-to-one because if x_1 and x_2 are any elements in A with $i_A(x_1) = i_A(x_2)$, then, by definition of i_A, $x_1 = x_2$. The identity function is also onto because if y is any element of A, then $y = i_A(y)$ by definition of i_A. Hence i_A is a one-to-one correspondence from A to A. [*So there exists a one-to-one correspondence from A to A, as was to be shown.*]

Part (b), Symmetry: Suppose A and B are any sets and A has the same cardinality as B. [*We must show that B has the same cardinality as A.*] Since A has the same cardinality as B, there is a function f from A to B that is one-to-one and onto. But then, by Theorems 7.2.1 and 7.2.2, there is a function f^{-1} from B to A that is also one-to-one and onto. Hence B has the same cardinality as A [*as was to be shown*].

Part (c), Transivity: Suppose A, B, and C are any sets and A has the same cardinality as B and B has the same cardinality as C. [*We must show that A has the same cardinality as C.*] Since A has the same cardinality as B, there is a function f from A to B that is one-to-one and onto, and since B has the same cardinality as C, there is a function g from B to C that is one-to-one and onto. But then, by Theorems 7.4.3 and 7.4.4, $g \circ f$ is a function from A to C that is one-to-one and onto. Hence A has the same cardinality as C [*as was to be shown*].

Note that Theorem 7.5.1(b) makes it possible to say simply that two sets have the same cardinality instead of always having to say that one set has the same cardinality as another. That is, the following definition can be made.

• Definition

A and B **have the same cardinality** if, and only if, A has the same cardinality as B or B has the same cardinality as A.

The following example illustrates a very important property of infinite sets—namely, that an infinite set can have the same cardinality as a proper subset of itself. This property is sometimes taken as the definition of infinite set. The example shows that even though it may seem reasonable to say that there are twice as many integers as there are even integers, the elements of the two sets can be matched up exactly, and so, according to the definition, the two sets have the same cardinality.

Example 7.5.1 An Infinite Set and a Proper Subset Can Have the Same Cardinality

Let 2**Z** be the set of all even integers. Prove that 2**Z** and **Z** have the same cardinality.

Solution   Consider the function H from **Z** to 2**Z** defined as follows:

$$H(n) = 2n \quad \text{for all } n \in \mathbf{Z}.$$

A (partial) arrow diagram for H is shown below.

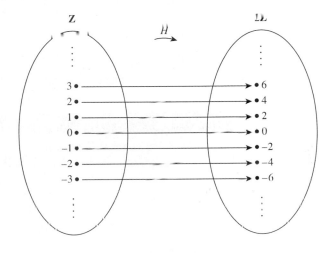

To show that H is one-to-one, suppose $H(n_1) = H(n_2)$ for some integers n_1 and n_2. Then $2n_1 = 2n_2$ by definition of H, and dividing both sides by 2 gives $n_1 = n_2$. Hence h is one-to-one.

To show that H is onto, suppose m is any element of 2**Z**. Then m is an even integer, and so $m = 2k$ for some integer k. It follows that $H(k) = 2k = m$. Thus there exists k in **Z** with $H(k) = m$, and hence H is onto.

Therefore, by definition of cardinality, **Z** and 2**Z** have the same cardinality.   ∎

Theorem 7.3.2 states that a function from one finite set to another set of the same size is one-to-one if, and only if, it is onto. This result does not hold for infinite sets. Although it is true that for two infinite sets to have the same cardinality there must exist a function from one to the other that is both one-to-one and onto, it is also always the case that:

> If A and B are infinite sets with the same cardinality, then there exist functions from A to B that are one-to-one but not onto and functions from A to B that are onto but not one-to-one.

For instance, since the function H in Example 7.5.1 is one-to-one and onto, **Z** and 2**Z** have the same cardinality. But the "inclusion function" I from 2**Z** to **Z**, given by $I(n) = n$ for all even integers n, is one-to-one but not onto. And the function J from **Z** to 2**Z** defined by $J(n) = 2\lfloor n/2 \rfloor$, for all integers n, is onto but not one-to-one. (See exercise 6 at the end of this section.)

Countable Sets

The set $\mathbf{Z}^+$ of counting numbers $\{1, 2, 3, 4, \ldots\}$ is, in a sense, the most basic of all infinite sets. A set A having the same cardinality as this set is called *countably infinite*. The reason is that the one-to-one correspondence between the two sets can be used to "count" the elements of A: If F is a one-to-one and onto function from $\mathbf{Z}^+$ to A, then $F(1)$ can be designated as the first element of A, $F(2)$ as the second element of A, $F(3)$ as the third element of A, and so forth. This is illustrated graphically in Figure 7.5.1. Because F is one-to-one, no element is ever counted twice, and because it is onto, every element of A is counted eventually.

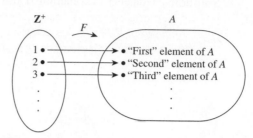

Figure 7.5.1 "Counting" a Countably Infinite Set

> **• Definition**
>
> A set is called **countably infinite** if, and only if, it has the same cardinality as the set of positive integers $\mathbf{Z}^+$. A set is called **countable** if, and only if, it is finite or countably infinite. A set that is not countable is called **uncountable.**

Example 7.5.2 Countability of Z, the Set of All Integers

Show that the set $\mathbf{Z}$ of all integers is countable.

Solution   The set $\mathbf{Z}$ of all integers is certainly not finite, so if it is countable, it must be because it is countably infinite. To show that $\mathbf{Z}$ is countably infinite, find a function from the positive integers $\mathbf{Z}^+$ to $\mathbf{Z}$ that is one-to-one and onto. Looked at in one light, this contradicts common sense; judging from the diagram below, there appear to be more than twice as many integers as there are positive integers.

$$\underbrace{\cdots -5 \quad -4 \quad -3 \quad -2 \quad -1 \quad 0 \quad \overbrace{1 \quad 2 \quad 3 \quad 4 \quad 5 \cdots}^{\text{positive integers}}}_{\text{all integers}}$$

But you were alerted that results in this section might be surprising. Try to think of a way to "count" the set of all integers anyway.

The trick is to start in the middle and work outward systematically. Let the first integer be 0, the second 1, the third -1, the fourth 2, the fifth -2, and so forth as shown in Figure 7.5.2, starting at 0 and swinging outward in back-and-forth arcs from positive to negative integers and back again, picking up one additional integer at each swing.

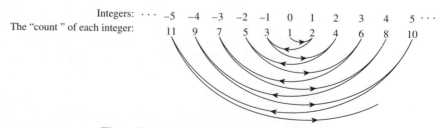

Figure 7.5.2 "Counting" the Set of All Integers

It is clear from the diagram that no integer is counted twice (so the function is one-to-one) and every integer is counted eventually (so the function is onto). Consequently, this diagram defines a function from $\mathbf{Z}^+$ to $\mathbf{Z}$ that is one-to-one and onto. Even though in one sense there seem to be more integers than positive integers, the elements of the two sets can be paired up one for one. It follows by definition of cardinality that $\mathbf{Z}^+$ has the same cardinality as $\mathbf{Z}$. Thus $\mathbf{Z}$ is countably infinite and hence countable.

The diagrammatic description of the above function is acceptable as given. You can check, however, that the function can also be described by the explicit formula

$$F(n) = \begin{cases} \dfrac{n}{2} & \text{if } n \text{ is an even positive integer} \\[2mm] -\dfrac{n-1}{2} & \text{if } n \text{ is an odd positive integer} \end{cases}.$$

■

Example 7.5.3 Countability of 2Z, the Set of All Even Integers

Show that the set $2\mathbf{Z}$ of all even integers is countable.

Solution   Example 7.5.2 showed that $\mathbf{Z}^+$ has the same cardinality as $\mathbf{Z}$, and Example 7.5.1 showed that $\mathbf{Z}$ has the same cardinality as $2\mathbf{Z}$. Thus, by the transitive property of cardinality, $\mathbf{Z}^+$ has the same cardinality as $2\mathbf{Z}$. It follows by definition of countably infinite that $2\mathbf{Z}$ is countably infinite and hence countable. ■

The Search for Larger Infinities: The Cantor Diagonalization Process

Every infinite set we have discussed so far has been countably infinite. Do any larger infinities exist? Are there uncountable sets? Here is one candidate.

Imagine the number line as shown below.

As noted in Section 2.1, the integers are spread along the number line at discrete intervals. The rational numbers, on the other hand, are *dense*: Between any two rational numbers (no matter how close) lies another rational number (the average of the two numbers, for instance; see exercise 17). This suggests the conjecture that the infinity of the set of rational numbers is larger than the infinity of the set of integers.

Amazingly, this conjecture is false. Despite the fact that the rational numbers are crowded onto the number line whereas the integers are quite separated, the set of all rational numbers can be put into one-to-one correspondence with the set of integers. The next example gives a partial proof of this fact. It shows that the set of all positive rational numbers can be put into one-to-one correspondence with the set of all positive integers. In exercise 16 at the end of this section you are asked to use this result, together with a technique similar to that of Example 7.5.2, to show that the set of *all* rational numbers is countable.

Example 7.5.4 The Set of All Positive Rational Numbers Is Countable

Show that the set $\mathbf{Q}^+$ of all positive rational numbers is countable.

Solution   Display the elements of the set $\mathbf{Q}^+$ of positive rational numbers in a grid as shown in Figure 7.5.3.

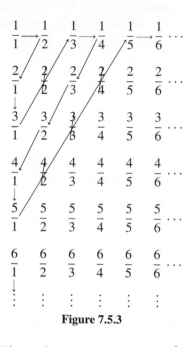

Figure 7.5.3

Define a function F from $\mathbf{Z}^+$ to $\mathbf{Q}^+$ by starting to count at $\frac{1}{1}$ and following the arrows as indicated, skipping over any number that has already been counted.

To be specific: Set $F(1) = \frac{1}{1}$, $F(2) = \frac{1}{2}$, and $F(3) = \frac{2}{1}$. Then skip $\frac{2}{2}$ since $\frac{2}{2} = \frac{1}{1}$, which was counted first. After that, set $F(4) = \frac{3}{1}$, $F(5) = \frac{1}{3}$, $F(6) = \frac{1}{4}$, $F(7) = \frac{2}{3}$, $F(8) = \frac{3}{2}$, $F(9) = \frac{4}{1}$, and $F(10) = \frac{5}{1}$. Then skip $\frac{4}{2}$, $\frac{3}{3}$, and $\frac{2}{4}$ (since $\frac{4}{2} = \frac{2}{1}$, $\frac{3}{3} = \frac{1}{1}$, and $\frac{2}{4} = \frac{1}{2}$) and set $F(11) = \frac{1}{5}$. Continue in this way, defining $F(n)$ for each positive integer n.

Note that every positive rational number appears somewhere in the grid, and the counting procedure is set up so that every point in the grid is reached eventually. Thus the function F is onto. Also, skipping numbers that have already been counted ensures that no number is counted twice. Thus F is one-to-one. Consequently, F is a function from $\mathbf{Z}^+$ to $\mathbf{Q}^+$ that is one-to-one and onto, and so $\mathbf{Q}^+$ is countably infinite and hence countable. ■

In 1874 the German mathematician Georg Cantor achieved success in the search for a larger infinity by showing that the set of all real numbers is uncountable. His method of proof was somewhat complicated, however. We give a proof of the uncountability of the set of all real numbers between 0 and 1 using a simpler technique introduced by Cantor in 1891 and now called the **Cantor diagonalization process.** Over the intervening years, this technique and variations on it have been used to establish a number of important results in logic and the theory of computation.

Before stating and proving Cantor's theorem, we note that every real number, which is a measure of location on a number line, can be represented by a decimal expansion of the form

$$a_0.a_1a_2a_3\ldots,$$

where a_0 is an integer (positive, negative, or zero) and for each $i \geq 1$, a_i is an integer from 0 through 9.

This way of thinking about numbers was developed over several centuries by mathematicians in the Chinese, Hindu, and Islamic worlds, culminating in the work of Ghiyath al-Dīn Jamshīd al-Kashi in 1427. In Europe it was first clearly formulated and successfully promoted by the Flemish mathematician Simon Stevin in 1585. We illustrate the concept with an example.

Consider the point P in Figure 7.5.4. Figure 7.5.4(a) shows P located between 1 and 2. When the interval from 1 to 2 is divided into ten equal subintervals (see Figure 7.5.4(b)) P is seen to lie between 1.6 and 1.7. If the interval from 1.6 to 1.7 is itself divided into ten equal subintervals (see Figure 7.5.4(c)), the P is seen to lie between 1.62 and 1.63 but closer to 1.62 than to 1.63. So to two-decimal-place accuracy, the decimal expansion for P is 1.62.

*al-Kashi
(1380–1429)*

*Simon Stevin
(1548–1620)*

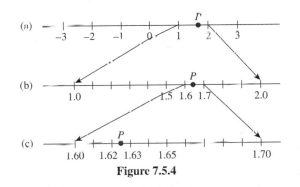

Figure 7.5.4

Assuming that any interval of real numbers, no matter how small, can be divided into ten equal subintervals, the process of obtaining additional digits in the decimal expansion for P can, in theory, be repeated indefinitely. If at any stage P is seen to be a subdivision point, then all further digits in the expansion may be taken to be 0. If not, then the process gives an expansion with an infinite number of digits.

The resulting decimal representation for P is unique except for numbers that end in infinitely repeating 9's or infinitely repeating 0's. For example (see exercise 25 at the end of this section),

$$0.199999\ldots = 0.200000\ldots.$$

Let us agree to express any such decimal in the form that ends in all 0's.

Theorem 7.5.2 (Cantor)

The set of all real numbers between 0 and 1 is uncountable.

Proof (by contradiction):

Suppose the set of all real numbers between 0 and 1 is countable. Then the decimal representations of these numbers can be written in a list as follows:

$$0.a_{11}a_{12}a_{13} \cdots a_{1n} \cdots$$
$$0.a_{21}a_{22}a_{23} \cdots a_{2n} \cdots$$
$$0.a_{31}a_{32}a_{33} \cdots a_{3n} \cdots$$
$$\vdots$$
$$0.a_{n1}a_{n2}a_{n3} \cdots a_{nn} \cdots$$
$$\vdots$$

[*We will derive a contradiction by showing that there is a number between* 0 *and* 1 *that does not appear on this list.*]

   For each pair of positive integers i and j, the jth decimal digit of the ith number on the list is a_{ij}. In particular, the first decimal digit of the first number on the list is a_{11}, the second decimal digit of the second number on the list is a_{22}, and so forth. As an example, suppose the list of real numbers between 0 and 1 starts out as follows:

$$0.\,②\;\;0\;\;1\;\;4\;\;8\;\;8\;\;0\;\;2\ldots$$
$$0.\;1\;\;①\;\;6\;\;6\;\;6\;\;0\;\;2\;\;1\ldots$$
$$0.\;0\;\;3\;\;③\;\;5\;\;3\;\;3\;\;2\;\;0\ldots$$
$$0.\;9\;\;6\;\;7\;\;⑦\;\;6\;\;8\;\;0\;\;9\ldots$$
$$0.\;0\;\;0\;\;0\;\;3\;\;①\;\;0\;\;0\;\;2\ldots$$
$$\vdots$$

The diagonal elements are circled: a_{11} is 2, a_{22} is 1, a_{33} is 3, a_{44} is 7, a_{55} is 1, and so forth.

   Construct a new decimal number $d = 0.d_1d_2d_3 \cdots d_n \cdots$ as follows:

$$d_n = \begin{cases} 1 & \text{if } a_{nn} \neq 1 \\ 2 & \text{if } a_{nn} = 1 \end{cases}.$$

In the above example,

$$d_1 \text{ is 1 because } a_{11} = 2 \neq 1,$$
$$d_2 \text{ is 2 because } a_{22} = 1,$$
$$d_3 \text{ is 1 because } a_{33} = 3 \neq 1,$$
$$d_4 \text{ is 1 because } a_{44} = 7 \neq 1,$$
$$d_5 \text{ is 2 because } a_{55} = 1,$$

and so forth. Hence d would equal $0.12112\ldots$.

   The crucial observation is that for *each integer n, d differs in the nth decimal position from the nth number on the list.* But this implies that d is not on the list! In other words, d is a real number between 0 and 1 that is not on the list of *all* real numbers between 0 and 1. This contradiction shows the falseness of the supposition that the set of all numbers between 0 and 1 is countable. Hence the set of all real numbers between 0 and 1 is uncountable.

Along with demonstrating the existence of an uncountable set, Cantor developed a whole arithmetic theory of infinite sets of various sizes. One of the most basic theorems of the theory states that any subset of a countable set is countable.

Theorem 7.5.3

Any subset of any countable set is countable.

Proof:

Let A be a particular but arbitrarily chosen countable set and let B be any subset of A. [*We must show that B is countable.*] Either B is finite or it is infinite. If B is finite, then B is countable by definition of countable, and we are done. So suppose B is infinite. Since A is countable, the distinct elements of A can be represented as a sequence

$$a_1, a_2, a_3, \ldots.$$

Define a function $g \colon \mathbf{Z}^+ \to B$ inductively as follows:

1. Search sequentially through elements of $a_1, a_2, a_3, \ldots$ until an element of B is found. [*This must happen eventually since $B \subseteq A$ and $B \neq \emptyset$.*] Call that element $g(1)$.

2. For each integer $k \geq 2$, suppose $g(k-1)$ has been defined. Then $g(k-1) = a_i$ for some a_i in $\{a_1, a_2, a_3, \ldots\}$. Starting with a_{i+1}, search sequentially through $a_{i+1}, a_{i+2}, a_{i+3}, \ldots$ trying to find an element of B. One must be found eventually because B is infinite, and $\{g(1), g(2), \ldots, g(k-1)\}$ is a finite set. When an element of B is found, define it to be $g(k)$.

By (1) and (2) above, the function g is defined for each positive integer.

Since the elements of $a_1, a_2, a_3, \ldots$ are all distinct, g is one-to-one. Furthermore, the searches for elements of B are sequential: Each picks up where the previous one left off. Thus every element of A is reached during some search. But all the elements of B are located somewhere in the sequence $a_1, a_2, a_3, \ldots$, and so every element of B is eventually found and made the image of some integer. Hence g is onto. These remarks show that g is a one-to-one correspondence from $\mathbf{Z}^+$ to B. So B is countably infinite and thus countable.

An immediate consequence of Theorem 7.5.3 is the following corollary.

Corollary 7.5.4

Any set with an uncountable subset is uncountable.

Proof:

Consider the following equivalent phrasing of Theorem 7.5.3: For all sets S and for all subsets A of S, if S is countable, then A is countable. The contrapositive of this statement is logically equivalent to it and states: For all sets S and for all subsets A of S, if A is uncountable then S is uncountable. But this is an equivalent phrasing for the corollary. So the corollary is proved.

Corollary 7.5.4 implies that the set of all real numbers is uncountable because the subset of numbers between 0 and 1 is uncountable. In fact, as Example 7.5.5 shows, the set of all real numbers has the same cardinality as the set of all real numbers between 0 and 1! This fact is further explored in exercises 13 and 14 at the end of this section.

Example 7.5.5   The Cardinality of the Set of All Real Numbers

Show that the set of all real numbers has the same cardinality as the set of real numbers between 0 and 1.

Solution   Solution Let S be the open interval of real numbers between 0 and 1:

$$S = \{x \in \mathbf{R} \mid 0 < x < 1\}.$$

Imagine picking up S and bending it into a circle as shown below. Since S does not include either endpoint 0 or 1, the top-most point of the circle is omitted from the drawing.

Define a function $F\colon S \to \mathbf{R}$ as follows:

Draw a number line and place the interval, S, somewhat enlarged and bent into a circle, tangent to the line above the point 0. This is shown below.

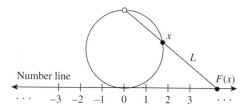

For each point x on the circle representing S, draw a straight line L through the top-most point of the circle and x. Let $F(x)$ be the point of intersection of L and the number line. ($F(x)$ is called the *projection* of x onto the number line.)

It is clear from the geometry of the situation that distinct points on the circle go to distinct points on the number line, so F is one-to-one. In addition, given any point y on the number line, a line can be drawn through y and the top-most point of the circle. This line must intersect the circle at some point x, and, by definition, $y = F(x)$. Thus F is onto. Hence F is a one-to-one correspondence from S to $\mathbf{R}$, and so S and $\mathbf{R}$ have the same cardinality. ∎

Application: Cardinality and Computability

Knowledge of the countability and uncountability of certain sets can be used to answer a question of computability. We begin by showing that a certain set is countable.

Example 7.5.6 Countability of the Set of Computer Programs in a Computer Language

Show that the set of all computer programs in a given computer language is countable.

Solution This result is a consequence of the fact that any computer program in any language can be regarded as a finite string of symbols in the (finite) alphabet of the language.

Given any computer language, let P be the set of all computer programs in the language. Either P is finite or P is infinite. If P is finite, then P is countable and we are done. If P is infinite, set up a binary code to translate the symbols of the alphabet of the language into strings of 0's and 1's. (For instance, either the seven-bit American Standard Code for Information Interchange, known as ASCII, or the eight-bit Extended Binary-Coded Decimal Interchange Code, known as EBCDIC, might be used.)

For each program in P, use the code to translate all the symbols in the program into 0's and 1's. Order these strings by length, putting shorter before longer, and order all strings of a given length by regarding each string as a binary number and writing the numbers in ascending order.

Define a function $F: \mathbf{Z}^+ \to P$ by specifying that

$$F(n) = \text{the } n\text{th program in the list} \quad \text{for each } n \in \mathbf{Z}^+.$$

By construction, F is one-to-one and onto, and so P is countably infinite and hence countable. As a simple example, suppose the following are all the programs in P that translate into bit strings of length less than or equal to 5:

$$10111, \ 11, \ 0010, \ 1011, \ 01, \ 00100, \ 1010, \ 00010.$$

Ordering these by length gives

$$length\ 2:\ 11, 01$$
$$length\ 4:\ 0010, 1011, 1010$$
$$length\ 5:\ 10111, 00100, 00010$$

And ordering those of each given length by the size of the binary number they represent gives

$$01 \quad = F(1)$$
$$11 \quad = F(2)$$
$$0010 = F(3)$$
$$1010 = F(4)$$
$$1011 = F(5)$$
$$00010 = F(6)$$
$$00100 = F(7)$$
$$10111 = F(8)$$

Note that when viewed purely as numbers, ignoring leading zeros, $0010 = 00010$. This shows the necessity of first ordering the strings by length before arranging them in ascending numeric order. ∎

The final example of this section shows that a certain set is uncountable and hence that there must exist a noncomputable function.

Example 7.5.7 The Cardinality of a Set of Functions and Computability

a. Let T be the set of all functions from the positive integers to the set $\{0, 1, 2, 3, 4, 5, 6, 7, 8, 9\}$. Show that T is uncountable.

b. Derive the consequence that there are noncomputable functions. Specifically, show that for any computer language there must be a function F from $\mathbf{Z}^+$ to $\{0, 1, 2, 3, 4, 5, 6, 7, 8, 9\}$ with the property that no computer program can be written in the language to take arbitrary values as input and output the corresponding function values.

Solution

a. Let S be the set of all real numbers between 0 and 1. As noted before, any number in S can be represented in the form

$$0.a_1a_2a_3 \ldots a_n \ldots,$$

where each a_i is an integer from 0 to 9. This representation is unique if decimals that end in all 9's are omitted.

Define a function F from S to a subset of T (the set of all functions from $\mathbf{Z}^+$ to $\{0, 1, 2, 3, 4, 5, 6, 7, 8, 9\}$) as follows:

$$F(0.a_1a_2a_3 \ldots a_n \ldots) = \text{the function that sends each}$$
$$\text{positive integer } n \text{ to } a_n.$$

Choose the co-domain of F to be exactly that subset of T that makes F onto. That is, define the co-domain of F to equal the image of F. Note that F is one-to-one because if $F(x_1) = F(x_2)$, then each decimal digit of x_1 equals the corresponding decimal digit of x_2, and so $x_1 = x_2$. Thus F is a one-to-one correspondence from S to a subset of T. But S is uncountable by Theorem 7.5.2. Hence T has an uncountable subset, and so, by Theorem 7.5.3, T is uncountable.

b. Part (a) shows that the set T of all functions from $\mathbf{Z}^+$ to $\{0, 1, 2, 3, 4, 5, 6, 7, 8, 9\}$ is uncountable. But Example 7.5.6 shows that given any computer language, the set of all programs in that language is countable. Consequently, in any computer language there are not enough programs to compute values of every function in T. There must exist functions that are not computable! ■

Exercise Set 7.5

1. When asked what it means to say that set A has the same cardinality as set B, a student replies, "A are B are one-to-one and onto." What *should* the student have replied? Why?

2. Show that "there are as many squares as there are numbers" by exhibiting a one-to-one correspondence from the positive integers, $\mathbf{Z}^+$, to the set S of all squares of positive integers:
$$S = \{n \in \mathbf{Z}^+ \mid n = k^2, \text{ for some positive integer } k\}.$$

3. Let $3\mathbf{Z} = \{n \in \mathbf{Z} \mid n = 3k, \text{ for some integer } k\}$. Prove that $\mathbf{Z}$ and $3\mathbf{Z}$ have the same cardinality.

4. Let $\mathbf{O}$ be the set of all odd integers. Prove that $\mathbf{O}$ has the same cardinality as $2\mathbf{Z}$, the set of all even integers.

5. Let $25\mathbf{Z}$ be the set of all integers that are multiples of 25. Prove that $25\mathbf{Z}$ has the same cardinality as $2\mathbf{Z}$, the set of all even integers.

H 6. Use the functions I and J defined in the paragraph following Example 7.5.1 to show that even though there is a one-to-one correspondence, H, from $2\mathbf{Z}$ to $\mathbf{Z}$, there is also a function from $2\mathbf{Z}$ to $\mathbf{Z}$ that is one-to-one but not onto and a function from $\mathbf{Z}$ to $2\mathbf{Z}$ that is onto but not one-to-one. In other words, show that I is one-to-one but not onto, and show that J is onto but not one-to-one.

7. a. Check that the formula for F given at the end of Example 7.5.2 produces the correct values for $n = 1, 2, 3$, and 4.
   b. Use the floor function to write a formula for F as a single algebraic expression for all positive integers n.

8. Use the result of exercise 3 to prove that $3\mathbf{Z}$ is countable.

9. Show that the set of all nonnegative integers is countable by exhibiting a one-to-one correspondence between $\mathbf{Z}^+$ and $\mathbf{Z}^{nonneg}$.

In 10–14, S denotes the set of real numbers strictly between 0 and 1. That is, $S = \{x \in \mathbf{R} \mid 0 < x < 1\}$.

10. Let $U = \{x \in \mathbf{R} \mid 0 < x < 2\}$. Prove that S and U have the same cardinality.

H 11. Let $V = \{x \in \mathbf{R} \mid 2 < x < 5\}$. Prove that S and V have the same cardinality.

12. Let a and b be real numbers with $a < b$, and suppose that $W = \{x \in \mathbf{R} \mid a < x < b\}$. Prove that S and W have the same cardinality.

13. Draw the graph of the function f defined by the following formula:

For all real numbers x with $0 < x < 1$,

$$f(x) = \tan\left(\pi x - \frac{\pi}{2}\right).$$

Use the graph to explain why S and $\mathbf{R}$ have the same cardinality.

✶ 14. Define a function g from the set of real numbers to S by the following formula:

For all real numbers x,

$$g(x) = \frac{1}{2} \cdot \left(\frac{x}{1 + |x|}\right) + \frac{1}{2}.$$

Prove that g is a one-to-one correspondence. (It is possible to prove this statement either with calculus or without it.) What conclusion can you draw from this fact?

15. Show that the set of all bit strings (strings of 0's and 1's) is countable.

16. Show that $\mathbf{Q}$, the set of all rational numbers, is countable.

17. Show that the set $\mathbf{Q}$ of all rational numbers is dense along the number line by showing that given any two rational numbers r_1 and r_2 with $r_1 < r_2$, there exists a rational number x such that $r_1 < x < r_2$.

H 18. Must the average of two irrational numbers always be irrational? Prove or give a counterexample.

H ✶ 19. Show that the set of all irrational numbers is dense along the number line by showing that given any two real numbers, there is an irrational number in between.

20. Give two examples of functions from $\mathbf{Z}$ to $\mathbf{Z}$ that are one-to-one but not onto.

21. Give two examples of functions from $\mathbf{Z}$ to $\mathbf{Z}$ that are onto but not one-to-one.

H 22. Define a function $g: \mathbf{Z}^+ \times \mathbf{Z}^+ \to \mathbf{Z}^+$ by the formula $g(m, n) = 2^m 3^n$ for all $(m, n) \in \mathbf{Z}^+ \times \mathbf{Z}^+$. Show that g is one-to-one and use this result to prove that $\mathbf{Z}^+ \times \mathbf{Z}^+$ is countable.

23. a. Explain how to use the following diagram to show that $\mathbf{Z}^{nonneg} \times \mathbf{Z}^{nonneg}$ and $\mathbf{Z}^{nonneg}$ have the same cardinality.

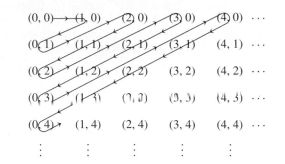

H ✶ b. Define a function $H: \mathbf{Z}^{nonneg} \times \mathbf{Z}^{nonneg} \to \mathbf{Z}^{nonneg}$ by the formula

$$H(m, n) = n + \frac{(m + n)(m + n + 1)}{2}$$

for all nonnegative integers m and n. Interpret the action of H geometrically using the diagram of part (a).

✶ 24. Prove that the function H defined in exercise 23 is a one-to-one correspondence.

H 25. Prove that $0.1999\ldots = 0.2$.

26. Prove that any infinite set contains a countably infinite subset.

27. If A is any countably infinite set, B is any set, and $g: A \to B$ is onto, then B is countable.

H 28. Prove that a union of any two countably infinite sets is countably infinite.

H 29. Use the result of exercise 28 to prove that the set of all irrational numbers is uncountable.

30. Prove that a union of any finite set and any countably infinite set is countably infinite.

31. Use the results of exercises 28 and 30 to prove that a union of any two countable sets is countable.

H 32. Prove that $\mathbf{Z} \times \mathbf{Z}$, the Cartesian product of the set of integers with itself, is countably infinite.

33. Use the results of exercises 27, 30, and 32 to prove the following: If R is the set of all solutions to all equations of the form $x^2 + bx + c = 0$, where b and c are integers, then R is countable.

H 34. Let $\mathscr{P}(S)$ be the set of all subsets of set S, and let T be the set of all functions from S to $\{0, 1\}$. Show that $\mathscr{P}(S)$ and T have the same cardinality.

H 35. Let S be a set and let $\mathscr{P}(S)$ be the set of all subsets of S. Show that S and $\mathscr{P}(S)$ do not have the same cardinality.

✱ 36. The Schroeder–Bernstein theorem states the following: If A and B are any sets with the property that there is a one-to-one function from A to B and a one-to-one function from B to A, then A and B have the same cardinality. Use this theorem to prove that there are as many functions from $\mathbf{Z}^+$ to $\{0, 1, 2, 3, 4, 5, 6, 7, 8, 9\}$ as there are functions from $\mathbf{Z}^+$ to $\{0, 1\}$.

H 37. Prove that if A and B are any countably infinite sets, then $A \times B$ is countably infinite.

✱ 38. Suppose $A_1, A_2, A_3, \ldots$ is an infinite sequence of countable sets. Let

$$\bigcup_{i=1}^{\infty} A_i = \{x \mid x \in A_i \text{ for some } i\}.$$

Prove that $\bigcup_{i=1}^{\infty} A_i$ is countable. (In other words, prove that a countably infinite union of countable sets is countable.)

CHAPTER 8

RECURSION

A sequence is said to be defined recursively if certain initial values are specified and later terms of the sequence are defined by relating them to a number of earlier terms. In the first section of this chapter, we give a variety of examples that show how to analyze certain kinds of problems by thinking recursively to obtain a recursively defined sequence. In the next two sections we address the problem of finding an explicit formula for a sequence that is defined recursively. And in the final section we discuss more general recursive definitions, such as the one used for the careful formulation of the concept of Boolean expression, and the idea of recursive function.

8.1 Recursively Defined Sequences

So, Nat'ralists observe, a Flea/Hath smaller Fleas that on him prey,/And these have smaller Fleas to bite 'em,/And so proceed ad infinitum. — Jonathan Swift, 1733

A sequence can be defined in a variety of different ways. One informal way is to write the first few terms with the expectation that the general pattern will be obvious. We might say, for instance, "consider the sequence $3, 5, 7, \ldots$." Unfortunately, misunderstandings can occur when this approach is used. The next term of the sequence could be 9 if we mean the sequence of odd integers, or it could be 11 if we mean the sequence of odd prime numbers.

A second way to define a sequence is to give an explicit formula for its nth term. For example, a sequence $a_0, a_1, a_2, \ldots$ can be specified by writing

$$a_n = \frac{(-1)^n}{n+1} \quad \text{for all integers } n \geq 0.$$

The advantage of defining a sequence by such an explicit formula is that each term of the sequence is uniquely determined and any term can be computed in a fixed, finite number of steps. In this case, for instance,

$$a_0 = \frac{(-1)^0}{0+1} = 1, \qquad a_1 = \frac{(-1)^1}{1+1} = -\frac{1}{2}, \qquad \text{and so forth.}$$

A third way to define a sequence is to use recursion. This requires giving both an equation, called a *recurrence relation*, that relates later terms in the sequence to earlier terms and a specification, called *initial conditions,* of the values of the first few terms of the sequence. The initial conditions are also called the *base* or *bottom* of the recursion. For instance, define a sequence $b_0, b_1, b_2, \ldots$ recursively as follows: For all integers $k \geq 2$,

$$(1) \quad b_k = b_{k-1} + b_{k-2} \qquad \text{recurrence relation}$$

$$(2) \quad b_0 = 1, \quad b_1 = 3 \qquad \text{initial conditions.}$$

Since b_0 and b_1 are given, b_2 can be computed using the recurrence relation.

$$\begin{aligned} b_2 &= b_1 + b_0 && \text{by substituting } k = 2 \text{ into (1)} \\ &= 3 + 1 && \text{since } b_1 = 3 \text{ and } b_0 = 1 \text{ by (2)} \end{aligned}$$

$$(3) \quad \therefore b_2 = 4$$

Then, since both b_1 and b_2 are now known, b_3 can be computed using the recurrence relation.

$$\begin{aligned} b_3 &= b_2 + b_1 && \text{by substituting } k = 3 \text{ into (1)} \\ &= 4 + 3 && \text{since } b_2 = 4 \text{ by (3) and } b_1 = 3 \text{ by (2)} \end{aligned}$$

$$(4) \quad \therefore b_3 = 7$$

In general, the recurrence relation says that any term of the sequence after a certain given term is the sum of the two preceding terms. Thus

$$b_4 = b_3 + b_2 = 7 + 4 = 11,$$
$$b_5 = b_4 + b_3 = 11 + 7 = 18,$$

and so forth. It should be clear that any later term of the sequence can be computed from this point by continuing in a step-by-step fashion.

Sometimes it is very difficult or impossible to find an explicit formula for a sequence, but it *is* possible to define the sequence using recursion. Note that defining sequences recursively is similar to proving theorems by mathematical induction. The recurrence relation is like the inductive step and the initial conditions are like the basis step. Indeed, the fact that sequences can be defined recursively is equivalent to the fact that mathematical induction works as a method of proof.

• Definition

A **recurrence relation** for a sequence $a_0, a_1, a_2, \ldots$ is a formula that relates each term a_k to certain of its predecessors $a_{k-1}, a_{k-2}, \ldots, a_{k-i}$, where i is an integer and k is any integer greater than or equal to i. The **initial conditions** for such a recurrence relation specify the values of $a_0, a_1, a_2, \ldots, a_{i-1}$, if i is a fixed integer, or $a_0, a_1, \ldots, a_m$, where m is an integer with $m \geq 0$, if i depends on k.

Example 8.1.1 Computing Terms of a Recursively Defined Sequence

Define a sequence $c_0, c_1, c_2, \ldots$ recursively as follows: For all integers $k \geq 2$,

$$(1) \quad c_k = c_{k-1} + kc_{k-2} + 1 \qquad \text{recurrence relation}$$

$$(2) \quad c_0 = 1 \quad \text{and} \quad c_1 = 2 \qquad \text{initial conditions.}$$

Find $c_2, c_3,$ and c_4.

Solution

$$c_2 = c_1 + 2c_0 + 1 \qquad \text{by substituting } k = 2 \text{ into (1)}$$
$$= 2 + 2 \cdot 1 + 1 \qquad \text{since } c_1 = 2 \text{ and } c_0 = 1 \text{ by (2)}$$
(3) $\therefore c_2 = 5$
$$c_3 = c_2 + 3c_1 + 1 \qquad \text{by substituting } k = 3 \text{ into (1)}$$
$$= 5 + 3 \cdot 2 + 1 \qquad \text{since } c_2 = 5 \text{ by (3) and } c_1 = 2 \text{ by (2)}$$
(4) $\therefore c_3 = 12$
$$c_4 = c_3 + 4c_2 + 1 \qquad \text{by substituting } k = 4 \text{ into (1)}$$
$$12 + 4 \cdot 5 + 1 \qquad \text{since } c_3 = 12 \text{ by (4) and } c_2 = 5 \text{ by (3)}$$
(5) $\therefore c_4 = 33$ ∎

A given recurrence relation may be expressed in several different ways.

Example 8.1.2 Writing a Recurrence Relation in More Than One Way

Let $s_0, s_1, s_2, \ldots$ be a sequence that satisfies the following recurrence relation:

$$\text{for all integers } k \geq 1, \quad s_k = 3s_{k-1} - 1.$$

Explain why the following statement is true:

$$\text{for all integers } k \geq 0, \quad s_{k+1} = 3s_k - 1.$$

Solution   In informal language, the recurrence relation says that any term of the sequence equals 3 times the previous term minus 1. Now for any integer $k \geq 0$, the term previous to s_{k+1} is s_k. Thus for any integer $k \geq 0$, $s_{k+1} = 3s_k - 1$. ∎

A sequence defined recursively need not start with a subscript of zero. Also, a given recurrence relation may be satisfied by many different sequences; the actual values of the sequence are determined by the initial conditions.

Example 8.1.3 Sequences That Satisfy the Same Recurrence Relation

Let $a_1, a_2, a_3, \ldots$ and $b_1, b_2, b_3, \ldots$ satisfy the recurrence relation that the kth term equals 3 times the $(k-1)$st term for all integers $k \geq 2$:

$$(1) \quad a_k = 3a_{k-1} \quad \text{and} \quad b_k = 3b_{k-1}.$$

But suppose that the initial conditions for the sequences are different:

$$(2) \quad a_1 = 2 \quad \text{and} \quad b_1 = 1.$$

Find (a) a_2, a_3, a_4 and (b) b_2, b_3, b_4.

Solution

a. $a_2 = 3a_1 = 3 \cdot 2 = 6$ b. $b_2 = 3b_1 = 3 \cdot 1 = 3$
   $a_3 = 3a_2 = 3 \cdot 6 = 18$ $b_3 = 3b_2 = 3 \cdot 3 = 9$
   $a_4 = 3a_3 = 3 \cdot 18 = 54$ $b_4 = 3b_3 = 3 \cdot 9 = 27$

Thus

$$a_1, a_2, a_3, \ldots \text{ begins } 2, 6, 18, 54, \ldots \text{ and}$$
$$b_1, b_2, b_3, \ldots \text{ begins } 1, 3, 9, 27, \ldots.$$ ∎

Example 8.1.4 Showing That a Sequence Given by an Explicit Formula Satisfies a Certain Recurrence Relation

Show that the sequence $1, -1!, 2!, -3!, 4!, \ldots, (-1)^n n!, \ldots$, for $n \geq 0$, satisfies the recurrence relation

$$s_k = (-k)s_{k-1} \quad \text{for all integers } k \geq 1.$$

Solution   The recurrence relation specifies that the kth term of the sequence equals $-k$ times the $(k-1)$st term. Call the general term of the sequence s_n starting with $s_0 = 1$. Then by definition of the sequence,

$$s_n = (-1)^n n! \quad \text{for each integer } n > 0.$$

Substitute k and $k - 1$ for n to get

$$s_k = (-1)^k k! \qquad\qquad 8.1.1$$
$$s_{k-1} = (-1)^{k-1}(k-1)! \qquad\qquad 8.1.2$$

It follows that

$$
\begin{aligned}
(-k)s_{k-1} &= (-k)[(-1)^{k-1}(k-1)!] && \text{by substitution from (8.1.2)} \\
&= (-1)k(-1)^{k-1}(k-1)! \\
&= (-1)(-1)^{k-1}k(k-1)! \\
&= (-1)^k k! && \text{by basic algebra} \\
&= s_k && \text{by substitution from (8.1.1).} \quad\blacksquare
\end{aligned}
$$

Examples of Recursively Defined Sequences

Recursion is one of the central ideas of computer science. To solve a problem recursively means to find a way to break it down into smaller subproblems each having the same form as the original problem—and to do this in such a way that when the process is repeated many times, the last of the subproblems are small and easy to solve and the solutions of the subproblems can be woven together to form a solution to the original problem.

Probably the most difficult part of solving problems recursively is to figure out how knowing the solution to smaller subproblems of the same type as the original problem will give you a solution to the problem as a whole. You *suppose* you know the solutions to smaller subproblems and ask yourself how you would best make use of that knowledge to solve the larger problem. The supposition that the smaller subproblems have already been solved has been called the *recursive paradigm* or the *recursive leap of faith*. Once you take this leap, you are right in the middle of the most difficult part of the problem, but generally, the path to a solution from this point, though difficult, is short. The recursive leap of faith is similar to the inductive hypothesis in a proof by mathematical induction.

Example 8.1.5 The Tower of Hanoi

In 1883 a French mathematician, Édouard Lucas, invented a puzzle that he called The Tower of Hanoi (La Tour D'Hanoï). The puzzle consisted of eight disks of wood with holes in their centers, which were piled in order of decreasing size on one pole in a row of three. A facsimile of the cover of the box is shown in Figure 8.1.1. Those who played the game were supposed to move all the disks one by one from one pole to another, never placing a larger disk on top of a smaller one. The directions to the puzzle claimed it was based on an old Indian legend:

*Édouard Lucas
(1842–1891)*

*On the steps of the altar in the temple of Benares, for many, many years
Brahmins have been moving a tower of sixty-four golden disks from one
pole to another, one by one, never placing a larger on top of a smaller.
When all the disks have been transferred the Tower and the Brahmins will
fall, and it will be the end of the world.*

Figure 8.1.1

The puzzle offered a prize of ten thousand francs (about $34,000 dollars today) to anyone
who could move a tower of 64 disks by hand while following the rules of the game. (See
Figure 8.1.2.) Assuming that you transferred the disks as efficiently as possible, how
many moves would be required to win the prize?

Figure 8.1.2

Solution   An elegant and efficient way to solve this problem is to think recursively. Suppose that you, somehow or other, have found the most efficient way possible to transfer a
tower of $k - 1$ disks one by one from one pole to another, obeying the restriction that you
never place a larger disk on top of a smaller one. What is the most efficient way to transfer a

tower of k disks from one pole to another? The answer is sketched in Figure 8.1.3, where pole A is the initial pole and pole C is the target pole, and is described below.

Step 1: Transfer the top $k - 1$ disks from pole A to pole B. If $k > 2$, execution of this step will require a number of moves of individual disks among the three poles. But the point of thinking recursively is not to get caught up in imagining the details of how those moves will occur.

Step 2: Move the bottom disk from pole A to pole C.

Step 3: Transfer the top $k - 1$ disks from pole B to pole C. (Again, if $k > 2$, execution of this step will require more than one move.)

To see that this sequence of moves is most efficient, observe that to move the bottom disk of a stack of k disks from one pole to another, you must first transfer the top $k - 1$ disks to a third pole to get them out of the way. Thus transferring the stack of k disks from pole A to pole C requires at least two transfers of the top $k - 1$ disks: one to transfer them off the bottom disk to free the disk so that it can be moved and another to transfer them back on top of the bottom disk after the bottom disk has been moved to pole C. If the bottom disk were not moved directly from pole A to pole C but were moved to pole B first, at least two additional transfers of the top $k - 1$ disks would be necessary: one to move them from pole A to pole C so that the bottom disk could be moved from pole A to pole B and another to move them off pole C so that the bottom disk could be moved onto pole C. This would increase the total number of moves and result in a less efficient transfer.

Thus the minimum sequence of moves must include going from the initial position (a) to position (b) to position (c) to position (d). If follows that

$$\begin{bmatrix} \text{the minimum} \\ \text{number of moves} \\ \text{needed to transfer} \\ \text{a tower of } k \text{ disks} \\ \text{from pole } A \text{ to} \\ \text{pole } C \end{bmatrix} = \begin{bmatrix} \text{the minimum} \\ \text{number of} \\ \text{moves needed} \\ \text{to go from} \\ \text{position (a)} \\ \text{to position (b)} \end{bmatrix} + \begin{bmatrix} \text{The minimum} \\ \text{number of} \\ \text{moves needed} \\ \text{to go from} \\ \text{position (b)} \\ \text{to position (c)} \end{bmatrix} + \begin{bmatrix} \text{the minimum} \\ \text{number of} \\ \text{moves needed} \\ \text{to go from} \\ \text{position (c)} \\ \text{to position (d)} \end{bmatrix} \quad 8.1.3$$

For each integer $n \geq 1$, let

$$m_n = \begin{bmatrix} \text{the minimum number of moves needed to transfer} \\ \text{a tower of } n \text{ disks from one pole to another} \end{bmatrix}$$

Note that the numbers m_n are independent of the labeling of the poles; it takes the same minimum number of moves to transfer n disks from pole A to pole C as to transfer n disks from pole A to pole B, for example. Also the values of m_n are independent of the number of larger disks that may lie below the top n, provided these remain stationary while the top n are moved. Because the disks on the bottom are all larger than the ones on the top, the top disks can be moved from pole to pole as though the bottom disks were not present.

Now going from position (a) to position (b) requires m_{k-1} moves, going from position (b) to position (c) requires just one move, and going from position (c) to position (d) requires m_{k-1} moves. By substitution into equation (8.1.3), therefore,

$$m_k = m_{k-1} + 1 + m_{k-1}$$
$$= 2m_{k-1} + 1 \qquad \text{for all integers } k \geq 2.$$

The initial condition, or base, of this recursion is found by using the definition of the sequence. Because just one move is needed to move one disk from one pole to another,

$$m_1 = \begin{bmatrix} \text{the minimum number of moves needed to move} \\ \text{a tower of one disk from one pole to another} \end{bmatrix} = 1.$$

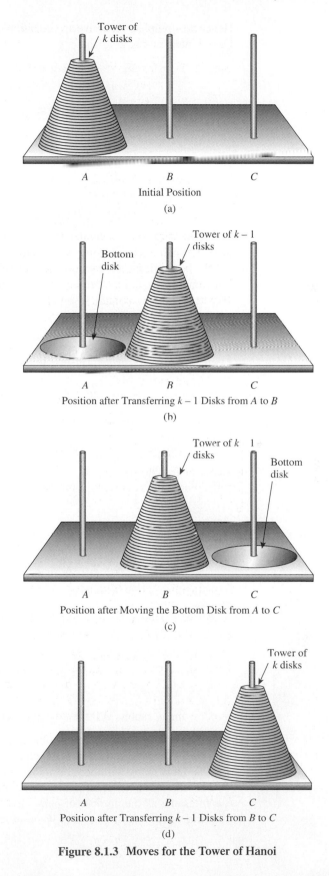

Tower of
k disks

A B C

Initial Position

(a)

Tower of $k - 1$
disks

Bottom
disk

A B C

Position after Transferring $k - 1$ Disks from A to B

(b)

Tower of k 1
disks

Bottom
disk

A B C

Position after Moving the Bottom Disk from A to C

(c)

Tower of
k disks

A B C

Position after Transferring $k - 1$ Disks from B to C

(d)

Figure 8.1.3 Moves for the Tower of Hanoi

Hence the complete recursive specification of the sequence $m_1, m_2, m_3, \ldots$ is as follows: For all integers $k \geq 2$,

$$(1) \quad m_k = 2m_{k-1} + 1 \qquad \text{recurrence relation}$$

$$(2) \quad m_1 = 1 \qquad \text{initial condition}$$

Here is a computation of the next five terms of the sequence:

$$(3) \quad m_2 = 2m_1 + 1 = 2 \cdot 1 + 1 = 3 \qquad \text{by (1) and (2)}$$

$$(4) \quad m_3 = 2m_2 + 1 = 2 \cdot 3 + 1 = 7 \qquad \text{by (1) and (3)}$$

$$(5) \quad m_4 = 2m_3 + 1 = 2 \cdot 7 + 1 = 15 \qquad \text{by (1) and (4)}$$

$$(6) \quad m_5 = 2m_4 + 1 = 2 \cdot 15 + 1 = 31 \qquad \text{by (1) and (5)}$$

$$(7) \quad m_6 = 2m_5 + 1 = 2 \cdot 31 + 1 = 63 \qquad \text{by (1) and (6)}$$

Going back to the legend, suppose the priests work rapidly and move one disk every second. Then the time from the beginning of creation to the end of the world would be m_{64} seconds. In the next section we derive an explicit formula for m_n. Meanwhile, we can compute m_{64} on a calculator or a computer by continuing the process started above (Try it!). The approximate result is

$$1.844674 \times 10^{19} \text{ seconds} \cong 5.84542 \times 10^{11} \text{ years}$$

$$\cong 584.5 \text{ billion years},$$

which is obtained by the estimate of

$$60 \cdot 60 \cdot 24 \cdot (365.25) = 31{,}557{,}600$$

seconds per minute	minutes per hour	hours per day	days per year	seconds per year

seconds in a year (figuring 365.25 days in a year to take leap years into account). Surprisingly, this figure is close to some scientific estimates of the life of the universe! ■

Example 8.1.6 The Fibonacci Numbers

Fibonacci (Leonardo of Pisa) (ca. 1175–1250)

One of the earliest examples of a recursively defined sequence arises in the writings of Leonardo of Pisa, commonly known as Fibonacci, who was the greatest European mathematician of the Middle Ages. In 1202 Fibonacci posed the following problem:

A single pair of rabbits (male and female) is born at the beginning of a year. Assume the following conditions:

1. Rabbit pairs are not fertile during their first month of life but thereafter give birth to one new male/female pair at the end of every month.

2. No rabbits die.

How many rabbits will there be at the end of the year?

Solution One way to solve this problem is to plunge right into the middle of it using recursion. Suppose you know how many rabbit pairs there were at the ends of previous months. How many will there be at the end of the current month?

The crucial observation is that the number of rabbit pairs born at the end of month k is the same as the number of pairs alive at the end of month $k - 2$. Why? Because it is exactly the rabbit pairs that were alive at the end of month $k - 2$ that were fertile during month k. The rabbits born at the end of month $k - 1$ were not.

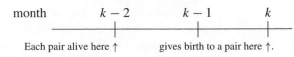

Each pair alive here ↑   gives birth to a pair here ↑.

Now the number of rabbit pairs alive at the end of month k equals the ones alive at the end of month $k - 1$ plus the pairs newly born at the end of the month. Thus

$$\begin{bmatrix} \text{the number} \\ \text{of rabbit} \\ \text{pairs alive} \\ \text{at the end} \\ \text{of month } k \end{bmatrix} = \begin{bmatrix} \text{the number} \\ \text{of rabbit} \\ \text{pairs alive} \\ \text{at the end} \\ \text{of month } k - 1 \end{bmatrix} + \begin{bmatrix} \text{the number} \\ \text{of rabbit} \\ \text{pairs born} \\ \text{at the end} \\ \text{of month } k \end{bmatrix}$$

$$= \begin{bmatrix} \text{the number} \\ \text{of rabbit} \\ \text{pairs alive} \\ \text{at the end} \\ \text{of month } k - 1 \end{bmatrix} + \begin{bmatrix} \text{the number} \\ \text{of rabbit} \\ \text{pairs alive} \\ \text{at the end} \\ \text{of month } k - 2 \end{bmatrix} \qquad 8.1.4$$

For each integer $n \geq 1$, let

$$F_n = \begin{bmatrix} \text{the number of rabbit pairs} \\ \text{alive at the end of month } n \end{bmatrix}$$

and let

$$F_0 = \text{the initial number of rabbit pairs}$$
$$= 1.$$

Then by substitution into equation (8.1.4), for all integers $k \geq 2$,

$$F_k = F_{k-1} + F_{k-2}.$$

Now $F_0 = 1$, as already noted, and $F_1 = 1$ also, because the first pair of rabbits is not fertile until the second month. Hence the complete specification of the Fibonacci sequence is as follows: For all integers $k \geq 2$,

(1) $F_k = F_{k-1} + F_{k-2}$ recurrence relation

(2) $F_0 = 1, \quad F_1 = 1$ initial conditions.

To answer Fibonacci's question, compute F_2, F_3, and so forth through F_{12}:

(3) $F_2 = F_1 + F_0 \quad = 1 + 1 \quad = 2$ by (1) and (2)

(4) $F_3 = F_2 + F_1 \quad = 2 + 1 \quad = 3$ by (1), (2) and (3)

(5) $F_4 = F_3 + F_2 \quad = 3 + 2 \quad = 5$ by (1), (3) and (4)

(6) $F_5 = F_4 + F_3 \quad = 5 + 3 \quad = 8$ by (1), (4) and (5)

(7) $F_6 = F_5 + F_4 \quad = 8 + 5 \quad = 13$ by (1), (5) and (6)

(8) $F_7 = F_6 + F_5 \quad = 13 + 8 \quad = 21$ by (1), (6) and (7)

(9) $F_8 = F_7 + F_6 \quad = 21 + 13 \quad = 34$ by (1), (7) and (8)

(10) $F_9 = F_8 + F_7 \quad = 34 + 21 \quad = 55$ by (1), (8) and (9)

(11) $F_{10} = F_9 + F_8 \quad = 55 + 34 \quad = 89$ by (1), (9) and (10)

(12) $F_{11} = F_{10} + F_9 \quad = 89 + 55 \quad = 144$ by (1), (10) and (11)

(13) $F_{12} = F_{11} + F_{10} = 144 + 89 = 233$ by (1), (11) and (12)

At the end of the twelfth month there are 233 rabbit pairs, or 466 rabbits in all. ■

Example 8.1.7 Compound Interest

On your twenty-first birthday you get a letter informing you that on the day you were born an eccentric rich aunt deposited $100,000 in a bank account earning 4% interest compounded annually and she now intends to turn the account over to you, provided you can figure out how much it is worth. What is the amount currently in the account?

Solution   To approach this problem recursively, observe that

$$\begin{bmatrix} \text{the amount in} \\ \text{the account at} \\ \text{the end of any} \\ \text{particular year} \end{bmatrix} = \begin{bmatrix} \text{the amount in} \\ \text{the account at} \\ \text{the end of the} \\ \text{previous year} \end{bmatrix} + \begin{bmatrix} \text{the interest} \\ \text{earned on the} \\ \text{account during} \\ \text{the year} \end{bmatrix}.$$

Now the interest earned during the year equals the interest rate, $4\% = 0.04$ times the amount in the account at the end of the previous year. Thus

$$\begin{bmatrix} \text{the amount in} \\ \text{the account at} \\ \text{the end of any} \\ \text{particular year} \end{bmatrix} = \begin{bmatrix} \text{the amount in} \\ \text{the account at} \\ \text{the end of the} \\ \text{previous year} \end{bmatrix} + (0.04) \cdot \begin{bmatrix} \text{the amount in} \\ \text{the account at} \\ \text{the end of the} \\ \text{previous year} \end{bmatrix}. \qquad 8.1.5$$

For each positive integer n, let

$$A_n = \begin{bmatrix} \text{the amount in the account} \\ \text{at the end of year } n \end{bmatrix}$$

and let

$$A_0 = \begin{bmatrix} \text{the initial amount} \\ \text{in the account} \end{bmatrix} = \$100,000.$$

Then for any particular year k, substitution into equation (8.1.5) gives

$$A_k = A_{k-1} + (0.04) \cdot A_{k-1}$$
$$= (1 + 0.04) \cdot A_{k-1} = (1.04) \cdot A_{k-1} \quad \text{by factoring out } A_{k-1}.$$

Consequently, the values of the sequence $A_0, A_1, A_2, \ldots$ are completely specified as follows: for all integers $k \geq 1$,

(1) $A_k = (1.04) \cdot A_{k-1}$ recurrence relation

(2) $A_0 = \$100,000$ initial condition.

The number 1.04 is called the *growth factor* of the sequence.

In the next section we derive an explicit formula for the value of the account in any year n. The value on your twenty-first birthday can also be computed by repeated substitution as follows:

(3) $A_1 = 1.04 \cdot A_0 = (1.04) \cdot \$100,000 = \$104,000$ by (1) and (2)

(4) $A_2 = 1.04 \cdot A_1 = (1.04) \cdot \$104,000 = \$108,160$ by (1) and (3)

(5) $A_3 = 1.04 \cdot A_2 = (1.04) \cdot \$108,160 = \$112,486.40$ by (1) and (4)

$\vdots$ $\vdots$

(22) $A_{20} = 1.04 \cdot A_{19} \cong (1.04) \cdot \$210,684.92 \cong \$219,112.31$ by (1) and (21)

(23) $A_{21} = 1.04 \cdot A_{20} \cong (1.04) \cdot \$219,112.31 \cong \$227,876.81$ by (1) and (22)

The amount in the account is $227,876.81 (to the nearest cent). Fill in the dots (to check the arithmetic) and collect your money!   ■

Example 8.1.8 Compound Interest with Compounding Several Times a Year

When an annual interest rate of i is compounded m times per year, the interest rate paid per period is i/m. For instance, if $3\% = 0.03$ annual interest is compounded quarterly, then the interest rate paid per quarter is $0.03/4 = 0.0075$.

For each integer $k \geq 1$, let $P_k =$ the amount on deposit at the end of the kth period, assuming no additional deposits or withdrawals. Then the interest earned during the kth period equals the amount on deposit at the end of the $(k-1)$st period times the interest rate for the period:

$$\text{interest earned during } k\text{th period } = P_{k-1}\left(\frac{i}{m}\right).$$

The amount on deposit at the end of the kth period, P_k, equals the amount at the end of the $(k-1)$st period, P_{k-1}, plus the interest earned during the kth period:

$$P_k = P_{k-1} + P_{k-1}\left(\frac{i}{m}\right) = P_{k-1}\left(1 + \frac{i}{m}\right). \qquad \text{8.1.6}$$

Suppose $10,000 is left on deposit at 3% compounded quarterly.

a. How much will the account be worth at the end of one year, assuming no additional deposits or withdrawals?

b. The **annual percentage rate (APR)** is the percentage increase in the value of the account over a one-year period. What is the APR for this account?

Solution

a. For each integer $n \geq 1$, let $P_n =$ the amount on deposit after n consecutive quarters, assuming no additional deposits or withdrawals, and let P_0 be the initial $10,000. Then by equation (8.1.6) with $i = 0.03$ and $m = 4$, a recurrence relation for the sequence $P_0, P_1, P_2, \ldots$ is

   (1) $P_k = P_{k-1}(1 + 0.0075) = (1.0075) \cdot P_{k-1}$ for all integers $k \geq 1$.

   The amount on deposit at the end of one year (four quarters), P_4, can be found by successive substitution:

   (2) $P_0 = \$10,000$

   (3) $P_1 = 1.0075 \cdot P_0 = (1.0075) \cdot \$10,000.00 = \$10,075.00$ by (1) and (2)

   (4) $P_2 = 1.0075 \cdot P_1 = (1.0075) \cdot \$10,075.00 = \$10,150.56$ by (1) and (3)

   (5) $P_3 = 1.0075 \cdot P_2 \cong (1.0075) \cdot \$10,150.56 = \$10,226.69$ by (1) and (4)

   (6) $P_4 = 1.0075 \cdot P_3 \cong (1.0075) \cdot \$10,226.69 = \$10,303.39$ by (1) and (5)

   Hence after one year there is $10,303.39 (to the nearest cent) in the account.

b. The percentage increase in the value of the account, or APR, is

$$\frac{10303.39 - 10000}{10000} = 0.03034 = 3.034\%. \qquad ■$$

Example 8.1.9 Number of Bit Strings with a Certain Property

a. Recall that a bit string is a sequence of 0's and 1's, with ϵ denoting the null string, or string with no characters. Make a list of all bit strings of lengths 0, 1, 2, and 3 that do not contain the bit pattern 11.

b. For each integer $n \geq 0$, let

$$s_n = \begin{bmatrix} \text{the number of bit strings of length } n \\ \text{that do not contain the pattern 11} \end{bmatrix}.$$

Find s_0, s_1, s_2, and s_3.

c. Find the number of bit strings of length ten that do not contain the pattern 11.

Solution

a. One way to solve this problem is to make a list of all bit strings of lengths 0, 1, 2, and 3 and to cross off all those that contain the pattern 11:

length 0: ϵ

length 1: 0, 1

length 2: 00, 01, 10, ~~11~~

length 3: 000, 001, 010, ~~011~~, 100, 101, ~~110~~, ~~111~~

b. Counting the number of strings of each length that are listed in part (a) gives

$$s_0 = 1, \quad s_1 = 2, \quad s_2 = 3, \quad \text{and} \quad s_3 = 5.$$

c. To find the number of strings of length ten that do not contain the pattern 11, you could list all $2^{10} = 1,024$ strings of length ten and cross off those that contain the pattern 11, as was done in part (a). However, this approach would be very time-consuming. A more efficient solution uses recursion.

Suppose you know the number of bit strings that have length *less* than some integer k and do not contain the pattern 11. To use recursion to find the number of bit strings that have length k and do not contain the pattern 11, you have to describe strings that do not contain the pattern 11 in terms of shorter strings that do not contain the pattern 11.

Consider the set of all bit strings of length k that do not contain the pattern 11. Any string in the set begins with either a 0 or a 1. If the string begins with a 0, the remaining $k-1$ characters can be any sequence of 0's and 1's except that the pattern 11 cannot appear. If the string begins with a 1, then the second character must be a 0, for otherwise the string would contain the pattern 11; the remaining $k-2$ characters can be any sequence of 0's and 1's that does not contain the pattern 11. Thus the set of all bit strings of length k that do not contain the pattern 11 can be partitioned into two mutually disjoint subsets as shown in Figure 8.1.4.

Set of All Bit Strings of Length k That Do Not Contain the Pattern 11

Set of all bit strings of the form
0 _ _ _ ⋯ _

These positions may be filled in with any string of $k-1$ 0's and 1's that does not contain the pattern 11.

Set of all bit strings of the form
10 _ _ _ ⋯ _

These positions may be filled in with any string of $k-2$ 0's and 1's that does not contain the pattern 11.

There are s_{k-1} strings that start with a 0.

There are s_{k-2} strings that start with a 1.

Figure 8.1.4 Partition of a Set of Bit Strings

By the addition rule, the number of elements in the entire set equals the sum of the numbers of elements in the two disjoint subsets:

$$
\begin{bmatrix}
\text{the number of} \\
\text{bit strings of} \\
\text{length } k \text{ that} \\
\text{do not contain} \\
\text{the pattern 11}
\end{bmatrix}
=
\begin{bmatrix}
\text{the number of} \\
\text{bit strings of} \\
\text{length } k - 1 \text{ that} \\
\text{do not contain} \\
\text{the pattern 11}
\end{bmatrix}
+
\begin{bmatrix}
\text{the number of} \\
\text{bit strings of} \\
\text{length } k - 2 \text{ that} \\
\text{do not contain} \\
\text{the pattern 11}
\end{bmatrix}. \qquad 8.1.7
$$

Thus by substitution into equation (8.1.7) for all integers $k \geq 2$,

$$(1) \quad s_k = s_{k-1} + s_{k-2} \qquad \text{recurrence relation}$$

and by part (b),

$$(2) \quad s_0 = 1 \quad s_1 = 2 \qquad \text{initial conditions.}$$

It follows that

$$(3) \quad s_2 = s_1 + s_0 = 2 + 1 = 3 \qquad \text{by (1) and (2)}$$

$$(4) \quad s_3 = s_2 + s_1 = 3 + 2 = 5 \qquad \text{by (1), (2) and (3)}$$

$$(5) \quad s_4 = s_3 + s_2 = 5 + 3 = 8 \qquad \text{by (1), (3) and (4)}$$

$$\vdots$$

$$(10) \quad s_{10} = s_9 + s_8 = 89 + 55 = 144 \qquad \text{by (1), (8) and (9).}$$

Hence there are 144 bit strings of length ten that do not contain the bit pattern 11.

Note that, because of the similarity in the defining relation, the sequence $s_0, s_1, s_2, \ldots$ has almost the same set of values as the Fibonacci sequence. ■

The Number of Partitions of a Set Into r Subsets

In an ordinary (or *singly indexed*) sequence, integers n are associated to numbers a_n. In a *doubly indexed* sequence, ordered pairs of integers (m, n) are associated to numbers $a_{m,n}$. For example, combinations can be thought of as terms of the doubly indexed sequence defined by $C_{n,r} = \binom{n}{r}$ for all integers n and r with $0 \leq r \leq n$.

An important example of a doubly indexed sequence is the sequence of *Stirling numbers of the second kind*. These numbers, named after the Scottish mathematician James Stirling (1692–1770), arise in a surprisingly large variety of counting problems. They are defined recursively and can be interpreted in terms of partitions of a set.

Observe that if a set of three elements $\{x_1, x_2, x_3\}$ is partitioned into two subsets, then one of the subsets has one element and the other has two elements. Therefore, there are three ways the set can be partitioned:

$$\{x_1, x_2\}\{x_3\} \qquad \text{put } x_3 \text{ by itself}$$
$$\{x_1, x_3\}\{x_2\} \qquad \text{put } x_2 \text{ by itself}$$
$$\{x_2, x_3\}\{x_1\} \qquad \text{put } x_1 \text{ by itself}$$

In general, let

$$
\boxed{
\begin{aligned}
S_{n,r} = \text{ } & \text{number of ways a set of size } n \\
& \text{can be partitioned into } r \text{ subsets}
\end{aligned}
}
$$

Then, by the above, $S_{3,2} = 3$. The numbers $S_{n,r}$ are called **Stirling numbers of the second kind.**

Example 8.1.10 Values of Stirling Numbers

Find $S_{4,1}$, $S_{4,2}$, $S_{4,3}$, and $S_{4,4}$.

Solution Given a set with four elements, denote it by $\{x_1, x_2, x_3, x_4\}$. The Stirling number $S_{4,1} = 1$ because a set of four elements can be partitioned into one subset in only one way:

$$\{x_1, x_2, x_3, x_4\}.$$

Similarly, $S_{4,4} = 1$ because there is only one way to partition a set of four elements into four subsets:

$$\{x_1\}\{x_2\}\{x_3\}\{x_4\}.$$

The number $S_{4,2} = 7$. The reason is that any partition of $\{x_1, x_2, x_3, x_4\}$ into two subsets must consist either of two subsets of size two or of one subset of size three and one subset of size one. The partitions for which both subsets have size two must pair x_1 with x_2, with x_3, or with x_4, which gives rise to these three partitions:

$$\{x_1, x_2\}\{x_3, x_4\} \qquad x_2 \text{ paired with } x_1$$
$$\{x_1, x_3\}\{x_2, x_4\} \qquad x_3 \text{ paired with } x_1$$
$$\{x_1, x_4\}\{x_2, x_3\} \qquad x_4 \text{ paired with } x_1$$

The partitions for which one subset has size one and the other has size three can have any one of the four elements in the subset of size one, which leads to these four partitions:

$$\{x_1\}\{x_2, x_3, x_4\} \qquad x_1 \text{ by itself}$$
$$\{x_2\}\{x_1, x_3, x_4\} \qquad x_2 \text{ by itself}$$
$$\{x_3\}\{x_1, x_2, x_4\} \qquad x_3 \text{ by itself}$$
$$\{x_4\}\{x_1, x_2, x_3\} \qquad x_4 \text{ by itself}$$

It follows that the total number of ways that the set $\{x_1, x_2, x_3, x_4\}$ can be partitioned into two subsets is $3 + 4 = 7$.

Finally, $S_{4,3} = 6$ because any partition of a set of four elements into three subsets must have two elements in one subset and the other two elements in subsets by themselves. There are $\binom{4}{2} = 6$ ways to choose the two elements to put together, which results in the following six possible partitions:

$$\{x_1, x_2\}\{x_3\}\{x_4\} \qquad \{x_2, x_3\}\{x_1\}\{x_4\}$$
$$\{x_1, x_3\}\{x_2\}\{x_4\} \qquad \{x_2, x_4\}\{x_1\}\{x_3\}$$
$$\{x_1, x_4\}\{x_2\}\{x_3\} \qquad \{x_3, x_4\}\{x_1\}\{x_2\}$$

∎

Example 8.1.11 Finding a Recurrence Relation for $S_{n,r}$

Find a recurrence relation relating $S_{n,r}$ to values of the sequence with lower indices than n and r, and give initial conditions for the recursion.

Solution To solve this problem recursively, suppose a procedure has been found to count both the number of ways to partition a set of $n - 1$ elements into $r - 1$ subsets and the number of ways to partition a set of $n - 1$ elements into r subsets. The partitions of a set of n elements $\{x_1, x_2, \ldots, x_n\}$ into r subsets can be divided, as shown in Figure 8.1.5, into those that contain the set $\{x_n\}$ and those that do not.

To obtain the result shown in Figure 8.1.5, first count the number of partitions of $\{x_1, x_2, \ldots, x_n\}$ into r subsets where one of the subsets is $\{x_n\}$. To do this, imagine taking any one of the $S_{n-1,r-1}$ partitions of $\{x_1, x_2, \ldots, x_{n-1}\}$ into $r - 1$ subsets and adding the

Partitions of $\{x_1, x_2, \ldots, x_n\}$ into r Subsets

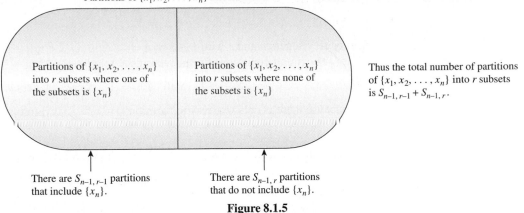

Partitions of $\{x_1, x_2, \ldots, x_n\}$ into r subsets where one of the subsets is $\{x_n\}$

Partitions of $\{x_1, x_2, \ldots, x_n\}$ into r subsets where none of the subsets is $\{x_n\}$

Thus the total number of partitions of $\{x_1, x_2, \ldots, x_n\}$ into r subsets is $S_{n-1, r-1} + S_{n-1, r}$.

There are $S_{n-1, r-1}$ partitions that include $\{x_n\}$.

There are $S_{n-1, r}$ partitions that do not include $\{x_n\}$.

Figure 8.1.5

subset $\{x_n\}$ to the partition. For example, if $n = 4$ and $r = 3$, you would take one of the three partitions of $\{x_1, x_2, x_3\}$ into two subsets, namely

$$\{x_1, x_2\}\{x_3\}, \quad \{x_1, x_3\}\{x_2\}, \quad \text{or} \quad \{x_2, x_3\}\{x_1\},$$

and add $\{x_4\}$. The result would be one of the partitions

$$\{x_1, x_2\}\{x_3\}\{x_4\}, \quad \{x_1, x_3\}\{x_2\}\{x_4\}, \quad \text{or} \quad \{x_2, x_3\}\{x_1\}\{x_4\}.$$

Clearly, any partition of $\{x_1, x_2, \ldots, x_n\}$ into r subsets with $\{x_n\}$ as one of the subsets can be obtained in this way. Hence $S_{n-1, r-1}$ is the number of partitions of $\{x_1, x_2, \ldots, x_n\}$ into r subsets of which one is $\{x_n\}$.

Next, count the number of partitions of $\{x_1, x_2, \ldots, x_n\}$ into r subsets where $\{x_n\}$ is *not* one of the subsets of the partition. Imagine taking any one of the $S_{n-1, r}$ partitions of $\{x_1, x_2, \ldots, x_{n-1}\}$ into r subsets. Now imagine choosing one of the r subsets of the partition and adding in the element x_n. The result is a partition of $\{x_1, x_2, \ldots, x_n\}$ into r subsets none of which is the singelton subset $\{x_n\}$. Since the element x_n could have been added to any one of the r subsets of the partition, it follows from the multiplication rule that there are $r S_{n-1, r}$ partitions of this type. For instance, if $n = 4$ and $r = 3$, you would take the (unique) partition of $\{x_1, x_2, x_3\}$ into three subsets, namely $\{x_1\}\{x_2\}\{x_3\}$, and add x_4 to one of these sets. The result would be one of the partitions

$$\{x_1, x_4\}\{x_2\}\{x_3\}, \quad \{x_1\}\{x_2, x_4\}\{x_3\}, \quad \text{or} \quad \{x_1\}\{x_2\}\{x_3, x_4\}.$$

↑	↑	↑
x_4 is added to $\{x_1\}$	x_4 is added to $\{x_2\}$	x_4 is added to $\{x_3\}$

Clearly, any partition of $\{x_1, x_2, \ldots, x_n\}$ into r subsets, none of which is $\{x_n\}$, can be obtained in the way described above, for when x_n is removed from whatever subset contains it in such a partition, the result is a partition of $\{x_1, x_2, \ldots, x_{n-1}\}$ into r subsets. Hence $r S_{n-1, r}$ is the number of partitions of $\{x_1, x_2, \ldots, x_n\}$ that do not contain $\{x_n\}$.

Since any partition of $\{x_1, x_2, \ldots, x_n\}$ either contains $\{x_n\}$ or does not,

$$\begin{bmatrix} \text{the number of partitions} \\ \text{of } \{x_1, x_2, \ldots, x_n\} \\ \text{into } r \text{ subsets} \end{bmatrix} = \begin{bmatrix} \text{the number of partitions of} \\ \{x_1, x_2, \ldots, x_n\} \text{ into } r \text{ subsets} \\ \text{of which } \{x_n\} \text{ is one} \end{bmatrix}$$

$$+ \begin{bmatrix} \text{the number of partitions of} \\ \{x_1, x_2, \ldots, x_n\} \text{ into } r \text{ subsets} \\ \text{none of which is } \{x_n\} \end{bmatrix}$$

Thus

$$S_{n,r} = S_{n-1,r-1} + rS_{n-1,r}$$

for all integers n and r with $1 \leq r \leq n$.

The initial conditions for the recurrence relation are

$$S_{n,1} = 1 \quad \text{and} \quad S_{n,n} = 1 \quad \text{for all integers } n \geq 1$$

because there is only one way to partition $\{x_1, x_2, \ldots, x_n\}$ into one subset, namely

$$\{x_1, x_2, \ldots, x_n\},$$

and only one way to partition $\{x_1, x_2, \ldots, x_n\}$ into n subsets, namely

$$\{x_1\}\{x_2\}, \ldots, \{x_n\}. \qquad \blacksquare$$

Exercise Set 8.1*

Find the first four terms of each of the recursively defined sequences in 1–8.

1. $a_k = 2a_{k-1} + k$, for all integers $k \geq 2$
   $a_1 = 1$

2. $b_k = b_{k-1} + 3k$, for all integers $k \geq 2$
   $b_1 = 1$

3. $c_k = k(c_{k-1})^2$, for all integers $k \geq 1$
   $c_0 = 1$

4. $d_k = k(d_{k-1})^2$, for all integers $k \geq 1$
   $d_0 = 3$

5. $s_k = s_{k-1} + 2s_{k-2}$, for all integers $k \geq 2$
   $s_0 = 1, s_1 = 1$

6. $t_k = t_{k-1} + 2t_{k-2}$, for all integers $k \geq 2$
   $t_0 = -1, t_1 = 2$

7. $u_k = ku_{k-1} - u_{k-2}$, for all integers $k \geq 3$
   $u_1 = 1, u_2 = 1$

8. $v_k = v_{k-1} + v_{k-2} + 1$, for all integers $k \geq 3$
   $v_1 = 1, v_2 = 3$

9. Let $a_0, a_1, a_2, \ldots$ be defined by the formula $a_n = 3n + 1$, for all integers $n \geq 0$. Show that this sequence satisfies the recurrence relation $a_k = a_{k-1} + 3$, for all integers $k \geq 1$.

10. Let $b_0, b_1, b_2, \ldots$ be defined by the formula $b_n = 4^n$, for all integers $n \geq 0$. Show that this sequence satisfies the recurrence relation $b_k = 4b_{k-1}$, for all integers $k \geq 1$.

11. Show that the sequence $0, 1, 3, 7, \ldots, 2^n - 1, \ldots$, defined for $n \geq 0$, satisfies the recurrence relation

$$c_k = 2c_{k-1} + 1 \quad \text{for all integers } k \geq 1.$$

12. Show that the sequence $1, -1, \dfrac{1}{2}, \dfrac{-1}{3!}, \ldots, \dfrac{(-1)^n}{n!}, \ldots$, defined for $n \geq 0$, satisfies the recurrence relation

$$s_k = \frac{-s_{k-1}}{k} \quad \text{for all integers } k \geq 1.$$

13. Show that the sequence $2, 3, 4, 5, \ldots, 2 + n, \ldots$, defined for $n \geq 0$, satisfies the recurrence relation

$$t_k = 2t_{k-1} - t_{k-2} \quad \text{for all integers } k \geq 2.$$

14. Show that the sequence $0, 1, 5, 19, \ldots, 3^n - 2^n, \ldots$, defined for $n \geq 0$, satisfies the recurrence relation

$$d_k = 5d_{k-1} - 6d_{k-2} \quad \text{for all integers } k \geq 2.$$

15. Define a sequence $a_0, a_1, a_2, \ldots$ by the formula

$$a_n = (-2)^{\lfloor n/2 \rfloor} = \begin{cases} (-2)^{n/2} & \text{if } n \text{ is even} \\ (-2)^{(n-1)/2} & \text{if } n \text{ is odd} \end{cases}$$

for all integers $n \geq 0$. Show that this sequence satisfies the recurrence relation $a_k = -2a_{k-2}$, for all integers $k \geq 2$.

16. The sequence of Catalan numbers was defined in Exercise Set 6.6 by the formula $C_n = \dfrac{1}{n+1}\dbinom{2n}{n}$, for each integer $n \geq 1$. Show that this sequence satisfies the recurrence relation $C_k = \dfrac{4k-2}{k+1}C_{k-1}$, for all integers $k \geq 2$.

17. Use the recurrence relation and values for the Tower of Hanoi sequence $m_1, m_2, m_3, \ldots$ discussed in Example 8.1.5 to compute m_7 and m_8.

18. *Tower of Hanoi with Adjacency Requirement:* Suppose that in addition to the requirement that they never move a larger disk on top of a smaller one, the priests who move the disks

---

of the Tower of Hanoi are also allowed only to move disks one by one from one pole to an *adjacent* pole. Assume poles A and C are at the two ends of the row and pole B is in the middle. Let

$$a_n = \begin{bmatrix} \text{the minimum number of moves} \\ \text{needed to transfer a tower of } n \\ \text{disks from pole } A \text{ to pole } C \end{bmatrix}.$$

a. Find a_1, a_2, and a_3.   **b.** Find a_4.
c. Find a recurrence relation for a_1, a_2, a_3.

19. *Tower of Hanoi with Adjacency Requirement:* Suppose the same situation as in exercise 18. Let

$$b_n = \begin{bmatrix} \text{the minimum number of moves} \\ \text{needed to transfer a tower of } n \\ \text{disks from pole } A \text{ to pole } B \end{bmatrix}.$$

a. Find b_1, b_2, and b_3.   **b.** Find b_4.
c. Show that $b_k = a_{k-1} + 1 + b_{k-1}$ for all integers $k \geq 2$, where a_1, a_2, a_3, ... is the sequence defined in exercise 18.
d. Show that $b_k < 3b_{k-1} + 1$ for all integers $k \geq 2$.
✱**H** e. Show that $b_k = 3b_{k-1} + 1$ for all integers $k > 2$.

20. *Four-Pole Tower of Hanoi:* Suppose that the Tower of Hanoi problem has four poles in a row instead of three. Disks can be transferred one by one from one pole to any other pole, but at no time may a larger disk be placed on top of a smaller disk. Let s_n be the minimum number of moves needed to transfer the entire tower of n disks from the left-most to the right-most pole.
a. Find s_1, s_2, and s_3.   **b.** Find s_4.
c. Show that $s_k < 2s_{k-2} + 3$ for all integers $k \geq 3$.

21. *Double Tower of Hanoi:* In this variation of the Tower of Hanoi there are three poles in a row and $2n$ disks, two of each of n different sizes, where n is any positive integer. Initially one of the poles contains all the disks placed on top of each other in pairs of decreasing size. Disks are transferred one by one from one pole to another, but at no time may a larger disk be placed on top of a smaller disk. However, a disk may be placed on top of one of the same size. Let t_n be the minimum number of moves needed to transfer a tower of $2n$ disks from one pole to another.
a. Find t_1 and t_2.   **b.** Find t_3.
c. Find a recurrence relation for t_1, t_2, t_3, ...

22. *Fibonacci Variation:* A single pair of rabbits (male and female) is born at the beginning of a year. Assume the following conditions (which are more realistic than Fibonacci's):
(1) Rabbit pairs are not fertile during their first month of life but thereafter give birth to four new male/female pairs at the end of every month.
(2) No rabbits die.
a. Let r_n = the number of pairs of rabbits alive at the end of month n, for each integer $n \geq 1$, and let $r_0 = 1$. Find a recurrence relation for r_0, r_1, r_2, ...
b. Compute r_0, r_1, r_2, r_3, r_4, r_5, and r_6.
c. How many rabbits will there be at the end of the year?

23. *Fibonacci Variation:* A single pair of rabbits (male and female) is born at the beginning of a year. Assume the following conditions:
(1) Rabbit pairs are not fertile during their first *two* months of life, but thereafter give birth to three new male/female pairs at the end of every month.
(2) No rabbits die.
a. Let s_n = the number of pairs of rabbits alive at the end of month n, for each integer $n \geq 1$, and let $s_0 = 1$. Find a recurrence relation for s_0, s_1, s_2, ...
b. Compute s_0, s_1, s_2, s_3, s_4, and s_5.
c. How many rabbits will there be at the end of the year?

In 24–32, F_0, F_1, F_2, ... is the Fibonacci sequence.

24. Use the recurrence relation and values for F_0, F_1, F_2, ... given in Example 8.1.6 to compute F_{13} and F_{14}.

25. The Fibonacci sequence satisfies the recurrence relation $F_k = F_{k-1} + F_{k-2}$, for all integers $k \geq 2$.
a. Explain why the following is true:
$$F_{k+1} = F_k + F_{k-1} \quad \text{for all integers } k \geq 1.$$
b. Write an equation expressing F_{k+2} in terms of F_{k+1} and F_k.
c. Write an equation expressing F_{k+3} in terms of F_{k+2} and F_{k+1}

26. Prove that $F_k = 3F_{k-3} + 2F_{k-4}$ for all integers $k \geq 4$.

27. Prove that $F_k^2 - F_{k-1}^2 = F_k F_{k+1} - F_{k+1} F_{k-1}$, for all integers $k \geq 1$.

28. Prove that $F_{k+1}^2 - F_k^2 - F_{k-1}^2 = 2F_k F_{k-1}$, for all integers $k \geq 1$.

29. Prove that $F_{k+1}^2 - F_k^2 = F_{k-1} F_{k+2}$, for all integers $k \geq 1$.

30. Use mathematical induction to prove that for all integers $n \geq 0$, $F_{n+2} F_n - F_{n+1}^2 = (-1)^n$.

31. (For students who have studied calculus) Find $\lim_{n \to \infty} \left(\dfrac{F_{n+1}}{F_n} \right)$, assuming that the limit exists.

H ✱ 32. (For students who have studied calculus) Prove that $\lim_{n \to \infty} \left(\dfrac{F_{n+1}}{F_n} \right)$ exists.

33. (For students who have studied calculus) Define x_0, x_1, x_2, ... as follows:
$$x_k = \sqrt{2 + x_{k-1}} \quad \text{for all integers } k \geq 1$$
$$x_0 = 0$$
Find $\lim_{n \to \infty} x_n$. (Assume that the limit exists.)

34. *Compound Interest:* Suppose a certain amount of money is deposited in an account paying 4% annual interest compounded quarterly. For each positive integer n, let R_n = the amount on deposit at the end of the nth quarter, assuming no additional deposits or withdrawals, and let R_0 be the initial amount deposited.

a. Find a recurrence relation for $R_0, R_1, R_2, \ldots$.
b. If $R_0 = \$5000$, find the amount of money on deposit at the end of one year.
c. Find the APR for the account.

35. *Compound Interest:* Suppose a certain amount of money is deposited in an account paying 3% annual interest compounded monthly. For each positive integer n, let $S_n = $ the amount on deposit at the end of the nth month, and let S_0 be the initial amount deposited.
   a. Find a recurrence relation for $S_0, S_1, S_2, \ldots$, assuming no additional deposits or withdrawals during the year.
   b. If $S_0 = \$10,000$, find the amount of money on deposit at the end of one year.
   c. Find the APR for the account.

36. *Counting Strings:*
   a. Make a list of all bit strings of lengths zero, one, two, three, and four that do not contain the bit pattern 111.
   b. For each integer $n \geq 0$, let $d_n = $ the number of bit strings of length n that do not contain the bit pattern 111. Find d_0, d_1, d_2, d_3, and d_4.
   c. Find a recurrence relation for $d_0, d_1, d_2, \ldots$.
   d. Use the results of parts (b) and (c) to find the number of bit strings of length five that do not contain the pattern 111.

37. *Counting Strings:* Consider the set of all strings of a's, b's, and c's.
   a. Make a list of all of these strings of lengths zero, one, two, and three that do not contain the pattern aa.
   b. For each integer $n \geq 0$, let $s_n = $ the number of strings of a's, b's, and c's of length n that do not contain the pattern aa. Find s_0, s_1, s_2, and s_3.
   H c. Find a recurrence relation for $s_0, s_1, s_2, \ldots$.
   d. Use the results of parts (b) and (c) to find the number of strings of a's, b's, and c's of length four that do not contain the pattern aa.

38. For each integer $n \geq 0$, let a_n be the number of bit strings of length n that do not contain the pattern 101.
   a. Show that $a_k = a_{k-1} + a_{k-3} + \cdots + a_0 + 2$, for all integers $k \geq 3$.
   b. Use the result of part (a) to show that if $k \geq 3$, then $a_k = 2a_{k-1} - a_{k-2} + a_{k-3}$.

39. With each step you take when climbing a staircase, you can move up either one stair or two stairs. As a result, you can climb the entire staircase taking one stair at a time, taking two at a time, or taking a combination of one- and two-stair increments. For each integer $n \geq 1$, if the staircase consists of n stairs, let c_n be the number of different ways to climb the staircase. Find a recurrence relation for $c_1, c_2, c_3, \ldots$.

40. A set of blocks contains blocks of heights 1, 2, and 4 inches. Imagine constructing towers by piling blocks of different heights directly on top of one another. (A tower of height 6 inches could be obtained using six 1-inch blocks, three

2-inch blocks, one 2-inch block with one 4-inch block on top, one 4-inch block with one 2-inch block on top, and so forth.) Let t_n be the number of ways to construct a tower of height n inches using blocks from the set. (Assume an infinite supply of block of each size.) Find a recurrence relation for $t_1, t_2, t_3, \ldots$.

✶ 41. For each integer $n \geq 2$ let a_n be the number of permutations of $\{1, 2, 3, \ldots, n\}$ in which no number is more than one place removed from its "natural" position. Thus $a_1 = 1$ since the one permutation of $\{1\}$, namely 1, does not move 1 from its natural position. Also $a_2 = 2$ since neither of the two permutations of $\{1, 2\}$, namely 12 and 21, moves either number more than one place from its natural position.
   a. Find a_3.
   b. Find a recurrence relation for $a_1, a_2, a_3, \ldots$.

✶ 42. A row in a classroom has n seats. Let s_n be the number of ways nonempty sets of students can sit in the row so that no student is seated directly adjacent to any other student. (For instance, a row of three seats could contain a single student in any of the seats or a pair of students in the two outer seats. Thus $s_3 = 4$.) Find a recurrence relation for $s_1, s_2, s_3, \ldots$.

✶ 43. Let P_n be the number of partitions of a set with n elements. Show that

$$P_n = \binom{n-1}{0} P_{n-1} + \binom{n-1}{1} P_{n-2} + \cdots + \binom{n-1}{n-1} P_0$$

for all integers $n \geq 1$.

Exercises 44–50 refer to the sequence of Stirling numbers of the second kind.

44. Find $S_{5,4}$ by exhibiting all the partitions of $\{x_1, x_2, x_3, x_4, x_5\}$ into four subsets.

45. Use the values computed in Example 8.1.10 and the recurrence relation and initial conditions found in Example 8.1.11 to compute $S_{5,2}$.

46. Use the values computed in Example 8.1.10 and the recurrence relation and initial conditions found in Example 8.1.11 to compute $S_{5,3}$.

47. Find the total number of different partitions of a set with five elements.

48. Use mathematical induction and the recurrence relation found in Example 8.1.11 to prove that for all integers $n \geq 2$, $S_{n,2} = 2^{n-1} - 1$.

49. Use mathematical induction and the recurrence relation found in Example 8.1.11 to prove that for all integers $n \geq 2$, $\sum_{k=1}^{n} (3^{n-k} S_{k,2}) = S_{n+1,3}$.

H 50. If X is a set with n elements and Y is a set with m elements, express the number of onto functions from X and Y using Stirling numbers of the second kind. Justify your answer.

In 51 and 52, assume that $F_0, F_1, F_2, \ldots$ is the Fibonacci sequence.

✱ **51.** Use strong mathematical induction to prove that $F_n < 2^n$ for all integers $n \geq 1$.

H ✱ **52.** Prove that for all integers $n \geq 0$, $\gcd(F_{n+1}, F_n) = 1$.

53. A gambler decides to play successive games of blackjack until he loses three times in a row. (Thus the gambler could play five games by losing the first, winning the second, and losing the final three or by winning the first two and losing the final three. These possibilities can be symbolized as $LWLLL$ and $WWLLL$.) Let g_n be the number of ways the gambler can play n games.
  a. Find g_3, g_4, and g_5.
  b. Find g_6.
  H c. Find a recurrence relation for $g_3, g_4, g_5, \ldots$.

✱ **54.** A *derangement* of the set $\{1, 2, \ldots, n\}$ is a permutation that moves every element of the set away from its "natural" posi-

tion. Thus 21 is a derangement of $\{1, 2\}$, and 231 and 312 are derangements of $\{1, 2, 3\}$. For each positive integer n, let d_n be the number of derangements of the set $\{1, 2, \ldots, n\}$.
  a. Find d_1, d_2, and d_3.
  b. Find d_4.
  H c. Find a recurrence relation for $d_1, d_2, d_3, \ldots$.

55. Note that a product $x_1 x_2 x_3$ may be parenthesized in two different ways: $(x_1 x_2)x_3$ and $x_1(x_2 x_3)$. Similarly, there are several different ways to parenthesize $x_1 x_2 x_3 x_4$. Two such ways are $(x_1 x_2)(x_3 x_4)$ and $x_1((x_2 x_3)x_4)$. Let P_n be the number of different ways to parenthesize the product $x_1 x_2 \cdots x_n$. Show that if $P_1 = 1$, then

$$P_n = \sum_{k=1}^{n-1} P_k P_{n-k} \quad \text{for all integers } n \geq 2.$$

(It turns out that the sequence $P_1, P_2, P_3, \ldots$ is the same as the sequence of Catalan numbers.)

8.2 Solving Recurrence Relations by Iteration

The keener one's sense of logical deduction, the less often one makes hard and fast inferences. — Bertrand Russell, 1872–1970

Suppose you have a sequence that satisfies a certain recurrence relation and initial conditions. It is often helpful to know an explicit formula for the sequence, especially if you need to compute terms with very large subscripts or if you need to examine general properties of the sequence. Such an explicit formula is called a **solution** to the recurrence relation. In this section and the next, we discuss methods for solving recurrence relations. In the text and exercises of this section, we will show that the Tower of Hanoi sequence of Example 8.1.5 satisfies the formula

$$m_n = 2^n - 1,$$

and that the compound interest sequence of Example 8.1.7 satisfies

$$A_n = (1.04)^n \cdot \$100{,}000.$$

In Section 8.3 we will show that the Fibonacci sequence of Example 8.1.6 satisfies the formula

$$F_n = \frac{1}{\sqrt{5}} \left[\left(\frac{1 + \sqrt{5}}{2} \right)^{n+1} - \left(\frac{1 - \sqrt{5}}{2} \right)^{n+1} \right].$$

The Method of Iteration

The most basic method for finding an explicit formula for a recursively defined sequence is **iteration**. Iteration works as follows: Given a sequence $a_0, a_1, a_2, \ldots$ defined by a recurrence relation and initial conditions, you start from the initial conditions and calculate successive terms of the sequence until you see a pattern developing. At that point you guess an explicit formula.

Example 8.2.1 Finding an Explicit Formula

Let $a_0, a_1, a_2, \ldots$ be the sequence defined recursively as follows: For all integers $k \geq 1$,

(1) $a_k = a_{k-1} + 2$ recurrence relation

(2) $a_0 = 1$ initial condition.

Use iteration to guess an explicit formula for the sequence.

Solution   Recall that to say

$$a_k = a_{k-1} + 2 \quad \text{for all integers } k \geq 1$$

means

$$a_\square = a_{\square - 1} + 2 \quad \text{no matter what positive integer is placed into the box } \square.$$

In particular,

$$a_1 = a_0 + 2,$$
$$a_2 = a_1 + 2,$$
$$a_3 = a_2 + 2,$$

and so forth. Now use the initial condition to begin a process of successive substitutions into these equations, not just of numbers (as was done in Section 8.1) but of *numerical expressions*.

The reason for using numerical expressions rather than numbers is that in these problems you are seeking a numerical pattern that underlies a general formula. The secret of success is to leave most of the arithmetic undone. However, you do need to eliminate parentheses as you go from one step to the next. Otherwise, you will soon end up with a bewilderingly large nest of parentheses. Also, it is nearly always helpful to use shorthand notations for regrouping additions, subtractions, and multiplications. Thus, for instance, you would write

$$5 \cdot 2 \quad \text{instead of} \quad 2 + 2 + 2 + 2 + 2$$

and

$$2^5 \quad \text{instead of} \quad 2 \cdot 2 \cdot 2 \cdot 2 \cdot 2.$$

Notice that you don't lose any information about the number patterns when you use these shorthand notations.

Here's how the process works for the given sequence:

$a_0 = 1$ the initial condition

$a_1 = a_0 + 2 = 1 + 2$ by substitution

$a_2 = a_1 + 2 = (1 + 2) + 2 \qquad = 1 + 2 + 2$ eliminate parentheses

$a_3 = a_2 + 2 = (1 + 2 + 2) + 2 \qquad = 1 + 2 + 2 + 2$ eliminate parentheses again; write $3 \cdot 2$ instead of $2 + 2 + 2$?

$a_4 = a_3 + 2 = (1 + 2 + 2 + 2) + 2 = 1 + 2 + 2 + 2 + 2$ eliminate parentheses again; definitely write $4 \cdot 2$ instead of $2 + 2 + 2 + 2$—the length of the string of 2's is getting out of hand.

Since it appears helpful to use the shorthand $k \cdot 2$ in place of $2 + 2 + \cdots + 2$ (k times), we do so, starting again from a_0.

$$a_0 = 1 \qquad\qquad\qquad\quad = 1 + 0 \cdot 2 \quad \text{the initial condition}$$

$$a_1 = a_0 + 2 = \overbrace{1 + 2} \qquad = 1 + 1 \cdot 2 \quad \text{by substitution}$$

$$a_2 = a_1 + 2 = \overbrace{(1 + 2)} + 2 \quad = \underbrace{1 + 2 \cdot 2}$$

$$a_3 = a_2 + 2 = \overbrace{(1 + 2 \cdot 2)} + 2 = \underbrace{1 + 3 \cdot 2}$$

$$a_4 = a_3 + 2 = \overbrace{(1 + 3 \cdot 2)} + 2 = \underbrace{1 + 4 \cdot 2}$$

At this point it certainly seems likely that the general pattern is $1 + n \cdot 2$; check whether the next calculation supports this.

$$a_5 = a_4 + 2 = \overbrace{(1 + 4 \cdot 2)} + 2 = 1 + 5 \cdot 2$$

It does! So go ahead and write an answer. It's only a guess, after all.

$$\vdots$$

Guess: $\quad a_n = 1 + n \cdot 2 = 1 + 2n$

The answer obtained for this problem is just a guess. To be sure of the correctness of this guess, you will need to check it by mathematical induction. Later in this section, we will show how to do this. ∎

A sequence like the one in Example 8.2.1, in which each term equals the previous term plus a fixed constant, is called an *arithmetic sequence*. In the exercises at the end of this section you are asked to show that the $(n + 1)$st term of an arithmetic sequence always equals the initial value of the sequence plus n times the fixed constant.

• Definition

A sequence $a_0, a_1, a_2, \ldots$ is called an **arithmetic sequence** if, and only if, there is a constant d such that

$$a_k = a_{k-1} + d \quad \text{for all integers } k \geq 1.$$

Or, equivalently,

$$a_n = a_0 + dn \quad \text{for all integers } n \geq 0.$$

Example 8.2.2 An Arithmetic Sequence

Under the force of gravity, an object falling in a vacuum falls about 9.8 meters farther each second than it fell the second before. Thus, neglecting air resistance, a skydiver leaving an airplane falls approximately 9.8 meters between 0 and 1 seconds after departure, $9.8 + 9.8 = 19.6$ meters between 1 and 2 seconds after departure, and so forth. If air resistance is neglected, how many meters would the diver fall between 60 and 61 seconds after leaving the airplane?

Solution Let d_n be the distance the skydiver would fall between n and $n + 1$ seconds after exiting the airplane if there were no air resistance. Thus d_0 is the distance fallen between 0 and 1 seconds after exiting, d_1 is the distance fallen between 1 and 2 seconds after exiting, and so forth. Then $d_0 = 9.8$, and since the diver would fall 9.8 meters farther each second than the second before,

$$d_k = d_{k-1} + 9.8 \text{ meters} \quad \text{for all integers } k \geq 1.$$

It follows that $d_0, d_1, d_2, \ldots$ is an arithmetic sequence with a constant adder of 9.8 and that

$$d_n = d_0 + n \cdot (9.8) \quad \text{for each integer } n \geq 0.$$

Hence between the 60th and the 61st seconds after exiting, the diver would fall

$$d_{60} = 9.8 + 60 \cdot (9.8) = 597.8 \text{ meters}.$$

Note that 597.8 meters is approximately equal to 1,961 feet or about three or four city blocks, which is a long way to fall in one second. Of course, this result was obtained by neglecting air resistance, which in fact cuts the diver's speed considerably. ∎

In an arithmetic sequence, each term equals the previous term plus a fixed constant. In a geometric sequence, each term equals the previous term *times* a fixed constant. Geometric sequences arise in a large variety of applications, such as compound interest, certain models of population growth, radioactive decay, and the number of operations needed to execute certain computer algorithms.

Example 8.2.3 The Explicit Formula for a Geometric Sequence

Let r be a fixed nonzero constant, and suppose a sequence $a_0, a_1, a_2, \ldots$ is defined recursively as follows:

$$a_k = ra_{k-1} \quad \text{for all integers } k \geq 1,$$
$$a_0 = a.$$

Use iteration to guess an explicit formula for this sequence.

Solution

$$a_0 = a$$

$$a_1 = ra_0 = ra$$

$$a_2 = ra_1 = r(ra) = r^2 a$$

$$a_3 = ra_2 = r(r^2 a) = r^3 a$$

$$a_4 = ra_3 = r(r^3 a) = r^4 a$$

$$\vdots$$

Guess: $\quad a_n = r^n a = ar^n \quad$ for any arbitrary integer $n \geq 0$

In the exercises at the end of this section, you are asked to prove that this formula is correct. ∎

> **• Definition**
>
> A sequence $a_0, a_1, a_2, \ldots$ is called a **geometric sequence** if, and only if, there is a constant r such that
>
> $$a_k = ra_{k-1} \quad \text{for all integers } k \geq 1.$$
>
> Or, equivalently,
>
> $$a_n = a_0 r^n \quad \text{for all integers } n \geq 0.$$

Example 8.2.4 A Geometric Sequence

As shown in Example 8.1.7, if a bank pays interest at a rate of 4% per year compounded annually and A_n denotes the amount in the account at the end of year n, then $A_k = (1.04)A_{k-1}$, for all integers $k \geq 1$, assuming no deposits or withdrawals during the year. Suppose the initial amount deposited is $100,000, and assume that no additional deposits or withdrawals are made.

a. How much will the account be worth at the end of 21 years?

b. In how many years will the account be worth $1,000,000?

Solution

a. $A_0, A_1, A_2, \ldots$ is a geometric sequence with initial value 100,000 and constant multiplier 1.04. Hence,

$$A_n = \$100,000 \cdot (1.04)^n \quad \text{for all integers } n \geq 0.$$

After 21 years, the amount in the account will be

$$A_{21} = \$100,000 \cdot (1.04)^{21} \cong \$227,876.81.$$

This is the same answer as that obtained in Example 8.1.7 but is computed much more easily (at least if a calculator with a powering key, such as $\boxed{\wedge}$ or $\boxed{x^y}$, is used).

b. Let t be the number of years needed for the account to grow to $1,000,000. Then

$$\$1,000,000 = \$100,000 \cdot (1.04)^t.$$

Dividing both sides by 100,000 gives

$$10 = (1.04)^t,$$

and taking natural logarithms of both sides results in

$$\ln(10) = \ln(1.04)^t.$$

Then

$$\ln(10) \cong t \ln(1.04) \qquad \begin{array}{l} \text{because } \log_b(x^a) = a \log_b(x) \\ \text{(see exercise 30 of Section 7.2)} \end{array}$$

and so

$$t = \frac{\ln(10)}{\ln(1.04)} \cong 58.7$$

Hence the account will grow to $1,000,000 in approximately 58.7 years. ∎

An important property of a geometric sequence with constant multiplier greater than 1 is that its terms increase very rapidly in size as the subscripts get larger and larger. For instance, the first ten terms of a geometric sequence with a constant multiplier of 10 are

$$1, 10, 10^2, 10^3, 10^4, 10^5, 10^6, 10^7, 10^8, 10^9.$$

Thus, by its tenth term, the sequence already has the value $10^9 = 1,000,000,000 = 1$ billion. The following box indicates some quantities that are approximately equal to certain powers of 10.

$10^7 \cong$ number of seconds in a year

$10^{11} \cong$ number of neurons in a human brain

$10^{17} \cong$ age of the universe in seconds (according to one theory)

$10^{30} \cong$ number of bytes of memory in a personal computer

$10^{31} \cong$ number of seconds to process all possible positions of a checkers game if moves are processed at a rate of 1 per billionth of a second

$10^{81} \cong$ number of atoms in the universe

$10^{111} \cong$ number of seconds to process all possible positions of a chess game if moves are processed at a rate of 1 per billionth of a second

Using Formulas to Simplify Solutions Obtained by Iteration

Explicit formulas obtained by iteration can often be simplified by using formulas such as those developed in Section 4.2. For instance, according to the formula for the sum of a geometric sequence with initial term 1 (Theorem 4.2.3), for each real number r except $r = 1$,

$$1 + r + r^2 + \cdots + r^n = \frac{r^{n+1} - 1}{r - 1} \quad \text{for all integers } n \geq 0.$$

And according to the formula for the sum of the first n integers (Theorem 4.2.2),

$$1 + 2 + 3 + \cdots + n = \frac{n(n + 1)}{2} \quad \text{for all integers } n \geq 1.$$

Example 8.2.5 An Explicit Formula for the Tower of Hanoi Sequence

Recall that the Tower of Hanoi sequence $m_1, m_2, m_3, \ldots$ of Example 8.1.5 satisfies the recurrence relation

$$m_k = 2m_{k-1} + 1 \quad \text{for all integers } k \geq 2$$

and has the initial condition

$$m_1 = 1.$$

Use iteration to guess an explicit formula for this sequence, and make use of a formula from Section 4.2 to simplify the answer.

Solution   By iteration

$$m_1 = 1$$

$$m_{②} = 2m_1 + 1 = 2 \cdot 1 + 1 \qquad\qquad = 2^{①} + 1,$$

$$m_{③} = 2m_2 + 1 = 2\overbrace{(2+1)} + 1 \qquad\qquad = 2^{②} + 2 + 1,$$

$$m_{④} = 2m_3 + 1 = 2\overbrace{(2^2 + 2 + 1)} + 1 \qquad = 2^{③} + 2^2 + 2 + 1,$$

$$m_{⑤} = 2m_4 + 1 = 2\overbrace{(2^3 + 2^2 + 2 + 1)} + 1 = 2^{④} + 2^3 + 2^2 + 2 + 1.$$

These calculations show that each term up to m_5 is a sum of successive powers of 2, starting with $2^0 = 1$ and going up to 2^k, where k is 1 less than the subscript of the term. The pattern would seem to continue to higher terms because each term is obtained from the preceding one by multiplying by 2 and adding 1; multiplying by 2 raises the exponent of each component of the sum by 1, and adding 1 adds back the 1 that was lost when the previous 1 was multiplied by 2. For instance, for $n = 6$,

$$m_6 = 2m_5 + 1 = 2(2^4 + 2^3 + 2^2 + 2 + 1) + 1 = 2^5 + 2^4 + 2^3 + 2^2 + 2 + 1.$$

Thus it seems that, in general,

$$m_n = 2^{n-1} + 2^{n-2} + \cdots + 2^2 + 2 + 1.$$

By the formula for the sum of a geometric sequence (Theorem 4.2.3),

$$2^{n-1} + 2^{n-2} + \cdots + 2^2 + 2 + 1 = \frac{2^n - 1}{2 - 1} = 2^n - 1.$$

Hence the explicit formula seems to be

$$m_n = 2^n - 1 \quad \text{for all integers } n \geq 1. \qquad\blacksquare$$

⚠ ***Caution!*** It is not true that

~~$2 \cdot (2+1) + 1 = 2^2 + 1 + 1$~~ ← This is false.

A common mistake people make when doing problems such as this is to misuse the laws of algebra. For instance, by the distributive law,

$$a \cdot (b + c) = a \cdot b + a \cdot c \quad \text{for all real numbers } a, b, \text{ and } c.$$

Thus, in particular, for $a = 2, b = 2$, and $c = 1$,

$$2 \cdot (2 + 1) = 2 \cdot 2 + 2 \cdot 1 = 2^2 + 2.$$

It follows that

$$2 \cdot (2 + 1) + 1 = (2^2 + 2) + 1 = 2^2 + 2 + 1.$$

Example 8.2.6 Using the Formula for the Sum of the First n Integers

Let K_n be the picture obtained by drawing n dots (which we call *vertices*) and joining each pair of vertices by a line segment (which we call an *edge*). (In Chapter 11 we discuss these objects in a more general context.) Then $K_1, K_2, K_3,$ and K_4 are as follows:

$K_1 \qquad K_2 \qquad K_3 \qquad K_4$

Observe that K_5 may be obtained from K_4 by adding one vertex and drawing edges between this new vertex and all the vertices of K_4 (the old vertices). The reason this procedure gives the correct result is that each pair of old vertices is already joined by an edge, and adding the new edges joins each pair of vertices consisting of an old one and the new one.

New vertex

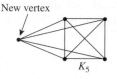

K_5

Thus

$$\text{the number of edges of } K_5 = 4 + \text{the number of edges of } K_4.$$

By the same reasoning, for all integers $k \geq 2$, the number of edges of K_k is $k - 1$ more than the number of edges of K_{k-1}. That is, if for each integer $n \geq 1$

$$s_n = \text{the number of edges of } K_n,$$

then

$$s_k = s_{k-1} + (k - 1) \quad \text{for all integers } k \geq 2.$$

Use iteration to find an explicit formula for $s_1, s_2, s_3, \ldots$.

Solution Because

$$s_k = s_{k-1} + (k - 1) \quad \text{for } all \text{ integers } k \geq 2$$

and

$$s_① = ⓪ \overset{1-1}{\nwarrow}$$

then, in particular,

$$s_② = s_1 + 1 = \overbrace{0 + ①}^{\quad} \overset{2-1}{\nwarrow},$$

$$s_③ = s_2 + 2 = \overbrace{(0 + 1)} + 2 = \underbrace{0 + 1 + ②}\overset{3-1}{\nwarrow},$$

$$s_④ = s_3 + 3 = \overbrace{(0 + 1 + 2)} + 3 = \underbrace{0 + 1 + 2 + ③}\overset{4-1}{\nwarrow},$$

$$s_⑤ = s_4 + 4 = \overbrace{(0 + 1 + 2 + 3)} + 4 = 0 + 1 + 2 + 3 + ④\overset{5-1}{\nwarrow},$$

$$\vdots$$

Guess: $\quad s_ⓝ = 0 + 1 + 2 + \cdots + (n - 1).$

But by Theorem 4.2.2,

$$0 + 1 + 2 + 3 + \cdots + (n-1) = \frac{(n-1)n}{2} = \frac{n(n-1)}{2}.$$

Hence it appears that

$$s_n = \frac{n(n-1)}{2}. \qquad \blacksquare$$

Checking the Correctness of a Formula by Mathematical Induction

As you can see from some of the previous examples, the process of solving a recurrence relation by iteration can involve complicated calculations. It is all too easy to make a mistake and come up with the wrong formula. That is why it is important to confirm your calculations by checking the correctness of your formula. The most common way to do this is to use mathematical induction.

Example 8.2.7 Using Mathematical Induction to Verify the Correctness of a Solution to a Recurrence Relation

In Example 8.2.5 we obtained a formula for the Tower of Hanoi sequence. Use mathematical induction to show that this formula is correct.

Solution   What does it mean to show the correctness of a formula for a recursively defined sequence? You are given a sequence of numbers that satisfies a certain recurrence relation and initial condition. Your job is to show that each term of the sequence satisfies the proposed explicit formula. To do this, you need to prove the following statement:

> If $m_1, m_2, m_3, \ldots$ is the sequence defined by
> $$m_k = 2m_{k-1} + 1 \quad \text{for all integers } k \geq 2, \text{ and}$$
> $$m_1 = 1,$$
> then $m_n = 2^n - 1 \quad$ for all integers $n \geq 1$.

To prove this by mathematical induction, you prove a basis step (that the formula holds for $n = 1$) and an inductive step (that if the formula holds for an integer $n = k$, then it holds for $n = k + 1$). In other words, you show that

1. $m_1 = 2^1 - 1$

2. If $m_k = 2^k - 1$, for some integer $k \geq 1$, then $m_{k+1} = 2^{k+1} - 1$.

Proof of Correctness:

Show that the formula holds for $n = 1$: Observe that $m_1 = 1$ by definition of the sequence $m_1, m_2, m_3, \ldots$. And $2^1 - 1 = 1$ by basic algebra. Hence $m_1 = 2^1 - 1$, and so the formula holds for $n = 1$.

Show that for all integers $k \geq 1$, if the formula holds for $n = k$, then it holds for $n = k + 1$: Suppose that

$$m_k = 2^k - 1 \quad \text{for some integer } k \geq 1. \qquad \text{This is the inductive hypothesis.}$$

We must show that

$$m_{k+1} = 2^{k+1} - 1.$$

But the left-hand side of this equation is

$$m_{k+1} = 2m_{(k+1)-1} + 1 \quad \text{by definition of } m_1, m_2, m_3, \ldots$$
$$= 2m_k + 1$$
$$= 2(2^k - 1) + 1 \quad \text{by substitution from the inductive hypothesis}$$
$$= 2^{k+1} - 2 + 1 \quad \text{by the distributive law and the fact that } 2 \cdot 2^k = 2^{k+1}$$
$$= 2^{k+1} - 1 \quad \text{by basic algebra}$$

which equals the right-hand side of the equation. [*Since the basis and inductive steps have been proved, it follows by mathematical induction that the given formula holds for all integers $n \geq 1$.*] ■

Discovering That an Explicit Formula Is Incorrect

The following example shows how the process of trying to verify a formula by mathematical induction may reveal a mistake.

Example 8.2.8 Using Verification by Mathematical Induction to Find a Mistake

Let $c_0, c_1, c_2, \ldots$ be the sequence defined as follows:

$$c_k = 2c_{k-1} + k \quad \text{for all integers } k \geq 1,$$
$$c_0 = 1.$$

Suppose your calculations suggest that $c_0, c_1, c_2, \ldots$ satisfies the following explicit formula:

$$c_n = 2^n + n \quad \text{for all integers } n \geq 0.$$

Is this formula correct?

Solution   Start to prove the statement by mathematical induction and see what develops. The proposed formula passes the basis step of the inductive proof with no trouble, for on the one hand, $c_0 = 1$ by definition of $c_0, c_1, c_2, \ldots$, and on the other hand, $2^0 + 0 = 1 + 0 = 1$ also.

In the inductive step, you suppose

$$c_k = 2^k + k \quad \text{for some integer } k \geq 0 \qquad \text{This is the inductive hypothesis.}$$

and then you must show that

$$c_{k+1} = 2^{k+1} + (k + 1).$$

To do this, you start with c_{k+1}, substitute from the recurrence relation, and then use the inductive hypothesis as follows:

$$c_{k+1} = 2c_k + (k + 1) \qquad \text{by the recurrence relation}$$
$$= 2(2^k + k) + (k + 1) \quad \text{by substitution from the inductive hypothesis}$$
$$= 2^{k+1} + 3k + 1 \qquad \text{by basic algebra}$$

To finish the verification, therefore, you need to show that

$$2^{k+1} + 3k + 1 = 2^{k+1} + (k + 1).$$

Now this equation is equivalent to

$$2k = 0 \quad \text{by subtracting } 2^{k+1} + k + 1 \text{ from both sides,}$$

which is equivalent to

$$k = 0 \quad \text{by dividing both sides by 2.}$$

But this is false since k may be *any* nonnegative integer. Hence the sequence $c_0, c_1, c_2, \ldots$ does not satisfy the proposed formula. ∎

Once you have found a proposed formula to be false, you should look back at your calculations to see where you made a mistake, correct it, and try again.

Exercise Set 8.2

1. The formula

$$1 + 2 + 3 + \cdots + n = \frac{n(n+1)}{2}$$

is true for all integers $n \geq 1$. Use this fact to solve each of the following problems:
   a. If k is an integer and $k \geq 2$, find a formula for the expression $1 + 2 + 3 + \cdots + (k - 1)$.
   b. If n is an integer and $n \geq 1$, find a formula for the expression $3 + 2 + 4 + 6 + 8 + \cdots + 2n$.
   c. If n is an integer and $n \geq 1$, find a formula for the expression $3 + 3 \cdot 2 + 3 \cdot 3 + \cdots + 3 \cdot n + n$.

2. The formula

$$1 + r + r^2 + \cdots + r^n = \frac{r^{n+1} - 1}{r - 1}$$

is true for all real numbers r except $r = 1$ and for all integers $n \geq 0$. Use this fact to solve each of the following problems:
   a. If i is an integer and $i \geq 1$, find a formula for the expression $1 + 2 + 2^2 + \cdots + 2^{i-1}$.
   b. If n is an integer and $n \geq 1$, find a formula for the expression $3^{n-1} + 3^{n-2} + \cdots + 3^2 + 3 + 1$.
   c. If n is an integer and $n \geq 2$, find a formula for the expression $2^n + 2^{n-2} \cdot 3 + 2^{n-3} \cdot 3 + \cdots + 2^2 \cdot 3 + 2 \cdot 3 + 3$.
   d. If n is an integer and $n \geq 1$, find a formula for the expression $2^n - 2^{n-1} + 2^{n-2} - 2^{n-3} + \cdots + (-1)^{n-1} \cdot 2 + (-1)^n$.

In each of 3–15 a sequence is defined recursively. Use iteration to guess an explicit formula for the sequence. Use the formulas from Section 4.2 to simplify your answers whenever possible.

3. $a_k = ka_{k-1}$, for all integers $k \geq 1$
   $a_0 = 1$

4. $b_k = \dfrac{b_{k-1}}{1 + b_{k-1}}$, for all integers $k \geq 1$
   $b_0 = 1$

5. $c_k = 3c_{k-1} + 1$, for all integers $k \geq 2$
   $c_1 = 1$

H **6.** $d_k = 2d_{k-1} + 3$, for all integers $k \geq 2$
   $d_1 = 2$

7. $e_k = 4e_{k-1} + 5$, for all integers $k \geq 1$
   $e_0 = 2$

8. $f_k = f_{k-1} + 2^k$, for all integers $k \geq 2$
   $f_1 = 1$

H **9.** $g_k = \dfrac{g_{k-1}}{g_{k-1} + 2}$, for all integers $k \geq 2$
   $g_1 = 1$

10. $h_k = 2^k - h_{k-1}$, for all integers $k \geq 1$
    $h_0 = 1$

11. $p_k = 2p_{k-1} + 3^k$
    $p_1 = 2$

12. $s_k = s_{k-1} + 2k$, for all integers $k \geq 1$
    $s_0 = 3$

13. $t_k = t_{k-1} + 3k + 1$, for all integers $k \geq 1$
    $t_0 = 0$

✶ **14.** $x_k = 3x_{k-1} + k$, for all integers $k \geq 2$
    $x_1 = 1$

15. $y_k = y_{k-1} + k^2$, for all integers $k \geq 2$
    $y_1 = 1$

16. Solve the recurrence relation obtained as the answer to exercise 18(c) of Section 8.1.

17. Solve the recurrence relation obtained as the answer to exercise 21(c) of Section 8.1.

18. Suppose d is a fixed constant and $a_0, a_1, a_2, \ldots$ is a sequence that satisfies the recurrence relation $a_k = a_{k-1} + d$, for all integers $k \geq 1$. Use mathematical induction to prove that $a_n = a_0 + nd$, for all integers $n \geq 0$.

19. A worker is promised a bonus if he can increase his productivity by 2 units a day every day for a period of 30 days. If on day 0 he produces 170 units, how many units must he produce on day 30 to qualify for the bonus?

20. A runner targets herself to improve her time on a certain course by 3 seconds a day. If on day 0 she runs the course in 3 minutes, how fast must she run it on day 14 to stay on target?

21. Suppose r is a fixed constant and $a_0, a_1, a_2 \ldots$ is a sequence that satisfies the recurrence relation $a_k = ra_{k-1}$, for all integers $k \geq 1$ and $a_0 = a$. Use mathematical induction to prove that $a_n = ar^n$, for all integers $n \geq 0$.

22. As shown in Example 8.1.8, if a bank pays interest at a rate of i compounded m times a year, then the amount of money P_k at the end of k time periods (where one time period $= 1/m$th of a year) satisfies the recurrence relation $P_k = [1 + (i/m)]P_{k-1}$ with initial condition $P_0 =$ the initial amount deposited. Find an explicit formula for P_n.

23. Suppose the population of a country increases at a steady rate of 3% per year. If the population is 50 million at a certain time, what will it be 25 years later?

24. A chain letter works as follows: One person sends a copy of the letter to five friends, each of whom sends a copy to five friends, each of whom sends a copy to five friends, and so forth. How many people will have received copies of the letter after the twentieth repetition of this process, assuming no person receives more than one copy?

25. A certain computer algorithm executes twice as many operations when it is run with an input of size k as when it is run with an input of size $k - 1$ (where k in an integer that is greater than 1). When the algorithm is run with an input of size 1, it executes seven operations. How many operations does it execute when it is run with an input of size 25?

26. A person saving for retirement makes an initial deposit of $1,000 to a bank account earning interest at a rate of 3% per year compounded monthly, and each month she adds an additional $200 to the account.
    a. For each nonnegative integer n, let A_n be the amount in the account at the end of n months. Find a recurrence relation relating A_k to A_{k-1}.
    H b. Use iteration to find an explicit formula for A_n.
    c. Use mathematical induction to prove the correctness of the formula you obtained in part (b).
    d. How much will the account be worth at the end of 20 years? At the end of 40 years?
    H e. In how many years will the account be worth $10,000?

27. A person borrows $3,000 on a bank credit card at a nominal rate of 18% per year, which is actually charged at a rate of 1.5% per month.
    a. What is the annual percentage rate (APR) for the card? (See Example 8.1.8 for a definition of APR.)
    b. Assume that the person does not place any additional charges on the card and pays the bank $150 each month to pay off the loan. Let B_n be the balance owed on the card after n months. Find an explicit formula for B_n.
    c. How long will be required to pay off the debt?
    d. What is the total amount of money the person will have paid for the loan?

In 28–42 use mathematical induction to verify the correctness of the formula you obtained in the referenced exercise.

28. Exercise 3   29. Exercise 4   **30.** Exercise 5

31. Exercise 6   32. Exercise 7   33. Exercise 8

34. Exercise 9   *H* **35.** Exercise 10   36. Exercise 11

H **37.** Exercise 12   38. Exercise 13   **39.** Exercise 14

40. Exercise 15   41. Exercise 16   42. Exercise 17

In each of 43–49 a sequence is defined recursively. (a) Use iteration to guess an explicit formula for the sequence. (b) Use strong mathematical induction to verify that the formula of part (a) is correct.

43. $a_k = \dfrac{a_{k-1}}{2a_{k-1} - 1}$, for all integers $k \geq 1$
$a_0 = 2$

44. $b_k = \dfrac{2}{b_{k-1}}$, for all integers $k \geq 2$
$b_1 = 1$

45. $v_k = v_{\lfloor k/2 \rfloor} + v_{\lfloor (k+1)/2 \rfloor} + 2$, for all integers $k \geq 2$,
$v_1 = 1$.

H **46.** $s_k = 2s_{k-2}$, for all integers $k \geq 2$,
$s_0 = 1, s_1 = 2$.

47. $t_k = k - t_{k-1}$, for all integers $k \geq 1$,
$t_0 = 0$.

H **48.** $w_k = w_{k-2} + k$, for all integers $k \geq 3$,
$w_1 = 1, w_2 = 2$.

H **49.** $u_k = u_{k-2} \cdot u_{k-1}$, for all integers $k \geq 2$,
$u_0 = u_1 = 2$.

In 50 and 51 determine whether the given recursively defined sequence satisfies the explicit formula $a_n = (n - 1)^2$, for all integers $n \geq 1$.

50. $a_k = 2a_{k-1} + k - 1$, for all integers $k \geq 2$
$a_1 = 0$

51. $a_k = (a_{k-1} + 1)^2$, for all integers $k \geq 2$
$a_1 = 0$

52. A single line divides a plane into two regions. Two lines (by crossing) can divide a plane into four regions; three lines can divide it into seven regions (see the figure). Let P_n be the maximum number of regions into which n lines divide a plane, where n is a positive integer.

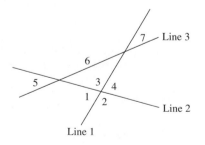

H **a.** Derive a recurrence relation for P_k in terms of P_{k-1}, for all integers $k \geq 2$.
  b. Use iteration to guess an explicit formula for P_n.

H **53.** Compute $\begin{bmatrix} 1 & 1 \\ 1 & 0 \end{bmatrix}^n$ for small values of n (up to about 5 or 6). Conjecture explicit formulas for the entries in this matrix, and prove your conjecture using mathematical induction.

54. In economics the behavior of an economy from one period to another is often modeled by recurrence relations. Let Y_k be the income in period k and C_k be the consumption in period k. In one economic model, income in any period is assumed to be the sum of consumption in that period plus investment and government expenditures (which are assumed to be constant from period to period), and consumption in each period is assumed to be a linear function of the income

of the preceding period. That is,

$$Y_k = C_k + E \qquad \text{where } E \text{ is the sum of investment plus government expenditures}$$

$$C_k = c + mY_{k-1} \qquad \text{where } c \text{ and } m \text{ are constants.}$$

Substituting the second equation into the first gives $Y_k = E + c + mY_{k-1}$.

a. Use iteration on the above recurrence relation to obtain

$$Y_n = (E + c)\left(\frac{m^n - 1}{m - 1}\right) + m^n Y_0$$

for all integers $n \geq 1$.

b. (For students who have studied calculus) Show that if $0 < m < 1$, then $\lim_{n \to \infty} Y_n = \dfrac{E + c}{1 - m}$.

8.3 Second-Order Linear Homogeneous Recurrence Relations with Constant Coefficients

Genius is 1% inspiration and 99% perspiration. — Thomas Alva Edison, 1932

In section 8.2 we discussed finding explicit formulas for recursively defined sequences using iteration. This is a basic technique that does not require any special tools beyond the ability to discern patterns. In many cases, however, a pattern is not readily discernible and other methods must be used. A variety of techniques are available for finding explicit formulas for special classes of recursively defined sequences. The method explained in this section is one that works for the Fibonacci and other similarly defined sequences.

> **• Definition**
>
> A **second-order linear homogeneous recurrence relation with constant coefficients** is a recurrence relation of the form
>
> $$a_k = Aa_{k-1} + Ba_{k-2} \quad \text{for all integers } k \geq \text{ some fixed integer,}$$
>
> where A and B are fixed real numbers with $B \neq 0$.

"Second-order" refers to the fact that the expression for a_k contains the two previous terms a_{k-1} and a_{k-2}, "linear" to the fact that a_{k-1} and a_{k-2} appear in separate terms and to the first power, "homogeneous" to the fact that the total degree of each term is the same (thus there is no constant term), and "constant coefficients" to the fact that A and B are fixed real numbers that do not depend on k.

Example 8.3.1 Second-Order Linear Homogeneous Recurrence Relations with Constant Coefficients

State whether each of the following is a second-order linear homogeneous recurrence relation with constant coefficients:

a. $a_k = 3a_{k-1} + 2a_{k-2}$

b. $b_k = b_{k-1} + b_{k-2} + b_{k-3}$

c. $c_k = \frac{1}{2}c_{k-1} - \frac{3}{7}c_{k-2}$

d. $d_k = d_{k-1}^2 + d_{k-1} \cdot d_{k-2}$

e. $e_k = 2e_{k-2}$

f. $f_k = 2f_{k-1} + 1$

g. $g_k = g_{k-1} + g_{k-2}$

h. $h_k = (-1)h_{k-1} + (k-1)h_{k-2}$

Solution

a. Yes; $A = 3$ and $B = 2$

b. No; not second-order

c. Yes; $A = \frac{1}{2}$ and $B = -\frac{3}{7}$

d. No; not linear

e. Yes; $A = 0$ and $B = 2$

f. No; not homogeneous

g. Yes; $A = 1$ and $B = 1$

h. No; nonconstant coefficients ■

The Distinct-Roots Case

Consider a second-order linear homogeneous recurrence relation with constant coefficients:

$$a_k = Aa_{k-1} + Ba_{k-2} \quad \text{for all integers } k \geq 2, \qquad 8.3.1$$

where A and B are fixed real numbers. *Suppose* that for some number t with $t \neq 0$, the sequence

$$1, t, t^2, t^3, \ldots, t^n, \ldots$$

satisfies relation (8.3.1). This means that each term of the sequence equals A times the previous term plus B times the term before that. So for all integers $k \geq 2$,

$$t^k = At^{k-1} + Bt^{k-2}.$$

Since $t \neq 0$, this equation may be divided by t^{k-2} to obtain

$$t^2 = At + B,$$

or, equivalently,

$$t^2 - At - B = 0. \qquad 8.3.2$$

This is a quadratic equation, and the values of t that make it true can be found either by factoring or by using the quadratic formula.

Now work backward. *Suppose* t is any number that satisfies equation (8.3.2). Does the sequence $1, t, t^2, t^3, \ldots, t^n, \ldots$ satisfy relation (8.3.1)? To answer this question, multiply equation (8.3.2) by t^{k-2} to obtain

$$t^{k-2} \cdot t^2 - t^{k-2} \cdot At - t^{k-2} \cdot B = 0.$$

This is equivalent to

$$t^k - At^{k-1} - Bt^{k-2} = 0$$

or

$$t^k = At^{k-1} + Bt^{k-2}.$$

Hence the answer is yes: $1, t, t^2, t^3, \ldots, t^n, \ldots$ satisfies relation (8.3.1).

This discussion proves the following lemma.

Lemma 8.3.1

Let A and B be real numbers. A recurrence relation of the form

$$a_k = Aa_{k-1} + Ba_{k-2} \qquad \text{8.3.1}$$

is satisfied by the sequence

$$1, t, t^2, t^3, \ldots, t^n, \ldots,$$

where t is a nonzero real number, if, and only if, t satisfies the equation

$$t^2 - At - B = 0. \qquad \text{8.3.2}$$

Equation (8.3.2) is called the *characteristic equation* of the recurrence relation.

• Definition

Given a second-order linear homogeneous recurrence relation with constant coefficients:

$$a_k = Aa_{k-1} + Ba_{k-2} \quad \text{for all integers } k \geq 2, \qquad \text{8.3.1}$$

the **characteristic equation of the relation** is

$$t^2 - At - B = 0. \qquad \text{8.3.2}$$

Example 8.3.2 Using the Characteristic Equation to Find Solutions to a Recurrence Relation

Consider the recurrence relation that specifies that the kth term of a sequence equals the sum of the $(k-1)$st term plus twice the $(k-2)$nd term. That is,

$$a_k = a_{k-1} + 2a_{k-2} \quad \text{for all integers } k \geq 2. \qquad \text{8.3.3}$$

Find all sequences that satisfy relation (8.3.3) and have the form

$$1, t, t^2, t^3, \ldots, t^n, \ldots,$$

where t is nonzero.

Solution   By Lemma 8.3.1, relation (8.3.3) is satisfied by a sequence $1, t, t^2, t^3, \ldots, t^n, \ldots$ if, and only if, t satisfies the characteristic equation

$$t^2 - t - 2 = 0.$$

Since

$$t^2 - t - 2 = (t - 2)(t + 1),$$

the only possible values of t are 2 and -1. It follows that the sequences

$$1, 2, 2^2, 2^3, \ldots, 2^n, \ldots \quad \text{and} \quad 1, -1, (-1)^2, (-1)^3, \ldots, (-1)^n, \ldots.$$

are both solutions for relation (8.3.3) and there are no other solutions of this form. Note that these sequences can be rewritten more simply as

$$1, 2, 2^2, 2^3, \ldots, 2^n, \ldots \quad \text{and} \quad 1, -1, 1, -1, \ldots, (-1)^n, \ldots. \qquad \blacksquare$$

The example above shows how to find two distinct sequences that satisfy a given second-order linear homogeneous recurrence relation with constant coefficients. It turns out that any linear combination of such sequences produces another sequence that also satisfies the relation.

Lemma 8.3.2

If $r_0, r_1, r_2, \ldots$ and $s_0, s_1, s_2, \ldots$ are sequences that satisfy the same second-order linear homogeneous recurrence relation with constant coefficients, and if C and D are *any* numbers, then the sequence $a_0, a_1, a_2, \ldots$ defined by the formula

$$a_n = Cr_n + Ds_n \quad \text{for all integers } n \geq 0$$

also satisfies the same recurrence relation.

Proof:

Suppose $r_0, r_1, r_2, \ldots$ and $s_0, s_1, s_2, \ldots$ are sequences that satisfy the same second-order linear homogeneous recurrence relation with constant coefficients. In other words, suppose that for some real numbers A and B,

$$r_k = Ar_{k-1} + Br_{k-2} \quad \text{and} \quad s_k = As_{k-1} + Bs_{k-2} \qquad 8.3.4$$

for all integers $k \geq 2$. Suppose also that C and D are any numbers. Let $a_0, a_1, a_2, \ldots$ be the sequence defined by

$$a_n = Cr_n + Ds_n \quad \text{for all integers } n \geq 0. \qquad 8.3.5$$

[*We must show that $a_0, a_1, a_2, \ldots$ satisfies the same recurrence relation as $r_0, r_1, r_2, \ldots$ and $s_0, s_1, s_2, \ldots$. That is, we must show that $a_k = Aa_{k-1} + Ba_{k-2}$, for all integers $k \geq 2$.*]

For all integers $k \geq 2$,

$$
\begin{aligned}
Aa_{k-1} + Ba_{k-2} &= A(Cr_{k-1} + Ds_{k-1}) + B(Cr_{k-2} + Ds_{k-2}) && \text{by substitution from (8.3.5)} \\
&= C(Ar_{k-1} + Br_{k-2}) + D(As_{k-1} + Bs_{k-2}) && \text{by basic algebra} \\
&= Cr_k + Ds_k && \text{by substitution from (8.3.4)} \\
&= a_k && \text{by substitution from (8.3.5).}
\end{aligned}
$$

Hence $a_0, a_1, a_2, \ldots$ satisfies the same recurrence relation as $r_0, r_1, r_2, \ldots$ and $s_0, s_1, s_2, \ldots$ [*as was to be shown*].

Given a second-order linear homogeneous recurrence relation with constant coefficients, if the characteristic equation has two distinct roots, then Lemmas 8.3.1 and 8.3.2 can be used together to find a sequence that satisfies both the recurrence relation and the specific initial conditions.

Example 8.3.3 Finding the Linear Combination That Satisfies the Initial Conditions

Find a sequence that satisfies the recurrence relation of Example 8.3.2,

$$a_k = a_{k-1} + 2a_{k-2} \quad \text{for all integers } k \geq 2, \qquad 8.3.3$$

and that also satisfies the initial conditions

$$a_0 = 1 \quad \text{and} \quad a_1 = 8.$$

Solution   By Example 8.3.2, the sequences

$$1, 2, 2^2, 2^3, \ldots, 2^n, \ldots \quad \text{and} \quad 1, -1, 1, -1, \ldots, (-1)^n, \ldots$$

both satisfy relation (8.3.3) (though neither satisfies the given initial conditions). By Lemma 8.3.2, therefore, any sequence $a_0, a_1, a_2, \ldots$ that satisfies an explicit formula of the form

$$a_n = C \cdot 2^n + D(-1)^n, \qquad\qquad 8.3.6$$

where C and D are numbers, also satisfies relation (8.3.3). You can find C and D so that $a_0, a_1, a_2, \ldots$ satisfies the specified initial conditions by substituting $n = 0$ and $n = 1$ into equation (8.3.6) and solving for C and D:

$$a_0 = 1 = C \cdot 2^0 + D(-1)^0,$$
$$a_1 = 8 = C \cdot 2^1 + D(-1)^1.$$

When you simplify, you obtain the system

$$1 = C + D$$
$$8 = 2C - D,$$

which can be solved in various ways. For instance, if you add the two equations, you get

$$9 = 3C,$$

and so

$$C = 3.$$

Then, by substituting into $1 = C + D$, you get

$$D = -2.$$

If follows that the sequence $a_0, a_1, a_2, \ldots$ given by

$$a_n = 3 \cdot 2^n + (-2)(-1)^n = 3 \cdot 2^n - 2(-1)^n,$$

for integers $n \geq 0$, satisfies both the recurrence relation and the given initial conditions. ∎

The techniques of Examples 8.3.2 and 8.3.3 can be used to find an explicit formula for *any* sequence that satisfies a second-order linear homogeneous recurrence relation with constant coefficients for which the characteristic equation has distinct roots, provided that the first two terms of the sequence are known. This is made precise in the next theorem.

Theorem 8.3.3 Distincts-Roots Theorem

Suppose a sequence $a_0, a_1, a_2, \ldots$ satisfies a recurrence relation

$$a_k = A a_{k-1} + B a_{k-2} \qquad\qquad 8.3.1$$

for some real numbers A and B and all integers $k \geq 2$. If the characteristic equation

$$t^2 - At - B = 0 \qquad\qquad 8.3.2$$

has two distinct roots r and s, then $a_0, a_1, a_2, \ldots$ satisfies the explicit formula

$$a_n = Cr^n + Ds^n,$$

where C and D are the numbers whose values are determined by the values a_0 and a_1.

Note: To say "C and D are determined by the values of a_0 and a_1" means that C and D are the solutions to the system of simultaneous equations

$$a_0 = Cr^0 + Ds^0 \quad \text{and} \quad a_1 = Cr^1 + Ds^1,$$

or, equivalently,

$$a_0 = C + D \quad \text{and} \quad a_1 = Cr + Ds.$$

In exercise 19 at the end of this section you are asked to verify that this system always has a solution when $r \neq s$.

Proof:

Suppose that for some real numbers A and B, a sequence $a_0, a_1, a_2, \ldots$ satisfies the recurrence relation $a_k = Aa_{k-1} + Ba_{k-2}$, for all integers $k \geq 2$, and suppose the characteristic equation $t^2 - At - B = 0$ has two distinct roots r and s. We will show that

$$\text{for all integers } n \geq 0, \quad a_n = Cr^n + Ds^n,$$

where C and D are numbers such that

$$a_0 = Cr^0 + Ds^0 \quad \text{and} \quad a_1 = Cr^1 + Ds^1.$$

Consider the formula

$$a_n = Cr^n + Ds^n.$$

We use strong mathematical induction to prove that the formula holds for all integers $n \geq 0$. In the basis step, we prove not only that the formula holds for $n = 0$ but also that it holds for $n = 1$. The reason we do this is that in the inductive step, we need the formula to hold for $n = 0$ and $n = 1$ in order to prove that it holds for $n = 2$.

Show that the formula holds for $n = 0$ and $n = 1$: The truth of the formula for $n = 0$ and $n = 1$ is automatic because C and D are exactly those numbers that make the following equations true:

$$a_0 = Cr^0 + Ds^0 \quad \text{and} \quad a_1 = Cr^1 + Ds^1.$$

Show that for all integers k, if $k \geq 2$ and the formula holds for all integers i with $0 \leq i < k$, then it holds for k: Suppose that $k \geq 2$ and for all integers i with $0 \leq i < k$,

$$a_i = Cr^i + Ds^i. \quad \text{This is the inductive hypothesis.}$$

We must show that

$$a_k = Cr^k + Ds^k.$$

Now by the inductive hypothesis,

$$a_{k-1} = Cr^{k-1} + Ds^{k-1} \quad \text{and} \quad a_{k-2} = Cr^{k-2} + Ds^{k-2},$$

so

$$a_k = Aa_{k-1} + Ba_{k-2} \qquad \text{by definition of } a_0, a_1, a_2, \ldots$$

$$= A(Cr^{k-1} + Ds^{k-1}) + B(Cr^{k-2} + Ds^{k-2}) \qquad \text{by inductive hypothesis}$$

$$= C(Ar^{k-1} + Br^{k-2}) + D(As^{k-1} + Bs^{k-2}) \qquad \text{by combining terms involving } C \text{ and } D \text{ together}$$

$$= Cr^k + Ds^k \qquad \text{by Lemma 8.3.1.}$$

This is what was to be shown.

[*The reason the last equality follows from Lemma* 8.3.1 *is that since r and s satisfy the characteristic equation* (8.3.2), *the sequences* $r^0, r^1, r^2, \ldots$ *and* $s^0, s^1, s^2, \ldots$ *satisfy the recurrence relation* (8.3.1).]

Remark: The t of Lemma 8.3.1 and the C and D of Lemma 8.3.2 and Theorem 8.3.3 are referred to simply as numbers. This is to allow for the possibility of complex as well as real number values. If both roots of the characteristic equation of the recurrence relation are real numbers, then C and D will be real. But if the roots are nonreal complex numbers, then C and D will be nonreal complex numbers.

The next example shows how to use the distinct-roots theorem to find an explicit formula for the Fibonacci sequence.

Example 8.3.4 A Formula for the Fibonacci Sequence

The Fibonacci sequence $F_0, F_1, F_2, \ldots$ satisfies the recurrence relation

$$F_k = F_{k-1} + F_{k-2} \quad \text{for all integers } k \geq 2$$

with initial conditions

$$F_0 = F_1 = 1.$$

Find an explicit formula for this sequence.

Solution   The Fibonacci sequence satisfies part of the hypothesis of the distinct-roots theorem since the Fibonacci relation is a second-order linear homogeneous recurrence relation with constant coefficients ($A = 1$ and $B = 1$). Is the second part of the hypothesis also satisfied? Does the characteristic equation

$$t^2 - t - 1 = 0$$

have distinct roots? By the quadratic formula, the roots are

$$t = \frac{1 \pm \sqrt{1 - 4(-1)}}{2} = \begin{cases} \dfrac{1 + \sqrt{5}}{2} \\ \dfrac{1 - \sqrt{5}}{2} \end{cases}$$

and so the answer is yes. It follows from the distinct-roots theorem that the Fibonacci sequence satisfies the explicit formula

$$F_n = C\left(\frac{1 + \sqrt{5}}{2}\right)^n + D\left(\frac{1 - \sqrt{5}}{2}\right)^n \quad \text{for all integers } n \geq 0, \qquad 8.3.7$$

where C and D are the numbers whose values are determined by the fact that $F_0 = F_1 = 1$. To find C and D, write

$$F_0 = 1 = C\left(\frac{1 + \sqrt{5}}{2}\right)^0 + D\left(\frac{1 - \sqrt{5}}{2}\right)^0 = C \cdot 1 + D \cdot 1 = C + D$$

and

$$F_1 = 1 = C\left(\frac{1 + \sqrt{5}}{2}\right)^1 + D\left(\frac{1 - \sqrt{5}}{2}\right)^1$$

$$= C\left(\frac{1 + \sqrt{5}}{2}\right) + D\left(\frac{1 - \sqrt{5}}{2}\right)$$

Thus the problem is to find numbers C and D such that

$$C + D = 1$$

and

$$C\left(\frac{1 + \sqrt{5}}{2}\right) + D\left(\frac{1 - \sqrt{5}}{2}\right) = 1.$$

This may look complicated, but in fact it is just a system of two equations in two unknowns. In exercise 7 at the end of this section, you are asked to show that

$$C = \frac{1 + \sqrt{5}}{2\sqrt{5}} \quad \text{and} \quad D = \frac{-(1 - \sqrt{5})}{2\sqrt{5}}.$$

Substituting these values for C and D into formula (8.3.7) gives

$$F_n = \left(\frac{1 + \sqrt{5}}{2\sqrt{5}}\right)\left(\frac{1 + \sqrt{5}}{2}\right)^n + \left(\frac{-(1 - \sqrt{5})}{2\sqrt{5}}\right)\left(\frac{1 - \sqrt{5}}{2}\right)^n,$$

or, simplifying,

$$F_n = \frac{1}{\sqrt{5}}\left(\frac{1 + \sqrt{5}}{2}\right)^{n+1} - \frac{1}{\sqrt{5}}\left(\frac{1 - \sqrt{5}}{2}\right)^{n+1} \qquad 8.3.8$$

for all integers $n \geq 0$. Remarkably, even though the formula for F_n involves $\sqrt{5}$, all of the values of the Fibonacci sequence are integers. It is also interesting to note that the numbers $(1 + \sqrt{5})/2$ and $(1 - \sqrt{5})/2$ are related to the golden ratio of Greek mathematics. (See exercise 24 at the end of this section.) ∎

The Single-Root Case

Consider again the recurrence relation

$$a_k = Aa_{k-1} + Ba_{k-2} \quad \text{for all integers } k \geq 2, \qquad 8.3.1$$

where A and B are real numbers, but suppose now that the characteristic equation

$$t^2 - At - B = 0 \qquad 8.3.2$$

has a single real root r. By Lemma 8.3.1, one sequence that satisfies the recurrence relation is

$$1, r, r^2, r^3, \ldots, r^n, \ldots.$$

But another sequence that also satisfies the relation is

$$0, r, 2r^2, 3r^3, \ldots, nr^n, \ldots.$$

To see why this is so, observe that since r is the unique root of $t^2 - At - B = 0$, the left-hand side of the equation can be factored as $(t - r)^2$, and so

$$t^2 - At - B = (t - r)^2 = t^2 - 2rt + r^2. \qquad 8.3.9$$

Equating coefficients in equation (8.3.9) gives

$$A = 2r \quad \text{and} \quad B = -r^2. \qquad 8.3.10$$

Let $s_0, s_1, s_2, \ldots$ be the sequence defined by the formula

$$s_n = nr^n \quad \text{for all integers } n \geq 0.$$

Then

$$
\begin{aligned}
As_{k-1} + Bs_{k-2} &= A(k - 1)r^{k-1} + B(k - 2)r^{k-2} && \text{by definition} \\
&= 2r(k - 1)r^{k-1} - r^2(k - 2)r^{k-2} && \text{by substitution from (8.3.10)} \\
&= 2(k - 1)r^k - (k - 2)r^k \\
&= (2k - 2 - k + 2)r^k \\
&= kr^k && \text{by basic algebra} \\
&= s_k && \text{by definition.}
\end{aligned}
$$

Thus $s_0, s_1, s_2, \ldots$ satisfies the recurrence relation. This argument proves the following lemma.

Lemma 8.3.4

Let A and B be real numbers and suppose the characteristic equation

$$t^2 - At - B = 0$$

has a single root r. Then the sequences $1, r^1, r^2, r^3, \ldots, r^n, \ldots$ and $0, r, 2r^2, 3r^3, \ldots, nr^n, \ldots$ both satisfy the recurrence relation

$$a_k = Aa_{k-1} + Ba_{k-2}$$

for all integers $k \geq 2$.

Lemmas 8.3.2 and 8.3.4 can be used to establish the *single-root theorem,* which tells how to find an explicit formula for any recursively defined sequence satisfying a second-order linear homogeneous recurrence relation with constant coefficients for which the characteristic equation has just one root. Taken together, the distinct-roots and single-root theorems cover all second-order linear homogeneous recurrence relations with constant coefficients. The proof of the single-root theorem is very similar to that of the distinct-roots theorem and is left as an exercise.

Theorem 8.3.5 Single-Root Theorem

Suppose a sequence $a_0, a_1, a_2, \ldots$ satisfies a recurrence relation

$$a_k = Aa_{k-1} + Ba_{k-2}$$

for some real numbers A and B with $B \neq 0$ and for all integers $k \geq 2$. If the characteristic equation $t^2 - At - B = 0$ has a single (real) root r, then $a_0, a_1, a_2, \ldots$ satisfies the explicit formula

$$a_n = Cr^n + Dnr^n,$$

where C and D are the real numbers whose values are determined by the values of a_0 and any other known value of the sequence.

Example 8.3.5 Single-Root Case

Suppose a sequence $b_0, b_1, b_2, \ldots$ satisfies the recurrence relation

$$b_k = 4b_{k-1} - 4b_{k-2} \quad \text{for all integers } k \geq 2, \qquad 8.3.11$$

with initial conditions

$$b_0 = 1 \quad \text{and} \quad b_1 = 3.$$

Find an explicit formula for $b_0, b_1, b_2, \ldots$.

Solution This sequence satisfies part of the hypothesis of the single-root theorem because it satisfies a second-order linear homogeneous recurrence relation with constant coefficients ($A = 4$ and $B = -4$). The single-root condition is also met because the characteristic equation

$$t^2 - 4t + 4 = 0$$

has the unique root $r = 2$ [*since* $t^2 - 4t + 4 = (t-2)^2$].

It follows from the single-root theorem that $b_0, b_1, b_2, \ldots$ satisfies the explicit formula

$$b_n = C \cdot 2^n + Dn2^n \quad \text{for all integers } n \geq 0, \qquad 8.3.12$$

where C and D are the real numbers whose values are determined by the fact that $b_0 = 1$ and $b_1 = 3$. To find C and D, write

$$b_0 = 1 = C \cdot 2^0 + D \cdot 0 \cdot 2^0 = C$$

and

$$b_1 = 3 = C \cdot 2^1 + D \cdot 1 \cdot 2^1 = 2C + 2D.$$

Hence the problem is to find numbers C and D such that

$$C = 1$$

and

$$2C + 2D = 3.$$

Substitute $C = 1$ into the second equation to obtain

$$2 + 2D = 3,$$

and so

$$D = \frac{1}{2}.$$

Now substitute $C = 1$ and $D = \frac{1}{2}$ into formula (8.3.12) to conclude that

$$b_n = 2^n + \frac{1}{2}n2^n = 2^n \left(1 + \frac{n}{2}\right) \quad \text{for all integers } n \geq 0. \qquad \blacksquare$$

Example 8.3.6 Gambler's Ruin

A gambler repeatedly bets $1 that a coin will come up heads when tossed. Each time the coin comes up heads, the gambler wins $1; each time it comes up tails, he loses $1. The gambler will quit playing either when he is ruined (loses all his money) or when he has $M (where M is a positive number he has decided in advance). Let P_n be the probability that the gambler is ruined if he begins playing with $n. Then if the coin is fair (has an equal chance of coming up heads or tails),

$$P_{k-1} = \frac{1}{2}P_k + \frac{1}{2}P_{k-2} \quad \text{for each integer } k \text{ with } 2 \leq k \leq M.$$

(This follows from the fact that if the gambler has $(k-1)$, then he has equal chance of winning $1 or losing $1, and if he wins $1, then his chance of being ruined is P_k, whereas if he loses $1, then his chance of being ruined is P_{k-2}.) Also $P_0 = 1$ (because if he has $0, he is certain of being ruined) and $P_M = 0$ (because once he has $M, he quits and so stands no chance of being ruined). Find an explicit formula for P_n. How should the gambler choose M to minimize his chance of being ruined?

Solution Multiplying both sides of $P_{k-1} = \frac{1}{2}P_k + \frac{1}{2}P_{k-2}$ by 2 and subtracting P_{k-2} from both sides gives

$$P_k = 2P_{k-1} - P_{k-2},$$

which is a second-order homogeneous recurrence relation with constant coefficients. Its characteristic equation is

$$t^2 - 2t + 1 = 0,$$

which has the single root $r = 1$. Thus, by the single-root theorem,

$$P_n = Cr^n + Dnr^n = C + Dn$$

(since $r = 1$), where C and D are determined by two values of the sequence. But $P_0 = 1$ and $P_M = 0$. Hence

$$1 = P_0 = C + D \cdot 0 = C,$$
$$0 = P_M = C + DM = 1 + DM.$$

It follows that $C = 1$ and $D = -\dfrac{1}{M}$, and so

$$P_n = 1 - \frac{1}{M}n = \frac{M-n}{M} \quad \text{for each integer } n \text{ with } 0 \leq n \leq M.$$

For instance, a gambler who starts with $20 and decides to quit either if his total grows to $100 or if he goes broke has the following chance of going broke:

$$P_{20} = \frac{100 - 20}{100} = \frac{80}{100} = 80\%.$$

Observe that the larger M is relative to n, the closer P_n is to 1. In other words, the larger the amount of money the gambler sets himself as a target, the more likely he is to go broke. Conversely, the more modest he is in his goal, the more likely he is to reach it. $\blacksquare$

Exercise Set 8.3

1. Which of the following are second-order linear homogeneous recurrence relations with constant coefficients?
   a. $a_k = 2a_{k-1} - 5a_{k-2}$ b. $b_k = kb_{k-1} + b_{k-2}$
   c. $c_k = 3c_{k-1} \cdot c_{k-2}^2$ d. $d_k = 3d_{k-1} + d_{k-2}$
   e. $r_k = r_{k-1} - r_{k-2} - 2$ f. $s_k = 10s_{k-2}$

2. Which of the following are second-order linear homogeneous recurrence relations with constant coefficients?
   a. $a_k = (k-1)a_{k-1} + 2ka_{k-2}$
   b. $b_k = -b_{k-1} + 7b_{k-2}$
   c. $c_k = 3c_{k-1} + 1$
   d. $d_k = 3d_{k-1}^2 + d_{k-2}$
   e. $r_k = r_{k-1} - 6r_{k-3}$
   f. $s_k = s_{k-1} + 10s_{k-2}$

3. Let $a_0, a_1, a_2, \ldots$ be the sequence defined by the explicit formula
   $$a_n = C \cdot 2^n + D \quad \text{for all integers } n \geq 0,$$
   where C and D are real numbers.
   a. Find C and D so that $a_0 = 1$ and $a_1 = 3$. What is a_2 in this case?
   b. Find C and D so that $a_0 = 0$ and $a_1 = 2$. What is a_2 in this case?

4. Let $b_0, b_1, b_2, \ldots$ be the sequence defined by the explicit formula
   $$b_n = C \cdot 3^n + D(-2)^n \quad \text{for all integers } n \geq 0,$$
   where C and D are real numbers.
   a. Find C and D so that $b_0 = 0$ and $b_1 = 5$. What is b_2 in this case?
   b. Find C and D so that $b_0 = 3$ and $b_1 = 4$. What is b_2 in this case?

5. Let $a_0, a_1, a_2, \ldots$ be the sequence defined by the explicit formula
   $$a_n = C \cdot 2^n + D \quad \text{for all integers } n \geq 0,$$
   where C and D are real numbers. Show that for any choice of C and D,
   $$a_k = 3a_{k-1} - 2a_{k-2} \quad \text{for all integers } k \geq 2.$$

6. Let $b_0, b_1, b_2, \ldots$ be the sequence defined by the explicit formula
   $$b_n = C \cdot 3^n + D(-2)^n \quad \text{for all integers } n \geq 0,$$
   where C and D are real numbers. Show that for any choice of C and D,
   $$b_k = b_{k-1} + 6b_{k-2} \quad \text{for all integers } k \geq 2.$$

7. Solve the system of equations in Example 8.3.4 to obtain
   $$C = \frac{1 + \sqrt{5}}{2\sqrt{5}} \quad \text{and} \quad D = \frac{-(1 - \sqrt{5})}{2\sqrt{5}}.$$

In each of 8–10: (a) suppose a sequence of the form $1, t, t^2, t^3, \ldots, t^n, \ldots$, where $t \neq 0$, satisfies the given recurrence relation (but not necessarily the initial conditions), and find all possible values of t; (b) suppose a sequence satisfies the given initial conditions as well as the recurrence relation, and find an explicit formula for the sequence.

8. $a_k = 2a_{k-1} + 3a_{k-2}$, for all integers $k \geq 2$
   $a_0 = 1, a_1 = 2$

9. $b_k = 7b_{k-1} - 10b_{k-2}$, for all integers $k \geq 2$
   $b_0 = 2, b_1 = 2$

10. $c_k = c_{k-1} + 6c_{k-2}$, for all integers $k \geq 2$
    $c_0 = 0, c_1 = 3$

In each of 11–15 suppose a sequence satisfies the given recurrence relation and initial conditions. Find an explicit formula for the sequence.

11. $d_k = 4d_{k-2}$, for all integers $k \geq 2$
    $d_0 = 1, d_1 = -1$

12. $e_k = 9e_{k-2}$, for all integers $k \geq 2$
    $e_0 = 0, e_1 = 2$

13. $r_k = 2r_{k-1} - r_{k-2}$, for all integers $k \geq 2$
    $r_0 = 1, r_1 = 4$

14. $s_k = -4s_{k-1} - 4s_{k-2}$, for all integers $k \geq 2$
    $s_0 = 0, s_1 = -1$

15. $t_k = 6t_{k-1} - 9t_{k-2}$, for all integers $k \geq 2$
    $t_0 = 1, t_1 = 3$

H 16. Find an explicit formula for the sequence of exercise 37 in Section 8.1.

17. Find an explicit formula for the sequence of exercise 39 in Section 8.1.

18. Suppose that the sequences $s_0, s_1, s_2, \ldots$ and $t_0, t_1, t_2, \ldots$ both satisfy the same second-order linear homogeneous recurrence relation with constant coefficients:
    $$s_k = 5s_{k-1} - 4s_{k-2} \quad \text{for all integers } k \geq 2,$$
    $$t_k = 5t_{k-1} - 4t_{k-2} \quad \text{for all integers } k \geq 2.$$
    Show that the sequence $2s_0 + 3t_0, 2s_1 + 3t_1, 2s_2 + 3t_2, \ldots$ also satisfies the same relation. In other words, show that
    $$2s_k + 3t_k = 5(2s_{k-1} + 3t_{k-1}) - 4(2s_{k-2} + 3t_{k-2})$$
    for all integers $k \geq 2$. Do *not* use Lemma 8.3.2.

19. Show that if r, s, a_0, and a_1 are numbers with $r \neq s$, then there exist unique numbers C and D so that
    $$C + D = a_0$$
    $$Cr + Ds = a_1.$$

20. Show that if r is a nonzero real number, k and m are distinct integers, and a_k and a_m are any real numbers, then there exist unique real numbers C and D so that

$$Cr^k + KDr^k = a_k$$
$$Cr^m + lDr^m = a_m.$$

H 21. Prove Theorem 8.3.5 for the case where the values of C and D are determined by a_0 and a_1.

Exercises 22 and 23 are intended for students who are familiar with complex numbers.

22. Find an explicit formula for a sequence $a_0, a_1, a_2, \ldots$ that satisfies

$$a_k = 2a_{k-1} - 2a_{k-2} \quad \text{for all integers } k \geq 2$$

with initial conditions $a_0 = 1$ and $a_1 = 2$.

23. Find an explicit formula for a sequence $b_0, b_1, b_2, \ldots$ that satisfies

$$b_k = 2b_{k-1} - 5b_{k-2} \quad \text{for all integers } k \geq 2$$

with initial conditions $b_0 = 1$ and $b_1 = 1$.

24. The numbers $\dfrac{1 + \sqrt{5}}{2}$ and $\dfrac{1 - \sqrt{5}}{2}$ that appear in the explicit formula for the Fibonacci sequence are related to a quantity called the *golden ratio* in Greek mathematics. Consider a rectangle of length ϕ units and height 1, where $\phi > 1$.

Divide the rectangle into a rectangle and a square as shown in the preceding diagram. The square is 1 unit on each side, and the rectangle has sides of lengths 1 and $\phi - 1$. The ancient Greeks considered the outer rectangle to be perfectly proportioned (saying that the lengths of its sides were in a *golden ratio* to each other) if the ratio of the length to the width of the outer rectangle equaled the ratio of the length to the width of the inner rectangle. That is,

$$\frac{\psi}{1} = \frac{1}{\phi - 1}.$$

a. Show that ϕ satisfies the following quadratic equation: $t^2 - t - 1 = 0$.

b. Find the two solutions of $t^2 - t - 1 = 0$ and call them ϕ_1 and ϕ_2.

c. Express the explicit formula for the Fibonacci sequence in terms of ϕ_1 and ϕ_2.

H 25. A gambler repeatedly bets that a die will come up 6 when rolled. Each time the die comes up 6, the gambler wins \$1; each time it does not, the gambler loses \$1. He will quit playing either when he is ruined or when he wins \$300. If P_n is the probability that the gambler is ruined when he begins play with \$$n$, then $P_{k-1} = \frac{1}{6}P_k + \frac{5}{6}P_{k-2}$ for all integers k with $2 \leq k \leq 300$. Also $P_0 = 1$ and $P_{300} = 0$. Find an explicit formula for P_n and use it to calculate P_{20}.

✱ 26. A circular disk is cut into n distinct sectors, each shaped like a piece of pie and all meeting at the center point of the disk. Each sector is to be painted red, green, yellow, or blue in such a way that no two adjacent sectors are painted the same color. Let S_n be the number of ways to paint the disk.

H a. Find a recurrence relation for S_k in terms of S_{k-1} and S_{k-2} for each integer $k \geq 4$.

b. Find an explicit formula for S_n for $n \geq 2$.

8.4 General Recursive Definitions

GENIE: Oh, aren't you acquainted with recursive acronyms? I thought everybody knew about them. You see, "GOD" stands for "GOD Over Djinn"—which can be expanded as "GOD Over Djinn, Over Djinn"—and that can, in turn, be expanded to "GOD Over Djinn, Over Djinn, Over Djinn"—which can, in its turn, be further expanded.... You can go as far as you like.
ACHILLES: But I'll never finish!
GENIE: Of course not. You can never totally expand GOD.

— Douglas Hofstadter, *Gödel, Escher, Bach,* 1979

Sequences of numbers are not the only objects that can be defined recursively. In this section we discuss recursive definitions for sets, sums, products, unions, intersections, and functions.

Recursively Defined Sets

To define a set of objects recursively, you identify a few core objects as belonging to the set and give rules showing how to build new set elements from old. More formally, a recursive definition for a set consists of the following three components:

   I. BASE: A statement that certain objects belong to the set.

  II. RECURSION: A collection of rules indicating how to form new set objects from those already known to be in the set.

 III. RESTRICTION: A statement that no objects belong to the set other than those coming from I and II.

Example 8.4.1 Recursive Definition of Boolean Expressions

The set of Boolean expressions was introduced in Section 1.4 as "legal" expressions involving letters from the alphabet such as p, q, and r, and the symbols $\wedge$, $\vee$, and $\sim$ [*a legal expression being, for instance, $p \wedge (q \vee \sim r)$ and an illegal one being $\wedge \sim pqr \vee$*]. To make precise which expressions are legal, the set of Boolean expressions over a general alphabet is defined recursively.

   I. BASE: Each symbol of the alphabet is a Boolean expression.

  II. RECURSION: If P and Q are Boolean expressions, then so are

$$\text{(a)} \ (P \wedge Q) \quad \text{and} \quad \text{(b)} \ (P \vee Q) \quad \text{and} \quad \text{(c)} \ \sim P.$$

 III. RESTRICTION: There are no Boolean expressions over the alphabet other than those obtained from I and II.

Derive the fact that the following is a Boolean expression over the English alphabet $\{a, b, c, \ldots, x, y, z\}$:

$$(\sim(p \wedge q) \vee (\sim r \wedge p)).$$

Solution   (1) By I, p, q, and r are Boolean expressions.

     (2) By (1) and II(a) and (c), $(p \wedge q)$ and $\sim r$ are Boolean expressions.

     (3) By (2) and II(c) and (a), $\sim(p \wedge q)$ and $(\sim r \wedge p)$ are Boolean expressions.

     (4) By (3) and II(b), $(\sim(p \wedge q) \vee (\sim r \wedge p))$ is a Boolean expression. ∎

Example 8.4.2 The Set of Strings over an Alphabet

Consider the set S of all strings in a's and b's. S is defined recursively as follows:

   I. BASE: ϵ is in S, where ϵ is the null string.

  II. RECURSION: If $s \in S$, then

$$\text{(a)} \ sa \in S \quad \text{and} \quad \text{(b)} \ sb \in S,$$

where sa and sb are the concatenations of s with a and b respectively.

 III. RESTRICTION: Nothing is in S other than objects defined in I and II above.

Derive the fact that $ab \in S$.

Solution   (1) By I, $\epsilon \in S$.

(2) By (1) and II(a), $\epsilon a \in S$. But ϵa is the concatenation of the null string and a, which equals a. So $a \in S$.

(3) By (2) and II(b), $ab \in S$. ■

Example 8.4.3 Sets of Strings with Certain Properties

In *Gödel, Escher, Bach,* Douglas Hofstadter introduces the following recursively defined set of strings of M's, I's, and U's, which he calls the MIU-system*:

I. BASE: MI is in the MIU-system.

II. RECURSION:

a. If xI is in the MIU-system, where x is a string, then xIU is in the MIU-system. (In other words, you can add a U to any string that ends in I. For example, since MI is in the system, so is MIU.)

b. If Mx is in the MIU-system, where x is a string, then Mxx is in the MIU-system. (In other words, you can repeat all the characters in a string that follow an initial M. For example, if MUI is in the system, so is $MUIUI$.)

c. If $xIIIy$ is in the MIU-system, where x and y are strings (possibly null), then xUy is also in the MIU-system. (In other words, you can replace III by U. For example, if $MIIII$ is in the system, so are MIU and MUI.)

d. If $xUUy$ is in the MIU-system, where x and y are strings (possibly null), then xUy is also in the MIU-system. (In other words, you can replace UU by U. For example, if $MIIUU$ is in the system, so is $MIIU$.)

III. RESTRICTION: No strings other than those derived from I and II are in the MIU-system.

Derive the fact that $MUIU$ is in the MIU-system.

Solution   (1) By I, MI is in the MIU-system.

(2) By (1) and II(b), MII is in the MIU-system.

(3) By (2) and II(b), $MIIII$ is in the MIU-system.

(4) By (3) and II(c), MUI is in the MIU-system.

(5) By (4) and II(a), $MUIU$ is in the MIU-system. ■

Example 8.4.4 Parenthesis Structures

Certain configurations of parentheses in algebraic expressions are "legal" [*such as* (())() *and* ()()()], whereas others are not [*such as*)()))) *and* ()))((( ]. Here is a recursive definition to generate the set P of legal configurations of parentheses.

I. BASE: () is in P.

II. RECURSION:

a. If E is in P, so is (E).

b. If E and F are in P, so is EF.

*Douglas Hofstadter, *Gödel, Escher, Bach* (New York: Basic Books), pp. 33–35.

III. RESTRICTION: No configurations of parentheses are in P other than those derived from I and II above.

Derive the fact that $(())()$ is in P.

Solution　(1) By I, $()$ is in P.

(2) By (1) and II(a), $(())$ is in P.

(3) By (2), (1), and II(b), $(())()$ is in P.　■

Proving Properties about Recursively Defined Sets

When a set has been defined recursively, a version of mathematical induction, called **structural induction,** can be used to prove that every object in the set satisfies a given property.

> ### Structural Induction for Recursively Defined Sets
>
> Let S be a set that has been defined recursively, and consider a property that objects in S may or may not possess. To prove that every object in S possesses the property:
>
> 1. Show that each object in the BASE for S satisfies the property;
>
> 2. Show that for each rule in the RECURSION, if the rule is applied to an object in S that satisfies the property, then the object defined by the rule also satisfies the property.
>
> Because no objects other than those obtained through the BASE and RECURSION conditions are contained in S, it must be the case that every object in S possesses the property.

Example 8.4.5 A Property of the Set of Parenthesis Structures

Consider the set P of all grammatical configurations of parentheses defined in Example 8.4.4. Prove that every configuration in P contains an equal number of left and right parentheses.

Solution

Proof (by structural induction): Let the property be the following sentence: "The parenthesis configuration has an equal number of left and right parentheses."

Show that each object in the BASE for P satisfies the property: The only object in the base for P is $()$, which has one left parenthesis and one right parenthesis, so it has an equal number of left and right parentheses.

Show that for each rule in the RECURSION for P, if the rule is applied to an object in P that satisfies the property, then the object defined by the rule also satisfies the property: The recursion for P consists of two rules denoted II(a) and II(b).

Suppose E is a parenthesis configuration that has an equal number of left and right parentheses. When rule II(a) is applied to E, the result is (E), so both the number of left parentheses and the number of right parentheses are increased by one. Since these numbers were equal to start with, they remain equal when each is increased by one.

Suppose E and F are parenthesis configurations with equal numbers of left and right parentheses. Say E has m left and right parentheses, and F has n left and right parentheses. When rule II(b) is applied, the result is EF, which has an equal number, namely $m + n$, of left and right parentheses.

Thus when each rule in the RECURSION is applied to a configuration of parentheses in P with an equal number of left and right parentheses, the result is a configuration with an equal number of left and right parentheses.

Therefore, every structure in P has an equal number of left and right parentheses. ∎

Recursive Definitions of Sum, Product, Union, and Intersection

Addition and multiplication are called *binary operations* because only two numbers can be added or multiplied at a time. Careful definitions of sums and products of more than two numbers use recursion.

• Definition

Given numbers $a_1, a_2, \ldots, a_n$, where n is a positive integer, the **summation from $i = 1$ to n of the a_i**, denoted $\sum_{i=1}^{n} a_i$, is defined as follows:

$$\sum_{i=1}^{1} a_i = a_1 \quad \text{and} \quad \sum_{i=1}^{n} a_i = \left(\sum_{i=1}^{n-1} a_i \right) + a_n, \quad \text{if } n > 1.$$

The **product from $i = 1$ to n of the a_i**, denoted $\prod_{i=1}^{n} a_i$, is defined by

$$\prod_{i=1}^{1} a_i = a_1 \quad \text{and} \quad \prod_{i=1}^{n} a_i = \left(\prod_{i=1}^{n-1} a_i \right) \cdot a_n, \quad \text{if } n > 1.$$

The effect of these definitions is to specify an *order* in which sums and products of more than two numbers are computed. For example,

$$\sum_{i=1}^{4} a_i = \left(\sum_{i=1}^{3} a_i \right) + a_4 = \left(\left(\sum_{i=1}^{2} a_i \right) + a_3 \right) + a_4 = ((a_1 + a_2) + a_3) + a_4.$$

Sometimes these recursive definitions are started at $n = 0$ by decreeing that $\sum_{i=1}^{0} a_i = 0$ and $\prod_{i=1}^{0} a_i = 1$. Before rejecting these definitions as formalistic nonsense, observe that the usual computer algorithms to compute sums and products use them in a very natural way. For instance, to compute the sum of $a[1], a[2], \ldots, a[n]$, one normally writes

$$sum := 0$$
for $k := 1$ **to** n
$$sum := sum + a[k]$$
next k.

The recursive definitions are used with mathematical induction to establish various properties of general finite sums and products.

Example 8.4.6 A Sum of Sums

Prove that for any positive integer n, if $a_1, a_2, \ldots, a_n$ and $b_1, b_2, \ldots, b_n$ are real numbers, then

$$\sum_{i=1}^{n} (a_i + b_i) = \sum_{i=1}^{n} a_i + \sum_{i=1}^{n} b_i.$$

Solution   The proof is by mathematical induction.

Show that the formula holds for $n = 1$: Suppose a_1 and b_1 are real numbers. Then

$$\sum_{i=1}^{1}(a_i + b_i) = a_1 + b_1 \qquad \text{by definition of } \sum$$

$$= \sum_{i=1}^{1} a_i + \sum_{i=1}^{1} b_i \quad \text{also by definition of } \sum.$$

Show that for all integers $k \geq 1$, if the formula holds for $n = k$, then it holds for $n = k + 1$: Suppose $a_1, a_2, \ldots, a_k, a_{k+1}$ and $b_1, b_2, \ldots, b_k, b_{k+1}$ are real numbers and that for some $k \geq 1$

$$\sum_{i=1}^{k}(a_i + b_i) = \sum_{i=1}^{k} a_i + \sum_{i=1}^{k} b_i. \quad \text{This is the inductive hypothesis.}$$

We must show that

$$\sum_{i=1}^{k+1}(a_i + b_i) = \sum_{i=1}^{k+1} a_i + \sum_{i=1}^{k+1} b_i.$$

[*We will show that the left-hand side of this equation equals the right-hand side.*]

But the left-hand side of the equation is

$$\sum_{i=1}^{k+1}(a_i + b_i) = \sum_{i=1}^{k}(a_i + b_i) + (a_{k+1} + b_{k+1}) \qquad \text{by definition of } \sum$$

$$= \left(\sum_{i=1}^{k} a_i + \sum_{i=1}^{k} b_i\right) + (a_{k+1} + b_{k+1}) \quad \text{by inductive hypothesis}$$

$$= \left(\sum_{i=1}^{k} a_i + a_{k+1}\right) + \left(\sum_{i=1}^{k} b_i + b_{k+1}\right) \quad \begin{array}{l}\text{by the associative and cummutative}\\\text{laws of algebra}\end{array}$$

$$= \sum_{i=1}^{k+1} a_i + \sum_{i=1}^{k+1} b_i \qquad \text{by definition of } \sum$$

which equals the right-hand side of the equation. This is what was to be shown. ■

Like sum and product, union and intersection are binary operations, and unions and intersections of more than two sets can be defined recursively.

• Definition

Given sets $A_1, A_2, \ldots, A_n$, where n is a positive integer, the **union of the A_i from $i = 1$ to n,** denoted $\bigcup_{i=1}^{n} A_i$, is defined by

$$\bigcup_{i=1}^{1} A_i = A_1 \quad \text{and} \quad \bigcup_{i=1}^{n} A_i = \left(\bigcup_{i=1}^{n-1} A_i\right) \bigcup A_n.$$

The **intersection of the A_i from $i = 1$ to n,** denoted $\bigcap_{i=1}^{n} A_i$, is defined by

$$\bigcap_{i=1}^{1} A_i = A_1 \quad \text{and} \quad \bigcap_{i=1}^{n} A_i = \left(\bigcap_{i=1}^{n-1} A_i\right) \bigcap A_n.$$

Example 8.4.7 A Generalized De Morgan's Law

Prove that for all integers $n \geq 1$, if $A_1, A_2, \ldots, A_n$ are sets, then

$$\left(\bigcup_{i=1}^{n} A_i \right)^c = \bigcap_{i=1}^{n} (A_i)^c.$$

Solution   The proof is by mathematical induction.

Show that the formula holds for $n = 1$: We must show that

$$\left(\bigcup_{i=1}^{1} A_i \right)^c = \left(\bigcap_{i=1}^{1} (A_i)^c \right).$$

But

$$\left(\bigcup_{i=1}^{1} A_i \right)^c = (A_i)^c = \left(\bigcap_{i=1}^{1} A_i \right)^c.$$

Show that for all integers $k \geq 1$, if the formula is true for $n = k$, then it is true for $n = k + 1$: Suppose that for some integer $k \geq 1$,

$$\left(\bigcup_{i=1}^{k} A_i \right)^c = \bigcap_{i=1}^{k} (A_i)^c. \qquad \text{This is the inductive hypothesis.}$$

We must show that

$$\left(\bigcup_{i=1}^{k+1} A_i \right)^c = \bigcap_{i=1}^{k+1} (A_i)^c.$$

But

$$\left(\bigcup_{i=1}^{k+1} A_i \right)^c = \left(\left(\bigcup_{i=1}^{k} A_i \right) \bigcup A_n \right)^c \qquad \text{by the recursive definition of union}$$

$$= \left(\bigcup_{i=1}^{k} A_i \right)^c \bigcap (A_n)^c \qquad \text{by De Morgan's law for two sets}$$

$$= \left(\bigcap_{i=1}^{k} (A_i)^c \right) \bigcap (A_n)^c \qquad \text{by inductive hypothesis}$$

$$= \left(\bigcap_{i=1}^{k+1} (A_i)^c \right) \qquad \text{by the recursive definition of intersection.}$$

This is what was to be shown. ∎

Recursive Functions

A function is said to be **defined recursively** or to be a **recursive function** if its rule of definition refers to itself. Because of this self-reference, it is sometimes difficult to tell whether a given recursive function is well defined. Recursive functions are of great importance in the theory of computation in computer science.

Example 8.4.8 McCarthy's 91 Function

The following function $M: \mathbf{Z}^+ \to \mathbf{Z}$ was defined by John McCarthy, a pioneer in the theory of computation and in the study of artificial intelligence:

$$M(n) = \begin{cases} n - 10 & \text{if } n > 100 \\ M(M(n+11)) & \text{if } n \le 100 \end{cases}$$

for all positive integers n. Find $M(99)$.

Solution   By repeated use of the definition of M,

$$
\begin{aligned}
M(99) &= M(M(110)) && \text{since } 99 \le 100 \\
&= M(100) && \text{since } 110 > 100 \\
&= M(M(111)) && \text{since } 100 \le 100 \\
&= M(101) && \text{since } 111 > 100 \\
&= 91 && \text{since } 101 > 100
\end{aligned}
$$

John McCarthy
(born 1927)

The remarkable thing about this function is that it takes the value 91 for all positive integers less than or equal to 101. (You are asked to show this in exercise 27 at the end of this section.) Of course, for $n > 101$, $M(n)$ is well defined because it equals $n - 10$. ∎

Example 8.4.9 The Ackermann Function

In the 1920s the German logician and mathematician Wilhelm Ackermann first defined a version of the function that now bears his name. This function is important in computer science because it helps answer the question of what can and what cannot be computed on a computer. It is defined on the set of all pairs of nonnegative integers as follows:

$$
\begin{aligned}
&A(0, n) = n + 1 && \text{for all nonnegative integers } n && 8.4.1 \\
&A(m, 0) = A(m - 1, 1) && \text{for all positive integers } m && 8.4.2 \\
&A(m, n) = A(m - 1, A(m, n - 1)) && \text{for all positive integers } m \text{ and } n && 8.4.3
\end{aligned}
$$

Find $A(1, 2)$.

Wilhelm Ackermann
(1896–1962)

Solution

$$
\begin{aligned}
A(1, 2) &= A(0, A(1, 1)) && \text{by (8.4.3) with } m = 1 \text{ and } n = 2 \\
&= A(0, A(0, A(1, 0))) && \text{by (8.4.3) with } m = 1 \text{ and } n = 1 \\
&= A(0, A(0, A(0, 1))) && \text{by (8.4.2) with } m = 1 \\
&= A(0, A(0, 2)) && \text{by (8.4.1) with } n = 1 \\
&= A(0, 3) && \text{by (8.4.1) with } n = 2 \\
&= 4 && \text{by (8.4.1) with } n = 3.
\end{aligned}
$$

The special properties of the Ackermann function are a consequence of its phenomenal rate of growth. While the values of $A(0, 0) = 1$, $A(1, 1) = 3$, $A(2, 2) = 7$, and $A(3, 3) = 61$ are not especially impressive,

$$A(4, 4) \cong 2^{2^{2^{65536}}}$$

and the values of $A(n, n)$ continue to increase with extraordinary rapidity thereafter. ∎

The argument is somewhat technical, but it is not difficult to show that the Ackermann function is well defined. The following is an example of a recursive "definition" that does not define a function.

Example 8.4.10 A Recursive "Function" That Is Not Well Defined

Consider the following attempt to define a recursive function G from $\mathbf{Z}^+$ to $\mathbf{Z}$. For all integers $n \geq 1$,

$$G(n) = \begin{cases} 1 & \text{if } n \text{ is } 1 \\ 1 + G\left(\dfrac{n}{2}\right) & \text{if } n \text{ is even} \\ G(3n-1) & \text{if } n \text{ is odd and } n > 1 \end{cases}.$$

Is G well defined? Why?

Solution   Suppose G is a function. Then by definition of G,

$G(1) = 1,$

$G(2) = 1 + G(1) = 1 + 1 = 2,$

$G(3) = G(8) = 1 + G(4) = 1 + (1 + G(2)) = 1 + (1 + 2) = 4,$

$G(4) = 1 + G(2) = 1 + 2 = 3.$

However,

$G(5) = G(14) = 1 + G(7) = 1 + G(20)$

$\qquad = 1 + (1 + G(10)) = 1 + (1 + (1 + G(5))) = 3 + G(5).$

Subtracting $G(5)$ from both sides gives $0 = 3$, which is false. Since the supposition that G is a function leads logically to a false statement, it follows that G is not a function. ■

A slight modification of the formula of Example 8.4.10 produces a "function" whose status of definition is unknown. Consider the following formula: For all integers $n \geq 1$,

$$T(n) = \begin{cases} 1 & \text{if } n \text{ is } 1 \\ T\left(\dfrac{n}{2}\right) & \text{if } n \text{ is even} \\ T(3n+1) & \text{if } n \text{ is odd} \end{cases}.$$

In the 1930s, a student, Luther Collatz, became interested in the behavior of a related function g, which is defined as follows: $g(n) = n/2$ if n is even, and $g(n) = 3n + 1$ if n is odd. Collatz conjectured that for any initial positive number n, computation of successive values of $g(n)$, $g^2(n)$, $g^3(n), \ldots$ would eventually produce the number 1. Determining whether this conjecture is true or false is called the **$3n + 1$ problem** (or the **$3x + 1$ problem**). If Collatz's conjecture is true, the formula for T defines a function; if the conjecture is false, T is not well defined. As of the publication of this book the answer is not known, although computer calculation has established that it holds for extremely large values of n.

Exercise Set 8.4

1. Consider the set of Boolean expressions defined in Example 8.4.1. Give derivations showing that each of the following is a Boolean expression over the English alphabet $\{a, b, c, \ldots, x, y, z\}$.
   a. $(\sim p \vee (q \wedge (r \vee \sim s)))$
   b. $((p \vee q) \vee \sim ((p \wedge \sim s) \wedge r))$

2. Let S be defined as in Example 8.4.2. Give derivations showing that each of the following is in S.
   a. aab **b.** bb

3. Consider the MIU-system discussed in Example 8.4.3. Give derivations showing that each of the following is in the MIU-system.
   a. $MIUI$ **b.** $MUIIU$

$H \ast$ **4.** Is the string MU in the MIU-system?

5. Consider the set P of parenthesis structures defined in Example 8.4.4. Give derivations showing that each of the following is in P.
   a. $(\,)((\,))$ **b.** $((\,))((\,))$

$\ast$ **6.** Determine whether either of the following parenthesis structures is in the set P defined in Example 8.4.4.
   a. $(\,)((\,)$ **b.** $((\,)(\,)))((\,)$

7. The set of arithmetic expressions over the real numbers can be defined recursively as follows:

   I. BASE: Each real number r is an arithmetic expression.

   II. RECURSION: If u and v are arithmetic expressions, then the following are also arithmetic expressions:
     a. $(+u)$ **b.** $(-u)$ **c.** $(u+v)$
     d. $(u-v)$ **e.** $(u \cdot v)$ **f.** $\left(\dfrac{u}{v}\right)$

   III. RESTRICTION: There are no arithmetic expressions over the real numbers other than those obtained from I and II.

   (Note that the *expression* $\left(\dfrac{u}{v}\right)$ is legal even though the value of v may be 0.) Give derivations showing that each of the following is an arithmetic expression.
   a. $((2 \cdot (0.3 - 4.2)) + (-7))$ **b.** $\left(\dfrac{(9 \cdot (6.1 + 2))}{((4-7) \cdot 6)}\right)$

8. Define a set S recursively as follows:

   I. BASE: $1 \in S$

   II. RECURSION: If $s \in S$, then
     a. $0s \in S$ **b.** $1s \in S$

   III. RESTRICTION: Nothing is in S other than objects defined in I and II above.

   Use structural induction to prove that every string in S ends in a 1.

9. Define a set S recursively as follows:

   I. BASE: $a \in S$

   II. RECURSION: If $s \in S$, then,
     a. $sa \in S$ **b.** $sb \in S$

   III. RESTRICTION: Nothing is in S other than objects defined in I and II above.

   Use structural induction to prove that every string in S begins with an a.

10. Define a set S recursively as follows:

   I. BASE: $\epsilon \in S$

   II. RECURSION: If $s \in S$, then
     a. $bs \in S$ **b.** $sb \in S$
     c. $saa \in S$ **d.** $aas \in S$

   III. RESTRICTION: Nothing is in S other than objects defined in I and II above.

Use structural induction to prove that every string in S contains an even number of a's.

11. Define a set S recursively as follows:

   I. BASE: $1 \in S$, $2 \in S$, $3 \in S$, $4 \in S$, $5 \in S$, $6 \in S$, $7 \in S$, $8 \in S$, $9 \in S$

   II. RECURSION: If $s \in S$ and $t \in S$, then
     a. $s0 \in S$ **b.** $st \in S$

   III. RESTRICTION: Nothing is in S other than objects defined in I and II above.

   Use structural induction to prove that no string in S represents an integer with a leading zero.

H **12.** Define a set S recursively as follows:

   I. BASE: $1 \in S, 3 \in S, 5 \in S, 7 \in S, 9 \in S$

   II. RECURSION: If $s \in S$ and $t \in S$ then
     a. $st \in S$ **b.** $2s \in S$ **c.** $4s \in S$
     d. $6s \in S$ **e.** $8s \in S$

   III. RESTRICTION: Nothing is in S other than objects defined in I and II above.

   Use structural induction to prove that every string in S represents an odd integer.

H **13.** Define a set S recursively as follows:

   I. BASE: $0 \in S, 5 \in S$

   II. RECURSION: If $s \in S$ and $t \in S$ then
     a. $s+t \in S$ **b.** $s-t \in S$

   III. RESTRICTION: Nothing is in S other than objects defined in I and II above.

   Use structural induction to prove that every integer in S is divisible by 5.

14. Define a set S recursively as follows:

   I. BASE: $0 \in S$

   II. RECURSION: If $s \in S$, then
     a. $s+3 \in S$ **b.** $s-3 \in S$

   III. RESTRICTION: Nothing is in S other than objects defined in I and II above.

   Use structural induction to prove that every integer in S is divisible by 3.

15. Give a recursive definition for the set of all strings of 0's and 1's that have the same number of 0's as 1's.

16. Give a recursive definition for the set of all strings of 0's and 1's for which all the 0's precede all the 1's.

17. Give a recursive definition for the set of all strings of a's and b's that contain an odd number of a's.

18. Give a recursive definition for the set of all strings of a's and b's that contain exactly one a.

19. Use the recursive definition of summation, together with mathematical induction, to prove the generalized distributive law that for all positive integers n, if $a_1, a_2, \ldots, a_n$ and c are real numbers, then

$$\sum_{i=1}^{n} ca_i = c \left(\sum_{i=1}^{n} a_i \right).$$

20. Use the recursive definition of product, together with mathematical induction, to prove that for all positive integers n, if $a_1, a_2, \ldots, a_n$ and $b_1, b_2, \ldots, b_n$ are real numbers, then

$$\prod_{i=1}^{n} (a_i b_i) = \left(\prod_{i=1}^{n} a_i \right) \left(\prod_{i=1}^{n} b_i \right).$$

21. Use the recursive definition of product, together with mathematical induction, to prove that for all positive integers n, if $a_1, a_2, \ldots, a_n$ and c are real numbers, then

$$\prod_{i=1}^{n} (ca_i) = c^n \left(\prod_{i=1}^{n} a_i \right).$$

H **22.** The triangle inequality for absolute value states that for all real numbers a and b, $|a + b| \leq |a| + |b|$. Use the recursive definition of summation, the triangle inequality, the definition of absolute value, and mathematical induction to prove that for all positive integers n, if $a_1, a_2, \ldots, a_n$ are real numbers, then

$$\left| \sum_{i=1}^{n} a_i \right| \leq \sum_{i=1}^{n} |a_i|.$$

23. Use the recursive definitions of union and intersection to prove the following general distributive law: For all positive integers, n, if A and $B_1, B_2, \ldots, B_n$ are sets, then

$$A \cap \left(\bigcup_{i=1}^{n} B_i \right) = \bigcup_{i=1}^{n} (A \cap B_i).$$

24. Use the recursive definitions of union and intersection to prove the following general distributive law: For all positive integers n, if A and $B_1, B_2, \ldots, B_n$ are sets, then

$$A \cup \left(\bigcap_{i=1}^{n} B_i \right) = \bigcap_{i=1}^{n} (A \cup B_i).$$

25. Use the recursive definitions of union and intersection to prove the following general De Morgan's law: For all positive integers n, if $A_1, A_2, \ldots, A_n$ are sets, then

$$\left(\bigcap_{i=1}^{n} A_i \right)^c = \bigcup_{i=1}^{n} (A_i)^c.$$

26. Use the definition of McCarthy's 91 function in Example 8.4.8 to show the following:
   a. $M(86) = M(91)$ **b.** $M(91) = 91$

★ **27.** Prove that McCarthy's 91 function equals 91 for all positive integers less than or equal to 101.

28. Use the definition of the Ackermann function in Example 8.4.9 to compute the following:
   a. $A(1, 1)$ **b.** $A(2, 1)$

29. Use the definition of the Ackermann function to show the following:
   a. $A(1, n) = n + 2$, for all nonnegative integers n.
   b. $A(2, n) = 3 + 2n$, for all nonnegative integers n.
   c. $A(3, n) = 8 \cdot 2^n - 3$, for all nonnegative integers n.

30. Compute $T(2), T(3), T(4), T(5), T(6)$, and $T(7)$ for the "function" T defined after Example 8.4.10.

31. Student A tries to define a function $F: \mathbf{Z}^+ \to \mathbf{Z}$ by the rule

$$F(n) = \begin{cases} 1 & \text{if } n \text{ is 1} \\ F\left(\dfrac{n}{2}\right) & \text{if } n \text{ is even} \\ 1 + F(5n - 9) & \text{if } n \text{ is odd and } n > 1 \end{cases}$$

for all integers $n \geq 1$. Student B claims that F is not well defined. Justify student B's claim.

32. Student C tries to define a function $G: \mathbf{Z}^+ \to \mathbf{Z}$ by the rule

$$G(n) = \begin{cases} 1 & \text{if } n \text{ is 1} \\ G\left(\dfrac{n}{2}\right) & \text{if } n \text{ is even} \\ 2G(3n - 2) & \text{if } n \text{ is odd and } n > 1 \end{cases}$$

for all integers $n \geq 1$. Student D claims that G is not well defined. Justify student D's claim.

CHAPTER 9

THE EFFICIENCY
OF ALGORITHMS

*René Descartes
(1596–1650)*

In 1637 the French mathematician and philosopher René Descartes published his great philosophical work *Discourse on Method.* An appendix to this work, called "Geometry," laid the foundation for the subject of analytic geometry, in which geometric methods are applied to the study of algebraic objects, such as functions, equations, and inequalities, and algebraic methods are used to study geometric objects, such as straight lines, circles, and half-planes.

The analytic geometry of Descartes provides the foundation for the main topic of this chapter: the big-*O*, big-Omega, and big-Theta notations and their application to the analysis of algorithms. In Section 9.1 we briefly discuss certain properties of graphs of real-valued functions of a real variable that are needed to understand these notations. In Section 9.2 we define the notations and apply them to power and polynomial functions, and in Section 9.3 we show how the notations are used to study the efficiency of algorithms. Because the analysis of algorithms often involves logarithmic and exponential functions, we develop the needed properties of these functions in Section 9.4 and use them to analyze several algorithms in Section 9.5.

9.1 Real-Valued Functions of a Real Variable and Their Graphs

The first precept was never to accept a thing as true until I knew it as such without a single doubt — René Descartes, 1637

A **Cartesian plane** or **two-dimensional Cartesian coordinate system** is a pictorial representation of $\mathbf{R} \times \mathbf{R}$ obtained by setting up a one-to-one correspondence between ordered pairs of real numbers and points in a Euclidean plane. To obtain it, two perpendicular lines, called the **horizontal** and **vertical axes,** are drawn in the plane. Their point of intersection is called the **origin,** and a unit of distance is chosen for each axis. An ordered pair (x, y) of real numbers corresponds to the point P that lies $|x|$ units to the right or left of the vertical axis and $|y|$ units above or below the horizontal axis. On each axis the positive direction is marked with an arrow.

A **real-valued function of a real variable** is a function from one set of real numbers to another. If f is such a function, then for each real number x in the domain of f, there is a unique corresponding real number $f(x)$. Thus it is possible to define the *graph of f* as follows:

510

> **• Definition**
>
> Let f be a real-valued function of a real variable. The **graph of f** is the set of all points (x, y) in the Cartesian coordinate plane with the property that x is in the domain of f and $y = f(x)$.

The definition of graph (see Figure 9.1.1) means that for all x in the domain of f:

$$y = f(x) \quad \Leftrightarrow \quad \text{the point } (x, y) \text{ lies on the graph of } f.$$

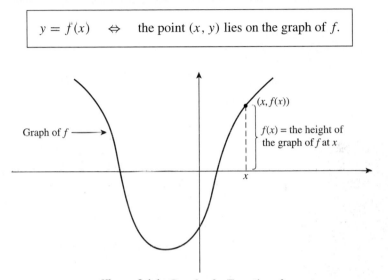

Figure 9.1.1 Graph of a Function f

Note that if $f(x)$ can be written as an algebraic expression in x, the graph of the function f is the same as the graph of the equation $y = f(x)$ where x is restricted to lie in the domain of f.

Power Functions

A function that sends a real number x to a particular power, x^a, is called a *power function*. For applications in computer science, we are almost invariably concerned with situations where x and a are nonnegative, and so we restrict our definition to these cases.

> **• Definition**
>
> Let a be any nonnegative number. Define p_a, the **power function with exponent a,** as follows:
>
> $$p_a(x) = x^a \quad \text{for each nonnegative real number } x.$$

Example 9.1.1 Graphs of Power Functions

Plot the graphs of the power functions p_0, $p_{1/2}$, p_1, and p_2 on the same coordinate axes.

Solution Because the power function with exponent zero satisfies $p_0(x) = x^0 = 1$ for all nonnegative numbers x,* all points of the form $(x, 1)$ lie on the graph of p_0 for all such x. So

*As in Section 6.7 (see page 364), we simply define $0^0 = 1$.

the graph is just a horizontal half-line of height 1 lying above the horizontal axis. Similarly, $p_1(x) = x$ for all nonnegative numbers x, and so the graph of p_1 consists of all points of the form (x, x) where x is nonnegative. The graph is therefore the half-line of slope 1 that emanates from $(0, 0)$.

Since for each nonnegative number x, $p_{1/2}(x) = x^{1/2} = \sqrt{x}$, any point with coordinates $(x, \sqrt{x})$, where x is nonnegative, is on the graph of $p_{1/2}$. For instance, the graph of $p_{1/2}$ contains the points $(0, 0)$, $(1, 1)$, $(4, 2)$, and $(9, 3)$. Similarly, since $p_2(x) = x^2$, any point with coordinates (x, x^2) lies on the graph of p_2. Thus, for instance, the graph of p_2 contains the points $(0, 0)$, $(1, 1)$, $(2, 4)$, and $(3, 9)$.

The graphs of all four functions are shown in Figure 9.1.2.

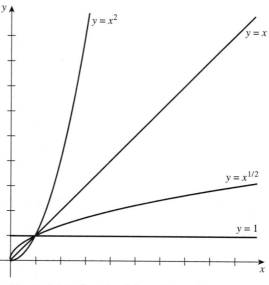

Figure 9.1.2 Graphs of Some Power Functions   ■

The Floor Function

The floor and ceiling functions arise in many computer science contexts. Example 9.1.2 illustrates the graph of the floor function. In exercise 5 at the end of this section you are asked to draw the graph of the ceiling function.

Example 9.1.2 Graph of the Floor Function

Recall that each real number either is an integer itself or sits between two consecutive integers: For each real number x, there exists a unique integer n such that $n \leq x < n + 1$. The floor of a number is the integer immediately to its left on the number line. More formally, the floor function F is defined by the rule

For each real number x,

$$F(x) = \lfloor x \rfloor$$

$$= \text{the greatest integer that is less than or equal to } x$$

$$= \text{the unique integer } n \text{ such that } n \leq x < n + 1.$$

Graph the floor function.

Solution   If n is any integer, then for each real number x in the interval $n \leq x < n + 1$, the floor of x, $\lfloor x \rfloor$, equals n. Thus on each such interval, the graph of the floor function is horizontal; for each x in the interval, the height of the graph is n.

It follows that the graph of the floor function consists of horizontal line segments, like a staircase, as shown in Figure 9.1.3. The open circles at the right-hand edge of each step are used to show that those points are *not* on the graph.

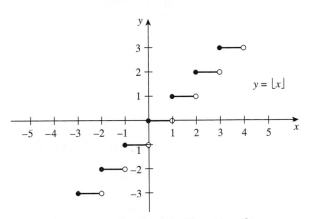

Figure 9.1.3 Graph of the Floor Function

Graphing Functions Defined on Sets of Integers

Many real-valued functions used in computer science are defined on sets of integers and not on intervals of real numbers. Suppose you know what the graph of a function looks like when it is given by a certain formula on an interval of real numbers. You can obtain the graph of the function defined by the same formula on the integers in the interval by selecting out only those points on the known graph with integers as their first coordinates. For instance, if f is the function defined by the same formula as the power function p_1 but having as its domain the set of nonnegative integers, then $f(n) = n$ for all nonnegative integers n. The graphs of p_1, reproduced from Example 9.1.2, and f are shown side-by-side below.

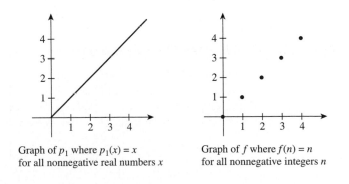

Graph of p_1 where $p_1(x) = x$
for all nonnegative real numbers x

Graph of f where $f(n) = n$
for all nonnegative integers n

Example 9.1.3 Graph of a Function Defined on a Set of Integers

Consider an integer version of the power function $p_{1/2}$. In other words, define a function g by the formula $g(n) = n^{1/2}$ for all nonnegative integers n. Draw the graph of g.

Solution   Look back at the graph of $p_{1/2}$ in Figure 9.1.2. Draw the graph of g by reproducing only those points on the graph of $p_{1/2}$ with integer first coordinates. Thus for each nonnegative integer n, the point $(n, n^{1/2})$ is on the graph of g.

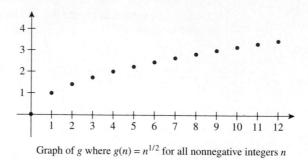

Graph of g where $g(n) = n^{1/2}$ for all nonnegative integers n

Graph of a Multiple of a Function

A *multiple* of a function is obtained by multiplying every value of the function by a fixed number. To understand the concept of O-notation, it is helpful to understand the relation between the graph of a function and the graph of a multiple of the function.

> • **Definition**
>
> Let f be a real-valued function of a real variable and let M be any real number. The function Mf, called the **multiple of f by M or M times f,** is the real-valued function with the same domain as f that is defined by the rule
>
> $$(Mf)(x) = Mf(x) \quad \text{for all } x \in \text{domain of } f.$$

If the graph of a function is known, the graph of any multiple can easily be deduced. Specifically, if f is a function and M is a real number, the height of the graph of Mf at any real number x is M times the quantity $f(x)$. To sketch the graph of Mf from the graph of f, you plot the heights $Mf(x)$ on the basis of knowledge of M and visual inspection of the heights $f(x)$.

Example 9.1.4 Graph of a Multiple of a Function

Let f be the function whose graph is shown below. Sketch the graph of $2f$.

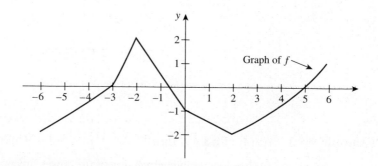

Solution   At each real number x, you obtain the height of the graph of $2f$ by measuring the height of the graph of f at x and multiplying that number by 2. The result is the following graph. Note that the general shapes of f and $2f$ are very similar, but the graph of $2f$ is "stretched out": the "highs" are twice as high and the "lows" are twice as low.

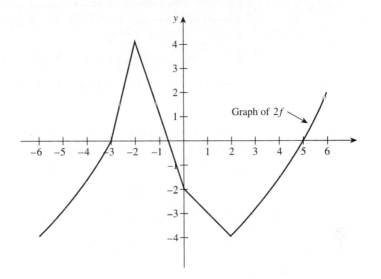

Increasing and Decreasing Functions

Consider the *absolute value function, A,* which is defined as follows:

$$A(x) = |x| = \begin{cases} x & \text{if } x \geq 0 \\ -x & \text{if } x < 0 \end{cases} \quad \text{for all real numbers } x.$$

When $x \geq 0$, the graph of A is the same as the graph of $y = x$, the straight line with slope 1 that passes through the origin $(0, 0)$. For $x < 0$, the graph of A is the same as the graph of $y = -x$, which is the straight line with slope -1 that passes through $(0, 0)$. (See Figure 9.1.4.)

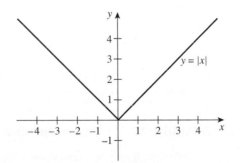

Figure 9.1.4 Graph of the Absolute Value Function

Note that as you trace from left to right along the graph to the left of the origin, the height of the graph continually *decreases*. For this reason, the absolute value function is said to be *decreasing* on the set of real numbers less than 0. On the other hand, as you trace from left to right along the graph to the right of the origin, the height of the graph continually *increases*. Consequently, the absolute value function is said to be *increasing* on the set of real numbers greater than 0.

Since the height of the graph of a function f at a point x is $f(x)$, these geometric concepts translate to the following analytic definition.

> **• Definition**
>
> Let f be a real-valued function defined on a set of real numbers, and suppose the domain of f contains a set S. We say that f is **increasing on the set S** if, and only if,
>
> for all real numbers x_1 and x_2 in S, if $x_1 < x_2$ then $f(x_1) < f(x_2)$.
>
> We say that f is **decreasing on the set S** if, and only if,
>
> for all real numbers x_1 and x_2 in S, if $x_1 < x_2$ then $f(x_1) > f(x_2)$.
>
> We say that f is an **increasing** (or **decreasing**) **function** if, and only if, f is increasing (or decreasing) on its entire domain.

Figure 9.1.5 illustrates the analytic definitions of increasing and decreasing.

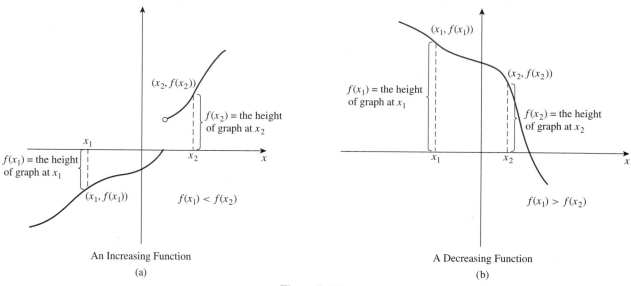

An Increasing Function
(a)

A Decreasing Function
(b)

Figure 9.1.5

It follows almost immediately from the definitions that both increasing functions and decreasing functions are one-to-one. You are asked to show this in the exercises.

Example 9.1.5 A Positive Multiple of an Increasing Function Is Increasing

Suppose that f is a real-valued function of a real variable that is increasing on a set S of real numbers, and suppose M is any positive real number. Show that Mf is also increasing on S.

Solution   Suppose x_1 and x_2 are particular but arbitrarily chosen elements of S such that

$$x_1 < x_2.$$

[*We must show that* $(Mf)(x_1) < (Mf)(x_2)$.] From the facts that $x_1 < x_2$ and f is increasing, it follows that

$$f(x_1) < f(x_2).$$

Then

$$Mf(x_1) < Mf(x_2),$$

since multiplying both sides of the inequality by a positive number does not change the direction of the inequality. Hence, by definition of Mf,

$$(Mf)(x_1) < (Mf)(x_2),$$

and, consequently, Mf is increasing on S. ∎

It is also true that a positive multiple of a decreasing function is decreasing, that a negative multiple of a increasing function is decreasing, and that a negative multiple of a decreasing function is increasing. The proofs of these facts are left to the exercises.

Exercise Set 9.1*

1. The graph of a function f is shown below.
 a. Is $f(0)$ positive or negative?
 b. For what values of x does $f(x) = 0$?
 c. As x increases from -3 to -1, do the values of f increase or decrease?
 d. As x increases from 0 to 4, do the values of f increase or decrease?

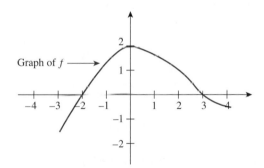

2. Draw the graphs of the power functions $p_{1/3}$ and $p_{1/4}$ on the same set of axes. When $0 < x < 1$, which is greater: $x^{1/3}$ or $x^{1/4}$? When $x > 1$, which is greater: $x^{1/3}$ or $x^{1/4}$?

3. Draw the graphs of the power functions p_3 and p_4 on the same set of axes. When $0 < x < 1$, which is greater: x^3 or x^4? When $x > 1$, which is greater: x^3 or x^4?

4. Draw the graphs of $y = 2\lfloor x \rfloor$ and $y = \lfloor 2x \rfloor$ for all real numbers x. What can you conclude from these graphs?

Graph each of the functions defined in 5–8 below.

5. $g(x) = \lceil x \rceil$ for all real numbers x (Recall that the ceiling of x, $\lceil x \rceil$, is the least integer that is greater than or equal to x. That is, $\lceil x \rceil$ = the unique integer n such that $n - 1 < x \le n$.)

6. $h(x) = \lceil x \rceil - \lfloor x \rfloor$ for all real numbers x

7. $F(x) = \lfloor x^{1/2} \rfloor$ for all real numbers x

8. $G(x) = x - \lfloor x \rfloor$ for all real numbers x

In each of 9–12 a function is defined on a set of integers. Graph each function.

9. $f(n) = |n|$ for each integer n

10. $g(n) = (n/2) + 1$ for each integer n

11. $h(n) = \lfloor n/2 \rfloor$ for each integer $n \ge 0$

12. $k(n) = \lfloor n^{1/2} \rfloor$ for each integer $n \ge 0$

13. The graph of a function f is shown below. Find the intervals on which f is increasing and the intervals on which f is decreasing.

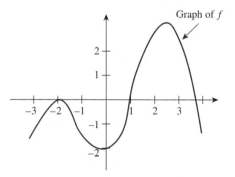

14. Show that the function $f: \mathbf{R} \to \mathbf{R}$ defined by the formula $f(x) = 2x - 3$ is increasing on the set of all real numbers.

15. Show that the function $g: \mathbf{R} \to \mathbf{R}$ defined by the formula $g(x) = -(x/3) + 1$ is decreasing on the set of all real numbers.

16. Let h be the function from $\mathbf{R}$ to $\mathbf{R}$ defined by the formula $h(x) = x^2$ for all real numbers x.
 a. Show that h is decreasing on the set of all real numbers less than zero.
 b. Show that h is increasing on the set of all real numbers greater than zero.

*For exercises with blue numbers or letters, solutions are given in Appendix B. The symbol **H** indicates that only a hint or a partial solution is given. The symbol ✳ signals that an exercise is more challenging than usual.

17. Let $k: \mathbf{R} \rightarrow \mathbf{R}$ be the function defined by the formula $k(x) = (x - 1)/x$ for all real numbers $x \neq 0$.
    a. Show that k is increasing for all real numbers $x > 0$.
    b. Is k increasing or decreasing for $x < 0$? Prove your answer.

18. Show that if a function $f: \mathbf{R} \rightarrow \mathbf{R}$ is increasing, then f is one-to-one.

19. Given real-valued functions f and g with the same domain D, the sum of f and g, denoted $f + g$, is defined as follows:

    For all real numbers x, $(f + g)(x) = f(x) + g(x)$.

    Show that if f and g are both increasing on a set S, then $f + g$ is also increasing on S.

20. **a.** Let m be any positive integer, and define $f(x) = x^m$ for all nonnegative real numbers x. Use the binomial theorem to show that f is an increasing function.
    b. Let m and n be any positive integers, and let $g(x) = x^{m/n}$ for all nonnegative real numbers x. Prove that g is an increasing function.
    The results of this exercise are used in the exercises for Sections 9.2 and 9.4.

21. Let f be the function whose graph is shown below. Draw the graph of $3f$.

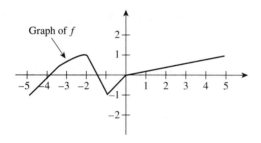

22. Let h be the function whose graph is shown below. Draw the graph of $2h$.

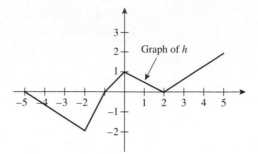

23. Let f be a real-valued function of a real variable. Show that if f is decreasing on a set S and if M is any positive real number, then Mf is decreasing on S.

24. Let f be a real-valued function of a real varaible. Show that if f is increasing on a set S and if M is any negative real number, then Mf is decreasing on S.

25. Let f be a real-valued function of a real variable. Show that if f is decreasing on a set S and if M is any negative real number, then Mf is increasing on S.

In 26 and 27, functions f and g are defined. In each case draw the graphs of f and $2g$ on the same set of axes and find a number x_0 so that $f(x) \leq 2g(x)$ for all $x > x_0$. You can find x_0 by solving a quadratic equation, or you can find an approximate value for x_0 by using the zoom and trace features on a graphing calculator.

26. $f(x) = x^2 + 10x + 11$ and $g(x) = x^2$ for all real numbers $x \geq 0$

27. $f(x) = 2x^2 + 125x + 254$ and $g(x) = 3x^2$ for all real numbers $x \geq 0$

9.2 O, Ω, and Θ Notations

Although this may seem a paradox, all exact science is dominated by the idea of approximation. — Bertrand Russell, 1872–1970

It often happens that any one of several algorithms could be used to do a certain job but the time or memory space they require varies dramatically. The O-, Ω-, and Θ-notations provide approximations that make it easy to evaluate large-scale differences in algorithm efficiency, while ignoring differences of a constant factor and differences that occur only for small sets of input data.

The oldest of the notations, O-notation (read "big-O notation"), was introduced by the German mathematician Paul Bachmann in 1894 in a book on analytic number theory. Both the Ω- (read "big-Omega") and Θ- (read "big-Theta") notations were developed by Donald Knuth, one of the pioneers of the science of computer programming.

The idea of the notations is this. Suppose f and g are real-valued functions of a real variable x.

1. If, for sufficiently large values of x, the values of $|f|$ are less than those of a multiple of $|g|$, then f is of order *at most* g, or $f(x)$ is $O(g(x))$.

2. If, for sufficiently large values of x, the values of $|f|$ are greater than those of a multiple of $|g|$, then f is of order *at least* g, or $f(x)$ is $\Omega(g(x))$.

3. If, for sufficiently large values of x, the values of $|f|$ are bounded both above and below by those of multiples of $|g|$, then f is of order g, or $f(x)$ is $\Theta(g(x))$.

These relationships are illustrated in Figure 9.2.1.

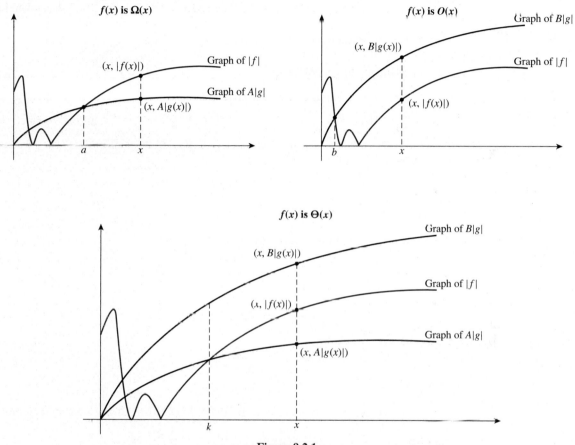

Figure 9.2.1

• Definition

Let f and g be real-valued functions defined on the same set of nonnegative real numbers. Then

1. **f is of order at least g,** written **$f(x)$ is $\Omega(g(x))$,** if, and only if, there exist positive real numbers A and a such that

$$A|g(x)| \leq |f(x)| \quad \text{for all real numbers } x > a.$$

2. **f is of order at most g,** written **$f(x)$ is $O(g(x))$,** if, and only if, there exist positive real numbers B and b such that

$$|f(x)| \leq B|g(x)| \quad \text{for all real numbers } x > b.$$

3. **f is of order g,** written **$f(x)$ is $\Theta(g(x))$,** if, and only if, there exist positive real numbers A, B, and k such that

$$A|g(x)| \leq |f(x)| \leq B|g(x)| \quad \text{for all real numbers } x > k.$$

Remark on Notation: In Section 7.1 we stated that we would generally make a careful distinction between a function f and its value $f(x)$. The traditional use of the order notation violates this general rule. For instance, in the statement "$f(x)$ is $\Theta(g(x))$," the symbols $f(x)$ and $g(x)$ are understood to refer to the functions f and g defined by the expressions $f(x)$ and $g(x)$, respectively. Thus the statement

$$3\sqrt{x} + 4 \text{ is } \Theta(x^{1/2})$$

means that f is of order g where f and g are defined by $f(x) = 3\sqrt{x} + 4$ and $g(x) = x^{1/2}$ with some common domain (usually the largest set of nonnegative real numbers for which both function formulas are defined).

Example 9.2.1 Translating to Θ-Notation

Use Θ-notation to express the statement

$$10|x^6| \leq |17x^6 - 45x^3 + 2x + 8| \leq 30|x^6| \quad \text{for all real numbers } x > 2.$$

Solution Let $A = 10$, $B = 30$, and $k = 2$. By definition of Θ-notation, the statement translates to

$$17x^6 - 45x^3 + 2x + 8 \text{ is } \Theta(x^6). \qquad \blacksquare$$

Example 9.2.2 Translating to O- and Ω-Notations

a. Use Ω and O notations to express the statements

(i) $15|\sqrt{x}| \leq \left| \dfrac{15\sqrt{x}(2x + 9)}{x + 1} \right|$, for all real numbers $x > 0$.

(ii) $\left| \dfrac{15\sqrt{x}(2x + 9)}{x + 1} \right| \leq 45|\sqrt{x}|$, for all real numbers $x > 7$.

b. Justify the statement $\dfrac{15\sqrt{x}(2x + 9)}{x + 1}$ is $\Theta(\sqrt{x})$.

Solution

a. (i) Let $A = 15$ and $a = 0$. By definition of Ω-notation, the statement translates to

$$\frac{15\sqrt{x}(2x + 9)}{x + 1} \text{ is } \Omega(\sqrt{x}).$$

(ii) Let $B = 45$ and $b = 7$. By definition of O-notation, the statement translates to

$$\frac{15\sqrt{x}(2x + 9)}{x + 1} \text{ is } O(\sqrt{x}).$$

b. Let $A = 10$, $B = 30$, and let k be the larger of 0 and 7. Then when $x > k$, both inequalities in a(i) and a(ii) are satisfied, and so

$$A|\sqrt{x}| \leq \left| \frac{15\sqrt{x}(2x + 9)}{x + 1} \right| \leq B|\sqrt{x}| \quad \text{for all real numbers } x > k.$$

Hence by definition of Θ-notation, $\dfrac{15\sqrt{x}(2x + 9)}{x + 1}$ is $\Theta(\sqrt{x})$. $\qquad \blacksquare$

Part (b) of Example 9.2.2 illustrates the fact that if you know both that f is of order at most g and that f is of order at least g, then you may take x_0 to be the larger of the numbers a and b promised in the definitions for big-Omega and big-O and conclude that f is of order g. Conversely, if f is of order g, then both a and b may be taken to be the number x_0 promised in the definition for big-Theta to show that f is of order at most g and f is of order at least g. These results are stated formally in the following theorem, along with a number of other useful properties of the notations.

Theorem 9.2.1 Properties of *O*-, Ω-, and Θ-Notations

Let f and g be real-valued functions defined on the same set of nonnegative real numbers.

1. $f(x)$ is $\Omega(g(x))$ and $f(x)$ is $O(g(x))$ if, and only if $f(x)$ is $\Theta(g(x))$.

2. $f(x)$ is $\Omega(g(x))$ if, and only if, $g(x)$ is $O(f(x))$.

3. $f(x)$ is $O(f(x))$, $f(x)$ is $\Omega(f(x))$, and $f(x)$ is $\Theta(f(x))$.

4. If $f(x)$ is $O(g(x))$ and $g(x)$ is $O(h(x))$, then $f(x)$ is $O(h(x))$.

5. If $f(x)$ is $O(g(x))$ and c is any nonzero real number, then $cf(x)$ is $O(g(x))$.

6. If $f(x)$ is $O(h(x))$ and $g(x)$ is $O(k(x))$, then $f(x) + g(x)$ is $O(G(x))$ where $G(x) = \max(|h(x)|, |k(x)|)$ for each x in the domain of the functions.

7. If $f(x)$ is $O(h(x))$ and $g(x)$ is $O(k(x))$, then $f(x)g(x)$ is $O(h(x)k(x))$.

Proof:

1. The proof of this property was given before the statement of the theorem.

2. We first show that if $f(x)$ is $\Omega(g(x))$, then $g(x)$ is $O(f(x))$. Thus, suppose $f(x)$ is $\Omega(g(x))$. By definition of Ω-notation, there exist positive real numbers A and a such that

$$A|g(x)| \leq |f(x)| \quad \text{for all real numbers } x > a.$$

Divide both sides by A to obtain

$$|g(x)| \leq \frac{1}{A}|f(x)| \quad \text{for all real numbers } x > a.$$

Let $B = 1/A$ and $b = a$. Then

$$|g(x)| \leq B|f(x)| \quad \text{for all real numbers } x > b,$$

and so $g(x)$ is $O(f(x))$ by definition of O-notation. The proof that if $g(x)$ is $O(f(x))$ then $f(x)$ is $\Omega(g(x))$ is left as an exercise.

4. Suppose $f(x)$ is $O(g(x))$ and $g(x)$ is $O(h(x))$. By definition of O-notation, there exist positive real numbers B_1, b_1, B_2, and b_2 such that

$$|f(x)| \leq B_1|g(x)| \quad \text{for all real numbers } x > b_1,$$

and

$$|g(x)| \leq B_2|h(x)| \quad \text{for all real numbers } x > b_2.$$

(If $B_2 < 1$, we can redefine it to equal 1 and the inequality will still hold.) Let $B = B_1 B_2$ and b be the greater of b_1 and b_2. Then if $x > b$,

$$|f(x)| \leq B_1|g(x)| \leq B_1(B_2|h(x)|) \leq B|h(x)|.$$

Thus, by definition of O-notation, $f(x)$ is $O(h(x))$.

3, 5–7. The proofs of these properties are left as exercises.

Orders of Power Functions

Observe that if

$$1 < x,$$

then

$$x < x^2 \qquad \text{multiplying both sides by } x \text{ (which is positive)}$$

and so

$$x^2 < x^3 \qquad \text{multiplying again by } x.$$

Thus if $1 < x$, then

$$1 < x < x^2 < x^3 \qquad \text{by transitivity of } <.$$

The following generalization of this result is developed in exercises 15 and 50 at the end of this section.

> For any rational numbers r and s,
> $$\text{if } x > 1 \text{ and } r < s, \text{ then } x^r < x^s. \qquad\qquad 9.2.1$$

Property (9.2.1) has the following consequence for orders.

> For any rational numbers r and s,
> $$\text{if } r < s, \text{ then } x^r \text{ is } O(x^s). \qquad\qquad 9.2.2$$

The relation among the graphs of various positive power functions of x for $x \geq 1$ is shown graphically in Figure 9.2.2.

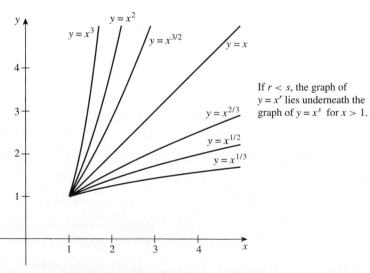

If $r < s$, the graph of $y = x^r$ lies underneath the graph of $y = x^s$ for $x > 1$.

Figure 9.2.2 Graphs of Powers of x for $x \geq 1$

Orders of Polynomial Functions

The following example shows how to use property (9.2.1) to derive a polynomial inequality.

Example 9.2.3 A Polynomial Inequality

Show that for any real number x,

$$\text{if } x > 1, \quad \text{then} \quad 3x^3 + 2x + 7 \le 12x^3.$$

Solution   Suppose x is a real number and $x > 1$. Then by property (9.2.1),

$$x < x^3 \quad \text{and} \quad 1 < x^3.$$

Multiply the left-hand inequality by 2 and the right-hand inequality by 7 to get

$$2x < 2x^3 \quad \text{and} \quad 7 < 7x^3.$$

Now add $3x^3 \le 3x^3$, $2x < 2x^3$, and $7 < 7x^3$ to obtain

$$3x^3 + 2x + 7 \le 3x^3 + 2x^3 + 7x^3 = 12x^3. \quad \blacksquare$$

The method of Example 9.2.3 is used in the next example (more compactly) to show that a polynomial function has a certain order.

Example 9.2.4 Using the Definitions to Show That a Polynomial Function with Positive Coefficients Has a Certain Order

Use the definitions of big-Omega, big-O, and big-Theta to show that $2x^4 + 3x^3 + 5$ is $\Theta(x^4)$.

Solution   Define functions f and g as follows. For all nonnegative real numbers x,

$$f(x) = 2x^4 + 3x^3 + 5, \text{ and}$$
$$g(x) = x^4.$$

Observe that for all real numbers $x > 0$,

$$2x^4 \le 2x^4 + 3x^3 + 5 \qquad \text{because } 3x^3 + 5 > 0 \text{ for } x > 0,$$

and so

$$2|x^4| \le |2x^4 + 3x^3 + 5| \qquad \begin{array}{l}\text{because all terms on both sides} \\ \text{of the inequality are positive.}\end{array}$$

Let $M_1 = 2$ and $x_1 = 0$. Then

$$A|x^4| \le |2x^4 + 3x^3 + 5| \quad \text{for all } x > a,$$

and so by definition of Ω-notation, $2x^4 + 3x^3 + 5$ is $\Omega(x^4)$.
Also for $x > 1$,

$$2x^4 + 3x^3 + 5 \le 2x^4 + 3x^4 + 5x^4 \qquad \begin{array}{l}\text{because by (9.2.1), } x^3 < x^4 \text{ and } 1 < x^4, \\ \text{and so } 3x^3 < 3x^4 \text{ and } 5 < 5x^4\end{array}$$

$$\Rightarrow \quad 2x^4 + 3x^3 + 5 \le 10x^4 \qquad \text{because } 2 + 3 + 5 = 10$$

$$\Rightarrow \quad |2x^4 + 3x^3 + 5| \le 10|x^4| \qquad \begin{array}{l}\text{because all terms on both sides} \\ \text{of the inequality are positive}\end{array}$$

(When the implication arrow, $\Rightarrow$, is placed at the beginning of a line, it means that the truth of the statement in that line is implied by the truth of the statement in the line above).

Let $B = 10$ and $b = 1$. Then

$$|2x^4 + 3x^3 + 5| \leq B|x^4| \quad \text{for all } x > b,$$

and so, by definition of O-notation, $2x^4 + 3x^3 + 5$ is $O(x^4)$.

Since $2x^4 + 3x^3 + 5$ is both $\Omega(x^4)$ and $O(x^4)$, by Theorem 9.2.1, it is $\Theta(x^4)$. ∎

The technique used in Example 9.2.4 can be generalized to show that any polynomial with nonnegative coefficients is big-Theta of its highest-power term. Taken together, the next two examples show that such a result can hold for a polynomial with negative as well as positive coefficients.

Example 9.2.5 A Big-O Approximation for a Polynomial with Some Negative Coefficients

a. Use the definition of O-notation to show that $3x^3 - 1000x - 200$ is $O(x^3)$.

b. Show that $3x^3 - 1000x - 200$ is $O(x^s)$ for all integers $s > 3$.

Solution

a. According to the triangle inequality for absolute value (see exercise 53 in Section 3.4),

$$\boxed{|a + b| \leq |a| + |b| \quad \text{for all real numbers } a \text{ and } b.} \qquad \text{triangle inequality}$$

If $-b$ is substituted in place of b, the result is

$$|a - b| = |a + (-b)| \leq |a| + |-b| = |a| + |b|, \quad \text{or}$$
$$|a - b| \leq |a| + |b|.$$

It follows that for all real numbers $x > 1$,

$$|3x^3 - 1000x - 200| \leq |3x^3| + |1000x| + |200|$$
$$\Rightarrow \quad |3x^3 - 1000x - 200| \leq |3x^3| + |1000x| + |200| \qquad \text{because all terms on the right side of the inequality are positive when } x > 1$$
$$\Rightarrow \quad |3x^3 - 1000x - 200| \leq 3x^3 + 1000x^3 + 200x^3 \qquad \text{because by (9.2.1), } x < x^3 \text{ and } 1 < x^3, \text{ and so } 1000x < 1000x^3 \text{ and } 200 < 200x^3$$
$$\Rightarrow \quad |3x^3 - 1000x - 200| \leq 1203x^3 \qquad \text{because } 3 + 1000 + 200 = 1203$$
$$\Rightarrow \quad |3x^3 - 1000x - 200| \leq 1203|x^3| \qquad \text{because } x^3 \text{ is positive.}$$

Let $b = 1$ and $B = 1203$. Then

$$|3x^3 - 1000x - 200| \leq B|x^3| \quad \text{for all real numbers } x > b.$$

So, by definition of O-notation, $3x^3 - 1000x - 200$ is $O(x^3)$.

b. Suppose s is an integer with $s > 3$. By property (9.2.1), $x^3 < x^s$ for all real numbers $x > 1$. So $B|x^3| < B|x^s|$ for all real numbers $x > b$ (because $b = 1$), and thus by part (a),

$$|3x^3 - 1000x - 200| \leq B|x^s| \quad \text{for all real numbers } x > b.$$

Hence, by definition of O-notation, $3x^3 - 1000x - 200$ is $O(x^s)$ for all integers $s > 3$. ∎

Example 9.2.6 A Big-Omega Approximation for a Polynomial with Some Negative Coefficients

    a. Use the definition of Ω-notation to show that $3x^3 - 1000x - 100$ is $\Omega(x^3)$.

    b. Show that $3x^3 - 1000x - 200$ is $\Omega(x^r)$ for all integers $r < 3$.

Solution

    a. To show that $3x^3 - 1000x - 200$ is $\Omega(x^3)$, you need to find numbers a and A so that $A|x^3| \le |3x^3 - 1000x - 100|$ for all real numbers $x > a$. Exercise 27 at the end of the section shows that the following procedure for choosing a will always produce an A that will give the desired result.

        Choose a as follows: Add up the absolute values of the coefficients of the lower-order terms of $3x^3 - 1000x - 100$, divide by the absolute value of the highest-power term, and multiply the result by 2. The result is $a = 2(1000 + 200)/3$, which equals 800. If you follow the steps below, you will see that when a is chosen in this way, A can be taken to be one-half of the absolute value of the highest power of the polynomial. Accordingly, assume that $x > a$. Then

$$x > 800$$

$\Rightarrow \qquad x > 2\left(\dfrac{1000 + 200}{3}\right)$ because $2(1000 + 200)/3 = 800$

$\Rightarrow \qquad x > \dfrac{2 \cdot 1000}{3} + \dfrac{2 \cdot 200}{3}$ by the rules for adding fractions

$\Rightarrow \qquad x > \dfrac{2 \cdot 1000}{3} \cdot \dfrac{1}{x} + \dfrac{2 \cdot 200}{3} \cdot \dfrac{1}{x^2}$ because $x > 800$ and so by (9.2.1), $1 > \dfrac{1}{x}$ and $1 > \dfrac{1}{x^2}$

$\Rightarrow \qquad \dfrac{3}{2}x^3 > 1000x + 200$ by multiplying both sides by $\frac{3}{2}x^2$

$\Rightarrow \qquad 3x^3 - \dfrac{3}{2}x^3 > 1000x + 200$ because $\frac{3}{2} = 3 - \frac{3}{2}$

$\Rightarrow \quad 3x^3 - 1000x - 200 > \dfrac{3}{2}x^3$ by adding $\frac{3}{2}x^3 - 1000x - 200$ to both sides

$\Rightarrow \quad |3x^3 - 1000x - 200| > \dfrac{3}{2}|x^3|$ because the expressions on both sides of the inequality are positive when $x > 800$.

Let $A = \frac{3}{2}$ and let $a = 800$. Then

$$A|x^3| \le |3x^3 - 1000x - 100| \quad \text{for all real numbers } x > a.$$

So, by definition of Ω-notation, $3x^3 - 1000x - 100$ is $\Omega(x^3)$.

    b. Suppose r is an integer with $r < 3$. By property (9.2.1), $x^r < x^3$ for all real numbers $x > 1$. So, since $a = 800 > 1$, $A|x^r| < A|x^3|$ for all real numbers $x > a$. Thus, by part (a),

$$A|x^r| \le |3x^3 - 1000x - 200| \quad \text{for all real numbers } x > a.$$

Hence, by definition of Ω-notation, $3x^3 - 1000x - 200$ is $\Omega(x^r)$ for all integers $r < 3$. ∎

    By Theorem 9.2.1, it follows immediately from Examples 9.2.5(a) and 9.2.6(a) that $3x^3 - 1000x - 100$ is big-Theta of x^3, and the techniques used in the examples can be generalized to show that every polynomial is big-Theta of the power function of its highest power. Moreover, the findings in parts (b) of the examples—that $3x^3 - 1000x - 100$ is also big-O of x^s for every integer s greater than 3 and is big-Omega of x^r for every integer r less than 3—can also be generalized to all polynomials. These facts are summarized in the next theorem.

Theorem 9.2.2 On Polynomial Orders

Suppose $a_0, a_1, a_2, \ldots, a_n$ are real numbers and $a_n \neq 0$.

1. $a_n x^n + a_{n-1} x^{n-1} + \cdots + a_1 x + a_0$ is $O(x^s)$ for all integers $s \geq n$.

2. $a_n x^n + a_{n-1} x^{n-1} + \cdots + a_1 x + a_0$ is $\Omega(x^r)$ for all integers $r \leq n$.

3. $a_n x^n + a_{n-1} x^{n-1} + \cdots + a_1 x + a_0$ is $\Theta(x^n)$.

Theorem 9.2.2 can easily be proved using calculus. As suggested by Examples 9.2.5 and 9.2.6, however, it can also be derived without calculus. (See exercises 26, 27, and 49 at the end of this section.)

Example 9.2.7 Calculating Polynomial Orders Using the Theorem on Polynomial Orders

Use the theorem on polynomial orders to find orders for the functions given by the following formulas.

a. $f(x) = 7x^5 + 5x^3 - x + 4$, for all real numbers x.

b. $g(x) = \dfrac{(x-1)(x+1)}{4}$, for all real numbers x.

Solution

a. By direct application of the theorem on polynomial orders, $7x^5 + 5x^3 - x + 4$ is $\Theta(x^5)$

b. $g(x) = \dfrac{(x-1)(x+1)}{4}$

$$= \frac{1}{4}(x^2 - 1)$$

$$= \frac{1}{4}x^2 - \frac{1}{4} \quad \text{by algebra}$$

Thus $g(x)$ is $\Theta(x^2)$ by the theorem on polynomial orders.   ■

Example 9.2.8 Showing That Two Power Functions Have Different Orders

Show that x^2 is not $O(x)$, and deduce that x^2 is not $\Theta(x)$.

Solution   [*Argue by contradiction.*] Suppose that x^2 is $O(x)$. [*Derive a contradiction.*] By the supposition that x^2 is $O(x)$, there exist a positive real number B and a real number b such that

$$|x^2| \leq B \quad \text{for all real numbers } x > b. \tag{$*$}$$

Let x be a positive real number that is greater than both B and b. Then

$$x \cdot x > B \cdot x \quad \text{by multiplying both sides of } x > B \text{ by } x \text{ which is positive}$$

$$\Rightarrow \quad |x^2| > B|x| \quad \text{because } b \text{ is positive.}$$

Thus there is a real number $x > b$ such that

$$|x^2| > B|x|.$$

This contradicts (∗). Hence the supposition is false, and so x^2 is not $O(x)$.

By Theorem 9.2.1, if x^2 is $\Theta(x)$, then x^2 is $O(x)$. But x^2 is not $O(x)$, and thus x^2 is not $\Theta(x)$. ∎

The technique used in Example 9.2.8 can be extended and generalized to prove that any polynomial function in x of degree n is not big-O (or big-Theta) of the mth power function x^m for any $m < n$. (See exercise 53 at the end of this section.)

> **Theorem 9.2.3 Limitation on Orders of Polynomial Functions**
>
> Let n be a positive integer, and let $a_0, a_1, a_2, \ldots, a_n$ be real numbers with $a_n \neq 0$. If m is any integer with $m < n$, then
>
> $$a_n x^n + a_{n-1} x^{n-1} + \cdots + a_1 x + a_0 \quad \text{is not} \quad O(x^m)$$
>
> and so also
>
> $$a_n x^n + a_{n-1} x^{n-1} + \cdots + a_1 x + a_0 \quad \text{is not} \quad \Theta(x^m).$$

It follows from Theorems 9.2.2 and 9.2.3 that integral power functions are convenient benchmarks for comparisons among general polynomial functions because every polynomial function has the same order as some power function, and no power function has the same order as any other.

Orders for Functions of Integer Variables

It is traditional to use the symbol x to denote a real number variable, whereas n is used to represent an integer variable. Thus, given a statement of the form

$$f(n) \text{ is } \Theta(g(n)),$$

we assume that f and g are functions defined on sets of *integers*. If it is true that

$$f(x) \text{ is } \Theta(g(x)),$$

where f and g are functions defined for *real numbers,* then it is certainly true that $f(n)$ is $\Theta(g(n))$. The reason is that if $f(x)$ is $\Theta(g(x))$, then an inequality $A|g(x)| \leq |f(x)| \leq B|g(x)|$ holds for all real numbers $x > k$. Hence, in particular, the inequality $A|g(n)| \leq |f(n)| \leq B|g(n)|$ holds for all integers $n > k$.

Example 9.2.9 An Order for the Sum of the First *n* Integers

Sums of the form $1 + 2 + 3 + \cdots + n$ arise in the analysis of computer algorithms such as selection sort. Show that for a positive integer variable n,

$$1 + 2 + 3 + \cdots + n \text{ is } \Theta(n^2).$$

Solution   By the formula for the sum of the first n integers (see Theorem 4.2.2), for all positive integers n,

$$1 + 2 + 3 + \cdots + n = \frac{n(n+1)}{2}.$$

But

$$\frac{n(n+1)}{2} = \frac{1}{2}n^2 + \frac{1}{2}n \quad \text{by basic algebra.}$$

And, by the theorem on polynomial orders,

$$\frac{1}{2}n^2 + \frac{1}{2}n \text{ is } \Theta(n^2).$$

Hence

$$1 + 2 + 3 + \cdots + n \text{ is } \Theta(n^2). \qquad \blacksquare$$

Extension to Functions Composed of Rational Power Functions

Consider a function of the form

$$\frac{(x^{3/2} + 3)(x - 2)^2}{x^{1/2}(2x^{1/2} + 1)} = \frac{x^{7/2} - 4x^{5/2} + 4x^{3/2} + 3x^2 - 12x + 12}{2x + x^{1/2}}.$$

When the numerator and denominator are expanded, each is a sum of terms of the form ax^r, where a is a real number and r is a positive rational number. The degree of such a sum can be taken to be the largest exponent of x that occurs in one of its terms. If the difference between the degree of the numerator and that of the denominator is called the degree of the function and denoted d, then it can be shown that $f(x)$ is $\Theta(x^d)$, that $f(x)$ is $O(x^c)$ for all real numbers $c > d$, and that $f(x)$ is not $O(x^c)$ for any real number $c < d$. For the example given above, this means that $d = 7/2 - 1 = 5/2$ and that

$$\frac{(x^{3/2} + 3)(x - 2)^2}{x^{1/2}(2x^{1/2} + 1)} \text{ is } \Theta(x^{5/2}),$$

$$\frac{(x^{3/2} + 3)(x - 2)^2}{x^{1/2}(2x^{1/2} + 1)} \text{ is } O(x^c) \quad \text{for all real numbers } c > 5/2,$$

and

$$\frac{(x^{3/2} + 3)(x - 2)^2}{x^{1/2}(2x^{1/2} + 1)} \text{ is not } O(x^c) \quad \text{for any real number } c < 5/2.$$

We state the general result as Theorem 9.2.4.

Theorem 9.2.4 Orders of Functions Composed of Rational Power Functions

Let m and n be positive integers, and let $r_0, r_1, r_2, \ldots, r_n$ and $s_0, s_1, s_2, \ldots, s_m$ be non-negative rational numbers with $r_0 < r_1 < r_2 < \cdots < r_n$ and $s_0 < s_1 < s_2 < \cdots < s_m$. Let $a_0, a_1, a_2, \ldots, a_n$ and $b_0, b_1, b_2, \ldots, b_m$ be real numbers with $a_n \neq 0$ and $b_m \neq 0$. Then

$$\frac{a_n^{r_n} + a_{n-1}^{r_{n-1}} + \cdots + a_1^{r_1} + a_0^{r_0}}{b_m^{s_m} + b_{m-1}^{s_{m-1}} + \cdots + b_1^{s_1} + b_0^{s_0}} \text{ is } \Theta(x^{r_n - s_m}).$$

$$\frac{a_n^{r_n} + a_{n-1}^{r_{n-1}} + \cdots + a_1^{r_1} + a_0^{r_0}}{b_m^{s_m} + b_{m-1}^{s_{m-1}} + \cdots + b_1^{s_1} + b_0^{s_0}} \text{ is } O(x^c) \quad \text{for all real numbers } c > r_n - s_m.$$

$$\frac{a_n^{r_n} + a_{n-1}^{r_{n-1}} + \cdots + a_1^{r_1} + a_0^{r_0}}{b_m^{s_m} + b_{m-1}^{s_{m-1}} + \cdots + b_1^{s_1} + b_0^{s_0}} \text{ is not } O(x^c) \quad \text{for any real number } c < r_n - s_m.$$

Exercise Set 9.2

1. The following is a formal definition for Ω-notation, written using quantifiers and variables: $f(x)$ is $\Omega(g(x))$ if, and only if, $\exists$ positive real numbers a and A such that $\forall x > a$,

$$A|g(x)| \leq |f(x)|.$$

   a. Write the formal negation for the definition using the symbols $\forall$ and $\exists$.
   b. Restate the negation less formally without using the symbols $\forall$ and $\exists$.

2. The following is a formal definition for *O*-notation, written using quantifiers and variables: $f(x)$ is $O(g(x))$ if, and only if, $\exists$ positive real numbers b and B such that $\forall x > b$,

$$|f(x)| \leq B|g(x)|.$$

   a. Write the formal negation for the definition using the symbols $\forall$ and $\exists$.
   b. Restate the negation less formally without using the symbols $\forall$ and $\exists$.

3. The following is a formal definition for Θ-notation, written using quantifiers and variables: $f(x)$ is $\Theta(g(x))$ if, and only if, $\exists$ positive real numbers k, A, and B such that $\forall x > k$,

$$A|g(x)| \leq |f(x)| \leq B|g(x)|.$$

   a. Write the formal negation for the definition using the symbols $\forall$ and $\exists$.
   b. Restate the negation less formally without using the symbols $\forall$ and $\exists$.

In 4–9, express each statement using Ω-, *O*-, or Θ-notation.

4. $|5x^8 - 9x^7 + 2x^5 + 3x - 1| \leq 6|x^8|$ for all real numbers $x > 3$. (Use *O*-notation.)

5. $|x| \leq \left| \dfrac{(x^2 - 1)(12x + 25)}{3x^2 + 4} \right| \leq 6|x|$ for all real numbers $x > 2$.

6. $|x^{7/2}| \leq \left| \dfrac{(x^2 - 7)^2(10x^{1/2} + 3)}{x + 1} \right|$ for all real numbers $x > 4$. (Use Ω-notation.)

7. $|3x^6 + 5x^4 - x^3| \leq 9|x^6|$ for all real numbers $x > 1$. (Use *O*-notation.)

8. $\frac{1}{2}x^4 \leq |x^4 - 50x^3 + 1|$ for all real numbers $x > 101$. (Use Ω-notation.)

9. $\frac{1}{2}x^2 \leq |3x^2 - 80x + 7| \leq 3|x^2|$ for all real numbers $x > 25$.

In each of 10–14 assume f and g are real-valued functions defined on the same set of nonnegative real numbers.

10. Prove that if $g(x)$ is $O(f(x))$, then $f(x)$ is $\Omega(g(x))$.

11. Prove that if $f(x)$ is $O(g(x))$ and c is any nonzero real number, then $cf(x)$ is $O(g(x))$.

12. Prove that if $f(x)$ is $O(h(x))$ and $g(x)$ is $O(k(x))$, then $f(x) + g(x)$ is $O(G(x))$, where, for each x in the domain, $G(x) = \max(|h(x)|, |k(x)|)$.

13. Prove that $f(x)$ is $\Theta(f(x))$.

H **14.** Prove that if $f(x)$ is $O(h(x))$ and $g(x)$ is $O(k(x))$, then $f(x)g(x)$ is $O(h(x)k(x))$.

15. a. Use mathematical induction to prove that if x is any real number with $x > 1$, then $x^n > 1$ for all integers $n \geq 1$.
   H b. Prove that if x is any real number with $x > 1$, then $x^m < x^n$ for any integers m and n with $m < n$.

16. a. Show that for any real number x,

$$\text{if } x > 1 \text{ then } |x^2| \leq |2x^2 + 15x + 4|.$$

   b. Show that for any real number x,

$$\text{if } x > 1 \text{ then } |2x^2 + 15x + 4| \leq 21|x^2|.$$

   c. Use the Ω- and *O*-notations to express the results of parts (a) and (b).
   d. What can you deduce about the order of $2x^2 + 15x + 4$?

17. a. Show that for any real number x,

$$\text{if } x > 1 \text{ then } |x^4| \leq |23x^4 + 8x^2 + 4x|.$$

   b. Show that for any real number x,

$$\text{if } x > 1 \text{ then } |23x^4 + 8x^2 + 4x| \leq 35|x^4|.$$

   c. Use the Ω- and *O*-notations to express the results of parts (a) and (b).
   d. What can you deduce about the order of $23x^4 + 8x^2 + 4x$?

18. Use the definition of Θ-notation to show that

$$5x^3 + 65x + 30 \text{ is } \Theta(x^3).$$

19. Use the definition of Θ-notation to show that

$$x^2 + 100x + 88 \text{ is } \Theta(x^2).$$

20. a. Show that for any real number x, if $x > 1$ then $|x^2| \leq |\lceil x^2 \rceil|$.
   b. Show that for any real number x, if $x > 1$ then $\frac{1}{2}|\lceil x^2 \rceil| \leq |x^2|$.
   c. Use the Ω- and *O*-notations to express the results of parts (a) and (b).
   d. What can you deduce about the order of $\lceil x^2 \rceil$?

21. a. Show that for any real number x, if $x > 1$ then $||\lfloor \sqrt{x} \rfloor|| \leq |\sqrt{x}|$.
   b. Show that for any real number x, if $x > 1$ then $\frac{1}{2}|\sqrt{x}| \leq |\lfloor x \rfloor|$.
   c. Use the Ω- and *O*-notation to express the results of parts (a) and (b).
   d. What can you deduce about the order of $\lfloor \sqrt{x} \rfloor$?

22. a. Show that for any real number x, if $x > 1$ then $|7x^4 - 95x^3 + 3| \leq 105|x^4|$.

b. Use O-notation to express the result of part (a).

23. a. Show that for any real number x, if $x > 1$ then $|\frac{1}{5}x^2 - 42x - 8| \leq 51|x^2|$.

b. Use O-notation to express the result of part (a).

24. a. Show that for any real number x, if $x > 1$ then $|\frac{1}{4}x^5 - 50x^3 + 3x + 12| \leq 66|x^5|$.

b. Use O-notation to express the result of part (a).

H 25. Show that x^5 is not $O(x^2)$.

26. Suppose $a_0, a_1, a_2, \ldots, a_n$ are real numbers and $a_n \neq 0$. Use the generalization of the triangle inequality to n integers (exercise 22, Section 8.4) to show that

$$a_n x^n + a_{n-1} x^{n-1} + \cdots + a_1 x + a_0 \text{ is } O(x^n).$$

27. Suppose $a_0, a_1, a_2, \ldots, a_n$ are real numbers and $a_n \neq 0$. Show that $a_n x^n + a_{n-1} x^{n-1} + \cdots + a_1 x + a_0$ is $\Omega(x^n)$ by letting

$$x_1 = 2 \left(\frac{|a_0| + |a_1| + |a_2| + \cdots + |a_{n-1}|}{|a_n|} \right).$$

In 28–30: (a) Let a be the number obtained by adding up the absolute values of the coefficients of the lower-order terms of the given polynomial, dividing by the absolute value of the highest-order term, and multiplying the result by 2, and let A be half the coefficient of the absolute value of the highest-order term of the polynomial. (b) Show that if $x > a$, the absolute value of the polynomial will be greater than the product of A and the absolute value of the highest-order term of the polynomial. (c) Deduce the result given in the exercise.

28. $7x^4 - 95x^3 + 3$ is $\Omega(x^4)$.

29. $\frac{1}{5}x^2 - 42x - 8$ is $\Omega(x^2)$.

30. $\frac{1}{4}x^5 - 50x^3 + 3x + 12$ is $\Omega(x^5)$.

31. Refer to the results of exercises 22 and 28 to find an order for $7x^4 - 95x^3 + 3$ from among the set of power functions.

32. Refer to the results of exercises 23 and 29 to find an order for $\frac{1}{5}x^2 - 42x - 8$ from among the set of power functions.

33. Refer to the results of exercises 24 and 30 to find an order for $\frac{1}{4}x^5 - 50x^3 + 3x + 12$ from among the set of power functions.

Use the theorem on polynomial orders to prove each of the statements in 34–39.

34. $\dfrac{(x + 1)(x - 2)}{4}$ is $\Theta(x^2)$.

35. $\dfrac{x}{3}(4x^2 - 1)$ is $\Theta(x^3)$.

36. $\dfrac{x(x - 1)}{2} + 3x$ is $\Theta(x^2)$.

37. $\dfrac{n(n + 1)(2n + 1)}{6}$ is $\Theta(n^3)$.

38. $\left[\dfrac{n(n + 1)}{2} \right]^2$ is $\Theta(n^4)$.

39. $2(n - 1) + \dfrac{n(n + 1)}{2} + 4 \left(\dfrac{n(n - 1)}{2} \right)$ is $\Theta(n^2)$.

Prove each of the statements in 40–47, assuming n is a variable that takes positive integer values. (Use formulas from the exercise set of Section 4.2 and the theorem on polynomial orders as appropriate.)

40. $1^2 + 2^2 + 3^2 + \cdots + n^2$ is $\Theta(n^3)$.

41. $1^3 + 2^3 + 3^3 + \cdots + n^3$ is $\Theta(n^4)$.

42. $2 + 4 + 6 + \cdots + 2n$ is $\Theta(n^2)$.

43. $5 + 10 + 15 + 20 + 25 + \cdots + 5n$ is $\Theta(n^2)$.

44. $\displaystyle\sum_{i=1}^{n} (4i - 9)$ is $\Theta(n^2)$.   **45.** $\displaystyle\sum_{k=1}^{n} (k + 3)$ is $\Theta(n^2)$.

H 46. $\displaystyle\sum_{i=1}^{n} i(i + 1)$ is $\Theta(n^3)$.   **47.** $\displaystyle\sum_{k=3}^{n} (k^2 - 2k)$ is $\Theta(n^3)$.

H 48. (Requires the concept of limit from calculus)

a. Let $a_0, a_1, a_2, \ldots, a_n$ be real numbers with $a_n \neq 0$. Prove that

$$\lim_{x \to \infty} \left| \frac{a_n x^n + a_{n-1} x^{n-1} + \cdots + a_1 x + a_0}{a_n x^n} \right| = 1.$$

b. Use the result of part (a) and the definition of limit to prove that

$$a_n x^n + a_{n-1} x^{n-1} + \cdots + a_1 x + a_0 \text{ is } \Theta(x^n).$$

49. Another approach to proving part of the theorem on polynomial orders uses properties of O-notation.

a. Show that if f, g, and h are functions from **R** to **R** and $f(x)$ is $O(h(x))$ and $g(x)$ is $O(f(x))$ then $f(x) + g(x)$ is $O(h(x))$.

b. How does it follow from part (a) and Theorem 9.2.1(4) that $x^4 + x^2$ is $O(x^4)$?

c. Show that if f is a function from **R** to **R**, $f(x)$ is $O(g(x))$, and c is any nonzero real number, then $cf(x)$ is $O(g(x))$.

d. How does it follow from parts (a) and (c) that $12x^5 - 34x^2 + 7$ is $O(x^5)$?

e. Use the results of parts (a) and (c) to show that if n is a positive integer and $a_1, a_2, \ldots, a_n$ are real numbers, then

$$a_n x^n + a_{n-1} x^{n-1} + \cdots + a_1 x + a_0 \text{ is } O(x^n).$$

50. a. Let x be any positive real number. Use mathematical induction to prove that for all integers $n \geq 1$, if $x \leq 1$ then $x^n \leq 1$.

b. Explain how it follows from part (a) that if x is any positive real number, then for all integers $n \geq 1$, if $x^n > 1$ then $x > 1$.

c. Explain how it follows from part (b) that if x is any positive real number, then for all integers $n \geq 1$, if $x > 1$ then $x^{1/n} > 1$.

H d. Let p, q, and s be positive integers, let r be a nonnegative integer, and suppose $p/q > r/s$. Use part (c) and the result of exercise 15 to show that for any real number x, if $x > 1$ then $x^{p/q} > x^{r/s}$.

Explain how each statement in 51 and 52 follows from exercise 50 and parts (a) and (c) of exercise 49.

51. $4x^{4/3} - 15x + 7$ is $O(x^{4/3})$.

52. $\sqrt{x}(38x^5 + 9)$ is $O(x^{11/2})$.

H 53. Prove that if r and s are rational numbers with $r > s$, then x^r is not $O(x^s)$.

In 54–56, use Theorem 9.2.4 to find an order for each of the given functions from among the set of rational power functions.

54. $f(x) = \dfrac{\sqrt{x}(3x + 5)}{2x + 1}$

55. $f(x) = \dfrac{(2x^{7/2} + 1)(x - 1)}{(x^{1/2} + 1)(x - 1)}$

56. $f(x) = \dfrac{(5x^2 + 1)(\sqrt{x} - 1)}{4x^{3/2} - 2x}$

★57. a. Use mathematical induction to prove that

$$\sqrt{1} + \sqrt{2} + \sqrt{3} + \cdots + \sqrt{n} \leq n^{3/2}$$

for all integers $n \geq 1$.

H b. Use mathematical induction to prove that

$$\frac{1}{2}n^{3/2} \leq \sqrt{1} + \sqrt{2} + \sqrt{3} + \cdots + \sqrt{n}.$$

c. What can you conclude from parts (a) and (b) about an order of $\sqrt{1} + \sqrt{2} + \sqrt{3} + \cdots + \sqrt{n}$?

★58. a. Use mathematical induction to prove that $1^{1/3} + 2^{1/3} + \cdots + n^{1/3} \leq n^{4/3}$, for all integers $n \geq 1$.

b. Use mathematical induction to prove that

$$\frac{1}{2}n^{4/3} \leq 1^{1/3} + 2^{1/3} + 3^{1/3} + \cdots + n^{1/3}.$$

c. What can you conclude from parts (a) and (b) about an order for $1^{1/3} + 2^{1/3} + 3^{1/3} + \cdots + n^{1/3}$?

Exercises 59–61 use the following definition, which requires the concept of limit from calculus.

Definition: If f and g are real-valued functions of a real variable and $\lim_{x \to \infty} g(x) \neq 0$, then

$$f(x) \text{ is } o(g(x)) \quad \Leftrightarrow \quad \lim_{x \to \infty} \frac{f(x)}{g(x)} = 0.$$

The notation $f(x)$ is $o(g(x))$ is read "$f(x)$ is little-oh of $g(x)$."

59. Prove that if $f(x)$ is $o(g(x))$, then $f(x)$ is $O(g(x))$.

60. Prove that if $f(x)$ and $g(x)$ are both $o(h(x))$, then for all real numbers a and b, $af(x) + bg(x)$ is $o(h(x))$.

61. Prove that for any positive real numbers a and b, if $a < b$ then x^a is $o(x^b)$.

9.3 Application: Efficiency of Algorithms I

Charles Babbage
(1792–1871)

As soon as an Analytical Engine exists, it will necessarily guide the future course of the science. Whenever any result is sought by its aid, the question will then arise—by what course of calculation can these results be arrived at by the machine in the shortest time?
— Charles Babbage, 1864

Charles Babbage's Analytical Engine was similar in concept to a modern computer. The above quotation suggests that Babbage anticipated the importance of analyzing the efficiencies of computer algorithms well over a hundred years ago. In the 1950s and 1960s, a number of mathematicians and computer scientists contributed to the development of algorithm analysis, especially Donald Knuth, whose multivolume work *The Art of Computer Programming* provides a foundation for the subject that is both elegant and mathematically rigorous.*

*Donald E. Knuth, *The Art of Computer Programming*, vol. 1: *Fundamental Algorithms,* 3rd ed. (1997); vol. 2: *Seminumerical Algorithms,* 3rd ed., (1997); vol. 3: *Searching and Sorting,* 2nd ed. (1998) (Reading, MA: Addison-Wesley).

Courtesy of Donald Knuth

Understanding the relative efficiencies of algorithms designed to do the same job is of much more than academic interest. In industrial and scientific settings, the choice of an efficient over an inefficient program may result in the saving of many thousands of dollars or may make the difference between being able or not being able to do a project at all.

Two aspects of algorithm efficiency are important: the amount of time required to execute the algorithm and the amount of memory space needed when it is run. In this chapter we introduce basic techniques for calculating time efficiency. Similar techniques exist for calculating space efficiency. Occasionally, one algorithm may make more efficient use of time but less efficient use of memory space than another, forcing a trade-off based on the resources available to the user.

Donald Knuth
(born 1938)

Time Efficiency of an Algorithm

How can the time efficiency of an algorithm be calculated? The answer depends on several factors. One is the size of the set of data that is input to the algorithm; for example, it takes longer for a sort algorithm to process 1,000,000 items than 100 items. Consequently, the execution time of an algorithm is generally expressed as a function of its input size.

Another factor that may affect the run time of an algorithm is the nature of the input data. For instance, a program that searches sequentially through a list of length n to find a data item requires only one step if the item is first on the list, but it uses n steps if the item is last on the list. Thus algorithms are frequently analyzed in terms of their "best case," "worst case," and "average case" performances for an input of size n.

Roughly speaking, the analysis of an algorithm for time efficiency begins by trying to count the number of elementary operations that must be performed when the algorithm is executed with an input of size n (in the best case, worst case, or average case). What is classified as an "elementary operation" may vary depending on the nature of the problem the algorithms being compared are designed to solve. For instance, to compare two algorithms for evaluating a polynomial, the crucial issue is the number of additions and multiplications that are needed, whereas to compare two algorithms for searching a list to find a particular element, the important distinction is the number of comparisons that are required. For simplicity, we will classify the following as **elementary operations:** addition, subtraction, multiplication, division, and comparison.

When algorithms are implemented in a particular programming language and run on a particular computer, some operations are executed faster than others, and, of course, there are differences in execution times from one machine to another. In certain practical situations these factors are taken into account when we decide which algorithm or which machine to use to solve a particular problem. In other cases, however, the machine is fixed, and rough estimates are all that we need to determine the clear superiority of one algorithm over another. Since each elementary operation is executed in time no longer than the slowest, the time efficiency of an algorithm is approximately proportional to the number of elementary operations required to execute the algorithm.

Consider the example of two algorithms, A and B, designed to do a certain job. Suppose that for an input of size n, the number of elementary operations needed to perform algorithm A is between $10n$ and $20n$ (at least for large n) and the number of elementary operations needed to perform algorithm B is between $2n^2$ and $4n^2$. Note that $20n < 2n^2$ whenever $n > 10$, which means that the maximum number of operations required to execute A is less than the *minimum* number of operations required to execute B whenever $n > 10$. In fact, $20n$ is very much less than $2n^2$ when n is large. For instance, if $n = 1000$, then $20n = 20,000$, whereas $2n^2 = 2,000,000$. We say that in the worst case, algorithm A is $\Theta(n)$ (or has worst-case order n) and that in the worst case, algorithm B is $\Theta(n^2)$ (or has worst-case order n^2).

> **• Definition**
>
> Let A be an algorithm.
>
> 1. Suppose the number of elementary operations performed when A is executed for an input of size n depends on n alone and not on the nature of the input data; say it equals $f(n)$. If $f(n)$ is $\Theta(g(n))$, we say that **A is $\Theta(g(n))$** or **A is of order $g(n)$.**
>
> 2. Suppose the number of elementary operations performed when A is executed for an input of size n depends on the nature of the input data as well as on n.
>
>   a. Let $b(n)$ be the *minimum* number of elementary operations required to execute A for all possible input sets of size n. If $b(n)$ is $\Theta(g(n))$, we say that **in the best case, A is $\Theta(g(n))$** or **A has a best case order of $g(n)$.**
>
>   b. Let $w(n)$ be the *maximum* number elementary operations required to execute A for all possible input sets of size n. If $w(n)$ is $\Theta(g(n))$, we say that **in the worst case, A is $\Theta(g(n))$** or **A has a worst case order of $g(n)$.**

Some of the orders most commonly used to describe algorithm efficiencies are shown in Table 9.3.1. As you see from the table, differences between the orders of various types of algorithms are more than astronomical. The time required for an algorithm of order 2^n to operate on a data set of size 100,000 is approximately $10^{29,976}$ times the estimated 15 billion years since the universe began (according to one theory of cosmology). On the other hand, an algorithm of order $\log_2 n$ needs at most a fraction of a second to process the same data set.

Table 9.3.1 Time Comparisons of Some Algorithm Orders

Approximate Time to Execute $f(n)$ Operations Assuming One Operation per Nanosecond*				
$f(n)$	$n = 10$	$n = 1{,}000$	$n = 100{,}000$	$n = 10{,}000{,}000$
$\log_2 n$	3.3×10^{-9} sec	10^{-8} sec	1.7×10^{-8} sec	2.3×10^{-8} sec
n	10^{-8} sec	10^{-6} sec	0.0001 sec	0.01 sec
$n \log_2 n$	3.3×10^{-8} sec	10^{-5} sec	0.0017 sec	0.23 sec
n^2	10^{-7} sec	0.001 sec	10 sec	27.8 min
n^3	10^{-6} sec	1 sec	11.6 days	$31{,}688$ yr
2^n	10^{-6} sec	3.4×10^{284} yr	3.2×10^{30095} yr	3.1×10^{3001022} yr

*one nanosecond $= 10^{-9}$ second

Example 9.3.1 Computing an Order of an Algorithm Segment

Assume n is a positive integer and consider the following algorithm segment:

$$p := 0, x := 2$$
$$\textbf{for } i := 2 \textbf{ to } n$$
$$p := (p + i) \cdot x$$
$$\textbf{next } i$$

a. Compute the actual number of additions and multiplications that must be performed when this algorithm segment is executed.

b. Find an order for this algorithm segment from among the set of power functions.

Solution

a. There are one multiplication and one addition for each iteration of the loop, so there are twice as many multiplications and additions as there are iterations of the loop. Now the number of iterations of the **for-next** loop equals the top index of the loop minus the bottom index plus 1; that is, $n - 2 + 1 = n - 1$. Hence there are $2(n - 1) = 2n - 2$ multiplications and additions.

b. By the theorem on polynomial orders,

$$2n - 2 \text{ is } \Theta(n),$$

and so this algorithm segment is $\Theta(n)$. ∎

The next example looks at an algorithm segment that contains a nested loop.

Example 9.3.2 An Order for an Algorithm with a Nested Loop

Assume n is a positive integer and consider the following algorithm segment:

$$s := 0$$
for $i := 1$ **to** n
    for $j := 1$ **to** i
        $s := s + j \cdot (i - j + 1)$
    next j
next i

a. Compute the actual number of additions, subtractions, and multiplications that must be performed when this algorithm segment is executed.

b. Find an order for this algorithm segment from among the set of power functions.

Solution

a. There are two additions, one multiplication, and one subtraction for each iteration of the inner loop, so the total number of additions, multiplications, and subtractions is four times the number of iterations of the inner loop. Now the inner loop is iterated

one time when $i = 1$,

two times when $i = 2$,

three times when $i = 3$,

$$\vdots$$

n times when $i = n$.

You can see this easily if you construct a table that shows the values of i and j for which the statements in the inner loop are executed. There is one iteration for each column in the table.

i	1	2		3			4				$\cdots$	n				
j	1	1	2	1	2	3	1	2	3	4	$\cdots$	1	2	3	$\cdots$	n

Hence the total number of iterations of the inner loop is

$$1 + 2 + 3 + \cdots + n = \frac{n(n+1)}{2} \quad \text{by Theorem 4.2.2,}$$

and so the number of additions, subtractions, and multiplications is

$$4 \cdot \frac{n(n+1)}{2} = 2n(n+1).$$

An alternative method for computing the number of columns of the table uses an approach discussed in Section 6.5. Observe that the number of columns in the table is the same as the number of ways to place two ×'s in n categories, $1, 2, \ldots, n$, where the location of the ×'s indicates the values of i and j. By Theorem 6.5.1, this number is

$$\binom{n-1+2}{2} = \binom{n+1}{2} = \frac{(n+1)!}{2!((n+1)-2)!} = \frac{(n+1)n(n-1)!}{2(n-1)!} = \frac{n(n+1)}{2}.$$

Although, for this example, the alternative method is more complicated than the one preceding it, it is simpler when the number of loop nestings exceeds two. (See exercise 19.)

b. By the theorem on polynomial orders, $2n(n+1) = 2n^2 + 2n$ is $\Theta(n^2)$, and so this algorithm segment is $\Theta(n^2)$. ∎

Example 9.3.3 When the Number of Iterations Depends on the Floor Function

Assume n is a positive integer and consider the following algorithm segment:

for $i := \lfloor n/2 \rfloor$ **to** n
$\qquad u := n - i$
next i

a. Compute the actual number of subtractions that must be performed when this algorithm segment is executed.

b. Find an order for this algorithm segment from among the set of power functions.

Solution

a. There is one subtraction for each iteration of the loop, and the loop is iterated $n - \left\lfloor \dfrac{n}{2} \right\rfloor + 1$ times. If n is even, then $\left\lfloor \dfrac{n}{2} \right\rfloor = \dfrac{n}{2}$, and so the number of subtractions is

$$n - \left\lfloor \frac{n}{2} \right\rfloor + 1 = n - \frac{n}{2} + 1 = \frac{n+2}{2}.$$

If n is odd, then $\left\lfloor \dfrac{n}{2} \right\rfloor = \dfrac{n-1}{2}$, and so the number of subtractions is

$$n - \left\lfloor \frac{n}{2} \right\rfloor + 1 = n - \frac{n-1}{2} + 1 = \frac{2n-(n-1)+2}{2} = \frac{n+3}{2}.$$

b. By the theorem on polynomial orders,

$$\frac{n+2}{2} \text{ is } \Theta(n) \quad \text{and} \quad \frac{n+3}{2} \text{ is } \Theta(n)$$

also. Hence, regardless of whether n is even or odd, this algorithm segment is $\Theta(n)$. ∎

The Sequential Search Algorithm

The object of a search algorithm is to hunt through an array of data in an attempt to find a particular item x. In a sequential search, x is compared to the first item in the array, then to the second, then to the third, and so on. The search is stopped if a match is found at any stage. On the other hand, if the entire array is processed without finding a match, then x is not in the array. An example of a sequential search is shown diagrammatically in Figure 9.3.1.

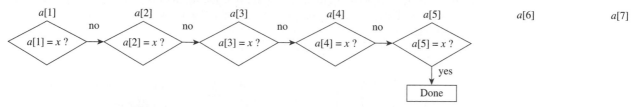

Figure 9.3.1 Sequential Search of $a[1], a[2], \ldots, a[7]$ for x where $x = a[5]$

Example 9.3.4 Best- and Worst-Case Orders for Sequential Search

Find best- and worst-case orders for the sequential search algorithm from among the set of power functions.

Solution   Suppose the sequential search algorithm is applied to an input array $a[1], a[2], \ldots, a[n]$ to find an item x. In the best case, the algorithm requires only one comparison between x and the items in $a[1], a[2], \ldots, a[n]$. This occurs when x is the first item in the array. Thus in the best case, the sequential search algorithm is $\Theta(1)$. (Note that $\Theta(1) = \Theta(n^0)$.) In the worst case, however, the algorithm requires n comparisons. This occurs when $x = a[n]$ or when x does not appear in the array at all. Thus in the worst case, the sequential search algorithm is $\Theta(n)$. ∎

The Insertion Sort Algorithm

Insertion sort is an algorithm for arranging the items in an array into ascending order. Initially, the second item is compared to the first. If the second item is less than the first, their values are interchanged, and as a result the first two array items are in ascending order. The idea of the algorithm is gradually to lengthen the section of the array that is known to be in ascending order by inserting each subsequent array item into its correct position relative to the preceding ones. When the last item has been placed, the entire array is in ascending order.

Figure 9.3.2 illustrates the action of step k of insertion sort on an array $a[1], a[2], a[3], \ldots, a[n]$.

Figure 9.3.2 Step k of Insertion Sort

The following is a formal algorithm for insertion sort.

Algorithm 9.3.1 Insertion Sort

[*The aim of this algorithm is to take an array* $a[1], a[2], a[3], \ldots, a[n]$, *where* $n \geq 1$, *and reorder it. The output array is also denoted* $a[1], a[2], a[3], \ldots, a[n]$. *It has the same values as the input array, but they are in ascending order. In the kth step,* $a[1], a[2], a[3], \ldots, a[k-1]$ *is in ascending order, and* $a[k]$ *is inserted into the correct position with respect to it.*]

Input: n [*a positive integer*], $a[1], a[2], a[3], \ldots, a[n]$ [*an array of data items capable of being ordered*]

Algorithm Body:

    for $k := 2$ **to** n
    [*Compare* $a[k]$ *to previous items in the array* $a[1], a[2], a[3], \ldots, a[k-1]$, *starting from the largest and moving downward. Whenever* $a[i-1] \leq a[k] < a[i]$, *interchange the values of* $a[k]$ *and* $a[i]$. *If* $a[k]$ *is greater than or equal to* $a[k-1]$, *then leave it as* $a[k]$.]

        do
        $x := a[k]$
        $j := k - 1$
            while $(j \neq 0$ and $a[j] > x)$
                do $a[j+1] := a[j]$
                    $j := j - 1$
            end do
            $a[j+1] := x$
        end while
        end do
    next k

Output: $a[1], a[2], a[3], \ldots, a[n]$ in ascending order

Figure 9.3.3 shows the result of each step when insertion sort is applied to the particular array

$$a[1] = 6, a[2] = 3, a[3] = 5, a[4] = 7, \quad \text{and} \quad a[5] = 2.$$

	$a[1]$	$a[2]$	$a[3]$	$a[4]$	$a[5]$
Initial	6	3	5	7	2
Result of step 1	3	6	5	7	2
Result of step 2	3	5	6	7	2
Result of step 3	3	5	6	7	2
Result of step 4	2	3	5	6	7

The top row of the table shows the initial values of the array, and the bottom row shows the final values. The result of each step is shown in a separate row. For each step, the sorted section of the array is shaded.

Figure 9.3.3 Action of Insertion Sort on an Array

Example 9.3.5 develops a trace table for the action of insertion sort on a particular array.

Example 9.3.5 A Trace Table for Insertion Sort

Construct a trace table showing the action of insertion sort on the array

$$a[1] = 6, a[2] = 3, a[3] = 5, a[4] = 7, a[5] = 2.$$

Solution   The first column shows the state of the variables before the first iteration of the **for-next** loop. When the **for-next** loop is first iterated, k is assigned the value 2; x the value of $a[2]$, which is 3; and j the value of $k - 1$, which is 1. Because $j \neq 0$ and $a[1] > x$, the **while** loop is entered. Then $a[2]$ is assigned the value of $a[1]$, which is 6 and j is assigned the value of $j - 1$, which is 0. The condition governing the **while** loop is tested again, but since $j = 0$, it is not satisfied, and so the **while** loop is not entered. Thus the value of k is incremented by 1 (so that it equals 3), and the **for-next** loop is entered a second time. This process continues until the value of k has been incremented to 6. Because 6 is greater than the top value in the **for-next** loop, execution of the algorithm ceases, and the array items are seen to be in ascending order.

n	5											
$a[1]$	6		3								2	
$a[2]$	3	6			5					2	3	
$a[3]$	5			6					2	5		
$a[4]$	7						2	6				
$a[5]$	2					7						
k		2		3		4	5					6
x		3		5		7	2					
j		1	0	2	1	3	4	3	2	1	0	

Example 9.3.6 Finding a Worst-Case Order for Insertion Sort

a. What is the maximum number of comparisons that are performed when insertion sort is applied to the array $a[1], a[2], a[3], \ldots, a[n]$?

b. Find a worst-case order for insertion sort from among the set of power functions.

Solution

a. In each attempted iteration of the **while** loop, two comparisons are made: one to test whether $j \neq 0$ and the other to test whether $a[j] > x$. During the time that $a[k]$ is put into position relative to $a[1], a[2], \ldots, a[k - 1]$, the maximum number of attempted iterations of the **while** loop is k. This happens when $a[k]$ is less than every $a[1], a[2], \ldots, a[k - 1]$; on the kth attempted iteration, the condition of the **while** loop is not satisfied because $j = 0$. Thus the maximum number of comparisons for a given value of k is $2k$. Because k goes from 2 to n, it follows that the maximum total number of comparisons occurs when the items in the array are in reverse order, and it equals

$$
\begin{aligned}
2 \cdot 2 + 2 \cdot 3 + \cdots + 2 \cdot n &= 2(2 + 3 + \cdots + n) && \text{by factoring out the 2} \\
&= 2[(1 + 2 + 3 + \cdots + n) - 1] && \text{by adding and subtracting 1} \\
&= 2\left(\frac{n(n+1)}{2} - 1\right) && \text{by Theorem 4.2.2} \\
&= n(n+1) - 2 \\
&= n^2 + n - 2 && \text{by algebra}
\end{aligned}
$$

b. By the theorem on polynomial orders, $n^2 + n - 2$ is $\Theta(n^2)$, and so the insertion sort algorithm has worst-case order $\Theta(n^2)$. ∎

The definition of expected value that was introduced in Section 6.8 can be used to find an average-case order for insertion sort.

Example 9.3.7 Finding an Average-Case Order for Insertion Sort

a. What is the average number of comparisons that are performed when insertion sort is applied to the array $a[1], a[2], a[3], \ldots, a[n]$?

b. Find an average case order for insertion sort from among the set of power functions.

Solution

a. Let E_n be the average, or expected, number of comparisons used to sort $a[1], a[2], \ldots, a[n]$ with insertion sort. Note that for each integer $k = 2, 3, \ldots, n$,

$$
\begin{bmatrix} \text{the expected number of} \\ \text{comparisons used to} \\ \text{sort } a[1], a[2], \ldots, a[k] \end{bmatrix}
$$

$$
= \begin{bmatrix} \text{the expected number of} \\ \text{comparisons used to} \\ \text{sort } a[1], a[2], \ldots, a[k-1] \end{bmatrix} + \begin{bmatrix} \text{the expected number of comparisons} \\ \text{used to place } a[k] \text{ into position} \\ \text{relative to } a[1], a[2], \ldots, a[k-1] \end{bmatrix}.
$$

Thus

$$
E_k = E_{k-1} + \begin{bmatrix} \text{the expected number of comparisons} \\ \text{used to place } a[k] \text{ into position} \\ \text{relative to } a[1], a[2], \ldots, a[k-1] \end{bmatrix}.
$$

Also, $E_1 = 0$ because when there is just one item in the array, $n = 1$ and no iterations of the outer loop are performed.

Now at the time $a[k]$ is placed relative to $a[1], a[2], \ldots, a[k-1]$, it is equally likely to belong in any one of the first k positions. Thus the probability of its belonging in any particular position is $1/k$. If it actually belongs in position j, then $2(k - j + 1)$ comparisons will be used in moving it, because there will be $k - j + 1$ attempted iterations of the **while** loop and there are 2 comparisons per attempted iteration.

According to the definition of expected value given in Section 6.8, the expected number of comparisons used to place $a[k]$ relative to $a[1], a[2], \ldots, a[k-1]$ is therefore

$$
\begin{aligned}
\sum_{j=1}^{k} \frac{1}{k} 2(k-j+1) &= \frac{2}{k}[k + (k-1) + \cdots + 3 + 2 + 1] && \text{by writing the summation} \\
&&& \text{in expanded form} \\
&= \frac{2}{k}\left(\frac{k(k+1)}{2}\right) && \text{by Theorem 4.2.2} \\
&= k+1 && \text{by algebra.}
\end{aligned}
$$

Hence

$$
E_k = E_{k-1} + k + 1 \quad \text{for all integers } k \geq 2, \quad \text{and}
$$
$$
E_1 = 0.
$$

Exercise 27 at the end of the section asks you to solve this recurrence relation to show that

$$
E_n = \frac{n^2 + 3n - 4}{2} \quad \text{for each integer } n \geq 1.
$$

b. By the theorem on polynomial orders, $\dfrac{n^2 + 3n - 4}{2} = \dfrac{1}{2}n^2 + \dfrac{3}{2}n - 2$ is $\Theta(n^2)$, and so the average-case order of insertion sort is also $\Theta(n^2)$. ∎

Exercise Set 9.3

1. Suppose a computer takes 1 nanosecond ($= 10^{-9}$ second) to execute each operation. Approximately how long will it take for the computer to execute the following numbers of operations? Convert your answers into seconds, minutes, hours, days, weeks, or years, as appropriate. For example, instead of 2^{50} nanoseconds, write 13 days.
   a. $\log_2 200$ b. 200 c. $200 \log_2 200$
   d. 200^2 e. 200^8 f. 2^{200}

2. Suppose an algorithm requires cn^2 operations when performed with an input of size n (where c is a constant).
   a. How many operations will be required when the input size is increased from m to $2m$ (where m is a positive integer)?
   b. By what factor will the number of operations increase when the input size is doubled?
   c. By what factor will the number of operations increase when the input size is increased by a factor of ten?

3. Suppose an algorithm requires cn^3 operations when performed with an input of size n (where c is a constant).
   a. How many operations will be required when the input size is increased from m to $2m$ (where m is a positive integer)?
   b. By what factor will the number of operations increase when the input size is doubled?
   c. By what factor will the number of operations increase when the input size is increased by a factor of ten?

Exercises 4–5 explore the fact that for relatively small values of n, algorithms with larger orders can be more efficient than algorithms with smaller orders.

4. Suppose that when run with an input of size n, algorithm A requires $2n^2$ operations and algorithm B requires $80n^{3/2}$ operations.
   a. What are orders for algorithms A and B from among the set of power functions?
   b. For what values of n is algorithm A more efficient than algorithm B?
   c. For what values of n is algorithm B at least 100 times more efficient than algorithm A?

5. Suppose that when run with an input of size n, algorithm A requires $10^6 n^2$ operations and algorithm B requires n^3 operations.
   a. What are orders for algorithms A and B from among the set of power functions?
   b. For what values of n is algorithm A more efficient than algorithm B?
   c. For what values of n is algorithm B at least 100 times more efficient than algorithm A?

For each of the algorithm segments in 6–14, assume that n is a positive integer. (a) Compute the actual number of additions, subtractions, multiplications, divisions, and comparisons that must be performed when the algorithm segment is executed. For simplicity, however, count only comparisons that occur within **if-then** statements; ignore those implied by **for-next** loops. (b) Find an order for the algorithm segment from among the set of power functions.

6. for $i := 3$ **to** $n - 1$
$\qquad a := 3 \cdot n + 2 \cdot i - 1$
next i

7. $max := a[1]$
$\quad$ **for** $i := 2$ **to** n
$\qquad$ **if** $max < a[i]$ **then** $max := a[i]$
$\quad$ **next** i

8. for $i := 1$ **to** $\lfloor n/2 \rfloor$
$\qquad a := n - i$
next i

9. for $i := 1$ **to** n
$\qquad$ **for** $j := 1$ **to** $2n$
$\qquad\qquad a := 2 \cdot n + i \cdot j$
$\qquad$ **next** j
next i

10. for $k := 2$ **to** n
$\qquad$ **for** $j := 1$ **to** $3n$
$\qquad\qquad x := a[k] - b[j]$
$\qquad$ **next** j
next k

11. for $k := 1$ **to** $n - 1$
$\qquad$ **for** $j := 1$ **to** $k + 1$
$\qquad\qquad x := a[k] + b[j]$
$\qquad$ **next** j
next k

12. for $k := 1$ **to** $n - 1$
$\qquad max := a[k]$
$\qquad$ **for** $i := k + 1$ **to** n
$\qquad\qquad$ **if** $max < a[i]$ **then** $max := a[i]$
$\qquad$ **next** i
$\qquad a[k] := max$
next k

13. for $i := 1$ **to** $n - 1$
$\qquad$ **for** $j := i + 1$ **to** n
$\qquad\qquad$ **if** $a[j] > a[i]$ **then do**
$\qquad\qquad\quad temp := a[i]$
$\qquad\qquad\quad a[i] := a[j]$
$\qquad\qquad\quad a[j] := temp$
$\qquad\qquad$ **end do**
$\qquad$ **next** j
next i

14. $t := 0$
$\quad$ **for** $i := 1$ **to** n
$\qquad s := 0$
$\qquad$ **for** $j := 1$ **to** i
$\qquad\qquad s := s + a[j]$
$\qquad$ **next** j
$\qquad t := t + s^2$
$\quad$ **next** i

15. $r := 0$
$\quad$ **for** $i := 1$ **to** n
$\qquad p := 1$
$\qquad q := 1$
$\qquad$ **for** $j := i + 1$ **to** n
$\qquad\qquad p := p \cdot c[j]$
$\qquad\qquad q := q \cdot c[j]^2$
$\qquad$ **next** j
$\qquad r := p + q$
$\quad$ **next** i

16. $t := 0$
$\quad$ **for** $k := 1$ **to** n
$\qquad s := 0$
$\qquad$ **for** $j := 1$ **to** $i - 1$
$\qquad\qquad s := s + j \cdot (i - j + 1)$
$\qquad$ **next** j
$\qquad r := s^2$
$\quad$ **next** k

17. for $i := 1$ **to** n
$\qquad$ **for** $j := 1$ **to** $\lfloor (i + 1)/2 \rfloor$
$\qquad\qquad a := (n - i) \cdot (n - j)$
$\qquad$ **next** j
next i

18. for $i := 1$ **to** n
$\qquad$ **for** $j := 1$ **to** $2n$
$\qquad\qquad$ **for** $k := 1$ **to** n
$\qquad\qquad\quad x := i \cdot j \cdot k$
$\qquad\qquad$ **next** k
$\qquad$ **next** j
next i

$H \ast$ **19. for** $i := 1$ **to** n
$\qquad$ **for** $j := 1$ **to** i
$\qquad\qquad$ **for** $k := 1$ **to** j
$\qquad\qquad\quad x := i \cdot j \cdot k$
$\qquad\qquad$ **next** k
$\qquad$ **next** j
next i

20. Construct a table showing the result of each step when insertion sort is applied to the array $a[1] = 6$, $a[2] = 2$, $a[3] = 1$, $a[4] = 8$, and $a[5] = 4$.

21. Construct a table showing the result of each step when insertion sort is applied to the array $a[1] = 7$, $a[2] = 3$, $a[3] = 6$, $a[4] = 9$, and $a[5] = 5$.

22. Construct a trace table showing the action of insertion sort on the array of exercise 20.

23. Construct a trace table showing the action of insertion sort on the array of exercise 21.

24. How many comparisons between values of $a[j]$ and x actually occur when insertion sort is applied to the array of exercise 20?

25. How many comparisons between values of $a[j]$ and x actually occur when insertion sort is applied to the array of exercise 21?

26. According to Example 9.3.6, the maximum number of comparisons needed to perform insertion sort on an array of length five is $5^2 - 5 + 2 = 22$. Find an array of length five that requires the maximum number of comparisons when insertion sort is applied to it.

H 27. Consider the recurrence relation that arose in Example 9.3.7: $E_1 = 0$ and $E_k = E_{k-1} + k + 1$, for all integers $k \geq 2$.
   a. Use iteration to find an explicit formula for the sequence.
   b. Use mathematical induction to verify the correctness of the formula.

Exercises 28–35 refer to *selection sort*, which is another algorithm to arrange the items in an array in ascending order.

Algorithm 9.3.2 Selection Sort

[*The aim of this algorithm is to take an array $a[1], a[2], a[3], \ldots, a[n]$ (where $n \geq 1$) and interchange its values if necessary to put them in ascending order. In the first step, the array item with the least value is found, and its value is assigned to $a[1]$. In general, in the kth step, $a[k]$ is compared to each $a[i]$ for $i = k + 1, 2, \ldots, n$. Whenever the value of $a[k]$ is greater than that of $a[i]$, the two values are interchanged. The process continues through the $(n-1)$st step, after which the array items are in ascending order.*]

Input: n [*a positive integer*], $a[1], a[2], a[3], \ldots, a[n]$ [*an array of data items capable of being ordered*]

Algorithm Body:

    for $k := 1$ **to** $n - 1$
        for $i := k + 1$ **to** n
            if $a[i] < a[k]$ **then**
                do $temp := a[k]$
                    $a[k] := a[i]$
                    $a[i] := temp$
                end do
        next i
    next k

Output: $a[1], a[2], a[3], \ldots, a[n]$ in ascending order

The action of selection sort can be represented pictorially as follows:

$$a[1]\, a[2] \cdots \boxed{a[k]}\, a[k+1] \cdots a[n]$$

kth step: Compare $a[k]$ to each of $a[k+1], \ldots, a[n]$ and interchange values each time $a[k]$ is greater.

28. Construct a table showing the interchanges that occur when selection sort is applied to the array $a[1] = 5$, $a[2] = 3$, $a[3] = 4$, $a[4] = 6$, and $a[5] = 2$.

29. Construct a table showing the interchanges that occur when selection sort is applied to the array $a[1] = 6$, $a[2] = 4$, $a[3] = 5$, $a[4] = 8$, and $a[5] = 1$.

30. Construct a trace table showing the action of selection sort on the array of exercise 28.

31. Construct a trace table showing the action of selection sort on the array of exercise 29.

32. When selection sort is applied to the array of exercise 28, how many times is the comparison in the **if-then** statement performed?

33. When selection sort is applied to the array of exercise 29, how many times is the comparison in the **if-then** statement performed?

34. When selection sort is applied to an array $a[1], a[2], a[3], a[4]$, how many times is the comparison in the **if-then** statement performed?

35. Consider applying selection sort to an array $a[1], a[2], a[3], \ldots, a[n]$.
   a. How many times is the comparison in the **if-then** statement performed when $a[1]$ is compared to each of $a[2], a[3], \ldots, a[n]$?
   b. How many times is the comparison in the **if-then** statement performed when $a[2]$ is compared to each of $a[3], a[4], \ldots, a[n]$?
   c. How many times is the comparison in the **if-then** statement performed when $a[k]$ is compared to each of $a[k+1], a[k+2], \ldots, a[n]$?
H d. Using the number of times the comparison in the **if-then** statement is performed as a measure of the time efficiency of selection sort, find an order for selection sort from among the set of power functions.

Exercises 36–39 refer to the following algorithm to compute the value of a real polynomial.

Algorithm 9.3.3 Term-by-Term Polynomial Evaluation

[*This algorithm computes the value of the real polynomial* $a[n]x^n + a[n-1]x^{n-1} + \cdots + a[2]x^2 + a[1]x + a[0]$ *by computing each term separately, starting with* $a[0]$, *and adding it on to an accumulating sum.*]

Input: n [*a nonnegative integer*], $a[0], a[1], a[2], \ldots, a[n]$ [*an array of real numbers*], x [*a real number*]

Algorithm Body:

$polyval := a[0]$

for $i := 1$ **to** n
$\quad term := a[i]$
$\quad$ **for** $j := 1$ **to** i
$\quad\quad term := term \cdot x$
$\quad$ **next** j
$\quad polyval := polyval + term$
next i
[*At this point*
$polyval = a[n]x^n + a[n-1]x^{n-1}$
$\quad\quad\quad + \cdots + a[2]x^2 + a[1]x + a[0].$]

Output: $polyval$ [*a real number*]

Algorithm 9.3.4 Horner's Rule

[*This algorithm computes the value of the real polynomial* $a[n]x^n + a[n-1]x^{n-1} + \cdots + a[2]x^2 + a[1]x + a[0]$ *by nesting successive additions and multiplications as indicated in the following parenthesization:*

$$((\cdots((a[n]x + a[n-1])x + a[n-2])x$$
$$+ \cdots + a[2])x + a[1])x + a[0].$$

At each stage, starting with $a[n]$, *the current value of polyval is multiplied by* x *and the next lower coefficient of the polynomial is added on.*]

Input: n [*a nonnegative integer*] $a[0], a[1], a[2], \ldots, a[n]$ [*an array of real numbers*], x [*a real number*]

Algorithm Body:

$polyval := a[n]$

for $i := 1$ **to** n
$\quad polyval := polyval \cdot x + a[n-i]$
next i
[*At this point*
$polyval = a[n]x^n + a[n-1]x^{n-1}$
$\quad\quad\quad + \cdots + a[2]x^2 + a[1]x + a[0].$]

Output: $polyval$ [*a real number*]

36. Trace Algorithm 9.3.3 for the input $n = 3$, $a[0] = 2$, $a[1] = 1$, $a[2] = -1$, $a[3] = 3$, and $x = 2$.

37. Trace Algorithm 9.3.3 for the input $n = 2$, $a[0] = 5$, $a[1] = -1$, $a[2] = 2$, and $x = 3$.

38. Let s_n = the number of additions and multiplications that must be performed when Algorithm 9.3.3 is executed for a polynomial of degree n. Express s_n as a function of n.

39. Find an order for Algorithm 9.3.3 from among the set of power functions.

Exercises 40–43 refer to another algorithm, known as Horner's rule, for finding the value of a real polynomial.

40. Trace Algorithm 9.3.4 for the input $n = 3$, $a[0] = 2$, $a[1] = 1$, $a[2] = -1$, $a[3] = 3$, and $x = 2$.

41. Trace Algorithm 9.3.4 for the input $n = 2$, $a[0] = 5$, $a[1] = -1$, $a[2] = 2$, and $x = 3$.

H 42. Let t_n = the number of additions and multiplications that must be performed when Algorithm 9.3.4 is executed for a polynomial of degree n. Express t_n as a function of n.

43. Find an order for Algorithm 9.3.4 from among the set of power functions. How does this order compare with that of Algorithm 9.3.3?

9.4 Exponential and Logarithmic Functions: Graphs and Orders

We ought never to allow ourselves to be persuaded of the truth of anything unless on the evidence of our own reason. — René Descartes, 1596–1650

Exponential and logarithmic functions are of great importance in mathematics in general and in computer science in particular. Several important computer algorithms have execution times that involve logarithmic functions of the size of the input data (which means they are relatively efficient for large data sets), and some have execution times that are exponential functions of the size of the input data (which means they are quite inefficient for large data sets). In addition, since exponential and logarithmic functions arise naturally in the descriptions of many growth and decay processes and in the computation of many kinds of probabilities, these functions are used in the analysis of computer operating systems, in queuing theory, and in the theory of information.

Graphs of Exponential Functions

As defined in Section 7.2, the exponential function with base $b > 0$ is the function that sends each real number x to b^x. The graph of the exponential function with base 2 (together with a partial table of its values) is shown in Figure 9.4.1. Note that the values of this function increase with extraordinary rapidity. If we tried to continue drawing the graph using the scale shown in Figure 9.4.1, we would have to plot the point $(10, 2^{10})$ more than 21 feet above the horizontal axis. And the point $(30, 2^{30})$ would be located more than 610,080 miles above the axis—well beyond the moon!

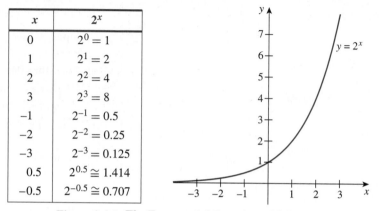

x	2^x
0	$2^0 = 1$
1	$2^1 = 2$
2	$2^2 = 4$
3	$2^3 = 8$
-1	$2^{-1} = 0.5$
-2	$2^{-2} = 0.25$
-3	$2^{-3} = 0.125$
0.5	$2^{0.5} \cong 1.414$
-0.5	$2^{-0.5} \cong 0.707$

Figure 9.4.1 The Exponential Function with Base 2

The graph of any exponential function with base $b > 1$ has a shape that is similar to the graph of the exponential function with base 2. If $0 < b < 1$, then $1/b > 0$ and the graph of the exponential function with base b is the reflection across the vertical axis of the exponential function with base $1/b$. These facts are illustrated in Figure 9.4.2.

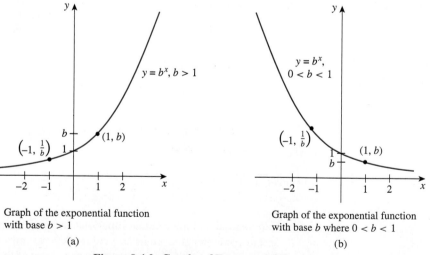

Graph of the exponential function
with base $b > 1$

(a)

Graph of the exponential function
with base b where $0 < b < 1$

(b)

Figure 9.4.2 Graphs of Exponential Functions

Graphs of Logarithmic Functions

Logarithms were first introduced by the Scotsman John Napier. Astronomers and navigators found them so useful for reducing the time needed to do multiplication and division that they quickly gained wide acceptance and played a crucial role in the remarkable development of those areas in the seventeenth century. Nowadays, however, electronic calculators and computers are available to handle most computations quickly and conveniently, and logarithms and logarithmic functions are used primarily as conceptual tools.

Recall the definition of the logarithmic function with base b from Section 7.1. We state it formally below.

John Napier (1550–1617)

• Definition

The **logarithmic function with base b, $\log_b$: $\mathbf{R}^+ \to \mathbf{R}$,** is the function that sends each positive real number x to the number $\log_b x$, which is the exponent to which b must be raised to obtain x.

If b is a positive real number not equal to 1, then the logarithmic function with base b is, in fact, the inverse of the exponential function with base b. (See exercise 10 at the end of this section.) It follows that the graphs of the two functions are symmetric with respect to the line $y = x$. The graph of the logarithmic function with base $b > 1$ is shown in Figure 9.4.3.

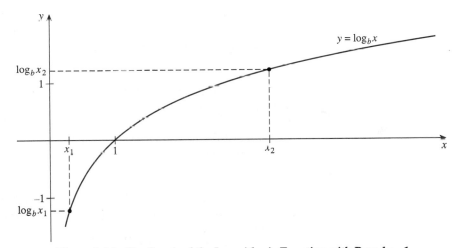

Figure 9.4.3 The Graph of the Logarithmic Function with Base $b > 1$

Observe that if its base b is greater than 1, the logarithmic function is increasing. Analytically, this means that

---

if $b > 1$, then for all positive numbers x_1 and x_2,

$$\text{if } x_1 < x_2, \text{ then } \log_b(x_1) < \log_b(x_2).$$ 9.4.1

Corresponding to the rapid growth of the exponential function, however, is the very slow growth of the logarithmic function. Thus you must go very far out on the horizontal axis to find points whose logarithms are large numbers. For instance, $\log_2(1,024)$ is only 10 and $\log_2(1,048,576)$ is just 20.

The following example shows how to make use of the increasing nature of the logarithmic function with base 2 to derive a remarkably useful property.

Example 9.4.1 Base 2 Logarithms of Numbers between Two Consecutive Powers of 2

Prove the following property:

a.

> If k is an integer and x is a real number with
> $$2^k \leq x < 2^{k+1}, \text{ then } \lfloor \log_2 x \rfloor = k. \qquad 9.4.2$$

b. Describe property (9.4.2) in words and give a graphical interpretation of the property for $x > 1$.

Solution

a. Suppose that k is an integer and x is a real number with

$$2^k \leq x < 2^{k+1}.$$

Because the logarithmic function with base 2 is increasing, this implies that

$$\log_2(2^k) \leq \log_2 x < \log_2(2^{k+1}).$$

But $\log_2(2^k) = k$ [*the exponent to which you must raise 2 to get 2^k is k*] and $\log_2(2^{k+1}) = k + 1$ [*for a similar reason*]. Hence

$$k \leq \log_2 x < k + 1.$$

By definition of the floor function, then,

$$\lfloor \log_2 x \rfloor = k.$$

b. Recall that the floor of a positive number is its integer part. For instance, $\lfloor 2.82 \rfloor = 2$. Hence property (9.4.2) can be described in words as follows:

> If x is a positive number that lies between two consecutive integer powers of 2, the floor of the logarithm with base 2 of x is the exponent of the smaller power of 2.

A graphical interpretation follows:

One consequence of property (9.4.2) does not appear particularly interesting in its own right but is frequently needed as a step in the analysis of algorithm efficiency.

Example 9.4.2 When $\lfloor \log_2(n-1) \rfloor = \lfloor \log_2 n \rfloor$

Prove the following property:

> For any odd integer $n > 1$, $\lfloor \log_2(n-1) \rfloor = \lfloor \log_2 n \rfloor$.   9.4.3

Solution   If n is an odd integer that is greater than 1, then n lies strictly between two successive powers of 2:

$$2^k < n < 2^{k+1} \quad \text{for some integer } k > 0. \qquad 9.4.4$$

It follows that $2^k \leq n - 1$ because $2^k < n$ and both 2^k and n are integers. Consequently,

$$2^k \leq n - 1 < 2^{k+1}. \qquad 9.4.5$$

Applying property (9.4.2) to both (9.4.4) and (9.4.5) gives

$$\lfloor \log_2 n \rfloor = k \quad \text{and also} \quad \lfloor \log_2(n-1) \rfloor = k.$$

Hence $\lfloor \log_2 n \rfloor = \lfloor \log_2(n-1) \rfloor$. ∎

Application: Number of Bits Needed to Represent an Integer in Binary Notation

Given a positive integer n, how many binary digits are needed to represent n? To answer this question, recall from Section 4.4 that any positive integer n can be written in a unique way as

$$n = 2^k + c_{k-1} \cdot 2^{k-1} + \cdots + c_2 \cdot 2^2 + c_1 \cdot 2 + c_0,$$

where k is a nonnegative integer and each $c_0, c_1, c_2, \ldots c_{k-1}$ is either 0 or 1. Then the binary representation of n is

$$1 c_{k-1} c_{k-2} \cdots c_2 c_1 c_0,$$

and so the number of binary digits needed to represent n is $k + 1$.

What is $k + 1$ as a function of n? Observe that since each $c_i \leq 1$,

$$n = 2^k + c_{k-1} \cdot 2^{k-1} + \cdots + c_2 \cdot 2^2 + c_1 \cdot 2 + c_0 \leq 2^k + 2^{k-1} + \cdots + 2^2 + 2 + 1.$$

But by the formula for the sum of a geometric sequence (Theorem 4.2.3),

$$2^k + 2^{k-1} + \cdots + 2^2 + 2 + 1 = \frac{2^{k+1} - 1}{2 - 1} = 2^{k+1} - 1.$$

Hence, by transitivity of order,

$$n \leq 2^{k+1} - 1 < 2^{k+1} \qquad 9.4.6$$

In addition, because each $c_i \geq 0$,

$$2^k \leq 2^k + c_{k-1} \cdot 2^{k-1} + \cdots + c_2 \cdot 2^2 + c_1 \cdot 2 + c_0 = n. \qquad 9.4.7$$

Putting inequalities (9.4.6) and (9.4.7) together gives the double inequality

$$2^k \leq n < 2^{k+1}.$$

But then, by property (9.4.2),

$$k = \lfloor \log_2 n \rfloor.$$

Thus the number of binary digits needed to represent n is $\lfloor \log_2 n \rfloor + 1$.

Example 9.4.3 Number of Bits in a Binary Representation

How many binary digits are needed to represent 52,837 in binary notation?

Solution If you compute the logarithm with base 2 using formula (7.3.6) in Example 7.3.6 and a calculator that gives you approximate values of logarithms with base 10, you find that

$$\log_2(52,837) \cong \frac{\log_{10}(52,837)}{\log_{10}(2)} \cong \frac{4.722938151}{0.3010299957} \cong 15.7.$$

Thus the binary representation of 52,837 has $\lfloor 15.7 \rfloor + 1 = 15 + 1 = 16$ binary digits. ■

Application: Using Logarithms to Solve Recurrence Relations

In Chapter 8 we discussed methods for solving recurrence relations. One class of recurrence relations that is very important in computer science has solutions that can be expressed in terms of logarithms. One such recurrence relation is discussed in the next example.

Example 9.4.4 A Recurrence Relation with a Logarithmic Solution

Define a sequence $a_1, a_2, a_3, \ldots$ recursively as follows:

$$a_1 = 1,$$
$$a_k = 2a_{\lfloor k/2 \rfloor} \quad \text{for all integers } k \geq 2.$$

a. Use iteration to guess an explicit formula for this sequence.

b. Use strong mathematical induction to confirm the correctness of the formula obtained in part (a).

Solution

a. Begin by iterating to find the values of the first few terms of the sequence.

$$\begin{aligned}
a_1 &= 1 & 1 &= 2^0 \\
a_2 &= 2a_{\lfloor 2/2 \rfloor} = 2a_1 = 2 \cdot 1 = 2 \\
a_3 &= 2a_{\lfloor 3/2 \rfloor} = 2a_1 = 2 \cdot 1 = 2 & 2 &= 2^1 \\
a_4 &= 2a_{\lfloor 4/2 \rfloor} = 2a_2 = 2 \cdot 2 = 4 & 4 &= 2^2 \\
a_5 &= 2a_{\lfloor 5/2 \rfloor} = 2a_2 = 2 \cdot 2 = 4 \\
a_6 &= 2a_{\lfloor 6/2 \rfloor} = 2a_3 = 2 \cdot 2 = 4 \\
a_7 &= 2a_{\lfloor 7/2 \rfloor} = 2a_3 = 2 \cdot 2 = 4 \\
a_8 &= 2a_{\lfloor 8/2 \rfloor} = 2a_4 = 2 \cdot 4 = 8 & 8 &= 2^3 \\
a_9 &= 2a_{\lfloor 9/2 \rfloor} = 2a_4 = 2 \cdot 4 = 8 \\
&\vdots & &\vdots \\
a_{15} &= 2a_{\lfloor 15/2 \rfloor} = 2a_7 = 2 \cdot 4 = 8 \\
a_{16} &= 2a_{\lfloor 16/2 \rfloor} = 2a_8 = 2 \cdot 8 = 16 & 16 &= 2^4 \\
&\vdots & &\vdots
\end{aligned}$$

Note that in each case when the subscript n is between two powers of 2, a_n equals the smaller power of 2. More precisely:

$$\text{If } 2^i \leq n < 2^{i+1}, \text{ then } a_n = 2^i. \qquad\qquad 9.4.8$$

But since n satisfies the inequality

$$2^i \leq n < 2^{i+1},$$

then (by property 9.4.2)

$$i = \lfloor \log_2 n \rfloor.$$

Substituting into statement (9.4.8) gives

$$a_n = 2^{\lfloor \log_2 n \rfloor}.$$

b. The following proof shows that if $a_1, a_2, a_3, \ldots$ is a sequence of numbers that satisfies

$$a_1 = 1,$$

and

$$a_k = 2a_{\lfloor k/2 \rfloor} \quad \text{for all } k \geq 2,$$

then the sequence satisfies the formula

$$a_n = 2^{\lfloor \log_2 n \rfloor} \quad \text{for all integers } n \geq 1.$$

Proof:

Show that the sequence satisfies the formula for $n = 1$: By definition of the sequence, $a_1 = 1$. And $2^{\lfloor \log_2 1 \rfloor} = 2^0 = 1$ also. Hence $a_1 = 2^{\lfloor \log_2 1 \rfloor}$, and so the formula holds for $n = 1$.

Show that for all integers $k \geq 2$, if the sequence satisfies the formula for all integers i with $1 \leq i < k$, then the sequence satisfies the formula for k: Let k be an integer that is greater than or equal to 2, and suppose $a_i = 2^{\lfloor \log_2 i \rfloor}$ for all integers i with $1 \leq i < k$. Now either k is odd or k is even.

Case 1 (k is odd): In this case,

$$
\begin{aligned}
a_k &= 2a_{\lfloor k/2 \rfloor} && \text{by definition of } a_1, a_2, a_3, \ldots \\
&= 2a_{(k-1)/2} && \text{because } \lfloor k/2 \rfloor = (k-1)/2 \text{ since } k \text{ is odd} \\
&= 2 \cdot 2^{\lfloor \log_2((k-1)/2) \rfloor} && \text{by inductive hypothesis (since } k \geq 2, 1 \leq k/2 < k) \\
&= 2^{\lfloor \log_2((k-1)/2) \rfloor + 1} && \text{by the laws of exponents from algebra (7.2.1)} \\
&= 2^{\lfloor \log_2(k-1) - \log_2 2 \rfloor + 1} && \text{by the identity } \log_b(x/y) = \log_b x - \log_b y \text{ derived} \\
& && \text{in exercise 29 of Section 7.2} \\
&= 2^{\lfloor \log_2(k-1) - 1 \rfloor + 1} && \text{since } \log_2 2 = 1 \\
&= 2^{\lfloor \log_2(k-1) \rfloor - 1 + 1} && \text{by substituting } x = \log_2(k-1) \text{ into the identity} \\
& && \lfloor x - 1 \rfloor = \lfloor x \rfloor - 1 \text{ derived in exercise 15 of Section 3.5} \\
&= 2^{\lfloor \log_2(k-1) \rfloor} && \\
&= 2^{\lfloor \log_2 k \rfloor} && \text{by property (9.4.3)}
\end{aligned}
$$

Case 2 (k is even): The analysis of this case is very similar to that of case 1 and is left as an exercise.

Thus in either case, $a_n = 2^{\lfloor \log_2 n \rfloor}$, as was to be shown. ■

Exponential and Logarithmic Orders

Now consider the question "How do graphs of logarithmic and exponential functions compare with graphs of power functions?" It turns out that for large enough values of x, the graph of the logarithmic function with any base $b > 1$ lies *below* the graph of any positive power function, and the graph of the exponential function with any base $b > 1$ lies *above* the graph of any positive power function. In analytic terms, this says the following:

For all real numbers b and r with $b > 1$ and $r > 0$,

$$\log_b x \le x^r \quad \text{for all sufficiently large values of } x. \qquad 9.4.9$$

and

$$x^r \le b^x \quad \text{for all sufficiently large values of } x. \qquad 9.4.10$$

These statements have the following implications for O-notation.

For all real numbers b and r with $b > 1$ and $r > 0$,

$$\log_b x \quad \text{is} \quad O(x^r) \qquad 9.4.11$$

and

$$x^r \quad \text{is} \quad O(b^x) \qquad 9.4.12$$

Another important function in the analysis of algorithms is the function f defined by the formula

$$f(x) = x \log_b x \quad \text{for all real numbers } x > 0.$$

For large values of x, the graph of this function fits in between the graph of the identity function and the graph of the squaring function. More precisely:

For all real numbers b with $b > 1$ and for all sufficiently large values of x,

$$x \le x \log_b x \le x^2. \qquad 9.4.13$$

The O-notation versions of these facts are as follows:

For all real numbers $b > 1$,

$$x \quad \text{is} \quad O(x \log_b x) \quad \text{and} \quad x \log_b x \quad \text{is} \quad O(x^2). \qquad 9.4.14$$

Although proofs of some of these facts require calculus, proofs of some cases can be obtained using the algebra of inequalities. (See the exercises at the end of this section.) Figure 9.4.4 illustrates the relationships among some power functions, the logarithmic function with base 2, the exponential function with base 2, and the function defined by the formula $x \rightarrow x \log_2 x$. Note that different scales are used on the horizontal and vertical axes.

Example 9.4.5 shows how to use inequalities such as (9.4.9), (9.4.10), and (9.4.13) to derive additional orders involving the logarithmic function.

Figure 9.4.4 Graphs of Some Logarithmic, Exponential, and Power Functions

Example 9.4.5 Deriving an Order from Logarithmic Inequalities

Show that $x + x \log_2 x$ is $\Theta(x \log_2 x)$.

Solution   First observe that $x + x \log_2 x$ is $\Omega(x)$ because for all real numbers $x > 1$,

$$x \log_2 x \ \leq \ x + x \log_2 x,$$

and since all quantities are positive,

$$|x \log_2 x| \ \leq \ |x + x \log_2 x|.$$

Let $A = 1$ and $a = 1$. Then

$$A|x \log_2 x| \ \leq \ |x + x \log_2 x| \quad \text{for all } x > a.$$

Hence, by definition of Ω-notation,

$$x + x \log_2 x \ \text{ is } \ \Omega(x \log_2 x).$$

To show that $x + x \log_2 x$ is $O(x)$, note that according to property (9.4.13) with $b = 2$, there is a number x_2 such that for all $x > x_2$,

$$x < x \log_2 x$$
$$\Rightarrow \quad x + x \log_2 x < 2x \log_2 x \quad \text{by adding } x \log_2 x \text{ to both sides}$$

Thus, if x_2 is taken to be greater than 2, then

$$|x + x \log_2 x| \ < \ 2|x \log_2 x| \qquad \begin{array}{l} \text{because when } x > 2, x \log_2 x > 0, \text{ and so} \\ |x + x \log_2 x| = x + x \log_2 x \text{ and} \\ \log_2 x = |x \log_2 x|. \end{array}$$

Let $B = 2$. Then

$$|x + x \log_2 x| \leq B|x \log_2 x| \quad \text{for all } x > b.$$

Hence, by definition of O-notation

$$x + x \log_2 x \quad \text{is} \quad O(x \log_2 x).$$

Therefore, since $x + x \log_2 x$ is $\Omega(x)$ and $x + x \log_2 x$ is $O(x)$, by Theorem 9.2.1,

$$x + x \log_2 x \quad \text{is} \quad \Theta(x). \qquad \blacksquare$$

Example 9.4.5 illustrates a special case of a useful general fact about O-notation: *If one function "dominates" another (in the sense of being larger for large values of the variable), then the sum of the two is big-O of the dominating function.* (See exercise 49a at the end of Section 9.2.)

Example 9.4.6 shows that any two logarithmic functions with bases greater than 1 have the same order.

Example 9.4.6 Logarithm with Base b Is Big-Theta of Logarithm with Base c

Show that if b and c are real numbers such that $b > 1$ and $c > 1$, then $\log_b x$ is $\Theta(\log_c x)$.

Solution   Suppose b and c are real numbers and $b > 1$ and $c > 1$. To show that $\log_b x$ is $\Theta(\log_c x)$, positive real numbers A, B, and k must be found such that

$$A|\log_c x| \leq |\log_b x| \leq B|\log_c x| \quad \text{for all real numbers } x > k.$$

By property (7.2.7) in Example 7.2.6,

$$\log_b x = \frac{\log_c x}{\log_c b} = \left(\frac{1}{\log_c b}\right)\log_c x. \qquad (*)$$

Since $b > 1$ and the logarithmic function with base c is strictly increasing, then $\log_c b > \log_c 1 = 0$, and so $\dfrac{1}{\log_c b} > 0$ also. Furthermore, if $x > 1$, then $\log_b x > 0$ and $\log_c x > 0$. It follows from equation $(*)$, therefore, that

$$\left(\frac{1}{2\log_c b}\right)\log_c x \leq \log_b x \leq \left(\frac{2}{\log_c b}\right)\log_c x \qquad (**)$$

for all real numbers $x > 1$. Accordingly, let $A = \dfrac{1}{2\log_c b}$, $B = \dfrac{2}{\log_c b}$, and $k = 1$. Then, since all quantities in $(**)$ are positive,

$$A|\log_c x| \leq |\log_b x| \leq B|\log_c x| \quad \text{for all real numbers } x > k.$$

Hence, by definition of Θ-notation,

$$\log_b x \quad \text{is} \quad \Theta(\log_c x). \qquad \blacksquare$$

Example 9.4.7 shows how a logarithmic order can arise from the computation of a certain kind of sum. It requires the following fact from calculus:

The area underneath the graph of $y = 1/x$ between $x = 1$ and $x = n$ equals $\ln n$, where $\ln n = \log_e n$. This fact is illustrated in Figure 9.4.5.

Figure 9.4.5 Area Under Graph of $y = \dfrac{1}{x}$ Between $x = 1$ and $x = n$

Example 9.4.7 Order of Harmonic Sum

Sums of the form $1 + \dfrac{1}{2} + \cdots + \dfrac{1}{n}$ are called *harmonic sums*. They occur in the analysis of various computer algorithms such as quick sort. Show that $1 + \dfrac{1}{2} + \dfrac{1}{3} + \cdots + \dfrac{1}{n}$ is $\Omega(\ln n)$ by performing the following steps:

a. Interpret Figure 9.4.6 to show that

$$\frac{1}{2} + \frac{1}{3} + \cdots + \frac{1}{n} \leq \ln n.$$

and

$$\ln n \leq 1 + \frac{1}{2} + \frac{1}{3} + \cdots + \frac{1}{n}.$$

b. Show that if n is an integer that is at least 3, then $1 \leq \ln n$.

c. Deduce from (a) and (b) that if the integer n is greater than or equal to 3, then

$$\ln n \leq 1 + \frac{1}{2} + \frac{1}{3} + \cdots + \frac{1}{n} \leq 2 \ln n.$$

d. Deduce from (c) that

$$1 + \frac{1}{2} + \frac{1}{3} + \cdots + \frac{1}{n} \text{ is } \Theta(\ln n).$$

Solution

a. Figure 9.4.6(a) shows rectangles whose bases are the intervals between each pair of integers from 1 to n and whose heights are the heights of the graph of $y = 1/x$ above the right-hand endpoints of the intervals. Figure 9.4.6(b) shows rectangles with the

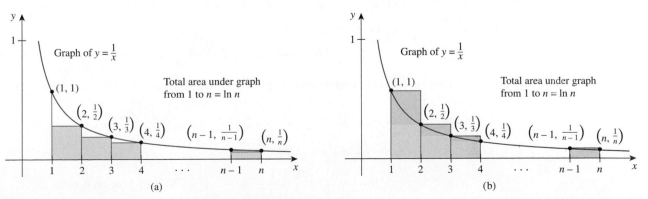

Figure 9.4.6

same bases but whose heights are the heights of the graph above the left-hand endpoints of the intervals.

Now the area of each rectangle is its base times its height. Since all the rectangles have base 1, the area of each rectangle equals its height. Thus in Figure 9.4.6(a),

$$\text{the area of the rectangle from 1 to 2 is } \frac{1}{2};$$

$$\text{the area of the rectangle from 2 to 3 is } \frac{1}{3};$$

$$\vdots$$

$$\text{the area of the rectangle from } n-1 \text{ to } n \text{ is } \frac{1}{n}.$$

So the sum of the areas of all the rectangles is $\frac{1}{2} + \frac{1}{3} + \cdots + \frac{1}{n}$. From the picture it is clear that this sum is no larger than the area underneath the graph of f between $x = 1$ and $x = n$, which is known to equal $\ln n$. Hence

$$\frac{1}{2} + \frac{1}{3} + \cdots + \frac{1}{n} \leq \ln n.$$

A similar analysis of the areas of the combined blue and gray rectangles in Figure 9.4.6(b) shows that

$$\ln n \leq 1 + \frac{1}{2} + \frac{1}{3} + \cdots + \frac{1}{n}.$$

b. Suppose n is an integer and $n \geq 3$. Since $e \cong 2.718$, then $n \geq e$. Now the logarithmic function with base e is strictly increasing. Thus since $e \leq n$, then $1 = \ln e \leq \ln n$.

c. By part (a),

$$\frac{1}{2} + \frac{1}{3} + \cdots + \frac{1}{n} \leq \ln n,$$

and by part (b),

$$1 \leq \ln n.$$

Adding these two inequalities together gives

$$1 + \frac{1}{2} + \frac{1}{3} + \cdots + \frac{1}{n} \leq 2 \ln n \quad \text{for any integer } n \geq 3.$$

d. Putting together the results of parts (a) and (c) leads to the conclusion that for all integers $n \geq 3$,

$$\ln n \leq 1 + \frac{1}{2} + \frac{1}{3} + \cdots + \frac{1}{n} \leq 2 \ln n.$$

And because all the quantities are positive for $n \geq 3$,

$$|\ln n| \leq \left| 1 + \frac{1}{2} + \frac{1}{3} + \cdots + \frac{1}{n} \right| \leq 2 |\ln n|.$$

Let $A = 1$, $B = 2$, and $k = 3$. Then

$$A |\ln n| \leq \left| 1 + \frac{1}{2} + \frac{1}{3} + \cdots + \frac{1}{n} \right| \leq B |\ln n| \quad \text{for all } n > k.$$

Hence by definition of Θ-notation,

$$1 + \frac{1}{2} + \frac{1}{3} + \cdots + \frac{1}{n} \quad \text{is} \quad \Theta(\ln n).$$ ∎

Exercise Set 9.4

Graph each function defined in 1–8 below.

1. $f(x) = 3^x$ for all real numbers x

2. $g(x) = \left(\frac{1}{3}\right)^x$ for all real numbers x

3. $h(x) = \log_{10} x$ for all positive real numbers x

4. $k(x) = \log_2 x$ for all positive real numbers x

5. $F(x) = \lfloor \log_2 x \rfloor$ for all positive real numbers x

6. $G(x) = \lceil \log_2 x \rceil$ for all positive real numbers x

7. $H(x) = x \log_2 x$ for all positive real numbers x

8. $K(x) = x \log_{10} x$ for all positive real numbers x

9. The scale of the graph shown in Figure 9.4.1 is one-fourth inch to each unit. If the point $(2, 2^{64})$ is plotted on the graph of $y = 2^x$, how many miles will it lie above the horizontal axis? What is the ratio of the height of the point to the distance of the earth from the sun? (There are 12 inches per foot and 5,280 feet per mile. The earth is approximately 93,000,000 miles from the sun on average.) ($\frac{1}{4}$ inch $\cong 0.635$ cm, 1 mile $\cong 0.62$ km)

10. a. Use the definition of logarithm to show that $\log_b b^x = x$ for all real numbers x.
    b. Use the definition of logarithm to show that $b^{\log_b x} = x$ for all positive real numbers x.
    c. By the result of exercise 26 in Section 7.4, if $f: X \to Y$ and $g: Y \to X$ are functions and $g \circ f = i_X$ and $f \circ g = i_Y$, then f and g are inverse functions. Use this result to show that $\log_b$ and $\exp_b$ (the exponential function with base b) are inverse functions.

11. Let $b > 1$.
    a. Use the fact that $u = \log_b v \Leftrightarrow v = b^u$ to show that a point (u, v) lies on the graph of the logarithmic function with base b if, and only if, (v, u) lies on the graph of the exponential function with base b.
    b. Plot several pairs of points of the form (u, v) and (v, u) on a coordinate system. Describe the geometric relationship between the locations of the points in each pair.
    c. Draw the graphs of $y = \log_2 x$ and $y = 2^x$. Describe the geometric relationship between these graphs.

12. Give a graphical interpretation for property (9.4.2) in Example 9.4.1(a) for $0 < x < 1$.

H 13. Suppose a positive real number x satisfies the inequality $10^m \le x < 10^{m+1}$ where m is an integer. What can be inferred about $\lfloor \log_{10} x \rfloor$? Justify your answer.

14. a. Prove that if x is a positive real number and k is a nonnegative integer such that $2^{k-1} < x \le 2^k$, then $\lceil \log_2 x \rceil = k$.
    b. Describe in words the statement proved in part (a).

15. If n is an odd integer and $n > 1$, is $\lceil \log_2(n-1) \rceil = \lceil \log_2(n) \rceil$? Justify your answer.

H 16. If n is an odd integer and $n > 1$, is $\lceil \log_2(n+1) \rceil = \lceil \log_2(n) \rceil$? Justify your answer.

17. If n is an odd integer and $n > 1$, is $\lfloor \log_2(n+1) \rfloor = \lfloor \log_2(n) \rfloor$? Justify your answer.

In 18 and 19, indicate how many binary digits are needed to represent the numbers in binary notation. Use the method shown in Example 9.4.3.

18. 148,206   19. 5,067,329

20. It was shown in the text that the number of binary digits needed to represent a positive integer n is $\lfloor \log_2 n \rfloor + 1$. Can this also be given as $\lceil \log_2 n \rceil$? Why or why not?

In each of 21 and 22, a sequence is specified by a recurrence relation and initial conditions. In each case, (a) use iteration to guess an explicit formula for the sequence; (b) use strong mathematical induction to confirm the correctness of the formula you obtained in part(a).

21. $a_k = a_{\lfloor k/2 \rfloor} + 2$, for all integers $k \ge 2$
    $a_1 = 1$

22. $b_k = b_{\lceil k/2 \rceil} + 1$, for all integers $k \ge 2$
    $b_1 = 1$.

H 23. Define a sequence $c_1, c_2, c_3, \ldots,$ recursively as follows:

$$c_1 = 0,$$
$$c_k = 2c_{\lfloor k/2 \rfloor} + k, \quad \text{for all integers } k \ge 2.$$

Use strong mathematical induction to show that $c_n \le n^2$ for all integers $n \ge 1$.

✶ H 24. Use strong mathematical induction to show that for the sequence of exercise 23, $c_n \le n \log_2 n$, for all integers $n \ge 4$.

Exercises 25–28 refer to properties 9.4.9 and 9.4.10. To solve them, think big!

25. Find a real number $x > 3$ such that $\log_2 x < x^{1/10}$.

26. Find a real number $x > 1$ such that $x^{50} < 2^x$.

27. Find a real number $x > 2$ such that $x < 1.0001^x$.

28. Use a graphing calculator or computer graphing program to find two distinct approximate values of x such that $x = 1.0001^x$. On what approximate intervals is $x > 1.0001^x$? On what approximate intervals is $x < 1.0001^x$?

29. Use Θ-notation to express the following statement:

$$|x^2| \leq |7x^2 + 3x \log_2 x| \leq 10|x^2|,$$

for all real numbers $x > 2$.

Derive each statement in 30–33.

30. $2x + \log_2 x$ is $\Theta(x)$.

31. $x^2 + 5x \log_2 x$ is $\Theta(x^2)$.

32. $n^2 + 2^n$ is $\Theta(2^n)$.

H 33. 2^{n+1} is $\Theta(2^n)$.

H 34. Show that 4^n is not $O(2^n)$.

Prove each of the statements in 35–40, assuming n is an integer variable that takes positive integer values. Use identities from Section 4.2 as needed.

35. $1 + 2 + 2^2 + 2^3 + \cdots + 2^n$ is $\Theta(2^n)$.

H 36. $4 + 4^2 + 4^3 + \cdots + 4^n$ is $\Theta(4^n)$.

37. $2 + 2 \cdot 3^2 + 2 \cdot 3^4 + \cdots + 2 \cdot 3^{2n}$ is $\Theta(3^{2n})$.

38. $\dfrac{1}{5} + \dfrac{4}{5^2} + \dfrac{4^2}{5^3} + \cdots + \dfrac{4^n}{5^{n+1}}$ is $\Theta\left(\left(\dfrac{4}{5}\right)^n\right)$.

39. $n + \dfrac{n}{2} + \dfrac{n}{4} + \cdots + \dfrac{n}{2^n}$ is $\Theta(n)$.

40. $\dfrac{2n}{3} + \dfrac{2n}{3^2} + \dfrac{2n}{3^3} + \cdots + \dfrac{2n}{3^n}$ is $\Theta(n)$.

41. Quantities of the form

$$kn + kn \log_2 n \quad \text{for positive integers } k \text{ and } n$$

arise in the analysis of the merge sort algorithm in computer science. Show that for any positive integer k,

$$kn + kn \log_2 n \quad \text{is} \quad \Theta(n \log_2 n).$$

42. Calculate the values of the harmonic sums

$$1 + \frac{1}{2} + \frac{1}{3} + \cdots + \frac{1}{n} \quad \text{for } n = 2, 3, 4, \text{ and } 5.$$

43. Use part (d) of Example 9.4.7 to show that

$$n + \frac{n}{2} + \frac{n}{3} + \cdots + \frac{n}{n} \quad \text{is} \quad \Theta(n \ln n).$$

44. Use the fact that $\log_2 x = \left(\dfrac{1}{\log_e 2}\right) \log_e x$ and $\log_e x = \ln x$, for all positive numbers x, and part (c) of Example 9.4.7 to show that

$$1 + \frac{1}{2} + \frac{1}{3} + \cdots + \frac{1}{n} \quad \text{is} \quad \Theta(\log_2 n).$$

45. a. Show that $\lfloor \log_2 n \rfloor$ is $\Theta(\log_2 n)$.
    b. Show that $\lfloor \log_2 n \rfloor + 1$ is $\Theta(\log_2 n)$.

46. Prove by mathematical induction that $n \leq 10^n$ for all integers $n \geq 1$.

H 47. Prove by mathematical induction that $\log_2 n \leq n$ for all integers $n \geq 1$.

H 48. Show that if n is a variable that takes positive integer values, then 2^n is $O(n!)$.

49. Let n be a variable that takes positive integer values.
    a. Show that $n!$ is $O(n^n)$.
    b. Use part (a) to show that $\log_2(n!)$ is $O(n \log_2 n)$.
    H c. Show that $n^n \leq (n!)^2$ for all integers $n \geq 2$.
    d. Use part (c) to show that $\log_2(n!)$ is $\Omega(n \log_2 n)$.
    e. Use parts (b) and (d) to find an order for $\log_2(n!)$.

★ 50. a. For all real numbers u, $\log_2 u < u$. Use this fact to show that for any positive integer n, $\log_2 x < nx^{1/n}$ for all real numbers $x > 0$.
    b. Interpret the statement of part (a) using O-notation.

51. a. For all positive real numbers x, $x < 2^x$. Use this fact to show that for any positive integer n, $x^n < n^n 2^x$ for all real numbers $x > 0$.
    b. Interpret the statement of part (a) using O-notation.

★ 52. For all positive real numbers u, $\log_2 u < u$. Use this fact and the result of exercise 20 in Section 9.1 to prove the following: For all integers $n \geq 1$, $\log_2 x < x^{1/n}$ for all real numbers $x > (2n)^{2n}$.

53. Use the result of exercise 52 above to prove the following: For all integers $n \geq 1$, $x^n < 2^x$ for all real numbers $x > (2n)^{2n}$.

Exercises 54 and 55 use L'Hôpital's rule from calculus.

54. a. Let b be any real number greater than 1. Use L'Hôpital's rule and mathematical induction to prove that for all integers $n \geq 1$,

$$\lim_{x \to \infty} \frac{x^n}{b^x} = 0.$$

    b. Use the result of part (a) and the definitions of limit and of O-notation to prove that x^n is $O(b^x)$ for any integer $n \geq 1$.

55. a. Let b be any real number greater than 1. Use L'Hôpital's rule to prove that for all integers $n \geq 1$,

$$\lim_{x \to \infty} \frac{\log_b x}{x^{1/n}} = 0.$$

    b. Use the result of part (a) and the definitions of limit and of O-notation to prove that $\log_b x$ is $O(x^{1/n})$ for any integer $n \geq 1$.

9.5 *Application: Efficiency of Algorithms II*

Have you ever played the "guess my number" game? A person thinks of a number between two other numbers, say 1 and 10 or 1 and 100 for example, and you try to figure out what it is, using the least possible number of guesses. Each time you guess a number, the person tells you whether you are correct, too low, or too high.

If you have ever played this game, you have probably already hit upon the most efficient strategy: Begin by guessing a number as close to the middle of the two given numbers as possible. If your guess is too high, then the number is between the lower of the two given numbers and the one you first chose. If your guess is too low, then the number is between the number you first chose and the higher of the two given numbers. In either case, you take as your next guess a number as close as possible to the middle of the new range in which you now know the number lies. You repeat this process as many times as necessary until you have found the person's number.

The technique described above is an example of a general strategy called **divide and conquer,** which works as follows: To solve a problem, reduce it to a fixed number of smaller problems of the same kind, which can themselves be reduced to the same fixed number of smaller problems of the same kind, and so forth until easily resolved problems are obtained. In this case, the problem of finding a particular number in a given range of numbers is reduced at each stage to finding a particular number in a range of numbers approximately half as long.

It turns out that algorithms using a divide-and-conquer strategy are generally quite efficient and nearly always have orders involving logarithmic functions. In this section we define the *binary search* algorithm, which is the formalization of the "guess my number" game described above, and we compare the efficiency of binary search to the sequential search discussed in Section 9.3. Then we develop a divide-and-conquer algorithm for sorting, *merge sort,* and compare its efficiency with that of insertion sort and selection sort, which were also discussed in Section 9.3.

Binary Search

Whereas a sequential search can be performed on an array whose elements are in any order, a binary search can be performed only on an array whose elements are arranged in ascending (or descending) order. Given an array $a[1], a[2], \ldots, a[n]$ of distinct elements arranged in ascending order, consider the problem of trying to find a particular element x in the array.

To use binary search, first compare x to the "middle element" of the array. If the two are equal, the search is successful. If the two are not equal, then because the array elements are in ascending order, comparing the values of x and the middle array element narrows the search either to the lower subarray (consisting of all the array elements below the middle element) or to the upper subarray (consisting of all array elements above the middle element).

The search continues by repeating this basic process over and over on successively smaller subarrays. It terminates either when a match occurs or when the subarray to which the search has been narrowed contains no elements. The efficiency of the algorithm is a result of the fact that at each step, the length of the subarray to be searched is roughly half the length of the array of the previous step. This process is illustrated in Figure 9.5.1.

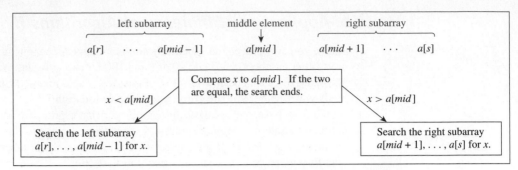

Figure 9.5.1 One Iteration of the Binary Search Process

To write down a formal algorithm for binary search, we introduce a variable *index* whose final value will tell us whether or not x is in the array and, if so, will indicate the location of x. Since the array goes from $a[1]$ to $a[n]$, we intialize *index* to be 0. If and when x is found, the value of *index* is changed to the subscript of the array element equaling x. If index still has the value 0 when the algorithm is complete, then x is not one of the elements in the array. Figure 9.5.2 shows the action of a particular binary search.

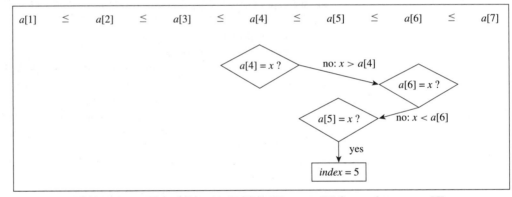

Figure 9.5.2 Binary Search of $a[1]$, $a[2]$, . . . , $a[7]$ for x where $x = a[5]$

Formalizing a binary search algorithm also requires that we be more precise about the meaning of the "middle element" of an array. (This issue was side-stepped by careful choice of n in Figure 9.5.2.) If the array consists of an even number of elements, there are two elements in the middle. For instance, both $a[6]$ and $a[7]$ are equally in the middle of the following array.

In a case such as this, the algorithm must choose which of the two middle elements to take, the smaller or the larger. The choice is arbitrary—either would do. We will write the algorithm to choose the smaller. The index of the smaller of the two middle elements is the floor of the average of the top and bottom indices of the array. That is, if

$$bot = \text{the bottom index of the array,}$$

$$top = \text{the top index of the array,} \quad \text{and}$$

$$mid = \text{the lower of the two middle indices of the array,}$$

then

$$mid = \left\lfloor \frac{bot + top}{2} \right\rfloor.$$

In this case, $bot = 3$ and $top = 10$, so the index of the "middle element" is

$$mid = \left\lfloor \frac{3 + 10}{2} \right\rfloor = \left\lfloor \frac{13}{2} \right\rfloor = \lfloor 6.5 \rfloor = 6.$$

The following is a formal algorithm for a binary search.

Algorithm 9.5.1 Binary Search

[The aim of this algorithm is to search for an element x in an ascending array of elements $a[1], a[2], \ldots, a[n]$. If x is found, the variable index is set equal to the index of the array element where x is located. If x is not found, index is not changed from its initial value, which is 0. The variables bot and top denote the bottom and top indices of the array currently being examined.]

Input: n *[a positive integer]*, $a[1], a[2], \ldots, a[n]$ *[an array of data items given in ascending order]*, x *[a data item of the same data type as the elements of the array]*

Algorithm Body:

$index := 0, bot := 1, top := n$

[Compute the middle index of the array, mid. Compare x to $a[mid]$. If the two are equal, the search is successful. If not, repeat the process either for the lower or for the upper subarray, either giving top the new value $mid - 1$ or giving bot the new value $mid + 1$. Each iteration of the loop either decreases the value of top or increases the value of bot. Thus, if the looping is not stopped by success in the search process, eventually the value of top will become less than the value of bot. This occurrence stops the looping process and indicates that x is not an element of the array.]

    while ($top \geq bot$ and $index = 0$)

$$mid := \left\lfloor \frac{bot + top}{2} \right\rfloor$$

      if $a[mid] = x$ **then** $index := mid$

      if $a[mid] > x$

          then $top := mid - 1$

          else $bot := mid + 1$

    end while

[If index has the value 0 at this point, then x is not in the array. Otherwise, index gives the index of the array where x is located.]

Output: $index$ *[a nonnegative integer]*.

Example 9.5.1 Tracing the Binary Search Algorithm

Trace the action of Algorithm 9.5.1 on the variables *index, bot, top, mid,* and the values of x given in (a) and (b) below for the input array

$$a[1] = \text{Ann}, a[2] = \text{Dawn}, a[3] = \text{Erik}, a[4] = \text{Gail}, a[5] = \text{Juan},$$
$$a[6] = \text{Matt}, a[7] = \text{Max}, a[8] = \text{Rita}, a[9] = \text{Tsuji}, a[10] = \text{Yuen}$$

where alphabetical ordering is used to campare elements of the array.

a. $x = \text{Max}$ b. $x = \text{Sara}$

Solution

a.

index	0				7
bot	1	6		7	
top	10		7		
mid		5	8	6	7

b.

index	0			
bot	1	6	9	
top	10			8
mid		5	8	9

The Efficiency of the Binary Search Algorithm

The idea of the derivation of the efficiency of the binary search algorithm is not difficult. Here it is in brief. At each stage of the binary search process, the length of the new subarray to be searched is approximately half that of the previous one, and in the worst case, every subarray down to a subarray with a single element must be searched. Consequently, in the worst case, the maximum number of iterations of the **while** loop in the binary search algorithm is 1 more than the number of times the original input array can be cut approximately in half. If the length n of this array is a power of 2 ($n = 2^k$ for some integer k), then n can be halved exactly $k = \log_2 n = \lfloor \log_2 n \rfloor$ times before an array of length 1 is reached. If n is not a power of 2, then $n = 2^k + m$ for some integer k (where $m < 2^k$), and so n can be split approximately in half k times also. So in this case, $k = \lfloor \log_2 n \rfloor$ also. Thus in the worst case, the number of iterations of the **while** loop in the binary search algorithm, which is proportional to the number of comparisons required to execute it, is $\lfloor \log_2 n \rfloor + 1$. The derivation is concluded by noting that $\lfloor \log_2 n \rfloor + 1$ is $O(\log_2 n)$.

The details of the derivation are developed in Examples 9.5.2–9.5.6. Throughout the derivation, for each integer $n \geq 1$, let

> $w_n = $ the number of iterations of the **while** loop
> in a *worst-case* execution of the binary search
> algorithm for an input array of length n.

The first issue to consider is this. If the length of the input array for one iteration of the **while** loop is known, what is the greatest possible length of the array input to the next iteration?

Example 9.5.2 The Length of the Input Array to the Next Iteration of the Loop

Prove that if an array of length k is input to the **while** loop of the binary search algorithm, then after one unsuccessful iteration of the loop, the input to the next iteration is an array of length at most $\lfloor k/2 \rfloor$.

Solution Consider what occurs when an array of length k is input to the **while** loop in the case where $x \neq a[mid]$:

$$\underbrace{a[bot], a[bot+1], \ldots, a[mid-1]}_{\substack{\text{new input to the while} \\ \text{loop if } x < a[mid]}}, \underset{\substack{\uparrow \\ \text{``middle} \\ \text{element''}}}{a[mid]}, \underbrace{a[mid+1], \ldots, a[top-1], a[top]}_{\substack{\text{new input to the while} \\ \text{loop if } x > a[mid]}}.$$

Since the input array has length k, the value of *mid* depends on whether k is odd or even. In both cases we match up the array elements with the integers from 1 to k and analyze the lengths of the left and right subarrays. In case k is odd, both the left and the right subarrays have length $\lfloor k/2 \rfloor$. In case k is even, the left subarray has length $\lfloor k/2 \rfloor - 1$ and the right subarray has length $\lfloor k/2 \rfloor$. The reasoning behind these results is shown in Figure 9.5.3.

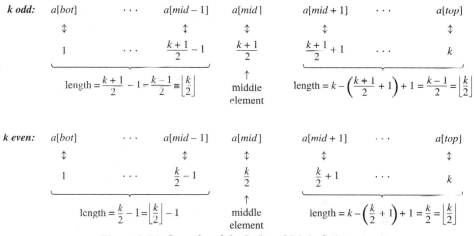

Figure 9.5.3 Lengths of the Left and Right Subarrays

Because the maximum of the numbers $\lfloor k/2 \rfloor$ and $\lfloor k/2 \rfloor - 1$ is $\lfloor k/2 \rfloor$, in the worst case this will be the length of the array input to the next iteration of the loop. ∎

To find the order of the algorithm, a formula for $w_1, w_2, w_3, \ldots$ is needed. The next example derives a recurrence relation for the sequence.

Example 9.5.3 A Recurrence Relation for $w_1, w_2, w_3, \ldots$

Prove that the sequence $w_1, w_2, \ldots, w_n, \ldots$ satisfies the recurrence relation and initial condition

$$w_1 = 1,$$
$$w_k = 1 + w_{\lfloor k/2 \rfloor} \quad \text{for all integers } k > 1.$$

Solution Example 9.5.2 showed that given an input array of length k to the **while** loop, the worst that can happen is that the next iteration of the loop will have to search an array of length $\lfloor k/2 \rfloor$. Hence the maximum number of iterations of the loop is 1 more than the maximum number necessary to execute it for an input array of length $\lfloor k/2 \rfloor$. In symbols,

$$w_k = 1 + w_{\lfloor k/2 \rfloor}.$$

Also

$$w_1 = 1$$

because for an input array of length 1 ($bot = top$), the **while** loop iterates only one time.

■

Now that a recurrence relation for $w_1, w_2, w_3, \ldots$ has been found, iteration can be used to come up with a good guess for an explicit formula.

Example 9.5.4 An Explicit Formula for $w_1, w_2, w_3, \ldots$

Apply iteration to the recurrence relation found in Example 9.5.3 to conjecture an explicit formula for $w_1, w_2, w_3, \ldots$.

Solution Begin by iterating to find the values of the first few terms of the sequence.

$$
\begin{aligned}
w_① &= ① & &1 = 2^0;\ 1 = 0 + 1 \\
w_② &= 1 + w_{\lfloor 2/2 \rfloor} = 1 + w_1 = 1 + 1 = ② & &2 = 2^1;\ 2 = 1 + 1 \\
w_3 &= 1 + w_{\lfloor 3/2 \rfloor} = 1 + w_1 = 1 + 1 = 2 & & \\
w_④ &= 1 + w_{\lfloor 4/2 \rfloor} = 1 + w_2 = 1 + 2 = ③ & &4 = 2^2;\ 3 = 2 + 1 \\
w_5 &= 1 + w_{\lfloor 5/2 \rfloor} = 1 + w_2 = 1 + 2 = 3 & & \\
w_6 &= 1 + w_{\lfloor 6/2 \rfloor} = 1 + w_3 = 1 + 2 = 3 & & \\
w_7 &= 1 + w_{\lfloor 7/2 \rfloor} = 1 + w_3 = 1 + 2 = 3 & & \\
w_⑧ &= 1 + w_{\lfloor 8/2 \rfloor} = 1 + w_4 = 1 + 3 = ④ & &8 = 2^3;\ 4 = 3 + 1 \\
w_9 &= 1 + w_{\lfloor 9/2 \rfloor} = 1 + w_4 = 1 + 3 = 4 & & \\
&\vdots & &\vdots \\
w_{15} &= 1 + w_{\lfloor 15/2 \rfloor} = 1 + w_7 = 1 + 3 = 4 & & \\
w_{⑯} &= 1 + w_{\lfloor 16/2 \rfloor} = 1 + w_8 = 1 + 4 = ⑤ & &16 = 2^4;\ 5 = 4 + 1 \\
&\vdots & &\vdots
\end{aligned}
$$

Note that in each case when the subscript n is between two powers of 2, w_n is 1 more than the exponent of the lower power of 2. In other words:

$$\text{If } 2^i \le n < 2^{i+1}, \text{ then } w_n = i + 1. \tag{9.5.1}$$

But if

$$2^i \le n < 2^{i+1},$$

then [*by property (9.4.2) of Example 9.4.1*]

$$i = \lfloor \log_2 n \rfloor.$$

Substitution into statement (9.5.1) gives the conjecture that

$$w_n = \lfloor \log_2 n \rfloor + 1.$$

■

Now mathematical induction can be used to verify the correctness of the formula found in Example 9.5.4.

Example 9.5.5 Verifying the Correctness of the Formula

Use strong mathematical induction to show that if $w_1, w_2, w_3, \ldots$ is a sequence of numbers that satisfies the recurrence relation and initial condition

$$w_1 = 1,$$
$$w_k = 1 + w_{\lfloor k/2 \rfloor} \quad \text{for all integers } k > 1,$$

then the sequence satisfies the formula

$$w_n = \lfloor \log_2 n \rfloor + 1 \quad \text{for all integers } n \geq 1.$$

Solution

Show that the sequence satisfies the formula for n = 1: For $n = 1$, $w_1 = 1$ and also $\lfloor \log_2 1 \rfloor + 1 = \lfloor 0 \rfloor + 1 = 1$. Hence the sequence satisfies the formula for $n = 1$.

Show that all integers k ≥ 2, if the sequence satisfies the formula for all integers i with 1 ≤ i < k, then it satisfies the formula for k: Suppose that for some integer $k \geq 2$,

$$w_i = \lfloor \log_2 i \rfloor + 1 \quad \text{for all integers } i \text{ with } 1 \leq i < k. \qquad \text{This is the inductive hypothesis.}$$

We must show that

$$w_k = \lfloor \log_2 k \rfloor + 1.$$

Consider the two cases: k is odd and k is even.

Case 1 (k is odd): In this case $\lfloor k/2 \rfloor = \dfrac{k-1}{2}$, and so

$$
\begin{aligned}
w_k &= 1 + w_{\lfloor k/2 \rfloor} & \text{by the recurrence relation} \\
&= 1 + w_{(k-1)/2} & \text{because } \lfloor k/2 \rfloor = (k-1)/2 \text{ since } k \text{ is odd} \\
&= 1 + \left(\left\lfloor \log_2 \left(\frac{k-1}{2} \right) \right\rfloor + 1 \right) & \text{by inductive hypothesis} \\
&= \lfloor \log_2 (k-1) - \log_2 2 \rfloor + 2 & \text{by substituting into the identity} \\
& & \log_b(x/y) = \log_b x - \log_b y \text{ derived in} \\
& & \text{exercise 28 of Section 7.2} \\
&= \lfloor \log_2 (k-1) - 1 \rfloor + 2 & \text{since } \log_2 2 = 1 \\
&= (\lfloor \log_2 (k-1) \rfloor - 1) + 2 & \text{by substituting } x = \log_2 (k-1) \text{ into the identity} \\
& & \lfloor x - 1 \rfloor = \lfloor x \rfloor - 1 \text{ derived in exercise 15 of Section 3.5} \\
&= \lfloor \log_2 k \rfloor + 1 & \text{by property (9.4.3) in Example 9.4.2}
\end{aligned}
$$

Case 2 (k is even): In this case, it can also be shown that $w_k = \lfloor \log_2 k \rfloor + 1$. The analysis is very similar to that of case 1 and is left as an exercise.

Hence regardless of whether k is odd or k is even,

$$w_k = \lfloor \log_2 k \rfloor + 1,$$

as was to be shown.

[*Since both the basis and the inductive steps have been demonstrated, the proof by strong mathematical induction is complete.*] ∎

The final example shows how to use the formula for $w_1, w_2, w_3, \ldots$ to find a worst-case order for the algorithm.

Example 9.5.6 The Binary Search Algorithm Is Logarithmic

Given that by Example 9.5.5, for all positive integers n,

$$w_n = \lfloor \log_2 n \rfloor + 1,$$

show that in the worst case, the binary search algorithm is $\Theta(\log_2 n)$.

Solution For any integer $n > 2$,

$$w_n = \lfloor \log_2 n \rfloor + 1 \qquad \text{by Example 9.5.5}$$

$$\Rightarrow \quad \frac{1}{2} \log_2 n \leq w_n \leq \log_2 n + 1 \qquad \text{because } \frac{x}{2} < \lfloor x \rfloor + 1 \text{ and } \lfloor x \rfloor \leq x \text{ for all real numbers } x$$

$$\Rightarrow \quad \frac{1}{2} \log_2 n \leq w_n \leq \log_2 n + \log_2 n \qquad \text{since the logorithm with base 2 is increasing, if } 2 < n, \text{ then } 1 = \log_2 2 < \log_2 n$$

$$\Rightarrow \quad \frac{1}{2} \log_2 n \leq w_n \leq 2 \log_2 n.$$

Both w_n and $\log_2 n$ are positive for $n > 2$. Therefore,

$$\frac{1}{2} |\log_2 n| \leq |w_n| \leq 2 |\log_2 n| \quad \text{for all integers } n > 2.$$

Let $A = \frac{1}{2}$, $B = 2$ and $k = 2$. Then

$$A |\log_2 n| \leq |w_n| \leq B |\log_2 n| \quad \text{for all integers } n > k.$$

Hence by definition of Θ-notation,

$$w_n \quad \text{is} \quad \Theta(\log_2 n).$$

But w_n, the number of iterations of the **while** loop, is proportional to the number of comparisons performed when the binary search algorithm is executed. Thus the binary search algorithm is $\Theta(\log_2 n)$. ∎

Examples 9.5.2–9.5.6 show that in the worst case, the binary search algorithm has order $\log_2 n$. As noted in Section 9.3, in the worst case the sequential search algorithm has order n. This difference in efficiency becomes increasingly more important as n gets larger and larger. Assuming one loop iteration is performed each nanosecond, then performing n iterations for $n = 100,000,000$ requires 0.1 second, whereas performing $\log_2 n$ iterations requires 0.000017 second. For $n = 100,000,000,000$ the times are 1.67 minutes and 0.000027 second, respectively. And for $n = 100,000,000,000,000$ the respective times are 2.78 hours and 0.000037 second.

Merge Sort

Note that it is much easier to write a detailed algorithm for sequential search than for binary search. Yet binary search is much more efficient than sequential search. Such trade-offs often occur in computer science. Frequently, the straightforward "obvious" solution to a problem is less efficient than a clever solution that is more complicated to describe.

In the text and exercises for Section 9.3, we gave two methods for sorting, insertion sort and selection sort, both of which are formalizations of methods human beings often use in ordinary situations. Can a divide-and-conquer approach be used to find a sorting method more efficient than these? It turns out that the answer is an emphatic "yes." In fact, over the past few decades, computer scientists have developed several divide-and-conquer sorting methods all of which are somewhat more complex to describe but are significantly more efficient than either insertion sort or selection sort.

One of these methods, **merge sort,** is obtained by thinking recursively. Imagine that an efficient way for sorting arrays of length less than k is already known. How can such knowledge be used to sort an array of length k? One way is to suppose the array of length

k is split into two roughly equal parts and each part is sorted using the known method. Is there an efficient way to combine the parts into a sorted array? Sure. Just "merge" them.

Figure 9.5.4 illustrates how a merge works. Imagine that the elements of two ordered subarrays, 2, 5, 6, 8 and 3, 6, 7, 9, are written on slips of paper (to make them easy to move around). Place the slips for each subarray in two columns on a tabletop, one at the left and one at the right. Along the bottom of the tabletop, set up eight positions into which the slips will be moved. Then, one-by-one, bring down the slips from the bottoms of the columns. At each stage compare the numbers on the slips currently at the column bottoms, and move the slip containing the smaller number down into the next position in the array as a whole. If at any stage the two numbers are equal, take, say, the slip on the left to move into the next position. And if one of the columns is empty at any stage, just move the slips from the other column into position one-by-one in order.

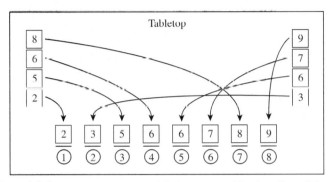

Figure 9.5.4 Merging Two Sorted Subarrays to Obtain a Sorted Array

One important observation about the merging algorithm described above: It requires memory space to move the array elements around. A second set of array positions as long as the original one is needed into which to place the elements of the two subarrays in order. In Figure 9.5.4 this second set of array positions is represented by the positions set up at the bottom of the tabletop. Of course, once the elements of the original array have been placed into this new array, they can be moved back in order into the original array positions.

In terms of time, however, merging is efficient because the total number of comparisons needed to merge two subarrays into an array of length k is just $k - 1$. You can see why by analyzing Figure 9.5.4. Observe that at each stage, the decision about which slip to move is made by comparing the numbers on the slips currently at the bottoms of the two columns—execpt when one of the columns is empty, in which case no comparisons are made at all. Thus in the worst case there will be one comparison for each of the k positions in the final array except the very last one (because when the last slip is placed into position, the other column is sure to be empty), or a total of $k - 1$ comparisons in all.

The merge sort algorithm is recursive: Its defining statements include references to itself. The algorithm is well defined, however, because at each stage the length of the array that is input to the algorithm is shorter than at the previous stage, so that, ultimately, the algorithm has to deal only with arrays of length 1, which are already sorted. Specifically, merge sort works as follows.

Given an array of elements that can be put into order, if the array consists of a single element, leave it as it is. It is already sorted. Otherwise:

1. Divide the array into two subarrays of as nearly equal length as possible.

2. Use merge sort to sort each subarray.

3. Merge the two subarrays together.

Figure 9.5.5 illustrates a merge sort in a particular case.

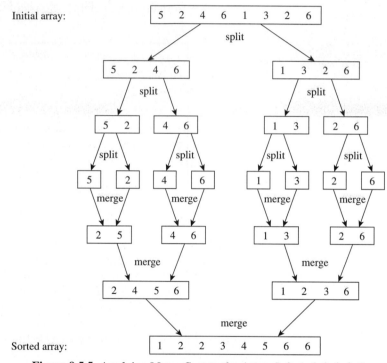

Figure 9.5.5 Applying Merge Sort to the Array 5, 2, 4, 6, 1, 3, 2, 6

As in the case of the binary search algorithm, in order to formalize merge sort we must decide at exactly what point to split each array. Given an array denoted by $a[bot]$, $a[bot + 1], \ldots, a[top]$, let $mid = \lfloor (bot + top)/2 \rfloor$. Take the left subarray to be $a[bot]$, $a[bot + 1], \ldots, a[mid]$ and the right subarray to be $a[mid + 1], a[mid + 2], \ldots, a[top]$. The following is a formal version of merge sort.

Algorithm 9.5.2 Merge Sort

[The aim of this algorithm is to take an array of elements $a[r]$, $a[r + 1], \ldots, a[s]$ (where $r \leq s$) and to order it. The output array is denoted $a[r]$, $a[r + 1], \ldots, a[s]$ also. It has the same values as the input array, but they are in ascending order. The input array is split into two nearly equal-length subarrays, each of which is ordered using merge sort. Then the two subarrays are merged together.]

Input: r and s [*positive integers with $r < s$*], $a[r], a[r + 1], \ldots, a[s]$ [*an array of data items that can be ordered*]

Algorithm Body:

$bot := r, top := s$

while $(bot < top)$

$$mid := \left\lfloor \frac{bot + top}{2} \right\rfloor$$

call **merge sort** with input bot, mid, and
$a[bot], a[bot + 1], \ldots, a[mid]$

call **merge sort** with input $mid + 1$, top and
$a[mid + 1], a[mid + 2], \ldots, a[top]$

[*After these steps are completed, the arrays $a[bot], a[bot + 1], \ldots, a[mid]$ and $a[mid + 1], a[mid + 2], \ldots, a[top]$ are both in order.*]

merge $a[bot], a[bot + 1], \ldots, a[mid]$ and
$a[mid + 1], a[mid + 2], \ldots, a[top]$

[*This step can be done with a call to a merge algorithm. To put the final array in ascending order, the merge algorithm must be written so as to take two arrays in ascending order and merge them into an array in ascending order.*]

end while

Output: $a[r], a[r + 1], \ldots, a[s]$ [*an array with the same elements as the input array but in ascending order*]

To derive the efficiency of merge sort, let

$m_n = $ the maximum number of comparisons used
when merge sort is applied to an array of length n.

Then $m_1 = 0$ because no comparisons are used when merge sort is applied to an array of length 1. Also for any integer $k > 1$, consider an array $a[bot], a[bot + 1], \ldots, a[top]$ of length k that is split into two subarrays, $a[bot], a[bot + 1], \ldots, a[mid]$ and $a[mid + 1], a[mid + 2], \ldots, a[top]$, where $mid = \lfloor (bot + top)/2 \rfloor$. In exercise 24 you are asked to show that the right subarray has length $\lfloor k/2 \rfloor$ and the left subarray has length $\lceil k/2 \rceil$. From the previous discussion of the merge process, it is known that to merge two subarrays into an array of length k, at most $k - 1$ comparisons are needed.

Consequently,

$$\begin{bmatrix} \text{the number of comparisons} \\ \text{when merge sort is applied} \\ \text{to an array of length } k \end{bmatrix} = \begin{bmatrix} \text{the number of comparisons} \\ \text{when merge sort is applied} \\ \text{to an array of length } \lfloor k/2 \rfloor \end{bmatrix}$$

$$+ \begin{bmatrix} \text{the number of comparisons} \\ \text{when merge sort is applied} \\ \text{to an array of length } \lceil k/2 \rceil \end{bmatrix} + \begin{bmatrix} \text{the number of comparisons} \\ \text{used to merge two subarrays} \\ \text{into an array of length } k \end{bmatrix}.$$

Or, in other words,

$$m_k = m_{\lfloor k/2 \rfloor} + m_{\lceil k/2 \rceil} + (k - 1) \quad \text{for all integers } k > 1.$$

In exercise 25 you are asked to use this recurrence relation to show that

$$n \log_2 n \leq m_n \leq 2n \log_2 n \quad \text{for all integers } n \geq 1.$$

It follows that merge sort is $\Theta(n \log_2 n)$.

In the text and exercises for Section 9.3, we showed that insertion sort and selection sort are both $\Theta(n^2)$. How much difference can it make that merge sort is $\Theta(n \log_2 n)$? If $n = 100,000,000$ and a computer is used that performs one operation each nanosecond, the time needed to perform $n \log_2 n$ operations is about 2.7 seconds, whereas the time needed to perform n^2 operations is over 115 days.

Tractable and Intractable Problems

At an opposite extreme from an algorithm such as binary search, which has logarithmic order, is an algorithm with exponential order. For example, consider an algorithm to direct the movement of each of the 64 disks in the Tower of Hanoi puzzle as they are transferred one by one from one pole to another. In Section 8.2 we showed that such a transfer requires $2^{64} - 1$ steps. If a computer took a nanosecond to calculate each transfer step, the total time to calculate all the steps would be

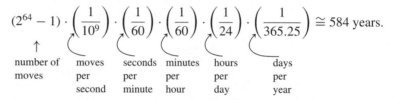

$$(2^{64} - 1) \cdot \left(\frac{1}{10^9}\right) \cdot \left(\frac{1}{60}\right) \cdot \left(\frac{1}{60}\right) \cdot \left(\frac{1}{24}\right) \cdot \left(\frac{1}{365.25}\right) \cong 584 \text{ years.}$$

number of moves — moves per second — seconds per minute — minutes per hour — hours per day — days per year

Problems whose solutions can be found with algorithms whose worst-case order with respect to time is a polynomial are said to belong to **class P.** They are called **polynomial-time algorithms** and are said to be **tractable.** Problems that cannot be solved in polynomial time are called **intractable.** For certain problems, it is possible to check the correctness of a proposed solution with a polynomial-time algorithm, but it may not be possible to find a solution in polynomial time. Such problems are said to belong to **class NP.**[*] The biggest open question in theoretical computer science is whether every problem in class NP belongs to class P. This is known as the **P vs. NP** problem. The Clay Institute, in Cambridge, Massachusetts, has offered a prize of $1,000,000 to anyone who can either prove or disprove that P = NP.

In recent years, computer scientists have identified a fairly large set of problems, called **NP-complete,** that all belong to class NP but are widely believed not to belong to class P. What is known for sure is that if any one of these problems is solvable in polynomial time, then so are all the others. One of the NP-complete problems, commonly known as the *traveling salesman problem,* is discussed in Section 11.2.

A Final Remark on Algorithm Efficiency

This section and the previous one on algorithm efficiency have offered only a partial view of what is involved in analyzing a computer algorithm. For one thing, it is assumed that searches and sorts take place in the memory of the computer. Searches and sorts on disk-based files require different algorithms, though the methods for their analysis are similar.

---

[*]Technically speaking, a problem whose solution can be verified on an ordinary computer (or *deterministic sequential machine*) with a polynomial-time algorithm can be solved on a *nondeterministic sequential machine* with a polynomial-time algorithm. This is the reason why such problems are called NP, which stands for *nondeterministic polynomial-time algorithm.*

For another thing, as mentioned at the beginning of Section 9.3, time efficiency is not the only factor that matters in the decision about which algorithm to choose. The amount of memory space required is also important, and there are mathematical techniques to estimate space efficiency very similar to those used to estimate time efficiency. Furthermore, as parallel processing of data becomes increasingly prevalent, current methods of algorithm analysis are being modified and extended to apply to algorithms designed for this new technology.

Exercise Set 9.5

1. Use the facts that $\log_2 10 \cong 3.32$ and that for all real numbers a, $\log_2(10^a) = a \log_2 10$ to find $\log_2(1,000)$, $\log_2(1,000,000)$, and $\log_2(1,000,000,000,000)$.

2. Suppose an algorithm requires $c\lfloor \log_2 n \rfloor$ operations when performed with an input of size n (where c is a constant).
   a. How many operations will be required when the input size is increased from m to m^2 (where m is a positive integer power of 2)?
   b. By what factor will the number of operations increase when the input size is increased from m to m^{10} (where m is a positive integer power of 2)?
   c. When n increases from $128\ (= 2^7)$ to $268,435,456$ $(= 2^{28})$, by what factor is $c\lfloor \log_2 n \rfloor$ increased?

Exercises 3 and 4 illustrate that for relatively small values of n, algorithms with larger orders can be more efficient than algorithms with smaller orders. Use a graphing calculator or computer to answer these questions.

3. For what values of n is an algorithm that requires n operations more efficient than an algorithm that requires $\lfloor 50 \log_2 n \rfloor$ operations?

4. For what values of n is an algorithm that requires $\lfloor n^2/10 \rfloor$ operations more efficient than an algorithm that requires $\lfloor n \log_2 n \rfloor$ operations?

In 5 and 6, trace the action of the binary search algorithm (Algorithm 9.5.1) on the variables *index, bot, top, mid*, and the given values of x for the input array $a[1] = $ Chia, $a[2] = $ Doug, $a[3] = $ Jan, $a[4] = $ Jim, $a[5] = $ José, $a[6] = $ Mary, $a[7] = $ Rob, $a[8] = $ Roy, $a[9] = $ Sue, $a[10] = $ Usha, where alphabetical ordering is used to compare elements of the array.

5. a. $x = $ Chia   b. $x = $ Max

6. a. $x = $ Amanda   b. $x = $ Roy

7. Suppose *bot* and *top* are positive integers with *bot* $\leq$ *top*. Consider the array
$$a[bot], a[bot + 1], \ldots, a[top].$$
   a. How many elements are in this array?
   b. Show that if the number of elements in the array is odd, then the quantity *bot* + *top* is even.
   c. Show that if the number of elements in the array is even, then the quantity *bot* + *top* is odd.

Exercises 8–11 refer to the following algorithm segment. For each positive integer n, let a_n be the number of iterations of the **while** loop.

$$\textbf{while } (n > 0)$$
$$n := n \ div \ 2$$
$$\textbf{end while}$$

8. Trace the action of this algorithm segment on n when the initial value of n is 27.

9. Find a recurrence relation for a_n.

10. Find an explicit formula for a_n.

11. Find an order for this algorithm segment.

Exercises 12–15 refer to the following algorithm segment. For each positive integer n, let b_n be the number of iterations of the **while** loop.

$$\textbf{while } (n > 0)$$
$$n := n \ div \ 3$$
$$\textbf{end while}$$

12. Trace the action of this algorithm segment on n when the initial value of n is 424.

13. Find a recurrence relation for b_n.

H 14. a. Use iteration to guess an explicit formula for b_n.
   b. Prove that if k is an integer and x is a real number with $3^k \leq x < 3^k$, then $\lfloor \log_3 x \rfloor = k$.
   c. Prove that for all integers $m \geq 1$,
   $$\lfloor \log_3(3m) \rfloor = \lfloor \log_3(3m + 1) \rfloor = \lfloor \log_3(3m + 2) \rfloor.$$
   d. Prove the correctness of the formula you found in part (a).

15. Find an order for the algorithm segment.

16. Complete the proof of case 2 of the strong induction argument in Example 9.5.5. In other words, show that if k is an even integer and $w_i = \lfloor \log_2 i \rfloor + 1$ for all integers i with $1 \leq i < k$, then $w_k = \lfloor \log_2 k \rfloor + 1$.

For 17–19, modify the binary search algorithm (Algorithm 9.5.1) to take the upper of the two middle array elements in case the input array has even length. In other words, in Algorithm 9.5.1 replace
$$mid := \left\lfloor \frac{bot + top}{2} \right\rfloor \text{ with } mid := \left\lceil \frac{bot + top}{2} \right\rceil.$$

17. Trace the modified binary search algorithm for the same input as was used in Example 9.5.1.

18. Suppose an array of length k is input to the **while** loop of the modified binary search algorithm. Show that after one iteration of the loop, if $a[mid] \neq x$, the input to the next iteration is an array of length at most $\lfloor k/2 \rfloor$.

19. Let w_n be the number of iterations of the **while** loop in a worst-case execution of the modified binary search algorithm for an input array of length n. Show that $w_k = 1 + w_{\lfloor k/2 \rfloor}$ for $k \geq 2$.

In 20 and 21, draw a diagram like Figure 9.5.4 to show how to merge the given subarrays into a single array in ascending order.

20. 3, 5, 6, 9, 12 and 2, 4, 7, 9, 11

21. F, K, L, R, U and C, E, L, P, W (alphabetical order)

In 22 and 23, draw a diagram like Figure 9.5.5 to show how merge sort works for the given input arrays.

22. R, G, B, U, C, F, H, G (alphabetical order)

23. 5, 2, 3, 9, 7, 4, 3, 2

24. Show that given an array $a[bot], a[bot +1], \ldots, a[top]$ of length k, if $mid = \lfloor (bot + top)/2 \rfloor$ then
    a. the subarray $a[mid + 1], a[mid + 2], \ldots, a[top]$ has length $\lfloor k/2 \rfloor$.
    b. the subarray $a[bot], a[bot + 1], \ldots, a[mid]$ has length $\lceil k/2 \rceil$.

H 25. The recurrence relation for $m_1, m_2, m_3, \ldots$, which arises in the calculation of the efficiency of merge sort, is

$$m_1 = 0$$

$$m_k = m_{\lfloor k/2 \rfloor} + m_{\lceil k/2 \rceil} + k - 1.$$

Show that for all integers $n \geq 1$,
a. $n \log_2 n \leq m_n$ b. $m_n \leq 2n \log_2 n$

26. You might think that $n - 1$ multiplications are needed to compute x^n, since

$$x^n = \underbrace{x \cdot x \cdots x.}_{n-1 \text{ multiplications}}$$

But observe that, for instance, since $6 = 4 + 2$,

$$x^6 = x^4 x^2 = (x^2)^2 x^2.$$

Thus x^6 can be computed using three multiplications: one to compute x^2, one to compute $(x^2)^2$, and one to multiply $(x^2)^2$ times x^2. Similarly, since $11 = 8 + 2 + 1$,

$$x^{11} = x^8 x^2 x^1 = ((x^2)^2)^2 x^2 x$$

and so x^{11} can be computed using five multiplications: one to compute x^2, one to compute $(x^2)^2$, one to compute $((x^2)^2)^2$, one to multiply $((x^2)^2)^2$ times x^2, and one to multiply that product by x.
a. Write an algorithm to take a real number x and a positive integer n and compute x^n by
   (i) calling Algorithm 4.1.1 to find the binary representation of n:

$$(r[k] \, r[k-1] \cdots r[0])_2,$$

   where each $r[i]$ is 0 or 1;
   (ii) computing $x^2, x^{2^2}, x^{2^3}, \ldots, x^{2^k}$ by squaring, then squaring again, and so forth,
   (iii) computing x^n using the fact that

$$x^n = x^{r[k]2^k + \cdots + r[2]2^2 + r[1]2^1 + r[0]2^0}$$

$$= x^{r[k]2^k} \cdots x^{r[2]2^2} \cdot x^{r[1]2^1} \cdot x^{r[0]2^0}$$

b. Show that the number of multiplications performed by the algorithm of part (a) is less than or equal to $2\lfloor \log_2 n \rfloor$.

CHAPTER 10

RELATIONS

There are many kinds of relationships in the world. For instance, we say that two people are related by blood if they share a common ancestor and that they are related by marriage if one shares a common ancestor with the spouse of the other. We also speak of the relationship between boyfriend and girlfriend, between student and teacher, between people who work for the same employer, and between people who share a common ethnic background.

Similarly, the objects of mathematics and computer science may be related in various ways. Two digital logic circuits may be said to be related if they have the same input/output table. A set A may be said to be related to a set B if A is a subset of B, or if A is not a subset of B, or if A is the complement of B. A number x may be said to be related to a number y if $x < y$, or if x divides y, or if $x^2 + y^2 = 1$. Two identifiers in a computer program may be said to be related if they have the same first eight characters, or if the same memory location is used to store their values when the program is executed. And the list could go on!

In this chapter we discuss the mathematics of relations defined on sets, focusing on ways to represent relations and exploring various properties they may have. The concept of equivalence relation is introduced in Section 10.3 and applied in Section 10.4 to modular arithmetic and cryptography. Partial order relations are discussed in Section 10.5, and an application is given showing how to use these relations to help coordinate and guide the flow of individual tasks that must be performed to accomplish a complex, large-scale project.

10.1 Relations on Sets

Strange as it may sound, the power of mathematics rests on its evasion of all unnecessary thought and on its wonderful saving of mental operations. — Ernst Mach, 1838–1916

Let $A = \{0, 1, 2\}$ and $B = \{1, 2, 3\}$. Let us say that an element x in A is related to an element y in B if, and only if, x is less than y. Let us use the notation $x \, R \, y$ as a shorthand for the sentence "x is related to y." Then

$$0 \, R \, 1 \quad \text{since} \quad 0 < 1,$$
$$0 \, R \, 2 \quad \text{since} \quad 0 < 2,$$
$$0 \, R \, 3 \quad \text{since} \quad 0 < 3,$$
$$1 \, R \, 2 \quad \text{since} \quad 1 < 2,$$
$$1 \, R \, 3 \quad \text{since} \quad 1 < 3, \quad \text{and}$$
$$2 \, R \, 3 \quad \text{since} \quad 2 < 3.$$

On the other hand, if the notation $x \, \mathcal{R} \, y$ represents the sentence "x is not related to y," then

$$0 \, \mathcal{R} \, 0 \quad \text{since} \quad 0 \nless 0,$$
$$1 \, \mathcal{R} \, 1 \quad \text{since} \quad 1 \nless 1,$$
$$2 \, \mathcal{R} \, 1 \quad \text{since} \quad 2 \nless 1, \quad \text{and}$$
$$2 \, \mathcal{R} \, 2 \quad \text{since} \quad 2 \nless 2.$$

Recall that the Cartesian product of A and B, $A \times B$, consists of all ordered pairs whose first element is in A and whose second element is in B:

$$A \times B = \{(x, y) \mid x \in A \text{ and } y \in B\}.$$

In this case,

$$A \times B = \{(0, 1), (0, 2), (0, 3), (1, 1), (1, 2), (1, 3), (2, 1), (2, 2), (2, 3)\}.$$

The elements of some ordered pairs in $A \times B$ are related, whereas the elements of other ordered pairs are not. Consider the set of all ordered pairs in $A \times B$ whose elements are related:

$$\{(0, 1), (0, 2), (0, 3), (1, 2), (1, 3), (2, 3)\}.$$

Observe that knowing which ordered pairs lie in this set is equivalent to knowing which elements are related to which. The relation itself can therefore be thought of as the totality of ordered pairs whose elements are related by the given condition. The formal mathematical definition of relation, based on this idea, was introduced by the American mathematician and logician C. S. Peirce in the nineteenth century.

• Definition

Let A and B be sets. A **(binary) relation R from A to B** is a subset of $A \times B$. Given an ordered pair (x, y) in $A \times B$, **x is related to y by R**, written $x \, R \, y$, if, and only if, (x, y) is in R.

The notation for relations may be written symbolically as follows:

$$x \, R \, y \quad \Leftrightarrow \quad (x, y) \in R$$

The notation $x \, \mathcal{R} \, y$ means that x is not related to y by R.

$$x \, \mathcal{R} \, y \quad \Leftrightarrow \quad (x, y) \notin R$$

The term *binary* is used in the definition above to refer to the fact that the relation is a subset of the Cartesian product of two sets. Because we mostly discuss binary relations in this text, when we use the term *relation* by itself, we will mean binary relation. A more general type of relation, called an *n-ary relation,* is defined later in this section.

Example 10.1.1 A Binary Relation as a Subset

Let $A = \{1, 2\}$ and $B = \{1, 2, 3\}$ and define a binary relation R from A to B as follows:

$$\text{Given any } (x, y) \in A \times B, \quad (x, y) \in R \quad \Leftrightarrow \quad x - y \text{ is even.}$$

a. State explicitly which ordered pairs are in $A \times B$ and which are in R.

b. Is $1 \, R \, 3$? Is $2 \, R \, 3$? Is $2 \, R \, 2$?

Solution

a. $A \times B = \{(1, 1), (1, 2), (1, 3), (2, 1), (2, 2), (2, 3)\}$. To determine explicitly the composition of R, examine each ordered pair in $A \times B$ to see whether its elements satisfy the defining condition for R.

$(1, 1) \in R$ because $1 - 1 = 0$ and 0 is even   since $0 = 2 \cdot 0$.

$(1, 2) \notin R$ because $1 - 2 = -1$ and -1 is not even   since $-1 \neq 2k$, for any integer k.

$(1, 3) \in R$ because $1 - 3 = -2$ and -2 is even.

$(2, 1) \notin R$ because $2 - 1 = 1$ and 1 is not even.

$(2, 2) \in R$ because $2 - 2 = 0$ and 0 is even.

$(2, 3) \notin R$ because $2 - 3 = -1$ and -1 is not even.

Thus

$$R = \{(1, 1), (1, 3), (2, 2)\}.$$

b. Yes, $1 \, R \, 3$ since $(1, 3) \in R$.

No, $2 \, \not\!R \, 3$ since $(2, 3) \notin R$.

Yes, $2 \, R \, 2$ since $(2, 2) \in R$.

∎

Example 10.1.2 The Congruence Modulo 2 Relation

Generalize the relation defined in Example 10.1.1 to the set of all integers $\mathbf{Z}$. That is, define a binary relation E from $\mathbf{Z}$ to $\mathbf{Z}$ as follows:

For all $(m, n) \in \mathbf{Z} \times \mathbf{Z}$, $m \, E \, n$ ⇔ $m - n$ is even.

a. Is $4 \, E \, 0$? Is $2 \, E \, 6$? Is $3 \, E \, (-3)$? Is $5 \, E \, 2$?

b. List five integers that are related by E to 1.

c. Prove that if n is any odd integer, then $n \, E \, 1$.

Solution

a. Yes, $4 \, E \, 0$ because $4 - 0 = 4$ and 4 is even.

Yes, $2 \, E \, 6$ because $2 - 6 = -4$ and -4 is even.

Yes, $3 \, E \, (-3)$ because $3 - (-3) = 6$ and 6 is even.

No, $5 \, \not\!E \, 2$ because $5 - 2 = 3$ and 3 is not even.

b. There are many such lists. One is

$$1 \quad \text{because } 1 - 1 = 0 \text{ is even,}$$
$$3 \quad \text{because } 3 - 1 = 2 \text{ is even,}$$
$$5 \quad \text{because } 5 - 1 = 4 \text{ is even,}$$
$$-1 \quad \text{because } -1 - 1 = -2 \text{ is even,}$$
$$-3 \quad \text{because } -3 - 1 = -4 \text{ is even.}$$

c. **Proof:** Suppose n is any odd integer. Then $n = 2k + 1$ for some integer k. Now by definition of E, $n \, E \, 1$ if, and only if, $n - 1$ is even. But by substitution,

$$n - 1 = (2k + 1) - 1 = 2k,$$

and since k is an integer, $2k$ is even. Hence $n \, E \, 1$ [*as was to be shown*].

It can be shown (see exercise 4 at the end of this section) that integers m and n are related by E if, and only if, $m \bmod 2 = n \bmod 2$ (that is, that both are even or both are odd). When this occurs m and n are said to be **congruent modulo 2**. ∎

Example 10.1.3 The Circle Relation

Define a binary relation C from $\mathbf{R}$ to $\mathbf{R}$ as follows:

$$\text{For any } (x, y) \in \mathbf{R} \times \mathbf{R}, \quad (x, y) \in C \iff x^2 + y^2 = 1.$$

a. Is $(1, 0) \in C$? Is $(0, 0) \in C$? Is $(-\frac{1}{2}, \frac{\sqrt{3}}{2}) \in C$? Is $-2 \, C \, 0$? Is $0 \, C \, (-1)$? Is $1 \, C \, 1$?

b. Draw a graph for C by plotting the points of C in the Cartesian plane.

Solution

a. Yes, $(1, 0) \in C$ because $1^2 + 0^2 = 1$.
   No, $(0, 0) \notin C$ because $0^2 + 0^2 = 0 \neq 1$.
   Yes, $(-\frac{1}{2}, \frac{\sqrt{3}}{2}) \in C$ because $(-\frac{1}{2})^2 + (\frac{\sqrt{3}}{2})^2 = \frac{1}{4} + \frac{3}{4} = 1$.
   No, $-2 \not{C} 0$ because $(-2)^2 + 0^2 = 4 \neq 1$.
   Yes, $0 \, C \, (-1)$ because $0^2 + (-1)^2 = 1$.
   No, $1 \not{C} 1$ because $1^2 + 1^2 = 2 \neq 1$.

b.

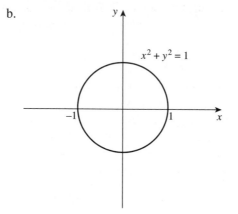

Example 10.1.4 A Relation on a Set of Strings

Let A be the set of all strings of length 6 consisting of x's and y's. Define a binary relation R from A to A as follows: For all strings s and t in A,

$$s \, R \, t \iff \text{the first four characters of } s \text{ equal} \\ \text{the first four characters of } t.$$

Is $xxyxyx \, R \, xxxyxy$? Is $yxyyyx \, R \, yxyyxy$? Is $xyxxxx \, R \, yxxxxx$?

Solution   No, $xxyxyx \not{R} xxxyxy$ because $xxyx \neq xxxy$.

Yes, $yxyyyx \, R \, yxyyxy$ because $yxyy = yxyy$.

No, $xyxxxx \not{R} yxxxxx$ because $xyxx \neq yxxx$. ∎

Arrow Diagram of a Relation

Suppose R is a relation from a set A to a set B. The **arrow diagram for R** is obtained as follows:

1. Represent the elements of A as points in one region and the elements of B as points in another region.

2. For each x in A and y in B, draw an arrow from x to y if, and only if, x is related to y by R. Symbolically:

$$\textbf{Draw an arrow from } \boldsymbol{x} \textbf{ to } \boldsymbol{y} \iff x \, R \, y \iff (x, y) \in R.$$

Example 10.1.5 Arrow Diagrams of Relations

Let $A = \{1, 2, 3\}$ and $B = \{1, 3, 5\}$ and define relations S and T from A to B as follows: For all $(x, y) \in A \times B$,

$$(x, y) \in S \quad \Leftrightarrow \quad x < y \qquad \text{\small S is a "less than" relation.}$$
$$T = \{(2, 1), (2, 5)\}.$$

Draw arrow diagrams for S and T.

Solution

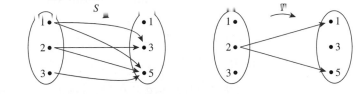

These example relations illustrate that it is possible to have several arrows coming out of the same element of A pointing in different directions. Also, it is quite possible to have an element of A that does not have an arrow coming out of it. ∎

Relations and Functions

With the introduction of Georg Cantor's set theory in the late nineteenth century, it began to seem possible to put mathematics on a firm logical foundation by developing all the different branches of mathematics from logic and set theory alone. In 1914, a crucial breakthrough in using sets to specify mathematical structures was made by Norbert Wiener (1894–1964), a young American who had recently received his Ph.D. from Harvard. What Wiener showed was that an ordered pair can be defined as a certain type of set. Unfortunately, his definition was somewhat awkward. At about the same time, the German mathematician Felix Hausdorff (1868–1942) offered another definition, but it turned out to have a slight flaw. Finally, in 1921, the Polish mathematician Kazimierez Kuratowski (1896–1980) published the version of the definition that has since become standard. It specifies that

$$(a, b) = \{\{a\}, \{a, b\}\}.$$

Note that this definition implies the fundamental property of ordered pairs:

$$(a, b) = (c, d) \quad \Leftrightarrow \quad a = c \text{ and } b = d.$$

The importance of this definition is that it makes it possible to define binary relations using nothing other than set theory, because Cartesian products are defined as sets of ordered pairs and binary relations are defined as subsets of Cartesian products. The concept of function is then defined as the following special kind of a binary relation.

• Definition

A **function** F **from a set** A **to a set** B is a relation from A to B that satisfies the following two properties:

1. For every element x in A, there is an element y in B such that $(x, y) \in F$.

2. For all elements x in A and y and z in B,

$$\text{if } (x, y) \in F \text{ and } (x, z) \in F, \text{ then } y = z.$$

If F is a function from A to B, we write

$$y = F(x) \quad \Leftrightarrow \quad (x, y) \in F.$$

Note that $y = F(x)$ if, and only if, y is the second element of an ordered pair in F whose first element is x. Note also that properties (1) and (2) can be stated less formally as follows: A binary relation F from A to B is a function if, and only if:

1. Every element of A is the first element of an ordered pair of F.

2. No two distinct ordered pairs in F have the same first element.

Example 10.1.6 Functions and Relations on Finite Sets

Let $A = \{2, 4, 6\}$ and $B = \{1, 3, 5\}$. Which of the relations R and S defined below are functions from A to B?

a. $R = \{(2, 5), (4, 1), (4, 3), (6, 5)\}$.

b. For all $(x, y) \in A \times B$, $(x, y) \in S \iff y = x + 1$.

Solution

a. R is not a function because it does not satisfy property (2). The ordered pairs (4, 1) and (4, 3) have the same first element but different second elements. You can see this graphically if you draw the arrow diagram for R.

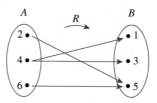

b. S is not a function because it does not satisfy property (1). It is not true that every element of A is the first element of an ordered pair in S. For example, $6 \in A$ but there is no y in B such that $y = 6 + 1 = 7$. You can also see this graphically by drawing the arrow diagram for S.

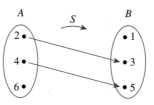

Example 10.1.7 Functions and Relations on Sets of Real Numbers

a. In Example 10.1.3 the circle relation C was defined as follows:

For all $(x, y) \in \mathbf{R} \times \mathbf{R}$, $(x, y) \in C \iff x^2 + y^2 = 1$.

Is C a function?

b. Define a relation from $\mathbf{R}$ to $\mathbf{R}$ as follows:

For all $(x, y) \in \mathbf{R} \times \mathbf{R}$, $(x, y) \in L \iff y = x - 1$.

Is L a function?

Solution

a. The graph of C, shown below, indicates that C does not satisfy either function property. To see why C does not satisfy property (1), observe that there are many real numbers x such that $(x, y) \notin C$ for any y.

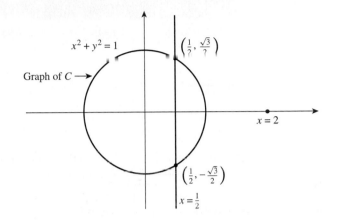

For instance, when $x = 2$, there is no real number y so that

$$x^2 + y^2 = 2^2 + y^2 = 4 + y^2 = 1$$

because if there were, then it would have to be true that

$$y^2 = -3,$$

which is not the case for any real number y.

To see why C does not satisfy property (2), note that for some values of x there are two distinct values of y so that $(x, y) \in C$. One way to see this graphically is to observe that there are vertical lines, such as $x = \frac{1}{2}$, that intersect the graph of C at two separate points: $(\frac{1}{2}, \frac{\sqrt{3}}{2})$ and $(\frac{1}{2}, -\frac{\sqrt{3}}{2})$.

b. L is a function. For each real number x, $y = x - 1$ is a real number, and so there is a real number y with $(x, y) \in L$. Also if $(x, y) \in L$ and $(x, z) \in L$, then $y = x - 1$ and $z = x - 1$, and so $y = z$.

You can also check these results by inspecting the graph of L, shown below. Note that for every real number x, the vertical line through $(x, 0)$ passes through the graph of L exactly once. This indicates both that every real number x is the first element of an ordered pair in L and also that no two distinct ordered pairs in L have the same first element.

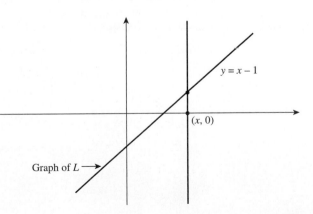

The Inverse of a Relation

If R is a relation from A to B, then a relation R^{-1} from B to A can be defined by interchanging the elements of all the ordered pairs of R.

> **• Definition**
>
> Let R be a relation from A to B. Define the inverse relation R^{-1} from B to A as follows:
>
> $$R^{-1} = \{(y, x) \in B \times A \mid (x, y) \in R\}.$$

This definition can be written operationally as follows:

> For all $x \in X$ and $y \in Y$, $(y, x) \in R^{-1} \Leftrightarrow (x, y) \in R$.

Example 10.1.8 The Inverse of a Finite Relation

Let $A = \{2, 3, 4\}$ and $B = \{2, 6, 8\}$ and let R be the "divides" relation from A to B:

For all $(x, y) \in A \times B$, $x \, R \, y \Leftrightarrow x \mid y$ x divides y.

a. State explicitly which ordered pairs are in R and R^{-1}, and draw arrow diagrams for R and R^{-1}.

b. Describe R^{-1} in words.

Solution

a. $R = \{(2, 2), (2, 6), (2, 8), (3, 6), (4, 8)\}$
   $R^{-1} = \{(2, 2), (6, 2), (8, 2), (6, 3), (8, 4)\}$

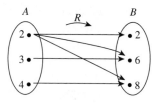

To draw the arrow diagram for R^{-1}, you can copy the arrow diagram for R but reverse the directions of the arrows.

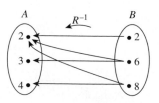

Or you can redraw the diagram so that B is on the left.

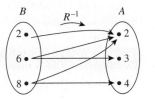

b. R^{-1} is defined in words as follows:

For all $(x, y) \in B \times A$, $y \, R^{-1} \, x \iff y$ is a multiple of x. ∎

Example 10.1.9 The Inverse of an Infinite Relation

Define a relation R from $\mathbf{R}$ to $\mathbf{R}$ as follows:

For all $(x, y) \in \mathbf{R} \times \mathbf{R}$, $x \, R \, y \iff y = 2|x|$.

Draw the graphs of R and R^{-1} in the Cartesian plane. Is R^{-1} a function?

Solution   A point (v, u) is on the graph of R^{-1} if, and only if, (u, v) is on the graph of R. Note that if $x \geq 0$, then the graph of $y = 2|x| = 2x$ is a straight line with slope 2. And if $x < 0$, then the graph of $y = 2|x| = 2(-x) = -2x$ is a straight line with slope -2. Some sample values are tabulated and the graphs are shown below.

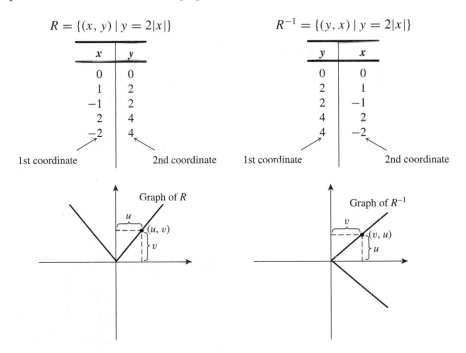

$R = \{(x, y) \mid y = 2|x|\}$

x	y
0	0
1	2
-1	2
2	4
-2	4

1st coordinate   2nd coordinate

$R^{-1} = \{(y, x) \mid y = 2|x|\}$

y	x
0	0
2	1
2	-1
4	2
4	-2

1st coordinate   2nd coordinate

Note that R^{-1} is not a function because, for instance, both $(2, 1)$ and $(2, -1)$ are in R^{-1}. ∎

Directed Graph of a Relation

In the remaining sections of this chapter, we discuss important properties of relations that are defined from a set to itself.

> ### • Definition
>
> A **binary relation on a set** A is a binary relation from A to A.

When a binary relation R is defined *on* a set A, the arrow diagram of the relation can be modified so that it becomes a **directed graph.** Instead of representing A as two separate sets of points, represent A only once, and draw an arrow from each point of A to each related point. As with an ordinary arrow diagram,

> For all points x and y in A,
>
> there is an arrow from x to y $\;\Leftrightarrow\;$ $x\,R\,y$ $\;\Leftrightarrow\;$ $(x, y) \in R$.

If a point is related to itself, a loop is drawn that extends out from the point and goes back to it.

Example 10.1.10 Directed Graph of a Relation

Let $A = \{3, 4, 5, 6, 7, 8\}$ and define a binary relation R on A as follows:

$$\text{For all } x, y \in A, \quad x\,R\,y \;\Leftrightarrow\; 2\,|\,(x - y).$$

Draw the directed graph of R.

Solution   Note that $3\,R\,3$ because $3 - 3 = 0$ and $2\,|\,0$ since $0 = 2 \cdot 0$. Thus there is a loop from 3 to itself. Similarly, there is a loop from 4 to itself, from 5 to itself, and so forth, since the difference of each integer with itself is 0, and $2\,|\,0$.

Note also that $3\,R\,5$ because $3 - 5 = -2 = 2 \cdot (-1)$. And $5\,R\,3$ because $5 - 3 = 2 = 2 \cdot 1$. Hence there is an arrow from 3 to 5 and also an arrow from 5 to 3. The other arrows in the directed graph, as shown below, are obtained by similar reasoning.

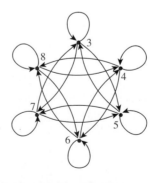

■

N-ary Relations and Relational Databases

N-ary relations form the mathematical foundation for relational database theory. A binary relation is a subset of the Cartesian product of two sets, similarly, an *n-ary* relation is a subset of the Cartesian product of n sets.

• Definition

Given sets $A_1, A_2, \ldots, A_n$, an n-ary relation R on $A_1 \times A_2 \times \cdots \times A_n$ is a subset of $A_1 \times A_2 \times \cdots \times A_n$. The special cases of 2-ary, 3-ary, and 4-ary relations are called **binary, ternary,** and **quaternary relations,** respectively,

---

Example 10.1.11 A Simple Database

The following is a radically simplified version of a database that might be used in a hospital. Let A_1 be a set of positive integers, A_2 a set of alphabetic character strings, A_3 a set of numeric character strings, and A_4 a set of alphabetic character strings. Define a quaternary relation R on $A_1 \times A_2 \times A_3 \times A_4$ as follows:

$$(a_1, a_2, a_3, a_4) \in R \quad \Leftrightarrow \quad \text{a patient with patient ID number } a_1, \text{ named } a_2, \text{ was admitted on date } a_3, \text{ with primary diagnosis } a_4.$$

At a particular hospital, this relation might contain the following 4-tuples:

(011985, John Schmidt, 020795, asthma)

(574329, Tak Kurosawa, 011495, pneumonia)

(466581, Mary Lazars, 010395, appendicitis)

(008352, Joan Kaplan, 112494, gastritis)

(011985, John Schmidt, 021795, pneumonia)

(244388, Sarah Wu, 010395, broken leg)

(778400, Jamal Baskers, 122794, appendicitis)

In discussions of relational databases, the tuples are normally thought of as being written in tables. Each row of the table corresponds to one tuple, and the header for each column gives the descriptive attribute for the elements in the column.

Operations within a database allow the data to be manipulated in many different ways. For example, in the database language SQL, if the above database is denoted S, the result of the query

SELECT Patient_ID#, Name FROM S WHERE

Admission_Date $= 010395$

would be a list of the ID numbers and names of all patients admitted on 01-03-95:

466581   Mary Lazars,

244388   Sarah Wu.

This is obtained by taking the intersection of the set $A_1 \times A_2 \times \{010395\} \times A_4$ with the database and then projecting onto the first two coordinates. (See exercise 20 of Section 7.1.) Similarly, SELECT can be used to obtain a list of all admission dates of a given patient. For John Schmidt this list is

02-07-95   and

02-17-95

Individual entries in a database can be added, deleted, or updated, and most databases can sort data entries in various ways. In addition, entire databases can be merged, and the entries common to two databases can be moved to a new database. ∎

Exercise Set 10.1*

1. Let $A = \{2, 3, 4\}$ and $B = \{6, 8, 10\}$ and define a binary relation R from A to B as follows:

$$\text{For all } (x, y) \in A \times B, \quad (x, y) \in R \quad \Leftrightarrow \quad x \mid y.$$

a. Is $4\,R\,6$? Is $4\,R\,8$? Is $(3, 8) \in R$? Is $(2, 10) \in R$?
b. Write R as a set of ordered pairs.

2. Let $C = \{2, 3, 4, 5\}$ and $D = \{3, 4\}$ and define a binary relation S from C to D as follows:

$$\text{For all } (x, y) \in C \times D, \quad (x, y) \in S \quad \Leftrightarrow \quad x \geq y.$$

a. Is $2\,S\,4$? Is $4\,S\,3$? Is $(4, 4) \in S$? Is $(3, 2) \in S$?
b. Write S as a set of ordered pairs.

3. As in Example 10.1.2, the **congruence modulo 2** relation E is defined from $\mathbf{Z}$ to $\mathbf{Z}$ as follows:

$$\text{For all integers } m \text{ and } n, \quad m\,E\,n \quad \Leftrightarrow \quad m - n \text{ is even.}$$

a. Is $0\,E\,0$? Is $5\,E\,2$? Is $(6, 6) \in E$? Is $(-1, 7) \in E$?
b. Prove that for any even integer n, $n\,E\,0$.

H 4. Prove that for all integers m and n, $m - n$ is even if, and only if, both m and n are even or both m and n are odd.

5. The **congruence modulo 3** relation, T, is defined from $\mathbf{Z}$ to $\mathbf{Z}$ as follows:

$$\text{For all integers } m \text{ and } n, \quad m\,T\,n \quad \Leftrightarrow \quad 3 \mid (m - n).$$

a. Is $10\,T\,1$? Is $1\,T\,10$? Is $(2, 2) \in T$? Is $(8, 1) \in T$?
b. List five integers n such that $n\,T\,0$.
c. List five integers n such that $n\,T\,1$.
d. List five integers n such that $n\,T\,2$.
H e. Make and prove a conjecture about which integers are related by T to 0, which integers are related by T to 1, and which integers are related by T to 2.

6. Define a binary relation S from $\mathbf{R}$ to $\mathbf{R}$ as follows:

$$\text{For all } (x, y) \in \mathbf{R} \times \mathbf{R}, \quad x\,S\,y \quad \Leftrightarrow \quad x \geq y.$$

a. Is $(2, 1) \in S$? Is $(2, 2) \in S$? Is $2\,S\,3$? Is $(-1)\,S\,(-2)$?
b. Draw the graph of S in the Cartesian plane.

7. Define a binary relation R from $\mathbf{R}$ to $\mathbf{R}$ as follows:

$$\text{For all } (x, y) \in \mathbf{R} \times \mathbf{R}, \quad x\,R\,y \quad \Leftrightarrow \quad y = x^2.$$

a. Is $(2, 4) \in R$? Is $(4, 2) \in R$? Is $(-3)\,R\,9$? Is $9\,R\,(-3)$?
b. Draw the graph of R in the Cartesian plane.

8. Define a binary relation P on $\mathbf{Z}$ as follows:

$$\text{For all } m, n \in \mathbf{Z},$$
$$m\,P\,n \quad \Leftrightarrow \quad m \text{ and } n \text{ have a common prime factor.}$$

a. Is $15\,P\,25$?   b. $22\,P\,27$?
c. Is $0\,P\,5$?   d. Is $8\,P\,8$?

9. Let $X = \{a, b, c\}$. Recall that $\mathscr{P}(X)$ is the power set of X. Define a binary relation $\mathscr{R}$ on $\mathscr{P}(X)$ as follows:

$$\text{For all } A, B \in \mathscr{P}(X),$$
$$A\,\mathscr{R}\,B \quad \Leftrightarrow \quad A \text{ has the same number of elements as } B.$$

a. Is $\{a, b\}\,\mathscr{R}\,\{b, c\}$?   b. Is $\{a\}\,\mathscr{R}\,\{a, b\}$?
c. Is $\{c\}\,\mathscr{R}\,\{b\}$?

10. Let $X = \{a, b, c\}$. Define a binary relation $\mathscr{J}$ on $\mathscr{P}(X)$ as follows:

$$\text{For all } A, B \in \mathscr{P}(X), \quad A\,\mathscr{J}\,B \quad \Leftrightarrow \quad A \cap B \neq \varnothing.$$

a. Is $\{a\}\,\mathscr{J}\,\{c\}$?   b. Is $\{a, b\}\,\mathscr{J}\,\{b, c\}$?
c. Is $\{a, b\}\,\mathscr{J}\,\{a, b, c\}$?

11. Let S be the set of all strings of a's and b's of length 4. Define a relation R on S as follows:

$$\text{For all } s, t \in S,$$
$$s\,R\,t \quad \Leftrightarrow \quad s \text{ has the same first two characters as } t.$$

a. Is *abaa R abba*?   b. Is *aabb R bbaa*?
c. Is *aaaa R aaab*?

H 12. Let $A = \{4, 5, 6\}$ and $B = \{5, 6, 7\}$ and define binary relations R, S, and T from A to B as follows:

$$\text{For all } (x, y) \in A \times B, \quad (x, y) \in R \quad \Leftrightarrow \quad x \geq y.$$
$$\text{For all } (x, y) \in A \times B, \quad x\,S\,y \quad \Leftrightarrow \quad 2 \mid (x - y).$$
$$T = \{(4, 7), (6, 5), (6, 7)\}.$$

a. Draw arrow diagrams for R, S, and T.
b. Indicate whether any of the relations R, S, and T are functions.

13. a. Find all binary relations from $\{0,1\}$ to $\{1\}$.
b. Find all functions from $\{0,1\}$ to $\{1\}$.
c. What fraction of the binary relations from $\{0, 1\}$ to $\{1\}$ are functions?

14. Find four binary relations from $\{a, b\}$ to $\{x, y\}$ that are not functions from $\{a, b\}$ to $\{x, y\}$.

H 15. Suppose A is a set with m elements and B is a set with n elements.
a. How many binary relations are there from A to B? Explain.
b. How many functions are there from A to B? Explain.
c. What fraction of the binary relations from A to B are functions?

16. Define a binary relation P from $\mathbf{R}$ to $\mathbf{R}$ as follows:

$$\text{For all real numbers } x \text{ and } y,$$
$$(x, y) \in P \quad \Leftrightarrow \quad x = y^2.$$

Is P a function? Explain.

*For exercises with blue numbers or letters, solutions are given in Appendix B. The symbol *H* indicates that only a hint or a partial solution is given. The symbol ✱ signals that an exercise is more challenging than usual.

17. Let $A = \{3, 4, 5\}$ and $B = \{4, 5, 6\}$ and let R be the "less than" relation. That is,

    For all $(x, y) \in A \times B$, $x \, R \, y \;\Leftrightarrow\; x < y$.

    State explicitly which ordered pairs are in R and R^{-1}.

18. Let $A = \{3, 4, 5\}$ and $B = \{4, 5, 6\}$ and let S be the "divides" relation. That is,

    For all $(x, y) \in A \times B$, $x \, S \, y \;\Leftrightarrow\; x \mid y$.

    State explicitly which ordered pairs are in S and S^{-1}.

19. Let S be the set of all strings in a's and b's. Define a relation T on S as follows:

    For all $s, t \in S$, $s \, T \, t \;\Leftrightarrow\; t = as$

    (that is, t is the concatenation of a with s).
    a. Is $ab \, T \, aab$?   **b.** Is $aab \, T \, ab$?
    c. Is $ba \, T \, aba$?   **d.** Is $aba \, T^{-1} \, ba$?
    e. Is $abb \, T^{-1} \, bba$?   **f.** Is $abba \, T^{-1} \, bba$?

20. Define a relation R from $\mathbf{R}$ to $\mathbf{R}$ as follows:

    For all $(x, y) \in \mathbf{R} \times \mathbf{R}$, $x \, R \, y \;\Leftrightarrow\; y = \lfloor x \rfloor$.

    Draw the graphs of R and R^{-1} in the Cartesian plane.

21. **a.** Rewrite the definition of one-to-one function using the notation of the definition of a function as a relation.
    b. Rewrite the definition of onto function using the notation of the definition of function as a relation.

22. **a.** Suppose a function $F: X \to Y$ is one-to-one but not onto. Is F^{-1} (the inverse relation for F) a function? Explain your answer.
    b. Suppose a function $F: X \to Y$ is onto but not one-to-one. Is F^{-1} (the inverse relation for F) a function? Explain your answer.

Draw the directed graphs of the binary relations defined in 23–27 below.

23. Define a binary relation R on $A = \{0, 1, 2, 3\}$ by $R = \{(0, 0), (1, 2), (2, 2)\}$.

24. Define a binary relation S on $B = \{a, b, c, d\}$ by $S = \{(a, b), (a, c), (b, c), (d, d)\}$.

25. Let $A = \{2, 3, 4, 5, 6, 7, 8\}$ and define a binary relation R on A as follows:

    For all $x, y \in A$, $x \, R \, y \;\Leftrightarrow\; x \mid y$.

H 26. Let $A = \{5, 6, 7, 8, 9, 10\}$ and define a binary relation S on A as follows:

    For all $x, y \in A$, $x \, S \, y \;\Leftrightarrow\; 2 \mid (x - y)$.

27. Let $A = \{2, 3, 4, 5, 6, 7, 8\}$ and define a binary relation T on A as follows:

    For all $x, y \in A$, $x \, T \, y \;\Leftrightarrow\; 3 \mid (x - y)$.

28. In Example 10.1.11 the result of the query SELECT Patient_ID#, Name FROM S WHERE Primary_Diagnosis $= X$ is the projection onto the first two coordinates of the intersection of the set $A_1 \times A_2 \times A_3 \times \{X\}$ with the database.
    a. Find the result of the query SELECT Patient_ID#, Name FROM S WHERE Primary_Diagnosis = pneumonia.
    b. Find the result of the query SELECT Patient_ID#, Name FROM S WHERE Primary_Diagnosis = appendicitis.

Exercises 29–33 refer to unions and intersections of relations. Since binary relations are subsets of Cartesian products, their unions and intersections can be calculated as for any subsets. Given two relations R and S from A to B,

$$R \cup S = \{(x, y) \in A \times B \mid (x, y) \in R \text{ or } (x, y) \in S\}$$
$$R \cap S = \{(x, y) \in A \times B \mid (x, y) \in R \text{ and } (x, y) \in S\}.$$

29. Let $A = \{2, 4\}$ and $B = \{6, 8, 10\}$ and define binary relations R and S from A to B as follows:

    For all $(x, y) \in A \times B$, $x \, R \, y \;\Leftrightarrow\; x \mid y$.
    For all $(x, y) \in A \times B$, $x \, S \, y \;\Leftrightarrow\; y - 4 = x$.

    State explicitly which ordered pairs are in $A \times B, R, S, R \cup S$, and $R \cap S$.

30. Let $A = \{-1, 1, 2, 4\}$ and $B = \{1, 2\}$ and define binary relations R and S from A to B as follows:

    For all $(x, y) \in A \times B$, $x \, R \, y \;\Leftrightarrow\; |x| = |y|$.
    For all $(x, y) \in A \times B$, $x \, S \, y \;\Leftrightarrow\; x - y$ is even.

    State explicitly which ordered pairs are in $A \times B, R, S, R \cup S$, and $R \cap S$.

31. Define R and S from $\mathbf{R}$ to $\mathbf{R}$ as follows:

    $$R = \{(x, y) \in \mathbf{R} \times \mathbf{R} \mid x < y\} \quad \text{and}$$
    $$S = \{(x, y) \in \mathbf{R} \times \mathbf{R} \mid x = y\}.$$

    That is, R is the "less than" relation and S is the "equals" relation from $\mathbf{R}$ to $\mathbf{R}$. Graph $R, S, R \cup S$, and $R \cap S$ in the Cartesian plane.

32. Define binary relations R and S from $\mathbf{R}$ to $\mathbf{R}$ as follows:

    $$R = \{(x, y) \in \mathbf{R} \times \mathbf{R} \mid x^2 + y^2 = 4\} \quad \text{and}$$
    $$S = \{(x, y) \in \mathbf{R} \times \mathbf{R} \mid x = y\}.$$

    Graph $R, S, R \cup S$, and $R \cap S$ in the Cartesian plane.

33. Define binary relations R and S from $\mathbf{R}$ to $\mathbf{R}$ as follows:

    $$R = \{(x, y) \in \mathbf{R} \times \mathbf{R} \mid y = |x|\} \quad \text{and}$$
    $$S = \{(x, y) \in \mathbf{R} \times \mathbf{R} \mid y = 1\}.$$

    Graph $R, S, R \cup S$, and $R \cap S$ in the Cartesian plane.

10.2 Reflexivity, Symmetry, and Transitivity

Mathematics is the tool specially suited for dealing with abstract concepts of any kind and there is no limit to its power in this field. — P. A. M. Dirac, 1902–1984

Let $A = \{2, 3, 4, 6, 7, 9\}$ and define a relation R on A as follows:

$$\text{For all } x, y \in A, \quad x \, R \, y \quad \Leftrightarrow \quad 3 \mid (x - y).$$

Then $2 \, R \, 2$ because $2 - 2 = 0$, and $3 \mid 0$. Similarly, $3 \, R \, 3, 4 \, R \, 4, 6 \, R \, 6, 7 \, R \, 7$, and $9 \, R \, 9$. Also $6 \, R \, 3$ because $6 - 3 = 3$, and $3 \mid 3$. And $3 \, R \, 6$ because $3 - 6 = -(6 - 3) = -3$, and $3 \mid (-3)$. Similarly, $3 \, R \, 9, 9 \, R \, 3, 6 \, R \, 9, 9 \, R \, 6, 4 \, R \, 7$, and $7 \, R \, 4$. Thus the directed graph for R has the appearance shown below.

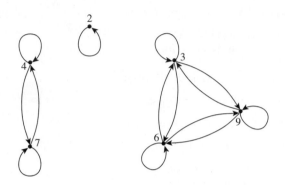

This graph has three important properties:

1. Each point of the graph has an arrow looping around from it back to itself.

2. In each case where there is an arrow going from one point to a second, there is an arrow going from the second point back to the first.

3. In each case where there is an arrow going from one point to a second and from a second point to a third, there is an arrow going from the first point to the third. That is, there are no "incomplete directed triangles" in the graph.

Properties (1), (2), and (3) correspond to properties of general binary relations called *reflexivity, symmetry,* and *transitivity*.

• Definition

Let R be a binary relation on a set A.

1. R is **reflexive** if, and only if, for all $x \in A$, $x \, R \, x$.

2. R is **symmetric** if, and only if, for all $x, y \in A$, *if* $x \, R \, y$ then $y \, R \, x$.

3. R is **transitive** if, and only if, for all $x, y, z \in A$, *if* $x \, R \, y$ and $y \, R \, z$ then $x \, R \, z$.

Because of the equivalence of the expressions $x \, R \, y$ and $(x, y) \in R$ for all x and y in A, the reflexive, symmetric, and transitive properties can also be written as follows:

> 1. R is reflexive $\Leftrightarrow$ for all x in A, $(x, x) \in R$.
>
> 2. R is symmetric $\Leftrightarrow$ for all x and y in A, ***if*** $(x, y) \in R$ then $(y, x) \in R$.
>
> 3. R is transitive $\Leftrightarrow$ for all x, y and z in A, ***if*** $(x, y) \in R$ and $(y, z) \in R$ then $(x, z) \in R$.

In informal terms, properties (1)–(3) say the following:

1. **Reflexive:** Each element is related to itself.

2. **Symmetric:** If any one element is related to any other element, then the second element is related to the first.

3. **Transitive:** If any one element is related to a second and that second element is related to a third, then the first element is related to the third.

Caution! One caution about the informal phrasing: The first, second, and third elements referred to need not all be distinct. This is the disadvantage of informal phrasing; it sometimes masks some nuances of the full formal definition.

Note that the definitions of reflexivity, symmetry, and transitivity are universal statements. This means that to prove a relation has one of the properties, you use either the method of exhaustion or the method of generalizing from the generic particular.

Now consider what it means for a relation *not* to have one of the properties defined above. Recall that the negation of a universal statement is existential. Hence if R is a binary relation on a set A, then

1. R is **not reflexive** $\Leftrightarrow$ there is an element x in A such that $x \not\!R\, x$ [*that is, such that* $(x, x) \notin R$].

2. R is **not symmetric** $\Leftrightarrow$ there are elements x and y in A such that $x\, R\, y$ but $y \not\!R\, x$ [*that is, such that* $(x, y) \in R$ *but* $(y, x) \notin R$].

3. R is **not transitive** $\Leftrightarrow$ there are elements x, y and z in A such that $x\, R\, y$ and $y\, R\, z$ but $x \not\!R\, z$ [*that is, such that* $(x, y) \in R$ *and* $(y, z) \in R$ *but* $(x, z) \notin R$].

It follows that you can show that a binary relation does *not* have one of the properties by finding a counterexample.

Example 10.2.1 Properties of Binary Relations on Finite Sets

Let $A = \{0, 1, 2, 3\}$ and define relations R, S, and T on A as follows:

$$R = \{(0, 0), (0, 1), (0, 3), (1, 0), (1, 1), (2, 2), (3, 0), (3, 3)\},$$
$$S = \{(0, 0), (0, 2), (0, 3), (2, 3)\},$$
$$T = \{(0, 1), (2, 3)\}.$$

a. Is R reflexive? symmetric? transitive?

b. Is S reflexive? symmetric? transitive?

c. Is T reflexive? symmetric? transitive?

Solution

a. The directed graph of R has the appearance shown below.

R is reflexive: There is a loop at each point of the directed graph. This means that each element of A is related to itself, so R is reflexive.

R is symmetric: In each case where there is an arrow going from one point of the graph to a second, there is an arrow going from the second point back to the first. This means that whenever one element of A is related by R to a second, then the second is related to the first. Hence R is symmetric.

R is not transitive: There is an arrow going from 1 to 0 and an arrow going from 0 to 3, but there is no arrow going from 1 to 3. This means that there are elements of A— 0, 1, and 3—such that $1\ R\ 0$ and $0\ R\ 3$ but $1\not{R}3$. Hence R is not transitive.

b. The directed graph of S has the appearance shown below.

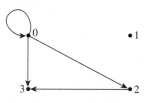

S is not reflexive: There is no loop at 1, for example. Thus $(1, 1) \notin S$, and so S is not reflexive.

S is not symmetric: There is an arrow from 0 to 2 but not from 2 to 0. Hence $(0, 2) \in S$ but $(2, 0) \notin S$, and so S is not symmetric.

S is transitive: There are three cases for which there is an arrow going from one point of the graph to a second and from the second point to a third: Namely, there are arrows going from 0 to 2 and from 2 to 3; there are arrows going from 0 to 0 and from 0 to 2; and there are arrows going from 0 to 0 and from 0 to 3. In each case there is an arrow going from the first point to the third. (Note again that the "first," "second," and "third" points need not be distinct.) This means that whenever $(x, y) \in S$ and $(y, z) \in S$, then $(x, z) \in S$, for all $x, y, z \in \{0, 1, 2, 3\}$, and so S is transitive.

c. The directed graph of T has the appearance shown below.

T is not reflexive: There is no loop at 0, for example. Thus $(0, 0) \notin T$, so T is not reflexive.

T is not symmetric: There is an arrow from 0 to 1 but not from 1 to 0. Thus $(0, 1) \in T$ but $(1, 0) \notin T$, and so T is not symmetric.

T is transitive: The transitivity condition is vacuously true for T. That is, T is transitive by default because it is *not* not transitive! To see this, observe that the transitivity condition says that

For all $x, y, z \in A$, if $(x, y) \in T$ and $(y, z) \in T$ then $(x, z) \subset T$.

The only way for this to be false would be for there to exist elements of A that make the hypothesis true and the conclusion false. That is, there would have to be elements x, y, and z in A such that

$$(x, y) \in T \quad \text{and} \quad (y, z) \in T \quad \text{and} \quad (x, z) \notin T.$$

In other words, there would have to be two ordered pairs in T that have the potential to "link up" by having the *second* element of one pair be the *first* element of the other pair. But the only elements in T are $(0, 1)$ and $(2, 3)$, and these do not have the potential to link up. Hence the hypothesis is never true. It follows that it is impossible for T *not* to be transitive, and thus T is transitive. ■

When a binary relation R is defined on a finite set A, it is possible to write computer algorithms to check whether R is reflexive, symmetric, and transitive. One way to do this is to represent A as a one-dimensional array, $(a[1], a[2], \ldots, a[n])$ and use the algorithm of exercise 33 in Section 5.1 to check whether an ordered pair in $A \times A$ is in R. Checking whether R is reflexive can be done with a loop that examines each element $a[i]$ of A in turn. If, for some i, $(a[i], a[i]) \notin R$, then R is not reflexive. Otherwise, R is reflexive. Checking for symmetry can be done with a nested loop that examines each pair $(a[i], a[j])$ of $A \times A$ in turn. If, for some i and j, $(a[i], a[j]) \in R$ and $(a[j], a[i]) \notin R$, then R is not symmetric. Otherwise, R is symmetric. Checking whether R is transitive can be done with a triply nested loop that examines each triple $(a[i], a[j], a[k])$ of $A \times A \times A$ in turn. If, for some triple, $(a[i], a[j]) \in R$, $(a[j], a[k]) \in R$, and $(a[i], a[k]) \notin R$, then R is not transitive. Otherwise, R is transitive. In the exercises for this section, you are asked to formalize these algorithms.

The Transitive Closure of a Relation

Generally speaking, a relation fails to be transitive because it fails to contain certain ordered pairs. For example, if $(1, 3)$ and $(3, 4)$ are in a relation R, then the pair $(1, 4)$ *must* be in R if R is to be transitive. To obtain a transitive relation from one that is not transitive, it is necessary to add ordered pairs. Roughly speaking, the relation obtained by adding the least number of ordered pairs to ensure transitivity is called the *transitive closure* of the relation. In a sense made precise by the formal definition, the transitive closure of a relation is the smallest transitive relation that contains the relation.

• Definition

Let A be a set and R a binary relation on A. The **transitive closure** of R is the binary relation R^t on A that satisfies the following three properties:

1. R^t is transitive.

2. $R \subseteq R^t$.

3. If S is any other transitive relation that contains R, then $R^t \subseteq S$.

Example 10.2.2 Transitive Closure of a Relation

Let $A = \{0, 1, 2, 3\}$ and consider the relation R defined on A as follows:

$$R = \{(0, 1), (1, 2), (2, 3)\}.$$

Find the transitive closure of R.

Solution   Every ordered pair in R is in R^t, so

$$\{(0, 1), (1, 2), (2, 3)\} \subseteq R^t.$$

Thus the directed graph of R contains the arrows shown below.

Since there are arrows going from 0 to 1 and from 1 to 2, R^t must have an arrow going from 0 to 2. Hence $(0, 2) \in R^t$. Then $(0, 2) \in R^t$ and $(2, 3) \in R^t$, so since R^t is transitive, $(0, 3) \in R^t$. Also, since $(1, 2) \in R^t$ and $(2, 3) \in R^t$, then $(1, 3) \in R^t$. Thus R^t contains at least the following ordered pairs:

$$\{(0, 1), (0, 2), (0, 3), (1, 2), (1, 3), (2, 3)\}.$$

But this relation *is* transitive; hence it equals R^t. Note that the directed graph of R^t is as shown below.

Properties of Relations on Infinite Sets

Suppose a binary relation R is defined on an infinite set A. To prove the relation is reflexive, symmetric, or transitive, first write down what is to be proved. For instance, for symmetry you need to prove that

$$\forall x, y \in A, \text{ if } x \, R \, y \text{ then } y \, R \, x.$$

Then use the definitions of A and R to rewrite the statement for the particular case in question. For instance, for the "equality" relation on the set of real numbers, the rewritten statement is

$$\forall x, y \in \mathbf{R}, \text{ if } x = y \text{ then } y = x.$$

Sometimes the truth of the rewritten statement will be immediately obvious (as it is here). At other times you will need to prove it using the method of generalizing from the generic particular. We give examples of both cases in this section. We begin with the relation of equality, one of the simplest and yet most important binary relations.

Example 10.2.3 Properties of Equality

Define a binary relation R on $\mathbf{R}$ (the set of all real numbers) as follows: For all real numbers x and y.

$$\boxed{x \, R \, y \quad \Leftrightarrow \quad x = y.}$$

a. Is R reflexive?   b. Is R symmetric?   c. Is R transitive?

Solution

a. **R is reflexive:** R is reflexive if, and only if, the following statement is true:

$$\text{For all } x \in \mathbf{R}, \quad x \, R \, x.$$

Since $x \, R \, x$ just means that $x = x$, this is the same as saying

$$\text{For all } x \in \mathbf{R}, \quad x = x.$$

But this statement is certainly true; every real number is equal to itself.

b. **R is symmetric:** R is symmetric if, and only if, the following statement is true:

$$\text{For all } x, y \in \mathbf{R}, \quad \text{if } x \, R \, y \text{ then } y \, R \, x.$$

By definition of R, $x \, R \, y$ means that $x = y$ and $y \, R \, x$ means that $y = x$. Hence R is symmetric if, and only if,

$$\text{For all } x, y \in \mathbf{R}, \quad \text{if } x = y \text{ then } y = x.$$

But this statement is certainly true; if one number is equal to a second, then the second is equal to the first.

c. **R is transitive:** R is transitive if, and only if, the following statement is true:

$$\text{For all } x, y, z \in \mathbf{R}, \quad \text{if } x \, R \, y \text{ and } y \, R \, z \text{ then } x \, R \, z.$$

By definition of R, $x \, R \, y$ means that $x = y$, $y \, R \, z$ means that $y = z$, and $x \, R \, z$ means that $x = z$. Hence R is transitive if, and only if, the following statement is true:

$$\text{For all } x, y, z \in \mathbf{R}, \quad \text{if } x = y \text{ and } y = z \text{ then } x = z.$$

But this statement is certainly true: If one real number equals a second and the second equals a third, then the first equals the third. ∎

Example 10.2.4 Properties of "Less Than"

Define a relation R on $\mathbf{R}$ (the set of all real numbers) as follows: For all $x, y \in R$,

$$\boxed{x \; R \; y \quad \Leftrightarrow \quad x < y.}$$

a. Is R reflexive?   b. Is R symmetric?   c. Is R transitive?

Solution

a. **R is not reflexive:** R is reflexive if, and only if, $\forall x \in \mathbf{R}, x \; R \; x$. By definition of R, this means that $\forall x \in \mathbf{R}, x < x$. But this is false: $\exists x \in \mathbf{R}$ such that $x \not< x$. As a counterexample, let $x = 0$ and note that $0 \not< 0$. Hence R is not reflexive.

b. **R is not symmetric:** R is symmetric if, and only if, $\forall x, y \in \mathbf{R}$, if $x \; R \; y$ then $y \; R \; x$. By definition of R, this means that $\forall x, y \in \mathbf{R}$, if $x < y$ then $y < x$. But this is false: $\exists x, y \in \mathbf{R}$ such that $x < y$ and $y \not< x$. As a counterexample, let $x = 0$ and $y = 1$ and note that $0 < 1$ but $1 \not< 0$. Hence R is not symmetric.

c. **R is transitive:** R is transitive if, and only if, for all $x, y, z \in \mathbf{R}$, if $x \; R \; y$ and $y \; R \; z$ then $x \; R \; z$. By definition of R, this means that for all $x, y, z \in \mathbf{R}$, if $x < y$ and $y < z$, then $x < z$. But this statement is true by the transitive law of order for real numbers (Appendix A, T17). Hence R is transitive. ∎

Sometimes a property is "universally false" in the sense that it is false for *every* element of its domain. It follows immediately, of course, that the property is false for each particular element of the domain and hence counterexamples abound. In such a case, it may seem more natural to prove the universal falseness of the property rather than to give a single counterexample. In the example above, for instance, you might find it natural to answer (a) and (b) as follows:

Alternative Answer to (a): R is not reflexive because $x \not< x$ for any real number x (by the trichotomy law—Appendix A, T16).

Alternative Answer to (b): R is not symmetric because for all x and y in A, if $x < y$, then $y \not< x$ (by the trichotomy law).

Example 10.2.5 Properties of Congruence Modulo 3

Define a relation R on $\mathbf{Z}$ (the set of all integers) as follows: For all integers m and n,

$$\boxed{m \; R \; n \quad \Leftrightarrow \quad 3 \mid (m - n).}$$

This relation is called **congruence modulo 3.**

a. Is R reflexive?   b. Is R symmetric?   c. Is R transitive?

Solution

a. **R is reflexive:** To show that R is reflexive, it is necessary to show that

$$\text{For all } m \in \mathbf{Z}, \quad m \, R \, m.$$

By definition of R, this means that

$$\text{For all } m \in \mathbf{Z}, \quad 3 \mid (m - m).$$

Or, since $m - m = 0$,

$$\text{For all } m \in \mathbf{Z}, \quad 3 \mid 0.$$

But this is true: $3 \mid 0$ since $0 = 3 \cdot 0$. Hence R is reflexive. This reasoning is formalized in the following proof.

Proof of Reflexivity: Suppose m is a particular but arbitrarily chosen integer. [*We must show that m R m.*] Now $m - m = 0$. But $3 \mid 0$ since $0 = 3 \cdot 0$. Hence $3 \mid (m - m)$. Thus, by definition of R, $m \, R \, m$ [*as was to be shown*].

b. **R is symmetric:** To show that R is symmetric, it is necessary to show that

$$\text{For all } m, n \in \mathbf{Z}, \quad \text{if } m \, R \, n \text{ then } n \, R \, m.$$

By definition of R this means that

$$\text{For all } m, n \in \mathbf{Z}, \quad \text{if } 3 \mid (m - n) \text{ then } 3 \mid (n - m).$$

Is this true? Suppose m and n are particular but arbitrarily chosen integers such that $3 \mid (m - n)$. Must it follow that $3 \mid (n - m)$? By definition of "divides," since

$$3 \mid (m - n),$$

then

$$m - n = 3k \quad \text{for some integer } k.$$

The crucial observation is that $n - m = -(m - n)$. Hence, you can multiply both sides of this equation by -1 to obtain

$$-(m - n) = -3k,$$

which is equivalent to

$$n - m = 3(-k).$$

Since $-k$ is an integer, this equation shows that

$$3 \mid (n - m).$$

It follows that R is symmetric.

The reasoning above is formalized in the following proof.

Proof of Symmetry: Suppose m and n are particular but arbitrarily chosen integers that satisfy the condition $m \, R \, n$. [*We must show that n R m.*] By definition of R, since $m \, R \, n$ then $3 \mid (m - n)$. By definition of "divides," this means that $m - n = 3k$, for some integer k. Multiplying both sides by -1 gives $n - m = 3(-k)$. Since $-k$ is an integer, this equation shows that $3 \mid (n - m)$. Hence, by definition of R, $n \, R \, m$ [*as was to be shown*].

c. **R is transitive:** To show that R is transitive, it is necessary to show that

$$\text{For all } m, n \in \mathbf{Z}, \quad \text{if } m \, R \, n \text{ and } n \, R \, p \text{ then } m \, R \, p.$$

By definition of R this means that

For all $m, n \in \mathbf{Z}$,   **if** $3 \mid (m - n)$ and $3 \mid (n - p)$ then $3 \mid (m - p)$.

Is this true? Suppose m, n, and p are particular but arbitrarily chosen integers such that $3 \mid (m - n)$ and $3 \mid (n - p)$. Must it follow that $3 \mid (m - p)$? By definition of "divides," since

$$3 \mid (m - n) \quad \text{and} \quad 3 \mid (n - p),$$

then

$$m - n = 3r \quad \text{for some integer } r,$$

and

$$n - p = 3s \quad \text{for some integer } s.$$

The crucial observation is that $(m - n) + (n - p) = m - p$. Add these two equations together to obtain

$$(m - n) + (n - p) = 3r + 3s,$$

which is equivalent to

$$m - p = 3(r + s).$$

Since r and s are integers, $r + s$ is an integer, and so this equation shows that

$$3 \mid (m - p).$$

It follows that R is transitive.

The reasoning above is formalized in the following proof.

Proof of Transitivity: Suppose m, n, and p are particular but arbitrarily chosen integers that satisfy the condition $m \, R \, n$ and $n \, R \, p$. [*We must show that $m \, R \, p$.*] By definition of R, since $m \, R \, n$ and $n \, R \, p$, then $3 \mid (m - n)$ and $3 \mid (n - p)$. By definition of "divides," this means that $m - n = 3r$ and $n - p = 3s$, for some integers r and s. Adding the two equations gives $(m - n) + (n - p) = 3r + 3s$, and simplifying gives that $m - p = 3(r + s)$. Since $r + s$ is an integer, this equation shows that $3 \mid (m - p)$. Hence, by definition of R, $m \, R \, p$ [*as was to be shown*]. ∎

Exercise Set 10.2

In 1–8 a number of binary relations are defined on the set $A = \{0, 1, 2, 3\}$. For each relation:
a. Draw the directed graph.
b. Determine whether the relation is reflexive.
c. Determine whether the relation is symmetric.
d. Determine whether the relation is transitive.
Give a counterexample in each case in which the relation does not satisfy one of the properties.

1. $R_1 = \{(0, 0), (0, 1), (0, 3), (1, 1), (1, 0), (2, 3), (3, 3)\}$

2. $R_2 = \{(0, 0), (0, 1), (1, 1), (1, 2), (2, 2), (2, 3)\}$

3. $R_3 = \{(2, 3), (3, 2)\}$

4. $R_4 = \{(1, 2), (2, 1), (1, 3), (3, 1)\}$

5. $R_5 = \{(0, 0), (0, 1), (0, 2), (1, 2)\}$

6. $R_6 = \{(0, 1), (0, 2)\}$ 7. $R_7 = \{(0, 3), (2, 3)\}$

8. $R_8 = \{(0, 0), (1, 1)\}$

In 9–11, R, S, and T are binary relations defined on $A = \{0, 1, 2, 3\}$.

9. Let $R = \{(0, 1), (0, 2), (1, 1), (1, 3), (2, 2), (3, 0)\}$.
Find R^t, the transitive closure of R.

10. Let $S = \{(0, 0), (0, 3), (1, 0), (1, 2), (2, 0), (3, 2)\}$.
Find S^t, the transitive closure of S.

11. Let $T = \{(0, 2), (1, 0), (2, 3), (3, 1)\}$. Find T^t, the transitive closure of T.

In 12–36 determine whether the given binary relation is reflexive, symmetric, transitive, or none of these. Justify your answers.

12. R is the "greater than or equal to" relation on the set of real numbers: For all $x, y \in \mathbf{R}$, $x \, R \, y \Leftrightarrow x \geq y$.

13. C is the circle relation on the set of real numbers: For all $x, y \in \mathbf{R}$, $x\,C\,y \Leftrightarrow x^2 + y^2 = 1$.

14. D is the binary relation defined on $\mathbf{R}$ as follows: For all $x, y \in \mathbf{R}$, $x\,D\,y \Leftrightarrow xy \geq 0$.

15. E is the congruence modulo 2 relation on $\mathbf{Z}$: For all $m, n \in \mathbf{Z}$, $m\,E\,n \Leftrightarrow 2\,|\,(m - n)$.

16. F is the congruence modulo 5 relation on $\mathbf{Z}$: For all $m, n \in \mathbf{Z}$, $m\,F\,n \Leftrightarrow 5\,|\,(m - n)$.

17. O is the binary relation defined on $\mathbf{Z}$ as follows: For all $m, n \in \mathbf{Z}$, $m\,O\,n \Leftrightarrow m - n$ is odd.

18. D is the "divides" relation on $\mathbf{Z}$: For all integers m and n, $m\,D\,n \Leftrightarrow m\,|\,n$.

19. A is the "absolute value" relation on $\mathbf{R}$: For all real numbers x and y, $x\,A\,y \Leftrightarrow |x| = |y|$.

20. Recall that a prime number is an integer that is greater than 1 and has no positive integer divisors other than 1 and itself. (In particular, 1 is not prime.) A binary relation P is defined on $\mathbf{Z}$ as follows: For all $m, n \in \mathbf{Z}$, $m\,P\,n \Leftrightarrow \exists$ a prime number p such that $p\,|\,m$ and $p\,|\,n$.

21. Let S be the set of all strings of a's and b's. A binary relation L is defined on S as follows: For all strings $s, t \in S$, $s\,L\,t \Leftrightarrow l(s) < l(t)$ where l is the length function (that is, the number of characters in s is less than the number of characters in t).

22. Let B be the set of all strings of 0's and 1's. A binary relation G is defined on B as follows: For all $s, t \in B$, $s\,G\,t \Leftrightarrow$ the number of 0's in s is greater than the number of 0's in t.

23. Let $X = \{a, b, c\}$ and $\mathscr{P}(X)$ be the power set of X (the set of all subsets of X). A binary relation $\#$ is defined on $\mathscr{P}(X)$ as follows: For all $A, B \in \mathscr{P}(X)$, $A\,\#\,B \Leftrightarrow$ the number of elements in A equals the number of elements in B.

24. Let $X = \{a, b, c\}$ and $\mathscr{P}(X)$ be the power set of X. A binary relation $\mathscr{R}$ is defined on $\mathscr{P}(X)$ as follows: For all $A, B \in \mathscr{P}(X)$, $A\,\mathscr{R}\,B \Leftrightarrow N(A) < N(B)$ (that is, the number of elements in A is less than the number of elements in B).

25. Let $X = \{a, b, c\}$ and $\mathscr{P}(X)$ be the power set of X. A binary relation $\mathscr{N}$ is defined on $\mathscr{P}(X)$ as follows: For all $A, B \in \mathscr{P}(X)$, $A\,\mathscr{N}\,B \Leftrightarrow N(A) \neq N(B)$ (that is, the number of elements in A is not equal to the number of elements in B).

26. Let A be a nonempty set and $\mathscr{P}(A)$ the power set of A. Define the "subset" relation $\mathscr{J}$ on $\mathscr{P}(A)$ as follows: For all $X, Y \in \mathscr{P}(A)$, $X\,\mathscr{J}\,Y \Leftrightarrow X \subseteq Y$.

27. Let A be a nonempty set and $\mathscr{P}(A)$ the power set of A. Define the "not equal to" relation $\mathscr{R}$ on $\mathscr{P}(A)$ as follows: For all $X, Y \in \mathscr{P}(A)$, $X\,\mathscr{R}\,Y \Leftrightarrow X \neq Y$.

28. Let A be a nonempty set and $\mathscr{P}(A)$ the power set of A. Define the "relative complement" relation $\mathscr{C}$ on $\mathscr{P}(A)$ as follows: For all $X, Y \in \mathscr{P}(A)$, $X\,\mathscr{C}\,Y \Leftrightarrow Y = A - X$.

29. Let A be a set with at least two elements and $\mathscr{P}(A)$ the power set of A. Define a relation $\mathscr{R}$ on $\mathscr{P}(A)$ as follows: For all $X, Y \in \mathscr{P}(A)$, $X\,\mathscr{R}\,Y \Leftrightarrow X \subseteq Y$ or $Y \subseteq X$.

30. Let A be the set of all English statements. A binary relation I is defined on A as follows: For all $p, q \in A$, $p\,I\,q \Leftrightarrow p \rightarrow q$ is true.

31. Let $A = \mathbf{R} \times \mathbf{R}$. A binary relation $\mathscr{R}$ is defined on A as follows: For all (x_1, y_1) and (x_2, y_2) in A, $(x_1, y_1)\,\mathscr{R}\,(x_2, y_2) \Leftrightarrow x_1 = x_2$.

32. Let $A = \mathbf{R} \times \mathbf{R}$. A binary relation $\mathscr{R}$ is defined on A as follows: For all (x_1, y_1) and (x_2, y_2) in A, $(x_1, y_1)\,\mathscr{R}\,(x_2, y_2) \Leftrightarrow y_1 = y_2$.

33. Let A be the "punctured plane"; that is, A is the set of all points in the Cartesian plane except the origin $(0, 0)$. A binary relation R is defined on A as follows: For all p_1 and p_2 in A, $p_1\,R\,p_2 \Leftrightarrow p_1$ and p_2 lie on the same half line emanating from the origin.

34. Let A be the set of people living in the world today. A binary relation R is defined on A as follows: For all $p, q \in A$, $p\,R\,q \Leftrightarrow p$ lives within 100 miles of q.

35. Let A be the set of all lines in the plane. A binary relation R is defined on A as follows: For all l_1 and l_2 in A, $l_1\,R\,l_2 \Leftrightarrow l_1$ is parallel to l_2. (Assume that a line is parallel to itself.)

36. Let A be the set of all lines in the plane. A binary relation R is defined on A as follows: For all l_1 and l_2 in A, $l_1\,R\,l_2 \Leftrightarrow l_1$ is perpendicular to l_2.

37. Let A be a set with eight elements.
    a. How many binary relations are there on A?
    b. How many binary relations on A are reflexive?
    c. How many binary relations on A are symmetric?
    d. How many binary relations on A are both reflexive and symmetric?

38. Write a computer algorithm to test whether a binary relation R defined on a finite set A is reflexive, where $A = \{a[1], a[2], \ldots, a[n]\}$.

39. Write a computer algorithm to test whether a binary relation R defined on a finite set A is symmetric, where $A = \{a[1], a[2], \ldots, a[n]\}$.

40. Write a computer algorithm to test whether a binary relation R defined on a finite set A is transitive, where $A = \{a[1], a[2], \ldots, a[n]\}$.

✱41. Let R be a binary relation on a set A and let R^t be the transitive closure of R. Prove that for all x and y in A, $x\, R^t\, y$ if, and only if, there is a sequence of elements of A, say $x_1, x_2, \ldots, x_n$, such that $x = x_1, x_1\, R\, x_2, x_2\, R\, x_3, \ldots, x_{n-1}\, R\, x_n$, and $x_n = y$.

✱42. Write a computer program to find the transitive closure of a binary relation R defined on a finite set $A = \{a[1], a[2], \ldots, a[n]\}$.

43. Suppose R and S are binary relations on a set A.
   a. If R and S are reflexive, is $R \cap S$ reflexive? Why?
   H b. If R and S are symmetric, is $R \cap S$ symmetric? Why?
   c. If R and S are transitive, is $R \cap S$ transitive? Why?

44. Suppose R and S are binary relations on a set A.
   a. If R and S are reflexive, is $R \cup S$ reflexive? Why?
   b. If R and S are symmetric, is $R \cup S$ symmetric? Why?
   c. If R and S are transitive, is $R \cup S$ transitive? Why?

In 45–52 the following definitions are used: A binary relation on a set A is defined to be

irreflexive if, and only if, for all $x \in A$, $x\, \cancel{R}\, x$;

asymmetric if, and only if, for all $x, y \in A$, if $x\, R\, y$ then $y\, \cancel{R}\, x$;

intransitive if, and only if, for all $x, y, z \in A$, if $x\, R\, y$ and $y\, R\, z$ then $x\, \cancel{R}\, z$.

For each of the binary relations in the referenced exercise, determine whether the relation is irreflexive, asymmetric, intransitive, or none of these.

45. Exercise 1   46. Exercise 2

47. Exercise 3   48. Exercise 4

49. Exercise 5   **50.** Exercise 6

51. Exercise 7   52. Exercise 8

10.3 Equivalence Relations

"You are sad" the Knight said in an anxious tone: "let me sing you a song to comfort you."

"Is it very long?" Alice asked, for she had heard a good deal of poetry that day.

"It's long," said the Knight, "but it's very, very beautiful. Everybody that hears me sing it—either it brings the tears into the eyes, or else—"

"Or else what?" said Alice, for the Knight had made a sudden pause.

"Or else it doesn't, you know. The name of the song is called 'Haddocks' Eyes.'"

"Oh, that's the name of the song, is it?" Alice said, trying to feel interested.

"No, you don't understand," the Knight said, looking a little vexed. "That's what the name is called. The name really is 'The Aged Aged Man.'"

"Then I ought to have said 'That's what the song is called'?" Alice corrected herself.

"No, you oughtn't: that's quite another thing! The song is called 'Ways and Means': but that's only what it's called, you know!"

"Well, what is the song, then?" said Alice, who was by this time completely bewildered.

"I was coming to that," the Knight said. "The song really is 'A-sitting on a Gate': and the tune's my own invention."

So saying, he stopped his horse and let the reins fall on its neck: then, slowly beating time with one hand, and with a faint smile lighting up his gentle foolish face, as if he enjoyed the music of his song, he began.

— Lewis Carroll, *Through the Looking Glass*, 1872

You know from your early study of fractions that each fraction has many equivalent forms. For example,

$$\frac{1}{2}, \frac{2}{4}, \frac{3}{6}, \frac{-1}{-2}, \frac{-3}{-6}, \frac{15}{30}, \ldots, \text{ and so on}$$

are all different ways to represent the same number. They may look different; they may be called different names; but they are all equal. The idea of grouping together things that "look different but are really the same" is the central idea of equivalence relations.

The Relation Induced by a Partition

A **partition** of a set A is a finite or infinite collection of nonempty, mutually disjoint subsets whose union is A. The diagram of Figure 10.3.1 illustrates a partition of a set A by subsets $A_1, A_2, \ldots, A_6$.

$$A_i \cap A_j = \emptyset \text{ whenever } i \neq j$$
$$A_i \cup A_2 \cup \cdots \cup A_6 = A$$

Figure 10.3.1 A Partition of a Set

• Definition

Given a partition of a set A, the **binary relation induced by the partition,** R, is defined on A as follows: For all $x, y \in A$,

$$x \, R \, y \quad \Leftrightarrow \quad \text{there is a subset } A \text{ of the partition}$$
$$\text{such that both } x \text{ and } y \text{ are in } A.$$

Example 10.3.1 Relation Induced by a Partition

Let $A = \{0, 1, 2, 3, 4\}$ and consider the following partition of A:

$$\{0, 3, 4\}, \{1\}, \{2\}.$$

Find the relation R induced by this partition.

Solution   Since $\{0, 3, 4\}$ is a subset of the partition.

$$0 \, R \, 3 \quad \text{because both 0 and 3 are in } \{0, 3, 4\},$$
$$3 \, R \, 0 \quad \text{because both 3 and 0 are in } \{0, 3, 4\},$$
$$0 \, R \, 4 \quad \text{because both 0 and 4 are in } \{0, 3, 4\},$$
$$4 \, R \, 0 \quad \text{because both 4 and 0 are in } \{0, 3, 4\},$$
$$3 \, R \, 4 \quad \text{because both 3 and 4 are in } \{0, 3, 4\}, \quad \text{and}$$
$$4 \, R \, 3 \quad \text{because both 4 and 3 are in } \{0, 3, 4\}.$$

Also,

$$0 \, R \, 0 \quad \text{because both 0 and 0 are in } \{0, 3, 4\}$$

[*This statement may seem strange, but, after all, it is not false!*],

$$3 \, R \, 3 \quad \text{because both 3 and 3 are in } \{0, 3, 4\}, \quad \text{and}$$
$$4 \, R \, 4 \quad \text{because both 4 and 4 are in } \{0, 3, 4\}.$$

Since $\{1\}$ is a subset of the partition,

$$1 \, R \, 1 \quad \text{because both 1 and 1 are in } \{1\},$$

and since $\{2\}$ is a subset of the partition,

$$2 \, R \, 2 \quad \text{because both 2 and 2 are in } \{2\}.$$

Hence

$$R = \{(0, 0), (0, 3), (0, 4), (1, 1), (2, 2), (3, 0), (3, 3), (3, 4), (4, 0), (4, 3), (4, 4)\}. \quad \blacksquare$$

The fact is that a relation induced by a partition of a set satisfies all three properties studied in Section 10.2: reflexivity, symmetry, and transitivity.

Theorem 10.3.1

Let A be a set with a partition and let R be the relation induced by the partition. Then R is reflexive, symmetric, and transitive.

Proof:

Suppose A is a set with a partition. In order to simplify notation, we assume that the partition consists of only a finite number of sets. The proof for an infinite partition is identical except for notation. Denote the partition subsets by

$$A_1, A_2, \ldots, A_n.$$

Then $A_i \cap A_j = \emptyset$ whenever $i \neq j$, and $A_1 \cup A_2 \cup \cdots \cup A_n = A$. The relation R induced by the partition is defined as follows: For all $x, y \in A$,

> $x \, R \, y \quad \Leftrightarrow \quad$ there is a set A_i of the partition
> such that $x \in A_i$ and $y \in A_i$.

R is reflexive: [*For R to be reflexive means that each element of A is related by R to itself. But by definition of R, for an element x to be related to itself means that x is in the same subset of the partition as itself. Well, if x is in some subset of the partition, then it is certainly in the same subset as itself. But x is in some subset of the partition because the union of the subsets of the partition is all of A. This reasoning is formalized as follows.*]

Suppose $x \in A$. Since $A_1, A_2, \ldots, A_n$ is a partition of A, it follows that $x \in A_i$ for some i. But then the statement

there is a set A_i of the partition such that $x \in A_i$ and $x \in A_i$

is true.* Thus, by definition of R, $x \, R \, x$.

R is symmetric: [*For R to be symmetric means that any time one element is related to a second, then the second is related to the first. Now for one element x to be related to a second element y means that x and y are in the same subset of the partition. But if this is the case, then y is in the same subset of the partition as x, so y is related to x by definition of R. This reasoning is formalized as follows.*]

Suppose x and y are elements of A such that $x \, R \, y$. Then

there is a subset A_i of the partition such that $x \in A_i$ and $y \in A_i$

by definition of R. It follows that the statement

there is a subset A_i of the partition such that $y \in A_i$ and $x \in A_i$

is also true.† Hence, by definition of R, $y \, R \, x$.

R is transitive: [*For R to be transitive means that any time one element of A is related by R to a second and that second is related to a third, then the first element*

---

*Since the statement forms p and $p \wedge p$ are logically equivalent, if p is true then $p \wedge p$ is also true.
†This follows from the fact that the statement forms $p \wedge q$ and $q \wedge p$ are logically equivalent.

is related to the third. But for one element to be related to another means that there is a subset of the partition that contains both. So suppose x, y, and z are elements such that x is in the same subset as y and y is in the same subset as z. Must x be in the same subset as z? Yes, because the subsets of the partition are mutually disjoint. Since the subset that contains x and y has an element in common with the subset that contains y and z (namely y), the two subsets are equal. But this means that x, y, and z are all in the same subset, and so in particular, x and z are in the same subset. Hence x is related by R to z. This reasoning is formalized as follows.]

Suppose x, y, and z are in A and x R y and y R z. By definition of R, there are subsets A_i and A_j of the partition such that

$$x \text{ and } y \text{ are in } A_i \quad \text{and} \quad y \text{ and } z \text{ are in } A_j.$$

Suppose $A_i \neq A_j$. [*We will deduce a contradiction.*] Then $A_i \cap A_j = \emptyset$ since $\{A_1, A_2, A_3, \ldots, A_n\}$ is a partition of A. But y is in A_i and y is in A_j also. Hence $A_i \cap A_j \neq \emptyset$. [*This contradicts the fact that $A_i \cap A_j = \emptyset$.*] Thus $A_i = A_j$. It follows that x, y, and z are all in A_i, and so in particular,

$$x \text{ and } z \text{ are in } A_i.$$

Thus, by definition of R, x R z.

Definition of an Equivalence Relation

A binary relation that satisfies the three properties of reflexivity, symmetry, and transitivity is called an *equivalence relation*.

> **• Definition**
>
> Let A be a set and R a binary relation on A. R is an **equivalence relation** if, and only if, R is reflexive, symmetric, and transitive.

Thus, according to Theorem 10.3.1, the relation induced by a partition is an equivalence relation. Another example is congruence modulo 3. In Example 10.2.5 it was shown that this relation is reflexive, symmetric, and transitive. Hence it, also, is an equivalence relation.

The following notation is used frequently when referring to congruence relations. It was introduced by Carl Friedrich Gauss in the first chapter of his book *Disquisitiones Arithmeticae*. This work, which was published when Gauss was only 24, laid the foundation for modern number theory.

*Carl Friedrich Gauss
(1777–1855)*

> **• Notation**
>
> Let m and n be integers and let d be a positive integer. The notation
>
> $$m \equiv n \;(\textbf{mod } d)$$
>
> is read "m is congruent to n modulo d" and means that
>
> $$d \mid (m - n).$$
>
> Symbolically,
>
> $$m \equiv n \;(\text{mod } d) \quad \Leftrightarrow \quad d \mid (m - n)$$

Exercise 15(b) at the end of this section asks you to show that $m \equiv n \pmod{d}$ if, and only if, $m \bmod d = n \bmod d$, where m, n, and d are integers and d is positive.

Example 10.3.2 Evaluating Congruences

Determine which of the following congruences are true and which are false.

a. $12 \equiv 7 \pmod{5}$ b. $6 \equiv -8 \pmod{4}$ c. $3 \equiv 3 \pmod{7}$

Solution

a. True. $12 - 7 = 5 = 5 \cdot 1$. Hence $5 \mid (12 - 7)$, and so $12 \equiv 7 \pmod{5}$.

b. False. $6 - (-8) = 14$, and $4 \nmid 14$ because $14 \neq 4 \cdot k$ for any integer k. Consequently, $6 \not\equiv -8 \pmod{4}$.

c. True. $3 - 3 = 0 = 7 \cdot 0$. Hence $7 \mid (3 - 3)$, and so $3 \equiv 3 \pmod{7}$. ∎

Example 10.3.3 Equivalence of Digital Logic Circuits Is an Equivalence Relation

Let S be the set of all digital logic circuits with a fixed number n of inputs. Define a binary relation $\mathscr{E}$ on S as follows: For all circuits C_1 and C_2 in S,

$$C_1 \, \mathscr{E} \, C_2 \quad \Leftrightarrow \quad C_1 \text{ has the same input/output table as } C_2.$$

If $C_1 \, \mathscr{E} \, C_2$, then circuit C_1 is said to be *equivalent* to circuit C_2. Prove that $\mathscr{E}$ is an equivalence relation on S.

Solution

$\mathscr{E}$ is reflexive: Suppose C is a digital logic circuit in S. [*We must show that $C \, \mathscr{E} \, C$.*] Certainly C has the same input/output table as itself. Thus, by definition of $\mathscr{E}$, $C \, \mathscr{E} \, C$ [*as was to be shown*].

$\mathscr{E}$ is symmetric: Suppose C_1 and C_2 are digital logic circuits in S such that $C_1 \, \mathscr{E} \, C_2$. [*We must show that $C_2 \, \mathscr{E} \, C_1$.*] By definition of $\mathscr{E}$, since $C_1 \, \mathscr{E} \, C_2$, then C_1 has the same input/output table as C_2. It follows that C_2 has the same input/output table as C_1. Hence, by definition of $\mathscr{E}$, $C_2 \, \mathscr{E} \, C_1$ [*as was to be shown*].

$\mathscr{E}$ is transitive: Suppose C_1, C_2, and C_3 are digital logic circuits in S such that $C_1 \, \mathscr{E} \, C_2$ and $C_2 \, \mathscr{E} \, C_3$. [*We must show that $C_1 \, \mathscr{E} \, C_3$.*] By definition of $\mathscr{E}$, since $C_1 \, \mathscr{E} \, C_2$ and $C_2 \, \mathscr{E} \, C_3$, then

$$C_1 \text{ has the same input/output table as } C_2$$

and

$$C_2 \text{ has the same input/output table as } C_3.$$

It follows that

$$C_1 \text{ has the same input/output table as } C_3.$$

Hence, by definition of $\mathscr{E}$, $C_1 \, \mathscr{E} \, C_3$ [*as was to be shown*].

Since $\mathscr{E}$ is reflexive, symmetric, and transitive, $\mathscr{E}$ is an equivalence relation on S. ∎

Certain implementations of computer languages do not place a limit on the allowable length of an identifier. This permits a programmer to be as precise as necessary in naming variables without having to worry about exceeding length limitations. However, compilers for such languages often ignore all but some specified number of initial characters: As far as the compiler is concerned, two identifiers are the same if they have the same initial characters, even though they may look different to a human reader of the program. For example, to a compiler than ignores all but the first eight characters of an identifier, the following identifiers would be the same:

NumberOfScrews   NumberOfBolts.

Obviously, in using such a language, the programmer has to be sure to avoid giving two distinct identifiers the same first eight characters. When a compiler lumps identifiers together in this way, it sets up an equivalence relation on the set of all possible identifiers in the language. Such a relation is described in the next example.

Example 10.3.4 A Binary Relation on a Set of Identifiers

Let L be the set of all allowable identifiers in a certain computer language, and define a relation R on L as follows: For all strings s and t in L,

$$s \; R \; t \quad \Leftrightarrow \quad \text{the first eight characters of } s \text{ equal the first eight characters of } t.$$

Prove that R is an equivalence relation on L.

Solution

R is reflexive: Let $s \in L$. [*We must show that $s \; R \; s$.*] Clearly s has the same first eight characters as itself. Thus, by definition of R, $s \; R \; s$ [*as was to be shown*].

R is symmetric: Let s and t be in L and suppose that $s \; R \; t$. [*We must show that $t \; R \; s$.*] By definition of R, since $s \; R \; t$, the first eight characters of s equal the first eight characters of t. But then the first eight characters of t equal the first eight characters of s. And so, by definition of R, $t \; R \; s$ [*as was to be shown*].

R is transitive: Let s, t, and u be in L and suppose that $s \; R \; t$ and $t \; R \; u$. [*We must show that $s \; R \; u$.*] By definition of R, since $s \; R \; t$ and $t \; R \; u$, the first eight characters of s equal the first eight characters of t, and the first eight characters of t equal the first eight characters of u. Hence the first eight characters of s equal the first eight characters of u. Thus, by definition of R, $s \; R \; u$ [*as was to be shown*].

Since R is reflexive, symmetric, and transitive, R is an equivalence relation on L. ■

Equivalence Classes of an Equivalence Relation

Suppose there is an equivalence relation on a certain set. If a is any particular element of the set, then one can ask, "What is the subset of all elements that are related to a?" This subset is called the *equivalence class* of a.

• Definition

Suppose A is a set and R is an equivalence relation on A. For each element a in A, the **equivalence class of a**, denoted **[a]** and called the **class of a** for short, is the set of all elements x in A such that x is related to a by R.

Written symbolically, this definition becomes:

$$[a] = \{x \in A \mid x \, R \, a\}$$

When several equivalence relations on a set are under discussion, the notation $[a]_R$ is often used to denote the equivalence class of a under R.

The procedural version of this definition is

$$\text{for all } x \in A, \quad x \in [a] \iff x \, R \, a.$$

Example 10.3.5 Equivalence Classes of a Relation Defined on a Finite Set

Let $A = \{0, 1, 2, 3, 4\}$ and define a binary relation R on A as follows:

$$R = \{(0, 0), (0, 4), (1, 1), (1, 3), (2, 2), (4, 0), (3, 3), (3, 1), (4, 4)\}.$$

The directed graph for R is as shown below. As can be seen by inspection, R is an equivalence relation on A. Find the distinct equivalence classes of R.

Solution   First find the equivalence class of every element of A.

$$[0] = \{x \in A \mid x \, R \, 0\} = \{0, 4\}$$
$$[1] = \{x \in A \mid x \, R \, 1\} = \{1, 3\}$$
$$[2] = \{x \in A \mid x \, R \, 2\} = \{2\}$$
$$[3] = \{x \in A \mid x \, R \, 3\} = \{1, 3\}$$
$$[4] = \{x \in A \mid x \, R \, 4\} = \{0, 4\}$$

Note that $[0] = [4]$ and $[1] = [3]$. Thus the *distinct* equivalence classes of the relation are

$$\{0, 4\}, \{1, 3\}, \text{ and } \{2\}. \qquad \blacksquare$$

When a problem asks you to find the *distinct* equivalence classes of an equivalence relation, you will generally solve the problem in two steps. In the first step you either explicitly construct (as in Example 10.3.5) or imagine constructing (as in infinite cases) the equivalence class for every element of the domain A of the relation. Usually several of the classes will contain exactly the same elements, so in the second step you must take a careful look at the classes to determine which are the same. You then indicate the distinct equivalence classes by describing them without duplication.

Example 10.3.6 Equivalence Classes of Identifiers

In Example 10.3.4 it was shown that the relation R of having the same first eight characters is an equivalence relation on the set L of allowable identifiers in a computer language. Describe the distinct equivalence classes of R.

Solution By definition of R, two strings in L are related by R if, and only if, they have the same first eight characters. Given any string s in L,

$$[s] = \{t \in L \mid t \ R \ s\}$$
$$= \{t \in L \mid \text{the first eight characters of } t \text{ equal the first eight characters of } s\}.$$

Thus the distinct equivalence classes of R are sets of strings such that (1) each class consists entirely of strings all of which have the same first eight characters, and (2) any two distinct classes contain strings that differ somewhere in their first eight characters. ∎

Example 10.3.7 Equivalence Classes of the Identity Relation

Let A be any set and define a relation R on A as follows: For all x and y in A,

$$x \ R \ y \quad \Leftrightarrow \quad x = y.$$

Then R is an equivalence relation. [*To prove this, just generalize the argument used in Example 10.2.3.*] Describe the distinct equivalence classes of R.

Solution Given any a in A, the class of a is

$$[a] = \{x \in A \mid x \ R \ a\}.$$

But by definition of R, $a \ R \ x$ if, and only if, $a = x$. So

$$[a] = \{x \in A \mid x = a\}$$
$$= \{a\} \qquad \text{since the only element of } A \text{ that equals } a \text{ is } a.$$

Hence, given any a in A,

$$[a] = \{a\},$$

and if $x \neq a$, then $\{x\} \neq \{a\}$. Consequently, all the classes of all the elements of A are distinct, and the distinct equivalence classes of R are all the single-element subsets of A. ∎

In each of Examples 10.3.5, 10.3.6, and 10.3.7, the set of distinct equivalence classes of the relation consists of mutually disjoint subsets whose union is the entire domain A of the relation. This means that the set of equivalence classes of the relation forms a partition of the domain A. In fact, it is always the case that the equivalence classes of an equivalence relation partition the domain of the relation into a union of mutually disjoint subsets. We establish the truth of this statement in stages, first proving two lemmas and then proving the main theorem.

The first lemma says that if two elements of A are related by an equivalence relation R, then their equivalence classes are the same.

Lemma 10.3.2

Suppose A is a set, R is an equivalence relation on A, and a and b are elements of A. If $a \ R \ b$, then $[a] = [b]$.

This lemma says that if a certain condition is satisfied, then $[a] = [b]$. Now $[a]$ and $[b]$ are *sets*, and two sets are equal if, and only if, each is a subset of the other. Hence the proof of the lemma consists of two parts: first, a proof that $[a] \subseteq [b]$ and second, a proof that $[b] \subseteq [a]$. To show each subset relation, it is necessary to show that every element in the left-hand set is an element of the right-hand set.

Proof of Lemma 10.3.2:

Let A be a set, let R be an equivalence relation on A, and suppose

$$a \text{ and } b \text{ are elements of } A \text{ such that } a \; R \; b.$$

[*We must show that* $[a] = [b]$.]

Proof that $[a] \subseteq [b]$: Let $x \in [a]$. [*We must shown that* $x \in [b]$.] Since

$$x \in [a]$$

then

$$x \; R \; a$$

by definition of class. But

$$a \; R \; b$$

by hypothesis. Thus, by transitivity of R,

$$x \; R \; b.$$

Hence

$$x \in [b]$$

by definition of class. [*This is what was to be shown.*]

Proof that $[b] \subseteq [a]$: Let $x \in [b]$. [*We must show that* $x \in [a]$.] Since

$$x \in [b]$$

then

$$x \; R \; b$$

by definition of class. Now

$$a \; R \; b$$

by hypothesis. Thus, since R is symmetric,

$$b \; R \; a$$

also. Then, since R is transitive and $x \; R \; b$ and $b \; R \; a$,

$$x \; R \; a.$$

Hence,

$$x \in [a]$$

by definition of class. [*This is what was to be shown.*]

Since $[a] \subseteq [b]$ and $[b] \subseteq [a]$, it follows that $[a] = [b]$ by definition of set equality.

The second lemma says that any two equivalence classes of an equivalence relation are either mutually disjoint or identical.

Lemma 10.3.3

If A is a set, R is an equivalence relation on A, and a and b are elements of A, then

$$\text{either} \quad [a] \cap [b] = \emptyset \quad \text{or} \quad [a] = [b].$$

The statement of Lemma 10.3.3 has the form

$$\text{if } p \text{ then } q \text{ or } r,$$

where p is the statement "A is a set, R is an equivalence relation on A, and a and b are elements of A," q is the statement "$[a] \cap [b] = \emptyset$," and r is the statement "$[a] = [b]$." To prove the lemma, we will prove the logically equivalent statement[*]

$$\text{if } p \text{ and not } q \text{ then } r.$$

That is, we will prove the following:

If A is a set, R is an equivalence relation on A, a and b are elements of A, and $[a] \cap [b] \neq \emptyset$, then $[a] = [b]$.

Proof of Lemma 10.3.3:

Suppose A is a set, R is an equivalence relation on A, a and b are elements of A, and

$$[a] \cap [b] \neq \emptyset.$$

[*We must show that* $[a] = [b]$.] Since $[a] \cap [b] \neq \emptyset$, there exists an element x in A such that $x \in [a] \cap [b]$. By definition of intersection,

$$x \in [a] \quad \text{and} \quad x \in [b]$$

and so

$$x \, R \, a \quad \text{and} \quad x \, R \, b$$

by definition of class. Since R is symmetric [*being an equivalence relation*] and $x \, R \, a$, then $a \, R \, x$. But R is also transitive [*since it is an equivalence relation*], and so, since $a \, R \, x$ and $x \, R \, b$,

$$a \, R \, b.$$

Now a and b satisfy the hypothesis of Lemma 10.3.2. Hence, by that lemma,

$$[a] = [b].$$

[*This is what was to be shown.*]

Theorem 10.3.4

If A is a set and R is an equivalence relation on A, then the distinct equivalence classes of R form a partition of A; that is, the union of the equivalence classes is all of A, and the intersection of any two distinct classes is empty.

[*]See exercise 14 in Section 1.2.

The proof of Theorem 10.3.4 is divided into two parts: first, a proof that A is the union of the equivalence classes of R and second, a proof that the intersection of any two distinct equivalence classes is empty. The proof of the first part follows from the fact that the relation is reflexive. The proof of the second part follows from Lemma 10.3.3.

Proof of Theorem 10.3.4:

Suppose A is a set and R is an equivalence relation on A. For notational simplicity, we assume that R has only a finite number of distinct equivalence classes, which we denote

$$A_1, A_2, \ldots, A_n,$$

where n is a positive integer. (When the number of classes is infinite, the proof is identical except for notation.)

Proof that $A = A_1 \cup A_2 \cup \cdots \cup A_n$: [*We must show that $A \subseteq A_1 \cup A_2 \cup \cdots \cup A_n$ and that $A_1 \cup A_2 \cup \cdots \cup A_n \subseteq A$.*]

To show that $A \subseteq A_1 \cup A_2 \cup \cdots \cup A_n$, suppose x is any element of A. [*We must show that $x \in A_1 \cup A_2 \cup \cdots \cup A_n$.*] By reflexivity of R, $x\, R\, x$. But this implies that $x \in [x]$ by definition of class. Since x is in *some* equivalence class, it must be in one of the distinct equivalence classes $A_1, A_2, \ldots,$ or A_n. Thus $x \in A_i$ for some index i, and hence $x \in A_1 \cup A_2 \cup \cdots \cup A_n$ by definition of union [*as was to be shown*].

To show that $A_1 \cup A_2 \cup \cdots \cup A_n \subseteq A$, suppose $x \in A_1 \cup A_2 \cup \cdots \cup A_n$. [*We must show that $x \in A$.*] Then $x \in A_i$ for some $i = 1, 2, \ldots,$ or n, by definition of union. But each A_i is an equivalence class of R. And equivalence classes are subsets of A. Hence $A_i \subseteq A$ and so $x \in A$ [*as was to be shown*].

Since $A \subseteq A_1 \cup A_2 \cup \cdots \cup A_n$ and $A_1 \cup A_2 \cup \cdots \cup A_n \subseteq A$, then by definition of set equality, $A = A_1 \cup A_2 \cup \cdots \cup A_n$.

Proof that the distinct classes of R are mutually disjoint: Suppose that A_i and A_j are any two distinct equivalence classes of R. [*We must show that A_i and A_j are disjoint.*] Since A_i and A_j are distinct, then $A_i \neq A_j$. And since A_i and A_j are equivalence classes of R, there must exist elements a and b in A such that $A_i = [a]$ and $A_j = [b]$. By Lemma 10.3.3,

$$\text{either} \quad [a] \cap [b] = \emptyset \quad \text{or} \quad [a] = [b].$$

But $[a] \neq [b]$ because $A_i \neq A_j$. Hence $[a] \cap [b] = \emptyset$. Thus $A_i \cap A_j = \emptyset$, and so A_i and A_j are disjoint [*as was to be shown*].

Example 10.3.8 Equivalence Classes of Congruence Modulo 3

Let R be the relation of congruence modulo 3 on the set $\mathbf{Z}$ of all integers. That is, for all integers m and n,

$$m\, R\, n \quad \Leftrightarrow \quad 3 \mid (m - n) \quad \Leftrightarrow \quad m \equiv n \pmod 3.$$

Describe the distinct equivalence classes of R.

Solution   For each integer a,

$$[a] = \{x \in \mathbf{Z} \mid x \, R \, a\}$$
$$= \{x \in \mathbf{Z} \mid 3 \mid (x - a)\}$$
$$= \{x \in \mathbf{Z} \mid x - a = 3k, \text{ for some integer } k\}.$$

Therefore,

$$[a] = \{x \in \mathbf{Z} \mid x = 3k + a, \text{ for some integer } k\}.$$

In particular,

$$[0] = \{x \in \mathbf{Z} \mid x = 3k + 0, \text{ for some integer } k\}$$
$$= \{x \in \mathbf{Z} \mid x = 3k, \text{ for some integer } k\}$$
$$= \{\ldots -9, -6, -3, 0, 3, 6, 9, \ldots\},$$
$$[1] = \{x \in \mathbf{Z} \mid x = 3k + 1, \text{ for some integer } k\}$$
$$= \{\ldots -8, -5, -2, 1, 4, 7, 10, \ldots\},$$
$$[2] = \{x \in \mathbf{Z} \mid x = 3k + 2, \text{ for some integer } k\}$$
$$= \{\ldots -7, -4, -1, 2, 5, 8, 11, \ldots\}.$$

Now since $3 \, R \, 0$, then by Lemma 10.3.2,

$$[3] = [0].$$

More generally, by the same reasoning,

$$[0] = [3] = [-3] = [6] = [-6] = \ldots, \text{ and so on.}$$

Similarly,

$$[1] = [4] = [-2] = [7] = [-5] = \ldots, \text{ and so on.}$$

And

$$[2] = [5] = [-1] = [8] = [-4] = \ldots, \text{ and so on.}$$

Notice that every integer is in class $[0]$, $[1]$, or $[2]$. Hence the distinct equivalence classes are

$$\{x \in \mathbf{Z} \mid x = 3k, \text{ for some integer } k\},$$
$$\{x \in \mathbf{Z} \mid x = 3k + 1, \text{ for some integer } k\}, \quad \text{and}$$
$$\{x \in \mathbf{Z} \mid x = 3k + 2, \text{ for some integer } k\}.$$

In words, the three classes of congruence modulo 3 are (1) the set of all integers that are divisible by 3, (2) the set of all integers that leave a remainder of 1 when divided by 3, and (3) the set of all integers that leave a remainder of 2 when divided by 3. ∎

Example 10.3.8 illustrates a very important property of equivalence classes, namely that an equivalence class may have many different names. In Example 10.3.8, for instance, the class of 0, $[0]$, may also be *called* the class of 3, $[3]$, or the class of -6, $[-6]$. But what the class *is* is the set

$$\{x \in \mathbf{Z} \mid x = 3k, \text{ for some integers } k\}.$$

(The quote at the beginning of this section refers in a humorous way to the philosophically interesting distinction between what things are *called* and what they *are*.)

> **• Definition**
>
> Suppose R is an equivalence relation on a set A and S is an equivalence class of R. A **representative** of the class S is any element a such that $[a] = S$.

In the exercises at the end of this section, you are asked to show that if x is any element of an equivalence class S, then $S = [x]$. Hence *any* element of an equivalence class is a representative of that class.

Example 10.3.9 Equivalence Classes of Digital Logic Circuits

In Example 10.3.3 it was shown that the relation of equivalence among circuits is an equivalence relation. Let S be the set of all digital logic circuits with exactly two inputs and one output. The binary relation $\mathscr{E}$ is defined on S as follows: For all C_1 and C_2 in S,

$$C_1 \mathscr{E} C_2 \quad \Leftrightarrow \quad C_1 \text{ has the same input/output table as } C_2.$$

Describe the equivalence classes of this relation. How many distinct equivalence classes are there? Write representative circuits for two of the distinct classes.

Solution Given a circuit C, the equivalence class of C is the set of all circuits with two input signals and one output signal that have the same input/output table as C. Now each input/output table has exactly four rows, corresponding to the four possible combinations of inputs: 11, 10, 01, and 00. A typical input/output table is the following:

Input		Output
P	**Q**	**R**
1	1	0
1	0	0
0	1	0
0	0	1

There are exactly as many such tables as there are binary strings of length 4. The reason is that distinct input/output tables can be formed by changing the pattern of the four 0's and 1's in the output column, and there are as many ways to do that as there are strings of four 0's and 1's. But the number of binary strings of length 4 is $2^4 = 16$. Hence there are 16 distinct input/output tables.

This implies that there are exactly 16 equivalence classes of circuits, one for each distinct input/output table. However, there are infinitely many circuits that give rise to each table. For instance, two representative circuits for the above input/output table are shown below.

Example 10.3.10 Rational Numbers Are Really Equivalence Classes

For a moment, forget what you know about fractional arithmetic and look at the numbers

$$\frac{1}{3} \quad \text{and} \quad \frac{2}{6}$$

as *symbols*. Considered as symbolic expressions, these *appear* quite different. In fact, if they were written as ordered pairs

$$(1, 3) \quad \text{and} \quad (2, 6)$$

they would *be* different. The fact that we regard them as "the same" is a specific instance of our general agreement to regard any two numbers

$$\frac{a}{b} \quad \text{and} \quad \frac{c}{d}$$

as equal provided the *cross products* are equal: $ad = bc$. This can be formalized as follows, using the language of equivalence relations.

Let A be the set of all ordered pairs of integers for which the second element of the pair is nonzero. Symbolically,

$$A = \mathbf{Z} \times (\mathbf{Z} - \{0\}).$$

Define a binary relation R on A as follows: For all $(a, b), (c, d) \in A$,

$$\boxed{(a, b) \ R \ (c, d) \quad \Leftrightarrow \quad ad = bc.}$$

The fact is that R is an equivalence relation.

a. Prove that R is transitive. (Proofs that R is reflexive and symmetric are left to the exercises.)

b. Describe the distinct equivalence classes of R.

Solution

a. [*We must show that for all* $(a, b), (c, d), (e, f) \in A$, *if* $(a, b) \ R \ (c, d)$ *and* $(c, d) \ R \ (e, f)$, *then* $(a, b) \ R \ (e, f)$.] Suppose $(a, b), (c, d)$, and (e, f) are particular but arbitrarily chosen elements of A such that $(a, b) \ R \ (c, d)$ and $(c, d) \ R \ (e, f)$. [*We must show that* $(a, b) \ R \ (e, f)$.] By definition of R,

$$(1) \ ad = bc \quad \text{and} \quad (2) \ cf = de.$$

Since the second elements of all ordered pairs in A are nonzero, $b \neq 0, d \neq 0$, and $f \neq 0$. Multiply (1) and (2) together to obtain

$$adcf = bcde,$$

which implies that

$$af(cd) = be(cd).$$

In case $c \neq 0$, by the cancellation law for multiplication (T7 in Appendix A), $af = be$. In case $c = 0$, it follows by the zero product property that $a = 0$ and $e = 0$ because $b \neq 0, d \neq 0, ad = bc$, and $cf = de$. So $af = be$ in this case also. Hence, in either case,

$$af = be.$$

It follows, by definition of R, that $(a, b) \ R \ (e, f)$ [*as was to be shown*].

b. There is one equivalence class for each distinct rational number. Each equivalence class consists of all ordered pairs (a, b) that, if written as fractions a/b, would equal each other. The reason for this is that the condition for two rational numbers to be equal is the same as the condition for two ordered pairs to be related. For instance, the class of $(1, 2)$ is

$$[(1, 2)] = \{(1, 2), (-1, -2), (2, 4), (-2, -4), (3, 6), (-3, -6), \ldots\}$$

since $\dfrac{1}{2} = \dfrac{-1}{-2} = \dfrac{2}{4} = \dfrac{-2}{-4} = \dfrac{3}{6} = \dfrac{-3}{-6}$ and so forth. ∎

It is possible to expand the result of Example 10.3.10 to define operations of addition and multiplication on the equivalence classes of R that satisfy all the same properties as the addition and multiplication of rational numbers. (See exercise 39.) It follows that the rational numbers can be defined as equivalence classes of ordered pairs of integers. Similarly (see exercise 40), it can be shown that all integers, negative and zero included, can be defined as equivalence classes of ordered pairs of positive integers. But in the late nineteenth century, F. L. G. Frege and Giuseppe Peano showed that the positive integers can be defined entirely in terms of sets. And just a little earlier, Richard Dedekind (1848–1916) showed that all real numbers can be defined as sets of rational numbers. All together, these results show that the real numbers can be defined using logic and set theory alone.

Exercise Set 10.3

1. Suppose that $S = \{a, b, c, d, e\}$ and R is a binary relation on S such that $a\,R\,b$, $b\,R\,c$, and $d\,R\,e$. List all of the following that must be true if R is (a) reflexive (but not symmetric or transitive), (b) symmetric (but not reflexive or transitive), (c) transitive (but not reflexive or symmetric), and (d) an equivalence relation.

$c\,R\,b \quad c\,R\,c \quad a\,R\,c \quad b\,R\,a \quad a\,R\,d \quad e\,R\,a \quad e\,R\,d \quad c\,R\,a$

2. Each of the following partitions of $\{0, 1, 2, 3, 4\}$ induces a relation R on $\{0, 1, 2, 3, 4\}$. In each case, find the ordered pairs in R.
a. $\{0, 2\}, \{1\}, \{3, 4\}$ b. $\{0\}, \{1, 3, 4\}, \{2\}$
c. $\{0\}, \{1, 2, 3, 4\}$

In 2–12, the relation R is an equivalence relation on the set A. Find the distinct equivalence classes of R.

3. $A = \{0, 1, 2, 3, 4\}$
$R = \{(0, 0), (0, 4), (1, 1), (1, 3), (2, 2), (3, 1), (3, 3),$
$(4, 0), (4, 4)\}$

4. $A = \{a, b, c, d\}$
$R = \{(a, a), (b, b), (b, d), (c, c), (d, b), (d, d)\}$

5. $A = \{1, 2, 3, 4, \ldots, 20\}$. R is defined on A as follows:

For all $x, y \in A$, $\quad x\,R\,y \quad \Leftrightarrow \quad 4\,|\,(x - y)$.

6. $A = \{-4, -3, -2, -1, 0, 1, 2, 3, 4, 5\}$. R is defined on A as follows:

For all $x, y \in A$, $\quad x\,R\,y \quad \Leftrightarrow \quad 3\,|\,(x - y)$.

7. $A = \{(1, 3), (2, 4), (-4, -8), (3, 9), (1, 5), (3, 6)\}$. R is defined on A as follows: For all $(a, b), (c, d) \in A$,

$$(a, b)\,R\,(c, d) \quad \Leftrightarrow \quad ad = bc.$$

8. $X = \{a, b, c\}$ and $A = \mathscr{P}(X)$. R is defined on A as follows:

For all sets U and V in $\mathscr{P}(X)$, $\quad U\,R\,V \quad \Leftrightarrow \quad N(U) = N(V)$.

(That is, the number of elements in U equals the number of elements in V.)

9. $X = \{-1, 0, 1\}$ and $A = \mathscr{P}(X)$. R is defined on $\mathscr{P}(X)$ as follows: For all sets S and T in $\mathscr{P}(X)$,

$S\,R\,T \quad \Leftrightarrow \quad$ the sum of the elements in S equals the sum of the elements in T.

10. A is the set of all strings of length 4 in a's and b's. R is defined on A as follows: For all strings s and t in A,

$s\,R\,t \quad \Leftrightarrow \quad$ the first two characters of s equal the first two characters of t.

11. A is the set of all strings of length 2 in 0's, 1's, and 2's. R is defined on A as follows: For all strings s and t in A,

$s\,R\,t \quad \Leftrightarrow \quad$ the sum of the characters in s equals the sum of the characters in t.

12. $A = \{-5, -4, -3, -2, -1, 0, 1, 2, 3, 4, 5\}$. R is defined on A as follows:

For all $m, n \in \mathbf{Z}$, $\quad m\,R\,n \quad \Leftrightarrow \quad 3\,|\,(m^2 - n^2)$.

13. Determine which of the following congruence relations are true and which are false.
    a. $17 \equiv 2 \pmod 5$ b. $4 \equiv -5 \pmod 7$
    c. $-2 \equiv -8 \pmod 3$ d. $-6 \equiv 22 \pmod 2$

14. a. Let R be the relation of congruence modulo 3. Which of the following equivalence classes are equal?
    $$[7], [-4], [-6], [17], [4], [27], [19]$$
    b. Let R be the relation of congruence modulo 7. Which of the following equivalence classes are equal?
    $$[35], [3], [-7], [12], [0], [-2], [17]$$

15. a. Prove that for all integers m and n, $m \equiv n \pmod 3$ if, and only if, $m \bmod 3 = n \bmod 3$.
    b. Prove that for all integers m and n and any positive integer d, $m \equiv n \pmod d$ if, and only if, $m \bmod d = n \bmod d$.

16. a. Give an example of two sets that are distinct but not disjoint.
    b. Find sets A_1 and A_2 and elements x, y and z such that x and y are in A_1 and y and z are in A_2 but x and z are not both in either of the sets A_1 or A_2.

In 17–28, (1) prove that the relation is an equivalence relation, and (2) describe the distinct equivalence classes of each relation.

17. A is the set of all students at your college.
    a. R is the relation defined on A as follows: For all x and y in A,
    $$x \, R \, y \iff x \text{ has the same major (or double major) as } y.$$
    (Assume "undeclared" is a major.)
    b. S is the relation defined on A as follows: For all $x, y \in A$,
    $$x \, S \, y \iff x \text{ is the same age as } y.$$

H 18. E is the relation defined on $\mathbf{Z}$ as follows:
    $$\text{For all } m, n \in \mathbf{Z}, \quad m \, E \, n \iff 2 \mid (m - n).$$

19. F is the relation defined on $\mathbf{Z}$ as follows:
    $$\text{For all } m, n \in \mathbf{Z}, \quad m \, F \, n \iff 4 \mid (m - n).$$

20. Let A be the set of all statement forms in three variables $p, q,$ and r. $\mathscr{R}$ is the relation defined on A as follows: For all P and Q in A,
    $$P \, \mathscr{R} \, Q \iff P \text{ and } Q \text{ have the same truth table.}$$

21. Let P be a set of parts shipped to a company from various suppliers. S is the relation defined on P as follows: For all $x, y \in P$,
    $$x \, S \, y \iff x \text{ has the same part number and is shipped from the same supplier as } y.$$

22. A is the "absolute value" relation defined on $\mathbf{R}$ as follows:
    $$\text{For all } x, y \in \mathbf{R}, \quad x \, A \, y \iff |x| = |y|.$$

23. I is the relation defined on $\mathbf{R}$ as follows:
    $$\text{For all } x, y \in \mathbf{R}, \quad x \, I \, y \iff x - y \text{ is an integer.}$$

24. D is the relation defined on $\mathbf{Z}$ as follows:
    $$\text{For all } m, n \in \mathbf{Z}, \quad m \, D \, n \iff 3 \mid (m^2 - n^2).$$

25. Define P on the set $\mathbf{R} \times \mathbf{R}$ of ordered pairs of real numbers as follows: For all $(w, x), (y, z) \in \mathbf{R} \times \mathbf{R}$,
    $$(w, x) \, P \, (y, z) \iff w = y.$$

26. Let A be the set of identifiers in a computer program. It is common for identifiers to be used for only a short part of the execution time of a program and not to be used again to execute other parts of the program. In such cases, arranging for identifiers to share memory locations makes efficient use of a computer's memory capacity. Define R on A as follows: For all identifiers x and y,
    $$x \, R \, y \iff \text{the values of } x \text{ and } y \text{ are stored in the same memory location during execution of the program.}$$

H 27. Let A be the set of all straight lines in the Cartesian plane. Define a relation $\|$ on A as follows:
    $$\text{For all } l_1 \text{ and } l_2 \text{ in } A, \quad l_1 \| l_2 \iff l_1 \text{ is parallel to } l_2.$$

28. Let P be the set of all points in the Cartesian plane except the origin. R is the relation defined on P as follows: For all p_1 and p_2 in P,
    $$p_1 \, R \, p_2 \iff p_1 \text{ and } p_2 \text{ lie on the same half-line emanating from the origin.}$$

29. Let A be the set of points in the rectangle with x and y coordinates between 0 and 1. That is,
    $$A = \{(x, y) \in \mathbf{R} \times \mathbf{R} \mid 0 \le x \le 1 \text{ and } 0 \le y \le 1\}.$$
    Define a relation R on A as follows: For all (x_1, y_1) and (x_2, y_2) in A,
    $$(x_1, y_1) \, R \, (x_2, y_2) \iff$$
    $$(x_1, y_1) = (x_2, y_2); \quad \text{or}$$
    $$x_1 = 0 \text{ and } x_2 = 1 \text{ and } y_1 = y_2; \quad \text{or}$$
    $$x_1 = 1 \text{ and } x_2 = 0 \text{ and } y_1 = y_2; \quad \text{or}$$
    $$y_1 = 0 \text{ and } y_2 = 1 \text{ and } x_1 = x_2; \quad \text{or}$$
    $$y_1 = 1 \text{ and } y_2 = 0 \text{ and } x_1 = x_2.$$

    In other words, all points along the top edge of the rectangle are related to the points along the bottom edge directly beneath them, and all points directly opposite each other along the left and right edges are related to each other. The points in the interior of the rectangle are not related to anything other than themselves. Then R is an equivalence relation on A. Imagine gluing together all the points that are in the same equivalence class. Describe the resulting figure.

30. The documentation for the computer language Java recommends that when an "equals method" is defined for an object, it be an equivalence relation. That is, if R is defined as follows:

$$x \; R \; y \quad \Leftrightarrow \quad x.equals(y) \quad \text{for all objects in the class,}$$

then R should be an equivalence relation. Suppose that in trying to optimize some of the mathematics of a graphics application, a programmer creates an object called a point, consisting of two coordinates in the plane. The programmer defines an equals method as follows: If p and q are any points, then

$$p.equals(q) \quad \Leftrightarrow \quad \text{the distance from } p \text{ to } q \text{ is} \\ \text{less than or equal to } c$$

where c is a small positive number that depends on the resolution of the computer display. Is the programmer's equals method an equivalence relation? Justify your answer.

Let R be an equivalence relation on a set A. Prove each of the statements in 31–36 directly from the definitions of equivalence relation and equivalence class without using the results of Lemma 10.3.2, Lemma 10.3.3, or Theorem 10.3.4.

31. For all a in A, $a \in [a]$.

32. For all a and b in A, if $b \in [a]$ then $a \; R \; b$.

33. For all a, b and c in A, if $b \; R \; c$ and $c \in [a]$ then $b \in [a]$.

34. For all a and b in A, if $[a] = [b]$ then $a \; R \; b$.

35. For all a, b, and x in A, if $a \; R \; b$ and $x \in [a]$, then $x \in [b]$.

H 36. For all a and b in A, if $a \in [b]$ then $[a] = [b]$.

37. Find an additional representative circuit for the input/output table of Example 10.3.9.

38. Let R be the binary relation defined in Example 10.3.10.
    a. Prove that R is reflexive.
    b. Prove that R is symmetric.
    c. List four distinct elements in $[(1, 3)]$.
    d. List four distinct elements in $[(2, 5)]$.

★ 39. In Example 10.3.10, define operations of addition $(+)$ and multiplication $(\cdot)$ as follows: For all (a, b), $(c, d) \in A$,

$$[(a, b)] + [(c, d)] = [(ad + bc, bd)]$$
$$[(a, b)] \cdot [(c, d)] = [(ac, bd)].$$

    a. Prove that this addition is well defined. That is, show that if $[(a, b)] = [(a', b')]$ and $[(c, d)] = [(c', d')]$, then $[(ad + bc, bd)] = [(a'd' + b'c', b'd')]$.

    b. Prove that this multiplication is well defined. That is, show that if $[(a, b)] = [(a', b')]$ and $[(c, d)] = [(c', d')]$, then $[(ac, bd)] = [(a'c', b'd')]$.

    c. Show that $[(0, 1)]$ is an identity element for addition. That is, show that for any $(a, b) \in A$,
    $$[(a, b)] + [(0, 1)] = [(a, b)].$$

    d. Find an identity element for multiplication.

    e. For any $(a, b) \in A$, show that $[(-a, b)]$ is an inverse for $[(a, b)]$ for addition. That is, show that $[(-a, b)] + [(a, b)] = [(0, 1)]$.

    f. Given any $(a, b) \in A$ with $a \neq 0$, find an inverse for $[(a, b)]$ for multiplication.

40. Let $A = \mathbf{Z}^+ \times \mathbf{Z}^+$. Define a binary relation R on A as follows: For all (a, b) and (c, d) in A,

$$(a, b) \; R \; (c, d) \quad \Leftrightarrow \quad a + d = c + b.$$

    a. Prove that R is reflexive.
    b. Prove that R is symmetric.
    H c. Prove that R is transitive.
    d. List five elements in $[(1, 1)]$.
    e. List five elements in $[(3, 1)]$.
    f. List five elements in $[(1, 2)]$.
    g. Describe the distinct equivalence classes of R.

41. The following argument claims to prove that the requirement that an equivalence relation be reflexive is redundant. In other words, it claims to show that if a relation is symmetric and transitive, then it is reflexive. Find the mistake in the argument.

    "**Proof:** Let R be a binary relation on a set A and suppose R is symmetric and transitive. For any two elements x and y in A, if $x \; R \; y$ then $y \; R \; x$ since R is symmetric. But then it follows by transitivity that $x \; R \; x$. Hence R is reflexive."

42. Let R be a binary relation on a set A and suppose R is symmetric and transitive. Prove the following: If for every x in A there is a y in A such that $x \; R \; y$, then R is an equivalence relation.

43. Refer to the quote at the beginning of this section to answer the following questions.
    a. What is the name of the Knight's song called?
    b. What is the name of the Knight's song?
    c. What is the Knight's song called?
    d. What *is* the Knight's song?
    e. What is your (full, legal) name?
    f. What are you called?
    g. What *are* you? (Do not answer this on paper; just think about it.)

10.4 *Modular Arithmetic with Applications to Cryptography*

The "real" mathematics of the "real" mathematicians, the mathematics of Fermat and Euler and Gauss and Abel and Riemann, is almost wholly "useless." . . . It is not possible to justify the life of any genuine professional mathematician on the ground of the "utility" of his work. — G. H. Hardy, *A Mathematician's Apology*, 1941

Cryptography is the study of methods for sending secret messages. It involves **encryption,** in which a message, called **plaintext,** is converted into a form, called **ciphertext,** that may be sent over channels possibly open to view by outside parties. The receiver of the ciphertext uses **decryption** to convert the ciphertext back into plaintext.

In the past the primary use of cryptography was for government and military intelligence, and this use continues to be important. In fact, the National Security Agency, whose main business is cryptography, is the largest employer of mathematicians in the United States. With the rise of electronic communication systems, however, especially the Internet, an extremely important current use of cryptography is to make it possible to send private information, such as credit card numbers, banking data, medical records, and so forth, over electronic channels.

Many systems for sending secret messages require both the sender and the receiver to know both the encryption and the decryption procedures. For instance, an encryption system once used by Julius Caesar, and now called the **Caesar cipher,** encrypts messages by changing each letter of the alphabet to the one three places farther along, with X wrapping around to A, Y to B, and Z to C. In other words, say each letter of the alphabet is coded by its position relative to the others—so that $A = 01$, $B = 02, \ldots, Z = 26$. If the numerical version of the plaintext for a letter is denoted M and the numeric version of the ciphertext is denoted C, then

$$C = (M + 3) \bmod 26.$$

The receiver of such a message can easily decrypt it by using the formula

$$M = (C - 3) \bmod 26.$$

For reference, here are the letters of the alphabet, together with their numeric equivalents:

A	B	C	D	E	F	G	H	I	J	K	L	M
01	02	03	04	05	06	07	08	09	10	11	12	13
N	O	P	Q	R	S	T	U	V	W	X	Y	Z
14	15	16	17	18	19	20	21	22	23	24	25	26

Example 10.4.1 Encrypting and Decrypting with the Caesar Cipher

a. Use the Caesar cipher to encrypt the message HOW ARE YOU.

b. Use the Caesar cipher to decrypt the message L DP ILQH.

Solution

a. First translate the letters of HOW ARE YOU into their numeric equivalents:

$$08 \quad 15 \quad 23 \qquad 01 \quad 18 \quad 05 \qquad 25 \quad 15 \quad 21.$$

Next encrypt the message by adding 3 to each number. The result is

$$11 \quad 18 \quad 26 \qquad 04 \quad 21 \quad 08 \qquad 02 \quad 18 \quad 24.$$

Finally, substitute the letters that correspond to these numbers. The encrypted message becomes

$$\text{KRZ} \quad \text{DUH} \quad \text{BRX.}$$

b. First translate the letters of L DP ILQH into their numeric equivalents:

$$12 \qquad 04 \quad 16 \qquad 09 \quad 12 \quad 17 \quad 08.$$

Next decrypt the message by subtracting 3 from each number:

$$09 \qquad 01 \quad 13 \qquad 06 \quad 09 \quad 14 \quad 05.$$

Then translate back into letters to obtain the original message: I AM FINE. ■

One problem with the Caesar cipher is that given a sufficient amount of ciphertext a person with knowledge of letter frequencies in the language can easily figure out the cipher. Partly for this reason, even Caesar himself did not make extensive use of it. Another problem with a system like the Caesar cipher is that knowledge of how to encrypt a message automatically gives knowledge of how to decrypt it. When a potential recipient of messages passes the encryption information to a potential sender of messages, the channel over which the information is passed may itself be insecure. Thus the information may leak out, enabling an outside party to decrypt messages intended to be kept secret.

With public-key cryptography, a potential recipient of encrypted messages openly distributes a public key containing the encryption information. However, knowledge of the public key provides virtually no clue about how messages are decrypted. Only the recipient has that knowledge. Regardless of how many people learn the encryption information, only the recipient should be able to decrypt messages that are sent.

The first public-key cryptography system was developed in 1976–1977 by three young mathematician/computer scientists working at M.I.T.: Ronald Rivest, Adi Shamir, and Leonard Adleman. In their honor it is called the RSA cipher. In order for you to learn how it works, you need to know some additional properties of congruence modulo n.

From left to right: Ronald Rivest (born 1948), Adi Shamir (born 1952), and Leonard Adleman (born 1945)

Properties of Congruence Modulo n

The first theorem in this section brings together a variety of equivalent ways of expressing the same basic arithmetic fact. Sometimes one way is most convenient; sometimes another way is best. You need to be comfortable moving from one to another, depending on the nature of the problem you are trying to solve.

Theorem 10.4.1 Modular Equivalences

Let a, b, and n be any integers and suppose $n > 1$. The following statements are all equivalent:

1. $n \mid (a - b)$

2. $a \equiv b \pmod{n}$

3. $a = b + kn$ for some integer k

4. a and b have the same (nonnegative) remainder when divided by n

5. $a \bmod n = b \bmod n$

Proof:

We will show that $(1) \Rightarrow (2) \Rightarrow (3) \Rightarrow (4) \Rightarrow (5) \Rightarrow (1)$. It will follow by the transitivity of if-then that all five statements are equivalent.

So let a, b, and n be any integers with $n > 1$.

Proof that (1) $\Rightarrow$ (2): Suppose that $n \mid (a - b)$. By definition of congruence modulo n, we can immediately conclude that $a \equiv b \pmod{n}$.

Proof that (2) $\Rightarrow$ (3): Suppose that $a \equiv b \pmod{n}$. By definition of congruence modulo n, $n \mid (a - b)$. Thus, by definition of divisibility, $a - b = kn$, for some integer k. Adding b to both sides gives that $a = b + kn$.

Proof that (3) $\Rightarrow$ (4): Suppose that $a = b + kn$, for some integer k. Use the quotient-remainder theorem to divide a by n to obtain

$$a = qn + r \quad \text{where } q \text{ and } r \text{ are integers and } 0 \leq r < n.$$

Substituting $b + kn$ for a in this equation gives that

$$b + kn = qn + r$$

and subtracting kn from both sides and factoring out n yields

$$b = (q - k)n + r.$$

But since $0 \leq r < n$, the uniqueness property of the quotient-remainder theorem guarantees that r is also the remainder obtained when b is divided by n. Thus a and b have the same remainder when divided by n.

Proof that (4) $\Rightarrow$ (5): Suppose that a and b have the same remainder when divided by n. It follows immediately from the definition of the *mod* function that $a \bmod n = b \bmod n$.

continued on page 614

Proof that $(5) \Rightarrow (1)$: Suppose that $a \bmod n = b \bmod n$. By definition of the *mod* function, a and b have the same remainder when divided by n. Thus, by the quotient-remainder theorem, we can write

$$a = q_1 n + r \quad \text{and} \quad b = q_2 n + r \quad \text{where } q_1, q_2, \text{ and } r \text{ are integers and } 0 \leq r < n.$$

It follows that

$$a - b = (q_1 n + r) - (q_2 n + r) = (q_1 - q_2)n.$$

Therefore, since $q_1 - q_2$ is an integer, $n \mid (a - b)$.

Another consequence of the quotient-remainder theorem is this: When an integer a is divided by an integer n, a unique quotient q and remainder r are obtained with the property that $a = nq + r$ and $0 \leq r < n$. Because there are exactly n integers that satisfy the inequality $0 \leq r < n$ (the numbers from 0 through $n - 1$), there are exactly n possible remainders that can occur. These are called the *least nonnegative residues modulo n* or simply the *residues modulo n*.

• Definition

Given integers a and n with $n > 1$, **the residue of a modulo n** is $a \bmod n$, the non-negative remainder obtained when a is divided by n. The numbers $0, 1, 2, \ldots, n - 1$ are called a **complete set of residues modulo n**. To **reduce a number modulo n** means to set it equal to its residue modulo n. If a modulus $n > 1$ is fixed throughout a discussion and an integer a is given, the words "modulo n" are often dropped and we simply speak of **the residue of a**.

The following theorem generalizes several examples from Section 10.3.

Theorem 10.4.2 Congruence Modulo n Is an Equivalence Relation

If n is any integer with $n > 1$, congruence modulo n is an equivalence relation on the set of all integers. The distinct equivalence classes of the relation are the sets $[0], [1], [2], \ldots, [n - 1]$, where for each $a = 0, 1, 2, \ldots, n - 1$,

$$[a] = \{m \in Z \mid m \equiv a \,(\mathrm{mod}\, n)\},$$

or, equivalently,

$$[a] = \{m \in Z \mid m = a + kn \text{ for some integer } k\}.$$

Proof:

Suppose n is any integer with $n > 1$. We must show that congruence modulo n is reflexive, symmetric, and transitive.

Proof of reflexivity: Suppose a is any integer. To show that $a \equiv a \,(\mathrm{mod}\, n)$, we must show that $n \mid (a - a)$. But $a - a = 0$, and $n \mid 0$ because $0 = n \cdot 0$. Therefore $a \equiv a \,(\mathrm{mod}\, n)$.

***Proof of symmetry*:** Suppose a and b are any integers such that $a \equiv b \pmod{n}$. We must show that $b \equiv a \pmod{n}$. But since $a \equiv b \pmod{n}$, then $n \mid (a - b)$. Thus, by definition of divisibility, $a - b = nk$, for some integer k. Multiply both sides of this equation by -1 to obtain

$$-(a - b) = -nk,$$

or, equivalently,

$$b - a = n(-k).$$

Thus, by definition of divisibility $n \mid (b - a)$, and so, by definition of congruence modulo n, $b \equiv a \pmod{n}$.

***Proof of transitivity*:** This is left as exercise 5 at the end of the section.

Proof that the distinct equivalence classes are $[0], [1], [2], \ldots, [n-1]$**:** This is left as exercise 6 at the end of the section.

Observe that there is a one-to-one correspondence between the distinct equivalence classes for congruence modulo n and the elements of a complete set of residues modulo n.

Modular Arithmetic

A fundamental fact about congruence modulo n is that if you first perform an addition, subtraction, or multiplication on integers and then reduce the result modulo n, you will obtain the same answer as if you had first reduced each of the numbers modulo n, performed the operation, and then reduced the result modulo n. For instance, instead of computing

$$(5 \cdot 8) = 40 \equiv 1 \pmod{3}$$

you will obtain the same answer if you compute

$$(5 \bmod 3)(8 \bmod 3) = 2 \cdot 2 = 4 \equiv 1 \pmod{3}.$$

The fact that this process works is a result of the following theorem.

Theorem 10.4.3 Modular Arithmetic

Let a, b, c, d, and n be integers with $n > 1$, and suppose $a \equiv c \pmod{n}$ and $b \equiv d \pmod{n}$. Then

1. $(a + b) \equiv (c + d) \pmod{n}$

2. $(a - b) \equiv (c - d) \pmod{n}$

3. $ab \equiv cd \pmod{n}$

4. $a^m \equiv c^m \pmod{n}$ for all integers m.

Proof:

Because we will make greatest use of part 3 of this theorem, we prove it here and leave the proofs of the remaining parts of the theorem to exercises 9–11 at the end of the section.

continued on page 616

Proof of Part 3: Suppose a, b, c, d, and n are integers with $n > 1$, and suppose $a \equiv b \pmod{n}$ and $c \equiv d \pmod{n}$. By Theorem 10.4.1, there exist integers s and t such that

$$a = c + sn \quad \text{and} \quad b = d + tn.$$

Then

$$
\begin{aligned}
ab &= (c + sn)(d + tn) &&\text{by substitution}\\
&= cd + ctn + snd + sntn\\
&= cd + n(ct + sd + stn) &&\text{by algebra.}
\end{aligned}
$$

Let $k = ct + sd + stn$. Then k is an integer and $ab = cd + nk$. Thus by Theorem 10.4.1, $ab \equiv cd \pmod{n}$.

Example 10.4.2 Getting Started with Modular Arithmetic

The most practical use of modular arithmetic is to reduce computations involving large integers to computations involving smaller ones. For instance, note that $55 \equiv 3 \pmod{4}$ because $55 - 3 = 52$, which is divisible by 4, and $26 \equiv 2 \pmod{4}$ because $26 - 2 = 24$, which is also divisible by 4. Verify the following statements.

a. $55 + 26 \equiv (3 + 2) \pmod{4}$ b. $55 - 26 \equiv (3 - 2) \pmod{4}$

c. $55 \cdot 26 \equiv (3 \cdot 2) \pmod{4}$ d. $55^2 \equiv 3^2 \pmod{4}$

Solution

a. Compute $55 + 26 = 81$ and $3 + 2 = 5$. By definition of congruence modulo n, to show that $81 \equiv 5 \pmod{4}$, you need to show that $4 \mid (81 - 5)$. But this is true because $81 - 5 = 76$, and $4 \mid 76$ since $76 = 4 \cdot 19$.

b. Compute $55 - 26 = 29$ and $3 - 2 = 1$. By definition of congruence modulo n, to show that $29 \equiv 1 \pmod{4}$, you need to show that $4 \mid (29 - 1)$. But this is true because $29 - 1 = 28$, and $4 \mid 28$ since $28 = 4 \cdot 7$.

c. Compute $55 \cdot 26 = 1430$ and $3 \cdot 2 = 6$. By definition of congruence modulo n, to show that $1430 \equiv 6 \pmod{4}$, you need to show that $4 \mid (1430 - 6)$. But this is true because $1430 - 6 = 1424$, and $4 \mid 1424$ since $1424 = 4 \cdot 356$.

d. Compute $55^2 = 3025$ and $3^2 = 9$. By definition of congruence modulo n, to show that $3025 \equiv 9 \pmod{4}$, you need to show that $4 \mid (3025 - 9)$. But this is true because $3025 - 9 = 3016$, and $4 \mid 3016$ since $3016 = 4 \cdot 754$. ■

In order to facilitate the computations performed in this section, it is convenient to express part 3 of Theorem 10.4.3 in a slightly differently form.

Corollary 10.4.4

Let a, b, and n be integers with $n > 1$. Then

$$ab \equiv [(a \bmod n)(b \bmod n)] \pmod{n},$$

or, equivalently,

$$ab \bmod n = [(a \bmod n)(b \bmod n)] \bmod n.$$

In particular, if m is a positive integer, then

$$a^m \equiv [(a \bmod n)^m] \pmod{n}.$$

Example 10.4.3 Computing a Product Modulo n

As in Example 10.4.2, note that $55 \equiv 3 \pmod 4$ and $26 \equiv 2 \pmod 4$. Because both 3 and 2 are less than 4, each of these numbers is a least nonnegative residue modulo 4. Therefore, $55 \bmod 4 = 3$ and $26 \bmod 4 = 2$. Use the notation of Corollary 10.4.4 to find the residue of $55 \cdot 26$ modulo 4.

Solution   Recall that to use a calculator to compute remainders, you can use the formula $n \bmod d = n - d \cdot \lfloor n/d \rfloor$. If you are using a hand calculator with an "integer part" feature and both n and d are positive, then $\lfloor n/d \rfloor$ is the integer part of the division of n by d. When you divide a positive integer n by a positive integer d with a more basic calculator, you can see $\lfloor n/d \rfloor$ on the calculator display by simply ignoring the digits that follow the decimal point.

By Corollary 10.4.4,

$$
\begin{aligned}
(55 \cdot 26) \bmod 4 &= [(55 \bmod 4)(26 \bmod 4)] \bmod 4 \\
&\equiv (3 \cdot 2) \bmod 4 \qquad \text{because } 55 \bmod 4 = 3 \\
&\qquad\qquad\qquad\qquad\quad \text{and } 26 \bmod 4 = 2 \\
&\equiv 6 \bmod 4 \\
&\equiv 2 \qquad\qquad\qquad \text{because } 4 \mid (6 - 2) \text{ and } 2 < 4. \quad \blacksquare
\end{aligned}
$$

When modular arithmetic is performed with very large numbers, as is the case for RSA crytography, computations are facilitated by using two properties of exponents. The first is

$$
x^{2a} = (x^a)^2 \quad \text{for all real numbers } x \text{ and } a \text{ with } x \geq 0. \tag{10.4.1}
$$

Thus, for instance, if x is any positive real number, then

$$
\begin{aligned}
x^4 \bmod n &= (x^2)^2 \bmod n \qquad \text{because } (x^2)^2 = x^4 \\
&= (x^2 \bmod n)^2 \bmod n \qquad \text{by Corollary 10.4.4.}
\end{aligned}
$$

Hence you can reduce x^4 modulo n by reducing x^2 modulo n and then reducing the square of the result modulo n. Because all the residues are less than n, this process limits the size of the computations to numbers that are less than n^2, which makes them easier to work with, both for humans (when the numbers are relatively small) and for computers (when the numbers are very large).

A second useful property of exponents is

$$
x^{a+b} = x^a x^b \quad \text{for all real numbers } x, a, \text{ and } b \text{ with } x \geq 0. \tag{10.4.2}
$$

For instance, because $7 = 4 + 2 + 1$,

$$
x^7 = x^4 x^2 x^1.
$$

Thus, by Corollary 10.4.4,

$$
x^7 \bmod n = [(x^4 \bmod n)(x^2 \bmod n)(x^1 \bmod n)] \bmod n.
$$

We first show an example that illustrates the application of formula (10.4.1) and then an example that uses both (10.4.1) and (10.4.2).

Example 10.4.4 Computing a^k *mod* n When k Is a Power of 2

Find 144^4 *mod* 713.

Solution Use property (10.4.1) to write $144^4 = (144^2)^2$. Then

$$
\begin{aligned}
144^4 \ mod \ 713 &= (144^2)^2 \ mod \ 713 \\
&= (144^2 \ mod \ 713)^2 \ mod \ 713 \\
&= (20736 \ mod \ 713)^2 \ mod \ 713 && \text{because } 144^2 = 20736 \\
&= 59^2 \ mod \ 713 && \text{because } 20736 \ mod \ 713 = 59 \\
&= 3481 \ mod \ 713 && \text{because } 59^2 = 3481 \\
&= 629 && \text{because } 3481 \ mod \ 713 = 629. \quad \blacksquare
\end{aligned}
$$

Example 10.4.5 Computing a^k *mod* n When k Is Not a Power of 2

Find 12^{43} *mod* 713.

Solution First write the exponent as a sum of powers of 2:

$$43 = 2^5 + 2^3 + 2 + 1 = 32 + 8 + 2 + 1.$$

Next compute 12^{2^k} for $k = 1, 2, 3, 4, 5$.

$$
\begin{aligned}
12 \ mod \ 713 &= 12 \\
12^2 \ mod \ 713 &= 144 \\
12^4 \ mod \ 713 &= 144^2 \ mod \ 713 &= 59 && \text{by Example 10.4.4} \\
12^8 \ mod \ 713 &= 59^2 \ mod \ 713 &= 629 && \text{by Example 10.4.4} \\
12^{16} \ mod \ 713 &= 629^2 \ mod \ 713 &= 639 && \text{by the method of Example 10.4.4} \\
12^{32} \ mod \ 713 &= 639^2 \ mod \ 713 &= 485 && \text{by the method of Example 10.4.4}
\end{aligned}
$$

By property (10.4.2),

$$12^{43} = 12^{32+8+2+1} = 12^{32} \cdot 12^8 \cdot 12^2 \cdot 12^1.$$

Thus, by Corollary 10.4.4,

$12^{43} \ mod \ 713$

$$= [(12^{32} \ mod \ 713) \cdot (12^8 \ mod \ 713) \cdot (12^2 \ mod \ 713) \cdot (12 \ mod \ 713)] \ mod \ 713.$$

By substitution,

$$
\begin{aligned}
12^{43} \ mod \ 713 &= (485 \cdot 629 \cdot 144 \cdot 12) \ mod \ 713 \\
&= 527152320 \ mod \ 713 \\
&= 48. \quad \blacksquare
\end{aligned}
$$

It is important to understand how to do the computations in Example 10.4.5 by hand using only a simple electronic calculator, but if you are computing a lot of residues, especially ones involving large numbers, you may want to write a short computer or calculator program to do the computations for you.

Extending the Euclidean Algorithm

An extended version of the Euclidean algorithm can be used to find a concrete expression for the greatest common divisor of integers a and b.

> • **Definition**
>
> An integer d is said to be a **linear combination of integers** a and b if, and only if, there exist integers s and t such that $as + bt = d$.

Theorem 10.4.5 Writing a Greatest Common Divisor as a Linear Combination

For all integers a and b, not both zero, if $d = \gcd(a, b)$, then there exist integers s and t such that $as + bt = d$.

Proof:

Given integers a and b, not both zero, and given $d = \gcd(a, b)$, let

$S = \{x \mid x$ is a positive integer and $x = as + bt$ for some integers s and $t\}$.

Note that S is a nonempty set because (1) if $a > 0$ then $1 \cdot a + 0 \cdot b \in S$, (2) if $a < 0$ then $(-1) \cdot a + 0 \cdot b \in S$, and (3) if $a = 0$, then by assumption $b \neq 0$, and hence $0 \cdot a + 1 \cdot b \in S$ or $0 \cdot a + (-1) \cdot b \in S$. Thus, because S consists entirely of positive integers, by the well-ordering principle there is a least element c in S. By definition of S,

$$c = as + bt \quad \text{for some integers } s \text{ and } t. \qquad 10.4.3$$

We will show that (1) $c \geq d$ and (2) $c \leq d$ and will therefore be able to conclude that $c = d = \gcd(a, b)$.

(1) *Proof that $c \geq d$*:
[*In this part of the proof, we show that d is a divisor of c and thus that $d \mid c$.*] Because $d = \gcd(a, b)$, by definition of greatest common divisor, $d \mid a$ and $d \mid b$. Hence $a = dx$ and $b = dy$ for some integers x and y. Then

$$
\begin{aligned}
c &= as + bt &&\text{by (10.4.3)} \\
&= (dx)s + (dy)t &&\text{by substitution} \\
&= d(xs + yt) &&\text{by factoring out the } d.
\end{aligned}
$$

But $xs + yt$ is an integer because it is a sum of products of integers. Thus, by definition of divisibility, $d \mid c$. Both c and d are positive, and hence, by Example 3.3.3, $c \geq d$.

(2) *Proof that $c \leq d$*:
[*In this part of the proof, we show that c is a divisor of both a and b and therefore that c is less than or equal to the greatest common divisor of a and b, which is d.*] Apply the quotient-remainder theorem to the division of a by c to obtain

$$a = cq + r \quad \text{for some integers } q \text{ and } r \text{ with } 0 \leq r < c. \qquad 10.4.4$$

Thus for some integers q and r with $0 \leq r < c$,

$$r = a - cq$$

Now $c = as + bt$. Therefore, for some integers q and r with $0 \leq r < c$,

$$
\begin{aligned}
r &= a - (as + bt)q &&\text{by substitution} \\
&= a(1 - sq) - btq.
\end{aligned}
$$

continued on page 620

Thus r is a linear combination of a and b, and hence $r \in S$. If $r > 0$, then r would be a smaller element of S than c, which would contradict the fact that c is the least element of S. Hence $r = 0$. By substitution into (10.4.4),

$$a = cq$$

and therefore $c \mid a$.

An almost identical argument establishes that $c \mid b$ and is left as exercise 30 at the end of the section.

Because $c \mid a$ and $c \mid b$, c is a common divisor of a and b. Hence it is less than the greatest common divisor of a and b. In other words, $c \leq d$.

From (1) and (2), we conclude that $c = d$. It follows that d, the greatest common divisor of a and b, is equal to $as + bt$.

The following example shows a practical method for expressing the greatest common divisor of two integers as a linear combination of the two.

Example 10.4.6 Expressing a Greatest Common Divisor as a Linear Combination

In Example 3.8.6 we showed how to use the Euclidean algorithm to find that the greatest common divisor of 330 and 156 is 6. Use the results of those calculations to express gcd(330, 156) as a linear combination of 330 and 156.

Solution   The first four steps of the solution restate and extend results from Example 3.8.6, which were obtained by successive applications of the quotient-remainder theorem. The fifth step shows how to find the coefficients of the linear combination by substituting back through the results of the previous steps.

Step 1: $330 = 156 \cdot 2 + 18$, which implies that $18 = 330 - 156 \cdot 2$.

Step 2: $156 = 18 \cdot 8 + 12$, which implies that $12 = 156 - 18 \cdot 8$.

Step 3: $18 = 12 \cdot 1 + 6$, which implies that $6 = 18 - 12 \cdot 1$.

Step 4: $12 = 6 \cdot 2 + 0$, which implies that gcd(330, 156) = 6.

Step 5: By substituting back through steps 3 to 1:

$$
\begin{aligned}
6 &= 18 - 12 \cdot 1 && \text{from step 3} \\
&= 18 - (156 - 8 \cdot 18) \cdot 1 && \text{by substitution from step 2} \\
&= 9 \cdot 18 + (-1) \cdot 156 && \text{by algebra} \\
&= 9 \cdot (330 - 156 \cdot 2) + (-1) \cdot 156 && \text{by substitution from step 1} \\
&= 9 \cdot 330 + (-19) \cdot 156 && \text{by algebra.}
\end{aligned}
$$

Thus gcd(330, 156) $= 9 \cdot 330 + (-19) \cdot 156$. (It is always a good idea to check the result of a calculation like this to be sure you did not make a mistake. In this case, you find that $9 \cdot 330 + (-19) \cdot 156$ does indeed equal 6.) ∎

The Euclidean algorithm given in Section 3.8 can be adapted so as to compute the coefficients of the linear combination of the gcd at the same time as it computes the gcd itself. This extended Euclidean algorithm is described in the exercises at the end of the section.

Finding an Inverse Modulo *n*

Suppose you want to solve the following congruence for x:

$$2x \equiv 3 \,(\text{mod } 5)$$

Note that $3 \cdot 2 = 6 \equiv 1 \,(\text{mod } 5)$. So you can think of 3 as a kind of inverse for 2 modulo 5 and multiply both sides of the congruence to be solved by 3 to obtain

$$6x = 3 \cdot 2x \equiv 3 \cdot 3 \,(\text{mod } 5) \equiv 9 \,(\text{mod } 5) \equiv 4 \,(\text{mod } 5),$$

But $6 \equiv 1 \,(\text{mod } 5)$, and so by Theorem 10.4.3(3), $6x \equiv 1x \,(\text{mod } 5) \equiv x \,(\text{mod } 5)$. Thus, by the symmetric and transitive properties of modular congruence,

$$x \equiv 4 \,(\text{mod } 5),$$

and hence a solution is $x = 4$. (You can check that $2 \cdot 4 = 8 \equiv 3 \,(\text{mod } 5)$.)

Unfortunately, it is not always possible to find an "inverse" modulo an integer n. For instance, observe that

$$2 \cdot 1 \equiv 2 \,(\text{mod } 4)$$
$$2 \cdot 2 \equiv 0 \,(\text{mod } 4)$$
$$2 \cdot 3 \equiv 2 \,(\text{mod } 4).$$

By Theorem 10.4.3, these calculations suffice for us to conclude that the number 2 does not have an inverse modulo 4.

Describing the circumstances in which inverses exist in modular arithmetic requires the concept of relative primeness.

> **• Definition**
>
> Integers a and b are **relatively prime** if, and only if, $\gcd(a, b) = 1$. Integers $a_1, a_2, a_3, \ldots, a_n$ are **pairwise relatively prime** if, and only if, $\gcd(a_i, a_j) = 1$ for all integers i and j with $1 \le i, j \le n$, and $i \ne j$.

Given the definition of relatively prime integers, the following corollary is an immediate consequence of Theorem 10.4.5.

> **Corollary 10.4.6**
>
> If a and b are relatively prime integers, then there exist integers s and t such that $as + bt = 1$.

Example 10.4.7 Expressing 1 as a Linear Combination of Relatively Prime Integers

Show that 660 and 43 are relatively prime, and find a linear combination of 660 and 43 that equals 1.

Solution

Step 1: Divide 660 by 43 to obtain $660 = 43 \cdot 15 + 15$, which implies that $15 = 660 - 43 \cdot 15$.

Step 2: Divide 43 by 15 to obtain $43 = 15 \cdot 2 + 13$, which implies that $13 = 43 - 15 \cdot 2$.

Step 3: Divide 15 by 13 to obtain $15 = 13 \cdot 1 + 2$, which implies that $2 = 15 - 13$.

Step 4: Divide 13 by 2 to obtain $13 = 2 \cdot 6 + 1$, which implies that $1 = 13 - 2 \cdot 6$.

Step 5: Divide 2 by 1 to obtain $2 = 1 \cdot 2 + 0$, which implies that $\gcd(660, 43) = 1$ and so 660 and 43 are relatively prime.

Step 6: To express 1 as a linear combination of 660 and 43, substitute back through steps 4 to 1:

$$
\begin{aligned}
1 &= 13 - 2 \cdot 6 && \text{from step 4}\\
&= 13 - (15 - 13) \cdot 6 && \text{by substitution from step 3}\\
&= 7 \cdot 13 - 6 \cdot 15 && \text{by algebra}\\
&= 7 \cdot (43 - 15 \cdot 2) - 6 \cdot 15 && \text{by substitution from step 2}\\
&= 7 \cdot 43 - 20 \cdot 15 && \text{by algebra}\\
&= 7 \cdot 43 - 20 \cdot (660 - 43 \cdot 15) && \text{by substitution from step 1}\\
&= 307 \cdot 43 - 20 \cdot 660 && \text{by algebra.}
\end{aligned}
$$

Thus $\gcd(660, 43) = 1 = 307 \cdot 43 - 20 \cdot 660$. (And a check by direct computation confirms that $307 \cdot 43 - 20 \cdot 660$ does indeed equal 1.) ∎

A consequence of Corollary 10.4.6 is that under certain circumstances, it is possible to find an inverse for an integer modulo n.

Corollary 10.4.7 Existence of Inverses Modulo n

For all integers a and n, if $\gcd(a, n) = 1$, then there exists an integer s such that $as \equiv 1 \pmod{n}$. The integer s is called the **inverse of a modulo n.**

Proof:

Suppose a and n are integers and $\gcd(a, n) = 1$. By Corollary 10.4.6, there exist integers s and t such that

$$as + nt = 1.$$

Subtracting nt from both sides gives that

$$as = 1 - nt = 1 + (-t)n.$$

Thus, by definition of congruence modulo n,

$$as \equiv 1 \pmod{n}.$$

Example 10.4.8 Finding an Inverse Modulo n

a. Find an inverse for 43 modulo 660. That is, find an integer s such that $43s \equiv 1 \pmod{660}$.

b. Find a positive inverse for 3 modulo 40. That is, find a positive integer s such that $3s \equiv 1 \pmod{40}$.

Solution

a. By Example 10.4.7,

$$307 \cdot 43 - 20 \cdot 660 = 1.$$

Adding $20 \cdot 660$ to both sides gives that

$$307 \cdot 43 = 1 + 20 \cdot 660.$$

Thus, by definition of congruence modulo 660,

$$307 \cdot 43 \equiv 1 \pmod{660},$$

so 307 is an inverse for 43 modulo 660.

b. Use the technique of Example 10.4.7 to find a linear combination of 3 and 40 that equals 1.

Step 1: Divide 40 by 3 to obtain $40 = 3 \cdot 13 + 1$. This implies that $1 = 40 - 3 \cdot 13$.

Step 2: Divide 3 by 1 to obtain $3 = 3 \cdot 1 + 0$. This implies that $\gcd(3, 40) = 1$.

Step 3: Use the result of step 1 to write

$$3 \cdot (-13) = 1 + (-1)40.$$

This result implies that -13 is an inverse for 3 modulo 40. In symbols, $3 \cdot (-13) \equiv 1 \pmod{40}$. To find a positive inverse, compute $40 - 13$. The result is 27, and

$$27 \equiv -13 \pmod{40}$$

because $27 - (-13) = -40 = (-1)40$. So, by Theorem 10.4.3(3),

$$3 \cdot 27 \equiv 3 \cdot (-13) \equiv 1 \pmod{40},$$

and thus by the transitive property of congruence modulo n, 27 is a positive integer that is an inverse for 3 modulo 40. ■

RSA Cryptography

At this point we have developed enough number theory to explain how to encrypt and decrypt messages using the RSA cipher. The effectiveness of the system is based on the fact that although modern computer algorithms make it quite easy to find large numbers p and q—say on the order of several hundred digits each—that are virtually certain to be prime, even the fastest computers are not currently able to factor their product, a number with approximately twice that many digits. In order to encrypt a message using the RSA cipher, a person needs to know the value of pq and of another number e, both of which are made publicly available. But only a person who knows the individual values of p and q can decrypt an encrypted message.

We first give an example to show *how* the cipher works and then discuss some of the theory to explain *why* it works. The example is unrealistic in the sense that because p and q are so small, it would be easy to figure out what they are just by knowing their product. But working with small numbers conveys the idea of the system, while keeping the computations in a range that can be performed with a hand calculator.

Suppose Alice decides to set up an RSA cipher. She chooses two prime numbers, say $p = 5$ and $q = 11$, and computes $pq = 55$. She then chooses a positive integer e that is relatively prime to $(p - 1)(q - 1)$. In this case, $(p - 1)(q - 1) = 4 \cdot 10 = 40$, so she may take $e = 3$ because 3 is relatively prime to 40. (In practice, taking e to be small could compromise the secrecy of the cipher, so she would take a larger number than 3. However, the mathematics of the cipher works as well for 3 as for a larger number, and the smaller number makes for easier calculations.)

The two numbers $pq = 55$ and $e = 3$ are the **public key,** which she may distribute widely. Because the RSA cipher works only on numbers, Alice also informs people how she will interpret the numbers in the messages they send her. Let us suppose that she encodes letters of the alphabet the same way as was done for the Caesar cipher:

$$A = 1, B = 2, C = 3, \ldots, Z = 26.$$

Let us also assume that the messages Alice receives consist of blocks, each of which, for simplicity, is taken to be a single, numerically encoded letter of the alphabet.

Someone who wants to send Alice a message breaks the message into blocks, each consisting of a single letter, and finds the numeric equivalent for each block. The plaintext, M, in a block is converted into ciphertext, C, according to the following formula:

$$C = M^e \bmod pq.$$

10.4.5

Note that because both pq and e are public keys, anyone who is given the keys and knows modular arithmetic can encrypt a message to send to Alice.

Example 10.4.9 Encrypting a Message Using RSA Cryptography

Bob wants to send Alice the message HI. What is the ciphertext for his message?

Solution   Bob will send his message in two blocks, one for the H and another for the I. Because H is the eighth letter in the alphabet, it is encoded as 08, or 8. The corresponding ciphertext is computed using formula (10.4.5) as follows:

$$\begin{aligned} C &= 8^3 \bmod 55 \\ &= 512 \bmod 55 \\ &= 17. \end{aligned}$$

Because I is the ninth letter in the alphabet, it is encoded as 09, or 9. The corresponding ciphertext is

$$\begin{aligned} C &= 9^3 \bmod 55 \\ &= 729 \bmod 55 \\ &= 14. \end{aligned}$$

Accordingly, Bob sends Alice the message: 17 14.   ■

To decrypt the message, Alice needs to compute the decryption key, a number d that is a positive inverse to e modulo $(p - 1)(q - 1)$. She obtains the plaintext M from the ciphertext C by the formula

$$M = C^d \bmod pq.$$

10.4.6

Note that because $M + kpq \equiv M \pmod{pq}$, M must be taken to be less than pq, as in the above example, in order for the decryption to be guaranteed to produce the original message. But because p and q are normally taken to be so large, this requirement does not cause problems. A long message can be broken into sections, if necessary, to meet the restriction.

Example 10.4.10 Decrypting a Message Using RSA Cryptography

Imagine that Alice has hired you to help her decrypt messages and has shared with you the values of p and q. Decrypt the following ciphertext for her: 17 14.

Solution   Because $p = 5$ and $q = 11$, $(p - 1)(q - 1) = 40$, and so you first need to find the decryption key, which is a positive inverse for 3 modulo 40. Knowing that you would be needing this number, we computed it in Example 10.4.4(b) and found it to be 27. Thus you need to compute $M = 17^{27} \bmod 55$. To do so, note that $27 = 16 + 8 + 2 + 1 = 2^4 + 2^3 + 2 + 1$. Thus you will find the residues obtained when 17 is raised to successively higher powers of 2, up to $2^4 = 16$.

$$
\begin{aligned}
17 \, mod \, 55 &= 17 \, mod \, 55 \\
17^2 \, mod \, 55 &= 17^2 \, mod \, 55 &= 14 \\
17^4 \, mod \, 55 &= (17^2)^2 \, mod \, 55 &= 14^2 \, mod \, 55 &= 31 \\
17^8 \, mod \, 55 &= (17^4)^2 \, mod \, 55 &= 31^2 \, mod \, 55 &= 26 \\
17^{16} \, mod \, 55 &= (17^8)^2 \, mod \, 55 &= 26^2 \, mod \, 55 &= 16
\end{aligned}
$$

Then you will use the fact that

$$17^{27} = 17^{16+8+2+1} = 17^{16} \cdot 17^8 \cdot 17^2 \cdot 17^1$$

to write

$$
\begin{aligned}
17^{27} \, mod \, 55 &= (17^{16} \cdot 17^8 \cdot 17^2 \cdot 17) \, mod \, 55 \\
&\equiv [(17^{16} \, mod \, 55)(17^8 \, mod \, 55)(17^2 \, mod \, 55)(17 \, mod \, 55)] \, (\text{mod } 55) \\
&\qquad\qquad\qquad\qquad\qquad\qquad\qquad\qquad\qquad\qquad \text{by Corollary 10.4.4} \\
&\equiv (16 \cdot 26 \cdot 14 \cdot 17) \, (\text{mod } 55) \\
&\equiv 99008 \, (\text{mod } 55) \\
&\equiv 8 \, (\text{mod } 55).
\end{aligned}
$$

Hence $17^{27} \, mod \, 55 = 8$, and thus the plaintext of the first part of Bob's message is 8, or 08. In the last step, you find the letter corresponding to 08, which is H. In exercises 14 and 15 at the end of this section, you are asked to show that when you decrypt 14, the result is 9, which corresponds to the letter I, so you can tell Alice that Bob's message is HI. ∎

Euclid's Lemma

Another consequence of Theorem 10.4.5 is known as *Euclid's lemma*. It is the crucial fact behind the unique factorization theorem for the integers and is also of great importance in many other parts of number theory.

Theorem 10.4.8 Euclid's Lemma

For all integers a, b, and c, if $\gcd(a, c) = 1$ and $a \mid bc$, then $a \mid b$.

Proof:

Suppose a, b and c are integers, $\gcd(a, c) = 1$, and $a \mid bc$. [*We must show that $a \mid b$.*] By Theorem 10.4.5, there exist integers s and t so that

$$as + ct = 1.$$

Multiply both sides of this equation by b to obtain

$$bas + bct = b. \qquad\qquad 10.4.7$$

Since $a \mid bc$, by definition of divisibility there exists an integer k such that

$$bc = ak. \qquad\qquad 10.4.8$$

Substituting (10.4.8) into (10.4.7), rewriting, and factoring out an a gives that

$$b = bas + (ak)t = a(bs + kt).$$

Let $r = bs + kt$. Then r is an integer (because b, s, k, and t are all integers), and $b = ar$. Thus $a \mid b$ by definition of divisibility.

The unique factorization theorem for the integers states that any integer greater than 1 has a unique representation as a product of prime numbers, except possibly for the order in which the numbers are written. The hint for exercise 11 of Section 4.4 outlined a proof of the existence part of the proof, and the uniqueness of the representation follows quickly from Euclid's lemma. In exercise 41 at the end of this section, we outline a proof for you to complete.

Another application of Euclid's lemma is a cancellation theorem for congruence modulo n. This theorem allows us—under certain circumstances—to divide out a common factor in a congruence relation.

Theorem 10.4.9 Cancellation Theorem for Modular Congruence

For all integers a, b, c, and n, if $\gcd(c, n) = 1$ and $ac \equiv bc \,(\text{mod}\, n)$, then $a \equiv b \,(\text{mod}\, n)$.

Proof:

Suppose a, b, c, and n are any integers, $\gcd(c, n) = 1$, and $ac \equiv bc \,(\text{mod}\, n)$. [*We must show that $a \equiv b \,(\text{mod}\, n)$.*] By Theorem 10.4.3(3), $ac - bc \equiv 0 \,(\text{mod}\, n)$, or, equivalently,

$$(a - b)c \equiv 0 \,(\text{mod}\, n).$$

Thus by definition of congruence modulo n,

$$n \mid (a - b)c.$$

Because $\gcd(c, n) = 1$, we may apply Euclid's lemma to obtain

$$n \mid (a - b),$$

and so, by definition of congruence modulo n,

$$a \equiv b \,(\text{mod}\, n).$$

An alternative proof for Theorem 10.4.9 uses Corollary 10.4.7. Because $\gcd(c, n) = 1$, the corollary guarantees an inverse for c modulo n. In the proof of Theorem 10.4.9, let d denote an inverse for c. Apply Theorem 10.4.3(3) repeatedly, first to multiply both sides of $ac \equiv bc \,(\text{mod}\, n)$ by d to obtain $(ac)d \equiv (bd)d \,(\text{mod}\, n)$, and then to use the fact that $cd \equiv 1 \,(\text{mod}\, n)$ to simplify the congruence and conclude that $a \equiv b \,(\text{mod}\, n)$.

Fermat's Little Theorem and the Chinese Remainder Theorem

Fermat's little theorem was given that name to distinguish it from Fermat's last theorem, which we discussed in Section 3.1. Together with the Chinese remainder theorem, it provides the theoretical underpinning for RSA cryptography.

Theorem 10.4.10 Fermat's Little Theorem

If p is any prime number and a is any integer such that $p \nmid a$, then $a^{p-1} \equiv 1 \,(\text{mod}\, p)$.

Proof:

Suppose p is any prime number and a is any integer such that $p \nmid a$. Note that $a \neq 0$ because otherwise p would divide a. Consider the set of integers

$$S = \{a, 2a, 3a, \ldots, (p - 1)a\}.$$

We claim that no two elements of S are congruent modulo p. For suppose $sa \equiv ra \pmod{p}$ for some integers s and r with $1 \leq r < s \leq p - 1$. Then, by definition of congruence modulo p,

$$p \mid (sa - ra), \quad \text{or, equivalently,} \quad p \mid (s - r)a.$$

Now $p \nmid a$ by hypothesis, and because p is prime, $\gcd(a, p) = 1$. Thus, by Euclid's lemma, $p \mid (s - r)$. But this is impossible because $0 < s - r < p$.

Consider the function F from S to the set $T = \{1, 2, 3, \ldots, (p - 1)\}$ that sends each element of S to its residue modulo p. Then F is one-to-one because no two elements of S are congruent modulo p. But Theorem 7.3.3 states that if a function from one finite set to another is one-to-one, then it is also onto. Hence F is onto, and so the $p - 1$ residues of the $p - 1$ elements of S are exactly the numbers $1, 2, 3, \ldots, (p - 1)$.

It follows by Theorem 10.4.3(3) that

$$a \cdot 2a \cdot 3a \cdots (p - 1)a \equiv [1 \cdot 2 \cdot 3 \cdots (p - 1)] \pmod{p},$$

or equivalently,

$$a^{p-1}(p - 1)! \equiv (p - 1)! \pmod{p}.$$

But because p is prime, p and $(p - 1)!$ are relatively prime. Thus, by the cancellation theorem for modular congruence (Theorem 10.4.9),

$$a^{p-1} \equiv 1 \pmod{p}.$$

The Chinese remainder theorem was given its name because of the work of ancient Chinese mathematicians, notably Sun Tsu, who posed and solved the following problem in the first century A.D.: Find a number with a remainder of 2 when divided by 3, a remainder of 3 when divided by 5, and a remainder of 2 when divided by 7. The calculations in his solution are essentially those that would come from the construction in the proof of the theorem. It is interesting to note that a Greek mathematician, Nicomachus, posed and solved the same problem at about the same time and in about the same way.

Theorem 10.4.11 The Chinese Remainder Theorem

If $n_1, n_2, \ldots, n_k$ are pairwise relatively prime positive integers and $a_1, a_2, \ldots, a_k$ are any integers, then the congruences

$$x \equiv a_1 \pmod{n_1}$$
$$x \equiv a_2 \pmod{n_2}$$
$$\vdots$$
$$x \equiv a_k \pmod{n_k}$$

have a simultaneous solution x that is unique modulo n, where $n = n_1 n_2 \cdots n_k$.

Proof:

We restrict the proof to the case $k = 3$ because the proof in this case embodies all the ideas of the general proof. Suppose $n_1, n_2,$ and n_3 are pairwise relatively prime positive integers and $a_1, a_2,$ and a_3 are any integers. Let

$$N_1 = n_2 n_3, \quad N_2 = n_1 n_3, \quad \text{and} \quad N_3 = n_1 n_2.$$

continued on page 628

Because $n_1, n_2,$ and n_3 are pairwise relatively prime, so is each pair N_i and n_i for $i = 1, 2, 3$. Thus, by Corollary 10.4.7 there exist integers $x_1, x_2,$ and x_3 such that x_i is an inverse for N_i for each $i = 1, 2, 3$. That is,

$$N_1 x_1 \equiv 1 \,(\mathrm{mod}\ n_1), \quad N_2 x_2 \equiv 1 \,(\mathrm{mod}\ n_2), \quad \text{and} \quad N_3 x_3 \equiv 1 \,(\mathrm{mod}\ n_3).$$

For each $i = 1, 2, 3$, multiply the ith congruence by a_i. The result is

$$a_1 N_1 x_1 \equiv a_1 \,(\mathrm{mod}\ n_1), \quad a_2 N_2 x_2 \equiv a_2 \,(\mathrm{mod}\ n_2), \quad \text{and} \quad a_3 N_3 x_3 \equiv a_3 \,(\mathrm{mod}\ n_3).$$

Let

$$x = a_1 N_1 x_1 + a_2 N_2 x_2 + a_3 N_3 x_3.$$

Note that

$$N_2 = n_1 n_3 \equiv 0 \,(\mathrm{mod}\ n_1) \quad \text{and} \quad N_3 = n_1 n_2 \equiv 0 \,(\mathrm{mod}\ n_1)$$

because $n_1 \mid (n_1 n_3 - 0)$ and $n_1 \mid (n_1 n_2 - 0)$. Thus, by Theorem 10.4.3(3),

$$a_2 N_2 x_2 \equiv 0 \,(\mathrm{mod}\ n_1) \quad \text{and} \quad a_2 N_2 x_2 \equiv 0 \,(\mathrm{mod}\ n_1).$$

It follows by Theorem 10.4.3(1) that

$$x = a_1 N_1 x_1 + a_2 N_2 x_2 + a_3 N_3 x_3 \equiv a_1 N_1 x_1 \,(\mathrm{mod}\ n_1).$$

But $N_1 x_1 \equiv 1 \,(\mathrm{mod}\ n_1)$, and so by Theorem 10.4.3(3),

$$a_1 N_1 x_1 \equiv a_1 \,(\mathrm{mod}\ n_1).$$

Therefore, by transitivity of modular congruence,

$$x \equiv a_1 \,(\mathrm{mod}\ n_1).$$

Similarly,

$$N_1 \equiv 0 \,(\mathrm{mod}\ n_2), \quad N_3 \equiv 0 \,(\mathrm{mod}\ n_2), \quad N_1 \equiv 0 \,(\mathrm{mod}\ n_3), \quad \text{and} \quad N_2 \equiv 0 \,(\mathrm{mod}\ n_3),$$

and so, by the same reasoning,

$$x = a_1 N_1 x_1 + a_2 N_2 x_2 + a_3 N_3 x_3 \equiv a_2 N_2 x_2 \equiv a_2 \,(\mathrm{mod}\ n_2)$$

and

$$x = a_1 N_1 x_1 + a_2 N_2 x_2 + a_3 N_3 x_3 \equiv a_3 N_3 x_3 \equiv a_3 \,(\mathrm{mod}\ n_3).$$

Thus x is a solution for the set of congruences. Proving the uniqueness of the solution requires proving that if x' is any other solution, then $x' = x \,(\mathrm{mod}\ n)$, where $n = n_1 n_2 n_3$. This proof is left as exercise 47 at the end of the section.

Exercise 44 at the end of the section outlines the steps of the solution to Sun-Tsu's problem.

Why Does the RSA Cipher Work?

The crucial part of the RSA cipher is the formula

$$M = C^d \bmod pq,$$

which claims to produce the original plaintext message. How can we be sure that it always does so? We know that $C = M^e \bmod pq$, and so, by substitution,

$$M = (M^e \bmod pq)^d \bmod pq.$$

And by Theorem 10.4.3(4),

$$(M^e \bmod pq)^d \equiv M^{ed} \pmod{pq}.$$

Thus it suffices to show that

$$M = M^{ed} \pmod{pq}.$$

Recall that d was chosen to be an inverse for e modulo $(p-1)(q-1)$, which exists because $\gcd(e, (p-1)(q-1)) = 1$. In other words,

$$ed \equiv 1 \pmod{(p-1)(q-1)},$$

or, equivalently,

$$ed = 1 + k(p-1)(q-1) \quad \text{for some integer } k.$$

Therefore,

$$M^{ed} = M^{1+k(p-1)(q-1)} = M(M^{p-1})^{k(q-1)} = M(M^{q-1})^{k(p-1)}$$

If $p \nmid M$, then by Fermat's little theorem, $M^{p-1} \equiv 1 \pmod{p}$, and so

$$M^{ed} = M(M^{p-1})^{k(q-1)} \equiv M(1)^{k(q-1)} \pmod{p} \equiv M \pmod{p}.$$

Similarly, if $q \nmid M$, then by Fermat's little theorem, $M^{q-1} \equiv 1 \pmod{q}$, and so

$$M^{ed} = M(M^{q-1})^{k(p-1)} \equiv M(1)^{k(p-1)} \equiv M \pmod{q}.$$

Thus, if M is relatively prime to pq,

$$M^{ed} = M \pmod{p} \quad \text{and} \quad M^{ed} \equiv M \pmod{q}.$$

It follows from the Chinese remainder theorem that

$$M^{ed} \equiv M \pmod{pq},$$

or, equivalently,

$$M = M^{ed} \pmod{pq},$$

as was to be shown.

This proof has shown that the RSA cipher gives the correct result provided $M < pq$ (so that the solution is unique) and M is relatively prime to pq. The restriction of relative primeness is not significant because it is easy to add a step to an encrypting algorithm to check that this condition is satisfied. In any case, because $M < pq$, it is only when $M = p$ or $M = q$ that the restriction is not satisfied, and this is extremely unlikely when p and q are very large.

Additional Remarks on Number Theory and Cryptography

The famous British mathematician G. H. Hardy (1877–1947) was fond of comparing the beauty of pure mathematics, especially number theory, to the beauty of art. Indeed, the theorems in this section have many beautiful and striking consequences beyond those we have had the space to describe, and the subject of number theory extends far beyond these theorems. Hardy also enjoyed describing pure mathematics as useless. Hence it is ironic that there are now whole books devoted to applications of number theory to computer science, RSA cryptography being just one such application. Furthermore, as the need for public-key cryptography has developed, techniques from other areas of mathematics, such as abstract algebra and algebraic geometry, have been used to develop additional cryptosystems.

Exercise Set 10.4

1. a. Use the Caesar cipher to encrypt the message WHERE SHALL WE MEET.
   b. Use the Caesar cipher to decrypt the message LQ WKH FDIHWHULD.

2. a. Use the Caesar cipher to encrypt the message AN APPLE A DAY.
   b. Use the Caesar cipher to decrypt the message NHHSV WKH GRFWRU DZDB.

3. Let $a = 25, b = 19$, and $n = 3$.
   a. Verify that $3 \mid (25 - 19)$.
   b. Explain why $25 \equiv 19 \pmod{3}$.
   c. What value of k has the property that $25 = 19 + 3k$?
   d. What is the (nonnegative) remainder obtained when 25 is divided by 3? When 19 is divided by 3?
   e. Explain why $25 \bmod 3 = 19 \bmod 3$.

4. Let $a = 68, b = 33$, and $n = 7$.
   a. Verify that $7 \mid (68 - 33)$.
   b. Explain why $68 \equiv 33 \pmod{7}$.
   c. What value of k has the property that $68 = 33 + 7k$?
   d. What is the (nonnegative) remainder obtained when 68 is divided by 7? When 33 is divided by 7?
   e. Explain why $68 \bmod 7 = 33 \bmod 7$.

5. Prove the transitivity of modular congruence. That is, prove that for all integers a, b, c, and n with $n > 1$, if $a \equiv b \pmod{n}$ and $b \equiv c \pmod{n}$ then $a \equiv c \pmod{n}$.

H 6. Prove that the distinct equivalence classes of the relation of congruence modulo n are the sets $[0], [1], [2], \ldots, [n - 1]$, where for each $a = 0, 1, 2, \ldots, n - 1$,

$$[a] = \{m \in Z \mid m \equiv a \pmod{n}\}.$$

7. Verify the following statements.
   a. $128 \equiv 2 \pmod{7}$ and $61 \equiv 5 \pmod{7}$
   b. $(128 + 61) \equiv (2 + 5) \pmod{7}$
   c. $(128 - 61) \equiv (2 - 5) \pmod{7}$
   d. $(128 \cdot 61) \equiv (2 \cdot 5) \pmod{7}$
   e. $128^2 \equiv 2^2 \pmod{7}$

8. Verify the following statements.
   a. $45 \equiv 3 \pmod{6}$ and $104 \equiv 2 \pmod{6}$
   b. $(45 + 104) \equiv (3 + 2) \pmod{6}$
   c. $(45 - 104) \equiv (3 - 2) \pmod{6}$
   d. $(45 \cdot 104) \equiv (3 \cdot 2) \pmod{6}$
   e. $45^2 \equiv 2^2 \pmod{6}$

In 9–11, prove each of the given statements, assuming that a, b, c, d, and n are integers with $n > 1$ and that $a \equiv c \pmod{n}$ and $b \equiv d \pmod{n}$.

9. $(a + b) \equiv (c + d) \pmod{n}$

10. $(a - b) \equiv (c - d) \pmod{n}$

11. $a^m \equiv c^m \pmod{n}$

12. a. Use mathematical induction and modular arithmetic to prove that for all integers $n \geq 0, 10^n \equiv 1 \pmod{9}$.
    b. Use part (a) to prove that a positive integer is divisible by 9 if, and only if, the sum of its digits is divisible by 9.

13. a. Use strong mathematical induction and modular arithmetic to prove that for all integers $n \geq 1, 10^n \equiv (-1)^n \pmod{11}$.
    b. Use part (a) to prove that a positive integer is divisible by 11 if, and only if, the alternating sum of its digits is divisible by 11. (For instance, the alternating sum of the digits of 82,379 is $8 - 2 + 3 - 7 + 9 = 11$ and $82,379 = 11 \cdot 7489$.)

14. Use the technique of Example 10.4.4 to find $14^2 \bmod 55$, $14^4 \bmod 55$, $14^8 \bmod 55$, and $14^{16} \bmod 55$.

15. Use the result of exercise 14 and the technique of Example 10.4.5 to find $14^{27} \bmod 55$.

In 16–18, use the techniques of Example 10.4.4 and Example 10.4.5 to find the given numbers.

16. $675^{307} \bmod 713$ 17. $89^{307} \bmod 713$

18. $48^{307} \bmod 713$

In 19–24, use the RSA cipher from Examples 10.4.9 and 10.4.10. In 19–21, translate the message into its numeric equivalent and encrypt it. In 22–24, decrypt the ciphertext and translate the result into letters of the alphabet to discover the message.

19. HELLO   20. WELCOME   21. EXCELLENT

22. 13 20 20 09   23. 08 05 15   24. 51 14 49 15

H 25. Use Theorem 4.2.3 to prove that if a and n are positive integers and $a^n - 1$ is prime, then $a = 2$ and n is prime.

In 26 and 27, use the extended Euclidean algorithm to find the greatest common divisor of the given numbers and express it as a linear combination of the two numbers.

26. 6664 and 765   27. 4158 and 1568

Exercises 28 and 29 refer to the following formal version of the extended Euclidean algorithm.

Algorithm 10.4.1 Extended Euclidean Algorithm

[*Given integers A and B with $A > B > 0$, this algorithm computes* $\gcd(A, B)$ *and finds integers s and t such that* $sA + tB = \gcd(A, B)$.]

Input: A, B [*integers with $A > B > 0$*]

Algorithm Body:

$q := A$, $b := B$, $s := 1$, $t := 0$, $u := 0$, $v := 1$,
[*pre-condition: $a = sA + tB$ and $b = uA + vB$*]

while ($b \neq 0$)

    [*loop invariant: $a = sA + tB$ and $b = uA + vB$, $\gcd(a, b) = \gcd(A, B)$*]

    $r := a \bmod b$, $q := a \operatorname{div} b$

    $a := b$, $b := r$

    if ($b \neq 0$) **then do**

        $newu := s - uq$, $newv := t - vq$

        $s := u$, $t := v$

        $u := newu$, $v := newv$

    end do

end while

$\gcd := a$
[*post-condition: $\gcd(A, B) = a$, $sA + tB$*]

Output: $\gcd$[*a positive integer*], s, t [*integers*]

In 28 and 29, for the given values of A and B, make a table showing the value of $sa + tb$ before the start of the **while** loop and after each iteration of the loop.

28. $A = 330, B = 156$ **29.** $A = 284, B = 168$

30. Finish the proof of Theorem 10.4.5 by proving that if a, b, d, and c are as in the proof, then $c \mid b$.

31. a. Find an inverse for 210 modulo 13.
    b. Find a positive inverse for 210 modulo 13.
    c. Find a partial congruence for the congruence $210x \equiv 8 \pmod{13}$.

32. a. Find an inverse for 41 modulo 660.
    b. Find the least positive solution for the following congruence: $41x \equiv 125 \pmod{660}$.

H **33.** Use Corollary 10.4.6 to prove that for all integers a, b, and c, if $\gcd(a, b) = 1$ and $a \mid c$ and $b \mid c$, then $ab \mid c$.

34. Give a counterexample to show that the converse of exercise 33 is false.

35. Corollary 10.4.7 guarantees the existence of an inverse modulo n for an integer a when a and n are relatively prime. Use Euclid's lemma to prove that the inverse is unique modulo n. In other words, show that any two integers whose product with a is congruent to 1 modulo n are congruent to each other modulo n.

In 36, 37, 39, and 40, use the RSA cipher with public key $n = 713 = 23 \cdot 31$ and $e = 43$. In 36 and 37, encode the messages into their numeric equivalents and encrypt them. In 39 and 40, decrypt the given ciphertext and find the original messages.

36. HELP   **37.** COME

38. Show that 307 is an inverse for 43 modulo 660.

39. 675 089 089 048

40. 028 018 675 129

H **41.** a. Use mathematical induction and Euclid's lemma to prove that for all positive integers s, if p and $q_1, q_2, \ldots, q_s$ are prime numbers and $p \mid q_1 q_2 \cdots q_s$, then $p = q_i$ for some i with $1 \leq i \leq s$.

    b. The uniqueness part of the unique factorization theorem for the integers says that given any integer n, if

$$n = p_1 p_2 \cdots p_r = q_1 q_2 \cdots q_s$$

for some positive integers r and s and prime numbers $p_1 \leq p_2 \leq \cdots \leq p_r$ and $q_1 \leq q_2 \leq \cdots \leq q_s$, then $r = s$ and $p_i = q_i$ for all integers i with $1 \leq i \leq r$.

    Use the result of part (a) to fill in the details of the following sketch of a proof. Suppose that n is an integer with two different prime factorizations: $n = p_1 p_2 \cdots p_t = q_1 q_2 \cdots q_u$. All the prime factors that appear on both sides can be cancelled (as many times as they appear on both sides) to arrive at the situation where $p_1 p_2 \cdots p_r = q_1 q_2 \cdots q_s$, $p_1 \leq p_2 \leq \cdots \leq p_r$, $q_1 \leq q_2 \leq \cdots \leq q_s$, and $p_i \neq q_j$ for any integers i and j. Then $r = s$ and $p_i = q_i$ for all i with $1 \leq i \leq r$, and so the prime factorization of n is unique except, possibly, for the order in which the prime factors are written.

42. According to Fermat's little theorem, if p is a prime number and a and p are relatively prime, then $a^{p-1} \equiv 1 \pmod{p}$. Verify that this theorem gives correct results for
    a. $a = 15$ and $p = 7$ **b.** $a = 8$ and $p = 11$

43. Fermat's little theorem can be used to show that a number is not prime by finding a number a relatively prime to p with the property that $a^{p-1} \not\equiv 1 \pmod{p}$. However, it cannot be used to show that a number *is* prime. Find an example to illustrate this fact. That is, find integers a and p such that a and p are relatively prime and $a^{p-1} \equiv 1 \pmod{p}$ but p is not prime.

44. The parts of this exercise are steps in the solution of Sun-Tsu's problem using the construction in the proof of the Chinese remainder theorem. Sun-Tsu's problem is to find an integer whose remainders upon division by $n_1 = 3, n_2 = 5$, and $n_3 = 7$ are $a_1 = 2, a_2 = 3$, and $a_3 = 2$, respectively. Let $N = n_1 n_2 n_3$, $N_1 = n_2 n_3$, $N_2 = n_1 n_3$, and $N_3 = n_2 n_3$.
    a. Find an inverse x_1 for N_1 modulo n_1.
    b. Find an inverse x_2 for N_2 modulo n_2.
    c. Find an inverse x_3 for N_3 modulo n_3.

d. Compute $x = a_1 N_1 x_1 + a_2 N_2 x_2 + a_3 N_3 x_3$ and verify that $x \equiv a_1 \pmod{n_1}$, $x \equiv a_2 \pmod{n_2}$, and $x \equiv a_3 \pmod{n_3}$.

e. Find $x \bmod N$, the least nonnegative solution for the congruence.

45. Another problem attributed to ancient China is this: Fifteen pirates steal a sack of identical gold coins. When they try to divide them evenly, two coins are left over. A fight results in which one pirate is killed. The remaining pirates again try to divide the coins evenly, but this time one coin is left over. A second fight breaks out, and a second pirate is killed. When the remaining pirates again divide the coins, they find that each gets an equal number. Use the Chinese remainder theorem to find the smallest number of gold coins that could have been in the sack at the beginning.

H 46. The Indian mathematician Brahmagupta, born in A.D. 598, posed the following problem: There are n eggs in a basket. When eggs are removed 2, 3, 4, 5, or 6 at a time, the number left over is 1, 2, 3, 4, or 5, respectively. Only when eggs are removed 7 at a time is none left over. What is the least number of eggs that could be in the basket? Use the Chinese remainder theorem to answer this question.

47. Finish the proof of the Chinese remainder theorem by proving the uniqueness of the solution modulo n.

10.5 Partial Order Relations

There is no branch of mathematics, however abstract, which may not some day be applied to phenomena of the real world. — Nicolai Ivanovitch Lobachevsky, 1792–1856

In order to obtain a degree in computer science at a certain university, a student must take a specified set of required courses, some of which must be completed before others can be started. Given the prerequistite structure of the program, one might ask what is the least number of school terms needed to fulfill the degree requirements, or what is the maximum number of courses that can be taken in the same term, or whether there is a sequence in which a part-time student can take the courses one per term. Later in this section, we will show how representing the prerequisite structure of the program as a partial order relation makes it relatively easy to answer such questions.

Antisymmetry

In Section 10.2 we defined three properties of relations: reflexivity, symmetry, and transitivity. A fourth property of relations is called *antisymmetry*. In terms of the arrow diagram of a relation, saying that a relation is antisymmetric is the same as saying that whenever there is an arrow going from one element to another *distinct* element, there is *not* an arrow going back from the second to the first.

• Definition

Let R be a relation on a set A. R is **antisymmetric** if, and only if,

for all a and b in A,　if $a\,R\,b$ and $b\,R\,a$ then $a = b$.

By taking the negation of the definition, you can see that a relation R is **not** **antisymmetric** if, and only if,

there are elements a and b in A such that $a\,R\,b$ and $b\,R\,a$ but $a \neq b$.

Example 10.5.1 Testing for Antisymmetry of Finite Relations

Let R_1 and R_2 be the relations on $\{0, 1, 2\}$ defined as follows: Draw the directed graphs for R_1 and R_2 and indicate which relations are antisymmetric.

a. $R_1 = \{(0, 2), (1, 2), (2, 0)\}$

b. $R_2 = \{(0, 0), (0, 1), (0, 2), (1, 1), (1, 2)\}$

Solution

a. R_1 is not antisymmetric.

Since $0\ R_1\ 2$ and $2\ R_1\ 0$ but $0 \neq 2$, R_1 is not antisymmetric.

b. R_2 is antisymmetric.

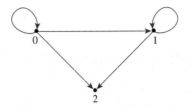

In order for R_2 not to be antisymmetric, there would have to exist a pair of distinct elements of A such that each is related to the other by R_2. But you can see by inspection that no such pair exists.

Example 10.5.2 Testing for Antisymmetry of "Divides" Relations

Let R_1 be the "divides" relation on the set of all positive integers, and let R_2 be the "divides" relation on the set of all integers.

$$
\begin{aligned}
\text{For all } a, b \in Z^+, & \quad a\ R_1\ b \iff a \mid b. \\
\text{For all } a, b \in Z, & \quad a\ R_2\ b \iff a \mid b.
\end{aligned}
$$

a. Is R_1 antisymmetric? Prove or give a counterexample.

b. Is R_2 antisymmetric? Prove or give a counterexample.

Solution

a. R_1 is antisymmetric.

Proof:

Suppose a and b are positive integers such that $a\ R_1\ b$ and $b\ R_1\ a$. [*We must show that $a = b$.*] By definition of R_1, $a \mid b$ and $b \mid a$. Thus, by definition of divides, there are integers k_1 and k_2 with $b = k_1 a$ and $a = k_2 b$. It follows that

$$b = k_1 a = k_1(k_2 b) = (k_1 k_2)b.$$

Dividing both sides by b gives

$$k_1 k_2 = 1.$$

Now since a and b are both integers k_1 and k_2 are both positive integers also. But the only product of two positive integers that equals 1 is $1 \cdot 1$. Thus

$$k_1 = k_2 = 1$$

and so

$$a = k_2 b = 1 \cdot b = b.$$

[*This is what was to be shown.*]

b. R_2 is not antisymmetric.

Counterexample:

Let $a = 2$ and $b = -2$. Then $a \mid b$ [*since* $-2 = (-1) \cdot 2$] and $b \mid a$ [*since* $2 = (-1)(-2)$]. Hence a R_2 b and b R_2 a but $a \neq b$. ■

Example 10.5.2 illustrates the fact that a relation may be antisymmetric on a subset of a set but not antisymmetric on the set itself.

Partial Order Relations

A relation that is reflexive, antisymmetric, and transitive is called a *partial order*.

> • **Definition**
>
> Let R be a binary relation defined on a set A. R is a **partial order relation** if, and only if, R is reflexive, antisymmetric, and transitive.

Two fundamental partial order relations are the "less than or equal to" relation on a set of real numbers and the "subset" relation on a set of sets. These can be thought of as models, or paradigms, for general partial order relations.

Example 10.5.3 The "Subset" Relation

Let $\mathscr{A}$ be any collection of sets and define the "subset" relation, $\subseteq$, on $\mathscr{A}$ as follows: For all $U, V \in \mathscr{A}$,

$$ U \subseteq V \quad \Leftrightarrow \quad \text{for all } x, \text{ if } x \in U \text{ then } x \in V. $$

By an argument almost identical to that of the solution for exercise 26 of Section 10.2, $\subseteq$ is reflexive and transitive. Finish the proof that $\subseteq$ is a partial order relation by proving that $\subseteq$ is antisymmetric.

Solution For $\subseteq$ to be antisymmetric means that for all sets U and V in $\mathscr{A}$ if $U \subseteq V$ and $V \subseteq U$ then $U = V$. But this is true by definition of equality of sets. ■

Example 10.5.4 A "Divides" Relation on a Set of Positive Integers

Let $\mid$ be the "divides" relation on a set A of positive integers. That is, for all $a, b \in A$,

$$ a \mid b \quad \Leftrightarrow \quad b = ka \text{ for some integer } k. $$

Prove that $\mid$ is a partial order relation on A.

Solution

$\mid$ *is reflexive:* [*We must show that for all* $a \in A, a \mid a$.] Suppose $a \in A$. Then $a = 1 \cdot a$, so $a \mid a$ by definition of divisibility.

| *is antisymmetric*: [*We must show that for all $a, b \in A$, if $a \mid b$ and $b \mid a$ then $a = b$.*] The proof of this is virtually identical to that of Example 10.5.2(a).

| *is transitive*: To show transitivity means to show that for all $a, b, c \in A$, if $a \mid b$ and $b \mid c$ then $a \mid c$. But this was proved as Theorem 3.3.1.

Since | is reflexive, antisymmetric, and transitive, | is a partial order relation on A.

∎

Example 10.5.5 The "Less Than or Equal to" Relation

Let S be a set of real numbers and define the "less than or equal to" relation, $\leq$, on S as follows: For all real numbers x and y in S,

$$x \leq y \quad \Leftrightarrow \quad x < y \text{ or } x = y.$$

Show that $\leq$ is a partial order relation.

Solution

$\leq$ *is reflexive*: For $\leq$ to be reflexive means that $x \leq x$ for all real numbers x in S. But $x \leq x$ means that $x < x$ or $x = x$, and $x = x$ is always true.

$\leq$ *is antisymmetric*: For $\leq$ to be antisymmetric means that for all real numbers x and y in S, if $x \leq y$ and $y \leq x$ then $x = y$. This follows immediately from the definition of $\leq$ and the trichotomy property (see Appendix A, T16), which says that given any real numbers, x and y, exactly one of the following holds: $x < y$ or $x = y$ or $x > y$.

$\leq$ *is transitive*: For $\leq$ to be transitive means that for all real numbers x, y, and z in S if $x \leq y$ and $y \leq z$ then $x \leq z$. This follows from the definition of $\leq$ and the transitivity property of order (see Appendix A, T17), which says that given any real numbers x, y, and z, if $x < y$ and $y < z$ then $x < z$.

Because $\leq$ is reflexive, antisymmetric, and transitive, it is a partial order relation.

∎

• Notation

Because of the special paradigmatic role played by the $\leq$ relation in the study of partial order relations, the symbol $\preceq$ is often used to refer to a general partial order relation, and the notation $x \preceq y$ is read "x is less than or equal to y" or "y is greater than or equal to x."

Lexicographic Order

To figure out which of two words comes first in an English dictionary, you compare their letters one by one from left to right. If all letters have been the same to a certain point and one word runs out of letters, that word comes first in the dictionary. For example, *play* comes before *playhouse*. If all letters up to a certain point are the same and the next letters differ, then the word whose next letter is located earlier in the alphabet comes first in the dictionary. For instance, *playhouse* comes before *playmate*.

More generally, if A is any set with a partial order relation, then a *dictionary* or *lexicographic* order can be defined on a set of strings over A as indicated in the following theorem.

Theorem 10.5.1

Let A be a set with a partial order relation R, and let S be a set of strings over A. Define a relation $\preceq$ on S as follows:

For any positive integers m and n and $a_1 a_2 \cdots a_m$ and $b_1 b_2 \cdots b_n$ in S,

1. If $m \leq n$ and $a_i = b_i$ for all $i = 1, 2, \ldots, m$, then

$$a_1 a_2 \cdots a_m \preceq b_1 b_2 \cdots b_n.$$

2. If for some integer k with $k \leq m$, $k \leq n$, and $k \geq 1$, $a_i = b_i$ for all $i = 1$, $2, \ldots, k - 1$, and $a_k \, R \, b_k$ but $a_k \neq b_k$, then

$$a_1 a_2 \cdots a_m \preceq b_1 b_2 \cdots b_n.$$

3. If ε is the null string and s is any string in S, then $\epsilon \preceq s$.

If no strings are related other than by these three conditions, then $\preceq$ is a partial order relation.

The proof of Theorem 10.5.1 is technical but straightforward. It is left for the exercises.

• Definition

The partial order relation of Theorem 10.5.1 is called the **lexicographic order for S** that corresponds to the partial order R on A.

Example 10.5.6 A Lexicographic Order

Let $A = \{x, y\}$ and let R be the following partial order relation on A:

$$R = \{(x, x), (x, y), (y, y)\}.$$

Let S be the set of all strings over A, and denote by $\preceq$ the lexicographic order for S that corresponds to R.

a. Is $x \preceq xx$? $x \preceq xy$? $xx \preceq xxx$? $yxy \preceq yxyxxx$?

b. Is $x \preceq y$? $xx \preceq xyx$? $xxxy \preceq xy$? $yxyxxyy \preceq yxyxy$?

c. Is $\epsilon \preceq x$? $\epsilon \preceq xy$? $\epsilon \preceq yyxy$?

Solution

a. Yes in all cases, by property (1) of the definition of $\preceq$.

b. Yes in all cases, by property (2) of the definition of $\preceq$.

c. Yes in all cases, by property (3) of the definition of $\preceq$. ∎

Hasse Diagrams

Let $A = \{1, 2, 3, 9, 18\}$ and consider the "divides" relation on A: For all $a, b \in A$,

$$a \mid b \quad \Leftrightarrow \quad b = ka \text{ for some integer } k.$$

The directed graph of this relation has the following appearance:

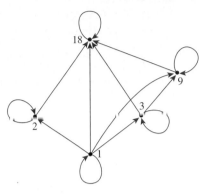

Note that there is a loop at every vertex, all other arrows point in the same direction (upward), and any time there is an arrow from one point to a second and from the second point to a third, there is an arrow from the first point to the third. Given any partial order relation defined on a finite set, it is possible to draw the directed graph in such a way that all of these properties are satisfied. This makes it possible to associate a somewhat simpler graph, called a **Hasse diagram** (after Helmut Hasse, a twentieth-century German number theorist), with a partial order relation defined on a finite set. To obtain a Hasse diagram, proceed as follows:

Start with a directed graph of the relation in which all arrows point upward. Then eliminate

1. the loops at all the vertices,

2. all arrows whose existence is implied by the transitive property,

3. the direction indicators on the arrows.

For the relation given above, the Hasse diagram is as follows:

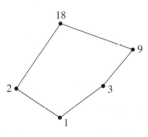

Example 10.5.7 Constructing a Hasse Diagram

Consider the "subset" relation, $\subseteq$, on the set $\mathscr{P}(\{a, b, c\})$. That is, for all sets U and V in $\mathscr{P}(\{a, b, c\})$,

$$U \subseteq V \quad \Leftrightarrow \quad \forall x, \text{ if } x \in U \text{ then } x \in V.$$

Construct the Hasse diagram for this relation.

Solution   Draw the directed graph of the relation in such a way that all arrows except loops point upward.

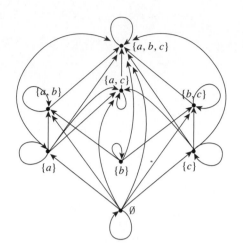

Then strip away all loops, unnecessary arrows, and direction indicators to obtain the Hasse diagram.

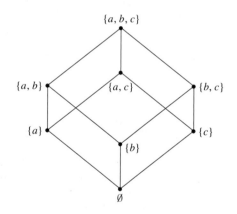

To recover the directed graph of a relation from the Hasse diagram, just reverse the instructions given above, using the knowledge that the original directed graph was sketched so that all arrows pointed upward:

1. Reinsert the direction markers on the arrows making all arrows point upward.

2. Add loops at each vertex.

3. For each sequence of arrows from one point to a second and from that second point to a third, add an arrow from the first point to the third.

Example 10.5.8 Obtaining the Directed Graph of a Partial Order Relation from the Hasse Diagram of the Relation

A partial order relation R has the following Hasse diagram. Find the directed graph of R.

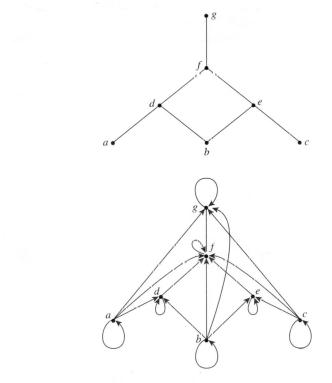

Solution

Partially and Totally Ordered Sets

Given any two real numbers x and y, either $x \leq y$ or $y \leq x$. In a situation like this, the elements x and y are said to be *comparable*. On the other hand, given two subsets A and B of $\{a, b, c\}$, it may be the case that neither $A \subseteq B$ nor $B \subseteq A$. For instance, let $A = \{a, b\}$ and $B = \{b, c\}$. Then $A \nsubseteq B$ and $B \nsubseteq A$. In such a case, A and B are said to be *noncomparable*.

• Definition

Suppose R is a partial order relation on a set A. Elements a and b of A are said to be **comparable** if, and only if, either $a \, R \, b$ or $b \, R \, a$. Otherwise, a and b are called **noncomparable.**

When all the elements of a partial order relation are comparable, the relation is called a *total order*.

• Definition

If R is a partial order relation on a set A, and for any two elements a and b in A either $a \, R \, b$ or $b \, R \, a$, then R is a **total order relation** on A.

Both the "less than or equal to" relation on sets of real numbers and the lexicographic order of the set of words in a dictionary are total order relations. Note that the Hasse diagram for a total order relation can be drawn as a single vertical "chain."

Many important partial order relations have elements that are not comparable and are, therefore, not total order relations. For instance, the subset relation on $\mathscr{P}(\{a, b, c\})$ is not a total order relation because, as shown above, the subsets $\{a, b\}$ and $\{a, c\}$ of $\{a, b, c\}$ are not comparable. In addition, a "divides" relation is not a total order relation unless the elements are all powers of a single integer. (See exercise 21 at the end of this section.)

A set A is called a **partially ordered set** (or **poset**) with respect to a relation $\preceq$ if, and only if, $\preceq$ is a partial order relation on A. For instance, the set of real numbers is a partially ordered set with respect to the "less than or equal to" relation $\leq$, and a set of sets is partially ordered with respect to the "subset" relation $\subseteq$. It is entirely straightforward to show that *any subset of a partially ordered set is partially ordered.* (See exercise 35 at the end of this section.) This, of course, assumes the "same definition" for the relation on the subset as for the set as a whole. A set A is called a **totally ordered set** with respect to a relation $\preceq$ if, and only if, A is partially ordered with respect to $\preceq$ and $\preceq$ is a total order.

A set that is partially ordered but not totally ordered may have totally ordered subsets. Such subsets are called *chains*.

• **Definition**

Let A be a set that is partially ordered with respect to a relation $\preceq$. A subset B of A is called a **chain** if, and only if, each pair of elements in B is comparable. In other words, $a \preceq b$ or $b \preceq a$ for all a and b in A. The **length of a chain** is one less than the number of elements in the chain.

Observe that if B is a chain in A, then B is a totally ordered set with respect to the "restriction" of $\preceq$ to B. (See exercise 35 at the end of this section.)

Example 10.5.9 A Chain of Subsets

The set $\mathscr{P}(\{a, b, c\})$ is partially ordered with respect to the subset relation. Find a chain of length 3 in $\mathscr{P}(\{a, b, c\})$.

Solution   Since $\emptyset \subseteq \{a\} \subseteq \{a, b, \} \subseteq \{a, b, c\}$, the set

$$S = \{\emptyset, \{a\}, \{a, b\}, \{a, b, c\}\}$$

is a chain of length 3 in $\mathscr{P}(\{a, b, c\})$. ∎

In exercise 39 at the end of this section, you are asked to show that a set that is partially ordered with respect to a relation $\preceq$ is totally ordered with respect to $\preceq$ if, and only if, it is a chain.

A *maximal element* in a partially ordered set is an element that is greater than or equal to every element to *which it is comparable.* (There may be many elements to which it is *not* comparable.) A *greatest element* in a partially ordered set is an element that is greater than or equal to *every* element in the set (so it is comparable to every element in the set). Minimal and least elements are defined similarly.

> • **Definition**
>
> Let a set A be partially ordered with respect to a relation $\preceq$.
>
> 1. An element a in A is called a **maximal element of A** if, and only if, for all b in A, either $b \preceq a$ or b and a are not comparable.
>
> 2. An element a in A is called a **greatest element of A** if, and only if, for all b in A, $b \preceq a$.
>
> 3. An element a in A is called a **minimal element of A** if, and only if, for all b in A, either $a \preceq b$ or b and a are not comparable.
>
> 4. An element a in A is called a **least element of A** if, and only if, for all b in A, $a \preceq b$.

A greatest element is maximal, but a maximal element need not be a greatest element. However, every finite subset of a totally ordered set has both a least element and a greatest element. (See exercise 40 at the end of the section.) Similarly, a least element is minimal, but a minimal element need not be a least element. Furthermore, a set that is partially ordered with respect to a relation can have at most one greatest element and one least element (see exercise 42 at the end of the section), but it may have more than one maximal or minimal element. The next example illustrates some of these facts.

Example 10.5.10 Maximal, Minimal, Greatest, and Least Elements

Let $A = \{a, b, c, d, e, f, g, h, i\}$ have the partial ordering $\preceq$ defined by the following Hasse diagram. Find all maximal, minimal, greatest, and least elements of A.

Solution   There is just one maximal element, g, which is also the greatest element. The minimal elements are c, d, and i, and there is no least element. ∎

Topological Sorting

Is it possible to input the sets of $\mathcal{P}(\{a, b, c\})$ into a computer in a way that is *compatible* with the subset relation $\subseteq$ in the sense that if set U is a subset of set V, then U is input before V? The answer, as it turns out, is yes. For instance, the following input order satisfies the given condition:

$$\emptyset, \{a\}, \{b\}, \{c\}, \{a, b\}, \{a, c\}, \{b, c\}, \{a, b, c\}.$$

Another input order that satisfies the condition is

$$\emptyset, \{a\}, \{b\}, \{a, b\}, \{c\}, \{b, c\}, \{a, c\}, \{a, b, c\}.$$

• **Definition**

Given partial order relations $\preceq$ and $\preceq'$ on a set A, $\preceq'$ is **compatible** with $\preceq$ if, and only if, for all a and b in A, if $a \preceq b$ then $a \preceq' b$.

Given an arbitrary partial order relation $\preceq$ on a set A, is there a total order $\preceq'$ on A that is compatible with $\preceq$? If the set on which the partial order is defined is finite, then the answer is yes. A total order that is compatible with a given order is called a *topological sorting*.

• **Definition**

Given partial order relations $\preceq$ and $\preceq'$ on a set A, $\preceq'$ is a **topological sorting** for $\preceq$ if, and only if, $\preceq'$ is a total order that is compatible with $\preceq$.

The construction of a topological sorting for a general finite partially ordered set is based on the fact that *any partially ordered set that is finite and nonempty has a minimal element.* (See exercise 41 at the end of the section.) To create a total order for a partially ordered set, simply pick any minimal element and make it number one. Then consider the set obtained when this element is removed. Since the new set is a subset of a partially ordered set, it is partially ordered. If it is empty, stop the process. If not, pick a minimal element from it and call that element number two. Then consider the set obtained when this element also is removed. If this set is empty, stop the process. If not, pick a minimal element and call it number three. Continue in this way until all the elements of the set have been used up.

Here is a somewhat more formal version of the algorithm:

Constructing a Topological Sorting

Let $\preceq$ be a partial order relation on a nonempty finite set A. To construct a topological sorting,

1. Pick any minimal element x in A. [*Such an element exists since A is nonempty.*]

2. Set $A' := A - \{x\}$.

3. Repeat steps a–c while $A' \neq \emptyset$.
   a. Pick any minimal element y in A'.
   b. Define $x \preceq' y$.
   c. Set $A' := A' - \{y\}$ and $x := y$.
   [*Completion of steps 1–3 of this algorithm gives enough information to construct the Hasse diagram for the total ordering $\preceq'$. We have already shown how to use the Hasse diagram to obtain a complete directed graph for a relation.*]

Example 10.5.11 A Topological Sorting

Consider the set $A = \{2, 3, 4, 6, 18, 24\}$ ordered by the "divides" relation $|$. The Hasse diagram of this relation is the following:

The ordinary "less than or equal to" relation $\leq$ on this set is a topological sorting for it since for positive integers a and b, if $a \mid b$ then $a \leq b$. Find another topological sorting for this set.

Solution   The set has two minimal elements: 2 and 3. Either one may be chosen; say you pick 3. The beginning of the total order is

$$\text{total order: } 3.$$

Set $A' = A - \{3\}$. You can indicate this by removing 3 from the Hasse diagram as shown below.

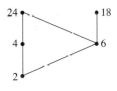

Next choose minimal element from $A' - \{3\}$. Only 2 is minimal, so you must pick it. The total order thus far is

$$\text{total order: } 3 \preceq 2.$$

Set $A' = (A - \{3\}) - \{2\} = A - \{3, 2\}$. You can indicate this by removing 2 from the Hasse diagram, as is shown below.

Choose a minimal element from $A' - \{3, 2\}$. Again you have two choices: 4 and 6. Say you pick 6. The total order for the elements chosen thus far is

$$\text{total order: } 3 \preceq 2 \preceq 6.$$

You continue in this way until every element of A has been picked. One possible sequence of choices gives

$$\text{total order: } 3 \preceq 2 \preceq 6 \preceq 18 \preceq 4 \preceq 24.$$

You can verify that this order is compatible with the "divides" partial order by checking that for each pair of elements a and b in A such that $a \mid b$, then $a \preceq b$. Note that it is *not* the case that if $a \preceq b$ then $a \mid b$. ■

An Application

To return to the example that introduced this section, note that the following defines a partial order relation on the set of courses required for a university degree: For all required courses x and y,

$$x \preceq y \quad \Leftrightarrow \quad x = y \quad \text{or} \quad x \text{ is a prerequisite for } y$$

If the Hasse diagram for the relation is drawn, then the questions raised at the beginning of this section can be answered easily. For instance, consider the Hasse diagram for the requirements at a particular university, which is shown in Figure 10.5.1.

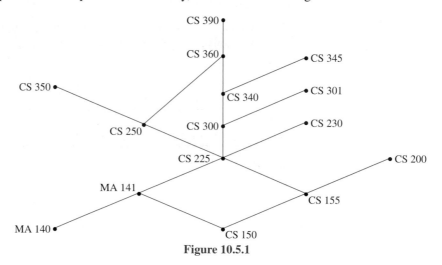

Figure 10.5.1

The minimum number of school terms needed to complete the requirements is the length of the longest chain, which is 7 (150, 155, 225, 300, 340, 360, 390, for example). The maximum number of courses that could be taken in the same term (assuming the university allows it) is the maximum number of noncomparable courses, which is 6 (350, 360, 345, 301, 230, 200, for example). A part-time student could take the courses in a sequence determined by constructing a topological sorting for the set. (One such sorting is 140, 150, 141, 155, 200, 225, 230, 300, 250, 301, 340, 345, 350, 360, 390. There are many others.)

PERT and CPM

Two important and widely used applications of partial order relations are **PERT** (Program Evaluation and Review Technique) and **CPM** (Critical Path Method). These techniques came into being in the 1950s as planners came to grips with the complexities of scheduling the individual activities needed to complete very large projects, and although they are very similar, their developments were independent. PERT was developed by the U.S. Navy to help organize the construction of the Polaris submarine, and CPM was developed by the E. I. Du Pont de Nemours company for scheduling chemical plant maintenance. Here is a somewhat simplified example of the way the techniques work.

Example 10.5.12 A Job Scheduling Problem

At an automobile assembly plant, the job of assembling an automobile can be broken down into these tasks:

1. Build frame.

2. Install engine, power train components, gas tank.

3. Install brakes, wheels, tires.

4. Install dashboard, floor, seats.

5. Install electrical lines.

6. Install gas lines.

7. Install brake lines.

8. Attach body panels to frame.

9. Paint body.

Certain of these tasks can be carried out at the same time, whereas some cannot be started until other tasks are finished. Table 10.5.1 summarizes the order in which tasks can be performed and the time required to perform each task.

Table 10.5.1

Task	Immediately Preceding Tasks	Time Needed to Perform Task
1		7 hours
2	1	6 hours
3	1	3 hours
4	2	6 hours
5	2, 3	3 hours
6	4	1 hour
7	2, 3	1 hour
8	4, 5	2 hours
9	6, 7, 8	5 hours

Let T be the set of all tasks, and consider the partial order relation $\preceq$ defined on T as follows: For all tasks x and y in T,

$$x \preceq y \quad \Leftrightarrow \quad x = y \text{ or } x \text{ precedes } y.$$

If the Hasse diagram of this relation is turned sideways (as is customary in PERT and CPM analysis), it has the appearance shown below.

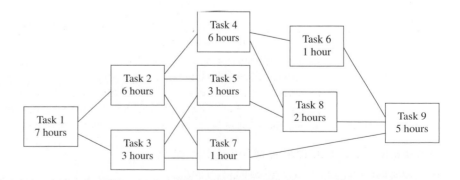

What is the minimum time required to assemble a car? You can determine this by working from left to right across the diagram, noting for each task (say, just above the box representing that task) the minimum time needed to complete that task starting from the beginning of the assembly process. For instance, you can put a 7 above the box for task 1 because task 1 requires 7 hours. Task 2 requires completion of task 1 (7 hours) plus 6 hours for itself, so the minimum time required to complete task 2, starting at the beginning of the assembly process, is $7 + 6 = 13$ hours. You can put a 13 above the box for task 2. Similarly, you can put a 10 above the box for task 3 because $7 + 3 = 10$. Now consider what number you should write above the box for task 5. The minimum times to complete tasks 2 and 3, starting from the beginning of the assembly process, are 13 and 10 hours respectively. Since *both* tasks must be completed before task 5 can be started, the minimum time to complete task 5, starting from the beginning, is the time needed for task 5 itself (3 hours) plus the *maximum* of the times to complete tasks 2 and 3 (13 hours), and this equals $3 + 13 = 16$ hours. Thus you should place the number 16 above the box for task 5. The same reasoning leads you to place a 14 above the box for task 7. Similarly, you can place a 19 above the box for task 4, a 20 above the box for task 6, a 21 above the box for task 8, and a 26 above the box for task 9, as shown below.

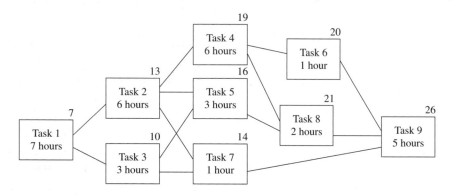

This analysis shows that at least 26 hours are required to complete task 9 starting from the beginning of the assembly process. When task 9 is finished, the assembly is complete, so 26 hours is the minimum time needed to accomplish the whole process.

Note that the minimum time required to complete tasks 1, 2, 4, 8, and 9 in sequence is exactly 26 hours. This means that a delay in performing any one of these tasks causes a delay in the total time required for assembly of the car. For this reason, the path through tasks 1, 2, 4, 8, and 9 is called a **critical path.** ∎

Exercise Set 10.5

1. Each of the following is a relation on $\{0, 1, 2, 3\}$. Draw directed graphs for each relation, and indicate which relations are antisymmetric.
   a. $R_1 = \{(0, 0), (0, 2), (1, 0), (1, 3), (2, 2), (3, 0), (3, 1)\}$
   b. $R_2 = \{(0, 1), (0, 2), (1, 1), (1, 2), (1, 3), (2, 2), (3, 2)\}$
   c. $R_3 = \{(0, 0), (0, 3), (1, 0), (1, 3), (2, 2), (3, 3), (3, 2)\}$
   d. $R_4 = \{(0, 0), (1, 0), (1, 2), (1, 3), (2, 0), (2, 1), (3, 2), (3, 0)\}$

2. Let P be the set of all people in the world and define a relation R on P as follows: For all $x, y \in P$,
$$x\, R\, y \quad \Leftrightarrow \quad x \text{ is no older then } y.$$
Is R antisymmetric? Prove or give a counterexample.

3. Let S be the set of all strings of a's and b's. Define a relation R on S as follows: For all $t \in S$,
$$s\, R\, t \quad \Leftrightarrow \quad l(s) \leq l(t),$$

where $l(x)$ denotes the length of a string x. Is R antisymmetric? Prove or give a counterexample.

4. Let R be the "less than" relation on the set $\mathbf{R}$ of all real numbers: For all $x, y \in \mathbf{R}$,
$$x\, R\, y \quad \Leftrightarrow \quad x < y.$$
Is R antisymmetric? Prove or give a counterexample.

5. Let $\mathbf{R}$ be the set of all real numbers and define a relation R on $\mathbf{R} \times \mathbf{R}$ as follows: For all (a, b) and (c, d) in $\mathbf{R} \times \mathbf{R}$,
$$(a, b)\, R\, (c, d) \quad \Leftrightarrow \quad \text{either } a < c \text{ or both } a = c \text{ and } b \leq d.$$
Is R a partial order relation? Prove or give a counterexample.

6. Let P be the set of all people who have ever lived and define a relation R on P as follows: For all $r, s \in P$,

$$r \, R \, s \quad \Leftrightarrow \quad r \text{ is an ancestor of } s \text{ or } r = s.$$

Is R a partial order relation? Prove or give a counterexample.

7. Define a relation R on the set $\mathbf{Z}$ of all integers as follows: For all $m, n \in \mathbf{Z}$,

$$m \, R \, n \quad \Leftrightarrow \quad \text{every prime factor of } m$$
$$\text{is a prime factor of } n.$$

Is R a partial order relation? Prove or give a counterexample.

8. Define a relation R on the set $\mathbf{Z}$ of all integers as follows: For all $m, n \in \mathbf{Z}$,

$$m \, R \, n \quad \Leftrightarrow \quad m + n \text{ is even.}$$

Is R a partial order relation? Prove or give a counterexample.

9. Define a relation R on the set of all real numbers $\mathbf{R}$ as follows: For all $x, y \in \mathbf{R}$,

$$x \, R \, y \quad \Leftrightarrow \quad x^2 \leq y^2.$$

Is R a partial order relation? Prove or give a counterexample.

10. Suppose R and S are antisymmetric relations on a set A. Must $R \cup S$ also be antisymmetric? Explain.

11. Let $A = \{a, b\}$, and suppose A has the partial order relation R where $R = \{(a, a), (a, b), (b, b)\}$. Let S be the set of all strings in a's and b's and let $\preceq$ be the corresponding lexicographic order on S. Indicate which of the following statements are true, and for each true statement cite as a reason part (1), (2), or (3) of the definition of lexicographic order given in Theorem 10.5.1.
    a. $aab \preceq aaba$
    b. $bbab \preceq bba$
    c. $\epsilon \preceq aba$
    d. $aba \preceq abb$
    e. $bbab \preceq bbaa$
    f. $ababa \preceq ababaa$
    g. $bbaba \preceq bbabb$

12. Prove Theorem 10.5.1.

13. Let $A = \{a, b\}$. Describe all partial order relations on A.

14. Let $A = \{a, b, c\}$.
    a. Describe all partial order relations on A for which a is a maximal element.
    b. Describe all partial order relations on A for which a is a minimal element.

H 15. Suppose a relation R on a set A is reflexive, symmetric, transitive, and antisymmetric. What can you conclude about R? Prove your answer.

16. Consider the "divides" relation on each of the following sets A. Draw the Hasse diagram for each relation.
    a. $A = \{1, 2, 4, 5, 10, 15, 20\}$
    b. $A = \{2, 3, 4, 6, 8, 9, 12, 18\}$

17. Consider the "subset" relation on $\mathscr{P}(S)$ for each of the following sets S. Draw the Hasse diagram for each relation.
    a. $S = \{0, 1\}$
    b. $S = \{0, 1, 2\}$

18. Let $S = \{0, 1\}$ and consider the partial order relation R defined on $S \times S$ as follows: For all ordered pairs (a, b) and (c, d) in $S \times S$,

$$(a, b) \, R \, (c, d) \quad \Leftrightarrow \quad \text{either } a < c \text{ or } a = c \text{ and } b \leq d,$$

where $<$ denotes the usual "less than" and $\leq$ denotes the usual "less than or equal to" relation for real numbers. Draw the Hasse diagram for R.

19. Let $S = \{0, 1\}$ and consider the partial order relation R defined on $S \times S$ as follows: For all ordered pairs (a, b) and (c, d) in $S \times S$,

$$(a, b) \, R \, (c, d) \quad \Leftrightarrow \quad a \leq c \text{ and } b \leq d,$$

where $\leq$ denotes the usual "less than or equal to" relation for real numbers. Draw the Hasse diagram for R.

20. Let $S = \{0, 1\}$ and consider the partial order relation R defined on $S \times S \times S$ as follows: For all ordered triples (a, b, c) and (d, e, f) in $S \times S \times S$,

$$(a, b, c) \, R \, (d, e, f) \quad \Leftrightarrow \quad a \leq d, b \leq e, \text{ and } c \leq f,$$

where $\leq$ denotes the usual "less than or equal to" relation for real numbers. Draw the Hasse diagram for R.

21. Consider the "divides" relation defined on the set $A = \{1, 2, 2^2, 2^3, \ldots, 2^n\}$, where n is a nonnegative integer.
    a. Prove that this relation is a total order relation on A.
    b. Draw the Hasse diagram for this relation for $n = 4$.

In 22–29, find all greatest, least, maximal, and minimal elements for the relations in each of the referenced exercises.

22. Exercise 16(a)   23. Exercise 16(b)

24. Exercise 17(a)   25. Exercise 17(b)

26. Exercise 18   27. Exercise 19

28. Exercise 20   29. Exercise 21

30. Each of the following sets is partially ordered with respect to the "less than or equal to" relation, $\leq$, for real numbers. In each case, determine whether the set has a greatest or least element.
    a. $\mathbf{R}$
    b. $\{x \in \mathbf{R} \mid 0 \leq x \leq 1\}$
    c. $\{x \in \mathbf{R} \mid 0 < x < 1\}$
    d. $\{x \in \mathbf{Z} \mid 0 < x < 10\}$

31. Let $A = \{a, b, c, d\}$, and let R be the relation

$$R = \{(a, a), (b, b), (c, c), (d, d), (c, a), (a, d),$$
$$(c, d), (b, c), (b, d), (b, a)\}.$$

Is R a total order on A? Justify your answer.

32. Let $A = \{a, b, c, d\}$, and let R be the relation

$$R = \{(a, a), (b, b), (c, c), (d, d), (c, b), (a, d),$$
$$(b, a), (b, d), (c, d), (c, a)\}.$$

Is R a total order on A? Justify your answer.

33. Consider the set $A = \{12, 24, 48, 3, 9\}$ ordered by the "divides" relation. Is A totally ordered with respect to the relation? Justify your answer.

34. How many total orderings are there on a set with n elements? Explain your answer.

H 35. Suppose that R is a partial order relation on a set A and that B is a subset of A. The **restriction of R to B** is defined as follows:

The restriction of R to B
$$= \{(x, y) \mid x \in B, y \in B, \text{ and } (x, y) \in R\}.$$

In other words, two elements of B are related by the restriction of R to B if, and only if, they are related by R. Prove that the restriction of R to B is a partial order relation on B. (In less formal language, this says that a subset of a partially ordered set is partially ordered.)

36. The set $\mathscr{P}(\{w, x, y, z\})$ is partially ordered with respect to the "subset" relation $\subseteq$. Find a chain of length 4 in $\mathscr{P}(\{w, x, y, z\})$.

37. The set $A = \{2, 4, 3, 6, 12, 18, 24\}$ is partially ordered with respect to the "divides" relation. Find a chain of length 3 in A.

38. Find a chain of length 2 for the relation defined in exercise 19.

39. Prove that a partially ordered set is totally ordered if, and only if, it is a chain.

40. Suppose that A is a totally ordered set. Use mathematical induction to prove that for any integer $n \geq 1$, every subset of A with n elements has both a least element and a greatest element.

41. Prove that a nonempty finite partially ordered set has
    a. at least one minimal element,
    b. at least one maximal element.

42. Prove that a finite partially ordered set has
    a. at most one greatest element,
    b. at most one least element.

43. Draw a Hasse diagram for a partially ordered set that has two maximal elements and two minimal elements and is such that each element is comparable to exactly two other elements.

44. Draw a Hasse diagram for a partially ordered set that has three maximal elements and three minimal elements and is such that each element is either greater than or less than exactly two other elements.

45. Use the algorithm given in the text to find a topological sorting for the relation of exercise 16(a) that is different from the "less than or equal to" relation $\leq$.

46. Use the algorithm given in the text to find a topological sorting for the relation of exercise 16(b) that is different from the "less than or equal to" relation $\leq$.

47. Use the algorithm given in the text to find a topological sorting for the relation of exercise 19.

48. Use the algorithm given in the text to find a topological sorting for the relation of exercise 20.

49. Use the algorithm given in the text to find a topological sorting for the "subset" relation on $\mathscr{P}(\{a, b, c, d\})$.

50. Refer to the prerequisite structure shown in Figure 10.5.1.
    a. Find a list of six noncomparable courses that is different from the list given in the text.
    b. Find two topological sortings that are different from the one given in the text.

51. A set S of jobs can be ordered by writing $x \preceq y$ to mean that either $x = y$ or x must be done before y, for all x and y in S. The following is a Hasse diagram for this relation for a particular set S of jobs.

    a. If one person is to perform all the jobs, one after another, find an order in which the jobs can be done.
    b. Suppose enough people are available to perform any number of jobs simultaneously.
       (i) If each job requires one day to perform, what is the least number of days needed to perform all ten jobs?
       (ii) What is the maximum number of jobs that can be performed at the same time?

52. Suppose the tasks described in Example 10.5.12 require the following performance times:

Task	Time Needed to Perform Task
1	9 hours
2	7 hours
3	4 hours
4	5 hours
5	7 hours
6	3 hours
7	2 hours
8	4 hours
9	6 hours

    a. What is the minimum time required to assemble a car?
    b. Find a critical path for the assembly process.

CHAPTER 11

GRAPHS AND TREES

Graphs and trees have already appeared in this book as convenient visualizations to use in a variety of situations. For instance, a possibility tree shows all possible outcomes of a multistep operation with a finite number of outcomes for each step, the directed graph of a relation on a set shows which elements of the set are related to which, a Hasse diagram illustrates the relations among elements in a partially ordered set, and a PERT diagram shows which tasks must precede which in executing a project.

In this chapter we present some of the mathematics of graphs and trees, discussing concepts such as the degree of a vertex, connectedness, Euler and Hamiltonian circuits, representation of graphs by matrices, isomorphisms of graphs, the relation between the number of vertices and the number of edges of a tree, rooted trees, and the spanning tree of a graph. Applications include uses of graphs and trees in the study of artificial intelligence, chemistry, scheduling problems, and transportation systems.

11.1 Graphs: An Introduction

The whole of mathematics consists in the organization of a series of aids to the imagination in the process of reasoning. — Alfred North Whitehead, 1861–1947

Imagine an organization that has acquired six different computers in recent years. In an effort to upgrade computer services, the organization proposes to connect the computers to form an integrated system. It is not necessary that every computer be linked with every other computer, however. In fact, analysis shows that the following connections are optimal:

Connect computer A with B, C, D, and E;

connect computer B with A and C;

connect computer C with A, B, D, and E;

connect computer D with A and C;

connect computer E with A, C, and F;

connect computer F with E.

This information can be conveniently displayed in the diagram shown below.

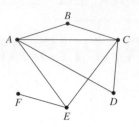

A drawing such as this is an illustration of a *graph*. The dots are called *vertices* (plural of *vertex*) and the line segments joining vertices are called *edges*. As you can see from the drawing, it is possible for two edges to cross at a point that is not a vertex. Note also that the type of graph described here is quite different from the "graph of an equation" or the "graph of a function."

In general, a graph consists of a set of vertices and a set of edges connecting various vertices. The edges may be straight or curved and should either connect one vertex to another or a vertex to itself, as shown below.

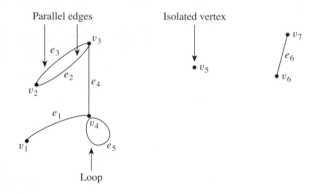

In this drawing, the vertices have been labeled with v's and the edges with e's. When an edge connects a vertex to itself (as e_5 does), it is called a *loop*. When two edges connect the same pair of vertices (as e_2 and e_3 do), they are said to be *parallel*. It is quite possible for a vertex to be unconnected by an edge to any other vertex in the graph (as v_5 is), and in that case the vertex is said to be *isolated*. The formal definition of a graph follows.

• Definition

A **graph** G consists of two finite sets: a set $V(G)$ of **vertices** and a set $E(G)$ of **edges,** where each edge is associated with a set consisting of either one or two vertices called its **endpoints.** The correspondence from edges to endpoints is called the **edge-endpoint function.** An edge with just one endpoint is called a **loop,** and two distinct edges with the same set of endpoints are said to be **parallel.** An edge is said to **connect** its endpoints; two vertices that are connected by an edge are called **adjacent;** and a vertex that is an endpoint of a loop is said to be **adjacent to itself.** An edge is said to be **incident on** each of its endpoints, and two edges incident on the same endpoint are called **adjacent.** A vertex on which no edges are incident is called **isolated.** A graph with no vertices is called **empty,** and one with at least one vertex is called **nonempty.**

Graphs have pictorial representations in which the vertices are represented by dots and the edges by line segments. A given pictorial representation uniquely determines a graph.

Example 11.1.1 Terminology

Consider the following graph:

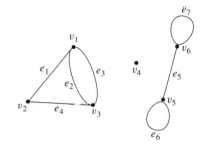

a. Write the vertex set and the edge set, and give a table showing the edge-endpoint function.

b. Find all edges that are incident on v_1, all vertices that are adjacent to v_1, all edges that are adjacent to e_1, all loops, all parallel edges, all vertices that are adjacent to themselves, and all isolated vertices.

Solution

a. vertex set $= \{v_1, v_2, v_3, v_4, v_5, v_6\}$
   edge set $= \{e_1, e_2, e_3, e_4, e_5, e_6, e_7\}$
   edge-endpoint function:

Edge	Endpoints
e_1	$\{v_1, v_2\}$
e_2	$\{v_1, v_3\}$
e_3	$\{v_1, v_3\}$
e_4	$\{v_2, v_3\}$
e_5	$\{v_5, v_6\}$
e_6	$\{v_5\}$
e_7	$\{v_6\}$

Note that the isolated vertex v_4 does not appear in this table. Although each edge must have either one or two endpoints, a vertex need not be an endpoint of an edge.

b. e_1, e_2, and e_3 are incident on v_1.
   v_2 and v_3 are adjacent to v_1.
   e_2, e_3, and e_4 are adjacent to e_1.
   e_6 and e_7 are loops.
   e_2 and e_3 are parallel.
   v_5 and v_6 are adjacent to themselves.
   v_4 is an isolated vertex.

■

As noted earlier, a given pictorial representation uniquely determines a graph. However, a given graph may have more than one pictorial representation. Such things as the lengths or curvatures of the edges and the relative position of the vertices on the page may vary from one pictorial representation to another.

Example 11.1.2 Drawing More Than One Picture for a Graph

Consider the graph specified as follows:

$$\text{vertex set} = \{v_1, v_2, v_3, v_4\}$$
$$\text{edge set} = \{e_1, e_2, e_3, e_4\}$$
edge-endpoint function:

Edge	Endpoints
e_1	$\{v_1, v_3\}$
e_2	$\{v_2, v_4\}$
e_3	$\{v_2, v_4\}$
e_4	$\{v_3\}$

Both drawings (a) and (b) shown below are pictorial representations of this graph.

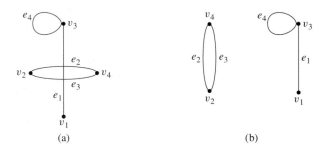

(a)   (b)   ■

Example 11.1.3 Labeling Drawings to Show They Represent the Same Graph

Consider the two drawings shown in Figure 11.1.1. Label vertices and edges in such a way that both drawings represent the same graph.

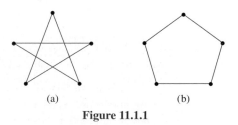

(a)   (b)

Figure 11.1.1

Solution   Imagine putting one end of a piece of string at the top vertex of Figure 11.1.1(a) (call this vertex v_1), then laying the string to the next adjacent vertex on the lower right (call this vertex v_2), then laying it to the next adjacent vertex on the upper left (v_3), and so forth, returning finally to the top vertex v_1. Call the first edge e_1, the second e_2, and so forth, as shown below.

Now imagine picking up the piece of string, together with its labels, and repositioning it as follows:

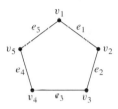

This is the same as Figure 11.1.1(b), so both drawings are representations of the graph with vertex set $\{v_1, v_2, v_3, v_4, v_5\}$, edge set $\{e_1, e_2, e_3, e_4, e_5\}$, and edge-endpoint function as follows:

Edge	Endpoints
e_1	$\{v_1, v_2\}$
e_2	$\{v_2, v_3\}$
e_3	$\{v_3, v_4\}$
e_4	$\{v_4, v_5\}$
e_5	$\{v_5, v_1\}$

■

In Chapter 10 we discussed the directed graph of a binary relation on a set. The general definition of directed graph is similar to the definition of graph, except that one associates an *ordered pair* of vertices with each edge instead of a *set* of vertices. Thus each edge of a directed graph can be drawn as an arrow going from the first vertex to the second vertex of the ordered pair.

• Definition

A **directed graph,** or **digraph,** consists of two finite sets: a set $V(G)$ of vertices and a set $D(G)$ of directed edges, where each is associated with an ordered pair of vertices called its **endpoints.** If edge e is associated with the pair (v, w) of vertices, then e is said to be the **(directed) edge** from v to w.

Note that each directed graph has an associated ordinary (undirected) graph, which is obtained by ignoring the directions of the edges.

Examples of Graphs

You have already seen a number of examples of directed and undirected graphs in this book. Flowcharts, possibility trees, Hasse diagrams, and PERT diagrams can all be viewed as graphs if additional structure is stripped away. The following examples illustrate some other situations in which graphs are used.

Example 11.1.4 Using a Graph to Represent a Communication System

Computer networks, the Internet, and telephone, electric power, gas pipeline, and air transport systems can all be represented by graphs. Questions that arise in the design of such systems involve choosing connecting edges to minimize cost, optimize a certain type of service, and so forth. A typical communication system is shown below.

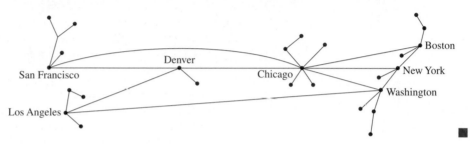

Example 11.1.5 Using a Graph to Represent Knowledge

In many applications of artifical intelligence, a knowledge base of information is collected and represented inside a computer. Because of the way the knowledge is represented and because of the properties that govern the artificial intelligence program, the computer is not limited to retrieving data in the same form as it was entered; it can also derive new facts from the knowledge base by using certain built-in rules of inference. For example, from the knowledge that the *Los Angeles Times* is a big-city daily and that a big-city daily contains national news, an artifical intelligence program could infer that the *Los Angeles Times* contains national news. The directed graph shown in Figure 11.1.2 is a pictorial representation for a simplified knowledge base about periodical publications.

According to this knowledge base, what paper finish does the *New York Times* use?

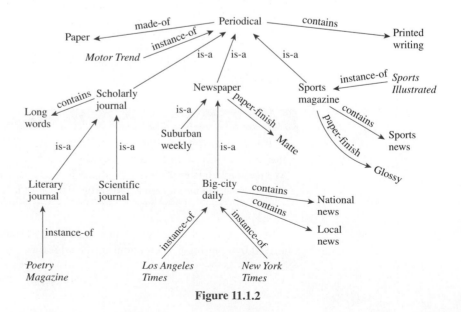

Figure 11.1.2

Solution   The arrow going from *New York Times* to big-city daily (labeled "instance-of") shows that the *New York Times* is a big-city daily. The arrow going from big-city daily to newspaper (labeled "is-a") shows that a big-city daily is a newspaper. The arrow going from newspaper to matte (labeled "paper-finish") indicates that the paper finish on a newspaper is matte. Hence it can be inferred that the paper finish on the *New York Times* is matte. ∎

Example 11.1.6 Using a Graph to Solve a Problem: Vegetarians and Cannibals

The following is a variation of a famous puzzle often used as an example in the study of artificial intelligence. It concerns an island on which all the people are of one of two types, either vegetarians or cannibals. Initially, two vegetarians and two cannibals are on the left bank of a river. With them is a boat that can hold a maximum of two people. The aim of the puzzle is to find a way to transport all the vegetarians and cannibals to the right bank of the river. What makes this difficult is that at no time can the number of cannibals on either bank outnumber the number of vegetarians. Otherwise, disaster befalls the vegetarians!

Solution   A systematic way to approach this problem is to introduce a notation that can indicate all possible arrangements of vegetarians, cannibals, and the boat on the banks of the river. For example, you could write (vvc/Bc) to indicate that there are two vegetarians and one cannibal on the left bank and one cannibal and the boat on the right bank. Then $(vvccB/)$ would indicate the initial position in which both vegetarians, both cannibals, and the boat are on the left bank of the river. The aim of the puzzle is to figure out a sequence of moves to reach the position $(/Bvvcc)$ in which both vegetarians, both cannibals, and the boat are on the right bank of the river.

Construct a graph whose vertices are the various arrangements that can be reached in a sequence of legal moves starting from the initial position. Connect vertex x to vertex y if it is possible to reach vertex y in one legal move from vertex x. For instance, from the initial position there are four legal moves: one vegetarian and one cannibal can take the boat to the right bank; two cannibals can take the boat to the right bank; one cannibal can take the boat to the right bank; or two vegetarians can take the boat to the right bank. You can show these by drawing edges connecting vertex $(vvccB/)$ to vertices (vc/Bvc), (vv/Bcc), $(vvcBc)$, and (cc/Bvv). (It might seem natural to draw directed edges rather than undirected edges from one vertex to another. The rationale for drawing undirected edges is that each legal move is reversible.) From the position (vc/Bvc), the only legal moves are to go back to $(vvccB/)$ or to go to $(vvcB/c)$. You can also show these by drawing in edges. Continue this process until finally you reach $(/Bvvcc)$. From Figure 11.1.3 it is apparent that one successful sequence of moves is $(vvccB/) \rightarrow (vc/Bvc) \rightarrow (vvcB/c) \rightarrow (c/Bvvc) \rightarrow (ccB/vv) \rightarrow (/Bvvcc)$.

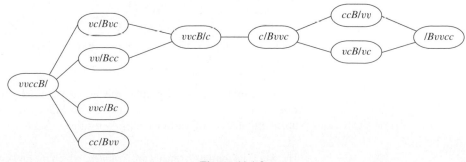

Figure 11.1.3 ∎

Special Graphs

One important class of graphs consists of those that do not have any loops or parallel edges. Such graphs are called *simple*. In a simple graph, no two edges share the same set of endpoints, so specifying two endpoints is sufficient to determine an edge.

> **• Definition and Notation**
>
> A **simple graph** is a graph that does not have any loops or parallel edges. In a simple graph, an edge with endpoints v and w is denoted $\{v, w\}$.

Example 11.1.7 A Simple Graph

Draw all simple graphs with the four vertices $\{u, v, w, x\}$ and two edges, one of which is $\{u, v\}$.

Solution   Each possible edge of a simple graph corresponds to a subset of two vertices. Given four vertices, there are $\binom{4}{2} = 6$ such subsets in all: $\{u, v\}$, $\{u, w\}$, $\{u, x\}$, $\{v, w\}$, $\{v, x\}$, and $\{w, x\}$. Now one edge of the graph is specified to be $\{u, v\}$, so any of the remaining five from this list can be chosen to be the second edge. Thus the possibilities are as follows:

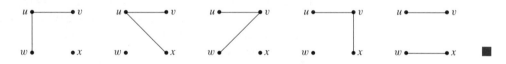

Another important class of graphs consists of those that are "complete" in the sense that all pairs of vertices are connected by edges.

> **• Definition**
>
> Let n be a positive integer. A **complete graph on n vertices,** denoted K_n, is a simple graph with n vertices $v_1, v_2, \ldots, v_n$ whose set of edges contains exactly one edge for each pair of distinct vertices.*

Example 11.1.8 Complete Graphs on n Vertices: K_1, K_2, K_3, K_4, K_5

The complete graphs $K_1, K_2, K_3, K_4,$ and K_5 can be drawn as follows:

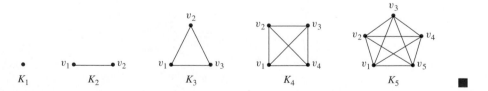

In yet another class of graphs, the vertex set can be separated into two subsets: Each vertex in one of the subsets is connected by exactly one edge to each vertex in the other subset, but not to any vertices in its own subset. Such a graph is called *complete bipartite*.

*The K stands for the German word *komplett,* which means "complete."

> **• Definition**
>
> Let m and n be positive integers. A **complete bipartite graph on (m, n) vertices,** denoted $K_{m,n}$, is a simple graph with vertices $v_1, v_2, \ldots, v_m$ and $w_1, w_2, \ldots, w_n$ that satisfies the following properties:
>
> $$\text{For all } i, k = 1, 2, \ldots, m \text{ and for all } j, l = 1, 2, \ldots, n,$$
>
> 1. There is an edge from each vertex v_i to each vertex w_j.
> 2. There is not an edge from any vertex v_i to any other vertex v_k.
> 3. There is not an edge from any vertex w_j to any other vertex w_l.

Example 11.1.9 Complete Bipartite Graphs: $K_{3,2}$ and $K_{3,3}$

The bipartite graphs $K_{3,2}$ and $K_{3,3}$ are illustrated below.

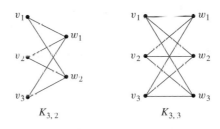

$K_{3,2}$ $K_{3,3}$

> **• Definition**
>
> A graph H is said to be a **subgraph** of a graph G if, and only if, every vertex in H is also a vertex in G, every edge in H is also an edge in G, and every edge in H has the same endpoints as in G.

Example 11.1.10 Subgraphs

List all nonempty subgraphs of the graph G with vertex set $\{v_1, v_2\}$ and edge set $\{e_1, e_2, e_3\}$, where the endpoints of e_1 are v_1 and v_2, the endpoints of e_2 are v_1 and v_2, and e_3 is a loop at v_1.

Solution G can be drawn as shown below.

There are 11 nonempty subgraphs of G, which can be grouped according to those that do not have any edges, those that have one edge, those that have two edges, and those that have three edges. The 11 nonempty subgraphs are shown in Figure 11.1.4.

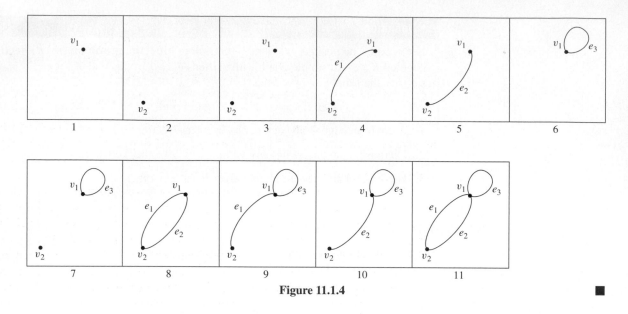

Figure 11.1.4 ■

The Concept of Degree

The *degree of a vertex* is the number of edges that "stick out of" the vertex. We will show that the sum of the degrees of all the vertices in a graph is twice the number of edges of the graph.

> **• Definition**
>
> Let G be a graph and v a vertex of G. **The degree of v,** denoted **deg(v),** equals the number of edges that are incident on v, with an edge that is a loop counted twice. The **total degree of G** is the sum of the degrees of all the vertices of G.

Since an edge that is a loop is counted twice, the degree of a vertex can be obtained from the drawing of a graph by counting how many end segments of edges are incident on the vertex. This is illustrated below.

The degree of this vertex equals 5.

Example 11.1.11 Degree of a Vertex and Total Degree of a Graph

Find the degree of each vertex of the graph G shown below. Then find the total degree of G.

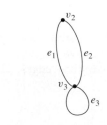

Solution $\deg(v_1) = 0$ since no edge is incident on v_1 (v_1 is isolated).

$\deg(v_2) = 2$ since both e_1 and e_2 are incident on v_2.

$\deg(v_3) = 4$ since e_1 and e_2 are incident on v_3 and the loop e_3 is also incident on v_3 (and contributes 2 to the degree of v_3).

total degree of $G = \deg(v_1) + \deg(v_2) + \deg(v_3) = 0 + 2 + 4 = 6$.   ■

Note that the total degree of the graph G of Example 11.1.11, which is 6, equals twice the number of edges of G, which is 3. Roughly speaking, this is true because each edge has two end segments, and each end segment is counted once toward the degree of some vertex. This result generalizes to any graph.

In fact, for any graph without loops, the general result can be explained as follows: Imagine a group of people at a party. Depending on how social they are, each person shakes hands with various other people. So each person participates in a certain number of handshakes—perhaps many, perhaps none —but because each handshake is experienced by two different people, if the numbers experienced by each person are added together, the sum will equal twice the total number of handshakes. This is such an attractive way of understanding the situation that the following theorem is often called the *handshake lemma* or the *handshake theorem*. As the proof demonstrates, the conclusion is true even if the graph contains loops.

Theorem 11.1.1 The Handshake Theorem

If G is any graph, then the sum of the degrees of all the vertices of G equals twice the number of edges of G. Specifically, if the vertices of G are $v_1, v_2, \ldots, v_n$, where n is a nonnegative integer, then

$$\text{the total degree of } G = \deg(v_1) + \deg(v_2) + \cdots + \deg(v_n)$$
$$= 2(\text{the number of edges of } G).$$

Proof:

Let G be a particular but arbitrarily chosen graph. If G does not have any vertices, than it does not have any edges, and so its total degree, which is 0, is twice the number of its edges, which is also 0. If G has n vertices $v_1, v_2, \ldots, v_n$ and m edges, where n is a positive integer and m is a nonnegative integer, we claim that each edge of G contributes 2 to the total degree of G. For suppose e is an arbitrarily chosen edge with endpoints v_i and v_j. This edge contributes 1 to the degree of v_i and 1 to the degree v_j. As shown below, this is true even if $i = j$, because an edge that is a loop is counted twice in computing the degree of the vertex on which it is incident.

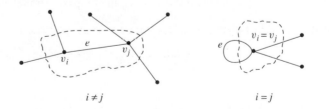

Therefore, e contributes 2 to the total degree of G. Since e was arbitrarily chosen, this shows that *each* edge of G contributes 2 to the total degree of G. Thus

$$\text{the total degree of } G = 2(\text{the number of edges of } G).$$

The following corollary is an immediate consequence of Theorem 11.1.1.

Corollary 11.1.2

The total degree of a graph is even.

Proof:

Since the total degree of G equals 2 times the number of edges, which is an integer, the total degree of G is even.

Example 11.1.12 Determining Whether Certain Graphs Exist

Draw a graph with the specified properties or show that no such graph exists.

a. A graph with four vertices of degrees 1, 1, 2, and 3

b. A graph with four vertices of degrees 1, 1, 3, and 3

c. A simple graph with four vertices of degrees 1, 1, 3, and 3

Solution

a. No such graph is possible. By Corollary 11.1.2, the total degree of a graph is even. But a graph with four vertices of degrees 1, 1, 2, and 3 would have a total degree of $1 + 1 + 2 + 3 = 7$, which is odd.

b. Let G be any of the graphs shown below.

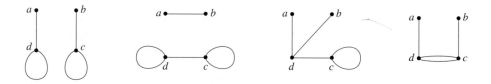

In each case, no matter how the edges are labeled, $\deg(a) = 1, \deg(b) = 1, \deg(c) = 3$, and $\deg(d) = 3$.

c. There is no simple graph with four vertices of degrees 1, 1, 3, and 3.

Proof (by contradiction):

Suppose there were a simple graph G with four vertices of degrees 1, 1, 3, and 3. Call a and b the vertices of degree 1, and call c and d the vertices of degree 3. Since $\deg(c) = 3$ and G does not have any loops or parallel edges (because it is simple), there must be edges that connect c to a, b, and d.

By the same reasoning, there must be edges connecting d to a, b and c.

But then $\deg(a) \geq 2$ and $\deg(b) \geq 2$, which contradicts the supposition that these vertices have degree 1. Hence the supposition is false, and consequenlty there is no simple graph with four vertices of degrees 1, 1, 3, and 3. ∎

Example 11.1.13 Application to an Acquaintance Graph

Is it possible in a group of nine people for each to be friends with exactly five others?

Solution The answer is no. Imagine constructing an "acquaintance graph" in which each of the nine people represented by a vertex and two vertices are joined by an edge if, and only if, the people they represent are friends. Suppose each of the people were friends with exactly five others. Then the degree of each of the nine vertices of the graph would be five, and so the total degree of the graph would be 45. But this contradicts Corollary 11.1.2, which says that the total degree of a graph is even. This contradiction shows that the supposition is false, and hence it is impossible for each person in a group of nine people to be friends with exactly five others. ∎

The following proposition is easily deduced from Corollary 11.1.2 using properties of even and odd integers.

Proposition 11.1.3

In any graph there are an even number of vertices of odd degree.

Proof:

Suppose G is any graph, and suppose G has n vertices of odd degree and m vertices of even degree, where n and m are nonnegative integers. [*We must show that n is even.*] If n is 0, then, since 0 is even, G has an even number of vertices of odd degree. So suppose that $n \geq 1$. Let E be the sum of the degrees of all the vertices of even degree, O the sum of the degrees of all the vertices of odd degree, and T the total degree of G. If $u_1, u_2, \ldots, u_m$ are the vertices of even degree and $v_1, v_2, \ldots, v_n$ are the vertices of odd degree, then

$$E = \deg(u_1) + \deg(u_2) + \cdots + \deg(u_m),$$
$$O = \deg(v_1) + \deg(v_2) + \cdots + \deg(v_n), \quad \text{and}$$
$$T = \deg(u_1) + \cdots + \deg(u_m) + \deg(v_1) + \cdots + \deg(v_n) = E + O.$$

Now T, the total degree of G, is an even integer by Corollary 11.1.2. Also E is even since either E is zero, which is even, or E is a sum of the numbers $\deg(u_i)$, each of which is even. But

$$T = E + O,$$

and therefore

$$O = T - E.$$

Hence O is a difference of two even integers, and so O is even.

By assumption, $\deg(v_i)$ is odd for all $i = 1, 2, \ldots, n$. Thus O, an even integer, is a sum of the n odd integers $\deg(v_1), \deg(v_2), \ldots, \deg(v_n)$. But if a sum of n odd integers is even, then n is even. (See exercise 32 at the end of this section.) Therefore, n is even [*as was to be shown*].

Example 11.1.14 Applying the Fact That the Number of Vertices with Odd Degree Is Even

Is there a graph with ten vertices of degrees 1, 1, 2, 2, 2, 3, 4, 4, 4, and 6?

Solution   No. Such a graph would have three vertices of odd degree, which is impossible by Proposition 11.1.3.

Note that this same result could have been deduced directly from Corollary 11.1.2 by computing the total degree $(1 + 1 + 2 + 2 + 2 + 3 + 4 + 4 + 4 + 6 = 29)$ and noting that it is odd. However, use of Proposition 11.1.3 gives the result without the need to perform this addition.   ■

Exercise Set 11.1*

In 1 and 2, graphs are represented by drawings. Define each graph formally by specifying its vertex set, its edge set, and a table giving the edge-endpoint function.

In 5–7, show that the two drawings represent the same graph by labeling the vertices and edges of the right-hand drawing to correspond to those of the left-hand drawing.

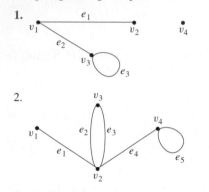

1.

2.

In 3 and 4, draw pictures of the specified graphs.

3. Graph G has vertex set $\{v_1, v_2, v_3, v_4, v_5\}$ and edge set $\{e_1, e_2, e_3, e_4\}$, with edge-endpoint function as follows:

Edge	Endpoints
e_1	$\{v_1, v_2\}$
e_2	$\{v_1, v_2\}$
e_3	$\{v_2, v_3\}$
e_4	$\{v_2\}$

4. Graph H has vertex set $\{v_1, v_2, v_3, v_4, v_5\}$ and edge set $\{e_1, e_2, e_3, e_4\}$ with edge-endpoint function as follows:

Edge	Endpoints
e_1	$\{v_1\}$
e_2	$\{v_2, v_3\}$
e_3	$\{v_2, v_3\}$
e_4	$\{v_1, v_5\}$

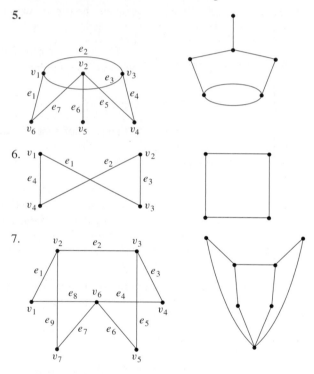

5.

6.

7.

For each of the graphs in 8 and 9:
  (i) Find all edges that are incident on v_1.
  (ii) Find all vertices that are adjacent to v_3.
  (iii) Find all edges that are adjacent to e_1.
  (iv) Find all loops.
  (v) Find all parallel edges.
  (vi) Find all isolated vertices.
  (vii) Find the degree of v_3.
  (viii) Find the total degree of the graph.

8.

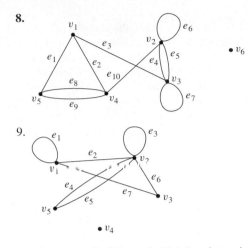

9.

10. Use the graph of Example 11.1.5 to determine
   a. whether *Sports Illustrated* contains printed writing;
   b. whether *Poetry Magazine* contains long words.

11. Find three other winning sequences of moves for the vegetarians and the cannibals in Example 11.1.6.

12. Another famous puzzle used as an example in the study of
artificial intelligence seems first to have appeared in a collection of problems, *Problems for the Quickening of the Mind*,
which was compiled about A.D. 775. It involves a wolf, a
goat, a bag of cabbage, and a ferryman. From an initial position on the left bank of a river, the ferryman is to transport
the wolf, the goat, and the cabbage to the right bank. The
difficulty is that the ferryman's boat is only big enough for
him to transport one object at a time, other than himself.
Yet, for obvious reasons, the wolf cannot be left alone with
the goat, and the goat cannot be left alone with the cabbage.
How should the ferryman proceed?

13. Solve the vegetarians-and-cannibals puzzle for the case
where there are three vegetarians and three cannibals to be
transported from one side of a river to the other.

H 14. Two jugs A and B have capacities of 3 quarts and 5 quarts,
respectively. Can you use the jugs to measure out exactly
1 quart of water, while obeying the following restrictions:
You may fill either jug to capacity from a water tap; you
may empty the contents of either jug into a drain; and you
may pour water from either jug into the other.

In each of 15–23, either draw a graph with the specified properties or explain why no such graph exists.

15. Graph with five vertices of degrees 1, 2, 3, 3, and 5.

16. Graph with four vertices of degrees 1, 2, 3, and 3.

17. Graph with four vertices of degrees 1, 1, 1, and 4.

18. Graph with four vertices of degrees 1, 2, 3, and 4.

19. Simple graph with four vertices of degrees 1, 2, 3, and 4.

20. Simple graph with five vertices of degrees 2, 3, 3, 3, and 5.

21. Simple graph with five vertices of degrees 1, 1, 1, 2, and 3.

22. Simple graph with six edges and all vertices of degree 3.

23. Simple graph with nine edges and all vertices of degree 3.

24. Find all nonempty subgraphs of each of the following
graphs.

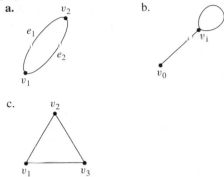

25. a. In a group of 15 people, is it possible for each person to
have exactly 3 friends? Explain. (Assume that friend-
ship is a symmetric relationship: If x is a friend of y,
then y is a friend of x.)
   b. In a group of 4 people, is it possible for each person to
have exactly 3 friends? Why?

26. In a group of 25 people, is it possible for each to shake hands
with exactly 3 other people? Explain.

27. Is there a simple graph, each of whose vertices has even
degree? Explain.

28. Suppose a graph has vertices of degrees 0, 2, 2, 3, and 9.
How many edges does the graph have?

29. Suppose a graph has vertices of degrees 1, 1, 4, 4, and 6.
How many edges does the graph have?

30. Suppose that G is a graph with v vertices and e edges and
that the degree of each vertex is at least $d_{\min}$ and at most
$d_{\max}$. Show that
$$\frac{1}{2}d_{\min} \cdot v \le e \le \frac{1}{2}d_{\max} \cdot v.$$

31. Prove that any sum of an odd number of odd integers is odd.

H 32. Deduce from exercise 31 that for any positive integer n, if
a sum of n odd integers is even, then n is even.

33. Recall that K_n denotes a complete graph on n vertices.
   a. Draw K_6.
   H b. Show that for all integers $n \ge 1$, the number of edges of
K_n is $\dfrac{n(n-1)}{2}$.

34. Use the result of exercise 33 to show that the number of
edges of a simple graph with n vertices is less than or equal
to $\dfrac{n(n-1)}{2}$.

35. Is there a nonempty simple graph with twice as many edges as vertices? Explain. (You may find it helpful to use the result of exercise 34.)

36. Recall that $K_{m,n}$ denotes a complete bipartite graph on (m, n) vertices.
   a. Draw $K_{4,2}$
   b. Draw $K_{1,3}$
   c. Draw $K_{3,4}$
   d. How many vertices of $K_{m,n}$ have degree m? degree n?
   e. What is the total degree of $K_{m,n}$?
   f. Find a formula in terms of m and n for the number of edges of $K_{m,n}$. Explain.

37. A **bipartite graph** G is a simple graph whose vertex set can be partitioned into two mutually disjoint nonempty subsets V_1 and V_2 such that vertices in V_1 may be connected to vertices in V_2, but no vertices in V_1 are connected to other vertices in V_1 and no vertices in V_2 are connected to other vertices in V_2. For example, the graph G illustrated in (i) can be redrawn as shown in (ii). From the drawing in (ii), you can see that G is bipartite with mutually disjoint vertex sets $V_1 = \{v_1, v_3, v_5\}$ and $V_2 = \{v_2, v_4, v_6\}$.

Find which of the following graphs are bipartite. Redraw the bipartite graphs so that their bipartite nature is evident.

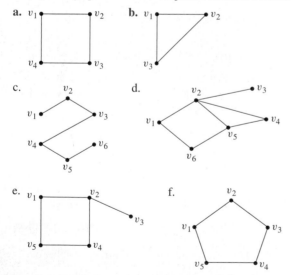

38. Suppose r and s are any positive integers. Does there exist a graph G with the property that G has vertices of degrees r and s and of no other degrees? Explain.

Definition: If G is a simple graph, the **complement of G,** denoted G', is obtained as follows: The vertex set of G' is identical to the vertex set of G. However, two distinct vertices v and w of G' are connected by an edge if, and only if, v and w are not connected by an edge in G. For example, if G is the graph

then G' is

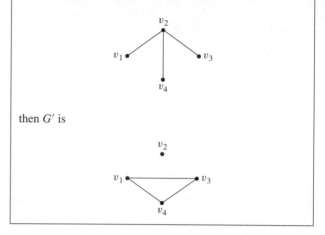

39. Find the complement of each of the following graphs.

40. **a.** Find the complement of the graph K_4, the complete graph on four vertices. (See Example 11.1.8.)
   b. Find the complement of the graph $K_{3,2}$, the complete bipartite graph on $(3, 2)$ vertices. (See Example 11.1.9.)

41. Suppose that in a group of five people A, B, C, D, and E the following pairs of people are acquainted with each other: A and C, A and D, B and C, C and D, C and E.
   a. Draw a graph to represent this situation.
   b. Draw a graph that illustrates who among these five people are *not* acquainted. That is, draw an edge between two people if, and only if, they are not acquainted.

H **42.** Let G be a simple graph with n vertices. What is the relation between the number of edges of G and the number of edges of the complement G'?

43. Show that at a party with at least two people, there are at least two mutual acquaintances or at least two mutual strangers.

44. **a.** In a simple graph, must every vertex have degree that is less than the number of vertices in the graph? Why?
   b. Can there be a simple graph that has four vertices each of different degrees?

H ✱ **c.** Can there be a simple graph that has n vertices all of different degrees?

H ✳ **45.** In a group of two or more people, must there always be at least two people who are acquainted with the same number of people within the group? Why?

H **46.** In this exercise a graph is used to help solve a scheduling problem. Eleven faculty members in a mathematics department serve on the following committees:

Undergraduate Education: Bergen, Jones, Kashina, Cohen

Graduate Education: Gatto, Moussa, Cohen, Catoiu

Colloquium: Sahin, Goldman, Ash

Library: Cortzen, Bergen, Sahin

Hiring: Gatto, Goldman, Moussa, Jones

Personnel: Moussa, Wang, Cortzen

The committees must all meet during the first week of classes, but there are only three time slots available. Find a schedule that will allow all faculty members to attend the meetings of all committees on which they serve. To do this, represent each committee as the vertex of a graph, and draw an edge between two vertices if the two committees have a common member.

11.2 Paths and Circuits

One can begin to reason only when a clear picture has been formed in the imagination.
— W. W. Sawyer, *Mathematician's Delight*, 1943

The subject of graph theory began in the year 1736 when the great mathematician Leonhard Euler published a paper giving the solution to the following puzzle:

> The town of Königsberg in Prussia (now Kaliningrad in Russia) was built at a point where two branches of the Pregel River came together. It consisted of an island and some land along the river banks. These were connected by seven bridges as shown in Figure 11.2.1.
>
> The question is this: Is it possible for a person to take a walk around town, starting and ending at the same location and crossing each of the seven bridges exactly once?*

Figure 11.2.1 The Seven Bridges of Königsberg

To solve this puzzle, Euler translated it into a graph theory problem. He noticed that all points of a given land mass can be identified with each other since a person can travel from any one point to any other point of the same land mass without crossing a bridge. Thus for the purpose of solving the puzzle, the map of Königsberg can be identified with the graph shown in Figure 11.2.2, in which the vertices *A*, *B*, *C*, and *D* represent land masses and the seven edges represent the seven bridges.

*In his original paper, Euler did not require the walk to start and end at the same point. The analysis of the problem is simplified, however, by adding this condition. Later in the section, we discuss walks that start and end at different points.

*Leonhard Euler
(1707–1783)*

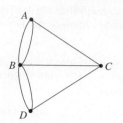

Figure 11.2.2 Graph Version of Königsberg Map

In terms of this graph, the question becomes the following:

Is it possible to find a route through the graph that starts and ends at some vertex, one of A, B, C, or D, and traverses each edge exactly once?

Equivalently:

Is it possible to trace this graph, starting and ending at the same point, without ever lifting your pencil from the paper?

Take a few minutes to think about the question yourself. Can you find a route that meets the requirements? Try it!

Looking for a route is frustrating because you continually find yourself at a vertex that does not have an unused edge on which to leave, while elsewhere there are unused edges that must still be traversed. If you start at vertex A, for example, each time you pass through vertex B, C, or D, you use up two edges because you arrive on one edge and depart on a different one. So, if it is possible to find a route that uses all the edges of the graph and starts and ends at A, then the total number of arrivals and departures from each vertex B, C, and D must be multiple of 2. Or, in other words, the degrees of the vertices B, C, and D must be even. But they are not: $\deg(B) = 5$, $\deg(C) = 3$, and $\deg(D) = 3$. Hence there is no route that solves the puzzle by starting and ending at A. Similar reasoning can be used to show that there are no routes that solve the puzzle by starting and ending at B, C, or D. Therefore, it is impossible to travel all around the city crossing each bridge exactly once.

Definitions

Travel in a graph is accomplished by moving from one vertex to another along a sequence of adjacent edges. In the graph below, for instance, you can go from u_1 to u_4 by taking f_1 to u_2 and then f_7 to u_4. This is represented by writing

$$u_1 f_1 u_2 f_7 u_4.$$

Or you could take the roundabout route

$$u_1 f_1 u_2 f_3 u_3 f_4 u_2 f_3 u_3 f_5 u_4 f_6 u_4 f_7 u_2 f_3 u_3 f_5 u_4.$$

Certain types of sequences of adjacent vertices and edges are of special importance in graph theory: those that do not have a repeated edge, those that do not have a repeated vertex, and those that start and end at the same vertex.

• Definition

Let G be a graph, and let v and w be vertices in G.

A **walk from v to w** is a finite alternating sequence of adjacent vertices and edges of G. Thus a walk has the form

$$v_0 e_1 v_1 e_2 \cdots v_{n-1} e_n v_n,$$

where the v's represent vertices, the e's represent edges, $v_0 = v$, $v_n = w$, and for all $i = 1, 2, \ldots n$, v_{i-1} and v_i are the endpoints of e_i. The **trivial walk from v to v** consists of the single vertex v.

A **path from v to w** is a walk from v to w that does not contain a repeated edge. Thus a path from v to w is a walk of the form

$$v = v_0 e_1 v_1 e_2 \cdots v_{n-1} e_n v_n = w,$$

where all the e_i are distinct (that is, $e_i \neq e_k$ for any $i \neq k$).

A **simple path from v to w** is a path that does not contain a repeated vertex. Thus a simple path is a walk of the form

$$v = v_0 e_1 v_1 e_2 \cdots v_{n-1} e_n v_n = w,$$

where all the e_i are distinct and all the v_j are also distinct (that is, $v_j \neq v_m$ for any $j \neq m$).

A **closed walk** is a walk that starts and ends at the same vertex.

A **circuit** is a closed walk that does not contain a repeated edge. Thus a circuit is a walk of the form

$$v_0 e_1 v_1 e_2 \cdots v_{n-1} e_n v_n,$$

where $v_0 = v_n$ and all the e_i are distinct.

A **simple circuit** is a circuit that does not have any other repeated vertex except the first and last. Thus a simple circuit is a walk of the form

$$v_0 e_1 v_1 e_2 \cdots v_{n-1} e_n v_n,$$

where all the e_i are distinct and all the v_j are distinct except that $v_0 = v_n$.

For ease of reference, these definitions are summarized in the following table:

	Repeated Edge?	Repeated Vertex?	Starts and Ends at Same Point?
Walk	allowed	allowed	allowed
Path	no	allowed	allowed
Simple path	no	no	no
Closed walk	allowed	allowed	yes
Circuit	no	allowed	yes
Simple circuit	no	first and last only	yes

Often a walk can be specified unambiguously by giving either a sequence of edges or a sequence of vertices. The next two examples show how this is done.

Example 11.2.1 Notation for Walks

a. In the graph below, the notation $e_1e_2e_4e_3$ refers unambiguously to the following walk: $v_1e_1v_2e_2v_3e_4v_3e_3v_2$. On the other hand, the notation e_1 is ambiguous if used to refer to a walk. It could mean either $v_1e_1v_2$ or $v_2e_1v_1$.

b. In the graph of part (a), the notation v_2v_3 is ambiguous if used to refer to a walk. It could mean $v_2e_2v_3$ or $v_2e_3v_3$. On the other hand, in the graph below, the notation $v_1v_2v_2v_3$ refers unambiguously to the walk $v_1e_1v_2e_2v_2e_3v_3$.

Note that if a graph G does not have any parallel edges, then any walk in G is uniquely determined by its sequence of vertices.

Example 11.2.2 Walks, Paths, and Circuits

In the graph below, determine which of the following walks are paths, simple paths, circuits, and simple circuits.

a. $v_1e_1v_2e_3v_3e_4v_3e_5v_4$ b. $e_1e_3e_5e_5e_6$ c. $v_2v_3v_4v_5v_3v_6v_2$
d. $v_2v_3v_4v_5v_6v_2$ e. $v_2v_3v_4v_5v_6v_3v_2$ f. v_1

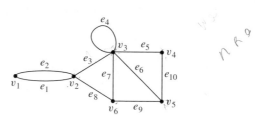

Solution

a. This walk has a repeated vertex but does not have a repeated edge, so it is a path from v_1 to v_4 but not a simple path.

b. This is just a walk from v_1 to v_5. It is not a path because it has a repeated edge.

c. This walk starts and ends at v_2 and does not have a repeated edge, so it is a circuit. Since the vertex v_3 is repeated in the middle, it is not a simple circuit.

d. This walk starts and ends at v_2, does not have a repeated edge, and does not have a repeated vertex. Thus it is a simple circuit.

e. This is just a closed walk starting and ending at v_2. It is neither a circuit nor a simple circuit because edge e_3 and vertex v_3 are repeated.

f. The first vertex of this walk is the same as its last vertex. (Try to disprove this statement if you are inclined not to believe it!) Also, this walk has neither a repeated vertex nor a repeated edge. Thus it is a simple circuit. A circuit such as this is called a *trivial circuit*. ■

> **• Definition**
>
> A **trivial circuit** is a walk consisting of a single vertex and no edge. A **nontrivial circuit** is a circuit with at least one edge.

Connectedness

It is easy to understand the concept of connectedness on an intuitive level. Roughly speaking, a graph is connected if it is possible to travel from any vertex to any other vertex along a sequence of adjacent edges of the graph. The formal definition of connectedness is stated in terms of walks.

> **• Definition**
>
> Let G be a graph. Two **vertices v and w of G are connected** if, and only if, there is a walk from v to w. The **graph G is connected** if, and only if, given *any* two vertices v and w in G, there is a walk from v to w. Symbolically,
>
> $$G \text{ is connected} \quad \Leftrightarrow \quad \forall \text{ vertices } v, w \in V(G), \exists \text{ a walk from } v \text{ to } w.$$

If you take the negation of this definition, you will see that a graph G is *not connected* if, and only if, there are two vertices of G that are not connected by any walk.

Example 11.2.3 Connected and Disconnected Graphs

Which of the following graphs are connected?

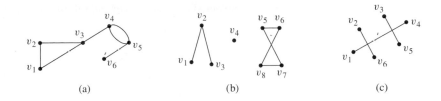

(a)   (b)   (c)

Solution   The graph represented in (a) is connected, whereas those of (b) and (c) are not. To understand why (c) is not connected, recall that in a drawing of a graph, two edges may cross at a point that is not a vertex. Thus the graph in (c) can be redrawn as follows:

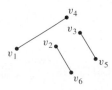

■

Some useful facts relating circuits and connectedness are collected in the following lemma. Proofs of (a) and (b) are left for the exercises. The proof of (c) is in Section 11.5.

Lemma 11.2.1

Let G be a graph.

a. If G is connected, then any two distinct vertices of G can be connected by a simple path.

b. If vertices v and w are part of a circuit in G and one edge is removed from the circuit, then there still exists a path from v to w in G.

c. If G is connected and G contains a nontrivial circuit, then an edge of the circuit can be removed without disconnecting G.

Look back at Example 11.2.3. The graphs in (b) and (c) are both made up of three pieces, each of which is itself a connected graph. A *connected component* of a graph is a connected subgraph of largest possible size.

• Definition

A graph H is a **connected component** of a graph G if, and only if,

1. H is subgraph of G;

2. H is connected; and

3. no connected subgraph of G has H as a subgraph and contains vertices or edges that are not in H.

The fact is that any graph is a kind of union of its connected components.

Example 11.2.4 Connected Components

Find all connected components of the following graph G.

Solution G has three connected components: H_1, H_2, and H_3 with vertex sets V_1, V_2, and V_3 and edge sets E_1, E_2, and E_3, where

$$V_1 = \{v_1, v_2, v_3\}, \qquad E_1 = \{e_1, e_2\},$$
$$V_2 = \{v_4\}, \qquad E_2 = \emptyset,$$
$$V_3 = \{v_5, v_6, v_7, v_8\}, \qquad E_3 = \{e_3, e_4, e_5\}.$$ ∎

Euler Circuits

Now we return to consider general problems similar to the puzzle of the Königsberg bridges. The following definition is made in honor of Euler.

> **• Definition**
>
> Let G be a graph. An **Euler circuit** for G is a circuit that contains every vertex and every edge of G. That is, an Euler circuit for G is a sequence of adjacent vertices and edges in G that starts and ends at the same vertex, uses every vertex of G at least once, and uses every edge of G exactly once.

The analysis used earlier to solve the puzzle of the Königsberg bridges generalizes to prove the following theorem:

> **Theorem 11.2.2**
>
> If a graph has an Euler circuit, then every vertex of the graph has even degree.
>
> **Proof:**
>
> Suppose G is a graph that has an Euler circuit. [*We must show that given any vertex v of G, the degree of v is even.*] Let v be any particular but arbitrarily chosen vertex of G. Since the Euler circuit contains every edge of G, it contains all edges incident on v. Now imagine taking a journey that begins in the middle of one of the edges adjacent to the start of the Euler circuit and continues around the Euler circuit to end in the middle of the starting edge. (See Figure 11.2.3.) Each time v is entered by traveling along one edge, it is immediately exited by traveling along another edge (since the journey ends in the *middle* of an edge).
>
>
>
> In this example, the Euler circuit is $v_0v_1v_2v_3v_4v_2v_5v_0$, and v is v_2. Each time v_2 is entered by one edge, it is exited by another edge.
>
> **Figure 11.2.3 Example for the Proof of Theorem 11.2.2**
>
> Because the Euler circuit uses every edge of G exactly once, every edge incident on v is traversed exactly once in this process. Hence the edges incident on v occur in entry/exit pairs, and consequently the degree of v must be a multiple of 2. But that means that the degree of v is even [*as was to be shown*].

Recall that the contrapositive of a statement is logically equivalent to the statement. The contrapositive of Theorem 11.2.2 is as follows:

> **Contrapositive Version of Theorem 11.2.2**
>
> If some vertex of a graph has odd degree, then the graph does not have an Euler circuit.

This version of Theorem 11.2.2 is useful for showing that a given graph does *not* have an Euler circuit.

Example 11.2.5 Showing That a Graph Does Not Have an Euler Circuit

Show that the graph below does not have an Euler circuit.

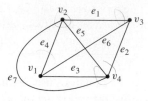

Solution Vertices v_1 and v_3 both have degree 3, which is odd. Hence by (the contrapositive form of) Theorem 11.2.2, this graph does not have an Euler circuit. ■

Now consider the converse of Theorem 11.2.2: If every vertex of a graph has even degree, then the graph has an Euler circuit. Is this true? The answer is no. There is a graph G such that every vertex of G has even degree but G does not have an Euler circuit. In fact, there are many such graphs. The illustration below shows one example.

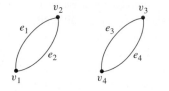

Every vertex has even degree, but the graph does not have an Euler circuit.

Note that the graph in the preceding drawing is not connected. It turns out that although the converse of Theorem 11.2.2 is false, a modified converse is true: If every vertex of a graph has even degree *and* if the graph is connected, then the graph has an Euler circuit. The proof of this fact is constructive: It contains an algorithm to find an Euler circuit for any connected graph in which every vertex has even degree.

Theorem 11.2.3

If every vertex of a nonempty graph has even degree and if the graph is connected, then the graph has an Euler circuit.

Proof:

Suppose that G is any nonempty connected graph and that every vertex of G has even degree. [*We must find an Euler circuit for G.*] If G consists of a single vertex v, the trivial walk from v to v is an Euler circuit. Otherwise, construct a circuit C by the following algorithm:

Step 1: Pick any vertex v of G at which to start.
[*This step can be accomplished because the vertex set of G is nonempty by assumption.*]

Step 2: Pick any sequence of adjacent vertices and edges, starting and ending at v and never repeating an edge. Call the resulting circuit C.
[*This step can be performed for the following reasons: Since the degree of each vertex of G is even, as each vertex other than v is entered by traveling on one edge, it can be exited by traveling on another previously unused edge. Thus a sequence of distinct adjacent edges can be produced indefinitely as long as v is not reached. But since the number of edges of the graph is finite*]

(by definition of graph), the sequence of distinct edges cannot go on forever. Thus the sequence must eventually return to the starting vertex v.]

Step 3:   Check whether C contains every edge and vertex of G. If so, C is an Euler circuit, and we are finished. If not, perform the following steps.

Step 3a: Remove all edges of C from G and also any vertices that become isolated when the edges of C are removed. Call the resulting subgraph G'. *[Note that G' may not be connected (as illustrated in Figure 11.2.4), but every vertex of G' has even degree (since removing the edges of C removes an even number of edges from each vertex, and the difference of two even integers is even).]*

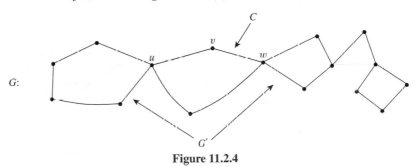

Figure 11.2.4

Step 3b: Pick any vertex w common to both C and G'.
[There must be at least one such vertex since G is connected. (See exercise 44.) (In Figure 11.2.4 there are two such vertices: u and w.)]

Step 3c: Pick any sequence of adjacent vertices and edges of G', starting and ending at w and never repeating an edge. Call the resulting circuit C'. *[This can be done since the degree of each vertex of G' is even and G' is finite. See the justification for step 2.]*

Step 3d: Patch C and C' together to create a new circuit C'' as follows: Start at v and follow C all the way to w. Then follow C' all the way back to w. After that, continue along the untraveled portion of C to return to v. *[The effect of executing steps 3c and 3d for the graph of Figure 11.2.4 is shown in Figure 11.2.5.]*

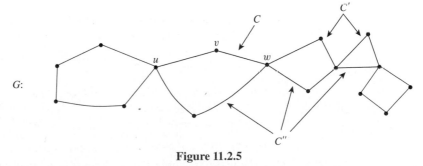

Figure 11.2.5

Step 3e: Let $C = C''$ and go back to step 3.

Since the graph G is finite, execution of the steps outlined in this algorithm must eventually terminate. At that point an Euler circuit for G will have been constructed. (Note that because of the element of choice in steps 1, 2, 3b, and 3c, a variety of different Euler circuits can be produced by using this algorithm.)

Example 11.2.6 Finding an Euler Circuit

Use Theorem 11.2.3 to check that the graph below has an Euler circuit. Then use the algorithm from the proof of the theorem to find an Euler circuit for the graph.

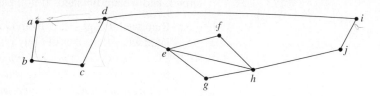

Solution   Observe that

$$\deg(a) = \deg(b) = \deg(c) = \deg(f) = \deg(g) = \deg(i) = \deg(j) = 2$$

and that $\deg(d) = \deg(e) = \deg(h) = 4$. Hence all vertices have even degree. Also, the graph is connected. Thus, by Theorem 11.2.3, the graph has an Euler circuit.

To construct an Euler circuit using the algorithm of Theorem 11.2.3, let $v = a$ and let C be

$$C: abcda.$$

C is represented by the labeled edges shown below.

Observe that C is not an Euler circuit for the graph but that C intersects the rest of the graph at d. Let C' be

$$C': deghjid.$$

Patch C' into C to obtain

$$C'': abcdeghjida.$$

Set $C = C''$. Then C is represented by the labeled edges shown below.

Observe that C is not an Euler circuit for the graph but that it intersects the rest of the graph at e. Let C' be

$$C': efhe.$$

Patch C' into C to obtain

$$C'': abcdefheghjida.$$

Set $C = C''$. Then C is represented by the labeled edges shown below.

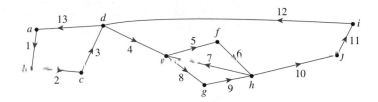

Since C includes every edge of the graph exactly once, C is an Euler circuit for the graph. ∎

In exercise 45 at the end of this section you are asked to show that any graph with an Euler circuit is connected. This result can be combined with Theorems 11.2.2 and 11.2.3 to give a complete characterization of graphs that have Euler circuits, as stated in Theorem 11.2.4.

Theorem 11.2.4

A graph G has an Euler circuit if, and only if, G is connected and every vertex of G has even degree.

A corollary to Theorem 11.2.4 gives a criterion for determining when it is possible to find a walk from one vertex of a graph to another, passing through every vertex of the graph at least once and every edge of the graph exactly once.

• Definition

Let G be a graph, and let v and w be two distinct vertices of G. An **Euler path from v to w** is a sequence of adjacent edges and vertices that starts at v, ends at w, passes through every vertex of G at least once, and traverses every edge of G exactly once.

Corollary 11.2.5

Let G be a graph, and let v and w be two distinct vertices of G. There is an Euler path from v to w if, and only if, G is connected, v and w have odd degree, and all other vertices of G have even degree.

The proof of this corollary is left as an exercise.

Example 11.2.7 Finding an Euler Path

The floor plan shown below is for a house that is open for public viewing. Is it possible to find a path that starts in room A, ends in room B, and passes through every interior doorway of the house exactly once? If so, find such a path.

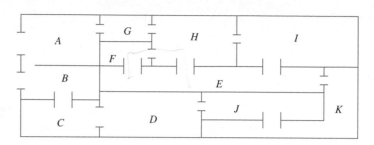

Solution   Let the floor plan of the house be represented by the graph below.

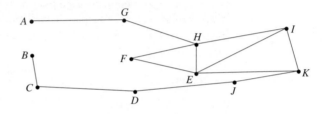

Each vertex of this graph has even degree except for A and B, each of which has degree 1. Hence by Corollary 11.2.5, there is an Euler path from A to B. One such path is

$$AGHFEIHEKJDCB.$$ ■

Hamiltonian Circuits

Theorem 11.2.4 completely answers the following question: Given a graph G, is it possible to find a circuit for G in which all the *edges* of G appear exactly once? A related question is this: Given a graph G, is it possible to find a circuit for G in which all the *vertices* of G (except the first and the last) appear exactly once?

In 1859 the Irish mathematician Sir William Rowan Hamilton introduced a puzzle in the shape of a dodecahedron (DOH-dek-a-HEE-dron). (Figure 11.2.6 contains a drawing of a dodecahedron, which is a solid figure with 12 identical pentagonal faces.)

Sir Wm. Hamilton (1805–1865)

Figure 11.2.6 Dodecahedron

Each vertex was labeled with the name of a city—London, Paris, Hong Kong, New York, and so on. The problem Hamilton posed was to start at one city and tour the world by visiting each other city exactly once and returning to the starting city. One way to solve the puzzle is to imagine the surface of the dodecahedron stretched out and laid flat in the plane, as follows:

The circuit denoted with black lines is one solution. Note that although every city is visited, many edges are omitted from the circuit. (More difficult versions of the puzzle required that certain cities be visited in a certain order.)

The following definition is made in honor of Hamilton.

> **• Definition**
>
> Given a graph G, a **Hamiltonian circuit** for G is a simple circuit that includes every vertex of G. That is, a Hamiltonian circuit for G is a sequence of adjacent vertices and distinct edges in which every vertex of G appears exactly once, except for the first and the last, which are the same.

Note that although an Euler circuit for a graph G must include every vertex of G, it may visit some vertices more than once and hence may not be a Hamiltonian circuit. On the other hand, a Hamiltonian circuit for G does not need to include all the edges of G and hence may not be an Euler circuit.

Despite the analogous-sounding definitions of Euler and Hamiltonian circuits, the mathematics of the two are very different. Theorem 11.2.4 gives a simple criterion for determining whether a given graph has an Euler circuit. Unfortunately, there is no analogous criterion for determining whether a given graph has a Hamiltonian circuit, nor is there even an efficient algorithm for finding such a circuit. There is, however, a simple technique that can be used in many cases to show that a graph does *not* have a Hamiltonian circuit. This follows from the following considerations:

Suppose a graph G with at least two vertices has a Hamiltonian circuit C given concretely as

$$C: v_0 e_1 v_1 e_2 \cdots v_{n-1} e_n v_n.$$

Since C is a simple circuit, all the e_i are distinct and all the v_j are distinct except that $v_0 = v_n$. Let H be the subgraph of G that is formed using the vertices and edges of C. An example of such an H is shown below.

H is indicated by the black lines.

Note that H has the same number of edges as it has vertices since all its n edges are distinct and so are its n vertices $v_1, v_2, \ldots, v_n$. Also, by definition of Hamiltonian circuit, every vertex of G is a vertex of H, and H is connected since any two of its vertices lie on a circuit. In addition, every vertex of H has degree 2. The reason for this is that there are exactly two edges incident on any vertex. These are e_i and e_{i+1} for any vertex v_i except $v_0 = v_n$, and they are e_1 and e_n for $v_0 (= v_n)$. These observations have established the truth of the following proposition in all cases where G has at least two vertices.

Proposition 11.2.6

If a graph G has a nontrivial Hamiltonian circuit, then G has a subgraph H with the following properties:

1. H contains every vertex of G.

2. H is connected.

3. H has the same number of edges as vertices.

4. Every vertex of H has degree 2.

Note that if G contains only one vertex and G has a nontrivial Hamiltonian circuit, then the circuit has the form $v\ e\ v$, where v is the vertex of G and e is an edge incident on v. In this case, the subgraph H consisting of v and e satisfies conditions (1)–(4) of Proposition 11.2.6.

Recall that the contrapositive of a statement is logically equivalent to the statement. The contrapositive of Proposition 11.2.6 says that if a graph G does *not* have a subgraph H with properties (1)–(4), then G does *not* have a Hamiltonian circuit, except the trivial Hamiltonian circuit in the case where C contains only one vertex.

Example 11.2.8 Showing That a Graph Does Not Have a Hamiltonian Circuit

Prove that the graph G shown below does not have a Hamiltonian circuit.

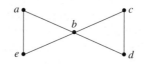

Solution   If G has a Hamiltonian circuit, then by Proposition 11.2.6, G has a subgraph H that (1) contains every vertex of G, (2) is connected, (3) has the same number of edges as vertices, and (4) is such that every vertex has degree 2. Suppose such a subgraph H exists. In other words, suppose there is a connected subgraph H of G such that H has five vertices (a, b, c, d, e) and five edges and such that every vertex of H has degree 2. Since the degree of b in G is 4 and every vertex of H has degree 2, two edges incident on b must be removed from G to create H. Edge $\{a, b\}$ cannot be removed because if it were, vertex a would have degree less than 2 in H. Similar reasoning shows that edges $\{e, b\}$, $\{b, a\}$, and $\{b, d\}$ cannot be removed either. It follows that the degree of b in H must be 4, which contradicts the condition that every vertex in H has degree 2 in H. Hence no such subgraph H exists, and so G does not have a Hamiltonian circuit. ∎

The next example illustrates a type of problem known as a **traveling salesman problem.** It is a variation of the problem of finding a Hamiltonian circuit for a graph.

Example 11.2.9 A Traveling Salesman Problem

Imagine that the drawing below is a map showing four cities and the distances in kilometers between them. Suppose that a salesman must travel to each city exactly once, starting and ending in city A. Which route from city to city will minimize the total distance that must be traveled?

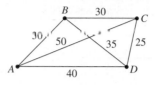

Solution This problem can be solved by writing all possible Hamiltonian circuits starting and ending at A and calculating the total distance traveled for each.

Route	Total Distance (In Kilometers)	
$ABCDA$	$30 + 30 + 25 + 40 = 125$	
$ABDCA$	$30 + 35 + 25 + 50 = 140$	
$ACBDA$	$50 + 30 + 35 + 40 = 155$	
$ACDBA$	140	[$ABDCA$ backwards]
$ADBCA$	155	[$ACBDA$ backwards]
$ADCBA$	125	[$ABCDA$ backwards]

Thus either route $ABCDA$ or $ADCBA$ gives a minimum total distance of 125 kilometers. ∎

The general traveling salesman problem involves finding a Hamiltonian circuit to minimize the total distance traveled for an arbitrary graph with n vertices in which each edge is marked with a distance. One way to solve the general problem is to use the method of Example 11.2.9: Write down all Hamiltonian circuits starting and ending at a particular vertex, compute the total distance for each, and pick one for which this total is minimal. However, even for medium-sized values of n this method is impractical. In the language of Chapter 9, any algorithm to implement this method has exponential order. Observe that for a complete graph with 30 vertices, there would be $29! \cong 8.84 \times 10^{30}$ different Hamiltonian circuits starting and ending at a particular vertex to check. Even if each circuit could be found and its total distance computed in just one nanosecond, it would require approximately 2.8×10^{14} years to finish the computation. At present, there is no known algorithm for solving the general traveling salesman problem that is more efficient. However, there are efficient algorithms that find "pretty good" solutions—that is, circuits that, while not necessarily having the least possible total distances, have smaller total distances than most other Hamiltonian circuits.

Exercise Set 11.2

1. In the graph at right, determine whether the following walks are paths, simple paths, closed walks, circuits, simple circuits, or just walks.

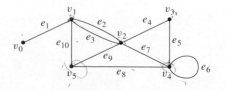

   a. $v_0 e_1 v_1 e_{10} v_5 e_9 v_2 e_2 v_1$
   b. $v_4 e_7 v_2 e_9 v_5 e_{10} v_1 e_3 v_2 e_9 v_5$
   c. v_2
   d. $v_5 v_2 v_3 v_4 v_4 v_5$
   e. $v_2 v_3 v_4 v_5 v_2 v_4 v_3 v_2$
   f. $e_5 e_8 e_{10} e_3$

2. In the graph below, determine whether the following walks are paths, simple paths, closed walks, circuits, simple circuits, or just walks.

   a. $v_1e_2v_2e_3v_3e_4v_4e_5v_2e_2v_1e_1v_0$ b. $v_2v_3v_4v_5v_2$
   c. $v_4v_2v_3v_4v_5v_2v_4$ d. $v_2v_1v_5v_2v_3v_4v_2$
   e. $v_0v_5v_2v_3v_4v_2v_1$ f. $v_5v_4v_2v_1$

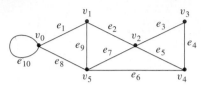

3. Let G be the graph

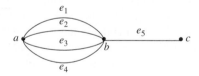

   and consider the walk $v_1e_1v_2e_2v_1$.
   a. Can this walk be written unambiguously as $v_1v_2v_1$? Why?
   b. Can this walk be written unambiguously as e_1e_2? Why?

4. Consider the following graph.

   a. How many simple paths are there from v_1 to v_4?
   b. How many paths are there from v_1 to v_4?
   c. How many walks are there from v_1 to v_4?

5. Consider the following graph.

   a. How many simple paths are there from a to c?
   b. How many paths are there from a to c?
   c. How many walks are there from a to c?

6. An edge whose removal disconnects the graph of which it is a part is called a **bridge.** Find all bridges for each of the following graphs.

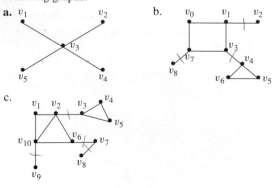

7. Given any positive integer n, (a) find a connected graph with n edges such that removal of just one edge disconnects the graph; (b) find a connected graph with n edges that cannot be disconnected by the removal of any single edge.

8. Find the number of connected components for each of the following graphs.

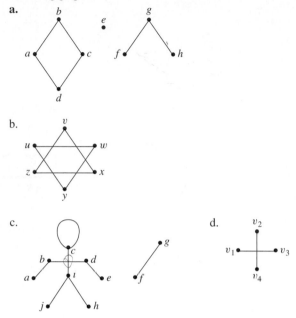

9. Each of (a)–(c) describes a graph. In each case answer *yes, no,* or *not necessarily* to this question: Does the graph have an Euler circuit? Justify your answers.
   a. G is a connected graph with five vertices of degrees 2, 2, 3, 3, and 4.
   b. G is a connected graph with five vertices of degrees 2, 2, 4, 4, and 6.
   c. G is a graph with five vertices of degrees 2, 2, 4, 4, and 6.

10. The solution for Example 11.2.5 shows a graph for which every vertex has even degree but which does not have an Euler circuit. Give another example of a graph satisfying these properties.

11. Is it possible for a citizen of Königsberg to make a tour of the city and cross each bridge exactly twice? (See Figure 11.2.1.) Why?

Determine which of the graphs in 12–17 have Euler circuits. If the graph does not have an Euler circuit, explain why not. If it does have an Euler circuit, describe one.

12.

13.

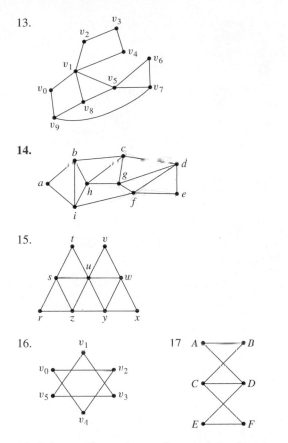

14.

15.

16.   17 A • — • B

                                C • — • D

                                E • — • F

18. Is it possible to take a walk around the city whose map is shown below, starting and ending at the same point and crossing each bridge exactly once? If so, how can this be done?

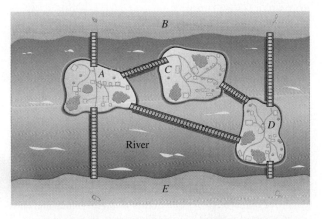

For each of the graphs in 19–21, determine whether there is an Euler path from u to w. If there is, find such a path.

19.

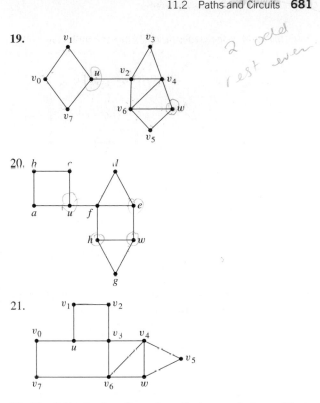

2 odd
rest even

20.

21.

22. The following is a floor plan of a house. Is it possible to enter the house in room A, travel through every interior doorway of the house exactly once, and exit out of room E? If so, how can this be done?

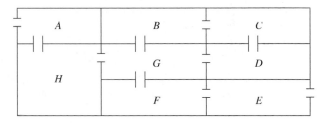

Find Hamiltonian circuits for each of the graphs in 23 and 24.

23.

24.

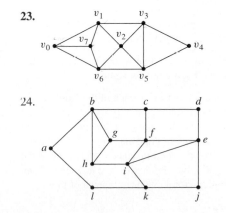

Show that none of the graphs in 25–27 has a Hamiltonian circuit.

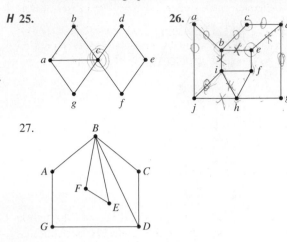

H 25.

26.

27.

In 28–31 find Hamiltonian circuits for those graphs that have them. Explain why the other graphs do not.

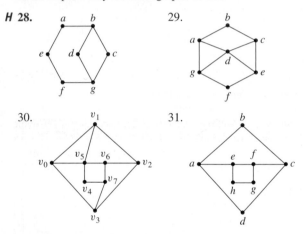

H 28.

29.

30.

31.

H 32. Give two examples of graphs that have Euler circuits but not Hamiltonian circuits.

H 33. Give two examples of graphs that have Hamiltonian circuits but not Euler circuits.

H 34. Give two examples of graphs that have circuits that are both Euler circuits and Hamiltonian circuits.

H 35. Give two examples of graphs that have Euler circuits and Hamiltonian circuits that are not the same.

36. A traveler in Europe wants to visit each of the cities shown on the map in the next column exactly once, starting and ending in Brussels. The distance (in kilometers) between each pair of cities is given in the table. Find a Hamiltonian circuit that minimizes the total distance traveled. (Use the map to narrow the possible circuits down to just a few. Then use the table to find the total distance for each of those.)

	Berlin	Brussels	Düsseldorf	Luxembourg	Munich
Brussels	783				
Düsseldorf	564	223			
Luxembourg	764	219	224		
Munich	585	771	613	517	
Paris	1,057	308	497	375	832

37. **a.** Prove that if a walk in a graph contains a repeated edge, then the walk contains a repeated vertex.
   b. Explain how it follows from part (a) that any walk with no repeated vertex has no repeated edge.

38. Prove Lemma 11.2.1(a): If G is a connected graph, then any two distinct vertices of G can be connected by a simple path.

39. Prove Lemma 11.2.1(b): If vertices v and w are part of a circuit in a graph G and one edge is removed from the circuit, then there still exists a path from v to w in G.

40. Draw a picture to illustrate Lemma 11.2.1(c): If a graph G is connected and G contains a circuit, then an edge of the circuit can be removed without disconnecting G.

41. Prove that if there is a path in a graph G from a vertex v to a vertex w, then there is a path from w to v.

42. If a graph contains a circuit that starts and ends at a vertex v, does the graph contain a simple circuit that starts and ends at v? Why?

43. Prove that if there is a circuit in a graph that starts and ends at a vertex v and if w is another vertex in the circuit, then there is a circuit in the graph that starts and ends at w.

44. Let G be a connected graph, and let C be a circuit in G. Let G' be the subgraph obtained by removing all the edges of C from G and also any vertices that become isolated when the edges of C are removed. Prove that if G' is nonempty, then there exists a vertex v such that v is in both C and G'.

45. Prove that any graph with an Euler circuit is connected.

46. Prove Corollary 11.2.5.

47. For what values of n does the complete graph K_n with n vertices have (a) an Euler circuit? (b) a Hamiltonian circuit?

✱**48.** For what values of m and n does the complete bipartite graph on (m, n) vertices have (a) an Euler circuit? (b) a Hamiltonian circuit?

✱**49.** What is the maximum number of edges a simple disconnected graph with n vertices can have? Prove your answer.

✱**50.** Show that a graph is bipartite if, and only if, it does not have a circuit with an odd number of edges. (See exercise 37 of Section 11.1 for the definition of bipartite graph.)

11.3 Matrix Representations of Graphs

Order and simplification are the first steps toward the mastery of a subject.
— Thomas Mann, *The Magic Mountain, 1924*

How can graphs be represented inside a computer? It happens that all the information needed to specify a graph can be conveyed by a structure called a *matrix,* and matrices (*matrices* is the plural of *matrix*) are easy to represent inside computers. This section contains some basic definitions about matrices and matrix operations, a description of the relation between graphs and matrices, and some applications.

Matrices

Matrices are two-dimensional analogues of sequences. They are also called two-dimensional arrays.

> **• Definition**
>
> An $m \times n$ (read "m by n") **matrix A over a set** S is a rectangular array of elements of S arranged into m rows and n columns:
>
> $$\mathbf{A} = \begin{bmatrix} a_{11} & a_{12} & \cdots & a_{1j} & \cdots & a_{1n} \\ a_{21} & a_{22} & \cdots & a_{2j} & \cdots & a_{2n} \\ \vdots & \vdots & & \vdots & & \vdots \\ a_{i1} & a_{i2} & \cdots & a_{ij} & \cdots & a_{in} \\ \vdots & \vdots & & \vdots & & \vdots \\ a_{m1} & a_{m2} & \cdots & a_{mj} & \cdots & a_{mn} \end{bmatrix} \leftarrow i\text{th row of } \mathbf{A}$$
>
> $\uparrow$
> jth column of **A**
>
> We write $\mathbf{A} = (a_{ij})$.

The *i***th row of** A is

$$[a_{i1} \quad a_{i2} \quad \cdots \quad a_{in}]$$

and the *j***th column of** A is

$$\begin{bmatrix} a_{1j} \\ a_{2j} \\ \vdots \\ a_{mj} \end{bmatrix}.$$

The entry a_{ij} in the ith row and jth column of $\mathbf{A}$ is called the ***ij*th entry of A.** An $m \times n$ matrix is said to have **size $m \times n$.** If $\mathbf{A}$ and $\mathbf{B}$ are matrices, then $\mathbf{A} = \mathbf{B}$ if, and only if, $\mathbf{A}$ and $\mathbf{B}$ have the same size and the corresponding entries of $\mathbf{A}$ and $\mathbf{B}$ are all equal; that is,

$$a_{ij} = b_{ij} \quad \text{for all } i = 1, 2, \ldots, m \text{ and } j = 1, 2 \ldots, n.$$

A matrix for which the numbers of rows and columns are equal is called a **square matrix.** If $\mathbf{A}$ is a square matrix of size $n \times n$, then the **main diagonal of A** consists of all the entries $a_{11}, a_{22}, \ldots, a_{nn}$:

main diagonal of $\mathbf{A}$

Example 11.3.1 Matrix Terminology

The following is a 3×3 matrix over the set of integers.

$$\begin{bmatrix} 1 & 0 & -3 \\ 4 & -1 & 5 \\ -2 & 2 & 0 \end{bmatrix}$$

a. What is the entry in row 2, column 3?

b. What is the second column of $\mathbf{A}$?

c. What are the entries in the main diagonal of $\mathbf{A}$?

Solution

a. 5   b. $\begin{bmatrix} 0 \\ -1 \\ 2 \end{bmatrix}$ c. 1, -1, and 0   ∎

Matrices and Directed Graphs

Consider the directed graph shown in Figure 11.3.1. This graph can be represented by the matrix $\mathbf{A} = (a_{ij})$ for which $a_{ij} = $ the number of arrows from v_i to v_j, for all $i = 1, 2, 3$ and $j = 1, 2, 3$. Thus $a_{11} = 1$ because there is one arrow from v_1 to v_1, $a_{12} = 0$ because there is no arrow from v_1 to v_2, $a_{23} = 2$ because there are two arrows from v_2 to v_3, and so forth. $\mathbf{A}$ is called the *adjacency matrix* of the directed graph. For convenient reference, the rows and columns of $\mathbf{A}$ are often labeled with the vertices of the graph G.

Directed Graph G

(a)

Adjacency Matrix

(b)

Figure 11.3.1 A Directed Graph and Its Adjacency Matrix

> **• Definition**
>
> Let G be a directed graph with ordered vertices $v_1, v_2, \ldots, v_n$. The **adjacency matrix of** G is the matrix $A = (a_{ij})$ over the set of nonnegative integers such that
>
> $$a_{ij} = \text{the number of arrows from } v_i \text{ to } v_j \quad \text{for all } i, j = 1, 2, \ldots, n.$$

Note that nonzero entries along the main diagonal of an adjacency matrix indicate the presence of loops, and entries larger than 1 correspond to parallel edges. Moreover, if the vertices of a directed graph are reordered, then the entries in the rows and columns of the corresponding adjacency matrix are moved around.

Example 11.3.2 The Adjacency Matrix of a Graph

The two directed graphs shown below differ only in the ordering of their vertices. Find their adjacency matrices.

(a)   (b)

Solution   Since both graphs have three vertices, both adjacency matrices are 3×3 matrices. For (a), all entries in the first row are 0 since there are no arrows from v_1 to any other vertex. For (b), the first two entries in the first row are 1 and the third entry is 0 since from v_1 there are single arrows to v_1 and to v_2 and no arrows to v_3. Continuing the analysis in this way, you obtain the following two adjacency matrices:

$$
\begin{array}{c}
\quad\; v_1 \; v_2 \; v_3 \\
\begin{array}{c} v_1 \\ v_2 \\ v_3 \end{array}
\left[\begin{array}{ccc} 0 & 0 & 0 \\ 0 & 1 & 1 \\ 2 & 1 & 0 \end{array} \right]
\end{array}
\qquad
\begin{array}{c}
\quad\; v_1 \; v_2 \; v_3 \\
\begin{array}{c} v_1 \\ v_2 \\ v_3 \end{array}
\left[\begin{array}{ccc} 1 & 1 & 0 \\ 1 & 0 & 2 \\ 0 & 0 & 0 \end{array} \right]
\end{array}
$$

(a)   (b)   ■

If you are given a square matrix with nonnegative integer entries, you can construct a directed graph with that matrix as its adjacency matrix. However, the matrix does not tell you how to label the edges, so the directed graph is not uniquely determined.

Example 11.3.3 Obtaining a Directed Graph from a Matrix

Let

$$
\mathbf{A} = \begin{bmatrix} 0 & 1 & 1 & 0 \\ 1 & 1 & 0 & 2 \\ 0 & 0 & 1 & 1 \\ 2 & 1 & 0 & 0 \end{bmatrix}.
$$

Draw a directed graph that has $\mathbf{A}$ as its adjacency matrix.

Solution   Let G be the graph corresponding to $\mathbf{A}$, and let v_1, v_2, v_3, v_4 be the vertices of G. Label $\mathbf{A}$ across the top and down the left side with these vertex names, as shown below.

$$
\mathbf{A} = \begin{array}{c} \\ v_1 \\ v_2 \\ v_3 \\ v_4 \end{array}
\begin{array}{cccc}
v_1 & v_2 & v_3 & v_4 \\
\left[\begin{array}{cccc}
0 & 1 & 1 & 0 \\
1 & 1 & 0 & 2 \\
0 & 0 & 1 & 1 \\
2 & 1 & 0 & 0
\end{array}\right]
\end{array}
$$

Then, for instance, the 2 in the fourth row and the first column means that there are two arrows from v_4 to v_1. The 0 in the first row and the fourth column means that there is no arrow from v_1 to v_4. A corresponding directed graph is shown below (without edge labels because the matrix does not determine those).

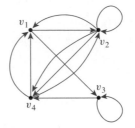

Matrices and (Undirected) Graphs

Once you know how to associate a matrix with a directed graph, the definition of the matrix corresponding to an undirected graph should seem natural to you. As before, you must order the vertices of the graph, but in this case you simply set the ijth entry of the adjacency matrix equal to the number of edges connecting the ith and jth vertices of the graph.

> **• Definition**
>
> Let G be an (undirected) graph with ordered vertices $v_1, v_2, \ldots, v_n$. The **adjacency matrix of G** is the matrix $\mathbf{A} = (a_{ij})$ over the set of nonnegative integers such that
>
> $$a_{ij} = \text{the number of edges connecting } v_i \text{ and } v_j$$
>
> for all $i, j = 1, 2, \ldots, n$.

Example 11.3.4 Finding the Adjacency Matrix of a Graph

Find the adjacency matrix for the graph G shown below.

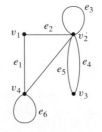

Solution

$$\begin{array}{c} \\ \mathbf{A} = \begin{array}{c} v_1 \\ v_2 \\ v_3 \\ v_4 \end{array} \begin{array}{cccc} v_1 & v_2 & v_3 & v_4 \\ \left[\begin{array}{cccc} 0 & 1 & 0 & 1 \\ 1 & 1 & 2 & 1 \\ 0 & 2 & 0 & 0 \\ 1 & 1 & 0 & 1 \end{array}\right] \end{array} \end{array}$$

∎

Note that if the matrix $\mathbf{A} = (a_{ij})$ in Example 11.3.4 is flipped across its main diagonal, it looks the same: $a_{ij} = a_{ji}$, for $i, j = 1, 2, \ldots, n$. Such a matrix is said to be *symmetric*.

• Definition

An $n \times n$ square matrix $\mathbf{A} = (a_{ij})$ is called **symmetric** if, and only if, for all $i, j = 1, 2, \ldots, n,$

$$a_{ij} = a_{ji}.$$

Example 11.3.5 Symmetric Matrices

Which of the following matrices are symmetric?

a. $\begin{bmatrix} 1 & 0 \\ 1 & 2 \end{bmatrix}$

b. $\begin{bmatrix} 0 & 1 & 2 \\ 1 & 1 & 0 \\ 2 & 0 & 3 \end{bmatrix}$

c. $\begin{bmatrix} 2 & 0 & 0 \\ 0 & 1 & 0 \end{bmatrix}$

Solution   Only (b) is symmetric. In (a) the entry in the first row and the second column differs from the entry in the second row and the first column; the matrix in (c) is not even square. ∎

It is easy to see that the matrix of *any* undirected graph is symmetric since it is always the case that the number of edges joining v_i and v_j equals the number of edges joining v_j and v_i for all $i, j = 1, 2, \ldots, n$.

Matrices and Connected Components

Consider a graph G, as shown below, that consists of several connected components.

The adjacency matrix of G is

$$\mathbf{A} = \begin{bmatrix} 1 & 0 & 1 & 0 & 0 & 0 & 0 \\ 0 & 0 & 2 & 0 & 0 & 0 & 0 \\ 1 & 2 & 0 & 0 & 0 & 0 & 0 \\ 0 & 0 & 0 & 0 & 1 & 0 & 0 \\ 0 & 0 & 0 & 1 & 1 & 0 & 0 \\ 0 & 0 & 0 & 0 & 0 & 0 & 2 \\ 0 & 0 & 0 & 0 & 0 & 2 & 0 \end{bmatrix}.$$

As you can see, $\mathbf{A}$ consists of square matrix blocks (of different sizes) down its diagonal and blocks of 0's everywhere else. The reason is that vertices in each connected component share no edges with vertices in other connected components. For instance, since v_1, v_2, and v_3 share no edges with v_4, v_5, v_6, or v_7, all entries in the top three rows to the right of the third column are 0 and all entries in the left three columns below the third row are also 0. Sometimes matrices whose entries are all 0's are themselves denoted 0. If this convention is followed here, $\mathbf{A}$ is written as

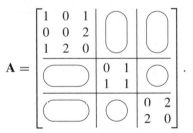

The above reasoning can be generalized to prove the following theorem:

Theorem 11.3.1

Let G be a graph with connected components $G_1, G_2, \ldots, G_k$. If there are n_i vertices in each connected component G_i and these vertices are numbered consecutively, then the adjacency matrix of G has the form

$$\begin{bmatrix} A_1 & O & O & \cdots & O & O \\ O & A_2 & O & \cdots & O & O \\ O & O & A_3 & \cdots & O & O \\ \vdots & \vdots & \vdots & & \vdots & \vdots \\ O & O & O & \cdots & O & A_k \end{bmatrix}$$

where each A_i is the $n_i \times n_i$ adjacency matrix of G_i, for all $i = 1, 2, \ldots, k$, and the O's represent matrices whose entries are all 0.

Matrix Multiplication

Matrix multiplication is an enormously useful operation that arises in many contexts, including the investigation of walks in graphs. Although matrix multiplication can be defined in quite abstract settings, the definition for matrices whose entries are real numbers will be sufficient for our applications. The product of two matrices is built up of *scalar* or *dot* products of their individual rows and columns.

• **Definition**

Suppose that all entries in matrices **A** and **B** are real numbers. If the number of elements, n, in the ith row of **A** equals the number of elements in the jth column of **B**, then the **scalar product** or **dot product** of the ith row of **A** and the jth column of **B** is the real number obtained as follows:

$$[a_{i1} \quad a_{i2} \quad \cdots \quad a_{in}] \begin{bmatrix} b_{1j} \\ b_{2j} \\ \vdots \\ b_{nj} \end{bmatrix} = a_{i1}b_{1j} + a_{i2}b_{2j} + \cdots + a_{in}b_{nj}.$$

Example 11.3.6 Multiplying a Row and a Column

$$[3 \quad 0 \quad -1 \quad 2] \begin{bmatrix} -1 \\ 2 \\ 3 \\ 0 \end{bmatrix} = 3 \cdot (-1) + 0 \cdot 2 + (-1) \cdot 3 + 2 \cdot 0$$

$$= -3 + 0 - 3 + 0 = -6 \qquad \blacksquare$$

More generally, if **A** and **B** are matrices whose entries are real numbers and if **A** and **B** have *compatible sizes* in the sense that the number of columns of **A** equals the number of rows of **B**, then the product **AB** is defined. It is the matrix whose ijth entry is the scalar product of the ith row of **A** times the jth column of **B**, for all possible values of i and j.

• **Definition**

Let $\mathbf{A} = (a_{ij})$ be an $m \times k$ matrix and $\mathbf{B} = (b_{ij})$ a $k \times n$ matrix with real entries. The (matrix) product of **A** times **B**, denoted **AB**, is that matrix (c_{ij}) defined as follows:

where

$$c_{ij} = a_{i1}b_{1j} + a_{i2}b_{2j} + \cdots + a_{ik}b_{kj} = \sum_{r=1}^{k} a_{ir}b_{rj},$$

for all $i = 1, 2, \ldots, m$ and $j = 1, 2, \ldots, n$.

Example 11.3.7 Computing a Matrix Product

Let $\mathbf{A} = \begin{bmatrix} 2 & 0 & 3 \\ -1 & 1 & 0 \end{bmatrix}$ and $\mathbf{B} = \begin{bmatrix} 4 & 3 \\ 2 & 2 \\ -2 & -1 \end{bmatrix}$. Compute $\mathbf{AB}$.

Solution $\mathbf{A}$ has size 2×3 and $\mathbf{B}$ has size 3×2, so the number of columns of $\mathbf{A}$ equals the number of rows of $\mathbf{B}$ and the matrix product of $\mathbf{A}$ and $\mathbf{B}$ can be computed. Then

$$\begin{bmatrix} 2 & 0 & 3 \\ -1 & 1 & 0 \end{bmatrix} \begin{bmatrix} 4 & 3 \\ 2 & 2 \\ -2 & -1 \end{bmatrix} = \begin{bmatrix} c_{11} & c_{12} \\ c_{21} & c_{22} \end{bmatrix},$$

where

$$c_{11} = 2 \cdot 4 + 0 \cdot 2 + 3 \cdot (-2) = 2 \qquad \begin{bmatrix} \boxed{2 \quad 0 \quad 3} \\ -1 \quad 1 \quad 0 \end{bmatrix} \begin{bmatrix} 4 & 3 \\ 2 & 2 \\ -2 & -1 \end{bmatrix}$$

$$c_{12} = 2 \cdot 3 + 0 \cdot 2 + 3 \cdot (-1) = 3 \qquad \begin{bmatrix} \boxed{2 \quad 0 \quad 3} \\ -1 \quad 1 \quad 0 \end{bmatrix} \begin{bmatrix} 4 & 3 \\ 2 & 2 \\ -2 & -1 \end{bmatrix}$$

$$c_{21} = (-1) \cdot 4 + 1 \cdot 2 + 0 \cdot (-2) = 2 \qquad \begin{bmatrix} 2 \quad 0 \quad 3 \\ \boxed{-1 \quad 1 \quad 0} \end{bmatrix} \begin{bmatrix} 4 & 3 \\ 2 & 2 \\ -2 & -1 \end{bmatrix}$$

$$c_{22} = (-1) \cdot 3 + 0 \cdot 2 + 3 \cdot (-1) = -1 \qquad \begin{bmatrix} 2 \quad 0 \quad 3 \\ \boxed{-1 \quad 1 \quad 0} \end{bmatrix} \begin{bmatrix} 4 & 3 \\ 2 & 2 \\ -2 & -1 \end{bmatrix}.$$

Hence

$$\mathbf{AB} = \begin{bmatrix} 2 & 3 \\ -2 & -1 \end{bmatrix}. \qquad \blacksquare$$

Matrix multiplication is both similar to and different from multiplication of real numbers. One difference is that although the product of any two numbers can be formed, only matrices with compatible sizes can be multiplied. Also, multiplication of real numbers is commutative (for all real numbers a and b, $a \cdot b = b \cdot a$), whereas matrix multiplication is not. For instance,

$$\begin{bmatrix} 1 & 1 \\ 0 & 1 \end{bmatrix} \begin{bmatrix} 0 & 1 \\ 0 & 1 \end{bmatrix} = \begin{bmatrix} 0 & 2 \\ 0 & 1 \end{bmatrix}, \quad \text{but} \quad \begin{bmatrix} 0 & 1 \\ 0 & 1 \end{bmatrix} \begin{bmatrix} 1 & 1 \\ 0 & 1 \end{bmatrix} = \begin{bmatrix} 0 & 1 \\ 0 & 1 \end{bmatrix}.$$

On the other hand, both real number and matrix multiplications are associative ($(ab)c = a(bc)$, for all elements a, b and c for which the products are defined). This is proved in Example 11.3.8 for products of 2×2 matrices. Additional exploration of matrix multiplication is offered in the exercises.

Example 11.3.8 Associativity of Matrix Multiplication for 2×2 Matrices

Prove that if $\mathbf{A}$, $\mathbf{B}$, and $\mathbf{C}$ are 2×2 matrices over the set of real numbers, then $(\mathbf{AB})\mathbf{C} = \mathbf{A}(\mathbf{BC})$.

Solution Suppose $\mathbf{A} = (a_{ij})$, $\mathbf{B} = (b_{ij})$, and $\mathbf{C} = (c_{ij})$ are particular but arbitrarily chosen 2×2 matrices with real entries. Since the numbers of rows and columns are all the same, $\mathbf{AB}$, $\mathbf{BC}$, $(\mathbf{AB})\mathbf{C}$, and $\mathbf{A}(\mathbf{BC})$ are defined. Let $\mathbf{AB} = (d_{ij})$ and $\mathbf{BC} = (e_{ij})$. Then for all integers $i = 1, 2$ and $j = 1, 2$,

$$\text{the } ij\text{th entry of } (\mathbf{AB})\mathbf{C} = \sum_{r=1}^{2} d_{ir} c_{rj} \qquad \text{by definition of the product of } \mathbf{AB} \text{ and } \mathbf{C}$$

$$= d_{i1} c_{1j} + d_{i2} c_{2j} \qquad \text{by definition of } \Sigma$$

$$= \left(\sum_{r=1}^{2} a_{ir} b_{r1} \right) c_{1j} + \left(\sum_{r=1}^{2} a_{ir} b_{r2} \right) c_{2j} \qquad \text{by definition of the product of } \mathbf{A} \text{ and } \mathbf{B}$$

$$= (a_{i1} b_{11} + a_{i2} b_{21}) c_{1j} \qquad \text{by definition of } \Sigma$$
$$+ (a_{i1} b_{12} + a_{i2} b_{22}) c_{2j}$$

$$= a_{i1} b_{11} c_{1j} + a_{i2} b_{21} c_{1j} + a_{i1} b_{12} c_{2j} + a_{i2} b_{22} c_{2j}.$$

Similarly, the ijth entry of $\mathbf{A}(\mathbf{BC})$ is

$$(\mathbf{A}(\mathbf{BC}))_{ij} = \sum_{r=1}^{2} a_{ir} e_{rj}$$

$$= a_{i1} e_{1j} + a_{i2} e_{2j}$$

$$= a_{i1} \left(\sum_{r=1}^{2} b_{1r} c_{rj} \right) + a_{i2} \left(\sum_{r=1}^{2} b_{2r} c_{rj} \right)$$

$$= a_{i1} (b_{11} c_{1j} + b_{12} c_{2j}) + a_{i2} (b_{21} c_{1j} + b_{22} c_{2j})$$

$$= a_{i1} b_{11} c_{1j} + a_{i1} b_{12} c_{2j} + a_{i2} b_{21} c_{1j} + a_{i2} b_{22} c_{2j}$$

$$= a_{i1} b_{11} c_{1j} + b_{i2} b_{21} c_{1j} + a_{i1} b_{12} c_{2j} + a_{i2} b_{22} c_{2j}.$$

Comparing the results of the two computations shows that for all i and j,

the ijth entry of $(\mathbf{AB})\mathbf{C}$ = the ijth entry of $\mathbf{A}(\mathbf{BC})$.

Since all corresponding entries are equal, $(\mathbf{AB})\mathbf{C} = \mathbf{A}(\mathbf{BC})$, as was to be shown. ∎

As far as multiplicative identities are concerned, there are both similarities and differences between real numbers and matrices. You know that the number 1 acts as a multiplicative identity for products of real numbers. It turns out that there are certain matrices, called *identity matrices,* that act as multiplicative identities for certain matrix products. For instance, mentally perform the following matrix multiplications to check that for any real numbers a, b, c, d, e, f, g, h and i,

$$\begin{bmatrix} 1 & 0 \\ 0 & 1 \end{bmatrix} \begin{bmatrix} a & b & c \\ d & e & f \end{bmatrix} = \begin{bmatrix} a & b & c \\ d & e & f \end{bmatrix}$$

and

$$\begin{bmatrix} a & b & c \\ d & e & f \\ g & h & i \end{bmatrix} \begin{bmatrix} 1 & 0 & 0 \\ 0 & 1 & 0 \\ 0 & 0 & 1 \end{bmatrix} \begin{bmatrix} a & b & c \\ d & e & f \\ g & h & i \end{bmatrix}.$$

These computations show that $\begin{bmatrix} 1 & 0 \\ 0 & 1 \end{bmatrix}$ acts as an identity on the left side for multiplication with 2×3 matrices and that $\begin{bmatrix} 1 & 0 & 0 \\ 0 & 1 & 0 \\ 0 & 0 & 1 \end{bmatrix}$ acts as an identity on the right side for multiplication with 3×3 matrices. Note that $\begin{bmatrix} 1 & 0 \\ 0 & 1 \end{bmatrix}$ cannot act as an identity on the right side for multiplication with 2×3 matrices because the sizes are not compatible.

Leopold Kronecker
(1823–1891)

• Definition

For each positive integer n, the $n \times n$ **identity matrix,** denoted $\mathbf{I}_n = (\delta_{ij})$ or just $\mathbf{I}$ (if the size of the matrix is obvious from context), is the $n \times n$ matrix in which all the entries in the main diagonal are 1's and all other entries are 0's. In other words,

$$\delta_{ij} = \begin{cases} 1 & \text{if } i = j \\ 0 & \text{if } i \neq j \end{cases}, \quad \text{for all } i, j = 1, 2, \ldots, n.$$

The German mathematician Leopold Kronecker introduced the symbol δ_{ij} to make matrix computations more convenient. In his honor, this symbol is called the *Kronecker delta*.

Example 11.3.9 An Identity Matrix Acts As an Identity

Prove that if $\mathbf{A}$ is any $m \times n$ matrix and $\mathbf{I}$ is the $n \times n$ identity matrix, then $\mathbf{AI} = \mathbf{A}$. (In exercise 14 at the end of this section you are asked to show that if $\mathbf{I}$ is the $m \times m$ identity matrix, then $\mathbf{IA} = \mathbf{A}$.)

Proof:

Let $\mathbf{A}$ be any $n \times n$ matrix and let A_{ij} be the ijth entry of $\mathbf{A}$ for all integers $i = 1, 2, \ldots, m$ and $j = 1, 2, \ldots, n$. Consider the product $\mathbf{AI}$, where $\mathbf{I}$ is the $n \times n$ identity matrix. Observe that

$$\begin{bmatrix} a_{11} & a_{12} & \cdots & a_{1n} \\ a_{21} & a_{22} & \cdots & 2_{2n} \\ \vdots & \vdots & & \vdots \\ a_{m1} & a_{m2} & \cdots & a_{mn} \end{bmatrix} \begin{bmatrix} 1 & 0 & \cdots & 0 \\ 0 & 1 & \cdots & 0 \\ \vdots & \vdots & & \vdots \\ 0 & 0 & \cdots & 1 \end{bmatrix} = \begin{bmatrix} a_{11} & a_{12} & \cdots & a_{1n} \\ a_{21} & a_{22} & \cdots & a_{2n} \\ \vdots & \vdots & & \vdots \\ a_{m1} & a_{m2} & \cdots & a_{mn} \end{bmatrix}$$

because

$$\text{the } ij\text{th entry of } \mathbf{AI} = \sum_{r=1}^{n} a_{ir} \delta_{rj} \qquad \text{by definition of 1}$$

$$= a_{i1}\delta_{1j} + a_{i2}\delta_{2j} + \cdots \qquad \text{by definition of } \Sigma$$
$$\quad + a_{ij}\delta_{jj} + \cdots + a_{in}\delta_{nj}$$

$$= a_{ij}\delta_{jj} \qquad \text{since } \delta_{kj} = 0 \text{ whenever } k \neq j \text{ and } \delta_{jj} = 1$$

$$= a_{ij}$$

$$= \text{the } ij\text{th entry of } \mathbf{A}.$$

Thus $\mathbf{AI} = \mathbf{A}$, as was to be shown. ∎

There are also similarities and differences between real numbers and matrices with respect to the computation of powers. Any number can be raised to a nonnegative integer power, but a matrix can be multiplied by itself only if it has the same number of rows as columns. As for real numbers, however, the definition of matrix powers is recursive. Just as any number to the zero power is defined to be 1, so any $n \times n$ matrix to the zero power is defined to be the $n \times n$ identity matrix. The nth power of an $n \times n$ matrix $\mathbf{A}$ is defined to be the product of $\mathbf{A}$ with its $(n - 1)$st power.

> **• Definition**
>
> For any $n \times n$ matrix A, the **powers of A** are defined as follows:
>
> $$\mathbf{A}^0 = \mathbf{I} \quad \text{where } \mathbf{I} \text{ is the } n \times n \text{ identity matrix}$$
>
> $$\mathbf{A}^n = \mathbf{A}\mathbf{A}^{n-1} \quad \text{for all integers } n \geq 1$$

Example 11.3.10   Powers of a Matrix

Let $\mathbf{A} = \begin{bmatrix} 1 & 2 \\ 2 & 0 \end{bmatrix}$. Compute $\mathbf{A}^0$, $\mathbf{A}^1$, $\mathbf{A}^2$, and $\mathbf{A}^3$.

Solution

$$\mathbf{A}^0 = \text{the } 2 \times 2 \text{ identity matrix} = \begin{bmatrix} 1 & 0 \\ 0 & 1 \end{bmatrix}$$

$$\mathbf{A}^1 = \mathbf{A}\mathbf{A}^0 = \mathbf{A}\mathbf{I} = \mathbf{A}$$

$$\mathbf{A}^2 = \mathbf{A}\mathbf{A}^1 = \mathbf{A}\mathbf{A} = \begin{bmatrix} 1 & 2 \\ 2 & 0 \end{bmatrix}\begin{bmatrix} 1 & 2 \\ 2 & 0 \end{bmatrix} = \begin{bmatrix} 5 & 2 \\ 2 & 4 \end{bmatrix}$$

$$\mathbf{A}^3 = \mathbf{A}\mathbf{A}^2 = \begin{bmatrix} 1 & 2 \\ 2 & 0 \end{bmatrix}\begin{bmatrix} 5 & 2 \\ 2 & 4 \end{bmatrix} = \begin{bmatrix} 9 & 10 \\ 10 & 4 \end{bmatrix}$$

■

Counting Walks of Length N

A walk in a graph consists of an alternating sequence of vertices and edges. If repeated edges are counted each time they occur, then the number of edges in the sequence is called the **length** of the walk. For instance, the walk $v_2 e_3 v_3 e_4 v_2 e_2 v_2 e_3 v_3$ has length 4 (counting e_3 twice). Consider the following graph G:

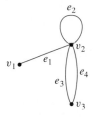

How many distinct walks of length 2 connect v_2 and v_2? Your can list the possibilities systematically as follows: From v_1, the first edge of the walk must go to *some* vertex of G: v_1, v_2, or v_3. There is one walk of length 2 from v_2 to v_2 that starts by going from v_2 to v_1:

$$v_2 e_1 v_1 e_1 v_2.$$

There is one walk of length 2 from v_2 to v_2 that starts by going from v_2 to v_2:

$$v_2 e_2 v_2 e_2 v_2.$$

And there are four walks of length 2 from v_2 to v_2 that start by going from v_2 to v_3:

$$v_2 e_3 v_3 e_4 v_2,$$

$$v_2 e_4 v_3 e_3 v_2,$$

$$v_2 e_3 v_3 e_3 v_2,$$

$$v_2 e_4 v_3 e_4 v_2.$$

Thus the answer is six.

The general question of finding the number of walks that have a given length and connect two particular vertices of a graph can easily be answered using matrix multiplication. Consider the adjacency matrix $\mathbf{A}$ of the graph G on the previous page:

$$\mathbf{A} = \begin{array}{c} \\ v_1 \\ v_2 \\ v_3 \end{array} \overset{\begin{array}{ccc} v_1 & v_2 & v_3 \end{array}}{\begin{bmatrix} 0 & 1 & 0 \\ 1 & 1 & 2 \\ 0 & 2 & 0 \end{bmatrix}}.$$

Compute $\mathbf{A}^2$ as follows:

$$\begin{bmatrix} 0 & 1 & 0 \\ 1 & 1 & 2 \\ 0 & 2 & 0 \end{bmatrix} \begin{bmatrix} 0 & 1 & 0 \\ 1 & 1 & 2 \\ 0 & 2 & 0 \end{bmatrix} = \begin{bmatrix} 1 & 1 & 2 \\ 1 & 6 & 2 \\ 2 & 2 & 4 \end{bmatrix}.$$

Note that the entry in the second row and the second column is 6, which equals the number of walks of length 2 from v_2 to v_2. This is no accident! To compute a_{22}, you multiply the second row of $\mathbf{A}$ times the second column of $\mathbf{A}$ to obtain a sum of three terms:

$$\begin{bmatrix} 1 & 1 & 2 \end{bmatrix} \begin{bmatrix} 1 \\ 1 \\ 2 \end{bmatrix} = 1 \cdot 1 + 1 \cdot 1 + 2 \cdot 2.$$

Observe that

$$\begin{bmatrix} \text{the first term} \\ \text{of this sum} \end{bmatrix} = \begin{bmatrix} \text{number of} \\ \text{edges from} \\ v_2 \text{ to } v_1 \end{bmatrix} \cdot \begin{bmatrix} \text{number of} \\ \text{edges from} \\ v_1 \text{ to } v_2 \end{bmatrix} = \begin{bmatrix} \text{number of pairs} \\ \text{of edges from} \\ v_2 \text{ to } v_1 \text{ and } v_1 \text{ to } v_2 \end{bmatrix}.$$

Now consider the ith term of this sum, for each $i = 1, 2$, and 3. It equals the number of edges from v_2 to v_i times the number of edges from v_i to v_2. By the multiplication rule this equals the number of pairs of edges from v_2 to v_i and from v_i back to v_2. But this equals the number of walks of length 2 that start and end at v_2 and pass through v_i. Since this analysis holds for each term of the sum for $i = 1, 2$, and 3, the sum as a whole equals the total number of walks of length 2 that start and end at v_2:

$$1 \cdot 1 + 1 \cdot 1 + 2 \cdot 2 = 1 + 1 + 4 = 6.$$

More generally, if $\mathbf{A}$ is the adjacency matrix of a graph G, the ijth entry of $\mathbf{A}^2$ equals the number of walks of length 2 connecting the ith vertex to the jth vertex of G. Even more generally, if n is any positive integer, the ijth entry of $\mathbf{A}^n$ equals the number of walks of length n connecting the ith and the jth vertices of G.

Theorem 11.3.2

If G is a graph with vertices $v_1, v_2, \ldots, v_m$ and $\mathbf{A}$ is the adjacency matrix of G, then for each positive integer n,

the ijth entry of $\mathbf{A}^n$ = the number of walks of length n from v_i to v_j
for all integers $i, j = 1, 2, \ldots, m$.

Proof:

Suppose G is a graph with vertices $v_1, v_2, \ldots, v_m$ and $\mathbf{A}$ is the adjacency matrix of G. We use mathematical induction to show that the following property holds for each positive integer n: For all integers $i, j = 1, 2, \ldots, m$,

the ijth entry of $\mathbf{A}^n$ = the number of walks of length n from v_i to v_j.

Show that the property holds for n = 1:

The ijth entry of $\mathbf{A}^1$ = the ijth entry of $\mathbf{A}$ because $\mathbf{A}^1 = \mathbf{A}$

                 = the number of edges   by definition of adjacency matrix
connecting v_i to v_j

                 = the number of walks of   because a walk of length 1
length 1 from v_i to v_j contains a single edge.

Show that for all integers k with k > 1, if the property holds for n = k, then it holds for n = k + 1: Suppose that for some integer $k \geq 1$, the ijth entry of $\mathbf{A}^k$ = the number of walks of length k from v_i to v_j. [*This is the inductive hypothesis.*] We must show that the ijth entry of $\mathbf{A}^{k+1}$ = the number of walks of length $k + 1$ from v_i to v_j.

Let $\mathbf{A} = (a_{ij})$ and $\mathbf{A}^k = (b_{ij})$. Since $\mathbf{A}^{k+1} = \mathbf{A}\mathbf{A}^k$, the ijth entry of $\mathbf{A}^{k+1}$ is obtained by multiplying the ith row of $\mathbf{A}$ by the jth column of $\mathbf{A}^k$:

$$\text{the } ij\text{th entry of } \mathbf{A}^{k+1} = a_{i1}b_{1j} + a_{i2}b_{2j} + \cdots + a_{im}b_{mj} \qquad \text{11.3.1}$$

for all $i, j = 1, 2, \ldots, m$. Now consider the individual terms of this sum: a_{i1} is the number of edges from v_i to v_1; and, by inductive hypothesis, b_{1j} is the number of walks of length k from v_1 to v_j. But any edge from v_i to v_1 can be joined with any walk of length k from v_1 to v_j to create a walk of length $k + 1$ from v_i to v_j with v_1 as its second vertex. Thus, by the multiplication rule,

$$a_{i1}b_{1j} = \begin{bmatrix} \text{the number of walks of length } k + 1 \text{ from} \\ v_i \text{ to } v_j \text{ that have } v_1 \text{ as their second vertex} \end{bmatrix}.$$

More generally, for each integer $r = 1, 2, \ldots, m$,

$$a_{ir}b_{rj} = \begin{bmatrix} \text{the number of walks of length } k + 1 \text{ from} \\ v_i \text{ to } v_j \text{ that have } v_r \text{ as their second vertex} \end{bmatrix}.$$

Since any walk of length $k + 1$ from v_i to v_j must have *one* of the vertices $v_1, v_2, \ldots,$ v_m as its second vertex, the total number of walks of length $k + 1$ from v_i to v_j equals the sum in (11.3.1), which equals the ijth entry of $\mathbf{A}^{k+1}$. Hence

the ijth entry of $\mathbf{A}^{k+1}$ = the number of walks of length $k + 1$ from v_i to v_j.

[*as was to be shown*].

Since both the basis step and the inductive step have been proved, the given equality is true for all integers $n \geq 1$.

Exercise Set 11.3

1. Find real numbers $a, b,$ and c such that the following are true.

    a. $\begin{bmatrix} a+b & a-c \\ c & b-a \end{bmatrix} = \begin{bmatrix} 1 & 0 \\ -1 & 3 \end{bmatrix}$

    b. $\begin{bmatrix} 2a & b+c \\ c-a & 2b-a \end{bmatrix} = \begin{bmatrix} 4 & 3 \\ 1 & -2 \end{bmatrix}$

2. Find the adjacency matrices for the following directed graphs.

3. Find directed graphs that have the following adjacency matrices:

a.
$$\begin{bmatrix} 1 & 0 & 1 & 2 \\ 0 & 0 & 1 & 0 \\ 0 & 2 & 1 & 1 \\ 0 & 1 & 1 & 0 \end{bmatrix}$$
b.
$$\begin{bmatrix} 0 & 1 & 0 & 0 \\ 2 & 0 & 1 & 0 \\ 1 & 2 & 1 & 0 \\ 0 & 0 & 1 & 0 \end{bmatrix}$$

4. Find adjacency matrices for the following (undirected) graphs.

a.

b.

c. K_4, the complete graph on four vertices
d. $K_{2,3}$, the complete bipartite graph on (2, 3) vertices

5. Find graphs that have the following adjacency matrices.

a.
$$\begin{bmatrix} 1 & 0 & 1 \\ 0 & 1 & 2 \\ 1 & 2 & 0 \end{bmatrix}$$
b.
$$\begin{bmatrix} 0 & 2 & 0 \\ 2 & 1 & 0 \\ 0 & 0 & 1 \end{bmatrix}$$

6. The following are adjacency matrices for graphs. In each case determine whether the graph is connected by analyzing the matrix without drawing the graph.

a.
$$\begin{bmatrix} 0 & 1 & 1 \\ 1 & 1 & 0 \\ 1 & 0 & 0 \end{bmatrix}$$
b.
$$\begin{bmatrix} 0 & 2 & 0 & 0 \\ 2 & 0 & 0 & 0 \\ 0 & 0 & 1 & 1 \\ 0 & 0 & 1 & 1 \end{bmatrix}$$

7. Suppose that for all i, all the entries in the ith row and ith column of the adjacency matrix of a graph are 0. What can you conclude about the graph?

8. Find each of the following products.

a. $\begin{bmatrix} 2 & -1 \end{bmatrix} \begin{bmatrix} 1 \\ 3 \end{bmatrix}$ b. $\begin{bmatrix} 4 & -1 & 7 \end{bmatrix} \begin{bmatrix} 1 \\ 2 \\ 0 \end{bmatrix}$

9. Find each of the following products.

a. $\begin{bmatrix} 3 & 0 \\ 1 & -2 \end{bmatrix} \begin{bmatrix} 1 & -1 & 4 \\ 0 & 2 & 1 \end{bmatrix}$

b. $\begin{bmatrix} 2 & 0 & 1 \\ 0 & -1 & 0 \end{bmatrix} \begin{bmatrix} 1 & 3 \\ 5 & -4 \\ -2 & 2 \end{bmatrix}$

c. $\begin{bmatrix} -1 \\ 2 \end{bmatrix} \begin{bmatrix} 2 & 3 \end{bmatrix}$

10. Let $\mathbf{A} = \begin{bmatrix} 1 & 1 & -1 \\ 0 & -2 & 1 \end{bmatrix}$, $\mathbf{B} = \begin{bmatrix} -2 & 0 \\ 1 & 3 \end{bmatrix}$, and $\mathbf{C} = \begin{bmatrix} 0 & -2 \\ 3 & 1 \\ 1 & 0 \end{bmatrix}$.

For each of the following, determine whether the indicated product exists, and compute it if it does.

a. **AB**   b. **BA**   c. $\mathbf{A}^2$ d. **BC**   e. **CB**
f. $\mathbf{B}^2$ g. $\mathbf{B}^3$ h. $\mathbf{C}^2$ i. **AC**   j. **CA**

11. Give an example different from that in the text to show that matrix multiplication is not commutative. That is, find 2×2 matrices **A** and **B** such that **AB** and **BA** both exist but $\mathbf{AB} \neq \mathbf{BA}$.

12. Let O denote the matrix $\begin{bmatrix} 0 & 0 \\ 0 & 0 \end{bmatrix}$. Find 2×2 matrices **A** and **B** such that $\mathbf{A} \neq \mathrm{O}$ and $\mathbf{B} \neq \mathrm{O}$, but $\mathbf{AB} = \mathrm{O}$.

13. Let O denote the matrix $\begin{bmatrix} 0 & 0 \\ 0 & 0 \end{bmatrix}$. Find 2×2 matrices **A** and **B** such that $\mathbf{A} \neq \mathbf{B}$, $\mathbf{B} \neq \mathrm{O}$, and $\mathbf{AB} \neq \mathrm{O}$, but $\mathbf{BA} = \mathrm{O}$.

In 14–18 assume that the entries of all matrices are real numbers.

H 14. Prove that if **I** is the $m \times m$ identity matrix and **A** is any $m \times n$ matrix, then $\mathbf{IA} = \mathbf{A}$.

15. Prove that if **A** is an $m \times m$ symmetric matrix, then $\mathbf{A}^2$ is symmetric.

16. Prove that matrix multiplication is associative: If **A**, **B**, and **C** are any $m \times k$, $k \times r$, and $r \times n$ matrices, respectively, then $(\mathbf{AB})\mathbf{C} = \mathbf{A}(\mathbf{BC})$.

17. Use mathematical induction to prove that if **A** is any $m \times m$ matrix, then $\mathbf{A}^n \mathbf{A} = \mathbf{A} \mathbf{A}^n$ for all integers $n \geq 1$. (You will need to use the result of exercise 16.)

18. Use mathematical induction to prove that if **A** is an $m \times m$ symmetric matrix, then for any integer $n \geq 1$, $\mathbf{A}^n$ is also symmetric.

19. a. Let $\mathbf{A} = \begin{bmatrix} 1 & 1 & 2 \\ 1 & 0 & 1 \\ 2 & 1 & 0 \end{bmatrix}$. Find $\mathbf{A}^2$ and $\mathbf{A}^3$.

b. Let G be the graph with vertices v_1, v_2, and v_3 and with **A** as its adjacency matrix. Use the answers to part (a) to find the number of walks of length 2 from v_1 to v_3 and the number of walks of length 3 from v_1 to v_3. Do not draw G to solve this problem.

c. Examine the calculations you performed in answering part (a) to find five walks of length 2 from v_3 to v_3. Then draw G and find the walks by visual inspection.

20. The following is an adjacency matrix for a graph:

	v_1	v_2	v_3	v_4
v_1	0	1	1	0
v_2	1	0	2	1
v_3	1	2	0	1
v_4	0	1	1	1

Answer the following questions by examining the matrix and its powers only, not by drawing the graph:

a. How many walks of length 2 are there from v_2 to v_3?
b. How many walks of length 2 are there from v_3 to v_4?
c. How many walks of length 3 are there from v_1 to v_4?
d. How many walks of length 3 are there from v_2 to v_3?

21. Let **A** be the adjacent matrix for K_3, the complete graph on three vertices. Use mathematical induction to prove that for each positive integer n, all the entries along the main diag-

onal of $\mathbf{A}^n$ are equal to each other and that all the entries that do not lie along the main diagonal are equal to each other.

22. a. Draw a graph that has

$$\begin{bmatrix} 0 & 0 & 0 & 1 & 2 \\ 0 & 0 & 0 & 1 & 1 \\ 0 & 0 & 0 & 2 & 1 \\ 1 & 1 & 2 & 0 & 0 \\ 2 & 1 & 1 & 0 & 0 \end{bmatrix}$$

as its adjacency matrix. Is this graph bipartite? (For a definition of bipartite, see exercise 37 in Section 11.1.)

Definition: Given an $m \times n$ matrix $\mathbf{A}$ whose ijth entry is denoted a_{ij}, the **transpose of A** is the matrix $\mathbf{A}^t$ whose ijth entry is a_{ji}, for all $i = 1, 2, \ldots, m$ and $j = 1, 2, \ldots, n$.

Note that the first row of $\mathbf{A}$ becomes the first column of $\mathbf{A}^t$, the second row of $\mathbf{A}$ becomes the second column of $\mathbf{A}^t$, and so forth. For instance,

$$\text{if } \mathbf{A} = \begin{bmatrix} 0 & 2 & 1 \\ 1 & 2 & 3 \end{bmatrix}, \text{ then } \mathbf{A}^t = \begin{bmatrix} 0 & 1 \\ 2 & 2 \\ 1 & 3 \end{bmatrix}.$$

H **b.** Show that a graph with n vertices is bipartite if, and only if, for some labeling of its vertices, its adjacency matrix has the form

$$\begin{bmatrix} \mathbf{O} & \mathbf{A} \\ \mathbf{A}^t & \mathbf{O} \end{bmatrix}$$

where $\mathbf{A}$ is a $k \times (n-k)$ matrix for some integer k such that $0 < k < n$, the top left O represents a $k \times k$ matrix all of whose entries are 0, $\mathbf{A}^t$ is the transpose of $\mathbf{A}$, and the bottom right O represents an $(n-k) \times (n-k)$ matrix all of whose entries are 0.

23. a. Let G be a graph with n vertices, and let v and w be distinct vertices of G. Prove that if there is a walk from v to w, then there is a walk from v to w that has length less than or equal to $n - 1$.

H **b.** If $\mathbf{A} = (a_{ij})$ and $\mathbf{B} = (b_{ij})$ are any $m \times n$ matrices, the matrix $\mathbf{A} + \mathbf{B}$ is the $m \times n$ matrix whose ijth entry is $a_{ij} + b_{ij}$ for all $i = 1, 2, \ldots, m$ and $j = 1, 2, \ldots, n$. Let G be a graph with n vertices where $n > 1$, and let $\mathbf{A}$ be the adjacency matrix of G. Prove that G is connected if, and only if, every entry of $\mathbf{A} + \mathbf{A}^2 + \cdots + \mathbf{A}^{n-1}$ is positive.

11.4 Isomorphisms of Graphs

Thinking is a momentary dismissal of irrelevancies. — R. Buckminster Fuller, 1969

Recall from Section 11.1 that the two drawings shown in Figure 11.4.1 both represent the same graph: Their vertex and edge sets are identical, and their edge-endpoint functions are the same.

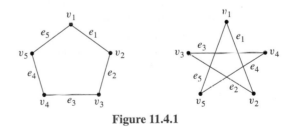

Figure 11.4.1

Call this graph G. Now consider the graph G' represented in Figure 11.4.2.

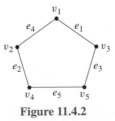

Figure 11.4.2

Observe that G' is a different graph from G (for instance, in G the endpoints of e_1 are v_1 and v_2, whereas in G' the endpoints of e_1 are v_1 and v_3). Yet G' is certainly very similar to G. In fact, if the vertices and edges of G' are relabeled by the functions shown in Figure 11.4.3, then G' becomes the same as G.

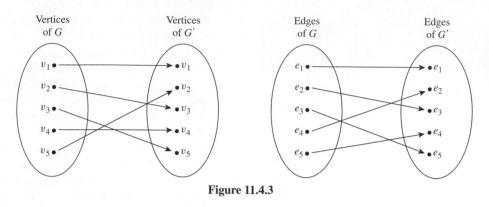

Figure 11.4.3

Note that these relabeling functions are one-to-one and onto.

Two graphs that are the same except for the labeling of their vertices and edges are called *isomorphic*. The word *isomorphism* comes from the Greek meaning "same form." Isomorphic graphs are those that have essentially the same form.

• Definition

Let G and G' be graphs with vertex sets $V(G)$ and $V(G')$ and edge sets $E(G)$ and $E(G')$, respectively. **G is isomorphic to G'** if, and only if, there exist one-to-one correspondences $g: V(G) \rightarrow V(G')$ and $h: E(G) \rightarrow E(G')$ that preserve the edge-endpoint functions of G and G' in the sense that for all $v \in V(G)$ and $e \in E(G)$,

$$v \text{ is an endpoint of } e \quad \Leftrightarrow \quad g(v) \text{ is an endpoint of } h(e).$$ 11.4.1

In words, G is isomorphic to G' if, and only if, the vertices and edges of G and G' can be matched up by one-to-one, onto functions such that the edges between corresponding vertices correspond to each other.

It is common in mathematics to identify objects that are isomorphic. For instance, if we are given a graph G with five vertices such that each pair of vertices is connected by an edge, then we may identify G with K_5, saying that G is K_5 rather than that G is isomorphic to K_5.

Example 11.4.1 Showing That Two Graphs Are Isomorphic

Show that the following two graphs are isomorphic.

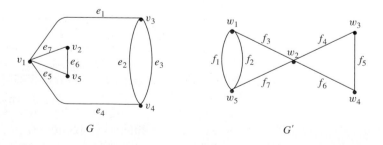

Solution   To solve this problem, you must find functions $g\colon V(G) \to V(G')$ and $h\colon E(G) \to E(G')$ such that for all $v \in V(G)$ and $e \in E(G)$, v is an endpoint of e if, and only if, $g(v)$ is an endpoint of $h(e)$. Setting up such functions is partly a matter of trial and error and partly a matter of deduction. For instance, since e_2 and e_3 are parallel (have the same endpoints), $h(e_2)$ and $h(e_3)$ must be parallel also. So $h(e_2) = f_1$ and $h(e_3) = f_2$ or $h(e_2) = f_2$ and $h(e_3) = f_1$. Also, the endpoints of e_2 and e_3 must correspond to the endpoints of f_1 and f_2, and so $g(v_3) = w_1$ and $g(v_4) = w_5$ or $g(v_3) = w_5$ and $g(v_4) = w_1$.

Similarly, since v_1 is the endpoint of four distinct edges ($e_1, e_7, e_5,$ and e_4), $g(v_1)$ must also be the endpoint of four distinct edges (because every edge incident on $g(v_1)$ is the image under h of an edge incident on v_1 and h is one-to-one and onto). But the only vertex in G' that has four edges coming out of it is w_2, and so $g(v_1) = w_2$. Now if $g(v_3) = w_1$, then since v_1 and v_3 are endpoints of e_1 in G, $g(v_1) = w_2$ and $g(v_3) = w_1$ must be endpoints of $h(e_1)$ in G'. This implies that $h(e_1) = f_3$.

By continuing in this way, possibly making some arbitrary choices as you go, you eventually can find functions g and h to define the isomorphism between G and G'. One pair of functions (there are several) is the following:

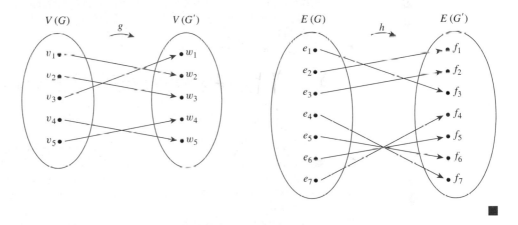

It is not hard to show that graph isomorphism is an equivalence relation on a set of graphs; in other words, it is reflexive, symmetric, and transitive. To prove the reflexive property, it must be shown that any graph is isomorphic to itself. Such an isomorphism can be defined using the identity functions on the set of vertices and on the set of edges.

To prove that graph isomorphism is symmetric, it must be shown that if a graph G is isomorphic to a graph G', then G' is isomorphic to G. But this is true because if g and h are vertex and edge correspondences from G to G' that preserve the edge-endpoint functions, then g^{-1} and h^{-1} are vertex and edge correspondences from G' to G that preserve the edge-endpoint functions. Note that as a consequence of the symmetry property, you can simply say "G and G' are isomorphic" instead of "G is isomorphic to G'" or "G' is isomorphic to G."

Finally, to establish that graph isomorphism is transitive, it must be shown that if a graph G is isomorphic to a graph G' and if G' is isomorphic to G'', then G is isomorphic to G''. But this follows from the fact that if g_1 and h_1 are vertex and edge correspondences from G to G' that preserve the edge-endpoint functions of G and G' and g_2 and h_2 are vertex and edge correspondences from G' to G'' that preserve the edge-endpoint functions of G' and G'', then $g_2 \circ g_1$ and $h_2 \circ h_1$ are vertex and edge correspondences from G to G'' that preserve the edge-endpoint functions of G and G''.

Example 11.4.2 Finding Representatives of Isomorphism Classes

Find all nonisomorphic graphs that have two vertices and two edges. In other words, find a collection of representative graphs with two vertices and two edges such that every such graph is isomorphic to one in the collection.

Solution   There are four nonisomorphic graphs that have two vertices and two edges. These can be drawn without vertex and edge labels because any two labelings give isomorphic graphs.

(a)   (b)   (c)   (d)

To see that these four drawings show all the nonisomorphic graphs that have two vertices and two edges, first note whether one of the edges joins the two vertices or not. If it does, there are two possibilities: The other edge can also join the two vertices (as in (a)) or it can be a loop incident on one of them (as in (b))—it makes no difference *which* vertex is chosen to have the loop because interchanging the two vertex labels gives isomorphic graphs). If neither edge joins the two vertices, then both edges are loops. In this case, there are only two possibilities: Either both loops are incident on the same vertex (as in (c)) or the two loops are incident on separate vertices (as in (d)). There are no other possibilities for placing the edges, so the listing is complete. ■

Now consider the question, "Is there a general method to figure out whether graphs G and G' are isomorphic?" In other words, is there some algorithm that will accept graphs G and G' as input and produce a statement as to whether they are isomorphic? In fact, there is such an algorithm. It consists of generating all one-to-one, onto functions from the set of vertices of G to the set of vertices of G' and from the set of edges of G to the set of edges of G' and checking each pair to determine whether it preserves the edge-endpoint functions of G and G'. The problem with this algorithm is that it takes an unreasonably long time to perform, even on a high-speed computer. If G and G' each have n vertices and m edges, the number of one-to-one correspondences from vertices to vertices is $n!$ and the number of one-to-one correspondences from edges to edges is $m!$, so the total number of pairs of functions to check is $n! \cdot m!$. For instance, if $m = n = 20$, there would be $20! \cdot 20! \cong 5.9 \times 10^{36}$ pairs to check. Assuming that each check takes just 1 nanosecond, the total time would be approximately 1.9×10^{20} years!

Unfortunately, there is no more efficient general method known for checking whether two graphs are isomorphic. However, there are some simple tests that can be used to show that certain pairs of graphs are *not* isomorphic. For instance, if two graphs are isomorphic, then they have the same number of vertices (because there is a one-to-one correspondence from the vertex set of one graph to the vertex set of the other). It follows that if you are given two graphs, one with 16 vertices and the other with 17, you can immediately conclude that the two are not isomorphic. More generally, a property that is preserved by graph isomorphism is called an *isomorphic invariant*. For instance, "having 16 vertices" is an isomorphic invariant: If one graph has 16 vertices, then so does any graph that is isomorphic to it.

• **Definition**

A property P is called an **isomorphic invariant** if, and only if, given any graphs G and G', if G has property P and G' is isomorphic to G, then G' has property P.

Theorem 11.4.1

Each of the following properties is an invariant for graph isomorphism, where n, m, and k are all nonnegative integers:

1. has n vertices;
2. has m edges;
3. has a vertex of degree k;
4. has m vertices of degree k;
5. has a circuit of length k;
6. has a simple circuit of length k;
7. has m simple circuits of length k;
8. is connected;
9. has an Euler circuit;
10. has a Hamiltonian circuit.

Example 11.4.3 Showing That Two Graph Are Not Isomorphic

Show that the following pairs of graphs are not isomorphic by finding an isomorphic invariant that they do not share.

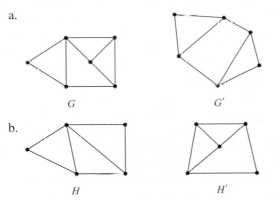

a.

G G'

b.

H H'

Solution

a. G has nine edges; G' has only eight.

b. H has a vertex of degree 4; H' does not. ∎

We prove part (3) of Theorem 11.4.1 below and leave the proofs of the other parts as exercises.

Example 11.4.4 Proof of Theorem 11.4.1, Part (3)

Prove that if G is a graph that has a vertex of degree k and G' is isomorphic to G, then G' has a vertex of degree k.

Proof:

Suppose G and G' are isomorphic graphs and G has a vertex v of degree k, where k is a nonnegative integer. [*We must show that G' has a vertex of degree k.*] Since G and G' are isomorphic, there are one-to-one, onto functions g and h from the vertices of G to the vertices of G' and from the edges of G to the edges of G' that preserve the edge-endpoint functions in the sense that for all edges e and all vertices u of G, u is an endpoint of e if, and only if, $g(u)$ is an endpoint of $h(e)$. An example for a particular vertex v is shown below.

Degree $v = 3 + 2 \cdot 2 = 7$ Degree $g(v) = 3 + 2 \cdot 2 = 7$

Let $e_1, e_2, \ldots, e_m$ be the m distinct edges that are incident on a vertex v in G, where m is a nonnegative integer. Then $h(e_1), h(e_2), \ldots, h(e_m)$ are m distinct edges that are incident on $g(v)$ in G'. [*The reason why $h(e_1), h(e_2), \ldots, h(e_m)$ are distinct is that h is one-to-one and $e_1, e_2, \ldots, e_m$ are distinct. And the reason why $h(e_1), h(e_2), \ldots, h(e_m)$ are incident on $g(v)$ is that g and h preserve the edge-endpoint functions of G and G' and $e_1, e_2, \ldots, e_m$ are incident on v.*]

Also, there are no edges incident on $g(v)$ other than the ones that are images under g of edges incident on v [*because g is onto and g and h preserve the edge-endpoint functions of G and G'*]. Thus the number of edges incident on v equals the number of edges incident on $g(v)$.

Finally, an edge e is a loop at v if, and only if, $h(e)$ is a loop at $g(v)$, so the number of loops incident on v equals the number of loops incident on $g(v)$. [*For since g and h preserve the edge-endpoint functions of G and G', a vertex w is an endpoint of e in G if, and only if, $g(w)$ is an endpoint of $h(e)$ in G'. It follows that v is the only endpoint of e in G if, and only if, $g(v)$ is the only endpoint of $h(e)$ in G'.*]

Now the degree of v, which is k, equals the number of edges incident on v plus the number of edges incident on v that are loops (since each loop contributes 2 to the degree of v). But we have already shown that the number of edges incident on v equals the number of edges incident on $g(v)$ and that the number of loops incident on v equals the number of loops incident on $g(v)$. Hence $g(v)$ also has degree k. ■

Graph Isomorphism for Simple Graphs

When graphs G and G' are both simple, the definition of G being isomorphic to G' can be written without referring to the correspondence between the edges of G and the edges of G'.

> **• Definition**
>
> If G and G' are simple graphs, then **G is isomorphic to G'** if, and only if, there exists a one-to-one correspondence g from the vertex set $V(G)$ of G to the vertex set $V(G')$ of G' that preserves the edge-endpoint functions of G and G' in the sense that for all vertices u and v of G,
>
> $$\{u, v\} \text{ is an edge in } G \quad \Leftrightarrow \quad \{g(u), g(v)\} \text{ is an edge in } G'. \qquad \text{11.4.2}$$

Example 11.4.5 Isomorphism of Simple Graphs

Are the two graphs shown below isomorphic? If so, define an isomorphism.

$$G \qquad\qquad G'$$

Solution   Yes. Define $f\colon V(G) \to V(G')$ by the arrow diagram shown below.

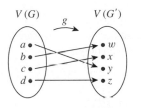

Then g is one-to-one and onto by inspection. The fact that g preserves the edge-endpoint functions of G and G' is shown by the following table:

Edges of G	Edges of G'
$\{a, b\}$	$\{y, w\} = \{g(a), g(b)\}$
$\{a, c\}$	$\{y, x\} = \{g(a), g(c)\}$
$\{a, d\}$	$\{y, z\} = \{g(a), g(d)\}$
$\{c, d\}$	$\{x, z\} = \{g(c), g(d)\}$

Exercise Set 11.4

For each pair of graphs G and G' in 1–5, determine whether G and G' are isomorphic. If they are, give functions $g\colon V(G) \to V(G')$ and $h\colon E(G) \to E(G')$ that define the isomorphism. If they are not, give an isomorphic invariant that they do not share.

1.

2.

3.

$$G \qquad\qquad G'$$

4.

5.

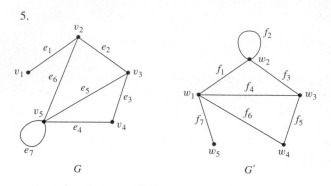

For each pair of simple graphs G and G' in 6–13, determine whether G and G' are isomorphic. If they are, give a function $g: V(G) \to V(G')$ that defines the isomorphism. If they are not, give an isomorphic invariant that they do not share.

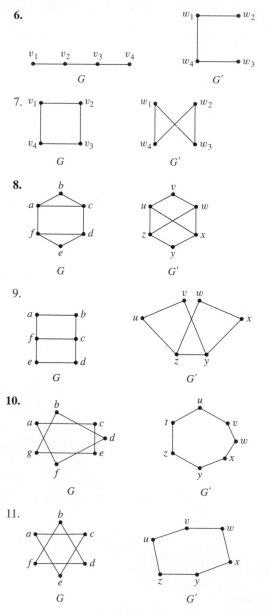

6.

7.

8.

9.

10.

11.

12.

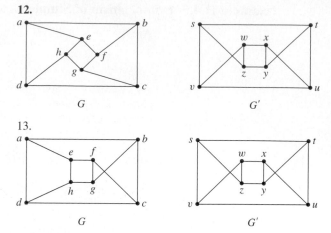

13.

14. Draw all nonisomorphic simple graphs with three vertices.

15. Draw all nonisomorphic simple graphs with four vertices.

16. Draw all nonisomorphic graphs with three vertices and no more than two edges.

17. Draw all nonisomorphic graphs with four vertices and no more than two edges.

H 18. Draw all nonisomorphic graphs with four vertices and three edges.

19. Draw all nonisomorphic graphs with six vertices, all having degree 2.

20. Draw four nonisomorphic graphs with six vertices, two of degree 4 and four of degree 3.

Prove that each of the properties in 21–29 is an invariant for graph isomorphism. Assume that n, m, and k are all nonnegative integers.

21. Has n vertices

22. Has m edges

23. Has a circuit of length k

24. Has a simple circuit of length k

H 25. Has m vertices of degree k

26. Has m simple circuits of length k

H 27. Is connected

28. Has an Euler circuit

29. Has a Hamiltonian circuit

30. Show that the following two graphs are not isomorphic by supposing they are isomorphic and deriving a contradiction.

11.5 Trees

A fool sees not the same tree that a wise man sees. — William Blake, 1757–1827

If a friend asks what you are studying and you answer "trees," your friend is likely to infer you are taking a course in botany. But trees are also a subject for mathematical investigation. In mathematics, a tree is a connected graph that does not contain any circuits except trivial ones. (Recall that a trivial circuit is one that consists of a single vertex.) Despite the formality of the definition, mathematical trees are similar in certain ways to their botanical namesakes.

> **• Definition**
>
> A graph is said to be **circuit-free** if, and only if, it has no nontrivial circuits. A graph is called a **tree** if, and only if, it is circuit-free and connected. A **trivial tree** is a graph that consists of a single vertex, and an **empty tree** is a tree that does not have any vertices or edges. A graph is called a **forest** if, and only if, it is circuit-free.

Example 11.5.1 Trees and Non-Trees

All the graphs shown in Figure 11.5.1 are trees, whereas those in Figure 11.5.2 are not.

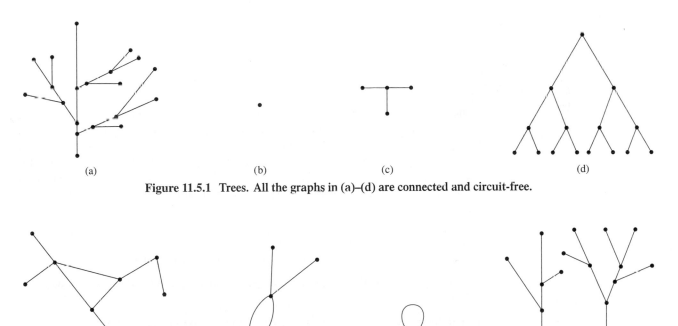

(a)　　　　　　(b)　　　　　(c)　　　　　　　　(d)

Figure 11.5.1 Trees. All the graphs in (a)–(d) are connected and circuit-free.

(a)　　　　　　(b)　　　　(c)　　　　　　(d)

Figure 11.5.2 Non-Trees. The graphs in (a), (b), and (c) all have circuits, and the graph in (d) is not connected. ■

Examples of Trees

The following examples illustrate just a few of the many and varied situations in which mathematical trees arise.

Example 11.5.2 A Decision Tree

During orientation week, a college administers an exam to all entering students to determine placement in the mathematics curriculum. The exam consists of two parts, and placement recommendations are made as indicated by the tree shown in Figure 11.5.3. Read the tree from left to right to decide what course should be recommended for a student who scored 9 on part I and 7 on part II.

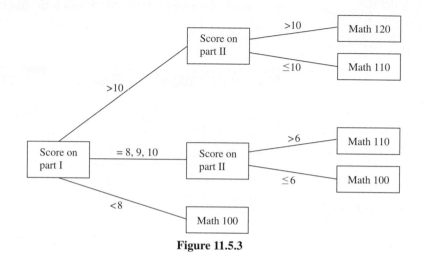

Figure 11.5.3

Solution   Since the student scored 9 on part I, the score on part II is checked. Since it is greater than 6, the student should be advised to take math 110.   ■

Example 11.5.3 A Parse Tree

In the last 30 years, Noam Chomsky and others have developed new ways to describe the syntax (or grammatical structure) of natural languages such as English. As is discussed briefly in Chapter 12, this work has proved useful in constructing compilers for high-level computer languages. In the study of grammars, trees are often used to show the derivation of grammatically correct sentences from certain basic rules. Such trees are called **syntactic derivation trees** or **parse trees**. A very small subset of English grammar, for example, specifies that

1. a sentence can be produced by writing first a noun phrase and then a verb phrase;

2. a noun phrase can be produced by writing an article and then a noun;

3. a noun phrase can also be produced by writing an article, then an adjective, and then a noun;

4. a verb phrase can be produced by writing a verb and then a noun phrase;

5. one article is "the";

6. one adjective is "young";

7. one verb is "caught";

8. one noun is "man";

9. one (other) noun is "ball."

*John Backus
(1924–1998)*

*Peter Naur
(born 1928)*

The rules of a grammar are called **productions.** It is customary to express them using the shorthand notation illustrated below. This notation, introduced by John Backus in 1959 and modified by Peter Naur in 1960, was used to describe the computer language Algol and is called the **Backus-Naur notation.** In the notation, the symbol | represents the word *or,* and angle brackets ⟨ ⟩ are used to enclose terms to be defined (such as a sentence or noun phrase).

1. ⟨sentence⟩ → ⟨noun phrase⟩⟨verb phrase⟩

2, 3. ⟨noun phrase⟩ → ⟨article⟩⟨noun⟩ | ⟨article⟩⟨adjective⟩⟨noun⟩

4. ⟨verb phrase⟩ → ⟨verb⟩⟨noun phrase⟩

5. ⟨article⟩ → the

6. ⟨adjective⟩ → young

7, 8. ⟨noun⟩ → man | ball

9. ⟨verb⟩ → caught

The derivation of the sentence "The young man caught the ball" from the above rules is described by the tree shown below.

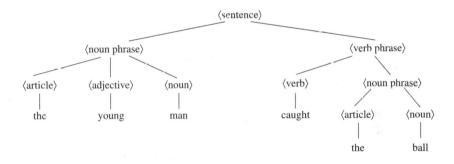

In the study of linguistics, **syntax** refers to the grammatical structure of sentences, and **semantics** refers to the meanings of words and their interrelations. A sentence can be syntactically correct but semantically incorrect, as in the nonsensical sentence "The young ball caught the man," which can be derived from the rules given above. Or a sentence can contain syntactic errors but not semantic ones, as, for instance, when a two-year-old child says, "Me hungry!" ■

Example 11.5.4 Structure of Hydrocarbon Molecules

The German physicist Gustav Kirchhoff (1824–1887) was the first to analyze the behavior of mathematical trees in connection with the investigation of electrical circuits. Soon after (and independently), the English mathematician Arthur Cayley used the mathematics of trees to enumerate all isomers for certain hydrocarbons. Hydrocarbon molecules are composed of carbon and hydrogen; each carbon atom can form up to four chemical bonds with other atoms, and each hydrogen atom can form one bond with another atom. Thus the structure of hydrocarbon molecules can be represented by graphs such as those shown following, in which the vertices represent atoms of hydrogen and carbon, denoted H and C, and the edges represent the chemical bonds between them.

Arthur Cayley (1821–1895)

Butane

Isobutane

Note that each of these graphs has four carbon atoms and ten hydrogen atoms, but the two graphs show different configurations of atoms. When two molecules have the same chemical formulae (in this case C_4H_{10}) but different chemical bonds, they are called *isomers*.

Certain *saturated hydrocarbon* molecules contain the maximum number of hydrogen atoms for a given number of carbon atoms. Cayley showed that if such a saturated hydrocarbon molecule has k carbon atoms, then it has $2k + 2$ hydrogen atoms. The first step in doing so is to prove that the graph of such a saturated hydrocarbon molecule is a tree. Prove this using proof by contradiction. (You are asked to finish the derivation of Cayley's result in exercise 4 at the end of this section.)

Solution   Suppose there is a hydrocarbon molecule that contains the maximum number of hydrogen atoms for the number of its carbon atoms and whose graph G is not a tree. [*We must derive a contradiction.*] Since G is not a tree, G is not connected or G has a nontrivial circuit. But the graph of any molecule is connected (all the atoms in a molecule must be connected to each other), and so G must have a nontrivial circuit. Now the edges of the circuit can link only carbon atoms because every vertex of a circuit has degree at least 2 and a hydrogen atom vertex has degree 1. Delete one edge of the circuit and add two new edges to join each of the newly disconnected carbon atom vertices to a hydrogen atom vertex as shown below.

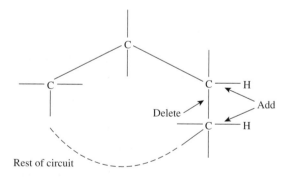

The resulting molecule has two more hydrogen atoms than the given molecule, but the number of carbon atoms is unchanged. This contradicts the supposition that the given molecule has the maximum number of hydrogen atoms for the given number of carbon atoms. Hence the supposition is false, and so G is a tree. ∎

Characterizing Trees

There is a somewhat surprising relation between the number of vertices and the number of edges of a tree. It turns out that if n is a positive integer, then any tree with n vertices (no matter what its shape) has $n - 1$ edges. Perhaps even more surprisingly, a partial converse to this fact is also true—namely, any *connected* graph with n vertices and $n - 1$ edges is a tree. It follows from these facts that if even one new edge (but no new vertex) is added to a tree, the resulting graph must contain a nontrivial circuit. Also, from the fact that removing an edge from a circuit does not disconnect a graph, it can be shown that every connected graph has a subgraph that is a tree. It follows that if n is a positive integer, any graph with n vertices and *fewer* than $n - 1$ edges is not connected.

A small but very important fact necessary to derive the first main theorem about trees is that any nontrivial tree must have at least one vertex of degree 1.

Lemma 11.5.1

Any tree that has more than one vertex has at least one vertex of degree 1.

A constructive way to understand this lemma is to imagine being given a tree T with more than one vertex. You pick a vertex v at random and then search outward along a path from v looking for a vertex of degree 1. As you reach each new vertex, you check whether it has degree 1. If it does, you are finished. If it does not, you exit from the vertex along a different edge from the one you entered on. Because T is circuit-free, the vertices included in the path never repeat. And since the number of vertices of T is finite, the process of building a path must eventually terminate. When that happens, the final vertex v' of the path must have degree 1. This process is illustrated below.

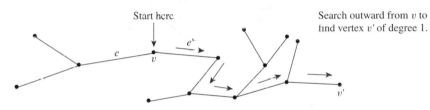

This discussion is made precise in the following proof.

Proof:

Let T be a particular but arbitrarily chosen tree that has more than one vertex, and consider the following algorithm:

Step 1: Pick a vertex v of T and let e be an edge incident on v.
[*If there were no edge incident on v, then v would be an isolated vertex. But this would contradict the assumption that T is connected (since it is a tree) and has at least two vertices.*]

Step 2: While $\deg(v) > 1$, repeat steps 2a, 2b, and 2c:

Step 2a: Choose e' to be an edge incident on v such that $e' \neq e$. [*Such an edge exists because $\deg(v) > 1$ and so there are at least two edges incident on v.*]

Step 2b: Let v' be the vertex at the other end of e' from v. [*Since T is a tree, e' cannot be a loop and therefore e' has two distinct endpoints.*]

Step 2c: Let $e = e'$ and $v = v'$. [*This is just a renaming process in preparation for a repetition of step 2.*]

The algorithm just described must eventually terminate because the set of vertices of the tree T is finite and T is circuit-free. When it does, a vertex v of degree 1 will have been found.

Using Lemma 11.5.1 it is not difficult to show that, in fact, any tree that has more than one vertex has at least *two* vertices of degree 1. This extension of Lemma 11.5.1 is left to the exercises at the end of this section.

• Definition

Let T be a tree. If T has only one or two vertices, then each is called a **terminal vertex.** If T has at least three vertices, then a vertex of degree 1 in T is called a **terminal vertex** (or a **leaf**), and a vertex of degree greater than 1 in T is called an **internal vertex** (or a **branch vertex**).

Example 11.5.5 Terminal and Internal Vertices

Find all terminal vertices and all internal vertices in the following tree:

Solution   The terminal vertices are v_0, v_2, v_4, v_5, v_7, and v_8. The internal vertices are v_6, v_1, and v_3. ∎

The following is the first of the two main theorems about trees:

Theorem 11.5.2

For any positive integer n, any tree with n vertices has $n - 1$ edges.

The proof is by mathematical induction. To do the inductive step, you assume the theorem is true for a positive integer k and then show it is true for $k + 1$. Thus you assume you have a tree T with $k + 1$ vertices, and you must show that T has $(k + 1) - 1 = k$ edges. As you do this, you are free to use the inductive hypothesis that *any* tree with k vertices has $k - 1$ edges. To make use of the inductive hypothesis, you need to reduce the tree T with $k + 1$ vertices to a tree with just k vertices. But by Lemma 11.5.1, T has a vertex v of degree 1, and since T is connected, v is attached to the rest of T by a single edge e as sketched below.

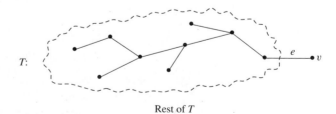

Rest of T

Now if e and v are removed from T, what remains is a tree T' with $(k+1) - 1 = k$ vertices. By inductive hypothesis, then, T' has $k - 1$ edges. But the original tree T has one more vertex and one more edge than T'. Hence T must have $(k-1) + 1 = k$ edges, as was to be shown. A formal version of this argument is given below.

Proof (by mathematical induction):

Let the property $P(n)$ be the sentence

Any tree with n vertices has $n - 1$ edges.

We use mathematical induction to show that this property is true for all integers $n \geq 1$.

Show that the property is true for $n = 1$: Let T be any tree with one vertex. Then T has zero edges (since it contains no loops). But $0 = 1 - 1$, so the property is true for $n = 1$.

Show that for all integers $k \geq 1$, if the property is true for k then it is true for $k + 1$: Suppose k is a positive integer and the property is true for k. In other words, suppose that any tree with k vertices has $k - 1$ edges. [*This is the inductive hypothesis.*] We must show that the property is true for $k + 1$. In other words, we must show that any tree with $k + 1$ vertices has $(k + 1) - 1 = k$ edges.

Let T be a particular but arbitrarily chosen tree with $k + 1$ vertices. [*We must show that T has k edges.*] Since k is a positive integer, $(k + 1) \geq 2$, and so T has more than one vertex. Hence by Lemma 11.5.1, T has a vertex v of degree 1. Also, since T has more than one vertex, there is at least one other vertex in T besides v. Thus there is an edge e connecting v to the rest of T. Define a subgraph T' of T so that

$$V(T') = V(T) - \{v\}$$
$$E(T') = E(T) - \{e\}.$$

Then

1. The number of vertices of T' is $(k + 1) - 1 = k$.

2. T' is circuit-free (since T is circuit-free, and removing an edge and a vertex cannot create a circuit).

3. T' is connected (see exercise 24 at the end of this section).

Hence, by the definition of tree, T' is a tree. Since T' has k vertices, by inductive hypothesis

$$\text{the number of edges of } T' = (\text{the number of vertices of } T') - 1$$
$$= k - 1.$$

But then

$$\text{the number of edges of } T = (\text{the number of edges of } T') + 1$$
$$= (k - 1) + 1$$
$$= k.$$

[*This is what was to be shown.*]

Example 11.5.6 Determining Whether a Graph Is a Tree

A graph G has ten vertices and twelve edges. Is it a tree?

Solution   No. By Theorem 11.5.2, any tree with ten vertices has nine edges, not twelve.

■

Example 11.5.7 Finding Nonisomorphic Trees

Find all nonisomorphic trees with four vertices.

Solution   By Theorem 11.5.2, any tree with four vertices has three edges. Thus the total degree of a tree with four vertices must be 6. Also, every tree with more than one vertex has at least two vertices of degree 1 (see the comment following Lemma 11.5.1 and exercise 29 at the end of this section). Thus the following combinations of degrees for the vertices are the only ones possible:

$$1, 1, 1, 3 \quad \text{and} \quad 1, 1, 2, 2.$$

There are nonisomorphic trees corresponding to both of these possibilities, as shown below.

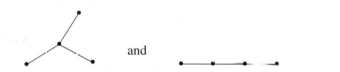

and

■

To prove the second major theorem about trees, we need another lemma.

Lemma 11.5.3

If G is any connected graph, C is any nontrivial circuit in G, and one of the edges of C is removed from G, then the graph that remains is connected.

Essentially, the reason why Lemma 11.5.3 is true is that any two vertices in a circuit are connected by two distinct paths. It is possible to draw the graph so that one of these goes "clockwise" and the other goes "counterclockwise" around the circuit. For example, in the circuit shown below, the clockwise path from v_2 to v_3 is

$$v_2 e_3 v_3$$

and the counterclockwise path from v_2 to v_3 is

$$v_2 e_2 v_1 e_1 v_0 e_6 v_5 e_5 v_4 e_4 v_3.$$

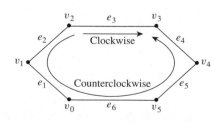

Proof:

Suppose G is a connected graph, C is a circuit in G, and e is an edge of C. Form a subgraph G' of G by removing e from G. Thus

$$V(G') = V(G)$$
$$E(G') = E(G) - \{e\}.$$

We must show that G' is connected. [*To show a graph is connected, we must show that if u and w are any vertices of the graph, then there exists a walk in G' from u to w.*] Suppose u and w are any two vertices of G'. [*We must find a walk from u to w.*] Since the vertex sets of G and G' are the same, u and w are both vertices of G, and since G is connected, there is a walk W in G from u to w.

Case 1 (e is not an edge of W): The only edge in G that is not in G' is e, so in this case W is also a walk in G'. Hence u is connected to w by a walk in G'.

Case 2 (e is an edge of W): In this case the walk W from u to w includes a section of the circuit C that contains e. Let C be denoted as follows:

$$C: v_0 e_1 v_1 e_2 v_2 \cdots e_n v_n (= v_0).$$

Now e equals one of the edges of C, so, to be specific, let $e = e_k$. Then the walk W contains either the sequence

$$v_{k-1} e_k v_k \quad \text{or} \quad v_k e_k v_{k-1}.$$

If W contains $v_{k-1} e_k v_k$, connect v_{k-1} to v_k by taking the "counterclockwise" walk W' defined as follows:

$$W': v_{k-1} e_{k-1} v_{k-2} \cdots v_0 e_n v_{n-1} \cdots e_{k+1} v_k.$$

An example showing how to obtain W' from W is given in Figure 11.5.4.

To go from v_{k-1} to v_k while avoiding e_k, go counterclockwise around the circuit.

Figure 11.5.4 An Example of a Walk from v_{k-1} to v_k That Does Not Include Edge e_k

If W contains $v_k e_k v_{k-1}$, connect v_k to v_{k-1} by taking the "clockwise" walk W'' defined as follows:

$$W'': v_k e_{k+1} v_{k+1} \cdots v_n e_1 v_1 e_2 \cdots e_{k-1} v_{k-1}.$$

Now patch either W' or W'' into W to form a new walk from u to w. For instance, to patch W' into W, start with the section of W from u to v_{k-1}, then take W' from v_{k-1} to v_k, and finally take the section of W from v_k to w. If this new walk still contains an occurrence of e, just repeat the process above until all occurrences are eliminated. [*This must happen eventually since the number of occurrences of e in G is finite.*] The result is a walk from u to w that does not contain e and hence is a walk in G'.

The arguments above show that both in case 1 and in case 2 there is a walk in G' from u to w. Since the choice of u and w was arbitrary, G' is connected.

The second major theorem about trees is modified converse to Theorem 11.5.2.

Theorem 11.5.4

For any positive integer n, if G is a connected graph with n vertices and $n - 1$ edges, then G is a tree.

Proof:

Let n be a positive integer and suppose G is a particular but arbitrarily chosen graph that is connected and has n vertices and $n - 1$ edges. [*We must show that G is a tree. Now a tree is a connected, circuit-free graph. Since we already know G is connected, it suffices to show that G is circuit-free.*] Suppose G is not circuit-free. That is, suppose G has a nontrivial circuit C. [*We must derive a contradiction.*] By Lemma 11.5.3, an edge of C can be removed from G to obtain a graph G' that is connected. If G' has a nontrivial circuit, then repeat this process: Remove an edge of the circuit from G' to form a new connected graph. Continue repeating the process of removing edges from circuits until eventually a graph G'' is obtained that is connected and is circuit-free. By definition, G'' is a tree. Since no vertices were removed from G to form G'', G'' has n vertices just as G does. Thus, by Theorem 11.5.2, G'' has $n - 1$ edges. But the supposition that G has a nontrivial circuit implies that at least one edge of G is removed to form G''. Hence G'' has no more than $(n - 1) - 1 = n - 2$ edges, which contradicts its having $n - 1$ edges. So the supposition is false. Hence G is circuit-free, and therefore G is a tree [*as was to be shown*].

Theorem 11.5.4 is not a full converse of Theorem 11.5.2. Although it is true that every *connected* graph with n vertices and $n - 1$ edges (where n is a positive integer) is a tree, it is not true that *every* graph with n vertices and $n - 1$ edges is a tree.

Example 11.5.8 A Graph with n Vertices and $n - 1$ Edges That Is Not a Tree

Give an example of a graph with five vertices and four edges that is not a tree.

Solution   By Theorem 11.5.4, such a graph cannot be connected. One example of such an unconnected graph is shown below.

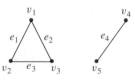

Rooted Trees

An outdoor tree is rooted and so is the kind of family tree that shows all the descendants of one particular person. The terminology and notation of rooted trees blends the language of botanical trees and that of family trees. In mathematics, a rooted tree is either an empty tree or a tree in which one vertex has been distinguished from the others and is designated the *root*. Given any other vertex v in the tree, there is a unique path from the root to v. (After all, if there were two distinct paths, a circuit could be constructed.) The number of edges in such a path is called the *level* of v, and the *height* of the tree is the length of the longest such path. (The height of the empty tree is defined to be 0.) It is traditional in drawing rooted trees to place the root at the top (as is done in family trees) and show the branches descending from it.

• **Definition**

A **rooted tree** is a tree in which one vertex is distinguished from the others and is called the **root.** The **level** of a vertex is the number of edges along the unique path between it and the root. The **height** of a rooted tree is the maximum level to any vertex of the tree. Given the root or any internal vertex v of a rooted tree, the **children** of v are all those vertices that are adjacent to v and are one level farther away from the root than v. If w is a child of v, then v is called the **parent** of w, and two vertices that are both children of the same parent are called **siblings.** Given vertices v and w, if v lies on the unique path between w and the root, then v is an **ancestor** of w and w is a **descendant** of v.

These terms are illustrated in Figure 11.5.5.

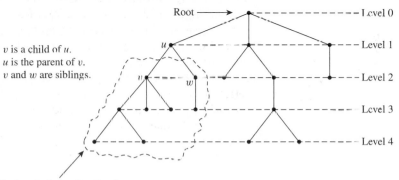

Figure 11.5.5 A Rooted Tree

Example 11.5.9 Rooted Trees

Consider the tree with root v_0 shown below.

a. What is the level of v_5? b. What is the level of v_0?

c. What is the height of this rooted tree? d. What are the children of v_3?

e. What is the parent of v_2? f. What are the siblings of v_8?

g. What are the descendants of v_3?

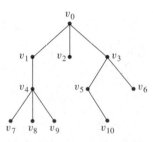

Solution

a. 2 b. 0 c. 3 d. v_5 and v_6 e. v_0 f. v_7 and v_9 g. v_5, v_6, v_{10}

∎

Note that in the tree with root v_0 shown below, v_1 has level 1 and is the child of v_0, and both v_0 and v_1 are terminal vertices.

$\bullet\, v_0$

$\bullet\, v_1$

Binary Trees

When every vertex in a rooted tree has at most two children and each child is designated either the (unique) left child or the (unique) right child, the result is a *binary tree*.

• Definition

A **binary tree** is a rooted tree in which every parent has at most two children. Each child in a binary tree is designated either a **left child** or a **right child** (but not both), and every parent has at most one left child and one right child. A **full binary tree** is a binary tree in which each parent has exactly two children.

Given any parent v in a binary tree T, the **left subtree** of v is the binary tree whose root is the left child of v, whose vertices consist of the left child of v and all its descendants, and whose edges consist of all those edges of T that connect the vertices of the left subtree. The **right subtree** of v is defined analogously.

These terms are illustrated in Figure 11.5.6.

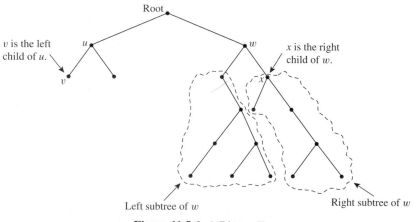

Figure 11.5.6 A Binary Tree

Example 11.5.10 Representation of Algebraic Expressions

Binary trees are used in computer science to represent algebraic expressions with arbitrary nesting of balanced parentheses. For instance, the following (labeled) binary tree represents the expression a/b: The operator is at the root and acts on the left and right children of the root in left-right order.

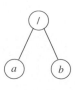

More generally, the binary tree shown below represents the expression $a/(c + d)$. In such a representation, the internal vertices are arithmetic operators, the terminal vertices are variables, and the operator at each vertex acts on its left and right subtrees in left-right order.

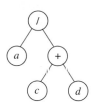

Draw a binary tree to represent the expression $((a - b) \cdot c) + (d/e)$.

Solution

One interesting theorem about binary trees says that if you know the number of internal vertices of a full binary tree, then you can calculate the total number of vertices and also the number of terminal vertices, and conversely. More specifically, a full binary tree with k internal vertices has a total of $2k + 1$ vertices of which $k + 1$ are terminal vertices.

Theorem 11.5.5

If k is a positive integer and T is a full binary tree with k internal vertices, then T has a total of $2k + 1$ vertices and has $k + 1$ terminal vertices.

Proof:

Suppose k is a positive integer and T is a full binary tree with k internal vertices. Observe that the set of all vertices of T can be partitioned into two disjoint subsets: the set of all vertices that have a parent and the set of all vertices that do not have a parent. Now there is just one vertex that does not have a parent, namely the root. Also, since every internal vertex of a full binary tree has exactly two children, the number of vertices that have a parent is twice the number of parents, or $2k$, since each parent is an internal vertex. Hence

$$\begin{bmatrix} \text{the total number} \\ \text{of vertices of } T \end{bmatrix} = \begin{bmatrix} \text{the number of} \\ \text{vertices that} \\ \text{have a parent} \end{bmatrix} + \begin{bmatrix} \text{the number of} \\ \text{vertices that do} \\ \text{not have a parent} \end{bmatrix}$$

$$= \quad 2k \quad + \quad 1.$$

continued on page 718

But it is also true that the total number of vertices of T equals the number of internal vertices plus the number of terminal vertices. Thus

$$\begin{bmatrix} \text{the total number} \\ \text{of vertices of } T \end{bmatrix} = \begin{bmatrix} \text{the number of} \\ \text{internal vertices} \end{bmatrix} + \begin{bmatrix} \text{the number of} \\ \text{terminal vertices} \end{bmatrix}$$

$$= \qquad k \qquad + \begin{bmatrix} \text{the number of} \\ \text{terminal vertices} \end{bmatrix}$$

Now equate the two expressions for the total number of vertices of T:

$$2k + 1 = k + \begin{bmatrix} \text{the number of} \\ \text{terminal vertices} \end{bmatrix}$$

Solving this equation gives

$$\begin{bmatrix} \text{the number of} \\ \text{terminal vertices} \end{bmatrix} = (2k + 1) - k = k + 1.$$

Thus the total number of vertices is $2k + 1$ and the number of terminal vertices is $k + 1$ [*as was to be shown*].

Example 11.5.11 Determining Whether a Certain Full Binary Tree Exists

Is there a full binary tree that has 10 internal vertices and 13 terminal vertices?

Solution   No. By Theorem 11.5.5, a full binary tree with 10 internal vertices has $10 + 1 = 11$ terminal vertices, not 13.   ∎

Another interesting theorem about binary trees specifies the maximum number of terminal vertices of a binary tree of a given height. Specifically, the maximum number of terminal vertices of a binary tree of height h is 2^h. Another way to say this is that a binary tree with t terminal vertices has height of at least $\log_2 t$.

Theorem 11.5.6

If T is a binary tree that has t terminal vertices and height h, then

$$t \le 2^h.$$

Equivalently,

$$\log_2 t \le h.$$

Proof:

We will use the strong form of mathematical induction on h to prove the truth of the following statement:

For all integers $h \ge 0$, if T is any binary tree of height h, then the number of terminal vertices of T is at most 2^h.

Let $P(h)$ be the property

If T is any binary tree of height h, then the number of terminal vertices of T is at most 2^h.

To go by strong induction, we will first show that the property is true for $h = 0$. Second, we will show that if the property is true for all nonnegative integers $k < h$, then it is true for h.

The property is true for h = 0: We must show that if T is any binary tree of height 0, then the number of terminal vertices of T is at most 2^0. Suppose T is a tree of height 0. Then T is the empty tree or T consists of a single vertex (the root) only. Let t be the number of terminal vertices of T. In the case where T is the empty tree, $t = 0$ and $h = 0$. Since $0 \leq 2^0$, then $t \leq 2^h$. In the case were T consists of a single vertex, $t = 1$ and $h = 0$. Since $1 = 2^0$, then in this case also $t \leq 2^h$. Hence in either case $t \leq 2^h$ [*as was to be shown*].

If h ≥ 1 and the property is true for all nonnegative integers k < h, then it is true for h: Suppose $h \geq 1$ is an integer and for all nonnegative integers $k < h$, $P(k)$ is true. In other words, suppose that for all integers $k < h$, any binary tree of height k has at most 2^k terminal vertices.

We must show that $P(h)$ is true. In other words, we must show that any binary tree of height h has at most 2^h terminal vertices.

Let T be a binary tree of height h and root v. Since $h \geq 1$, v has at least one child.

Case 1 (v has only one child): In this case we may assume without loss of generality that v's child is a left child and denote it by v_L. Let T_L be the left subtree of v. Then v_L is the root of T_L. (This situation is illustrated in Figure 11.5.7.) Because v has only one child, v is itself a terminal vertex, so the total number of terminal vertices in T equals the number of terminal vertices in T_L plus 1. Thus if t_L is the number of terminal vertices in T_L, then $t = t_L + 1$.

Now by inductive hypothesis, $t_L \leq 2^{h-1}$ because the height of T_L is $h - 1$, one less than the height of T. Also, because v has a child, $h \geq 1$ and so $2^{h-1} \geq 2^0 = 1$. Therefore,

$$t = t_L + 1 \leq 2^{h-1} + 1 \leq 2^{h-1} + 2^{h-1} = 2^h.$$

Left subtree T_L

Figure 11.5.7 A Binary Tree Whose Root Has One Child

Case 2 (v has two children): In this case, v has both a left child, v_L, and a right child, v_R, and v_L, and v_R are roots of a left subtree T_L and a right subtree T_R. Note that T_L and T_R are binary trees because T is a binary tree. (This situation is illustrated in Figure 11.5.8.)

continued on page 720

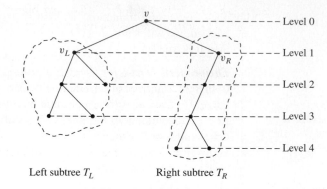

Left subtree T_L Right subtree T_R

Figure 11.5.8 A Binary Tree Whose Root Has Two Children

Now v_L and v_R are the roots of the left and right subtrees of v, denoted T_L and T_R, respectively. Note that T_L and T_R are binary trees because T is a binary tree. Let h_L and h_R be the heights of T_L and T_R, respectively. Then $h_L \leq h - 1$ and $h_R \leq h - 1$ since T is obtained by joining T_L and T_R and adding a level. Let t_L and t_R be the numbers of terminal vertices of T_L and T_R, respectively. Then, since both T_L and T_R have heights less than h, by inductive hypothesis

$$t_L \leq 2^{h_L} \quad \text{and} \quad t_R \leq 2^{h_R}.$$

But the terminal vertices of T consist exactly of the terminal vertices of T_L together with the terminal vertices of T_R. Therefore,

$$t = t_L + t_R \leq 2^{h_L} + 2^{h_R} \quad \text{by inductive hypothesis}$$
$$\text{since } h_L \leq h - 1 \text{ and } h_R \leq h - 1$$
$$\Rightarrow \quad t \leq 2^{h-1} + 2^{h-1}$$
$$\Rightarrow \quad t \leq 2(2^{h-1})$$
$$\Rightarrow \quad t \leq 2^h \qquad \text{by basic algebra.}$$

Thus the number of terminal vertices is at most 2^h [*as was to be shown*].

Since both the basis step and the inductive step have been proved, the first version of the theorem is proved.

The equivalent inequality $\log_2 t \leq h$ follows immediately from the fact that the logarithmic function with base 2 is increasing and from the definition of logarithm, for if

$$t \leq 2^h,$$

then applying the logarithm with base 2 function to both sides gives

$$\log_2 t \leq \log_2(2^h).$$

It follows from the definition of logarithm that $\log_2(2^h) = h$ [*because $\log_2(2^h)$ is the exponent to which 2 must be raised to obtain 2^h*]. Hence

$$\log_2 t \leq h$$

[*as was to be shown*].

Example 11.5.12 Determining Whether a Certain Binary Tree Exists

Is there a binary tree that has height 5 and 38 terminal vertices?

Solution   No. By Theorem 11.5.6, any binary tree T with height 5 has at most $2^5 = 32$ terminal vertices, so such a tree cannot have 38 terminal vertices. ∎

From Theorem 11.5.6 it can be deduced that any algorithm to sort a set of n data items has worst-case order of at least $n \log_2 n$. This result is obtained by analyzing a decision tree whose terminal vertices are all the $n!$ arrangements of the set to be sorted. The height of this tree is the minimum number of operations needed to sort the set. By Theorem 11.5.6 this height is at least $\log_2(n!)$. But $\log_2(n!) \geq M(n \log_2 n)$, where M is a positive constant (see exercise 49 of Section 9.4). It follows that the worst-case order for an algorithm to sort the set cannot be less than $n \log_2 n$.

Exercise Set 11.5

1. Read the tree in Example 11.5.2 from left to right to answer the following questions:
   a. What course should a student who scored 12 on part I and 4 on part II take?
   b. What course should a student who scored 8 on part I and 9 on part II take? ·

2. Draw trees to show the derivations of the following sentences from the rules given in Example 11.5.3.
   a. The young ball caught the man.
   b. The man caught the young ball.

H **3.** What is the total degree of a tree with n vertices? Why?

4. Let G be the graph of a hydrocarbon molecule with the maximum number of hydrogen atoms for the number of its carbon atoms.
   a. Draw the graph of G if G has three carbon atoms and eight hydrogen atoms.
   b. Draw the graphs of three isomers of C_5H_{12}.
   c. Use Example 11.5.4 and exercise 3 to prove that if the vertices of G consist of k carbon atoms and m hydrogen atoms, then G has a total degree of $2k + 2m - 2$.
   H **d.** Prove that if the vertices of G consist of k carbon atoms and m hydrogen atoms, then G has a total degree of $4k + m$.
   e. Equate the results of (c) and (d) to prove Cayley's result that a saturated hydrocarbon molecule with k carbon atoms and a maximum number of hydrogen atoms has $2k + 2$ hydrogen atoms.

H **5.** Extend the argument given in the proof of Lemma 11.5.1 to show that a tree with more than one vertex has at least two vertices of degree 1.

6. If graphs are allowed to have an infinite number of vertices and edges, then Lemma 11.5.1 is false. Give a counterexample that shows this. In other words, give an example of an "infinite tree" (a connected, circuit-free graph with an infinite number of vertices and edges) that has no vertex of degree 1.

7. Find all terminal vertices and all internal vertices for the following trees.

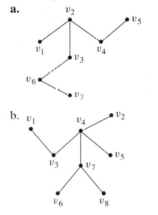

a.

b.

In each of 8–21, either draw a graph with the given specifications or explain why no such graph exists.

8. Tree, nine vertices, nine edges

9. Graph, connected, nine vertices, nine edges

10. Graph, circuit-free, nine vertices, six edges

11. Tree, six vertices, total degree 14

12. Tree, five vertices, total degree 8

13. Graph, connected, six vertices, five edges, has a nontrivial circuit

14. Graph, two vertices, one edge, not a tree

15. Graph, circuit-free, seven vertices, four edges

16. Tree, twelve vertices, fifteen edges

17. Graph, six vertices, five edges, not a tree

18. Tree, five vertices, total degree 10

19. Graph, connected, ten vertices, nine edges, has a nontrivial circuit

20. Simple graph, connected, six vertices, six edges

21. Tree, ten vertices, total degree 24

22. A connected graph has twelve vertices and eleven edges. Does it have a vertex of degree 1? Why?

23. A connected graph has nine vertices and twelve edges. Does it have a nontrivial circuit? Why?

24. Suppose that v is a vertex of degree 1 in a connected graph G and that e is the edge incident on v. Let G' be the subgraph of G obtained by removing v and e from G. Must G' be connected? Why?

25. A graph has eight vertices and six edges. Is it connected? Why?

H 26. If a graph has n vertices and $n - 2$ or fewer edges, can it be connected? Why?

27. A circuit-free graph has ten vertices and nine edges. Is it connected? Why?

H 28. Is a circuit-free graph with n vertices and at least $n - 1$ edges connected? Why?

29. Prove that every nonempty, nontrivial tree has at least two vertices of degree 1 by filling in the details and completing the following argument: Let T be a nontrivial tree and let S be the set of all paths from one vertex to another of T. Among all the paths in S, choose a path P with the most edges. (Why is it possible to find such a P?) What can you say about the initial and final vertices of P? Why?

30. Find all nonisomorphic trees with five vertices.

31. a. Prove that the following is an invariant for graph isomorphism: A vertex of degree i is adjacent to a vertex of degree j.
    H b. Find all nonisomorphic trees with six vertices.

32. Consider the tree shown below with root a.
    a. What is the level of n?
    b. What is the level of a?
    c. What is the height of this rooted tree?
    d. What are the children of n?
    e. What is the parent of g?
    f. What are the siblings of j?
    g. What are the descendants of f?

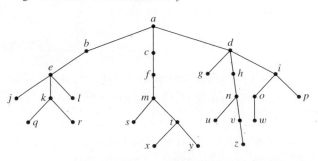

33. Consider the tree shown below with root v_0.
    a. What is the level of v_8?
    b. What is the level of v_0?
    c. What is the height of this rooted tree?
    d. What are the children of v_{10}?
    e. What is the parent of v_5?
    f. What are the siblings of v_1?
    g. What are the descendants of v_{12}?

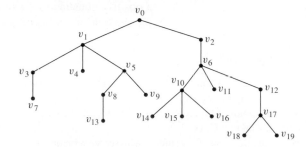

34. Draw binary trees to represent the following expressions:
    a. $a \cdot b - (c/(d + e))$ **b.** $a/(b - c \cdot d)$

In each of 35–50 either draw a graph with the given specifications or explain why no such graph exists.

35. Full binary tree, five internal vertices

36. Full binary tree, five internal vertices, seven terminal vertices

37. Full binary tree, seven vertices, of which four are internal vertices

38. Full binary tree, twelve vertices

39. Full binary tree, nine vertices

40. Binary tree, height 3, seven terminal vertices

41. Full binary tree, height 3, six terminal vertices

42. Binary tree, height 3, nine terminal vertices

43. Full binary tree, eight internal vertices, seven terminal vertices.

44. Binary tree, height 4, eight terminal vertices

45. Full binary tree, seven vertices

46. Full binary tree, nine vertices, five internal vertices

47. Full binary tree, four internal vertices

48. Binary tree, height 4, eighteen terminal vertices

49. Full binary tree, sixteen vertices

50. Full binary tree, height 3, seven terminal vertices

51. What can you deduce about the height of a binary tree if you know that it has the following properties?
    a. Twenty-five terminal vertices
    b. Forty terminal vertices
    c. Sixty terminal vertices

11.6 *Spanning Trees*

I contend that each science is a real science insofar as it is mathematics.
— Immanuel Kant, 1724–1804

An East Coast airline company wants to expand service to the Midwest and has received permission from the Federal Aviation Authority to fly any of the routes shown in Figure 11.6.1.

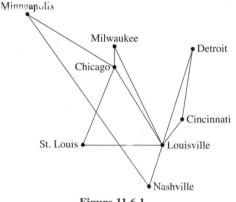

Figure 11.6.1

The company wishes to legitimately advertise service to all the cities shown but, for reasons of economy, wants to use the least possible number of individual routes to connect them. One possible route system is given in Figure 11.6.2.

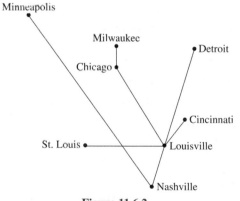

Figure 11.6.2

Clearly this system joins all the cities. Is the number of individual routes minimal? The answer is yes, and the reason may surprise you.

The fact is that the graph of any system of routes that satisfies the company's wishes is a tree, because if the graph were to contain a circuit, then one of the routes in the circuit could be removed without disconnecting the graph (by Lemma 11.5.3), and that would give a smaller total number of routes. But any tree with eight vertices has seven edges. Therefore, any system of routes that connects all eight vertices and yet minimizes the total number of routes consists of seven routes.

> **• Definition**
>
> A **spanning tree** for a graph G is a subgraph of G that contains every vertex of G and is a tree.

The preceding discussion contains the essence of the proof of the following proposition:

> **Proposition 11.6.1**
>
> 1. Every connected graph has a spanning tree.
>
> 2. Any two spanning trees for a graph have the same number of edges.
>
> **Proof of (1):**
>
> Suppose G is a connected graph. If G is circuit-free, then G is its own spanning tree and we are done. If not, then G has at least one circuit C_1. By Lemma 11.5.3, the subgraph of G obtained by removing an edge from C_1 is connected. If this subgraph is circuit-free, then it is a spanning tree and we are done. If not, then it has at least one circuit C_2, and, as above, an edge can be removed from C_2 to obtain a connected subgraph. Continuing in this way, we can remove successive edges from circuits, until eventually we obtain a connected, circuit-free subgraph T of G. [*This must happen at some point because the number of edges of G is finite, and at no stage does removal of an edge disconnect the subgraph.*] Also, T contains every vertex of G because no vertices of G were removed in constructing it. Thus T is a spanning tree for G.

The proof of part (2) is left as an exercise.

Example 11.6.1 Spanning Trees

Find all spanning trees for the graph G pictured below.

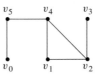

Solution   The graph G has one circuit $v_2 v_1 v_4 v_2$, and removal of any edge of the circuit gives a tree. Thus, as shown below, there are three spanning trees for G.

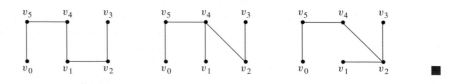

Mininum Spanning Trees

The graph of the routes allowed by the Federal Aviation Authority shown in Figure 11.6.1 can be annotated by adding the distances (in miles) between each pair of cities. This is done in Figure 11.6.3.

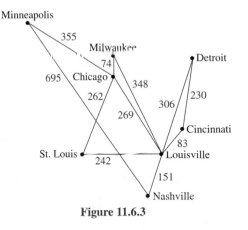

Figure 11.6.3

Now suppose the airline company wants to serve all the cities shown, but with a route system that minimizes the total mileage. Note that such a system is a tree, because if the system contained a circuit, removal of an edge from the circuit would not affect a person's ability to reach every city in the system from every other (again, by Lemma 11.5.3), but it would reduce the total mileage of the system.

More generally, a graph whose edges are labeled with numbers (known as *weights*) is called a *weighed graph*. A *minimum-weight spanning tree,* or simply a *minimum spanning tree,* is a spanning tree for which the sum of the weights of all the edges is as small as possible.

• Definition

A **weighted graph** is a graph for which each edge has an associated real number **weight.** The sum of the weights of all the edges is the **total weight** of the graph. A **minimum spanning tree** for a weighted graph is a spanning tree that has the least possible total weight compared to all other spanning trees for the graph.

If G is a weighed graph and e is an edge of G, then $w(e)$ denotes the weight of e and $w(G)$ denotes the total weight of G.

The problem of finding a minimum spanning tree for a graph is certainly solvable. One solution is to list all spanning trees for the graph, compute the total weight of each, and choose one for which this total is a minimum. (Note that the well-ordering principle for the integers guarantees the existence of such a minimum total.) This solution, however, is inefficient in its use of computing time because the number of distinct spanning trees is so large. For instance, a complete graph with n vertices has n^{n-2} spanning trees.

In 1956 and 1957 Joseph B. Kruskal and Robert C. Prim each described much more efficient algorithms to construct minimum spanning trees. For graphs with n vertices and m edges, Kruskal's and Prim's algorithms can be implemented so as to have worst-case orders of $m \log_2 m$ and n^2, respectively.

Kruskal's Algorithm

Joseph Kruskal
(born 1928)

In Kruskal's algorithm, the edges of a weighed graph are examined one by one in order of increasing weight. At each stage the edge being examined is added to what will become the minimum spanning tree, provided that this addition does not create a circuit. After $n - 1$ edges have been added (where n is the number of vertices of the graph), these edges, together with the vertices of the graph, form a minimum spanning tree for the graph.

Algorithm 11.6.1 Kruskal

Input: G [*a weighed graph with n vertices*]

Algorithm Body:
[*Build a subgraph T of G to consist of all the vertices of G with edges added in order of increasing weight. At each stage, let m be the number of edges of T.*]

1. Initialize T to have all the vertices of G and no edges.

2. Let E be the set of all edges of G, and let $m := 0$.
   [*pre-condition: G is connected.*]

3. **while** $(m < n - 1)$
   3a. Find an edge e in E of least weight.
   3b. Delete e from E.
   3c. if addition of e to the edge set of T does not produce a circuit
       then add e to the edge set of T and set $m := m + 1$

   end while
   [*post-condition: T is a minimum spanning tree for G.*]

Output: T

The following example shows how Kruskal's algorithm works for the graph of the airline route system.

Example 11.6.2 Action of Kruskal's Algorithm

Describe the action of Kruskal's algorithm for the graph shown in Figure 11.6.4.

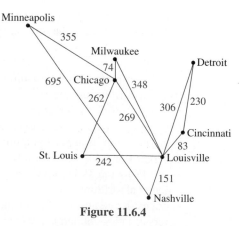

Figure 11.6.4

Solution

Iteration Number	Edge Considered	Weight	Action Taken
1	Chicago–Milwaukee	74	added
2	Louisville–Cincinnati	83	added
3	Louisville–Nashville	151	added
4	Cincinnati–Detroit	230	added
5	St. Louis–Louisville	242	added
6	St. Louis–Chicago	262	added
7	Chicago–Louisville	269	not added
8	Louisville–Detroit	306	not added
9	Louisville–Milwaukee	348	not added
10	Minneapolis–Chicago	355	added

The tree produced by Kruskal's algorithm is shown in Figure 11.6.5.

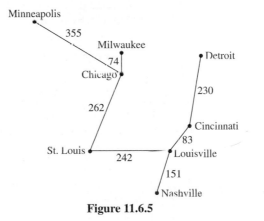

Figure 11.6.5

It is not obvious from the description of Kruskal's algorithm that it does what it is supposed to do. To be specific, what guarantees that it is possible at each stage to find an edge of least weight whose addition does not produce a circuit? And if such edges can be found, what guarantees that they will all eventually connect? And if they do connect, what guarantees that the resulting tree has minimum weight? Of course, the mere fact that Kruskal's algorithm is printed in this book may lead you to believe that everything works out. But the questions above are real, and they deserve serious answers.

Theorem 11.6.2 Correctness of Kruskal's Algorithm

When a connected, weighted graph is input to Kruskal's algorithm, the output is a minimum spanning tree.

Proof:

Suppose that G is a connected, weighted graph with n vertices and that T is a subgraph of G produced when G is input to Kruskal's algorithm. Clearly T is circuit-free [*since no edge that completes a circuit is ever added to T*]. Also T is connected. For as long as T has more than one connected component, the set of edges of G that can

continued on page 728

be added to T without creating a circuit is nonempty. [*The reason is that since G is connected, given any vertex v_1 in one connected component C_1 of T and any vertex v_2 in another connected component C_2, there is a path in G from v_1 to v_2. Since C_1 and C_2 are distinct, there is an edge e of this path that is not in T. Adding e to T does not create a circuit in T, because deletion of an edge from a circuit does not disconnect a graph and deletion of e would.*] The preceding arguments show that T is circuit-free and connected. Since by construction T contains every vertex of G, T is a spanning tree for G.

Now we show that T has minimum weight. Let T_1 be any minimum spanning tree for G. If $T = T_1$, then T is a minimum spanning tree for G and we are done. If $T \neq T_1$, then there is an edge e in T that is not an edge of T_1. [*Since T and T_1 both have the same vertex set, if they differ at all, they must have different edge sets.*] Now adding e to T_1 produces a graph with a unique nontrivial circuit (see exercise 14 at the end of this section). Let e' be an edge of this circuit such that e' is not in T. [*Such an edge must exist because T is a tree and hence circuit free.*] Let T_2 be the graph obtained from T_1 by removing e' and adding e. This situation is illustrated below.

e' (remove from T_1 to form T_2.)

e (add to T_1 to form T_2.)

The entire graph is G. T_1 has black edges. e is in T but not T_1. e' is in T_1 but not T.

Note that T_2 has $n - 1$ edges and n vertices and that T_2 is connected [*since by Lemma 11.5.3 the subgraph obtained by removing an edge from a circuit in a connected graph is connected*]. Consequently, T_2 is a spanning tree for G. In addition,

$$w(T_2) = w(T_1) - w(e') + w(e).$$

Now $w(e) \leq w(e')$ because at the stage in Kruskal's algorithm when e was added to T, e' was available to be added [*since it was not already in T, and at that stage its addition could not produce a circuit since e was not in T*], and e' would have been added had its weight been smaller than that of e. Thus

$$w(T_2) = w(T_1) - \underbrace{(w(e') - w(e))}_{\geq 0}$$

$$\leq w(T_1).$$

But T_1 is a minimum spanning tree. Since T_2 is a spanning tree with weight less than or equal to the weight of T_1, T_2 is also a minimum spanning tree for G.

Finally, note that by construction, T_2 has one more edge in common with T than does T_1. If T now equals T_2, then T is a minimum spanning tree and we are done. If not, then we can repeat the process described above to find a minimum spanning tree T_3 that has one more edge in common with T than T_2. Continuing in this way produces a sequence of minimum spanning trees $T_1, T_2, T_3, \ldots$ each of which has one more edge in common with T than the preceding tree. Since T has only a finite number of edges, this sequence is finite and so there is a minimum spanning tree, T_k, which is identical to T. It follows that T is, itself, a minimum spanning tree for G.

Prim's Algorithm

Prim's algorithm works differently from Kruskal's. It builds a minimum spanning tree T by expanding outward in connected links from some vertex. One edge and one vertex are added at each stage. The edge added is the one of least weight that connects the vertices already in T with those not in T, and the vertex is the endpoint of this edge that is not already in T.

Algorithm 11.6.2 Prim

Input: G [*a weighted graph with n vertices*]

Algorithm Body:
[*Build a subgraph T of G by starting with any vertex v of G and attaching edges (with their endpoints) one by one to an as-yet-unconnected vertex of T, each time choosing an adjacent edge of least weight.*]

1. Pick a vertex v of G and let T be the graph with one vertex, v, and no edges.

2. Let V be the set of all vertices of G except v.
   [*pre-condition: G is connected.*]

3. **for** $i := 1$ **to** $n - 1$

   3a. Find an edge e of G such that (1) e connects T to one of the vertices in V, and (2) e has the least weight of all edges connecting T to a vertex in V. Let w be the endpoint of e that is in V.

   3b. Add e and w to the edge and vertex sets of T, and delete w from V.

   next i
   [*post-condition: T is a minimum spanning tree for G.*]

Output: T

The following example shows how Prim's algorithm works for the graph of the airline route system.

Example 11.6.3 Action of Prim's Algorithm

Describe the action of Prim's algorithm for the graph in Figure 11.6.6 using the Minneapolis vertex as a starting point.

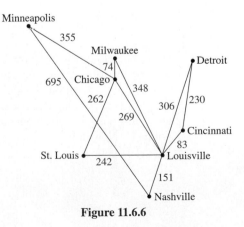

Figure 11.6.6

Caption at left of photo:
Robert Prim
(*born 1921*)

Courtesy of Lucent Technologies, Inc.

Solution

Iteration Number	Vertex Added	Edge Added	Weight
0	Minneapolis		
1	Chicago	Minneapolis–Chicago	355
2	Milwaukee	Chicago–Milwaukee	74
3	St. Louis	Chicago–St. Louis	262
4	Louisville	St. Louis–Louisville	242
5	Cincinnati	Louisville–Cincinnati	83
6	Nashville	Louisville–Nashville	151
7	Detroit	Cincinnati–Detroit	230

■

Note that the tree obtained is the same as that obtained by Kruskal's algorithm, but the edges are added in a different order. It is not hard to see that when a connected graph is input to Prim's algorithm, the result is a spanning tree. What is not so clear is that this spanning tree is a minimum. The proof of the following theorem establishes that it is.

Theorem 11.6.3 Correctness of Prim's Algorithm

When a connected, weighted graph G is input to Prim's algorithm, the output is a minimum spanning tree for G.

Proof:

Let G be a connected, weighted graph, and suppose G is input to Prim's algorithm. At each stage of execution of the algorithm, an edge must be found that connects a vertex in a subgraph to a vertex outside the subgraph. As long as there are vertices outside the subgraph, the connectedness of G ensures that such an edge can always be found. [*For if one vertex in the subgraph and one vertex outside it are chosen, then by the connectedness of G there is a walk in G linking the two. As one travels along this walk, at some point one moves along an edge from a vertex inside the subgraph to a vertex outside the subgraph.*]

Now it is clear that the output T of Prim's algorithm is a tree because the edge and vertex added to T at each stage are connected to other edges and vertices of T and because at no stage is a circuit created since each edge added connects vertices in two disconnected sets. [*Consequently, removal of a newly added edge produces a disconnected graph, whereas by Lemma 11.5.3, removal of an edge from a circuit produces a connected graph.*] Also, T includes every vertex of G because T, being a tree with $n - 1$ edges, has n vertices [*and that is all G has*]. Thus T is a spanning tree for G.

Next we prove that T has minimum weight. Let T_1 be *any* minimum spanning tree for G. If $T = T_1$, then we are done. If not, then there is an edge in T that is not in T_1. Of all edges in T and not T_1, let e be the first that was added when T was constructed using Prim's algorithm. Let V be the set of vertices of T just before the addition of e. Then one endpoint, say v, of e is in T, and the other, say w, is not. Since T_1 is a spanning tree for G, there is a path in T_1 joining v to w. As one travels along this path, one must encounter an edge e' joining a vertex in V to one that is not in V. Now at the stage when e was added to T, e' could also have been added, and it *would* have been added instead of e had its weight been less than that of e. Since e' was not added at that stage, we conclude that

$$w(e') \geq w(e).$$

Let T_2 be the graph obtained from T_1 by removing e' and adding e. [*Thus T_2 has one more edge in common with T than T_1 does.*] Note that T_2 is a tree. The reason is that since e' is part of a path in T_1 from v to w, and e connects v and w, adding e to T_1 creates a circuit. When e' is removed from this circuit, the resulting subgraph remains connected. In fact, T_2 is a spanning tree for G since no vertices were removed in forming T_2 from T_1. The argument showing that $w(T_2) \leq w(T_1)$ is left as an exercise. [*It is virtually identical to part of the proof of Theorem 11.6.2*].

It follows that T_2 is a minimum spanning tree for G with one more edge in common with T than T_1 has. If T now equals T_2, then we are done. If not, then, as above, we can find another minimum spanning tree T_3 having one more edge in common with T than T_2. Continuing in this way produces a sequence of minimum spanning trees $T_1, T_2, T_3, \ldots$ each of which has one more edge in common with T than does the preceding tree. Since T has only a finite number of edges, this sequence is finite, so there is a tree, T_k, that is identical to T. This shows that T is itself a minimum spanning tree.

Example 11.6.4 Finding Minimum Spanning Trees

Find all minimum spanning trees for the following graph. Use Kruskal's algorithm and Prim's algorithm starting at vertex a. Indicate the order in which edges are added to form each tree.

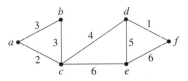

Solution When Kruskal's algorithm is applied, edges are added in one of the following two orders:

1. $\{d, f\}, \{a, c\}, \{a, b\}, \{c, d\}, \{d, e\}$

2. $\{d, f\}, \{a, c\}, \{b, c\}, \{c, d\}, \{d, e\}$

When Prim's algorithm is applied starting at a, edges are added in one of the following two orders:

1. $\{a, c\}, \{a, b\}, \{c, d\}, \{d, f\}, \{d, e\}$

2. $\{a, c\}, \{b, c\}, \{c, d\}, \{d, f\}, \{d, e\}$

Thus, as shown below, there are two distinct minimum spanning trees for this graph.

(a)   (b)

Exercise Set 11.6

Find all possible spanning trees for each of the graphs in 1 and 2.

1.

2.

Find a spanning tree for each of the graphs in 3 and 4.

3.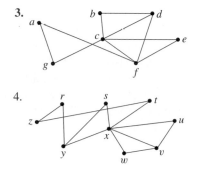

4.

Use Kruskal's algorithm to find a minimum spanning tree for each of the graphs in 5 and 6. Indicate the order in which edges are added to form each tree.

5.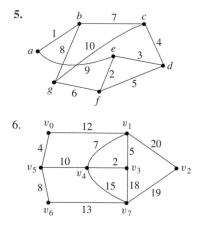

6.

Use Prim's algorithm starting with vertex a or v_0 to find a minimum spanning tree for each of the graphs in 7 and 8. Indicate the order in which edges are added to form each tree.

7. The graph of exercise 5.   **8.** The graph of exercise 6.

For each of the graphs in 9 and 10, find all minimum spanning trees that can be obtained using (a) Kruskal's algorithm and (b) Prim's algorithm starting with vertex a or t. Indicate the order in which edges are added to form each tree.

9.

10.

11. A pipeline is to be built that will link six cities. The cost (in hundreds of millions of dollars) of constructing each potential link depends on distance and terrain and is shown in the weighted graph below. Find a system of pipelines to connect all the cities and yet minimize the total cost.

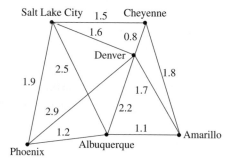

12. Prove part (2) of Proposition 11.6.1: Any two spanning trees for a graph have the same number of edges.

13. Given any two distinct vertices of a tree, there exists a unique path from one to the other.
  a. Give an informal justification for the above statement.
  ✶ b. Write a formal proof of the above statement.

14. Prove that if G is a graph with spanning tree T and e is an edge of G that is not in T, then the graph obtained by adding e to T contains one and only one set of edges that forms a nontrivial circuit.

15. Suppose G is a connected graph and T is a circuit-free subgraph of G. Suppose also that if any edge e of G not in T is added to T, the resulting graph contains a circuit. Prove that T is a spanning tree for G.

H **16. a.** Suppose T_1 and T_2 are two different spanning trees for a graph G. Must T_1 and T_2 have an edge in common? Prove or give a counterexample.
  b. Suppose that the graph G in part (a) is simple. Must T_1 and T_2 have an edge in common? Prove or give a counterexample.

H **17.** Prove that an edge e is contained in every spanning tree for a connected graph G if, and only if, removal of e disconnects G.

18. Consider the spanning trees T_1 and T_2 in the proof of Theorem 11.6.3. Prove that $w(T_2) \leq w(T_1)$.

19. Suppose that T is a minimum spanning tree for a connected, weighted graph G and that G contains an edge e (not a loop) that is not in T. Let v and w be the endpoints of e. By exercise 13 there is a unique path in T from v to w. Let e' be any edge of this path. Prove that $w(e') \leq w(e)$.

H **20.** Prove that if G is a connected, weighted graph and e is an edge of G (not a loop) that has smaller weight than any other edge of G, then e is in every minimum spanning tree for G.

✳ **21.** If G is a connected, weighted graph and no two edges of G have the same weight, does there exist a unique minimum spanning tree for G? Justify your answer.

✳ **22.** Prove that if G is a connected, weighted graph and e is an edge of G that (1) has greater weight than any other edge of G and (2) is in a circuit of G, then there is no minimum spanning tree T for G such that e is in T.

23. Suppose a disconnected graph is input to Kruskal's algorithm. What will be the output?

24. Suppose a disconnected graph is input to Prim's algorithm. What will be the output?

25. Prove that if a connected, weighted graph G is input to Algorithm 11.6.3 (shown at right), the output is a minimum spanning tree for G.

Algorithm 11.6.3

Input: G [*a graph*]

Algorithm Body:

1. $T := G$.

2. $E :=$ the set of all edges of G, $m :=$ the number of edges of G.
[*pre-condition: G is connected.*]

3. while ($m > 0$)

    3a. Find an edge e in E that has maximal weight.

    3b. Remove e from E and set $m := m - 1$.

    3c. if the subgraph obtained when e is removed from the edge set of T is connected **then** remove e from the edge set of T

    end while
[*post-condition: T is a minimum spanning tree for G.*]

Output: T [*a graph*]

CHAPTER 12

REGULAR EXPRESSIONS AND FINITE-STATE AUTOMATA

The theoretical foundations of computer science were derived from several disciplines: logic (the foundations of mathematics), electrical engineering (the design of switching circuits), brain research (models of neurons), and linguistics (the formal specification of languages).

As discussed briefly in Sections 5.4 and 7.5, the 1930s saw the development of mathematical treatments of basic questions concerning what can be proved in mathematics and what can be computed by means of a finite sequence of mechanized operations. Although the first digital computers were not built until the early 1940s, ten years earlier Alan Turing developed a simple abstract model of a machine, now called a Turing machine, by means of which he defined what it would mean for a function to be computable.

Around the same time, somewhat similar models of computation were developed by the American logicians Alonzo Church, Stephen C. Kleene, and Emil Post (who was born in Poland but came to the United States as a child), but Church and others showed these all to be equivalent. As a result, Church formulated a conjecture, now known as the **Church-Turing thesis,** asserting that the Turing machine is universal in the sense that anything that can ever be computed on a machine can be computed with a Turing machine. If this thesis is correct—which is widely believed—then all computers that have been or will ever be constructed are theoretically equivalent in what they can do, although they may differ widely in speed and storage capacity. For instance, quantum computers may have the capability to compute certain quantities enormously faster than classical computers. But Church's thesis implies that the theory of computation is likely to remain fundamentally the same, even though the enabling technology is subject to constant change.

In the early 1940s, Warren S. McCulloch and Walter Pitts, working at the Massachusetts Institute of Technology (M.I.T.), developed a model of how the neurons in the brain might work and how models of neurons could be combined to make "circuits" or "automata" capable of more complicated computations. To a certain extent, they were influenced by the results of Claude Shannon, who also worked at M.I.T. and had in the 1930s developed the foundations of a theory that implemented Boolean functions as switching circuits. In the 1950s, Kleene analyzed the work of McCulloch and Pitts and connected it with versions of the machine models introduced by Turing and others.

Another development of the 1950s was the introduction of high-level computer languages. During the same years, linguist Noam Chomsky's attempts to understand the underlying principles by means of which human beings generate speech led him to develop a theory of formal languages, which he defined using sets of abstract rules, called *grammars,* of varying levels of complexity. It soon became apparent that Chomsky's theory was of great utility in the analysis and construction of computer languages. For computer science, the most useful of Chomsky's language classifications are also the two simplest: the *regular languages* and the *context-free languages.*

Regular languages, which are defined by *regular expressions,* are used extensively for matching patterns within text (as in word processing or Internet searches) and for lexical analysis in computer language compilers. They are part of sophisticated text editors and a number of UNIX* utilities, and they are also used in transforming XML[†]documents.

Through use of the Backus-Naur notation (introduced in Section 11.5), context-free languages are able to describe many of the more complex aspects of modern high-level computer languages, and they form the basis for the main part of compilers, which translate programs written in a high-level language into machine code suitable for execution.

A remarkable fact is that all of the subjects referred to above are related. Each context-free grammar turns out to be equivalent to a type of automaton called a *push-down automaton,* and each regular expression turns out to be equivalent to a type of automaton called a *finite-state automaton.* In this chapter, we focus on the study of regular languages and finite-state automata, leaving the subject of context-free grammars and their equivalent automata to a later course in compiler construction or automata theory.

12.1 *Formal Languages and Regular Expressions*

The mind has finite means but it makes unbounded use of them and in very specific and organized ways. That's the core problem of language that it became possible to face [by the mid-twentieth century]. — Noam Chomsky, circa 1998

Noam Chomsky (born 1928)

An English sentence can be regarded as a string of words, and an English word can be regarded as a string of letters. Not every string of letters is a legitimate word, and not every string of words is a grammatical sentence. We could say that a word is legitimate if it can be found in an unabridged English dictionary and that a sentence is grammatical if it satisfies the rules in a standard English grammar book.

Computer languages are similar to English in that certain strings of characters are legitimate words of the language and certain strings of words can be put together according to certain rules to form syntactically correct programs. A compiler for a computer language analyzes the stream of characters in a program—first to recognize individual word and sentence units (this part of the compiler is called a lexical scanner), then to analyze the syntax, or grammar, of the sentences (this part is called a syntactic analyzer), and finally to translate the sentences into machine code (this part is called a code generator).

In computer science it has proved useful to look at languages from a very abstract point of view as strings of certain fundamental units and allow any finite set of symbols to be used as an alphabet. It is common to denote an alphabet by a capital Greek sigma: Σ. (This just happens to be the same symbol as the one used for summation, but the two concepts have no other connection.)

---

*UNIX is an operating system that was developed in 1969 by Kenneth Thompson at Bell Laboratories. It was later rewritten in Dennis Ritchie's C language, which was also developed at Bell Laboratories.

[†]XML is a standard for defining markup languages used for Internet applications.

A *string of characters of an alphabet* Σ (or a *string over* Σ) is either (1) an ordered n-tuple of elements of Σ written without parentheses or commas, or (2) the null string ϵ, which has no characters. As we indicated when strings were introduced earlier, the *length of a string of characters* is the number of characters that make up the string, with the null string having length 0. A *formal language over an alphabet* is any set of strings of characters of the alphabet. These definitions are summarized below.

Alphabet Σ:	a finite set of characters
String over Σ:	an ordered n-tuple of elements of Σ, written without parentheses or commas, or (2) the null string ϵ
Formal language over Σ:	a set of strings over the alphabet

Note that the empty set satisfies the criteria for being a formal language. Allowing the empty set to be a formal language turns out to be convenient in certain technical situations.

Example 12.1.1 Examples of Formal Languages

Let the alphabet $\Sigma = \{a, b\}$.

a. Define a language L_1 over Σ to be the set of all strings that begin with the character a and have length at most three characters. Find L_1.

b. A **palindrome** is a string that looks the same if the order of its characters is reversed. For instance, *aba* and *baab* are palindromes. Define a language L_2 over Σ to be the set of all palindromes obtained using the characters of Σ. Write ten elements of L_2.

Solution

a. $L_1 = \{a, aa, ab, aaa, aab, aba, abb\}$

b. L_2 contains the following ten strings (among infinitely many others):

$$\epsilon, a, b, aa, bb, aaa, bab, abba, babaabab, abaabbbbbaaba \qquad \blacksquare$$

Stephen C. Kleene
(1909–1994)

University of Wisconsin

• **Notation**

Let Σ be an alphabet. For each nonnegative integer n, let

Σ^n = the set of all strings over Σ that have length n,

Σ^+ = the set of all strings over Σ that have length at least 1, and

Σ^* = the set of all strings over Σ.

Note that Σ^n is essentially the Cartesian product of n copies of Σ. The language Σ^* is called the **Kleene closure of Σ**, in honor of Stephen C. Kleene (pronounced CLAY-knee). Σ^+ is the set of all strings over Σ except for ϵ and is called the **positive closure of Σ**.

Example 12.1.2 The Languages Σ^n, Σ^+, and Σ^*

Let $\Sigma = \{a, b\}$.

a. Find Σ^0, Σ^1, Σ^2, and Σ^3.

b. Let $A = \Sigma^0 \cup \Sigma^1$ and $B = \Sigma^2 \cup \Sigma^3$. Use words to describe A, B, and $A \cup B$.

c. Describe a systematic way of writing the elements of Σ^+. What change needs to be made to obtain the elements of Σ^*?

Solution

a. $\Sigma^0 = \{\epsilon\}$, $\Sigma^1 = \{a, b\}$, $\Sigma^2 = \{aa, ab, ba, bb\}$, and $\Sigma^3 = \{aaa, aab, aba, abb, baa, bab, bba, bbb\}$

b. A is the set of all strings over Σ of length at most 1.

B is the set of all strings over Σ of length 2 or 3.

$A \cup B$ is the set of all strings over Σ of length at most 3.

c. Elements of Σ^+ can be written systematically by writing all the strings of length 1, then all the strings of length 2, and so forth.

$$\Sigma^+: a, b, aa, ab, ba, bb, aaa, aab, aba, abb, baa, bab, bba, bbb, aaaa, \ldots$$

Of course the process of writing the strings in Σ^+ would continue forever, because Σ^+ is an infinite set. The only change that needs to be made to obtain Σ^* is to place the null string at the beginning of the list. ∎

Example 12.1.3 Polish Notation: A Language Consisting of Postfix Expressions

An expression such as $a + b$, in which a binary operator such as $+$ sits between the two quantities on which it acts, is said to be written in infix notation. Alternative notations are called prefix notation (in which the binary operator precedes the quantities on which it acts) and postfix notation (in which the binary operator follows the quantities on which it acts). In prefix notation, $a + b$ is written $+ab$. In postfix notation, $a + b$ is written $ab+$.

Prefix and postfix notations were introduced in 1920 by the Polish mathematician Jan Łukasiewicz (pronounced Wu-cash-AY-vich). In his honor—and because some people have difficulty pronouncing his name—they are often referred to as *Polish notation* and *reverse Polish notation,* respectively. A great advantage of these notations is that they eliminate the need for parentheses in writing arithmetic expressions. For instance, in postfix (or reverse Polish) notation, the expression 8 4 + 6 / is evaluated from left to right as follows: Add 8 and 4 to obtain 12, and then divide 12 by 6 to obtain 2.

Let $\Sigma = \{4, 1, +, -\}$, and let $L =$ the set of all strings over Σ obtained by writing either a 4 or a 1 first, then either a 4 or a 1, and finally either a $+$ or a $-$. List all elements of L between braces, and evaluate the resulting expressions.

Solution

$$L = \{4\,1\,+,\ 4\,1\,-,\ 1\,4\,+,\ 1\,4\,-\}$$

$4\,1\,+$ equals $4 + 1 = 5$, $\quad 4\,1\,-$ equals $4 - 1 = 3$,

$1\,4\,+$ equals $1 + 4 = 5$, $\quad 1\,4\,-$ equals $1 - 4 = -3$ ∎

The following definition describes ways in which languages can be combined to form new languages.

> **• Definition**
>
> Let Σ be a finite alphabet. Given any strings x and y over Σ, the **concatenation of x and y** is the string obtained by juxtaposing the characters of x and y. For any languages L and L' over Σ, three new languages can be defined as follows:
>
> The **concatenation of L and L'**, denoted LL', is
>
> $$LL' = \{xy \mid x \in L \text{ and } y \in L'\}.$$
>
> The **union of L and L'**, denoted $L \cup L'$, is
>
> $$L \cup L' = \{x \mid x \in L \text{ or } x \in L'\}.$$
>
> The **Kleene closure of L**, denoted L^*, is
>
> $$L^* = \{x \mid x \text{ is a concatenation of any finite number of strings in } L\}.$$
>
> Note that ϵ is in L^* because it is regarded as a concatenation of zero strings in L.

Example 12.1.4 New Languages from Old

Let L_1 be the set of all strings consisting of an even number of a's (namely, ϵ, aa, $aaaa$, $aaaaaa$, ...), and let $L_2 = \{b, bb, bbb\}$. Find $L_1 L_2$, $L_1 \cup L_2$, and $(L_1 \cup L_2)^*$. Note that the null string ϵ is in L_1 because 0 is an even number.

Solution

$L_1 L_2 = $ the set of all strings that consist of an even number of a's followed by b or by bb or by bbb.

$L_1 \cup L_2 = $ the set that includes the strings b, bb, bbb and any strings consisting of an even number of a's.

$(L_1 \cup L_2)^* = $ the set of all strings of a's and b's in which every occurrence of a consists of an even number of a's. ∎

The Language Defined by a Regular Expression

One of the most useful ways to define a language is by means of a *regular expression,* a concept first introduced by Kleene. We give a recursive definition for generating the set of all regular expressions over an alphabet.

> **• Definition**
>
> Given a finite alphabet Σ, the following are **regular expressions over Σ:**
>
> I. BASE: $\emptyset$, ϵ, and each individual symbol in Σ are regular expressions over Σ.
>
> II. RECURSION: If r and s are regular expressions over Σ, then the following are also regular expressions over Σ:
>
> $$\text{(i) } (rs) \qquad \text{(ii) } (r \mid s) \qquad \text{(iii) } (r^*)$$
>
> To agree with the terminology for strings, rs is called the **concatenation of r and s,** r^* is called the **Kleene closure of r,** and $r \mid s$ is read "r or s."
>
> III. RESTRICTION: Nothing is a regular expression over Σ except for objects defined in (I) and (II) above.

As an example, one regular expression over $\Sigma = \{a, b\}$ is

$$a \mid (b \mid c)^* \mid (ab)^*.$$

If the alphabet Σ happens to include symbols such as (, |,), or *, special provisions have to be made to avoid ambiguity. An *escape character,* usually a backslash, is added before the potentially ambiguous symbol. For instance, a left parenthesis would be written as \(and the backslash itself would be written as \\.

To eliminate parentheses, an order of precedence for the operations used to define regular expressions has been introduced. The highest is *, concatenation is next, and | is the lowest. It is also customary to eliminate the outer set of parentheses in a regular expression, because doing so does not produce ambiguity. Thus

$$(a((bc)^*)) = a(bc)^* \quad \text{and} \quad (a \mid (bc)) = a \mid bc.$$

Example 12.1.5 Order of Precedence for the Operations in a Regular Expression

a. Add parentheses to make the order of precedence clear in the following expression: $ab^* \mid b^*a$.

b. Use the convention about order of precedence to eliminate the parentheses in the following expression: $((a \mid ((b^*)c))(a^*))$.

Solution

a. $((a(b^*)) \mid ((b^*)a))$

b. $(a \mid b^*c)a^*$ ■

Given a finite alphabet, every regular expression r over the alphabet defines a formal language $L(r)$. The function L is defined recursively.

> • **Definition**
>
> For any finite alphabet Σ, the function L that associates a language to each regular expression over Σ is defined by (I) and (II) below. For each such regular expression r, $L(r)$ is called the **language defined by** r.
>
> I. BASE: $L(\emptyset) = \emptyset$, $L(\epsilon) = \{\epsilon\}$, $L(a) = \{a\}$ for every $a \in \Sigma$.
>
> II. RECURSION: If $L(r)$ and $L(r')$ are the languages defined by the regular expressions r and r' over Σ, then
>
> (i) $L(rr') = L(r)L(r')$ (ii) $L(r \mid r') = L(r) \cup L(r')$ (iii) $L(r^*) = (L(r))^*$

Note that any finite language can be defined by a regular expression. For instance, the language {cat, dog, bird} is defined by the regular expression (cat | dog | bird). An important example is the following.

Example 12.1.6 Using Set Notation to Describe the Language Defined by a Regular Expression

Let $\Sigma = \{a, b\}$, and consider the language defined by the regular expression $(a \mid b)^*$. Use set notation to find this language, and describe it in words.

Solution   The language defined by $(a\,|\,b)^*$ is

$$
\begin{aligned}
(L(a\,|\,b)^*) &= (L(a\,|\,b))^* \\
&= (L(a) \cup L(b))^* \\
&= (\{a\} \cup \{b\})^* \\
&= \{a, b\}^* \qquad\qquad \text{by definition of operations on languages} \\
&= \text{the set of all strings of } a\text{'s and } b\text{'s} \\
&= \Sigma^*.
\end{aligned}
$$ ∎

Note that concatenating strings and taking unions of sets are both associative operations. Thus for any regular expressions r, s and t,

$$L((rs)t) = L(r(st)).$$

Moreover,

$$
\begin{aligned}
L\,((r\,|\,s)\,|\,t) &= (L(r\,|\,s)) \cup L(t) && \text{by definition of } | \\
&= (L(r) \cup L(s)) \cup L(t) && \text{by definition of } | \\
&= L(r) \cup (L(s) \cup L(t)) && \text{by the associativity of union for sets} \\
&= L(r) \cup (L(s\,|\,t)) && \text{by definition of } | \\
&= L(r\,|\,(s\,|\,t)) && \text{by definition of } |.
\end{aligned}
$$

Because of these relationships, it is customary to drop the parentheses in "associative" situations and write

$$rst = (rs)t = r(st)$$

and

$$r\,|\,s\,|\,t = (r\,|\,s)\,|\,t = r\,|\,(s\,|\,t).$$

As you become accustomed to working with regular expressions, you will find that you do not need to go through a formal derivation in order to determine the language defined by an expression.

Example 12.1.7 The Language Defined by a Regular Expression

Let $\Sigma = \{0, 1\}$. Use words to describe the languages defined by the following regular expressions over Σ.

a. $0^*1^*\,|\,1^*0^*$ b. $0(0\,|\,1)^*$

Solution

a. The strings in this language consist either of a string of 0's followed by a string of 1's or of a string of 1's followed by a string of 0's. However, in either case the strings could be empty, which means that ϵ is also in the language.

b. The strings in this language have to start with a 0. The 0 may be followed by any finite number (including zero) of 0's and 1's in any order. Thus the language is the set of all strings of 0's and 1's that start with a 0. ∎

Example 12.1.8 Individual Strings in the Language Defined by a Regular Expression

In each of (a) and (b), let $\Sigma = \{a, b\}$ and consider the language L over Σ defined by the given regular expression.

a. The regular expression is $a^*b(a \mid b)^*$. Write five strings that belong to L.

b. The regular expression is $a^* \mid (ab)^*$. Indicate which of the following strings belong to L:

$$a \quad b \quad aaaa \quad abba \quad ababab$$

Solution

a. The strings b, ab, $abbb$, $abaaa$, and $ababba$ are five strings from the infinitely many in L.

b. The following strings are the only ones listed that belong to L: a, $aaaa$, and $ababab$. The string b does not belong to L because it is neither a string of a's nor a string of possibly repeated ab's. The string $abba$ does not belong to L because any two b's that might occur in a string of L are separated by one or more a's. ∎

Example 12.1.9 A Regular Expression That Defines a Language

Let $\Sigma = \{0, 1\}$. Find regular expressions over Σ that define that following languages.

a. The language consisting of all strings of 0's and 1's in which the 0's and 1's alternate.

b. The language consisting of all strings of 0's and 1's with an even number of 1's. Such strings are said to have *even parity*.

c. The language consisting of all strings of 0's and 1's in which every 1 has all its neighbors equal to 0. Note, however, that a string beginning with a 1 only has a neighbor to its right, and a string ending with a 1 only has a neighbor to its left.

Solution

a. Any string in the language starts with a 1 or a 0. If it starts with a 1, the pattern 10 must continue for the length of the string. If it starts with 0, the pattern 01 must continue for the length of the string. Also, the null string satisfies the condition by default. Thus an answer is

$$(10)^* \mid (01)^*.$$

b. Basic strings with even parity are ϵ, 0, and 10^*1. Concatenation of strings with even parity also have even parity. Because such a string may start or end with a string of 0's, an answer is

$$(0 \mid 10^*1)^*.$$

c. If the string starts or ends with a 1, then only the character to the right or left of the 1 has to be a 0, but if the string has length greater than 2, then every 1 between the two ends must be surrounded by 0's. Note that the null string satisfies the condition to be in the language, and so does any string consisting entirely of 0's. Thus an answer is

$$(\epsilon \mid 1)((101)^* \mid 0)^*(\epsilon \mid 1). \qquad \blacksquare$$

Note that a given language may be defined by more than one regular expression. For example, both

$$(a^* \mid b^*)^* \quad \text{and} \quad (a \mid b)^*$$

define the language consisting of the set of all strings of a's and b's.

Example 12.1.10 Deciding Whether Regular Expressions Define the Same Language

In (a) and (b), determine whether the given regular expressions define the same language. If they do, describe the language. If they do not, give an example of a string that is in one of the languages but not the other.

a. $(a \mid \epsilon)^*$ and a^*

b. $0^* \mid 1^*$ and $(01)^*$

Solution

a. Note that because the null string ϵ has no characters, when it is concatenated with any other string x, the result is just x: for all strings x, $x\epsilon = \epsilon x = x$. Now $L((a \mid \epsilon)^*)$ is the set of strings formed using a and ϵ in any order, and so, because $a\epsilon = \epsilon a = a$, this is the same as the set of strings consisting of zero or more a's. Thus $L((a \mid \epsilon)^*) = L(a^*)$.

b. The two languages defined by the given regular expressions are not the same: 0101 is in the second language but not the first. ∎

Practical Uses of Regular Expressions

Many applications of computers involve performing operations on pieces of text. For instance, word and text processing programs allow us to find certain words or phrases in a document and possibly replace them with others. A compiler for a computer language analyzes an incoming stream of characters to find groupings that represent aspects of the computer language such as keywords, constants, identifiers, and operators. And in bioinformatics, pattern matching and flexible searching techniques are used extensively to analyze the long sequences of the characters A, C, G, and T that occur in DNA.

Through their connection with finite-state automata, which we discuss in the next section, regular expressions provide an extremely useful way to describe a pattern in order to identify a string or a collection of strings within a piece of text. Regular expressions make it possible to replace a long, complicated set of if-then-else statements with code that is easy both to produce and to understand. Because of their convenience, regular expressions were introduced into a number of UNIX utilities, such as *grep* (short for *g*lobally *s*earch for *r*egular *e*xpression and *p*rint) and *egrep* (*e*xtended *grep*), in text editors, such as *QED* (short for *Q*uick *ED*itor, the first text editor to use regular expressions), *vi* (short for *v*isual *i*nterface), *sed* (short for *s*tream *ed*itor and originally developed for UNIX but now used by many systems), and *Emacs* (short for *E*ditor *mac*ros), and in the lexical scanner component of a compiler. The computer language Perl has a particularly powerful implementation for regular expressions, which has become a de facto standard. The implementations used in Java and .NET are similar.

A number of shorthand notations have been developed to facilitate working with regular expressions in text processing. When characters in an alphabet or in a part of an alphabet are understood to occur in a standard order, the notation [*beginning character–ending character*] is commonly used to represent the regular expression that consists of a single character in the range from the beginning to the ending character. It is called a **character class.** Thus

$$[A - C] \quad \text{stands for} \quad (A \mid B \mid C)$$

and

$$[0 - 9] \quad \text{stands for} \quad (0 \mid 1 \mid 2 \mid 3 \mid 4 \mid 5 \mid 6 \mid 7 \mid 8 \mid 9).$$

Character classes are also allowed to include more than one range of characters. For instance,

$$[A - C \ x - z] \quad \text{stands for} \quad (A \mid B \mid C \mid x \mid y \mid z)$$

As an example, consider the language defined by the regular expression

$$[A - Z \ a - z]([A - Z \ a - z] \mid [0 - 9])^*.$$

The following are some strings in the language:

Account Number,   *z23*,   *jsmith*109,   *Draft2rev*.

In general, the language is the set of all strings that start with a letter followed by a sequence of digits or letters. This set is the same as the set of allowable identifiers in a number of computer languages.

Other commonly used shorthands are

$$[ABC] \text{ to stand for } (A \mid B \mid C)$$

and a single dot

$$. \text{ to stand for an arbitrary character.}$$

Thus, for instance, if $\Sigma = \{A, B, C\}$, then

$$A.C \quad \text{stands for} \quad (AAC \mid ABC \mid ACC).$$

When the symbol ^ is placed at the beginning of a character class, it indicates that a character of the same type as those in the range of the class, but not any of the characters following the ^, is to occur at that point in the string. For instance,

$$[^{\wedge}D - Z][0 - 9][0 - 9]^*$$

stands for any string starting with a letter of the alphabet different from D to Z, followed by any positive number of digits from 0 to 9. Examples are $B3097$, $C0046$, and so forth. The symbol $+$ following a regular expression r means that the string contains at least one occurrence of r. In symbols,

$$r^+ = rr^*.$$

For example,

$$[A - Z]^+$$

represents any nonempty string of capital letters. If r is a regular expression, then

$$r? = (\epsilon \mid r).$$

That is, $r?$ denotes either zero occurrences or exactly one occurrence of r. Finally,

$$r\{n\} \quad \text{means that} \quad r \text{ is concatenated } n \text{ times,}$$

and

$$r\{m, n\} \quad \text{means that} \quad r \text{ is concatenated between } m \text{ and } n \text{ times.}$$

Thus a check to help determine whether a given string is a local telephone number in the United States is to see whether it has the form

$$[0 - 9][0 - 9][0 - 9] \text{ - } [0 - 9][0 - 9][0 - 9][0 - 9],$$

or, equivalently, whether it has the form

$$[0 - 9]\{3\} \text{ - } [0 - 9]\{4\}.$$

Example 12.1.11 A Regular Expression for a Date

People often write dates in a variety of formats. For instance, in the United States the following all represent the second of February of 1983:

$$2/5/83 \qquad 2\text{-}5\text{-}83 \qquad 02/05/83 \qquad 2/05/83 \qquad 02/5/83$$

Write a regular expression that would help check whether a given string might be a valid date written in one of these forms.

Solution   The language defined by following regular expression consists of all strings that begin with one or two digits followed by either a hyphen or a slash, followed by either one or two digits, followed by either a hyphen or a slash, followed by two digits.

$$[0-9]\{2\}[\text{ - }/]([0-9] \mid [0-9]\{2\})[\text{ - }/][0-9]\{2\}$$

All valid dates of the given format are elements of the language defined by this expression, but the language also includes strings that are not valid dates. For instance, 09/54/83 is in the language, but it is not a valid date because September does not have 54 days, and 38/12/83 is not valid because there is no 38th month. It is possible to write a more complicated regular expression that could be used to check all aspects of the validity of a date (see exercise 40 at the end of the section), but the kind of simpler expression given above is nonetheless useful. For instance, it provides an easy way to notify a user of an interactive program that a certain kind of mistake was made and that information should be reentered.

■

Exercise Set 12.1*

In 1 and 2 let $\Sigma = \{x, y\}$ be an alphabet.

1. a. Let L_1 be the language consisting of all strings over Σ that are palindromes and have length ≤ 4. List the elements of L_1 between braces.

b. Let L_2 be the language consisting of all strings over Σ that begin with an x and have length ≤ 3. List the elements of L_2.

2. a. Let L_3 be the language consisting of all strings over Σ of length ≤ 3 in which all the x's appear to the left of all the y's. List the elements of L_3 between braces.

b. List between braces the elements of Σ^4, the set of strings of length 4 over Σ.

c. Let $A = \Sigma^1 \cup \Sigma^2$ and $B = \Sigma^3 \cup \Sigma^4$. Describe A, B, and $A \cup B$ in words.

H 3. Let $\Sigma = \{1, 2, {}^*, /\}$ and let L be the set of all strings over Σ obtained by writing first a number (1 or 2), then a second number (1 or 2), which can be the same as the first one, and finally an operation (* or / where * indicates multiplication and / indicates division). Then L is a set of postfix, or reverse Polish, expressions. List all the elements of L between braces, and evaluate the resulting expressions.

In 4–6, describe $L_1 L_2$, $L_1 \cup L_2$, and $(L_1 \cup L_2)^*$ for the given languages L_1 and L_2.

4. L_1 is the set of all strings of a's and b's that start with an a and contain only that one a; L_2 is the set of all strings of a's and b's that contain an even number of a's.

5. L_1 is the set of all strings of a's, b's, and c's that contain no c's and have the same number of a's as b's; L_2 is the set of all strings of a's, b's, and c's that contain no a's or b's.

6. L_1 is the set of all strings of 0's and 1's that start with a 0, and L_2 is the set of all strings of 0's and 1's that end with a 0.

In 7–9, add parentheses to make the order of precedence clear in the given expressions.

7. $(a \mid b^* b)(a^* \mid ab)$ **8.** $0^* 1 \mid 0(0^* 1)^*$

9. $(x \mid yz^*)^*(yx \mid (yz)^* z)$

In 10–12 use the convention about order of precedence to eliminate the parentheses in the given regular expression.

10. $((a(b^*)) \mid (c(b^*)))\,((ac) \mid (bc))$

11. $(1(1^*)) \mid ((1(0^*)) \mid ((1^*)1))$

12. $(xy)(((x^*)y)^*) \mid (((yx) \mid y)(y^*))$

*For exercises with blue numbers or letters, solutions are given in Appendix B. The symbol **H** indicates that only a hint or a partial solution is given. The symbol ✶ signals that an exercise is more challenging than usual.

In 13–15 use set notation to derive the language defined by the given regular expression. Assume $\Sigma = \{a, b, c\}$.

13. $\epsilon \mid ab$ **14.** $\emptyset \mid \epsilon$ **15.** $(a \mid b)c$

In 16–18 write five strings that belong to the language defined by the given regular expression.

16. $0^*1(0^*1^*)^*$ **17.** $b^* \mid b^*ab^*$ **18.** $x^*(yxxy \mid x)^*$

In 19–21 use words to describe the language defined by the given regular expression.

19. $b^*ab^*ab^*a$ **20.** $1(0 \mid 1)^*00$ **21.** $(x \mid y)y(x \mid y)^*$

In 22–24 indicate whether the given strings belong to the language defined by the given regular expression. Briefly justify your answers.

22. Expression: $(b \mid \epsilon)a(a \mid b)^*a(b \mid \epsilon)$, strings: $aaaba, baabb$

23. Expression: $(x^*y \mid zy^*)^*$, strings: $zyyxz, zyyzy$

24. Expression: $(01^*2)^*$, strings: $120, 01202$

In 25–27 find a regular expression that defines the given language.

25. The language consisting of all strings of 0's and 1's with an odd number of 1's. (Such a string is said to have *odd parity*.)

26. The language consisting of all strings of a's and b's in which the third character from the end is a b.

27. The language consisting of strings of x's and y's in which the elements in every pair of x's are separated by at least one y.

Let r, s, and t be regular expressions over $\Sigma = \{a, b\}$. In 28–30 determine whether the two regular expressions define the same language. If they do, describe the language. If they do not, give an example of a string that is in one of the languages but not the other.

28. $(r \mid s)t$ and $rt \mid st$ **29.** $(rs)^*$ and r^*s^*

30. $(rs)^*$ and $((rs)^*)^*$

In 31–39 write a regular expression to define the given set of strings. Use the shorthand notations given in the section whenever convenient. In most cases, your expression will describe other strings in addition to the given ones, but try to make your answer fit the given strings as closely as possible within reasonable space limitations.

31. All words that are written in lower-case letters and start with the letters *pre* but do not consist of *pre* all by itself.

32. All words that are written in upper-case letters, and contain the letters *BIO* (as a unit) or *INFO* (as a unit).

33. All words that are written in lower-case letters, end in *ly*, and contain at least five letters.

34. All words that are written in lower-case letters and contain at least one of the vowels a, e, i, o, or u.

35. All words that are written in lower-case letters and contain exactly one of the vowels a, e, i, o, or u.

36. All words that are written in upper-case letters and do not start with one of the vowels A, E, I, O, or U but contain exactly two of these vowels next to each other.

37. All United States social security numbers (which consist of three digits, a hyphen, two digits, another hyphen, and finally four more digits), where the final four digits start with a 3 and end with a 6.

38. All telephone numbers that have three digits, then a hyphen, then three more digits, then a hyphen, and then four digits, where the first three digits are either 800 or 888 and the last four digits start and end with a 2.

39. All signed or unsigned numbers with or without a decimal point. A signed number has one of the prefixes + or –, and an unsigned number does not have a prefix. Represent the decimal point as \\. to distinguish it from the single dot symbol for an arbitrary character.

H **40.** Write a regular expression to perform a complete check to determine whether a given string represents a valid date from 1980 to 2079 written in one of the formats of Example 12.1.11. (During this period, leap years occur every four years starting in 1980.)

$\bigstar$ **41.** Write a regular expression to define the set of strings of 0's and 1's with an even number of 0's and even number of 1's.

12.2 Finite-State Automata

The world of the future will be an ever more demanding struggle against the limitations of our intelligence, not a comfortable hammock in which we can lie down to be waited upon by our robot slaves. — Norbert Wiener, 1964

The kind of circuit discussed in Section 1.4 is called a *combinational circuit*. Such a circuit is characterized by the fact that its output is completely determined by its input/output table, or, in other words, by a Boolean function. Its output does not depend in any way on the history of previous inputs to the circuit. For this reason, a combinational circuit is said to have no memory.

Combinational circuits are very important in computer design, but they are not the only type of circuits used. Equally important are *sequential circuits*. For sequential circuits one cannot predict the output corresponding to a particular input unless one also knows something about the prior history of the circuit, or, more technically, unless one knows the state the circuit was in before receiving the input. The behavior of a sequential circuit is a function not only of the input to the circuit but also of the state the circuit is in when the input is received. A computer memory circuit is a type of sequential circuit.

A **finite-state automaton** (aw-TAHM-uh-tahn) is an idealized machine that embodies the essential idea of a sequential circuit. Each piece of input to a finite-state automaton leads to a change in the state of the automaton, which in turn affects how subsequent input is processed. Imagine, for example, the act of dialing a telephone number. Dialing 1–800 puts the telephone circuit in a state of readiness to receive the final seven digits of a toll-free call, whereas dialing 328 leads to a state of expectation for the four digits of a local call. Vending machines operate similarly. Just knowing that you put a quarter into a vending machine is not enough for you to be able to predict what the behavior of the machine will be. You also have to know the state the machine was in when the quarter was inserted. If 75¢ had already been deposited, you might get a beverage or some candy, but if the quarter was the first coin deposited, you would probably get nothing at all.

Example 12.2.1 A Simple Vending Machine

A simple vending machine dispenses bottles of juice that cost $1 each. The machine accepts quarters and half-dollars only and does not give change. As soon as the amount deposited equals or exceeds $1 the machine releases a bottle of juice. The next coin deposited starts the process over again. The operation of the machine is represented by the diagram of Figure 12.2.1.

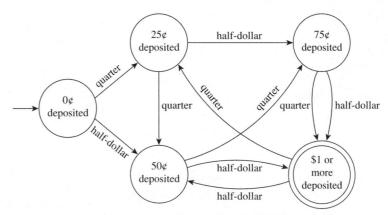

Figure 12.2.1 A Simple Vending Machine

Each circle represents a state of the machine: the state in which 0¢ has been deposited 25¢, 50¢, 75¢, and $1 or more. The unlabeled arrow pointing to "0¢ deposited" indicates that this is the initial state of the machine. The double circle around "$1 or more deposited" indicates that a bottle of juice is released when the machine has reached this state. (It is called an *accepting state* of the machine because when the machine is in this state, it has accepted the input sequence of coins as payment for juice.) The arrows that link the states indicate what happens when a particular input is made to the machine in each of its various states. For instance, the arrow labeled "quarter" that goes from "0¢ deposited" to "25¢ deposited" indicates that when the machine is in the state "0¢ deposited" and a quarter is inserted, the machine goes to the state "25¢ deposited." The arrow labeled "half-dollar" that goes from "75¢ deposited" to "$1 or more deposited" indicates that when the machine is in the state "75¢ deposited" and a half-dollar is inserted, the machine goes to the state "$1 or more deposited" and juice is dispensed. (In this case the purchaser would pay $1.25 for the juice because the machine does not return change.) The arrow labeled "quarter" that goes from "$1 or more deposited" to "25¢ deposited" indicates that when the machine is in the state "$1 or more deposited" and a quater is inserted, the machine goes back to the state "25¢ deposited." (This corresponds to the fact that after the machine has dispensed a bottle of juice, it starts operation all over again.)

Equivalently, the operation of the vending machine can be represented by a *next-state table* as shown in Table 12.2.1.

Table 12.2.1 Next-State Table

		Input	
		Quarter	**Half-Dollar**
→	0¢ deposited	25¢ deposited	50¢ deposited
	25¢ deposited	50¢ deposited	75¢ deposited
State	50¢ deposited	75¢ deposited	$1 or more deposited
	75¢ deposited	$1 or more deposited	$1 or more deposited
⊙	$1 or more deposited	25¢ deposited	50¢ deposited

The arrow pointing to "0¢ deposited" in the table indicates that the machine begins operation in this state. The double circle next to "$1 or more deposited" indicates that a bottle of juice is released when the machine has reached this state. Entries in the body of the table are interpreted in the obvious way. For instance, the entry in the third row of the column labeled *Half-Dollar* shows that when the machine is in state "50¢ deposited" and a half-dollar is deposited, it goes to state "$1 or more deposited."

Note that Table 12.2.1 conveys exactly the same information as the diagram of Figure 12.2.1. If the diagram is given, the table can be constructed; and if the table is given, the diagram can be drawn. ■

*David Hilbert
(1862–1943)*

David Eugene Smith Collection, Columbia University

*Alan M. Turing
(1912–1954)*

Godfrey Argent

Observe that the vending machine described in Example 12.2.1 can be thought of as having a primitive memory: It "remembers" how much money has been deposited (within limits) by referring to the state it is in. This capability for storing information and acting upon it is what gives finite-state automata* their tremendous power.

The most important finite-state automata are digital computers. Each computer consists of several subsystems: input devices, a processor, and output devices. A processor typically consists of a central processing unit and a finite number of memory locations. At any given time, the state of the processor is determined by the locations and values of all the bits stored within its memory. A computer that has n different locations for storing a single bit can therefore exist in 2^n different states. For a modern computer, n is many billions or even trillions, so the total number of states is enormous. But it *is* finite. Therefore, despite the complexity of a computer, just as for a vending machine, it is possible to predict the next state given knowledge of the current state and the input. Indeed, this is essentially what programmers try to do every time they write a program. Fortunately, modern, high-level computer languages provide a lot of help.

The basic theory of automata was developed to answer very theoretical questions about the foundations of mathematics posed by the great German mathematician David Hillbert in 1900. The ground-breaking work on automata was done in the mid-1930s by the English mathematician and logician Alan M. Turing. In the 1940s and 1950s, Turing's work played an important role in the development of real-world automatic computers.

**Automata is the plural of *automaton.*

Definition of a Finite-State Automaton

A general *finite-state automaton* is completely described by giving a set of states, together with an indication about which is the initial state and which are the accepting states (when something special happens), a list of all input elements, and specification for a *next-state function* that defines which state is produced by each input in each state. This is formalized in the following definition:

• Definition

A **finite-state automaton** A consists of five objects:

1. A set I, called the **input alphabet,** of input symbols;

2. A set S of **states** the automaton can be in;

3. A designated state s_0 called the **initial state;**

4. A designated set of states called the set of **accepting states;**

5. A **next-state function** $N: S \times I \rightarrow S$ that associates a "next-state" to each ordered pair consisting of a "current state" and a "current input." For each state s in S and input symbol m in I, $N(s, m)$ is called the state to which A goes if m is input to A when A is in state s.

The operation of a finite-state automaton is commonly described by a diagram called a **(state-)transition diagram,** similar to that of Figure 12.2.1. It is called a *transition diagram* because it shows the transitions the machine makes from one state to another in response to various inputs. In a transition diagram, states are represented by circles and accepting states by double circles. There is one arrow that points to the initial state and other arrows that are labeled with input symbols and point from each state to other states to indicate the action of the next-state function. Specifically, an arrow from state s to state t labeled m means that $N(s, m) = t$.

The **next-state table** for an automaton shows the values of the next-state function N for all possible states s and input symbols i. In the **annotated next-state table,** the initial state is indicated by an arrow and the accepting states are marked by double circles.

Example 12.2.2 A Finite-State Automaton Given by a Transition Diagram

Consider the finite-state automaton A defined by the transition diagram shown in Figure 12.2.2.

a. What are the states of A?

b. What are the input symbols of A?

c. What is the initial state of A?

d. What are the accepting states of A?

e. Find $N(s_1, 1)$.

f. Find the annotated next-state table for A.

Figure 12.2.2

Solution

a. The states of A are s_0, s_1, and s_2 [*since these are the labels of the circles*].

b. The input symbols of A are 0 and 1 [*since these are the labels of the arrows*].

c. The initial state of A is s_0 [*since the unlabeled arrow points to s_0*].

d. The only accepting state of A is s_2 [*since this is the only state marked by a double circle*].

e. $N(s_1, 1) = s_2$ [*since there is an arrow from s_1 to s_2 labeled 1*].

f.

		Input	
		0	1
$\rightarrow$	s_0	s_1	s_0
State $\quad$	s_1	s_1	s_2
$\odot$	s_2	s_1	s_0

■

Example 12.2.3 A Finite-State Automaton Given by an Annotated Next-State Table

Consider the finite-state automaton A defined by the following annotated next-state table:

		Input		
		a	b	c
$\rightarrow$	U	Z	Y	Y
$\odot$	V	V	V	V
State $\quad$	Y	Z	V	Y
$\odot$	Z	Z	Z	Z

a. What are the states of A?

b. What are the input symbols of A?

c. What is the initial state of A?

d. What are the accepting states of A?

e. Find $N(U, c)$.

f. Draw the transition diagram for A.

Solution

a. The states of A are U, V, Y, and Z.

b. The input symbols of A are a, b, and c.

c. The initial state of A is U [*since the arrow points to U*].

d. The accepting states of A are V and Z [*since these are marked with double circles*].

e. $N(U, c) = Y$ [*since the entry in the row labeled U and the column labeled c of the next-state table is Y*].

f. The transition diagram for A is shown in Figure 12.2.3. It can be drawn more compactly by labeling arrows with multiple-input symbols where appropriate. This is illustrated in Figure 12.2.4.

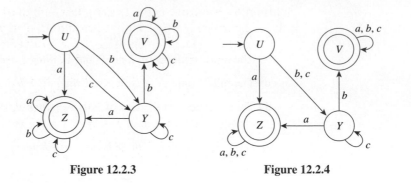

Figure 12.2.3   Figure 12.2.4   ■

The Language Accepted by an Automaton

Now suppose a string of input symbols is fed into a finite-state automaton in sequence. At the end of the process, after each successive input symbol has changed the state of the automaton, the automaton ends up in a certain state, which may be either an accepting state or a nonaccepting state. In this way, the action of a finite-state automaton separates the set of all strings of input symbols into two subsets: those that send the automaton to an accepting state and those that do not. Those strings that send the automaton to an accepting state are said to be *accepted* by the automaton.

> • **Definition**
>
> Let A be a finite-state automaton with set of input symbols I. Let I^* be the set of all strings over I, and let w be a string in I^*. Then **w is accepted by** A if, and only if, A goes to an accepting state when the symbols of w are input to A in sequence from left to right, starting when A is in its initial state. The **language accepted by** A, denoted $L(A)$, is the set of all strings that are accepted by A.

Example 12.2.4 Finding the Language Accepted by an Automaton

Consider the finite-state automaton A defined in Example 12.2.2 and shown again below.

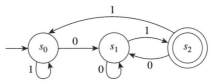

a. To what states does A go if the symbols of the following strings are input to A in sequence, starting from the initial state?

　　　　　　(i) 01　　(ii) 0011　　(iii) 0101100　　(iv) 10101

b. Which of the strings in part (a) send A to an accepting state?

c. What is the language accepted by A?

d. Is there a regular expression that defines the same language?

Solution

a. (i) s_2　　(ii) s_0　　(iii) s_1　　(iv) s_2

b. The strings 01 and 10101 send A to an accepting state.

c. Observe that if w is any string that ends in 01, then w is accepted by A. For if w is any string of length $n \geq 2$, then after the first $n-2$ symbols of w have been input, A is in one of its three states: s_0, s_1, or s_2. But from any of these three states, input of the symbols 01 in sequence sends A first to s_1 and then to the accepting state s_2. Hence any string that ends in 01 is accepted by A.

Also note that the only strings accepted by A are those that end in 01. (That is, no other strings besides those ending in 01 are accepted by A.) The reason for this is that the only accepting state of A is s_2, and the only arrow pointing to s_2 comes from s_1 and is labeled 1. Thus in order for an input string w of length n to send A to an accepting state, the last symbol of w must be a 1 and the first $n-1$ symbols of w must send A to state s_1. Now three arrows point to s_1, one from each of the three states of A, and all are labeled 0. Thus the last of the first $n-1$ symbols of w must be 0, or, in other words, the next-to-the-last symbol of w must be 0. Hence the last two symbols of w must be 01, and thus

$$L(A) = \text{the set of all strings of 0's and 1's that end in 01.}$$

d. Yes. One regular expression that defines $L(A)$ is $(0\,|\,1)^*01$. ∎

A finite-state automaton with multiple accepting states can have output devices attached to each one so that the automaton can classify input strings into a variety of different categories, one for each accepting state. This is how finite-state automata are used in the lexical scanner component of a computer compiler to group the symbols from a stream of input characters into identifiers, keywords, and so forth.

The Eventual-State Function

Now suppose a finite-state automaton is in one of its states (not necessarily the initial state) and a string of input symbols is fed into it in sequence. To what state will the automaton eventually go? The function that gives the answer to this question for every possible combination of input strings and states of the automaton is called the *eventual-state function*.

> **• Definition**
>
> Let A be a finite-state automaton with set of states S, set of input symbols I, and next-state function $N\colon S \times I \to S$. Let I^* be the set of all strings over I, and define the **eventual-state function** $N^*\colon S \times I^* \to S$ as follows:
>
> For any state s and for any input string w,
>
> $$N^*(s, w) = \begin{bmatrix} \text{the state to which } A \text{ goes if the} \\ \text{symbols of } w \text{ are input to } A \text{ in sequence,} \\ \text{starting when } A \text{ is in state } s \end{bmatrix}.$$

Example 12.2.5 Computing Values of the Eventual-State Function

Consider again the finite-state automaton of Example 12.2.2 shown below for convenience. Find $N^*(s_1, 10110)$.

Solution By definition of the eventual-state function,

$$N^*(s_1, 10110) = \begin{bmatrix} \text{the state to which } A \text{ goes if the} \\ \text{symbols of 10110 are input to } A \text{ in} \\ \text{sequence, starting when } A \text{ is in state } s_1 \end{bmatrix}.$$

By referring to the transition diagram for A, you can see that starting from s_1, when a 1 is input, A goes to s_2; then when a 0 is input, A goes back to s_1; after that, when a 1 is input, A goes to s_2; from there, when a 1 is input, A goes to s_0; and finally, when a 0 is input, A goes back to s_1. This sequence of state transitions can be written as follows:

$$s_1 \ \xrightarrow{1} \ s_2 \ \xrightarrow{0} \ s_1 \ \xrightarrow{1} \ s_2 \ \xrightarrow{1} \ s_0 \ \xrightarrow{0} \ s_1.$$

Thus, after all the symbols of 10110 have been input in sequence, the eventual state of A is s_1, so

$$N^*(s_1, 10110) = s_1. \qquad \blacksquare$$

The definitions of string and language accepted by an automaton can be restated symbolically using the eventual-state function. Suppose A is a finite-state automaton with set of input symbols I and next-state function N, and suppose that I^* is the set of all strings over I and that w is a string in I^*.

$$w \text{ is accepted by } A \quad \Leftrightarrow \quad N^*(s_0, w) \text{ is an accepting state of } A$$

$$L(A) = \{w \in I^* \mid N^*(s_0, w) \text{ is an accepting state of } A\}$$

Designing a Finite-State Automaton

Now consider the problem of starting with a description of a language and designing an automaton to accept exactly that language.

Example 12.2.6 A Finite-State Automaton That Accepts the Set of Strings of 0's and 1's for Which the Number of 1's Is Divisible by 3

a. Design a finite-state automaton A that accepts the set of all strings of 0's and 1's such that the number of 1's in the string is divisible by 3.

b. Is there a regular expression that defines this set?

Solution

a. Let s_0 be the initial state of A, s_1 its state after one 1 has been input, and s_2 its state after two 1's have been input. Note that s_0 is the state of A after zero 1's have been input, and since zero is divisible by 3 ($0 = 0 \cdot 3$), s_0 must be an accepting state. The states s_0, s_1, and s_2 must be different from one another because from state s_0 three 1's are needed to reach a new total divisible by 3, whereas from state s_1 two additional 1's are necessary, and from state s_2 just one more 1 is required.

Now the state of A after three 1's have been input can also be taken to be s_0 because after three 1's have been input, three more are needed to reach a new total divisible by 3.

More generally, if $3k$ 1's have been input to A, where k is any nonnegative integer, then three more are needed for the total again to be divisible by 3 (since $3k + 3 = 3(k + 1)$). Thus the state in which $3k$ 1's have been input, for any nonnegative integer k, can be taken to be the initial state s_0.

By similar reasoning, the states in which $(3k + 1)$ 1's and $(3k + 2)$ 1's have been input, where k is a nonnegative integer, can be taken to be s_1 and s_2, respectively.

Now every nonnegative integer can be written in one of the three forms $3k$, $3k + 1$, or $3k + 2$ (see Section 3.4), so the three states $s_0, s_1,$ and s_2 are all that is needed to create A. Thus the states of A can be drawn and labeled as shown below.

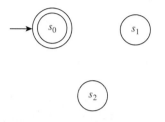

Next consider the possible inputs to A in each of its states. No matter what state A is in, if a 0 is input the total number of 1's in the input string remains unchanged. Thus there is a loop at each state labeled 0.

Now suppose a 1 is input to A when it is in state s_0. Then A goes to state s_1 (since the total number of 1's in the input string has changed from $3k$ to $3k + 1$). Similarly, if a 1 is input to A when it is in state s_1, then A goes to state s_2 (since the total number of 1's in the input string has changed from $3k + 1$ to $3k + 2$). Finally, if a 1 is input to A when it is in state s_2, then it goes to state s_0 (since the total number of 1's in the input string becomes $(3k + 2) + 1 = 3k + 3 = 3(k + 1)$, which is a multiple of 3.)

It follows that the transition diagram for A has the appearance shown below.

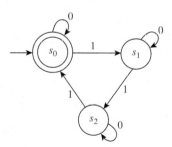

This automaton accepts the set of strings of 0's and 1's for which the number of 1's is divisible by 3.

b. A regular expression that defines the given set is $0^* \mid (0^*10^*10^*10^*)^*$. ■

Example 12.2.7 A Finite-State Automaton That Accepts the Set of All Strings of 0's and 1's Containing Exactly One 1

a. Design a finite-state automaton A to accept the set of all strings of 0's and 1's that contain exactly one 1.

b. Is there a regular expression that defines this set?

Solution

a. The automaton A must have at least two distinct states:

s_0: initial state;

s_1: state to which A goes when the input string contains exactly one 1.

If A is in state s_0 and a 0 is input, A may as well stay in state s_0 (since it still needs to wait for a 1 to move to state s_1), but as soon as a 1 is input, A moves to state s_1. Thus a partial drawing of the transition diagram is as shown below.

Now consider what happens when A is in state s_1. If a 0 is input, the input string still has a single 1, so A stays in state s_1. But if a 1 is input, then the input string contains more than one 1, so A must leave s_1 (since no string with more than one 1 is to be accepted by A). It cannot go back to state s_0 because there is a way to get from s_0 to s_1, and after input of the second 1, A can never return to state s_1. Hence A must go to a third state, s_2, from which there is no return to s_1. Thus from s_2 every input may as well leave A in state s_2. It follows that the completed transition diagram for A has the appearance shown below.

This automaton accepts the set of strings 0's and 1's, with exactly one 1.

b. A regular expression that defines the given set is 0^*10^*. ■

Simulating a Finite-State Automaton Using Software

Suppose items have been coded with strings of 0's and 1's. A program is to be written to govern the processing of items coded with strings that end 011; items coded any other way are to be ignored. This situation can be modeled by the finite-state automaton shown in Figure 12.2.5.

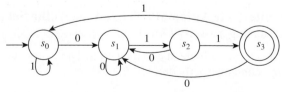

This automaton recognizes strings that end 011.

Figure 12.2.5

The symbols of the code for the item are fed into this automaton in sequence, and every string of symbols in a given code sends the automaton to one of the four states s_0, s_1, s_2, or s_3. If state s_3 is reached, the item is processed; if not, the item is ignored.

The action of this finite-state automaton can be simulated by a computer algorithm as given in Algorithm 12.2.1.

Algorithm 12.2.1 A Finite-State Automaton

[*This algorithm simulates the action of the finite-state automaton of Figure 12.2.5 by mimicking the functioning of the transition diagram. The states are denoted* 0, 1, 2, *and* 3.]

Input: *string* [*a string of* 0*'s and* 1*'s plus an end marker e*]

Algorithm Body:

$state := 0$
$symbol :=$ first symbol in the input string
while ($symbol \neq e$)
    if $state = 0$ **then if** $symbol = 0$
                      then $state := 1$
                      else $state := 0$
    else if $state = 1$ **then if** $symbol = 0$
                      then $state := 1$
                      else $state := 2$
    else if $state = 2$ **then if** $symbol = 0$
                      then $state := 1$
                      else $state := 3$
    else if $state = 3$ **then if** $symbol = 0$
                      then $state := 1$
                      else $state := 0$
    $symbol :=$ next symbol in the input string
end while
[*After execution of the* **while** *loop, the value of state is* 3 *if, and only if, the input string ends in* 011*e*.]

Output: *state*

Note how use of the finite-state automaton allows the creator of the algorithm to focus on each step of the analysis of the input string independently of the other steps.

An alternative way to program this automaton is to enter the values of the next-state function directly as a two-dimensional array. This is done in Algorithm 12.2.2.

> **Algorithm 12.2.2 A Finite-State Automaton**
>
> [*This algorithm simulates the action of the finite-state automaton of Figure 12.2.5 by repeated application of the next-state function. The states are denoted* 0, 1, 2, *and* 3.]
>
> **Input:** *string* [*a string of* 0*'s and* 1*'s plus an end marker e*]
>
> **Algorithm Body:**
>
> $N(0, 0) := 1, N(0, 1) := 0, N(1, 0) := 1, N(1, 1) := 2,$
> $N(2, 0) := 1, N(2, 1) := 3, N(3, 0) := 1, N(3, 1) := 0$
> *state* : $= 0$
> *symbol* : $=$ first symbol in the input string
>
> **while** (*symbol* $\neq$ *e*)
>   *state* := N (*state, symbol*)
>   *symbol* := next symbol in the input string
>
> **end while**
> [*After execution of the* **while** *loop, the value of state is* 3 *if, and only if, the input string ends in* 011*e*.]
>
> **Output:** *state*

Finite-State Automata and Regular Expressions

In the previous sections, each time we considered a language accepted by a finite-state automaton, we found a regular expression that defined the same language. Stephen Kleene showed that our ability to do this is not sheer coincidence. He proved that any language accepted by a finite-state automaton can be defined by a regular expression and that, conversely, any language defined by a regular expression is accepted by a finite-state automaton. Thus for the many applications of regular expressions discussed in Section 12.1, it is theoretically possible to find a corresponding finite-state automaton, which can then be simulated using the kinds of computer algorithms described in the previous subsection.

In practice, it is often of interest to retain only pieces of the patterns sought. For instance, to obtain a reference in an HTML document, one would specify a regular expression defining the full HTML tag, <a href= "the desired URL">, but one would be interested in retrieving only the string between the quotation marks. Because of these kinds of considerations, actual implementations of finite-state automata include additional features.*

We break the statement of Kleene's theorem into two parts.

> **Kleene's Theorem, Part 1**
>
> Given any language that is accepted by a finite-state automaton, there is a regular expression that defines the same language.

> **Proof:**
>
> Suppose A is a finite-state automaton with a set I of input symbols, a set S of n states, and a next-state function $N: S \times I \to S$. Let I^* denote the set of all strings

---

*For more information, see *Mastering Regular Expressions*, 2nd ed., by Jeffrey E. F. Friedl, (Sebastopol, CA: O'Reilly & Associates, 2002).

over I. Number the states $s_1, s_2, s_3, \ldots, s_n$, using s_1 to denote the initial state, and for each integer $k = 1, 2, 3, \ldots, n$, let

$$L_{i,j}^k = \left\{ x \in I^* \,\middle|\, \begin{array}{l} \text{when the symbols of } x \text{ are input to } A \text{ in sequence, } A \\ \text{goes from state } s_i \text{ to state } s_j \text{ without traveling through} \\ \text{an intermediate state } s_h \text{ for which } h > k \end{array} \right\}.$$

Note that either index i or index j in $L_{i,j}^k$ could be greater than k; the only restriction is that the symbols of a string in $L_{i,j}^k$ cannot make A both enter and exit an intermediate state with index greater than k.

If s_j is an accepting state and if $k = n$ and $i = 1$, then $L_{1,j}^n$ is the set of all strings that send A to s_j when the symbols of the string are input to A in sequence starting from s_1. Thus

$$L_{1,j}^n \subseteq L(A).$$

Moreover, because the sequence of symbols in every string in $L(A)$ sends A to *some* accepting state s_j,

$L(A)$ is the union of all the sets $L_{1,j}^n$, where s_j is an accepting state.

We use a version of mathematical induction to build up a set of regular expressions over I.

BASE: For each pair of integers i and j with $1 \leq i, j \leq n$, $L_{i,j}^0$ is the set of all strings that send A from s_i to s_j without sending it through any intermediate state s_h for which $h > 0$. Because the subscript of every state in A is greater than zero, the strings in $L_{i,j}^0$ do not send A through any intermediate states at all, and so each is a single input symbol from I. In other words, for all integers i and j with $1 \leq i, j \leq n$,

$$L_{i,j}^0 = \{a \in I \mid N(s_i, a) = s_j\}.$$

Hence $L_{i,j}^0$ is a subset of I, and so (because I is finite) we may denote the elements of $L_{i,j}^0$ as follows:

$$L_{i,j}^0 = \{a_1, a_2, a_3, \ldots, a_M\} \subseteq I.$$

Now, by definition of regular expression, each single input symbol of I is a regular expression over I; thus every element of $L_{i,j}^0$ is a regular expression over I. The result is that for all integers i and j with $1 \leq i, j \leq n$, the following regular expression defines $L_{i,j}^0$:

$$a_1 \mid a_2 \mid a_3 \mid \cdots \mid a_M$$

RECURSION: Let i and j be any pair of integers with $1 \leq i, j \leq n$, and suppose that for each pair of integers p and q with $1 \leq p, q \leq k$, there is a regular expression that defines $L_{p,q}^k$. Call it $r_{p,q}^k$. We will construct a regular expression $r_{i,j}^{k+1}$ that defines $L_{i,j}^{k+1}$. Observe that any string in $L_{i,j}^{k+1}$ sends A from s_i to s_j, either by a route that makes A pass through s_{k+1} or by a route that does not make A pass through s_{k+1}. Now each string that sends A from s_i to s_j and makes A pass through s_{k+1} one or more times can be broken into segments. The symbols in the first segment send A from s_i to s_{k+1} without making A pass through s_{k+1}; those in each of the intermediate segments send s_{k+1} to itself without making A pass through

continued on page 758

s_{k+1}; and those in the final segment send A from s_{k+1} to s_j without making A pass through s_{k+1}. (The intermediate segment could be the null string.) A typical path showing two intermediate segments is illustrated below.

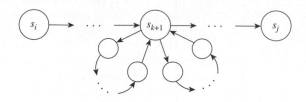

Note that each intermediate segment of the string is in $L^k_{k+1,k+1}$, and by assumption the regular expression $r^k_{k+1,k+1}$ defines this set. By the same reasoning, $r^k_{i,k+1}$ defines the set of all possible first segments of the string, and $r^k_{k+1,j}$ defines the set of all possible final segments of the string. In addition, $r^k_{i,j}$ defines the set of all strings that send A from s_i to s_j without making it pass through a state s_m with $m > k$. Thus we may define the regular expression $r^{k+1}_{i,j}$ as follows:

$$r^{k+1}_{i,j} = r^k_{i,j} \mid r^k_{i,k+1} \left(r^k_{k+1,k+1} \right)^* r^k_{k+1,j}.$$

Then $r^{k+1}_{i,j}$ defines the set of all strings that send A from s_i to s_j without making it pass through any states s_m with $m > k + 1$. Note that because i and j were allowed to be any pair of integers with $1 \le i, j \le n$, then $r^{k+1}_{i,j}$ is defined for all possible such pairs, which allows the next stage of the recursion to go forward.

To complete the proof, let s_j be any accepting state of A, and let r_j be the regular expression

$$r_j = r^0_{1,j} \mid r^1_{1,j} \mid r^2_{1,j} \mid \cdots \mid r^n_{1,j}.$$

Then r_j defines the set of all strings that send A to s_j when the symbols in the string are input to A in sequence, starting with s_1. Thus if the entire set of accepting states of A is $\{s_{j_1}, s_{j_2}, s_{j_3}, \ldots, s_{j_h}\}$, then the regular expression

$$r = r_{j_1} \mid r_{j_2} \mid r_{j_3} \mid \cdots \mid r_{j_h}$$

defines the set of all strings that send A to an accepting state. In other words, r defines $L(A)$.

Kleene's Theorem, Part 2

Given any language defined by a regular expression, there is a finite-state automaton that accepts the same language.

The most common way to prove part 2 of Kleene's theorem is to introduce a new category of automata called *nondeterministic finite-state automata*. These are similar to the (deterministic) finite-state automata we have been discussing, except that for any given state and input symbol, the next state is a subset of the set of states of the automaton, possibly even the empty set. Thus the next state of the automaton is not uniquely determined by the combination of a current state and an input symbol. A string is accepted by

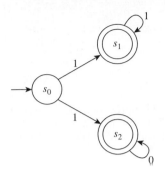

a nondeterministic finite-state automaton if, and only if, when the symbols in the string are input to the automaton in sequence, starting from an initial state, there is *some* sequence of next states through which the automaton could travel that would send it to an accepting state. For instance, the transition diagram at the left is an example of a very simple nondeterministic finite-state automaton that accepts the set of all strings beginning with a 1. Observe that $N(s_0, 1) = \{s_1, s_2\}$ and $N(s_0, 0) = \emptyset$.

Given a language defined by any regular expression, there is a straightforward recursive algorithm for finding a nondeterministic finite-state automaton that defines the same language. The proof of Kleene's theorem is completed by showing that for any such nondeterministic finite-state automaton, there is a (deterministic) finite-state automaton that defines the same language. We leave the details of the proof to a course in automata theory.

Regular Languages

According to Kleene's theorem, the set of languages defined by regular expressions is identical to the set of languages accepted by finite-state automata. Any such language is called a **regular language.** The brief allusions we made earlier to context-free languages and Chomsky's classification of languages suggest that not every language is regular. We will prove this by giving an example of a nonregular language.

To construct the example, note that because a finite-state automaton can assume only a finite number of states and because there are infinitely many input sequences, by the pigeonhole principle there must be at least one state to which the automaton returns over and over again. This is the essential feature of an automaton that makes it possible to find a nonregular language.

Example 12.2.8 Showing That a Language is Not Regular

Let the language L consist of all strings of the form $a^k b^k$, where k is a positive integer. Symbolically, L is the language over the alphabet $\Sigma = \{a, b\}$ defined by

$$L = \{s \in \Sigma^* \mid s = a^k b^k, \text{ where } k \text{ is a positive integer}\}.$$

Use the pigeonhole principle to show that L is not regular. In other words, show that there is no finite-state automaton that accepts L.

Solution [*Use a proof by contradiction.*] Suppose not. That is, suppose there is a finite-state automaton A that accepts L. [*A contradiction will be derived.*] Since A has only a finite number of states, these states can be denoted $s_1, s_2, s_3, \ldots, s_n$, where n is a positive integer. Consider all input strings that consist entirely of a's: $a, a^2, a^3, a^4, \ldots$. Now there are infinitely many such strings and only finitely many states. Thus, by the pigeonhole principle, there must be a state s_m and two input strings a^p and a^q with $p \neq q$ such that when either a^p or a^q is input to A, A goes to state s_m. (See Figure 12.2.6.) [*The pigeons are the strings of a's, the pigeonholes are the states, and the correspondence associates each string with the state to which A goes when the string is input.*]

Now, by supposition, A accepts L. Hence A accepts the string

$$a^p b^p.$$

This means that after p a's have been input, at which point A is in state s_m, inputting p additional b's sends A into an accepting state, say s_a. But that implies that

$$a^q b^p$$

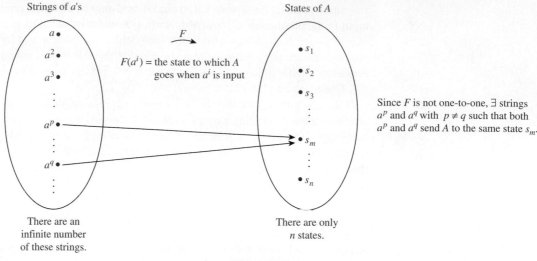

Strings of a's

States of A

$F(a^i) =$ the state to which A goes when a^i is input

Since F is not one-to-one, $\exists$ strings a^p and a^q with $p \neq q$ such that both a^p and a^q send A to the same state s_m.

There are an infinite number of these strings.

There are only n states.

Figure 12.2.6

also sends A to the accepting state s_a, and so $a^q b^p$ is accepted by A. The reason is that after q a's have been input, A is also in state s_m, and from that point, inputting p additional b's sends A to state s_a, which is an accepting state. Pictorially, if $p < q$, then

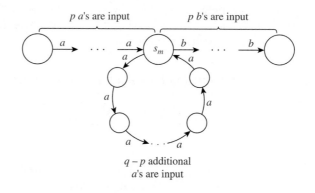

p a's are input

p b's are input

$q - p$ additional a's are input

Now, by supposition, L is the language accepted by A. Thus since s is accepted by A, $s \in L$. But by definition of L, L consists only of strings with equal numbers of a's and b's. So since $p \neq q$, $s \notin L$. Hence $s \in L$ and $s \notin L$, which is a contradiction.

It follows that the supposition is false, and so there is no finite-state automaton that accepts L. ∎

Exercise Set 12.2

1. Find the state of the vending machine in Example 12.2.1 after each of the following sequences of coins have been input.
   a. Quarter, half-dollar, quarter
   b. Quarter, half-dollar, half-dollar
   c. Half-dollar, quarter, quarter, quarter, half-dollar

In 2–7 a finite-state automaton is given by a transition diagram. For each automaton:
a. Find its states.
b. Find its input symbols.
c. Find its initial state.
d. Find its accepting states.
e. Write its annotated next-state table.

2.

3.

4.

5.

6.

7.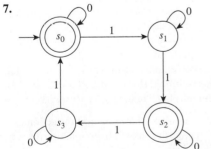

In 8 and 9 a finite-state automaton is given by an annotated next-state table. For each automaton:

a. Find its states.
b. Find its input symbols.
c. Find its initial state.
d. Find its accepting states.
e. Draw its transition diagram.

8. Next-State Table

	State	Input	
		0	**1**
→	s_0	s_1	s_2
	s_1	s_1	s_2
⊙	s_2	s_1	s_2

9. Next-State Table

	State	Input	
		0	**1**
→	s_0	s_0	s_1
⊗	s_1	s_1	s_2
	s_2	s_2	s_3
	s_3	s_3	s_0

10. A finite-state automaton A, given by the transition diagram below, has next-state function N and eventual-state function N^*.

a. Find $N(s_1, 1)$ and $N(s_0, 1)$.
b. Find $N(s_2, 0)$ and $N(s_1, 0)$.
c. Find $N^*(s_0, 10011)$ and $N^*(s_1, 01001)$.
d. Find $N^*(s_2, 11010)$ and $N^*(s_0, 01000)$.

11. A finite-state automaton A, given by the transition diagram below, has next-state function N and eventual-state function N^*.

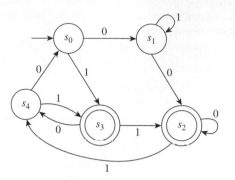

   a. Find $N(s_3, 0)$ and $N(s_2, 1)$.
   b. Find $N(s_0, 0)$ and $N(s_4, 1)$.
   c. Find $N^*(s_0, 010011)$ and $N^*(s_3, 01101)$.
   d. Find $N^*(s_0, 1111)$ and $N^*(s_2, 00111)$.

12. Consider again the finite-state automaton of exercise 2.
   a. To what state does the automaton go when the symbols of the following strings are input to it in sequence, starting from the initial state?
     (i) 1110001   (ii) 0001000   (iii) 11110000
   b. Which of the strings in part (a) send the automaton to an accepting state?
   c. What is the language accepted by the automaton?
   d. Find a regular expression that defines the language.

13. Consider again the finite-state automaton of exercise 3.
   a. To what state does the automaton go when the symbols of the following strings are input to it in sequence, starting from the initial state?
     (i) *bb*   (ii) *aabbbaba*   (iii) *babbbbbabaa*
     (iv) *bbaaaabaa*
   b. Which of the strings in part (a) send the automaton to an accepting state?
   c. What is the language accepted by the automaton?
   d. Find a regular expression that defines the language.

In each of 14–19, (a) find the language accepted by the automaton in the referenced exercise, and (b) find a regular expression that defines the same language.

14. Exercise 4   **15.** Exercise 5   16. Exercise 6

17. Exercise 7   **18.** Exercise 8   19. Exercise 9

In each of 20–29, (a) design an automaton with the given input alphabet that accepts the given set of strings, and (b) find a regular expression that defines the language accepted by the automaton.

20. Input alphabet $= \{0, 1\}$; Accepts the set of all strings for which the final three input symbols are 1.

H **21.** Input alphabet $= \{a, b\}$; Accepts the set of all strings of length at least 2 for which the final two input symbols are the same.

22. Input alphabet $= \{0, 1\}$; Accepts the set of all strings that start with 01 or 10.

23. Input alphabet $= \{0, 1\}$; Accepts the set of all strings that start with 01.

24. Input alphabet $= \{0, 1\}$; Accepts the set of all strings that start with 101.

25. Input alphabet $= \{0, 1\}$; Accepts the set of all strings that end in 10.

26. Input alphabet $= \{a, b\}$; Accepts the set of all strings that contain exactly two b's.

27. Input alphabet $= \{0, 1\}$; Accepts the set of all strings that start with 0 and contain exactly one 1.

28. Input alphabet $= \{0, 1\}$; Accepts the set of all strings that contain the pattern 010.

In 29–47, design a finite-state automaton to accept the language defined by the regular expression in the referenced exercise from Section 12.1.

29. Exercise 16   30. Exercise 17   **31.** Exercise 18

32. Exercise 19   **33.** Exercise 20   34. Exercise 21

35. Exercise 24   **36.** Exercise 25   37. Exercise 26

38. Exercise 27   **39.** Exercise 31   40. Exercise 32

41. Exercise 33   **42.** Exercise 34   43. Exercise 35

44. Exercise 36   **45.** Exercise 37   46. Exercise 38

47. Exercise 39

48. A simplified telephone switching system allows the following strings as legal telephone numbers:
   a. A string of seven digits that does not start with 00, 01, 10 or 11 (*a local call string*).
   b. A 1 followed by a three-digit *area code string* (any digit except 0 or 1 followed by a 0 or 1 followed by any digit) followed by a seven-digit local call string.
   c. A 0 alone or followed by a three-digit area code string plus a seven-digit local call string.
Design a finite-state automaton to recognize all the legal telephone numbers in (a), (b) and (c). Include an "error state" for invalid telephone numbers.

49. Write a computer algorithm that simulates the action of the finite-state automaton of exercise 2 by mimicking the action of the transition diagram.

50. Write a computer algorithm that simulates the action of the finite-state automaton of exercise 8 by repeated application of the next-state function.

H **51.** Let L be the language consisting of all strings of the form

$a^m b^n$, where m and n are positive integers and $m \geq n$.

Show that there is no finite-state automaton that accepts L.

52. Let L be the language consisting of all strings of the form

$a^m b^n$, where m and n are positive integers and $m \leq n$.

Show that there is no finite-state automaton that accepts L.

H **53.** Let L be the language consisting of all strings of the form

a^n, where $n = m^2$, for some positive integer m.

Show that there is no finite-state automaton that accepts L.

54. a. Let A be a finite-state automaton with input alphabet Σ, and suppose $L(A)$ is the language accepted by A. The complement of $L(A)$ is the set of all strings over Σ that are not in $L(A)$. Show that the complement of a regular language is regular by proving the following: If $L(A)$ is the language accepted by a finite-state automaton A, then there is a finite-state automaton A' that accepts the complement of $L(A)$.

b. Show that the intersection of any two regular languages is regular as follows: First prove that if $L(A_1)$ and $L(A_2)$ are languages accepted by automata A_1 and A_2, respectively, then there is an automaton A that accepts $(L(A_1))^c \cup (L(A_2))^c$. Then use one of De Morgan's laws for sets, the double complement law for sets, and the result of part (a) to prove that there is an automaton that accepts $L(A_1) \cap L(A_2)$.

12.3 Simplifying Finite-State Automata

Our life is frittered away by detail. . . . Simplify, simplify.
— Henry David Thoreau, *Walden,* 1854

Any string input to a finite-state automaton either sends the automaton to an accepting state or not, and the set of all strings accepted by an automaton is the language accepted by the automaton. It often happens that when an automaton is created to do a certain job (as in compiler construction, for example), the automaton that emerges "naturally" from the development process is unnecessarily complicated; that is, there may be an automaton with fewer states that accepts exactly the same language. It is desirable to find such an automaton because the memory space required to store an automaton with n states is approximately proportional to n^2. Thus approximately 10,000 memory spaces are required to store an automaton with 100 states, whereas only about 100 memory spaces are needed to store an automaton with 10 states. In addition, the fewer states an automaton has, the easier it is to write a computer algorithm based on it; and to see that two automata both accept the same language, it is easiest to simplify each to a minimal number of states and compare the simplified automata. In this section we show how to take a given automaton and simplify it in the sense of finding an automaton with fewer states that accepts the same language.

Example 12.3.1 An Overview

Consider the finite-state automata A and A' in Figure 12.3.1. A moment's thought should convince you that A' accepts all those strings, and only those strings, that contain an

A

Figure 12.3.1 Two Equivalent Automata

even number of 1's. But A, although it appears more complicated, accepts exactly those strings also. Thus the two automata are "equivalent" in the sense that they accept the same language, even though A' has fewer states than A.

Roughly speaking, the reason for the equivalence of these automata is that some of the states of A can be combined without affecting the acceptance or nonacceptance of any input string. It turns out that s_2 can be combined with state s_0 and that s_3 can be combined with state s_1. (How to figure out which states can be combined is explained later in this section.) The automaton with the two combined states $\{s_0, s_2\}$ and $\{s_1, s_3\}$ is called the *quotient automaton* of A and is denoted $\overline{A}$. Its transition diagram is obtained by combining the circles for s_0 and s_2 and for s_1 and s_3 and by replacing any arrow from a state s to a state t by an arrow from the combined state containing s to the combined state containing t. For instance, since there is an arrow labeled 1 from s_1 to s_2 in A, there is an arrow labeled 1 from $\{s_1, s_3\}$ to $\{s_0, s_2\}$ in $\overline{A}$. The complete transition diagram for $\overline{A}$ is shown in Figure 12.3.2. As you can see, except for labeling the names of the states, it is identical to the diagram for A'.

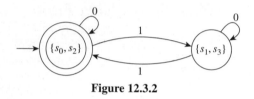

Figure 12.3.2   ∎

In general, simplification of a finite-state automaton involves identifying "equivalent states" that can be combined without affecting the action of the automaton on input strings. Mathematically speaking, this means defining an equivalence relation on the set of states of the automaton and forming a new automaton whose states are the equivalence classes of the relation. The rest of this section is devoted to developing an algorithm to carry out this process in a practical way.

*-Equivalence of States

Two states of a finite-state automaton are said to be *-*equivalent* (this is read "star equivalent") if any string accepted by the automaton when it starts from one of the states is accepted by the automaton when it starts from the other state. Recall that the value of the eventual-state function, N^*, for a state s and input string w is the state to which the automaton goes if the characters of w are input in sequence when the automaton is in state s.

> **• Definition**
>
> Let A be a finite-state automaton with next-state function N and eventual-state function N^*. Define a binary relation on the set of states of A as follows: Given any states s and t of A, we say that **s and t are *-equivalent** and write $s\ R_* \ t$ if, and only if, for all input strings w,
>
> > either both $N^*(s, w)$ and $N^*(t, w)$ are accepting states or both are nonaccepting states.

In other words, states s and t are *-equivalent if, and only if, for all input strings w,

$$N^*(s, w) \text{ is an accepting state} \quad \Leftrightarrow \quad N^*(t, w) \text{ is an accepting state.}$$

Or, more simply, for all input strings w,

$$\begin{bmatrix} A \text{ goes to an accepting state if} \\ w \text{ is input when } A \text{ is in state } s \end{bmatrix} \Leftrightarrow \begin{bmatrix} A \text{ goes to an accepting state if} \\ w \text{ is input when } A \text{ is in state } t \end{bmatrix}.$$

It follows immediately, by substitution into the definition, that

> R_* is an equivalence relation on S, the set of states of A.   12.3.1

You are asked to prove this formally in the exercises at the end of this section.

k-Equivalence of States

From a procedural point of view, it is difficult to determine the $*$-equivalence of two states using the definition directly. According to the definition, you must know the action of the automaton starting in states s and t on *all* input strings in order to tell whether s and t are equivalent. But since most languages have infinitely many input strings, you cannot check individually the effect of every string that is input to an automaton. As a practical matter, you can tell whether or not two states s and t are $*$-equivalent by using an iterative procedure based on a simpler kind of equivalence of states called *k-equivalence*. Two states are *k-equivalent* if any string *of length less than or equal to k* that is accepted by the automaton when it starts from one of the states is accepted by the automaton when it starts from the other state.

> **• Definition**
>
> Let A be a finite-state automaton with next-state function N and eventual-state function N^*. Define a binary relation on the set of states of A as follows: Given any states s and t of A and an integer $k \geq 0$, we say that s is **k-equivalent** to t and write $s \; R_k \; t$ if, and only if, for all input strings w *of length less than or equal to k*, either $N^*(s, w)$ and $N^*(t, w)$ are both accepting states or they are both nonaccepting states.

Certain useful facts follows quickly from the definition of k-equivalence:

> For each integer $k \geq 0$, k-equivalence is an equivalence relation.   12.3.2
>
> For each integer $k \geq 0$, the k-equivalence classes partition the set of all states of the automaton into a union of mutually disjoint subsets.   12.3.3
>
> For each integer $k \geq 1$, if two states are k-equivalent, then they are also $(k - 1)$ equivalent.   12.3.4
>
> For each integer $k \geq 1$, each k-equivalence class is a subset of a $(k - 1)$-equivalence class.   12.3.5
>
> Any two states that are k-equivalent for all integers $k \geq 0$ are $*$-equivalent.   12.3.6

Proofs of these facts are left for the exercises.

The following theorem gives a recursive description of k-equivalence of states. It says, first, that any two states are 0-equivalent if, and only if, either both are accepting states or both are nonaccepting states and, second, that any two states are k-equivalent (for $k \geq 1$) if, and only if, they are $(k - 1)$-equivalent and for any input symbols their next-states are also $(k - 1)$-equivalent.

Theorem 12.3.1

Let A be a finite-state automaton with next-state function N. Given any states s and t in A,

1. s is 0-equivalent to t $\Leftrightarrow$ $\begin{bmatrix} \text{either } s \text{ and } t \text{ are both accepting states} \\ \text{or they are both nonaccepting states} \end{bmatrix}$

2. for every integer $k \geq 1$, $\Leftrightarrow$ $\begin{bmatrix} s \text{ and } t \text{ are } (k-1)\text{-equivalent, and} \\ \text{for any input symbol } m, N(s, m) \text{ and} \\ N(t, m) \text{ are also } (k-1)\text{-equivalent} \end{bmatrix}$.
   s is k-equivalent to t

The truth of Theorem 12.3.1 follows from the fact that inputting a string w of length k has the same effect as inputting the first symbol of w and then the remaining $k - 1$ symbols of w. A detailed proof is somewhat technical.

Theorem 12.3.1 implies that if you know which states are $(k - 1)$-equivalent (where k is a positive integer) and if you know the action of the next-state function, then you can figure out which states are k-equivalent. Specifically, if s and t are $(k - 1)$-equivalent states whose next-states are $(k - 1)$-equivalent for any input symbol m, then s and t are k-equivalent. Thus the k-equivalence classes are obtained by subdividing the $(k - 1)$-equivalence classes according to the action of the next-state function on the members of the classes. An example should make this procedure clear.

Example 12.3.2 Finding k-Equivalence Classes

Find the 0-equivalence classes, the 1-equivalence classes, and the 2-equivalence classes for the states of the automaton shown below.

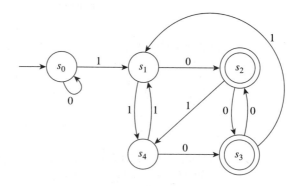

Solution

1. **0-equivalence classes:** By Theorem 12.3.1 two states are 0-equivalent if, and only if, both are accepting states or both are nonaccepting states. Thus there are two sets of 0-equivalent states:

   $\{s_0, s_1, s_4\}$ (the nonaccepting states) and $\{s_2, s_3\}$ (the accepting states),

   and so

   the 0-equivalence classes are $\{s_0, s_1, s_4\}$ and $\{s_2, s_3\}$.

2. *1-equivalence classes*: By Theorem 12.3.1, two states are 1-equivalent if, and only if, they are 0-equivalent and, after input of any input symbol, their next-states are 0-equivalent. Thus s_1 is not 1-equivalent to s_0 because when a 0 is input to the automaton in state s_1 it goes to state s_2, whereas when a 0 is input to the automaton in state s_0 it goes to state s_0, and s_2 and s_0 are not 0-equivalent. On the other hand, s_1 *is* 1-equivalent to s_4 because when a 0 is input to the automaton in state s_1 or s_4 the next-states are s_2 and s_3, which are 0-equivalent; and when a 1 is input to the automaton in state s_1 or s_4 the next-states are s_4 and s_1, which are 0-equivalent. By a similar argument, s_2 is 1-equivalent to s_3. Since 1-equivalent states must also be 0-equivalent [*by property* (12.3.4)], no other pairs of states can be 1-equivalent. Hence

the 1-equivalence classes are $\{s_0\}$, $\{s_1, s_4\}$, and $\{s_2, s_3\}$.

3. *2-equivalence classes*: By Theorem 12.3.1, two states are 2-equivalent if, and only if, they are 1-equivalent and, after input of any input symbol, their next-states are 1-equivalent. Now s_1 is 2-equivalent to s_4 because they are 1-equivalent; and when a 1 is input to the automaton in state s_1 or s_4 the next-states are s_4 and s_1, which are 1-equivalent; and when a 0 is input to the automaton in state s_1 or s_4 the next-states are s_2 and s_3, which are 1-equivalent. Similarly, s_2 is 2-equivalent to s_3. Since 2-equivalent states must also be 1-equivalent [*by property* (12.3.4)], no other pairs of states can be 2-equivalent. Hence

the 2-equivalence classes are $\{s_0\}$, $\{s_1, s_4\}$, and $\{s_2, s_3\}$.

Note that the set of 2-equivalence classes equals the set of 1-equivalence classes. ∎

Finding the ∗-Equivalence Classes

Example 12.3.2 illustrates the relative ease with which the sets of k-equivalence classes of states can be found. But to simplify a finite-state automaton, you need to find the set of ∗-equivalence classes of states. The next theorem says that for some integer K, the set of ∗-equivalence classes equals the set of K-equivalence classes.

Theorem 12.3.2

If A is a finite-state automaton, then for some integer, $K \geq 0$, the set of K-equivalence classes of states of A equals the set of $(K + 1)$-equivalence classes of states of A, and for all such K these are both equal to the set of ∗-equivalence classes of states of A.

The detailed proof of Theorem 12.3.2 is somewhat technical, but the idea of the proof is not hard to understand. Theorem 12.3.2 follows from the fact that for each positive integer k, the k-equivalence classes are obtained by subdividing the $(k - 1)$-equivalence classes according to a certain rule that is the same for each k. Since the number of states of the automaton is finite, this subdivision process cannot continue forever, and so for some integer $K \geq 0$, the set of K-equivalence classes equals the set of $(K + 1)$-equivalence classes. Moreover, the set of m-equivalence classes equals the set of K-equivalence classes for every integer $m \geq K$. But this implies that the set of ∗-equivalence classes equals the set of K-equivalence classes.

Example 12.3.3 Finding ∗-Equivalence Classes of R

Let A be the finite-state automaton defined in Example 12.3.2. Find the ∗-equivalence classes of states of A.

Solution   According to Example 12.3.2, the set of 1-equivalence classes for A equals the set of 2-equivalence classes. By Theorem 12.3.2, then, the set of ∗-equivalence classes also equals the set of 1-equivalence classes. Hence

the ∗-equivalence classes are $\{s_0\}$, $\{s_1, s_4\}$ and $\{s_2, s_3\}$.

In the notation of Section 10.3, the equivalence classes are denoted

$$[s_0] = \{s_0\} \quad [s_1] = \{s_1, s_4\} = [s_4] \quad [s_2] = \{s_2, s_3\} = [s_3].$$ ■

The Quotient Automaton

We next define the *quotient automaton* $\overline{A}$ of an automaton A. However, in order for all parts of the definition to make sense, we must point out two facts.

> No ∗-equivalence class of states of A can contain both accepting and nonaccepting states.   12.3.7

The reason this is true is that the 0-equivalence classes divide the set of states of A into accepting and nonaccepting states, and the ∗-equivalence classes are subsets of 0-equivalence classes.

> If two states are ∗-equivalent, then their next-states are also ∗-equivalent for any input symbol m.   12.3.8

This is true for the following reason. Suppose states s and t are ∗-equivalent. Then any input string that sends A to an accepting state when A is in state s sends A to an accepting state when A is in state t. Now suppose m is any input symbol, and consider the next-states $N(s, m)$ and $N(t, m)$. Inputting a string of length k to A when A is in state $N(s, m)$ or $N(t, m)$ produces the same effect as inputting a certain string of length $k + 1$ to A when A is in state s or t (namely the concatenation of m with the string of length k). Hence any string that sends A to an accepting state when A is in state $N(s, m)$ also sends A to an accepting state when A is in state $N(t, m)$. It follows that $N(s, m)$ and $N(t, m)$ are ∗-equivalent. Complete proofs of properties (12.3.7) and (12.3.8) are left to the exercises.

Now we can define the quotient automaton $\overline{A}$ of A. It is the finite-state automaton whose states are the ∗-equivalence classes of states of A, whose initial state is the ∗-equivalence class containing the initial state of A, whose accepting states are of the form $[s]$ where s is an accepting state of A, whose input symbols are the same as the input symbols of A, and whose next-state function is derived from the next-state function for A in the following way: To find the next-state of $\overline{A}$ for a state s and an input symbol m, pick any state t in $[s]$ and look to see what next-state A goes to if m is input when A is in state t; the equivalence class of this state is the next-state of $\overline{A}$.

> **• Definition**
>
> Let A be a finite-state automaton with set of states S, set of input symbols I, and next-state function N. The **quotient automaton** $\overline{A}$ is defined as follows:
>
> 1. The set of states, $\overline{S}$, of $\overline{A}$ is the set of $*$-equivalence classes of states of A.
>
> 2. The set of input symbols, $\overline{I}$, of $\overline{A}$ equals I.
>
> 3. The initial state of $\overline{A}$ is $[s_0]$, where s_0 is the initial state of A.
>
> 4. The accepting states of A are the states of the form $[s]$, where s is an accepting state of A.
>
> 5. The next-state function $\overline{N}\colon \overline{S} \times I \to \overline{S}$ is defined as follows:
>
>   For all states $[s]$ in $\overline{S}$ and input symbols m in I, $\quad \overline{N}([s], m) = [N(s, m)]$.
>
>   (That is, if m is input to $\overline{A}$ when $\overline{A}$ is in state $[s]$, then $\overline{A}$ goes to the state that is the $*$-equivalence class of $N(s, m)$.)

Note that since the states of $\overline{A}$ are *sets* of states of A, $\overline{A}$ generally has fewer states than A. (A and $\overline{A}$ have the same number of states only in the case where each $*$-equivalence class of states contains just one element.) Also, by property (12.3.7), each accepting state of $\overline{A}$ consists entirely of accepting states of A. Furthermore, property (12.3.8) guarantees that the next-state function $\overline{N}$ is well defined.

By construction, a quotient automaton $\overline{A}$ accepts exactly the same strings as A. We state this formally as Theorem 12.3.3. We leave the details to a more advanced course in automata theory.

> **Theorem 12.3.3**
>
> If A is a finite-state automaton, then the quotient automaton $\overline{A}$ accepts exactly the same languages as A. In other words, if $L(A)$ denotes the language accepted by A and $L(\overline{A})$ denotes the language accepted by $\overline{A}$, then
>
> $$L(A) = L(\overline{A}).$$

Constructing the Quotient Automaton

Let A be a finite-state automaton with set of states S, next-state function N, relation R_* of $*$-equivalence of states, and relation R_k of k-equivalence of states. It follows from Theorem 12.3.2 and 12.3.3 and from the definition of quotient automaton that to find the quotient automaton $\overline{A}$ of A, you can proceed as follows:

1. Find the set of 0-equivalence classes of S.

2. For each integer $k \geq 1$, subdivide the $(k-1)$-equivalence classes of S (as described earlier) to find the k-equivalence classes of S. Stop subdividing when you observe that for some integer K the set of $(K+1)$-equivalence classes equals the set of K-equivalence classes. At this point, conclude that the set of K-equivalence classes equals the set of $*$-equivalence classes.

3. Construct the quotient automaton $\overline{A}$ whose states are the *-equivalence classes of states of A and whose next-state function $\overline{N}$ is given by

$$\overline{N}([s], m) = [N(s, m)] \quad \text{for any state of } \overline{A} \text{ and any input symbol } m,$$

where s is any state in $[s]$. [*That is, to see where $\overline{A}$ goes if m is input to $\overline{A}$ when it is in state s, look to see where A goes if m is input to A when it is in state s. The *-equivalence class of that state is the answer.*]

Example 12.3.4 Constructing a Quotient Automaton

Consider the automaton A of Examples 12.3.2 and 12.3.3. This automaton is shown again below for reference. Find the quotient automaton of A.

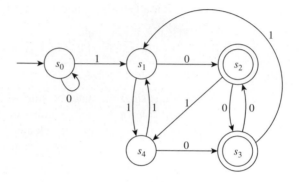

Solution According to Example 12.3.3 the *-equivalence classes of the states of A are

$$\{s_0\}, \quad \{s_1, s_4\}, \quad \text{and} \quad \{s_2, s_3\}.$$

Hence the states of the quotient automaton $\overline{A}$ are

$$[s_0] = \{s_0\}, \quad [s_1] = \{s_1, s_4\} = [s_4], \quad [s_2] = \{s_2, s_3\} = [s_3].$$

The accepting states of A are s_2 and s_3, so the accepting state of $\overline{A}$ is $[s_2] = [s_3]$. The next-state function $\overline{N}$ of $\overline{A}$ is defined as follows: for all states $[s]$ and input symbols m of $\overline{A}$,

$$\overline{N}([s], m) = [N(s, m)] = \text{the *-equivalence class of } N(s, m).$$

Thus,

$$\overline{N}([s_0], 0) = [N(s_0, 0)] = \text{the *-equivalence class of } N(s_0, 0).$$

But $N(s_0, 0) = s_0$, so

$$\overline{N}([s_0], 0) = \text{the *-equivalence class of } s_0 = [s_0].$$

Similarly,

$$\overline{N}([s_0], 1) = [N(s_0, 1)] = [s_1]$$
$$\overline{N}([s_1], 0) = [N(s_1, 0)] = [s_2]$$
$$\overline{N}([s_1], 1) = [N(s_1, 1)] = [s_4] = [s_1]$$
$$\overline{N}([s_2], 0) = [N(s_2, 0)] = [s_3] = [s_2]$$
$$\overline{N}([s_2], 1) = [N(s_2, 1)] = [s_4] = [s_1].$$

The transition diagram for $\overline{A}$ is, therefore, as shown below.

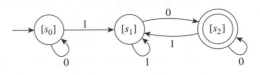

By Theorem 12.3.3, this automaton accepts the same language as the original automaton.

■

Equivalent Automata

Output devices may be attached to the states of finite-state automata to indicate whether they are accepting or nonaccepting states. For example, accepting states might produce an output of 1 and nonaccepting states an output of 0. Then a finite-state automaton can be thought of as an input/output device whose input consists of strings and whose output consists of 0's and 1's. Recall that a circuit can be thought of as a black box that transforms combinations of input signals into output signals. Two circuits that produce identical output signals for each combination of input signals are called *equivalent*. Similarly, a finite-state automaton can be regarded as a black box that processes input strings and produces output signals (indicating whether or not the strings are accepted). Two finite-state automata are called *equivalent* if they produce identical output signals for each input string. But this means that two finite-state automata are equivalent if, and only if, they accept the same language.

> **• Definition**
>
> Let A and A' be finite-state automata with the same set of input symbols I, Let $L(A)$ denote the language accepted by A and $L(A')$ the language accepted by A'. Then A is said to be **equivalent** to A' if, and only if, $L(A) = L(A')$.

Example 12.3.5 Showing That Two Automata Are Equivalent

Show that the automata A and A' that follow are equivalent.

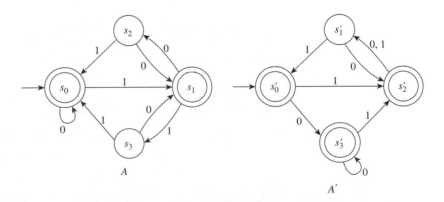

(The label 0, 1 on an arrow of a transition diagram means that for either input 0 or 1, the next-state of the automaton is the state to which the arrow points.)

Solution

For the automaton A: The 0-equivalence classes are

$$\{s_0, s_1\} \quad \text{and} \quad \{s_2, s_3\} \qquad \text{since } s_0 \text{ and } s_1 \text{ are accepting states and}$$
$$s_2 \text{ and } s_3 \text{ are nonaccepting states.}$$

The 1-equivalence classes are

$$\{s_0\}, \quad \{s_1\}, \quad \text{and} \quad \{s_2, s_3\} \qquad \text{since } s_0 \text{ and } s_1 \text{ are not 1-equivalent}$$
$$\text{(because } N(s_0, 1) = s_1 \text{, whereas}$$
$$N(s_1, 1) = s_3 \text{ and } s_1 \text{ is not 0-equivalent}$$
$$\text{to } s_3 \text{) but } s_2 \text{ and } s_3 \text{ are 1-equivalent.}$$

The 2-equivalence classes are

$$\{s_0\}, \quad \{s_1\}, \quad \text{and} \quad \{s_2, s_3\} \qquad \text{since } s_2 \text{ and } s_3 \text{ are 1-equivalent.}$$

This discussion shows that the set of 1-equivalence classes equals the set of 2-equivalence classes, so by Theorem 10.4.2 this is equal to the set of ∗-equivalence classes. Hence the ∗-equivalence classes are

$$\{s_0\}, \quad \{s_1\}, \quad \text{and} \quad \{s_2, s_3\}.$$

For the automaton A′: By reasoning similar to that above, the 0-equivalence classes are

$$\{s_0', s_2', s_3'\} \quad \text{and} \quad \{s_1'\}.$$

The 1-equivalence classes are

$$\{s_0', s_3'\}, \quad \{s_2'\}, \quad \text{and} \quad \{s_1'\}.$$

The 2-equivalence classes are the same as the 1-equivalence classes, which are therefore equal to the ∗-equivalence classes. Thus the ∗-equivalence classes are

$$\{s_0', s_3'\} \quad \{s_2'\}, \quad \text{and} \quad \{s_1'\}.$$

To calculate the next-state functions for $\overline{A}$ and $\overline{A'}$, you repeatedly use the fact that in the quotient automaton, the next-state of $[s]$ and m is the class of the next-state of s and m. For instance,

$$\overline{N}([s_1], 1) = [N(s_1, 1)] = [s_3] = [s_2]$$

and

$$\overline{N'}([s_0'], 0) = [N'(s_0', 0)] = [s_3'] = [s_0']$$

where N is the next-state function for A and N' is the next-state function for A'.

The complete transition diagrams for the quotient automata $\overline{A}$ and $\overline{A'}$ are shown below.

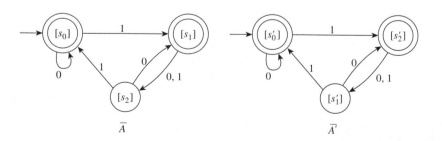

As you can see, except for the labeling of the names of the states, $\overline{A}$ and $\overline{A'}$ are identical and hence accept the same language. But by Theorem 12.3.3, each original automaton accepts the same language as its quotient automaton. Thus A and A' accept the same language, and so they are equivalent. ∎

In mathematics an object such as a finite-state automaton is called a *structure*. In general, when two mathematical structures are the same in all respects except for the labeling given to their elements, they are called **isomorphic,** which comes from the Greek words *isos,* meaning "same" or "equal," and *morphe,* meaning "from." It can be shown that two automata are equivalent if, and only if, their quotient automata are isomorphic, provided that "inaccessible states" have first been removed. (Inaccessible states are those that cannot be reached by inputting any string of symbols to the automaton when it is in its initial state.)

Exercise Set 12.3

1. Consider the finite-state automaton A given by the following transition diagram:

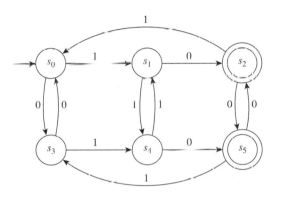

a. Find the 0-, 1-, and 2-equivalence classes of states of A.
b. Draw the transition diagram for A, the quotient automaton of A.

2. Consider the finite-state automaton A given by the following transition diagram:

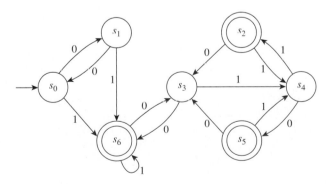

a. Find the 0-, 1-, and 2-equivalence classes of states of A.
b. Draw the transition diagram for $\overline{A}$, the quotient automaton of A.

3. Consider the finite-state automaton A discussed in Example 10.4.1:

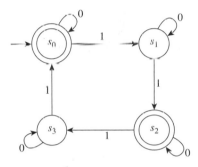

a. Find the 0- and 1-equivalence classes of states of A.
b. Draw the transition diagram of $\overline{A}$, the quotient automaton of A.

4. Consider the finite-state automaton given by the following transition diagram:

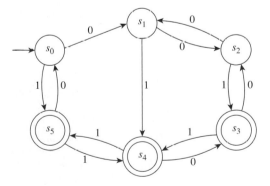

a. Find the 0-,1-, 2-, and 3-equivalence classes of states of A.
b. Draw the transition diagram for $\overline{A}$, the quotient automaton of A.

5. Consider the finite-state automaton given by the following transition diagram:

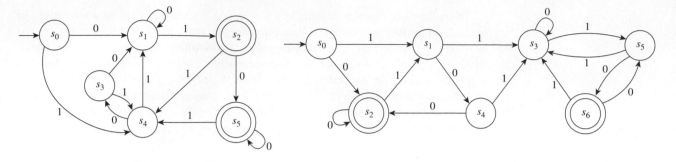

a. Find the 0-, 1-, 2-, and 3-equivalence classes of states of A.

b. Draw the transition diagram for $\overline{A}$, the quotient automaton of A.

6. Consider the finite-state automaton given by the following transition diagram:

H a. Find the 0-, 1-, 2-, and 3-equivalence classes of states of A.

b. Draw the transition diagram for $\overline{A}$, the quotient automaton of A.

7. Are the automata A and A' shown below equivalent?

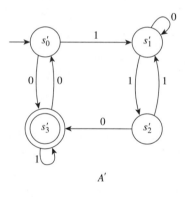

A

A'

8. Are the automata A and A' shown below equivalent?

A

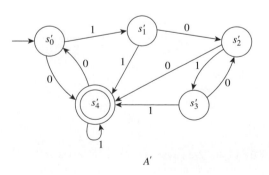

A'

9. Are the automata A and A' shown below equivalent?

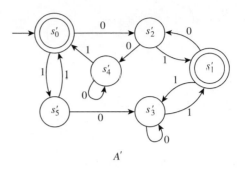

A

A'

10. Are the automata A and A' shown below equivalent?

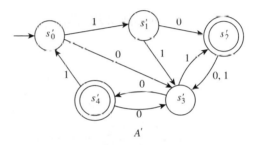

A

A'

_H_ 11. Prove property (12.3.1).

12. How should the proof of property (12.3.1) be modified to prove property (12.3.2)?

13. Prove property (12.3.3).   14. Prove property (12.3.4).

_H_ 15. Prove property (12.3.5).   16. Prove property (12.3.6).

_H_ 17. Prove that if two states of a finite-state automaton are k-equivalent for some integer k, then those states are m-equivalent for all nonnegative integers $m < k$.

18. Write a complete proof of property (12.3.7).

_H_ 19. Write a complete proof of property (12.3.8).

APPENDIX A

PROPERTIES OF THE REAL NUMBERS*

In this text we take the real numbers and their basic properties as our starting point. We give a core set of properties, called axioms, which the real numbers are assumed to satisfy, and we state some useful properties that can be deduced from these axioms.

We assume that there are two binary operations defined on the set of real numbers, called **addition** and **multiplication,** such that if a and b are any two real numbers, the **sum** of a and b, denoted $a + b$, and the **product** of a and b, denoted $a \cdot b$ or ab, are also real numbers. These operations satisfy properties F1–F6, which are called the **field axioms.**

F1. *Commutative Laws*   For all real numbers a and b,

$$a + b = b + a \quad \text{and} \quad ab = ba.$$

F2. *Associative Laws*   For all real numbers a, b, and c,

$$(a + b) + c = a + (b + c) \quad \text{and} \quad (ab)c = a(bc).$$

F3. *Distributive Laws*   For all real numbers a, b, and c,

$$a(b + c) = ab + ac \quad \text{and} \quad (b + c)a = ba + ca.$$

F4. *Existence of Identity Elements*   There exist two distinct real numbers, denoted 0 and 1, such that for every real number a,

$$0 + a = a + 0 = a \quad \text{and} \quad 1 \cdot a = a \cdot 1 = a.$$

F5. *Existence of Additive Inverses*   For every real number a, there is a real number, denoted $-a$ and called the **additive inverse** of a, such that

$$a + (-a) = (-a) + a = 0.$$

F6. *Existence of Reciprocals*   For every real number $a \neq 0$, there is a real number, denoted $1/a$ or a^{-1}, called the **reciprocal** of a, such that

$$a \cdot \left(\frac{1}{a}\right) = \left(\frac{1}{a}\right) \cdot a = 1.$$

All the usual algebraic properties of the real numbers that do not involve order can be derived from the field axioms. The most important are collected as theorems T1–T15 as follows. In all these theorems the symbols a, b, c, and d represent arbitrary real numbers.

*Adapted from Tom M. Apostol, *Calculus, Volume I* (New York: Blaisdell, 1961), pp. 13–19.

T1. *Cancellation Law for Addition*   If $a + b = a + c$, then $b = c$. (In particular, this shows that the number 0 of Axiom F4 is unique.)

T2. *Possibility of Subtraction*   Given a and b, there is exactly one x such that $a + x = b$. This x is denoted by $b - a$. In particular, $0 - a$ is the additive inverse of a, $-a$.

T3. $b - a = b + (-a)$.

T4. $-(-a) = a$.

T5. $a(b - c) = ab - ac$.

T6. $0 \cdot a = a \cdot 0 = 0$.

T7. *Cancellation Law for Multiplication*   If $ab = ac$ and $a \neq 0$, then $b = c$. (In particular, this shows that the number 1 of Axiom F4 is unique.)

T8. *Possibility of Division*   Given a and b with $a \neq 0$, there is exactly one x such that $ax = b$. This x is denoted by b/a and is called the **quotient** of b and a. In particular, $1/a$ is the reciprocal of a.

T9. If $a \neq 0$, then $b/a = b \cdot a^{-1}$.

T10. If $a \neq 0$, then $(a^{-1})^{-1} = a$.

T11. *Zero Product Property*   If $ab = 0$, then $a = 0$ or $b = 0$,

T12. *Rule for Multiplication with Negative Signs*

$$(-a)b = a(-b) = -(ab), \quad (-a)(-b) = ab,$$

and

$$-\frac{a}{b} = \frac{-a}{b} = \frac{a}{-b}.$$

T13. *Rule for Addition of Fractions*

$$\frac{a}{b} + \frac{c}{d} = \frac{ad + bc}{bd}, \quad \text{if } b \neq 0 \text{ and } d \neq 0.$$

T14. *Rule for Multiplication of Fractions*

$$\frac{a}{b} \cdot \frac{c}{d} = \frac{ac}{bd}, \quad \text{if } b \neq 0 \text{ and } d \neq 0.$$

T15. *Rule for Division of Fractions*

$$\frac{\dfrac{a}{b}}{\dfrac{c}{d}} = \frac{ad}{bc}, \quad \text{if } b \neq 0,\, c \neq 0,\, \text{and } d \neq 0.$$

The real numbers also satisfy the following axioms, called the **order axioms.** It is assumed that among all real numbers there are certain ones, called the **positive real numbers,** that satisfy properties Ord1–Ord3.

Ord1. For any real numbers a and b, if a and b are positive, so are $a + b$ and ab.

Ord2. For every real number $a \neq 0$, either a is positive or $-a$ is positive but not both.

Ord3. The number 0 is not positive.

The symbols $<, >, \leq$, and $\geq$, and negative numbers are defined in terms of positive numbers.

• Definition

Given real numbers a and b,

$a < b$ means $b + (-a)$ is positive.	$b > a$ means $a < b$.
$a \leq b$ means $a < b$ or $a = b$.	$b \geq a$ means $a \leq b$.
If $a < 0$, we say that a is **negative.**	If $a \geq 0$, we say that a is **nonnegative.**

From the order axioms Ord1–Ord3 and the above definition, all the usual rules for calculating with inequalities can be derived. The most important are collected as theorems T16–T25 as follows. In all these theorems the symbols a, b, c, and d represent arbitrary real numbers.

T16. *Trichotomy Law*   For arbitrary real numbers a and b, exactly one of the three relations $a < b, b < a$, or $a = b$ holds.

T17. *Transitive Law*   If $a < b$ and $b < c$, then $a < c$.

T18. If $a < b$, then $a + c < b + c$.

T19. If $a < b$ and $c > 0$, then $ac < bc$.

T20. If $a \neq 0$, then $a^2 > 0$.

T21. $1 > 0$.

T22. If $a < b$ and $c < 0$, then $ac > bc$.

T23. If $a < b$, then $-a > -b$. In particular, if $a < 0$, then $-a > 0$.

T24. If $ab > 0$, then both a and b are positive or both are negative.

T25. If $a < c$ and $b < d$, then $a + b < c + d$.

One final axiom distinguishes the set of real numbers from the set of rational numbers. It is called the **least upper bound axiom.**

LUB. Any nonempty set S of real numbers that is bounded above has a least upper bound. That is, if B is the set of all real numbers x such that $x \geq s$ for all s in S and if B has at least one element, then B has a smallest element. This element is called the **least upper bound** of S.

The least upper bound axiom holds for the set of real numbers but not for the set of rational numbers. For example, the set of all rational numbers that are less than $\sqrt{2}$ has upper bounds but not a least upper bound.

SOLUTIONS AND HINTS TO SELECTED EXERCISES

Section 1.1

1. Common form: If p then q.

   p.

   Therefore, q.

   $(a + 2b)(a^2 - b)$ can be written in prefix notation. All algebraic expressions can be written in prefix notation.

3. Common form: $p \vee q$.

   $\sim p$.

   Therefore, q.

   My mind is shot. Logic is confusing.

5. a. It is a statement because it is a true sentence. 1,024 is a perfect square because $1{,}024 = 32^2$, and the next smaller perfect square is $31^2 = 961$, which has less than four digits.

6. a. $s \wedge i$ **b.** $\sim s \wedge \sim i$

8. a. $(h \wedge w) \wedge \sim s$ **d.** $(\sim w \wedge \sim s) \wedge h$

10. a. $p \wedge q \wedge r$ **c.** $p \wedge (\sim q \vee \sim r)$

11. Inclusive or. For instance, a team could win the playoff by winning games 1, 3, and 4 and losing game 2. Such a team would satisfy both conditions.

12. "United States President" AND (14th OR fourteenth) AND NOT amendment.

14.

p	q	$\sim p$	$\sim p \wedge q$
T	T	F	F
T	F	F	F
F	T	T	T
F	F	T	F

16.

p	q	r	$q \wedge r$	$p \wedge (q \wedge r)$
T	T	T	T	T
T	T	F	F	F
T	F	T	F	F
T	F	F	F	F
F	T	T	T	F
F	T	F	F	F
F	F	T	F	F
F	F	F	F	F

18. *Hint:* The following is a partial truth table.

p	q	r	$(p \vee (\sim p \vee q)) \wedge \sim(q \wedge \sim r)$
T	T	T	T
T	T	F	F
T	F	T	T
T	F	F	T
F	T	T	
F	T	F	
F	F	T	
F	F	F	

19.

p	q	$p \wedge q$	$p \vee (p \wedge q)$	p
T	T	T	T	T
T	F	F	T	T
F	T	F	F	F
F	F	F	F	F

↑   ↑

$p \vee (p \wedge q)$ and p always have the same truth values, so they are logically equivalent. (This proves one of the absorption laws.)

21.

p	t	p ∨ t
T	T	T
F	T	T

↑   ↑

$p \vee t$ and t always have the same truth values, so they are logically equivalent. (This proves one of the universal bound laws.)

23.

p	q	r	p ∧ q	q ∧ r	(p ∧ q) ∧ r	p ∧ (q ∧ r)
T	T	T	T	T	T	T
T	T	F	T	F	F	F
T	F	T	F	F	F	F
T	F	F	F	F	F	F
F	T	T	F	T	F	F
F	T	F	F	F	F	F
F	F	T	F	F	F	F
F	F	F	F	F	F	F

↑   ↑

$(p \wedge q) \wedge r$ and $p \wedge (q \wedge r)$ always have the same truth values, so they are logically equivalent. (This proves the associative law for $\wedge$.)

25.

p	q	r	p ∧ q	q ∨ r	(p ∧ q) ∨ r	p ∧ (q ∨ r)
T	T	T	T	T	T	T
T	T	F	T	T	T	T
T	F	T	F	T	T	T
T	F	F	F	F	F	F
F	T	T	F	T	T	F
F	T	F	F	T	F	F
F	F	T	F	T	T	F
F	F	F	F	F	F	F

↑   ↑

$(p \wedge q) \vee r$ and $p \wedge (q \vee r)$ have different truth values in the fifth and seventh rows, so they are not logically equivalent. (This proves that parentheses are needed with $\wedge$ and $\vee$.)

27.

p	q	r	~p	~q	~r	~p ∨ q	p ∨ ~r	(~p ∨ q) ∧ (p ∨ ~r)	~p ∨ ~q	((~p ∨ q) ∧ (p ∨ ~r)) ∧ (~p ∨ ~q)	~(p ∨ r)
T	T	T	F	F	F	T	T	T	F	F	F
T	T	F	F	F	T	T	T	T	F	F	F
T	F	T	F	T	F	F	T	F	T	F	F
T	F	F	F	T	T	F	T	F	T	F	F
F	T	T	T	F	F	T	F	F	T	F	F
F	T	F	T	F	T	T	T	T	T	T	T
F	F	T	T	T	F	T	F	F	T	F	F
F	F	F	T	T	T	T	T	T	T	T	T

$((\sim p \lor q) \land (p \lor \sim r)) \land (\sim p \lor \sim q)$
and $\sim(p \lor r)$ always have the same truth
values, so they are logically equivalent.

29. Hal is not a math major or Hal's sister is not a computer science major.

31. The connector is not loose and the machine is not unplugged.

35. $-2 \geq x$ or $x \geq 7$ **37.** $1 \leq x$ or $x < -3$

39. This statement's logical form is $(p \land q) \lor r$, so its negation has the form $\sim((p \land q) \lor r) \equiv \sim(p \land q) \land \sim r \equiv (\sim p \lor \sim q) \land \sim r$. Thus a negation for the statement is (*num_orders* $\leq$ 100 or *num_instock* > 500) and *num_instock* > 200

41.

p	q	~p	~q	p ∧ q	p ∧ ~q	~p ∨ (p ∧ ~q)	(p ∧ q) ∨ (~p ∨ (p ∧ ~q))
T	T	F	F	T	F	F	T
T	F	F	T	F	T	T	T
F	T	T	F	F	F	T	T
F	F	T	T	F	F	T	T

↑
Its truth values are all T's, so
$(p \land q) \lor (\sim p \lor (p \land \sim q))$
is a tautology.

42.

p	q	~p	~q	p ∧ ~q	~p ∨ q	(p ∧ ~q) ∧ (~p ∨ q)
T	T	F	F	F	T	F
T	F	F	T	T	F	F
F	T	T	F	F	T	F
F	F	T	T	F	T	F

↑
Its truth values are all F's,
so $(p \land \sim q) \land (\sim p \lor q)$
is a contradiction.

45. a. The distributive law
   b. The commutative law for ∨
   c. The negation law for ∨
   d. The identity law for ∧

47. $(p \land \sim q) \lor p \equiv p \lor (p \land \sim q)$ by the commutative law for ∨

            $\equiv p$ by the absorption law (with $\sim q$ in place of q)

50. $\sim((\sim p \land q) \lor (\sim p \land \sim q)) \lor (p \land q)$

$\equiv \sim[\sim p \land (q \lor \sim q)] \lor (p \land q)$ by the distributive law

$\equiv \sim(\sim p \land \mathbf{t}) \lor (p \land q)$ by the negation law for ∨

$\equiv \sim(\sim p) \lor (p \land q)$ by the identity law for ∧

$\equiv p \lor (p \land q)$ by the double negative law

$\equiv p$ by the absorption law

52. a. *Solution 1:* Construct a truth table for $p \oplus p$ using the truth values for *exclusive or*.

p	$p \oplus p$
T	F
F	F

because an *exclusive or* statement is false when both components are true and when both components are false.

Since all its truth values are false, $p \oplus p \equiv \mathbf{c}$, a contradiction.

Solution 2: Replace q by p in the logical equivalence $p \oplus q \equiv (p \vee q) \wedge \sim(p \wedge q)$, and simplify the result.

$$p \oplus p \equiv (p \vee p) \wedge \sim(p \wedge p) \quad \text{by defintion of } \oplus$$
$$\equiv p \wedge \sim p \quad \text{by the identity laws}$$
$$\equiv \mathbf{c} \quad \text{by the negation law for } \wedge$$

53. There is a famous story about a philosopher who once gave a talk in which he observed that whereas in English and many other languages a double negative is equivalent to a positive, there is no language in which a double positive is equivalent to a negative. To this, a person in the back row responded sarcastically, "Yeah, yeah."

[*Strictly speaking, sarcasm functions like negation. When spoken sarcastically, the words "Yeah, yeah" are not a true double positive; they just mean "no."*]

Section 1.2

1. If this loop does not contain a **stop** or a **go to,** then it will repeat exactly N times.

3. If you do not freeze, then I'll shoot.

5.

				conclusion	hypothesis	
p	q	$\sim p$	$\sim q$	$\sim p \vee q$	$\sim p \vee q \to \sim q$	
T	T	F	F	T	F	
T	F	F	T	F	T	
F	T	T	F	T	F	
F	F	T	T	T	T	

7.

				conclusion	hypothesis	
p	q	r	$\sim q$	$p \wedge \sim q$	$p \wedge \sim q \to r$	
T	T	T	F	F	T	
T	T	F	F	F	T	
T	F	T	T	T	T	
T	F	F	T	T	F	
F	T	T	F	F	T	
F	T	F	F	F	T	
F	F	T	T	F	T	
F	F	F	T	F	T	

9.

p	q	r	$\sim r$	$p \wedge \sim r$	$q \vee r$	$p \wedge \sim r \leftrightarrow q \vee r$
T	T	T	F	F	T	F
T	T	F	T	T	T	T
T	F	T	F	F	T	F
T	F	F	T	T	F	F
F	T	T	F	F	T	F
F	T	F	T	F	T	F
F	F	T	F	F	T	F
F	F	F	T	F	F	T

12. If $x > 2$ then $x^2 > 4$, and if $x < -2$ then $x^2 > 4$.

13. a.

p	q	$\sim p$	$p \to q$	$\sim p \vee q$
T	T	F	T	T
T	F	F	F	F
F	T	T	T	T
F	F	T	T	T

$p \to q$ and $\sim p \vee q$ always have the same truth values, so they are logically equivalent.

14. a. *Hint:* $p \to q \vee r$ is true in all cases except when p is true and both q and r are false.

16. Let p represent "You paid full price" and q represent "You didn't buy it at Crown Books." Thus, "If you paid full price, you didn't buy it at Crown Books" has the form $p \to q$. And "You didn't buy it at Crown Books or you paid full price" has the form $q \vee p$.

p	q	$p \to q$	$q \vee p$
T	T	T	T
T	F	F	T
F	T	T	T
F	F	T	F

These two statements are not logically equivalent because their forms have different truth values in rows 2 and 4.

(An alternative representation for the forms of the two statements is $p \to \sim q$ and $\sim q \vee p$. In this case, the truth values differ in rows 1 and 3.)

19. False. The negation of an if-then statement is not an if-then statement. It is an *and* statement.

20. a. P is a square and P is not a rectangle.
   d. n is prime and both n is not odd and n is not 2.
   Or: n is prime and n is neither odd nor 2.
   f. Tom is Ann's father and either Jim is not her uncle or Sue is not her aunt.

21. a. Because $p \rightarrow q$ is false, p is true and q is false. Hence $\sim p$ is false, and so $\sim p \rightarrow q$ is true.

22. a. If P is not a rectangle, then P is not a square.
   d. If n is not odd and n is not 2, then n is not prime.
   f. If either Jim is not Ann's uncle or Sue is not her aunt, then Tom is not her father.

23. a. *Converse:* If P is a rectangle, then P is a square.
   Inverse: If P is not a square, then P is not a rectangle.
   d. *Converse:* If n is odd or n is 2, then n is prime.
   Inverse: If n is not prime, then n is not odd and n is not 2.
   f. *Converse:* If Jim is Ann's uncle and Sue is her aunt, then Tom is her father.
   Inverse: If Tom is not Ann's father, then Jim is not her uncle or Sue is not her aunt.

24.

p	q	$p \rightarrow q$	$q \rightarrow p$
T	T	T	T
T	F	F	T
F	T	T	F
F	F	T	T

$\quad\quad\quad\quad\uparrow\quad\quad\quad\quad\uparrow$

$p \rightarrow q$ and $q \rightarrow p$ have different truth values in the second and third rows, so they are not logically equivalent.

29. $(p \rightarrow (q \vee r)) \leftrightarrow ((p \wedge \sim q) \rightarrow r)$

p	q	r	$\sim q$	$q \vee r$	$p \wedge \sim q$	$p \rightarrow (q \vee r)$	$p \wedge \sim q \rightarrow r$	$(p \rightarrow (q \vee r)) \leftrightarrow ((p \wedge \sim q) \rightarrow r)$
T	T	T	F	T	F	T	T	T
T	T	F	F	T	F	T	T	T
T	F	T	T	T	T	T	T	T
T	F	F	T	F	T	F	F	T
F	T	T	F	T	F	T	T	T
F	T	F	F	T	F	T	T	T
F	F	T	T	T	F	T	T	T
F	F	F	T	F	F	T	T	T

$\uparrow$

$(p \rightarrow (q \vee r)) \leftrightarrow ((p \wedge \sim q) \rightarrow r)$ is a tautology because all of its truth values are T.

26.

p	q	$\sim q$	$\sim p$	$\sim q \rightarrow \sim p$	$p \rightarrow q$
T	T	F	F	T	T
T	F	T	F	F	F
F	T	F	T	T	T
F	F	T	T	T	T

$\quad\quad\quad\quad\quad\uparrow\quad\quad\quad\quad\quad\uparrow$

$\sim q \rightarrow \sim p$ and $p \rightarrow q$ have the same truth values, so they are logically equivalent.

28. *Hint:* A person who says "I mean what I say" claims to speak sincerely. A person who says "I say what I mean" claims to speak with precision.

32. If the Cubs do not win tomorrow's game, then they will not win the pennant.
   If the Cubs win the pennant, then they will have won tomorrow's game.

35. a. If a new hearing is not granted, payment will be made on the fifth.

38. a. $p \wedge \sim q \rightarrow r \equiv \sim(p \wedge \sim q) \vee r$
   b. Result of $(a) \equiv \sim[\sim(\sim(p \wedge \sim q)) \wedge \sim r]$
   an acceptable answer
   $\equiv \sim[(p \wedge \sim q) \wedge \sim r]$
   by the double negative law (another acceptable answer)

40. a. $(p \rightarrow r) \leftrightarrow (q \rightarrow r) \equiv (\sim p \vee r) \leftrightarrow (\sim q \vee r)$

$\equiv \sim (\sim p \vee r) \vee (\sim q \vee r)] \wedge [\sim (\sim q \vee r) \vee (\sim p \vee r)]$

an acceptable answer

$\equiv [(p \wedge \sim r) \vee (\sim q \vee r)] \wedge [(q \wedge \sim r) \vee (\sim p \vee r)]$

by De Morgan's law
(another acceptable answer)

b. Result of (a) $\equiv \sim [\sim (p \wedge \sim r) \wedge \sim (\sim q \vee r)] \wedge$

$\sim [\sim (q \wedge \sim r) \wedge \sim (\sim p \vee r)]$

by De Morgan's law

$\equiv \sim [\sim (p \wedge \sim r) \wedge (q \wedge \sim r)] \wedge$

$\sim [\sim (q \wedge \sim r) \wedge (p \wedge \sim r)]$

by De Morgan's law

43. If I catch the 8:05 bus, then I am on time for work.

45. If this number is not divisible by 3, then it is not divisible by 9.

If this number is divisible by 9, then it is divisible by 3.

47. If Jon's team wins the rest of its games, then it will win the championship.

49. a. This statement is the converse of the given statement, and so it is not necessarily true. For instance, if the actual boiling point of compound X were 200°C, then the given statement would be true but this statement would be false.

b. This statement must be true. It is the contrapositive of the given statement.

Section 1.3

1. $\sqrt{2}$ is not rational.   **3.** Logic is not easy.

6.

		premises		conclusion
p	q	$p \rightarrow q$	$q \rightarrow p$	$p \vee q$
T	T	T	T	T
T	F	F	T	T
F	T	T	F	T
F	F	T	T	F

This row shows that it is possible for an argument of this form to have true premises and a false conclusion. Thus this argument form is invalid.

7.

					premises		conclusion
p	q	r	$\sim q$	p	$p \rightarrow q$	$\sim q \vee r$	r
T	T	T	F	T	T	T	T
T	T	F	F	T	T	F	F
T	F	T	T	T	F	T	T
T	F	F	T	T	F	T	F
F	T	T	F	F	T	T	T
F	T	F	F	F	T	F	F
F	F	T	T	F	T	T	T
F	F	F	T	F	T	T	F

This row describes the only situation in which all the premises are true. Because the conclusion is also true here, the argument form is valid.

8.

				premises			conclusion
p	q	r	$\sim q$	$p \vee q$	$p \rightarrow \sim q$	$p \rightarrow r$	r
T	T	T	F	T	F	T	T
T	T	F	F	T	F	F	F
T	F	T	T	T	T	T	T
T	F	F	T	T	T	F	F
F	T	T	F	T	T	T	T
F	T	F	F	T	T	T	F
F	F	T	T	F	T	T	T
F	F	F	T	F	T	T	F

This row shows that it is possible for an argument of this form to have true premises and a false conclusion. Thus this argument form is invalid.

13. a.

		premises		conclusion
p	q	$p \rightarrow q$	q	p
T	T	T	T	T
T	F	F	F	T
F	T	T	T	F
F	F	T	F	F

⟵ This row shows that it is possible for an argument of this form to have true premises and a false conclusion. Thus this argument form is invalid.

14.

		premise	conclusion
p	q	p	$p \vee q$
T	T	T	T
T	F	T	T
F	T	F	T
F	F	F	F

⟵ These two rows show that in all situations where the premise is true, the conclusion is also true. Thus the argument form is valid.

18.

		premises		conclusion
p	q	$p \vee q$	$\sim q$	p
T	T	T	F	T
T	F	T	T	T
F	T	T	F	F
F	F	F	T	F

⟵ This row represents the only situation in which both premises are true. Because the conclusion is also true here the argument form is valid.

22. Let p represent "Tom is on team A" and q represent "Hua is on team B." Then the argument has the form

$$\sim p \rightarrow q$$
$$\sim q \rightarrow p$$
$$\therefore \sim p \vee \sim q$$

				premises		conclusion
p	q	$\sim p$	$\sim q$	$\sim p \rightarrow q$	$\sim q \rightarrow p$	$\sim p \vee \sim q$
T	T	F	F	T	T	F
T	F	F	T	T	T	T
F	T	T	F	T	T	T
F	F	T	T	F	F	T

⟵ This row shows that it is possible for an argument of this form to have true premises and a false conclusion. Thus this argument form is invalid.

24. $p \rightarrow q$ invalid: converse error
      q
     $\therefore p$

25. $p \vee q$ valid: elimination
      $\sim p$
     $\therefore q$

26. $p \rightarrow q$ valid: transitivity
      $q \rightarrow r$
     $\therefore p \rightarrow r$

27. $p \rightarrow q$ invalid: inverse error
      $\sim p$
     $\therefore \sim q$

36. The program contains an undeclared variable.
One explanation:
1. There is not a missing semicolon and there is not a misspelled variable name. *(by (c) and (d) and definition of ∧)*
2. It is not the case that there is a missing semicolon or a misspelled variable name. *(by (1) and De Morgan's laws)*
3. There is not a syntax error in the first five lines. *(by (b) and (2) and modus tollens)*
4. There is an undeclared variable. *(by (a) and (3) and elimination)*

37. The treasure is buried under the flagpole.

One explanation:

1. The treasure is not in the kitchen. *(by (c) and (a) and modus ponens)*
2. The tree in the front yard is not an elm. *(by (b) and (1) and modus tollens)*
3. The treasure is buried under the flagpole. *(by (d) and (2) and elimination)*

38. a. A is a knave and B is a knight.

One explanation:

1. Suppose A is a knight.
2. ∴ What A says is true. *(by definition of knight)*
3. ∴ B is a knight also. *(That's what A said.)*
4. ∴ What B says is true. *(by definition of knight)*
5. ∴ A is a knave. *(That's what B said.)*
6. ∴ We have a contradiction: A is a knight and a knave. *(by (1) and (5))*
7. ∴ The supposition that A is a knight is false. *(by the contradiction rule)*
8. ∴ A is a knave. *(negation of supposition)*
9. ∴ What B says is true. *(B said A was a knave, which we now know to be true.)*
10. ∴ B is a knight. *(by definition of knight)*

d. *Hint:* W and Y are knights; the rest are knaves.

39. The chauffeur killed Lord Hazelton.

One explanation:

1. Suppose the cook was in the kitchen at the time of the murder.
2. ∴ The butler killed Lord Hazelton with strychnine. *(by (c) and (1) and modus ponens)*
3. ∴ We have a contradiction: Lord Hazelton was killed by strychnine and a blow on the head. *(by (2) and (a))*
4. ∴ The supposition that the cook was in the kitchen is false. *(by the contradiction rule)*
5. ∴ The cook was not in the kitchen at the time of the murder. *(negation of supposition)*
6. ∴ Sara was not in the dining room when the murder was committed. *(by (e) and (5) and modus ponens)*
7. ∴ Lady Hazelton was in the dining room when the murder was committed. *(by (b) and (6) and elimination)*
8. ∴ The chauffeur killed Lord Hazelton. *(by (d) and (7) and modus ponens)*

41. (1) $p \rightarrow t$ by premise (d)
    $\sim t$ by premise (c)
    ∴ $\sim p$ by modus tollens
  (2) $\sim p$ by (1)
    ∴ $\sim p \vee q$ by generalization
  (3) $\sim p \vee q \rightarrow r$ by premise (a)
    $\sim p \vee q$ by (2)
    ∴ r by modus ponens
  (4) $\sim p$ by (1)
    r by (3)
    ∴ $\sim p \wedge r$ by conjunction

(5) $\sim p \wedge r \rightarrow \sim s$ by premise (e)
    $\sim p \wedge r$ by (4)
    ∴ $\sim s$ by modus ponens
(6) $s \vee \sim q$ by premise (b)
    $\sim s$ by (5)
    ∴ $\sim q$ by elimination

43. (1) $\sim w$ by premise (d)
    $u \vee w$ by premise (e)
    ∴ u by elimination
  (2) $u \rightarrow \sim p$ by premise (c)
    u by (1)
    ∴ $\sim p$ by modus ponens
  (3) $\sim p \rightarrow r \wedge \sim s$ by premise (a)
    $\sim p$ by (2)
    ∴ $r \wedge \sim s$ by modus ponens
  (4) $r \wedge \sim s$ by (3)
    ∴ $\sim s$ by specialization
  (5) $t \rightarrow s$ by premise (b)
    $\sim s$ by (4)
    ∴ $\sim t$ by modus tollens

Section 1.4

1. $R = 1$ **3.** $S = 1$

5.

Input		Output
P	Q	R
1	1	1
1	0	1
0	1	0
0	0	1

7.

Input			Output
P	Q	R	S
1	1	1	1
1	1	0	0
1	0	1	1
1	0	0	1
0	1	1	1
0	1	0	0
0	0	1	1
0	0	0	0

9. $P \vee \sim Q$ **11.** $(P \wedge \sim Q) \vee R$

13.

16.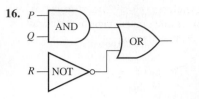

18. a. $(P \wedge Q \wedge \sim R) \vee (\sim P \wedge Q \wedge R)$

b.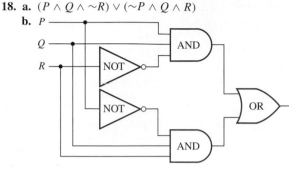

20. a. $(P \wedge Q \wedge R) \vee (P \wedge \sim Q \wedge R) \vee (\sim P \wedge \sim Q \wedge \sim R)$

b.

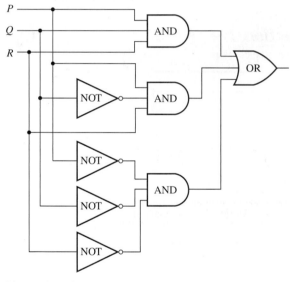

22. The input/output table is

Input			Output
P	**Q**	**R**	**S**
1	1	1	0
1	1	0	1
1	0	1	0
1	0	0	0
0	1	1	0
0	1	0	0
0	0	1	1
0	0	0	0

One circuit (among many) having this input/output table is shown below.

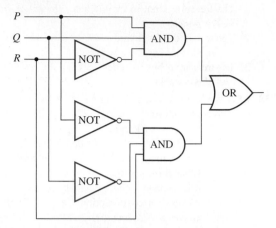

24. Let P and Q represent the positions of the switches in the classroom, with 0 being "down" and 1 being "up." Let R represent the condition of the light, with 0 being "off" and 1 being "on." Initially, $P = Q = 0$ and $R = 0$. If either P or Q (but not both) is changed to 1, the light turns on. So when $P = 1$ and $Q = 0$, then $R = 1$, and when $P = 0$ and $Q = 1$, then $R = 1$. Thus when one switch is up and the other is down the light is on, and hence moving the switch that is down to the up position turns the light off. So when $P = 1$ and $Q = 1$, then $R = 0$. It follows that the input/output table has the following appearance:

Input		Output
P	**Q**	**R**
1	1	0
1	0	1
0	1	1
0	0	0

One circuit (among many) having this input/output table is the following:

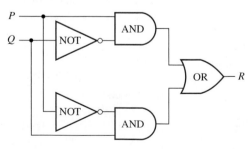

26. The Boolean expression for (a) is $(P \wedge Q) \vee Q$, and for (b) it is $(P \vee Q) \wedge Q$. We must show that if these expressions are regarded as statement forms, then they are logically equivalent. But

$(P \wedge Q) \vee Q$

$\equiv Q \vee (P \wedge Q)$	by the commutative law for $\vee$
$\equiv (Q \vee P) \wedge (Q \vee Q)$	by the distributive law
$= (Q \vee P) \wedge Q$	by the idempotent law
$\equiv (P \vee Q) \wedge Q$	by the commutative law for $\wedge$

Alternatively, by the absorption laws, both statement forms are logically equivalent to Q.

28. The Boolean expression for (a) is

$$(P \wedge Q) \vee (P \wedge \sim Q) \vee (\sim P \wedge \sim Q)$$

and for (b) it is $P \vee \sim Q$. We must show that if these expressions are regarded as statement forms, then they are logically equivalent. But

$(P \wedge Q) \vee (P \wedge \sim Q) \vee (\sim P \wedge \sim Q)$

$\equiv ((P \wedge Q) \vee (P \wedge \sim Q)) \vee (\sim P \wedge \sim Q)$	
	by inserting parentheses (which is legal by the associative law)
$\equiv (P \wedge (Q \vee \sim Q)) \vee (\sim P \wedge \sim Q)$	
	by the distributive law
$\equiv (P \wedge \mathbf{t}) \vee (\sim P \wedge \sim Q)$	by the negation law for $\vee$
$\equiv P \vee (\sim P \wedge \sim Q)$	by the identity law for $\wedge$
$\equiv (P \vee \sim P) \wedge (P \vee \sim Q)$	by the distibutive law
$\equiv \mathbf{t} \wedge (P \vee \sim Q)$	by the negation law for $\vee$
$\equiv (P \vee \sim Q) \wedge \mathbf{t}$	by the commutative law for $\wedge$
$\equiv P \vee \sim Q$	by the identity law for $\wedge$

30. $(P \wedge Q) \vee (\sim P \wedge Q) \vee (\sim P \wedge \sim Q)$

$\equiv (P \wedge Q) \vee ((\sim P \wedge Q) \vee (\sim P \wedge \sim Q))$	
	by inserting parentheses (which is legal by the associative law)
$\equiv (P \wedge Q) \vee (\sim P \wedge (Q \vee \sim Q))$	
	by the distributive law
$\equiv (P \wedge Q) \vee (\sim P \wedge \mathbf{t})$	by the negation law for $\vee$
$\equiv (P \wedge Q) \vee \sim P$	by the identity law for $\wedge$
$\equiv \sim P \vee (P \wedge Q)$	by the commutative law for $\vee$
$\equiv (\sim P \vee P) \wedge (\sim P \vee Q)$	by the distributive law
$\equiv (P \vee \sim P) \wedge (\sim P \vee Q)$	
	by the commutative law for $\vee$
$\equiv \mathbf{t} \wedge (\sim P \vee Q)$	by the negation law for $\vee$
$\equiv (\sim P \vee Q) \wedge \mathbf{t}$	by the commutative law for $\wedge$
$\equiv \sim P \vee Q$	by the identity law for $\wedge$

The following is, therefore, a circuit with at most two logic gates that has the same input/output table as the circuit corresponding to the given expression.

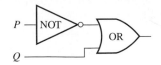

34. h. $(P \downarrow Q) \downarrow (P \downarrow Q)$

$\equiv \sim(P \downarrow Q)$	by part (a)
$\equiv \sim[\sim(P \vee Q)]$	by definition of $\downarrow$
$\equiv P \vee Q$	by the double negative law

d. *Hint:* Use the results of exercise 13 of Section 1.2 and part (a) and (c) of this exercise.

Section 1.5

1. $19_{10} = 16 + 2 + 1 = 10011_2$

4. $458_{10} = 256 + 128 + 64 + 8 + 2 = 111001010_2$

7. $1110_2 = 8 + 4 + 2 = 14_{10}$

10. $1100101_2 = 64 + 32 + 4 + 1 = 101_{10}$

13.
$$\begin{array}{r} 1 \ 1 \ 1 \\ 1 \ 0 \ 1 \ 1_2 \\ + 1 \ 0 \ 1_2 \\ \hline 1 \ 0 \ 1 \ 0 \ 0_2 \end{array}$$

15.
$$\begin{array}{r} 1 \ 1 \ 1 1 \\ 1 \ 0 \ 1 \ 1 \ 0 \ 1_2 \\ + 1 \ 1 \ 1 \ 0 \ 1_2 \\ \hline 1 \ 0 \ 0 \ 1 \ 0 \ 1 \ 0_2 \end{array}$$

17.
$$\begin{array}{r} 1 \\ 1 \ 10 \ \cancel{10} \ 1 \\ \cancel{1 \ 0} \ \cancel{1} \ 0 \ 0_2 \\ - 1 \ 1 \ 0 \ 1_2 \\ \hline 1 \ 1 \ 1_2 \end{array}$$

19.
$$\begin{array}{r} 0 \ \cancel{10} \\ 1 \ 0 \ 1 \ \cancel{1} \ 0 \ 1_2 \\ - 1 \ 0 \ 0 \ 1 \ 1_2 \\ \hline 1 \ 1 \ 0 \ 1 \ 0_2 \end{array}$$

21. a. $S = 0, T = 1$

23. $23_{10} = (16 + 4 + 2 + 1)_{10} = 00010111_2 \rightarrow 11101000 \rightarrow 11101001$. So the answer is 11101001.

25. $4_{10} = 00000100_2 \rightarrow 11111011 \rightarrow 11111100$. So the answer is 11111100.

27. Because the leading bit is 1, this is the 8-bit representation of a negative integer. $11010011 \rightarrow 00101100 \rightarrow 00101101_2 \leftrightarrow -(32 + 8 + 4 + 1)_{10} = -45_{10}$. So the answer is -45_{10}.

29. Because the leading bit is 1, this is the 8-bit representation of a negative integer. $11110010 \rightarrow 00001101 \rightarrow 00001110_2 \leftrightarrow -(8 + 4 + 2)_{10} = -14_{10}$. So the answer is -14_{10}.

31. $57_{10} = (32 + 16 + 8 + 1)_{10} = 111001_2 \rightarrow 00111001 -$
$118_{10} = -(64 + 32 + 16 + 4 + 2)_{10} = -1110110 \rightarrow$
$01110110 \rightarrow 10001001 \rightarrow 10001010$. So the 8-bit representations of 57 and -118 are 00111001 and 10001010. Adding the 8-bit representations gives

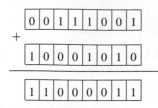

Since the leading bit of this number is a 1, the answer is negative. Converting back to decimal form gives

$11000011 \rightarrow 00111100 \rightarrow -00111101_2$
$$= -(32 + 16 + 8 + 4 + 1)_{10} = -61_{10}.$$

So the answer is -61.

32. $62_{10} = (32 + 16 + 8 + 4 + 2)_{10}$
$$= 111110_2 \rightarrow 00111110$$
$-18_{10} = -(16 + 2)_{10}$
$$= -10010_2 \rightarrow 00010010 \rightarrow 11101101 \rightarrow 11101110$$

Thus the 8-bit representations of 62 and -18 are 00111110 and 11101110. Adding the 8-bit representations gives

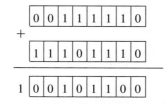

Truncating the 1 in the 2^8th position gives 00101100. Since the leading bit of this number is a 0, the answer is positive. Converting back to decimal form gives

$$00101100 \rightarrow 101100_2 = (32 + 8 + 4)_{10} = 44_{10}.$$

So the answer is 44.

33. $-6_{10} = -(4 + 2)_{10}$
$$= -110_2 \rightarrow 00000110 \rightarrow 11111001 \rightarrow 11111010$$
$-73_{10} = -(64 + 8 + 1)_{10} =$
$$-1001001_2 \rightarrow 01001001 \rightarrow 10110110 \rightarrow 10110111$$

Thus the 8-bit representations of -6 and -73 are 11111010 and 10110111. Adding the 8-bit representations gives

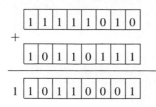

Truncating the 1 in the 2^8th position gives 10110001. Since the leading bit of this number is a 1, the answer is negative. Converting back to decimal form gives

$$10110001 \rightarrow 01001110 \rightarrow -01001111_2$$
$$= -(64 + 8 + 4 + 2 + 1)_{10} = -79_{10}.$$

So the answer is -79.

38. $A2BC_{16} = 10 \cdot 16^3 + 2 \cdot 16^2 + 11 \cdot 16 + 12$
$$= 41660_{10}$$

41. $0001110000001010101111110_2$

44. $2E_{16}$

47. a. $6 \cdot 8^4 + 1 \cdot 8^3 + 5 \cdot 8^2 + 0 \cdot 8 + 2 \cdot 1 = 25{,}410_{10}$

Section 2.1

1. a. False **b.** True

2. a. The statement is true. The integers correspond to certain of the points on a number line, and the real numbers correspond to all the points on the number line.

  b. The statement is false; 0 is neither positive nor negative.

  c. The statement is false. For instance, let $r = -2$. Then $-r = -(-2) = 2$, which is positive.

  d. The statement is false. For instance, the number $\frac{1}{2}$ is a real number, but it is not an integer.

3. a. $P(2)$ is "$2 > \frac{1}{2}$," which is true.
  $P(\frac{1}{2})$ is "$\frac{1}{2} > \frac{1}{\frac{1}{2}}$." This is false because $\frac{1}{\frac{1}{2}} = 2$, and $\frac{1}{2} \not> 2$.
  $P(-1)$ is "$-1 > \frac{1}{-1}$." This is false because $\frac{1}{-1} = -1$, and $-1 \not> -1$.
  $P(-\frac{1}{2})$ is "$-\frac{1}{2} > \frac{1}{-\frac{1}{2}}$." This is true because $\frac{1}{-\frac{1}{2}} = -2$ and $-\frac{1}{2} > -2$.
  $P(-8)$ is "$-8 > \frac{1}{-8}$." This is false because $\frac{1}{-8} = -\frac{1}{8}$ and $-8 \not> -\frac{1}{8}$.

  b. If the domain of $P(x)$ is the set of all real numbers, then its truth set is the set of all real numbers x for which either $x > 1$ or $-1 < x < 0$.

  c. If the domain of $P(x)$ is the set of all positive real numbers, then its truth set is the set of all real numbers x for which $x > 1$.

4. b. If the domain of $Q(n)$ is the set of all integers, then its truth set is $\{-5, -4, -3, -2, -1, 0, 1, 2, 3, 4, 5\}$.

5. a. $Q(-2, 1)$ is the statement "If $-2 < 1$ then $(-2)^2 < 1^2$." The hypothesis of this statement is $-2 < 1$, which is true. The conclusion is $(-2)^2 < 1^2$, which is false because $(-2)^2 = 4$ and $1^2 = 1$ and $4 \not< 1$. Thus $Q(-2, 1)$ is a conditional statement with a true hypothesis and a false conclusion. So $Q(-2, 1)$ is false.

  c. $Q(3, 8)$ is the statement "If $3 < 8$ then $3^2 < 8^2$." The hypothesis of this statement is $3 < 8$, which is true. The conclusion is $3^2 < 8^2$, which is also true because $3^2 = 9$ and $8^2 = 64$ and $9 < 64$. Thus $Q(3, 8)$ is a conditional statement with a true hypothesis and a true conclusion. So $Q(3, 8)$ is true.

7. a. The truth set is the set of all integers d such that $6/d$ is an integer, so the truth set is $\{-6, -3, -2, -1, 1, 2, 3, 6\}$.

   c. The truth set is the set of all real numbers x with the property that $1 \leq x^2 \leq 4$, so the truth set is $\{x \in \mathbf{R} \mid -2 \leq x \leq -1 \text{ or } 1 \leq x \leq 2\}$. In other words, the truth set is the set of all real numbers between -2 and -1 inclusive and between 1 and 2 inclusive.

8. a. $\{-9, -8, -7, -6, -5, -4, -3, -2, -1, 0, 1, 2, 3, 4, 5, 6, 7, 8, 9\}$

9. *Counterexample:* Let $x = 1 : 1 \not> \frac{1}{1}$. *(This is one counterexample among many.)*

11. *Counterexample:* Let $m = 1$ and $n = 1$. Then $m \cdot n = 1 \cdot 1 = 1$ and $m + n = 1 + 1 = 2$. But $1 \not\geq 2$, and so $m \cdot n \not\geq m + n$. *(This is one counterexample among many.)*

13. (a), (e), (f)   **14.** (b), (c), (e), (f)

16. a. $\forall$ dinosaurs x, x is extinct.

   c. $\forall$ irrational numbers x, x is not an integer.

   e. $\forall$ integers x, x^2 does not equal $2, 147, 581, 953$.

17. a. $\exists$ an exercise x such that x has an answer.

18. a. $\exists s \in D$ such that $E(s)$ and $M(s)$. (Or: $\exists s \in D$ such that $E(s) \wedge M(s)$.)

   b. $\forall s \in D$, if $C(s)$ then $E(s)$. (Or: $\forall s \in D$, $C(s) \rightarrow E(s)$.)

   e. $(\exists s \in D$ such that $C(s) \wedge E(s)) \wedge (\exists s \in D$ such that $C(s) \wedge \sim E(s))$.

19. (b), (d), (e)

21. a. $\forall x$, if x is a Java program, then x has at least 5 lines.

   c. $\forall$ integers x and y if x and y are even, then $x + y$ is even.

22. a. $\forall x$, if x is an even integer, then x^2 is even. $\forall$ even integers x, x^2 is even.

23. a. $\exists$ a hatter x such that x is mad. $\exists x$ such that x is a hatter and x is mad.

24. b. $\forall x (\text{Int}(x) \rightarrow \text{Ratl}(x)) \wedge \exists x (\text{Ratl}(x) \wedge \sim \text{Int}(x))$

25. a. False. Figure b is a circle that is not gray.

   b. True. All the gray figures are circles.

26. b. *One answer among many:* If a real number is negative, then when its opposite is computed, the result is a positive real number.

     This statement is true because for all real numbers x, $-(-|x|) = |x|$ (and any negative real number can be represented as $-|x|$, for some real number x).

   d. *One answer among many:* There is a real number that is not an integer. This statement is true. For instance, $\frac{1}{2}$ is a real number that is not an integer.

28. b. *One answer among many:* If an integer is prime, then it is not a perfect square.

     This statement is true because a prime number is an integer greater than 1 that is not a product of two smaller positive integers. So a prime number cannot be a perfect square because if it were, it would be a product of two smaller positive integers.

29. *Hint:* Your answer should have the appearance shown in the following made-up example:

   Statement: "If a function is differentiable, then it is continuous."

Formal version: $\forall$ functions f, if f is differentiable, then f is continuous.

Citation: Calculus by D. R. Mathematician, Best Publishing Company, 2004, page 263.

30. a. True: Any real number that is greater than 2 is greater than 1.

   c. False: $(-3)^2 > 4$ but $-3 \not> 2$.

31. a. True. Whenever both a and b are positive, so is their product.

   b. False. Let $a = -2$ and $b = -3$. Then $ab = 6$, which is not less than zero.

Section 2.2

1. (a) and (e) are negations.

3. a. $\exists$ a fish x such that x does not have gills.

   c. $\forall$ movies m, m is less than or equal to 6 hours long. (Or, $\forall$ movies m, m is no more than 6 hours long.)

In 4–6 there are other correct answers in addition to those shown.

4. a. Some pots do not have lids.

   c. No pigs can fly.

5. a. *Formal negation:* $\exists$ a dinosaur x such that x is not extinct.

     Informal negation: Some dinosaurs are not extinct.

   c. *Formal negation:* $\exists$ an irrational number x such that x is an integer.

     Informal negation: Some irrational numbers are integers.

6. a. *Formal negation:* $\forall$ exercises x, x does not have an answer.

     Informal negation: No exercises have answers.

7. The statement is not existential.

   Informal negation: There is at least one order from store A for item B.

   Formal version of statement: $\forall$ orders x, if x is from store A, then x is not for item B.

9. $\exists$ a real number x such that $x > 3$ and $x^2 \leq 9$.

11. The proposed negation is not correct. Consider the given statement: "The sum of any two irrational numbers is irrational." For this to be false means that it is possible to find at least one pair of irrational numbers whose sum is rational. On the other hand, the negation proposed in the exercise ("The sum of any two irrational numbers is rational") means that given any two irrational numbers, their sum is rational. This is a much stronger statement than the actual negation: The truth of this statement implies the truth of the negation (assuming that there are at least two irrational numbers), but the negation can be true without having this statement be true.

Correct negation: There are at least two irrational numbers whose sum is rational.

Or: The sum of some two irrational numbers is rational.

13. The proposed negation is not correct. There are two mistakes: The negation of a "for all" statement is not a "for all" statement; and the negation of an if-then statement is not an if-then statement.
    Correct negation: There exists an integer n such that n^2 is even and n is not even.

15. **a.** True: All the odd numbers in D are positive.
    c. False: $x = 16, x = 26, x = 32,$ and $x = 36$ are all counterexamples.

16. **a.** ∃ a Java program that does not have at least 5 lines.
    Or: ∃ a Java program that has fewer than 5 lines.
    c. ∃ integers m and n such that m and n are even and $m + n$ is not even.
    Or: ∃ even integers m and n such that $m + n$ is not even.

17. **a.** There exists an even integer whose square is not even.
    Or: ∃ an integer n such that n is even and n^2 is not even.

18. ∃ a real number x such that $x^2 \geq 1$ and $x \not> 0$. In other words, ∃ a real number x such that $x^2 \geq 1$ and $x \leq 0$.

20. ∃ a real number x such that $x(x + 1) > 0$ and both $x \leq 0$ and $x \geq -1$.

22. ∃ integers $a, b,$ and c such that $a - b$ is even and $b - c$ is even and $a - c$ is not even.

24. ∃ an integer n such that n is divisible by 2 and n is not even.

26. *One possible answer:* Let $P(x)$ be "$2x \neq 1$." The statement "$\forall x \in \mathbf{Z}, 2x \neq 1$" is true, but the statements "$\forall x \in \mathbf{Q}, 2x \neq 1$" and "$\forall x \in \mathbf{R}, 2x \neq 1$" are both false.

27. The claim is "$\forall x$, if $x = 1$ and x is a character in the string 0204, then x is to the left of all the 0's in the string."
    The negation is "$\exists x$ such that $x = 1$ and x is a character in the string 0204 and x is to the left of all the 0's in the string."
    The negation is false because the string does not contain the character 1. So the claim is vacuously true (or true by default).

29. *Contrapositive:* $\forall$ real numbers x, if $x \leq 0$ then $x^2 < 1$.
    Converse: $\forall$ real numbers x, if $x^2 \geq 1$ then $x > 0$.
    Inverse: $\forall$ real numbers x, if $x^2 < 1$ then $x \leq 0$.

31. *Contrapositive:* $\forall x \in \mathbf{R}$, if $x \leq 0$ and $x \geq -1$ then $x(x + 1) \leq 0$.
    Converse: $\forall x \in \mathbf{R}$, if $x > 0$ or $x < -1$ then $x(x + 1) > 0$.
    Inverse: $\forall x \in \mathbf{R}$, if $x(x + 1) \leq 0$, then $x \leq 0$ and $x \geq -1$.

33. *Contrapositive:* $\forall$ integers $a, b,$ and c, if $a - c$ is not even, then either $a - b$ is not even or $b - c$ is not even.
    Converse: $\forall$ integers $a, b,$ and c, if $a - c$ is even then both $a - b$ is even and $b - c$ is even.
    Inverse: $\forall$ integers a, b and c, if $a - b$ is not even or $b - c$ is not even, then $a - c$ is not even.

35. *Contrapositive:* $\forall$ integers n, if n is not even, then n is not divisible by 2.
    Converse: $\forall$ integers n, if n is even, then n is divisible by 2.
    Inverse: $\forall$ integers n, if n is not divisible by 2, then n is not even.

38. If a person earns a grade of C⁻ in this course, then the course counts toward graduation.

40. If a person is not on time each day, then the person will not keep this job.

42. It is not the case that if a number is divisible by 4, then that number is divisible by 8. In other words, there is a number that is divisible by 4 and is not divisible by 8.

44. It is not the case that if a person has a large income, then that person is happy. In other words, there is a person who has a large income and is not happy.

47. No. Interpreted formally, the statement says, "If carriers do not offer the same lowest fare, then you may not select among them," or, equivalently, "If you may select among carriers, then they offer the same lowest fare."

Section 2.3

1. **a.** True: Tokyo is the capital of Japan.
   b. False: Athens is not the capital of Egypt.

2. **a.** True: $2^2 > 3$ **b.** False: $1^2 \not> 1$

3. **a.** $y = \frac{1}{2}$ **b.** $y = -1$

4. **a.** Let $n = 16$. Then $n > x$ because $16 > 15.83$.

5. The statement says that no matter what circle anyone might give you, you can find a square of the same color. This is true because the only circles are $a, c,$ and b, and given a or c, which are blue, square j is also blue, and given b, which is gray, squares g and h are also gray.

7. This is true because triangle d is above every square.

9. **a.** There are five elements in D. For each, an element in E must be found so that the sum of the two equals 0. So: if $x = -2$, take $y = 2$; if $x = -1$, take $y = 1$; if $x = 0$, take $y = 0$; if $x = 1$, take $y = -1$; if $x = 2$, take $y = -2$.
   Alternatively, note that for each integer x in D, the integer $-x$ is also in D, including 0 (because $-0 = 0$), and for all integers x, $x + (-x) = 0$.

10. **a.** True. Every student chose at least one dessert: Uta chose pie, Tim chose both pie and cake, and Yuen chose pie.
    c. This statement says that some particular dessert was chosen by every student. This is true: Every student chose pie.

11. **a.** The statement has the form "∃ a student S in this class such that $\forall$ residence halls R at this school, S has dated at least one person from R." To determine whether this is true, you could present all the students in the class with a complete list of residence halls, asking them to check off all residence halls containing a person they have dated. Assuming all the students respond truthfully, if some student checks off every residence hall, then the statement is true. Otherwise, the statement is false.

12. **a.** There is a student who has seen *Casablanca*.
    c. Every student has seen at least one movie.
    d. There is a movie that has been seen by every student. (There are many other acceptable ways to state these answers.)

13. a. Negation: $\exists x$ in D such that $\forall y$ in E, $x + y \neq 1$.

The negation is true. When $x = -2$, the only number y with the property that $x + y = 1$ is $y = 3$, and 3 is not in E.

In 14–19 there are other correct answers in addition to those shown.

14. a. *Statement:* For every color, there is an animal of that color.

There are animals of every color.

b. *Negation:* $\exists$ a color C such that $\forall$ animals A, A is not colored C.

For some color, there is no animal of that color.

16. a. *Statement:* For every odd integer n, there is an integer k such that $n = 2k + 1$.

Given any odd integer, there is another integer for which the given integer equals twice the other integer plus 1.

Given any odd integer n, we can find another integer k so that $n = 2k + 1$.

An odd integer is equal to twice some other integer plus 1.

Every odd integer has the form $2k + 1$ for some integer k.

b. *Negation:* $\exists$ an odd integer n such that $\forall$ integers k, $n \neq 2k + 1$.

There is an odd integer that is not equal to $2k + 1$ for any integer k.

Some odd integer does not have the form $2k + 1$ for any integer k.

18. a. *Statement:* For every real number x, there is a real number y such that $x + y = 0$.

Given any real number x, there exists a real number y such that $x + y = 0$.

Given any real number, we can find another real number (possibly the same) such that the sum of the given number plus the other number equals 0.

Every real number can be added to some other number (possibly itself) to obtain 0.

b. *Negation:* $\exists$ a real number x such that $\forall$ real numbers y, $x + y \neq 0$.

There is a real number x for which there is no real number y with $x + y = 0$.

There is a real number x with the property that $x + y \neq 0$ for any real number y.

Some real number has the property that its sum with any other real number is nonzero.

20. a. Statement (1) says that no matter what circle anyone might give you, you can find a triangle of a different color. This is true because the only circles are a, c, and b, and given a and c, which are blue, any of triangles d, f, and i, which are gray or black, could be taken, and given circle b, which is gray, d, which is black, could be taken. In each case the chosen triangle would have a different color from the given circle.

Statement (2) says that there is a triangle that has a different color from every circle. This is also true. Triangle d is black, and all the circles are either blue or gray, so triangle d has a different color from all the circles.

21. a. Given any real number, you can find a real number so that the sum of the two is zero. In other words, every real number has an additive inverse. This statement is true.

b. There is a real number with the following property: No mattter what real number is given, the sum of the two will be zero. In other words, there is one particular real number whose sum with any real number is zero. This statement is false; no one number will work for all numbers. For instance, if $x + 0 = 0$, then $x = 0$, but in that case $x + 1 = 1 \neq 0$.

23. a. $\sim(\forall x \in D(\forall y \in E(P(x, y))))$
$$\equiv \exists x \in D(\sim(\forall y \in E(P(x, y))))$$
$$\equiv \exists x \in D(\exists y \in E(\sim P(x, y)))$$

24. This statement says that all of the circles are above all of the squares. This statement is true because the circles are a, b and c, and the squares are e, g, and h, and all of a, b, and c lie above all of e, g, and h.

Negation: There is a circle x and a square y such that x is not above y. In other words, at least one of the circles is not above at least one of the squares.

26. The statement says that there are a circle and a square with the property that the circle is above the square and has a different color from the square. This statement is true. For example, circle a lies above square e and is differently colored from e. (Several other examples could also be given.)

28. a. *Version with interchanged quantifiers:* $\exists x \in \mathbf{R}$ such that $\forall y \in \mathbf{R}$, $x < y$.

b. The given statement says that given any real number x, there is a real number y that is greater than x. This is true: For any real number x, let $y = x + 1$. Then $x < y$. The version with interchanged quantifiers says that there is a real number that is less than every other real number. This is false.

30. $\forall$ people x, $\exists$ a person y such that x is older than y.

31. $\exists$ a person x such that $\forall$ people y, x is older than y.

32. a. *Formal version:* $\forall$ people x, $\exists$ a person y such that x loves y.

b. *Negation:* $\exists$ a person x such that $\forall$ people y, x does not love y. In other words, there is someone who does not love anyone.

33. a. *Formal version:* $\exists$ a person x such that $\forall$ people y, x loves y.

b. *Negation:* $\forall$ people x, $\exists$ a person y such that x does not love y. In other words, everyone has someone whom they love.

36. a. *Statement:* $\forall$ even integers n, $\exists$ an integer k such that $n = 2k$.

b. *Negation:* $\exists$ an even integer n such that $\forall$ integers k, $n \neq 2k$.

There is some even integer that is not equal to twice any other integer.

38. a. *Statement:* ∃ a program P such that ∀ questions Q posed to P, P gives the correct answer to Q.

b. *Negation:* ∀ programs P, there is a question Q that can be posed to P such that P does not give the correct answer to Q.

39. ∀ minutes m, ∃ a sucker s such that s was born in minute m.

40. a. This statement says that given any positive integer, there is a positive integer such that the first integer is one more than the second integer. This is false. Given the positive integer $x = 1$, the only integer with the property that $x = y + 1$ is $y = 0$, and 0 is not a positive integer.

b. This statement says that given any integer, there is an integer such that the first integer is one more than the second integer. This is true. Given any integer x, take $y = x - 1$. Then y is an integer, and $y + 1 = (x - 1) + 1 = x$.

e. This statement says that given any real number, there is a real number such that the product of the two is equal to 1. This is false because $0 \cdot y = 0 \neq 1$ for every number y. So when $x = 0$, there is no real number y with the property that $xy = 1$.

41. ∃$\varepsilon > 0$ such that ∀ integers N, ∃ an integer n such that $n > N$ and either $L - \varepsilon \geq a_n$ or $a_n \geq L + \varepsilon$. In other words, there is a positive number ε such that for all integers N, it is possible to find an integer n that is greater than N and has the property that a_n does not lie between $L - \varepsilon$ and $L + \varepsilon$.

43. a. This statement is true. The unique real number with the given property is 1. Note that

$$1 \cdot y = y \quad \text{for all real numbers } y,$$

and if x is any real number such that for instance, $x \cdot 2 = 2$, then dividing both sides by 2 gives $x = \frac{2}{2} = 1$.

45. a. True. Both triangles a and c lie above all the squares.

b. *Formal version:* ∃x(Triangle$(x) \wedge$ (∀y(Square$(y) \rightarrow$ Above(x, y))))

c. *Formal negation:* ∀x($\sim$Triangle$(x) \vee$ (∃y(Square$(y) \wedge$ $\sim$Above(x, y))))

47. a. False. There is no square to the right of circle k.

b. *Formal version:* ∀x(Circle$(x) \rightarrow$ (∃y(Square$(y) \wedge$ RightOf(y, x))))

c. *Formal negation:* ∃x(Circle$(x) \wedge$ (∀y($\sim$Square$(y) \vee$ $\sim$RightOf(y, x))))

49. a. False. There is no object that has a different color from every other object.

b. *Formal version:* ∃y(∀x($x \neq y \rightarrow$ $\sim$SameColor(x, y)))

c. *Formal negation:* ∀y(∃x($x \neq y \wedge$ SameColor(x, y)))

51. a. False

b. *Formal version:* ∃x(Circle$(x) \wedge$ (∃y(Square$(y) \wedge$ SameColor(x, y))))

c. *Formal negation:* ∀x($\sim$Circle$(x) \vee$ (∀y($\sim$Square$(y) \vee$ $\sim$SameColor(x, y))))

53. a. No matter what the domain D or the predicates $P(x)$ and $Q(x)$ are, the given statements have the same truth value. If the statement "∀x in D, $(P(x) \wedge Q(x))$" is true, then $P(x) \wedge Q(x)$ is true for every x in D, which implies that both $P(x)$ and $Q(x)$ are true for every x in D. But then $P(x)$ is true for every x in D, and also $Q(x)$ is true for every x in D. So the statement "(∀x in D, $P(x)$) ∧ (∀x, in D, $Q(x)$)" is true. Conversely, if the statement "(∀x in D, $P(x)$) ∧ (∀x in D, $Q(x)$)" is true, then $P(x)$ is true for every x in D, and also $Q(x)$ is true for every x in D. This implies that both $P(x)$ and $Q(x)$ are true for every x in D, and so $P(x) \wedge Q(x)$ is true for every x in D. Hence the statement "∀x in D, $(P(x) \wedge Q(x))$" is true.

57. a. Yes **b.** $X = w_1$, $X = w_2$ **c.** $X = b_2$, $X = w_2$

Section 2.4

1. b. $(f_i + f_j)^2 = f_i^2 + 2f_if_j + f_j^2$

c. $(3u + 5v)^2 = (3u)^2 + 2(3u)(5v) + (5v)^2$
$(= 9u^2 + 30uv + 25v^2)$

d. $(g(r) + g(s))^2 = (g(r))^2 + 2g(r)g(s) + (g(s))^2$

2. 0 is even.

3. $\frac{2}{3} + \frac{4}{5} = \frac{2 \cdot 5 + 3 \cdot 4}{3 \cdot 5} (= \frac{22}{15})$

5. Adster is not a healthy person.

7. Invalid; converse error

8. Valid by universal modus ponens (or universal instantiation)

9. Invalid; inverse error

10. Valid by universal modus tollens

16. Invalid; converse error.

19. ∀x, if x is a good car, then x is not cheap.

a. Valid, universal modus ponens (or universal instantiation)

b. Invalid, converse error

21. Valid. (A valid argument can have false premises and a true conclusion!)

The major premise says the set of people is included in the set of mice. The minor premise says the set of mice is included in the set of mortals. Assuming both of these premises are true, it must follow that the set of people is included in the set of mortals. Since it is impossible for the conclusion to be false if the premises are true, the argument is valid.

23. Valid. The major and minor premises can be diagrammed as follows:

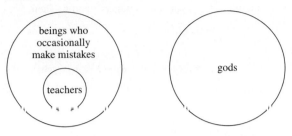

According to the diagram, the set of teachers and the set of gods can have no common elements. Hence, if the premises are true, then the conclusion must also be true, and so the argument is valid.

25. Invalid. Let C represent the set of all college cafeteria food, G the set of all good food, and W the set of all wasted food. Then any one of the following diagrams could represent the given premises.

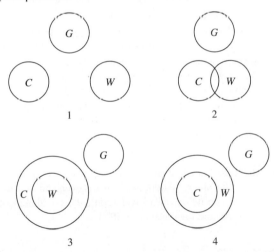

Only in drawing (1) is the conclusion true. Hence it is possible for the premises to be true while the conclusion is false, and so the argument is invalid.

28. (3) *Contrapositive form:* If an object is gray, then it is a circle.
(2) If an object is a circle, then it is to the right of all the blue objects.
(1) If an object is to right of all the blue objects, then it is above all the triangles.
∴ If an object is gray, then it is above all the triangles.

31. 4. If an animal is in the yard, then it is mine.
1. If an animal belongs to me, then I trust it.
5. If I trust an animal, then I admit it into my study.
3. If I admit an animal into my study, then it will beg when told to do so.
6. If an animal begs when told to do so, then that animal is a dog.
2. If an animal is a dog, then that animal gnaws bones.
∴ If an animal is in the yard, then that animal gnaws bones; that is, all the animals in the yard gnaw bones.

33. 2. If a bird is in this aviary, then it belongs to me.
4. If a bird belongs to me, then it is at least 9 feet high.
1. If a bird is at least 9 feet high, then it is an ostrich.
3. If a bird lives on mince pies, then it is not an ostrich.
Contrapositive: If a bird is an ostrich, then it does not live on mince pies.
∴ If a bird is in this aviary, then it does not live on mince pies; that is, no bird in this aviary lives on mince pies.

Section 3.1

1. a. Yes: $-17 = 2(-9) + 1$ **b.** Yes: $0 = 2 \cdot 0$
  c. Yes: $2k - 1 = 2(k - 1) + 1$ and $k - 1$ is an integer because it is a difference of integers.

2. a. Yes: $6m + 8n = 2(3m + 4n)$ and $(3m + 4n)$ is an integer because 3, 4, m and n are integers, and products and sums of integers are integers.
  b. Yes: $10mn + 7 = 2(5mn + 3) + 1$ and $5mn + 3$ is an integer because 3, 5, m, and n are integers, and products and sums of integers are integers.
  c. Not necessarily. For instance, if $m = 3$ and $n = 2$, then $m^2 - n^2 = 9 - 4 = 5$, which is prime. (Note that $m^2 - n^2$ is composite for many values of m and n because of the identity $m^2 - n^2 = (m - n)(m + n)$.)

4. For example, let $m = n = 2$. Then m and n are integers such that $m > 0$ and $n > 0$ and $\frac{1}{m} + \frac{1}{n} = \frac{1}{2} + \frac{1}{2} = 1$, which is an integer.

7. For example, let $n = 7$. Then n is an integer such that $n > 5$ and $2^n - 1 = 127$, which is prime.

9. For example, 25, 9, and 16 are all perfect squares, because $25 = 5^2, 9 = 3^2$, and $16 = 4^2$, and $25 = 9 + 16$. Thus 25 is a perfect square that can be written as a sum of two other perfect squares.

11. *Counterexample:* Let $a = -2$ and $b = -1$. Then $a < b$ because $-2 < -1$, but $a^2 \not< b^2$ because $(-2)^2 = 4$ and $(-1)^2 = 1$ and $4 \not< 1$.

14. This property is true for some integers and false for other integers. For instance, if $a = 0$ and $b = 1$, the property is true because $(0 + 1)^2 = 0^2 + 1^2$, but if $a = 1$ and $b = 1$, the property is false because $(1 + 1)^2 = 4$ and $1^2 + 1^2 = 2$ and $4 \neq 2$.

17. $2 = 1^2 + 1^2, 4 = 2^2, 6 = 2^2 + 1^2 + 1^2,$
  $8 = 2^2 + 2^2, 10 = 3^2 + 1^2, 12 = 2^2 + 2^2 + 2^2,$
  $14 = 3^2 + 2^2 + 1^2, 16 = 4^2,$
  $18 = 3^2 + 3^2 = 4^2 + 1^2 + 1^2, 20 = 4^2 + 2^2,$
  $22 = 3^2 + 3^2 + 2^2, 24 = 4^2 + 2^2 + 2^2$

19. *Theorem:* ∀ integers m and n, if m is even and n is odd, then $m + n$ is odd.
  a. any odd integer   **b.** integer r
  c. $2r + (2s + 1)$ **d.** $m + n$ is odd

20. *Start of proof:* Suppose m is an integer such that $m > 1$.
  Conclusion to be shown: $0 < 1/m < 1$.

22. *Start of proof:* Suppose m and n are integers and $mn = 1$.
  Conclusion to be shown: $m = n = 1$ or $m = n = -1$.

24. *Two versions of a correct proof are given below to illustrate some of the variety that is possible.*

Proof 1: Suppose n is any [*particular but arbitrarily chosen*] even integer. [*We must show that $-n$ is even.*] By definition of even, $n = 2k$ for some integer k. Multiplying both sides by -1 gives that $-n = -(2k) = 2(-k)$. Let $r = -k$. Then r is an integer because a product of two integers is an integer, $r = -k = (-1)k$, and -1 and k are integers. Hence $-n = 2r$ for some integer r, and so $-n$ is even [*as was to be shown.*]

Proof 2: Suppose n is any even integer. By definition of even, $n = 2k$ for some integer k. Then $-n = -2k = 2(-k)$. But $-k$ is an integer because it is a product of integers -1 and k. Thus $-n = 2 \cdot$ (some integer), and so $-n$ is even by definition of even.

25. *Proof:* Suppose a is any even integer and b is any odd integer. [*We must show that $a - b$ is odd.*] By definition of even and odd, $a = 2r$ and $b = 2s + 1$ for some integers r and s. By substitution and algebra, $a - b = 2r - (2s + 1) = 2r - 2s - 1 = 2(r - s - 1) + 1$. Let $t = r - s - 1$. Then t is an integer because sums and differences of integers are integers. Thus $a - b = 2t + 1$, where t is an integer, and so, by definition of odd, $a - b$ is odd [*as was to be shown*].

29. *Proof:* Suppose n is any even integer. Then $n = 2k$ for some integer k. Hence $(-1)^n = (-1)^{2k} = ((-1)^2)^k = 1^k = 1$ [*by the laws of exponents from algebra*]. This is what was to be shown.

31. The negation of the statement is "For all integers $m \geq 3$, $m^2 - 1$ is not prime."

Proof of the negation: Suppose m is any integer with $m \geq 3$. By basic algebra, $m^2 - 1 = (m - 1)(m + 1)$. Because $m \geq 3$, both $m - 1$ and $m + 1$ are positive integers greater than 1, and each is smaller than $m^2 - 1$. So $m^2 - 1$ is a product of two smaller positive integers, each greater than 1, and hence $m^2 - 1$ is not prime.

34. The incorrect proof just shows the theorem to be true in the one case where $k = 2$. A real proof must show that it is true for *all* integers $k > 0$.

35. The mistake in the "proof" is that the same symbol, k, is used to represent two different quantities. By setting $m = 2k$ and $n = 2k + 1$, the proof implies that $n = m + 1$, and thus it deduces the conclusion only for this one situation. When $m = 4$ and $n = 17$, for instance, the computations in the proof indicate that $n - m = 1$, but actually $n - m = 13$. In other words, the proof does not deduce the conclusion for an arbitrarily chosen even integer m and odd integer n, and hence it is invalid.

36. This incorrect proof begs the question. The word *since* in the third sentence is completely unjustified. The second sentence tells only what happens *if* $k^2 + 2k + 1$ is composite. But at that point in the proof, it has not been established that $k^2 + 2k + 1$ *is* composite. In fact, that is exactly what is to be proved.

39. True. *Proof:* Suppose m and n are any odd integers. [*We must show that mn is odd.*] By definition of odd, $n = 2r + 1$ and $m = 2s + 1$ for some integers r and s. Then

$$
\begin{aligned}
mn &= (2r + 1)(2s + 1) && \text{by subsitution} \\
&= 4rs + 2r + 2s + 1 \\
&= 2(2rs + r + s) + 1 && \text{by algebra.}
\end{aligned}
$$

Now $2rs + r + s$ is an integer because products and sums of integers are integers and 2, r, and s are all integers. Hence $mn = 2 \cdot$ (some integer) $+ 1$, and so, by definition of odd, mn is odd.

40. True. *Proof:* Suppose n is any odd integer. [*We must show that $-n$ is odd.*] By definition of odd, $n = 2k + 1$ for some integer k. By substitution and algebra, $-n = -(2k + 1) = -2k - 1 = 2(-k - 1) + 1$. Let $t = -k - 1$. Then t is an integer because differences of integers are integers. Thus $-n = 2t + 1$, where t is an integer, and so, by definition of odd, $-n$ is odd [*as was to be shown*].

41. False. *Counterexample:* Both 3 and 1 are odd, but their difference is $3 - 1 = 2$, which is even.

43. *Counterexample:* Let $m = 1$ and $n = 3$. Then $m + n = 4$ is even, but neither summand m nor summand n is even.

50. *Proof:* Suppose n is any integer. Then $4(n^2 + n + 1) - 3n^2 = 4n^2 + 4n + 4 - 3n^2 = n^2 + 4n + 4 = (n + 2)^2$ (by algebra). But $(n + 2)^2$ is a perfect square because $n + 2$ is an integer (being a sum of n and 2). Hence $4(n^2 + n + 1) - 3n^2$ is a perfect square, as was to be shown.

52. *Hint:* This is true.

58. *Hint:* The answer is no.

60. a. *Hint:* Note that $(x - r)(x - s) = x^2 - (r + s) + rs$. If both r and s are odd, then $r + s$ is even and rs is odd. So the coefficient of x^2 is 1 (odd), the coefficient of x is even, and the constant coefficient is odd.

Section 3.2

1. $-\frac{35}{6} = \frac{-35}{6}$ **3.** $\frac{4}{5} + \frac{2}{9} = \frac{4 \cdot 9 + 2 \cdot 5}{45} = \frac{46}{45}$

4. Let $x = 0.3737373737\ldots$.
Then $100x = 37.37373737\ldots$, and so
$100x - x = 37.37373737\ldots - 0.3737373737\ldots$.
Thus $99x = 37$, and hence $x = \frac{37}{99}$.

6. Let $x = 320.5492492492\ldots$.
Then $10000x = 3205492.492492\ldots$, and
$10x = 3205.492492492\ldots$, and so
$10000x - 10x = 3205492 - 3205$.
Thus $9990x = 3202287$, and hence $x = \frac{3202287}{9990}$.

8. b. $\forall$ real numbers x and y, if $x \neq 0$ and $y \neq 0$ then $xy \neq 0$.

9. Because a and b are integers, $b - a$ and ab^2 are both integers (since differences and products of integers are integers). Also, by the zero product property, $ab^2 \neq 0$ because neither a nor b is zero. Hence $(b - a)/ab^2$ is a quotient of two integers with nonzero denominator, and so it is rational.

11. *Proof:* Suppose n is any [*particular but arbitrarily chosen*] integer. Then $n = n \cdot 1$, and so $n = n/1$ by dividing both sides by 1. Now n and 1 are both integers, and $1 \neq 0$. Hence n can be written as a quotient of integers with a nonzero denominator, and so n is rational.

12. a. any [*particular but arbitrarily chosen*] rational number
b. integers a and b **c.** $(a/b)^2$ **d.** b^2
e. zero product property **f.** r^2 is rational

13. *Proof:* Suppose r and s are rational numbers. By definition of rational, $r = a/b$ and $s = c/d$ for some integers $a, b, c,$ and d with $b \neq 0$ and $d \neq 0$. Then

$$rs = \frac{a}{b} \cdot \frac{c}{d} \qquad \text{by substitution}$$

$$-\frac{ac}{bd} \qquad \begin{array}{l} \text{by the rules of algebra for} \\ \text{multiplying fractions.} \end{array}$$

Now ac and bd are both integers (being products of integers) and $bd \neq 0$ (by the zero product property). Hence rs is a quotient of integers with a nonzero denominator, and so, by definition of rational, rs is rational.

14. *Hint: Counterexample:* Let r be any rational number and $s = 0$. Then r and s are both rational, but the quotient of r divided by s is undefined and therefore is not a rational number.
Revised statement to be proved: For all rational numbers r and s, if $s \neq 0$ then r/s is rational.

17. *Hint:* $\dfrac{a/b + c/d}{2} = \dfrac{(ad + bc)/(bd)}{2} = \dfrac{ad + bc}{2bd}$

18. *Hint:* If $a < b$ then $a + a < a + b$ (by T18 of Appendix A), or equivalently $2a < a + b$. Thus $a < \frac{a+b}{2}$ (by T19 Appendix A).

20. True. *Proof:* Suppose m is any even integer and n is any odd integer. [*We must show that $m^2 + 3n$ is odd.*] By properties 1 and 3 of Example 3.2.2, m^2 is even (because $m^2 = m \cdot m$) and $3n$ is odd (because both 3 and n are odd). It follows from property 5 [*and the commutative law for addition*] that $m^2 + 3n$ is odd [*as was to be shown*].

23. *Proof:* Suppose r and s are any rational numbers. By Theorem 3.2.1, both 2 and 3 are rational, and so, by Exercise 13, both $2r$ and $3s$ are rational. Hence, by Theorem 3.2.2, $2r + 3s$ is rational.

26. Let

$$x = \frac{1 - \dfrac{1}{2^{n+1}}}{1 - \dfrac{1}{2}} = \frac{1 - \dfrac{1}{2^{n+1}}}{\dfrac{1}{2}} = \frac{1 - \dfrac{1}{2^{n+1}}}{\dfrac{1}{2}} \cdot \frac{2^{n+1}}{2^{n+1}} = \frac{2^{n+1} - 1}{2^n}.$$

But $2^{n+1} - 1$ and 2^n are both integers (since n is a nonnegative integer) and $2^n \neq 0$ by the zero product property. Therefore, x is rational.

30. *Proof:* Suppose c is a real number such that

$$r_3 c^3 + r_2 c^2 + r_1 c + r_0 = 0,$$

where $r_0, r_1, r_2,$ and r_3 are rational numbers. By definition of rational, $r_0 = a_0/b_0, r_1 = a_1/b_1, r_2 = a_2/b_2,$ and

$r_3 = a_3/b_3$ for some integers, $a_0, a_1, a_2, a_3,$ and nonzero integers $b_0, b_1, b_2,$ and b_3. By substitution,

$$r_3 c^3 + r_2 c^2 + r_1 c + r_0$$

$$= \frac{a_3}{b_3} c^3 + \frac{a_2}{b_2} c^2 + \frac{a_1}{b_1} c + \frac{a_0}{b_0}$$

$$= \frac{b_0 b_1 b_2 a_3}{b_0 b_1 b_2 b_3} c^3 + \frac{b_0 b_1 b_3 a_2}{b_0 b_1 b_2 b_3} c^2 + \frac{b_0 b_2 b_3 a_1}{b_0 b_1 b_2 b_3} c + \frac{b_1 b_2 b_3 a_0}{b_0 b_1 b_2 b_3}$$

$$= 0.$$

Multiplying both sides by $b_0 b_1 b_2 b_3$ gives

$$b_0 b_1 b_2 a_3 \cdot c^3 + b_0 b_1 b_3 a_2 \cdot c^2 + b_0 b_2 b_3 a_1 \cdot c + b_1 b_2 b_3 a_0 = 0.$$

Let $n_3 = b_0 b_1 b_3 a_3, \; n_2 = b_0 b_1 b_3 a_2, \; n_1 = b_0 b_2 b_3 a_1,$ and $n_0 = b_1 b_2 b_3 a_0$. Then $n_0, n_1, n_2,$ and n_3 are all integers (being products of integers). Hence c satisfies the equation

$$n_3 c^3 + n_2 c^2 + n_1 c + n_0 = 0,$$

where $n_0, n_1, n_2,$ and n_3 are all integers. This is what was to be shown.

33. This "proof" begs the question by assuming what is to be proved.

34. By setting both r and s equal to a/b, this incorrect proof violates the requirement that r and s be arbitrarily chosen rational numbers. If both r and s equal a/b, then $r - s$.

Section 3.3

1. Yes, $52 = 13 \cdot 4$

4. Yes, $(3k + 1)(3k + 2)(3k + 3) = 3[(3k + 1)(3k + 2)(k + 1)]$, and $(3k + 1)(3k + 2)(k + 1)$ is an integer because k is an integer and sums and products of integers are integers.

6. No, $29/3 \cong 9.67$, which is not an integer.

7. Yes, $66 = (-3)(-22)$.

8. Yes, $6a(a + b) = 3a[2(a + b)]$, and $2(a + b)$ is an integer because a and b are integers and sums and products of integers are integers.

10. No, $34/7 \cong 4.86$, which is not an integer.

12. Yes, $n^2 - 1 = (4k + 1)^2 - 1 = (16k^2 + 8k + 1) - 1 = 16k^2 + 8k = 8(2k^2 + k),$ and $2k^2 + k$ is an integer because k is an integer and sums and products of integers are integers.

14. a. $a \mid b$ **b.** $a \cdot k$ **c.** integer
d. $-(a \cdot k)$ **e.** $a \mid (-b)$

15. *Proof:* Suppose $a, b,$ and c are any integers such that $a \mid b$ and $a \mid c$. [*We must show that $a \mid (b + c)$.*] By definition of divides, $b = ar$ and $c = as$ for some integers r and s. Then $b + c = ar + as = a(r + s)$ (by algebra).

Let $t = r + s$. Then t is an integer (being a sum of integers), and thus $b + c = at$ where t is an integer. By definition of divides, then, $a \mid (b + c)$ [*as was to be shown.*]

17. *Proof:* Suppose $n, n + 1,$ and $n + 2$ are any three consecutive integers. Then $n + (n + 1) + (n + 2) = 3n + 3 =$

$3(n + 1)$. This is divisible by 3 because $n + 1$ is an integer (since n is an integer, and a sum of integers is an integer).

19. The given statement can be rewritten formally as "$\forall$ integers n, if n is divisible by 6, then n is divisible by 2." This statement is true.

Proof 1: Suppose n is any integer that is divisible by 6. By definition of divisibility, $n = 6k$ for some integer k. But $6k = 2 \cdot (3k)$, and $3k$ is an integer because k is. Hence $n = 2 \cdot$ (some integer), and so n is divisible by 2.

Proof 2: Suppose n is any integer that is divisible by 6. We know that 6 is divisible by 2 because $6 = 2 \cdot 3$. So $2 \mid 6$ and $6 \mid n$. Hence by transitivity of divisibility (Theorem 3.3.1), $2 \mid n$, or, in other words, n is divisible by 2.

21. The statement is true. *Proof:* Suppose a, b, and c are any integers such that $a \mid b$ and $a \mid c$. [*We must show that $a \mid (2b - 3c)$.*] By definition of divisibility, we know that $b = am$ and $c = an$ for some integers m and n. It follows that $2b - 3c = 2(am) - 3(an)$ (*by substitution*) $= a(2m - 3n)$ (*by basic algebra*). Let $t = 2m - 3n$. Then t is an integer because it is a difference of products of integers. Hence $2b - 3c = at$, where t is an integer, and so $a \mid (2b - 3c)$ by definition of divisibility [*as was to be shown*].

22. *Hint:* The statement is true.

23. The statement is false. *Counterexample:* Let $a = 2$, $b = 3$, and $c = 8$. Then $a \mid c$ because 2 divides 8, but $ab \nmid c$ because $ab = 6$ and 6 does not divide 8.

24. *Hint:* The statement is false.

29. No. Each of these numbers is divisible by 3, and so their sum is also divisible by 3. But 100 is not divisible by 3. Thus the sum cannot equal $100.

33. a. The sum of the digits is 54, which is divisible by 9. Therefore, 637,425,403,705,125 is divisible by 9 and hence also divisible by 3 (by transitivity of divisibility). Because the rightmost digit is 5, then 637,425,403,705,125 is also divisible by 5. And because the two rightmost digits are 25, which is not divisible by 4, then 637,425,403,705,125 is not divisible by 4.

34. $1176 = 2^3 \cdot 3 \cdot 7^2$

35. a. $p_1^{2e_1} p_2^{2e_2} \ldots p_k^{2e_k}$
   b. $n = 42$, $2^5 \cdot 3 \cdot 5^2 \cdot 7^3 \cdot n = 5.880^2$

37. a. Because $12a = 25b$, the unique factorization theorem guarantees that the standard factored forms of $12a$ and $25b$ must be the same. Thus $25b$ contains the factors $2^2 \cdot 3 \,(= 12)$. But since neither 2 nor 3 divide 25, the factors $2^2 \cdot 3$ must all occur in b, and hence $12 \mid b$. Similarly, $12a$ contains the factors $5^2 = 25$, and since 5 is not a factor of 12, the factors 5^2 must occur in a. So $25 \mid a$.

39. a. $6! = 6 \cdot 5 \cdot 4 \cdot 3 \cdot 2 \cdot 1 = 2 \cdot 3 \cdot 5 \cdot 2 \cdot 2 \cdot 3 \cdot 2$
    $= 2^4 \cdot 3^2 \cdot 5$

41. *Proof:* Suppose n is a nonnegative integer whose decimal representation ends in 0. Then $n = 10m + 0 = 10m$ for some integer m. Factoring out a 5 yields $n = 10m = 5(2m)$, and $2m$ is an integer since m is an integer. Hence $10m$ is divisible by 5, which is what was to be shown.

44. *Hint:* You may take it as a fact that for any positive integer k, $10^k = \underbrace{99\ldots 9}_{k \text{ of these}} + 1$; that is,

$$10^k = 9 \cdot 10^{k-1} + 9 \cdot 10^{k-2} + \cdots + 9 \cdot 10^1 + 9 \cdot 10^0 + 1.$$

Section 3.4

1. $q = 7, r = 7$ **3.** $q = 0, r = 36$

5. $q = -5, r = 10$ **7. a.** 4   **b.** 7

11. a. When today is Saturday, 15 days from today is two weeks (which is Saturday) plus one day (which is Sunday). Hence $DayN$ should be 0. According to the formula, when today is Saturday, $DayT = 6$, and so when $N = 15$,

$$\begin{aligned} DayN &= (DayT + N) \bmod 7 \\ &= (6 + 15) \bmod 7 \\ &= 21 \bmod 7 = 0, \text{ which agrees.} \end{aligned}$$

13. *Solution 1:* $30 = 4 \cdot 7 + 2$. Hence the answer is two days after Monday, or Wednesday.

Solution 2: By the formula, the answer is $(1 + 30) \bmod 7 = 31 \bmod 7 = 3$, which is Wednesday.

14. *Hint:* There are two ways to solve this problem. One is to find that $1,000 = 7 \cdot 142 + 6$ and note that if today is Tuesday, then 1,000 days from today is 142 weeks plus 6 days from today. The other way is to use the formula $DayN = (DayT + N) \bmod T$, with $DayT = 2$ (Tuesday) and $N = 1000$.

16. *Hint:* By the quotient-remainder theorem, $0 \le r < d$. Thus $r - d < 0$. But $n\%d = r - d$, so it is false that $0 \le n\%d < d$.

17. Because the remainder obtained when a is divided by 7 is 4, we know that $a = 7q + 4$ for some integer q. Multiplying this equation through by 5 gives that $5a = 35q + 20 = 35q + 14 + 6 = 7(5q + 2) + 6$. Because q is an integer, $5q + 2$ is also an integer, and so $5a = 7 \cdot$ (an integer) $+ 6$. Thus, because $0 \le 6 < 7$, the remainder obtained when $5a$ is divided by 7 is 6.

20. Because $d \mid n$, $n = dq + 0$ for some integer q. Thus the remainder is 0.

21. *Hint:* You need to show that (1) for all nonnegative integers n and positive integers d, if n is divisible by d then $n \bmod d = 0$; and (2) for all nonnegative integers n and positive integers d, if $n \bmod d = 0$ then n is divisible by d.

22. $7609 + 5 = 7614$

25. *Proof:* Suppose n is any odd integer. By definition of odd, $n = 2q + 1$ for some integer q. Then $n^2 = (2q + 1)^2 = 4q^2 + 4q + 1 = 4(q^2 + q) + 1 = 4q(q + 1) + 1$. By the result of exercise 24, the product $q(q + 1)$ is even, so $q(q + 1) = 2m$ for some integer m. Then, by substitution, $n^2 = 4 \cdot 2m + 1 = 8m + 1$.

27. *Proof:* Suppose n is any integer. By the quotient-remainder theorem with $d = 3$, there exist integers q and r such that $n = 3q + r$ and $0 \le r < 3$. But the only nonnegative integers r that are less than 3 are 0, 1, and 2. Therefore, $n = 3q + 0 = 3q$, or $n = 3q + 1$, or $n = 3q + 2$ for some integer q.

28. a. *Proof:* Suppose $n, n + 1$, and $n + 2$ are any three consecutive integers. [*We must show that $n(n + 1)(n + 2)$ is divisible by 3.*] By the result of exercise 27, n can be written in one of the three forms, $3q, 3q + 1$, or $3q + 2$ for some integer q. We divide into cases accordingly.

Case 1 ($n = 3q$ for some integer q): In this case,

$n(n + 1)(n + 2)$

$= 3q(3q + 1)(3q + 2)$ by substitution

$= 3 \cdot [q(3q + 1)(3q + 2)]$ by factoring out a 3.

Let $m = q(3q + 1)(3q + 2)$. Then m is an integer because q is an integer, and sums and products of integers are integers. By substitution,

$n(n + 1)(n + 2) = 3m$ where m is an integer.

And so, by definition of divisible, $n(n + 1)(n + 2)$ is divisible by 3.

Case 2 ($n = 3q + 1$ for some integer q): In this case,

$n(n + 1)(n + 2)$

$= (3q + 1)((3q + 1) + 1)((3q + 1) + 2)$

by substitution

$= (3q + 1)(3q + 2)(3q + 3)$

$= (3q + 1)(3q + 2)3(q + 1)$

$= 3 \cdot [(3q + 1)(3q + 2)(q + 1)]$ by algebra

Let $m = (3q + 1)(3q + 2)(q + 1)$. Then m is an integer because q is an integer, and sums and products of integers are integers. By substitution,

$n(n + 1)(n + 2) = 3m$ where m is an integer.

And so, by definition of divisible, $n(n + 1)(n + 2)$ is divisible by 3.

Case 3 ($n = 3q + 2$ for some integer q): In this case,

$n(n + 1)(n + 2)$

$= (3q + 2)((3q + 2) + 1)((3q + 2) + 2)$

by substitution

$= (3q + 2)(3q + 3)(3q + 4)$

$= (3q + 2)3(q + 1)(3q + 4)$

$= 3 \cdot [(3q + 2)(q + 1)(3q + 4)]$ by algebra

Let $m = (3q + 2)(q + 1)(3q + 4)$. Then m is an integer because q is an integer, and sums and products of integers are integers. By substitution,

$n(n + 1)(n + 2) = 3m$ where m is an integer.

And so, by definition of divisible, $n(n + 1)(n + 2)$ is divisible by 3.

In each of the three cases, $n(n + 1)(n + 2)$ was seen to be divisible by 3. But by the quotient-remainder theorem,

one of these cases must occur. Therefore, the product of *any* three consecutive integers is divisible by 3.

b. For all integers n, $n(n + 1)(n + 2) \bmod 3 = 0$.

29. *Hint:* Given any integer n, begin by using the quotient-remainder theorem to say that n can be written in one of the three forms: $n = 3q$, or $n = 3q + 1$, or $n = 3q + 2$ for some integer q. Then divide into three cases according to these three possibilities. Show that in each case either $n^2 = 3k$ for some integer k, or $n^2 = 3k + 1$ for some integer k. For instance, when $n = 3q + 2$, then $n^2 = (3q + 2)^2 = 9q^2 + 6q + 4 = 3(3q^2 + 2q + 1) + 1$, and $3q^2 + 2q + 1$ is an integer because it is a sum of products of integers.

31. b. If $m^2 - n^2 = 56$, then $56 = (m + n)(m - n)$. Now $56 = 2^3 \cdot 7$, and by the unique factorization theorem, this factorization is unique. Hence the only representations of 56 as a product of two positive integers are $56 = 7 \cdot 8 = 14 \cdot 4 = 28 \cdot 2 = 56 \cdot 1$. By part (a), m and n must both be odd or both be even. Thus the only solutions are either $m + n = 14$ and $m - n = 4$ or $m + n = 28$ and $m - n = 2$. This gives either $m = 9$ and $n = 5$ or $m = 15$ and $n = 13$ as the only solutions.

32. Under the given conditions, $2a - (b + c)$ is even. *Proof:* Suppose a, b, and c are integers and $a - b$ is even and $b - c$ is even. [*We must show that $2a - (b + c)$ is even.*] By properties 6 and 7 in Example 3.2.3, it is impossible for one of a and b to be even and the other odd, and so either both are even or both are odd. In the case where both are even, c must also be even because otherwise $b - c$ would be odd (by property 7). Hence in this case $2a - (b + c)$ is even (by property 1). In the case where both a and b are odd, c must also be odd because otherwise $b - c$ would be odd (by property 7). Hence in this case $2a - (b + c)$ is even (by properties 4, 2, and 1).

34. *Hint:* Express n using the quotient-remainder theorem with $d = 3$.

36. *Hint:* Use the quotient-remainder theorem (as in Example 3.4.5) to say that $n = 4q, n = 4q + 1, n = 4q + 2$, or $n = 4q + 3$ and divide into cases accordingly.

38. *Hint:* Given any integer n, consider the two cases where n is even and where n is odd.

39. *Hint:* Given any integer n, analyze the sum $n + (n + 1) + (n + 2) + (n + 3)$.

42. *Hint:* Use the quotient-remainder theorem to say that n must have one of the forms $6q, 6q + 1, 6q + 2, 6q + 3, 6q + 4$, or $6q + 5$ for some integer q.

44. *Answer to first question:* No. *Counterexample:* Let $m = 1$, $n = 3$, and $d = 2$. Then $m \bmod d = 1$ and $n \bmod d = 1$ but $m \neq n$.

Answer to second question: Yes. *Proof:* Suppose m, n, and d are integers such that $m \bmod d = n \bmod d$. Let $r = m \bmod d = n \bmod d$. By definition of *mod*, $m = dp + r$ and $n = dq + r$ for some integers p and q. Then $m - n = (dp + r) - (dq + r) = d(p - q)$. But $p - q$ is an integer (being a difference of integers), and so $m - n$ is divisible by d by definition of divisible.

49. *Proof:* Suppose x is any real number. [*We must show that $|-x| = |x|$.*]

Case 1 ($x > 0$): In this case $-x < 0$, and so, by definition of absolute value, $|-x| = -(-x) = x$. In addition, because $x > 0$, by definition of absolute value $|x| = x$. Thus $|-x| = |x|$.

Case 2 ($x = 0$): In this case $-x$ also equals 0, and so, by definition of absolute value, both $|-x|$ and $|x|$ equal 0. Thus $|-x| = |x|$.

Case 3 ($x < 0$): In this case $-x > 0$ and so, by definition of absolute value, $|-x| = -x$. In addition, because $x < 0$, by definition of absolute value $|x| = -x$. Thus $|-x| = |x|$.

Hence, in all three cases $|-x| = |x|$ [*as was to be shown*].

51. *Proof:* Let x be any real number. We consider two cases.

Case 1 $x \geq 0$: Then by definition $|x| = x$. Since $|x| \geq 0$, we have $-|x| \leq 0$. Hence $-|x| \leq x \leq |x|$.

Case 2 $x < 0$: Then by definition $|x| = -x$, which gives $-|x| = x$. So $-|x| = x < 0 \leq |x|$. Hence $-|x| \leq x \leq |x|$.

Thus, in either case, $-|x| \leq x \leq |x|$.

Section 3.5

1. $\lfloor 37.999 \rfloor = 37$, $\lceil 37.999 \rceil = 38$

3. $\lfloor -14.00001 \rfloor = -15$, $\lceil -14.00001 \rceil = -14$

8. $\lfloor n/7 \rfloor$. The floor notation is more appropriate. If the ceiling notation is used, two different formulas are needed, depending on whether $n/7$ is an integer or not. (What are they?)

10. a. (i) $\left(2050 + \lfloor \frac{2049}{4} \rfloor - \lfloor \frac{2049}{100} \rfloor + \lfloor \frac{2049}{400} \rfloor \right) mod\ 7$

$= (2050 + 512 - 20 + 5)\ mod\ 7 = 2547\ mod\ 7$

$= 6$, which corresponds to a Saturday

b. *Hint:* One day is added every four years, except that each century the day is not added unless the century is a multiple of 400.

12. *Proof:* Suppose n is any even integer. By definition of even, $n = 2k$ for some integer k. Then

$$\left\lfloor \frac{n}{2} \right\rfloor = \left\lfloor \frac{2k}{2} \right\rfloor = \lfloor k \rfloor = k \qquad \text{because } k \text{ is an integer} \\ \text{and } k \leq k < k + 1.$$

But $\qquad k = \dfrac{n}{2} \qquad$ because $n = 2k$.

Thus, on the one hand, $\lfloor \frac{n}{2} \rfloor = k$, and on the other hand, $k = \frac{n}{2}$. It follows that $\lfloor \frac{n}{2} \rfloor = \frac{n}{2}$ [*as was to be shown*].

14. False. *Counterexample:* Let $x = 2$ and $y = 1.9$. Then $\lfloor x - y \rfloor = \lfloor 2 - 1.9 \rfloor = \lfloor 0.1 \rfloor = 0$, whereas $\lfloor x \rfloor - \lfloor y \rfloor = \lfloor 2 \rfloor - \lfloor 1.9 \rfloor = 2 - 1 = 1$.

15. True. *Proof:* Suppose x is any real number. Let $m = \lfloor x \rfloor$. By definition of floor, $m \leq x < m + 1$. Subtracting 1 from all parts of the inequality gives that $m - 1 \leq x - 1 < m$, and so, by definition of floor, $\lfloor x - 1 \rfloor = m - 1$. It follows by substitution that $\lfloor x - 1 \rfloor = \lfloor x \rfloor - 1$.

17. *Proof for the case where $n\ mod\ 3 = 2$:*

In the case where $n\ mod\ 3 = 2$, then $n = 3q + 2$ for some integer q by definition of mod. By substitution,

$$\left\lfloor \frac{n}{3} \right\rfloor = \left\lfloor \frac{3q + 2}{3} \right\rfloor$$

$$= \left\lfloor \frac{3q}{3} + \frac{2}{3} \right\rfloor$$

$$= \left\lfloor q + \frac{2}{3} \right\rfloor = q \qquad \text{because } q \text{ is an integer and} \\ q \leq q + 2/3 < q + 1.$$

But

$$q = \frac{n - 2}{3} \qquad \text{by solving } n = 3q + 2 \text{ for } q.$$

Thus, on the one hand, $\lfloor \frac{n}{3} \rfloor = q$, and on the other hand, $q = \frac{n-2}{3}$. It follows that $\lfloor \frac{n}{3} \rfloor = \frac{n-2}{3}$.

18. *Hint:* This is false. **19.** *Hint:* This is true.

23. *Proof:* Suppose x is a real number that is not an integer. Let $\lfloor x \rfloor = n$. Then, by definition of floor and because n is not an integer, $n < x < n + 1$. Multiplying both sides by -1 gives $-n > -x > -n - 1$, or equivalently, $-n - 1 < -x < -n$. Since $-n - 1$ is an integer, it follows by definition of floor that $\lfloor -x \rfloor = -n - 1$. Hence $\lfloor x \rfloor + \lfloor -x \rfloor = n + (-n - 1) = n - n - 1 = -1$, as was to be shown.

25. *Hint:* Let $n = \lfloor \frac{x}{2} \rfloor$ and consider the two cases: n is even and n is odd.

26. *Proof:* Suppose x is any real number such that $x - \lfloor x \rfloor < \frac{1}{2}$. Multiplying both sides by 2 gives $2x - 2\lfloor x \rfloor < 1$, or $2x < 2\lfloor x \rfloor + 1$. Now, by definition of floor, $\lfloor x \rfloor \leq x$. Hence, $2\lfloor x \rfloor \leq 2x$. Putting the two inequalities involving $2x$ together gives $2\lfloor x \rfloor \leq 2x < 2\lfloor x \rfloor + 1$. Thus, by definition of floor (and because $2\lfloor x \rfloor$ is an integer), $\lfloor 2x \rfloor = 2\lfloor x \rfloor$. This is what was to be shown.

30. This incorrect proof begs the question. The equality $\lfloor \frac{n}{2} \rfloor = \frac{n-1}{2}$ is what is to be shown. By substituting $2k + 1$ for n into both sides of the equality and working from the result as though it were known to be true, the proof assumes the truth of the conclusion to be proved.

Section 3.6

1. a. $x \leq y$ **b.** a contradiction

c. $\frac{x}{2}$ is a positive real number

d. multiplying both sides of the inequality $\frac{1}{2} < 1$ by x, which is positive, gives $\frac{x}{2} < x$

e. $\frac{x}{2}$ is a positive real number that is less than the least positive real number

3. *Proof:* Suppose not. That is, suppose there is an integer n such that $3n + 2$ is divisible by 3. [*We must derive a contradiction.*] By definition of divisibility, $3n + 2 = 3k$ for some integer k. Subtracting $3n$ from both sides gives that $2 = 3k - 3n = 3(k - n)$. So, by definition of divisibility, $3 \mid 2$. But by Example 3.3.3 this implies that $3 \leq 2$, which contradicts the fact that $3 > 2$. [*Thus for all integers n, $3n + 2$ is not divisible by 3.*]

77777777

77777777

7777777777

71777777777777

77777777

5. *Negation of statement:* There is a greatest even integer.

Proof of statement: Suppose not. That is, suppose there is a greatest even integer; call it N. Then N is an even integer, and $N \geq n$ for every even integer n. [*We must deduce a contradiction.*] Let $M = N + 2$. Then M is an even integer since it is a sum of even integers, and $M > N$ since $M = N + 2$. This contradicts the supposition that $N \geq n$ for *every* even integer n. [*Hence the supposition is false and the statement is true.*]

8. a. The mistake in this proof occurs in the second sentence where the negation written by the student is incorrect: Instead of being existential, it is universal. The problem is that if the student proceeds in a logically correct manner, all that is needed to reach a contradiction is one example of a rational and an irrational number whose sum is irrational. To prove the given statement, however, it is necessary to show that there is *no* rational number and *no* irrational number whose sum is rational.

b. *Proof (by contradiction):* Suppose not. That is, suppose $\exists$ a rational number x and an irrational number y such that $x - y$ is rational. [*We must derive a contradiction.*] By definition of rational, $x = a/b$ and $x - y = c/d$ for some integers $a, b, c,$ and d with $b \neq 0$ and $d \neq 0$. By substitution,

$$\frac{c}{d} = x - y = \frac{a}{b} - y.$$

Solving for y gives

$$y = \frac{a}{b} - \frac{c}{d} = \frac{ad - bc}{bd}.$$

But $ad - bc$ and bd are integers (because $a, b, c,$ and d are integers, and products and differences of integers are integers), and $bd \neq 0$ (by the zero product property). Thus, by definition of rational, y is rational. This contradicts the supposition that y is irrational. [*Hence the supposition is false and the given statement is true.*]

10. *Proof:* Suppose not. That is, suppose $\exists$ a nonzero rational number x and an irrational number y such that xy is rational. [*We must derive a contradiction.*] By definition of rational, $x = a/b$ and $xy = c/d$ for some integers a, b, c and d with $b \neq 0$ and $d \neq 0$. Also $a \neq 0$ because x is nonzero. By substitution, $xy = (a/b)y = c/d$. Solving for y gives $y = bc/ad$. Now bc and ad are integers (being products of integers) and $ad \neq 0$ (by the zero product property). Thus, by definition of rational, y is rational, which contradicts the supposition that y is irrational. [*Hence the supposition is false and the statement is true.*]

12. *Hint:* Suppose $n^2 - 2$ is divisible by 4, and consider the two cases where n is even and n is odd.

13. *Hint:* $a^2 = c^2 - b^2 = (c - b)(c + b)$.

14. *Hint:* (1) For any integer c, if 2 divides c, then 4 divides c^2. (2) The result of exercise 12 may be helpful.

15. *Hint:* Suppose $a, b,$ and c are odd integers, z is a solution to $ax^2 + bx + c = 0$, and z is rational. Then $z = p/q$ for some integers p and q with $q \neq 0$. We may assume p and q have no common factor. (Why? If p and q do have a common factor, we can divide out their greatest common factor to obtain two integers p' and q' that (1) have no common factor and (2) satisfy the equation $z = p'/q'$. Then we can redefine $q = q'$ and $p = p'$.) Note that because p and q have no common factor, they are not both even. Substitute p/q into $ax^2 + bx + c = 0$, and multiply through by q^2. Show that (1) the assumption that p is even leads to a contradiction, (2) the assumption that q is even leads to a contradiction, and (3) the assumption that both p and q are odd leads to a contradiction. The only remaining possibility is that both p and q are even, which has been ruled out.

16. a. $5 \mid n$ **b.** $5 \mid n^2$ **c.** $5k$ **d.** $(5k)^2$ **e.** $5 \mid n^2$

17. *Proof (by contraposition):* [*To go by contraposition, we must prove that $\forall$ positive real numbers, r and s, if $r \leq 10$ and $s \leq 10$, then $rs \leq 100$.*] Suppose r and s are positive real numbers and $r \leq 10$ and $s \leq 10$. By the algebra of inequalities, $rs \leq 100$. [*To derive this fact, multiply both sides of $r \leq 10$ by s to obtain $rs \leq 10s$. And multiply both sides of $s \leq 10$ by 10 to obtain $10s \leq 10 \cdot 10 = 100$. By transitivity of $\leq$, then, $rs \leq 100$.*] But this is what was to be shown.

19. a. *Proof by contradiction:* Suppose not. That is, suppose there is an integer n such that n^2 is odd and n is even. Show that this supposition leads logically to a contradiction.

b. *Proof by contraposition:* Suppose n is any integer such that n is odd. Show that n^2 is odd.

21. a. The contrapositive is the statement "$\forall$ real numbers x, if $-x$ is not irrational, then x is not irrational." Equivalently (because $-(-x) = x$), "$\forall$ real numbers x, if x is rational then $-x$ is rational."

Proof by contraposition: Suppose x is any rational number. [*We must show that $-x$ is also rational.*] By definition of rational, $x = a/b$ for some integers a and b with $b \neq 0$. Then $-x = -(a/b) = (-a)/b$. Since both $-a$ and b are integers and $b \neq 0$, $-x$ is rational [*as was to be shown.*]

b. *Proof:* Suppose not. [*We take the negation and suppose it to be true*]. That is, suppose $\exists$ an irrational number x such that $-x$ is rational. [*We must derive a contradiction.*] By definition of rational, $-x = a/b$ for some integers a and b with $b \neq 0$. Multiplying both sides by -1 gives $x = -(a/b) = -a/b$. But $-a$ and b are integers (since a and b are) and $b \neq 0$. Thus x is a ratio of the two integers $-a$ and b with $b \neq 0$. Hence x is rational (by definition of rational), which is a contradiction. [*This contradiction shows that the supposition is false, and so the given statement is true.*]

23. *Hints:* See the answer to exercise 19 and look carefully at the two proofs for Proposition 3.6.4.

24. a. *Proof by contraposition:* Suppose a, b, and c are any [*particular but arbitrarily chosen*] integers such that $a \mid b$. [*We must show that $a \mid bc$.*] By definition of divides, $b = ak$ for some integer k. Then $bc = (ak)c = a(kc)$. But kc is an integer (because it is a product of the integers k and c). Hence $a \mid bc$ by definition of divisibility [*as was to be shown.*]

b. *Proof by contradiction:* Suppose not. [*We take the negation and suppose it to be true.*] Suppose $\exists$ integers a, b, and c such that $a \nmid bc$ and $a \mid b$. Since $a \mid b$, there exists an integer k such that $b = ak$ by definition of divides. Then $bc = (ak)c = a(kc)$ [*by the associative law of algebra*]. But kc is an integer (being a product of integers), and so $a \mid bc$ by definition of divides. Thus $a \nmid bc$ and $a \mid bc$, which is a contradiction. [*This contradiction shows that the supposition is false, and hence the given statement is true.*]

25. a. *Hint:* The contrapositive is "For all integers m and n, if m and n are not both even and m and n are not both odd, then $m + n$ is not even." Equivalently: "For all integers m and n, if one of m and n is even and the other is odd, then $m + n$ is odd."

b. *Hint:* The negation of the given statement is the following: $\exists$ integers m and n such that $m + n$ is even, and either m is even and n is odd, or m is odd and n is even.

27. The negation of "Every integer is rational" is "There is at least one integer that is irrational" not "Every integer is irrational." Deriving a contradiction from an incorrect negation of a statement does not prove the statement is true.

28. a. *Proof:* Suppose r, s, and n are integers and $r > \sqrt{n}$ and $s > \sqrt{n}$. Note that r and s are both positive because $\sqrt{n}$ cannot be negative. By multiplying both sides of the first inequality by s and both sides of the second inequality by $\sqrt{n}$ (Appendix A, T19), we have that $rs > \sqrt{n}s$ and $\sqrt{n}s = \sqrt{n}\sqrt{n} = n$. Thus, by the transitive law for inequality (Appendix A, T17), $rs > n$.

b. *Hint:* The contrapositive is as follows: For all integers $n > 1$, if n is not prime, then n is divisible by any positive integer that is greater than 1 and less than or equal to $\sqrt{n}$.

29. a. $\sqrt{667} \cong 25.8$, and so the possible prime factors are 2, 3, 4, 7, 11, 13, 17, 19, and 23. Testing each in turn shows that 667 is not prime because $667 = 23 \cdot 29$.

b. $\sqrt{557} \cong 23.6$, and so the possible prime factors are 2, 3, 4, 7, 11, 13, 17, 19, and 23. Testing each in turn shows that none divides 557. Therefore, 557 is prime.

31. a. $\sqrt{9269} \cong 96.3$, and so the possible prime factors are all among those you found for exercise 30. Testing each in turn shows that 9,269 is not prime because $9,269 = 13 \cdot 713$.

b. $\sqrt{9103} \cong 95.4$, and so the possible prime factors are all among those you found for exercise 30. Testing each in turn shows that none divides 9,103. Therefore, 9,103 is prime.

32. *Hint:* Is it possible for all three of $n - 4, n - 6$, and $n - 8$ to be prime?

Section 3.7

1. The value of $\sqrt{2}$ given by a calculator is an approximation. Calculators can give exact values only for numbers that can be represented using at most the number of decimal digits in the calculator display. In particular, every number in a calculator display is rational, but even many rational numbers cannot be represented exactly. For instance, consider the number formed by writing a decimal point and following it with the first 1 million digits of $\sqrt{2}$. By the discussion in Section 3.2, this number is rational, but you could not infer this from the calculator display.

3. *Proof by contradiction:* Suppose not. That is, suppose $6 - 7\sqrt{2}$ is rational. [*We must prove a contradiction.*] By definition of rational, there exist integers a and $b \neq 0$ with

$$6 - 7\sqrt{2} = \frac{a}{b}.$$

Then $\sqrt{2} = \dfrac{1}{-7}\left(\dfrac{a}{b} - 6\right)$ by subtracting 6 from both sides and dividing both sides by -7

and so $\sqrt{2} = \dfrac{a - 6b}{-7b}$ by the rules of algebra.

But $a - 6b$ and $-7b$ are both integers (since a and b are integers and products and differences of integers are integers), and $-7b \neq 0$ by the zero product property. Hence $\sqrt{2}$ is a ratio of the two integers $a - 6b$ and $-7b$ with $-7b \neq 0$, so $\sqrt{2}$ is a rational number (by definition of rational). This contradicts the fact that $\sqrt{2}$ is irrational, and so the supposition is false and $6 - 7\sqrt{2}$ is irrational.

5. This is false. $\sqrt{4} = 2 = 2/1$, which is rational.

7. *Counterexample:* Let $x = \sqrt{2}$ and let $y = -\sqrt{2}$. Then x and y are irrational, but $x + y = 0 = 0/1$, which is rational.

9. True.

Formal version of the statement: $\forall$ real numbers r, if r is irrational, then $\sqrt{r}$ is irrational.

Proof by contraposition: Suppose r is any real number such that $\sqrt{r}$ is rational. [*We must show that r is rational.*] By definition of rational, $\sqrt{r} = \frac{a}{b}$ for some integers a and b with $b \neq 0$. Then $r = (\sqrt{r})^2 = (\frac{a}{b})^2 = \frac{a^2}{b^2}$. But both a^2 and b^2 are integers because they are products of integers, and $b^2 \neq 0$ by the zero product property. Thus r is rational [*as was to be shown*].

(The statement may also be proved by contradiction.)

13. *Hint:* Can you think of any "nice" integers x and y that are greater than 1 and have the property that $x^2 = y^3$?

16. a. *Proof by contradiction:* Suppose not. That is, suppose there is an integer n such that $n = 3q_1 + r_1 = 3q_2 + r_2$, where q_1, q_2, r_1, and r_2 are integers, $0 \leq r_1 < 3, 0 \leq r_2 < 3$, and $r_1 \neq r_2$. By interchanging the labels for r_1 and r_2 if necessary, we may assume that $r_2 > r_1$. Then $3(q_1 - q_2) = r_2 - r_1 > 0$, and because both r_1 and r_2 are less than 3, either $r_2 - r_1 = 1$ or $r_2 - r_1 = 2$. So either

$3(q_1 - q_2) = 1$ or $3(q_1 - q_2) = 2$. The first case implies that $3 \mid 1$, and hence, by Example 3.3.3, that $3 \leq 1$, and the second case implies that $3 \mid 2$, and hence, by Example 3.3.3, that $3 \leq 2$. These results contradict the fact that 3 is greater than both 1 and 2. Thus in either case we have reached a contradiction, which shows that the supposition is false and the given statement is true.

b. *Proof by contradiction:* Suppose not. That is, suppose there is an integer n such that n^2 is divisible by 3 and n is not divisible by 3. [*We must deduce a contradiction.*] By definition of divisible, $n^2 = 3q$ for some integer q, and by the quotient-remainder theorem and part (a), $n = 3k + 1$ or $n = 3k + 2$ for some integer k.

Case 1 ($n = 3k + 1$) *for some integer* k: In this case $n^2 = (3k + 1)^2 = 9k^2 + 6k + 1 = 3(3k^2 + 2k) + 1$. Let $s = 3k^2 + 2k$. Then $n^2 = 3s + 1$, and s is an integer because it is a sum of products of integers. It follows that $n^2 = 3q = 3s + 1$ for some integers q and s, which contradicts the result of part (a).

Case 2 ($n = 3k + 2$) *for some integer* k: In this case $n^2 = (3k + 2)^2 = 9k^2 + 12k + 4 = 3(3k^2 + 6k + 1) + 1$. Let $t = 3k^2 + 6k + 1$. Then $n^2 = 3t + 1$, and t is an integer because it is a sum of products of integers. It follows that $n^2 = 3q = 3t + 1$ for some integers q and t, which contradicts the result of part (a).

Thus, in either case, a contradiction is reached, which shows that the supposition is false and the given statement is true.

c. *Proof by contradiction:* Suppose not. That is, suppose $\sqrt{3}$ is rational. By definition of rational, $\sqrt{3} = \frac{a}{b}$ for some integers a and b with $b \neq 0$. Without loss of generality, assume that a and b have no common factor. (If not, divide both a and b by their greatest common factor to obtain integers a' and b' with the property that a' and b' have no common factor and $\sqrt{3} = \frac{a'}{b'}$. Then redefine $a = a'$ and $b = b'$.) Squaring both sides of $\sqrt{3} = \frac{a}{b}$ gives $3 = \frac{a^2}{b^2}$, and multiplying both sides by b^2 gives $3b^2 = a^2$ (*). Thus a^2 is divisible by 3, and so, by part (b), a is also divisible by 3. By definition of divisibility, then, $a = 3k$ for some integer k, and so $a^2 = 9k^2$ (**). Substituting equation (**) into equation (*) gives $3b^2 = 9k^2$, and dividing both sides by 3 yields $b^2 = 3k^2$. Hence b^2 is divisible by 3, and so, by part (b), b is also divisible by 3. Consequently, both a and b are divisible by 3, which contradicts the assumption that a and b have no common factor. Thus the supposition is false, and so $\sqrt{3}$ is irrational.

18. *Hint:* The proof is a generalization of the one given in the solution for exercise 16(a).

19. *Hints:* (1) The parts of the proof are similar to those in exercise 16(b) and 16(c). (2) Use the result of exercise 18.

20. *Hint:* This statement is true. If $a^2 - 3 = 9b$, then $a^2 = 9b + 3 = 3(3b + 1)$, and so a^2 is divisible by 3. Hence, by exercise 16(a), a is divisible by 3. Thus, $a^2 = (3c)^2$ for some integer c.

23. *Hint:* By the result of exercise 22, $\sqrt{6}$ is irrational.

25. *Hint:* $\frac{2 \cdot 3 \cdot 5 \cdot 7 + 1}{2} = 3 \cdot 5 \cdot 7 + \frac{1}{2}$ and $\frac{2 \cdot 3 \cdot 5 \cdot 7 + 1}{3} = 2 \cdot 5 \cdot 7 + \frac{1}{3}$.

26. *Hint:* You can deduce that $p = 3$.

27. a. *Hint:* For example, $N_4 = 2 \cdot 3 \cdot 5 \cdot 7 + 1 = 211$.

29. *Hint:* Every odd integer can be written as $4k + 1$ or as $4k + 3$ for some integer k. (Why?) If $p_1 p_2 \cdots p_n + 1 = 4k + 1$, then $4 \mid p_1 p_2 \cdots p_n$. Is this possible?

30. *Hint:* By Theorem 3.3.2 (divisibility by a prime) there is a prime number p such that $p \mid (n! - 1)$. Show that the supposition that $p \leq n$ leads to a contradiction. It will then follow that $n < p < n!$.

31. a. *Hint:* Prove the contrapositive: If for some integer $n > 2$ that is not a power of 2, $x^n + y^n = z^n$ has a positive integer solution, then for some prime number $p > 2$, $x^p + y^p = z^p$ has a positive integer solution. Note that if $n = kp$, then $x^n = x^{kp} = (x^k)^p$.

32. *Existence proof:* When $n = 2$, then $n^2 - 1 = 3$, which is prime. Hence there exists a prime number of the form $n^2 - 1$, where n is an integer and $n \geq 2$.

Uniqueness proof (by contradiction): Suppose to the contrary that m is another integer satisfying the given conditions. That is, $m > 2$ and $m^2 - 1$ is prime. [*We must derive a contradiction.*] Factor $m^2 - 1$ to obtain $m^2 - 1 = (m - 1)(m + 1)$. But $m > 2$, and so $m - 1 > 1$ and $m + 1 > 1$. Hence $m^2 - 1$ is not prime, which is a contradiction. [*This contradiction shows that the supposition is false, and so there is no other integer $m > 2$ such that $n^2 - 1$ is prime.*]

Uniqueness proof (direct): Suppose m is any integer such that $m \geq 2$ and $m^2 - 1$ is prime. [*We must show that $m = 2$.*] By factoring, $m^2 - 1 = (m - 1)(m + 1)$. Since $m^2 - 1$ is prime, either $m - 1 = 1$ or $m + 1 = 1$. But $m + 1 \geq 2 + 1 = 3$. Hence, by elimination, $m - 1 = 1$, and so $m = 2$.

34. *Proof (by contradiction):* Suppose not. That is, suppose there are two distinct real numbers a_1 and a_2 such that for all real numbers r,

$$(1) \quad a_1 + r = r \quad \text{and} \quad (2) \quad a_2 + r = r$$

Then

$$a_1 + a_2 = a_2 \quad \text{by (1) with } r = a_2$$

and

$$a_2 + a_1 = a_1 \quad \text{by (2) with } r = a_1.$$

It follows that

$$a_2 = a_1 + a_2 = a_2 + a_1 = a_1$$

which implies that $a_2 = a_1$. But this contradicts the supposition that a_1 and a_2 are distinct. [*Thus the supposition is false and there is at most one real number a such that $a + r = r$ for all real numbers r.*]

Proof (direct): Suppose a_1 and a_2 are real numbers such that for all real numbers r,

$$(1)\ a_1 + r = r \quad \text{and} \quad (2)\ a_2 + r = r$$

Then

$$a_1 + a_2 = a_2 \quad \text{by (1) with } r = a_2$$

and

$$a_2 + a_1 = a_1 \quad \text{by (2) with } r = a_1.$$

It follows that

$$a_2 = a_1 + a_2 = a_2 + a_1 = a_1.$$

Hence $a_2 = a_1$. [*Thus there is at most one real number a such that $a + r = r$ for all real numbers r.*]

Section 3.8

1. $z = 0$ **3. a.** $z = 18$ **4.** $a = \frac{17}{12}$

6.

Iteration Number

	0	1	2	3
a	26			
d	7			
q	0	1	2	3
r	26	19	12	5

8. a.

A	69	19	9	
q	2			
d		1		
n			1	
p				4

9. $\gcd(27, 72) = 9$ **10.** $\gcd(5, 9) = 1$

13. Divide the larger number, 1,188, by the smaller, 385, to obtain a quotient of 3 and a remainder of 33. Next divide 385 by 33 to obtain a quotient of 11 and a remainder of 22. Then divide 33 by 22 to obtain a quotient of 1 and a remainder of 11. Finally, divide 22 by 11 to obtain a quotient of 2 and a remainder of 0. Thus, by Lemma 3.8.1, $\gcd(1188, 385) = \gcd(385, 33) = \gcd(33, 22) = \gcd(22, 11) = \gcd(11, 0)$, and by Lemma 3.8.2, $\gcd(11, 0) = 11$. So $\gcd(1188, 385) = 11$.

14. Divide the larger number, 1,177, by the smaller, 509, to obtain a quotient of 2 and a remainder of 159. Next divide 509 by 159 to obtain a quotient of 3 and a remainder of 32. Next divide 159 by 32 to obtain a quotient of 4 and a remainder of 31. Then divide 32 by 31 to obtain a quotient of 1 and a remainder of 1. Finally, divide 31 by 1 to obtain a quotient of 31 and a remainder of 0. Thus, by Lemma 3.8.1,

$\gcd(1177, 509) = \gcd(509, 159) = \gcd(159, 32) = \gcd(32, 31) = \gcd(31, 1) = \gcd(1, 0)$, and by Lemma 3.8.2, $\gcd(1, 0) = 1$. So $\gcd(1177, 509) = 1$.

17.

A	1,001					
B	871					
r		130	91	39	13	0
b	871	130	91	39	13	0
a	1,001	871	130	91	39	13
gcd						13

19. *Hint:* Divide the proof into two parts. In part 1 suppose a and b are any positive integers such that $a \mid b$, and derive the conclusion that $\gcd(a, b) = a$. To do this, note that because it is also the case that $a \mid a$, a is a common divisor of a and b. Thus, by definition of greatest common divisor, a is less than or equal to the greatest common divisor of a and b. In symbols, $a \leq \gcd(a, b)$. Then show that $a \geq \gcd(a, b)$ by using the fact (from Example 3.3.3) that any integer that divides the positive integer a is less than or equal to a. In part 2 of the proof, suppose a and b are any positive integers such that $\gcd(a, b) = a$, and deduce that $a \mid b$.

22. a. *Hint 1:* If $a = dq - r$, then $-a = -dq + r = -dq - d + d - r = d(-q - 1) + (d - r)$.

    Hint 2: If $0 \leq r < d$, then $0 \geq -r > -d$. Add d to all parts of this inequality and see what results.

23. *Proof:* Suppose a, d, q, and r are integers such that $a = dq + r$ and $0 \leq r < d$. [*We must show that $q = \lfloor \frac{a}{d} \rfloor$ and $r = a - \lfloor \frac{a}{d} \rfloor \cdot d$.*] Solving $a = dq + r$ for r gives $r = a - dq$, and substituting into $0 \leq r < d$ gives $0 \leq a - dq < d$. Add dq to both sides to obtain $dq \leq a < d + dq = d(q + 1)$. Then divide through by d to obtain $q \leq \frac{a}{d} < q + 1$. Therefore, by definition of floor, $\lfloor \frac{a}{d} \rfloor = q$. Finally, substitution into $a = dq + r$ gives $a = d\lfloor \frac{a}{d} \rfloor + r$, and subtracting $d\lfloor \frac{a}{d} \rfloor$ from both sides yields $r = a - d\lfloor \frac{a}{d} \rfloor$ [*as was to be shown*].

24. b.

Iteration Number

	0	1	2	3	4
a	630	294	294	252	210
b	336	336	42	42	42
gcd					

Iteration Number

	5	6	7	8	9
a	168	126	84	42	0
b	42	42	42	42	42
gcd					42

25. a. $\text{lcm}(12, 18) = 36$

26. *Partial proof:* Let a and b be positive integers, and suppose $d = \gcd(a, b) = \text{lcm}(a, b)$. By definition of greatest common divisor and least common multiple, $d > 0$, $d \mid a$, $d \mid b$, $a \mid d$, and $b \mid d$. Thus, in particular, $a = dm$ and $d = an$ for some integers m and n. By substitution, $a = dm = (an)m = anm$. Dividing both sides by a gives $1 = nm$. But the only divisors of 1 are 1 and -1 (Example 3.3.4), and so $m = n = \pm 1$. Since both a and d are positive, $m = n = 1$, and hence $a = d$. Similar reasoning shows that $b = d$ also, and so $a = b$.

28. *Hint:* Divide the proof into two parts. In part 1, suppose a and b are any positive integers, and deduce that

$$\gcd(a, b) \cdot \text{lcm}(a, b) \le ab.$$

Derive this result by showing that $\text{lcm}(a, b) \le \frac{ab}{\gcd(a,b)}$. To do this, show that $\frac{ab}{\gcd(a,b)}$ is a multiple of both a and b. For instance, to see that $\frac{ab}{\gcd(a,b)}$ is a multiple of b, note that because $\gcd(a, b) \mid a$, $a = \gcd(a, b) \cdot k$ for some integer k, and thus $ab = \gcd(a, b) \cdot kb$. Divide both sides by $\gcd(a, b)$ to obtain $\frac{ab}{\gcd(a,b)} = kb$. But since k is an integer, this equation implies that $\frac{ab}{\gcd(a,b)}$ is a multiple of b. The argument that $\frac{ab}{\gcd(a,b)}$ is a multiple of a is almost identical. In part 2 of the proof, use the definition of least common multiple to show that $\frac{ab}{\text{lcm}(a,b)} \mid a$ and $\frac{ab}{\text{lcm}(a,b)} \mid b$ and thus that $\frac{ab}{\text{lcm}(a,b)} \mid \gcd(a, b)$. Conclude that $\frac{ab}{\text{lcm}(a,b)} \le \gcd(a, b)$ and hence that $ab \le \gcd(a, b) \cdot \text{lcm}(a, b)$.

Section 4.1

1. $\frac{1}{11}, \frac{2}{12}, \frac{3}{13}, \frac{4}{14}$ **3.** $1, -\frac{1}{3}, \frac{1}{9}, -\frac{1}{27}$ **5.** $0, 0, 2, 2$

8. $g_1 = \lfloor \log_2 1 \rfloor = 0$

$g_2 = \lfloor \log_2 2 \rfloor = 1$, $g_3 = \lfloor \log_2 3 \rfloor = 1$

$g_4 = \lfloor \log_2 4 \rfloor = 2$, $g_5 = \lfloor \log_2 5 \rfloor = 2$

$g_6 = \lfloor \log_2 6 \rfloor = 2$, $g_7 = \lfloor \log_2 7 \rfloor = 2$

$g_8 = \lfloor \log_2 8 \rfloor = 3$, $g_9 = \lfloor \log_2 9 \rfloor = 3$

$g_{10} = \lfloor \log_2 10 \rfloor = 3$, $g_{11} = \lfloor \log_2 11 \rfloor = 3$

$g_{12} = \lfloor \log_2 12 \rfloor = 3$, $g_{13} = \lfloor \log_2 13 \rfloor = 3$

$g_{14} = \lfloor \log_2 14 \rfloor = 3$, $g_{15} = \lfloor \log_2 15 \rfloor = 3$

When n is an integral power of 2, g_n is the exponent of that power. For instance, $8 = 2^3$ and $g_8 = 3$. More generally, if $n = 2^k$, where k is an integer, then $g_n = k$. All terms of the sequence from g_n up to g_m, where $m = 2^{k+1}$ is the next integral power of 2, have the same value as g_n, namely k. For instance, all terms of the sequence from g_8 through g_{15} have the value 3.

Exercises 10–16 have more than one correct answer.

10. $a_n = (-1)^n$, where n is an integer and $n \ge 1$.

11. $a_n = (n - 1)(-1)^n$, where n is an integer and $n \ge 1$.

12. $a_n = \dfrac{n}{(n + 1)^2}$, where n is an integer and $n \ge 1$

14. $a_n = \dfrac{n}{3^n}$, where n is an integer and $n \ge 1$

18. a. $2 + 3 + (-2) + 1 + 0 + (-1) + (-2) = 1$

b. $a_0 = 2$

c. $a_2 + a_4 + a_6 = -2 + 0 + (-2) = -4$

d. $2 \cdot 3 \cdot (-2) \cdot 1 \cdot 0 \cdot (-1) \cdot (-2) = 0$

19. $2 + 3 + 4 + 5 + 6 = 20$ **20.** $2^2 \cdot 3^2 \cdot 4^2 = 576$

23. $1(1 + 1) = 2$

27. $\left(\frac{1}{1} - \frac{1}{2}\right) + \left(\frac{1}{2} - \frac{1}{3}\right) + \left(\frac{1}{3} - \frac{1}{4}\right) + \left(\frac{1}{4} - \frac{1}{5}\right)$

$+ \left(\frac{1}{5} - \frac{1}{6}\right) + \left(\frac{1}{6} - \frac{1}{7}\right) + \left(\frac{1}{7} - \frac{1}{8}\right) + \left(\frac{1}{8} - \frac{1}{9}\right)$

$+ \left(\frac{1}{9} - \frac{1}{10}\right) + \left(\frac{1}{10} - \frac{1}{11}\right) = 1 - \frac{1}{11} = \frac{10}{11}$

29. $(-2)^1 + (-2)^2 + (-2)^3 + \cdots + (-2)^n$

$$= -2 + 2^2 - 2^3 + \cdots + (-1)^n 2^n$$

Exercises 32–41 have more than one correct answer.

32. $\displaystyle\sum_{k=1}^{7} (-1)^{k+1} k^2$ or $\displaystyle\sum_{k=0}^{6} (-1)^k (k + 1)^2$

35. $\displaystyle\sum_{j=2}^{6} \frac{(-1)^j j}{(j + 1)(j + 2)}$ or $\displaystyle\sum_{k=3}^{7} \frac{(-1)^{k+1}(k - 1)}{k(k + 1)}$

36. $\displaystyle\sum_{i=0}^{5} (-1)^i r^i$ **38.** $\displaystyle\sum_{k=1}^{n} k^3$

40. $\displaystyle\sum_{i=0}^{n-1} (n - i)$ **42.** $\dfrac{4 \cdot 3 \cdot 2 \cdot 1}{3 \cdot 2 \cdot 1} = 4$

45. $\dfrac{n(n - 1)(n - 2) \cdots 3 \cdot 2 \cdot 1}{(n - 1)(n - 2) \cdots 3 \cdot 2 \cdot 1} = n$

46. $\dfrac{(n - 1)(n - 2) \cdots 3 \cdot 2 \cdot 1}{(n + 1)n(n - 1)(n - 2) \cdots 3 \cdot 2 \cdot 1} = \dfrac{1}{n(n + 1)}$

48. $\dfrac{[(n + 1)n(n - 1)(n - 2) \cdots 3 \cdot 2 \cdot 1]^2}{(n(n - 1)(n - 2) \cdots 3 \cdot 2 \cdot 1)^2} = (n + 1)^2$

49.
$\dfrac{n(n - 1)(n - 2) \cdots (n - k + 1)(n - k)(n - k - 1) \cdots 2 \cdot 1}{(n - k)(n - k - 1) \cdots 2 \cdot 1}$

$$= n(n - 1)(n - 2) \cdots (n - k + 1)$$

51. a. *Proof:* Let n be an integer such that $n \ge 2$. By definition of factorial,

$$n! = \begin{cases} 2 \cdot 1 & \text{if } n = 2 \\ 3 \cdot 2 \cdot 1 & \text{if } n = 3 \\ n \cdot (n - 1) \cdots 2 \cdot 1 & \text{if } n > 3. \end{cases}$$

In each case, $n!$ has a factor of 2, and so $n! = 2k$ for some integer k. Then

$$n! + 2 = 2k + 2 \qquad \text{by substitution}$$
$$= 2(k + 1) \qquad \text{by factoring out the 2.}$$

Since $k + 1$ is an integer, $n! + 2$ is divisible by 2 [*as was to be shown*].

c. *Hint:* Consider the sequence $m! + 2, m! + 3, m! + 4, \ldots, m! + m$.

52. When $k = 0$, then $i = 1$. When $k = 5$, then $i = 6$. Since $i = k + 1$, then $k = i - 1$. Thus,

$$k(k - 1) = (i - 1)((i - 1) - 1) = (i - 1)(i - 2),$$

and so

$$\sum_{k=0}^{5} k(k - 1) = \sum_{i=1}^{6} (i - 1)(i - 2)$$

54. When $i = 1$, then $j = 0$. When $i = n + 1$, then $j = n$. Since $j = i - 1$, then $i = j + 1$. Thus

$$\frac{(i - 1)^2}{i \cdot n} = \frac{((j + 1) - 1)^2}{(j + 1) \cdot n} = \frac{j^2}{jn + n}.$$

(Note that n is constant as far as the sum is concerned.)

So $\sum_{i=1}^{n+1} \dfrac{(i - 1)^2}{i \cdot n} = \sum_{j=0}^{n} \dfrac{j^2}{jn + n}$.

55. When $i = 3$, then $j = 2$. When $i = n$ then $j = n - 1$. Since $j = i - 1$, then $i = j + 1$. Thus,

$$\sum_{i=3}^{n} \frac{i}{i + n - 1} = \sum_{j=2}^{n} \frac{j + 1}{(j + 1) + n - 1}$$

$$= \sum_{j=2}^{n} \frac{j + 1}{j + n}.$$

58. $\displaystyle\sum_{k=1}^{n} [3(2k - 3) + (4 - 5k)]$

$$= \sum_{k=1}^{n} [(6k - 9) + (4 - 5k)] = \sum_{k=1}^{n} (k - 5)$$

62. a. $m - 1$, $sum + a[i + 1]$

63.

Hence $90_{10} = 1011010_2$.

66.

a	23					
i	0	1	2	3	4	5
g	23	11	5	2	1	0
$r[0]$		1				
$r[1]$			1			
$r[2]$				1		
$r[3]$					0	
$r[4]$						1

70.

Hence $287_{10} = 11F_{16}$.

Section 4.2

1. *Proof:* Let $P(n)$ be the property "n cents can be obtained by using 3-cent and 8-cent coins."

Show that the property is true for $n = 15$:

Fifteen cents can be obtained by using five 3-cent coins.

Show that for all integers $k \geq 15$, if the property is true for $n = k$, then it is true for $n = k + 1$:

Suppose k cents (where $k \geq 15$) can be obtained using 3-cent and 8-cent coins. [*Inductive hypothesis*] We must show that $k + 1$ cents can be obtained using 3-cent and 8-cent coins. If the k cents includes an 8-cent coin, replace it by three 3-cent coins to obtain a total of $k + 1$ cents. If the k cents consists of 3-cent coins exclusively, then there must be least five 3-cent coins (since the total amount is at least 15 cents). In this case, replace five of the 3-cent coins by two 8-cent coins to obtain a total of $k + 1$ cents. Thus, in either case, $k + 1$ cents can be obtained using 3-cent and 8-cent coins. [*This is what we needed to show.*]

[*Since we have proved the basis step and the inductive step, we conclude that the given statement is true for all integers $n \geq 15$.*]

3. a. $P(1)$ is "$1^2 = \frac{1 \cdot (1+1) \cdot (2 \cdot 1 + 1)}{6}$." $P(1)$ is true because $1^2 = 1$ and $\frac{1 \cdot (1+1) \cdot (2 \cdot 1 + 1)}{6} = \frac{2 \cdot 3}{6} = 1$ also.

b. $P(k)$ is "$1^2 + 2^2 + \cdots + k^2 = \frac{k(k+1)(2k+1)}{6}$."

c. $P(k + 1)$ is "$1^2 + 2^2 + \cdots + (k + 1)^2$
$= \frac{(k+1)((k+1)+1)(2 \cdot (k+1)+1)}{6}$."

d. *Must show:* If for some integer $k \geq 1$,
$1^2 + 2^2 + \cdots + k^2 = \frac{k(k+1)(2k+1)}{6}$, then
$1^2 + 2^2 + \cdots + (k + 1)^2 = \frac{(k+1)[(k+1)+1][(2(k+1)+1)]}{6}$.

5. a. 1^2 **b.** k^2

c. $1 + 3 + 5 + \cdots + [2(k + 1) - 1]$

d. $(k + 1)^2$

e. the odd integer just before $2k + 1$ is $2k - 1$

f. inductive hypothesis

6. *Proof:* For the given statement, the property is the equation

$$2 + 4 + 6 + \cdots + 2n = n^2 + n. \quad \text{the property}$$

Show that the property is true for $n = 1$:

To prove the property for $n = 1$, we must show that when 1 is substituted into the equation in place of n, the left-hand side equals the right-hand side. But when 1 is substituted for n, the left-hand side is the sum of all the even integers from 2 to $2 \cdot 1$, which is just 2, and the right-hand side is $1^2 + 1$, which also equals 2. Thus the property is true for $n = 1$.

Show that for all integers $k \geq 1$, if the property is true for $n = k$ then it is true for $n = k + 1$:

Let k be any integer with $k \geq 1$, and suppose the property is true for $n = k$. That is, suppose

$$2 + 4 + 6 + \cdots + 2k = k^2 + k. \quad \text{inductive hypothesis}$$

We must show that the property is true for $n = k + 1$. That is, we must show that

$$2 + 4 + 6 + \cdots + 2(k + 1) = (k + 1)^2 + (k + 1).$$

Because $(k+1)^2 + (k+1) = k^2 + 2k + 1 + k + 1 = k^2 + 3k + 2$, this is equivalent to showing that

$$2 + 4 + 6 + \cdots + 2(k+1) = k^2 + 3k + 2. \qquad (*)$$

But the left-hand side of equation $(*)$ is

$$2 + 4 + 6 + \cdots + 2(k+1)$$

$$= 2 + 4 + 6 + \cdots + 2k + 2(k+1)$$

by making the next-to-last term explicit

$$= (k^2 + k) + 2(k+1) \qquad \text{by substitution from the inductive hypothesis}$$

$$= k^2 + 3k + 2, \qquad \text{by algebra}$$

and this is the right-hand side of equation $(*)$. Hence the property is true for $n = k + 1$.

[*Since both the basis step and the inductive step have been proved, the property is true for all integers $n \geq 0$.*]

8. *Proof:* For the given statement, the property is the equation

$$1 + 2 + 2^2 + \cdots + 2^n = 2^{n+1} - 1. \qquad \text{the property}$$

Show that the property is true for $n = 0$.

When $n = 0$, the left-hand side of the equation is 1, and the right-hand side is $2^{0+1} - 1 = 2 - 1 = 1$ also. Thus the property is true for $n = 0$.

Show that for all integers $k \geq 0$, if the property is true for $n = k$ then it is true for $n = k + 1$:

Let k be any integer with $k \geq 0$, and suppose the property is true for $n = k$. That is, suppose

$$1 + 2 + 2^2 + \cdots + 2^k = 2^{k+1} - 1. \qquad \text{inductive hypothesis}$$

We must show that the property is true for $n = k + 1$. That is, we must show that

$$1 + 2 + 2^2 + \cdots + 2^{k+1} = 2^{(k+1)+1} - 1,$$

or, equivalently,

$$1 + 2 + 2^2 + \cdots + 2^{k+1} = 2^{k+2} - 1. \qquad (*)$$

But the left-hand side of equation $(*)$ is

$$1 + 2 + 2^2 + \cdots + 2^{k+1}$$

$$= 1 + 2 + 2^2 + \cdots + 2^k + 2^{k+1}$$

by making the next-to-last term explicit

$$= (2^{k+1} - 1) + 2^{k+1} \qquad \text{by substitution from the inductive hypothesis}$$

$$= 2 \cdot 2^{k+1} - 1 \qquad \text{by combining like terms}$$

$$= 2^{k+2} - 1, \qquad \text{by the laws of exponents}$$

and this is the right-hand side of equation $(*)$. Hence the property is true for $n = k + 1$.

[*Since both the basis step and the inductive step have been proved, the property is true for all integers $n \geq 0$.*]

10. *Proof:* For the given statement, the property is the equation

$$1^2 + 2^2 + 3^2 + \cdots + n^2 = \frac{n(n+1)(2n+1)}{6}. \qquad \text{the property}$$

Show that the property is true for $n = 1$:

When $n = 1$, the left-hand side of the equation is $1^2 = 1$, and the right-hand side is $\frac{1(1+1)(2 \cdot 1+1)}{6} = \frac{2 \cdot 3}{6} = 1$ also. Thus the property is true for $n = 1$.

Show that for all integers $k \geq 1$, if the property is true for $n = k$ then it is true for $n = k + 1$:

Let k be any integer with $k \geq 1$, and suppose the property is true for $n = k$. That is, suppose

$$1^2 + 2^2 + 3^2 + \cdots + k^2 = \frac{k(k+1)(2k+1)}{6}.$$

inductive hypothesis

We must show that the property is true for $n = k + 1$. That is, we must show that

$$1^2 + 2^2 + 3^2 + \cdots + (k+1)^2$$

$$= \frac{(k+1)((k+1)+1)(2(k+1)+1)}{6},$$

or, equivalently,

$$1^2 + 2^2 + 3^2 + \cdots + (k+1)^2$$

$$= \frac{(k+1)(k+2)(2k+3)}{6}. \qquad (*)$$

But the left-hand side of equation $(*)$ is

$$1^2 + 2^2 + 3^2 + \cdots + (k+1)^2$$

$$= 1^2 + 2^2 + 3^2 + \cdots + k^2 + (k+1)^2$$

by making the next-to-last term explicit

$$= \frac{k(k+1)(2k+1)}{6} + (k+1)^2 \qquad \text{by substitution from the inductive hypothesis}$$

$$= \frac{k(k+1)(2k+1)}{6} + \frac{6(k+1)^2}{6} \qquad \text{because } \frac{6}{6} = 1$$

$$= \frac{k(k+1)(2k+1) + 6(k+1)^2}{6} \qquad \text{by adding fractions}$$

$$= \frac{(k+1)[k(2k+1) + 6(k+1)]}{6} \qquad \text{by factoring out } (k+1)$$

$$= \frac{(k+1)(2k^2 + 7k + 6)}{6} \qquad \text{by multiplying out and combining like terms}$$

$$= \frac{(k+1)(k+2)(2k+3)}{6},$$

because $(k+2)(2k+3) = 2k^2 + 7k + 6$

and this is the right-hand side of equation $(*)$. Hence the property is true for $n = k + 1$.

[*Since both the basis step and the inductive step have been proved, the property is true for all integers $n \geq 1$.*]

13. *Proof:* For the given statement, the property is the equation

$$\sum_{i=1}^{n-1} i(i+1) = \frac{n(n-1)(n+1)}{3}. \quad \text{the property}$$

Show that the property is true for n = 2:

When $n = 2$, the left-hand side of the equation is $\sum_{i=1}^{1} i(i+1) = 1 \cdot (1+1) = 2$, and the right-hand side is $\frac{2(2-1)(2+1)}{3} = \frac{6}{3} = 2$ also. Thus the property is true for $n = 2$.

Show that for all integers $k \geq 2$, if the property is true for $n = k$ then it is true for $n = k + 1$:

Let k be any integer with $k \geq 2$, and suppose the property is true for $n = k$. That is, suppose

$$\sum_{i=1}^{k-1} i(i+1) = \frac{k(k-1)(k+1)}{3} \quad \text{inductive hypothesis}$$

We must show that the property is true for $n = k + 1$. That is, we must show that

$$\sum_{i=1}^{(k+1)-1} i(i+1) = \frac{(k+1)((k+1)-1)((k+1)+1)}{3},$$

or, equivalently,

$$\sum_{i=1}^{k} i(i+1) = \frac{(k+1)k(k+2)}{3}. \quad (*)$$

But the left-hand side of equation (*) is

$$\sum_{i=1}^{k} i(i+1)$$

$$= \sum_{i=1}^{k-1} i(i+1) + k(k+1) \quad \begin{array}{l}\text{by writing the last}\\ \text{term separately}\end{array}$$

$$= \frac{k(k-1)(k+1)}{3} + k(k+1) \quad \begin{array}{l}\text{by substitution from the}\\ \text{inductive hypothesis}\end{array}$$

$$= \frac{k(k-1)(k+1)}{3} + \frac{3k(k+1)}{3} \quad \text{because } \frac{3}{3} = 1$$

$$= \frac{k(k-1)(k+1) + 3k(k+1)}{3} \quad \text{by adding the fractions}$$

$$= \frac{k(k+1)[(k-1)+3]}{3} \quad \text{by factoring out } k(k+1)$$

$$= \frac{k(k+1)(k+2)}{3}, \quad \text{by algebra}$$

and this is the right-hand side of equation (*). Hence the property is true for $n = k + 1$.

[*Since both the basis step and the inductive step have been proved, the property is true for all integers $n \geq 0$.*]

15. *Hint:* To prove the basis step, show that $\sum_{i=1}^{1} i(i!) = (1+1)! - 1$. To prove the inductive step, suppose that $\sum_{i=1}^{k} i(i!) = (k+1)! - 1$ for some integer $k \geq 1$ and show

that $\sum_{i=1}^{k+1} i(i!) = (k+2)! - 1$. Note that $[(k+1)! - 1] + (k+1)[(k+1)!] = (k+1)![1 + (k+1)] - 1$.

18. *Hints:* $\sin^2 x + \cos^2 x = 1$, $\cos(2x) = \cos^2 x - \sin^2 x = 1 - 2\sin^2 x$, $\sin(a+b) = \sin a \cos b + \cos a \sin b$, $\sin(2x) = 2 \sin x \cos x$.

19. $4 + 8 + 12 + 16 + \cdots + 200 = 4(1 + 2 + 3 + \cdots + 50) = 4\left(\frac{50 \cdot 51}{2}\right) = 5100$

21. $3 + 4 + 5 + 6 + \cdots + 1000 = (1 + 2 + 3 + 4 + \cdots + 1000) - (1 + 2) = \left(\frac{1000 \cdot 1001}{2}\right) - 3 = 500{,}497$

23. $\frac{(k-1)((k-1)+1)}{2} = \frac{k(k-1)}{2}$

24. a. $\dfrac{2^{26} - 1}{2 - 1} = 2^{26} - 1 = 67{,}108{,}863$

    b. $2 + 2^2 + 2^3 + \cdots + 2^{26}$

$$= 2(1 + 2 + 2^2 + \cdots + 2^{25})$$

$$= 2 \cdot (67{,}108{,}863) \quad \text{by part (a)}$$

$$= 134{,}217{,}726$$

27. $\dfrac{\left(\dfrac{1}{2}\right)^{n+1} - 1}{\dfrac{1}{2} - 1} = \dfrac{\dfrac{1}{2^{n+1}} - 1}{-\dfrac{1}{2}} = \left(\dfrac{1}{2^{n+1}} - 1\right)(-2)$

$$= -\frac{2}{2^{n+1}} + 2 = 2 - \frac{1}{2^n}$$

29. *Hint:* $a + (a+d) + (a+2d) + \cdots + (a+nd) = (n+1)a + d \cdot \frac{n(n+1)}{2}$.

32. In the inductive step, both the inductive hypothesis and what is to be shown are wrong. The inductive hypothesis should be

Suppose that for some integer $k \geq 1$,

$$1^2 + 2^2 + \cdots + k^2 = \frac{k(k+1)(2k+1)}{6}.$$

And what is to be shown should be

$$1^2 + 2^2 + \cdots + (k+1)^2$$

$$= \frac{(k+1)((k+1)+1)(2(k+1)+1)}{6}.$$

34. *Hint:* Form the sum $n^2 + (n+1)^2 + (n+2)^2 + \cdots + (n + (p-1))^2$, and show that it equals

$$pn^2 + 2n(1 + 2 + 3 + \cdots + (p-1))$$

$$+ (1 + 4 + 9 + 16 + \cdots + (p-1)^2).$$

Section 4.3

1. *General formula:* $\prod_{i=2}^{n} \left(1 - \frac{1}{i}\right) = \frac{1}{n}$ for all integers $n \geq 2$.
Proof (by mathematical induction): The property is the equation

$$\prod_{i=2}^{n} \left(1 - \frac{1}{i}\right) = \frac{1}{n}.$$

Show that the property is true for n = 2:

The property holds because for $n = 2$, the left-hand side of the equation is $\prod_{i=2}^{2}\left(1 - \frac{1}{i}\right) = 1 - \frac{1}{2} = \frac{1}{2}$, which equals the right-hand side.

Show that for all integers $k \geq 2$, if the property is true for $n = k$, then it is true for $n = k + 1$:

Suppose $\prod_{i=2}^{k}\left(1 - \frac{1}{i}\right) = \frac{1}{k}$ for some integer $k \geq 2$. [*Inductive hypothesis*] We must show that $\prod_{i=2}^{k+1}\left(1 - \frac{1}{i}\right) = \frac{1}{k+1}$. (*) But by the laws of algebra and substitution from the inductive hypothesis, the left-hand side of equation (*) is

$$\prod_{i=2}^{k+1}\left(1 - \frac{1}{i}\right)$$

$$= \prod_{i=2}^{k}\left(1 - \frac{1}{i}\right) \cdot \left(1 - \frac{1}{k+1}\right)$$

$$= \left(\frac{1}{k}\right) \cdot \left(1 - \frac{1}{k+1}\right) = \left(\frac{1}{k}\right) \cdot \left(\frac{(k+1) - 1}{k+1}\right)$$

$$= \frac{1}{k+1}, \text{ which is the right-hand side of equation (*)}$$

[*as was to be shown*].

3. *General formula:* $\frac{1}{1\cdot 3} + \frac{1}{3\cdot 5} + \cdots + \frac{1}{(2n-1)(2n+1)} = \frac{n}{2n+1}$ for all integers $n \geq 1$.

Proof (by mathematical induction): The property is the equation

$$\frac{1}{1\cdot 3} + \frac{1}{3\cdot 5} + \cdots + \frac{1}{(2n-1)(2n+1)} = \frac{n}{2n+1}.$$

Show that the property is true for n = 1:

When $n = 1$, the left-hand side of the equation equals $\frac{1}{1\cdot 3}$, and the right-hand side equals $\frac{1}{2\cdot 1+1}$. But both of these equal $\frac{1}{3}$, so the property holds for $n = 1$.

Show that for any integer $k \geq 1$, if the property holds for $n = k$ then it holds for $n = k + 1$:

Suppose that for some integer $k \geq 1$,

$$\frac{1}{1\cdot 3} + \frac{1}{3\cdot 5} + \cdots + \frac{1}{(2k-1)(2k+1)} = \frac{k}{2k+1}$$

inductive hypothesis

We must show that

$$\frac{1}{1\cdot 3} + \frac{1}{3\cdot 5} + \cdots + \frac{1}{(2(k+1)-1)(2(k+1)+1)}$$
$$= \frac{k+1}{2(k+1)+1},$$

or, equivalently,

$$\frac{1}{1\cdot 3} + \frac{1}{3\cdot 5} + \cdots + \frac{1}{(2k+1)(2k+3)} = \frac{k+1}{2k+3}. \quad (*)$$

But the left-hand side of equation (*) is

$$\frac{1}{1\cdot 3} + \frac{1}{3\cdot 5} + \cdots + \frac{1}{(2k+1)(2k+3)}$$

$$= \frac{1}{1\cdot 3} + \frac{1}{3\cdot 5} + \cdots + \frac{1}{(2k-1)(2k+1)}$$

$$\qquad\qquad + \frac{1}{(2k+1)(2k+3)}$$

$$= \frac{k}{2k+1} + \frac{1}{(2k+1)(2k+3)} \quad \text{by inductive hypothesis}$$

$$= \frac{k(2k+3)}{(2k+1)(2k+3)} + \frac{1}{(2k+1)(2k+3)}$$

$$= \frac{2k^2 + 3k + 1}{(2k+1)(2k+3)}$$

$$= \frac{(2k+1)(k+1)}{(2k+1)(2k+3)}$$

$$= \frac{k+1}{2k+3}, \quad \text{by algebra}$$

and this is the right-hand side of equation (*) [*as was to be shown*].

4. *Hint 1:* The general formula is

$$1 - 4 + 9 - 16 + \cdots + (-1)^{n-1}n^2$$
$$= (-1)^{n-1}(1 + 2 + 3 + \cdots + n) \quad \text{in expanded form}$$

$$\sum_{i=1}^{n}(-1)^{i-1}i^2 = (-1)^{n-1}\left(\sum_{i=1}^{n}i\right) \quad \begin{array}{l}\text{in summation}\\\text{notation.}\end{array}$$

Hint 2: In the proof, use the fact that

$$1 + 2 + 3 + \cdots + n = \sum_{i=1}^{n}i = \frac{n(n+1)}{2}.$$

6. **a.** $P(0)$ is "$5^0 - 1$ is divisible by 4." $P(0)$ is true because $5^0 - 1 = 0$, which is divisible by 4.
   b. $P(k)$ is "$5^k - 1$ is divisible by 4."
   c. $P(k+1)$ is "$5^{k+1} - 1$ is divisible by 4."
   d. *Must show:* If for some integer $k \geq 0$, $5^k - 1$ is divisible by 4, then $5^{k+1} - 1$ is divisible by 4.

8. *Proof (by mathematical induction):* For the given statement, the property is the sentence "$5^n - 1$ is divisible by 4."

Show that the property is true for n = 0:

When $n = 0$, the property is the sentence "$5^0 - 1$ is divisible by 4." But $5^0 - 1 = 1 - 1 = 0$, and 0 is divisible by 4 because $0 = 4 \cdot 0$. Thus the property is true for $n = 0$.

Show that for all integers $n \geq 0$, if the property is true for $n = k$ then it is true for $n = k + 1$:

Let k be an integer with $k \geq 0$, and suppose the property is true for $n = k$. That is, suppose $5^k - 1$ is divisible by 4.

[*This is the inductive hypothesis.*] We must show that the property is true for $n = k + 1$. That is, we must show that $5^{k+1} - 1$ is divisible by 4. Now

$$5^{k+1} - 1 = 5^k \cdot 5 - 1$$
$$= 5^k \cdot (4 + 1) - 1 = 5^k \cdot 4 + (5^k - 1). \quad (*)$$

By the inductive hypothesis $5^k - 1$ is divisible by 4, and so $5^k - 1 = 4r$ for some integer r. By substitution into equation (*),

$$5^{k+1} - 1 = 5^k \cdot 4 + 4r = 4(5^k + r).$$

But $5^k + r$ is an integer because k and r are integers. Hence, by definition of divisibility, $5^{k+1} - 1$ is divisible by 4 [*as was to be shown*].

An alternative proof of the inductive step goes as follows: Suppose that for some integer $k \geq 0$, $5^k - 1$ is divisible by 4. Then $5^k - 1 = 4r$ for some integer r, and hence $5^k = 4r + 1$.
It follows that $5^{k+1} = 5^k \cdot 5 = (4r + 1) \cdot 5 = 20r + 5$. Subtracting 1 from both sides gives that $5^{k+1} - 1 = 20r + 4 = 4(5r + 1)$. But $5r + 1$ is an integer, and so, by definition of divisibility, $5^{k+1} - 1$ is divisible by 4.

11. *Proof (by mathematical induction):* For the given statement, the property is the sentence "$3^{2n} - 1$ is divisible by 8."

Show that the property is true for $n = 0$:

When $n = 0$, the property is the sentence "$3^{2 \cdot 0} - 1$ is divisible by 8." But $3^{2 \cdot 0} - 1 = 1 - 1 = 0$, and 0 is divisible by 8 because $0 = 8 \cdot 0$. Thus the property is true for $n = 0$.

Show that for all integers $n \geq 0$, if the property is true for $n = k$ then it is true for $n = k + 1$:

Let k be an integer with $k \geq 0$, and suppose the property is true for $n = k$. That is, suppose $3^{2k} - 1$ is divisible by 8. [*This is the inductive hypothesis.*] We must show that the property is true for $n = k + 1$. That is, we must show that $3^{2(k+1)} - 1$ is divisible by 8, or equivalently, $3^{2k+2} - 1$ is divisible by 8. Now

$$3^{2k+2} - 1 = 3^{2k} \cdot 3^2 - 1 = 3^{2k} \cdot 9 - 1$$
$$= 3^{2k} \cdot (8 + 1) - 1 = 3^{2k} \cdot 8 + (3^{2k} - 1) \cdot \quad (*)$$

By the inductive hypothesis $3^{2k} - 1$ is divisible by 8, and so $3^{2k} - 1 = 8r$ for some integer r. By substitution into equation (*),

$$3^{2k+2} - 1 = 3^{2k} \cdot 8 + 8r = 8(3^{2k} + r).$$

But $3^{2k} + r$ is an integer because k and r are integers. Hence, by definition of divisibility, $3^{2k+2} - 1$ is divisible by 8 [*as was to be shown*].

13. *Hint:* $x^{k+1} - y^{k+1} = x^{k+1} - x \cdot y^k + x \cdot y^k - y^{k+1}$
$$= x \cdot (x^k - y^k) + y^k \cdot (x - y)$$

14. *Hint 1:* $(k + 1)^3 - (k + 1) = k^3 + 3k^2 + 3k + 1 - k - 1$
$$= (k^3 - k) + 3k^2 + 3k$$
$$= (k^3 - k) + 3k(k + 1)$$

Hint 2: $k(k + 1)$ is a product of two consecutive integers. By Theorem 3.4.3, one of these must be even.

16. *Proof (by mathematical induction):* For the given statement, the property is the inequality $2^n < (n + 1)!$.

Show that the property is true for $n = 2$:

When $n = 2$, the property says that $2^2 < (2 + 1)!$. The left-hand side is $2^2 = 4$ and the right-hand side is $3! = 6$. So because $4 < 6$ the property is true for $n = 2$.

Show that for all integers $n \geq 2$, if the property is true for $n = k$ then it is true for $n = k + 1$:

Let k be an integer with $k \geq 2$, and suppose the property is true for $n = k$. That is, suppose $2^k < (k + 1)!$. [*This is the inductive hypothesis.*] We must show that the property is true for $n = k + 1$. That is, we must show that $2^{k+1} < ((k + 1) + 1)$, or, equivalently, $2^{k+1} < (k + 2)!$. By the laws of exponents and the inductive hypothesis,

$$2^{k+1} = 2 \cdot 2^k < 2(k + 1)!. \quad (*)$$

Since $k \geq 2$, then $2 < k + 2$, and so

$$2(k + 1)! < (k + 2)(k + 1)! = (k + 2)!. \quad (**)$$

Combining inequalities (*) and (**) gives

$$2^{k+1} < (k + 2)!$$

[*as was to be shown*].

19. *Proof (by mathematical induction):* For the given statement, the property is the inequality $n^2 < 2^n$.

Show that the property is true for $n = 5$:

When $n = 5$, the property says that $5^2 < 2^5$. But $5^2 = 25$ and $2^5 = 32$, and $25 < 32$. Hence the property is true for $n = 5$.

Show that for any integer $k \geq 5$, if the property is true for $n = k$ then it is true for $n = k + 1$:

Let k be an integer with $k \geq 5$, and suppose the property is true for k. That is, suppose $k^2 < 2^k$. [*This is the inductive hypothesis.*] We must show that the property is true for $n = k + 1$. That is, we must show that $(k + 1)^2 < 2^{k+1}$. But

$$(k + 1)^2 = k^2 + 2k + 1 < 2^k + 2k + 1 \quad \text{by inductive hypothesis}$$

Also, by Proposition 4.3.2,

$$2k + 1 < 2^k \quad \text{Prop. 4.3.2 applies since } k \geq 5 \geq 3.$$

Putting these inequalities together gives

$$(k + 1)^2 < 2^k + 2k + 1 < 2^k + 2^k = 2^{k+1}$$

[*as was to be shown*].

24. *Proof (by mathematical induction):* For the given statement, the property is the equation $a_n = 3 \cdot 7^{n-1}$.

Show that the property is true for $n = 1$:

When $n = 1$, the left-hand side of the equation is a_1, which equals 3 by definition of the sequence. The right-hand side is $3 \cdot 7^{1-1} = 3$ also. Thus the property is true for $n = 1$.

Show that for all integers $k \geq 1$, if the property is true for $n = k$ then it is true for $n = k + 1$:

Let k be any integer with $k \geq 1$, and suppose the property is true for $n = k$. That is, suppose $a_k = 3 \cdot 7^{k-1}$. [*This is the inductive hypothesis.*] We must show that the property is true for $n = k + 1$. That is, we must show that $a_{k+1} = 3 \cdot 7^{(k+1)-1}$, or, equivalently, $a_{k+1} = 3 \cdot 7^k$. (*) But the left-hand side of equation (*) is

$$
\begin{aligned}
a_{k+1} &= 7a_k && \text{by definition of the sequence } a_1, a_2, a_3, \ldots \\
&= 7(3 \cdot 7^{k-1}) && \text{by inductive hypothesis} \\
&= 3 \cdot 7^k && \text{by the laws of exponents,}
\end{aligned}
$$

and this is the right-hand side of equation (*) [*as was to be shown*].

30. The inductive step fails for going from $n = 1$ to $n = 2$, because when $k = 1$,

$$A = \{a_1, a_2\} \quad \text{and} \quad B = \{a_1\},$$

and no set C can be defined to have the properties claimed for the C in the proof. The reason is that if $C = \{a_1\}$, then B and C share a common element; but one element of A, a_2, is not in either B or C. On the other hand, if $C = \{a_2\}$, then B and C have no common element.

Since the inductive step fails for going from $n = 1$ to $n = 2$, the truth of the following statement is never proved: "All the numbers in a set of two numbers are equal to each other." This breaks the sequence of inductive steps, and so none of the statements for $n > 2$ is proved true either.

Here is an explanation for what happens in terms of the domino analogy. The first domino *is* tipped backward (the basis step is proved). Also, if any domino from the second onward tips backward, then it tips the one behind it backward (the inductive step works for $n \geq 2$). However, when the first domino is tipped backward, it does *not* tip the second one backward. So only the first domino falls down; the rest remain standing.

31. The basis step is not proved, and in fact it is false because for $n = 1$, $3^n - 2 = 3^1 - 2 = 1$, which is odd.

34. *Hint:* Use proof by contradiction. If the statement is false, then there exists some ordering of the integers from 1 to 30, say $x_1, x_2, \ldots, x_{30}$, such that $x_1 + x_2 + x_3 < 45, x_2 + x_3 + x_4 < 45, \ldots$, and $x_{30} + x_1 + x_2 < 45$. Evaluate the sum of all these inequalities using the fact that $\sum_{i=1}^{30} x_i = \sum_{i=1}^{30} i$ and Theorem 4.2.2.

Section 4.4

1. The property "a_n is odd" holds for all integers $n \geq 1$.
Proof (by strong mathematical induction):

Show that the property is true for $n = 1$ and $n = 2$:
Observe that $a_1 = 1$ and $a_2 = 3$ and both 1 and 3 are odd. Thus the property is true for $n = 1$ and $n = 2$.

Show that for any integer $k > 2$, if the property is true for all integers i with $1 \leq i < k$, then it is true for k:

Let $k > 2$ be an integer, and suppose a_i is odd for all integers i with $1 \leq i < k$. [*This is the inductive hypothesis.*] We must show that a_k is odd. We know that $a_k = a_{k-2} + 2a_{k-1}$ by definition of $a_1, a_2, a_3, \ldots$. Moreover, $k - 2$ is less than k and is greater than or equal to 1 (because $k > 2$). Thus, by inductive hypothesis, a_{k-2} is odd. Also, every term of the sequence is an integer (being a sum of products of integers), and so $2a_{k-1}$ is even by definition of even. Hence a_k is the sum of an odd integer and an even integer and hence is odd (by exercise 19, in Section 3.1). [*This is what was to be shown.*]

4. *Proof (by strong mathematical induction):* Let the property $P(n)$ be the inequality $d_n \leq 1$.

Show that the property is true for $n = 1$ and $n = 2$:
Observe that $d_1 = \frac{9}{10}$ and $d_2 = \frac{10}{11}$ and both $\frac{9}{10} \leq 1$ and $\frac{10}{11} \leq 1$. Thus the property is true for $n = 1$ and $n = 2$.

Show that for any integer $k > 2$, if the property is true for all integers i with $1 \leq i < k$, then it is true for k:

Let $k > 2$ be an integer, and suppose $d_i \leq 1$ for all integers i with $1 \leq i < k$. [*This is the inductive hypothesis.*] We must show that $d_k < 1$. But, by definition of $d_1, d_2, d_3, \ldots, d_k = d_{k-1} \cdot d_{k-2}$. Now $d_{k-1} \leq 1$ and $d_{k-2} \leq 1$ by inductive hypothesis [*since $1 \leq k - 1 < k$ and $1 \leq k - 2 < k$ because $k > 2$*]. Consequently, $d_k = d_{k-1} \cdot d_{k-2} \leq 1$ because if two positive numbers are each less than or equal to 1, then their product is less than or equal to 1. [*If $0 < a \leq 1$ and $0 < b \leq 1$, then multiplying $a \leq 1$ by b gives $ab \leq b$, and since $b \leq 1$, then by transitivity of order, $ab \leq 1$.*] This is what was to be shown. [*Since we have proved both the basis step and the inductive step, we conclude that $d_n \leq 1$ for all integers $n \leq 1$.*]

7. *Proof (by strong mathematical induction):* Let the property $P(n)$ be the equation $g_n = 5 \cdot 3^n + 7 \cdot 2^n$.

Show that the property is true for $n = 0$ and $n = 1$:
We must show that $g_0 = 5 \cdot 3^0 + 7 \cdot 2^0$ and $g_1 = 5 \cdot 3^1 + 7 \cdot 2^1$. The left-hand side of the first equation is 12 (by definition of $g_0, g_1, g_2, \ldots$), and its right-hand side is $5 \cdot 1 + 7 \cdot 1 = 12$ also. The left-hand side of the second equation is 29 (by definition of $g_0, g_1, g_2, \ldots$), and its right-hand side is $5 \cdot 3 + 7 \cdot 2 = 29$ also. Thus the property is true for $n = 0$ and $n = 1$.

Show that for any integer $k \geq 2$, if the property is true for all integers i with $0 \leq i < k$, then it is true for k:

Let $k \geq 2$ be an integer, and suppose $g_i = 5 \cdot 3^i + 7 \cdot 2^i$ for all integers i with $0 \leq i < k$. [*Inductive hypothesis*] We must show that $5 \cdot 3^k + 7 \cdot 2^k$.

But

$$g_k = 5g_{k-1} - 6g_{k-2} \qquad \text{by definition of } g_0, g_1, g_2, \ldots$$
$$= 5(5 \cdot 3^{k-1} + 7 \cdot 2^{k-1}) - 6(5 \cdot 3^{k-2} + 7 \cdot 2^{k-2})$$
$$\text{by inductive hypothesis}$$
$$= 25 \cdot 3^{k-1} + 35 \cdot 2^{k-1} - 30 \cdot 3^{k-2} - 42 \cdot 2^{k-2}$$
$$= 25 \cdot 3^{k-1} + 35 \cdot 2^{k-1} - 10 \cdot 3 \cdot 3^{k-2} - 21 \cdot 2 \cdot 2^{k-2}$$
$$= 25 \cdot 3^{k-1} + 35 \cdot 2^{k-1} - 10 \cdot 3^{k-1} - 21 \cdot 2^{k-1}$$
$$= (25 - 10) \cdot 3^{k-1} + (35 - 21) \cdot 2^{k-1}$$
$$= 15 \cdot 3^{k-1} + 14 \cdot 2^{k-1}$$
$$= 5 \cdot 3 \cdot 3^{k-1} + 7 \cdot 2 \cdot 2^{k-1}$$
$$= 5 \cdot 3^k + 7 \cdot 2^k \qquad \text{by algebra.}$$

[*This is what was to be shown.*]

10. *Proof (by strong mathematical induction):* Let the property $P(n)$ be the sentence

> "*A* jigsaw puzzle consisting of n pieces takes $n - 1$ steps to put together."

Show that the property is true for $n = 1$:

A jigsaw puzzle consisting of just one piece does not take any steps to put together. Hence it is correct to say that it takes zero steps to put together.

Show that for any integer $k > 1$, if the property is true for all integers i with $1 \leq i < k$ then it is true for k:

Let $k > 1$ be an integer and suppose that for all integers i with $1 \leq i < k$, a jigsaw puzzle consisting of i pieces takes $i - 1$ steps to put together. [*This is the inductive hypothesis.*] We must show that a jigsaw puzzle consisting of k pieces takes $k - 1$ steps to put together. Consider assembling a jigsaw puzzle consisting of k pieces. The last step involves fitting together two blocks. Suppose one of the blocks consists of r pieces and the other consists of s pieces. Then $r + s = k$, and $1 \leq r < k$ and $1 \leq s < k$. Thus by inductive hypothesis, the numbers of steps required to assemble the blocks are $r - 1$ and $s - 1$, respectively. Then the total number of steps required to assemble the puzzle is $(r - 1) + (s - 1) + 1 = (r + s) - 1 = k - 1$ [*as was to be shown*].

11. *Sketch of proof:* Given any integer $k > 1$, either k is prime or k is a product of two smaller positive integers, each greater than 1. In the former case, the property is true. In the latter case, the inductive hypothesis ensures that both factors of k are products of primes and hence that k is also a product of primes.

12. *Proof (by strong mathematical induction):* Let the property $P(n)$ be the sentence "Any product of n odd integers is odd."

Show that the property is true for $n = 2$:

We must show that any product of two odd integers is odd. But this was established in Chapter 3 (exercise 39 of Section 3.1).

Show that for any integer $k > 2$, if the property is true for all integers i with $2 \leq i < k$ then it is true for k:

Let k be any integer with $k > 2$, and suppose that for all integers i with $2 \leq i < k$, any product of i odd integers is odd. [*Inductive hypothesis*] Consider any product P of k odd integers. Some multiplication is the final one that is used to obtain P. Thus there are integers A and B such that $P = AB$, and each of A and B is a product of between 1 and $k - 1$ odd integers. (For instance, if $P = ((a_1 a_2)a_3)a_4$, then $A = (a_1 a_2)a_3$ and $B = a_4$.) By inductive hypothesis, each of A and B is odd, and, as in the basis step, we know that any product of two odd integers is odd. Hence $P = AB$ is odd.

14. *Hint:* Let the property $P(n)$ be the sentence "If n is even, then any sum of n odd integers is even, and if n is odd, then any sum of n odd integers is odd." For the inductive step, consider any sum S of k odd integers. Some addition is the final one that is used to obtain S. Thus there are integers A and B such that $S = A + B$, and A is a sum of r odd integers and B is a sum of $k - r$ odd integers. Consider the two cases where k is even and k is odd, and for each case consider the two subcases where r is even and where r is odd.

15. $4^1 = 4, 4^2 = 16, 4^3 = 64, 4^4 = 256, 4^5 = 1024$, $4^6 = 4096, 4^7 = 16384$, and $4^8 = 65536$.

Conjecture: The units digit of 4^n equals 4 if n is odd and equals 6 if n is even.

Proof by strong mathematical induction: Let the property $P(n)$ be the sentence "The units digit of 4^n equals 4 if n is odd and equals 6 if n is even."

Show that the property is true for $n = 1$ and $n = 2$:

When $n = 1$, $4^n = 4^1 = 4$, and the units digit is 4. When $n = 2$, then $4^n = 4^2 = 16$, and the units digits is 6. Thus the property is true for $n = 1$ and $n = 2$.

Show that for any integer $k > 2$, if the property is true for all integers i with $1 \leq i < k$ then it is true for k:

Let k by any integer with $k > 2$, and suppose that for all integers i with $0 \leq i < k$, the units digit of 4^i equals 4 if i is odd and equals 6 if i is even. [*Inductive hypothesis*] We must show that the units digit of 4^k equals 4 if k is odd and equals 6 if k is even.

Case 1 (k is odd): In this case, $k - 1$ is even, and so, by inductive hypothesis, the units digits of 4^{k-1} is 6. Thus $4^{k-1} = 10q + 6$ for some nonnegative integer q. It follows that $4^k = 4^{k-1} \cdot 4 = (10q + 6) \cdot 4 = 40q + 24 = 10(4q + 2) + 4$. Thus the units digit of 4^k is 4 [*as was to be shown*].

Case 2 (k is even): In this case, $k - 1$ is odd, and so, by inductive hypothesis, the units digit of 4^{k-1} is 4. Thus $4^{k-1} = 10q + 4$ for some nonnegative integer q. It follows that $4^k = 4^{k-1} \cdot 4 = (10q + 4) \cdot 4 = 40q + 16 = 10(4q + 1) + 6$. Thus the units digit of 4^k is 6 [*as was to be shown*].

18. *Proof:* Let n be any integer greater than 1. Consider the set S of all positive integers other than 1 that divide n. Since $n \mid n$ and $n > 1$, there is at least one element in S. Hence, by the well-ordering principle, S has a smallest element; call it p. We claim that p is prime. For suppose p is not prime. Then there are integers a and b with $1 < a < p$, $1 < b < p$, and $p = ab$. By definition of divides, $a \mid p$. Also $p \mid n$ because p is in S and every element in S divides n. Therefore, $a \mid p$ and $p \mid n$, and so, by transitivity of divisibility, $a \mid n$. Consequently, $a \in S$. But this contradicts the fact that $a < p$, and p is the smallest element of S. [*This contradiction shows that the supposition that p is not prime is false.*] Hence p is prime, and we have shown the existence of a prime number that divides n.

20. *Proof:* Suppose r is any rational number. [*We need to show that there is an integer n such that $r < n$.*]

Case 1 ($r \leq 0$): In this case, take $n = 1$. Then $r < n$.

Case 2 ($r > 0$): In this case, $r = \frac{a}{b}$ for some positive integers a and b (by definition of rational and because r is positive). Note that $r = \frac{a}{b} < n$ if, and only if, $a < nb$. Let $n = 2a$. Multiply both sides of the inequality $1 < 2$ by a to obtain $a < 2a$, and multiply both sides of the inequality $1 < b$ by $2a$ to obtain $2a < 2ab = nb$. Thus $a < 2a < nb$, and so, by transitivity of order, $a < nb$. Dividing both sides by b gives that $\frac{a}{b} < n$, or, equivalently, that $r < n$. Hence, in both cases, $r < n$ [*as was to be shown*].

21. *Hint:* Given any rational number r, divide into three cases: $r > 0$, $r = 0$, and $r < 0$. For the case where $r > 0$, represent r as a quotient of integers, use the result of exercise 20, and use the well-ordering principle. For the case where $r < 0$, modify the result for the case where $r > 0$.

22. *Proof:* Let S be the set of all integers r such that $n = 2^i \cdot r$ for some integer i. Then $n \in S$ because $n = 2^0 \cdot n$, and so $S \neq \emptyset$. Also, since $n \geq 1$, each r in S is positive, and so, by the well-ordering principle, S has a least element m. This means that $n = 2^k \cdot m$ (*) for some nonnegative integer k and $m \leq r$ for every r in S. We claim that m is odd. The reason is that if m were even, then $m = 2p$ for some integer p. Substituting into equation (*) gives

$$n = 2^k \cdot m = 2^k \cdot 2p = (2^k \cdot 2)p = 2^{k+1} \cdot p.$$

It follows that $p \in S$ and $p < m$, which contradicts the fact that m is the *least* element of S. Hence m is odd, and so $n = m \cdot 2^k$ for some odd integer m and nonnegative integer k.

27. *Hint:* In the inductive step, divide into cases depending upon whether k can be written as $k = 3x$ or $k = 3x + 1$ or $k = 3x + 2$ for some integer x.

28. *Hint:* In the inductive step, let an integer $k \geq 0$ be given and suppose that there exist integers q' and r' such that $k = dq' + r'$ and $0 \leq r' < d$. You must show that there exist integers q and r such that

$$k + 1 = dq + r \quad \text{and} \quad 0 \leq r < d.$$

To do this, consider the two cases $r' < d - 1$ and $r' = d - 1$.

29. *Hint:* Given a predicate $P(n)$ that satisfies conditions (1) and (2) of the principle of mathematical induction, let S be the set of all integers greater than or equal to a for which $P(n)$ is false. Suppose that S has one or more elements, and use the well-ordering principle to derive a contradiction.

Section 4.5

1. *Proof:* Suppose the predicate $m + n = 100$ is true before entry to the loop. Then

$$m_{\text{old}} + n_{\text{old}} = 100.$$

After execution of the loop,

$$m_{\text{new}} = m_{\text{old}} + 1 \quad \text{and} \quad n_{\text{new}} = n_{\text{old}} - 1,$$

so

$$m_{\text{new}} + n_{\text{new}} = (m_{\text{old}} + 1) + (n_{\text{old}} - 1)$$
$$= m_{\text{old}} + n_{\text{old}} = 100.$$

3. *Proof:* Suppose the predicate $m^3 > n^2$ is true before entry to the loop. Then

$$m_{\text{old}}^3 > n_{\text{old}}^2.$$

After execution of the loop,

$$m_{\text{new}} = 3 \cdot m_{\text{old}} \quad \text{and} \quad n_{\text{new}} = 5 \cdot n_{\text{old}},$$

so

$$m_{\text{new}}^3 = (3 \cdot m_{\text{old}})^3 = 27 \cdot m_{\text{old}}^3 > 27 \cdot n_{\text{old}}^2.$$

But since $n_{\text{new}} = 5 \cdot n_{\text{old}}$, then $n_{\text{old}} = \frac{1}{5} n_{\text{new}}$. Hence

$$m_{\text{new}}^3 > 27 \cdot n_{\text{old}}^2 = 27 \cdot \left(\frac{1}{5} n_{\text{new}}\right)^2 = 27 \cdot \frac{1}{25} n_{\text{new}}^2$$
$$= \frac{27}{25} \cdot n_{\text{new}}^2 > n_{\text{new}}^2.$$

6. *Proof:* [*The wording of this proof is almost the same as that of Example 4.5.2.*]

I. Basis Property: [*$I(0)$ is true before the first iteration of the loop.*]

$I(0)$ is "$exp = x^0$ and $i = 0$." According to the precondition, before the first iteration of the loop $exp = 1$ and $i = 0$. Since $x^0 = 1$, $I(0)$ is evidently true.

II. Inductive Property: [*If $G \wedge I(k)$ is true before a loop iteration (where $k \geq 0$), then $I(k + 1)$ is true after the loop iteration.*]

Suppose k is a nonnegative integer such that $G \wedge I(k)$ is true before an iteration of the loop. Then as execution reaches the top of the loop, $i \neq m$, $exp = x^k$, and $i = k$. Since $i \neq m$, the guard is passed and statement 1 is executed. Now before execution of statement 1,

$$exp_{\text{old}} = x^k,$$

so execution of statement 1 has the following effect:

$$exp_{\text{new}} = exp_{\text{old}} \cdot x = x^k \cdot x = x^{k+1}.$$

Similarly, before statement 2 is executed,

$$i_{\text{old}} = k,$$

so after execution of statement 2,

$$i_{new} = i_{old} + 1 = k + 1.$$

Hence after the loop iteration, the two statements $exp = x^{k+1}$ and $i = k + 1$ are true, and so $I(k + 1)$ is true.

III. Eventual Falsity of Guard: [*After a finite number of iterations of the loop, G becomes false.*]

The guard G is the condition $i \neq m$, and m is a nonnegative integer. By I and II, it is known that

for *all* integers $n \geq 0$, if the loop is iterated n times, then $exp = x^n$ and $i = n$.

So after m iterations of the loop, $i = m$. Thus G becomes false after m iterations of the loop.

IV. Correctness of the Post-Condition: [*If N is the least number of iterations after which G is false and I(N) is true, then the value of the algorithm variables will be as specified in the post-condition of the loop.*]

According to the post-condition, the value of exp after execution of the loop should be x^m. But when G is false, $i = m$. And when $I(N)$ is true, $i = N$ and $exp = x^N$. Since *both* conditions (G false and $I(N)$ true) are satisfied, $m = i = N$ and $exp = x^m$, as required.

8. *Proof:*

I. Basis Property: $I(0)$ is "$i = 1$ and $sum = A[1]$." According to the pre-condition, this statement is true.

II. Inductive Property: Suppose k is a nonnegative integer such that $G \wedge I(k)$ is true before an iteration of the loop. Then as execution reaches the top of the loop, $i \neq m, i = k + 1$, and $sum = A[1] + A[2] + \cdots + A[k + 1]$. Since $i \neq m$, the guard is passed and statement 1 is executed. Now before execution of statement, 1, $i_{old} = k + 1$. So after execution of statement 1, $i_{new} = i_{old} + 1 = (k + 1) + 1 = k + 2$. Also before statement 2 is executed, $sum_{old} = A[1] + A[2] + \cdots + A[k + 1]$. Execution of statement 2 adds $A[k + 2]$ to this sum, and so after statement 2 is executed, $sum_{new} = A[1] + A[2] + \cdots + A[k + 1] + A[k + 2]$. Thus after the loop iteration, $I(k + 1)$ is true.

III. Eventual Falsity of Guard: The guard G is the condition $i \neq m$. By I and II, it is known that for all integers $n \geq 1$, after n iterations of the loop, $I(n)$ is true. Hence, after $m - 1$ iterations of the loop, $I(m)$ is true, which implies that $i = m$ and G is false.

IV. Correctness of the Post-Condition: Suppose that N is the least number of iterations after which G is false and $I(N)$ is true. Then (since G is false) $i = m$ and (since $I(N)$ is true) $i = N + 1$ and $sum = A[1] + A[2] + \cdots + A[N + 1]$. Putting these together gives $m = N + 1$, and so $sum = A[1] + A[2] + \cdots + A[m]$, which is the post-condition.

10. *Hint:* Assume $G \wedge I(k)$ is true for a nonnegative integer k. Then $a_{old} \neq 0$ and $b_{old} \neq 0$ and
(1) a_{old} and b_{old} are nonnegative integers with $\gcd(a_{old}, b_{old}) = \gcd(A, B)$.
(2) At most one of a_{old} and b_{old} equals 0.
(3) $0 \leq a_{old} + b_{old} \leq A + B - k$.

It must be shown that $I(k + 1)$ is true after the loop iteration. That means it is necessary to show that
(1) a_{new} and b_{new} are nonnegative integers with $\gcd(a_{new}, b_{new}) = \gcd(A, B)$.
(2) At most one of a_{new} and b_{new} equals 0.
(3) $0 \leq a_{new} + b_{new} \leq A + B - (k + 1)$.

To show (3), observe that

$$a_{new} + b_{new} = \begin{cases} a_{old} - b_{old} + b_{old} & \text{if } a_{old} \geq b_{old} \\ b_{old} - a_{old} + a_{old} & \text{if } a_{old} < b_{old} \end{cases}$$

[*The reason for this is that when $a_{old} \geq b_{old}$, then $a_{new} = a_{old} - b_{old}$ and $b_{new} = b_{old}$, and when $a_{old} < b_{old}$, then $b_{new} = b_{old} - a_{old}$ and $a_{new} = a_{old}$.*]
Thus

$$a_{new} + b_{new} = \begin{cases} a_{old} & \text{if } a_{old} \geq b_{old} \\ b_{old} & \text{if } a_{old} < b_{old} \end{cases}$$

But since $a_{old} \neq 0$ and $b_{old} \neq 0$ and a_{old} and b_{old} are nonnegative integers, then $a_{old} \geq 1$ and $b_{old} \geq 1$. Hence $a_{old} - 1 \geq 0$ and $b_{old} - 1 \geq 0$ and $a_{old} \leq a_{old} + b_{old} - 1$ and $b_{old} \leq b_{old} + a_{old} - 1$. It follows that $a_{new} + b_{new} \leq a_{old} + b_{old} - 1 \leq (A + B - k) - 1$ by the truth of (3) going into the kth iteration. Hence $a_{new} + b_{new} < A + B - (k + 1)$ by algebraic simplification.

Section 5.1

1. $A = C$ and $B = D$

4. a. $\{1, -1\}$
  c. $\emptyset$ (the set has no elements)
  d. $\mathbf{Z}$ (every integer is in the set)

5. a. The number 0 is not in $\emptyset$ because $\emptyset$ has no elements.
  b. No. The left-hand set is the empty set; it does not have any elements. The right-hand set is a set with one element, namely $\emptyset$.

6. a. The set of all x in U such that x is in A and x is in B. The shorthand notation is $A \cap B$.

7. a. No. $j \in B$ and $j \notin A$.
  d. Yes. Both elements of C are in A, but A contains elements (namely c and f) that are not in C.

8. a. Yes **b.** No **f.** No **i.** Yes

9. a. $\{1, 3, 5, 6, 7, 9\}$ **b.** $\{3, 9\}$
  c. $\{1, 2, 3, 4, 5, 6, 7, 8, 9\}$ **d.** $\emptyset$ **e.** $\{1, 5, 7\}$

10. a. $A \cup B = \{x \in \mathbf{R} \mid 0 < x < 4\}$
  b. $A \cap B = \{x \in \mathbf{R} \mid 1 \leq x \leq 2\}$
  c. $A^c = \{x \in \mathbf{R} \mid x \leq 0 \text{ or } x > 2\}$
  d. $A \cup C = \{x \in \mathbf{R} \mid 0 < x \leq 2 \text{ or } 3 \leq x < 9\}$

e. $A \cap C = \emptyset$

f. $B^c = \{x \in \mathbf{R} \mid x < 1 \text{ or } x \geq 4\}$

g. $A^c \cap B^c = \{x \in \mathbf{R} \mid x \leq 0 \text{ or } x \geq 4\}$

h. $A^c \cup B^c = \{x \in \mathbf{R} \mid x < 1 \text{ or } x > 2\}$

i. $(A \cap B)^c = \{x \in \mathbf{R} \mid x < 1 \text{ or } x > 2\}$

j. $(A \cup B)^c = \{x \in \mathbf{R} \mid x \leq 0 \text{ or } x \geq 4\}$

12. b. False. Many negative real numbers are not rational. For example, $-\sqrt{2} \in \mathbf{R}$ but $-\sqrt{2} \notin \mathbf{Q}$.

   d. False. $0 \in \mathbf{Z}$ but $0 \notin \mathbf{Z}^- \cup \mathbf{Z}^+$.

13. a. *Negation:* $\exists$ a set A such that $A \subseteq \mathbf{R}$ and $A \not\subseteq \mathbf{Z}$. This statement is true. As an example, take $A = \mathbf{R}$.

14. a. No, $R \not\subseteq T$. $2 \in R$ but $2 \notin T$.

   b. Yes, $T \subseteq R$. Every integer that is divisible by 6 is also divisible by 2.

16. a. $C \subseteq D$ because every element of C is in D. For if n is any element of C, then $n = 6r - 5$ for some integer r. Let $s = 2r - 2$. Then s is an integer (because products and differences of integers are integers), and $3s + 1 = 3(2r - 2) + 1 = 6r - 6 + 1 = 6r - 5$, which equals n. Thus n satisfies the condition for being in D. Hence, every element in C is in D.

   b. $D \not\subseteq C$ because there are elements of D that are not in C. For example, 4 is in D because $4 = 3 \cdot 1 + 1$. But 4 is not in C because if it were, then $4 = 6r - 5$ for some integer r, which would imply that $9 = 6r$, or, equivalently, that $r = \frac{3}{2}$, and this contradicts the fact that r is an integer.

17. a. $A \neq B$ because, for example, $4 \in A$ and $4 \notin B$.

   b. $A = C$. To see this, note that for any integer i, $5i - 1 = 5(i - 1) + 4$. Hence any integer that can be written in the form $5i - 1$, for some integer i, can also be written in the form $5r + 4$, where $r = i - 1$. Thus $A \subseteq C$. Conversely, observe that for any integer r, $5r + 4 = 5(r + 1) - 1$, and so any integer that can be written in the form $5r + 4$, for some integer r, can also be written in the form $5i - 1$, where $i = r + 1$. Thus $C \subseteq A$. Since $A \subseteq C$ and $C \subseteq A$, we conclude that $A = C$.

18. a.

19.

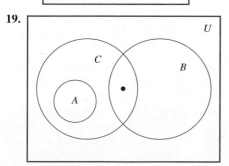

20. a. $A \cup (B \cap C) = \{a, b, c\}$, $(A \cup B) \cap C = \{b, c\}$, and $(A \cup B) \cap (A \cup C) = \{a, b, c, d\} \cap \{a, b, c, e\} = \{a, b, c\}$.

   Hence $A \cup (B \cap C) = (A \cup B) \cap (A \cup C)$.

21. a.

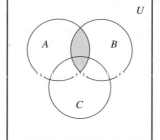

22. a. No. The element d is in two of the sets.

   d. No. None of the sets contains 6.

23. Yes. Every integer is either even or odd, and no integer is both even and odd.

26. a. $A \cap B = \{2\}$, so $\mathscr{P}(A \cap B) = \{\emptyset, \{2\}\}$.

   b. $A = \{1, 2\}$, so $\mathscr{P}(A) = \{\emptyset, \{1\}, \{2\}, \{1, 2\}\}$.

   c. $A \cup B = \{1, 2, 3\}$, so $\mathscr{P}(A \cup B) = \{\emptyset, \{1\}, \{2\}, \{3\}, \{1, 2\}, \{1, 3\}, \{2, 3\}, \{1, 2, 3\}\}$.

   d. $A \times B = \{(1, 2), (1, 3), (2, 2), (2, 3)\}$, so
   $\mathscr{P}(A \times B) = \{\emptyset, \{(1, 2)\}, \{(1, 3)\}, \{(2, 2)\}, \{(2, 3)\},$
   $\{(1, 2), (1, 3)\}, \{(1, 2), (2, 2)\},$
   $\{(1, 2), (2, 3)\}, \{(1, 3), (2, 2)\}, \{(1, 3), (2, 3)\},$
   $\{(2, 2), (2, 3)\}, \{(1, 2), (1, 3), (2, 2)\},$
   $\{(1, 2), (1, 3), (2, 3)\},$
   $\{(1, 2), (2, 2), (2, 3)\}, \{(1, 3), (2, 2), (2, 3)\},$
   $\{(1, 2), (1, 3), (2, 2), (2, 3)\}\}$.

27. a. $\mathscr{P}(A \times B) = \{\emptyset, \{(1, u)\}, \{(1, v)\}, \{(1, u), (1, v)\}\}$

28. b. $\mathscr{P}(\mathscr{P}(\emptyset)) = \mathscr{P}(\{\emptyset\}) = \{\emptyset, \{\emptyset\}\}$

29. a. $A \times B = \{(x, a), (y, a), (z, a), (w, a), (x, b), (y, b), (z, b), (w, b)\}$

30. a. $A \times (B \times C) = \{(1, (u, m)), (2, (u, m)), (3, (u, m)), (1, (u, n)), (2, (u, n)), (3, (u, n)), (1, (v, m)), (2, (v, m)), (3, (v, m)), (1, (v, n)), (2, (v, n)), (3, (v, n))\}$

31.

i	1 →			2 →				3 →		4
j		1	2	3	1	2	3	4	1 →	2
found		no	yes		no		yes		no	yes
answer	$A \subseteq B$									

Section 5.2

1. a. (1) A (2) $B \cup C$

   b. (1) $A \cap B$ (2) C

2. a. (1) $A - B$ (2) A (3) A (4) B

   b. (1) $x \in A$ (2) A (3) B (4) A

3. a. A **b.** C **c.** B **d.** C **e.** $B \subseteq C$

5. *Proof:* Suppose A and B are sets.

$B - A \subseteq B \cap A^c$: Suppose $x \in B - A$. By definition of set difference, $x \in B$ and $x \notin A$. But then by definition of complement, $x \in B$ and $x \in A^c$, and so by definition of intersection, $x \in B \cap A^c$. [*Thus $B - A \subseteq B \cap A^c$ by definition of subset.*]

$B \cap A^c \subseteq B - A$: Suppose $x \in B \cap A^c$. By definition of intersection, $x \in B$ and $x \in A^c$. But then by definition of complement, $x \in B$ and $x \notin A$, and so by definition of set difference, $x \in B - A$. [*Thus $B \cap A^c \subseteq B - A$ by definition of subset.*]

[*Since both set containments have been proved, $B - A = B \cap A^c$ by definition of set equality.*]

6. (1) **a.** $(A \cap B) \cup (A \cap C)$ **b.** A **c.** $B \cup C$
  d. $x \in C$ **e.** $A \cap B$ **f.** by definition of intersection, $x \in A \cap C$, and so by definition of union, $x \in (A \cap B) \cup (A \cap C)$.
  (2) **a.** $x \in A \cap (B \cup C)$ **b.** $A \cap B$ **c.** $A \cap C$
  d. $x \in A$ **e.** $x \in B$ **f.** $A \cap (B \cup C)$ **g.** by definition of intersection, $x \in A$ and $x \in C$. Since $x \in C$, by definition of union, $x \in B \cup C$. Hence $x \in A$ and $x \in B \cup C$, and so, by definition of intersection, $x \in A \cap (B \cup C)$.
  (3) $A \cap (B \cup C) = (A \cap B) \cup (A \cap C)$

7. *Hint:* This is somewhat similar to the proof in Example 5.2.3.

8. *Proof:* Suppose A, B, and C are any sets. To show that $(A - B) \cup (C - B) = (A \cup C) - B$, we must show that $(A - B) \cup (C - B) \subseteq (A \cup C) - B$ and that $(A \cup C) - B \subseteq (A - B) \cup (C - B)$.

$(A - B) \cup (C - B) \subseteq (A \cup C) - B$: Suppose that x is any element in $(A - B) \cup (C - B)$. [*We must show that $x \in (A \cup C) - B$.*] By definition of union, $x \in A - B$ or $x \in C - B$.

Case 1 ($x \in A - B$): Then, by definition of set difference, $x \in A$ and $x \notin B$. But because $x \in A$, we have that $x \in A \cup C$ by definition of union. Hence $x \in A \cup C$ and $x \notin B$, and so, by definition of set difference, $x \in (A \cup C) - B$.

Case 2 ($x \in C - B$): Then, by definition of set difference, $x \in C$ and $x \notin B$. But because $x \in C$, we have that $x \in A \cup C$ by definition of union. Hence $x \in A \cup C$ and $x \notin B$, and so, by definition of set difference, $x \in (A \cup C) - B$.

Thus, in both cases, $x \in (A \cup C) - B$ [*as was to be shown*]. So $(A - B) \cup (C - B) \subseteq (A \cup C) - B$.

$(A \cup C) - B \subseteq (A - B) \cup (C - B)$: Suppose that x is any element in $(A \cup C) - B$. [*We must show that $x \in (A - B) \cup (C - B)$.*] By definition of set difference, $x \in A \cup C$ and $x \notin B$. By definition of union, $x \in A$ or $x \in C$, and in both cases $x \notin B$.

Case 1 ($x \in A$ and $x \notin B$): Then, by definition of set difference $x \in A - B$, and so by definition of union, $x \in (A - B) \cup (C - B)$.

Case 2 ($x \in C$ and $x \notin B$): Then, by definition of set difference $x \in C - B$, and so by definition of union, $x \in (A - B) \cup (C - B)$.

Thus, in both cases, $x \in (A - B) \cup (C - B)$ [*as was to be shown*]. So $(A \cup C) - B \subseteq (A - B) \cup (C - B)$.

Because both subset relations have been proved, we conclude that $(A - B) \cup (C - B) = (A \cup C) - B$.

10. *Partial proof:* Suppose A and B are any sets. We will show that $A \cup (A \cap B) \subseteq A$. Suppose x is any element in $A \cup (A \cap B)$. [*We must show that $x \in A$.*] By definition of union, $x \in A$ or $x \in A \cap B$. In the case where $x \in A$, clearly $x \in A$. In the case where $x \in A \cap B$, $x \in A$ and $x \in B$ (by definition of intersection). Thus, in particular, $x \in A$. Hence, in both cases $x \in A$ [*as was to be shown*].

11. *Proof:* Let A be a set. [*We must show that $A \cup \emptyset = A$.*]

$A \cup \emptyset \subseteq A$: Suppose $x \in A \cup \emptyset$. Then $x \in A$ or $x \in \emptyset$ by definition of union. But $x \notin \emptyset$ since $\emptyset$ has no elements. Hence $x \in A$.

$A \subseteq A \cup \emptyset$: Suppose $x \in A$. Then the statement "$x \in A$ or $x \in \emptyset$" is true. Hence $x \in A \cup \emptyset$ by definition of union. [*Alternatively, $A \subseteq A \cup \emptyset$ by the inclusion in union property.*]

Since $A \cup \emptyset \subseteq A$ and $A \subseteq A \cup \emptyset$, then $A \cup \emptyset = A$ by definition of set equality.

12. *Proof:* Suppose A, B, and C are sets and $A \subseteq B$. Let $x \in A \cap C$. By definition of intersection, $x \in A$ and $x \in C$. But since $A \subseteq B$ and $x \in A$, then $x \in B$. Hence $x \in B$ and $x \in C$, and so, by definition of intersection, $x \in B \cap C$. [*Thus $A \cap C \subseteq B \cap C$ by definition of subset.*]

15. *Hint:* The proof has the following outline:
Suppose A, B, and C are any sets such that $A \subseteq B$ and $A \subseteq C$.
$\vdots$
Therefore, $A \subseteq B \cap C$.

16. *Proof:* Suppose A, B, and C are arbitrarily chosen sets.

$A \times (B \cup C) \subseteq (A \times B) \cup (A \times C)$: Suppose $(x, y) \in A \times (B \cup C)$. [*We must show that $(x, y) \in (A \times B) \cup (A \times C)$.*] Then $x \in A$ and $y \in B \cup C$. By definition of union, this means that $y \in B$ or $y \in C$.

Case 1 ($y \in B$): Then, since $x \in A$, $(x, y) \in A \times B$ by definition of Cartesian product. Hence $(x, y) \in (A \times B) \cup (A \times C)$ by the inclusion in union property.

Case 2 ($y \in C$): Then, since $x \in A$, $(x, y) \in A \times C$ by definition of Cartesian product. Hence $(x, y) \in (A \times B) \cup (A \times C)$ by the inclusion in union property.

Hence, in either case, $(x, y) \in (A \times B) \cup (A \times C)$ [*as was to be shown*].

Thus $A \times (B \cup C) \subseteq (A \times B) \cup (A \times C)$ by definition of subset.

$(A \times B) \cup (A \times C) \subseteq A \times (B \cup C)$: Suppose $(x, y) \in (A \times B) \cup (A \times C)$. Then $(x, y) \in A \times B$ or $(x, y) \in A \times C$.

Case 1 ($(x, y) \in A \times B$): In this case, $x \in A$ and $y \in B$. By definition of union, since $y \in B$, then $y \in B \cup C$. Hence $x \in A$ and $y \in B \cup C$, and so, by definition of Cartesian product, $(x, y) \in A \times (B \cup C)$.

Case 2 ($(x, y) \in A \times C$): In this case, $x \in A$ and $y \in C$. By definition of union, since $y \in C$, then $y \in B \cup C$. Hence $x \in A$ and $y \in B \cup C$, and so, by definition of Cartesian product, $(x, y) \in A \times (B \cup C)$.

Thus, in either case, $(x, y) \in A \times (B \cup C)$. [*Hence, by definition of subset, $(A \times B) \cup (A \times C) \subseteq A \times (B \cup C)$.*]

[*Since both subset relations have been proved, we can conclude that $A \times (B \cup C) = (A \times B) \cup (A \times C)$ by definition of set equality.*]

10. There is more than one error in this "proof." The most serious is the misuse of the definition of subset. To say that A is a subset of B means that for all x, **if** $x \in A$ **then** $x \in B$. It does not mean that there exists an element of A that is also an element of B. The second error in the proof occurs in the last sentence. Just because there is an element in A that is in B and an element in B that is in C, it does not follow that there is an element in A that is in C. For instance, suppose $A = \{1, 2\}$, $B = \{2, 3\}$, and $C = \{3, 4\}$. Then there is an element in A that is in B (namely 2) and there is an element in B that is in C (namely 3), but there is no element in A that is in C.

19. *Hint:* The statement "since $x \notin A$ or $x \notin B$, $x \notin A \cup B$" is fallacious. Try to think of an example of sets A and B and an element x such that the statement "$x \notin A$ or $x \notin B$" is true and the statement "$x \notin A \cup B$" is false.

21. a.

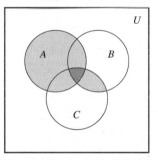

Entire shaded region is $A \cup (B \cap C)$.

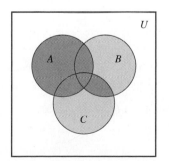

Darkly shaded region is $(A \cup B) \cap (A \cup C)$.

22. a. $(A - B) \cap (B - A)$ **b.** intersection   **c.** $B - A$
d. B **e.** A **f.** A **g.** $(A - B) \cap (B - A) = \emptyset$

23. *Proof:* Suppose $(A - B) \cap (A \cap B) \neq \emptyset$. That is, suppose there were an element x in $(A - B) \cap (A \cap B)$. Then,

by definition of intersection, $x \in A - B$ and $x \in A \cap B$. By definition of set difference, then, $x \in A$ and $x \notin B$, and by definition of intersection, $x \in A$ and $x \in B$. But then $x \in B$ and $x \notin B$, which is a contradiction. [*Hence the supposition is false, and so $(A - B) \cap (A \cap B) = \emptyset$.*]

25. *Proof:* Let A be a subset of a universal set U. Suppose $A \cap A^c \neq \emptyset$, that is, suppose there is an element x such that $x \in A \cap A^c$. Then by definition of intersection, $x \in A$ and $x \in A^c$, and so by definition of complement, $x \in A$ and $x \notin A$. This is a contradiction. [*Hence the supposition is false, and we conclude that $A \cap A^c = \emptyset$.*]

27. *Proof:* Let A be a set. Suppose $A \times \emptyset \neq \emptyset$. Then there would be an element (x, y) in $A \times \emptyset$. By definition of Cartesian product, $x \in A$ and $y \in \emptyset$. But there are no elements y such that $y \in \emptyset$. Hence there are no elements (x, y) such that $x \in A$ and $y \in \emptyset$. Consequently, $(x, y) \notin A \times \emptyset$. [*Thus the supposition is false, and so $A \times \emptyset = \emptyset$.*]

28. *Proof:* Let A and B be sets such that $A \subseteq B$. [*We must show that $A \cap B^c = \emptyset$.*] Suppose $A \cap B^c \neq \emptyset$; that is, suppose there were an element x such that $x \in A \cap B^c$. Then $x \in A$ and $x \in B^c$ by definition of intersection. So $x \in A$ and $x \notin B$ by definition of complement. But $A \subseteq B$ by hypothesis. So since $x \in A$, $x \in B$ by definition of subset. Thus $x \notin B$ and also $x \in B$, which is a contradiction. Hence the supposition that $A \cap B^c \neq \emptyset$ is false, and so $A \cap B^c = \emptyset$.

32. *Proof:* Let A, B, and C be any sets such that $C \subseteq B - A$. Suppose $A \cap C \neq \emptyset$. Then there is an element x such that $x \in A \cap C$. By definition of intersection, $x \in A$ and $x \in C$. Since $C \subseteq B - A$, then $x \in B$ and $x \notin A$. So $x \in A$ and $x \notin A$, which is a contradiction. Hence the supposition is false, and thus $A \cap C = \emptyset$.

35. *Proof (by mathematical induction):*

The formula holds for $n = 1$:

For $n = 1$ the formula is $A_1 \cap B = A_1 \cap B$, which is clearly true.

If the formula holds for $n = k$, then it holds for $n = k + 1$:

Let k be an integer with $k \geq 1$, and suppose the formula holds for $n = k$. We must show that the formula holds for $n = k + 1$; that is, for any sets $A_1, A_2, \ldots A_{k+1}$, and B,

$$(A_1 \cap B) \cup (A_2 \cap B) \cup \cdots \cup (A_{k+1} \cap B)$$
$$= (A_1 \cup A_2 \cup \cdots \cup A_{k+1}) \cap B.$$

But

$$(A_1 \cap B) \cup (A_2 \cap B) \cup \cdots \cup (A_{k+1} \cap B)$$
$$= [(A_1 \cap B) \cup (A_2 \cap B) \cup \cdots (A_k \cap B)] \cup (A_{k+1} \cap B)$$

by assumption

$$= [(A_1 \cup A_2 \cup \cdots \cup A_k) \cap B] \cup (A_{k+1} \cap B)$$

by the inductive hypothesis

$$= [(A_1 \cup A_2 \cup \cdots \cup A_k) \cup A_{k+1}] \cap B$$ by Example 5.2.5

$$= (A_1 \cup A_2 \cup \cdots \cup A_k \cup A_{k+1}) \cap B.$$ by assumption

Section 5.3

1. *Counterexample:* Any sets A, B, and C where C contains elements that are not in A will serve as a counterexample. For instance, let $A = \{1, 3\}$, $B = \{2, 3\}$, and $C = \{4\}$. Then $(A \cap B) \cup C = \{3\} \cup \{4\} = \{3, 4\}$, whereas $A \cap (B \cup C) = \{1, 3\} \cap \{2, 3, 4\} = \{3\}$. Since $\{3, 4\} \neq \{3\}$, $(A \cap B) \cup C \neq A \cap (B \cup C)$.

3. *Counterexample:* Any sets, A, B, and C where $A \subseteq C$ and B contains at least one element that is not in either A or C will serve as a counterexample. For instance, let $A = \{1\}$, $B = \{2\}$, and $C = \{1, 3\}$. Then $A \not\subseteq B$ and $B \not\subseteq C$ but $A \subseteq C$.

5. False. *Counterexample:* Any sets A, B, and C where A and C have elements in common that are not in B will serve as a counterexample. For instance, let $A = \{1, 2, 3\}$, $B = \{2, 3\}$, and $C = \{3\}$. Then $B - C = \{2\}$, and so $A - (B - C) = \{1, 2, 3\} - \{2\} = \{1, 3\}$. On the other hand $A - B = \{1, 2, 3\} - \{2, 3\} = \{1\}$, and so $(A - B) - C = \{1\} - \{3\} = \{1\}$. Since $\{1, 3\} \neq \{1\}$, $A - (B - C) \neq (A - B) - C$.

6. True. *Proof:* Let A and B be any sets.

$A \cap (A \cup B) \subseteq A$: Suppose $x \in A \cap (A \cup B)$. By definition of intersection, $x \in A$ and $x \in A \cup B$. In particular $x \in A$. Thus, by definition of subset, $A \cap (A \cup B) \subseteq A$.

$A \subseteq A \cap (A \cup B)$: Suppose $x \in A$. Then by definition of union, $x \in A \cup B$. Hence $x \in A$ and $x \in A \cup B$, and so, by definition of intersection $x \in A \cap (A \cup B)$. Thus, by definition of subset, $A \subseteq A \cap (A \cup B)$.

Because both $A \cap (A \cup B) \subseteq A$ and $A \subseteq A \cap (A \cup B)$ have been proved, we conclude that $A \cap (A \cup B) = A$.

9. True. *Proof:* Suppose A, B, and C are sets and $A \subseteq C$ and $B \subseteq C$. Let $x \in A \cup B$. By definition of union, $x \in A$ or $x \in B$. But if $x \in A$ then $x \in C$ (because $A \subseteq C$), and if $x \in B$ then $x \in C$ (because $B \subseteq C$). Hence, in either case, $x \in C$. [*So, by definition of subset, $A \cup B \subseteq C$.*]

11. *Hint:* The statement is true. To prove it, suppose A, B, and C are any sets with $A \subseteq C$ and $B \cap C = \emptyset$. Then use proof by contradiction to show that $A \cap C = \emptyset$.

13. True. *Proof:* Suppose A and B are any sets with $A \subseteq B$. [*We must show that $\mathscr{P}(A) \subseteq \mathscr{P}(B)$.*] So suppose $X \in \mathscr{P}(A)$. Then $X \subseteq A$ by definition of power set. But because $A \subseteq B$, we also have that $X \subseteq B$ by the transitive property for subsets, and thus, by definition of power set, $X \in \mathscr{P}(B)$. This proves that for all X, if $X \in \mathscr{P}(A)$ then $X \in \mathscr{P}(B)$, and so $\mathscr{P}(A) \subseteq \mathscr{P}(B)$ [*as was to be shown*].

14. False. *Counterexample:* For any sets A and B, $\mathscr{P}(A) \cup \mathscr{P}(B)$ contains only sets that are subsets of either A or B, whereas the sets in $\mathscr{P}(A \cup B)$ can contain elements of both A and B. Thus, if at least one of A or B contains elements that are not in the other set, $\mathscr{P}(A) \cup \mathscr{P}(B)$ and $\mathscr{P}(A \cup B)$ will not be equal. For instance, let $A = \{1\}$ and $B = \{2\}$. Then $\{1, 2\} \in \mathscr{P}(A \cup B)$ but $\{1, 2\} \notin \mathscr{P}(A) \cup \mathscr{P}(B)$.

15. *Hint:* The statement is true. To prove it, suppose A and B are any sets, and suppose $X \in \mathscr{P}(A) \cup \mathscr{P}(B)$. Show that $X \subseteq A \cup B$, and deduce the conclusion from this result.

18. **a.** *Statement:* $\forall$ sets S, $\exists$ a set T such that $S \cap T = \emptyset$.
Negation: $\exists$ a set S such that $\forall$ sets T, $S \cap T \neq \emptyset$.
The statement is true. Given any set S, take $T = S^c$. Then $S \cap T = S \cap S^c = \emptyset$ by the complement law for $\cap$. Alternatively, T could be taken to be $\emptyset$.

21. **a.** $S_1 = \{\emptyset, \{t\}, \{u\}, \{v\}, \{t, u\}, \{t, v\}, \{u, v\}, \{t, u, v\}\}$
b. $S_2 = \{\{w\}, \{t, w\}, \{u, w\}, \{v, w\}, \{t, u, w\}, \{t, v, w\}, \{u, v, w\}, \{t, u, v, w\}\}$
c. Yes

22. *Hint:* Use mathematical induction. In the inductive step, you will consider the set of all nonempty subsets of $\{2, \ldots, k\}$ and the set of all nonempty subsets of $\{2, \ldots, k + 1\}$. Any subset of $\{2, \ldots, k + 1\}$ either contains $k + 1$ or does not contain $k + 1$. Thus

$$\begin{bmatrix} \text{the sum of all products} \\ \text{of elements of nonempty} \\ \text{subsets of } \{2, \ldots, k + 1\} \end{bmatrix}$$

$$= \begin{bmatrix} \text{the sum of all products} \\ \text{of elements of nonempty} \\ \text{subsets of } \{2, \ldots, k + 1\} \\ \text{that do not contain } k + 1 \end{bmatrix} + \begin{bmatrix} \text{the sum of all products} \\ \text{of elements of nonempty} \\ \text{subsets of } \{2, \ldots, k + 1\} \\ \text{that contain } k + 1 \end{bmatrix}$$

But any subset of $\{2, \ldots, k + 1\}$ that does not contain $k + 1$ is a subset of $\{2, \ldots, k\}$. And any subset of $\{2, \ldots, k + 1\}$ that contains $k + 1$ is the union of a subset of $\{2, \ldots, k\}$ and $\{k + 1\}$.

23. **a.** commutative law for $\cap$
b. distributive law
c. commutative law for $\cap$

24. Partial answer:
a. set difference law
b. set difference law
c. commutative law for $\cap$
d. De Morgan's law

25. *Proof:* Let sets A, B, and C be given. Then

$(A \cap B) \cup C$
$= C \cup (A \cap B)$ by the commutative law for $\cup$
$= (C \cup A) \cap (C \cup B)$ by the distributive law
$= (A \cup C) \cap (B \cup C)$ by the commutative law for $\cup$.

26. *Proof:* Suppose A and B are sets. Then

$A \cup (B - A)$
$= A \cup (B \cap A^c)$ by the set difference law
$= (A \cup B) \cap (A \cup A^c)$ by the distributive law
$= (A \cup B) \cap U$ by the complement law for $\cup$
$= A \cup B$ by the identity law for $\cap$.

29. *Proof:* Let A, B, and C be any sets. Then

$((A^c \cup B^c) - A)^c$

$= ((A^c \cup B^c) \cap A^c)^c$	by the set difference law
$= (A^c \cup B^c)^c \cup (A^c)^c$	by De Morgan's law
$= ((A^c)^c \cap (B^c)^c) \cup (A^c)^c$	by De Morgan's law
$= (A \cap B) \cup A$	by the double complement law
$= A \cup (A \cap B)$	by the commutative law for $\cup$
$= A$	by the absorption law

32. *Partial proof:* Let A and B be any sets. Then

$(A - B) \cup (B - A)$

$= (A \cap B^c) \cup (B \cap A^c)$	by the set difference law
$= [(A \cap B^c) \cup B] \cap [(A \cap B^c) \cup A^c)]$	by the distributive law
$= [B \cup (A \cap B^c)] \cap [A^c \cup (A \cap B^c)]$	by the commutative law for $\cup$
$= [(B \cup A) \cap (B \cup B^c)] \cap [(A^c \cup A) \cap (A^c \cup B^c)]$	by the distributive law
$= [(A \cup B) \cap (B \cup B^c)] \cap [(A \cup A^c) \cap (A^c \cup B^c)]$	by the commutative law for $\cup$

34. *Hint:* The answer is $\emptyset$.

37. a. *Proof:* Suppose not. That is, suppose there exist sets A and B such that $A - B$ and B are not disjoint. [*We must derive a contradiction.*] Then $(A - B) \cap B \neq \emptyset$, and so there is an element x in $(A - B) \cap B$. By definition of intersection, $x \in A - B$ and $x \in B$, and by definition of difference, $x \in A$ and $x \notin B$. Hence $x \in B$ and also $x \notin B$, which is a contradiction. Thus the supposition is false, and we conclude that $A - B$ and B are disjoint.

b. Let A and B be any sets. Then

$(A - B) \cap B$

$= (A \cap B^c) \cap B$	by the set difference law
$= A \cap (B^c \cap B)$	by the associative law for $\cap$
$= A \cap (B \cap B^c)$	by the commutative law for $\cap$
$= A \cap \emptyset$	by the complement law for $\cap$
$= \emptyset$	by the universal bound law for $\cap$.

39. a. $A \triangle B = (A - B) \cup (B - A) = \{1, 2\} \cup \{5, 6\} = \{1, 2, 5, 6\}$

40. *Proof:* Let A and B be any sets. By definition of $\triangle$, showing that $A \triangle B = B \triangle A$ is equivalent to showing that $(A - B) \cup (B - A) = (B - A) \cup (A - B)$. But this follows immediately from the commutative law for $\cup$.

41. *Proof:* Let A be any set. Then

$A \triangle \emptyset$

$= (A - \emptyset) \cup (\emptyset - A)$	by definition of $\triangle$
$= (A \cap \emptyset^c) \cup (\emptyset \cap A^c)$	by the set difference law
$= (A \cap U) \cup (A^c \cap \emptyset)$	by the complement of U law and the commutative law for $\cap$
$= A \cup \emptyset$	by the identity law for $\cap$ and the universal bound law for $\cap$
$= A.$	by the identity law for $\cup$

44. *Hint:* First show that for any sets A and B and for any element x,

$$x \in A \triangle B \Leftrightarrow (x \in A \text{ and } x \notin B) \text{ or } (x \in B \text{ and } x \notin A),$$

and

$$x \notin A \triangle B \Leftrightarrow (x \notin A \text{ and } x \notin B) \text{ or } (x \in B \text{ and } x \in A).$$

45. Same hint as for exercise 44.

48. a. because 1 is an identity for $\cdot$
b. by the complement law for $+$
c. by the distributive law for $+$ over $\cdot$
d. by the complement law for $\cdot$
e. because 0 is an identity for $+$

51. *Proof:* For all elements a in B,

$a \cdot 0 = a \cdot (a \cdot \overline{a})$	by the complement law for $\cdot$
$= (a \cdot a) \cdot \overline{a}$	by the associative law for $\cdot$
$= a \cdot \overline{a}$	by exercise 48
$= 0.$	by the complement law for $\cdot$

53. *Proof:* $0 \cdot 1 = 0$ because 1 is an identity for $\cdot$, and $0 + 1 = 1 + 0 = 1$ because $+$ is commutative and 0 is an identity for $+$. Thus, by the uniqueness of the complement law, $\overline{0} = 1$.

55. *Proof:* Suppose B is a Boolean algebra and a and b are any elements of B. We first prove that $(a \cdot b) + (\overline{a} + \overline{b}) = 1$.

$a \cdot b + (\overline{a} + \overline{b})$

$= (\overline{a} + \overline{b}) + (a \cdot b)$	by the commutative law for $+$
$= ((\overline{a} + \overline{b}) + a) \cdot ((\overline{a} + \overline{b}) + b)$	by the distributive law of $+$ over $\cdot$
$= ((\overline{b} + \overline{a}) + a) \cdot (\overline{a} + (\overline{b} + b))$	by the commutative and associative laws for $+$
$= (\overline{b} + (\overline{a} + a)) \cdot (\overline{a} + (b + \overline{b}))$	by the associative and commutative laws for $+$
$= (\overline{b} + (a + \overline{a})) \cdot (\overline{a} + 1)$	by the commutative and complement laws for $+$
$= (\overline{b} + 1) \cdot 1$	by the complement and universal bound laws for $+$
$= 1 \cdot 1$	by the universal bound law for $+$
$= 1$	by the identity law for $\cdot$.

Next we prove that $(a \cdot b) \cdot (\overline{a} + \overline{b}) = 0$.

$(a \cdot b) \cdot (\overline{a} + \overline{b})$

$\quad = ((a \cdot b) \cdot \overline{a}) + (((a \cdot b) \cdot \overline{b})$

$\qquad$ by the distributive law of $\cdot$ over $+$

$\quad = ((b \cdot a) \cdot \overline{a}) + ((a \cdot (b \cdot \overline{b}))$

$\qquad$ by the commutative and associative laws for $\cdot$

$\quad = (b \cdot (a \cdot \overline{a})) + (a \cdot 0)$

$\qquad$ by the associative and complement laws for $\cdot$

$\quad = (b \cdot 0) + 0$

$\qquad$ by the complement and universal bound laws for $\cdot$

$\quad = 0 + 0 \quad$ by the universal bound law for $\cdot$

$\quad = 0 \qquad$ by the identity law for $+$.

Because both $(a \cdot b) + (\overline{a} + \overline{b}) = 1$ and $(a \cdot b) \cdot (\overline{a} + \overline{b}) = 0$, it follows, by the uniqueness of the complement law, that $\overline{a \cdot b} = \overline{a} + \overline{b}$.

57. *Hint:* One way to prove the statement is to use the result of exercise 50. Some stages in the proof are the following:

$$y = (y + x) \cdot y = (x \cdot y) + (z \cdot y) = z \cdot (x + y) = z.$$

58. a. (i) Because S has only two distinct elements, 0 and 1, we only need to check that $0 + 1 = 1 + 0$. But this is true because both sums equal 1.

(v) *Partial answer:*

$0 + (0 \cdot 0) = 0 + 0 = 0$ and $(0 + 0) \cdot (0 + 0) = 0 \cdot 0 = 0$ also

$0 + (0 \cdot 1) = 0 + 0 = 0$ and $(0 + 0) \cdot (0 + 1) = 0 \cdot 1 = 0$ also

$0 + (1 \cdot 0) = 0 + 0 = 0$ and $(0 + 1) \cdot (0 + 0) = 1 \cdot 0 = 0$ also

$0 + (1 \cdot 1) = 0 + 1 = 1$ and $(0 + 1) \cdot (0 + 1) = 1 \cdot 1 = 1$ also

b. *Hint:* Verify that $0 + x = x$ and that $1 \cdot x = x$ for all $x \in S$.

59. *Hints:* (1) Because the proofs of the absorption laws do not use the associative laws, the absorption laws may be used at any stage of the derivation.
(2) Show that for all x, y, and z in B, $x(x + (y + z)) = x$ and $x((x + y) + z)) = x$.
(3) Show that for all a, b, and c in B, both $a + (b + c)$ and $(a + b) + c$ equal $((a + b) + c)(a + (b + c))$.
(4) Use De Morgan's laws and the double complement law to deduce the associative law for $\cdot$.

Section 5.4

1. The sentence is not a statement because it is neither true nor false. If the sentence were true, then because it declares itself to be false, the sentence would be false. Therefore, the sentence is not true. On the other hand, if the sentence were false, then it would be false that "This sentence is false," and so the sentence would be true. Consequently, the sentence is not false.

2. This sentence is a statement because it is true. Recall that the only way for an if-then statement to be false is for the hypothesis to be true and the conclusion false. In this case the hypothesis is not true. So regardless of what the con-

clusion states, the sentence is true. (This is an example of a statement that is vacuously true, or true by default.)

5. This sentence is not a statement because it is neither true nor false. If the sentence were true, then either the sentence is false or $1 + 1 = 3$. But $1 + 1 \neq 3$, and so the sentence is false. Therefore, the sentence is not true. On the other hand, if the sentence were false, then it would be true that "This sentence is false or $1 + 1 = 3$," and so the sentence would be true. Consequently, the sentence is not false.

8. *Hint:* Suppose that apart from statement (ii), all of Nixon's other assertions about Watergate are evenly split between true and false.

9. No. Suppose there were a computer program P that had as output a list of all computer programs that do not list themselves in their output. If P lists itself as output, then it would be on the output list of P, which consists of all computer programs that do not list themselves in their output. Hence P would not list itself as output. But if P does not list itself as output, then P would be a member of the list of all computer programs that do not list themselves in their output, and this list is exactly the output of P. Hence P would list itself as output. This analysis shows that the assumption of the existence of such a program P is contradictory, and so no such program exists.

13. *Hint:* Show that any algorithm that solves the printing problem can be adapted to produce an algorithm that solves the halting problem.

Section 6.1

2. $3/4$, $1/2$, $1/2$

3. $\{1\blacklozenge, 2\blacklozenge, 3\blacklozenge, 4\blacklozenge, 5\blacklozenge, 6\blacklozenge, 7\blacklozenge, 8\blacklozenge, 9\blacklozenge, 10\blacklozenge, 1\heartsuit, 2\heartsuit, 3\heartsuit, 4\heartsuit, 5\heartsuit, 6\heartsuit, 7\heartsuit, 8\heartsuit, 9\heartsuit, 10\heartsuit\}$, probability $= 20/52 \cong 38.5\%$

5. $\{10\clubsuit, J\clubsuit, Q\clubsuit, K\clubsuit, A\clubsuit, 10\blacklozenge, J\blacklozenge, Q\blacklozenge, K\blacklozenge, A\blacklozenge, 10\heartsuit, J\heartsuit, Q\heartsuit, K\heartsuit, A\heartsuit, 10\spadesuit, J\spadesuit, Q\spadesuit, K\spadesuit, K\spadesuit, A\spadesuit\}$ probability $= 20/52 = 5/13 \cong 38.5\%$.

7. $\{26, 35, 44, 53, 62\}$, probability $= 5/36 \cong 13.9\%$

9. $\{11, 12, 13, 14, 15, 21, 22, 23, 24, 31, 32, 33, 41, 42, 51\}$ probability $= 15/36 = 41\frac{2}{3}\%$

11. a. $\{HHH, HHT, HTH, HTT, THH, THT, TTH, TTT\}$
b. (i) $\{HTT, THT, TTH\}$, probability $= 3/8 \cong 37.5\%$

12. a. $\{BBB, BBG, BGB, BGG, GBB, GBG, GGB, GGG\}$
b. (i) $\{GBB, BGB, BBG\}$ probability $= 3/8 = 37.5\%$

13. a. $\{CCC, CCW, CWC, CWW, WCC, WCW, WWC, WWW\}$
b. (i) $\{CWW, WCW, WWC\}$, probability $= 3/8 = 37.5\%$

14. a. probability $= 3/8 = 37.5\%$

16. a. $\{RRR, RRB, RRY, RBR, RBB, RBY, RYR, RYB, RYY, BRR, BRB, BRY, BBR, BBB, BBY, BYR, BYB, BYY, YRR, YRB, YRY, YBR, YBB, YBY, YYR, YYB, YYY\}$

b. $\{RBY, RYB, YBR, BRY, BYR, YRB\}$, probability $= 6/27 = 2/9 \cong 22.2\%$

c. $\{RRB, RBR, BRR, RRY, RYR, YRR, BBR, BRB, RBB, BBY, BYB, YBB, YYR, YRY, RYY, YYB, YBY, BYY\}$ probability $= 18/27 = 2/3 = 66\frac{2}{3}\%$

18. a. $\{B_1B_1, B_1B_2, B_1W, B_2B_1, B_2B_2, B_2W, WB_1, WB_2, WW\}$

b. $\{B_1B_1, B_1B_2, B_2B_1, B_2B_2\}$ probability $= \frac{4}{9} \cong 44.4\%$

c. $\{B_1W, B_2W, WB_1, WB_2\}$ probability $= \frac{4}{9} \cong 44.4\%$

21. a. 10 11 12 13 14 15 16 17 18 ... 96 97 98 99

$\updownarrow$ $\quad$ $\updownarrow$ $\quad$ $\updownarrow$ $\quad$ $\updownarrow$ $\quad$ $\updownarrow$

$3 \cdot 4$ $\quad$ $3 \cdot 5$ $\quad$ $3 \cdot 6$ $\;$ $3 \cdot 32$ $\quad$ $3 \cdot 33$

The above diagram shows that there are as many positive two-digit integers that are multiples of 3 as there are integers from 4 to 33 inclusive. By Theorem 6.1.1, there are $33 - 4 + 1$, or 30, such integers.

b. There are $99 - 10 + 1 = 90$ positive two-digit integers in all, and by part (a), 30 of these are multiples of 3. So the probability that a randomly chosen positive two-digit integer is a multiple of 3 is $30/90 = 1/3 = 33\frac{1}{3}\%$.

23. c. Probability $= \frac{m-3+1}{n} = \frac{m-2}{n}$

d. Because $\lfloor \frac{39}{2} \rfloor = 19$, the probability is $\frac{39-19+1}{39} = \frac{21}{39}$.

32. a. M Tu W Th F Sa Su M Tu W Th F Sa Su $\cdots$ F Sa Su M

$\;$ 1 $\;$ 2 $\;$ 3 $\;$ 4 $\;$ 5 $\;$ 6 $\;$ 7 $\;$ 8 $\;$ 9 $\;$ 10 $\;$ 11 $\;$ 12 $\;$ 13 $\;$ 14 $\quad$ 362 363 364 365

$\updownarrow$ $\qquad\qquad\qquad$ $\updownarrow$ $\qquad\qquad$ $\updownarrow$

$7 \cdot 1$ $\qquad\qquad\qquad$ $7 \cdot 2$ $\qquad\qquad$ $7 \cdot 52$

Sundays occur on the 7th day of the year, the 14th day of the year, and in fact on all days that are multiples of 7. There are 52 multiples of 7 between 1 and 365, and so there are 52 Sundays in the year.

Section 6.2

1.

There are five ways to complete the series:
A, B–A, B–B–A, B–B–B–A, and B–B–B–B.

3. Four ways: A–A–A–A, B–A–A–A–A, B–B–A–A–A–A, and B–B–B–A–A–A–A.

4. Two ways: A–B–A–B–A–B–A and B–A–B–A–B–A–B

24. a. (i) If n is even, there are $\lfloor \frac{n}{2} \rfloor = \frac{n}{2}$ elements in the sub-array.

(ii) If n is odd, there are $\lfloor \frac{n}{2} \rfloor = \frac{n-1}{2}$ elements in the sub-array.

b. There are n elements in the array, so

(i) The probability that an element is in the given sub-array when n is even is $\frac{\frac{n}{2}}{n} = \frac{1}{2}$,

(ii) The probability that an element is in the given sub-array when n is odd is $\frac{\frac{n-1}{2}}{n} = \frac{n-1}{2n}$.

26. Let k be the 27th element in the array. By Theorem 6.1.1, $k - 42 + 1 = 27$, and so $k = 42 + 27 - 1 = 68$. Thus the 27th element in the array is $A[68]$.

28. Let m be the smallest of the integers. By Theorem 6.1.1, $279 - m + 1 = 56$, and so $m = 279 - 56 + 1 = 224$. Thus the smallest of the integers is 224.

31. 1 2 3 4 5 6 7 8 9 ... 999 1000 1001

$\updownarrow$ $\quad$ $\updownarrow$ $\quad$ $\updownarrow$ $\qquad$ $\updownarrow$

$3 \cdot 1$ $\;$ $3 \cdot 2$ $\;$ $3 \cdot 3$ $\;$ $3 \cdot 333$

Thus there are 333 multiples of 3 between 1 and 1001.

6. a.

b. There are 12 equally likely outcomes of the experiment.

c. $2/12 = 1/6 = 16\frac{2}{3}\%$ **d.** $8/12 = 2/3 = 66\frac{2}{3}\%$

8. By the multiplication rule, the answer is $3 \cdot 2 \cdot 2 = 12$.

9. a. In going from city A to city B, one may take any of the 3 roads. In going from city B to city C, one may take any of the 5 roads. So, by the multiplication rule, there are $3 \cdot 5 = 15$ ways to travel from city A to city C via city B.

b. A round-trip journey can be thought of as a four-step operation:

Step 1: Go from A to B.

Step 3: Go from B to C.

Step 2: Go from C to B.

Step 4: Go from B to A.

Since there are 3 ways to perform step 1, 5 ways to perform step 2, 5 ways to perform step 3, and 3 ways to perform step 4, by the multiplication rule, there are $3 \cdot 5 \cdot 5 \cdot 3 = 225$ round-trip routes.

c. In this case the steps for making a round-trip journey are the same as in part (b), but since no route segment may be repeated, there are only 4 ways to perform step 3 and only 2 ways to perform step 4. So, by the multiplication rule, there are $3 \cdot 5 \cdot 4 \cdot 2 = 120$ round-trip routes in which no road is traversed twice.

11. a. Imagine constructing a bit string of length 8 as an eight-step process:

Step 1: Choose either a 0 or a 1 for the left-most position,

Step 2: Choose either a 0 or a 1 for the next position to the right.

Step 3: Choose either a 0 or a 1 for the next position to the right.

Since there are 2 ways to perform each step, the total number of ways to accomplish the entire operation, which is the number of different bit strings of length 8, is $2 \cdot 2 \cdot 2 \cdot 2 \cdot 2 \cdot 2 \cdot 2 \cdot 2 = 2^8 = 256$.

b. Imagine placing a 1 in the left-most position of an 8-bit string, and then imagine filling in the remaining seven positions as an operation with seven steps, where step i is to fill in the $(i+1)$st position. Since there are 2 ways to perform each of seven steps, there are 2^7 ways to perform the entire operation. So there are 2^7, or 128, 8-bit strings that begin with a 1.

12. a. There are 9 hexadecimal digits from 3 through B and 11 hexadecimal digits from 5 through F. Thus the answer is $8 \cdot 16 \cdot 16 \cdot 16 \cdot 11 = 405,504$.

13. a. In each of the four tosses there are two possible results: Either a head (H) or a tail (T) is obtained. Thus, by the multiplication rule, the number of outcomes is $2 \cdot 2 \cdot 2 \cdot 2 = 2^4 = 16$.

b. There are six outcomes with two heads: $HHTT, HTHT, HTTH, THHT, THTH, TTHH$. Thus the probability of obtaining exactly two heads is $6/16 = 3/8$.

14. a. Let each of steps 1–3 be to choose a letter of the alphabet to put in positions 1–3, and let each of steps 4–6 be to choose a digit to put in positions 4–6. Since there are 26

letters and 10 digits (0–9), the number of license plates is

$$26 \cdot 26 \cdot 26 \cdot 10 \cdot 10 \cdot 10 = 17,576,000.$$

b. In this case there is only one way to perform step 1 (because the first letter must be an A) and only one way to perform step 6 (because the last digit must be a 0). Therefore, the number of license plates is $26 \cdot 26 \cdot 10 \cdot 10 = 67,600$.

d. In this case there are 26 ways to perform step 1, 25 ways to perform step 2, 24 ways to perform step 3, 10 ways to perform step 4, 9 ways to perform step 5, and 8 ways to perform step 6, so the number of license plates is $26 \cdot 25 \cdot 24 \cdot 10 \cdot 9 \cdot 8 = 11,232,000$.

16. a. Let step 1 be to choose either the number 2 or one of the letters corresponding to the number 2 on the keypad, let step 2 be to choose either the number 1 or one of the letters corresponding to the number 1 on the keypad, and let steps 3 and 4 be to choose either the number 3 or one of the letters corresponding to the number 3 on the keypad. There are 4 ways to perform step 1, 3 ways to perform step 2, and 4 ways to perform each of steps 3 and 4. So by the multiplication rule, there are $4 \cdot 3 \cdot 4 \cdot 4 = 192$ ways to perform the entire operation. Thus there are 192 different PINs that are keyed the same as 2133. Note that on a computer keyboard, these PINs would not be keyed the same way.

17.

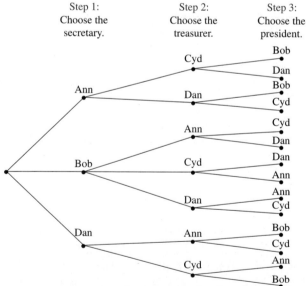

There are 14 different paths from "root" to "leaf" of this possibility tree, and so there are 14 ways the officers can be chosen. Because $14 = 2 \cdot 7$, reordering the steps will not make it possible to use the multiplication rule alone to solve this problem.

18. a. The number of ways to perform step 4 is not constant; it depends on how the previous steps were performed. For instance, if 3 digits had been chosen in steps 1–3, then there would be $10 - 3 = 7$ ways to perform step 4, but if

3 letters had been chosen in steps 1–3, then there would be 10 ways to perform step 4.

19. a. Two solutions:

(i) number of integers

$$= \begin{bmatrix} \text{number of} \\ \text{ways to pick} \\ \text{first digit} \end{bmatrix} \begin{bmatrix} \text{number of} \\ \text{ways to pick} \\ \text{second digit} \end{bmatrix} = 9 \cdot 10 = 90$$

(ii) Using Theorem 6.1.1, number of integers $=$
$99 - 10 + 1 = 90$

b. Odd integers end in 1, 3, 5, 7, or 9.
number of odd integers

$$= \begin{bmatrix} \text{number of} \\ \text{ways to pick} \\ \text{first digit} \end{bmatrix} \begin{bmatrix} \text{number of} \\ \text{ways to pick} \\ \text{second digit} \end{bmatrix} = 9 \cdot 5 = 45$$

Alternative solution: Use the listing method shown in the solution for Example 6.1.4.

c. $\begin{bmatrix} \text{number of integers} \\ \text{with distinct digits} \end{bmatrix}$

$$= \begin{bmatrix} \text{number of} \\ \text{ways to pick} \\ \text{first digit} \end{bmatrix} \begin{bmatrix} \text{number of} \\ \text{ways to pick} \\ \text{second digit} \end{bmatrix}$$

$$= 9 \cdot 9 = 81$$

d. $\begin{bmatrix} \text{number of odd integers} \\ \text{with distinct digits} \end{bmatrix}$

$$= \begin{bmatrix} \text{number of} \\ \text{ways to pick} \\ \text{second digit} \end{bmatrix} \begin{bmatrix} \text{number of} \\ \text{ways to pick} \\ \text{first digit} \end{bmatrix}$$

$$= 5 \cdot 8 = 40 \quad \text{because the first digit} \\ \text{can't equal 0, nor can it} \\ \text{equal the second digit}$$

e. $81/90 = 9/10$, $40/90 = 4/9$

21. The outer loop is iterated 30 times, and during each iteration of the outer loop there are 15 iterations of the inner loop. Hence, by the multiplication rule, the total number of iterations of the inner loop is $30 \cdot 15 = 450$.

24. The outer loop is iterated $50 - 5 + 1 = 46$ times, and during each iteration of the outer loop there are $20 - 10 + 1 = 11$ iterations of the inner loop. Hence, by the multiplication rule, the total number of iterations of the inner loop is $46 \cdot 11 = 506$.

26. *Hints:* One solution is to add leading zeros as needed to make each number five digits long. For instance, write 1 as 00001. Let some of the steps be to choose positions for the given digits. The answer is 720. Another solution is to consider separately the cases of four-digit and five-digit numbers.

28. a. There are $a + 1$ divisors: $1, p, p^2, \ldots, p^a$.

b. A divisor is a product of any one of the $a + 1$ numbers listed in part (a) times any one of the $b + 1$ numbers $1, q, q^2, \ldots, q^b$. So, by the multiplication rule, there are $(a + 1)(b + 1)$ divisors in all.

29. a. Since the nine letters of the word *ALGORITHM* are all distinct, there are as many arrangements of these letters in a row as there are permutations of a set with nine elements: $9! = 362,880$.

b. In this case there are effectively eight symbols to be permuted (because $\boxed{AL}$ may be regarded as a single symbol). So the number of arrangements is $8! = 40,320$.

31. The same reasoning as in Example 6.2.9 gives an answer of $4! = 24$.

32. $WX, WY, WZ, XW, XY, XZ, YW, YX, YZ, ZW, ZX, ZY$

34. a. $P(6, 4) = \dfrac{6!}{(6 - 4)!} = \dfrac{6 \cdot 5 \cdot 4 \cdot 3 \cdot \cancel{2 \cdot 1}}{\cancel{2 \cdot 1}} = 360$

35. a. $P(5, 3) = \dfrac{5 \cdot 4 \cdot 3 \cdot \cancel{2!}}{\cancel{2!}} = 60$

36. a. $P(9, 3) = \dfrac{9 \cdot 8 \cdot 7 \cdot \cancel{6!}}{\cancel{6!}} = 504$

c. $P(8, 5) = \dfrac{8 \cdot 7 \cdot 6 \cdot 5 \cdot 4 \cdot \cancel{3!}}{\cancel{3!}} = 6,720$

38. *Proof:* Let n be an integer and $n \geq 2$. Then

$$P(n + 1, 2) - P(n, 2)$$

$$= \frac{(n + 1)!}{[(n + 1) - 2]!} - \frac{n!}{(n - 2)!} = \frac{(n + 1)!}{(n - 1)!} - \frac{n!}{(n - 2)!}$$

$$= \frac{(n + 1) \cdot n \cdot \cancel{(n - 1)!}}{\cancel{(n - 1)!}} - \frac{n \cdot (n - 1) \cdot \cancel{(n - 2)!}}{\cancel{(n - 2)!}}$$

$$= n^2 + n - (n^2 - n) = 2n = 2 \cdot \frac{n \cdot (n - 1)!}{(n - 1)!}$$

$$= 2 \cdot \frac{n!}{(n - 1)!} = 2P(n, 1)$$

This is what was to be proved.

42. *Hint:* In the inductive step, suppose there exist $k!$ permutations of a set with k elements. Let X be a set with $k + 1$ elements. The process of forming a permutation of the elements of X can be considered a two-step operation where step 1 is to choose the element to write first. Step 2 is to write the remaining elements of X in some order.

Section 6.3

1. a. Set of Bit Strings Consisting of from 1 through 4 Bits

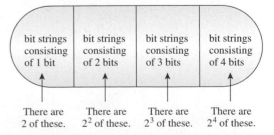

Applying the addition rule to the figure above shows that there are $2 + 2^2 + 2^3 + 2^4 = 30$ bit strings consisting of from one through four bits.

b. By reasoning similar to that of part (a), there are $2^5 + 2^6 + 2^7 + 2^8 = 480$ bit strings of from five through eight bits.

3. a. $\begin{bmatrix} \text{number of integers from 1 through 999} \\ \text{with no repeated digits} \end{bmatrix}$

$= \begin{bmatrix} \text{number of integers} \\ \text{from 1 through 9} \\ \text{with no repeated digits} \end{bmatrix} + \begin{bmatrix} \text{number of integers} \\ \text{from 10 through 99} \\ \text{with no repeated digits} \end{bmatrix}$

$+ \begin{bmatrix} \text{number of integers from} \\ \text{100 through 999 with} \\ \text{no repeated digits} \end{bmatrix}$

$= 9 + 9 \cdot 9 + 9 \cdot 9 \cdot 8 = 738$

b. $\begin{bmatrix} \text{number of integers from 1 through 999} \\ \text{with at least one repeated digit} \end{bmatrix}$

$= \begin{bmatrix} \text{total number of} \\ \text{integers from} \\ \text{1 through 999} \end{bmatrix} - \begin{bmatrix} \text{number of integers} \\ \text{from 1 through 999} \\ \text{with no repeated digits} \end{bmatrix}$

$= 999 - 738 = 261$

The probability that an integer chosen at random has at least one repeated digit is $261/999 \cong 26.1\%$.

4.

Set of Arrangements (without repetition)
or No More Than 3 Letters of NETWORK

| arrangements of no more than 1 letter | arrangements of no more than 2 letters | arrangements of no more than 3 letters |

There are 7 of these.   There are $7 \cdot 6$ of these.   There are $7 \cdot 6 \cdot 5$ of these.

Applying the addition rule to the figure above shows that there are $7 + 7 \cdot 6 + 7 \cdot 6 \cdot 5 = 259$ arrangements of three letters of the word *NETWORK* if repetition of letters is not permitted.

6. a. There are $1 + 26 + 26^2 + 26^3$ arrangements of from 0 through 3 letters of the alphabet. Any of these may be paired with all but one arrangement of from 0 through 4 digits, and there are $1 + 10 + 10^2 + 10^3 + 10^4$ arrangements of from 0 through 4 digits. So, by the multiplication rule and the difference rule, the number of license plates is

$(1 + 26 + 26^2 + 26^3)$
$\cdot (1 + 10 + 10^2 + 10^3 + 10^4) - 1 = 203{,}097{,}968$
    ↑
    the blank plate

b. $(1 + 26 + 26^2 + 26^3 - 85)$
$\cdot (1 + 10 + 10^2 + 10^3 + 10^4) - 1 = 202{,}153{,}533$

8. a. Each column of the table below corresponds to a pair of values of i and j for which the inner loop will be iterated.

i	1	2 →		3		→	4			→
j	1	1	2	1	2	3	1	2	3	4
	1	2		3			4			

Since there are $1 + 2 + 3 + 4 = 10$ columns, the inner loop will be iterated ten times.

9. a. The answer is the number of permutations of the five letters in *QUICK*, which equals $5! = 120$.

b. Because *QU* (in order) is to be considered as a single unit, the answer is the number of permutations of the four symbols $\boxed{QU}$, I, C, K. This is $4! = 24$.

c. By part (b), there are 4! arrangements of $\boxed{QU}$, I, C, K. Similarly, there are 4! arrangements of $\boxed{UQ}$, I, C, K. Therefore, by the addition rule, there are $4! + 4! = 48$ arrangements in all.

11. a. $\begin{bmatrix} \text{number of ways to place eight people} \\ \text{in a row keeping } A \text{ and } B \text{ together} \end{bmatrix}$

$= \begin{bmatrix} \text{number of ways to arrange} \\ \boxed{AB} \; CDEFGH \end{bmatrix}$

$+ \begin{bmatrix} \text{number of ways to arrange} \\ \boxed{BA} \; CDEFGH \end{bmatrix}$

$= 7! + 7! = 5{,}040 + 5{,}040 = 10{,}080$

b. $\begin{bmatrix} \text{number of ways to arrange the eight} \\ \text{people in a row keeping } A \text{ and } B \text{ apart} \end{bmatrix}$

$= \begin{bmatrix} \text{total number of ways} \\ \text{to place the eight} \\ \text{people in a row} \end{bmatrix} - \begin{bmatrix} \text{number of ways} \\ \text{to place the eight} \\ \text{people in a row} \\ \text{keeping } A \text{ and } B \\ \text{together} \end{bmatrix}$

$= 8! - 10{,}080 = 40{,}320 - 10{,}080$
$= 30{,}240$

12. number of variable names

$= \begin{bmatrix} \text{number of numeric} \\ \text{variable names} \end{bmatrix} + \begin{bmatrix} \text{number of string} \\ \text{variable names} \end{bmatrix}$

$= (26 + 26 \cdot 36) + (26 + 26 \cdot 36) = 1{,}924$

13. *Hint:* In exercise 12 note that

$$26 + 26 \cdot 36 = 26 \sum_{k=0}^{1} \cdot 36^k.$$

Generalize this idea here. Use Theorem 4.2.3 to evaluate the expression you obtain.

14. a. $10 \cdot 9 \cdot 8 \cdot 7 \cdot 6 \cdot 5 \cdot 4 = 604{,}800$

b. $\begin{bmatrix} \text{number of phone numbers with} \\ \text{at least one repeated digit} \end{bmatrix}$

$= \begin{bmatrix} \text{total number of} \\ \text{phone numbers} \end{bmatrix} - \begin{bmatrix} \text{number of phone numbers} \\ \text{with no repeated digits} \end{bmatrix}$

$= 10^7 - 604{,}800 = 9{,}395{,}200$

c. $9{,}395{,}200/10^7 \cong 93.95\%$

16. a. *Proof:* Let A and B be mutually disjoint events in a sample space S. By the addition rule, $N(A \cup B) = N(A) + N(B)$. Therefore, by the equally likely probability formula,

$$P(A \cup B) = \frac{N(A \cup B)}{N(S)} = \frac{N(A) + N(B)}{N(S)}$$
$$= \frac{N(A)}{N(S)} + \frac{N(B)}{N(S)} = P(A) + P(B).$$

17. *Hint:* Justify the following answer: $39 \cdot 38 \cdot 38$.

18. a. Identify the integers from 1 to 100,000 that contain the digit 6 exactly once with strings of five digits. Thus, for example, 306 would be identified with 00306. It is not necessary to use strings of six digits, because 100,000 does not contain the digit 6. Imagine the process of constructing a five-digit string that contains the digit 6 exactly once as a five-step operation that consists of filling in the five digit positions $\underset{1}{\underline{\quad}}\ \underset{2}{\underline{\quad}}\ \underset{3}{\underline{\quad}}\ \underset{4}{\underline{\quad}}\ \underset{5}{\underline{\quad}}$.

Step 1: Choose one of the five positions for the 6.

Step 2: Choose a digit for the left-most remaining position.

Step 3: Choose a digit for the next remaining position to the right.

Step 4: Choose a digit for the next remaining position to the right.

Step 5: Choose a digit for the right-most position.

Since there are 5 choices for step 1 (any one of the five positions) and 9 choices for each of steps 2–5 (any digit except 6), by the multiplication rule, the number of ways to perform this operation is $5 \cdot 9 \cdot 9 \cdot 9 \cdot 9 = 32,805$. Hence there are 32,805 integers from 1 to 100,000 that contain the digit 6 exactly once.

19. *Hint:* The answer is 2/3.

21. a. Let A = the set of integers that are multiples of 4 and B = the set of integers that are multiples of 7. Then $A \cap B$ = the set of integers that are multiples of 28.

But $n(A) = 250$ since $1\ 2\ 3\ 4\ 5\ 6\ 7\ 8 \ldots 999\ 1000$,
$$4 \cdot 1 \quad 4 \cdot 2 \ldots \quad 4 \cdot 250$$
or, equivalently, since $1{,}000 = 4 \cdot 250$.

Also $n(B) = 142$ since $1\ 2\ 3\ 4\ 5\ 6\ 7 \ldots 14 \ldots \quad 994\ 995 \ldots 1000$,
$$7 \cdot 1\ \ 7 \cdot 2 \ldots \ 7 \cdot 142$$
or, equivalently, since $1{,}000 = 7 \cdot 142 + 6$.

and $n(A \cap B) = 35$ since $1\ 2\ 3 \ldots 28 \ldots 56 \ldots \quad 980 \ldots 1000$,
$$28 \cdot 1\ \ 28 \cdot 2 \ldots\ 28 \cdot 35$$
or, equivalently, since $1{,}000 = 28 \cdot 35 + 20$.

So $n(A \cup B) = 250 + 142 - 35 = 357$.

23. a. $11001010_2 = 2 + 2^3 + 2^6 + 2^7 = 202$,
$00111000_2 = 2^3 + 2^4 + 2^5 = 56$,
$01101011_2 = 1 + 2 + 2^3 + 2^5 + 2^6 = 107$,
$11101110_2 = 2 + 2^2 + 2^3 + 2^5 + 2^6 + 2^7 = 238$
So the answer is 202.56.107.238.

b. The network ID for a Class A network consists of 8 bits and begins with 0. If all possible combinations of eight 0's and 1's that start with a 0 were allowed, there would be 2 choices (0 or 1) for each of the 7 positions from the second through the eighth. This would give $2^7 = 128$ possible ID's. But because neither 00000000 nor 01111111 is allowed, the total is reduced by 2, so there are 126 possible Class A networks.

c. Let $w.x.y.z$ be the dotted decimal form of the IP address for a computer in a Class A network. Because the network IDs for a Class A network go from 00000001 ($= 1$) through 01111110 ($= 126$), w can be any integer from 1 through 126. In addition, each of x, y, and z can be any integer from 0 ($= 00000000$) through 255 ($= 11111111$), except that x, y, and z cannot all be 0 simultaneously and cannot all be 255 simultaneously.

d. Twenty-four positions are allocated for the host ID in a Class A network. If each could be either 0 or 1, there would be $2^{24} = 16{,}777{,}216$ possible host IDs. But neither all 0's nor all 1's is allowed, which reduces the total by 2. Thus there are 16,777,214 possible host IDs in a Class A network.

i. Observe that $140 = 128 + 8 + 4 = 10001100_2$, which begins with 10. Thus the IP address comes from a Class B network. An alternative solution uses the result of Example 6.3.5: Network IDs for Class B networks range from 128 through 191. Thus, since $128 \leq 140 \leq 191$, the given IP address is from a Class B network.

24. a. There are 12 possible birth months for A, 12 for B, 12 for C, and 12 for D, so the total is $12^4 = 20{,}736$.

b. If no two people share the same birth month, there are 12 possible birth months for A, 11 for B, 10 for C, and 9 for D. Thus the total is $12 \cdot 11 \cdot 10 \cdot 9 = 11{,}880$.

c. If at least two people share the same birth month, the total number of ways birth months could be associated with A, B, C, and D is $20{,}736 - 11{,}880 = 8{,}856$.

d. The probability that at least two of the four people share the same birth month is $\frac{8856}{20736} \cong 42.7\%$.

e. When there are five people, the probability that at least two share the same birth month is $\frac{12^5 - 12 \cdot 11 \cdot 10 \cdot 9 \cdot 8}{12^5} \cong 61.8\%$, and when there are more than five people, the probability is even greater. Thus, since the probability for four people is less than 50%, the group must contain five or more people for the probability to be at least 50% that two or more share the same birth month.

25. *Hint:* Analyze the solution to exercise 24.

26. a. The number of students who checked at least one of the magazines is $N(T \cup N \cup U) = N(T) + N(N) + N(U) - N(T \cap N) - N(T \cap U) - N(N \cap U) + N(T \cap N \cap U) = 28 + 26 + 14 - 8 - 4 - 3 + 2 = 55.$

b. By the difference rule, the number of students who checked none of the magazines is the total number of students minus the number who checked at least one magazine. This is $100 - 55 = 45.$

d. The number of students who read *Time* and *Newsweek* but not *U.S. News* is

$$N((T \cap N) - N(T \cap N \cap U)) = 8 - 2 = 6.$$

28. Let

$M =$ the set of married people in the sample,

$Y =$ the set of people between 20 and 30 in the sample, and

$F =$ the set of females in the sample.

Then the number of people in the set $M \cup Y \cup F$ is less than or equal to the size of the sample. And so

$1,200 \geq N(M \cup Y \cup F)$

$= N(M) + N(Y) + N(F) - N(M \cap Y)$

$\qquad - N(M \cap F) - N(Y \cap F) + N(M \cap Y \cap F)$

$= 675 + 682 + 684 - 195 - 467 - 318 + 165$

$= 1,226.$

This is impossible since $1,200 < 1,226$, so the polltaker's figures are inconsistent. They could not have occurred as a result of an actual sample survey.

30. Let A be the set of all positive integers less than 1,000 that are not multiples of 2, and let B be the set of all positive integers less than 1,000 that are not multiples of 5. Since the only prime factors of 1,000 are 2 and 5, the number of positive integers that have no common factors with 1,000 is $N(A \cap B)$. Let the universe U be the set of all positive integers less than 1,000. Then A^c is the set of positive integers less than 1,000 that are multiples of 2, B^c is the set of positive integers less than 1,000 that are multiples of 5, and $A^c \cap B^c$ is the set of positive integers less than 1,000 that are multiples of 10. By one of the procedures discussed in Section 6.1 or 6.2, it is easily found that $N(A^c) = 499$, $N(B^c) = 199$, and $N(A^c \cap B^c) = 99$. Thus, by the inclusion/exclusion rule,

$$N(A^c \cup B^c) = N(A^c) + N(B^c) - N(A^c \cap B^c)$$
$$= 499 + 199 - 99 = 599.$$

But by De Morgan's law, $N(A^c \cup B^c) = N((A \cap B)^c)$, and so

$$N((A \cap B)^c) = 599. \qquad (*)$$

Now since $(A \cap B)^c = U - (A \cap B)$, by the difference rule we have

$$N((A \cap B)^c) = N(U) - N(A \cap B). \qquad (**)$$

Equating the right-hand sides of $(*)$ and $(**)$ gives $N(U) - N(A \cap B) = 599$. And because $N(U) = 999$, we conclude

that $999 - N(A \cap B) = 599$, or, equivalently, $N(A \cap B) = 999 - 599 = 400$. So there are 400 positive integers less than 1,000 that have no common factor with 1,000.

36. *Hint:* Use the generalized distributive law for sets from exercise 35, Section 5.2.

Section 6.4

1. a. 2-combinations: $\{x_1, x_2\}, \{x_1, x_3\}, \{x_2, x_3\}$.

Hence, $\dbinom{3}{2} = 3.$

b. Unordered selections: $\{a, b, c, d\}, \{a, b, c, e\},$ $\{a, b, d, e\}, \{a, c, d, e\}, \{b, c, d, e\}.$

Hence, $\dbinom{5}{4} = 5.$

3. $P(7, 2) = \dbinom{7}{2} \cdot 2!$

5. a. $\dbinom{5}{0} = \dfrac{5!}{0!(5 - 0)!} = \dfrac{\not{5}!}{1 \cdot \not{5}!} = 1$

b. $\dbinom{5}{1} = \dfrac{5!}{1!(5 - 1)!} = \dfrac{5 \cdot \not{4} \cdot \not{3} \cdot \not{2} \cdot 1}{1 \cdot \not{4} \cdot \not{3} \cdot \not{2} \cdot 1} = 5$

6. a. number of committees of 6

$$= \dbinom{15}{6} = \dfrac{15!}{(15 - 6)!6!}$$

$$= \dfrac{\overset{7}{\not{15}} \cdot \overset{}{\not{14}} \cdot 13 \cdot \not{12} \cdot 11 \cdot \overset{5}{\not{10}} \cdot \not{9}!}{\not{9}! \cdot \not{6} \cdot \not{5} \cdot \not{4} \cdot \not{3} \cdot \not{2}} = 5,005$$

b. $\begin{bmatrix} \text{number of committees} \\ \text{that don't contain } A \\ \text{and } B \text{ together} \end{bmatrix}$

$= \begin{bmatrix} \text{number of} \\ \text{committees with } A \\ \text{and five others—} \\ \text{none of them } B \end{bmatrix} + \begin{bmatrix} \text{number of} \\ \text{committees with } B \\ \text{and five others—} \\ \text{none of them } A \end{bmatrix}$

$\qquad + \begin{bmatrix} \text{number of committees} \\ \text{with neither } A \text{ nor } B \end{bmatrix}$

$= \dbinom{13}{5} + \dbinom{13}{5} + \dbinom{13}{6}$

$= 1,287 + 1,287 + 1,716 = 4,290$

Alternative solution:

$\begin{bmatrix} \text{number of committees} \\ \text{that don't contain } A \\ \text{and } B \text{ together} \end{bmatrix}$

$= \begin{bmatrix} \text{total number} \\ \text{of committees} \end{bmatrix} - \begin{bmatrix} \text{number of committees} \\ \text{that contain both } A \text{ and } B \end{bmatrix}$

$= \dbinom{15}{6} - \dbinom{13}{4}$

$= 5,005 - 715 = 4,290$

c. $\begin{bmatrix} \text{number of} \\ \text{committees with} \\ \text{both } A \text{ and } B \end{bmatrix} + \begin{bmatrix} \text{number of} \\ \text{committees with} \\ \text{neither } A \text{ and } B \end{bmatrix}$

$= \binom{13}{4} + \binom{13}{6} = 715 + 1{,}716 = 2{,}431$

d. (i) $\begin{bmatrix} \text{number of subsets} \\ \text{of three men} \\ \text{chosen from eight} \end{bmatrix} \cdot \begin{bmatrix} \text{number of subsets} \\ \text{of three women} \\ \text{chosen from seven} \end{bmatrix}$

$= \binom{8}{3}\binom{7}{3} = 56 \cdot 35 = 1{,}960$

(ii) $\begin{bmatrix} \text{number of committees} \\ \text{with at least one woman} \end{bmatrix}$

$= \begin{bmatrix} \text{total number of} \\ \text{committees} \end{bmatrix} - \begin{bmatrix} \text{number of all-male} \\ \text{committees} \end{bmatrix}$

$= \binom{15}{6} - \binom{8}{6} = 5{,}005 - 28$

$= 4{,}977$

e. $\begin{bmatrix} \text{number of} \\ \text{ways to choose} \\ \text{two freshmen} \end{bmatrix} \cdot \begin{bmatrix} \text{number of} \\ \text{ways to choose two} \\ \text{sophomores} \end{bmatrix}$

$\cdot \begin{bmatrix} \text{number of ways} \\ \text{to choose two juniors} \end{bmatrix} \cdot \begin{bmatrix} \text{number of ways} \\ \text{to choose two seniors} \end{bmatrix}$

$= \binom{3}{2}\binom{4}{2}\binom{3}{2}\binom{5}{2}$

$= 540$

8. *Hint:* The answers are: **a.** 66   **b.** (i) 35   (ii) 66   (iii) 10
   c. 21   **d.** 46

9. b. $\binom{24}{3}\binom{16}{3} + \binom{24}{4}\binom{16}{2} + \binom{24}{5}\binom{16}{1} + \binom{24}{6}\binom{16}{0} = 3{,}223{,}220$

11. a. (1) 4 (because there are as many royal flushes as there are suits)

(2) $\frac{4}{\binom{52}{5}} = \frac{4}{2{,}598{,}960} \cong 0.0000015$

c. (1) $13 \cdot \binom{48}{1} = 624$ (because one can first choose the denomination of the four-of-a-kind and then choose one additional card from the 48 remaining)

(2) $\frac{624}{\binom{52}{5}} = \frac{624}{2{,}598{,}960} = 0.00024$

f. (1) Imagine constructing a straight (including a straight flush and a royal flush) as a six-step process: step 1 is to choose the lowest denomination of any card of the five (which can be any one of $A, 2, \ldots, 10$), step 2 is to choose a card of that denomination, step 3 is to choose a card of the next higher denomination, and so forth until all five cards have been selected. By the multiplication rule, the number of ways to perform this process is

$10 \cdot \binom{4}{1}\binom{4}{1}\binom{4}{1}\binom{4}{1}\binom{4}{1} = 10 \cdot 4^5 = 10{,}240.$

By parts (a) and (b), 40 of these numbers represent royal or straight flushes, so there are $10{,}240 - 40 = 10{,}200$ straights in all.

(2) $\frac{10{,}200}{\binom{52}{5}} = \frac{10{,}200}{2{,}598{,}960} \cong 0.0039$

13. a. $2^{10} = 1{,}024$

d. $\begin{bmatrix} \text{number of outcomes} \\ \text{with at least one head} \end{bmatrix}$

$= \begin{bmatrix} \text{total number} \\ \text{of outcomes} \end{bmatrix} - \begin{bmatrix} \text{number of outcomes} \\ \text{with no heads} \end{bmatrix}$

$= 1{,}024 - 1 = 1{,}023$

15. a. 50   **b.** 50

c. To get an even sum, both numbers must be even or both must be odd. Hence

$\begin{bmatrix} \text{number of subsets of two integers from} \\ 1 \text{ to } 100 \text{ inclusive whose sum is even} \end{bmatrix}$

$= \begin{bmatrix} \text{number of subsets} \\ \text{of two even} \\ \text{integers chosen from} \\ \text{the 50 possible} \end{bmatrix} + \begin{bmatrix} \text{number of subsets} \\ \text{of two odd} \\ \text{integers chosen from} \\ \text{the 50 possible} \end{bmatrix}$

$= \binom{50}{2} + \binom{50}{2} = 2{,}450.$

d. To obtain an odd sum, one of the numbers must be even and the other odd. Hence the answer is $\binom{50}{1} \cdot \binom{50}{1} = 2{,}500$. Alternatively, note that the answer equals the total number of subsets of two integers chosen from 1 through 100 minus the number of such subsets for which the sum of the elements is even. Thus the answer is $\binom{100}{2} - 2{,}450 = 2{,}500.$

17. a. Two points determine a line. Hence

$\begin{bmatrix} \text{number of straight} \\ \text{lines determined} \\ \text{by the ten points} \end{bmatrix} = \begin{bmatrix} \text{number of subsets} \\ \text{of two points} \\ \text{chosen from ten} \end{bmatrix}$

$= \binom{10}{2} = 45.$

19. a. $\frac{10!}{2!1!1!3!2!1!} = 151{,}200$ since there are 2 A's, 1 B, 1 H, 3 L's, 2 O's, and 1 U

b. $\frac{8!}{2!1!1!2!2!} = 5{,}040$ **c.** $\frac{9!}{1!2!1!3!2!} = 15{,}120$

21. a. There are two choices for each of four positions in the string, so the answer is $2 \cdot 2 \cdot 2 \cdot 2 = 2^4 = 16.$

b. $\begin{bmatrix} \text{number of strings} \\ \text{with three } a\text{'s} \\ \text{and one } b \end{bmatrix} = \begin{bmatrix} \text{number of ways to pick} \\ \text{a subset of three positions} \\ \text{out of four into} \\ \text{which to place the } a\text{'s} \end{bmatrix}$

$= \binom{4}{3} = 4$

24. Rook must move seven squares to the right and seven squares up, so

$\begin{bmatrix} \text{the number of} \\ \text{paths the rook} \\ \text{can take} \end{bmatrix} = \begin{bmatrix} \text{the number} \\ \text{of orderings} \\ \text{of seven R's} \\ \text{and seven U's} \end{bmatrix}$ where R stands for "right" and U stands for "up"

$= \frac{14!}{7!7!} = 3{,}432.$

25. b. *Solution 1:* One factor can be 1, and the other factor can be the product of all the primes. (This gives 1 factorization.) One factor can be one of the primes, and the other factor can be the product of the other three. (This gives $\binom{4}{1} = 4$ factorizations.) One factor can be a product of two of the primes, and the other factor can be a product of the two other primes. The number $\binom{4}{2} = 6$ counts all possible sets of two primes chosen from the four primes, and each set of two primes corresponds to a factorization. Note, however, that the set $\{p_1, p_2\}$ corresponds to the same factorization as the set $\{p_3, p_4\}$, namely, $p_1 p_2 p_3 p_4$ (just written in a different order). In general, each choice of two primes corresponds to the same factorization as one other choice of two primes. Thus the number of factorizations in which each factor is a product of two primes is $\frac{\binom{4}{2}}{2} = 3$. (This gives 3 factorizations.) The foregoing cases account for all the possibilities, so the answer is $4 + 3 + 1 = 8$.

Solution 2: Let $S = \{p_1, p_2, p_3, p_4\}$. Let $p_1 p_2 p_3 p_4 = P$, and let $f_1 f_2$ be any factorization of P. The product of the numbers in any subset $A \subseteq S$ can be used for f_1, with the product of the numbers in A^c being f_2. There are as many ways to write $f_1 f_2$ as there are subsets of S, namely $2^4 = 16$ (by Theorem 5.3.1). But given any factors f_1 and f_2, $f_1 f_2 = f_2 f_1$. Thus counting the number of ways to write $f_1 f_2$ counts each factorization twice, so the answer is $\frac{16}{2} = 8$.

26. *Hint:* Use the difference rule and the generalization of the inclusion/exclusion rule for 4 sets. (See exercise 36 in Section 6.3.)

Section 6.5

1. a. $\binom{5+3-1}{5} = \binom{7}{5} = \frac{7 \cdot 6}{2} = 21$.

b. The three elements of the set are 1, 2 and 3. The 5-combinations are [1, 1, 1, 1, 1], [1, 1, 1, 1, 2], [1, 1, 1, 1, 3], [1, 1, 1, 2, 2], [1, 1, 1, 2, 3], [1, 1, 1, 3, 3], [1, 1, 2, 2, 2], [1, 1, 2, 2, 3], [1, 1, 2, 3, 3], [1, 1, 3, 3, 3], [1, 2, 2, 2, 2], [1, 2, 2, 2, 3], [1, 2, 2, 3, 3], [1, 2, 3, 3, 3], [1, 3, 3, 3, 3], [2, 2, 2, 2, 2], [2, 2, 2, 2, 3], [2, 2, 2, 3, 3], [2, 2, 3, 3, 3], [2, 3, 3, 3, 3], and [3, 3, 3, 3, 3].

2. a. $\binom{4+3-1}{4} = \binom{6}{4} = \frac{6 \cdot 5}{2} = 15$

3. a. $\binom{20+6-1}{20} = \binom{25}{20} = 53,130$

b. If at least three are eclairs, then 17 additional pastries are selected from six kinds. The number of selections is $\binom{17+6-1}{17} = \binom{22}{17} = 26,334$.

Note: In parts (a) and (b), it is assumed that the selections being counted are unordered.

c. By parts (a) and (b), the probability that at least three eclairs are among the pastries selected is $26334/53130 \cong 0.496 = 49.6\%$.

d. If exactly three of the pastries are eclairs, then 17 additional pastries are selected from five kinds. The number of selections is

$$\binom{17 + 5 - 1}{17} = \binom{21}{17} = 5,985.$$

Hence the probability that a random selection includes exactly three eclairs is $5985/53130 \cong 0.113 = 11.3\%$.

5. The answer equals the number of 4-combinations with repetition allowed that can be formed from a set of n elements. It is

$$\binom{4 + n - 1}{4} = \binom{n + 3}{4}$$
$$= \frac{(n + 3)(n + 2)(n + 1)n(n - 1)!}{4!(n - 1)!}$$
$$= \frac{n(n + 1)(n + 2)(n + 3)}{24}.$$

8. As in Example 6.5.4, the answer is the same as the number of quadruples of integers (i, j, k, m) for which $1 \leq i \leq j \leq k \leq m \leq n$. By exercise 5, this number is $\binom{n+3}{4} = \frac{n(n+1)(n+2)(n+3)}{24}$.

10. Think of the number 20 as divided into 20 individual units and the variables x_1, x_2, and x_3 as three categories into which these units are placed. The number of units in category x_i indicates the value of x_i in a solution of the equation. By Theorem 6.5.1, the number of ways to select 20 objects from the three categories is $\binom{20+3-1}{20} = \binom{22}{20} = \frac{22 \cdot 21}{2} = 231$, so there are 231 nonnegative integer solutions to the equation.

11. The analysis for this exercise is the same as for exercise 10 except that since each $x_i \geq 1$, we can imagine taking 3 of the 20 units, placing one in each category x_1, x_2, and x_3, and then distributing the remaining 17 units among the three categories. The number of ways to do this is $\binom{17+3-1}{17} = \binom{19}{17} = \frac{19 \cdot 18}{2} = 171$, so there are 171 positive integer solutions to the equation.

18. a. Because only ten eclairs are available, any selection of 20 pastries contains k eclairs, where $0 \leq k \leq 10$. Since such a selection includes $20 - k$ of the other five kinds of pastry, the number of such selections is $\binom{(20-k)+5-1}{20-k} = \binom{24-k}{20-k}$ (by Theorem 6.5.1.) Therefore, by the addition rule, the total number of selections is $\sum_{k=0}^{10} \binom{24-k}{20-k}$. The numerical value of this expression is 51,128, which can be obtained using a calculator that automatically computes values of $\binom{n}{r}$ or using a symbolic manipulation computer program such as Derive, Maple, or Mathematica.

b. For each combination of k eclairs and m napoleon slices, choose $20 - (k + m)$ pastries from the remaining four types. By Theorem 6.5.1, the number of ways to make such a selection is

$$\binom{(20-(k+m))+4-1}{20-(k+m)} = \binom{23-(k+m)}{20-(k+m)}.$$

Since $0 \le k \le 10$ and $0 \le m \le 8$, by the addition rule the total number of selections of 20 pastries is $\sum_{m=0}^{8} \sum_{k=0}^{10} \binom{23-(k+m)}{20-(k+m)}$. The numerical value of this expression is 46,761.

Section 6.6

1. $\binom{n}{0} = \dfrac{n!}{0!(n-0)!} = \dfrac{\not{n!}}{1 \cdot \not{n!}} = 1$

3. $\binom{n}{2} = \dfrac{n!}{(n-2)! \cdot 2!} = \dfrac{n \cdot (n-1) \cdot (n-2)!}{(n-2)! \cdot 2!}$

$= \dfrac{n(n-1)}{2}$

5. *Proof:* Suppose n and r are nonnegative integers and $r \le n$. Then

$$\binom{n}{r} = \frac{n!}{r!(n-r)!} \qquad \text{by Theorem 6.4.1}$$

$$= \frac{n!}{(n-(n-r))!(n-r)!} \qquad \begin{array}{l}\text{since } n-(n-r) = \\ n-n+r = r\end{array}$$

$$= \frac{n!}{(n-r)!(n-(n-r))!} \qquad \begin{array}{l}\text{by interchanging the} \\ \text{factors in the denominator}\end{array}$$

$$= \binom{n}{n-r} \qquad \text{by Theorem 6.4.1.}$$

6. Apply formula (6.6.2) with $n+k$ in place of n. This is legal since $n+k \ge 1$. The result is $\binom{n+k}{n+k-1} = n+k$.

9. $\binom{6}{4} = \binom{5}{3} + \binom{5}{4} = 10 + 5 = 15$

$\binom{6}{5} = \binom{5}{4} + \binom{5}{5} = 5 + 1 = 6$

14. *Proof by mathematical induction:* Let the property $P(n)$ be the formula

$$\sum_{i=2}^{n+1} \binom{i}{2} = \binom{n+2}{3}.$$

Show that the property is true for $n = 1$:

To prove the property for $n = 1$, we must show that

$$\sum_{i=2}^{1+1} \binom{i}{2} = \binom{1+2}{3}.$$

But

$$\sum_{i=2}^{1+1} \binom{i}{2} = \sum_{i=2}^{2} \binom{i}{2} = \binom{2}{2} = 1 = \binom{3}{3} = \binom{1+2}{3},$$

so the property is true for $n = 1$.

Show that for all integers $k \ge 1$, if the property is true for $n = k$, then it is true for $n = k+1$:

Suppose the property

$$\sum_{i=2}^{n+1} \binom{i}{2} = \binom{n+2}{3}$$

is true when an integer $k \ge 1$ is substituted for n. That is, suppose

$$\sum_{i=2}^{k+1} \binom{i}{2} = \binom{k+2}{3} \qquad \begin{array}{l}\text{for some integer } k \ge 1. \\ \text{Inductive hypothesis}\end{array}$$

[*We must show that the property* $\sum_{i=2}^{n+1} \binom{i}{2} = \binom{n+2}{3}$ *is true when $k+1$ is substituted for n.*] We must show that

$$\sum_{i=2}^{(k+1)+1} \binom{i}{2} = \binom{(k+1)+2}{3},$$

or, equivalently,

$$\sum_{i=2}^{k+2} \binom{i}{2} = \binom{k+3}{3}. \qquad (*)$$

But the left-hand side of equation (*) is

$$\sum_{i=2}^{k+2} \binom{i}{2} = \sum_{i=1}^{k+1} \binom{i}{2} + \binom{k+2}{2} \qquad \begin{array}{l}\text{by writing the last} \\ \text{term separately}\end{array}$$

$$= \binom{k+2}{3} + \binom{k+2}{2} \qquad \text{by inductive hypothesis}$$

$$= \binom{(k+2)+1}{3} \qquad \text{by Pascal's formula}$$

$$= \binom{k+3}{3},$$

which is the right-hand side of equation (*) [*as was to be shown*].

[*Since we have proved the basis step and the inductive step, we conclude that the property is true for all $n \ge 1$*].

15. *Hint:* Use the results of exercises 3 and 14.

17. a. $C_1 = \dfrac{1}{2}\binom{2}{1} = \dfrac{1}{2} \cdot 2 = 1,$

$C_2 = \dfrac{1}{3}\binom{4}{2} = \dfrac{1}{3} \cdot 6 = 2,$

$C_3 = \dfrac{1}{4}\binom{6}{3} = \dfrac{1}{4} \cdot 20 = 5$

19. *Hint:* This follows by letting $m = n = r$ in exercise 18 and using the result of Example 6.6.2.

Section 6.7

1. $1 + 7x + \binom{7}{2}x^2 + \binom{7}{3}x^3 + \binom{7}{4}x^4 + \binom{7}{5}x^5 + \binom{7}{6}x^6 + x^7 = 1 + 7x + 21x^2 + 35x^3 + 35x^4 + 21x^5 + 7x^6 + x^7$

3. $1 + 6(-x) + \binom{6}{2}(-x)^2 + \binom{6}{3}(-x)^3 + \binom{6}{4}(-x)^4 + \binom{6}{5}(-x)^5 + (-x)^6 = 1 - 6x + 15x^2 - 20x^3 + 15x^4 - 6x^5 + x^6$

5. $(p - 2q)^4 = \sum_{k=0}^{4} \binom{4}{k} p^{4-k}(-2q)^k$

$= \binom{4}{0} p^4(-2q)^0 + \binom{4}{1} p^3(-2q)^1$

$\quad + \binom{4}{2} p^2(-2q)^2 + \binom{4}{3} p^1(-2q)^3$

$\quad + \binom{4}{4} p^0(-2q)^4$

$= p^4 - 8p^3 q + 24 p^2 q^2 - 32 pq^3 + 16 q^4$

7. $\left(x + \dfrac{1}{x} \right)^5 = \sum_{k=0}^{5} \binom{5}{k} x^{5-k} \left(\dfrac{1}{x} \right)^k$

$= \binom{5}{0} x^5 \left(\dfrac{1}{x} \right)^0 + \binom{5}{1} x^4 \left(\dfrac{1}{x} \right)^1$

$\quad + \binom{5}{2} x^3 \left(\dfrac{1}{x} \right)^2 + \binom{5}{3} x^2 \left(\dfrac{1}{x} \right)^3$

$\quad + \binom{5}{4} x^1 \left(\dfrac{1}{x} \right)^4 + \binom{5}{5} x^0 \left(\dfrac{1}{x} \right)^5$

$= x^5 + 5x^3 + 10x + \dfrac{10}{x} + \dfrac{5}{x^3} + \dfrac{1}{x^5}$

11. The term is $\binom{9}{3}x^6 y^3 = 84 x^6 y^3$, so the coefficient is 84.

13. The term is $\binom{12}{7}a^5(-2b)^7 = 792 a^5(-128)b^7 = -101{,}376 a^5 b^7$, so the coefficient is $-101{,}376$.

15. The term is $\binom{15}{8}(3p^2)^8(-2q)^7 = \binom{15}{8}3^8(-2)^7 p^{16}q^7$, so the coefficient is $\binom{15}{8}3^8(-2)^7 = -5{,}404{,}164{,}480$.

17. *Proof:* Let $a = 1$, let $b = -1$, and let n be a positive integer. Substitute into the binomial theorem to obtain

$$(1 + (-1))^n = \sum_{k=0}^{n} \binom{n}{k} \cdot 1^{n-k} \cdot (-1)^k$$

$$= \sum_{k=0}^{n} \binom{n}{k} (-1)^k \qquad \text{since } 1^{n-k} = 1.$$

But $(1 + (-1))^n = 0^n = 0$, so

$$0 = \sum_{k=0}^{n} \binom{n}{k} (-1)^k$$

$$= \binom{n}{0} - \binom{n}{1} + \binom{n}{2} - \binom{n}{3} + \cdots + (-1)^n \binom{n}{n}.$$

18. *Hint:* $3 = 1 + 2$

19. *Proof:* Let m be any integer with $m \geq 0$, and apply the binomial theorem with $a = 2$ and $b = -1$. The result is

$$1 = 1^m = (2 + (-1))^m = \sum_{i=0}^{m} \binom{m}{i} 2^{m-i}(-1)^i$$

$$= \sum_{i=0}^{m} (-1)^i \binom{m}{i} 2^{m-i}.$$

22. *Hint:* Apply the binomial theorem with $a = -\frac{1}{2}$ and $b = 1$, and analyze the resulting equation when n is even and when n is odd.

24. $\displaystyle\sum_{k=0}^{n} \binom{n}{k} 5^k = \sum_{k=0}^{n} \binom{n}{k} 1^{n-k} 5^k = (1 + 5)^n = 6^n$

26. $\displaystyle\sum_{i=0}^{n} \binom{n}{i} x^i = \sum_{i=0}^{n} \binom{n}{i} 1^{n-i} x^i = (1 + x)^n$

28. $\displaystyle\sum_{j=0}^{2n} (-1)^j \binom{2n}{j} x^j = \sum_{j=0}^{2n} \binom{2n}{j} 1^{2n-j}(-x)^j = (1 - x)^{2n}$

32. $\displaystyle\sum_{i=0}^{m} (-1)^i \binom{m}{i} \frac{1}{2^i} = \sum_{i=0}^{m} \binom{m}{i} 1^{m-i} \left(-\frac{1}{2} \right)^i$

$$= \left(1 - \frac{1}{2} \right)^m = \frac{1}{2^m}$$

34. $\displaystyle\sum_{i=0}^{n} (-1)^i \binom{n}{i} 5^{n-i} 2^i = \sum_{i=0}^{n} \binom{n}{i} 5^{n-i}(-2)^i = (5 - 2)^n = 3^n$

36. b. $n(1 + x)^{n-1} = \displaystyle\sum_{k=1}^{n} \binom{n}{k} k x^{k-1}$.

[*The term corresponding to* $k = 0$ *is zero because* $\frac{d}{dx}(x^0) = 0.$]

c. (i) Substitute $x = 1$ in part (b) above to obtain

$$n(1 + 1)^{n-1} = \sum_{k=1}^{n} \binom{n}{k} k \cdot 1^{k-1} = \sum_{k=1}^{n} \binom{n}{k} k$$

$$= \binom{n}{1} \cdot 1 + \binom{n}{2} \cdot 2 + \binom{n}{3} \cdot 3 + \cdots + \binom{n}{n} \cdot n.$$

Dividing both sides by n and simplifying gives

$$2^{n-1} = \frac{1}{n}\left[\binom{n}{1} + 2\binom{n}{2} + 3\binom{n}{3} + \cdots + n\binom{n}{n} \right].$$

Section 6.8

1. By probability axiom 2, $P(\emptyset) = 0$.

2. a. By probability axiom 3, $P(A \cup B) = P(A) + P(B) = 0.3 + 0.5 = 0.8$.

b. Because $A \cup B \cup C = S$, $C = S - (A \cup B)$. Thus, by the formula for the probability of the complement of an event, $P(C) = P((A \cup B)^c) = 1 - P(A \cup B) = 1 - 0.8 = 0.2$.

4. By the formula for the probability of a general union of two events, $P(A \cup B) = P(A) + P(B) - P(A \cap B) = 0.8 + 0.7 - 0.6 = 0.9$.

7. a. $P(A \cup B) = 0.4 + 0.3 = 0.7$

b. $P(C) = P((A \cup B)^c) = 1 - P(A \cup B) = 1 - 0.7 = 0.3$

c. $P(A \cup C) = 0.4 + 0.3 = 0.7$

d. $P(A^c) = 1 - P(A) = 1 - 0.4 = 0.6$

e. $P(A^c \cap B^c) = P((A \cup B)^c) = 1 - P(A \cup B) = 1 - 0.7 = 0.3$

f. $P(A^c \cup B^c) = P((A \cap B)^c) = P(\emptyset^c) = P(S) = 1$

9. a. $P(A \cup B) = P(A) + P(B) - P(A \cap B) = 0.4 + 0.5 - 0.2 = 0.7$

d. $P(A^c \cap B^c) = P((A \cup B)^c) = 1 - P(A \cup B) = 1 - 0.7 = 0.3$

11. *Hint:* $V = (U \cup (V - U))$

12. *Hint:* Use the fact that for all sets U and V, $U \cup (V - U) = U \cup V$.

13. *Hint:* $(A_1 \cup A_2 \cup \cdots \cup A_k) \cap A_{k+1} = \emptyset$ and
$A_1 \cup A_2 \cup \cdots \cup A_k \cup A_{k+1} = (A_1 \cup A_2 \cup \cdots \cup A_k) \cup A_{k+1}$.

14. *Solution 1:* The net gain of the grand prize winner is $2,000,000 - \$2 = \$1,999,998$. Each of the 10,000 second prize winners has a net gain of $\$20 - \$2 = \$18$, and each of the 50,000 third prize winners has a net gain of $\$4 - \$2 = \$2$. The number of people who do not win anything is $1,500,000 - 1 - 10,000 - 50,000 = 1,439,999$, and each of these people has a net loss of $\$2$. Because all of the 1,500,000 tickets have an equal chance of winning a prize, the expected gain or loss of a ticket is

$$\frac{1}{1500000} (\$1,999,998 \cdot 1 + \$18 \cdot 10000$$
$$+ \$2 \cdot 50000 + (-\$2) \cdot 1,439,999) = -\$0.40.$$

Solution 2: The total income to the lottery organizer is $\$2$ (per ticket) $\cdot 1,500,000$ (tickets) $= \$3,000,000$. The payout the lottery organizer must make is $\$2,000,000 + (\$20)(10,000) + (\$4)(50,000) = \$2,400,000$, so the net gain to the lottery organizer is $\$600,000$, which amounts to $\frac{\$600,000}{1,500,000} = \$0.40$ per ticket. Thus the expected net loss to a purchaser of a ticket is $\$0.40$.

16. Let 2_1 and 2_2 denote the two balls with the number 2, and let 5 and 6 denote the other two balls. There are $\binom{6}{2} = 4$ subsets of 2 balls that can be chosen from the urn. The following table shows the sums of the numbers on the balls in each set and the corresponding probabilities:

Subset	Sum s	Probability that the sum $= s$
$\{2_1, 2_2\}$	4	$\frac{1}{6}$
$\{2_1, 5\}, \{2_2, 5\}$	7	$\frac{2}{6}$
$\{2_1, 6\}\{2_2, 6\}$	8	$\frac{2}{6}$
$\{5, 6\}$	11	$\frac{1}{6}$

So the expected value is $4 \cdot \frac{1}{6} + 7 \cdot \frac{2}{6} + 8 \cdot \frac{2}{6} + 11 \cdot \frac{1}{6} = 7.5$.

19. The following table displays the sum of the numbers showing face up on the dice:

	1	2	3	4	5	6
1	2	3	4	5	6	7
2	3	4	5	6	7	8
3	4	5	6	7	8	9
4	5	6	7	8	9	10
5	6	7	8	9	10	11
6	7	8	9	10	11	12

Each cell in the table represents an outcome whose probability is $\frac{1}{36}$. Thus the expected value of the sum is

$$2\left(\tfrac{1}{36}\right) + 3\left(\tfrac{2}{36}\right) + 4\left(\tfrac{3}{36}\right) + 5\left(\tfrac{4}{36}\right) + 6\left(\tfrac{5}{36}\right) + 7\left(\tfrac{6}{36}\right)$$
$$+ 8\left(\tfrac{5}{36}\right) + 9\left(\tfrac{4}{36}\right) + 10\left(\tfrac{3}{36}\right) + 11\left(\tfrac{2}{36}\right) + 12\left(\tfrac{1}{36}\right) = \tfrac{252}{36} = 7.$$

20. *Hint:* The answer is about 7.7 cents.

22. *Hint:* The answer is 1.875.

Section 6.9

1. $P(B) = \dfrac{P(A \cap B)}{P(A \mid B)} = \dfrac{1/6}{1/2} = \dfrac{1}{3}$

3. a. *Proof:* Suppose S is any sample space and A and B are any events in S such that $P(B) \neq 0$. Note that

(1) $A \cup A^c = S$ by the complement law for $\cup$.
(2) $B \cap S = B$ by the identity law for $\cap$.
(3) $B \cap (A \cup A^c) = (A \cap B) \cup (A^c \cap B)$ by the distributive law and commutative laws for sets.
(4) $(A \cap B) \cup (A^c \cap B) = \emptyset$ by the complement law for $\cap$ and the commutative and associative laws for sets.

Thus $B = (A \cap B) \cup (A^c \cap B)$, and, by probability axiom 3, $P(B) = P(A \cap B) + P(A^c \cap B)$. Therefore, $P(A^c \cap B) = P(B) - P(A \cap B)$. By definition of conditional probability, it follows that

$$P(A^c \mid B) = \frac{P(A^c \cap B)}{P(B)} = \frac{P(B) - P(A \cap B)}{P(B)}$$
$$= 1 - \frac{P(A \cap B)}{P(B)} = 1 - P(A \mid B).$$

4. *Hints:* (1) $A = (A \cap B) \cup (A \cap B^c)$.
(2) The answer is $P(A \mid B^c) = \dfrac{P(A) - P(A \mid B)P(B)}{1 - P(B)}$.

5. Let R_1 be the event that the first ball is red,
R_2 be the event that the second ball is red,
B_1 be the event that the first ball is blue,
B_2 be the event that the second ball is blue.
Then $P(R_1) = \frac{25}{40} = \frac{5}{8}$, $P(B_1) = \frac{15}{40} = \frac{3}{8}$, $P(R_2 \mid R_1) = \frac{24}{39} = \frac{8}{13}$, $P(R_2 \mid B_1) = \frac{25}{39}$, $P(B_2 \mid R_1) = \frac{15}{39}$, $P(B_2 \mid B_1) = \frac{14}{39}$.

a. The probability that both balls are red is $P(R_1 \cap R_2) = P(R_2 \mid R_1)P(R_1) = \frac{8}{13} \cdot \frac{5}{8} = \frac{5}{13} \cong 38.5\%$.

b. The probability that the second ball is red but the first ball is not is $P(R_2 \cap B_1) = P(R_2 \mid B_1)P(B_1) = \frac{25}{39} \cdot \frac{3}{8} \cong 24.0\%$.

c. The probability that the second ball is red is
$$P(R_2) = P((R_2 \cap R_1) \cup (R_2 \cap B_1))$$
$$= P(R_2 \mid R_1)P(R_1) + P(R_2 \mid B_1)P(B_1)$$
$$= \tfrac{8}{13} \cdot \tfrac{5}{8} + \tfrac{25}{39} \cdot \tfrac{3}{8} = \tfrac{5}{8} = 62.5\%.$$

d. The probability that at least one of the balls is red is
$$P(R_1 \cup R_2) = P(R_1) \cup P(R_2) - P(R_1 \cap R_2)$$
$$= \tfrac{5}{8} + \tfrac{5}{8} - \tfrac{5}{13} \cong 86.5\%.$$

7. a. Let W_1 be the event that a woman is chosen on the first draw,

W_2 be the event that a woman is chosen on the second draw,

M_1 be the event that a man is chosen on the first draw,

M_2 be the event that a man is chosen on the second draw. Then $P(W_1) = \frac{3}{10}$ and $P(W_2 \mid W_1) = \frac{2}{9}$, and thus

$P(W_1 \cap W_2) = P(W_2 \mid W_1)P(W_1) = \frac{2}{9} \cdot \frac{3}{10} = \frac{1}{15} = 6\frac{2}{3}\%$.

c. *Hint:* The answer is $\frac{7}{15} = 46\frac{2}{3}\%$.

8. *Hint:* Use the facts that $P(B_k \mid A) = \dfrac{P(B_k \cap A)}{P(A)}$ and that $(A \cap B_1) \cup (A \cap B_2) = A$.

9. *Hint:* For the inductive step, note that given sets $B_1, B_2, \ldots B_k, B_{k+1}$ that partition the sample space S, and given a set A with $P(A) \neq 0$, you can let $B'_k = B_k \cup B_{k+1}$. Then $B_1, B_2, \ldots, B_{k-1}, B'_k$ will also be a partition of S, and you can apply the inductive hypothesis to these sets and A.

10. Let U_1 be the event that the first urn is chosen, U_2 the event that the second urn is chosen, and B the event that the chosen ball is blue. Then

$$P(B|U_1) = \tfrac{12}{19} \quad \text{and} \quad P(B|U_2) = \tfrac{8}{27}.$$

$$P(B \cap U_1) = P(B|U_1)P(U_1) = \tfrac{12}{19} \cdot \tfrac{1}{2} = \tfrac{12}{38}.$$

Also

$$P(A \cap U_2) = P(B|U_2)P(U_2) = \tfrac{8}{27} \cdot \tfrac{1}{2} = \tfrac{8}{54}.$$

Now B is the disjoint union of $B \cap U_1$ and $B \cap U_2$. So

$$P(B) = P(B \cap U_1) + P(B \cap U_2) = \tfrac{12}{38} + \tfrac{8}{54} \cong 46.4\%.$$

Thus the probability that the chosen ball is blue is approximately 46.4%.

12. *Hint:* The answers to parts (a) and (b) are approximately 52.9% and 54.0%, respectively.

13. Let A be the event that a randomly chosen person tests positive for drugs, let B_1 be the event that a randomly chosen person uses drugs, and let B_2 be the event that a randomly chosen person does not use drugs. Then A^c is the event that a randomly chosen person does not test positive for drugs, and $P(B_1) = 0.04$, $P(B_2) = 0.96$, $P(A \mid B_2) = 0.03$, and $P(A^c \mid B_1) = 0.02$. Hence $P(A|B_1) = 0.97$ and $P(A^c|B_2) = 0.98$.

a. $P(B_1|A) = \dfrac{P(A|B_1)P(B_1)}{P(A|B_1)P(B_1) + P(A|B_2)P(B_2)}$

$= \dfrac{(0.97)(0.04)}{(0.97)(0.04) + (0.03)(0.96)} \cong 57.4\%$

b. $P(B_2|A^c) = \dfrac{P(A^c|B_2)P(B_2)}{P(A^c|B_1)P(B_1) + P(A^c|B_2)P(B_2)}$

$= \dfrac{(0.98)(0.96)}{(0.02)(0.04) + (0.98)(0.96)} \cong 99.9\%$

15. *Hint:* The answers to parts (a) and (b) are 11.25% and $21\frac{1}{3}\%$, respectively.

16. *Proof:* Suppose A and B are events in a sample space S, and $P(A|B) = P(A) \neq 0$. Then

$$P(B|A) = \frac{P(B \cap A)}{P(A)} = \frac{P(A|B)P(B)}{P(A)}$$
$$= \frac{P(A)P(B)}{P(A)} = P(B).$$

18. As in Example 6.9.1, the sample space is the set of all 36 outcomes obtained from rolling the two dice and noting the numbers showing face up on each. Let A be the event that the number on the blue die is 2 and B the event that the number on the gray die is 4 or 5. Then

$$A = \{21, 22, 23, 24, 25, 26\},$$
$$B = \{14, 24, 34, 44, 54, 64, 15, 25, 35, 45, 55, 65\}, \quad \text{and}$$
$$A \cap B = \{24, 25\}.$$

Since the dice are fair (so all outcomes are equally likely), $P(A) = \frac{6}{36}$, $P(B) = \frac{12}{36}$ and $P(A \cap B) = \frac{2}{36}$. By definition of conditional probability,

$$P(A \mid B) = \frac{P(A \cap B)}{P(B)} = \frac{\frac{2}{36}}{\frac{12}{36}} = \frac{1}{6} \quad \text{and}$$

$$P(B|A) = \frac{P(A \cap B)}{P(A)} = \frac{\frac{2}{36}}{\frac{6}{36}} = \frac{1}{3}.$$

But $P(A) = \frac{6}{36} = \frac{1}{6}$ and $P(B) = \frac{12}{36} = \frac{1}{3}$. Hence $P(A|B) = P(A)$ and $P(B \mid A) = P(B)$.

21. *Proof:* Suppose A and B are independent events in a sample space S. By definition of independence, $P(A \cap B) = P(A)P(B)$. We must show that A^c and B are also independent. By definition of independence, this means that we must show that $P(A^c \cap B) = P(A^c)P(B)$. As in the solution to exercise 3,

(1) $A \cup A^c = S$ by the complement law for $\cup$.

(2) $B \cap S = B$ by the identity law for $\cap$.

(3) $B \cap (A \cup A^c) = (A \cap B) \cup (A^c \cap B)$ by the distributive and commutative laws for sets.

(4) $(A \cap B) \cap (A^c \cap B) = \emptyset$ by the complement law for $\cap$ and the commutative and associative laws for sets.

Thus $B = (A \cap B) \cup (A^c \cap B)$, and, by probability axiom 3, $P(B) = P(A \cap B) + P(A^c \cap B)$. Therefore, $P(A^c \cap B) = P(B) - P(A \cap B) = P(B) - P(A)P(B)$ because A and B are independent. Factoring out $P(B)$ gives $P(A^c \cap B) = (1 - P(A))P(B) = P(A^c)P(B)$, as was to be shown.

Note: An alternative proof could make direct use of the result of exercise 3.

23. Let A be the event that the student answers the first question correctly, and let B be the event that the student answers the second answer correctly. Because two choices can be eliminated on the first question, $P(A) = \frac{1}{3}$, and because no choices can be eliminated on the second question, $P(B) = \frac{1}{5}$. Thus $P(A^c) = \frac{2}{3}$ and $P(B^c) = \frac{4}{5}$.

a. The probability that the student answers both questions correctly is

$$P(A \cap B) = P(A)P(B) = \tfrac{1}{3} \cdot \tfrac{1}{5} = \tfrac{1}{15} = 6\tfrac{2}{3}\%.$$

b. The probability that the student answers exactly one question correctly is

$$P((A \cap B^c) \cup (A^c \cap B))$$
$$= P(A \cap B^c) + P(A^c \cap B)$$
$$= P(A)P(B^c) + P(A^c)P(B)$$
$$= \tfrac{1}{3} \cdot \tfrac{4}{5} + \tfrac{2}{3} \cdot \tfrac{1}{5} = \tfrac{6}{15} = \tfrac{2}{5} = 40\%.$$

c. One solution is to say that the probability that the student answers both questions incorrectly is $P(A^c \cap B^c)$, and $P(A^c \cap B^c) = P(A^c)P(B^c)$ by the result of exercise 22. Thus the answer is

$$P(A^c)P(B^c) = \tfrac{2}{3} \cdot \tfrac{4}{5} = \tfrac{8}{15} = 53\tfrac{1}{3}\%.$$

Another solution uses the fact that the event that the student answers both questions incorrectly is the complement of the event that the student answers at least one question correctly. Thus, by the results of parts (a) and (b), the answer is $1 - \left(\tfrac{1}{15} + \tfrac{2}{5}\right) = \tfrac{8}{15} = 53\tfrac{1}{3}\%.$

25. Let H_i be the event that the result of toss i is heads, and let T_i be the event that the result of toss i is tails. Then $P(H_i) = 0.7$ and $P(T_i) = 0.3$ for $i = 1, 2$.

b. The probability of obtaining exactly one head is

$$P((H_1 \cap T_2) \cup (T_1 \cap H_2))$$
$$= P(H_1 \cap T_2) + P(T_1 \cap H_2)$$
$$= P(H_1)P(T_2) + P(T_1)P(H_2)$$
$$= (0.7)(0.3) + (0.3)(0.7) - 42\%.$$

27. *Hint:* The answer is $\tfrac{1}{2}$.

28. a. $P(\text{seven heads})$

$$= \begin{bmatrix} \text{the number of different} \\ \text{ways seven heads can} \\ \text{be obtained in ten tosses} \end{bmatrix} (0.7)^7(0.3)^3$$
$$= 120(0.7)^7(0.3)^3 \cong 0.267 = 26.7\%.$$

29. a. $P(\text{none is defective})$

$$= \begin{bmatrix} \text{the number of different} \\ \text{ways of having 0 defective} \\ \text{items in the sample of 10} \end{bmatrix} (0.03)^0(0.97)^{10}$$
$$= 1 \cdot (0.3.)^0(0.97)^{10} \cong 0.737 = 73.7\%$$

30. b. The probability that a woman will have at least one false positive result over a period of ten years is $1 - (0.96)^{10} \cong 33.5\%.$

31. a. $P(\text{none is male}) \cong 1.3\%$

b. $P(\text{at least one is male}) = 1 - P(\text{none is male}) \cong 1 - 0.013 = 98.7\%$

Section 7.1

1. a. domain of $f = \{1, 3, 5\}$, co-domain of $f = \{s, t, u, v\}$

b. $f(1) = v, f(3) = s, f(5) = v$

c. range of $f = \{s, v\}$

d. yes, no

e. inverse image of $s = \{3\}$, inverse image of $u = \emptyset$, inverse image of $v = \{1, 5\}$

f. $\{(1, v), (3, s), (5, v)\}$

3. a. This arrow-diagram does not define a function because there are two arrows coming out of the 2.

b. This diagram does not define a function because the element 5 in the domain is not related to any element in the co-domain. (There is no arrow coming out of the 5.)

4. a. True. The definition of function says that for any input there is one and only one output, so if two inputs are equal, their outputs must also be equal.

c. True. The definition of function does not prohibit this occurrence.

5. a. There are four functions from X to Y as shown below.

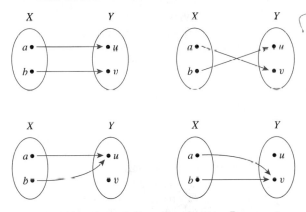

6. a. The answer is $4 \cdot 4 \cdot 4 = 4^3 = 64$. Imagine creating a function from a 3-element set to a 4-element set as a three-step process: Step 1 is to send the first element of the 3-element set to an element of the 4-element set (there are four ways to perform this step); step 2 is to send the second element of the 3-element set to an element of the 4-element set (there are also four ways to perform this step); and step 3 is to send the third element of the 3-element set to an element of the 4-element set (there are four ways to perform this step). Thus the entire process can be performed in $4 \cdot 4 \cdot 4$ different ways.

7. For all $x \in \mathbf{R}$,

$$g(x) = \frac{2x^3 + 2x}{x^2 + 1} = \frac{2x(x^2 + 1)}{x^2 + 1}$$
$$= 2x = f(x) \quad \text{since } x^2 + 1 \neq 0 \text{ for any real number } x.$$

Hence $f = g$.

9. $F \cdot G$ and $G \cdot F$ are equal because for all real numbers x,

$$(F \cdot G)(x) = F(x) \cdot G(x) \quad \text{by definition of } F \cdot G$$
$$= G(x) \cdot F(x) \quad \text{by the commutative law of multiplication of real numbers}$$
$$= (G \cdot F)(x) \quad \text{by definition of } G \cdot F.$$

11. a. e **b.** b_i^{jk}

12. a. The sequence is given by the function $f: \mathbf{Z}^{nonneg} \to \mathbf{R}$ defined by the rule

$$f(n) = \frac{(-1)^n}{2n+1} \quad \text{for all nonnegative integers } n.$$

13. a. 1 [*because there is an odd number of elements in* $\{1, 3, 4\}$]

  c. 0 [*because there is an even number of elements in* $\{2, 3\}$]

14. $f(aba) = 0$ [*because there are no b's to the left of the left-most a in aba*]

  $f(bbab) = 2$ [*because there are two b's to the left of the left-most a in bbab*]

  $f(b) = 0$ [*because the string b contains no a's*]

  range of $f = \mathbf{Z}^{nonneg}$

15. a. $2^3 = 8$ **c.** $4^1 = 4$

16. a. $\log_3 81 = 4$ because $3^4 = 81$

  c. $\log_3(\frac{1}{27}) = -3$ because $3^{-3} = \frac{1}{27}$

17. Let b be any positive real number with $b \neq 1$. Since $b^1 = b$, by definition of logarithm, $\log_b b = 1$.

19. *Proof:* Suppose b and u are any positive real numbers. [*We must show that* $\log_b(\frac{1}{u}) = -\log_b(u)$.] Let $v = \log_b(\frac{1}{u})$. By definition of logarithm, $b^v = \frac{1}{u}$. Multiplying both sides by u and dividing by b^v gives $u = b^{-v}$, and thus, by definition of logarithm, $-v = \log_b(u)$. Now multiply both sides of this equation by -1 to obtain $v = -\log_b(u)$. Therefore, $\log_b(\frac{1}{u}) = -\log_b(u)$ because both expressions equal v. [*This is what was to be shown.*]

20. *Hint:* Use a proof by contradiction. Suppose $\log_3 7$ is rational. Then $\log_3 7 = \frac{a}{b}$ for some integers a and b with $b \neq 0$. Apply the definition of logarithm to rewrite $\log_3 7 = \frac{a}{b}$ in exponential form.

21. Suppose b and y are positive real numbers with $\log_b y = 3$. By definition of logarithm, this implies that $b^3 = y$. Then

$$y = b^3 = \frac{1}{\frac{1}{b^3}} = \frac{1}{\left(\frac{1}{b}\right)^3} = \left(\frac{1}{b}\right)^{-3}.$$

Thus, by definition of logarithm (with base $1/b$), $\log_{1/b}(y) = -3$.

23. a. $p_1(2, y) = 2$, $p_1(5, x) = 5$, range of $p_1 = \{2, 3, 5\}$

24. a. $mod(67, 10) = 7$ and $div(67, 10) = 6$ since $67 = 10 \cdot 6 + 7$.

25. a. $E(0110) = 000111111000$ and $D(111111000111) = 1101$

26. a. $H(10101, 00011) = 3$

27. a.

```
1 2 3    1 2 3    1 2 3
↓ ↓ ↓    ↓ ↓ ↓    ↓ ↓ ↓
1 2 3    2 1 3    3 2 1

1 2 3    1 2 3    1 2 3
↓ ↓ ↓    ↓ ↓ ↓    ↓ ↓ ↓
1 3 2    2 3 1    3 1 2
```

c.

```
1 2 3    1 2 3
↓ ↓ ↓    ↓ ↓ ↓
2 3 1    3 1 2
```

28. a. Domain of f Co-domain of f

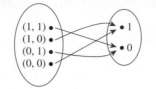

30. a. $f(1, 1, 1) = (4 \cdot 1 + 3 \cdot 1 + 2 \cdot 1) \bmod 2 = 9 \bmod 2 = 1$

  $f(0, 0, 1) = (4 \cdot 0 + 3 \cdot 0 + 2 \cdot 1) \bmod 2 = 2 \bmod 2 = 0$

31. g is not well-defined. Suppose g were well-defined. Then $g(1/2) = g(2/4)$ since $1/2 = 2/4$, but also $g(1/2) \neq g(2/4)$ because $g(1/2) = 1 - 2 = -1$ and $g(2/4) = 2 - 4 = -2$. This contradiction shows that the supposition that g is well-defined is false. Hence g is not well-defined.

35. a. $\phi(15) = 8$ [*because* 1, 2, 4, 7, 8, 11, 13, *and* 14 *have no common factors with* 15 *other than* ± 1]

  b. $\phi(2) = 1$ [*because the only positive integer less than or equal to 2 having no common factors with 2 other than* ± 1 *is* 1]

  c. $\phi(5) = 4$ [*because* 1, 2, 3, *and* 4 *have no common factors with 5 other than* ± 1]

36. *Proof:* Let p be any prime number and n any integer with $n \geq 1$. There are p^{n-1} positive integers less than or equal to p^n that have a common factor other than ± 1 with p^n, namely $p, 2p, 3p, \ldots, (p^{n-1})p$. Hence, by the difference rule, there are $p^n - p^{n-1}$ positive integers less than or equal to p^n that have no common factor with p^n except ± 1.

37. *Hint:* Use the result of exercise 36 with $p = 2$.

38. *Hint:* Let A and B be the sets of all positive integers less than or equal to n that are divisible by p and q, respectively. Then $\phi(n) = n - (N(A \cup B))$.

40. The statement is true. *Proof:* Let f be a function from X to Y, and suppose $A \subseteq X$, $B \subseteq X$, and $A \subseteq B$. Let $y \in f(A)$. [*We must show that* $y \in f(B)$]. Then, by definition of image of a set, $y = f(x)$ for some $x \in A$. Since $A \subseteq B$, $x \in B$, and so $y = f(x)$ for some $x \in B$. Hence $y \in f(B)$ [*as was to be shown*].

42. The statement is false. *Counterexample:* Let $X = \{1, 2, 3\}$, let $Y = \{a, b\}$, and define a function $f: X \to Y$ by the arrow diagram shown below.

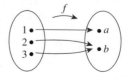

Let $A = \{1, 2\}$ and $B = \{1, 3\}$. Then $f(A) = \{a, b\} = f(B)$, and so $f(A) \cap f(B) = \{a, b\}$. But $f(A \cap B) = f(\{1\}) = \{a\} \neq \{a, b\}$. And so $f(A) \cap f(B) \neq f(A \cap B)$. (This is just one of many possible counterexamples.)

44. The statement is true. *Proof:* Let f be a function from a set X to a set Y, and suppose $C \subseteq Y$, $D \subseteq Y$, and $C \subseteq D$. [*We must show that* $f^{-1}(C) \subseteq f^{-1}(D)$]. Suppose $x \in f^{-1}(C)$. Then $f(x) \in C$. Since $C \subseteq D$, $f(x) \in D$ also. Hence by definition of inverse image, $x \in f^{-1}(D)$. [*So* $f^{-1}(C) \subseteq f^{-1}(D)$].

45. *Hint:* $x \in f^{-1}(C \cup D) \Leftrightarrow f(x) \in C \cup D \Leftrightarrow f(x) \in C$ or $f(x) \in D$

Section 7.2

1. The second statement is the contrapositive of the first.

2. a. most **b.** least

3. Consider the function defined by the arrow diagram shown below:

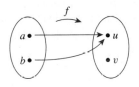

Observe that a is sent to exactly one element of Y, namely, u, and b is also sent to exactly one element of Y, namely, u also. So it is true that every element of X is sent to exactly one element of Y. But f is not one-to-one because $f(a) = f(b)$ but $a \neq b$. Note that to say, "Every element of X is sent to exactly one element of Y" is just another way of saying that in the arrow diagram for the function there is only one arrow coming out of each element of X. But this statement is part of the definition of *any* function, not just of a one-to-one function.

4. *Hint:* The statement is true.

5. *Hint:* One of the incorrect ways is (b).

6. a. f is not one-to-one because $f(1) = 4 = f(9)$ and $1 \neq 9$. f is not onto because $f(x) \neq 3$ for any x in X.

b. g is one-to-one because $g(1) \neq g(5)$, $g(1) \neq g(9)$, and $g(5) \neq g(9)$. g is onto because each element of Y is the image of some element of X: $3 = g(5)$, $4 = g(9)$, and $7 = g(1)$.

7. a. F is not one-to-one because $F(c) = x = F(d)$ and $c \neq d$. F is onto because each element of Y is the image of some element of X: $x = F(c) = F(d)$, $y = F(a)$, and $z = F(b)$.

9. a. One example of many is the following:

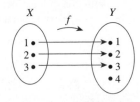

10. a. There are four choices for where to send the first element of the domain (any element of the co-domain may be chosen), three choices for where to send the second (since the function is one-to-one, the second element of the domain must go to a different element of the co-domain from the one to which the first element went), and two choices for where to send the third element (again since the function is one-to-one). Thus the answer is $4 \cdot 3 \cdot 2 = 24$.

b. none

e. *Hint:* The answer is

$$n(n-1)\cdots(n-m+1).$$

11. a. Let the elements of the domain be called a, b, and c and the elements of the co-domain be called u and v. In order for a function from $\{a, b, c\}$ to $\{u, v\}$ to be onto, two elements of the domain must be sent to u and one to v, or two elements must be sent to v and one to u. There are as many ways to send two elements of the domain to u and one to v as there are ways to choose which elements of $\{a, b, c\}$ to send to u, namely, $\binom{3}{2} = 3$. Similarly, there are $\binom{3}{2} = 3$ ways to send two elements of the domain to v and one to u. Therefore, there are $3 + 3 = 6$ onto functions from a set with three elements to a set with two elements.

c. *Hint:* The answer is 6.

d. Consider functions from a set with four elements to a set with two elements. Denote the set of four elements by $X = \{a, b, c, d\}$ and the set of two elements by $Y = \{u, v\}$. Divide the set of all onto functions from X to Y into two categories. The first category consists of all those that send the three elements in $\{a, b, c\}$ onto $\{u, v\}$ and that send d to either u or v. The functions in this category can be defined by the following two-step process:

Step 1: Construct an onto function from $\{a, b, c\}$ to $\{u, v\}$.

Step 2: Choose whether to send d to u or to v.

By part (a), there are six ways to perform step 1, and, because there are two choices for where to send d, there are two ways to perform step 2. Thus, by the multiplication rule, there are $6 \cdot 2 = 12$ ways to define the functions in the first category.

The second category consists of all those onto functions from X to Y that send all three elements in $\{a, b, c\}$ to either u or v and that send d to whichever of u or v is not the image of the others. Because there are only two choices for where to send the elements in $\{a, b, c\}$, and because d is simply sent to wherever the others do not go, there are just two functions in the second category.

Every onto function from X to Y either sends d to a single element of Y or it does not. If it sends d to a single element of Y, then it is in the second category. If it does not, then the image of $\{a, b, c\}$ is $\{u, v\}$ and so the "restriction" of the function to $\{a, b, c\}$ is onto. Therefore, the function is one of those included in the first category. Thus all onto functions from X to Y are in one of the two categories and no function is in both categories, and so the total number of onto functions is $12 + 2 = 14$.

f. *Hint:* Let X be a set with m elements and Y be a set with n elements, where $m \geq n \geq 1$, and let $x \in X$. Then

$$c_{m,n} = \left[\begin{array}{c} \text{the number of onto} \\ \text{functions from } X \text{ to } Y \end{array}\right]$$

$$= \left[\begin{array}{c} \text{the number of onto} \\ \text{functions for which} \\ f^{-1}(f(x)) \text{ has more} \\ \text{than one element} \end{array}\right]$$

$$+ \left[\begin{array}{c} \text{the number of onto} \\ \text{functions for which} \\ f^{-1}(f(x)) \text{ has one} \\ \text{element} \end{array}\right] \quad [Why?]$$

12. a. (i) *f is one-to-one:* Suppose $f(n_1) = f(n_2)$ for some integers n_1 and n_2. [*We must show that $n_1 = n_2$*]. By definition of f, $2n_1 = 2n_2$, and dividing both sides by 2 gives $n_1 = n_2$, as was to be shown.

(ii) *f is not onto:* Consider $1 \in \mathbf{Z}$. We claim that $1 \neq f(n)$, for any integer n, because if there were an integer n such that $1 = f(n)$, then, by definition of f, $1 = 2n$. Dividing both sides by 2 would give $n = 1/2$. But $1/2$ is not an integer. Hence $1 \neq f(n)$ for any integer n, and so f is not onto.

b. *h is onto:* Suppose $m \in 2\mathbf{Z}$. [*We must show that there exists an integer n such that $h(n) = m$*]. *Since* $m \in 2\mathbf{Z}$, $m = 2k$ for some integer k. Let $n = k$. Then $h(n) = 2n = 2k = m$. Hence there exists an integer (namely, n) such that $h(n) = m$. This is what was to be shown.

14. a. (i) *H is not one-to-one:* $H(1) = 1 = H(-1)$ but $1 \neq -1$.

(ii) *H is not onto:* $H(x) \neq -1$ for any real number x (since no real numbers have negative squares).

15. The "proof" claims that f is one-to-one because for each integer n there is only one possible value for $f(n)$. But to say that for each integer n there is only one possible value for $f(n)$ is just another way of saying that f satisfies one of the conditions necessary for it to be a function. To show that f is one-to-one, one must show that any integer n has a *different* function value from that of the integer m whenever $n \neq m$.

16. *f is one-to-one. Proof:* Suppose $f(x_1) = f(x_2)$ where x_1 and x_2 are nonzero real numbers. [*We must show that $x_1 = x_2$*]. By definition of f,

$$\frac{x_1 + 1}{x_1} = \frac{x_2 + 1}{x_2}$$

cross-multiplying gives

$$x_1 x_2 + x_2 = x_1 x_2 + x_1,$$

and so

$$x_1 = x_2 \quad \begin{array}{l} \text{by subtracting } x_1 x_2 \\ \text{from both sides} \end{array}$$

[*This is what was to be shown.*]

17. f is not one-to-one. Note that

$$\frac{x_1}{x_1^2 + 1} = \frac{x_2}{x_2^2 + 1} \Rightarrow x_1 x_2^2 + x_1 = x_2 x_1^2 + x_2$$

$$\Rightarrow x_1 x_2^2 - x_2 x_1^2 = x_2 - x_1$$

$$\Rightarrow x_1 x_2(x_2 - x_1) = x_2 - x_1$$

$$\Rightarrow x_1 = x_2 \text{ or } x_1 x_2 = 1.$$

Thus for a counterexample take any x_1 and x_2 with $x_1 \neq x_2$ but $x_1 x_2 = 1$. For instance, take $x_1 = 2$ and $x_2 = 1/2$. Then $f(x_1) = f(2) = 2/5$ and $f(x_2) = f(1/2) = 2/5$, but $2 \neq 1/2$.

20. a. Note that because $\frac{417302072}{7} \cong 59614581.7$ and $417302072 - 7 \cdot 59614581 = 5$, $h(417\text{-}30\text{-}2072) = 5$. But position 5 is already occupied, so the next position is checked. It is free, and thus the record is placed in position 6.

21. Recall that $\lfloor x \rfloor = $ that unique integer n such that $n \leq x < n + 1$.

a. *Floor is not one-to-one:*
Floor$(0) = 0 = $ Floor $(1/2)$ but $0 \neq 1/2$.

b. *Floor is onto:* Suppose $m \in \mathbf{Z}$. [*We must show that there exists a real number y such that Floor$(y) = m$*.] Let $y = m$. Then Floor$(y) = $ Floor$(m) = m$ since m is an integer. (Actually, Floor takes the value m for *all* real numbers in the interval $m \leq x < m + 1$.) Hence there exists a real number y such that Floor$(y) = m$. This is what was to be shown.

22. a. *l is not one-to-one:* $l(0) = l(1) = 1$ but $1 \neq 0$.

b. *l is onto:* Suppose n is a nonnegative integer. [*We must show that there exists a string s in S such that $l(s) = n$.*] Let

$$s = \begin{cases} \epsilon \text{ (the null string)} & \text{if } n = 0 \\ \underbrace{00\ldots 0}_{n \text{ 0's}} & \text{if } n > 0 \end{cases}.$$

Then $l(s) = $ the length of $s = n$. This is what was to be shown.

24. a. *F is not one-to-one:* Let $A = \{a\}$ and $B = \{b\}$. Then $F(A) = F(B) = 1$ but $A \neq B$.

25. b. *N is not onto:* The number -1 is in $\mathbf{Z}$ but $N(s) \neq -1$ for any string s in S because no string has a negative number of a's.

28. a. Let $x = \log_8 27$ and $y = \log_2 3$. [*The question is: Is $x = y$?*] By definition of logarithm, both of these equations can be written in exponential form as

$$8^x = 27 \quad \text{and} \quad 2^y = 3.$$

Now $8 = 2^3$. So

$$8^x = (2^3)^x = 2^{3x}.$$

Also $27 = 3^3$ and $3 = 2^y$. So

$$27 = 3^3 = (2^y)^3 = 2^{3y}.$$

Hence, since $8^x = 27$,

$$2^{3x} = 2^{3y}.$$

By (7.2.5), then,

$$3x = 3y,$$

and so

$$x = y.$$

But $x = \log_8 27$ and $y = \log_2 3$, and so $\log_8 27 = y = \log_2 3$ and the answer to the question is yes.

29. *Proof:* Suppose that $b, x,$ and y are positive real numbers and $b \neq 1$. Let $u = \log_b(x)$ and $v = \log_b(y)$. By definition of logarithm, $b^u = x$ and $b^v = y$. By substitution, $\frac{x}{y} = \frac{b^u}{b^v} = b^{u-v}$ by (7.2.3) *and the fact that* $b^{-v} = \frac{1}{b^v}$]. Translating $\frac{x}{y} = b^{u-v}$ into logarithmic form gives $\log_b(\frac{x}{y}) = u - v$, and so, by substitution, $\log_b(\frac{x}{y}) = \log_b(x) - \log_b(y)$ [*as was to be shown*].

31. *Proof:* Suppose $a, b,$ and x are [*particular but arbitrarily chosen*] real numbers such that b and x are positive and $b \neq 1$. [*We must show that* $\log_b(x^a) = a \log_b x$.] Let

$$r = \log_b(x^a) \quad \text{and} \quad s = \log_b x$$

By definition of logarithm, these equations may be written in exponential form as

$$(*) \ \ b^r = x^a \quad \text{and} \quad (**) \ \ b^s = x$$

Then

$$b^r = x^a = (b^s)^a \quad \text{by substituting } (**) \text{ into } (*)$$
$$= b^{sa} \quad \text{by (7.2.2.)}$$

It follows from property (7.2.5) that $r = sa$, which equals as. Substituting the values of r and s into this equation yields $\log_b(x^a) = a \log_b x$ [*as was to be shown.*]

32. No. *Counterexample:* Define $f: \mathbf{R} \to \mathbf{R}$ and $g: \mathbf{R} \to \mathbf{R}$ as follows: $f(x) = x$ and $g(x) = -x$ for all real numbers x. Then f and g are both one-to-one [*because for all real number x_1 and x_2, if $f(x_1) = f(x_2)$ then $x_1 = x_2$, and if $g(x_1) = g(x_2)$ then $-x_1 = -x_2$ and so $x_1 = x_2$ also*], but $f + g$ is not one-to-one [*because $f + g$ satisfies the equation $(f + g)(x) = x + (-x) = 0$ for all real numbers x, and so, for instance, $(f + g)(1) = (f + g)(2)$ but $1 \neq 2$*].

34. Yes. *Proof:* Let b be a one-to-one function from $\mathbf{R}$ to $\mathbf{R}$, and let c be any nonzero real number. Suppose $(cf)(x_1) = (cf)(x_2)$. [*We must show that $x_1 = x_2$.*] It follows by definition of cf that $cf(x_1) = cf(x_2)$. Since $c \neq 0$, we may divide both sides of the equation by c to obtain $f(x_1) = f(x_2)$. But since f is one-to-one, this implies that $x_1 = x_2$ [*as was to be shown.*]

36.

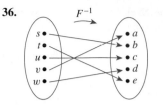

38. The function is not onto. Hence it is not a one-to-one correspondence.

39. The answer to exercise 12(b) shows that h is onto. To show that h is one-to-one, suppose $h(n_1) = h(n_2)$. By definition of h, this implies that $2n_1 = 2n_2$. Dividing both sides by 2 gives $n_1 = n_2$. Hence h is one-to-one.

Given any even integer m, if $m = h(n)$, then by definition of h, $m = 2n$, and so $n = m/2$. Thus

$$h^{-1}(m) = \frac{m}{2} \quad \text{for all } m \in 2\mathbf{Z}.$$

40. The function g is not a one-to-one correspondence because it is not onto. For instance, if $m = 2$, it is impossible to find an integer n such that $g(n) = m$. (This is because if $g(n) = m$, then $4n - 5 = 2$, and so $n = \frac{7}{4}$. Thus the only number n with the property that $g(n) = m$ is $\frac{7}{4}$. But $\frac{7}{4}$ is not an integer.)

43. The function is not one-to-one. Hence it is not a one-to-one correspondence.

44. The function is not one-to-one. Hence it is not a one-to-one correspondence.

48. The answer to exercise 16 shows that f is one-to-one, and if the co-domain is taken to be the set of all real numbers not equal to 1, then f is also onto. [*The reason is that given any real number $y \neq 1$, if we take $x = \frac{1}{y-1}$, then*

$$f(x) = f\left(\frac{1}{y-1}\right) = \frac{\frac{1}{y-1} + 1}{\frac{1}{y-1}} = \frac{1 + (y-1)}{1} = y.]$$

$$f^{-1}(y) = \frac{1}{y-1} \quad \text{for each real number } y \neq 1.$$

49. *Hint:* Is there a real number x such that $f(x) = 1$?

53. *Hint:* Let a function F be given and suppose the domain of F is represented as a one-dimensional array $a[1], a[2], \ldots, a[n]$. Introduce a variable *answer* whose initial value is "one-to-one." The main part of the body of the algorithm could be written as follows:

```
while (i ≤ n − 1 and answer = "one-to-one")
    j := i + 1
    while (j ≤ n and answer = "one-to-one")
        if (F(a[i]) = F(a[j]) and a[i] ≠ a[j])
        then answer := "not one-to-one"
        j := j + 1
    end while
    i := i + 1
end while
```

What can you say if execution reaches this point?

54. *Hint:* Let a function F be given and suppose the domain and co-domain of F are represented by the one-dimensional arrays $a[1], a[2], \ldots, a[n]$ and $b[1], b[2], \ldots, b[m]$, respectively. Introduce a variable *answer* whose initial value is "onto." For each $b[i]$ from $i = 1$ to m, make a search through $a[1], a[2], \ldots, a[n]$ to check whether $b[i] = F(a[j])$ for some $a[j]$. Introduce a Boolean variable to indicate whether a search has been successful. (Set the variable equal to 0 before the start of each search, and let it have the value 1 if the search is successful.) At the end of each search, check the value of the Boolean variable. If it is 0, then F is not onto. If all searches are successful, then F is onto.

Section 7.3

1. a. No. For instance, the aces of the four different suits could be selected.
  b. Yes. Let x_1, x_2, x_3, x_4, x_5 be the five cards. Consider the function S that sends each card to its suit.

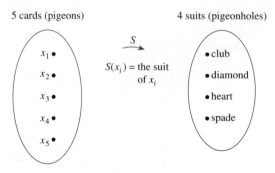

By the pigeonhole principle, S is not one-to-one: $S(x_i) = S(x_j)$ for some two cards x_i and x_j. Hence at least two cards have the same suit.

3. Yes. Denote the residents by $x_1, x_2, \ldots, x_{500}$. Consider the function B from residents to birthdays that sends each resident to his or her birthday:

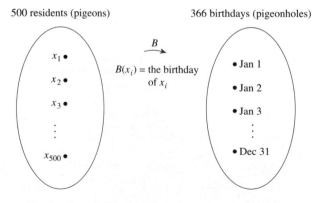

By the pigeonhole principle, B is not one-to-one: $B(x_i) = B(x_j)$ for some two residents x_i and x_j. Hence at least two residents have the same birthday.

5. a. Yes. There are only three possible remainders that can be obtained when an integer is divided by 3: 0, 1, and

2. Thus, by the pigeonhole principle, if four integers are each divided by 3, then at least two of them must have the same remainder.

More formally, call the integers n_1, n_2, n_3, and n_4, and consider the function R that sends each integer to the remainder obtained when that integer is divided by 3:

By the pigeonhole principle, R is not one-to-one, $R(n_i) = R(n_j)$ for some two integers n_i and n_j. Hence at least two integers must have the same remainder.
  b. No. For instance, $\{0, 1, 2\}$ is a set of three integers no two of which have the same remainder when divided by 3.

7. *Hint:* Look at Example 7.3.3.

9. a. Yes.

Solution 1: Only six of the numbers from 1 to 12 are even (namely, 2, 4, 6, 8, 10, 12), so at most six even numbers can be chosen from between 1 and 12 inclusive. Hence if seven numbers are chosen, at least one must be odd.

Solution 2: Partition the set of all integers from 1 through 12 into six subsets (the pigeonholes), each consisting of an odd and an even number: $\{1, 2\}, \{3, 4\}, \{5, 6\}, \{7, 8\}, \{9, 10\}, \{11, 12\}$. If seven integers (the pigeons) are chosen from among 1 through 12, then, by the pigeonhole principle, at least two must be from the same subset. But each subset contains one odd and one even number. Hence at least one of the seven numbers is odd.

Solution 3: Let $S = \{x_1, x_2, x_3, x_4, x_5, x_6, x_7\}$ be a set of seven numbers chosen from the set $T = \{1, 2, 3, 4, 5, 6, 7, 8, 9, 10, 11, 12\}$, and let P be the following partition of T: $\{1, 2\}, \{3, 4\}, \{5, 6\}, \{7, 8\}, \{9, 10\}$, and $\{11, 12\}$. Since each element of S lies in exactly one subset of the partition, we can define a function F from S to P by letting $F(x_i)$ be the subset that contains x_i.

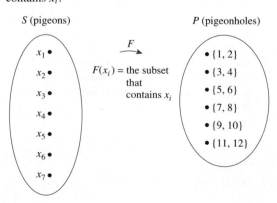

Since S has 7 elements and P has 6 elements, by the pigeonhole principle, F is not one-to-one. Thus two distinct numbers of the seven are sent to the same subset, which implies that these two numbers are the two distinct elements of the subset. Therefore, since each pair consists of one odd and one even integer, one of the seven numbers is odd.

b. No. For instance, none of the 10 numbers 1, 3, 5, 7, 9, 11, 13, 15, 17, 19 is even.

10. Yes. There are n even integers in the set $\{1, 2, 3, \ldots, 2n\}$, namely $2(= 2 \cdot 1), 4(= 2 \cdot 2), 6(= 2 \cdot 3), \ldots, 2n(= 2 \cdot n)$. So the maximum number of even integers that can be chosen is n. Thus if $n + 1$ integers are chosen, at least one of them must be odd.

12. The answer is 27. There are only 26 black cards in a standard 52-card deck, so at most 26 black cards can be chosen. Hence if 27 are taken, at least one must be red.

14. There are 61 integers from 0 to 60 inclusive. Of these, 31 are even $(0 = 2 \cdot 0, 2 = 2 \cdot 1, 4 = 2 \cdot 2, \ldots, 60 = 2 \cdot 30)$ and so 30 are odd. Hence if 32 integers are chosen, at least one must be odd, and if 31 integers are chosen, at least one must be even.

17. The answer is 8. (There are only seven possible remainders for division by 7: 0, 1, 2, 3, 4, 5, 6.)

20. The answer is 20,483 [*namely*, 0, 1, 2, …, 20482].

22. This number is irrational; the decimal expansion neither terminates nor repeats.

24. Let A be the set of the thirteen chosen numbers, and let B be the set of all prime numbers between 1 and 40. Note that $B = \{2, 3, 5, 7, 11, 13, 17, 19, 23, 29, 31, 37\}$. For each x in A, let $F(x)$ be the smallest prime number that divides x. Since A has 13 elements and B has 12 elements, by the pigeonhole principle F is not one-to-one. Thus $F(x_1) = F(x_2)$ for some $x_1 \neq x_2$ in A. By definition of F, this means that the smallest prime number that divides x_1 equals the smallest prime number that divides x_2. Therefore, two numbers in A, namely x_1 and x_2, have a common divisor greater than 1.

25. Yes. This follows from the generalized pigeonhole principle with 30 pigeons, 12 pigeonholes, and $k = 2$, using the fact that $30 > 2 \cdot 12$.

26. No. For instance, the birthdays of the 30 people could be distributed as follows: three birthdays in each of the six months January through June and two birthdays in each of the six months July through December.

29. The answer is $x = 3$. There are 18 years from 17 through 34. Now $40 > 18 \cdot 2$, so by the generalized pigeonhole principle, you can be sure that there are at least $x = 3$ students of the same age. However, since $18 \cdot 3 > 40$, you cannot be sure of having more than three students with the same age. (For instance, three students could be each of the ages 17 through 20, and two could be each of the ages from 21 through 34.) So x cannot be taken to be greater than 3.

31. *Hint:* Use the same type of reasoning as in Example 7.3.6.

32. *Hints:* (1) The number of subsets of the six integers is $2^6 = 64$. (2) Since each integer is less than 13, the largest possible sum is 57. (Why? What gives this sum?)

34. *Hint:* Let X be the set consisting of the given 52 positive integers, and let Y be the set containing the following elements: $\{00\}$, $\{50\}$, $\{01, 99\}$, $\{02, 98\}$, $\{03, 97\}$, …, $\{48, 52\}$, $\{49, 51\}$. Define a function F from X to Y by the rule $F(x) =$ the set containing the last two digits of x. Use the pigeonhole principle to argue that F is not one-to-one, and show how the desired conclusion follows.

35. *Hint:* Represent each of the 101 integers x_i as $a_i 2^{k_i}$ where a_i is odd and $k_i \geq 0$. Now $1 \leq x_i \leq 200$, and so $1 \leq a_i \leq 199$ for all i. There are only 100 odd integers from 1 to 199 inclusive.

36. b. *Hint:* For each $k = 1, 2, \ldots, n$, let $a_k = x_1 + x_2 + \cdots + x_k$. If some a_k is divisible by n, then the problem is solved: the consecutive subsequence is $x_1, x_2, \ldots, x_k$. If no a_k is divisible by n, then $a_1, a_2, a_3, \ldots, a_n$ satisfies the hypothesis of part (a). Hence $a_j - a_i$ is divisible by n for some integers i and j with $j > i$. Write $a_j - a_i$ in terms of the x_i's to derive the given conclusion.

37. *Hint:* Let $a_1, a_2, \ldots, a_{n^2+1}$ be any sequence of $n^2 + 1$ distinct real numbers, and suppose that this sequence contains neither a strictly increasing subsequence of length $n + 1$ nor a strictly decreasing subsequence of length $n + 1$. Let S be the set of all ordered pairs of integers (i, d), where $1 \leq i \leq n$ and $1 \leq d \leq n$. For each term a_k in the sequence, let $F(a_k) = (i_k, d_k)$, where i_k is the length of the longest increasing sequence starting at a_k, and d_k is the length of the longest decreasing sequence starting at a_k. Suppose that F is one-to-one and derive a contradiction.

Section 7.4

1. $g \circ f$ is defined as follows:
$$(g \circ f)(1) = g(f(1)) = g(5) = 1,$$
$$(g \circ f)(3) = g(f(3)) = g(3) = 5,$$
$$(g \circ f)(5) = g(f(5)) = g(1) = 3.$$

$f \circ g$ is defined as follows:
$$(f \circ g)(1) = f(g(1)) = f(3) = 3,$$
$$(f \circ g)(3) = f(g(3)) = f(5) = 1,$$
$$(f \circ g)(5) = f(g(5)) = f(1) = 5.$$

Then $g \circ f \neq f \circ g$ because, for example, $(g \circ f)(1) \neq (f \circ g)(1)$.

3. $(G \circ F)(x) = G(F(x)) = G(x^3) = x^3 - 1$ for all real numbers x.
$(F \circ G)(x) = F(G(x)) = F(x - 1) = (x - 1)^3$ for all real numbers x.
$G \circ F \neq F \circ G$ because, for instance, $(G \circ F)(2) = 2^3 - 1 = 7$, whereas $(F \circ G)(2) = (2 - 1)^3 = 1$.

5. $G \circ F$ is defined by $(G \circ F)(n) = n$, for all integers n, because for any integer n, $(G \circ F)(n) = G(F(n)) = G(2n) = \lfloor \frac{2n}{2} \rfloor = \lfloor n \rfloor = n$. $F \circ G$ is defined by

$(F \circ G)(n) = 2 \lfloor \frac{n}{2} \rfloor$, for all integers n, because for any integer n, $(F \circ G)(n) = F(G(n)) = F(\lfloor \frac{n}{2} \rfloor) = 2 \lfloor \frac{n}{2} \rfloor$. Then $G \circ F \neq F \circ G$ because, for instance, $(G \circ F)(3) = 3$, whereas $(F \circ G)(3) = 2 \lfloor \frac{3}{2} \rfloor = 2 \cdot 1 = 2$.

7. $(T \circ L)(abaa) = T(L(abaa)) = T(4) = 4 \bmod 3 = 1$
$(T \circ L)(baaab) = T(L(baaab)) = T(5) = 5 \bmod 3 = 2$
$(T \circ L)(aaa) = T(L(aaa)) = T(3) = 3 \bmod 3 = 0$

9. $(F^{-1} \circ F)(x) = F^{-1}(F(x)) = F^{-1}(3x + 2)$

$$= \frac{(3x + 2) - 2}{3} = \frac{3x}{3} = x = i_{\mathbf{R}}(x)$$

for all x in $\mathbf{R}$. Hence $F^{-1} \circ F = i_{\mathbf{R}}$ by definition of equality of functions.

$$(F \circ F^{-1})(y) = F(F^{-1}(y)) = F\left(\frac{y-2}{3}\right)$$

$$= 3\left(\frac{y-2}{3}\right) + 2 = (y - 2) + 2$$

$$= y = i_{\mathbf{R}}(y)$$

for all y in $\mathbf{R}$. Hence $F \circ F^{-1} = i_{\mathbf{R}}$ by definition of equality of functions.

12. **a.** By definition of logarithm with base b, for each real number x, $\log_b(b^x)$ is the exponent to which b must be raised to obtain b^x. But this exponent is just x. So $\log_b(b^x) = x$.

13. *Hint:* Suppose f is any function from a set X to a set Y, and show that for all x in X, $(i_Y \circ f)(x) = f(x)$.

15. **a.** $s_k = s_m$

16. No. *Counterexample:* Define f and g by the arrow diagrams below.

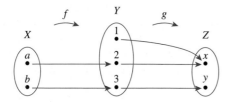

Then $g \circ f$ is one-to-one but g is not one-to-one. (So it is false that *both* f and g are one-to-one by De Morgan's law!) (This is one counterexample among many. Can you construct a different one?)

18. *Hint:* Suppose $f: X \to Y$ and $g: Y \to Z$ are functions and $g \circ f$ is one-to-one. Given x_1 and x_2 in X, if $f(x_1) = f(x_2)$ then $(g \circ f)(x_1) = (g \circ f)(x_2)$. (Why?) Then use the fact that $g \circ f$ is one-to-one.

19. *Hint:* Suppose $f: X \to Y$ and $g: Y \to Z$ are functions and $g \circ f$ is onto. Given $z \in Z$, there is an element x in X such that $(g \circ f)(x) = z$. (Why?) Let $y = f(x)$. Then $g(y) = z$.

21. True. *Proof:* Suppose X is any set and f, g, and h are functions from X to X such that h is one-to-one and $h \circ f = h \circ g$. [*We must show that for all x in X, $f(x) = g(x)$.*] Suppose x is any element in X. Because $h \circ f = h \circ g$, we have that $(h \circ f)(x) = (h \circ g)(x)$ by definition of equality

of functions. Then, by definition of composition of functions, $h(f(x)) = h(g(x))$. But since h is one-to-one, this implies that $f(x) = g(x)$ [*as was to be shown.*]

23.

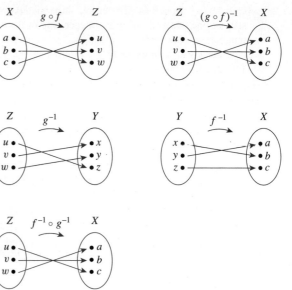

The functions $(g \circ f)^{-1}$ and $f^{-1} \circ g^{-1}$ are equal.

26. *Hints:* (1) Theorems 7.4.3 and 7.4.4 taken together ensure that $g \circ f$ is one-to-one and onto. (2) Use the inverse function property: $F^{-1}(b) = a \Leftrightarrow F(a) = b$, for all a in the domain of F and b in the domain of F^{-1}.

27. *Counterexample:* Let $X = \{1, 2\}$, $Y = \{3\}$, and $A = \{1\}$, and define $f: X \to Y$ by specifying that $f(1) = f(2) = 3$. Then $F(A) = \{3\}$, and $f^{-1}(f(A)) = f^{-1}(\{3\}) = \{1, 2\}$. Thus $f^{-1}(f(A)) \neq A$. (This is one counterexample among many. Can you construct a different one?)

29. *Proof:* Let $f: X \to Y$ be any function, and let C be any subset of Y. Suppose $y \in f(f^{-1}(C))$. [*We must show that $y \in C$.*] Then, by definition of image of a set, $y = f(x)$ for some $x \in f^{-1}(C)$, and so, by definition of inverse image of a set, $f(x) \in C$. Hence $y \in C$ [*as was to be shown.*]

Section 7.5

1. The student should have replied that for A to have the same cardinality as B means that there is a function from A to B that is one-to-one and onto. A set cannot have the property of being one-to-one or onto another set; only a function can have these properties.

2. Define a function $f: \mathbf{Z}^+ \to S$ as follows: For all positive integers k, $f(k) = k^2$.

f is *one-to-one:* [*We must show that for all $k_1, k_2 \in \mathbf{Z}^+$, if $f(k_1) = f(k_2)$ then $k_1 = k_2$.*] Suppose k_1 and k_2 are positive integers and $f(k_1) = f(k_2)$. By definition of f, $(k_1)^2 = (k_2)^2$, so $k_1 = \pm k_2$. But k_1 and k_2 are *positive.* Hence $k_1 = k_2$.

f is *onto:* [*We must show that for all $n \in S$, there exists $k \in \mathbf{Z}^+$ such that $n = f(k)$.*] Suppose $n \in S$. By definition of S, $n = k^2$ for some positive integer k. But then by definition of f, $n = f(k)$.

Since there is a one-to-one, onto function (namely, f) from $\mathbf{Z}^+$ to S, the two sets have the same cardinality.

3. Define $f: \mathbf{Z} \to 3\mathbf{Z}$ by the rule $f(n) = 3n$ for all integers n. The function f is one-to-one because for any integers n_1 and n_2, if $f(n_1) = f(n_2)$ then $3n_1 = 3n_2$ and so $n_1 = n_2$. Also f is onto because if m is any element in $3\mathbf{Z}$, then $m = 3k$ for some integer k. But then $f(k) = 3k = m$ by definition of f. Thus, since there is a function $f: \mathbf{Z} \to 3\mathbf{Z}$ that is one-to-one and onto, $\mathbf{Z}$ has the same cardinality as $3\mathbf{Z}$.

6. *Hint:* If $m \in 2\mathbf{Z}$, show that $J(m) = J(m+1) = m$.

7. b. For each positive integer n, $F(n) = (-1)^n \left\lfloor \frac{n}{2} \right\rfloor$.

8. It was shown in Example 7.5.2 that $\mathbf{Z}$ is countably infinite, which means that $\mathbf{Z}^+$ has the same cardinality as $\mathbf{Z}$. By exercise 3, $\mathbf{Z}$ has the same cardinality as $3\mathbf{Z}$. It follows by the transitive property of cardinality (Theorem 7.5.1 (c)) that $\mathbf{Z}^+$ has the same cardinality as $3\mathbf{Z}$. Thus $3\mathbf{Z}$ is countably infinite [*by definition of countably infinite*], and hence $3\mathbf{Z}$ is countable [*by definition of countable*].

10. *Proof:* Define $f: S \to U$ by the rule $f(x) = 2x$ for all real numbers x in S. Then f is one-to-one by the same argument as in exercise 12a of Section 7.2 with $\mathbf{R}$ in place of $\mathbf{Z}$. Furthermore, f is onto because if y is any element in U, then $0 < y < 2$ and so $0 < y/2 < 1$. Consequently, $y/2 \in S$ and $f(y/2) = 2(y/2) = y$. Hence f is a one-to-one correspondence, and so S and U have the same cardinality

11. *Hint:* Define $h: S \to V$ as follows: $h(x) = 3x + 2$, for all $x \in S$.

13.

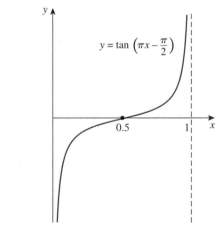

$$y = \tan\left(\pi x - \frac{\pi}{2}\right)$$

It is clear from the graph that f is one-to-one (since it is increasing) and that the image of f is all of $\mathbf{R}$ (since the lines $x = 0$ and $x = 1$ are vertical asymptotes). Thus S and $\mathbf{R}$ have the same cardinality.

16. In Example 7.5.4 it was shown that there is a one-to-one correspondence from $\mathbf{Z}^+$ to $\mathbf{Q}^+$. This implies that the positive rational numbers can be written as an infinite sequence: $r_1, r_2, r_3, r_4, \dots$. Now the set $\mathbf{Q}$ of all rational numbers consists of the numbers in this sequence together with 0 and the negative rational numbers: $-r_1, -r_2, -r_3, -r_4, \dots$. Let $r_0 = 0$. Then the elements of the set of all rational numbers

can be "counted" as follows:

$$r_0, r_1, -r_1, r_2, -r_2, r_3, -r_3, r_4, -r_4, \dots.$$

In other words, we can define a one-to-one correspondence

$$G(n) = \begin{cases} r_{n/2} & \text{if } n \text{ is even} \\ -r_{(n-1)/2} & \text{if } n \text{ is odd} \end{cases} \quad \text{for all integers } n \geq 1.$$

Therefore, $\mathbf{Q}$ is countably infinite and hence countable.

18. *Hint:* No.

19. *Hint:* Suppose r and s are real numbers with $s > r > 0$. Let n be an integer such that $n > \dfrac{\sqrt{2}}{s - r}$. Then $s - r > \dfrac{\sqrt{2}}{n}$. Let $m = \left\lfloor \dfrac{nr}{\sqrt{2}} \right\rfloor + 1$. Then $m > \dfrac{nr}{\sqrt{2}} \geq m - 1$. Use the fact that $s = r + (s - r)$ to show that $r < \dfrac{\sqrt{2}m}{n} < s$.

22. *Hint:* Use the unique factorization theorem (Theorem 3.3.3) and Theorem 7.5.3.

23. a. Define a function $G: \mathbf{Z}^{nonneg} \to \mathbf{Z}^{nonneg} \times \mathbf{Z}^{nonneg}$ as follows: Let $G(0) = (0, 0)$, and then follow the arrows in the diagram, letting each successive ordered pair of integers be the value of G for the next successive integer. Thus, for instance, $G(1) = (1, 0)$, $G(2) = (0, 1)$, $G(3) = (2, 0)$, $G(3) = (1, 1)$, $G(4) = (0, 2)$, $G(5) = (3, 0)$, $G(6) = (2, 1)$, $G(7) = (1, 2)$, and so forth.

b. *Hint:* Observe that if the top ordered pair of any given diagonal is $(k, 0)$, the entire diagonal (moving from top to bottom) consists of $(k, 0)$, $(k - 1, 1)$, $(k - 2, 2)$, \dots, $(2, k - 2)$, $(1, k - 1)$, $(0, k)$. Thus for all the ordered pairs (m, n) within any given diagonal, the value of $m + n$ is constant, and as you move down the ordered pairs in the diagonal, starting at the top, the value of the second element of the pair keeps increasing by 1.

25. *Hint:* There are at least two different approaches to this problem. One is to use the method discussed in Section 3.2. Another is to suppose that $1.999999\dots < 2$ and derive a contradiction. (Show that the difference between 2 and $1.999999\dots$ can be made smaller than any given positive number.)

26. Let A be an infinite set. Construct a countably infinite subset $a_1, a_2, a_3, \dots$ of A, by letting a_1 be any element of A, letting a_2 be any element of A other than a_1, letting a_3 be any element of A other than a_1 or a_2, and so forth. This process never stops (and hence $a_1, a_2, a_3, \dots$ is an infinite sequence) because A is an infinite set. More formally,

1. Let a_1 be any element of A.
2. For each integer $n \geq 2$, let a_n be any element of $A - \{a_1, a_2, a_3, \dots, a_{n-1}\}$. Such an element exists, for otherwise $A - \{a_1, a_2, a_3, \dots, a_{n-1}\}$ would be empty and A would be finite.

27. *Proof:* Suppose A is any countably infinite set, B is any set, and $g: A \to B$ is onto. Since A is countably infinite, there is a one-to-one correspondence $f: \mathbf{Z}^+ \to A$. Then, in particular, f is onto, and so by Theorem 7.4.4, $g \circ f$ is an onto function from $\mathbf{Z}^+$ to B. Define a function $h: B \to \mathbf{Z}^+$ as follows: Suppose x is any element of B. Since $g \circ f$ is onto, $\{m \in \mathbf{Z}^+ \mid (g \circ f)(m) = x\} \neq \emptyset$. Thus, by the well-ordering principle for the integers, this set has a least element. In other words, there is a least positive integer n with $(g \circ f)(n) = x$. Let $h(x)$ be this integer.

We claim that h is a one-to-one. For suppose $h(x_1) = h(x_2) = n$. By definition of h, n is the least positive integer with $(g \circ f)(n) = x_1$. But also by definition of h, n is the least positive integer with $(g \circ f)(n) = x_2$. Hence $x_1 = (g \circ f)(n) = x_2$.

Thus h is a one-to-one correspondence between B and a subset S of positive integers (the range of h). Since any subset of a countable set is countable (Theorem 7.5.3), S is countable, and so there is a one-to-one correspondence between B and a countable set. Hence, by the transitive property of cardinality, B is countable.

28. *Hint:* Suppose that A and B are countably infinite. Then there are one-to-one correspondences $f_A: \mathbf{Z}^+ \to A$ and $f_B: \mathbf{Z}^+ \to B$. Define a function $g: \mathbf{Z}^+ \to A \cup B$ as follows:

$$g(n) = \begin{cases} f_A\left(\frac{n}{2}\right) & \text{if } n \text{ is even} \\ f_B\left(\frac{n+1}{2}\right) & \text{if } n \text{ is odd} \end{cases}.$$

Show that g is onto, and finish by using the result of exercise 27.

29. *Hint:* Use proof by contradiction and the fact that the set of all real numbers is uncountable.

32. *Hint:* Use the one-to-one correspondence $F: \mathbf{Z}^+ \to \mathbf{Z}$ of Example 7.5.2 to define a function $G: \mathbf{Z}^+ \times \mathbf{Z}^+ \to \mathbf{Z} \times \mathbf{Z}$ by the formula $G(m, n) = (F(m), F(n))$. Show that G is a one-to-one correspondence, and use the result of exercise 22 and the transitive property of cardinality.

34. *Hint for Solution 1:* Define a function $f: \mathscr{P}(S) \to T$ as follows: For each subset A of S, let $f(A) = \chi_A$, the *characteristic function* of A, where $\chi_A: S \to \{0, 1\}$ is defined by the rule

$$\chi_A(x) = \begin{cases} 1 & \text{if } x \in A \\ 0 & \text{if } x \notin A \end{cases} \text{ for all } x \in S.$$

Show that f is one-to-one (for all $A_1, A_2 \subseteq S$, if $\chi_{A_1} = \chi_{A_2}$ then $A_1 = A_2$) and that f is onto (given any function $g: S \to \{0, 1\}$, there is a subset A of S such that $g = \chi_A$).

Hint for Solution 2: Define $H: T \to \mathscr{P}(S)$ by letting $H(f) = \{x \in S \mid f(x) = 1\}$. Show that H is a one-to-one correspondence?

35. *Partial proof (by contradiction):* Suppose not. Suppose there is a one-to-one, onto function $f: S \to \mathscr{P}(S)$. Let

$$A = \{x \in S \mid x \notin f(x)\}.$$

Then $A \in \mathscr{P}(S)$ and since f is onto, there is a $z \in S$ such that $A = f(z)$. [*Now derive a contradiction!*]

37. *Hint:* Since A and B are countable, their elements can be listed as

$$A: a_1, a_2, a_3, \ldots \quad \text{and} \quad B: b_1, b_2, b_3, \ldots$$

Represent the elements of $A \times B$ in a grid:

$$\begin{array}{lll} (a_1, b_1) & (a_1, b_2) & (a_1, b_3) \ldots \\ (a_2, b_1) & (a_2, b_2) & (a_2, b_3) \ldots \\ (a_3, b_1) & (a_3, b_2) & (a_3, b_3) \ldots \\ \vdots & \vdots & \vdots \end{array}$$

Now use a counting method similar to that of Example 7.5.4.

Section 8.1

1. $a_1 = 1, a_2 = 2a_1 + 2 = 2 \cdot 1 + 2 = 4,$
$a_3 = 2a_2 + 3 = 2 \cdot 4 + 3 = 11,$
$a_4 = 2a_3 + 4 = 2 \cdot 11 + 4 = 26$

3. $c_0 = 1, c_1 = 1 \cdot (c_0)^2 = 1 \cdot (1)^2 = 1,$
$c_2 = 2(c_1)^2 = 2 \cdot (1)^2 = 2,$
$c_3 = 3(c_2)^2 = 3 \cdot (2)^2 = 12$

5. $s_0 = 1, s_1 = 1, s_2 = s_1 + 2s_0 = 1 + 2 \cdot 1 = 3,$
$s_3 = s_2 + 2s_1 = 3 + 2 \cdot 1 = 5$

7. $u_1 = 1, u_2 = 1, u_3 = 3u_2 - u_1 = 3 \cdot 1 - 1 = 2,$
$u_4 = 4u_3 - u_2 = 4 \cdot 2 - 1 = 7$

9. By definition of $a_0, a_1, a_2, \ldots$, for each integer $k \geq 1$,

$$(*) \qquad a_k = 3k + 1 \quad \text{and}$$
$$(**) \qquad a_{k-1} = 3(k-1) + 1.$$

Then $a_{k-1} + 3$

$$\begin{aligned} &= 3(k-1) + 1 + 3 \\ &= 3k - 3 + 1 + 3 \\ &= 3k + 1 \\ &= a_k \end{aligned}$$

11. Call the nth term of the sequence c_n. Then, by definition, $c_n = 2^n - 1$, for each integer $n \geq 0$. Substitute k and $k-1$ in place of n to get

$$(*) \qquad c_k = 2^k - 1 \quad \text{and}$$
$$(**) \qquad c_{k-1} = 2^{k-1} - 1$$

for all integers $k \geq 1$. Then

$$\begin{aligned} 2c_{k-1} + 1 &= 2(2^{k-1} - 1) + 1 && \text{by substitution from } (**) \\ &= 2^k - 2 + 1 \\ &= 2^k - 1 && \text{by basic algebra} \\ &= c_k && \text{by substitution from } (*) \end{aligned}$$

13. Call the nth term of the sequence t_n. Then, by definition, $t_n = 2 + n$, for each integer $n \geq 0$. Substitute $k, k - 1$, and $k - 2$ in place of n to get

$$(*) \qquad\qquad t_k = 2 + k,$$

$$(**) \qquad\qquad t_{k-1} = 2 + (k - 1), \quad \text{and}$$

$$(***) \qquad\qquad t_{k-2} = 2 + (k - 2)$$

for each integer $k \geq 2$. Then

$$2t_{k-1} - t_{k-2}$$

$$= 2(2 + (k - 1) - (2 + (k - 2)) \qquad \text{by substitution from } (**) \text{ and } (***)$$

$$= 2(k + 1) - k$$

$$= 2 + k \qquad\qquad\qquad \text{by basic algebra}$$

$$= t_k \qquad\qquad\qquad\qquad \text{by substitution from } (*).$$

15. Let k be an integer and $k \geq 2$.

Case 1 (k is even): Then

$$a_k = (-2)^{k/2} \quad \text{and} \quad a_{k-2} = (-2)^{(k-2)/2}.$$

So

$$-2a_{k-2} = -2 \cdot (-2)^{(k-2)/2}$$

$$= (-2)^{1+(k-2)/2}$$

$$= (-2)^{k/2} \quad \text{since} \quad 1 + \frac{k-2}{2} = \frac{2}{2} + \frac{k-2}{2}$$

$$= \frac{2 + k - 2}{2}$$

$$= \frac{k}{2}$$

$$= a_k.$$

Case 2 (k is odd): Then

$$a_k = (-2)^{(k-1)/2}$$

and

$$a_{k-2} = (-2)^{(k-3)/2} \quad \text{since} \quad \frac{(k-2)-1}{2} = \frac{k-3}{2}$$

So

$$-2a_{k-2} = -2 \cdot (-2)^{(k-3)/2}$$

$$= (-2)^{1+(k-3)/2}$$

$$= (-2)^{(k-1)/2} \quad \text{since} \quad 1 + \frac{k-3}{2} = \frac{2}{2} + \frac{k-3}{2}$$

$$= \frac{2 + k - 3}{2}$$

$$= \frac{k-1}{2}$$

$$= a_k.$$

Hence in either case, $a_k = -2a_{k-2}$, as was to be shown.

18. a. $a_1 = 2$

$a_2 = 2$ (moves to move the top disk from pole A to pole C)

$+ 1$ (move to move the bottom disk from pole A to pole B)

$+ 2$ (moves to move the top disk from pole C to pole A)

$+ 1$ (move to move the bottom disk from pole B to pole C)

$+ 2$ (moves to move top disk from pole A to pole C)

$= 8$

$a_3 = 8 + 1 + 8 + 1 + 8 = 26$

c. For all integers $k \geq 2$.

$a_k = a_{k-1}$ (moves to move the top $k - 1$ disks from pole A to pole C)

$+ 1$ (move to move the bottom disk from pole A to pole B)

$+ a_{k-1}$ (moves to move the top disk from pole C to pole A)

$+ 1$ (move to move the bottom disks from pole B to pole C)

$+ a_{k-1}$ (moves to move the top disks from pole A to pole C)

$= 3a_{k-1} + 2.$

19. b. $b_4 = 40$

e. *Hint:* One solution is to use mathematical induction and apply the formula from part (c). Another solution is to prove by mathematical induction that when a most efficient transfer of n disks from one end pole to the other end pole is performed, at some point all the disks are on the middle pole.

20. a. $s_1 = 1$, $s_2 = 1 + 1 + 1 = 3$,
$s_3 = s_1 + (1 + 1 + 1) + s_1 = 5$

b. $s_4 = s_2 + (1 + 1 + 1) + s_2 = 9$

21. b. $t_3 = 14$

22. b. $r_0 = 1$, $r_1 = 1$, $r_2 = 1 + 4 \cdot 1 = 5$, $r_3 = 5 + 4 \cdot 1 = 9$,
$r_4 = 9 + 4 \cdot 5 = 29$, $r_5 = 29 + 4 \cdot 9 = 65$,
$r_6 = 65 + 4 \cdot 29 = 181$

23. c. There are 904 rabbit pairs, or 1,808 rabbits, after 12 months.

25. a. Each term of the Fibonacci sequence beyond the second equals the sum of the previous two. For any integer $k \geq 1$, the two terms previous to F_{k+1} are F_k and F_{k-1}. Hence, for all integers $k \geq 1$, $F_{k+1} = F_k + F_{k-1}$.

26. By repeated use of definition of the Fibonacci sequence, for all integers $k \geq 4$,

$$F_k = F_{k-1} + F_{k-2} = (F_{k-2} + F_{k-3}) + (F_{k-3} + F_{k-4})$$
$$= ((F_{k-3} + F_{k-4}) + F_{k-3}) + (F_{k-3} + F_{k-4})$$
$$= 3F_{k-3} + 2F_{k-4}.$$

27. For all integers $k \geq 1$,

$$F_k^2 - F_{k-1}^2$$
$$= (F_k - F_{k-1})(F_k + F_{k-1}) \quad \text{by basic algebra (difference of two squares)}$$
$$= (F_k - F_{k-1})F_{k+1} \quad \text{by definition of the Fibonacci sequence}$$
$$= F_k F_{k+1} - F_{k-1}F_{k+1}$$

31. Let $L = \lim\limits_{n \to \infty} \dfrac{F_{n+1}}{F_n}$. Since each $F_{n+1} > F_n > 0$, $L > 0$. Then, by definition of the Fibonacci sequence,

$$L = \lim_{n \to \infty} \left(\frac{F_{n-1} + F_n}{F_n} \right) = \lim_{n \to \infty} \left(\frac{F_{n-1}}{F_n} + \frac{F_n}{F_n} \right)$$
$$= \lim_{n \to \infty} \left(\frac{1}{\frac{F_n}{F_{n-1}}} + 1 \right) = \frac{1}{\lim_{n \to \infty} \left(\frac{F_n}{F_{n-1}} \right)} + 1$$
$$= \frac{1}{L} + 1.$$

Hence $L = \frac{1}{L} + 1$. Multiply both sides by L to obtain $L^2 = 1 + L$, or, equivalently, $L^2 - L - 1 = 0$. By the quadratic formula, then, $L = \frac{1 \pm \sqrt{5}}{2}$. But one of these numbers, $\frac{1-\sqrt{5}}{2}$, is less than zero, and $L > 0$. Hence $L = \frac{1+\sqrt{5}}{2}$.

32. *Hint:* Use the result of exercise 30 to prove that the infinite sequence $\frac{F_0}{F_1}, \frac{F_2}{F_3}, \frac{F_4}{F_5}, \ldots$ is strictly decreasing and that the infinite sequence $\frac{F_1}{F_2}, \frac{F_3}{F_4}, \frac{F_5}{F_6}, \ldots$ is strictly increasing. The first sequence is bounded below by 0, and the second sequence is bounded above by 1. Deduce that the limits of both sequences exist, and show that they are equal.

34. a. Because the 4% annual interest is compounded quarterly, the quarterly interest rate is $(4\%)/4 = 1\%$. Then $R_k = R_{k-1} + 0.01R_{k-1} = 1.01R_{k-1}$.

b. Because one year equals four quarters, the amount on deposit at the end of one year is $R_4 = \$5203.02$ (rounded to the nearest cent).

c. The annual percentage rate (APR) for the account is $\frac{\$5203.02 - \$5000.00}{\$5000.00} = 4.0604\%$.

36. a. Length 0: ϵ
 Length 1: 0, 1
 Length 2: 00, 01, 10, 11
 Length 3: 000, 001, 010, 011, 100, 101, 110
 Length 4: 0000, 0001, 0010, 0011, 0100, 0101, 0110,
  1000, 1001, 1010, 1011, 1100, 1101

b. By part (a), $d_0 = 1$, $d_1 = 2$, $d_2 = 4$, $d_3 = 7$, and $d_4 = 13$.

c. Let k be an integer with $k \geq 3$. Any string of length k that does not contain the bit pattern 111 starts either with a 0 or with a 1. If it starts with a 0, this can

be followed by any string of $k - 1$ bits that does not contain the pattern 111. There are d_{k-1} of these. If the string starts with a 1, then the first two bits are 10 or 11. If the first two bits are 10, then these can be followed by any string of $k - 2$ bits that does not contain the pattern 111. There are d_{k-2} of these. If the string starts with a 11, then the third bit must be 0 (because the string does not contain 111), and these three bits can be followed by any string of $k - 3$ bits that does not contain the pattern 111. There are d_{k-3} of these. Therefore, for all integers $k \geq 3$, $d_k = d_{k-1} + d_{k-2} + d_{k-3}$.

d. By parts (b) and (c), $d_5 = d_4 + d_3 + d_2 = 13 + 7 + 4 = 24$. This is the number of bit strings of length five that do not contain the pattern 111.

37. c. *Hint:* $s_k = 2s_{k-1} + 2s_{k-2}$

39. When one is climbing a staircase consisting of n stairs, the last step taken is either a single stair or two stairs together. The number of ways to climb the staircase and have the final step be a single stair is c_{n-1}; the number of ways to climb the staircase and have the final step be two stairs is c_{n-2}. Therefore, by the addition rule, $c_n = c_{n-1} + c_{n-2}$. Note also that $c_1 = 1$ and $c_2 = 2$ [*because either the two stairs can be climbed one by one or they can be climbed as a unit*].

41. a. $a_3 = 3$ (The three permutations that do not move more than one place from their "natural" positions are 213, 132, and 123.)

43. Call the set X, and suppose that $X = \{x_1, x_2, \ldots, x_n\}$. For each integer $i = 0, 1, 2, \ldots, n - 1$, we can consider the set of all partitions of X (let's call them *partitions of type i*) where one of the subsets of the partition is an $(i + 1)$-element set that contains x_n and i elements chosen from $\{x_1, \ldots, x_{n-1}\}$. The remaining subsets of the partition will be a partition of the remaining $(n - 1) - i$ elements of $\{x_1, \ldots, x_{n-1}\}$. For instance, if $X = \{x_1, x_2, x_3\}$, there are five partitions of the various types, namely,

Type 0: two partitions where one set is a 1-element set containing x_3: [$\{x_3\}, \{x_1\}, \{x_2\}$], [$\{x_3\}, \{x_1, x_2\}$]

Type 1: two partitions where one set is a 2-element set containing x_3: [$\{x_1, x_3\}, \{x_2\}$], [$\{x_2, x_3\}, \{x_1\}$]

Type 2: one partition where one set is a 3-element set containing x_3: $\{x_1, x_2, x_3\}$

In general, we can imagine constructing a partition of type i as a two-step process:

Step 1: Select out the i elements of $\{x_1, \ldots, x_{n-1}\}$ to put together with x_n,

Step 2: Choose any partition of the remaining $(n - 1) - i$ elements of $\{x_1, \ldots, x_{n-1}\}$ to put with the set formed in step 1.

There are $\binom{n-1}{i}$ ways to perform step 1 and $P_{(n-1)-i}$ ways to perform step 2. Therefore, by the multiplication rule, there are $\binom{n-1}{i} \cdot P_{(n-1)-i}$ partitions of type i. Because any partition of X is of type i for some $i = 0, 1, 2, \ldots, n - 1$, it follows from the addition rule that the total number of partitions is

$$\binom{n-1}{0}P_{n-1} + \binom{n-1}{1}P_{n-2}$$
$$+ \binom{n-1}{2}P_{n-3} + \cdots + \binom{n-1}{n-1}P_0.$$

45. $S_{5,2} = S_{4,1} + 2S_{4,2} = 1 + 2 \cdot 7 = 15$

48. *Proof* (*by mathematical induction*): Let the property $P(n)$ be the formula $S_{n,2} = 2^{n-1} - 1$.

Show that the property is true for n = 2:

We must show that $S_{2,2} = 2^{2-1} - 1$. By Example 8.1.11, $S_{2,2} = 1$, and $2^{2-1} - 1 = 2 - 1 = 1$ also. So the property is true for $n = 2$.

Show that for all integers k ≥ 2, if the property is true for n = k, then it is true for n = k + 1:

Suppose that for some integer $k \geq 2$, $S_{k,2} = 2^{k-1} - 1$. [*Inductive hypothesis.*] We must show that $S_{k+1,2} = 2^{(k+1)-1} - 1 = 2^k - 1$. But according to Example 8.1.11, $S_{k+1,2} = S_{k,1} + 2S_{k,2}$ and $S_{k,1} = 1$. So by substitution and the inductive hypothesis,

$$S_{k+1,2} = 1 + 2S_{k,2} = 1 + 2(2^{k-1} - 1)$$
$$= 1 + 2^k - 2 = 2^k - 1$$

[*as was to be shown*].

50. *Hint:* Observe that the number of onto functions from $X = \{x_1, x_2, x_3, x_4\}$ to $Y = \{y_1, y_2, y_3\}$ is $S_{4,3} \cdot 3!$ because the construction of an onto function can be thought of as a two-step process where step 1 is to choose a partition of X into three subsets and step 2 is to choose, for each subset of the partition, an element of Y for the elements of the subset to be sent to.

52. *Hint:* Use mathematical induction. In the inductive step, use Lemma 3.8.2 and the fact that $F_{k+2} = F_{k+1} + F_k$ to deduce that

$$\gcd(F_{k+2}, F_{k+1}) = \gcd(F_{k+1}, F_k).$$

53. c. *Hint:* If $k \geq 6$, any sequence of k games must begin with W, LW, or LLW, where L stands for "lose" and W stands for "win."

54. c. *Hint:* Divide the set of all derangements into two subsets: one subset consists of all derangements in which the number 1 changes places with another number, and the other subset consists of all derangements in which the number 1 goes to position $i \neq 1$ but i does not go to position 1. The answer is $d_k = (k-1)d_{k-1} + (k-1)d_{k-2}$. Can you justify it?

Section 8.2

1. a. $1 + 2 + 3 + \cdots + (k-1)$

$$= \frac{(k-1)((k-1)+1)}{2} = \frac{(k-1)k}{2}$$

b. $3 + 2 + 4 + 6 + 8 + \cdots + 2n$

$$= 3 + 2(1 + 2 + 3 + \cdots + n)$$
$$= 3 + 2\frac{n(n+1)}{2} = 3 + n(n+1)$$
$$= n^2 + n + 3$$

2. a. $1 + 2 + 2^2 + \cdots + 2^{i-1} = \frac{2^{(i-1)+1} - 1}{2 - 1} = 2^i - 1$

c. $2^n + 2^{n-2} \cdot 3 + 2^{n-3} \cdot 3 + \cdots + 2^2 \cdot 3 + 2 \cdot 3 + 3$

$$= 2^n + 3(2^{n-2} + 2^{n-3} + \cdots + 2^2 + 2 + 1)$$
$$= 2^n + 3(1 + 2 + 2^2 + \cdots + 2^{n-3} + 2^{n-2})$$
$$= 2^n + 3\left(\frac{2^{(n-2)+1} - 1}{2 - 1}\right)$$
$$= 2^n + 3(2^{n-1} - 1)$$
$$= 2 \cdot 2^{n-1} + 3 \cdot 2^{n-1} - 3$$
$$= 5 \cdot 2^{n-1} - 3$$

3. $a_0 = 1$

$a_1 = 1 \cdot a_0 = 1 \cdot 1 = 1$

$a_2 = 2a_1 = 2 \cdot 1$

$a_3 = 3a_2 = 3 \cdot 2 \cdot 1$

$a_4 = 4a_3 = 4 \cdot 3 \cdot 2 \cdot 1$

$\vdots$

Guess:

$a_n = n(n-1) \cdots 3 \cdot 2 \cdot 1 = n!$

5. $c_1 = 1$

$c_2 = 3c_1 + 1 = 3 \cdot 1 + 1 = 3 + 1$

$c_3 = 3c_2 + 1 = 3 \cdot (3 + 1) + 1 = 3^2 + 3 + 1$

$c_4 = 3c_3 + 1 = 3 \cdot (3^2 + 3 + 1) + 1$

$\quad = 3^3 + 3^2 + 3 + 1$

$\vdots$

Guess:

$c_n = 3^{n-1} + 3^{n-2} + \cdots + 3^3 + 3^2 + 3 + 1$

$\quad = \frac{3^n - 1}{3 - 1}$ by Theorem 4.2.3 with $r = 3$

$\quad = \frac{3^n - 1}{2}$

6. *Hint:*

$d_n = 2^n + 2^{n-2} \cdot 3 + 2^{n-3} \cdot 3 + \cdots + 2^2 \cdot 3 + 2 \cdot 3 + 3$

$\quad = 5 \cdot 2^{n-1} - 3$ for all integers $n \geq 1$

9. *Hint:* For any positive real numbers a and b,

$$\frac{\frac{a}{b}}{\frac{a}{b} + 2} = \frac{\frac{a}{b}}{\frac{a}{b} + 2} \cdot \frac{b}{b} = \frac{a}{a + 2b}.$$

10. $h_0 = 1$

$$h_1 = 2^1 - h_0 = 2^1 - 1$$

$$h_2 = 2^2 - h_1 = 2^2 - (2^1 - 1) = 2^2 - 2^1 + 1$$

$$h_3 = 2^3 - h_2 = 2^3 - (2^2 - 2^1 + 1)$$
$$= 2^3 - 2^2 + 2^1 - 1$$

$$h_4 = 2^4 - h_3 = 2^4 - (2^3 - 2^2 + 2^2 - 1)$$
$$= 2^4 - 2^3 + 2^2 - 2^1 + 1$$

$$\vdots$$

Guess:

$$h_n = 2^n - 2^{n-1} + \cdots + (-1)^n \cdot 1$$
$$= (-1)^n [1 - 2 + 2^2 - \cdots + (-1)^n \cdot 2^n]$$
$$= (-1)^n [1 + (-2)$$
$$\qquad + (-2)^2 - \cdots + (-2)^n] \quad \text{by basic algebra}$$
$$= (-1)^n \left[\frac{(-2)^{n+1} - 1}{(-2) - 1} \right] \quad \text{by Theorem 4.2.3}$$
$$= \frac{(-1)^{n+1} \cdot [(-2)^{n+1} - 1]}{(-1) \cdot (-3)}$$
$$= \frac{2^{n+1} - (-1)^{n+1}}{3} \quad \text{by basic algebra}$$

12. $s_0 = 3$

$$s_1 = s_0 + 2 \cdot 1 = 3 + 2 \cdot 1$$

$$s_2 = s_1 + 2 \cdot 2 = [3 + 2 \cdot 1] + 2 \cdot 2$$
$$= 3 + 2 \cdot (1 + 2)$$

$$s_3 = s_2 + 2 \cdot 3 = [3 + 2 \cdot (1 + 2)] + 2 \cdot 3$$
$$= 3 + 2 \cdot (1 + 2 + 3)$$

$$s_4 = s_3 + 2 \cdot 4 = [3 + 2 \cdot (1 + 2 + 3)] + 2 \cdot 4$$
$$= 3 + 2 \cdot (1 + 2 + 3 + 4)$$

$$\vdots$$

Guess:

$$s_n = 3 + 2 \cdot (1 + 2 + 3 + \cdots + (n - 1) + n)$$
$$= 3 + 2 \cdot \frac{n(n + 1)}{2} \quad \text{by Theorem 4.2.2}$$
$$= 3 + n(n + 1) \quad \text{by basic algebra}$$

14. $x_1 = 1$

$$x_2 = 3x_1 + 2 = 3 + 2$$

$$x_3 = 3x_2 + 3 = 3(3 + 2) + 3 = 3^2 + 3 \cdot 2 + 3$$

$$x_4 = 3x_3 + 4 = 3(3^2 + 3 \cdot 2 + 3) + 4$$
$$= 3^3 + 3^2 \cdot 2 + 3 \cdot 3 + 4$$

$$x_5 = 3x_4 + 5 = 3(3^3 + 3^2 \cdot 2 + 3 \cdot 3 + 4) + 5$$
$$= 3^4 + 3^3 \cdot 2 + 3^2 \cdot 3 + 3 \cdot 4 + 5$$

$$x_6 = 3x_5 + 6$$
$$= 3(3^4 + 3^3 \cdot 2 + 3^2 \cdot 3 + 4 \cdot 3 + 5) + 6$$
$$= 3^5 + 3^4 \cdot 2 + 3^3 \cdot 3 + 3^2 \cdot 4 + 3 \cdot 5 + 6$$

$$\vdots$$

Guess:

$$x_n = 3^{n-1} + 3^{n-2} \cdot 2 + 3^{n-3} \cdot 3 + \cdots + 3(n - 1) + n$$
$$= 3^{n-1} + \underbrace{3^{n-2} + 3^{n-2}}_{2 \text{ times}} + \underbrace{3^{n-3} + 3^{n-3} + 3^{n-3}}_{3 \text{ times}} +$$
$$\qquad + \underbrace{3 + 3 + \cdots + 3}_{(n-1) \text{ times}} + \underbrace{1 + 1 + \cdots + 1}_{n \text{ times}}$$
$$= (3^{n-1} + 3^{n-2} + \cdots + 3^2 + 3 + 1)$$
$$\qquad + (3^{n-2} + 3^{n-3} + \cdots + 3^2 + 3 + 1) + \cdots$$
$$\qquad\qquad + (3^2 + 3 + 1) + (3 + 1) + 1$$
$$= \frac{3^n - 1}{2} + \frac{3^{n-1} - 1}{2} + \cdots + \frac{3^3 - 1}{2}$$
$$\qquad\qquad + \frac{3^2 - 1}{2} + \frac{3 - 1}{2}$$
$$= \tfrac{1}{2}[(3^n + 3^{n-1} + \cdots + 3^2 + 3) - n]$$
$$= \tfrac{1}{2}[3(3^{n-1} + 3^{n-2} + \cdots + 3 + 1) - n]$$
$$= \tfrac{1}{2}\left(3\left(\tfrac{3^n - 1}{3 - 1}\right) - n\right)$$
$$= \tfrac{1}{4}(3^{n+1} - 3 - 2n)$$

18. *Proof:* Let d be any fixed constant, and let $a_0, a_1, a_2, \ldots$ be the sequence defined recursively by $a_k = a_{k-1} + d$ for all integers $k \geq 1$. The property, $P(n)$, is the equation $a_n = a_0 + nd$. We show by mathematical induction that this property is true for all integers $n \geq 0$.

Show that the property is true for $n = 0$:

When $n = 0$, the left-hand side of the equation is a_0, and the right-hand side is $a_0 + 0 \cdot d = a_0$, which equals the left-hand side. Thus the property is true for $n = 0$.

Show that for all integers $k \geq 0$, if the property is true for $n = k$, then it is true for $n = k + 1$:

Suppose

$$a_k = a_0 + kd, \quad \text{for some integer } k \geq 0.$$

$$[\textit{This is the inductive hypothesis.}]$$

We must show that $a_{k+1} = a_0 + (k + 1)d$. But

$$
\begin{aligned}
a_{k+1} &= a_k + d && \text{by definition of } a_0, a_1, a_2, \ldots \\
&= [a_0 + kd] + d && \text{by substitution from the} \\
& && \text{inductive hypothesis} \\
&= a_0 + (k + 1)d && \text{by basic algebra}
\end{aligned}
$$

19. Let $U_n = $ the number of units produced on day n. Then

$$U_k = U_{k-1} + 2 \quad \text{for all integers } k \geq 1,$$
$$U_0 = 170.$$

Hence $U_0, U_1, U_2, \ldots$ is an arithmetic sequence with constant adder 2. It follows that when $n = 30$,

$$U_n = U_0 + n \cdot 2 = 170 + 2n = 170 + 2 \cdot 30$$
$$= 230 \text{ units.}$$

Thus the worker must produce 230 units on day 30.

24. $\sum_{k=0}^{20} 5^k = \dfrac{5^{21} - 1}{4} \cong 1.192 \times 10^{14} \cong$

119,200,000,000,000 $\cong$ 119 trillion people (This is about 20,000 times the current population of the earth!)

26. b. *Hint:* Before simplification,

$A_n = 1000(1.0025)^n + 200[(1.0025)^{n-1} + (1.0025)^{n-1} + \cdots + (1.0025)^2 + 1.0025 + 1].$

d. $A_{240} \cong \$67{,}481.15,\quad A_{480} \cong \$185{,}215.22$

e. *Hint:* Use logarithms to solve the equation $A_n = 10{,}000$, where A_n is the expression found (after simplification) in part (b).

27. a. APR $\cong 19.6\%$

c. approximately two years

28. *Proof:* Let $a_0, a_1, a_2, \ldots$ be the sequence defined recursively by $a_0 = 1$ and $a_k = k a_{k-1}$ for all integers $k \geq 1$. The property, $P(n)$, is the equation $a_n = n!$. We show by mathematical induction that this property is true for all integers $n \geq 0$.

Show that the property is true for $n = 0$:

When $n = 0$, the right-hand side of the equation is $0! = 1$, and by definition of $a_0, a_1, a_2, \ldots$, the left-hand side of the equation, a_0, is also 1. Thus the property is true for $n = 0$.

Show that for all integers $k \geq 0$, if the property is true for $n = k$, then it is true for $n = k + 1$:

Suppose

$a_k = k!$ for some integer $k \geq 0$.

[This is the inductive hypothesis.]

We must show that $a_{k+1} = (k+1)!$. But

$\begin{aligned} a_{k+1} &= (k+1) \cdot a_k && \text{by definition of } a_0, a_1, a_2, \ldots \\ &= (k+1) \cdot k! && \text{by substitution from the inductive hypotheses} \\ &= (k+1)! && \text{by definition of factorial.} \end{aligned}$

[Hence if the property is true for $n = k$, then it is true for $n = k + 1$.]

30. *Proof:* Let $c_1, c_2, c_3, \ldots$ be the sequence defined recursively by $c_1 = 1$ and $c_k = 3c_{k-1} + 1$ for all integers $k \geq 2$. The property, $P(n)$, is the equation $c_n = \dfrac{3^n - 1}{2}$. We show by mathematical induction that this property is true for all integers $n \geq 1$.

Show that the property is true for $n = 1$:

When $n = 1$, the right-hand side of the equation is $\dfrac{3^1 - 1}{2} = \dfrac{3 - 1}{2} = 1$, and by definition of $c_1, c_2, c_3, \ldots$, the left-hand side of the equation, c_1, is also 1. Thus the property is true for $n = 1$.

Show that for all integers $k \geq 1$, if the property is true for $n = k$, then it is true for $n = k + 1$:

Suppose that

$c_k = \dfrac{3^k - 1}{2}$ for some integer $k \geq 1$.

[This is the inductive hypothesis.]

We must show that $c_{k+1} = \dfrac{3^{k+1} - 1}{2}$. But

$\begin{aligned} c_{k+1} &= 3c_k + 1 && \text{by definition of } c_1, c_2, c_3, \ldots \\ &= 3\left(\dfrac{3^k - 1}{2}\right) + 1 && \text{by substitution from the inductive hypothesis} \\ &= \dfrac{3^{k+1} - 3}{2} + \dfrac{2}{2} \\ &= \dfrac{3^{k+1} - 1}{2} && \text{by basic algebra.} \end{aligned}$

35. *Hint:* $2^{k+1} - \dfrac{2^{k+1} - (-1)^{k+1}}{3}$

$= \dfrac{3 \cdot 2^{k+1}}{3} - \dfrac{2^{k+1} - (-1)^{k+1}}{3}$

$= \dfrac{2 \cdot 2^{k+1} + (-1)^{k+1}}{3} = \dfrac{2^{k+2} - (-1)^{k+2}}{3}$

37. *Hint:* $[3 + k(k+1)] + 2(k+1)$

$= 3 + k^2 + k + 2k + 2 = 3 + [k^2 + 3k + 2]$

$= 3 + (k+1)(k+2)$

$= 3 + (k+1)[(k+1) + 1]$

39. *Proof:* Let $x_1, x_2, x_3, \ldots$ be the sequence defined recursively by $x_1 = 1$ and $x_k = 3x_{k-1} + k$ for all integers $k \geq 2$. The property, $P(n)$, is the equation $x_n = \dfrac{3^{n+1} - 2n - 3}{4}$. We show by mathematical induction that this property is true for all integers $n \geq 1$.

Show that the property is true for $n = 1$:

When $n = 1$, the right-hand side of the equation is $\dfrac{3^{1+1} - 2 \cdot 1 - 3}{4} = \dfrac{3^2 - 2 - 3}{4} = 1$, and by definition of $x_1, x_2, x_3, \ldots$, the left-hand side of the equation, x_1, is also 1. Thus the property is true for $n = 1$.

Show that for all integers $k \geq 1$, if the property is true for $n = k$, then it is true for $n = k + 1$:

Suppose that for some integer $k \geq 0$, $x_k = \dfrac{3^{k+1} - 2k - 3}{4}$. *[Inductive hypothesis]* We must show that

$x_{k+1} = \dfrac{3^{(k+1)+1} - 2(k+1) - 3}{4}$, or, equivalently,

$x_{k+1} = \dfrac{3^{k+2} - 2k - 5}{4}$. But

$\begin{aligned} x_{k+1} &= 3x_k + k && \text{by definition of } x_1, x_2, x_3, \\ &= 3\left(\dfrac{3^{k+1} - 2k - 3}{4}\right) + k + 1 && \text{by inductive hypothesis} \\ &= \dfrac{3 \cdot 3^{k+1} - 3 \cdot 2k - 3 \cdot 3}{4} + \dfrac{4(k+1)}{4} \\ &= \dfrac{3^{k+2} - 6k - 9 + 4k + 4}{4} \\ &= \dfrac{3^{k+2} - 2k - 5}{4} && \text{by algebra.} \end{aligned}$

[This is what was to be shown.]

43. a. $a_0 = 2$

$$a_1 = \frac{a_0}{2a_0 - 1} = \frac{2}{2 \cdot 2 - 1} = \frac{2}{3}$$

$$a_2 = \frac{a_1}{2a_1 - 1} = \frac{\frac{2}{3}}{2 \cdot \frac{2}{3} - \frac{3}{3}} = \frac{\frac{2}{3}}{\frac{1}{3}} = 2$$

$$a_3 = \frac{a_2}{2a_2 - 1} = \frac{2}{2 \cdot 2 - 1} = \frac{2}{3}$$

$$a_4 = \frac{a_3}{2a_3 - 1} = \frac{\frac{2}{3}}{2 \cdot \frac{2}{3} - \frac{3}{3}} = \frac{\frac{2}{3}}{\frac{1}{3}} = 2$$

Guess: $a_n = \begin{cases} 2 & \text{if } n \text{ is even} \\ \frac{2}{3} & \text{if } n \text{ is odd} \end{cases}$.

b. *Proof:* Let $a_0, a_1, a_2, \ldots$ be the sequence defined recursively by $x_0 = 2$ and $a_k = \dfrac{a_{k-1}}{2a_{k-1} - 1}$ for all integers $k \geq 1$. The property, $P(n)$, is the formula

$$a_n = \begin{cases} 2 & \text{if } n \text{ is even} \\ \frac{2}{3} & \text{if } n \text{ is odd} \end{cases}.$$

We show by mathematical induction that this property is true for all integers $n \geq 1$.

Show that the property is true for $n = 0$ and $n = 1$:
When $n = 0$ and $n = 1$, the results of part (a) show that the property is true.

Show that for all integers $k > 0$, if the property is true for all integers i with $0 \leq i < k$, then the property is true for k:
Let k be an integer with $k > 0$, and suppose that for all integers i with $0 \leq i < k$,

$$a_i = \begin{cases} 2 & \text{if } i \text{ is even} \\ \frac{2}{3} & \text{if } i \text{ is odd} \end{cases}. \quad \textit{[Inductive hypothesis.]}$$

We must show that

$$a_k = \begin{cases} 2 & \text{if } k \text{ is even} \\ \frac{2}{3} & \text{if } k \text{ is odd} \end{cases}.$$

But

$$a_k = \frac{a_{k-1}}{2a_{k-1} - 1} \qquad \text{by definition of } a_0, a_1, a_2, \ldots$$

$$= \begin{cases} \frac{2}{2 \cdot 2 - 1} & \text{if } k - 1 \text{ is even} \\ \frac{\frac{2}{3}}{2 \cdot \frac{2}{3} - 1} & \text{if } k - 1 \text{ is odd} \end{cases} \qquad \text{by inductive hypothesis}$$

$$= \begin{cases} \frac{2}{3} & \text{if } k - 1 \text{ is even} \\ \frac{\frac{2}{3}}{\frac{1}{3}} & \text{if } k - 1 \text{ is odd} \end{cases}$$

$$= \begin{cases} \frac{2}{3} & \text{if } k \text{ is odd} \\ 2 & \text{if } k \text{ is even} \end{cases} \qquad \begin{array}{l} \text{because } k \text{ is odd when} \\ k - 1 \text{ is even} \\ \text{and } k \text{ is even when} \\ k - 1 \text{ is odd.} \end{array}$$

[This is what was to be shown.]

45. $v_1 = 1$

$$v_2 = v_{\lfloor 2/2 \rfloor} + v_{\lfloor 3/2 \rfloor} + 2 = v_1 + v_1 + 2$$
$$= 1 + 1 + 2$$

$$v_3 = v_{\lfloor 3/2 \rfloor} + v_{\lfloor 4/2 \rfloor} + 2 = v_1 + v_2 + 2$$
$$= 1 + (1 + 1 + 2) + 2 = 3 + 2 \cdot 2$$

$$v_4 = v_{\lfloor 4/2 \rfloor} + v_{\lfloor 5/2 \rfloor} + 2 = v_2 + v_2 + 2$$
$$= (1 + 1 + 2) + (1 + 1 + 2) + 2$$
$$= 4 + 3 \cdot 2$$

$$v_5 = v_{\lfloor 5/2 \rfloor} + v_{\lfloor 6/2 \rfloor} + 2 = v_2 + v_3 + 2$$
$$= (3 + 2 \cdot 2) + (1 + 1 + 2) + 2$$
$$= 5 + 4 \cdot 2$$

$$v_6 = v_{\lfloor 6/2 \rfloor} + v_{\lfloor 7/2 \rfloor} + 2 = v_3 + v_3 + 2$$
$$= (3 + 2 \cdot 2) + (3 + 2 \cdot 2) + 2$$
$$= 6 + 5 \cdot 2$$

$$\vdots$$

Guess:

$$v_n = n + 2(n - 1) = 3n - 2 \text{ for all integers } n \geq 1$$

b. *Proof:* Let $v_1, v_2, v_3, \ldots$ be the sequence defined recursively by $v_1 = 1$ and $v_k = v_{\lfloor k/2 \rfloor} + v_{\lfloor (k+1)/2 \rfloor} + 2$ for all integers $k \geq 1$. The property, $P(n)$, is the equation

$$v_n = 3n - 2.$$

We show by mathematical induction that this property is true for all integers $n \geq 1$.

Show that the property is true for $n = 1$:
When $n = 1$, the right-hand side of the equation is $3 \cdot 1 - 2 = 1$, which equals v_1 by definition of $v_1, v_2, v_3, \ldots$.

Show that for all integers $k > 1$, if the property is true for all integers i with $0 \leq i < k$, then the property is true for k:
Let k be an integer with $k > 1$, and suppose that for all integers i with $1 \leq i < k$, $v_i = 3i - 2$.
[This is the inductive hypothesis.] We must show that $v_k = 3k - 2$. But

$$v_k = v_{\lfloor k/2 \rfloor} + v_{\lfloor (k+1)/2 \rfloor} + 2 \qquad \begin{array}{l} \text{by definition of} \\ v_1, v_2, v_3, \ldots \end{array}$$

$$= (3\lfloor k/2 \rfloor - 2) + (3\lfloor (k+1)/2 \rfloor - 2) + 2 \qquad \begin{array}{l} \text{by substitution from the} \\ \text{inductive hypotheses} \end{array}$$

$$= 3(\lfloor k/2 \rfloor + \lfloor (k+1)/2 \rfloor) - 2$$

$$= \begin{cases} 3\left(\frac{k}{2} + \frac{k}{2}\right) - 2 & \text{if } k \text{ is even} \\ 3\left(\frac{k-1}{2} + \frac{k+1}{2}\right) - 2 & \text{if } k \text{ is odd} \end{cases}$$

$$= \begin{cases} 3k - 2 & \text{if } k \text{ is even} \\ 3\left(\frac{2k-2+2k+2}{2}\right) - 2 & \text{if } k \text{ is odd} \end{cases}$$

$$= 3k - 2 \qquad \text{by the laws of algebra.}$$

[This is what was to be shown.]

46. *Hint:* Show that for all integers $n \geq 0$, $s_{2n} = 2^n$ and $s_{2n+1} = 2^{n+1}$. Then combine these formulas using the ceiling function to obtain $s_n = 2^{\lceil n/2 \rceil}$.

48. a. *Hint:*
$$w_n = \begin{cases} \left(\dfrac{n+1}{2}\right)^2 & \text{if } n \text{ is odd} \\ \dfrac{n}{2}\left(\dfrac{n}{2}+1\right) & \text{if } n \text{ is even} \end{cases}$$

49. a. *Hint:* Express the answer using the Fibonacci sequence.

50. The sequence does not satisfy the formula. According to the formula, $a_4 = (4-1)^2 = 9$. But by definition of the sequence, $a_1 = 0$, $a_2 = 2 \cdot 0 + (2+1) = 1$, $a_3 = 2 \cdot 1 + (3-1) = 4$, and so $a_4 = 2 \cdot 4 + (4-1) = 11$. Hence the sequence does not satisfy the formula for $n = 4$.

52. a. *Hint:* The maximum number of regions is obtained when each additional line crosses all the previous lines, but not at any point that is already the intersection of two lines. When a new line is added, it divides each region through which it passes into two pieces. The number of regions a newly added line passes through is one more than the number of lines it crosses.

53. *Hint:* The answer involves the Fibonacci numbers!

Section 8.3

1. (a), (d), and (f)

3. a. $\left.\begin{array}{l} a_0 = C \cdot 2^0 + D = C + D = 1 \\ a_1 = C \cdot 2^1 + D = 2C + D = 3 \end{array}\right\}$

$\Leftrightarrow \left\{\begin{array}{l} D = 1 - C \\ 2C + (1-C) = 3 \end{array}\right\} \Leftrightarrow \left\{\begin{array}{l} C = 2 \\ D = -1 \end{array}\right.$

$a_2 = 2 \cdot 2^2 + (-1) = 7$

4. a. $\left.\begin{array}{l} b_0 = C \cdot 3^0 + D \cdot (-2)^0 = C + D = 0 \\ b_1 = C \cdot 3^1 + D \cdot (-2)^1 = 3C - 2D = 5 \end{array}\right\}$

$\Leftrightarrow \left\{\begin{array}{l} D = -C \\ 3C - 2(-C) = 5 \end{array}\right\} \Leftrightarrow \left\{\begin{array}{l} C = 1 \\ D = -1 \end{array}\right.$

$b_2 = 3^2 + (-1)(-2)^2 = 9 - 4 = 5$

5. *Proof:* Given that $a_n = C \cdot 2^n + D$, then for any choice of C and D and integer $k > 2$,

$$a_k = C \cdot 2^k + D,$$
$$a_{k-1} = C \cdot 2^{k-1} + D,$$
$$a_{k-2} = C \cdot 2^{k-2} + D.$$

Hence

$$3a_{k-1} - 2a_{k-2} = 3(C \cdot 2^{k-1} + D) - 2(C \cdot 2^{k-2} + D)$$
$$= 3C \cdot 2^{k-1} + 3D - 2C \cdot 2^{k-2} - 2D$$
$$= 3C \cdot 2^{k-1} - C \cdot 2^{k-1} + D$$
$$= 2C \cdot 2^{k-1} + D$$
$$= C \cdot 2^k + D = a_k.$$

8. a. If for all $k > 2$, $t^k = 2t^{k-1} + 3t^{k-2}$ and $t \neq 0$ then $t^2 = 2t + 3$ [*by dividing by t^{k-2}*], and so $t^2 - 2t - 3 = 0$. But $t^2 - 2t - 3 = (t-3)(t+1)$; hence $t = 3$ or $t = -1$.

b. It follows from (a) and the distinct roots theorem that for some constants C and D, $a_0, a_1, a_2, \ldots$ satisfies the equation

$$a_n = C \cdot 3^n + D \cdot (-1)^n \quad \text{for all integers } n \geq 0.$$

Since $a_0 = 1$ and $a_1 = 2$, then

$$\left.\begin{array}{l} a_0 = C \cdot 3^0 + D \cdot (-1)^0 = C + D = 1 \\ a_1 = C \cdot 3^1 + D \cdot (-1)^1 = 3C - D = 2 \end{array}\right\}$$

$$\Leftrightarrow \left\{\begin{array}{l} D = 1 - C \\ 3C - (1-C) = 2 \end{array}\right\}$$

$$\Leftrightarrow \left\{\begin{array}{l} D = 1 - C \\ 4C - 1 = 2 \end{array}\right\}$$

$$\Leftrightarrow \left\{\begin{array}{l} C = 3/4 \\ D = 1/4 \end{array}\right.$$

Thus

$$a_n = \tfrac{3}{4}(3^n) + \tfrac{1}{4}(-1)^n \quad \text{for all integers } n \geq 0.$$

11. *Characteristic equation:* $t^2 - 4 = 0$. Since $t^2 - 4 = (t-2)(t+2)$, $t = 2$ and $t = -2$ are the roots. By the distinct roots theorem, for some constants C and D

$$d_n = C \cdot (2^n) + D \cdot (-2)^n \quad \text{for all integers } n \geq 0.$$

Since $d_0 = 1$ and $d_1 = -1$, then

$$\left.\begin{array}{l} d_0 = C \cdot 2^0 + D \cdot (-2)^0 = C + D = 1 \\ d_1 = C \cdot 2^1 + D \cdot (-2)^1 = 2C - 2D = -1 \end{array}\right\}$$

$$\Leftrightarrow \left\{\begin{array}{l} D = 1 - C \\ 2C - 2(1-C) = -1 \end{array}\right\}$$

$$\Leftrightarrow \left\{\begin{array}{l} D = 1 - C \\ 4C - 2 = -1 \end{array}\right\}$$

$$\Leftrightarrow \left\{\begin{array}{l} C = \tfrac{1}{4} \\ D = \tfrac{3}{4} \end{array}\right.$$

Thus

$$d_n = \tfrac{1}{4}(2^n) + \tfrac{3}{4}(-2)^n \quad \text{for all integers } n \geq 0.$$

13. *Characteristic equation:* $t^2 - 2t + 1 = 0$. By the quadratic formula,

$$t = \frac{2 \pm \sqrt{4 - 4 \cdot 1}}{2} = \frac{2}{2} = 1.$$

By the single root theorem, for some constants C and D

$$r_n = C \cdot (1^n) + Dn \cdot (1^n)$$
$$= C + nD \quad \text{for all integers } n \geq 0.$$

Since $r_0 = 1$ and $r_1 = 4$, then

$$\left.\begin{array}{l} r_0 = C + 0 \cdot D = C = 1 \\ r_1 = C + 1 \cdot D = C + D = 4 \end{array}\right\} \Leftrightarrow \left\{\begin{array}{l} C = 1 \\ 1 + D = 4 \end{array}\right\}$$

$$\Leftrightarrow \left\{\begin{array}{l} C = 1 \\ D = 3 \end{array}\right.$$

Thus $r_n = 1 + 3n$ for all integers $n \geq 0$.

16. *Hint:* For all integers $n \geq 0$,

$$s_n = \frac{\sqrt{3}+2}{2\sqrt{3}}(1+\sqrt{3})^n + \frac{\sqrt{3}-2}{2\sqrt{3}}(1-\sqrt{3})^n.$$

19. *Proof:* Suppose r, s, a_0 and a_1 are numbers with $r \neq s$. Consider the system of equations

$$C + D = a_0$$
$$Cr + Ds = a_1.$$

By solving for D and substituting, we find that

$$D = a_0 - C$$
$$Cr + (a_0 - C)s = a_1.$$

Hence

$$C(r - s) = a_1 - a_0 s.$$

Since $r \neq s$, both sides may be divided by $r - s$. Thus the given system of equations has the unique solution

$$C = \frac{a_1 - a_0 s}{r - s}$$

and

$$D = a_0 - C = a_0 - \frac{a_1 - a_0 s}{r - s}$$

$$= \frac{a_0 r - a_0 s - a_1 + a_0 s}{r - s} = \frac{a_0 r - a_1}{r - s}.$$

Alternative solution: Since the determinant of the system is $1 \cdot s - r \cdot 1 = s - r$ and since $r \neq s$, the given system has a nonzero determinant and therefore has a unique solution.

21. *Hint:* Use strong mathematical induction. First note that the formula holds for $n = 0$ and $n = 1$. To prove the inductive step, suppose that for some $k \geq 2$, the formula holds for all i with $0 \leq i < k$. Then show that the formula holds for k. Use the proof of Theorem 8.3.3 (the distinct roots theorem) as a model.

22. The characteristic equation is $t^2 - 2t + 2 = 0$. By the quadratic formula, its roots are

$$t = \frac{2 \pm \sqrt{4-8}}{2} = \frac{2 \pm 2i}{2} = \begin{cases} 1+i \\ 1-i \end{cases}.$$

By the distinct roots theorem, for some constants C and D

$$a_n = C(1+i)^n + D(1-i)^n$$

for all integers $n \geq 0$.

Since $a_0 = 1$ and $a_1 = 2$, then

$$a_0 = C(1+i)^0 + D(1-i)^0 = C + D = 1$$
$$a_1 = C(1+i)^1 + D(1-i)^1$$
$$= C(1+i) + D(1-i) = 2$$

$$\Leftrightarrow \begin{cases} D = 1 - C \\ C(1+i) + (1-C)(1-i) = 2 \end{cases}$$

$$\Leftrightarrow \begin{cases} D = 1 - C \\ C(1+i-1+i) + 1 - i = 2 \end{cases}$$

$$\Leftrightarrow \begin{cases} D = 1 - C \\ C(2i) = 1 + i \end{cases}$$

$$\Leftrightarrow \begin{cases} D = 1 - C \\ C = \dfrac{1+i}{2i} = \dfrac{1+i}{2i} \cdot \dfrac{i}{i} = \dfrac{i-1}{-2} = \dfrac{1-i}{2} \end{cases}$$

$$\Leftrightarrow \begin{cases} D = 1 - \dfrac{1-i}{2} = \dfrac{2-1+i}{2} = \dfrac{1+i}{2} \\ C = \dfrac{1-i}{2} \end{cases}$$

Thus for all integers $n \geq 0$,

$$a_n = \left(\frac{1-i}{2}\right)(1+i)^n + \left(\frac{1+i}{2}\right)(1-i)^n.$$

25. *Hint:* $P_{20} = \dfrac{5^{300} - 5^{20}}{5^{300} - 1} \cong 1$

26. a. *Hint:* The answer is $s_k = 2s_{k-1} + 3s_{k-2}$ for $k \geq 4$.

Section 8.4

1. a. (1) p, q, r, and s are Boolean expressions by I.
   (2) $\sim s$ is a Boolean expression by (1) and II(c).
   (3) $(r \vee \sim s)$ is a Boolean expression by (1), (2), and II(b).
   (4) $(q \wedge (r \vee \sim s))$ is a Boolean expression by (1), (3), and II(a).
   (5) $\sim p$ is a Boolean expression by (1) and II(c).
   (6) $(\sim p \vee (q \wedge (r \vee \sim s)))$ is a Boolean expression by (4), (5), and II(b).

2. a. (1) $\epsilon \in S$ by I.
   (2) $a = \epsilon a \in S$ by (1) and II(a).
   (3) $aa \in S$ by (2) and II(a).
   (4) $aab \in S$ by (3) and II(b).

3. a. (1) MI is in the MIU system by I.
   (2) MII is in the MIU system by (1) and II(b).
   (3) $MIIII$ is in the MIU system by (3) and II(b).
   (4) $MIIIIIIII$ is in the MIU system by (3) and II(b).
   (5) $MIUIIII$ is in the MIU system by (4) and II(c).
   (6) $MIUUI$ is in the MIU system by (5) and II(c).
   (7) $MIUI$ is in the MIU system by (6) and II(d).

4. *Hint:* Can the number of I's in a string in the MIU system be a multiple of 3? How do rules II(a)–(d) affect the number of I's in a string?

5. a. (1) $(\,)$ is in P by I.
   (2) $((\,))$ is in P by (1) and II(a).
   (3) $(\,)((\,))$ is in P by (1), (2), and II(b).

6. a. This structure is not in P. Define a function $f: P \to Z$ as follows: For each parenthesis structure S in P, let

$$f(S) = \begin{bmatrix} \text{the number of left} \\ \text{parentheses in } S \end{bmatrix} - \begin{bmatrix} \text{the number of right} \\ \text{parentheses in } S \end{bmatrix}.$$

Observe that for all S in P, $f(S) = 0$. To see why, use the reasoning of structural induction:

1. The base element of P is sent by f to 0: $f[()] = 0$ [*because there is one left and one right parenthesis in ( )*].
2. For all $S \in P$, if $f[S] = 0$ then $f[(S)] = 0$ [*because if $k - m = 0$ then $(k + 1) - (m + 1) = 0$*].
3. For all S and T in P, if $f[S] = 0$ and $f[T] = 0$, then $f[ST] = 0$ [*because if $k - m = 0$ and $n - p = 0$, then $(k + n) - (m + p) = 0$*].

Items (1), (2), and (3) show that all parenthesis structures obtainable from the base structure ( ) by repeated application of II(a) and II(b) are sent to 0 by f. But by III (the restriction condition), there are no other elements of P besides those obtainable from the base element by applying II(a) and II(b). Hence $f(S) = 0$ for all $S \in P$. Now if ( )(( ) were in P, then it would be sent to 0 by f. But $f[()(()] = 3 - 2 = 1 \neq 0$. Thus ( )(( ) $\notin P$.

7. a. (1) 2, 0.3, 4.2, and 7 are arithmetic expressions by I.
(2) $(0.3 - 4.2)$ is an arithmetic expression by (1) and II(d).
(3) $(2 \cdot (0.3 - 4.2))$ is an arithemetic expression by (1), (2), and II(e).
(4) (-7) is an arithmetic expression by (1) and II(b).
(5) $((2 \cdot (0.3 - 4.2)) + (-7))$ is an arithmetic expression by (3), (4), and II(c).

8. *Proof:* Let the property be the following sentence: The string ends in a 1.

Show that each object in the BASE for S satisfies the property:

The only object in the base is 1, and the string 1 ends in a 1.

Show that for each rule in the RECURSION for S, if the rule is applied to an object in S that satisfies the property, then the objects defined by the rule also satisfy the property:

The recursion for S consists of two rules denoted II(a) and II(b). Suppose s is a string in S that ends in a 1. In the case where rule II(a) is applied to s, the result is the string $0s$, which also ends in a 1. In the case where rule II(b) is applied to s, the result is the string $s1$, which also ends in a 1. Thus when each rule in the RECURSION is applied to a string in S that ends in a 1, the result is also a string that ends in a 1.

10. *Proof:* Let the property be the following sentence: The string contains an even number of a's.

Show that each object in the BASE for S satisfies the property:

The only object in the base is ϵ, which contains 0 a's. Because 0 is an even number, ϵ contains an even number of a's.

Show that for each rule in the RECURSION for S, if the rule is applied to an object in S that satisfies the property, then the objects defined by the rule also satisfy the property:

The recursion for S consists of four rules denoted II(a), II(b), II(c), and II(d). Suppose s is a string in S that contains an even number of a's. In the case where either rule II(a) or rule II(b) is applied to s, the result is the string bs or the string sb, each of which contain the same number of a's as s and hence an even number of a's. In the case where either rule II(c) or rule II(d) is applied to s, the result is the string aas or the string saa, each of which contain two more a's than the number of a's in s. Because two more than any even integer is an even integer, both aas and saa contain an even number of a's. Thus when each rule in 'the RECURSION is applied to a string in S that contains an even number of a's, the result is also a string that contains even number of a's.

12. *Hint:* Let the property be the following sentence: The string represents an odd integer. In the decimal notation, a string represents an odd integer if, and only if, it ends in 1, 3, 5, 7 or 9.

13. *Hint:* By divisibility results from Chapter 3 (exercises 15 and 16 of Section 3.3), if both s and t are divisible by 5, then so are $s + t$ and $s - t$.

15. Let S be the set of all strings of 0's and 1's with the same number of 0's and 1's. The following is a recursive definition of S.
   I. BASE: The null string $\epsilon \in S$.
   II. RECURSION: If $s \in S$, then
   a. $01s \in S$ **b.** $s01 \in S$ **c.** $10s \in S$
   d. $s10 \in S$ **e.** $0s1 \in S$ **f.** $1s0 \in S$
   III. RESTRICTION: There are no elements of S other that those obtained from I and II.

17. Let T be the set of all strings of a's and b's that contain an odd number of a's. The following is a recursive definition of T.
   I. BASE: The $a \in T$.
   II. RECURSION: If $t \in T$, then
   a. $bt \in T$ **b.** $tb \in T$ **c.** $aat \in T$
   d. $ata \in T$ **e.** $taa \in T$
   III. RESTRICTION: There are no elements of T other than those obtained from I and II.

19. *Proof (by mathematical induction):* Let the property, $P(n)$, be the equation $\sum_{i=1}^{n} ca_i = c \sum_{i=1}^{n} a_i$.

Show that the property is true for $n = 1$:

Let a_1 and c be any real numbers. By the recursive definition of sum, $\sum_{i=1}^{1} (ca_i) = ca_1$ and $\sum_{i=1}^{1} a_i = a_1$. Therefore, $\sum_{i=1}^{1} (ca_i) = c \sum_{i=1}^{1} a_i$, and so the property is true for $n = 1$.

Show that for all integers $k \geq 1$, if the property is true for $n = k$, then it is true for $n = k + 1$:

Let k be an integer with $k \geq 1$. Suppose that for any real numbers $a_1, a_2, a_3, \ldots, a_k$ and c, $\sum_{i=1}^{k} (ca_i) = c \sum_{i=1}^{k} a_i$. [*This is the inductive hypothesis*] [*We must show that for* any real numbers $a_1, a_2, a_3, \ldots a_{k+1}$ and c, $\sum_{i=1}^{k+1}(ca_i) = c\sum_{i=1}^{k+1} a_i$.] Let $a_1, a_2, a_3, \ldots, a_{k+1}$ and c be any real numbers. Then

$$\sum_{i=1}^{k+1} ca_i = \sum_{i=1}^{k} ca_i + ca_{k+1} \qquad \text{by the recursive definition of } \Sigma$$

$$= c\sum_{i=1}^{k} a_i + ca_{k+1} \qquad \text{by inductive hypothesis}$$

$$= c\left(\sum_{i=1}^{k} a_i + a_{k+1}\right) \qquad \text{by the distributive law for the real numbers}$$

$$= c\sum_{i=1}^{k+1} a_i \qquad \text{by the recursive definition of } \Sigma.$$

22. *Hint:* Let the property be the inequality

$$\left|\sum_{i=1}^{n} a_i\right| \le \sum_{i=1}^{n} |a_i|.$$

To prove the inductive step, note that because $\left|\sum_{i=1}^{k+1} a_i\right| = \left|\sum_{i=1}^{k} a_i + a_{k+1}\right|$, you can use the triangle inequality for absolute value (exercise 53 in Section 3.4) to deduce $\left|\sum_{i=1}^{k} a_i + a_{k+1}\right| \le \left|\sum_{i=1}^{k} a_i\right| + |a_{k+1}|$.

23. *Proof (by mathematical induction):* Let the property, $P(n)$, be the following sentence:

For all sets A and $B_1, B_2, \ldots B_n$, $\bigcup_{i=1}^{n}(A \cap B_i) = A \cap \bigcup_{i=1}^{n} B_i$.

Show that the property is true for $n = 1$:
Let B_1 and A be any sets. By the recursive definition of union, $\bigcup_{i=1}^{1} B_i = B_1$ and $\bigcup_{i=1}^{1}(A \cap B_i) = A \cap B_1$. Therefore, $A \cap (\bigcup_{i=1}^{1} B_i) = A \cap B_1 = \bigcup_{i=1}^{1}(A \cap B_i)$, and so the property is true for $n = 1$.

Show that for all integers $k \ge 1$, if the property is true for $n = k$ then it is true for $n = k + 1$:
Suppose that for any sets $B_1, B_2, B_3, \ldots, B_k$ and A, $A \cap (\bigcup_{i=1}^{k} B_i) = \bigcup_{i=1}^{k}(A \cap B_i)$. [*This is the inductive hypothesis.*] [*We must show that for any sets $B_1, B_2, B_3, \ldots, B_{k+1}$ and A, $A \cap (\bigcup_{i=1}^{k+1} B_i) = \bigcup_{i=1}^{k+1}(A \cap B_i)$.*] Let $B_1, B_2, B_3, \ldots, B_{k+1}$, and A be any sets. Then

$$A \cap \left(\bigcup_{i=1}^{k+1} B_i\right)$$

$$= A \cap \left[\left(\bigcup_{i=1}^{k} B_i\right) \cup B_{k+1}\right] \qquad \text{by the recursive definition of union}$$

$$= \left(A \cap \left(\bigcup_{i=1}^{k} B_i\right)\right) \cup (A \cap B_{k+1}) \qquad \text{by one of the distributive laws for sets (Theorem 5.2.2(3)).}$$

$$= \left(\bigcup_{i=1}^{k}(A \cap B_i)\right) \cup (A \cap B_{k+1}) \qquad \text{by inductive hypothesis}$$

$$= \bigcup_{i=1}^{k+1}(A \cap B_i) \qquad \text{by the recursive definition of union.}$$

26. a.
$$\begin{aligned} M(86) &= M(M(97)) && \text{since } 86 \le 100 \\ &= M(M(M(108))) && \text{since } 97 \le 100 \\ &= M(M(98)) && \text{since } 108 > 100 \\ &= M(M(M(109))) && \text{since } 98 < 100 \\ &= M(M(99)) && \text{since } 109 > 100 \\ &= M(91) && \text{by Example 8.4.8} \end{aligned}$$

28. a.
$$\begin{aligned} A(1, 1) &= A(0, A(1, 0)) && \text{by (8.4.3) with } m = 1 \text{ and } n = 1 \\ &= A(1, 0) + 1 && \text{by (8.4.1) with } n = A(1, 0) \\ &= A(0, 1) + 1 && \text{by (8.4.2) with } m = 1 \\ &= (1 + 1) + 1 && \text{by (8.4.1) with } n = 1 \\ &= 3 \end{aligned}$$

Alternative solution:

$$\begin{aligned} A(1, 1) &= A(0, A(1, 0)) && \text{by (8.4.3) with } m = 1 \text{ and } n = 1 \\ &= A(0, A(0, 1)) && \text{by (8.4.2) with } m = 1 \\ &= A(0, 2) && \text{by (8.4.1) with } n = 1 \\ &= 3 && \text{by (8.4.1) with } n = 2 \end{aligned}$$

29. a. *Proof (by mathematical induction):* Let the property, $P(n)$, be the equation $A(1, n) = n + 2$.

Show that the property is true for $n = 0$:
When $n = 0$,

$$\begin{aligned} A(1, n) &= A(1, 0) && \text{by substitution} \\ &= A(0, 1) && \text{by (8.4.2)} \\ &= 1 + 1 && \text{by (8.4.1)} \\ &= 2. \end{aligned}$$

On the other hand, $n + 2 = 0 + 2$ also. Thus $A(1, n) = n + 2$ for $n = 0$.

Show that for all integers $k \ge 0$, if the property is true for $n = k$, then it is true for $n = k + 1$:
Let k be an integer with $k \ge 1$ and suppose the property is true for $n = k$. In other words, suppose $A(1, k) = k + 2$. [*This is the inductive hypothesis.*] We must show that the formula holds for $n = k + 1$. In other words, we must show that $A(1, k + 1) = (k + 1) + 2 = k + 3$. But

$$\begin{aligned} A(1, k + 1) &= A(0, A(1, k)) && \text{by (8.1.3)} \\ &= A(1, k) + 1 && \text{by (8.4.1)} \\ &= (k + 2) + 1 && \text{by inductive hypothesis} \\ &= k + 3. \end{aligned}$$

[*This is what was to be shown.*]
[*Since both the basis and the inductive steps have been proved, we conclude that the formula holds for all non-negative integers n.*]

31. Suppose F is a function. Then $F(1) = 1$, $F(2) = F(1) = 1$, $F(3) = 1 + F(5 \cdot 3 - 9) = 1 + F(6) = 1 + F(3)$. Subtracting $F(3)$ from both sides gives $1 = 0$, which is false. Hence F is not a function.

Section 9.1

1. **a.** $f(0)$ is positive.
   b. $f(x) = 0$ when $x = -2$ and $x = 3$ (approximately)
   c. increase
   d. decrease

2.
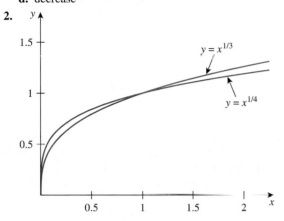

When $0 < x < 1$, $x^{1/3} < x^{1/4}$. When $x > 1$, $x^{1/3} > x^{1/4}$.

4.

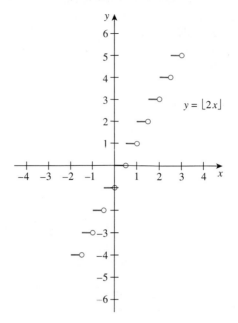

The graphs show that $2\lfloor x \rfloor \neq \lfloor 2x \rfloor$ for many values of x.

5.
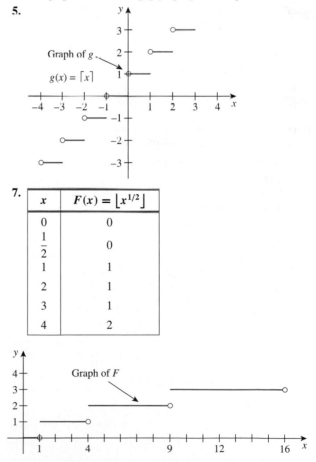

7.

x	$F(x) = \lfloor x^{1/2} \rfloor$
0	0
$\dfrac{1}{2}$	0
1	1
2	1
3	1
4	2

9.

n	$f(n) = \lvert n \rvert$
0	0
1	1
2	2
3	3
-1	1
-2	2
-3	3

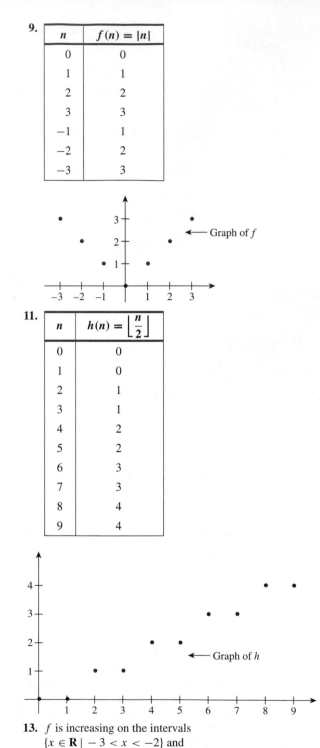

← Graph of f

11.

n	$h(n) = \left\lfloor \dfrac{n}{2} \right\rfloor$
0	0
1	0
2	1
3	1
4	2
5	2
6	3
7	3
8	4
9	4

← Graph of h

13. f is increasing on the intervals
$\{x \in \mathbf{R} \mid -3 < x < -2\}$ and
$\{x \in \mathbf{R} \mid 0 < x < 2.5\}$, and f is decreasing on
$\{x \in \mathbf{R} \mid -2 < x < 0\}$ and $\{x \in \mathbf{R} \mid 2.5 < x < 4\}$ (approximately).

14. *Proof:* Suppose x_1 and x_2 are particular but arbitrarily chosen real numbers such that $x_1 < x_2$. [*We must show that* $f(x_1) < f(x_2)$.] Since

$$x_1 < x_2$$

then

$$2x_1 < 2x_2$$

and

$$2x_1 - 3 < 2x_2 - 3$$

by basic properties of inequalities. But then, by definition of f,

$$f(x_1) < f(x_2)$$

[*as was to be shown*]. Hence f is increasing on the set of all real numbers.

16. a. *Proof:* Suppose x_1 and x_2 are real numbers with $x_1 < x_2 < 0$. [*We must show that* $h(x_1) > h(x_2)$.] Multiply both sides of $x_1 < x_2$ by x_1 to obtain $(x_1)^2 > x_1 x_2$ [*by T22 of Appendix A since* $x_1 < 0$], and multiply both sides of $x_1 < x_2$ by x_2 to obtain $x_1 x_2 > (x_2)^2$ [*by T22 of Appendix A since* $x_2 < 0$]. By transitivity of order [*Appendix A, T17*] $(x_2)^2 < (x_1)^2$, and so, by definition of h, $h(x_2) < h(x_1)$.

17. a. *Preliminaries:* If both x_1 and x_2 are positive, then by the rules for working with inequalities (see Appendix A),

$$\frac{x_1 - 1}{x_1} < \frac{x_2 - 1}{x_2} \Rightarrow x_2(x_1 - 1) < x_1(x_2 - 1)$$

by multiplying both sides by $x_1 x_2$ (which is positive)

$$\Rightarrow x_1 x_2 - x_2 < x_1 x_2 - x_1$$

by multiplying out

$$\Rightarrow -x_2 < -x_1$$

by subtracting $x_1 x_2$ from both sides

$$\Rightarrow x_2 > x_1 \quad \text{by multiplying by } -1.$$

Are these steps reversible? Yes!

Proof: Suppose that x_1 and x_2 are positive real numbers and $x_1 < x_2$. [*We must show that* $k(x_1) < k(x_2)$.] Then

$$x_1 < x_2$$

$$\Rightarrow -x_2 < -x_1 \qquad \text{by multiplying by } -1$$

$$\Rightarrow x_1 x_2 - x_2 < x_1 x_2 - x_1 \qquad \text{by adding } x_1 x_2 \text{ to both sides}$$

$$\Rightarrow x_2(x_1 - 1) < x_1(x_2 - 1) \qquad \text{by factoring both sides}$$

$$\Rightarrow \frac{x_1 - 1}{x_1} < \frac{x_2 - 1}{x_2} \qquad \text{by dividing both sides by the positive number } x_1 x_2$$

$$\Rightarrow k(x_1) < k(x_2) \qquad \text{by definition of } k.$$

[*This is what was to be shown.*]

18. *Proof:* Suppose $f : \mathbf{R} \to \mathbf{R}$ is increasing. [*We must show that f is one-to-one. In other words, we must show that for all real numbers x_1 and x_2, if $x_1 \neq x_2$ then $f(x_1) \neq f(x_2)$.*] Suppose x_1 and x_2 are real numbers and $x_1 \neq x_2$. By the trichotomy law [*Appendix A, T16*] $x_1 < x_2$, or $x_1 > x_2$. In case $x_1 < x_2$, then since f is increasing, $f(x_1) < f(x_2)$ and so $f(x_1) \neq f(x_2)$. Similarly in case $x_1 > x_2$, then $f(x_1) > f(x_2)$ and so $f(x_1) \neq f(x_2)$. Thus in either case, $f(x_1) \neq f(x_2)$ [as was to be shown].

20. a. *Proof:* Suppose u and v are nonnegative real numbers with $u < v$. [*We must show that $f(u) < f(v)$.*] Note that $v = u + h$ for some positive real number h. By substitution and the binomial theorem,

$$v^m = (u + h)^m$$

$$= u^m + \left[\binom{m}{1} u^{m-1}h + \binom{m}{2} u^{m-2}h^2 + \cdots \right.$$

$$\left. + \binom{m}{m-1} uh^{m-1} + h^m \right].$$

The bracketed sum is positive because $u \geq 0$ and $h > 0$, and a sum of nonnegative terms that includes at least one positive term is positive. Hence

$$v^m = u^m + \text{a positive number,}$$

and so $f(u) = u^m < v^m = f(v)$ [as was to be shown].

21.

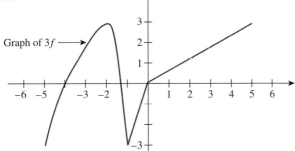

Graph of $3f$ →

23. *Proof:* Suppose that f is a real-valued function of a real variable, f is decreasing on a set S, and M is any positive real number. [*We must show that Mf is decreasing on S. In other words, we must show that for all x_1 and x_2 in S, if $x_1 < x_2$ then $(Mf)(x_1) > (Mf)(x_2)$.*] Suppose x_1 and x_2 are in S and $x_1 < x_2$. Since f is decreasing on S, $f(x_1) > f(x_2)$, and since M is positive, $Mf(x_1) > Mf(x_2)$ [*because when both sides of an inequality are multiplied by a positive number, the direction of the inequality is unchanged*]. It follows by definition of Mf that $(Mf)(x_1) > (Mf)(x_2)$ [as was to be shown].

26. To find the answer algebraically, solve the equation $2x^2 = x^2 + 10x + 11$ for x. Subtracting x^2 from both sides gives $x^2 - 10x - 11 = 0$, and either factoring $x^2 - 10x - 11 = (x - 11)(x + 1)$ or using the quadratic formula gives $x = 11$ (since $x > 0$). To find an approximate answer with a graphing calculator, plot both $f(x) = x^2 + 10x + 11$ and $2g(x) = 2x^2$ for $x > 0$, as shown in the figure, and find that $2g(x) > f(x)$ when $x > 11$ (approximately). You can obtain only an approximate answer from a graphing calculator

because the calculator computes values only to an accuracy of a finite number of decimal places.

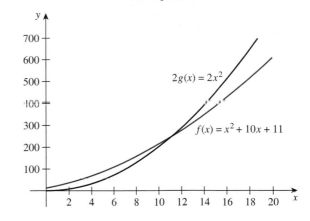

Section 9.2

1. a. $\forall$ positive real numbers a and A, $\exists x > a$ such that $A|g(x)| > |f(x)|$.

b. No matter what positive real numbers a and A might be chosen, it is possible to find a number x greater than a with the property that $A|g(x)| > |f(x)|$.

4. $5x^8 - 9x^7 + 2x^5 + 3x - 1$ is $O(x^8)$

5. $\dfrac{(x^2 - 1)(12x + 25)}{3x^2 + 4}$ is $\Theta(x)$

6. $\dfrac{(x^2 - 7)^2(10x^{1/2} + 3)}{x + 1}$ is $\Omega(x^{7/2})$

10. *Proof:* Suppose f and g are real-valued functions of a real variable that are defined on the same set of nonnegative real numbers, and suppose $g(x)$ is $O(f(x))$. By definition of O-notation, there exist positive real numbers b and B such that $|g(x)| \leq B|f(x)|$ for all real numbers $x > b$. Divide both sides of the inequality by B to obtain $\frac{1}{B}|g(x)| \leq |f(x)|$. Let $A = \frac{1}{B}$ and let $a = b$. Then $A|g(x)| \leq |f(x)|$ for all real numbers $x > a$, and so, by definition of Ω-notation, $f(x)$ is $\Omega(g(x))$.

12. *Proof:* Suppose $f, g, h,$ and k are real-valued functions of a real variable that are defined on the same set D of nonnegative real numbers, and suppose $f(x)$ is $O(h(x))$ and $g(x)$ is $O(k(x))$. By definition of O-notation, there exist positive real numbers $b_1, B_1, b_2,$ and B_2 such that $|f(x)| \leq B_1|h(x)|$ for all real numbers $x > b_1$, and $|g(x)| \leq B_2|k(x)|$ for all real numbers $x > b_2$. For each x in D, define $G(x) = \max(|h(x)|, |k(x)|)$, and let $b = \max(b_1, b_2)$ and $B = B_1 + B_2$. Note that the triangle inequality for absolute value (exercise 53, Section 3.4) implies that

$$|f(x) + g(x)| \leq |f(x)| + |g(x)|$$

for all real numbers x in D. Suppose that $x > b$. Then

because b is greater than both b_1 and b_2,

$$|f(x)| \le B_1|h(x)| \quad \text{and} \quad |g(x)| \le B_2|h(x)|,$$

and so, by adding the inequalities (Appendix A, T25), we get

$$|f(x)| + |g(x)| \le B_1|h(x)| + B_2|k(x)|.$$

Thus, by the transitive law for inequalities (Appendix A, T17),

$$|f(x)| + g(x)| \le B_1|h(x)| + B_2|k(x)|.$$

Now, because each value of $G(x) = |G(x)|$ is greater than or equal to $|h(x)|$ and $|k(x)|$,

$$B_1|h(x)| + B_2|k(x)| \le B_1|G(x)|$$
$$+ B_2|G(x)| \le (B_1 + B_2)|G(x)|.$$

Hence, again by transitivity and because $B = B_1 + B_2$,

$$|f(x) + g(x)| \le B|G(x)| \quad \text{for all real numbers } x > b.$$

Therefore, by definition of O-notation, $f(x) + g(x)$ is $O(G(x))$.

14. *Start of proof:* Suppose $f, g, h,$ and k are real-valued functions of a real variable that are defined on the same set D of nonnegative real numbers, and suppose $f(x)$ is $O(h(x))$ and $g(x)$ is $O(k(x))$. By definition of O-notation, there exist positive real numbers $b_1, B_1, b_2,$ and B_2 such that $|f(x)| \le B_1|h(x)|$ for all real numbers $x > b_1$, and $|g(x)| \le B_2|k(x)|$ for all real numbers $x > b_2$. Let $B = B_1 B_2$ and let $b = \max(b_1, b_2)$.

15. **b.** *Hint:* By the laws of exponents, $x^{n-m} = \dfrac{x^n}{x^m}$. Thus if $x^{n-m} > 1$, then $\dfrac{x^n}{x^m} > 1$.

16. **a.** For all real numbers $x > 1$, $x^2 + 15x + 4 \ge 0$ because all terms are nonnegative. Adding x^2 to both sides gives $2x^2 + 15x + 4 \ge x^2$. By the nonnegativity of all terms when $x > 1$, absolute value signs may be added to both sides of the inequality. Thus $|x^2| \le |2x^2 + 15x + 4|$ for all real numbers $x > 1$.

b. For all real numbers $x > 1$,

$$|2x^2 + 15x + 4| = 2x^2 + 15x + 4$$
$$\text{because } 2x^2 + 15x + 4$$
$$\text{is positive (since } x > 1)$$
$$\Rightarrow |2x^2 + 15x + 4| \le 2x^2 + 15x^2 + 4x^2$$
$$\text{because since } x > 1,$$
$$\text{then } x < x^2 \text{ and } 1 < x^2$$
$$\Rightarrow |2x^2 + 15x + 4| \le 21x^2 \quad \text{because } 2 + 15 + 4 = 21$$
$$\Rightarrow |2x^2 + 15x + 4| \le 21|x^2| \quad \text{because } x^2 \text{ is positive.}$$

c. Let $A = 1$ and $a = 1$. Then by part (a), $A|x^2| \le |2x^2 + 15x + 4|$ for all real numbers $x > a$, and so, by definition of Ω-notation, $2x^2 + 15x + 4$ is $\Omega(x^2)$.

Let $B = 21$ and $b = 1$. Then, by part (b), $|2x^2 + 15x + 4| \le B|x^2|$ for all real numbers $x > b$, and so, by definition of O-notation, $2x^2 + 15x + 4$ is $O(x^2)$.

d. Let $k = 1$, $A = 1$, and $B = 21$. By parts (a) and (b), for all real numbers $x > k$,

$$A|x^2| \le |2x^2 + 15x + 4| \le B|x^2|$$

and thus, by definition of Θ-notation, $2x^2 + 15x + 4$ is $\Theta(x^2)$. In other words, $2x^2 + 15x + 4$ has order x^2. (Alternatively, Theorem 9.2.1(1) could be used to derive this result.)

18. First observe that for all real numbers $x > 1$, $4x^3 + 65x + 30 \ge 0$ because all terms are nonnegative. Adding x^3 to both sides gives $5x^3 + 65x + 30 \ge x^3$. By the nonnegativity of the terms when $x > 1$, absolute value signs may be added to both sides of the inequality to obtain $|x^3| \le |5x^3 + 65x + 30|$ for all real numbers $x > 1$. Let $a = 1$ and $A = 1$. Then $A|x^3| \le |5x^3 + 65x + 30|$ (*) for all real numbers $x > a$.

Second, note that when $x > 1$,

$$|5x^3 + 65x + 30| \le 5x^3 + 65x + 30$$
$$\text{because all the terms are}$$
$$\text{positive since } x > 1.$$
$$\Rightarrow |5x^3 + 65x + 30| \le 5x^3 + 65x^3 + 30x^3$$
$$\text{because since } x > 1, \text{ then}$$
$$65x \le 65x^3 \text{ and } 30 \le 30x^3$$
$$\Rightarrow |5x^3 + 65x + 30| \le 100x^3$$
$$\text{because } 5 + 65 + 30 = 100$$
$$\Rightarrow |5x^3 + 65x + 30| \le 100|x^3|$$
$$\text{because } x^3 \text{ is positive since } x > 1.$$

Let $b = 1$ and $B = 100$. Then $|5x^3 + 65x + 30| \le B|x^3|$ (**) for all real numbers $x > b$.

Let $k = \max(a, b)$. Putting inequalities (*) and (**) together gives that for all real numbers $x > k$,

$$A|x^3| \le |5x^3 + 65x + 30| \le B|x^3|.$$

Hence, by definition of Θ-notation, $5x^3 + 65x + 30$ is $\Theta(x^3)$; in other words, $5x^3 + 65x + 30$ has order x^3.

20. **a.** By definition of ceiling, for any real number x, $\lceil x^2 \rceil$ is that integer n such that $n - 1 < x^2 \le n$, and thus, by substitution, $x^2 \le \lceil x^2 \rceil$. Since $x > 1$, both sides of the inequality are positive, and so $|x^2| \le |\lceil x^2 \rceil|$.

b. As in part (a), $\lceil x^2 \rceil$ is that integer n such that $n - 1 < x^2 \le n$. Adding 1 to all parts of this inequality gives $n < x^2 + 1 \le n + 1$, so $\lceil x^2 \rceil < x^2 + 1$. Thus if x is any real number with $x > 1$, then

$$|\lceil x^2 \rceil| \le \lceil x^2 \rceil \quad \text{because } \lceil x^2 \rceil \text{ is positive}$$
$$\Rightarrow |\lceil x^2 \rceil| \le x^2 + 1 \quad \text{by the argument above}$$
$$\Rightarrow |\lceil x^2 \rceil| \le x^2 + x^2 \quad \text{because } 1 < x^2 \text{ since } x > 1$$
$$\Rightarrow |\lceil x^2 \rceil| \le 2x^2$$
$$\Rightarrow |\lceil x^2 \rceil| \le 2|x^2| \quad \text{because } x^2 \text{ is positive.}$$

c. Let $A = 1$ and $a = 1$. Then, by part (a), $|x^2| \le A|\lceil x^2 \rceil|$ for all real numbers $x > a$, and thus, by definition of Ω-notation, $\lceil x^2 \rceil$ is $\Omega(x^2)$.

Let $B = 2$ and $b = 1$. Then, by part (b), $|x^2| \le B\lceil \lceil x^2 \rceil \rceil$ for all real numbers $x > b$, and thus, by definition of O-notation, $\lceil x^2 \rceil$ is $O(x^2)$.

d. We conclude that $\lceil x^2 \rceil$ is $\Theta(x^2)$ by part (c) and Theorem 9.2.1(1). Alternatively, we can use the results of parts (a) and (b), letting $k = \max(a, b)$, to obtain the result that for all real numbers $x > k$,

$$A|x^2| \le |\lceil x^2 \rceil| \le D|x^2|$$

and conclude directly from the definition of Θ-notation that $\lceil x^2 \rceil$ is $\Theta(x^2)$.

22. a. For all real numbers $x > 1$,

$$|7x^4 - 95x^3 + 3| \le |7x^4| + |95x^3| + |3|$$

$$\text{by the triangle inequality}$$

$$\Rightarrow \quad |7x^4 - 95x^3 + 3| \le 7x^4 + 95x^3 + 3$$

because all terms are positive since $x > 1$

$$\Rightarrow \quad |7x^4 - 95x^3 + 3| \le 7x^4 + 95x^4 + 3x^4$$

because $x > 1$ implies that $x^3 \le x^4$ and $1 \le x^4$

$$\Rightarrow \quad |7x^4 - 95x^3 + 3| \le 105|x^4|$$

because $7 + 95 + 3 = 105$ and $x^4 > 0$.

b. $7x^4 - 95x^3 + 3$ is $O(x^4)$

25. *Hint:* Use an argument by contradiction similar to the one in Example 9.2.8.

26. *Proof:* Suppose $a_0, a_1, a_2, \ldots, a_n$ are real numbers and $a_n \ne 0$. By the generalized triangle inequality,

$$|a_n x^n + a_{n-1}x^{n-1} + \cdots + a_1 x + a_0|$$
$$\le |a_n x^n| + |a_{n-1}x^{n-1}| + \cdots + |a_1 x| + |a_0|,$$

and because the absolute value of a product is the product of the absolute values (exercise 50, Section 3.4),

$$|a_n x^n| + |a_{n-1}x^{n-1}| + \cdots + |a_1 x| + |a_0|$$
$$\le |a_n||x^n| + |a_{n-1}||x^{n-1}| + \cdots + |a_1||x| + |a_0|.$$

In addition, when $x > 1$, property (9.2.1) implies that

$$x^n \le x^n, \quad x^{n-1} \le x^n, \ldots, x^2 \le x^n, \quad x \le x^n, \quad 1 \le x^n,$$

and also $x^n = |x^n|$ because $x > 1$. Thus

$$|a_n x^n + a_{n-1}x^{n-1} + \cdots + a_1 x + a_0|$$
$$\le |a_n||x^n| + |a_{n-1}||x^n| + \cdots + |a_1||x^n| + |a_0||x^n|$$
$$\le (|a_n| + |a_{n-1}| + \cdots + |a_1| + |a_0|)|x^n|.$$

Let $b = 1$ and $B = |a_n| + |a_{n-1}| + \cdots + |a_1| + |a_0|$. Then for all real numbers $x > b$,

$$|a_n x^n + a_{n-1}x^{n-1} + \cdots + a_1 x + a_0| \le B|x^n|$$

and so, by definition of O-notation,

$$a_n x^n + a_{n-1}x^{n-1} + \cdots + a_1 x + a_0 \text{ is } O(x^n).$$

28. Let $a = \left(\frac{95+3}{7}\right) \cdot 2 = 28$, and let $A = \frac{7}{2}$. If $x > a$, then

$$x \ge \left(\frac{95+3}{7}\right) \cdot 2$$

$$\Rightarrow \quad x \ge \frac{95}{7} \cdot 2 + \frac{3}{7} \cdot 2$$

$$\Rightarrow \quad x \ge \frac{95}{7} \cdot 2 + \frac{3}{7} \cdot 2 \frac{1}{x^3}$$

because $\frac{1}{x^3} < 1$ since $x > 28$

$$\Rightarrow \quad \frac{7}{2}x^4 \ge 95x^3 + 3$$

by multiplying both sides by $\frac{7x^3}{2}$

$$\Rightarrow \quad \left(7 - \frac{7}{2}\right)x^4 \ge 95x^3 - 3$$

because $95x^3 + 3 \ge 95x^3 - 3$ and $7 - \frac{7}{2} = \frac{7}{2}$

$$\Rightarrow \quad 7x^4 - \frac{7}{2}x^4 \ge 95x^3 - 3$$

by multiplying out

$$\Rightarrow \quad 7x^4 - 95x^3 + 3 \ge \frac{7}{2}x^4$$

by adding $\frac{7}{2}x^4 - 95x^3 + 3$ to both sides

$$\Rightarrow \quad 7x^4 - 95x^3 + 3 \ge Ax^4$$

because $A = \frac{7}{2}$

$$\Rightarrow \quad |7x^4 - 95x^3 + 3| \ge A|x^4|$$

because both sides are nonnegative.

Hence, by definition of Ω-notation, $7x^4 - 95x^3 + 3$ is $\Omega(x^4)$.

31. By exercise 22, $7x^4 - 95x^3 + 3$ is $O(x^4)$, and by exercise 28, $7x^4 - 95x^3 + 3$ is $\Omega(x^4)$. Thus, by Theorem 9.2.1 (1), $7x^4 - 95x^3 + 3$ is $\Theta(x^4)$.

34. $\frac{(x+1)(x-2)}{4} = \frac{x^2 - x - 2}{4} = \frac{1}{4}x^2 - \frac{1}{4}x - \frac{1}{2}$ is $\Theta(x^2)$ by the theorem on polynomial orders.

37. $\frac{n(n+1)(2n+1)}{6} = \frac{2n^3 + 3n^2 + n}{6} = \frac{1}{3}n^3 + \frac{1}{2}n^2 + \frac{1}{6}n$, which is $\Theta(n^3)$ by the theorem on polynomial orders.

40. By exercise 10 of Section 4.2, $1^2 + 2^2 + 3^2 + \cdots + n^2 = \frac{n(n+1)(2n+1)}{6}$, and, by exercise 37 above, $\frac{n(n+1)(2n+1)}{6}$ is $\Theta(n^3)$. Hence $1^2 + 2^2 + 3^2 + \cdots + n^2$ is $\Theta(n^3)$.

42. By Theorem 4.2.2, $2 + 4 + 6 + \cdots + 2n = 2\left(\frac{n(n+1)}{2}\right) = n^2 + n$, and by the theorem on polynomial orders, $n^2 + n$ is $\Theta(n^2)$. Thus $2 + 4 + 6 + \cdots + 2n$ is $\Theta(n^2)$.

44. By direct calculation or by Theorem 4.1.1, $\sum_{i=1}^n (4i - 9) = 4\sum_{i=1}^n i - \sum_{i=1}^n 9 = 4\left(\frac{n(n+1)}{2}\right) - 9n$. The last equality holds because of Theorem 4.2.2 and the fact that $\sum_{i=1}^n 9 = 9 + 9 + \cdots + 9 \, (n \text{ summands}) = 9n$. Then $4\left(\frac{n(n+1)}{2}\right) - 9n = 2n^2 + 2n - 9n = 2n^2 - 7n$, and hence $\sum_{i=1}^n (4i - 9) = 2n^2 - 7n$. But $2n^2 - 7n$ is $\Theta(n^2)$ by the theorem on polynomial orders. Thus $\sum_{i=1}^n (4i - 9)$ is $\Theta(n^2)$.

46. *Hint:* Use the result of exercise 13 from Section 4.2.

48. *Hints:*
a. $\dfrac{a_n x^n + a_{n-1}x^{n-1} + \cdots + a_1 x + a_0}{a_n x^n}$

$$= 1 + \frac{a_{n-1}}{a_n} \cdot \frac{1}{x} + \frac{a_{n-2}}{a_n} \cdot \frac{1}{x^2} + \cdots + \frac{a_1}{a_n} \cdot \frac{1}{x^{n-1}} + \frac{a_0}{a_n} \cdot \frac{1}{x^n}.$$

b. $\lim_{n \to \infty} f(x) = L$ means that given any real number $\varepsilon > 0$, there is a real number $M > 0$ such that $L - \varepsilon < f(x) < L + \varepsilon$ for all real numbers $x > M$. Apply the definition of limit to the result of part (a), using $\varepsilon = \frac{1}{2}$.

49. a. Let f, g, and h be functions from **R** to **R**, and suppose $f(x)$ is $O(h(x))$ and $g(x)$ is $O(h(x))$. Then there exist real numbers b_1, b_2, B_1, and B_2 such that $|f(x)| \le B_1|h(x)|$ for all $x > b_1$ and $|g(x)| \le B_2|h(x)|$ for all $x > b_2$. Let $B = B_1 + B_2$, and let b be the greater of b_1 and b_2. Then, for all $x > b$,

$$|f(x) + g(x)| < |f(x)| + |g(x)|$$

  by the triangle inequality

$$\Rightarrow \quad |f(x) + g(x)| \le B_1|h(x)| + B_2|h(x)|$$

  by hypothesis

$$\Rightarrow \quad |f(x) + g(x)| \le (B_1 + B_2)|h(x)|$$

  by algebra

$$\Rightarrow \quad |f(x) + g(x)| \le B|h(x)|$$

  because $B = B_1 + B_2$.

Hence, by definition of O-notation, $f(x) + g(x)$ is $O(h(x))$.

b. By exercise 15, for all $x > 1$, $x^2 < x^4$. Hence $|x^2| \le 1 \cdot |x^4|$ for all $x > 1$. Thus, by definition of O-notation, x^2 is $O(x^4)$. Clearly also, $|x^4| \le 1 \cdot |x^4|$ for all x, and so x^4 is $O(x^4)$. It follows by part (a) that $x^2 + x^4$ is $O(x^4)$.

50. d. *Hint:* If p, q, and s are positive integers, r is a nonnegative integer, and $\frac{p}{q} > \frac{r}{s}$, then $ps > qr$ and so $ps - qr > 0$. Also $\dfrac{x^{p/q}}{x^{r/s}} = x^{(p/q - r/s)} = x^{(pq - rs)/qs}$. Apply part (c) to $x^{1/qs}$, and use the fact that $ps - qr$ is an integer and $ps - qr > 0$ to make use of the result of exercise 15.

51. By part (d) of exercise 50, for all $x > 1$, $x \le x^{4/3}$ and $1 = x^0 \le x^{4/3}$. Hence, by definition of O-notation (since all expressions are positive), x is $O(x^{4/3})$ and 1 is $O(x^{4/3})$. By part (c) of exercise 49, then, $-15x = (-15)x$ is $O(x^{4/3})$ and $7 = 7 \cdot 1$ is $O(x^{4/3})$. It follows, by part (a) of exercise 49 (applied twice), that $4x^{4/3} - 15x + 7 = 4x^{4/3} + (-15x) + 7$ is $O(x^{4/3})$.

53. *Hint:* The proof is similar to the solution in Example 9.2.8. (Choose a real number x so that $x > B^{1/(r-s)}$ and $x > b$.)

54. $f(x) = \dfrac{\sqrt{x}(3x + 5)}{2x + 1} = \dfrac{3x^{3/2} + 5x^{1/2}}{2x + 1}$. The numerator of $f(x)$ is a sum of rational power functions with highest power $3/2$, and the denominator is a sum of rational power functions with highest power 1. Because $3/2 - 1 = 1/2$, Theorem 9.2.4 implies that $f(x)$ is $\Theta(x^{1/2})$.

57. a. *Proof (by mathematical induction):* Let the property $P(n)$ be the inequality

$$\sqrt{1} + \sqrt{2} + \sqrt{3} + \cdots + \sqrt{n} \le n^{3/2}.$$

Show that the property is true for $n = 1$:

When $n = 1$, the left-hand side of the inequality is 1, and the right-hand side is $1^{3/2}$, which is also 1. Thus the property is true for $n = 1$.

Show that for all integers $k \ge 1$, if the property is true for $n = k$, then it is true for $n = k + 1$:

Let k be any integer with $k \ge 1$, and suppose

$$\sqrt{1} + \sqrt{2} + \sqrt{3} + \cdots + \sqrt{k} \le k^{3/2}.$$

  [*Inductive hypothesis*]

We must show that

$$\sqrt{1} + \sqrt{2} + \sqrt{3} + \cdots + \sqrt{k+1} \le (k+1)^{3/2}.$$

But

$$\sqrt{1} + \sqrt{2} + \sqrt{3} + \cdots + \sqrt{k+1}$$
$$= \sqrt{1} + \sqrt{2} + \sqrt{3} + \cdots + \sqrt{k} + \sqrt{k+1}$$

  by making the next-to-last term explicit

$$\Rightarrow \quad \sqrt{1} + \sqrt{2} + \sqrt{3} + \cdots + \sqrt{k+1} \le k^{3/2} + \sqrt{k+1}$$

  by inductive hypothesis

$$\Rightarrow \quad \sqrt{1} + \sqrt{2} + \sqrt{3} + \cdots + \sqrt{k+1} \le k\sqrt{k} + \sqrt{k+1}$$

  because $k^{3/2} = k\sqrt{k}$

$$\Rightarrow \quad \sqrt{1} + \sqrt{2} + \sqrt{3} + \cdots + \sqrt{k+1}$$
$$\le k\sqrt{k+1} + \sqrt{k+1}$$

  because $\sqrt{k} < \sqrt{k+1}$

$$\Rightarrow \quad \sqrt{1} + \sqrt{2} + \sqrt{3} + \cdots + \sqrt{k+1} \le (k+1)\sqrt{k+1}$$

  by factoring out $\sqrt{k+1}$

$$\Rightarrow \quad \sqrt{1} + \sqrt{2} + \sqrt{3} + \cdots + \sqrt{k+1} \le (k+1)^{3/2}.$$

b. *Hint:* When $k \ge 1, k^2 \ge k^2 - 1$. Use the fact that $k^2 - 1 = (k-1)(k+1)$ and divide both sides by $k(k-1)$ to obtain $\frac{k}{k-1} \ge \frac{k+1}{k}$. But $\frac{k+1}{k} \ge 1$, and any number greater than or equal to 1 is greater than or equal to its square root. Thus $\frac{k}{k-1} \ge \frac{k+1}{k} \ge \sqrt{\frac{k+1}{k}} = \frac{\sqrt{k+1}}{\sqrt{k}}$. Hence $k\sqrt{k} \ge (k-1)\sqrt{k+1} = (k+1-2)\sqrt{k+1} = (k+1)\sqrt{k+1} - 2\sqrt{k+1}$, and so $k\sqrt{k} + 2\sqrt{k+1} \ge (k+1)\sqrt{k+1}$.

c. $\sqrt{1} + \sqrt{2} + \sqrt{3} + \cdots + \sqrt{n}$ is $\Theta(x^{3/2})$.

59. *Proof:* Suppose $f(x)$ is $o(g(x))$. By definition of o-notation, $\lim_{x \to \infty} \frac{f(x)}{g(x)} = 0$. By definition of limit, this implies that given any real number $\varepsilon > 0$, there exists a real number x_0 such that $\left| \frac{f(x)}{g(x)} - 0 \right| < \varepsilon$ for all $x > x_0$. Let $b = \max(x_0, 1)$. Then $|f(x)| \le \varepsilon|g(x)|$ for all $x > b$. Choose $\varepsilon = 1$, and set $B = 1$. Then there exists a real number b such that $|f(x)| \le B|g(x)|$ for all $x > b$. Hence, by definition of O-notation, $f(x)$ is $O(g(x))$.

Section 9.3

1. a. $\log_2(200) = \frac{\ln 200}{\ln 2} \cong 7.6$ nanoseconds $=$ 0.0000000076 second

d. $200^2 = 40{,}000$ nanoseconds $= 0.00004$ second

e. $200^8 = 2.56 \times 10^{18}$ nanoseconds $\cong$ $\frac{2.56 \times 10^{18}}{10^9 \cdot 60 \cdot 60 \cdot 24 \cdot (365.25)}$ years $\cong 81{,}121.5$ years

  [*because there are 10^9 nanoseconds in a second, 60 seconds in a minute, 60 minutes in an hour, 24 hours in a day and approximately 365.25 days in a year on average*].

2. a. When the input size is increased from m to $2m$, the number of operations increases from cm^2 to $c(2m)^2 = 4cm^2$.

b. By part (a), the number of operations increases by a factor of $(4cm^2)/cm^2 = 4$.

c. When the input size is increased by a factor of 10 (from m to $10m$), the number of operations increases by a factor of $(c(10m)^2)/(cm^2) = (100cm^2)/cm^2 = 100$.

4. a. Algorithm A has order n^2 and algorithm B has order $n^{3/2}$.

b. Algorithm A is more efficient than algorithm B when $2n^2 < 80n^{3/2}$. This occurs exactly when

$$n^2 < 40n^{3/2} \Leftrightarrow \frac{n^2}{n^{3/2}} < 40 \Leftrightarrow n^{1/2} < 40 \Leftrightarrow n < 40^2.$$

Thus, algorithm A is more efficient than algorithm B when $n < 1,600$.

c. Algorithm B is at least 100 times more efficient than algorithm A for values of n with $100(80n^{3/2}) \leq 2n^2$. This occurs exactly when $8,000n^{3/2} \leq 2n^2 \Leftrightarrow 4,000 \leq \frac{n^2}{n^{3/2}} \Leftrightarrow 4,000 \leq \sqrt{n} \Leftrightarrow 16,000,000 \leq n$. Thus, algorithm B is at least 100 times more efficient than algorithm A when $n > 16,000,000$.

6. a. There are two multiplications, one addition, and one subtraction for each iteration of the loop, so there are four times as many operations as there are iterations of the loop. The loop is iterated $(n-1) - 3 + 1 = n - 3$ times (since the number of iterations equals the top minus the bottom index plus 1). Thus the total number of operations is $4(n-3) = 4n - 12$.

b. By the theorem on polynomial orders, $4n - 12$ is $\Theta(n)$, so the algorithm segment has order n.

8. a. There is one subtraction for each iteration of the loop, and there are $\lfloor n/2 \rfloor$ iterations of the loop.

b. $\lfloor n/2 \rfloor = \begin{cases} n/2 & \text{if } n \text{ is even} \\ (n-1)/2 & \text{if } n \text{ is odd} \end{cases}$

is $\Theta(n)$ by theorem on polynomial orders, so the algorithm segment has order n.

9. a. For each iteration of the inner loop, there are two multiplications and one addition. There are $2n$ iterations of the inner loop for each iteration of the outer loop, and there are n iterations of the outer loop. Therefore, the number of iterations of the inner loop is $2n \cdot n = 2n^2$. It follows that the total number of elementary operations that must be performed when the algorithm is executed is $3 \cdot 2n^2 = 6n^2$.

b. Since $6n^2$ is $\Theta(n^2)$ (by the theorem on polynomial orders), the algorithm segment has order n^2.

11. a. There is one addition for each iteration of the inner loop. The number of iterations in the inner loop can be deduced from the following table, which shows the values of k and j for which the inner loop is executed.

Hence the total number of iterations of the inner loop is

$$2 + 3 + \cdots + n = (1 + 2 + 3 + \cdots + n) - 1$$
$$= \frac{n(n+1)}{2} - 1 = \frac{1}{2}n^2 + \frac{1}{2}n - 1$$

(by Theorem 4.2.2). Because one operation is performed for each iteration of the inner loop, the total number of operations is $\frac{1}{2}n^2 + \frac{1}{2}n - 1$.

b. By the theorem on polynomial orders, $\frac{1}{2}n^2 + \frac{1}{2}n - 1$ is $\Theta(n^2)$, and so the algorithm segment has order n^2.

14. a. There is one addition for each iteration of the inner loop, and there is one additional addition and one multiplication for each iteration of the outer loop. The number of iterations in the inner loop can be deduced from the following table, which shows the values of i and j for which the inner loop is executed.

i	1	2		3			$\cdots$	n			
j	1	1	2	1	2	3	$\cdots$	1	2	3	$\cdots$	n

$$\underbrace{\quad}_{1} \underbrace{\quad}_{2} \underbrace{\quad}_{3} \underbrace{\quad}_{n}$$

Hence the total number of iterations of the inner loop is

$$1 + 2 + 3 + \cdots + n = (1 + 2 + 3 + \cdots + n)$$
$$= \frac{n(n+1)}{2} = \frac{1}{2}n^2 + \frac{1}{2}n$$

(by Theorem 4.2.2). Because one addition is performed for each iteration of the inner loop, the number of operations performed when the inner loop is executed is $\frac{1}{2}n^2 + \frac{1}{2}$. Now an additional two operations are performed each time the outer loop is executed, and because the outer loop is executed n times, this gives an additional $2n$ operations. Therefore, the total number of operations is

$$\frac{1}{2}n^2 + \frac{1}{2}n + 2n = \frac{1}{2}n^2 + \frac{5}{2}n.$$

b. By the theorem on polynomial orders, $\frac{1}{2}n^2 + \frac{5}{2}n$ is $\Theta(n^2)$, and so the algorithm segment has order n^2.

17. a. There are two subtractions and one multiplication for each iteration of the inner loop.

If n is odd, the number of iterations of the inner loop can be deduced from the following table, which shows the values of i and j for which the inner loop is executed.

i	1	2	3		4		5		6			$\cdots$	$n-1$		$\cdots$		n		$\cdots$	
$\lfloor \frac{i+1}{2} \rfloor$	1	1	2		2		3		3			$\cdots$	$\frac{n-1}{2}$		$\cdots$		$\frac{n+1}{2}$		$\cdots$	
j	1	1	1	2	1	2	1	2	3	1	2	3	$\cdots$	1	2	$\cdots$	$\frac{n-1}{2}$	1	2	$\cdots$	$\frac{n+1}{2}$

$$\underbrace{}_{1}\ \underbrace{}_{1}\ \underbrace{}_{2}\ \underbrace{}_{2}\ \underbrace{}_{3}\ \underbrace{}_{3}\qquad \underbrace{}_{\frac{n-1}{2}}\qquad \underbrace{}_{\frac{n+1}{2}}$$

Thus the number of iterations of the inner loop is

$$1 + 1 + 2 + 2 + \cdots + \frac{n-1}{2} + \frac{n-1}{2} + \frac{n+1}{2}$$

$$= 2 \cdot \left(1 + 2 + 3 + \cdots + \frac{n-1}{2}\right) + \frac{n+1}{2}$$

$$= 2 \cdot \frac{\frac{n-1}{2}\left(\frac{n-1}{2}+1\right)}{2} + \frac{n+1}{2} \qquad \text{by Theorem 4.2.2}$$

$$= \frac{n^2 - 2n + 1}{4} + \frac{n-1}{2} + \frac{n+1}{2}$$

$$= \frac{1}{4}n^2 + \frac{1}{2}n + \frac{1}{4}.$$

By similar reasoning, if n is even, then the number of iterations of the inner loop is

$$1 + 1 + 2 + 2 + 3 + 3 + \cdots + \frac{n}{2} + \frac{n}{2}$$

$$= 2 \cdot \left(1 + 2 + 3 + \cdots + \frac{n}{2}\right)$$

$$= 2 \cdot \left(\frac{\frac{n}{2}\left(\frac{n}{2}+1\right)}{2}\right) \qquad \text{by Theorem 4.2.2}$$

$$= \frac{n^2}{4} + \frac{n}{2}.$$

Because three operations are performed for each iteration of the inner loop, the answer is $3\left(\frac{n^2}{4} + \frac{n}{2}\right)$ when n is even and $3\left(\frac{1}{4}n^2 + \frac{1}{2}n + \frac{1}{4}\right)$ when n is odd.

b. Since $3\left(\frac{n^2}{4} + \frac{n}{2}\right)$ is $\Theta(n^2)$ and $3\left(\frac{1}{4}n^2 + \frac{1}{2}n + \frac{1}{4}\right)$ is also $\Theta(n^2)$ (by the theorem on polynomial orders), this algorithm segment has order n^2.

19. *Hint:* See Section 6.5 for a discussion of how to count the number of iterations of the innermost loop.

20.

	$a[1]$	$a[2]$	$a[3]$	$a[4]$	$a[5]$
Initial order	6	2	1	8	4
Result of step 1	2	6	1	8	4
Result of step 2	1	2	6	8	4
Result of step 3	1	2	6	8	4
Final order	1	2	4	6	8

22.

n	5									
$a[1]$	6		2			1				
$a[2]$	2	6			1	2				
$a[3]$	1			6					4	
$a[4]$	8							4	6	
$a[5]$	4					8				
k		2		3			4	5		6
x		2		1			8	4		
j		1	0	2	1	0	3	4	3	2

24. *Solution 1:* The answer is 7, the same as the number of distinct nonzero values of j.

Solution 2: The answer is 7: 1 comparison from step 1, 1 from step 2, 3 from step 3, and 2 from step 4.

27. *Hint:* The answer to part (a) is $E_n = 3 + 4 + \cdots + (n+1)$, which equals $(1 + 2 + 3 + \cdots + (n+1)) - (1 + 2)$.

28. The top row of the table below shows the initial values of the array, and the bottom row shows the final values. The result of each interchange is shown in a separate row.

$a[1]$	$a[2]$	$a[3]$	$a[4]$	$a[5]$
5	3	4	6	2
3	5	4	6	2
2	5	4	6	3
2	4	5	6	3
2	3	5	6	4
2	3	4	6	5
2	3	4	5	6

30.

n	5										
a[1]	5	3		2							
a[2]	3	5			4	3					
a[3]	4				5			4			
a[4]	6										5
a[5]	2			3		4			5	6	
k		1			2			3			4
i		2	3	4	5	3	4	5	4	5	5
temp		5			3	5		4		5	6

32. There is one comparison for each combination of values of k and i: namely, $4 + 3 + 2 + 1 = 10$.

35. **b.** $n - 3 + 1 = n - 2$ **d.** *Hint:* The answer is n^2.

36.

n	3								
a[0]	2								
a[1]	1								
a[2]	−1								
a[3]	3								
x	2								
polyval	2	4			0			24	
i	1	2			3				
term	1	2	−1	−2		−4	3	6	12	24
j	1		1	2		1	2	3	

38. Number of multiplications

$$= \text{number of iterations of the inner loop}$$
$$= 1 + 2 + 3 + \cdots + n$$
$$= \frac{n(n + 1)}{2} \qquad \text{by Theorem 4.2.2}$$

number of additions

$$= \text{number of iterations of the outer loop}$$
$$= n$$

Hence the total number of multiplications and additions is

$$\frac{n(n + 1)}{2} + n = \frac{1}{2}n^2 + \frac{3}{2}n.$$

40.

n	3			
a[0]	2			
a[1]	1			
a[2]	−1			
a[3]	3			
x	2			
polyval	3	5	11	24
i	1	2	3	

42. *Hint:* The answer is $t_n = 2n$.

Section 9.4

1.

x	$f(x) = 3^x$
0	$3^0 = 1$
1	$3^1 = 3$
2	$3^2 = 9$
−1	$3^{-1} = 1/3$
−2	$3^{-2} = 1/9$
1/2	$3^{1/2} \cong 1.7$
−(1/2)	$3^{-(1/2)} \cong 0.6$

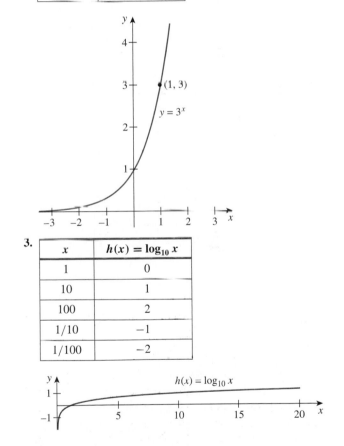

3.

x	$h(x) = \log_{10} x$
1	0
10	1
100	2
1/10	−1
1/100	−2

5.

x	$\lfloor \log_2 x \rfloor$
$1 \le x < 2$	0
$2 \le x < 4$	1
$4 \le x < 8$	2
$8 \le x < 16$	3
$1/2 \le x < 1$	-1
$1/4 \le x < 1/2$	-2

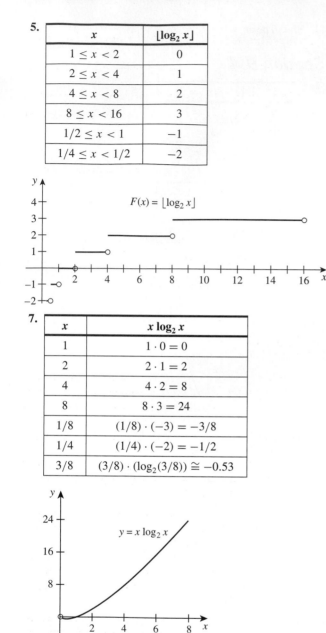

7.

x	$x \log_2 x$
1	$1 \cdot 0 = 0$
2	$2 \cdot 1 = 2$
4	$4 \cdot 2 = 8$
8	$8 \cdot 3 = 24$
1/8	$(1/8) \cdot (-3) = -3/8$
1/4	$(1/4) \cdot (-2) = -1/2$
3/8	$(3/8) \cdot (\log_2(3/8)) \cong -0.53$

9. The distance above the axis is $(2^{64} \text{ units}) \cdot (\frac{1}{4}\frac{\text{inch}}{\text{unit}}) = \frac{2^{64}}{4}$ inches $= \frac{2^{64}}{4 \cdot 12 \cdot 5280}$ miles $\cong 72{,}785{,}448{,}520{,}000$ miles. The ratio of the height of the point to the average distance of the earth to the sun is approximately $72785448520000/93000000 \cong 782{,}639$. (If you perform the computation using metric units and the approximation $0.635 \text{ cm} \cong 1/4$ inch, the ratio comes out to be approximately $780{,}912$.)

10. b. By definition of logarithm, $\log_b x$ is the exponent to which b must be raised to obtain x. Thus when b is actually raised to this exponent, x is obtained. That is, $b^{\log_b x} = x$.

11. b.

Each pair of points (u, v) and (v, u) are "mirror image reflections" across the line $y = x$.

13. *Hints:* (1) $\lfloor \log_{10} x \rfloor = m$, (2) See Example 9.4.1.

15. No. *Counterexample:* Let $n = 2$. Then $\lceil \log_2(n-1) \rceil = \lceil \log_2 1 \rceil = \lceil 0 \rceil = 0$, whereas $\lceil \log_2 n \rceil = \lceil \log_2 2 \rceil = \lceil 1 \rceil = 1$.

16. *Hint:* The statement is true.

18. $\lfloor \log_2 148206 \rfloor + 1 = 18$

21. a. $a_1 = 1$

$a_2 = a_{\lfloor 2/2 \rfloor} + 2 = a_1 + 2 = 1 + 2$

$a_3 = a_{\lfloor 3/2 \rfloor} + 2 = a_1 + 2 = 1 + 2$

$a_4 = a_{\lfloor 4/2 \rfloor} + 2 = a_2 + 2 = (1+2) + 2$
$= 1 + 2 \cdot 2$

$a_5 = a_{\lfloor 5/2 \rfloor} + 2 = a_2 + 2 = (1+2) + 2$
$= 1 + 2 \cdot 2$

$a_6 = a_{\lfloor 6/2 \rfloor} + 2 = a_3 + 2 = (1+2) + 2$
$= 1 + 2 \cdot 2$

$a_7 = a_{\lfloor 7/2 \rfloor} + 2 = a_3 + 2 = (1+2) + 2$
$= 1 + 2 \cdot 2$

$a_8 = a_{\lfloor 8/2 \rfloor} + 2 = a_4 + 2$
$= (1 + 2 \cdot 2) + 2 = 1 + 3 \cdot 2$

$a_9 = a_{\lfloor 9/2 \rfloor} + 2 = a_4 + 2$
$= (1 + 2 \cdot 2) + 2 = 1 + 3 \cdot 2$

$\vdots$

$a_{15} = a_{\lfloor 15/2 \rfloor} + 2 = a_7 + 2$
$= (1 + 2 \cdot 2) + 2 = 1 + 3 \cdot 2$

$a_{16} = a_{\lfloor 16/2 \rfloor} + 2 = a_8 + 2$
$= (1 + 3 \cdot 2) + 2 = 1 + 4 \cdot 2$

$\vdots$

Guess:

$a_n = 1 + 2\lfloor \log_2 n \rfloor$

b. *Proof:* Suppose the sequence $a_1, a_2, a_3, \ldots$ is defined recursively as follows: $a_1 = 1$ and $a_k = a_{\lfloor k/2 \rfloor} + 2$ for all integers $k \geq 2$. We will show by strong mathematical induction that the following property is true for all integers $n \geq 1$: $a_n = 1 + 2 \lfloor \log_2 n \rfloor$.

Show that the property is true for $n = 1$:

When $n = 1, 1 + 2 \lfloor \log_2 n \rfloor = 1 + 2 \lfloor \log_2 1 \rfloor = 1 + 2 \cdot 0 = 1$, which is the value of a_1.

Show that for any integer $k > 1$, if the property is true for all integers i with $1 \leq i < k$, then it is true for $n = k$:

Let k be an integer with $k > 1$ and suppose $a_i = 1 + 2 \lfloor \log_2 i \rfloor$ for all integers i with $1 \leq i < k$. [*This is the inductive hypothesis.*] We must show that $a_k = 1 + 2 \lfloor \log_2 k \rfloor$.

Case 1 (k is even):

$$a_k = a_{\lfloor k/2 \rfloor} + 2$$
by the recursive definition of $a_1, a_2, a_3, \ldots$

$$= a_{k/2} + 2$$
because k is even

$$= 1 + 2 \lfloor \log_2(k/2) \rfloor + 2$$
by inductive hypothesis

$$= 3 + 2 \lfloor \log_2 k - \log_2 2 \rfloor$$
because $\log_b(x/y) = \log_b x - \log_b y$
(exercise 29, Section 7.2)

$$= 3 + 2 \lfloor \log_2 k - 1 \rfloor$$
because $\log_2 2 = 1$

$$= 3 + 2(\lfloor \log_2 k \rfloor - 1)$$
because for all real numbers x $\lfloor x - 1 \rfloor = \lfloor x \rfloor - 1$
(by exercise 15, Section 3.5)

$$= 1 + 2 \lfloor \log_2 k \rfloor$$
by algebra

Case 2 (k is odd):

$$a_k = a_{\lfloor k/2 \rfloor} + 2$$
by the recursive definition of $a_1, a_2, a_3, \ldots$

$$= a_{\lfloor (k-1)/2 \rfloor} + 2 \quad \text{because } k \text{ is odd}$$

$$= 1 + 2 \lfloor \log_2((k-1)/2) \rfloor + 2$$
by inductive hypothesis

$$= 3 + 2 \lfloor \log_2(k-1) - \log_2 2 \rfloor$$
because $\log_b(x/y) = \log_b x - \log_b y$

$$= 3 + 2 \lfloor \log_2(k-1) - 1 \rfloor$$
because $\log_2 2 = 1$

$$= 3 + 2(\lfloor \log_2(k-1) \rfloor - 1)$$
because for all real numbers x, $\lfloor x - 1 \rfloor = \lfloor x \rfloor - 1$
(by exercise 15, Section 3.5)

$$= 1 + 2 \lfloor \log_2(k-1) \rfloor \quad \text{by algebra}$$

$$= 1 + 2 \lfloor \log_2 k \rfloor$$
by property (9.4.3)

Thus in either case, $a_k = 1 + 2 \lfloor \log_2 k \rfloor$ [*as was to be shown*].

23. *Hint:* When $k \geq 2$, then $k^2 \geq 2k$, and so $k \leq \dfrac{k^2}{2}$. Hence $\dfrac{k^2}{2} + k \leq \dfrac{k^2}{2} + \dfrac{k^2}{2} = k^2$. Also when $k \geq 2$, then $k^2 > 1$, and so $\dfrac{1}{2} < \dfrac{k^2}{2}$. Consequently, $\dfrac{k^2}{2} + \dfrac{1}{2} < \dfrac{k^2}{2} + \dfrac{k^2}{2} = k^2$.

24. *Hint:* Here is the argument for the inductive step in the case where k is even.

$$c_k = 2c_{\lfloor k/2 \rfloor} + k$$
by the recursive definition of $c_1, c_2, c_3, \ldots$

$$\Rightarrow \quad c_k = 2c_{k/2} + k$$
because k is even

$$\Rightarrow \quad c_k \leq 2 \left[\tfrac{k}{2} \log_2 \left(\tfrac{k}{2} \right) \right] + k$$
by inductive hypothesis

$$\Rightarrow \quad c_k \leq k(\log_2 k - \log_2 2) + k$$
by algebra and the fact that $\log_b(x/y) = \log_b x - \log_b y$

$$\Rightarrow \quad c_k \leq k(\log_2 k - 1) + k$$
because $\log_2 2 = 1$

$$\Rightarrow \quad c_k \leq k \log_2 k$$
by algebra

25. *Solution 1:* One way to solve this problem is to compare values for $\log_2 x$ and $x^{1/10}$ for conveniently chosen, large values of x. For instance, if powers of 10 are used, the following results are obtained: $\log_2(10^{10}) = 10 \log_2 10 \cong 33.2$ and $(10^{10})^{1/10} = 10^{10 \cdot (1/10)} = 10^1 = 10$. Thus the value $x = 10^{10}$ does not work.

However, since $\log_2(10^{20}) = 20 \log_2 10 \cong 66.4$ and $(10^{20})^{1/10} = 10^{20 \cdot (1/10)} = 10^2 = 100$, and since $66.4 < 100$, the value $x = 10^{20}$ works.

Solution 2: Another approach is to use a graphing calculator or computer to sketch graphs of $y = \log_2 x$ and $y = x^{1/10}$, taking seriously the hint to "think big" in choosing the interval size for the x's. A few tries and use of the zoom and trace features make it appear that the graph of $y = x^{1/10}$ crosses above the graph of $y = \log_2 x$ at about 4.9155×10^{17}. Thus, for values of x larger than this, $x^{1/10} > \log_2 x$.

27. As with exercise 25, you can solve this problem either by numerical exploration or by exploring with a graphing calculator or computer. For instance, if you raise 1.0001 to successive large powers of 10, you can find the solution $x = 10^6 = 1,000,000$. That is, $(1.0001)^{1000000} > 2.67 \times 10^{43} > 1,000,000$. (This is the first power of 10 that works.)

Alternatively, you can use a graphing calculator or computer to sketch graphs of $y_1 = (1.0001)^x$ and $y_2 = x$ and look to see where the graph of $y_1 = (1.0001)^x$ rises above the graph of $y_2 = x$. You will need to zoom in carefully to obtain an accurate answer. If you use this method, you will find that if $x > 116703$, then $(1.0001)^x > x$.

29. $7x^2 + 3x \log_2 x$ is $\Theta(x^2)$.

30. [*To show that $2x + \log_2 x$ is $\Theta(x)$, we must find positive real numbers A, B, and k such that $A|x| \leq |2x + \log_2 x| \leq B|x|$ for all $x > k$.*] Observe that if $x > 1$, $\log_2 x > 0$ and, by property (9.4.9), $\log_2 x \le x$. Adding $2x$ to both sides gives $2x + \log_2 x \le 3x$, or, because all terms are positive,

$$|2x + \log_2 x| \le 3|x|.$$

Also, by similar reasoning, when $x > 1$, then $x < x + (x + \log_2 x) = 2x + \log_2 x$. Thus when $x > 1$,

$$|x| \le |2x + \log_2 x|$$

Therefore, let $k = 1$, $A = 1$, and $B = 3$. Then for all real numbers $x > k$,

$$A|x| \le |2x + \log_2 x| \le B|x|$$

and hence, by definition of Θ-notation, $2x + \log_2 x$ is $\Theta(x)$.

32. For all integers n, $2^n \le n^2 + 2^n$. Also, by property (9.4.10), there is a real number k such that $n^2 \le 2^n$ for all $n > k$. Adding 2^n to both sides gives $n^2 + 2^n \le 2^n + 2^n = 2 \cdot 2^n$. Because all quantities are nonnegative, we can write

$$|2^n| \le |n^2 + 2^n| \le 2 \cdot |2^n| \quad \text{for all integers } n > k.$$

Let $A = 1$ and $B = 2$. Then

$$A|2^n| \le |n^2 + 2^n| \le B|2^n| \quad \text{for all integers } n > k,$$

and hence, by definition of Θ-notation, $n^2 + 2^n$ is $\Theta(2^n)$.

33. *Hint:* $2^{n+1} = 2 \cdot 2^n$

34. *Hint:* Use a proof by contradiction. Start by supposing that there are positive real numbers B and b such that $4^n \le B \cdot 2^n$ for all real numbers $n > b$, and use the fact that $\frac{4^n}{2^n} = \left(\frac{4}{2}\right)^n = 2^n$ to obtain a contradiction.

35. By Theorem 4.2.3, for all integers $n \ge 0$,

$$1 + 2 + 2^2 + \cdots + 2^n = \frac{2^{n+1} - 1}{2 - 1} = 2^{n+1} - 1.$$

Also

$$2^{n+1} - 1 \le 2^{n+1} = 2 \cdot 2^n.$$

Thus, by transitivity of order,

$$1 + 2 + 2^2 + \cdots + 2^n \le 2 \cdot 2^n. \qquad (*)$$

Moreover, if $n > 0$, then

$$2^n \le 1 + 2 + 2^2 + \cdots + 2^n. \qquad (**)$$

Combining (*) and (**) gives

$$1 \cdot 2^n \le 1 + 2 + 2^2 + \cdots + 2^n \le 2 \cdot 2^n,$$

and so, because all parts are positive,

$$1 \cdot |2^n| \le |1 + 2 + 2^2 + \cdots + 2^n| \le 2 \cdot |2^n|.$$

Let $A = 1$, $B = 2$, and $k = 1$. Then for all integers $n > k$,

$$A \cdot |2^n| \le |1 + 2 + 2^2 + \cdots + 2^n| \le B \cdot |2^n|.$$

Thus, by definition of Θ-notation, $1 + 2 + 2^2 + \cdots + 2^n$ is $\Theta(2^n)$.

36. *Hint:* This is similar to the solution for exercise 35. Use the fact that $4 + 4^2 + 4^3 + \cdots + 4^n = 4(1 + 4 + 4^2 + 4^3 + \cdots + 4^{n-1})$.

39. Factor out the n to obtain

$$n + \frac{n}{2} + \frac{n}{4} + \cdots + \frac{n}{2^n}$$

$$= n \left(1 + \frac{1}{2} + \frac{1}{4} + \cdots + \frac{1}{2^n}\right)$$

$$= n \left(\frac{\left(\frac{1}{2}\right)^{n+1} - 1}{\frac{1}{2} - 1}\right) \qquad \text{by Theorem 4.2.3}$$

$$= n \left(\frac{1 - 2^{n+1}}{2^n(1 - 2)}\right) \qquad \begin{array}{l}\text{by multiplying numerator} \\ \text{and denominator by } 2^{n+1}\end{array}$$

$$= n \left(\frac{2^{n+1} - 1}{2^n}\right)$$

$$= n \left(2 - \frac{1}{2^n}\right) \qquad \text{by algebra.}$$

Now $1 \le 2 - \dfrac{1}{2^n} \le 2$ when $n > 1$. Thus

$$1 \cdot n \le n \left(2 - \frac{1}{2^n}\right) \le 2 \cdot n,$$

and so, by substitution,

$$1 \cdot n \le n + \frac{n}{2} + \frac{n}{4} + \cdots + \frac{n}{2^n} \le 2 \cdot n.$$

Let $A = 1$, $B = 2$, and $k = 1$. Then, because all quantities are positive, for all integers $n > k$,

$$A \cdot |n| \le \left|n + \frac{n}{2} + \frac{n}{4} + \cdots + \frac{n}{2^n}\right| \le B \cdot |n|.$$

Hence, by definition of Θ-notation, $n + \frac{n}{2} + \frac{n}{4} + \cdots + \frac{n}{2^n}$ is $\Theta(n)$.

43. If n is any integer with $n \ge 3$, then

$$n + \frac{n}{2} + \frac{n}{3} + \cdots + \frac{n}{n} = n \left(1 + \frac{1}{2} + \frac{1}{3} + \cdots + \frac{1}{n}\right).$$

By Example 9.4.7,

$$\ln(n) \le 1 + \frac{1}{2} + \frac{1}{3} + \cdots + \frac{1}{n} \le 2 \ln(n).$$

If $n > 1$, then we may multiply through by n and use the fact that all quantities are positive to obtain

$$|n \ln(n)| \le \left|n + \frac{n}{2} + \frac{n}{3} + \cdots + \frac{n}{n}\right| \le 2 |n \ln(n)|.$$

Let $A = 1$, $B = 2$, and $k = 1$. Then for all integers $n > k$,

$$A \cdot |n \ln(n)| \le \left|n + \frac{n}{2} + \frac{n}{3} + \cdots + \frac{n}{n}\right| \le B \cdot |n \ln(n)|$$

and so, by definition of Θ-notation, $n + \frac{n}{2} + \frac{n}{3} + \cdots + \frac{n}{n}$ is $\Theta(n \ln(n))$.

46. *Proof (by mathematical induction):* Let the property $P(n)$ be the inequality $n \leq 10^n$.

Show that the property is true for $n = 1$:
When $n = 1$, the inequality is $1 \leq 10$, which is true.

Show that for all integers $k \geq 1$, if the property is true for $n = k$, then it is true for $n = k + 1$:

Let k be any integer with $k \geq 1$, and suppose $k \leq 10^k$. [*This is the inductive hypothesis.*] We must show that $k + 1 \leq 10^{k+1}$. By inductive hypothesis, $k \leq 10^k$. Adding 1 to both sides gives $k + 1 \leq 10^k + 1$. But when $k \geq 1$, $10^k + 1 \leq 10^k + 9 \cdot 10^k = 10 \cdot 10^k = 10^{k+1}$. Thus, by transitivity of order, $k + 1 \leq 10^{k+1}$ [*as was to be shown*].

47. *Hint:* To prove the inductive step, use the fact that if $k > 1$, then $k + 1 \leq 2k$. Apply the logarithmic function with base 2 to both sides of this inequality, and use properties of logarithms.

48. *Hint:* $\underbrace{2 \cdot 2 \cdot 2 \cdots 2}_{n \text{ factors}} \leq 2 \cdot (2 \cdot 3 \cdot 4 \cdots n) = 2 \cdot n!$

49. a. *Proof:* Suppose n is a variable that takes positive integer values. Then

$$n! = \underbrace{n \cdot (n-1) \cdot (n-2) \cdot \ldots \cdot 2 \cdot 1}_{n \text{ factors}}$$

$$\leq \underbrace{n \cdot n \cdot n \cdot n \cdot \ldots \cdot n}_{n \text{ factors}} = n^n$$

because $(n - 1) \leq n, (n - 2) \leq n, \ldots,$ and $1 \leq n$. Let $B = 1$ and $b = 1$. It follows from the displayed inequality and the fact that $n!$ and n^n are positive that $|n!| \leq B \cdot |n^n|$ for all integers $n > b$. Hence, by definition of O-notation, $n!$ is $O(n^n)$.

c. *Hint:* $(n!)^2 = n! \cdot n! = (1 \cdot 2 \cdot 3 \cdots n)(n \cdot (n-1) \cdots 3 \cdot 2 \cdot 1) = \left(\prod_{r=1}^{n} r \right) \left(\prod_{r=1}^{n} (n - r + 1) \right) = \prod_{r=1}^{n} r(n - r + 1)$.
Show that for all integers $r = 1, 2, \ldots, n, n^2 - nr + r \geq n$.

50. a. Let n be a positive integer. For any real number $x > 1$, properties of exponents and logarithms (see Section 7.2) imply that $0 \leq \log_2(x) = \log_2((x^{1/n})^n) = n \log_2(x^{1/n}) < nx^{1/n}$ (where the last inequality holds by substituting $x^{1/n}$ in place of u in $\log_2 u < u$).

b. Let $B = n$ and $b = 1$. Then if $x > x_0, |\log_2 x| = \log_2 x \leq B \cdot |x^{1/n}|$, and so $\log_2 x$ is $O(x^{1/n})$.

52. Let n be a positive integer, and suppose that $x > (2n)^{2n}$. By properties of logarithms,

$$\log_2 x = (2n)\left(\frac{1}{2n}\right)(\log_2 x)$$

$$= (2n) \log_2\left(x^{\frac{1}{2n}}\right) < 2nx^{\frac{1}{2n}} \qquad (*)$$

(where the last inequality holds by substituting $x^{\frac{1}{2n}}$ in place of u in $\log_2 u < u$). But raising both sides of $x > (2n)^{2n}$ to the 1/2 power gives $x^{1/2} > ((2n)^{2n})^{1/2} = (2n)^n$. When both sides are multiplied by $x^{1/2}$, the result is $x = x^{1/2}x^{1/2} >$

$x^{1/2}(2n)^n = x^{1/2}(2n)^n$, or, more compactly,

$$x^{1/2}(2n)^n < x.$$

Then, since the power function defined by $x \rightarrow x^{1/n}$ is increasing for all $x > 0$ (see exercise 20 of Section 9.1), we can take the nth root of both sides of the inequality and use the laws of exponents to obtain

$$(x^{1/2}(2n)^n)^{1/n} < x^{1/n}$$

or, equivalently,

$$2nx^{\frac{1}{2n}} < x^{1/n}. \qquad (**)$$

Now use transitivity of $<$ (Appendix A, T17) to combine $(*)$ and $(**)$ and conclude that $\log_2 x < x^{1/n}$ [*as was to be shown*].

54. *Proof (by mathematical induction):* Let b be a real number with $b > 1$, and let the property $P(n)$ be the equation

$$\lim_{x \to \infty} \left(\frac{x^n}{b^x}\right) = 0.$$

Show that the property is true for $n = 1$:

By L'Hôpital's rule, $\lim_{x \to \infty}\left(\frac{x^1}{b^x}\right) = \lim_{x \to \infty}\left(\frac{1}{b^x(\ln b)}\right) = 0$. Thus the property is true when $n = 1$.

Show that for all integers $k \geq 1$, if the property is true for $n = k$, then it is true for $n = k + 1$:

Let k be any integer with $k \geq 1$, and suppose $\lim_{x \to \infty}\left(\frac{x^k}{b^x}\right) = 0$. [*This is the inductive hypothesis.*] We must show that $\lim_{x \to \infty}\left(\frac{x^{k+1}}{b^x}\right) = 0$. But by L'Hôpital's rule, $\lim_{x \to \infty}\frac{x^{k+1}}{b^x} = \lim_{x \to \infty}\frac{(k+1)x^k}{(\ln b)b^x} = \frac{(k+1)}{(\ln b)}\lim_{x \to \infty}\frac{x^k}{b^x} = \frac{(k+1)}{(\ln b)} \cdot 0$ [*by inductive hypothesis*] $= 0$ [*This is what was to be shown.*]

b. By the result of part (a) and the definition of limit, given any real number $\varepsilon > 0$, there exists an integer N such that $|\frac{x^n}{b^n} - 0| < 1$ for all $x > N$. In this case take $\varepsilon = 1$. It follows that for all $x > N, |\frac{x^n}{b^x}| = |\frac{x^n}{b^x}| < 1$. Multiply both sides by $|b^x|$ to obtain $|x^n| < |b^x|$. Let $B = 1$ and $x_0 = N$. Then $|x^n| < B \cdot |b^x|$ for all $x > x_0$. Hence, by definition of O-notation, x^n is $O(b^x)$.

Section 9.5

1. $\log_2 1000 = \log_2(10^3) = 3\log_2 10 \cong 3(3.32) \cong 9.96$
$\log_2(1,000,000) = \log_2(10^6) = 6\log_2 10 \cong 6(3.32)$
$\cong 19.92$
$\log_2(1,000,000,000,000) = \log_2(10^{12}) = 12\log_2 10$
$\cong 12(3.32) = 39.84$

2. a. If $m = 2^k$, where k is a positive integer, then the algorithm requires $c\lfloor\log_2(2^k)\rfloor = c\lfloor k \rfloor = ck$ operations. If the input size is increased to $m^2 = (2^k)^2 = 2^{2k}$, then the number of operations required is $c\lfloor\log_2(2^{2k})\rfloor = c\lfloor 2k \rfloor = 2(ck)$. Hence the number of operations doubles.

b. As in part (a), for an input of size $m = 2^k$, where k is a positive integer, the algorithm requires ck operations. If the input size is increased to $m^{10} = (2^k)^{10} = 2^{10k}$, then the number of operations required is $c\lfloor \log_2(2^{10k})\rfloor = c\lfloor 10k\rfloor = 10(ck)$. Thus the number of operations increases by a factor of 10.

c. When the input size is increased from 2^7 to 2^{28}, the factor by which the number of operations increases is
$$\frac{c\lfloor \log_2(2^{28})\rfloor}{c\lfloor \log_2(2^7)\rfloor} = \frac{28c}{7c} = 4.$$

3. A little numerical exploration can help find an initial window to use to draw the graphs of $y = x$ and $y = \lfloor 50\log_2 x\rfloor$. Note that when $x = 2^8 = 256$, $\lfloor 50\log_2 x\rfloor = \lfloor 50\log_2(2^8)\rfloor = \lfloor 50 \cdot 8\rfloor = \lfloor 400\rfloor = 400 > 256 = x$. But when $x = 2^9 = 512$, $\lfloor 50\log_2 x\rfloor = \lfloor 50\log_2(2^9)\rfloor = \lfloor 50 \cdot 9\rfloor = \lfloor 450\rfloor = 450 < 512 = x$. So a good choice of initial window would be the interval from 256 to 512. Drawing the graphs, zooming if necessary, and using the trace feature reveal that when $n < 438$, $n < \lfloor 50\log_2 n\rfloor$.

5. a.

index	0			1
bot	1			
top	10	4	1	
mid		5	2	1

b.

index	0			
bot	1	6		7
top	10		7		6
mid		5	8	6	7

7. a. $top - bot + 1$

b. *Proof:* Suppose *top* and *bot* are particular but arbitrarily chosen positive integers such that $top - bot + 1$ is an odd number. Then, by definition of odd, there is an integer k such that
$$top - bot + 1 = 2k + 1$$
Adding $2 \cdot bot - 1$ to both sides gives
$$bot + top = 2 \cdot bot - 1 + 2k + 1$$
$$= 2(bot + k).$$
But $bot + k$ is an integer. Hence, by definition of even, $bot + top$ is even.

8.

n	27	13	6	3	1	0

9. For each positive integer n, n *div* $2 = \lfloor n/2\rfloor$. Thus when the algorithm segment is run for a particular n and the **while** loop has iterated one time, the input to the next iteration is $\lfloor n/2\rfloor$. It follows that the number of iterations of the loop for n is one more than the number of iterations for $\lfloor n/2\rfloor$. That is, $a_n = 1 + a_{\lfloor n/2\rfloor}$. Also $a_1 = 1$.

10. The recurrence relation and initial condition of $a_1, a_2, a_3, \ldots$ derived in exercise 9 are the same as those for the sequence $w_1, w_2, w_3, \ldots$ discussed in the worst-case analysis of the binary search algorithm. Thus the general formulas for the two sequences are the same. That is, $a_n = 1 + \lfloor \log_2 n\rfloor$, for all integers $n \geq 1$.

11. In the analysis of the binary search algorithm, it was shown that $1 + \lfloor \log_2 n\rfloor$ is $\Theta(\log_2 n)$. Thus the algorithm segment has order $\log_2 n$.

14. *Hint:* The formula is $b_n = 1 + \lfloor \log_3 n\rfloor$.

20.

22.

Initial array:

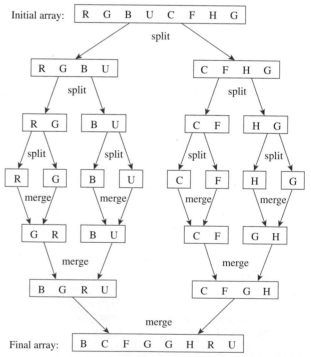

Final array:

24. b. Refer to Figure 9.5.3 and observe that when k is odd, the subarray $a[bot], a[bot + 1], \ldots, a[mid]$ has length $(k + 1)/2 = \lceil k/2\rceil$ and that when k is even, it also has length $k/2 = \lceil k/2\rceil$.

25. *Hint:* In part (a), applying the inductive hypothesis in the case where k is even gives

$$m_k \geq \frac{k}{2}\log_2\left(\frac{k}{2}\right) + \frac{k}{2}\log_2\left(\frac{k}{2}\right) + k + 1$$

$$\Rightarrow \quad m_k \geq k\log_2\left(\frac{k}{2}\right) + k + 1$$

$$\Rightarrow \quad m_k \geq k(\log_2 k - \log_2 2) + k + 1$$

$$\Rightarrow \quad m_k \geq k(\log_2 k - 1) + k + 1$$

$$\Rightarrow \quad m_k \geq k\log_2 k + 1 \geq k\log_2 k.$$

In part (b), applying the inductive hypothesis in the case where k is odd gives

$$\Rightarrow \quad m_k \leq 2\left(\frac{k-1}{2}\right)\log_2\left(\frac{k-1}{2}\right)$$

$$+ 2\left(\frac{k+1}{2}\right)\log_2\left(\frac{k+1}{2}\right) + k - 1$$

$$\Rightarrow \quad m_k \leq (k-1)(\log_2 k - 1) + (k+1)\log_2 k + k - 1$$

$$\Rightarrow \quad m_k \leq 2k\log_2 k.$$

Section 10.1

1. a. No. Yes. No. Yes

   b. $R = \{(2, 6), (2, 8), (2, 10), (3, 6), (4, 8)\}$

3. a. $0\ E\ 0$ because $0 - 0 = 0 = 2 \cdot 0$, so $2 \mid (0 - 0)$.

   $5\ \cancel{E}\ 2$ because $5 - 2 = 3$ and $3 \neq 2k$ for any integer k so $2 \nmid (5 - 2)$.

   $(6, 6) \in E$ because $6 - 6 = 0 = 2 \cdot 0$, so $2 \mid (6 - 6)$.

   $(-1, 7) \in E$ because $-1 - 7 = -8 = 2 \cdot (-4)$, so $2 \mid (-1 - 7)$.

4. *Hint:* To show a statement of the form $p \leftrightarrow (q \vee r)$, you need to show $p \rightarrow (q \vee r)$ and $(q \vee r) \rightarrow p$. To show a statement of the form $p \rightarrow (q \vee r)$, you can show $(p \wedge \sim q) \rightarrow r$ (since these two statement forms are logically equivalent). To show a statement of the form $(q \vee r) \rightarrow p$, you can show $(q \rightarrow p) \wedge (r \rightarrow p)$ (since these two statement forms are logically equivalent). In this case, suppose m and n are any integers, and let p be "$m - n$ is even," let q be "m and n are both even," and let r be "$m - n$ is even," let q be "m and n are both even," and let r be "m and n are both odd."

5. a. $10\ T\ 1$ because $10 - 1 = 9 = 3 \cdot 3$, so $3 \mid (10 - 1)$.

   $1\ T\ 10$ because $1 - 10 = -9 = 3 \cdot (-3)$, so $3 \mid (1 - 10)$.

   $2\ T\ 2$ because $2 - 2 = 0 = 3 \cdot 0$, so $3 \mid (2 - 2)$.

   $8\ \cancel{T}\ 1$ because $8 - 1 = 7 \neq 3k$, for any integer k. So $3 \nmid (8 - 1)$.

   b. *One possible answer:* $3, 6, 9, -3, -6$

   e. *Hint:* All integers of the form $3k + 1$, for some integer k, are related by T to 1.

6. b.

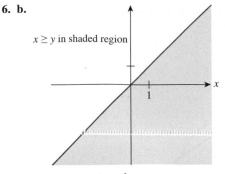

$x \geq y$ in shaded region

7. a. Yes, because $4 = 2^2$.

   No, because $2 \neq 4^2$.

   Yes, because $9 = (-3)^2$.

   No, because $-3 \neq 9^2$.

8. a. Yes, because 15 and 25 are both divisible by 5, which is prime.

   b. No, because 22 and 27 have no common prime factor.

9. a. Yes, because both $\{a, b\}$ and $\{b, c\}$ have two elements.

10. a. No, because $\{a\} \cap \{c\} = \emptyset$.

11. a. Yes, because both $abaa$ and $abba$ have the same first two characters ab.

12. *Hint:*

   a.

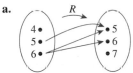

   b. R is not a function. It satisfies neither property (1) nor property (2). It fails property (1) because $(4, y) \notin R$ for any y in B. It fails property (2) because $(6, 5) \in R$ and $(6, 6) \in R$ and $5 \neq 6$.

13. a. $\emptyset, \{(0, 1)\}, \{(1, 1)\}, \{(0, 1), (1, 1)\}$

   b. $\{(0, 1), (1, 1)\}$

   c. $1/4$

15. *Hint:*

   a. The answer is 2^{mn}.   **b.** The answer is n^m.

16. No, because, for example, $(4, 2) \in P$ and $(4, -2) \in P$ but $2 \neq -2$.

17. $R = \{(3, 4), (3, 5), (3, 6), (4, 5), (4, 6), (5, 6)\}$

   $R^{-1} = \{(4, 3), (5, 3), (6, 3), (5, 4), (6, 4), (6, 5)\}$

19. a. Yes, because aab is the concatenation of a with ab.

   b. No, because ab is not the concatenation of a with aab.

   d. Yes, because $aba\ T^{-1}\ ba \Leftrightarrow ba\ T\ aba \Leftrightarrow aba$ is the concatenation of a with ba, which is true.

21. a. A function $F: X \rightarrow Y$ is one-to-one if, and only if, for all $x_1, x_2 \in X$, if $(x_1, y) \in F$ and $(x_2, y) \in F$, then $x_1 = x_2$.

22. a. No. If $F: X \rightarrow Y$ is not onto, then F^{-1} is not defined on all of Y. In other words, there is an element y in Y such that $(y, x) \notin F^{-1}$ for any $x \in X$. Consequently, F^{-1} does not satisfy property (1) of the definition of function.

23. **25.**

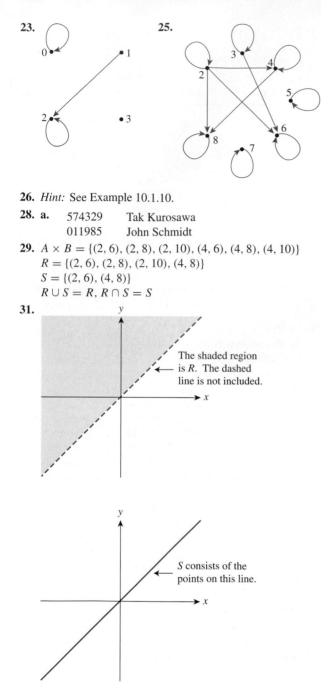

26. *Hint:* See Example 10.1.10.

28. a.   574329   Tak Kurosawa
          011985   John Schmidt

29. $A \times B = \{(2, 6), (2, 8), (2, 10), (4, 6), (4, 8), (4, 10)\}$
$R = \{(2, 6), (2, 8), (2, 10), (4, 8)\}$
$S = \{(2, 6), (4, 8)\}$
$R \cup S = R, R \cap S = S$

31.

The shaded region is R. The dashed line is not included.

S consists of the points on this line.

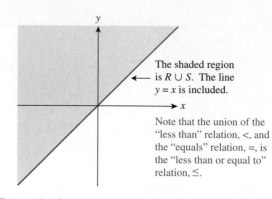

The shaded region is $R \cup S$. The line $y = x$ is included.

Note that the union of the "less than" relation, $<$, and the "equals" relation, $=$, is the "less than or equal to" relation, $\leq$.

The graph of the intersection of R and S is obtained by finding the set of all points common to both graphs. But there are no points for which both $x < y$ and $x = y$. Hence $R \cap S = \emptyset$ and the graph consists of no points at all.

Section 10.2

1. R_1:
a.

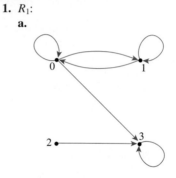

b. R_1 is not reflexive: $2 \not\!R_1 2$.
c. R_1 is not symmetric: $2\ R_1\ 3$ *but* $3 \not\!R_1\ 2$.
d. R_1 is not transitive: $1\ R_1\ 0$ and $0\ R_1\ 3$ but $1 \not\!R_1\ 3$.

3. R_3:
a. $0 \bullet$ $\bullet 1$

b. R_3 is not reflexive: $(0, 0) \notin R_3$
c. R_3 is symmetric. (If R_3 were not symmetric, there would be elements x and y in $A = \{0, 1, 2, 3\}$ such that $(x, y) \in R_3$ but $(y, x) \notin R_3$. It is clear by inspection that no such elements exist.)
d. R_3 is not transitive: $(2, 3) \in R_3$ and $(3, 2) \in R_3$ but $(2, 2) \notin R_3$

6. R_6:

a.

b. R_6 is not reflexive: $(0, 0) \notin R_6$
c. R_6 is not symmetric: $(0, 1) \in R_6$ but $(1, 0) \notin R_6$.
d. R_6 is transitive. (If R_6 were not transitive, there would be elements x, y, and z in $\{0, 1, 2, 3\}$ such that $(x, y) \in R_6$ and $(y, z) \in R_6$ but $(x, z) \notin R_6$. It is clear by inspection that no such elements exist.)

9. $R^t = R \cup \{(0, 0), (0, 3), (1, 0), (3, 1), (3, 2), (3, 3),$
$\qquad (0, 2), (1, 2)\}$
$\qquad = \{(0, 0), (0, 1), (0, 2), (0, 3), (1, 0), (1, 1), (1, 2),$
$\qquad (1, 3), (2, 2), (3, 0), (3, 1), (3, 2)(3, 3)\}$

12. *R is reflexive:* R is reflexive $\Leftrightarrow$ for all real numbers x, $x \, R \, x$. By definition of R, this means that for all real numbers x, $x \geq x$. In other words, for all real numbers x, $x > x$ or $x = x$. But this is true.
R is not symmetric: R is symmetric $\Leftrightarrow$ for all real numbers x and y, if $x \, R \, y$ then $y \, R \, x$. By definition of R, this means that for all real numbers x and y, if $x \geq y$ then $y \geq x$. But this is false. As a counterexample, take $x = 1$ and $y = 0$. Then $x \geq y$ but $y \ngeq x$ because $1 \geq 0$ but $0 \ngeq 1$.
R is transitive: R is transitive $\Leftrightarrow$ for all real numbers x, y, and z, if $x \, R \, y$ and $y \, R \, z$ then $x \, R \, z$. By definition of R, this means that for all real numbers x, y and z, if $x \geq y$ and $y \geq z$ then $x \geq z$. But this is true by definition of $\geq$ and the transitive property of order for the real numbers. (See Appendix A, T17.)

14. *D is reflexive:* For D to be reflexive means that for all real numbers x, $x \, D \, x$. But by definition of D, this means that for all real numbers x, $xx = x^2 \geq 0$, which is true.
D is symmetric: For D to be symmetric means that for all real numbers x and y, if $x \, D \, y$ then $y \, D \, x$. But by definition of D, this means that for all real numbers x and y, if $xy \geq 0$ then $yx \geq 0$, which is true by the commutative law of multiplication.
D is not transitive: For D to be transitive means that for all real numbers x, y, and z, if $x \, D \, y$ and $y \, D \, z$ then $x \, D \, z$. By definition of D, this means that for all real numbers x, y, and z, if $xy \geq 0$ and $yz \geq 0$ then $xz \geq 0$. But this is false: there exist real numbers x, y, and z such that $xy \geq 0$ and $yz \geq 0$ but $xz \ngeq 0$. As a counterexample, let $x = 1$, $y = 0$, and $z = -1$. Then $x \, D \, y$ and $y \, D \, z$ because $1 \cdot 0 \geq 0$ and $0 \cdot (-1) \geq 0$. But $x \not\!\!D z$ because $1 \cdot (-1) \ngeq 0$.

15. *E is reflexive:* [*We must show that for all integers m, $m \, E \, m$.*] Suppose m is any integer. Since $m - m = 0$ and $2 \,|\, 0$, we have that $2 \,|\, (m - m)$. Consequently, $m \, E \, m$ by definition of E.
E is symmetric: [*We must show that for all integers m and n, if $m \, E \, n$ then $n \, E \, m$.*] Suppose m and n are any integers such

that $m \, E \, n$. By definition of E, this means that $2 \,|\, (m - n)$, and so, by definition of divisibility, $m - n = 2k$ for some integer k. Now $n - m = -(m - n)$. Hence, by substitution, $n - m = -(2k) = 2(-k)$. It follows that $2 \,|\, (n - m)$ by definition of divisibility (since $-k$ is an integer), and thus $n \, E \, m$ by definition of E.
E is transitive: [*We must show that for all integers m, n and p if $m \, E \, n$ and $n \, E \, p$ then $m \, E \, p$.*] Suppose m, n, and p are any integers such that $m \, E \, n$ and $n \, E \, p$. By definition of E this means that $2 \,|\, (m - n)$ and $2 \,|\, (n - p)$, and so, by definition of divisibility, $m - n = 2k$ for some integer k and $n - p = 2l$ for some integer l. Now $m - p = (m - n) + (n - p)$. Hence, by substitution, $m - p = 2k + 2l = 2(k + l)$. It follows that $2 \,|\, m - p$ by definition of divisibility (since $k + l$ is an integer), and thus $m \, E \, p$ by definition of E.

18. *D is reflexive:* [*We must show that for all positive integers m, $m \, D \, m$.*] Suppose m is any positive integer. Since $m = m \cdot 1$, by definition of divisibility $m \,|\, m$. Hence $m \, D \, m$ by definition of D.
D is not symmetric: For D to be symmetric would mean that for all positive integers m and n, if $m \, D \, n$ then $n \, D \, m$. By definition of divisibility, this would mean that for all positive integers m and n, if $m \,|\, n$ then $n \,|\, m$. But this is false. As a counterexample, take $m = 2$ and $n = 4$. Then $m \,|\, n$ because $2 \,|\, 4$ but $n \nmid m$ because $4 \nmid 2$.
D is transitive: To prove transitivity of D, we must show that for all positive integers m, n, and p, if $m \, D \, n$ and $n \, D \, p$ then $m \, D \, p$. By definition of D, this means that for all positive integers m, n, and p, if $m \,|\, n$ and $n \,|\, p$ then $m \,|\, p$. But this is true by Theorem 3.3.1 (the transitivity of divisibility).

21. *L is not reflexive:* L is reflexive $\Leftrightarrow$ for all strings $s \in S$, $s \, L \, s$. By definition of L, this means that for all strings s in S, $\ell(s) < \ell(s)$, which means that the length of s is less than the length of s. But this is false for every string in S. For instance, let $s = \epsilon$. Then $\ell(s) = 0$ and $0 \not< 0$.
L is not symmetric: For L to be symmetric would mean that for all strings s and t in S, if $s \, L \, t$ then $t \, L \, s$. By definition of L, this would mean that for all strings s and t in S, if $\ell(s) < \ell(t)$ then $\ell(t) < \ell(s)$. But this is false for all strings s and t in S. For instance, take $s = 01$ and $t = 010$. Then $\ell(s) = 2$ and $\ell(t) = 3$, and so $\ell(s) < \ell(t)$ but $\ell(t) \not< \ell(s)$.
L is transitive: To prove transitivity of L, we must show that for all strings s, t, and u in S, if $s \, L \, t$ and $t \, L \, u$ then $s \, L \, u$. By definition of L, this means that for all strings s, t, and u in S, if $\ell(s) < \ell(t)$ and $\ell(t) < \ell(u)$ then $\ell(s) < \ell(u)$. But this is true by the transivity property of order (Appendix A, T17).

23. *# is reflexive:* # is reflexive $\Leftrightarrow$ for all subsets A of X, $A \,\#\, A$. By definition of #, this means that for all subsets A of X, A has the same number of elements as A. But this is true.
 # is symmetric: # is symmetric $\Leftrightarrow$ for all subsets A and B of X, if $A \,\#\, B$ then $B \,\#\, A$. By definition of #, this means that if A has the same number of elements as B, then B has the same number of elements as A. But this is true.
 # is transitive: # is transitive $\Leftrightarrow$ for all subsets A, B, and C

of X, if $A \# B$ and $B \# C$, then $A \# C$. By definition of $\#$, this means that for all subsets, A, B, and C of X, if A has the same number of elements as B and B has the number of elements as C, then A has the same number of elements as C. But this is true.

26. $\mathscr{J}$ *is reflexive:* $\mathscr{J}$ is reflexive $\Leftrightarrow$ for all subsets X of A, $X \mathscr{J} X$. By definition of $\mathscr{J}$, this means that for all subsets X of A, $X \subseteq X$. But this is true because every set is a subset of itself.

$\mathscr{J}$ *is not symmetric:* $\mathscr{J}$ is symmetric $\Leftrightarrow$ for all subsets X and Y of A, if $X \mathscr{J} Y$ then $Y \mathscr{J} X$. By definition of $\mathscr{J}$, this means that for all subsets X and Y of A, if $X \subseteq Y$ then $Y \subseteq X$. But this is false because $A \neq \emptyset$ and so there is an element a in A. As a counterexample, take $X = \emptyset$, and $Y = \{a\}$. Then $X \subseteq Y$ but $Y \nsubseteq X$.

$\mathscr{J}$ *is transitive:* $\mathscr{J}$ is transitive $\Leftrightarrow$ for all subsets X, Y, and Z of A, if $X \mathscr{J} Y$ and $Y \mathscr{J} Z$, then $X \mathscr{J} Z$. By definition of $\mathscr{J}$, this means that for all subsets X, Y, and Z of A, if $X \subseteq Y$ and $Y \subseteq Z$ then $X \subseteq Z$. But this is true by the transitive property of subsets (Theorem 5.2.1 (3)).

30. I *is reflexive:* [*We must show that for all statements p, $p I p$.*] Suppose p is a statement. The only way a conditional statement can be false is for its hypothesis to be true and its conclusion false. Consider the statement $p \rightarrow p$. Both the hypothesis and the conclusion have the same truth value. Thus it is impossible for $p \rightarrow p$ to be false, and so $p \rightarrow p$ must be true.

I *is not symmetric:* I is symmetric $\Leftrightarrow$ for all statements p and q, if $p I q$ then $q I p$. By definition of I, this means that for all statements p and q, if $p \rightarrow q$ then $q \rightarrow p$. But this false. As a counterexample, let p be the statement "10 is divisible by 4" and let q be "10 is divisible by 2." Then $p \rightarrow q$ is the statement "If 10 is divisible by 4, then 10 is divisible by 2." This is true because its hypothesis, p, is false. On the other hand, $q \rightarrow p$ is the statement "If 10 is divisible by 2, then 10 is divisible by 4." This is false because its hypothesis, q, is true and its conclusion, p, is false.

I *is transitive:* [*We must show that for all statements p, q, and r, if $p I q$ and $q I r$ then $p I r$.*] Suppose p, q, and r are statements such that $p I q$ and $q I r$. By definition of I, this means that $P \rightarrow q$ and $q \rightarrow r$ are both true. By transitivity of if-then (Example 1.3.7 and exercise 20 of Section 1.3), we can conclude that $p \rightarrow r$ is true. Hence, by definition of I, $p I r$.

31. $\mathscr{R}$ *is reflexive:* $\mathscr{R}$ is reflexive $\Leftrightarrow$ for all elements (x, y) in $\mathbf{R} \times \mathbf{R}$, $(x, y) \mathscr{R} (x, y)$. By definition of $\mathscr{R}$, this means that for all elements (x, y) in $\mathbf{R} \times \mathbf{R}$, $x = x$. But this is true.

$\mathscr{R}$ *is symmetric:* [*We must show that for all elements (x_1, y_1) and (x_2, y_2) in $\mathbf{R} \times \mathbf{R}$, if $(x_1, y_1) \mathscr{R} (x_2, y_2)$ then $(x_2, y_2) \mathscr{R} (x_1, y_1)$.*] Suppose (x_1, y_1) and (x_2, y_2) are elements of $\mathbf{R} \times \mathbf{R}$ such that (x_1, y_1), $\mathscr{R}(x_2, y_2)$. By definition of $\mathscr{R}$, this means that $x_1 = x_2$. By symmetry of equality, $x_2 = x_1$. Thus, by definition of $\mathscr{R}$, $(x_2, y_2) \mathscr{R}(x_1, y_1)$.

$\mathscr{R}$ *is transitive:* [*We must show that for all elements (x_1, y_1), (x_2, y_2) and (x_3, y_3) in $\mathbf{R} \times \mathbf{R}$, if $(x_1, y_1) \mathscr{R}(x_2, y_2)$*

and $(x_2, y_2) \mathscr{R}(x_3, y_3)$ then $(x_1, y_1) \mathscr{R}(x_3, y_3)$.] Suppose (x_1, y_1), (x_2, y_2), and (x_3, y_3) are elements of $\mathbf{R} \times \mathbf{R}$ such that $(x_1, y_1) \mathscr{R}(x_2, y_2)$ and $(x_2, y_2) \mathscr{R}(x_3, y_3)$. By definition of $\mathscr{R}$, this means that $x_1 = x_2$ and $x_2 = x_3$. By transitivity of equality, $x_1 = x_3$. Hence, by definition of $\mathscr{R}$, $(x_1, y_1) \mathscr{R}(x_3, y_3)$.

34. R *is reflexive:* R is reflexive $\Leftrightarrow$ for all people p in A, $p R p$. By definition of R, this means that for all people p living in the world today, p lives within 100 miles of p. But this is true.

R *is symmetric:* [*We must show that for all people p and q in A, if $p R q$ then $q R p$.*] Suppose p and q are people in A such that $p R q$. By definition of R, this means that p lives within 100 miles of q. But this implies that q lives within 100 miles of p. So, by definition of R, $q R p$.

R *is not transitive:* R is transitive $\Leftrightarrow$ for all people p, q and r, if $p R q$ and $q R r$ then $p R r$. But this is false. As a counterexample, take p to be an inhabitant of Chicago, Illinois, q an inhabitant of Kankakee, Illinois, and r an inhabitant of Champaign, Illinois. Then $p R q$ because Chicago is less then 100 miles from Kankakee, and $q R r$ because Kankakee is less than 100 miles from Champaign, but $p \not{R} r$ because Chicago is not less than 100 miles from Champaign.

37. a. A binary relation is any subset of $A \times A$, and $A \times A$ has $8^2 = 64$ elements. So there are 2^{64} binary relations on A.

c. Form a symmetric relation by a two-step process: (1) pick a set of elements of the form (a, a) (there are eight such elements, so 2^8 sets); (2) pick a set of pairs of elements of the form (a, b) and (b, a) where $a \neq b$ (there are $(64 - 8)/2 = 28$ such pairs, so 2^{28} such sets). The answer is therefore $2^8 \cdot 2^{28} = 2^{36}$.

38. Algorithm—Test for Reflexivity

[*The input for this algorithm consists of a binary relation R variable answer is initially set equal to "yes," and then each element $a[i]$ of A is examined in turn to see whether it is related by R defined on a set A that is represented as the one-dimensional array $a[1], a[2], \ldots, a[n]$. To test whether R is reflexive, the to itself. If any element is not related to itself by R, then answer is set equal to "no," the while loop is not repeated, and processing terminates.*]

Input: n [*a positive integer*], $a[1], a[2], \ldots, a[n]$ [*a one-dimensional array representing a set A*], R [*a subset of $A \times A$*]

Algorithm Body:

```
i := 1, answer := "yes"
while (answer = "yes" and i ≤ n)
    if (a[i], a[i]) ∉ R then answer := "no"
    i := i + 1
end while
```

Output: *answer* [*a string*]

43. a. $R \cap S$ *is reflexive:* Suppose R and S are reflexive. [*To show that $R \cap S$ is reflexive, we must show that $\forall x \in A$, $(x, x) \in R \cap S$.*] So suppose $x \in A$. Since R is reflexive, $(x, x) \in R$, and since S is reflexive, $(x, x) \in S$.

Thus, by definition of intersection, $(x, x) \in R \cap S$ [*as was to be shown*].

b. *Hint:* The answer is yes.

44. b. Yes. To prove this we must show that for all x and y in A, if $(x, y) \in R \cup S$ then $(y, x) \in R \cup S$. So suppose (x, y) is a particular but arbitrarily chosen element in $R \cup S$. [*We must show that $(y, x) \in R \cup S$.*] By definition of union, $(x, y) \in R$ or $(x, y) \in S$. If $(x, y) \in R$, then $(y, x) \in R$ because R is symmetric. Hence $(y, x) \in R \cup S$ by definition of union. But also, if $(x, y) \in S$ then $(y, x) \in S$ because S is symmetric. Hence $(y, x) \in R \cup S$ by definition of union. Thus, in either case, $(y, x) \in R \cup S$ [*as was to be shown*].

45. R_1 is not irreflexive because $(0, 0) \in R_1$. R_1 is not asymmetric because $(0, 1) \in R_1$ and $(1, 0) \in R_1$. R_1 is not intransitive because $(0, 1) \in R_1$ and $(1, 0) \in R_1$ and $(0, 0) \in R_1$.

47. R_3 is irreflexive. R_3 is not asymmetric because $(2, 3) \in R_3$ and $(3, 2) \subset R_3$. R_3 is intransitive.

50. R_6 is irreflexive. R_6 is asymmetric. R_6 is intransitive (by default).

Section 10.3

1. a. cRc **b.** bRa, cRb, eRd **c.** aRc
   d. $cRc, bRa, cRb, eRd, aRc, cRa$

2. a. $R = \{(0, 0), (0, 2), (1, 1), (2, 0), (2, 2), (3, 3), (3, 4), (4, 3), (4, 4)\}$

3. $\{0, 4\}, \{1, 3\}, \{2\}$

5. $\{1, 5, 9, 13, 17\}, \{2, 6, 10, 14, 18\}, \{3, 7, 11, 15, 19\}, \{4, 8, 12, 16, 20\}$

7. $\{(1, 3), (3, 9)\}, \{(2, 4), (-4, -8), (3, 6)\}, \{(1, 5)\}$

8. $\{\emptyset\}, \{\{a\}, \{b\}, \{c\}\}, \{\{a, b\}, \{a, c\}, \{b, c\}\}, \{\{a, b, c\}\}$

10. $\{aaaa, aaab, aaba, aabb\}$, $\{abaa, abab, abba, abbb\}$, $\{baaa, baab, baba, babb\}, \{bbaa, bbab, bbba, bbbb\}$

13. a. True. $17 - 2 = 15$ and $5 \mid 15$.

14. a. $[7] = [4] = [19], [-4] = [17], [-6] = [27]$

15. a. *Proof:* Suppose that m and n are integers such that $m \equiv n \pmod 3$. [*We must show that $m \bmod 3 = n \bmod 3$.*] By definition of congruence, $3 \mid (m - n)$, and so by definition of divisibility, $m - n = 3k$ for some integer k. Let $m \bmod 3 = r$. Then $m = 3l + r$ for some integer l. Since $m - n = 3k$, then by substitution, $(3l + r) - n = 3k$, or, equivalently, $n = 3(l - k) + r$. Since $l - k$ is an integer and $0 \le r < 3$, it follows, by definition of $\bmod$, that $n \bmod 3 = r$ also. So $m \bmod 3 = n \bmod 3$.

Suppose that m and n are integers such that $m \bmod 3 = n \bmod 3$. [*We must show that $m \equiv n \pmod 3$.*] Let $r = m \bmod 3 = n \bmod 3$. Then, by definition of $\bmod$, $m = 3p + r$ and $n = 3q + r$ for some integers p and q. By substitution, $m - n = (3p + r) - (3q + r) = 3(p - q)$. Since $p - q$ is an integer, it follows that $3 \mid (m - n)$, and so, by definition of congruence, $m \equiv n \pmod 3$.

16. a. For example, let $A = \{1, 2\}$ and $B = \{2, 3\}$. Then $A \ne B$, so A and B are distinct. But A and B are not disjoint since $2 \in A \cap B$.

17. a. (1) *Proof:* R is reflexive because it is true that for each student x at a college, x has the same major (or double major) as x.

R is symmetric because it is true that for all students x and y at a college, if x has the same major (or double major) as y, then y has the same major (or double major) as x.

R is transitive because it is true that for all students x, y, and z at a college, if x has the same major (or double major) as y and y has the same major (or double major) as z, then x has the same major (or double major) as z. R is an equivalence relation because it is reflexive, symmetric, and transitive.

(2) There is one equivalence class for each major and double major at the college. Each class consists of all students with that major.

18. (1) *Hint:* The proof is similar to the one in Example 10.2.5. (2) Two distinct classes: $\{x \in \mathbf{Z} \mid x = 2k, \text{ for some integer } k\}$ and $\{x \in \mathbf{Z} \mid x = 2k + 1, \text{ for some integer } k\}$.

22. (1) *Proof:* A is reflexive because each real number has the same absolute value as itself.

A is symmetric because for all real numbers x and y, if $|x| = |y|$ then $|y| = |x|$.

A is transitive because for all real numbers x, y, and z, if $|x| = |y|$ and $|y| = |z|$ then $|x| = |z|$.

A is an equivalence relation because it is reflexive, symmetric, and transitive.

(2) The distinct classes are all sets of the form $\{x, -x\}$, where x is a real number.

23. (1) *Proof:* I is reflexive because the difference between each real number and itself is 0, which is an integer.

I is symmetric because for all real numbers x and y, if $x - y$ is in integer, then $y - x = (-1)(x - y)$, which is also an integer.

I is transitive because for all real numbers x, y, and z, if $x - y$ is an integer and $y - z$ is an integer, then $x - z = (x - y) + (y - z)$ is the sum of two integers and thus an integer.

I is an equivalence relation because it is reflexive, symmetric, and transitive.

(2) There is one class for each real number x with $0 \le x < 1$. The distinct classes are all sets of the form $\{y \in \mathbf{R} \mid y = n + x, \text{ for some integer } n\}$, where x is a real number such that $0 \le x < 1$.

25. (1) *Proof:* P is reflexive because each ordered pair of real numbers has the same first element as itself.

P is symmetric for the following reason: Suppose (w, x) and (y, z) are ordered pairs of real numbers such that $(w, x) P (y, z)$. Then, by definition of P, $w = y$. But by the symmetric property of equality, this implies that $y = w$, and so, by definition of P, $(y, z) P (w, x)$.

P is transitive for the following reason: Suppose (u, v), (w, x), and (y, z) are ordered pairs of real numbers such that $(u, v) P (w, x)$ and $(w, x) P (y, z)$. Then, by definition of P, $u = w$ and $w = y$. But by the transitive property of equality, this implies that $u = w$, and so, by definition of P, $(u, v) P (w, x)$.

P is an equivalence relation because it is reflexive, symmetric, and transitive.

(2) There is one equivalence class for each real number. The distinct equivalence classes are all sets of ordered pairs $\{(x, y) \in \mathbf{R} \times \mathbf{R} \mid x = a\}$, for each real number a. Equivalently, the equivalence classes consist of all vertical lines in the Cartesian plane.

27. *Partial Solution:* There is one equivalence class for each real number t such that $0 \le t < \pi$. One line in each class goes through the origin, and that line makes an angle of t with the positive horizontal axis.

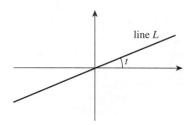

Alternatively, there is one equivalence class for every possible slope: all real numbers plus "undefined."

30. No. If points p, q, and r all lie on a straight line with q in the middle, and if p is c units from q and q is c units from r, than p is more then c units from r.

31. *Proof:* Suppose R is an equivalence relation on a set A and $a \in A$. Because R is an equivalence relation, R is reflexive, and because R is reflexive, each element of A is related to itself by R. In particular, $a \, R \, a$. Hence by definition of equivalence class, $a \in [a]$.

33. *Proof:* Suppose R is an equivalence relation on a set A and a, b, and c are elements of A with $b \, R \, c$ and $c \in [a]$. Since $c \in [a]$, then $c \, R \, a$ by definition of equivalence class. But R is transitive since R is an equivalence relation. Thus since $b \, R \, c$ and $c \, R \, a$, then $b \, R \, a$. It follows that $b \in [a]$ by definition of class.

35. *Proof:* Suppose a, b and x are in A, $a \, R \, b$, and $x \in [a]$. By definition of equivalence class, $x \, R \, a$. So $x \, R \, a$ and $a \, R \, b$, and thus, by transitivity, $x \, R \, b$. Hence $x \in [b]$.

36. *Hint:* To show that $[a] = [b]$, show that $[a] \subseteq [b]$ and $[b] \subseteq [a]$. To show that $[a] \subseteq [b]$, show that for all x in A, if $x \in [a]$ then $x \in [b]$.

38. **c.** For example $(2, 6), (-2, -6), (3, 9), (-3, -9)$.

39. **a.** Suppose that (a, b), (a', b'), (c, d) and (c', d') are any elements of A such that $[(a, b)] = [(a', b')]$ and $[(c, d)] = [(c', d')]$. By definition of the relation, $ab' = ba'$ (*) and $cd' = dc'$ (**). We must show that $[(a, b)] + [(c, d)] = [(a', b')] + [(c', d')]$. By defintion of the addition, this

equation is true if, and only if,

$$[(ad + bc, bd)] = [(a'd' + b'c', b'd')].$$

And, by definition of the relation, this equation is true if, and only if,

$$(ad + bc)b'd' = bd(a'd' + b'c'),$$

which is equivalent to

$$adb'd' + bcb'd' = bda'd' + bdb'c', \quad \text{by multiplying out.}$$

But this equation is equivalent to

$$(ab')(dd') + (cd')(bb')$$
$$= (ba')(dd') + (dc')(bb') \quad \text{by regrouping}$$

and, by substitution from (*) and (**), this last equation is true.

c. Suppose that (a, b) is any element of A. We must show that $[(a, b)] + [(0, 1)] = [(a, b)]$. By definition of the addition, this equation is true if, and only if,

$$[(a \cdot 1 + b \cdot 0, b \cdot 1)] = [(a, b)].$$

But this last equation is true because $a \cdot 1 + b \cdot 0 = a$ and $b \cdot 1 = b$.

e. Suppose that (a, b) is any element of A. We must show that $[(a, b)] + [(-a, b)] = [(0, 1)]$. By definition of the addition, this equation is true if, and only if,

$$[(ab + b(-a), bb)] = [(0, 1)],$$

or, equivalently,

$$[(0, bb)] = [(0, 1)]$$

By definition of the relation, this last equation is true if, and only if, $0 \cdot 1 = bb \cdot 0$, which is true.

40. **a.** Let (a, b) be any element of $\mathbf{Z}^+ \times \mathbf{Z}^+$. We must show that $(a, b) R (a, b)$. By definition of R, this relationship holds if, and only if, $a + b = b + a$. But this equation is true by the commutative law of addition for real numbers. Hence R is reflexive.

c. *Hint:* You will need to show that for any positive integers a, b, c, and d, if $a + d = c + b$ and $c + f = d + e$, then $a + f = b + e$.

d. *One possible answer:* $(1, 1), (2, 2), (3, 3), (4, 4), (5, 5)$

g. Observe that for any positive integers a and b, the equivalence class of (a, b) consists of all ordered pairs in $\mathbf{Z}^+ \times \mathbf{Z}^+$ for which the difference between the first and second coordinates equals $a - b$. Thus there is one equivalence class for each integer: positive, negative, and zero. Each positive integer n corresponds to the class of $(n + 1, 1)$; each negative integer $-n$ corresponds to the class of $(1, n + 1)$; and zero corresponds to the class $(1, 1)$.

43. **c.** "Ways and Means"

Section 10.4

1. **a.** ZKUHUH VKDOO ZH PHHW
   b. IN THE CAFETERIA

3. **a.** The relation $3 \mid (25 - 19)$ is true because $25 - 19 = 6$ and $3 \mid 6$ (since $6 = 3 \cdot 2$).
   b. By definition of congruence modulo n, to show that $25 \equiv 19 \pmod{3}$, one must show that $3 \mid (25 - 19)$. This was verified in part (a).
   c. To show that $25 = 19 + 3k$ for some integer k, one solves the equation for k and checks that the result is an integer. In this case, $k = (25 - 19)/3 = 2$, which is an integer. Thus $25 = 19 + 2 \cdot 3$.
   d. When 25 is divided by 3, the remainder is 1 because $25 = 3 \cdot 8 + 1$. When 19 is divided by 3, the remainder is also 1 because $19 = 3 \cdot 6 + 1$. Thus 25 and 19 have the same remainder when divided by 3.
   e. By definition, $25 \bmod 3$ is the remainder obtained when 25 is divided by 3, and $19 \bmod 3$ is the remainder obtained when 19 is divided by 3. In part (d) these two numbers were shown to be equal.

6. *Hints:* (1) Use the quotient-remainder theorem and Theorem 10.4.1 to show that given any integer a, a is in one of the classes $[0], [1], [2], \dots [n-1]$. (2) Use the result of Example 3.3.3 to prove that if $0 \le a < n$, $0 \le b < n$, and $a \equiv b \pmod{n}$, then $a = b$.

7. **a.** $128 \equiv 2 \pmod{7}$ because $128 - 2 = 126 = 7 \cdot 18$, and $61 \equiv 5 \pmod{7}$ because $61 - 5 = 56 = 7 \cdot 8$.
   b. $128 + 61 \equiv (2 + 5) \pmod{7}$ because $128 + 61 = 189$, $2 + 5 = 7$, and $189 - 7 = 182 = 7 \cdot 26$.
   c. $128 - 61 \equiv (2 - 5) \pmod{7}$ because $128 - 61 = 67$, $2 - 5 = -3$, and $67 - (-3) = 70 = 7 \cdot 10$.
   d. $128 \cdot 61 \equiv (2 \cdot 5) \pmod{7}$ because $128 \cdot 61 = 7808$, $2 \cdot 5 = 10$, and $7808 - (10) = 7798 = 7 \cdot 1114$.
   e. $128^2 \equiv 2^2 \pmod{7}$ because $128^2 = 16384$, $2^2 = 4$, and $16384 - 4 = 16380 = 7 \cdot 2340$.

9. *Proof:* Suppose a, b, c, d, and n are integers with $n > 1$, $a \equiv c \pmod{n}$, and $b \equiv d \pmod{n}$. By Theorem 10.4.1, $a - c = nr$ and $b - d = ns$ for some integers r and s. Then

$$(a + b) - (c + d) = (a - c) + (b - d) = nr + ns$$
$$= n(r + s).$$

But $r + s$ is an integer, and so, by Theorem 10.4.1, $a + b \equiv (c + d) \pmod{n}$.

12. **a.** *Proof (by mathematical induction):* Let the property $P(n)$ be the congruence $10^n \equiv 1 \pmod{9}$

   Show that the property is true for $n = 0$:

   When $n = 0$, the left-hand side of the congruence is $10^0 = 1$ and the right-hand side is also 1. Thus the property is true for $n = 0$.

   Show that for all integers $k \ge 0$, if the property is true for $n = k$, then it is true for $n = k + 1$:

   Let k be any integer with $k \ge 0$, and suppose the property is true for $n = k$. That is, suppose $10^k \equiv 1 \pmod{9}$. (*) [*This is the inductive hypothesis. We must show*

that $10^{k+1} \equiv 1 \pmod{9}$.] By Theorem 10.4.1, $10 \equiv 1 \pmod{9}$ (**) because $10 - 1 = 9 = 9 \cdot 1$. And by Theorem 10.4.3, we can multiply the left- and right-hand sides of (*) and (**) to obtain $10^k \cdot 10 \equiv 1 \cdot 1 \pmod{9}$, or, equivalently, $10^{k+1} \equiv 1 \pmod{9}$. [*This is what was to be shown.*]

14. $14^1 \bmod 55 = 14$
   $14^2 \bmod 55 = 195 \bmod 55 = 31$
   $14^4 \bmod 55 = (14^2 \bmod 55)^2 \bmod 55 = 31^2 \bmod 55 = 26$
   $14^8 \bmod 55 = (14^4 \bmod 55)^2 \bmod 55 = 26^2 \bmod 55 = 16$
   $14^{16} \bmod 55 = (14^8 \bmod 55)^2 \bmod 55 = 16^2 \bmod 55 = 36$

15. $4^{27} \bmod 55 = 14^{16+8+2+1} \bmod 55$
   $= \big[(14^{16} \bmod 55)(14^8 \bmod 55)(14^2 \bmod 55)$
   $\qquad\qquad (14^1 \bmod 55)\big] \bmod 55$
   $= (36 \cdot 16 \cdot 31 \cdot 14) \bmod 55 = 249984 \bmod 55 = 9$

16. Note that $307 = 256 + 32 + 16 + 2 + 1$.
   $675^1 \bmod 713 = 675$
   $675^2 \bmod 713 = 18$
   $675^4 \bmod 713 = 18^2 \bmod 713 = 324$
   $675^8 \bmod 713 = 324^2 \bmod 713 = 165$
   $675^{16} \bmod 713 = 165^2 \bmod 713 = 131$
   $675^{32} \bmod 713 = 131^2 \bmod 713 = 49$
   $675^{64} \bmod 713 = 49^2 \bmod 713 = 262$
   $675^{128} \bmod 713 = 262^2 \bmod 713 = 196$
   $675^{256} \bmod 713 = 196^2 \bmod 713 = 627$

   Thus
   $675^{307} \bmod 713 = 675^{256+32+16+2+1} \bmod 713$
   $\quad = (675^{256} \cdot 675^{32} \cdot 675^{16} \cdot 675^2 \cdot 675^1) \bmod 713$
   $\quad = (627 \cdot 49 \cdot 131 \cdot 18 \cdot 675) \bmod 713 = 3.$

19. The letters in HELLO translate numerically into 08, 05, 12, 12, and 15. By Example 10.4.9, the H is encrypted as 17. To encrypt E, we compute $5^3 \bmod 55 = 15$. To encrypt L, we compute $12^3 \bmod 55 = 23$. And to encrypt 0, we compute $15^3 \bmod 55 = 20$. Thus the ciphertext is 17 15 23 23 20. (In practice, individual letters of the alphabet are grouped together in blocks during encryption so that deciphering cannot be accomplished through knowledge of frequency patterns of letters or words.)

22. By Example 10.4.10, the decryption key is 27. Thus the residues modulo 55 for 13^{27}, 20^{27}, and 9^{27} must be found and then translated into letters of the alphabet. Because $27 = 16 + 8 + 2 + 1$, we first perform the following computations:

$13^1 \equiv 13 \pmod{55}$	$20^1 \equiv 20 \pmod{55}$
$13^2 \equiv 4 \pmod{55}$	$20^2 \equiv 15 \pmod{55}$
$13^4 \equiv 4^2 \equiv 16 \pmod{55}$	$20^4 \equiv 15^2 \equiv 5 \pmod{55}$
$13^8 \equiv 16^2 \equiv 36 \pmod{55}$	$20^8 \equiv 25^2 \equiv 5 \pmod{55}$
$13^{16} \equiv 36^2 \equiv 31 \pmod{55}$	$20^{16} \equiv 25^2 \equiv 20 \pmod{55}$

   $9^1 \equiv 9 \pmod{55}$
   $9^2 \equiv 26 \pmod{55}$
   $9^4 \equiv 26^2 \equiv 16 \pmod{55}$
   $9^8 \equiv 16^2 \equiv 36 \pmod{55}$
   $9^{16} \equiv 36^2 \equiv 31 \pmod{55}$

Then we compute

$$13^{27} \bmod 55 = (31 \cdot 36 \cdot 4 \cdot 13) \bmod 55 = 7,$$
$$20^{27} \bmod 55 = (20 \cdot 25 \cdot 15 \cdot 20) \bmod 55 = 15,$$
$$9^{27} \bmod 55 = (31 \cdot 36 \cdot 26 \cdot 9) \bmod 55 = 4.$$

Finally, because 7, 15, and 4 translate into letters as G, O, and D, we see that the message is GOOD.

25. *Hint:* By Theorem 4.2.3, using a in place of r and $n - 1$ in place of n, we have $1 + a + a^2 + \cdots + a^{n-1} = \frac{a^n - 1}{a - 1}$. Multiplying both sides by $a - 1$ gives $a^n - 1 = (a - 1)$ $(1 + a + a^2 + \cdots + a^{n-1})$.

26. Step 1: $6664 = 765 \cdot 8 + 544$, and so $544 = 6664 - 765 \cdot 8$
    Step 2: $765 = 544 \cdot 1 + 221$, and so $221 = 765 - 544$
    Step 3: $544 = 221 \cdot 2 + 102$, and so $102 = 544 - 221 \cdot 2$
    Step 4: $221 = 102 \cdot 2 + 17$, and so $17 = 221 - 102 \cdot 2$
    Step 5: $102 = 17 \cdot 6 + 0$

Thus $\gcd(6664, 765) = 17$ (which is the remainder obtained just before the final division). Substitute back through steps 4–1 to express 17 as a linear combination of 6664 and 765:

$$17 = 221 - 102 \cdot 2$$
$$= 221 - (544 - 221 \cdot 2) = 221 \cdot 5 - 544 \cdot 2$$
$$= (765 - 544) \cdot 5 - 544 \cdot 2 = 765 \cdot 5 - 544 \cdot 7$$
$$= 765 \cdot 5 - (6664 - 765 \cdot 8) \cdot 7 = (-7) \cdot 6664 + 61 \cdot 765.$$

(When you have finished this final step, it is wise to verify that you have not made a mistake by checking that the final expression really does equal the greatest common divisor.)

28.

a	330	156	18	12	6
b	156	18	12	6	0
r		18	12	6	0
q		2	8	1	2
s	1	0	1	−8	
t	0	1	−2	17	
u	0	1	−8	9	
v	1	−2	17	−19	
$newu$		1	−8	9	
$newv$		−2	17	−19	
$sa + tb$	330	18	−6	6	

31. **a.** Step 1: $210 = 13 \cdot 16 + 2$, and so $2 = 210 - 16 \cdot 13$
    Step 2: $13 = 2 \cdot 6 + 1$, and so $1 = 13 - 2 \cdot 6$
    Step 3: $6 = 1 \cdot 6 + 0$, and so $\gcd(210, 13) = 1$

Substitute back through steps 2–1:

$$1 = 13 - 2 \cdot 6$$
$$= 13 - (210 - 16 \cdot 13) \cdot 6 = (-6) \cdot 210 + 97 \cdot 13$$

Thus $210 \cdot (-6) \equiv 1 \pmod{13}$, and so -6 is an inverse for 210 modulo 13.

b. Compute $13 - 6 = 7$, and note that $7 \equiv -6 \pmod{13}$ because $7 - (-6) = 13 = 13 \cdot 1$. Thus, by Theorem 10.4.3(3), $210 \cdot 7 \equiv 210 \cdot (-6) \pmod{13}$. It fol-

lows, by the transitive property of congruence, that $210 \cdot 7 \equiv 1 \pmod{13}$, and so 7 is a positive inverse for 210 modulo 13.

c. This problem can be solved using either the result of part (a) or that of part (b). By part (b) $210 \cdot 7 \equiv 1 \pmod{13}$. Multiply both sides by 8 and apply Theorem 10.4.3(3) to obtain $210 \cdot 56 \equiv 8 \pmod{13}$. Thus a positive solution for $210x \equiv 8 \pmod{13}$ is $x = 56$. Note that the least positive residue corresponding to this solution is also a solution. By Theorem 10.4.1, $56 \equiv 4 \pmod{13}$ because $56 = 13 \cdot 4 + 4$, and so, by Theorem 10.4.3(3), $210 \cdot 56 \equiv 210 \cdot 4 \equiv 9 \pmod{13}$. This shows that 4 is also a solution for the congruence, and because $0 \le 4 < 13$, 4 is the least positive solution for the congruence.

33. *Hint:* If $as + bt = 1$ and $c = au = bv$, then $c = asc + btc = as(bv) + bt(au)$.

35. *Proof:* Suppose a, n, s and s' are integers such that $as \equiv as' \equiv 1 \pmod{n}$. Consider the quantity $as's$, and note that $as's = (as') \cdot s = (as) \cdot s'$. By Theorem 10.4.3(3), $(as') \cdot s \equiv 1 \cdot s = s \pmod{n}$ and $(as) \cdot s' \equiv 1 \cdot s' = s' \pmod{n}$. Thus by transitivity of congruence modulo n, $s \equiv s' \pmod{n}$. This shows that any two inverses for a are congruent modulo n.

36. The numeric equivalents of H, E, L, and P are 08, 05, 12 and 16. To encrypt these letters, the following quantities must be computed: $8^{43} \bmod 713$, $5^{43} \bmod 713$, $12^{43} \bmod 713$, and $16^{43} \bmod 713$. We use the fact that $43 = 32 + 8 + 2 + 1$.

H: $8 \equiv 8 \pmod{713}$
    $8^2 \equiv 64 \pmod{713}$
    $8^4 \equiv 64^2 \equiv 531 \pmod{713}$
    $8^8 \equiv 531^2 \equiv 326 \pmod{713}$
    $8^{16} \equiv 326^2 \equiv 39 \pmod{713}$
    $8^{32} \equiv 39^2 \equiv 95 \pmod{713}$
    Thus the ciphertext is
    $8^{43} \bmod 713$
    $= (95 \cdot 326 \cdot 64 \cdot 8) \bmod 713 = 233.$

E: $5 \equiv 5 \pmod{713}$
    $5^2 \equiv 25 \pmod{713}$
    $5^4 \equiv 625 \pmod{713}$
    $5^8 \equiv 625^2 \equiv 614 \pmod{713}$
    $5^{16} \equiv 614^2 \equiv 532 \pmod{713}$
    $5^{32} \equiv 532^2 \equiv 676 \pmod{713}$
    Thus the ciphertext is
    $5^{43} \bmod 713$
    $= (676 \cdot 614 \cdot 25 \cdot 5) \bmod 713 = 129.$

L: $12 \equiv 12 \pmod{713}$
    $12^2 \equiv 144 \pmod{713}$
    $12^4 \equiv 144^2 \equiv 59 \pmod{713}$
    $12^8 \equiv 59^2 \equiv 629 \pmod{713}$
    $12^{16} \equiv 629^2 \equiv 639 \pmod{713}$
    $12^{32} \equiv 639^2 \equiv 485 \pmod{713}$
    Thus the ciphertext is
    $12^{43} \bmod 713$
    $= (485 \cdot 629 \cdot 144 \cdot 12) \bmod 713 = 48.$

P: $16 \equiv 16 \pmod{713}$
$16^2 \equiv 256 \pmod{713}$
$16^4 \equiv 256^2 \equiv 653 \pmod{713}$
$16^8 \equiv 653^2 \equiv 35 \pmod{713}$
$16^{16} \equiv 35^2 \equiv 512 \pmod{713}$
$16^{32} \equiv 512^2 \equiv 473 \pmod{713}$

Thus the ciphertext is
$16^{43} \bmod 713$
$= (473 \cdot 35 \cdot 256 \cdot 16) \bmod 713 = 128.$

Therefore, the encrypted message is 233 129 048 128. (Again, note that in practice, individual letters of the alphabet are grouped together in blocks during encryption so that deciphering cannot be accomplished through knowledge of frequency patterns of letters or words. We kept them separate so that the numbers in the computations would be smaller and easier to work with.)

39. By exercise 38, the decryption key, d, is 307. Hence, to decrypt the message, the following quantities must be computed: $675^{307} \bmod 713$, $89^{307} \bmod 713$, and $48^{307} \bmod 713$. We use the fact that $307 = 256 + 32 + 16 + 2 + 1$.

$675 \equiv 675 \pmod{713}$
$675^2 \equiv 18 \pmod{713}$
$675^4 \equiv 18^2 \equiv 324 \pmod{713}$
$675^8 \equiv 324^2 \equiv 165 \pmod{713}$
$675^{16} \equiv 165^2 \equiv 131 \pmod{713}$
$675^{32} \equiv 131^2 \equiv 49 \pmod{713}$
$675^{64} \equiv 49^2 \equiv 262 \pmod{713}$
$675^{128} \equiv 262^2 \equiv 196 \pmod{713}$
$675^{256} \equiv 196^2 \equiv 627 \pmod{713}$

$89 \equiv 89 \pmod{713}$
$89^2 \equiv 78 \pmod{713}$
$89^4 \equiv 78^2 \equiv 380 \pmod{713}$
$89^8 \equiv 380^2 \equiv 374 \pmod{713}$
$89^{16} \equiv 374^2 \equiv 128 \pmod{713}$
$89^{32} \equiv 128^2 \equiv 698 \pmod{713}$
$89^{64} \equiv 698^2 \equiv 225 \pmod{713}$
$89^{128} \equiv 225^2 \equiv 2 \pmod{713}$
$89^{256} \equiv 2^2 \equiv 4 \pmod{713}$

$48 \equiv 48 \pmod{713}$
$48^2 \equiv 165 \pmod{713}$
$48^4 \equiv 131 \pmod{713}$
$48^8 \equiv 49 \pmod{713}$
$48^{16} \equiv 262 \pmod{713}$
$48^{32} \equiv 196 \pmod{713}$
$48^{64} \equiv 627 \pmod{713}$
$48^{128} \equiv 627^2 \equiv 266 \pmod{713}$
$48^{256} \equiv 266^2 \equiv 169 \pmod{713}$

Thus the decryption for 675 is
$675^{307} \bmod 713 = (675^{256+32+16+2+1}) \bmod 713$
$= (627 \cdot 49 \cdot 131 \cdot 18 \cdot 675) \bmod 713 = 3$, which corresponds to the letter C.

The decryption for 89 is
$89^{307} \bmod 713 = (89^{256+32+16+2+1}) \bmod 713$
$= (4 \cdot 698 \cdot 128 \cdot 78 \cdot 89) \bmod 713 = 15$, which corresponds to the letter O.

The decryption for 48 is
$48^{307} \bmod 713 = (48^{256+32+16+2+1}) \bmod 713$
$= (169 \cdot 196 \cdot 262 \cdot 165 \cdot 48) \bmod 713 = 12$, which corresponds to the letter L.

Thus the decrypted message is COOL.

41. a. *Hint:* For the inductive step, assume $p \mid q_1 q_2 \ldots q_{s+1}$ and let $a = q_1 q_2 \ldots q_s$. Then $p \mid a q_{s+1}$, and either $p = q_{s+1}$ or Euclid's lemma and the inductive hypothesis can be applied.

42. a. When $a = 15$ and $p = 7$, $a^{p-1} = 15^6 = 11390625 \equiv 1 \pmod{7}$ because $11390625 - 1 = 7 \cdot 1627232$.

44. For this problem, $n_1 = 3$, $n_2 = 5$, and $n_3 = 7$, so $N = 3 \cdot 5 \cdot 7 = 105$, $N_1 = 5 \cdot 7 = 35$, $N_2 = 3 \cdot 7 = 21$, and $N_3 = 3 \cdot 5 = 15$.

a. To find x_1, an inverse for 35 modulo 3 must be found; that is, the congruence $35x_1 \equiv 1 \pmod{3}$ must be solved.

$35 = 3 \cdot 11 + 2$, which implies that $2 = 35 - 3 \cdot 11$.
$3 = 2 \cdot 1 + 1$, which implies that $1 = 3 - 2$.

Back substitution gives $1 = 3 - 2 = 3 - (35 - 3 \cdot 11) = 35 \cdot (-1) + 3 \cdot 12$, which implies that $35(-1) \equiv 1 \pmod{3}$. Thus $x_1 = -1$ is an inverse for 35 modulo 3.

b. To find x_2, an inverse for 21 modulo 5 must be found; that is, the congruence $21x_2 \equiv 1 \pmod{5}$ must be solved. But $21 = 5 \cdot 4 + 1$, which implies that $1 = 21 - 5 \cdot 4 = 21 \cdot 1 + 5 \cdot (-4)$. Hence $21 \cdot 1 \equiv 1 \pmod{5}$, and thus $x_2 = 1$ is an inverse for 21 modulo 5.

c. To find x_3, an inverse for 15 modulo 7 must be found; that is, the congruence $15x_3 \equiv 1 \pmod{7}$ must be solved. But $15 = 2 \cdot 7 + 1$, which implies that $1 = 15 - 2 \cdot 7 = 15 \cdot 1 + 7 \cdot (-2)$. Hence $15 \cdot 1 \equiv 1 \pmod{7}$, and thus $x_3 = 1$ is an inverse for 15 modulo 7.

d. For this problem, $a_1 = 2$, $a_2 = 3$, and $a_3 = 2$.
Let $x = a_1 N_1 x_1 + a_2 N_2 x_2 + a_3 N_3 x_3$
$= 2 \cdot 35 \cdot (-1) + 3 \cdot 21 \cdot (1) + 2 \cdot 15 \cdot (1) = 23.$

Check:
$23 \equiv 2 \pmod{3}$ because $23 - 2 = 21$ and $21 = 3 \cdot 7$.
$23 \equiv 3 \pmod{5}$ because $23 - 3 = 20$ and $20 = 5 \cdot 4$.
$23 \equiv 2 \pmod{7}$ because $23 - 2 = 21$ and $21 = 7 \cdot 3$.

e. Because $23 < N = 105$, the least nonnegative solution for the congruence is 23.

46. *Hint:* Reduce the problem by showing that (1) if $x \equiv 2 \pmod{3}$ and $x \equiv 1 \pmod{2}$, then $x \equiv 6 \pmod{5}$, and (2) if $x \equiv 3 \pmod{4}$, then $x \equiv 1 \pmod{2}$. Thus you will need only to consider the congruences modulo 3, 4, 5, and 7, which are pairwise relatively prime.

Section 10.5

1. a.

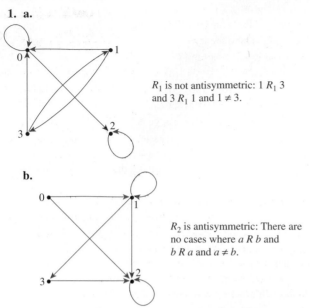

R_1 is not antisymmetric: $1 \, R_1 \, 3$ and $3 \, R_1 \, 1$ and $1 \neq 3$.

b.

R_2 is antisymmetric: There are no cases where $a \, R \, b$ and $b \, R \, a$ and $a \neq b$.

2. R is not antisymmetric. Let x and y be any two distinct people of the same age. Then $x \, R \, y$ and $y \, R \, x$ but $x \neq y$.

5. R is a partial order relation.

Proof:

R is reflexive: Suppose $(a, b) \in \mathbf{R} \times \mathbf{R}$. Then $(a, b) \, R \, (a, b)$ because $a = a$ and $b \leq b$.

R is antisymmetric: Suppose (a, b) and (c, d) are ordered pairs of real numbers such that $(a, b) \, R \, (c, d)$ and $(c, d) \, R \, (a, b)$. Then

$$\text{either } a < c \quad \text{or} \quad \text{both } a = c \text{ and } b \leq d$$

and

$$\text{either } c < a \quad \text{or} \quad \text{both } c = a \text{ and } d \leq b.$$

Thus

$$a \leq c \text{ and } c \leq a$$

and so

$$a = c.$$

Consequently,

$$b \leq d \quad \text{and} \quad d \leq b.$$

and so

$$b = d.$$

Hence $(a, b) = (c, d)$.

R is transitive: Suppose (a, b), (c, d), and (e, f) are ordered pairs of real numbers such that $(a, b) \, R \, (c, d)$ and $(c, d) \, R \, (e, f)$. Then

$$\text{either } a < c \quad \text{or} \quad \text{both } a = c \text{ and } b \leq d$$

and

$$\text{either } c < e \quad \text{or} \quad \text{both } c = e \text{ and } d \leq f.$$

It follows that one of the following cases must occur.

Case 1 ($a < c$ and $c < e$): Then by transitivity of $<$, $a < e$, and so $(a, b) \, R \, (e, f)$ by definition of R.

Case 2 ($a < c$ and $c = e$): Then by substitution, $a < e$, and so $(a, b) \, R \, (e, f)$ by definition of R.

Case 3 ($a = c$ and $c < e$): Then by substitution, $a < e$, and so $(a, b) \, R \, (e, f)$ by definition of R.

Case 4 ($a = c$ and $c = e$): Then by definition of R, $b \leq d$ and $d \leq f$, and so by transitivity of $\leq$, $b \leq f$. Hence $a = e$ and $b \leq f$, and so $(a, b) \, R \, (e, f)$ by definition of R.

In each case, $(a, b) \, R \, (e, f)$. Therefore, R is transitive. Since R is reflexive, antisymmetric, and transitive, R is a partial order relation.

8. R is not a partial order relation because R is not antisymmetric.

Counterexample: $1 \, R \, 3$ (because $1 + 3$ is even) and $3 \, R \, 1$ (because $3 + 1$ is even) but $1 \neq 3$.

10. No. *Counterexample:* Define relations R and S on the set $\{1, 2\}$ as follows: $R = \{(1, 2)\}$ and $S = \{(2, 1)\}$. Then both R and S are antisymmetric, but $R \cup S = \{(1, 2), (2, 1)\}$ is not antisymmetric because $(1, 2) \in R \cup S$ and $(2, 1) \in R \cup S$ but $1 \neq 2$.

11. a. This follows from (1).

b. False. By (1), $bba \preceq bbab$.

13. $R_1 = \{(a, a), (b, b)\}$, $R_2 = \{(a, a), (b, b), (a, b)\}$, $R_3 = \{(a, a), (b, b), (b, a)\}$

14. a. $R_1 = \{(a, a), (b, b), (c, c)\}$,

$R_2 = \{(a, a), (b, b), (c, c), (b, a)\}$,

$R_3 = \{(a, a), (b, b), (c, c), (c, a)\}$,

$R_4 = \{(a, a), (b, b), (c, c), (b, a), (c, a)\}$,

$R_5 = \{(a, a), (b, b), (c, c), (c, b), (c, a)\}$,

$R_6 = \{(a, a), (b, b), (c, c), (b, c), (b, a)\}$,

$R_7 = \{(a, a), (b, b), (c, c), (c, b), (b, a), (c, a)\}$

$R_8 = \{(a, a), (b, b), (c, c), (b, c), (b, a), (c, a)\}$,

$R_9 = \{(a, a), (b, b), (c, c), (b, c)\}$,

$R_{10} = \{(a, a), (b, b), (c, c), (c, b)\}$

15. *Hint:* R is the identity relation on A: $x \, R \, x$ for all $x \in A$ and $x \, R' \, y$ if $x \neq y$.

16. a.

17. a.

18.

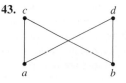

21. a. *Proof:* [*We must show that for all a and b in A, a | b or b | a.*] Let a and b be particular but arbitrarily chosen elements of A. By definition of A, there are nonnegative integers r and s such that $a = 2^r$ and $b = 2^s$. Now either $r \leq s$ or $s < r$. If $r \leq s$, then

$$b = 2^s = 2^r \cdot 2^{s-r} = a \cdot 2^{s-r},$$

where $s - r \geq 0$. It follows, by definition of divisibility, that $a \mid b$. By a similar argument, if $s < r$, then $b \mid a$. Hence either $a \mid b$ or $b \mid a$ [*as was to be shown*].

b.

```
●————————●————————●————————●————————●
1        2        2²       2³       2⁴
```

22. Greatest element: none; least element: 1;
Maximal elements: 15, 20; minimal element: 1

24. Greatest element: $\{0, 1\}$; least element: $\emptyset$;
Maximal elements: $\{0, 1\}$; minimal elements: $\emptyset$

26. Greatest element: $(1, 1)$; least element: $(0, 0)$;
Maximal elements: $(1, 1)$; minimal elements: $(0, 0)$

30. a. No greatest element, no least element
b. Least element is 0, greatest element is 1

31. R is a total order relation because it is reflexive, antisymmetric, and transitive (so it is a partial order) and because $\{b, a, c, d\}$ is a chain that contains every element of A: bRc, cRa, and aRd.

35. *Hint:* Let R' be the restriction of R to B and show that R' is reflexive, antisymmetric, and transitive. In each case, this follows almost immediately from the fact that R is reflexive, antisymmetric, and transitive.

36. $\emptyset \subseteq \{w\} \subseteq \{w, x\} \subseteq \{w, x, y\} \subseteq \{w, x, y, z\}$

39. *Proof:* Suppose A is a partially ordered set with respect to a relation $\preceq$. By definition of total order, A is totally ordered if, and only if, any two elements of A are comparable. By definition of chain, this is true if, and only if, A is a chain.

40. *Proof (by mathematical induction):* Let A be a set that is totally ordered with respect to a relation $\preceq$, and let the property $P(n)$ be the sentence "Every subset of A with n elements has both a least element and a greatest element."

Show that the property is true for $n = 1$:

If $A = \emptyset$, then the property is true by default. So assume that A has at least one element, and suppose $S = \{a_1\}$ is a subset of A with one element. Because $\preceq$ is reflexive, $a_1 \preceq a_1$.

So, by definition of least element and greatest element, a_1 is both a least element and a greatest element of S, and thus the property is true for $n = 1$.

Show that for all integers $k \geq 1$, if the property is true for $n = k$, then it is true for $n = k + 1$:

Let k be any integer with $k \geq 1$, and suppose that any subset of A with k elements has both a least element and a greatest element. We must show that any subset of A with $k + 1$ elements has both a least element and a greatest element. If A has fewer than $k + 1$ elements, then the statement is true by default. So assume that A has at least $k + 1$ elements and that $S = \{a_1, a_2, \ldots, a_{k+1}\}$ is a subset of A with $k + 1$ elements. By inductive hypothesis, $S - \{a_{k+1}\}$ has both a least element s and a greatest element b. Now because A is totally ordered, a_{k+1} and s are comparable. If $a_{k+1} \preceq s$, then, by transitivity of $\preceq$, a_{k+1} is the least element of S; otherwise, s remains the least element of S. And if $b \preceq a_{k+1}$, then, by transitivity of $\preceq$, a_{k+1} is the greatest element of S; otherwise, b remains the greatest element of S. Thus S has both a greatest element and a least element.

41. a. *Proof by contradiction:* Suppose not. Suppose A is a finite set that is partially ordered with respect to a relation $\preceq$ and A has no minimal element. Construct a sequence of elements $x_1, x_2, x_3, \ldots$ of A as follows:

1. Pick any element of A and call it x_1.
2. For each $i = 2, 3, 4, \ldots$, pick x_i to be an element of A for which $x_i \preceq x_{i-1}$ and $x_i \neq x_{i-1}$. [*Such an element must exist because otherwise x_{i-1} would be minimal, and we are supposing that no element of A is minimal.*] Now $x_i \neq x_j$ for any $i \neq j$. [*If $x_i = x_j$ where $i < j$, then on the one hand, $x_j \preceq x_{j-1} \preceq \cdots \preceq x_{i+1} \preceq x_i$ and so $x_i \preceq x_{i+1}$, and on the other hand, since $x_i = x_j$ then $x_j = x_i \succeq x_{i+1}$, and so $x_j \succeq x_{i+1}$. Hence by antisymmetry, $x_j = x_{i+1}$, and so $x_i = x_{i+1}$. But this contradicts the definition of the sequence $x_1, x_2, x_3, \ldots$.*] Thus $x_1, x_2, x_3, \ldots$ is an infinite sequence of distinct elements, and consequently $\{x_1, x_2, x_3, \ldots\}$ is an infinite subset of the finite set A. This is impossible. Hence the supposition is false and we conclude that any partially ordered subset of a finite set has a minimal element.

43.

```
c              d
 ●            ●
 │╲          ╱│
 │ ╲        ╱ │
 │  ╲      ╱  │
 │   ╲    ╱   │
 │    ╲  ╱    │
 │     ╲╱     │
 │     ╱╲     │
 │    ╱  ╲    │
 │   ╱    ╲   │
 │  ╱      ╲  │
 │ ╱        ╲ │
 ●            ●
 a              b
```

45. One such total order is 1, 5, 2, 15, 10, 4, 20.

47. One such total order is $(0, 0)$, $(1, 0)$, $(0, 1)$, $(1, 1)$.

51. a. *One possible answer:* 1, 6, 10, 9, 5, 7, 2, 4, 8, 3

52. b. Critical path: 1, 2, 5, 8, 9.

Section 11.1

1. $V(G) = \{v_1, v_2, v_3, v_4\}$, $E(G) = \{e_1, e_2, e_3\}$
Edge-endpoint function:

Edge	Endpoints
e_1	$\{v_1, v_2\}$
e_2	$\{v_1, v_3\}$
e_3	$\{v_3\}$

3.

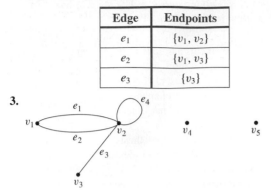

5. Imagine that the edges are strings and the vertices are knots. You can pick up the left-hand figure and lay it down again to form the right-hand figure as shown below.

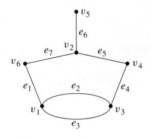

8. (i) e_1, e_2, and e_3 are incident on v_1.
(ii) v_1, v_2, and v_3 are adjacent to v_3.
(iii) e_2, e_8, e_9, and e_3 are adjacent to e_1.
(iv) Loops are e_6 and e_7.
(v) e_8 and e_9 are parallel; e_4 and e_5 are parallel.
(vi) v_6 is an isolated vertex.
(vii) degree of $v_3 = 5$
(viii) total degree $= 20$

10. a. Yes. According to the graph, *Sports Illustrated* is an instance of a sports magazine, a sports magazine is a periodical, and a periodical contains printed writing.

12. To solve this puzzle using a graph, introduce a notation in which, for example, wc/mg means that the wolf and the cabbage are on the left bank of the river and the man and the goat are on the right bank. Then draw those arrangements of wolf, cabbage, goat, and ferryman that can be reached from the initial arrangement $(wgcf/)$ and that are not arrangements to be avoided (such as (wg/fc)). At each stage ask yourself, "Where can I go from here?" and draw lines or arrows pointing to those arrangements. This method gives the graph shown at the top of the next column.

Start

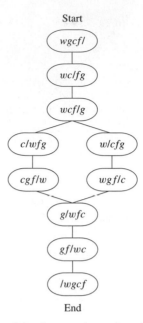

End

Examination of the diagram shows the solutions

$$(wgcf/) \rightarrow (wc/gf) \rightarrow (wcf/g) \rightarrow (w/gcf) \rightarrow$$
$$(wgf/c) \rightarrow (g/wcf) \rightarrow (gf/wc) \rightarrow (/wgcf)$$

and

$$(wgcf/) \rightarrow (wc/gf) \rightarrow (wcf/g) \rightarrow (c/wgf) \rightarrow$$
$$(gcf/w) \rightarrow (g/wcf) \rightarrow (gf/wc) \rightarrow (/wgcf)$$

14. *Hint:* The answer is yes. Represent possible amounts of water in jugs A and B by ordered pairs. For instance, the ordered pair $(1, 3)$ would indicate that there is one quart of water in jug A and three quarts in jug B. Starting with $(0, 0)$, draw arrows from one ordered pair to another if it is possible to go from the situation represented by one pair to that represented by the other by either filling a jug, emptying a jug, or transferring water from one jug to another. You need only draw arrows from states that have arrows pointing to them; the other states cannot be reached. Then find a directed path (sequence of directed edges) from the initial state $(0, 0)$ to a final state $(1, 0)$ or $(0, 1)$.

15. One such graph is

16. If there were a graph with four vertices of degrees 1, 2, 3, and 3, then its total degree would be 9, which is odd. But by Corollary 11.1.2, the total degree of the graph must be even. [*This is a contradiction.*] Hence there is no such graph.

(Alternatively, if there were such a graph, it would have an odd number of vertices of odd degree. But by Proposition 11.1.3 this is impossible.)

19. Suppose there were a simple graph with four vertices of degrees 1, 2, 3, and 4. Then the vertex of degree 4 would have to be connected by edges to four distinct vertices other than itself because of the assumption that the graph is simple (and hence has no loops or parallel edges.) This contradicts the assumption that the graph has four vertices in total. Hence there is no simple graph with four vertices of degrees 1, 2, 3, and 4.

22.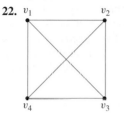

24. **a.** The nonempty subgraphs are as follows:

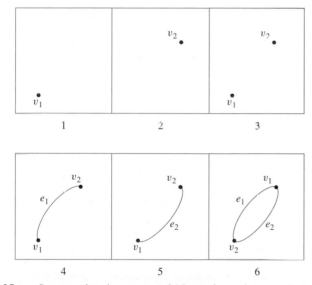

25. **a.** Suppose that, in a group of 15 people, each person had exactly three friends. Then you could draw a graph representing each person by a vertex and connecting two vertices by an edge if the corresponding people were friends. But such a graph would have 15 vertices, each of degree 3, for a total degree of 45. This would contradict the fact that the total degree of any graph is even. Hence the supposition must be false, and in a group of 15 people it is not possible for each to have exactly three friends.

28. The total degree of the graph is $0 + 2 + 2 + 3 + 9 = 16$, so by Theorem 11.1.1, the number of edges is $16/2 = 8$.

31. We give two proofs for the following statement, one less formal and the other more formal.

> For all integers $n \geq 0$, if $a_1, a_2, a_3, \ldots, a_{2n+1}$ are odd integers, then $\sum_{i=1}^{2n+1} a_i$ is odd.

Proof 1 (by mathematical induction): It is certainly true that the "sum" of one odd integer is odd. Suppose that for a certain positive odd integer r, the sum of r odd integers is odd. We must show that the sum of $r + 2$ odd integers is odd (because $r + 2$ is the next odd integer after r). But any sum of $r + 2$ odd integers equals a sum of r odd integers (which is odd by inductive hypothesis) plus a sum of two more odd integers (which is even). Thus the total sum is an odd integer plus an even integer, which is odd. [*This is what was to be shown.*]

Proof 2 (by mathematical induction): Let the property $P(n)$ be the following sentence: "If $a_1, a_2, a_3, \ldots, a_{2n+1}$ are odd integers, then $\sum_{i=1}^{2n+1} a_i$ is odd.

Show that the property is true for $n = 0$:

Suppose a_1 is an odd integer. Then $\sum_{i=1}^{2 \cdot 0 + 1} a_i = \sum_{i=1}^{1} a_i = a_1$, which is odd.

Show that for all integers $k \geq 0$, if the property is true for $n = k$, then it is true for $n = k + 1$:

Let k be an integer with $k > 0$, and suppose that

$$\text{if } a_2, a_3, \ldots, a_{2k+1} \text{ are odd integers, then } \sum_{i=1}^{2k+1} a_i \text{ is odd:}$$

[*This is the inductive hypothesis.*]

Suppose $a_1, a_2, a_3, \ldots, a_{2(k+1)+1}$ are odd integers. [*We must show that $\sum_{i=1}^{2(k+1)+1} a_i$ is odd, or, equivalently, that $\sum_{i=1}^{2k+3} a_i$ is odd.*] But

$$\sum_{i=1}^{2k+3} a_i = \sum_{i=1}^{2k+1} a_i + (a_{2k+2} + a_{2k+3}).$$

Since the sum of any two odd integers is even, $a_{2k+2} + a_{2k+3}$ is even, and, by inductive hypothesis, $\sum_{i=1}^{2k+1} a_i$ is odd. Therefore, $\sum_{i=1}^{2k+3} a_i$ is the sum of an odd integer and an even integer, which is odd. [*This is what was to be shown.*]

32. *Hint:* Use proof by contradiction.

33. **a.** K_6:

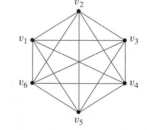

b. A proof of this fact was given in Section 8.2 using recursion. Try to find a different proof.

Hint for Proof 1: There are as many edges in K_n as there are subsets of two vertices (the endpoints) that can be chosen from a set of n vertices.

Hint for Proof 2: Use mathematical induction. A complete graph on $k + 1$ vertices can be obtained from a complete graph on k vertices by adding one vertex and connecting this vertex by k edges to each of the other vertices.

Hint for Proof 3: Use the fact that the number of edges of a graph is half the total degree. What is the degree of each vertex of K_n?

35. Suppose G is a nonempty simple graph with n vertices and $2n$ edges. By exercise 34, its number of edges cannot exceed $\frac{n(n-1)}{2}$. Thus $2n \le \frac{n(n-1)}{2}$, or $4n \le n^2 - n$. Equivalently, $n^2 - 5n \ge 0$, or $n(n-5) \ge 0$. This implies that $n \ge 5$ since G is nonempty. Hence a nonempty simple graph with twice as many edges as vertices must have at least five vertices. But a complete graph with five vertices has $\frac{5(5-1)}{2} = 10$ edges and $10 = 2 \cdot 5$. Consequently, the answer to the question is yes because K_5 is a graph with twice as many edges as vertices.

36. a. $K_{4,2}$:

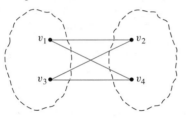

37. a. This graph is bipartite.

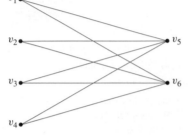

b. Suppose this graph is bipartite. Then the vertex set can be partitioned into two mutually disjoint subsets such that vertices in each subset are connected by edges only to vertices in the other subset and not to vertices in the same subset. Now v_1 is in one subset of the partition, say V_1. Since v_1 is connected by edges to v_2 and v_3, both v_2 and v_3 must be in the other subset, V_2. But v_2 and v_3 are connected by an edge to each other. This contradicts the fact that no vertices in V_2 are connected by edges to other vertices in V_2. Hence the supposition is false, and so the graph is not bipartite.

39. a.

41. b.

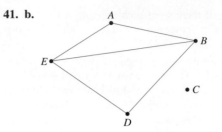

42. *Hint:* Consider the graph obtained by taking the vertices and edges of G plus all the edges of G'. Use exercise 33(b).

44. c. *Hint:* Suppose there were a simple graph with n vertices (where $n \ge 2$) each of which had a different degree. Then no vertex could have degree more than $n - 1$ (why?), so the degrees of the n vertices must be $0, 1, 2, \ldots, n - 1$ (why?). This is impossible (why?).

45. *Hint:* Use the result of exercise 44(c).

46. *Hint:* One solution is to begin by choosing a vertex of maximal degree and assigning the first time slot to it and to all other vertices that do not share an edge with it or with each other. Then choose a vertex of maximal degree from those remaining, and assign the second time slot to it and to all those still unassigned that do not share an edge with it or with each other. Continue in this way until all vertices have been assigned.

Section 11.2

1. a. path (no repeated edge), not a simple path (repeated vertex—v_1), not a circuit
b. walk, not a path (has repeated edge—e_9), not a circuit
c. simple circuit (no repeated edge, no repeated vertex, starts and ends at same vertex)
d. circuit (no repeated edge, starts and ends at same vertex), not a simple circuit (vertex v_4 is repeated)
e. closed walk (starts and ends at the same vertex but has repeated edges $-\{v_2, v_3\}$ and $\{v_3, v_4\}$)
f. simple path

3. a. No. The notation $v_1 v_2 v_1$ could equally well refer to $v_1 e_1 v_2 e_2 v_1$ or to $v_1 e_2 v_2 e_1 v_1$, which are different walks.

4. a. Three (There are three ways to choose the middle edge.)
b. $3! + 3 = 9$ (In addition to the three simple paths, there are $3!$ with vertices $v_1, v_2, v_3, v_2, v_3, v_4$. The reason is that from v_2 there are three choices of an edge to go to v_3, then two choices of different edges to go back to v_2, and then one choice of different edge to return to v_3. This makes $3!$ paths from v_2 to v_3.)
c. Infinitely many (Since a walk may have repeated edges, a walk from v_1 to v_4 may contain an arbitrarily large number of repetitions of edges joining a pair of vertices along the way.)

6. a. $\{v_1, v_3\}$, $\{v_2, v_3\}$, $\{v_4, v_3\}$, and $\{v_5, v_3\}$ are all bridges.

8. a. Three connected components.

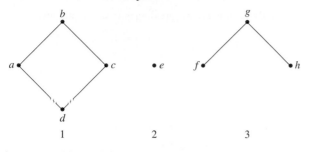

| 1 | 2 | 3 |

9. a. No. This graph has two vertices of odd degree, whereas all vertices of a graph with an Euler circuit have even degree.

12. One Euler circuit is $e_4 c_5 e_6 e_3 e_2 e_7 e_8 e_1$.

14. One Euler circuit is $i a b i h b c h g c d g f d e f i$.

19. There is an Euler path since $\deg(u)$ and $\deg(w)$ are odd, all other vertices have even degree, and the graph is connected. One Euler path is $u v_1 v_0 v_7 u v_2 v_3 v_4 v_2 v_6 v_4 w v_5 v_6 w$.

23. $v_0 v_7 v_1 v_2 v_3 v_4 v_5 v_6 v_0$

25. *Hint:* See the solution to Example 11.2.8.

26. Here is one sequence of reasoning you could use: Call the given graph G, and suppose G has a Hamiltonian circuit. Then G has a subgraph H that satisfies conditions (1)–(4) of Proposition 11.2.6. Since the degree of b in G is 4 and every vertex in H has degree 2, two edges incident on b must be removed from G to create H. Edge $\{a, b\}$ cannot be removed because doing so would result in vertex d having degree less than 2 in H. Similar reasoning shows that edge $\{b, c\}$ cannot be removed either. So edges $\{b, i\}$ and $\{b, e\}$ must be removed from G to create H. Because vertex e must have degree 2 in H and because edge $\{b, e\}$ is not in H, both edges $\{e, d\}$ and $\{e, f\}$ must be in H. Similarly, since both vertices c and g must have degree 2 in H, edges $\{c, d\}$ and $\{g, d\}$ must also be in H. But then three edges incident on d, namely $\{e, d\}$, $\{c, d\}$, and $\{g, d\}$, must be all in H, which contradicts the fact that vertex d must have degree 2 in H.

28. *Hint:* This graph does not have a Hamiltonian circuit.

32. *Partial answer:*

This graph has an Euler circuit $v_0 v_1 v_2 v_3 v_1 v_4 v_0$ but no Hamiltonian circuit.

33. *Partial answer:*

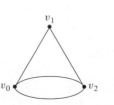

This graph has a Hamiltonian circuit $v_0 v_1 v_2 v_0$ but no Euler circuit.

34. *Partial answer:*

The walk $v_0 v_1 v_2 v_0$ is both an Euler circuit and a Hamiltonian circuit for this graph.

35. *Partial answer:*

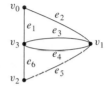

This graph has the Euler circuit $e_1 e_2 e_3 e_4 e_5 e_6$ and the Hamiltonian circuit $v_0 v_1 v_2 v_3 v_0$. These are not the same.

37. a. *Proof:* Suppose G is a graph and W is a walk in G that contains a repeated edge e. Let v and w be the endpoints of e. In case $v = w$, then v is a repeated vertex of W. In case $v \neq w$, then one of the following must occur: (1) W contains two copies of vew or of wev (for instance, W might contain a section of the form $vewe'vew$, as illustrated below); (2) W contains separate sections of the form vew and wev (for instance, W might contain a section of the form $vewe'wev$, as illustrated below); or (3) W contains a section of the form $vewev$ or of the form $wevew$ (as illustrated below). In cases (1) and (2), both vertices v and w are repeated, and in case (3), one of v or w is repeated. In all cases, there is at least one vertex in W that is repeated.

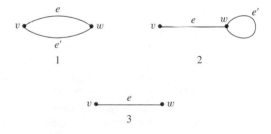

38. *Proof:* Suppose G is a connected graph. [*We must show that any two vertices of G can be connected by a simple path.*] Let v and w be any particular but arbitrarily chosen vertices of G. Since G is connected, there is a walk from v to w. If the walk contains a repeated vertex, then delete the portion of the walk from the first occurrence of the vertex to its next occurrence. (For example, in the walk $v e_1 v_2 e_5 v_7 e_6 v_2 e_3 w$, the vertex v_2 occurs twice. Deleting the portion of the walk from one occurrence to the next gives $v e_1 v_2 e_3 w$). If the resulting walk still contains a repeated vertex, do the above deletion process another time. Then check again for a repeated vertex. Continue in this way until all repeated vertices have been deleted. (This must occur eventually, since the total number of vertices is finite.) The resulting walk connects v to w but has no repeated vertex. By exercise 37(b), it has no repeated edge either. Hence it is a simple path from v to w.

40. The graph below contains a circuit, any edge of which can be removed without disconnecting the graph. For instance, if edge e is removed, then the following walk can be used to go from v_1 to v_2: $v_1 v_5 v_3 v_2$.

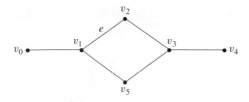

42. Yes. Suppose a graph contains a circuit that starts and ends at a vertex v. Successively delete sections of this circuit as follows: For each repeated vertex w in the circuit (excluding the first vertex if its only repetition is at the end of the circuit but including the first vertex if it is repeated in the middle of the circuit), there is a section of the circuit of the following form: $w e_1 v_1 e_2 v_2 \dots e_{n-1} v_{n-1} e_n w$. Replace this section of the circuit by the single vertex w. Because the circuit has finite length, only a finite number of such deletions can be made, after which a simple circuit starting and ending at v will remain.

44. *Proof:* Let G be a connected graph and let C be a circuit in G. Let G' be the nonempty subgraph obtained by removing all the edges of C from G and also any vertices that become isolated when the edges of C are removed. [*We must show that there exists a vertex v such that v is in both C and G'.*] Pick any vertex v of C and any vertex w of G'. Since G is connected, there is a simple path from v to w (by Lemma 11.2.1(a)):

$$v = v_0 e_1 v_1 e_2 v_2 \dots v_{i-1} e_i v_i e_{i+1} v_{i+1} \dots v_{n-1} e_n v_n = w.$$

$$\begin{matrix} \uparrow & & \uparrow & \uparrow & & \uparrow \\ \text{in } C & & \text{in } C & \text{not in } C & & \text{in } G' \end{matrix}$$

Let i be the largest subscript such that v_i is in C. If $i = n$, then $v_n = w$ is in C and also in G', and we are done. If $i < n$, then v_i is in C and v_{i+1} is not in C. This implies that e_{i+1} is not in C (for if it were, both endpoints would be in C by definition of circuit). Hence when G' is formed by removing the edges and resulting isolated vertices from G, then e_{i+1} is not removed. That means that v_i does not become an isolated vertex, so v_i is not removed either. Hence v_i is in G'. Consequently, v_i is in both C and G' [*as was to be shown*].

45. *Proof:* Suppose G is a graph with an Euler circuit. If v and w are any two vertices of G, then v and w each appear at least once in the Euler circuit (since an Euler circuit contains every vertex of the graph). The section of the circuit between the first occurrence of one of v or w and the first occurrence of the other is a walk from one of the two vertices to the other.

Section 11.3

1. a. By equating corresponding entries, we find that
$$a + b = 1,$$
$$a - c = 0,$$
$$c = -1,$$
$$b - a = 3.$$

Thus $a - c = a - (-1) = 0$, and so $a = -1$. Consequently, $a + b = (-1) + b = 1$, and hence $b = 2$. The last equation should be checked to make sure the answer is consistent: $b - a = 2 - (-1) = 3$, which agrees.

2. a.
$$\begin{array}{c} \\ v_1 \\ v_2 \\ v_3 \end{array}\begin{array}{ccc} v_1 & v_2 & v_3 \\ \left[\begin{array}{ccc} 0 & 1 & 1 \\ 1 & 0 & 0 \\ 0 & 0 & 0 \end{array}\right] \end{array}$$

3. a.

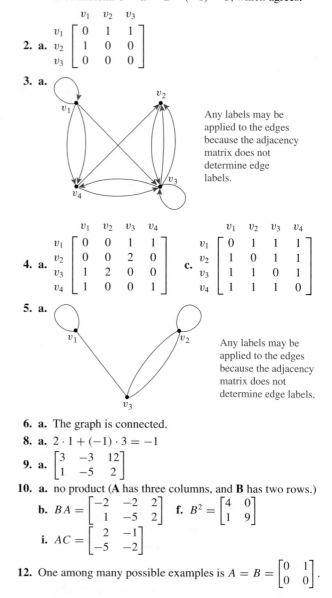

Any labels may be applied to the edges because the adjacency matrix does not determine edge labels.

4. a.
$$\begin{array}{c} \\ v_1 \\ v_2 \\ v_3 \\ v_4 \end{array}\begin{array}{cccc} v_1 & v_2 & v_3 & v_4 \\ \left[\begin{array}{cccc} 0 & 0 & 1 & 1 \\ 0 & 0 & 2 & 0 \\ 1 & 2 & 0 & 0 \\ 1 & 0 & 0 & 1 \end{array}\right] \end{array}$$

c.
$$\begin{array}{c} \\ v_1 \\ v_2 \\ v_3 \\ v_4 \end{array}\begin{array}{cccc} v_1 & v_2 & v_3 & v_4 \\ \left[\begin{array}{cccc} 0 & 1 & 1 & 1 \\ 1 & 0 & 1 & 1 \\ 1 & 1 & 0 & 1 \\ 1 & 1 & 1 & 0 \end{array}\right] \end{array}$$

5. a.

Any labels may be applied to the edges because the adjacency matrix does not determine edge labels.

6. a. The graph is connected.

8. a. $2 \cdot 1 + (-1) \cdot 3 = -1$

9. a. $\begin{bmatrix} 3 & -3 & 12 \\ 1 & -5 & 2 \end{bmatrix}$

10. a. no product (**A** has three columns, and **B** has two rows.)

b. $BA = \begin{bmatrix} -2 & -2 & 2 \\ 1 & -5 & 2 \end{bmatrix}$ **f.** $B^2 = \begin{bmatrix} 4 & 0 \\ 1 & 9 \end{bmatrix}$

i. $AC = \begin{bmatrix} 2 & -1 \\ -5 & -2 \end{bmatrix}$

12. One among many possible examples is $A = B = \begin{bmatrix} 0 & 1 \\ 0 & 0 \end{bmatrix}$.

14. *Hint:* If the entries of the $m \times m$ identity matrix are denoted by δ_{ik}, then $\delta_{ik} = \begin{cases} 0 & \text{if } i \neq k \\ 1 & \text{if } i = k \end{cases}$. The ijth entry of **IA** is $\sum_{k=1}^{m} \delta_{ik} A_{kj}$.

15. *Proof:* Suppose **A** is an $m \times m$ symmetric matrix. Then for all integers i and j with $1 \leq i, j \leq m$,

$$(A^2)_{ij} = \sum_{k=1}^{m} A_{ik} A_{kj} \quad \text{and} \quad (A^2)_{ji} = \sum_{k=1}^{m} A_{jk} A_{ki}.$$

But since **A** is symmetric, $A_{ik} = A_{ki}$ and $A_{kj} = A_{jk}$ for all i, j, and k, and thus $A_{ik} A_{kj} = A_{jk} A_{ki}$ [*by the commutative law for multiplication of real numbers*]. Hence $(A^2)_{ij} = (A^2)_{ji}$ for all integers i and j with $1 \leq i, j \leq m$.

17. *Proof (by mathematical induction):*
Let the property $P(n)$ be the equation $\mathbf{A}^n\mathbf{A} = \mathbf{A}\mathbf{A}^n$.

Show that the property is true for $n = 1$:
We must show that $\mathbf{A}^1\mathbf{A} = \mathbf{A}\mathbf{A}^1$. But this is true because $\mathbf{A}^1 = \mathbf{A}$ and $\mathbf{A}\mathbf{A} = \mathbf{A}\mathbf{A}$.

Show that for all integers $k \geq 1$, if the property is true for $n = k$, then it is true for $n = k + 1$:
Suppose that for some integer $k \geq 1$, $\mathbf{A}^k\mathbf{A} = \mathbf{A}\mathbf{A}^k$. We must show that $\mathbf{A}^{k+1}\mathbf{A} = \mathbf{A}\mathbf{A}^{k+1}$. But

$$\mathbf{A}^{k+1}\mathbf{A} = (\mathbf{A}\mathbf{A}^k)\mathbf{A} \qquad \text{by definition of matrix power}$$
$$= \mathbf{A}(\mathbf{A}^k\mathbf{A}) \qquad \text{by exercise 16}$$
$$= \mathbf{A}(\mathbf{A}\mathbf{A}^k) \qquad \text{by inductive hypothesis}$$
$$= \mathbf{A}\mathbf{A}^{k+1} \qquad \text{by definition of matrix power.}$$

19. a.
$$A^2 = \begin{bmatrix} 1 & 1 & 2 \\ 1 & 0 & 1 \\ 2 & 1 & 0 \end{bmatrix}\begin{bmatrix} 1 & 1 & 2 \\ 1 & 0 & 1 \\ 2 & 1 & 0 \end{bmatrix} = \begin{bmatrix} 6 & 3 & 3 \\ 3 & 2 & 2 \\ 3 & 2 & 5 \end{bmatrix}$$

$$A^3 = \begin{bmatrix} 1 & 1 & 2 \\ 1 & 0 & 1 \\ 2 & 1 & 0 \end{bmatrix}\begin{bmatrix} 6 & 3 & 3 \\ 3 & 2 & 2 \\ 3 & 2 & 5 \end{bmatrix} = \begin{bmatrix} 15 & 9 & 15 \\ 9 & 5 & 8 \\ 15 & 8 & 8 \end{bmatrix}$$

20. a. 2 since $(A^2)_{23} = 2$
b. 3 since $(A^2)_{34} = 3$
c. 6 since $(A^3)_{14} = 6$
d. 17 since $(A^3)_{23} = 17$

22. b. *Hint:* If G is bipartite, then its vertices can be partitioned into two sets V_1 and V_2 so that no vertices in V_1 are connected to each other by an edge and no vertices in V_2 are connected to each other by an edge. Label the vertices in V_1 as $v_1, v_2, \ldots, v_k$ and label the vertices in V_2 as $v_{k+1}, v_{k+2}, \ldots, v_n$. Now look at the matrix of G formed according to the given vertex labeling.

23. b. *Hint:* Consider the ijth entry of

$$\mathbf{A} + \mathbf{A}^2 + \mathbf{A}^3 + \cdots + \mathbf{A}^n.$$

If G is connected, then given the vertices v_i and v_j, there is a walk connecting v_i and v_j. If this walk has length

k, then by Theorem 11.3.2, the ijth entry of A^k is not equal to 0. Use the facts that all entries of each power of **A** are nonnegative and a sum of nonnegative numbers is positive provided that at least one of the numbers is positive.

Section 11.4

1. The graphs are isomorphic. One way to define the isomorphism is as follows:

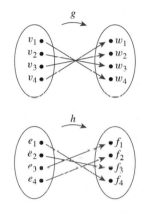

2. The graphs are not isomorphic. G has five vertices and G' has six.

6. The graphs are isomorphic. One isomorphism is the following:

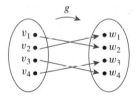

8. The graphs are not isomorphic. G has a simple circuit of length 3; G' does not.

10. The graphs are isomorphic. One way to define the isomorphism is as follows:

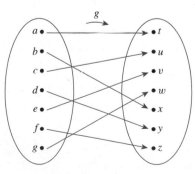

12. These graphs are isomorphic. One isomorphism is the following:

14.

16.

18. *Hint:* There are 19.

19.

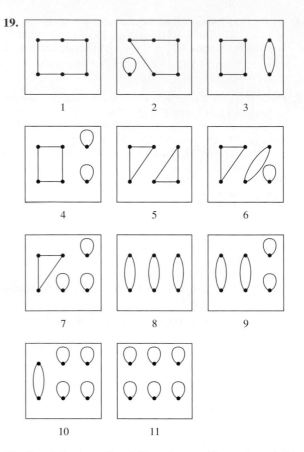

21. *Proof:* Suppose G and G' are isomorphic graphs and G has n vertices, where n is a nonnegative integer. [*We must show that G' has n vertices.*] By definition of graph isomorphism, there is a one-to-one correspondence $g: V(G) \rightarrow V(G')$ sending vertices of G to vertices of G'. Since $V(G)$ is a finite set and g is a one-to-one correspondence, the number of vertices in $V(G')$ equals the number of vertices in $V(G)$. Hence G' has n vertices [*as was to be shown*].

23. *Proof:* Suppose G and G' are isomorphic graphs and suppose G has a circuit C of length k, where k is a nonnegative integer. Let C be $v_0 e_1 v_1 e_2 \ldots e_k v_k (= v_0)$. By definition of graph isomorphism, there are one-to-one correspondences $g: V(G) \rightarrow V(G')$ and $h: E(G) \rightarrow E(G')$ that preserve the edge-endpoint functions in the sense that for all v in $V(G)$ and e in $E(G)$, v is an endpoint of $e \Leftrightarrow g(v)$ is an endpoint of $h(e)$. Let C' be $g(v_0)h(e_1)g(v_1)h(e_2) \ldots h(e_k)g(v_k)(= g(v_0))$. Then C' is a circuit of length k in G'. The reason is that (1) because g and h preserve the edge-endpoint functions, for all $i = 0, 1, \ldots, k-1$ both $g(v_i)$ and $g(v_{i+1})$ are incident on $h(e_{i+1})$ so that C' is a walk from $g(v_0)$ to $g(v_0)$, and (2) since C is a circuit, then $e_1, e_2, \ldots, e_k$ are distinct, and since h is a one-to-one correspondence, $h(e_1), h(e_2), \ldots, h(e_k)$ are also distinct, which implies that C' has k distinct edges. Therefore, G' has a circuit C of length k.

25. *Hint:* Suppose G and G' are isomorphic and G has m vertices of degree k; call them $v_1, v_2, \ldots, v_m$. Since G and G' are isomorphic, there are one-to-one correspondences $g: V(G) \to V(G')$ and $h: E(G) \to E(G')$. Show that $g(v_1), g(v_2), \ldots, g(v_m)$ are m distinct vertices of G' each of which has degree k.

27. *Hint:* Suppose G and G' are isomorphic and G is connected. To show that G' is connected, suppose w and x are any two vertices of G'. Show that there is a walk connecting w with x by finding a walk connecting the corresponding vertices in G.

Section 11.5

1. a. Math 110

2. a.

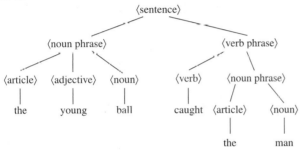

3. *Hint:* The answer is $2n - 2$. To obtain this result, use the relationship between the total degree of a graph and the number of edges of the graph.

4. a.

d. *Hint:* Each carbon atom in G is bonded to four other atoms in G, because otherwise an additional hydrogen atom could be bonded to it, and this would contradict the assumption that G has the maximum number of hydrogen atoms for its number of carbon atoms. Also each hydrogen atom is bonded to exactly one carbon atom in G, because otherwise G would not be connected.

5. *Hint:* Revise the algorithm given in the proof of Lemma 11.5.1 to keep track of which vertex and edge were chosen in step 1 (by, say, labeling them v_0 and e_0). Then after one vertex of degree 1 is found, return to v_0 and search for another vertex of degree 1 by moving along a path outward from v_0 starting with e_0.

7. a. Internal vertices: v_2, v_3, v_4, v_6
Terminal vertices: v_1, v_5, v_7

8. Any tree with nine vertices has eight edges, not nine. Thus there is no tree with nine vertices and nine edges.

9. One such graph is

10. One such graph is

11. There is no tree with six vertices and a total degree of 14. Any tree with six vertices has five edges and hence (by Theorem 11.1.1) a total degree of 10, not 14.

12. One such tree is shown.

13. No such graph exists. By Theorem 11.5.4, a connected graph with six vertices and five edges is a tree. Hence such a graph cannot have a nontrivial circuit.

14.

$\bullet$
v_1 v_2

22. Yes. Since it is connected and has 12 vertices and 11 edges, by Theorem 11.5.4 it is a tree. It follows from Lemma 11.5.1 that it has vertex of degree 1.

25. Suppose there were a connected graph with eight vertices and six edges. Either the graph itself would be a tree or edges could be eliminated from its circuits to obtain a tree. In either case, there would be a tree with eight vertices and six or fewer edges. But by Theorem 11.5.2, a tree with eight vertices has seven edges, not six or fewer. This contradiction shows that the supposition is false, so there is no connected graph with eight vertices and six edges.

26. *Hint:* See the answer to exercise 25.

27. Yes. Suppose G is a circuit-free graph with ten vertices and nine edges. Let $G_1, G_2, \ldots, G_k$ be the connected components of G [*To show that G is connected, we will show that $k = 1$.*] Each G_i is a tree since each G_i is connected and circuit-free. For each $i = 1, 2, \ldots, k$, let G_i have n_i vertices. Note that since G has ten vertices in all,

$$n_1 + n_2 + \cdots + n_k = 10.$$

By Theorem 11.5.2,

$$G_1 \text{ has } n_1 - 1 \text{ edges,}$$
$$G_2 \text{ has } n_2 - 1 \text{ edges,}$$
$$\vdots$$
$$G_k \text{ has } n_k - 1 \text{ edges.}$$

So the number of edges of G equals

$$(n_i - 1) + (n_2 - 1) + \cdots + (n_k - 1)$$
$$= (n_1 + n_2 + \cdots + n_k) - \underbrace{(1 + 1 + \cdots + 1)}_{k \text{ 1's}}$$
$$= 10 - k.$$

But we are given that G has nine edges. Hence $10 - k = 9$, and so $k = 1$. Thus G has just one connected component, G_1, and so G is connected.

28. *Hint:* See the answer to exercise 27.

31. b. *Hint:* There are six.

32. a. 3 **b.** 0 **c.** 5 **d.** u, v
 e. d **f.** k, l **g.** m, s, t, x, y

34. a.

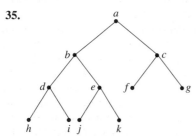

Exercise 35 and 39–41 have other answers in addition to the ones shown.

35.

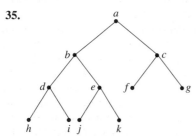

36. There is no full binary tree with the given properties because any full binary tree with five internal vertices has six terminal vertices, not seven.

37. Any full binary tree with four internal vertices has five terminal vertices for a total of nine, not seven, vertices in all. Thus there is no full binary tree with the given properties.

38. There is no full binary tree with 12 vertices because any full binary tree has $2k + 1$ vertices, where k is the number of internal vertices. But $2k + 1$ is always odd, and 12 is even.

39.

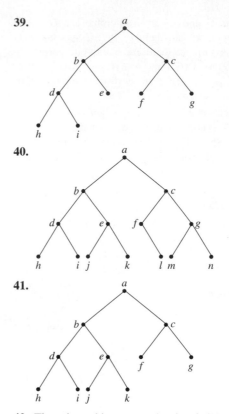

40.

41.

42. There is no binary tree that has height 3 and nine terminal vertices because any binary tree of height 3 has at most $2^3 = 8$ terminal vertices.

51. a. Height of tree $\geq \log_2 25 \cong 4.6$. Since the height of any tree is an integer, the height must be at least 5.

Section 11.6

1.

3. One of many spanning trees is as follows:

5. Minimum spanning tree:

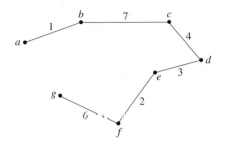

Order of adding the edges:
$\{a, b\}, \{e, f\}, \{e, d\}, \{d, c\}, \{g, f\}, \{b, c\}$

7. Minimum spanning tree: same as in exercise 5
Order of adding the edges:
$\{a, b\}, \{b, c\}, \{c, d\}, \{d, e\}, \{e, f\}, \{f, g\}$

9. There are four minimum spanning trees:

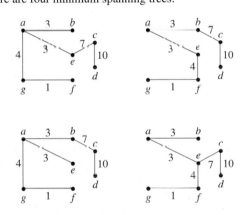

When Prim's algorithm is used, edges are added in any of the orders obtained by following one of the eight paths from left to right across the diagram below.

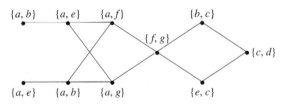

When Kruskal's algorithm is used, edges are added in any of the orders obtained by following one of the eight paths from left to right across the diagram below.

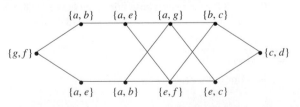

13. b. *Proof:* Suppose not. Suppose that for some tree T, u and v are distinct vertices of T, and P_1 and P_2 are two distinct paths joining u and v. [*We must deduce a contradiction.*

In fact, we will show that T contains a circuit.] Let P_1 be denoted $u = v_0, v_1, v_2, \ldots, v_m = v$, and let P_2 be denoted $u = w_0, w_1, w_2, \ldots, w_n = v$. Because P_1 and P_2 are distinct, and T has no parallel edges, the sequence of vertices in P_1 must diverge from the sequence of vertices in P_2 at some point. Let i be the least integer such that $v_i \neq w_i$. Then $v_{i-1} = w_{i-1}$. Let j and k be the least integers greater than i so that $v_j = w_k$. (There must be such integers because $v_m = w_n$.) Then

$$v_{i-1} v_i v_{i+1} \ldots v_j (= w_k) w_{k-1} \ldots w_i w_{i-1} (= v_{i-1})$$

is a circuit in T. The existence of such a circuit contradicts the fact that T is a tree. Hence the supposition must be false. That is, given any tree with vertices u and v, there is a unique path joining u and w.

15. *Proof:* Suppose G is a connected graph, T is a circuit free subgraph of G, and if any edge e of G not in T is added to T, the resulting graph contains a circuit. Suppose that T is not a spanning tree for G. [*We must derive a contradiction.*]

Case 1 (T is not connected): In this case, there are vertices u and v in T such that there is no walk in T from u to v. Now, since G is connected, there is a walk in G from u to v, and hence, by Lemma 11.2.1, there is a simple path in G from u to v. Let $e_1, e_2, \ldots, e_k$ be the edges of this path that are not in T. When these edges are added to T, the result is a graph T' in which u and v are connected by a path. In addition, by hypothesis, each of the edges e_i creates a circuit when added to T. Now remove these edges one by one from T'. By the same argument used in the proof of Lemma 11.5.3, each such removal leaves u and v connected since each e_i is an edge of a circuit when added to T. Hence, after all the e_i have been removed, u and v remain connected. But this contradicts the fact that there is no walk in T from u to v.

Case 2 (T is connected): In this case, since T is not a spanning tree and T is circuit-free, there is a vertex v in G such that v is not in T. [*For if T were connected, circuit-free, and contained every vertex in G, then T would be a spanning tree for G.*] Since G is connected, v is not isolated. Thus there is an edge e in G with v as an endpoint. Let T' be the graph obtained from T by adding e and v. [*Note that e is not already in T because if it were, its endpoint v would also be in T and it is not.*] Then T' contains a circuit because, by hypothesis, addition of any edge to T creates a circuit. Also T' is connected because T is and because when e is added to T, e becomes part of a circuit in T'. Now deletion of an edge from a circuit does not disconnect a graph, so if e is deleted from T' the result is a connected graph. But the resulting graph contains v, which means that there is an edge in T connecting v to another vertex of T. This implies that v is in T [*because both endpoints of any edge in a graph must be part of the vertex set of the graph,*] which contradicts the fact that v is not in T.

Thus, in either case, the supposition that T is not a spanning tree leads to a contradiction. Hence the supposition is false, and T is a spanning tree for G.

16. a. No. *Counterexample:* Let G be the following graph.

Then G has the spanning trees shown below.

These trees have no edge in common.

17. *Hint:* Suppose e is contained in every spanning tree of G and the graph obtained by removing e from G is connected. Let G' be the subgraph of G obtained by removing e, and let T' be a spanning tree for G'. How is T' related to G?

19. *Proof:* Suppose that $w(e') > w(e)$. Form a new graph T' by adding e to T and deleting e'. By exercise 15, addition of an edge to a spanning tree creates a circuit, and by Lemma 11.5.3, deletion of an edge from a circuit does not disconnect a graph. Consequently, T' is also a spanning tree for G. Furthermore, $w(T') < w(T)$ because $w(T') = w(T) - w(e') + w(e) = w(T) - (w(e') - w(e)) < w(T)$ [*since* $w(e') > w(e)$, *which implies that* $w(e') - w(e) > 0$.] But this contradicts the fact that T is a minimum spanning tree for G. Hence the supposition is false, and so $w(e') \leq w(e)$.

20. *Hint:* Suppose e is an edge that has smaller weight than any other edge of G, and suppose T is a minimum spanning tree for G that does not contain e. Create a new spanning tree T' by adding e to T and removing another edge of T (which one?). Then $w(T') < w(T)$.

21. Yes. *Proof by contradiction:* Suppose G is a weighted graph in which all the weights of all the edges are distinct, and suppose G has two distinct minimum spanning trees T_1 and T_2. Let e be the edge of least weight that is in one of the trees but not the other. Without loss of generality, we may say that e is in T_1. Add e to T_2 to obtain a graph G'. By exercise 14, G' contains a nontrivial circuit. At least one other edge f of this circuit is not in T_1 because otherwise T_1 would contain the complete circuit, which would contradict the fact that T_1 is a tree. Now f has weight greater than e because all edges have distinct weights, f is in T_2 and not in T_1, and e is the edge of least weight that is in one of the trees and not the other. Remove f from G' to obtain a tree T_3. Then $w(T_3) < w(T_2)$ because T_3 is the same as T_2 except that it contains e rather than f and $w(e) < w(f)$. Consequently, T_3 is a spanning tree for G of smaller weight than T_2. This contradicts the supposition that T_2 is a minimum spanning tree for G. Thus G cannot have more than one minimum spanning tree.

23. The output will be a "spanning forest" for the graph. It will contain one spanning tree for each connected component of the input graph.

Section 12.1

1. a. $L_1 = \{\epsilon, x, y, xx, yy, xxx, xyx, yxy, yyy, xxxx, xyyx, yxxy, yyyy\}$

b. $L_2 = \{x, xx, xy, xxx, xxy, xyx, xyy\}$

3. $11* = 1 \cdot 1 = 1, \quad 12* = 1 \cdot 2 = 2, \quad 21/ = 2/1 = 2$

4. $L_1 L_2$ is the set of all strings of a's and b's that start with an a and contain an odd number of a's.

$L_1 \cup L_2$ is the set of all strings of a's and b's that contain an even number of a's or that start with an a and contain only that one a. (Note that because 0 is an even number, both ϵ and b are in $L_1 \cup L_2$.)

$(L_1 \cup L_2)^*$ is the set of all strings of a's and b's. The reason is that a and b are both in $L_1 \cup L_2$, and thus every string in a and b is in $(L_1 \cup L_2)^*$.

7. $(a \,|\, ((b^*)b))((a^*) \,|\, (ab))$

10. $(ab^* \,|\, cb^*)(ac \,|\, bc)$

13. $L(\epsilon \,|\, ab) = L(\epsilon) \cup L(ab) = \{\epsilon\} \cup L(a)L(b)$
$= \{\epsilon\} \cup \{xy \,|\, x \in L(a) \text{ and } y \in L(b)\}$
$= \{\epsilon\} \cup \{xy \,|\, x \in \{a\} \text{ and } y \in \{b\}\}$
$= \{\epsilon\} \cup \{ab\} = \{\epsilon, ab\}$

16. Here are five strings out of infinitely many: 0101, 1, 01, 10000, and 011100.

19. The language consists of all strings of a's and b's that contain exactly three a's.

22. *aaaba* is in the language but *baabb* is not because if a string in the language contains a b to the right of the left-most a, then it must contain another a to the right of the all b's.

25. One solution is $0^*10^*(0^*10^*10^*)^*$.

28. $L((r \,|\, s)t) = L(r \,|\, s)L(t) = (L(r) \cup L(s))L(t)$
$= \{xy \,|\, x \in (L(r) \cup L(s)) \text{ and } y \in L(t)\}$
$= \{xy \,|\, (x \in L(r) \text{ or } x \in L(s)) \text{ and } y \in L(t)\}$
$= \{xy \,|\, (x \in L(r) \text{ and } y \in L(t)) \text{ or }$
$\qquad\qquad (x \in L(s) \text{ and } y \in L(t))\}$
$= \{xy \,|\, xy \in L(rt) \text{ or } xy \in L(st)\}$
$= L(rt) \cup L(st)$
$= L(rt \,|\, st)$

31. $pre[a - z]^+$

34. $[a - z]^*(a \,|\, e \,|\, i \,|\, o \,|\, u)[a - z]^*$

37. $[0 - 9]\{3\}$ - $[0 - 9]\{2\}$ - $[0 - 9]\{4\}$

39. $([+ -] \,|\, \epsilon)[0 - 9]^*(\backslash. \,|\, \epsilon)[0 - 9]^*$

40. *Hint:* Leap years from 1980 to 2079 are 1980, 1984, 1988, 1992, 1996, 2000, 2004, etc. Note that the fourth digit is 0, 4, or 8 for the ones whose third digit is even and that the fourth digit is 2 or 6 for those whose third digit is odd.

Section 12.2

1. a. \$1 or more deposited

2. a. s_0, s_1, s_2 **b.** 0, 1 **c.** s_0 **d.** s_2
e. Annotated next-state table:

		Input	
		0	1
→	s_0	s_1	s_0
State	s_1	s_1	s_2
⊙	s_2	s_2	s_2

5. a. A, B, C, D, E, F **b.** x, y **c.** A **d.** D, E
e. Annotated next-state table:

		Input	
		x	y
→	A	C	B
	B	F	D
State	C	E	F
⊙	D	F	D
⊙	E	E	F
	F	F	F

7. a. s_0, s_1, s_2, s_3 **b.** 0, 1 **c.** s_0 **d.** s_0, s_2
e. Annotated next-state table:

		Input	
		0	1
→ ⊙	s_0	s_0	s_1
State	s_1	s_1	s_2
⊙	s_2	s_2	s_3
	s_3	s_3	s_0

8. a. s_0, s_1, s_2 **b.** 0, 1 **c.** s_0 **d.** s_2
e.

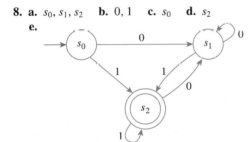

10. a. $N(s_1, 1) = s_2$, $N(s_0, 1) = s_3$
c. $N^*(s_0, 10011) = s_2$, $N^*(s_1, 01001) = s_2$

11. a. $N(s_3, 0) = s_4$, $N(s_2, 1) = s_4$
c. $N^*(s_0, 010011) = s_3$, $N^*(s_3, 01101) = s_4$

Note that multiple correct answers exist for part (d) of exercises 12 and 13, part (b) of exercises 14–19, and for exercises 20–48.

12. a. (i) s_2 (ii) s_2 (iii) s_1
b. those in (i) and (ii) but not (iii)
c. The language accepted by this automaton is the set of all strings of 0's and 1's that contain at least one 0 followed (not necessarily immediately) by at least one 1.
d. 1*00*1(0 | 1)*

14. a. The language accepted by this automaton is the set of all strings of 0's and 1's that end 00.
b. (0 | 1)*00

15. a. The language accepted by this automaton is the set of all strings of x's and y's of length at least two that consist either entirely of x's or entirely of y's.
b. $xxx^* | yyy^*$

17. a. The language accepted by this automaton is the set of all strings of 0's and 1's with the following property: If n is the number of 1's in the string, then $n \bmod 4 = 0$ or $n \bmod 4 = 2$. This is equivalent to saying that n is even.
b. $0^* | (0^*10^*10^*)^*$

18. a. The language accepted by this automaton is the set of all strings of 0's and 1's that end in 1.
b. $(0 | 1)^*1$

20. a. Call the automaton being constructed A. Acceptance of a string by A depends on the values of three consecutive inputs. Thus A requires at least four states:

s_0: initial state

s_1: state indicating that the last input character was a 1

s_2: state indicating that the last two input characters were 1's

s_3: state indicating that the last three input characters were 1's, the acceptance state

If a 0 is input to A when it is in state s_0, no progress is made toward achieving a string of three consecutive 1's. Hence A should remain in state s_0. If a 1 is input to A when it is in state s_0, it goes to state s_1, which indicates that the last input character of the string is a 1. From state s_1, A goes to state s_2 if a 1 is input. This indicates that the last two characters of the string are 1's. But if a 0 is input, A should return to s_0 because the wait for a string of three consecutive 1's must start all over again. When A is in state s_2 and a 1 is input, then a string of three consecutive 1's is achieved, so A should go to state s_3. If a 0 is input when A is in state s_2, then progress toward accumulating a sequence of three consecutive 1's is lost, so A should return to s_0. When A is in a state s_3 and a 1 is input, then the final three symbols of the input string are 1's, and so A should stay in state s_3. If a 0 is input when A is in state s_3, then A should return to state s_0 to await the input of more 1's. Thus the transition diagram is as follows:

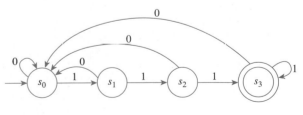

b. $(0 | 1)^*111$

21. *Hint:* Use five states: s_0 (the initial state), s_1 (the state indicating that the previous input symbol was an a), s_2 (the state indicating that the previous input symbol was a b), s_3

(the state indicating that the previous two input symbols were a's), and s_4 (the state indicating that the previous two input symbols were b's).

23. a.

b. $01(0\,|\,1)^*$

25. a.

b. $(0\,|\,1)^*10$

26. a.

b. $a^*ba^*ba^*$

28. a.

b. $(0\,|\,1)^*010(0\,|\,1)^*$

29.

31.

33.

36.

39. Let $\hat{P}$ denote a list of all letters of a lower-case alphabet except p, $\hat{R}$ denote a list of all the letters of a lower-case alphabet except r, and $\hat{E}$ denote a list of all the letters of a lower-case alphabet except e.

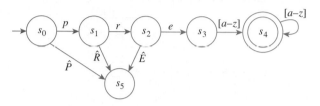

42. Let $\mathscr{C}$ denote a list of all the consonants in a lower-case alphabet.

45.

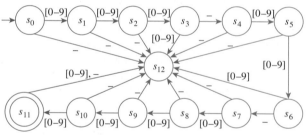

51. *Hint:* This proof is virtually identical to that of Example 12.2.8. Just take p and q in that proof so that $p > q$. From the fact that A accepts $a^p b^p$, you can deduce that A accepts $a^q b^p$. Since $p > q$, this string is not in L.

53. *Hint:* Suppose the automaton A has N states. Choose an integer m such that $(m + 1)^2 - m^2 > N$. Consider strings of a's of lengths between m^2 and $(m + 1)^2$.

Since there are more strings than states, at least two strings must send A to the same state s_i:

$$(m + 1)^2$$

$$\overbrace{aa \ldots aaa \ldots a}^{} aa \ldots aaa \ldots a$$

$$\underbrace{}_{m^2}$$

after both of these inputs, A is in state s_i

It follows (by removing the a's shown in color) that the automaton must accept a string of the form a^k, where $m^2 < k < (m + 1)^2$.

Section 12.3

1. a. 0-equivalence classes: $\{s_0, s_1, s_3, s_4\}, \{s_2, s_5\}$
1-equivalence classes: $\{s_0, s_3\}, \{s_1, s_4\}, \{s_2, s_5\}$
2-equivalence classes: $\{s_0, s_3\}, \{s_1, s_4\}, \{s_2, s_5\}$

b.

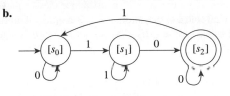

4. a. 0-equivalence classes: $\{s_0, s_1, s_2\}, \{s_3, s_4, s_5\}$
1-equivalence classes: $\{s_0, s_1, s_2\}, \{s_3, s_5\}, \{s_4\}$
2-equivalence classes: $\{s_0, s_2\}, \{s_1\}, \{s_3, s_5\}\{s_4\}$
3-equivalence classes: $\{s_0, s_2\}, \{s_1\}, \{s_3, s_5\}, \{s_4\}$

b.

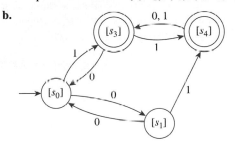

6. a. *Hint:* The 3-equivalence classes are $\{s_0\}, \{s_1\}, \{s_2\}, \{s_3\}, \{s_4\}, \{s_5\},$ and $\{s_6\}$.

7. Yes. For A:

0-equivalence classes: $\{s_0, s_2\}, \{s_1, s_3\}$
1-equivalence classes: $\{s_0\}, \{s_2\}, \{s_1, s_3\}$
2-equivalence classes: $\{s_0\}, \{s_2\}, \{s_1, s_3\}$

Transition diagram for $\overline{A}$:

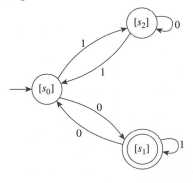

For A':

0-equivalence classes: $\{s_0', s_1', s_2'\}, \{s_3'\}$
1-equivalence classes: $\{s_0', s_2'\}, \{s_1'\}, \{s_3'\}$
2-equivalence classes: $\{s_0', s_2'\}, \{s_1'\}, \{s_3'\}$

Transition diagram for $\overline{A'}$:

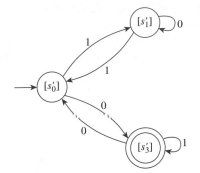

Except for the labeling of the states, the transition diagrams for $\overline{A}$ and $\overline{A'}$ are identical. Hence $\overline{A}$ and $\overline{A'}$ accept the same language, and so, by Theorem 12.3.3, A and A' also accept the same language. Thus A and A' are equivalent automata.

9. For A:

0-equivalence classes: $\{s_1, s_2, s_4, s_5\}, \{s_0, s_3\}$
1-equivalence classes: $\{s_1, s_2\}, \{s_4, s_5\}, \{s_0, s_3\}$
2-equivalence classes: $\{s_1\}, \{s_2\}, \{s_4, s_5\}, \{s_0, s_3\}$
3-equivalence classes: $\{s_1\}, \{s_2\}, \{s_4, s_5\}, \{s_0, s_3\}$

Therefore, the states of $\overline{A}$ are the 3-equivalence classes of A.

For A':

0-equivalence classes: $\{s_2', s_3', s_4', s_5'\}, \{s_0', s_1'\}$
1-equivalence classes: $\{s_2', s_3', s_4', s_5'\}, \{s_0', s_1'\}$

Therefore, the states of A' are the 1-equivalence classes of A'.

According to the text, two automata are equivalent if, and only if, their quotient automata are isomorphic, provided inaccessible states have first been removed. Now A and A' have no inaccessible states, and $\overline{A}$ has four states whereas $\overline{A'}$ has only two states. Therefore, A and A' are not equivalent.

This result can also be obtained by noting, for example, that the string 11 is accepted by A' but not by A.

11. *Partial answer:* Suppose A is a finite-state automaton with set of states S and relation R_* of *-equivalence of states. [*To show that R_* is an equivalence relation, we must show that R is reflexive, symmetric, and transitive.*]

Proof that R_* is symmetric:
[*We must show that for all states s and t, if $s \ R_* \ t$ then $t \ R_* \ s$.*] Suppose that s and t are states of A such that $s R_* t$. [*We must show that $t \ R_* \ s$*] Since $s \ R_* \ t$, then for all input strings w,

$$\begin{bmatrix} N^*(s, w) \text{ is an} \\ \text{accepting state} \end{bmatrix} \Leftrightarrow \begin{bmatrix} N^*(t, w) \text{ is an} \\ \text{accepting state} \end{bmatrix}$$

where N^* is the eventual-state function on A. But then, by symmetry of the $\Leftrightarrow$ relation, it is true that for all input strings w,

$$\begin{bmatrix} N^*(t, w) \text{ is an} \\ \text{accepting state} \end{bmatrix} \Leftrightarrow \begin{bmatrix} N^*(s, w) \text{ is an} \\ \text{accepting state} \end{bmatrix}$$

Hence $t \ R_* \ s$ [*as was to be shown*], so R_* is symmetric.

12. The proof is identical to the proof of property (12.3.1) given in the solution to exercise 11 provided each occurrence of "for all input strings w" is replaced by "for all input strings w of length less than or equal to k."

13. *Proof:* By property (12.3.2), for each integer $k \geq 0$, k-equivalence is an equivalence relation. But by Theorem 10.3.4, the distinct equivalence classes of an equivalence relation form a partition of the set on which the relation is defined. In this case, the relation is defined on the states of the automaton. So the k-equivalence classes form a partition of the set of all states of the automaton.

15. *Hint 1:* Suppose C_k is a particular but arbitrarily chosen k-equivalence class. You must show that there is a $(k-1)$-equivalence class C_{k-1} such that $C_k \subseteq C_{k-1}$.

   Hint 2: If s is any element in C_k, then s is a state of the automaton. Now the $(k-1)$-equivalence classes partition the set of all states of the automaton into a union of mutually disjoint subsets, so $s \in C_{k-1}$ for *some* $(k-1)$-equivalence class C_{k-1}.

   Hint 3: To show that $C_k \subseteq C_{k-1}$, you must show that for any state t, if $t \in C_k$, then $t \in C_{k-1}$.

17. *Hint:* If $m < k$, then every input string of length less than or equal to m has length less than or equal to k.

19. *Hint:* Suppose two states s and t are equivalent. You must show that for any input symbol m, the next-states $N(s, m)$ and $N(t, m)$ are equivalent. To do this, use the definition of equivalence and the fact that for any string w', input symbol m, and state s, $N^*(N(s, m), w') = N^*(s, mw')$.

INDEX

CREDITS

This page constitutes an extension of the copyright page. We have made every effort to trace the ownership of all copyrighted material and to secure permission from copyright holders. In the event of any question arising as to the use of any material, we will be pleased to make the necessary corrections in future printings. Thanks are due to the following authors, publishers, and agents for permission to use the material indicated.

Chapter 1   **1** Bettmann/CORBIS; **10** Culver Pictures; **39** Indiana University Archives; **43** M.I.T Museum; **44** Courtesy of Intel Corporation; **45** Courtesy of IBM; **48** CORBIS; **54** Courtesy of Harvard University Archives

Chapter 2   **78** (top) Culver Pictures; **78** (bottom) Friedrich Schiller, Universitat Jena/Thuringer Universitats/und Landesbibliothek Unfversitatsarchiv; **85** Public Domain; **102** Public Domain; **116** Culver Pictures

Chapter 3   **138** (top) CORBIS; **138** (bottom) Andrew Wiles/Princeton University; **180** CORBIS; **186** Getty Images; **190** Suleymaniye Kutuphanesi, Istanbul,Turkey

Chapter 4   **202** CORBIS; **244** (top) Courtesy of Christiane Floyd; **244** (bottom) Center for American History/The University of Texas at Austin; **247** Courtesy of Tony Hoare

Chapter 5   **255** David Eugene Smith Collection, Rare Book and Manuscript Library, Columbia University; **257** The President and Council of the Royal Society of London; **260** Stock Montage; **293** Bettmann/CORBIS; **294** Public Domain

Chapter 6   **298** Reprinted by permission of United Feature Syndicate, Inc.; **302** Bettmann/CORBIS; **358** Hulton-Deutch Collection/CORBIS; **362** Academic Royale de Belgique; **370** Andrei Nikolavich Kolmogorov—Yevgney Khaldei/CORBIS; **379** Public Domain

Chapter 7   **390** Stock Montage; **396** Courtesy of The U.S. Naval Academy; **443** Bettmann/Corbis; **449** (top) Public Domain; **449** (bottom) Getty Images

Chapter 8   **461** (top) Courtesy of Francis Lucas; **461** (middle) Courtesy of Paul Stockmeyer/www.cs.wm.edu/~pkstoc/toh.html; **464** CORBIS; **506** (top) Roger Russmeyer/CORBIS

Chapter 9   **510** CORBIS; **531** CORBIS; **532** Courtesy of Donald Knuth; **545** CORBIS

Chapter 10   **597** Bettmann/CORBIS; **612** RSA 2003/Courtesy of Leonard Adleman

Chapter 11   **666** Bettmann/CORBIS; **676** Bettmann/CORBIS; **707** (top) Courtesy of IBM Corporation © 2003; **707** (bottom) Courtesy of Peter Naur; **708** CORBIS; **726** Courtesy of Joseph Kruskal; **729** Courtesy of Lucent Technologies Inc. © 2003, Lucent Technologies Inc. All rights reserved.

Chapter 12   **735** Christopher Felver/CORBIS; **736** University of Wisconsin; **747** (top) David Eugene Smith Collection, Rare Book and Manuscript Library, Columbia University; **747** (bottom) Godfrey Argent

List of Symbols

Subject	Symbol	Meaning	Page
Formal Languages and Finite-State Automata	Σ	an alphabet of a language	736
	ϵ	the null string	310
	Σ^n	the set of all strings over Σ of length n	736
	Σ^*	the set of all strings over Σ	736
	Σ^+	the set of all strings over Σ with length at least 1	736
	LL'	the concatenation of languages L and L'	738
	L^*	the Kleene closure of L	738
	$(rs), (r \mid s), (r^*)$	regular expressions	738
	$[x_1 - x_n], [\wedge x_m - x_n]$	character classes	742, 743
	$x+, x?, x\{n\}, x\{m, n\}$	shorthand notations for regular expressions	743
	$N(s, m)$	the value of the next-state function for a state s and input symbol m	748
	$\rightarrow \widehat{s_0}$	initial state	748
	$\widehat{\widehat{s_a}}$	accepting state	748
	$L(A)$	language accepted by A	750
	$N^*(s, w)$	the value of the eventual-state function for a state s and input string w	751
	$s\, R_*\, t$	s and t are $*$-equivalent	764
	$s\, R_k\, t$	s and t are k-equivalent	765
	$\overline{A}$	the quotient automaton of A	768
Matrices	$\mathbf{A}$	matrix	683
	$\mathbf{I}$	identity matrix	692
	$\mathbf{A} + \mathbf{B}$	sum of matrices $\mathbf{A}$ and $\mathbf{B}$	697
	$\mathbf{AB}$	product of matrices $\mathbf{A}$ and $\mathbf{B}$	689
	$\mathbf{A}^n$	matrix $\mathbf{A}$ to the power n	693
Graphs and Trees	$V(G)$	the set of vertices of a graph G	650
	$E(G)$	the set of edges of a graph G	650
	$\{v, w\}$	the edge joining v and w in a simple graph	656
	K_n	complete graph on n vertices	656
	$K_{m,n}$	complete bipartite graph on (m, n) vertices	657
	$deg(v)$	degree of vertex v	658
	$v_0 e_1 v_1 e_2 \cdots e_n v_n$	a walk from v_0 to v_n	667
	$w(e)$	the weight of edge e	725
	$w(G)$	the total weight of graph G	725

Reference Formulas

Topic	Name	Formula	Page
Logic	De Morgan's law	$\sim(p \wedge q) \equiv \sim p \vee \sim q$	10
	De Morgan's law	$\sim(p \vee q) \equiv \sim p \wedge \sim q$	10
	Negation of $\rightarrow$	$\sim(p \rightarrow q) \equiv p \wedge \sim q$	20
	Equivalence of a conditional and its contrapositive	$p \rightarrow q \equiv \sim q \rightarrow \sim p$	21
	Nonequivalence of a conditional and its converse	$p \rightarrow q \not\equiv q \rightarrow p$	23
	Nonequivalence of a conditional and its inverse	$p \rightarrow q \not\equiv \sim p \rightarrow \sim q$	23
	Negation of a universal statement	$\sim(\forall x \text{ in } D, Q(x)) \equiv \exists x \text{ in } D \text{ such that } \sim Q(x)$	88
	Negation of an existential statement	$\sim(\exists x \text{ in } D \text{ such that } Q(x)) \equiv \forall x \text{ in } D, \sim Q(x)$	89
Sums	Sum of the first n integers	$1 + 2 + \cdots + n = \dfrac{n(n+1)}{2}$	218
	Sum of powers of r	$1 + r + r^2 + \cdots + r^n = \dfrac{r^{n+1} - 1}{r - 1}$	222
Counting and Probability	Probability in the equally likely case	$P(E) = \dfrac{N(E)}{N(S)}$	299
	Number of r-permutations of a set with n elements	$P(n, r) = \dfrac{n!}{(n-r)!}$	315
	Number of elements in a union	$N(A \cup B) = N(A) + N(B) - N(A \cap B)$	327
	Number of subsets of size r of a set with n elements	$\dbinom{n}{r} = \dfrac{n!}{r!(n-r)!}$	337
	Pascal's formula	$\dbinom{n+1}{r} = \dbinom{n}{r-1} + \dbinom{n}{r}$	358